LENGTH

scales for comparison of metric and U.S. units of measurement

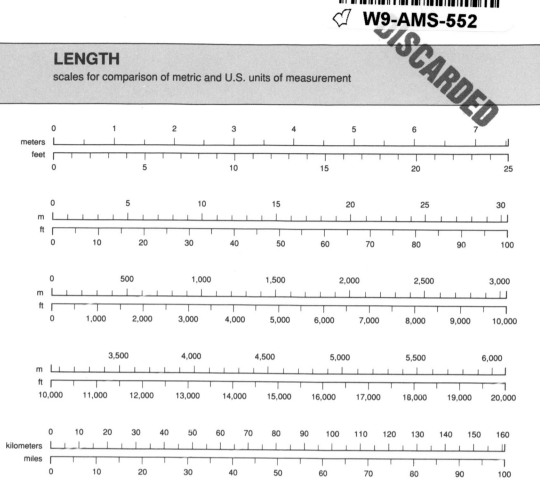

CONVERSION TABLES

	U.S. to Metric			Metric to U.S.	
	to convert	*multiply by*		*to convert*	*multiply by*
LENGTH	in. to mm.	25.4		mm. to in.	0.039
	in. to cm.	2.54		cm. to in.	0.394
	ft. to m.	0.305		m. to ft.	3.281
	yd. to m.	0.914		m. to yd.	1.094
	mi. to km.	1.609		km. to mi.	0.621
AREA	sq. in. to sq. cm.	6.452		sq. cm. to sq. in.	0.155
	sq. ft. to sq. m.	0.093		sq. m. to sq. ft.	10.764
	sq. yd. to sq. m.	0.836		sq. m. to sq. yd.	1.196
	sq. mi. to ha.	258.999		ha. to sq. mi.	0.004
VOLUME	cu. in. to cc.	16.387		cc. to cu. in.	0.061
	cu. ft. to cu. m.	0.028		cu. m. to cu. ft.	35.315
	cu. yd. to cu. m.	0.765		cu. m. to cu. yd.	1.308
CAPACITY	fl. oz. to liter	0.03		liter to fl. oz.	33.815
(liquid)	qt. to liter	0.946		liter to qt.	1.057
	gal. to liter	3.785		liter to gal.	0.264
MASS	oz. avdp. to g.	28.35		g. to oz. avdp.	0.035
(weight)	lb. avdp. to kg.	0.454		kg. to lb. avdp.	2.205
	ton to t.	0.907		t. to ton	1.102
	l. t. to t.	1.016		t. to l. t.	0.984

Abbreviations

avdp. avoirdupois
cc. cubic centimeter(s)
cm. centimeter(s)
cu. cubic
ft. foot, feet
g. gram(s)
gal. gallon(s)
ha. hectare(s)
in. inch(es)
kg. kilogram(s)
lb. pound(s)
l. t. long ton(s)
m. meter(s)
mi. mile(s)
mm. millimeter(s)
oz. ounce(s)
qt. quart(s)
sq. square
t. metric ton(s)
yd. yard(s)

Walker's
Mammals
of the World

Ernest P. Walker, 1891–1969

Walker's
Mammals of the World

Sixth Edition

Volume I

Ronald M. Nowak

The Johns Hopkins University Press
Baltimore and London   1999

NOTICE

Persons who find errors in this work or who know of better photographs or additional facts that should be incorporated in revisions are invited to communicate with The Johns Hopkins University Press, 2715 North Charles Street, Baltimore, Maryland 21218-4363.

Frontispiece photograph by Eugene Maliniak

The Johns Hopkins University Press
2715 North Charles Street
Baltimore, Maryland 21218-4363
www.press.jhu.edu

The first edition of this work appeared under the title *Mammals of the World* by Ernest P. Walker (senior author) and six coauthors: Florence Warnick, Sybil E. Hamlet, Kenneth I. Lange, Mary A. Davis, Howard E. Uible, and Patricia F. Wright. The second and third editions were revised by John L. Paradiso. The fourth edition was revised by Ronald M. Nowak and John L. Paradiso. The fifth edition was revised by Ronald M. Nowak.

Library of Congress Cataloging-in-Publication Data

Nowak, Ronald M.
    Walker's mammals of the world.— 6th ed. / Ronald M. Nowak.
        p.   cm.
    Includes bibliographical references (v. 1, p.   ) and index.
    ISBN 0-8018-5789-9 (alk. paper)
    1. Mammals.   2. Mammals—Classification.   I. Title.
QL703.N68   1999
599—dc21   98-23686 CIP

A catalog record for this book is available from the British Library.

To the MAMMALS, GREAT AND SMALL,
who contribute so much to the welfare and happiness
of man, another mammal, but receive so little in return,
except blame, abuse, and extermination.

Ernest P. Walker

Contents

Volume I

Volume II

Foreword

In the foreword to the first edition (1964) of *Mammals of the World*, Ernest P. Walker's monumental contribution to zoological literature, Fairfield Osborn, president of the New York Zoological Society, pointed out that no other single series of books dealt with all of the known genera of mammals on earth. Ronald M. Nowak has carried on the Walker legacy through the fourth and fifth editions and now continues the tradition with a greatly expanded and updated sixth edition. E. Raymond Hall, Nowak's major professor at the University of Kansas, explained in the foreword to the fourth edition that comparable works have been limited to selected groups of mammals, have dealt only with a particular region, or have provided only lists of names and distributions. And Durward Allen, longtime professor of wildlife ecology at Purdue University, noted in the foreword to the fifth edition that Walker's original intent was to provide a comprehensive reference that would be accessible to the general public but also serve the professional community of mammalogists. This latest edition meets those goals admirably.

The text has been expanded by 25 percent and now covers every genus of mammal known to have existed within the last 5,000 years. For instance, this edition includes an account of the woolly mammoth, recently demonstrated to have survived until less than 4,000 years ago. Detailed but readable accounts include myriad facts about the morphology, relationships, and natural history of all genera, including 81 additional ones, 21 of which have been described as new to science since preparation of the fifth edition. There is a useful introduction to the class Mammalia in this edition, as well as separate accounts for all seven orders of marsupials now recognized. Each generic account includes a complete list of currently recognized species, with an indication of their distribution. These species lists now correspond largely with those in the 1993 edition of *Mammal Species of the World*, the comprehensive work on taxonomy and distribution published by the Smithsonian Institution. All substantive differences from that work are explained by Nowak.

There are more than 2,700 new reference citations, nearly all of which were published in the last decade. The herculean task of assimilating this far-flung scientific literature has yielded a wealth of new data on systematics, ecology, behavior, reproduction, longevity, and conservation status. The *1996 IUCN Red List of Threatened Animals* was used to record the official classification of all species and subspecies in danger of extinction. Walker's original goal of presenting a quality photograph of a living representative of every genus of mammal continues, with coverage of such mammals as the recently discovered giant muntjac of Southeast Asia. The overall result is a newly updated and even better version of a contemporary classic.

Don E. Wilson
Director, Biodiversity Programs
Smithsonian Institution

Preface

Ernest P. Walker was involved for more than 30 years in preparing the first edition of *Mammals of the World*, published in 1964. He began the project, originally called "Genera of Recent Mammals of the World," shortly after becoming assistant director of the National Zoological Park, Washington, D.C., in 1930. In this capacity he accumulated extensive knowledge of mammals, both by direct observation and through contact with other zoo personnel, mammalogists, and naturalists. He was an accomplished photographer and made a point of taking a picture of each kind of mammal that arrived at the National Zoo. He loved all living things and personally worked with many different species of wild mammals, sometimes caring for them in his own office or home.

Walker formally retired at the end of 1956 but continued to maintain an office at the National Zoo. Through a research grant from the National Institutes of Health, and under sponsorship of the New York Zoological Society, he then was able to devote himself full-time to the compilation of data and the collection of additional facts and photographs needed for *Mammals of the World*. He employed a team of research and clerical personnel to assist him in assembling information, examining literature, and writing text. Six persons are considered coauthors of the first edition: Florence Warnick, Sybil E. Hamlet, Kenneth I. Lange, Mary A. Davis, Howard E. Uible, and Patricia F. Wright. During the last four years of the project, Walker was compelled, owing to failing health, to leave many details of the work to these coauthors.

Walker wished to present, in nontechnical language, a description of, and the basic facts of natural history about, each genus of living mammals. He wanted a book that could be easily understood and appreciated by the general public but also would serve the professional community. Although the latter group has received the book favorably, a consistent criticism of the first three editions was the lack of citation of specific references for most of the facts presented in the text. In fairness to Walker, it should here be pointed out (as it was in the preface to the first edition) that many of these facts were learned through direct observation by the authors or were taken from unpublished data received orally, through correspondence, or through addition to the manuscript by reviewers. At least 10 persons are thought to have examined the manuscript for each of the original generic accounts.

The first edition of *Mammals of the World* included an entire third volume comprising a bibliography of the literature of mammalogy through about the end of 1960. In the 35 years since the publication of the first edition there has been an enormous increase in mammalian research and in the resulting literature. An especially great amount of new information has been obtained through the systematic study of the ecology and behavior of mammals under natural conditions and through modern taxonomic approaches such as karyology, biochemistry, morphology of the reproductive system, and computer-based multivariate analysis of skeletal measurements. Many genera that were known during the preparation of the first edition only by casual observation or examination of a few specimens now have been studied in detail. The second and third editions of *Mammals of the World*, published in 1968 and 1975, included many new facts, several accounts of newly described genera, and more than 350 new photographs. There was, however, no attempt to extensively revise and expand the text.

The fourth edition incorporated major changes in format and content. The overall length of the text was increased by about 50 percent (the physical size of the volumes remained about the same, however, because much blank space was eliminated and type size was reduced). Approximately 90 percent of the generic accounts received substantive modification, and many were completely rewritten. There was a break with the original practice of restricting the information on each genus to a single page, and there was expansion of the accounts of those genera for which more data were available. In the fifth edition, text length was increased by about 22 percent, substantive changes were made in about 80 percent of the previously existing generic accounts, and 106 new generic accounts were added (mainly because of elevation of former subgenera and synonyms to generic rank). Respective figures for the present edition, which continues the objectives of the fourth and fifth, are a 25 percent increase in text length, 95 percent of the previous generic accounts substantively modified, and 81 new generic accounts.

Nonetheless, much text remains the same as in earlier editions, especially portions covering physical description. An effort also has been made to keep to Walker's basic objective of producing a book that can be useful to both scientists and the public in general. Ordinal and familial accounts usually precede the generic accounts and give information applicable to all members of the group involved. When possible, the following topics are covered for each genus: scientific and common name, number and distribution of species, measurements, physical description, habitat, locomotion, daily and seasonal activity, diet, population dynamics, home range, social life, reproduction, longevity, and relationship with people. There is little emphasis on such subjects as internal morphology, physiology, genetics, laboratory experimentation, parasitology, pathology, and paleontology. Walker's charts showing the world distribution of mammals have been retained, but they have been revised to reflect changes in taxonomy, distributional knowledge, and pagination. As in the earlier editions, the metric system is used throughout (conversion scales and tables appear on the endpapers of both volumes).

It was Walker's objective to include a satisfactory photograph of a living representative of every genus in *Mammals of the World*. In the preface to the first edition, he wrote:

> In selecting photographs we have kept in mind the idea of conveying a better image of the mammal than could be given in a picture. On the order and family pages

are photographs illustrating points mentioned in these general texts. The photographs included on almost every generic page are fairly representative of each group.

In some instances, several photographs are used when the members of the genus differ greatly in appearance or where more are needed to show structural peculiarities or poses. Many attractive and appealing photographs have been rejected because they did not show the characteristic form of the mammals to the best advantage.

Within some genera there is considerable variation in coloration, color pattern, or size of the species. The relative sizes of species within a genus cannot always be shown in the photographs, and only a few of the species with outstandingly different color patterns are shown, but information regarding such variations is given in each generic text.

Some of the photographs of mammals in captivity are of immature individuals which do not have fully developed adult characters, or old individuals that have been in captivity a long time and have become more heavy-bodied than those in the wild.

Many of the photographs have been retouched to eliminate distracting, extraneous objects; a few have been retouched to make anatomical characters visible that were obscure or to restore some that were missing in the original photograph. Fidelity to life has been followed in every instance, although we cannot vouch for the accuracy of drawings and paintings that have been copied. We must assume, however, that the author of the article tried to have the drawing or painting as accurate as information then available permitted.

Despite Walker's heroic efforts, the first edition did not actually contain a photograph of a living representative of every genus. And although additional photos were found subsequently, it was judged that the third edition still included nearly 400 accounts of living genera that were accompanied only by a photo of a museum specimen or dead individual, by a drawing, by an unsatisfactory photo of a live animal, or by no illustration at all. In the course of preparing the fourth and fifth editions, acceptable photos of living representatives were found for more than 200 generic accounts. For the present edition, approximately 50 more genera are newly represented by photos that I think would meet Walker's criteria. The total number of illustrations new to the present edition, either to replace or to supplement older material, is about 150. Of those genera still not adequately illustrated, many are little known, very rare, or long extinct. However, it is reasonable to expect that suitable photos of many others, including some of those newly recognized at the generic level in this edition, will become available. Persons who know of photos that would improve future editions of this book are encouraged to contact the Johns Hopkins University Press.

One of the problems with the first three editions of *Mammals of the World* was the inconsistent treatment of the names and distributions of the species within genera. To resolve this difficulty, and to perhaps contribute to the usefulness of the book, it was decided to list in the fourth edition the name and range of every species of every genus. For most genera with more than one species an attempt was made to arrange the list in systematic order, with more closely related species grouped together. The objective of systematic listing now has been extended to all genera. Admittedly, information still is far from complete for many genera, and taxonomic interpretations sometimes vary widely.

Even when such problems have been resolved, it may be difficult to adequately express relationships in a linear fashion.

Because of limited space, distributions of wide-ranging species can not be given in detail. The following examples of distributional information are taken from the text, with explanations given here in parentheses:

Z. [Zapus] hudsonius, southern Alaska to Labrador and northern Georgia, isolated populations in mountains of Arizona and New Mexico (the species occurs in the region bounded approximately by swaths connecting southern Alaska, Labrador, and northern Georgia and also occurs in isolated parts of Arizona and New Mexico);

B. [Bandicota] bengalensis, probably occurs naturally from Pakistan through India and Nepal to Burma and in Sri Lanka, also found on Penang Island off the west coast of the Malay Peninsula, northern Sumatra, and eastern Java (the species occurs naturally from Pakistan to Burma and on Sri Lanka but does not occur on the mainland east of Burma, its presence on Penang, Sumatra, and Java apparently resulting from introduction);

A. [Akodon] albiventer, Andes of southeastern Peru, western Bolivia, northern Chile, and northwestern Argentina (the species occurs in only the Andean mountainous parts of all four of the areas indicated).

In cases where more than one reference is cited in support of the systematic aspects of a generic account, each of the references was consulted in making a decision regarding the names, number, arrangement, and distributions of the species listed. All references do not necessarily agree with the information given in the accounts here. Major points of systematic controversy are discussed after the accepted list of species.

It would not have been possible to list all species had it not been for the existence of published sources providing the names and distributions of the mammals of major parts of the world. Especially helpful were Bannister et al. (1988), Cabrera (1957, 1961), Chasen (1940), Corbet (1978), Corbet and Hill (1991, 1992), Ellerman and Morrison-Scott (1966), Flannery (1995), E. R. Hall (1981), Laurie and Hill (1954), Meester and Setzer (1977), Rice (1977), Ride (1970), and Taylor (1934). The most extensive compilation of the world's mammalian species, edited recently by Wilson and Reeder (1993), comprises sections prepared by 20 authorities; much original information is included. It was used both as a primary source and to check the systematic lists given here. All points of disagreement with Wilson and Reeder except those involving minor differences in species distribution and more recently published information are discussed.

Wilson and Reeder (1993) was largely followed regarding the number and sequence of mammalian orders, resulting in major changes in the traditional arrangement accepted in the first edition of *Mammals of the World*. The Macroscelidea and the Scandentia are recognized as full and unrelated orders, and for the first time the marsupials are divided into seven separate orders. Also covered here is the extinct but historical order Bibymalagasia of Madagascar, which was described by MacPhee (1994). Substantive changes were made in the fourth and fifth editions with respect to number and arrangement of families, and the present edition contains many additional changes at this level. The names and contents of suborders and subfamilies are given at the beginning of, respectively, the ordinal and familial accounts. Accepted subgeneric designations appear at appropriate places within the lists of species in generic accounts.

The total number of taxonomic categories accepted in this edition are: 28 orders, 146 families, 1,192 genera, and 4,809 species. The genus remains the basic unit of treatment, though some generic texts contain separate species accounts. Of the genera, 21 were named subsequent to preparation of the fifth edition of *Mammals of the World: Gracilinanus* Gardner and Creighton, 1989; *Nesoscaptor* Abe, Shiraishi, and Arai, 1991; *Koopmania* Owen, 1991; *Pseudopotto* Schwartz, 1996; *Babakotia* Godfrey, Simons, Chatrath, and Rakotosamimanana, 1990; *Paralouatta* Rivero and Arredondo, 1991; *Antillothrix* MacPhee, Horovitz, Arredondo, and Jiménez Vasquez, 1995; *Megamuntiacus* Tuoc, Dung, Dawson, Arctander, and MacKinnon, 1994; *Pseudonovibos* Peter and Feiler, 1994; *Pseudoryx* Dung, Giao, Chinh, Tuoc, Arctander, and MacKinnon, 1993; *Amphinectomys* Malygin, Aniskin, Isaev, and Milishnikov, 1994; *Lundomys* Voss and Carleton, 1993; *Microakodontomys* Hershkovitz, 1993; *Pearsonomys* Patterson, 1992; *Salinomys* Braun and Mares, 1995; *Maresomys* Braun, 1993; *Monticolomys* Carleton and Goodman, 1996; *Volemys* Zagorodnyuk, 1990; *Pithecheirops* Emmons, 1993; *Coccymys* Menzies, 1990; and *Rhizoplagiodontia* Woods, 1989.

Also since preparation of the fifth edition came the extraordinary announcement that *Mammuthus* Burnett, 1829, previously thought to have become extinct shortly after the end of the Pleistocene, actually survived into historical time on Wrangel Island, off northeastern Siberia. Equally remarkable is that three of the above new genera—*Megamuntiacus, Pseudonovibos,* and *Pseudoryx*—represent large ungulates all evidently still living in the forests of Indochina. It seems that the age of great discovery in mammalogy is far from over.

There are altogether 81 new generic accounts in the present edition, and 5 accounts from the fifth edition have been deleted. The large net increase results mainly, not from newly described genera, but from a continuing trend by systematic mammalogists to elevate former subgenera and synonyms to full generic rank. In addition, some of the new accounts (including those of *Babakotia, Paralouatta, Antillothrix, Rhizoplagiodontia,* and *Mammuthus*) reflect an attempt, began in the fifth edition, to resolve the inconsistent treatment in earlier editions of genera that are known only by subfossil material but are thought to have lived in historical time. A separate account now is provided for every extinct genus known to have lived within the last 5,000 years. Most such genera inhabited the West Indies, islands of the Mediterranean, and Madagascar and are thought to have disappeared through human agency. Reference also is made to extinct species that lived in historical time and to the former distribution of extant species.

Work on the present edition began in mid-1993, and an attempt (not fully successful) was made to consult all relevant mammalogical literature covered through volume 131 (1994–95) of *Zoological Record* and number 251 (November 1995) of *Wildlife Review*. Many references, some published as recently as 1997, were obtained by other means. As in the fourth and fifth editions but unlike in the first edition of *Mammals of the World,* there are textual references for all new information except minor changes in measurements.

This edition contains substantial information on the relationship between people and other mammals, especially with regard to economic importance and conservation. Such information generally appears at the end of generic accounts. The distributional data provided at the beginning of accounts usually refer to the original situation in historical time. Changes in distribution caused by human influence are covered at the end of the accounts. Each account lists any species, subspecies, or population that at the time of preparation of the manuscript was classified in a category of concern by the World Conservation Union (also known as the International Union for Conservation of Nature and Natural Resources and commonly abbreviated as IUCN). The dates given in the text for certain IUCN classifications refer to dates of various individual data sheets in the original IUCN *Red Data Book*, issued through 1978, information from which is cited here. Many new IUCN *Red Data Books*, action plans, status surveys, and other publications have since appeared and greatly expanded our knowledge of declining mammals.

The most recently available *IUCN Red List of Threatened Animals* (Baillie and Groombridge 1996) made newspaper headlines with its shocking announcement that 1,096 full species of mammals, nearly one-fourth of all thought to exist, now qualified for classification as "critically endangered" (169 species), "endangered" (315), or "vulnerable" (612). Closer examination of the publication revealed that the situation is even worse, as there are another 75 mammal species in the category "conservation dependent" and 598 designated "near threatened." These latter two categories by definition cover species that come close to meeting the criteria for classification as vulnerable or endangered. Still another 86 species are considered "extinct," and 3 are regarded as "extinct in the wild." In addition, the IUCN classifies as critically endangered, endangered, vulnerable, conservation dependent, and near threatened 220 mammal subspecies that are *not* components of any of the full species of mammals that also are in those categories. When all of these data are considered, the actual number of mammals in jeopardy is much closer to *half* than to one-fourth of the world's total. In any case, all of the classifications given in Baillie and Groombridge (1996), together with any available supporting information, are reported here.

Also indicated here are species, subspecies, and populations of mammals that were classified as "endangered" or "threatened" by the United States Department of the Interior (USDI), based on a list issued in 1996, or that were on appendix 1 or 2 of the Convention on International Trade in Endangered Species of Wild Fauna and Flora (CITES), based on a list issued following the June 1997 conference of the latter organization. By definition, appendix 1 includes species threatened with extinction that are or may be affected by trade. Appendix 2 includes species that are not necessarily threatened with extinction now but may become threatened unless trade is subject to strict regulation. It should be noted that all species of the order Primates, the order Cetacea, and the family Felidae are on appendix 2, except for those species on appendix 1. It must be emphasized that these designations are subject to constant change.

Readers also should be cautioned that the USDI classification process has become hopelessly subject to delay and manipulation by bureaucratic, political, and commercial interests. Its resulting list is now almost meaningless as an expression of the extent and diversity of the world's declining mammals. For example, the total number of mammal species and subspecies in all IUCN categories of concern is now 2,078, which represents a net increase of 1,661 since late 1987. In the same period the USDI list had a net gain of only 17 mammals (21 added but 4 delisted). When John L. Paradiso and I signed the preface to the fourth edition in 1983, we proudly added our affiliation with the USDI's Office of Endangered Species (now abolished). There then was reason to hope that the United States would extend its role as a leader in the international movement to recognize and conserve endangered species. The utter failure of that hope recently led to my own retirement from the USDI.

Many persons contributed to the preparation of this edition. I am grateful to all of those who sent photographs, reprints, or other information and who pointed out errors and problems in the fifth edition. Among those who showed an unusual degree of kindness in this regard are: Tim Flannery, Pavel German, Jan M. Haft, Klaus-Gerhard Heller, J. R. Malcolm, J. C. Ray, Robert S. Voss, and the late Frederick A. Ulmer Jr. Credit for each photograph used is given in its legend. Linda M. Coley proved indispensable in locating many of the references used in compiling the text and provided other research assistance.

It again has been a great pleasure to work with the members of the staff of the Johns Hopkins University Press. Former science editors Richard T. O'Grady and Robert M. Harington were most helpful and supportive during early stages of preparation of the sixth edition. Much kindness and dedication also were shown by Sarah Cline, Wendy Harris, Michael Jensen, Barbara Lamb, and Lee Sioles. Joanne Allen, of Jonesboro, Arkansas, served as primary editor; she again found more errors and omissions than I would have imagined possible and made many changes that smooth the flow of the text. Despite her efforts, and those of the other participants in the project, I know that problems remain, and I ask readers to send corrections and suggestions for improvement to the Johns Hopkins University Press.

Finally, I again thank John L. Paradiso, colleague of Walker, editor of the second and third editions, and my coauthor for the fourth edition. It was he who recommended my original participation in the project and introduced me to many of the involved procedures. We both consider it to have been an honor and a challenge to try to update and modify the work of Ernest P. Walker, a person of remarkable experience, energy, and devotion to the wild mammals of the world. We continue to hope that he would not be disappointed.

World Distribution of Mammals

For maximum usefulness, it has been necessary to devise the simplest practicable outline of the approximate distribution of the genera in the sequence used in the text. The tabulation should be regarded as an index guide to groups of mammals or to geographic regions. At the same time it gives a good overall picture of the general distribution of mammals.

The major geographic distribution of the genera of Recent mammals that appears in the tabulation is designed to show their natural distribution at the present time or within comparatively recent times. It should be noted that most of the animals occupy only a portion of the geographic region that appears at the head of the column. Some are limited to the tropical regions, others to temperate zones, and still others to the colder areas. Mountain ranges and streams sometimes have been natural barriers preventing the spread of animals on the lands that are a part of the same continent. Also, many restricted ranges cannot be designated either by letters to show the general area or by footnotes because of limited space on the tabulation. *It therefore should not be assumed that a mark indicating that an animal occurs within a geographic region implies that it inhabits that entire area.* For more detailed outlines of the ranges of the respective genera, it is necessary to consult the generic texts.

Explanation of Geographic Column Headings
Europe and Asia constitute a single land mass, but this land mass comprises widely different types of zoogeographic areas created by high mountain ranges, plateaus, latitudes, and prevailing winds. The general distribution of Recent mammals can be shown much more accurately by two columns, headed "Europe" and "Asia," than by a single column headed "Eurasia."

Most islands are included with the major land masses nearby unless otherwise specified, though in many instances some of the mammals indicated for the continental mass do not occur on the islands.

With Europe are included the British Isles and other adjacent islands, including those in the Arctic.

With Asia are included the Japanese Islands, Taiwan, Hainan, Sri Lanka, and other adjacent islands, including those in the Arctic.

With North America are included Mexico and Central America south to Panama, adjacent islands, the Aleutian chain, the islands in the arctic region, and Greenland but not the West Indies.

With South America are included Trinidad, the Netherlands Antilles, and other small adjacent islands but not the Falkland and Galapagos Islands unless named in footnotes.

With Africa are included only Zanzibar Island and small islands close to the continent but not the Cape Verde or Canary Islands.

The island groups treated separately are:

Southeastern Asian islands, in which are included the Andamans, the Nicobars, the Mentawais, Sumatra, Java, the Lesser Sundas, Borneo, Sulawesi, the Moluccas, and the many other adjacent small islands;

New Guinea and small adjacent islands;

the Australian region, in which are included Australia, Tasmania, and adjacent small islands;

the Philippine Islands and small adjacent islands;

the West Indies;

Madagascar and small adjacent islands;

other islands that have only one or a few forms of mammals and are named in footnotes.

Seals, sea lions, walruses, and fresh-water dolphins frequent the water areas adjacent to the lands for which they are recorded. Whales, porpoises, and oceanic dolphins are designated by the water areas they inhabit.

Footnotes indicate the major easily definable deviations from the distribution indicated in the tables.

Symbols

†	The mammals are extinct.
■	The mammals occur on the land or in the water area.
N	Northern portion
S	Southern portion
E	Eastern portion
W	Western portion
Ne	Northeastern portion
Se	Southeastern portion
Sw	Southwestern portion
Nw	Northwestern portion
C	Central portion

Examples: "N, C" = northern and central; "Nc" = north-central. Numerals refer to footnotes indicating clearly defined limited ranges within the general area.

Genera of Recent Mammals	page	North America	West Indies	South America	Madagascar	Africa	Europe	Asia	Southeast Asia Islands	Philippine Islands	New Guinea	Australian Region	Antarctic Region	Arctic Region	Atlantic Ocean	Indian Ocean	Pacific Ocean
MONOTREMATA TACHYGLOSSIDAE																	
Tachyglossus	10										■	■1					
Zaglossus	11										■						
MONOTREMATA ORNITHORHYNCHIDAE																	
Ornithorhynchus	13											■E,1					
DIDELPHIMORPHIA MARMOSIDAE																	
Gracilinanus	19			■													
Marmosops	20	■S		■													
Marmosa	21	■S		■													
Micoureus	22	■S		■													
Thylamys	23			■													
Lestodelphys	23			■S													
Metachirus	24	■S		■													
Monodelphis	25	■S		■													
DIDELPHIMORPHIA CALUROMYIDAE																	
Caluromys	27	■S		■													
Caluromysiops	28			■Nw													
DIDELPHIMORPHIA GLIRONIIDAE																	
Glironia	29			■Wc													
DIDELPHIMORPHIA DIDELPHIDAE																	
Philander	30	■G		■													
Didelphis	31	■		■													
Chironectes	33	■S		■													
Lutreolina	34			■													
PAUCITUBERCULATA CAENOLESTIDAE																	
Caenolestes	36			■Nw													
Rhyncholestes	37			■2													
MICROBIOTHERIA MICROBIOTHERIIDAE																	
Dromiciops	39			■Sw													
DASYUROMORPHIA DASYURIDAE																	
Murexia	42										■						
Phascolosorex	43										■						
Neophascogale	44										■						
Phascogale	45											■					
Antechinus	47										■	■1					
Planigale	49										■S	■N,E					
Ningaui	50											■					
Sminthopsis	51										■S	■1					
Antechinomys	54											■					
Parantechinus	55											■N,W					
Dasykaluta	56											■Nw					
Pseudantechinus	56											■N,W					
Myoictis	57										■						
Dasyuroides	59											■Nc					
Dasycercus	60											■					

1. And Tasmania.   2. Chile only.

Genera of Recent Mammals	page	North America	West Indies	South America	Madagascar	Africa	Europe	Asia	Southeast Asia Islands	Philippine Islands	New Guinea	Australian Region	Antarctic Region	Arctic Region	Atlantic Ocean	Indian Ocean	Pacific Ocean
DASYUROMORPHIA DASYURIDAE Continued																	
Dasyurus	61										■	■1					
Sarcophilus	64											■2					
DASYUROMORPHIA MYRMECOBIIDAE																	
Myrmecobius	65											■S					
DASYUROMORPHIA THYLACINIDAE																	
Thylacinus	67											■2					
PERAMELEMORPHIA PERMELIDAE																	
Macrotis	71											■C,S					
Chaeropus	72											■					
Isoodon	73										■S	■1					
Perameles	75											■1					
PERAMELEMORPHIA PERORYCTIDAE																	
Echymipera	77										■3	■Ne					
Rhynchomeles	78								■4								
Microperoryctes	79										■						
Peroryctes	81										■						
NOTORYCTEMORPHIA NOTORYCTIDAE																	
Notoryctes	82											■					
DIPROTODONTIA PHASCOLARCTIDAE																	
Phascolarctos	84											■E					
DIPROTODONTIA VOMBATIDAE																	
Vombatus	87											■E,1					
Lasiorhinus	88											■E,S					
DIPROTODONTIA PHALANGERIDAE																	
Ailurops	90								■5								
Strigocuscus	91								■6		■						
Trichosurus	92											■1					
Wyulda	94											■Nw					
Spilocuscus	95								■4		■3	■Ne					
Phalanger	96								■7		■8	■Ne					
DIPROTODONTIA POTOROIDAE																	
Hypsiprymnodon	98											■Ne					
Potorous	99											■S,1					
Bettongia	100											■1					
Caloprymnus	102											■C					
Aepyprymnus	102											■E					
DIPROTODONTIA MACROPODIDAE																	
Lagostrophus	105											■W,S					
Dorcopsis	106										■						
Dorcopsulus	107										■						
Dendrolagus	108										■	■Ne					
Setonix	110											■Sw					

1. And Tasmania.   2. Tasmania only.   3. And Bismarck Archipelago.   4. Seram only.   5. Sulawesi and Talaud Islands only.
6. Sulawesi only.   7. Moluccas, Seram, and Timor.   8. And Bismarck Archipelago and Solomon Islands.

Genera of Recent Mammals	page	North America	West Indies	South America	Madagascar	Africa	Europe	Asia	Southeast Asia Islands	Philippine Islands	New Guinea	Australian Region	Antarctic Region	Arctic Region	Atlantic Ocean	Indian Ocean	Pacific Ocean
DIPROTODONTIA MACROPODIDAE Continued																	
Thylogale	111										■1	■E,2					
Petrogale	112											■					
Peradorcas	114											■N					
Lagorchestes	115											■					
Onychogalea	116											■					
Wallabia	117											■E					
Macropus	118										■S	■2					
DIPROTODONTIA BURRAMYIDAE																	
Cercartetus	129										■	■					
Burramys	130											■Se					
DIPROTODONTIA PSEUDOCHEIRIDAE																	
Pseudochirulus	132										■	■Ne					
Pseudocheirus	133											■E,2					
Pseudochirops	134										■	■Ne					
Petropseudes	135											■N					
Hemibelideus	136											■Ne					
Petauroides	136											■E					
DIPROTODONTIA PETAURIDAE																	
Petaurus	139								■3		■4	■					
Gymnobelideus	140											■Se					
Dactylopsila	142										■	■Ne					
DIPROTODONTIA TARSIPEDIDAE																	
Tarsipes	143											■Sw					
DIPROTODONTIA ACROBATIDAE																	
Distoechurus	144										■						
Acrobates	145											■E					
XENARTHRA MEGALONYCHIDAE																	
Synocnus†	149		■5														
Parocnus†	149		■5														
Choloepus	149	■S		■N													
XENARTHRA BRADYPODIDAE																	
Bradypus	152	■S		■													
XENARTHRA MYRMECOPHAGIDAE																	
Myrmecophaga	155	■S		■													
Tamandua	156	■S		■													
Cyclopes	157	■S		■													
XENARTHRA DASYPODIDAE																	
Chaetophractus	159			■S													
Euphractus	160			■													
Zaedyus	160			■S													
Priodontes	161			■													
Cabassous	162	■S		■													
Tolypeutes	163			■													

1. And Bismarck Archipelago.   2. And Tasmania.   3. Moluccas only.   4. And Bismarck and Louisiade Archipelagoes.
5. Hispaniola only.

Genera of Recent Mammals	page	North America	West Indies	South America	Madagascar	Africa	Europe	Asia	Southeast Asia Islands	Philippine Islands	New Guinea	Australian Region	Antarctic Region	Arctic Region	Atlantic Ocean	Indian Ocean	Pacific Ocean
XENARTHRA DASYPODIDAE Continued																	
Dasypus	165	■S		■													
Chlamyphorus	166			■S													
INSECTIVORA ERINACEIDAE																	
Hylomys	171							■Se	■								
Neotetracus	171							■Se									
Neohylomys	172							■1									
Echinosorex	173							■Se	■2								
Podogymnura	174									■S							
Erinaceus	174						■	■									
Atelerix	176					■											
Paraechinus	177					■N		■Sw									
Hemiechinus	178					■Ne	■Se	■									
Mesechinus	179							■Ec									
INSECTIVORA CHRYSOCHLORIDAE																	
Chrysospalax	180					■S											
Cryptochloris	181					■S											
Chrysochloris	181					■											
Eremitalpa	183					■S											
Calcochloris	184					■S											
Chlorotalpa	184					■											
Amblysomus	185					■											
INSECTIVORA TENRECIDAE																	
Potamogale	187					■W,C											
Micropotamogale	187					■W,C											
Geogale	188				■												
Oryzorictes	189				■												
Microgale	190				■												
Limnogale	192				■												
Tenrec	192				■												
Setifer	194				■												
Hemicentetes	196				■												
Echinops	198				■S												
INSECTIVORA SOLENODONTIDAE																	
Solenodon	199		■3														
INSECTIVORA NESOPHONTIDAE																	
Nesophontes†	201		■														
INSECTIVORA SORICIDAE																	
Sorex	204	■					■	■									
Blarinella	208							■E									
Cryptotis	208	■		■N													
Blarina	210	■C,E															
Notiosorex	211	■S															
Megasorex	212	■4															
Neomys	212						■	■									
Nesiotites†	212						■5											
Soriculus	214							■									

1. Hainan Island only.   2. Sumatra and Borneo only.   3. Cuba and Hispaniola only.   4. Mexico only.   5. Corsica, Sardinia, and Balearic Islands only.

Genera of Recent Mammals	page	North America	West Indies	South America	Madagascar	Africa	Europe	Asia	Southeast Asia Islands	Philippine Islands	New Guinea	Australian Region	Antarctic Region	Arctic Region	Atlantic Ocean	Indian Ocean	Pacific Ocean
INSECTIVORA SORICIDAE Continued																	
Nectogale	214							■Ec									
Chimarrogale	215							■	■1								
Anourosorex	216							■Se									
Myosorex	217					■											
Crocidura	218					■	■	■	■	■							
Paracrocidura	222					■Wc											
Suncus	223					■	■S	■S	■	■							
Sylvisorex	225					■											
Ruwenzorisorex	225						■c										
Feroculus	226							■2									
Solisorex	227							■2									
Diplomesodon	227							■Wc									
Scutisorex	228					■C											
INSECTIVORA TALPIDAE																	
Uropsilus	230							■Ec									
Desmana	230						■Se	■Wc									
Galemys	232						■Sw										
Scaptonyx	233							■Ec									
Talpa	234						■	■									
Scaptochirus	235							■Ne									
Euroscaptor	235							■E									
Mogera	236							■L									
Nesoscaptor	236							■3									
Parascaptor	237							■Se									
Urotrichus	237							■4									
Neurotrichus	238	■W															
Scapanulus	239							■Ec									
Parascalops	239	■E															
Scalopus	240	■E,S															
Scapanus	241	■W															
Condylura	242	■E															
SCANDENTIA TUPAIIDAE																	
Tupaia	244							■	■	■W							
Anathana	246							■Sc									
Dendrogale	247							■Se	■5								
Urogale	248									■S							
Ptilocercus	248							■6	■								
DERMOPTERA CYNOCEPHALIDAE																	
Cynocephalus	250							■Se	■	■S							
CHIROPTERA PTEROPODIDAE																	
Eidolon	260				■	■		■Sw									
Rousettus	261				■	■		■S	■	■	■7						
Boneia	263								■8								
Myonycteris	263					■											
Pteropus	264				■			■S	■	■	■7	■N,E				■	■9
Acerodon	271								■E	■							
Neopteryx	272								■8								

1. Sumatra and Borneo only. 2. Sri Lanka only. 3. Uotsuri-jima (near Taiwan) only. 4. Japan only. 5. Borneo only. 6. Malay Peninsula only. 7. And Bismarck Archipelago and Solomon Islands. 8. Sulawesi only. 9. East to Cook Islands.

Genera of Recent Mammals	page	North America	West Indies	South America	Madagascar	Africa	Europe	Asia	Southeast Asia Islands	Philippine Islands	New Guinea	Australian Region	Antarctic Region	Arctic Region	Atlantic Ocean	Indian Ocean	Pacific Ocean
CHIROPTERA PTEROPODIDAE Continued																	
Pteralopex	273										■1						■2
Styloctenium	273								■3								
Dobsonia	273								■E	■C	■4	■Ne					
Aproteles	276										■C						
Harpyionycteris	276								■3	■S							
Plerotes	277					■Sc											
Hypsignathus	278					■W,C											
Epomops	280					■											
Epomophorus	281					■											
Micropteropus	282					■											
Nanonycteris	283					■W											
Scotonycteris	284					■W											
Casinycteris	285					■C											
Cynopterus	286							■S	■	■							
Megaerops	287							■Se	■	■S							
Ptenochirus	288									■							
Dyacopterus	288								■5	■6	■N						
Chironax	289								■5	■							
Thoopterus	290									■7							
Sphaerias	290							■S									
Balionycteris	290								■5	■8							
Aethalops	291								■5	■							
Penthetor	292								■5	■6							
Haplonycteris	292									■							
Otopteropus	294									■N							
Alionycteris	295									■S							
Latidens	296							■Sc									
Nyctimene	296								■E	■C	■4	■Ne					■9
Paranyctimene	298										■						
Eonycteris	298							■Se	■	■							
Megaloglossus	301					■W,C											
Macroglossus	301							■Se	■	■S	■10	■N					
Syconycteris	302								■E		■11	■E					
Melonycteris	303										■10						
Notopteris	304																■Sw
CHIROPTERA RHINOPOMATIDAE																	
Rhinopoma	305					■N		■S	■12								
CHIROPTERA EMBALLONURIDAE																	
Taphozous	307				■	■		■	■	■	■	■				■	
Saccolaimus	309					■		■S	■	■	■13	■N					
Mosia	310								■E		■10						
Emballonura	310				■			■5	■	■	■10						■14
Coleura	312					■		■Sw								■15	
Rhynchonycteris	313	■S		■N													
Centronycteris	314	■S		■N													
Balantiopteryx	315	■S		■16													
Saccopteryx	316	■S		■													
Cormura	317	■S		■N													
Peropteryx	318	■S		■													
Cyttarops	319	■S		■N													

1. Solomon Islands only.   2. Fiji Islands.   3. Sulawesi only.   4. And Bismarck and Louisiade Archipelagoes and Solomon Islands.
5. Malay Peninsula only.   6. Sumatra and Borneo only.   7. Sulawesi and Morotai only.   8. Borneo only.   9. Santa Cruz Islands.
10. And Bismarck Archipelago and Solomon Islands.   11. And Bismarck and Louisiade Archipelagoes.   12. Sumatra only.
13. And Solomon Islands.   14. East to Samoa and Tonga.   15. Seychelles Islands only.   16. Ecuador and probably Colombia only.

Genera of Recent Mammals	page	North America	West Indies	South America	Madagascar	Africa	Europe	Asia	Southeast Asia Islands	Philippine Islands	New Guinea	Australian Region	Antarctic Region	Arctic Region	Atlantic Ocean	Indian Ocean	Pacific Ocean
CHIROPTERA EMBALLONURIDAE Continued																	
Diclidurus	320	■s		■N													
CHIROPTERA CRASEONYCTERIDAE																	
Craseonycteris	320							■1									
CHIROPTERA NYCTERIDAE																	
Nycteris	322				■	■		■s	■								
CHIROPTERA MEGADERMATIDAE																	
Macroderma	325											■					
Cardioderma	326					■E											
Megaderma	326							■s	■	■							
Lavia	327					■											
CHIROPTERA RHINOLOPHIDAE																	
Rhinolophus	328					■	■	■	■	■	■2	■E					
CHIROPTERA HIPPOSIDERIDAE																	
Hipposideros	333				■	■		■s	■	■	■3	■N					■4
Asellia	337					■N		■Sw									
Anthops	338										■5						
Aselliscus	338							■Se	■E		■3						■4
Rhinonycteris	339											■N					
Triaenops	340				■	■C,E		■Sw									
Cloeotis	341					■E,S											
Coelops	342							■Se	■	■s							
Paracoelops	343							■6									
CHIROPTERA MORMOOPIDAE																	
Pteronotus	344	■s	■	■													
Mormoops	346	■s	■	■N													
CHIROPTERA NOCTILIONIDAE																	
Noctilio	347	■s	■	■													
CHIROPTERA PHYLLOSTOMIDAE																	
Desmodus	353	■s		■													
Diaemus	355	■s		■													
Diphylla	356	■s		■													
Macrotus	356	■s	■														
Micronycteris	358	■s		■													
Vampyrum	359	■s		■													
Trachops	360	■s		■													
Chrotopterus	361	■s		■													
Phylloderma	362	■s		■													
Phyllostomus	362	■s		■													
Tonatia	364	■s		■													
Mimon	365	■s		■													
Lonchorhina	366	■s		■N													
Macrophyllum	367	■s		■													
Glossophaga	368	■s	■	■													
Monophyllus	370		■														

1. Thailand only.   2. And Bismarck and Louisiade Archipelagoes.   3. And Bismarck and Louisiade Archipelagoes and Solomon Islands.   4. East to New Hebrides.   5. Solomon Islands only.   6. Viet Nam only.

Genera of Recent Mammals	page	North America	West Indies	South America	Madagascar	Africa	Europe	Asia	Southeast Asia Islands	Philippine Islands	New Guinea	Australian Region	Antarctic Region	Arctic Region	Atlantic Ocean	Indian Ocean	Pacific Ocean
CHIROPTERA PHYLLOSTOMIDAE Continued																	
Leptonycteris	371	■s		■N													
Lonchophylla	372	■s		■													
Lionycteris	372	■1		■													
Anoura	373	■s		■													
Scleronycteris	374			■Nc													
Lichonycteris	375	■s		■													
Hylonycteris	376	■s															
Choeroniscus	377	■s		■N													
Choeronycteris	378	■s															
Musonycteris	378	■2															
Platalina	379			■3													
Brachyphylla	380		■														
Erophylla	381		■														
Phyllonycteris	382		■														
Carollia	384	■s	■4	■													
Rhinophylla	385			■													
Sturnira	385	■s	■	■													
Uroderma	388	■s		■													
Platyrrhinus	389	■s		■													
Vampyrodes	390	■s		■													
Vampyressa	391	■s		■													
Mesophylla	393	■s		■													
Chiroderma	394	■s	■4	■													
Ectophylla	395	■s		■N,5													
Enchisthenes	395	■s		■													
Dermanura	396	■s		■													
Koopmania	397			■N													
Artibeus	397	■s	■	■													
Ardops	400		■4														
Phyllops	400		■6														
Ariteus	401		■7														
Stenoderma	401		■8														
Pygoderma	401			■													
Ametrida	403	■1		■N													
Sphaeronycteris	403			■													
Centurio	404	■s		■N													
CHIROPTERA MYSTACINIDAE																	
Mystacina	405																■9
CHIROPTERA NATALIDAE																	
Natalus	408	■s	■	■N													
CHIROPTERA FURIPTERIDAE																	
Furipterus	410	■s		■N													
Amorphochilus	411			■w													
CHIROPTERA THYROPTERIDAE																	
Thyroptera	412	■s		■													

1. Panama only.   2. Mexico only.   3. Peru only.   4. Lesser Antilles only.   5. Colombia only.   6. Cuba and Hispaniola only.
7. Jamaica only.   8. Puerto Rico and Virgin Islands only.   9. New Zealand.

Genera of Recent Mammals	page	North America	West Indies	South America	Madagascar	Africa	Europe	Asia	Southeast Asia Islands	Philippine Islands	New Guinea	Australian Region	Antarctic Region	Arctic Region	Atlantic Ocean	Indian Ocean	Pacific Ocean
CHIROPTERA MYZOPODIDAE																	
Myzopoda	414				■												
CHIROPTERA VESPERTILIONIDAE																	
Cistugo	418					■S											
Myotis	418	■	■1	■	■2	■	■	■	■	■	■3	■					■4
Lasionycteris	423	■													■5		
Eudiscopus	425							■6									
Pipistrellus	425	■			■	■	■	■			■7	■8			■9		
Scotozous	429							■Sc									
Nyctalus	429						■	■							■10		
Glischropus	431							■Se	■	■S							
Laephotis	431					■											
Philetor	433							■Se	■	■	■11						
Hesperoptenus	433							■S	■								
Chalinolobus	434										■	■12					■13
Glauconycteris	436					■											
Nycticeius	437	■C,E	■14														
Nycticeinops	437					■		■Sw									
Scoteanax	438											■F					
Scotorepens	439								■15		■Se	■					
Scotoecus	440					■		■Sc									
Eptesicus	440	■	■	■		■	■	■									
Ia	442							■9									
Vespertilio	443						■	■									
Histiotus	444			■													
Tylonycteris	445							■S	■	■							
Mimetillus	446					■W,C											
Rhogeessa	446	■S		■													
Scotomanes	448							■Se									
Scotophilus	448				■	■		■S	■	■						■16	
Lasiurus	450	■	■	■17											■18		■19
Otonycteris	453					■N		■Sw									
Barbastella	454					■Nw	■	■							■20		
Plecotus	455					■N	■	■							■9		
Corynorhinus	455	■															
Idionycteris	457	■S															
Euderma	458	■Wc															
Antrozous	459	■W,S	■14														
Bauerus	461	■S															
Miniopterus	461			■21	■	■S	■	■	■	■	■7	■					■22
Murina	463							■	■	■S							
Harpiocephalus	465							■Se	■								
Kerivoula	465					■		■S	■	■	■Ne						
Nyctophilus	468								■23		■	■8					
Pharotis	469										■Se						
Tomopeas	469			■24													
CHIROPTERA MOLOSSIDAE																	
Mormopterus	472		■14	■W	■	■			■25		■	■				■26	
Sauromys	473					■S											
Platymops	474					■Ec											
Molossops	474	■S		■													

1. Lesser Antilles only.   2. And Comoro Islands.   3. And Solomon Islands.   4. East to New Hebrides.   5. Bermuda and Bahamas. 6. Burma and Laos only.   7. And Bismarck and Louisiade Archipelagoes and Solomon Islands.   8. And Tasmania.   9. Canary and Cape Verde Islands.   10. Azores, Madeira, and Canary Islands.   11. And Bismarck Archipelago.   12. And Tasmania.   13. New Caledonia and New Zealand.   14. Cuba only.   15. Timor only.   16. Reunion Island.   17. And Galapagos Islands.   18. Bermuda, Iceland, and Orkney Island.   19. Hawaii only.   20. Canary Islands.   21. And Comoro Islands.   22. East to New Hebrides and Loyalty Islands.   23. Lembata Island only.   24. Peru only.   25. Sumatra only.   26. Mauritius and Reunion.

Genera of Recent Mammals	page	North America	West Indies	South America	Madagascar	Africa	Europe	Asia	Southeast Asia Islands	Philippine Islands	New Guinea	Australian Region	Antarctic Region	Arctic Region	Atlantic Ocean	Indian Ocean	Pacific Ocean
CHIROPTERA MOLOSSIDAE Continued																	
Neoplatymops	475			■N,C													
Cabreramops	476			■1													
Myopterus	476					■W,C											
Tadarida	477	■C,S	■	■	■	■	■	■			■	■		■2			
Chaerephon	479				■	■		■S	■	■	■3	■N				■4	■5
Mops	481				■	■		■S	■		■S						
Otomops	482				■	■		■Sc	■6		■						
Nyctinomops	483	■C,S	■	■													
Eumops	484	■S	■	■													
Promops	485	■S		■													
Molossus	486	■S	■	■													
Cheiromeles	488							■7	■	■							
PRIMATES LORISIDAE																	
Pseudopotto	493					■C											
Arctocebus	495					■Wc											
Loris	496							■Sc									
Nycticebus	497							■Se	■		■S						
Perodicticus	498					■W,C											
Euoticus	500					■Wc											
Galago	500					■											
Otolemur	502					■E,S											
Galagoides	503					■											
PRIMATES CHEIROGALEIDAE																	
Microcebus	505				■												
Mirza	507				■W												
Cheirogaleus	507				■												
Allocebus	509				■E												
Phaner	510				■												
PRIMATES LEMURIDAE																	
Hapalemur	512				■												
Lemur	513				■S												
Varecia	515				■												
Pachylemur†	517				■												
Eulemur	517				■8												
PRIMATES MEGALADAPIDAE																	
Lepilemur	522				■												
Megaladapis†	523				■												
PRIMATES INDRIIDAE																	
Indri	524				■Ne												
Avahi	525				■												
Propithecus	526				■												
PRIMATES PALAEOPROPITHECIDAE																	
Mesopropithecus†	529				■												
Babakotia†	529				■N												
Palaeopropithecus†	530				■												
Archaeoindris†	531				■												

1. Ecuador only.   2. Madeira and Canary Islands.   3. And Solomon Islands.   4. Comoro and Seychelles Islands.   5. East to Fiji.
6. Java and Alor Island only.   7. Malay Peninsula only.   8. And Comoro Islands.

Genera of Recent Mammals	page	North America	West Indies	South America	Madagascar	Africa	Europe	Asia	Southeast Asia Islands	Philippine Islands	New Guinea	Australian Region	Antarctic Region	Arctic Region	Atlantic Ocean	Indian Ocean	Pacific Ocean
PRIMATES ARCHAEOLEMURIDAE																	
Archaeolemur†	531				■												
Hadropithecus†	532				■												
PRIMATES DAUBENTONIIDAE																	
Daubentonia	532				■												
PRIMATES TARSIIDAE																	
Tarsius	534								■	■S							
PRIMATES CEBIDAE																	
Lagothrix	538			■Nw													
Ateles	540	■S		■N,C													
Brachyteles	541			■E													
Alouatta	543	■S		■													
Pithecia	545			■N,C													
Chiropotes	546			■C													
Cacajao	547			■Nc													
Xenothrix†	548		■1														
Paralouatta†	549		■2														
Antillothrix†	549		■3														
Callicebus	549			■													
Aotus	551	■4		■													
Cebus	553	■S		■													
Saimiri	555	■S		■													
PRIMATES CALLITRICHIDAE																	
Callimico	559			■Wc													
Saguinus	560	■S		■N,C													
Callithrix	564			■C													
Cebuella	566			■Wc													
Leontopithecus	567			■E													
PRIMATES CERCOPITHECIDAE																	
Erythrocebus	570					■											
Chlorocebus	571					■											
Cercopithecus	573					■											
Miopithecus	575					■Wc											
Allenopithecus	576					■C											
Cercocebus	577					■											
Lophocebus	579					■W,C											
Macaca	580					■Nw		■S	■	■							
Papio	588					■		■Sw									
Mandrillus	591					■Wc											
Theropithecus	593					■5											
Nasalis	595								■6								
Simias	596								■7								
Pygathrix	597							■Se									
Rhinopithecus	597							■E									
Presbytis	599							■8	■								
Semnopithecus	600							■Sc									
Trachypithecus	602							■S	■								
Colobus	604					■											

1. Jamaica only. 2. Cuba only. 3. Hispaniola only. 4. Panama only. 5. Ethiopia only. 6. Borneo only. 7. Mentawai Islands only. 8. Malay Peninsula only.

Genera of Recent Mammals	page	North America	West Indies	South America	Madagascar	Africa	Europe	Asia	Southeast Asia Islands	Philippine Islands	New Guinea	Australian Region	Antarctic Region	Arctic Region	Atlantic Ocean	Indian Ocean	Pacific Ocean
PRIMATES CERCOPITHECIDAE Continued																	
Piliocolobus	606					■											
Procolobus	607					■W											
PRIMATES HYLOBATIDAE																	
Hylobates	608							■Se	■								
PRIMATES PONGIDAE																	
Pongo	615								■1								
Gorilla	618					■C											
Pan	622					■W,C											
PRIMATES HOMINIDAE																	
Homo	626	■	■	■	■	■	■	■	■	■	■	■					
CARNIVORA CANIDAE																	
Vulpes	636	■				■	■	■									
Fennecus	643					■N		■Sw									
Alopex	644	■N					■N	■N						■			
Urocyon	647	■		■N													
Lycalopex	648			■C													
Pseudalopex	648			■													
Dusicyon	650														■2		
Cerdocyon	651			■													
Nyctereutes	652							■E									
Atelocynus	653			■N													
Speothos	654	■3		■													
Canis	655	■				■	■	■									
Chrysocyon	672			■													
Otocyon	673					■E,S											
Cuon	674							■	■4								
Lycaon	676					■											
CARNIVORA URSIDAE																	
Tremarctos	680			■W													
Ursus	681	■				■Nw	■	■	■					■			
Ailuropoda	693							■Ec									
CARNIVORA PROCYONIDAE																	
Ailurus	695							■Ec									
Bassariscus	696	■W															
Procyon	698	■	■5	■													
Nasua	700	■S		■													
Nasuella	701			■Nw													
Potos	702	■S		■N													
Bassaricyon	703	■S		■N													
CARNIVORA MUSTELIDAE																	
Mustela	705	■		■N		■N	■	■	■								
Vormela	714						■Se	■W,C									
Martes	716	■					■	■	■								
Eira	720	■S		■													
Galictis	721	■S		■													

1. Sumatra and Borneo only.   2. Falkland Islands.   3. Panama only.   4. Sumatra and Java only.   5. Lesser Antilles and New Providence Island only.

Genera of Recent Mammals	page	North America	West Indies	South America	Madagascar	Africa	Europe	Asia	Southeast Asia Islands	Philippine Islands	New Guinea	Australian Region	Antarctic Region	Arctic Region	Atlantic Ocean	Indian Ocean	Pacific Ocean
CARNIVORA MUSTELIDAE Continued																	
Lyncodon	722			■S													
Ictonyx	723					■											
Poecilictis	724					■N											
Poecilogale	724					■C,S											
Gulo	726	■					■N	■N									
Mellivora	727					■		■S									
Meles	728						■	■									
Arctonyx	730							■C	■1								
Mydaus	730								■	■W							
Taxidea	731	■W,C															
Melogale	732							■Se	■2								
Spilogale	733	■															
Mephitis	735	■															
Conepatus	736	■		■													
Lutra	737					■	■	■	■								
Lutrogale	739							■S	■								
Lontra	740	■		■													
Pteronura	742			■													
Aonyx	743					■		■S	■	■W							
Enhydra	745	■W						■Ne									■N
CARNIVORA VIVERRIDAE																	
Viverra	749							■S	■	■							
Civettictis	750					■											
Viverricula	751							■S	■								
Genetta	751					■	■Sw	■Sw									
Osbornictis	754					■3											
Poiana	755					■W,C											
Prionodon	756							■Se	■								
Nandinia	756					■											
Arctogalidia	757							■Se	■								
Paradoxurus	758							■S	■	■							
Paguma	759							■S	■								
Macrogalidia	760								■4								
Arctictis	761							■Se	■	■W							
Hemigalus	762							■Se	■5								
Diplogale	763								■6								
Chrotogale	763							■Se									
Cynogale	764							■Se	■5								
Fossa	765				■												
Eupleres	766				■												
CARNIVORA HERPESTIDAE																	
Galidia	768				■												
Galidictis	769				■												
Mungotictis	769				■W												
Salanoia	771				■Ne												
Herpestes	771					■	■Sw	■S	■	■W							
Galerella	773					■											
Mungos	774					■											
Crossarchus	775					■W,C											
Liberiictis	776					■7											

1. Sumatra only.   2. Java, Bali, and Borneo only.   3. Zaire only.   4. Sulawesi only.   5. Sumatra and Borneo only.   6. Borneo only.
7. Liberia and probably Ivory Coast only.

Genera of Recent Mammals	page	North America	West Indies	South America	Madagascar	Africa	Europe	Asia	Southeast Asia Islands	Philippine Islands	New Guinea	Australian Region	Antarctic Region	Arctic Region	Atlantic Ocean	Indian Ocean	Pacific Ocean
CARNIVORA HERPESTIDAE Continued																	
Helogale	777					■E,S											
Dologale	778					■C											
Bdeogale	779					■		■1									
Rhynchogale	780					■E,S											
Ichneumia	780					■		■Sw									
Atilax	781					■											
Cynictis	782					■S											
Paracynictis	783					■S											
Suricata	783					■S											
Cryptoprocta	785				■												
CARNIVORA HYAENIDAE																	
Proteles	786					■E,S											
Parahyaena	787					■S											
Hyaena	790					■		■S									
Crocuta	791					■											
CARNIVORA FELIDAE																	
Felis	796	■		■		■	■	■	■	■							
Neofelis	820							■Se	■2								
Panthera	821	■S		■		■	■3	■	■								
Acinonyx	834					■		■S									
PINNIPEDIA OTARIIDAE																	
Callorhinus	840	■W						■Ne									■N
Arctocephalus	844	■Wc		■4		■S						■S	■		■S	■S	■S
Zalophus	852	■Wc		■5				■6									
Phocarctos	855																■7
Neophoca	857											■S					
Otaria	859			■8													
Eumetopias	860	■W						■Ne									
PINNIPEDIA ODOBENIDAE																	
Odobenus	862													■	■N		■N
PINNIPEDIA PHOCIDAE																	
Monachus	869	■Se	■			■N	■S	■Sw									■9
Lobodon	871			■S		■S						■S	■				■7
Hydrurga	873			■S		■S						■S	■		■S	■S	■S
Leptonychotes	874			■S		■							■		■S	■S	■S
Ommatophoca	875												■				
Mirounga	877	■W		■S									■		■S	■S	■
Erignathus	881	■N					■N	■N						■	■N		■N
Cystophora	882	■N					■N							■	■N		
Halichoerus	884	■Ne					■N								■N		
Phoca	886	■10					■10	■11						■	■N		■N
CETACEA PLATANISTIDAE																	
Platanista	900							■Sc									
CETACEA LIPOTIDAE																	
Lipotes	902							■E									

1. Yemen only.   2. Sumatra and Borneo only.   3. No longer present.   4. And Galapagos and Falkland Islands.   5. Galapagos Islands only.   6. Japan and Korea only.   7. New Zealand region.   8. And Falkland Islands.   9. Hawaii only.   10. Coastal waters. 11. Northern and eastern coastal waters.

Genera of Recent Mammals	page	North America	West Indies	South America	Madagascar	Africa	Europe	Asia	Southeast Asia Islands	Philippine Islands	New Guinea	Australian Region	Antarctic Region	Arctic Region	Atlantic Ocean	Indian Ocean	Pacific Ocean
CETACEA PONTOPORIIDAE																	
Pontoporia	903			■Se													
CETACEA INIIDAE																	
Inia	904			■N,C													
CETACEA MONODONTIDAE																	
Delphinapterus	907													■	■N		■N
Monodon	909													■			
CETACEA PHOCOENIDAE																	
Phocoena	911	■1		■1			■1	■I									
Neophocaena	913							■S,E	■								
Australophocaena	914			■1											■2	■3	■4
Phocoenoides	915																■N
CETACEA DELPHINIDAE																	
Steno	917														■	■	■
Sousa	920					■		■S	■			■					
Sotalia	920			■E													
Lagenorhynchus	923														■	■	■
Grampus	925														■	■	■
Tursiops	925														■	■	■
Stenella	929														■	■	■
Delphinus	932														■	■	■
Lagenodelphis	933														■	■	■
Lissodelphis	935														■S	■S	■
Orcaella	936							■Se	■		■	■N					
Cephalorhynchus	937			■S		■Sw									■2	■3	■4
Peponocephala	939														■	■	■
Feresa	940														■	■	■
Pseudorca	941														■	■	■
Orcinus	942												■	■	■	■	■
Globicephala	944														■	■	■
CETACEA ZIPHIIDAE																	
Berardius	946												■		■S	■S	■
Ziphius	947														■	■	■
Tasmacetus	948											■			■Sw		■S
Hyperoodon	949														■	■	■
Indopacetus	951															■	■
Mesoplodon	951														■	■	■
CETACEA PHYSETERIDAE																	
Kogia	953														■	■	■
Physeter	954												■	■	■	■	■
CETACEA ESCHRICHTIIDAE																	
Eschrichtius	959													■	■5		■N
CETACEA NEOBALAENIDAE																	
Caperea	962														■S	■S	■S

1. Coastal waters.   2. Falkland Islands and South Georgia.   3. Kerguelen Islands.   4. New Zealand region.   5. No longer present.

Genera of Recent Mammals	page	North America	West Indies	South America	Madagascar	Africa	Europe	Asia	Southeast Asia Islands	Philippine Islands	New Guinea	Australian Region	Antarctic Region	Arctic Region	Atlantic Ocean	Indian Ocean	Pacific Ocean
CETACEA BALAENIDAE																	
Eubalaena	964														■	■	■
Balaena	967													■	■N		■N
CETACEA BALAENOPTERIDAE																	
Balaenoptera	972												■	■	■	■	■
Megaptera	979												■	■	■	■	■
SIRENIA DUGONGIDAE																	
Dugong	983					■E		■S	■	■	■					■	■1
Hydrodamalis†	986							■2									
SIRENIA TRICHECHIDAE																	
Trichechus	987	■Se	■	■		■W											
PROBOSCIDEA ELEPHANTIDAE																	
Elephas	994							■S	■S								
Loxodonta	998					■											
Mammuthus†	1004							■3									
PERISSODACTYLA EQUIDAE																	
Equus	1008					■	■E	■									
PERISSODACTYLA TAPIRIDAE																	
Tapirus	1025	■S		■				■Se	■4								
PERISSODACTYLA RHINOCEROTIDAE																	
Dicerorhinus	1030							■Se	■5								
Rhinoceros	1032							■S	■6								
Diceros	1034					■											
Ceratotherium	1037					■											
HYRACOIDEA PROCAVIIDAE																	
Procavia	1043					■	■Sw										
Heterohyrax	1044					■											
Dendrohyrax	1046					■											
TUBULIDENTATA ORYCTEROPODIDAE																	
Orycteropus	1048					■											
BIBYMALAGASIA PLESIORYCTEROPODIDAE																	
Plesiorycteropus†	1050				■												
ARTIODACTYLA SUIDAE																	
Sus	1054					■N	■	■	■	■							
Potamochoerus	1058				■7	■											
Hylochoerus	1059					■W,C											
Phacochoerus	1060					■											
Babyrousa	1062								■8								
ARTIODACTYLA TAYASSUIDAE																	
Catagonus	1064			■Sc													
Pecari	1065	■S		■													

1. East to Carolines and New Hebrides.   2. Commander Islands only.   3. Wrangel Island only in historical times.   4. Sumatra only.
5. Sumatra and Borneo only.   6. Sumatra and Java only.   7. Probably introduced here and on Mayotte Island.   8. Sulawesi and nearby islands only.

Genera of Recent Mammals	page	North America	West Indies	South America	Madagascar	Africa	Europe	Asia	Southeast Asia Islands	Philippine Islands	New Guinea	Australian Region	Antarctic Region	Arctic Region	Atlantic Ocean	Indian Ocean	Pacific Ocean
ARTIODACTYLA TAYASSUIDAE Continued																	
Tayassu	1066	■s		■													
ARTIODACTYLA HIPPOPOTAMIDAE																	
Hippopotamus	1068				■1	■		■2									
Hexaprotodon	1071					■w											
ARTIODACTYLA CAMELIDAE																	
Lama	1075			■w,s													
Vicugna	1077			■Sw													
Camelus	1078							■C									
ARTIODACTYLA TRAGULIDAE																	
Hyemoschus	1081					■w,c											
Moschiola	1082							■Sc									
Tragulus	1082							■Se	■	■Sw							
ARTIODACTYLA GIRAFFIDAE																	
Okapia	1085					■3											
Giraffa	1086					■											
ARTIODACTYLA MOSCHIDAE																	
Moschus	1089							■E									
ARTIODACTYLA CERVIDAE																	
Hydropotes	1093							■Ec									
Elaphodus	1094							■Ec									
Muntiacus	1094							■Se	■								
Megamuntiacus	1096							■4									
Dama	1098					■N	■S	■Sw									
Axis	1100							■S	■5	■W							
Cervus	1102	■				■Nw	■	■	■	■							
Elaphurus	1113							■Ec									
Odocoileus	1114	■		■N													
Blastocerus	1118			■Sc													
Ozotoceros	1119			■													
Hippocamelus	1122			■w													
Mazama	1123	■s		■													
Pudu	1124			■w													
Alces	1126	■N					■N	■N									
Rangifer	1128	■N					■N	■N									
Capreolus	1131						■	■									
ARTIODACTYLA ANTILOCAPRIDAE																	
Antilocapra	1132	■Wc															
ARTIODACTYLA BOVIDAE																	
Tragelaphus	1135					■		■Sw									
Taurotragus	1143					■											
Boselaphus	1145							■Sc									
Tetracerus	1146							■Sc									
Bubalus	1147							■S	■	■W							
Syncerus	1151					■											

1. Until early historical times.   2. In Palestine in early historical times.   3. Zaire and possibly Uganda only.   4. Laos, Viet Nam, and possibly Cambodia only.   5. Bawean Island only.

Genera of Recent Mammals	page	North America	West Indies	South America	Madagascar	Africa	Europe	Asia	Southeast Asia Islands	Philippine Islands	New Guinea	Australian Region	Antarctic Region	Arctic Region	Atlantic Ocean	Indian Ocean	Pacific Ocean
ARTIODACTYLA BOVIDAE Continued																	
Bos	1153					■N1	■1	■C,S	■2								
Bison	1161	■						■									
Cephalophus	1164					■											
Sylvicapra	1165					■											
Kobus	1166					■											
Redunca	1170					■											
Pelea	1172					■S											
Hippotragus	1174					■											
Oryx	1175					■		■Sw									
Addax	1178					■N											
Damaliscus	1179					■											
Alcelaphus	1181					■											
Sigmoceros	1183					■E,S											
Connochaetes	1184					■E,S											
Oreotragus	1186					■											
Ourebia	1187					■											
Raphicerus	1188					■E,S											
Neotragus	1189					■											
Madoqua	1191					■E,S											
Dorcatragus	1193					■Ec											
Antilope	1193							■Sc									
Aepyceros	1194					■E,S											
Ammodorcas	1196					■Ec											
Litocranius	1197					■E											
Gazella	1199					■N,E		■C,S									
Antidorcas	1202					■S											
Procapra	1203							■Ec									
Pseudonovibos	1204							■3									
Pantholops	1205							■C									
Saiga	1206						■E	■C									
Pseudoryx	1207							■4									
Capricornis	1209							■E	■5								
Naemorhedus	1210							■Ec									
Oreamnos	1212	■Nw															
Rupicapra	1213						■	■Sw									
Myotragus†	1214						■6										
Budorcas	1215							■Ec									
Ovibos	1215	■N															
Hemitragus	1219							■S									
Capra	1220					■Ne	■S	■									
Pseudois	1228							■Ec									
Ammotragus	1229					■N											
Ovis	1231	■w					■7	■									
PHOLIDOTA MANIDAE																	
Manis	1239					■		■S	■	■w							
RODENTIA APLODONTIDAE																	
Aplodontia	1245	■w															
RODENTIA SCIURIDAE																	
Tamias	1247	■					■N	■N,E									

1. No longer present.   2. Java and Borneo only.   3. Cambodia and Viet Nam only.   4. Laos and Viet Nam only.   5. Sumatra only.
6. Balearic Islands only.   7. Corsica, Sardinia, and Cyprus (probably through introduction).

Genera of Recent Mammals	page	North America	West Indies	South America	Madagascar	Africa	Europe	Asia	Southeast Asia Islands	Philippine Islands	New Guinea	Australian Region	Antarctic Region	Arctic Region	Atlantic Ocean	Indian Ocean	Pacific Ocean
RODENTIA SCIURIDAE Continued																	
Marmota	1251	■					■	■N,C									
Ammospermophilus	1253	■W															
Spermophilus	1254	■					■E	■N,C									
Cynomys	1258	■Wc															
Sciurotamias	1260							■E									
Atlantoxerus	1261					■Nw											
Xerus	1262					■											
Spermophilopsis	1263							■Wc									
Sciurus	1265	■		■			■	■									
Syntheosciurus	1268	■S															
Microsciurus	1268	■S		■Nw													
Sciurillus	1269			■N													
Prosciurillus	1270								■1								
Rheithrosciurus	1271								■2								
Tamiasciurus	1271	■N,W															
Funambulus	1274							■Sc									
Ratufa	1274							■S	■								
Protoxerus	1275					■W,C											
Epixerus	1276					■W											
Funisciurus	1277					■											
Paraxerus	1279					■											
Heliosciurus	1281					■											
Hyosciurus	1282								■3								
Myosciurus	1282					■W											
Callosciurus	1284							■Se	■								
Rubrisciurus	1285								■3								
Sundasciurus	1286							■4	■	■							
Tamiops	1287							■Se									
Menetes	1287							■Se									
Rhinosciurus	1288							■4	■								
Lariscus	1289							■4	■								
Dremomys	1289							■Se	■2								
Glyphotes	1290								■2								
Nannosciurus	1290								■								
Exilisciurus	1291								■2	■S							
Petaurista	1291							■E,S	■								
Biswamoyopterus	1295							■5									
Aeromys	1295							■4	■6								
Eupetaurus	1296							■Sc									
Pteromys	1297						■N	■N,E									
Glaucomys	1297	■															
Eoglaucomys	1300							■7									
Hylopetes	1301							■S	■	■S							
Petinomys	1302							■S	■	■S							
Aeretes	1303							■Ec									
Trogopterus	1303							■Ec									
Belomys	1304							■Se									
Pteromyscus	1304							■4	■6								
Petaurillus	1305							■4	■2								
Iomys	1305							■	■								

1. Sulawesi and Sanghir Islands only.   2. Borneo only.   3. Sulawesi only.   4. Malay Peninsula only.   5. Northeastern India only.
6. Sumatra and Borneo only.   7. Kashmir and adjacent areas only.

Genera of Recent Mammals	page	North America	West Indies	South America	Madagascar	Africa	Europe	Asia	Southeast Asia Islands	Philippine Islands	New Guinea	Australian Region	Antarctic Region	Arctic Region	Atlantic Ocean	Indian Ocean	Pacific Ocean
RODENTIA CASTORIDAE																	
Castor	1306	■					■	■									
RODENTIA GEOMYIDAE																	
Thomomys	1311	■W															
Geomys	1312	■															
Orthogeomys	1314	■S		■Nw													
Zygogeomys	1315	■1															
Pappogeomys	1316	■S															
Cratogeomys	1316	■C,S															
RODENTIA HETEROMYIDAE																	
Heteromys	1319	■S		■N													
Liomys	1321	■S															
Perognathus	1322	■W,C															
Chaetodipus	1323	■W,C															
Microdipodops	1324	■Wc															
Dipodomys	1325	■W															
RODENTIA DIPODIDAE																	
Sicista	1329						■	■N,C									
Zapus	1331	■															
Eozapus	1332							■Ec									
Napaeozapus	1333	■E															
Cardiocranius	1333							■Ec									
Salpingotus	1334							■C									
Salpingotulus	1335							■Sc									
Paradipus	1335							■Wc									
Euchoreutes	1336							■Ec									
Dipus	1337							■C									
Eremodipus	1338							■C									
Jaculus	1339					■N		■W,C									
Stylodipus	1340						■Se	■C									
Allactaga	1341					■Ne	■E	■W,C									
Allactodipus	1342							■C									
Alactagulus	1343						■Se	■C									
Pygeretmus	1343							■C									
RODENTIA MURIDAE																	
Nyctomys	1346	■S															
Otonyctomys	1347	■S															
Tylomys	1348	■S		■Nw													
Ototylomys	1348	■S															
Nelsonia	1350	■1															
Neotoma	1350	■															
Hodomys	1353	■1															
Xenomys	1353	■1															
Ochrotomys	1354	■Se															
Scotinomys	1355	■S															
Baiomys	1355	■S															
Megadontomys	1356	■1															

1. Mexico only.

Genera of Recent Mammals	page	North America	West Indies	South America	Madagascar	Africa	Europe	Asia	Southeast Asia Islands	Philippine Islands	New Guinea	Australian Region	Antarctic Region	Arctic Region	Atlantic Ocean	Indian Ocean	Pacific Ocean
RODENTIA MURIDAE Continued																	
Isthmomys	1357	■1		■2													
Osgoodomys	1357	■3															
Onychomys	1357	■C,W															
Neotomodon	1358	■3															
Podomys	1359	■4															
Habromys	1360	■S															
Peromyscus	1360	■															
Reithrodontomys	1364	■C,S		■Nw													
Oryzomys	1366	■S	■5	■6													
Oligoryzomys	1368	■S	■7	■													
Microryzomys	1369			■W													
Melanomys	1369	■S		■Nw													
Sigmodontomys	1370	■S		■Nw													
Oecomys	1370	■S		■													
Nesoryzomys	1371		■8														
Megalomys†	1371		■9														
Neacomys	1372	■1		■N,C													
Scolomys	1373			■10													
Nectomys	1374			■N,C													
Amphinectomys	1375			■11													
Pseudoryzomys	1375			■S													
Zygodontomys	1376	■S		■N													
Lundomys	1376			■S													
Holochilus	1377			■													
Chilomys	1378			■Nw													
Rhagomys	1379			■12													
Microakodontomys	1379			■12													
Abrawayaomys	1380			■12													
Rhipidomys	1380	■1		■													
Thomasomys	1382			■													
Delomys	1383			■12													
Wifredomys	1384			■S													
Aepeomys	1384			■Nw													
Megaoryzomys†	1384			■8													
Phaenomys	1384			■12													
Wiedomys	1385			■Ec													
Akodon	1386			■													
Microxus	1387			■Nw													
Bolomys	1388			■													
Thalpomys	1390			■12													
Abrothrix	1390			■13													
Chroeomys	1391			■S													
Podoxymys	1391			■Ne													
Lenoxus	1392			■Wc													
Juscelinomys	1392			■12													
Oxymycterus	1393			■C,S													
Blarinomys	1394			■12													
Notiomys	1395			■14													
Pearsonomys	1395			■15													
Geoxus	1396			■S													
Chelemys	1397			■S													
Scapteromys	1398			■S													

1. Panama only.   2. Colombia only.   3. Mexico only.   4. Florida only.   5. Jamaica only.   6. And Galapagos Islands.
7. St. Vincent Island only.   8. Galapagos Islands only.   9. Lesser Antilles only.   10. Ecuador and Peru only.   11. Northeastern Peru only.   12. Southeastern Brazil only.   13. Chile, Argentina, and Tierra del Fuego only.   14. Southern Argentina only.
15. Southern Chile only

Genera of Recent Mammals	page	North America	West Indies	South America	Madagascar	Africa	Europe	Asia	Southeast Asia Islands	Philippine Islands	New Guinea	Australian Region	Antarctic Region	Arctic Region	Atlantic Ocean	Indian Ocean	Pacific Ocean
RODENTIA MURIDAE Continued																	
Kunsia	1398			■Sc													
Bibimys	1399			■Se													
Punomys	1400			■1													
Calomys	1401			■													
Salinomys	1403			■2													
Graomys	1403			■C,S													
Andalgalomys	1404			■Sc													
Eligmodontia	1404			■Sw													
Paralomys	1405			■1													
Phyllotis	1406			■W													
Auliscomys	1406			■W													
Galenomys	1407			■Wc													
Maresomys	1407			■Wc													
Loxodontomys	1408			■S													
Euneomys	1409			■S													
Reithrodon	1409			■S													
Neotomys	1410			■Wc													
Chinchillula	1411			■Wc													
Andinomys	1411			■Wc													
Irenomys	1411			■Sw													
Sigmodon	1412	■S		■N													
Neusticomys	1414			■N													
Chibchanomys	1414			■Nw													
Anotomys	1415			■3													
Ichthyomys	1417			■Nw													
Rheomys	1417	■S															
Calomyscus	1418							■Sw									
Phodopus	1419							■Ec									
Cricetus	1419						■	■C									
Allocricetulus	1421							■Ec									
Cricetulus	1421							■Se ■C,E									
Tscherskia	1422							■E									
Cansumys	1423							■4									
Mesocricetus	1423						■Se	■Sw									
Mystromys	1425					■S											
Spalax	1426						■Sw										
Nannospalax	1427					■Ne	■Se	■Sw									
Myospalax	1428							■Ec									
Lophiomys	1430					■Ec											
Platacanthomys	1431							■Sc									
Typhlomys	1432							■Se									
Monticolomys	1432				■C												
Macrotarsomys	1433				■W												
Nesomys	1434				■												
Brachytarsomys	1434				■E												
Eliurus	1435				■												
Gymnuromys	1436				■E												
Hypogeomys	1437				■W												
Brachyuromys	1437				■E												
Parotomys	1437					■S											
Otomys	1439					■											
Rhizomys	1441							■Se	■5								

1. Peru only.   2. Northwestern Argentina only.   3. Ecuador only.   4. Northeastern China only.   5. Sumatra only.

Genera of Recent Mammals	page	North America	West Indies	South America	Madagascar	Africa	Europe	Asia	Southeast Asia Islands	Philippine Islands	New Guinea	Australian Region	Antarctic Region	Arctic Region	Atlantic Ocean	Indian Ocean	Pacific Ocean
RODENTIA MURIDAE Continued																	
Cannomys	1442							■Se									
Tachyoryctes	1443					■Ec											
Gerbillus	1446					■N,E		■Sw									
Microdillus	1448					■1											
Gerbillurus	1448					■S											
Tatera	1449					■		■S									
Taterillus	1451					■											
Desmodillus	1451					■S											
Desmodilliscus	1452					■W,C											
Pachyuromys	1453					■N											
Ammodillus	1453					■Ec											
Sekeetamys	1454					■Ne		■Sw									
Meriones	1455					■N	■Se	■									
Brachiones	1457							■Ec									
Psammomys	1458					■N		■Sw									
Rhombomys	1458							■C									
Clethrionomys	1459	■N,W					■	■									
Phaulomys	1461							■2									
Eothenomys	1462							■E									
Alticola	1462							■C,Ne									
Hyperacrius	1463							■Sc									
Dinaromys	1464						■Se										
Arborimus	1465	■W															
Phenacomys	1466	■N,W															
Arvicola	1467						■	■N,W									
Volemys	1468							■E									
Proedromys	1468							■3									
Lasiopodomys	1468							■Ec									
Chionomys	1469						■	■Sw									
Blanfordimys	1469							■C									
Microtus	1470	■				■4	■	■									
Tyrrhenicola†	1474						■5										
Lemmiscus	1475	■Wc															
Lagurus	1475						■Se	■C									
Eolagurus	1476							■Ec									
Neofiber	1477	■Se															
Ondatra	1477	■															
Dicrostonyx	1479	■N					■Ne	■N									
Lemmus	1481	■N					■N	■N									
Myopus	1482						■N	■N									
Synaptomys	1483	■N															
Prometheomys	1484							■6									
Ellobius	1485						■Se	■									
Dendromus	1485					■											
Megadendromus	1488					■7											
Malacothrix	1488					■S											
Dendroprionomys	1488					■8											
Prionomys	1489					■C											
Steatomys	1489					■											
Deomys	1490					■C											
Leimacomys	1491					■9											
Petromyscus	1492					■S											

1. Somalia only.   2. Japan only.   3. Central China only.   4. Libya only.   5. Corsica and Sardinia only.   6. Caucasus region only.
7. Ethiopia only.   8. Congo only.   9. Togo only.

Genera of Recent Mammals	page	North America	West Indies	South America	Madagascar	Africa	Europe	Asia	Southeast Asia Islands	Philippine Islands	New Guinea	Australian Region	Antarctic Region	Arctic Region	Atlantic Ocean	Indian Ocean	Pacific Ocean
RODENTIA MURIDAE Continued																	
Delanymys	1492					■C											
Beamys	1493					■E											
Saccostomus	1494					■E,S											
Cricetomys	1495					■											
Lenothrix	1496							■1	■2								
Pithecheirops	1496								■3								
Pithecheir	1497							■1	■4								
Hapalomys	1498							■Se									
Apodemus	1499					■Nw	■	■									
Rhagamys†	1501					■5											
Tokudaia	1501							■6									
Chiropodomys	1502							■Se	■		■W						
Vernaya	1503							■Ec									
Vandeleuria	1503							■S									
Micromys	1504						■	■									
Haeromys	1505								■7	■8							
Anonymomys	1506									■C							
Sundamys	1506							■1	■		■W						
Kadarsanomys	1507								■4								
Diplothrix	1507							■6									
Margaretamys	1508								■9								
Lenomys	1508								■9								
Eropeplus	1510								■9								
Papagomys	1510								■10								
Komodomys	1510								■11								
Melasmothrix	1511								■9								
Tateomys	1512								■9								
Palawanomys	1514									■W							
Echiothrix	1514								■9								
Paulamys	1515								■10								
Bunomys	1516								■9								
Rattus	1517					■		■	■	■	■12	■13					
Nesoromys	1522								■14								
Stenomys	1522										■						
Taeromys	1523								■9								
Paruromys	1523								■9								
Abditomys	1524									■N							
Tryphomys	1525									■N							
Limnomys	1526									■S							
Tarsomys	1527									■S							
Bullimus	1528									■							
Apomys	1528									■							
Crateromys	1530									■							
Batomys	1531									■							
Carpomys	1533									■N							
Phloeomys	1533									■							
Crunomys	1535									■							
Archboldomys	1536								■9	■N							
Rhynchomys	1536									■N							
Chrotomys	1538									■							
Celaenomys	1538									■N							
Chiruromys	1539										■						

1. Malay Peninsula only.   2. Sarawak and Tuangku only.   3. Sabah only.   4. Java only.   5. Corsica and Sardinia only.   6. Ryukyu Islands only.   7. Borneo and Sulawesi only.   8. Palawan only.   9. Sulawesi only.   10. Flores only.   11. Lesser Sunda Islands only.   12. And Solomon Islands.   13. And Tasmania.   14. Seram only.

Genera of Recent Mammals	page	North America	West Indies	South America	Madagascar	Africa	Europe	Asia	Southeast Asia Islands	Philippine Islands	New Guinea	Australian Region	Antarctic Region	Arctic Region	Atlantic Ocean	Indian Ocean	Pacific Ocean
RODENTIA MURIDAE Continued																	
Pogonomys	1540										■						
Mallomys	1541										■						
Hyomys	1542										■						
Anisomys	1544										■						
Uromys	1545										■1	■Ne					
Solomys	1547										■2						
Melomys	1548								■3		■1	■N,E					
Coccymys	1550										■						
Pogonomelomys	1550										■						
Xenuromys	1551										■						
Spelaeomys†	1552								■4								
Coryphomys†	1552								■5								
Lorentzimys	1552										■						
Macruromys	1553										■						
Mesembriomys	1554											■N					
Conilurus	1554										■C	■					
Leporillus	1556											■S					
Zyzomys	1556											■N					
Pseudomys	1558											■6					
Mastacomys	1561											■7					
Leggadina	1562											■					
Notomys	1563											■					
Golunda	1564							■Sc									
Hadromys	1565							■Sc									
Bandicota	1566							■S	■								
Nesokia	1567					■Ne		■									
Erythronesokia	1568							■8									
Millardia	1568							■Sc									
Cremnomys	1569							■Sc									
Srilankamys	1570							■9									
Dacnomys	1570							■S									
Diomys	1570							■Sc									
Chiromyscus	1571							■Se									
Niviventer	1571							■E	■								
Maxomys	1573							■Se		■W							
Leopoldamys	1574							■Se	■								
Berylmys	1575							■Se	■10								
Mastomys	1576					■											
Myomys	1577					■		■Sw									
Praomys	1578					■											
Hylomyscus	1579					■											
Heimyscus	1579					■Wc											
Malacomys	1580					■											
Thallomys	1581					■E,S											
Thamnomys	1582					■C											
Grammomys	1583					■											
Stenocephalemys	1584					■11											
Oenomys	1585					■											
Lamottemys	1586					■12											
Lophuromys	1586					■											
Zelotomys	1588					■											
Colomys	1589					■W,C											

1. And Bismarck Archipelago and Solomon Islands.   2. Solomon Islands only.   3. Seram and Talaud Islands only.   4. Flores only.
5. Timor only.   6. And Tasmania.   7. Southeastern corner and Tasmania.   8. Iraq only.   9. Sri Lanka only.   10. Sumatra only.
11. Ethiopia only.   12. Cameroon only.

Genera of Recent Mammals	page	North America	West Indies	South America	Madagascar	Africa	Europe	Asia	Southeast Asia Islands	Philippine Islands	New Guinea	Australian Region	Antarctic Region	Arctic Region	Atlantic Ocean	Indian Ocean	Pacific Ocean
RODENTIA MURIDAE Continued																	
Nilopegamys	1589					■1											
Acomys	1590					■		■Sw									
Uranomys	1592					■											
Malpaisomys†	1593														■2		
Hybomys	1593					■W,C											
Typomys	1594					■W											
Dephomys	1595					■W											
Rhabdomys	1595					■E,S											
Lemniscomys	1596					■											
Pelomys	1597					■E,S											
Desmomys	1598					■1											
Mylomys	1599					■W,C											
Stochomys	1599					■W,C											
Aethomys	1600					■											
Canariomys†	1601														■2		
Dasymys	1602					■											
Arvicanthis	1603					■		■Sw									
Mus	1604					■	■	■	■								
Muriculus	1608					■1											
Leptomys	1608										■						
Paraleptomys	1609										■						
Xeromys	1609											■Ne					
Hydromys	1611										■3	■4					
Pseudohydromys	1612										■						
Microhydromys	1613										■						
Neohydromys	1613										■						
Parahydromys	1613										■						
Crossomys	1614										■						
Mayermys	1616										■E						
RODENTIA ANOMALURIDAE																	
Idiurus	1619					■W,C											
Zenkerella	1619					■Wc											
Anomalurus	1619					■											
RODENTIA PEDETIDAE																	
Pedetes	1620					■E,S											
RODENTIA CTENODACTYLIDAE																	
Pectinator	1622					■Ec											
Felovia	1623					■Nw											
Massoutiera	1623					■N											
Ctenodactylus	1624					■N											
RODENTIA MYOXIDAE																	
Graphiurus	1626					■											
Glirulus	1627							■5									
Muscardinus	1628						■	■W									
Myoxus	1629						■	■W									
Dryomys	1631						■	■C									
Chaetocauda	1632							■Ec									
Eliomys	1632					■N	■	■Sw									

1. Ethiopia only.   2. Canary Islands only.   3. And Bismarck Archipelago.   4. And Tasmania.   5. Japan only.

Genera of Recent Mammals	page	North America	West Indies	South America	Madagascar	Africa	Europe	Asia	Southeast Asia Islands	Philippine Islands	New Guinea	Australian Region	Antarctic Region	Arctic Region	Atlantic Ocean	Indian Ocean	Pacific Ocean
RODENTIA MYOXIDAE Continued																	
Hypnomys†	1633						■1										
Myomimus	1633						■Se	■Sw									
Selevinia	1634							■C									
RODENTIA BATHYERGIDAE																	
Georychus	1636					■S											
Cryptomys	1637					■											
Heliophobius	1639					■E											
Bathyergus	1640					■S											
Heterocephalus	1642					■Ec											
RODENTIA HYSTRICIDAE																	
Trichys	1644							■2	■3								
Atherurus	1644					■W,C		■Se									
Hystrix	1645					■	■S	■S	■	■W							
RODENTIA PETROMURIDAE																	
Petromus	1649					■S											
RODENTIA THRYONOMYIDAE																	
Thryonomys	1650					■											
RODENTIA ERETHIZONTIDAE																	
Echinoprocta	1652			■4													
Coendou	1652	■S		■													
Sphiggurus	1656			■													
Erethizon	1656	■															
RODENTIA CHINCHILLIDAE																	
Lagostomus	1659			■S													
Lagidium	1661			■W,S													
Chinchilla	1662			■W													
RODENTIA DINOMYIDAE																	
Dinomys	1663			■Nw													
RODENTIA CAVIIDAE																	
Microcavia	1665			■S													
Galea	1666			■C,S													
Cavia	1667			■													
Kerodon	1669			■5													
Dolichotis	1670			■S													
RODENTIA HYDROCHAERIDAE																	
Hydrochaeris	1672	■6		■													
RODENTIA DASYPROCTIDAE																	
Dasyprocta	1674	■S		■													
Myoprocta	1676			■N,C													
RODENTIA AGOUTIDAE																	
Agouti	1676	■S		■													

1. Balearic Islands only.   2. Malay Peninsula only.   3. Sumatra and Borneo only.   4. Colombia only.   5. Eastern Brazil only.
6. Panama only.

Genera of Recent Mammals	page	North America	West Indies	South America	Madagascar	Africa	Europe	Asia	Southeast Asia Islands	Philippine Islands	New Guinea	Australian Region	Antarctic Region	Arctic Region	Atlantic Ocean	Indian Ocean	Pacific Ocean
RODENTIA CTENOMYIDAE																	
Ctenomys	1678			■S													
RODENTIA OCTODONTIDAE																	
Octodon	1681			■1													
Octodontomys	1682			■Sw													
Octomys	1683			■2													
Tympanoctomys	1683			■2													
Spalacopus	1683			■1													
Aconaemys	1685			■S													
RODENTIA ABROCOMIDAE																	
Abrocoma	1686			■Sw													
RODENTIA ECHIMYIDAE																	
Proechimys	1688	■S		■													
Hoplomys	1689	■S		■Nw													
Euryzygomatomys	1690			■E													
Clyomys	1690			■C													
Carterodon	1692			■3													
Thrichomys	1693			■Ec													
Mesomys	1693			■N,C													
Lonchothrix	1694			■C													
Isothrix	1695			■C													
Diplomys	1695	■4		■Nw													
Echimys	1696			■N,C													
Makalata	1697			■N													
Dactylomys	1698			■Wc													
Kannabateomys	1699			■Se													
Olallamys	1700			■Nw													
Chaetomys	1700			■3													
Heteropsomys†	1701		■5														
Brotomys†	1702		■6														
Boromys†	1703		■7														
RODENTIA CAPROMYIDAE																	
Mesocapromys	1703		■7														
Mysateles	1704		■7														
Capromys	1706		■7														
Geocapromys	1706		■														
Hexolobodon†	1708		■6														
Rhizoplagiodontia†	1708		■6														
Plagiodontia	1708		■6														
Isolobodon†	1709		■														
RODENTIA HEPTAXODONTIDAE																	
Quemisia†	1710		■6														
Elasmodontomys†	1711		■5														
Amblyrhiza†	1711		■8														
Clidomys†	1712		■9														

1. Chile only.   2. Argentina only.   3. Southeastern Brazil only.   4. Panama only.   5. Puerto Rico only.   6. Hispaniola only. 7. Cuba only.   8. Lesser Antilles only.   9. Jamaica only.

Genera of Recent Mammals	page	North America	West Indies	South America	Madagascar	Africa	Europe	Asia	Southeast Asia Islands	Philippine Islands	New Guinea	Australian Region	Antarctic Region	Arctic Region	Atlantic Ocean	Indian Ocean	Pacific Ocean
RODENTIA MYOCASTORIDAE																	
Myocastor	1712		■S														
LAGOMORPHA OCHOTONIDAE																	
Ochotona	1717	■W					■Se	■									
Prolagus†	1720						■1										
LAGOMORPHA LEPORIDAE																	
Pentalagus	1721							■2									
Bunolagus	1722					■S											
Pronolagus	1723					■E,S											
Nesolagus	1723								■3								
Romerolagus	1723	■4															
Brachylagus	1726	■W															
Sylvilagus	1726	■		■													
Oryctolagus	1729					■Nw	■Sw										
Poelagus	1731					■C											
Caprolagus	1731							■Se									
Lepus	1733	■				■	■	■									
MACROSCELIDEA MACROSCELIDIDAE																	
Rhynchocyon	1741					■											
Petrodromus	1742					■											
Macroscelides	1743					■S											
Elephantulus	1744					■											

1. Corsica and Sardinia only.   2. Ryukyu Islands only.   3. Sumatra only.   4. Mexico only.

Walker's
Mammals
of the World

Class Mammalia

Mammals

The class Mammalia is here considered to comprise 28 orders that have lived during historical time (approximately the last 5,000 years) and, in turn, 146 families, 1,192 genera, and 4,809 species. Mammals are thought to have evolved from the class Reptilia, particularly from within the order Therapsida of the subclass Synapsida, or "mammal-like reptiles." The class Aves (birds) also seems to have evolved from the Reptilia, but from an entirely separate subclass. Of the three main groups of living mammals, the monotremes lay eggs as did their reptilian ancestors, the marsupials give birth to live but relatively undeveloped young, and the placentals retain the embryo until it reaches a more advanced stage.

E. R. Hall (1981) characterized the class Mammalia as follows: "Beings especially notable for possessing mammary glands (mammae) that permit the female to nourish the newborn young with milk; hair present although confined to early stages of development in most of the Cetacea; mandibular ramus of lower jaw made up of single bone (dentary); lower jaw articulating directly with skull without intervention of quadrate bone; two exoccipital condyles; differing from both Aves and Reptilia in possessing diaphragm and in having nonnucleated red blood corpuscles; resembling Aves and differing from Reptilia in having 'warm blood,' complete double circulation, and four-chambered heart; differing from Amphibia and Pisces in presence of amnion and allantois and in absence of gills."

Clemens (1989) noted that other definitions have been proposed, some of them involving complex sets of skeletal characters. These diagnoses serve mainly to distinguish the Mammalia not from other living classes but rather from the ancestral mammal-like reptiles. There has been intensive debate regarding the morphological and temporal boundary between reptiles and the first mammals. Recent fossil studies have revealed some specimens that do not clearly fall into either group and have challenged the significance of the direct articulation of the lower jaw and skull as the key indicator of mammalian origin. Rowe (1988, 1993) and Rowe and Gauthier (1992) reviewed the many proposed diagnoses and the differing views regarding the exact stage of the synapsid lineage from which mammals arose. Their own conclusion was that the class Mammalia originated with the most recent common ancestor of, on the one hand, the order Monotremata, and, on the other hand, the marsupial and placental mammals. Based on these criteria, the earliest known mammal was considered to be *Phascolotherium bucklandi* from the middle Jurassic of England, about 165 million years old.

Other authorities have extended the boundary farther back to include all lineages that achieved some particular morphological suite of characters regardless of whether such lineages are considered to be in the direct ancestral line of modern mammals. Perhaps most views on mammalian origin focus on or near the family Morganucodontidae, a group of small creatures that developed in the late Triassic and persisted through much of the Jurassic and did not fully meet the mammalian skeletal criteria as set forth by Hall. Lucas and Luo (1993), questioning such criteria, as well as the position of Rowe and Gauthier, argued that the earliest known mammal was *Adelobasileus cromptoni*, from the late Triassic of Texas, about 225 million years old.

One point on which a consensus now does seem to be forming is that all currently living groups of mammals—monotremes, marsupials, and placentals—are monophyletic, having originated from a common ancestor, and that these groups probably are more closely related to one another than to the Morganucodontidae and earlier mammal or mammal-like lineages. Until recently, a widely accepted view was that the monotremes had diverged at a very early stage of mammalian evolution or even emerged from a synapsid branch entirely separate from that giving rise to other living mammals and that they actually were more closely related to long-extinct lineages.

There is less agreement with respect to the affinity of the monotremes to the Multituberculata, an order of herbivorous mammals with some superficial resemblance to rodents, known from Jurassic through Oligocene deposits in Europe, Asia, North America, and South America (Simmons 1993b). Some authorities have suggested that the Multituberculata and the Monotremata are immediately related to one another and that the resulting assemblage is the closest relative of the assemblage formed by the marsupials and the placentals. Rowe (1993), while supporting a close relationship among all four of these groups, thought that the monotremes diverged from the line leading to the other modern mammals at an earlier stage than did the multituberculates. In contrast, Miao (1993) concluded that the Multituberculata independently evolved the cranial characters that suggest close relationships with the modern mammals and actually diverged from all other mammals even before the development of the Morganucodontidae. Detailed analyses of craniodental evidence (Wible 1991; Wible and Hopson 1993), especially from the basicranial region of the skull, suggested something of an intermediate position, with the multituberculates diverging prior to the monotremes but subsequent to the morganucodonts.

McKenna (1975) had divided the class Mammalia into two subclasses, Prototheria and Theria. The Prototheria were thought to comprise the extinct infraclass Eotheria, with various primitive mammals, and the infraclass Or-

1

A typical mammal, the otter civet *(Cynogale bennettii)*, photo from San Diego Zoological Society.

nithodelphia, with the orders Monotremata and Multituberculata. The Theria comprised all other mammals, there being a group of extinct forms but only one living infraclass, the Tribosphenida, containing the marsupials and placentals. Studies by Kemp (1982, 1983), however, suggested that the nontherian mammals are divided into a number of separate groups that all became extinct by the Cretaceous and that the monotremes belong in the same lineage as the therians. Most subsequent morphological study has supported assignment of the monotremes to a taxonomic position close to the Theria, possibly within the otherwise extinct group that preceded the Tribosphenida (Archer et al. 1993; Clemens 1989; Lucas and Luo 1993; Rowe 1988, 1993). In addition, recent analysis of mitochondrial DNA has indicated that the divergence of the monotremes from the marsupials and placentals occurred only 5–10 million years before the two latter groups separated from one another, which Gemmell and Westerman (1994) estimated to have been about 120 million years ago.

The above line of thought suggests close affinity of all living mammals, though there remains uncertainty whether the subclass Theria should include the monotremes together with the marsupials and the placentals. The latter two groups clearly are parts of a monophyletic unit, being united by a characteristic triangular arrangement of the cusps of the cheek teeth (Clemens 1989). The name Tribosphenida has been retained for this unit, which also includes various therians that existed prior to the marsupial-placental divergence (Cifelli 1993). Currently it seems to be generally recognized that both the marsupials and the placentals constitute multiordinal mammalian groupings, which many authorities recognize as infraclasses, their respective technical names being Metatheria and Eutheria. However, Aplin and Archer (1987) retained the Tribosphenida as an infraclass including the "supercohorts" Marsupialia and Placentalia.

Moving below the infraclass level, there seems no question that the monotremes consist of but a single order, and there also appears to be a very general consensus that the placentals comprise approximately 20 orders that have lived during historical time. The sequence of presentation of these orders adopted here is the same as that followed by Corbet and Hill (1991) and Wilson and Reeder (1993), which in turn stems from the work of McKenna (1975). Among the phylogenetic views reflected in this scheme are that (1) the Xenarthra were the first living placental order

An American marsupial, the thick-tailed opossum *(Lutreolina crassicaudata)*, photo by Michael Mares.

to diverge; (2) the Insectivora are also basically primitive but with questionable relationships; (3) the Scandentia, Dermoptera, Chiroptera, and Primates form a related assemblage, probably with some ties to the Insectivora; (4) the Carnivora and Pinnipedia are closely related and perhaps belong in the same order; (5) the Cetacea seem to have affinity with the Carnivora from certain perspectives but with the Artiodactyla from others; (6) there is an association of the Artiodactyla and the other members of the ungulate group, with especially close connections between the Sirenia and Proboscidea and between the Perissodactyla, Hyracoidea, and Tubulidentata; (7) the Rodentia are a highly diverse and generally advanced assemblage; and (8) the Lagomorpha seem to be related to both the Rodentia and the Macroscelidea.

There does remain considerable disagreement about the phylogenetic arrangement followed herein. For example, using a synthesis of recent information from various disciplines, Novacek (1992, 1993) indicated the following evolutionary sequence for the living placental orders: Xenarthra, Pholidota, Lagomorpha, Rodentia, Macroscelidea, Primates, Scandentia, Dermoptera, Chiroptera, Insectivora, Carnivora (including Pinnipedia), Artiodactyla, Cetacea, Tubulidentata, Perissodactyla, Hyracoidea, Proboscidea, and Sirenia. Novacek suggested an early divergence of the Rodentia, a closer affinity of that order with the Lagomorpha than with the Macroscelidea, and a possible relationship between the Insectivora and Carnivora. He maintained that the Scandentia, Dermoptera, Chiroptera, and Primates form a monophyletic group (Archonta) but noted questions about their interrelationships.

Novacek's sequence may serve as a basis of comparison with other views. Placing the Pholidota alongside the Xenarthra follows a long tradition, but the sum of morphological and biochemical evidence indicates that although the Pholidota are well isolated from other placentals, they have no close affinity to the Xenarthra (Rose and Emry 1993). According to Graur (1993), recent molecular (biochemical and DNA) studies do suggest that the Rodentia were among the earliest of the placental groups to diverge and probably should be divided into three full orders and also that the Cetacea should actually be placed within the Artiodactyla, which in turn are closely related to the Car-

nivora. These studies, however, do not support an association of the Lagomorpha with either the Rodentia or the Macroscelidea, the latter instead showing an immediate relationship to the Hyracoidea. Prothero (1993) reported both morphological and molecular evidence to support the monophyly of an ungulate group that includes the Cetacea, with the Artiodactyla having been the first living order to diverge and with the Sirenia, Proboscidea, Perissodactyla, and Hyracoidea forming a related and more advanced assemblage within the group. Based on myological data, Shoshani (1993) agreed that the same six orders are monophyletic and considered the closest living relative of the group to be the Tubulidentata. Additional disagreements about whether certain of the placental orders should be combined with others or split into further orders are discussed below in the appropriate ordinal accounts.

Fluctuation in the views on the placental orders pales in comparison with the revolution that has shaken the taxonomic treatment of the marsupials (or Metatheria) in the last two decades. Although many sources still follow Simpson (1945) in the traditional placement of all marsupials within a single order (Marsupialia), such an approach now seems to be adhered to mainly for convenience or in an attempt to maintain familiar terminology. Most primary works on marsupials, as well as compilations attempting to express modern and balanced phylogenetic accounts, divide the Metatheria into several orders. The latter procedure, along with acceptance of more marsupial families than formerly were recognized, is reasonable in that the marsupials include species that seem just as diverse in morphology and ecology as moles, rabbits, mice, flying squirrels, wolves, and antelopes. Marsupials are united primarily on the basis of a common means of reproduction. The placental mammals also share a common reproductive method, but no one questions their division into many orders.

The number of marsupial orders recognized by various zoologists steadily increased with mounting paleontological, morphological, and biochemical evidence (Aplin and Archer 1987; Archer 1984a, 1984b; Kirsch 1977b; Kirsch and Calaby 1977; L. G. Marshall 1981; Marshall, Case, and Woodburne 1990; Szalay 1982). The number (7) and sequence of living marsupial orders accepted herein follow that given in Wilson and Reeder (1993): Didelphimorphia,

An Australian carnivorous marsupial, the kultarr *(Antechinomys laniger)*, photo by P. A. Wooley and D. Walsh.

Paucituberculata, and Microbiotheria of the New World; Dasyuromorphia, Peramelemorphia, Notoryctemorphia, and Diprotodontia of Australasia. However, based on detailed osteological analyses of fossil and modern specimens, Szalay (1993, 1994) considered that the diversity of marsupials, in comparison with the placentals, is sufficiently reflected by recognition of only three living orders: Delphida, which would include the Didelphimorphia as a suborder and the Paucituberculata as an infraorder; Gondwanadelphia, which would include the Microbiotheria and the Dasyuromorphia as suborders; and Syndactyla, which would include the Peramelemorphia and Notoryctemorphia at just the familial level and the Diprotodontia as a semiorder.

Szalay retained a division of the living marsupials into two great cohorts, one including just the American order Didelphida (or the orders Didelphimorphia and Paucituberculata) and the other comprising the American group Microbiotheria plus all of the Australasian marsupials. Other authorities, such as Archer (1984a), have placed all the American groups into one cohort and all the Australasian groups into another. In contrast, Hershkovitz (1992a) put all marsupials into one cohort, except for the Microbiotheria, which he considered to have evolved independently and to show some affinity to both the monotremes and eutherians (additional details are presented below in the account of the order Microbiotheria).

A traditional view holds that marsupials developed initially in the Northern Hemisphere and might formerly have been widespread but became restricted to a few southerly regions through the expansion of superior placentals. Although Kirsch (1977c) and various other authorities came to support an origin in Australia, where most marsupials now occur, Martin (1977) still thought the safest hypothesis was that marsupials arose in North America, where the largest collections of early fossils have been made, spread to South America, and then traveled via Antarctica to Australia. These continents still were joined near the end of the Cretaceous period, and dispersal probably occurred at that time. More recently, agreement with a North American origin was suggested by both Szalay (1994) and Cifelli (1993), though because of differing interpretation of the morphological boundary between meta-

therians and earlier tribosphenids Szalay set the time as early Cretaceous, whereas Cifelli set it as late Cretaceous.

There are 20 known extinct families of marsupials (Szalay 1994). Geological range is Oligocene to Recent in Australia, early Cretaceous to Recent in North America, late Cretaceous to Recent in South America, early Eocene to middle Miocene in Europe, middle Miocene in South Asia, and Oligocene in Central Asia and North Africa (Bown and Simons 1984; Ducrocq et al. 1992; Gabuniya, Shevyreva, and Gabuniya 1985; Kirsch 1977a, 1977b; Szalay 1994). The reported record of a fossil marsupial on New Caledonia has been shown to be invalid (Rich et al. 1987). Recently the fossilized remains of a marsupial belonging to the extinct family Polydolopidae (the most closely related living family is the Caenolestidae) were found in late Eocene deposits on Seymour Island, off the northern Antarctic Peninsula (Woodburne and Zinsmeister 1982, 1984). This discovery of the first land mammal in Antarctica supports the theory of an early dispersal of marsupials across the southern continents.

Renfree (1993) suggested that when mammals originated they simultaneously evolved lactation and certain other of their characteristic reproductive features. Lactation may have begun as a maternal sebaceous secretion with antibacterial functions that served to protect the eggs and hatchlings and then gradually evolved into a flow of nutrients from specialized glands. From the late Triassic through part of the Cretaceous the basic mammalian reproductive mode remained almost unchanged because it was the most appropriate for small, nocturnal, insectivorous creatures such as the early mammals are thought to have been. The major differences in living mammals evolved during the divergence of the marsupials and placentals and the subsequent adaptive radiations of the two groups in response to the metabolic requirements of increasing body size and the ecological constraints imposed by new ecological niches and modes of life. All three living groups of mammals share certain basic reproductive characters, including Graafian follicles, functional corpora lutea, bilaminar blastocysts, uterine secretion, yolk-sac placentae, mammary glands, and lactation. The Monotremata are distinguished largely by factors associated with egg laying and incubation of the egg outside of the body. The specializations of the marsupials have to do mainly with extended lactation and endocrine controls thereof. The unique characters of the placentals are associated with extended embryonic gestation and include the universal precocious development of the chorioallantoic villous placenta and a tendency to develop a variety of luteotrophic and luteolytic controls of the corpora lutea.

Further details on the unique characters of the Monotremata are given in the following ordinal account, but because there are a number of metatherian orders to cover, a summary of the characters that are common to all marsupials and that serve to distinguish them from the placentals is provided at this point. Detailed discussions and comparisons of the morphology, physiology, and natural history of these two groups of mammals were provided by Bryden (1989), Dawson et al. (1989), Jarman, Lee, and Hall (1989), and Russell, Lee, and Wilson 1989).

Most female marsupials possess an abdominal pouch, referred to as a marsupium, within which the young are carried. The pouch, however, is neither the only nor the most diagnostic characteristic of the group. The best-developed pouches are found in marsupials that climb (phalangers), hop (kangaroos), dig (bandicoots and wombats), or swim (the yapok), but some small terrestrial marsupials have no pouch (Kirsch 1977c). In certain didelphids and dasyurids, among others, the pouch consists merely of folds of skin around the mammae that help to protect the attached young. Many marsupials develop pouches only during the reproductive season. When well developed, a pouch may open either to the front or to the rear, depending on the genus. The mammae usually are abdominal and located within the pouch if one is present.

All marsupials lack a complete placenta, the membranous structure facilitating passage of nutrients from the mother's body to the embryo in the uterus. Marsupials form a yolk-sac placenta through which uterine secretions are absorbed by the amnion, an outer membrane of the embryo. In the order Peramelemorphia, and also evidently in the genus *Phascolarctos* (Lee and Carrick 1989), there is a more developed placenta consisting of additional embryonic membranes, the chorion and the allantois. In all other therian mammals, however, this chorioallantoic placenta develops villi that penetrate uterine tissue and provide for the most efficient means of nourishment. Mammals with such a completely developed placenta are known as placentals.

In marsupials the female reproductive tract is bifid; that is, the vagina and uterus are double. The two lateral vaginae spread sufficiently to allow the urinary ducts to pass between them (Tyndale-Biscoe 1973). During birth, however, the young typically are extruded through a third canal, the birth canal, or median vagina, which passes from a point of medial fusion between the two uteri to the urogenital sinus. In the Didelphimorphia, Dasyuromorphia, Peramelemorphia, and most of the Diprotodontia the median vagina is transitory and must be formed before each birth. In the Macropodidae and the Tarsipedidae, however, the median vagina becomes lined with epithelium at the first birth and forms a permanent birth canal (Sharman 1970). The passage of the ureters medially between the vaginae, rather than laterally as in placentals, is a critical expression of the dichotomy of the two groups and is their essential reproductive distinction. Since the median vagina in marsupials is so constricted, the young must be very small at birth and hence must subsequently undergo an extended period of development and relatively longer lactation before independence (Renfree 1993).

In male marsupials the scrotum is in front of the penis, except in the Notoryctidae, and there is no baculum. Many males have a bifid penis, the right and left prongs appar-

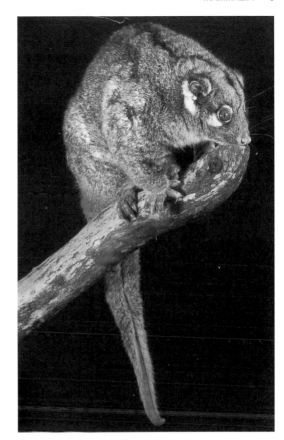

An Australian diprotodont marsupial, the coppery ring-tailed possum *(Pseudochirops archeri)*, photo from Queensland Museum.

ently being placed in the corresponding vaginal canals during mating. The vas deferens (ducts that carry sperm from the testes) pass laterally to the ureters instead of mesially as in placentals (Sharman 1970).

The gestation period in marsupials is short compared with that of placentals of equivalent size, and the tiny young are born in a practically embryonic state. The newborn crawl some distance anteriorly from the urogenital opening and take hold of nipples that expand in their mouths to ensure a firm attachment. Contrary to earlier belief, the young apparently are not assisted by the mother in their movement to the nipples. She does lick the fur through which the young crawl, but only after they have passed, in order to consume fluids and embryonic membranes (Sharman 1970).

In addition to reproduction, other characteristics help to set marsupials apart. The number of teeth, usually 40–50, often exceeds that of placental mammals. The Vombatidae, however, have only 24 teeth. The 7 or 8 cheek teeth, usually present on both sides, are divided into 3 premolars and 4 or 5 molars, in contrast with the 4 premolars and 3 molars of typical placental mammals. The lower jaws do not have the same number of incisors as the upper jaws. The skull has a large facial area and a relatively small cranial cavity; the brain, especially the telencephalon, is small compared with that of a placental mammal. The brain of the marsupials, like that of the monotremes, lacks a corpus callosum, a mass of tissue of many functions that connects the hemispheres of the brain in the placental mammals. The convo-

A marsupial mother and young, the tammar wallaby *(Macropus eugenii)*, photo by Wolfgang Dressen.

lutions of the cerebral hemispheres are generally simple. The nasal bones of the skull are large and expand posteriorly, and the zygomatic arches are complete, with the jugal bone extending backward below the zygomatic process of the squamosal bone as far as the glenoid fossa. The palate is usually imperfect, with spaces between the back molars. The angular process of the mandible is usually bent inward except in *Tarsipes.* Epipubic (marsupium) bones are associated with the pelvic girdle in both sexes of nearly all marsupials (and also monotremes and some multituberculates) but are small in the Notoryctidae and absent in the Thylacinidae (Dawson et al. 1989).

Marsupials traditionally have been viewed as primitive or second-class mammals. This line of thought now seems to be changing. Recent laboratory studies have demonstrated that the learning and problem-solving abilities of marsupials often equal or exceed those of some placental groups (Kirkby 1977). It is true that some Australian marsupials lost ground when placentals were introduced by human agency. The Virginia opossum, however, continues to thrive and even to extend its range in North America despite a host of placental competitors. Following its introduction on the West Coast it rapidly expanded its range over a large area. Other marsupials also have had a successful introduction, notably brush-tailed possums in New Zealand and wallabies in New Zealand, Great Britain, Germany, and Hawaii (Gilmore 1977). If marsupials did evolve mainly in the south, they would have had far less room to diversify than the placentals and much more difficulty in achieving a wide geographic distribution. These factors, rather than any inferiority on a one-to-one basis, may explain their current restricted range.

The reproductive process of marsupials, often considered less advanced and less efficient than that of placentals, may actually have advantages. A female marsupial invests relatively few resources during the brief gestation period of her young. Her major commitment comes later, during lactation, a phase that is more environmentally sensitive than that of internal nutrition and more easily terminated by adverse factors. The marsupial that loses her young is, therefore, able to make a second attempt at reproduction more quickly and in better condition than a placental in a comparable situation. This means of reproduction perhaps should be seen as a dynamic and highly derived alternative befitting the particular evolutionary history of the marsupials rather than as the survival of a primitive developmental stage (Kirsch 1977c; P. Parker 1977; Renfree 1983, 1993).

Order Monotremata

Monotremes

The monotremes, comprising two families, three genera, and three species of Australia and New Guinea, are the most distinctive of the 28 orders of living mammals here recognized. As already noted, some authorities have placed the monotremes in a subclass with long-extinct lineages and separate from all other extant mammals. Others have considered monotremes actually to be living therapsid reptiles, being nearly as remote from mammals as alligators are from birds.

The monotremes resemble reptiles and differ from all other mammals in that they lay shell-covered eggs that are incubated and hatched outside of the body of the mother. They also resemble reptiles in a number of anatomical details, such as the structure of the eye and the presence of certain bones in the skull. The pectoral or shoulder girdle of monotremes is reptilian, possessing distinct coracoid bones and an interclavicle, and some features of their ribs and vertebrae are reptilelike. The monotremes also exhibit similarities to reptiles in their digestive, reproductive, and excretory systems. In both sexes of all monotremes, as in all reptiles, the posterior end of the intestine, the ducts of the excretory system, and the genital ducts open into a common chamber known as the cloaca. Therefore, for the three systems there is only a single external opening, and this arrangement is the basis of the ordinal name, Monotremata. In male monotremes the penis is attached to the ventral wall of the cloaca and is divided at the tip into paired canals used only for the passage of sperm. The testes are abdominal. In female monotremes the oviducts, which conduct the eggs from the ovary, open separately into the cloaca. In the oviducts the eggs are fertilized and then covered with albumen and a flexible, sticky, leatherlike shell.

Although monotremes have many reptilian characteristics, in other ways they are typically mammalian. Like other mammals, they are furred, have a four-chambered heart (there is an incomplete right atrioventricular valve, however, and a single aortic arch on the left), nurse their young from milk secreted by specialized glands, and are warm-blooded (but the body temperature averages lower than that of other mammals, about 30°–32° C). In addition, the skeleton has mammalian features, such as the single bone of each side of the lower jaw (not several, as in reptiles) and the three middle ear bones (rather than one, as in reptiles). In the brain of monotremes, as in that of other mammals, the "pallial" region is the most strongly developed part of the cerebral hemispheres, not the "striatal" part as in reptiles and birds. Along with marsupials and reptiles, however, monotremes lack the corpus callosum, a multifunctional bridge of nervous tissue that connects the two hemispheres in other mammals.

One striking feature that monotremes possess in common with marsupials is epipubic, or "marsupium," bones associated with the pelvis. These bones are said to aid in supporting a pouch, but that function is doubtful as they are equally developed in both sexes. It seems more likely that they are a heritage from reptilian ancestors, associated with the attachment of abdominal muscles that support large hindquarters.

The skull of monotremes has a smooth, rounded cranial portion terminating in a long rostrum covered by rubbery, sensitive skin. The sutures of the skull tend to become obscured. Young Ornithorhynchidae have teeth that do not cut through the gums, but true functional teeth are not present in adult monotremes.

The males of all monotremes bear horny spurs on their ankles. In *Ornithorhynchus* these spurs are grooved to permit the passage of a poisonous glandular secretion.

The known geological range of the monotremes is early Cretaceous to Recent in Australia, Recent in New Guinea, and early Paleocene in South America (Archer et al. 1985; Griffiths 1978; Pascual et al. 1992). The presence of two distinctively different monotreme families in the early Cretaceous of Australia suggests that the order originated in that region long before the Cretaceous, when Australia still was joined to Antarctica. By the early Paleocene the order evidently had dispersed through Antarctica to South America (Flannery, Archer, et al. 1995).

MONOTREMATA; Family TACHYGLOSSIDAE

Echidnas, or Spiny Anteaters

This family of two Recent genera, each containing a single species, is found in Australia, Tasmania, and New Guinea.

Echidnas have a body covering of fur intermixed with dorsal and lateral barbless spines. The average size of males is larger than that of females. The body is robust and muscular yet supple. The limbs have broad, powerful feet modified for digging, each foot bearing three to five strong, curved claws. The second toe on the hind foot is elongated and is used for scratching and cleaning the fur and skin. The eyes are smaller than those of the Ornithorhynchidae. The pinna of the external ear is well developed, but it is partly concealed in the pelage. The sense of smell is acute. The tail is vestigial.

The braincase of the skull is large and rounded, and the

An echidna *(Tachyglossus aculeatus)* partially rolled up for self-protection. If it had been completely rolled up, as when danger threatens, the face, feet, and hands would be entirely enclosed in the spiny ball. Photo by John Warham.

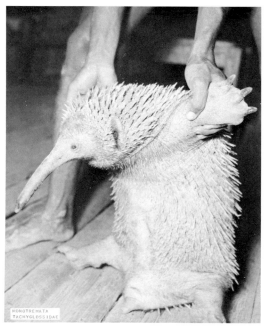

Long-nosed New Guinean echidna *(Zaglossus bruijni)*, photo by Sten Bergman.

brain is relatively large, with convoluted cerebral hemispheres. The rostrum is long, tubular, and tapering, and the lower jaw is represented by two slender bones. The mouth is small. Teeth are absent throughout life. The posterior part of the long, sticky tongue has horny serrations that work against hard ridges on the palate for grinding food.

Bollinger and Backhouse (1960) made a study of the hematology of blood samples of echidnas and concluded that the hemoglobin level was higher than in most mammals. Griffiths (1968), however, stated that wild marsupials and placentals exhibit higher hemoglobin levels than those found by Bollinger and Backhouse in echidnas. In other aspects, the blood of echidnas resembles that of other mammals.

Echidnas are powerful diggers. An individual in a horizontal position can burrow straight down with remarkable speed. Shelters include hollow logs, cavities under roots or rocks, and burrows. These animals can wedge themselves with their spines and feet into crevices and burrows in such a manner that it is practically impossible to dislodge them. The spines can be erected and the limbs withdrawn, as in a hedgehog *(Erinaceus).* Contrary to what has often been written, echidnas *(Tachyglossus* at least) do not hibernate. Improper diet may cause them to go into a state of torpor, but they do not hibernate in response to cold (Augee 1978). With their long, sticky tongue, provided with mucus from enlarged salivary glands, echidnas feed on termites, ants, other insects, and worms. Some individuals have thrived when fed milk, finely minced meat, hard-boiled eggs, meal-

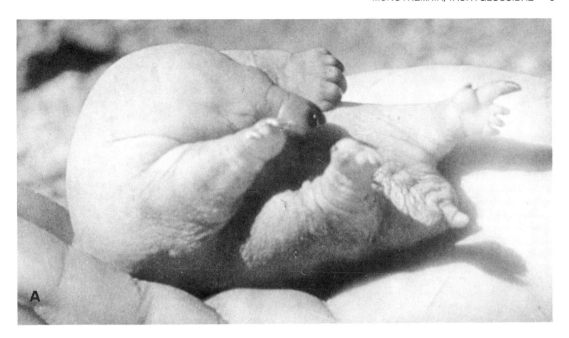

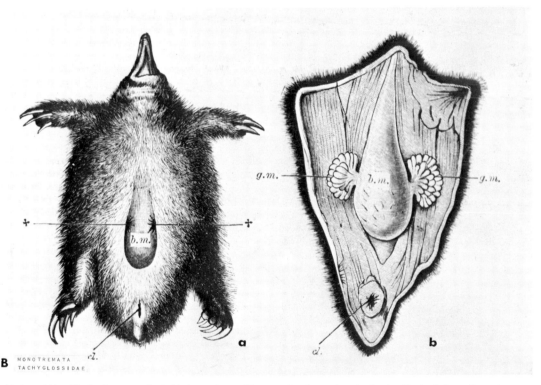

A. A baby echidna *(Tachyglossus aculeatus)* between two and four months of age, photo by Stephen Davis. B. Australian spiny anteater *(T. aculeatus):* a. Lower surface of brooding female; b. Dissection showing a dorsal view of the pouch, mammary glands, and two tufts of hair in the lateral folds of the mammary pouch from which the secretion flows; b.m. = pouch, c.l. = cloaca, g.m. = groups of mammary glands; photos from *Mammalia,* F. E. Beddard.

worms, and bread. The quantity of milk consumed by captives is remarkable. Echidnas have shown themselves able to exist without nourishment of any kind for at least a month.

Echidnas are generally solitary. Brattstrom (1973) studied the social behavior of *Tachyglossus* and found that many complicated behavioral postures, such as grooming, aggression, courtship, and maternal behavior, were missing. He concluded that echidnas seem to have a behavior that is not only simpler than that found in most mammals but perhaps also simpler than that of many lizards.

During the breeding season females develop a temporary pouch on the abdomen. The single egg, or rarely two or three, is transferred directly from the cloaca into this pouch. When hatched, the young is naked and measures about 12 mm in length. It sucks the thick, yellowish milk that flows from the mammary glands, which open into the pouch. The young remains in the pouch for six to eight weeks, or until its spines begin to develop. At this age it is about 90–100 mm long. It is then deposited in a sheltered spot by the mother, which continues to nurse it during periodic visits. Sexual maturity is reached at the end of one year of life.

The geological range of the family Tachyglossidae is Pliocene to Recent in Australia and Pleistocene to Recent in New Guinea and Tasmania (Marshall 1984).

MONOTREMATA; TACHYGLOSSIDAE; **Genus**
TACHYGLOSSUS
Illiger, 1811

Short-nosed Echidna

The single species, *T. aculeatus,* inhabits most of mainland Australia, many nearby islands, Tasmania, and central and southern New Guinea (Collins 1973; Griffiths 1968, 1978). A number of apparently well-marked subspecies have been described, based primarily on characters of the pelage and spines. The subspecies *T. a. setosus* of Tasmania is often considered a distinct species.

Head and body length is 350–530 mm, tail length is about 90 mm, and adult weight is 2.5–7.0 kg. The spines, which cover most of the body, measure up to 60 mm in length. They are usually yellow at the base and black at the tip, or, less frequently, entirely yellow. These specialized hairs are generally large, hollow, and thin-walled. In some individuals the brown or black body fur is almost concealed by the spines, but in the Tasmanian subspecies the short spines are largely hidden by the fur. The underparts lack spines but are covered with fur and thick bristles.

Like that of the platypus, the body of the short-nosed echidna is compressed dorsoventrally, but it lacks the streamlining so evident in the platypus. The back of *Tachyglossus* is domed, and the belly is flat, even concave, in outline. The head is small and seems to emerge from the body without any indication of a neck. At the posterior end of the body is a short, stubby tail, which is naked on its undersurface. The long and slender snout is about half the length of the entire head; it is nearly straight or curved slightly upward. According to Griffiths (1978), the snout is strong enough to be used as a tool to break open hollow logs and to plow up forest litter to get at the ants and termites the echidna catches with its long, sticky tongue. There also is evidence that the snout contains electroreceptors, which may assist the animal in moving about or locating prey (Augee and Gooden 1992). Each foot of *Tachyglossus* has five digits, with flat claws that are well adapted for digging. The eyes are small and situated well forward on the head, almost at the base of the snout; they look ahead more than sideways. After studying the vision of *Tachyglossus,* Gates (1978) concluded that its sight is somewhat better than "dismal." In fact, he stated that its visual discrimination is at least comparable to that of rats and compares favorably with that of "nonvisual" animals, such as bats.

There is no scrotum in males, the testes are internal, and in both sexes there is only a single opening for the passage of feces, urine, and reproductive products. During the

Echidna, or spiny anteater *(Tachyglossus aculeatus)*, photo by Bernhard Grzimek.

breeding season females develop on the abdomen a crescentic fold of skin that forms a pocket in which the single egg is deposited and incubated.

The short-nosed echidna frequents a variety of habitats, including forests, rocky areas, hilly tracts, and sandy plains. It shelters either in burrows or in crevices among rocks, and it emerges to forage in the afternoon or at night. It evidently retreats to its burrow when the air temperature exceeds 32° C and experiences severe heat stress if maintained above 35° C (Grant 1983). Abensperg-Traun and De Boer (1992) observed no foraging below 9° C or above 32° C, noting that winter activity generally began soon after midday, while during summer foraging usually started long after dark. When disturbed in its burrow, the echidna digs into the earth and clings with its claws and spines, or, if the substrate is too hard for digging, it rolls up into a spiny ball. The echidna walks with its legs fully extended, so that the stomach is relatively high off the ground, and with the hind toes directed outward and backward. It can run quite swiftly and can climb well.

According to Augee (1978), as long as the echidna is supplied with sufficient food it resists torpor regardless of the temperature. He stated that in each instance of reported torpor in the echidna there is evidence that the animal was in a condition of weight loss, generally because of an inappropriate diet. If food is available, the echidna will not become torpid at any time of the year. Likewise, if it does not have sufficient food it will eventually become torpid during any season. Tachyglossus has a body temperature that is lower than that of most other mammals and controlled by shivering thermogenesis rather than nonshivering thermogenesis, which is found in mammals other than monotremes.

Although Augee and other scientists once did not consider the echidna or any monotreme to be capable of true hibernation, recent investigations have demonstrated otherwise (Grigg, Beard, and Augee 1989). Studies in the Snowy Mountains of New South Wales (Beard, Grigg, and Augee 1992; Grigg, Augee, and Beard 1992) found the echidna to hibernate during the cold season, usually from about April to July, with sexually mature females mating immediately after arousal. Physiologically this hibernation resembled that of eutherian mammals, with body temperature falling as low as 3.7° C and weight losses of 2–3 percent per month. However, there were periodic arousals, up to 24 hours long, during which normal temperature was regained and individuals sometimes moved to a different retreat. As in other areas, animals also were found to enter brief periods of shallow torpor during the nonhibernating season.

Griffiths (1968) found Tachyglossus to be almost exclusively an ant and termite eater. For unknown reasons, in cool, moist areas ants compose the major portion of the diet, while in hot, dry areas termites are most often consumed. It seems not to be a question of availability since there are plenty of ants in both kinds of habitat.

In a radio-tracking study on Kangaroo Island, South Australia, Augee, Ealey, and Price (1975) found the short-nosed echidna to have a definite home range averaging about 800 meters in diameter. They also reported that the ranges of different individuals may overlap, that there is no specific nest site, that individuals are solitary for most of the year, and that there is no evidence of territorial behavior. Other radio-tracking studies have also found extensive overlap, with home range size being 24–76 ha. in the Snowy Mountains of New South Wales (Augee, Beard, and Grigg 1992) and 24–192 ha. for adults and 6–48 ha. for juveniles in the wheatbelt reserves of Western Australia (Abensperg-Traun 1991a). In a study of captives, Augee,

Bergin, and Morris (1978) found individuals to be mutually tolerant, but there was a dominance hierarchy among animals of the same sex.

Except as noted, the following information on reproduction in Tachyglossus was taken from Griffiths (1968, 1978). The height of the breeding season is July and August all over Australia. Females have been estimated to have an interbirth interval of 2 years and an estrous cycle of from 9 days to a month (Hayssen, Van Tienhoven, and Van Tienhoven 1993). The gestation period is brief, available evidence indicating a possible range from at least 9 days to 27 days. The single egg apparently is deposited into the pouch directly from the cloaca. This egg has a very large yolk and is enclosed in a flexible, leatherlike shell through which the baby breaks at birth with the help of an egg tooth. Incubation in the pouch, during which the greater part of embryogenesis occurs, lasts about 10–11 days. The young is ejected from the pouch when it is about 55 days old and 15–21 cm long. In the wild, after the female has ejected the young from her pouch she leaves it at a chosen spot while she hunts for food, returning occasionally to suckle it. According to Griffiths (1978), the young does not lick up milk from the mammary glands, but sucks as does any other mammal. He observed that the sucking is quite vigorous, involves movement of practically the entire body, and is clearly audible. Apparently, the young echidna is weaned when about 200 days old and 1,300 grams in weight (Abensperg-Traun 1989; Green, Griffiths, and Newgrain 1992). Dispersal from the natal home range occurs at about 12 months (Abensperg-Traun 1991a). Sexual maturity has been estimated to occur at 1–2 years (Hayssen, Van Tienhoven, and Van Tienhoven 1993). The short-nosed echidna may have an extremely long life span; wild individuals have been known to survive for about 20 years, and one captive lived to be more than 50 years old.

In contrast to many Australian mammals, Tachyglossus appears to have adapted well to the coming of European colonization, remaining common and widespread (Abensperg-Traun 1991b). Apparently, the echidna is not restricted by habitat and does not depend on vegetation for shelter. Its food source is abundant, readily available, and not subject to competition. Moreover, it is metabolically capable of tolerating low food supplies, as may result from fire or drought, and can go into torpor under extreme conditions. And because of its spines and other defensive strategies, the echidna is little affected by the introduced predators that have wiped out other native species. Nonetheless, the distinctive subspecies T. a. multiaculeatus, found only on Kangaroo Island off South Australia, is designated as near threatened by the IUCN.

MONOTREMATA; TACHYGLOSSIDAE; **Genus ZAGLOSSUS**
Gill, 1877

Long-nosed Echidna

At one time as many as four species of Zaglossus were recognized, but now these are generally considered to belong to the single species Z. bruijni (Griffiths 1978). This species currently is found throughout New Guinea except in the southern lowlands and along the north coast (Flannery 1990b). It also has been recorded with some question from Salawatti and Waigeo islands, just to the west (Flannery 1995).

Head and body length is about 450–775 mm, and the tail, like that of Tachyglossus, is only a slight projection. The

New Guinean long-nosed spiny anteater *(Zaglossus bruijni),* photo by Sten Bergman.

long-nosed echidna usually weighs 5–10 kg; one individual in the London Zoological Gardens weighed more than 16 kg, but it, like many zoo animals, may have been excessively fat. The hairs of *Zaglossus* are brownish or black and sometimes almost hide the spines on the back; the underparts are usually spineless. The spines range in color from white and light gray to solid black. The head is typically paler than the body and is almost white in one form that has a pale brown body.

In *Zaglossus* the long snout accounts for two-thirds of the length of the head and is curved downward, producing a pronounced convex profile. The three middle toes of each of the four feet usually have claws, while digits 1 and 5 are covered by a callosity. On the front feet, digits 1 and 5 are mere prominences in external aspect, but on the hind feet these digits are conspicuous.

Zaglossus is larger than *Tachyglossus* and has spines that are shorter, less numerous, blunter, and with smaller central cavities. The tongue of *Zaglossus* is longer, and the anterior third has a very deep groove on the dorsal surface; in the groove are three longitudinal rows of backwardly directed, sharp, keratinous "teeth" or spines.

In both *Zaglossus* and *Tachyglossus* the mammary glands open into the abdominal pocket of the breeding female. This pocket, or pouch, is not the same as the pouch of marsupials; it is a structure that occurs only temporarily in the female at the time of breeding. Also, in the males of both genera a spur is present on each of the hind legs, on the inner surface near the foot.

The long-nosed echidna generally is found in humid montane forests and may occur in alpine meadows at an elevation of 4,000 meters (Griffiths 1978). Its habits are not well known but probably are comparable to those of *Tachyglossus.* It is primarily nocturnal and forages on the forest floor. Dawson, Fanning, and Bergin (1978) found that specimens of *Zaglossus* had a body temperature of approximately 30° C.

The diet of *Zaglossus* differs from that of *Tachyglossus* in that earthworms apparently make up a major portion (Griffiths 1978). In fact, Collins (1973) reported that some captive specimens of *Zaglossus* would not eat ants at all. *Zaglossus* ingests earthworms buried in the forest litter by hooking them with the "teeth" within the groove on its tongue; the tongue protrudes only 2–3 cm beyond the end of the snout, and the groove opens by muscle flexion (Griffiths, Wells, and Barrie 1991). Griffiths (1978) speculated that the reason *Tachyglossus* is more widely distributed than *Zaglossus* may be that its principal foods, ants and termites, are found over a larger area and in many more kinds of habitat than are earthworms. Pleistocene fossils referable to *Zaglossus* have been found in Australia, but the genus no longer occurs there, possibly because climatic changes were unfavorable to the earthworms on which it preyed.

Data on reproduction in *Zaglossus* are limited, but they suggest that the breeding season may be centered in July (Griffiths 1978). *Zaglossus* almost certainly lays eggs, probably carries the small offspring in a pouch, and is said by local hunters to have 4–6 young at a time (Flannery 1990b). One captive specimen lived in the London Zoo for 30 years and 8 months, another lived at the Berlin Zoo for 30 years and 7 months, and both animals died within six days of each another in August 1943. An individual at the Taronga Zoo in Sydney that was captured in November 1963 may be still older (Marvin L. Jones, Zoological Society of San Diego, pers. comm., 1995).

The long-nosed echidna is designated as endangered by the IUCN and is on appendix 2 of the CITES. It is avidly hunted for food in New Guinea by natives using trained dogs. Because of such persecution, as well as loss of forest habitat to farming activity, *Zaglossus* is becoming rare in areas accessible to humans. According to Thornback and Jenkins (1982), the average population density of *Zaglossus* has been estimated at 1.6/sq km, which would indicate a total of about 300,000 animals in the remaining suitable habitat.

MONOTREMATA; **Family ORNITHORHYNCHIDAE;**
Genus ORNITHORHYNCHUS
Blumenbach, 1800

Duck-billed Platypus

The single living genus and species, *Ornithorhynchus anatinus*, inhabits freshwater streams, lakes, and lagoons of eastern Australia, in Queensland, New South Wales, Victoria, southeastern South Australia, and Tasmania (Grant 1992; Griffiths 1978).

Head and body length is 300–450 mm, tail length is 100–150 mm, and adult weight usually is 0.5–2.0 kg. Males are larger than females. Coloration is deep amber or blackish brown above and grayish white to yellowish chestnut below. Grant and Carrick (1978) stated that the fur, which is short and very dense, consists of a woolly underfur overlain with bladelike guard hairs. This coat traps a layer of air in the kinks of the fibers of the underfur and between the underfur and the blades of the guard hairs and thus contributes substantially to the insulation of the body.

The body of the platypus is streamlined and compressed dorsoventrally. The limbs are short and stout, and the webbed feet are broad. Each of the four feet has five clawed digits. The tail looks something like that of a beaver. The snout, which resembles a duck's bill, is elongated and covered with moist, soft, naked, leathery skin; it is perforated over its entire surface by openings to sensitive nerve endings. The nostrils also open through the snout, on its upper half, and are close together. The idea that the platypus has a horny bill like that of a bird arose from examination of dried skins only. The olfactory organs of the platypus are not as well developed as in echidnas. There is no ear pinna.

The young have small, calcified teeth with little enamel and numerous stubby roots; at least two pairs have replacement buds beneath. Adults lack functional teeth and instead have hornlike plates for most of the length of each jaw. Near the front of the mouth these plates are sharp ridges, but toward the back they are almost flat and function as crushing surfaces. They grow continuously and are rapidly worn by gritty food. The flattened tongue works against the palate and aids in mastication. The bones of the upper and lower jaws expand distally and support the bill. The brain of the platypus is relatively small, with smooth cerebral hemispheres.

The ankles of both hind limbs of the male have inwardly directed, hollow spurs that are connected with venom glands. The spur is found on both sexes when they are young, but it degenerates in the female. The gland secretes venom that is passed to the spur and can be injected into other animals by erection of the spur. Griffiths (1978) stated that the venom causes agonizing pain in humans and can kill a dog. Temple-Smith, cited by Griffiths (1978), surmised that the poison system of the platypus is used primarily in intraspecific fighting between males during the breeding season and that it is a potentially effective mechanism for ensuring spatial separation of males along riverine habitat.

The rarely heard voice of the platypus is a growl. This animal has keen sight and hearing, but when it submerges, the eyes and ears, which lie in a common furrow on each side of the head, are covered by skin folds that form the edges of the furrow. Thus rendered blind and deaf, the platypus relies mainly on the sense of touch, which is especially well developed in the bill.

Two kinds of burrows are constructed in banks of streams and ponds. One provides a shelter for both sexes and is retained by the male during the breeding season. The other, usually much deeper and more elaborate, is constructed only by the female and contains a nest for rearing the young. These burrows open above water, except during floods, and extend into the banks 1–7 meters above the water line and as far as 18 meters horizontally. A platypus, upon leaving the water, may enter its burrow and emerge moments later dry and glossy; the tunnels apparently squeeze out the water, which the soil absorbs like a sponge.

The platypus swims, dives, and digs well, using the forefeet more than the hind feet. It normally submerges for about a minute, but by holding onto some object it can stay underwater for as long as five minutes. The part of the web of the forefoot that extends beyond the nails when the platypus is swimming is folded under the palm when the animal is on land.

According to Collins (1973), the platypus is crepuscular in habit, usually confining its activities to the early morning and late evening hours. It then leaves its burrow and commences a unique quest for food. Swimming along the bottom of a freshwater stream or lake, a platypus probes the mud and gravel with the highly sensitive end of its rubbery bill. Until recently it was thought that edible material was located by the sense of touch alone, since the eyes and ears are closed when submerged, and that only those foodstuffs touched with the bill were snatched from the bottom. However, experiments by Scheich et al. (1986) indicate that the bill of the platypus is an electroreceptive as well as mechanoreceptive organ that allows the detection of muscle activity in prey animals. Proske, Gregory, and Iggo (1992) determined that weak electrical stimulus of the bill evoked a reaction in the brain.

When sufficient prey has been gathered and stored in the cheek pouches, or when breathing becomes necessary, the platypus proceeds to the surface and planes out with limbs outstretched and flattened tail trailing behind. Mastication of the food is accomplished by crushing and grinding it between the horny plates that act as teeth. Grit taken up with foodstuffs probably serves as an abrasive to aid in mastication. The diet consists of crayfish, shrimp, larvae of water insects, snails, tadpoles, worms, and small fish. In captivity the platypus has been known to eat food equivalent to about half its own weight in a day. The average daily intake of a 1.5-kg captive male included 450 grams of earthworms, 20–30 crayfish, 200 mealworms, 2 small frogs, and 2 coddled eggs.

Collins (1973) cited Fleay as stating that in Victoria the platypus undergoes periods of dormancy beginning in late May and ending before mid-September. However, Grant and Carrick (1978) indicated that the platypus is active year-round, even in the very cold water temperatures of winter, and responds to the thermal stress of cold by markedly declining in weight and tail volume.

Individuals may forage along several kilometers of a stream during a period of a few days (Grant et al. 1992). Apparently, a resident breeding population permanently occupies a particular area, though transient individuals also may be present (Gemmell et al. 1992). Radio-tracking studies in Victoria indicated a population density of 1.3–2.1 adults or subadults per kilometer of stream. Home range lengths of resident adult males were determined to be 2.9–7.0 km. Some of these ranges were mutually exclusive and others overlapped substantially, but in the latter case the animals avoided one another, spending most of their time in different parts of the shared area. All male ranges apparently overlapped those of two or more adult females, which in turn overlapped extensively (Gardner and Serena 1995; Serena 1994).

The platypus mates from July to November. Mating takes place in the water and is preceded by an unusual and

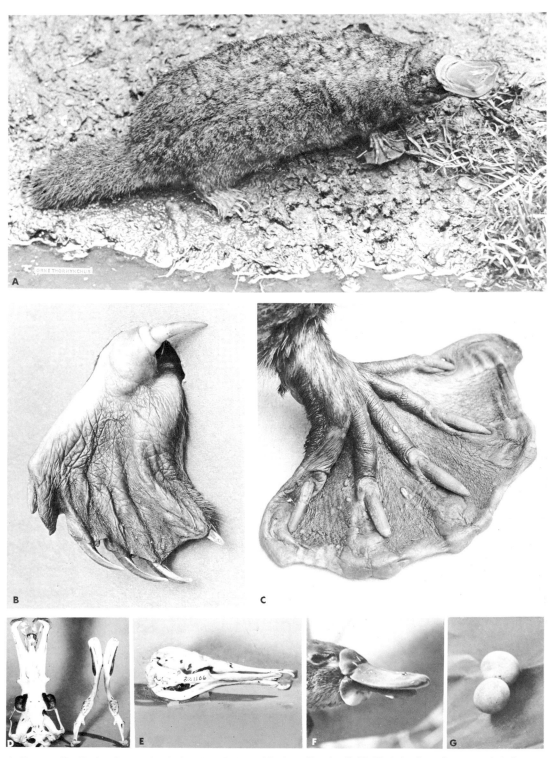

A. Platypus *(Ornithorhynchus anatinus),* photo from Zoological Society of London. B. Hind foot showing poison spur, photo from New York Zoological Society. C. Forefoot, photo from New York Zoological Society. D & E. Skull, photos by Howard E. Uible. F. Head and bill, photo from New York Zoological Society. G. Eggs, photo from New York Zoological Society.

intricate courtship during which, among other maneuvers, the male grasps the female's tail and the two animals then swim slowly in circles. The female carries bundles of wet leaves under her forwardly folded tail into the nest chamber. The wet leaves provide moisture and prevent the eggs from becoming too dry. The burrow used for incubation and rearing is plugged with earth at one or more places by the female. About 27 days after mating, one to three, usually two, eggs are laid in the nest. The young hatch after an incubation period of about 10 days. The incubating mother does not come out of the burrow for days at a time, and then only briefly to defecate, wash, and wet her fur. When leaving and returning to the nest chamber, she removes and then rebuilds the earth plugs in the tunnel. The eggs, 16–18 mm in length and 14–15 mm in diameter, are about the size

of those of a house sparrow though slightly more spherical. They adhere together after they are laid and thus do not roll away. The female curls about the eggs while incubating. No pouch is present in the female at any time during the reproductive process. When hatched, the young are about 25.4 mm long, blind, and naked, and the female curls around them. After about four months, when fully furred and about 335 mm in length, the young emerge from the burrow.

Grant and Griffiths (1992) captured 512 platypuses during an 18-year study in New South Wales. Lactation was observed to last three to four months, with most lactating females being found from November to February. Females did not begin to breed until their second or third year of life, and some did not breed in each consecutive year. Wild

Platypus *(Ornithorhynchus anatinus)*, photos by Bernhard Grzimek.

individuals up to 13 years old were found, and it was noted that captives have lived up to 21 years.

Until the turn of the century the platypus was intensively hunted for the fur trade, and until about 1950 it also was highly subject to accidental drowning in the nets of inland fisheries (Grant 1993). These factors probably suppressed platypus numbers, but the species now has made a successful comeback through effective government conservation efforts. Recent surveys indicate that it remains common in its original range except in South Australia; an introduced population is present on Kangaroo Island (Grant 1992). Nonetheless, Carrick (*in* Strahan 1983) regarded the platypus as vulnerable because of habitat disruption caused by dams, irrigation projects, fish netting and trapping, and pollution.

An early Cretaceous genus and species, *Steropodon galmani* from New South Wales, appears to represent an ornithorhynchidlike monotreme (Archer et al. 1985). An early Paleocene genus and species, *Monotrematum sudamericanum*, was recently discovered in southern Argentina and provides evidence of a former land bridge connecting Australia, Antarctica, and South America that allowed the dispersal of early mammals (Pascual et al. 1992). An Oligocene or Miocene genus, *Obdurodon*, with two species, *O. insignis* of South Australia and *O. dicksoni* of northwestern Queensland, has been assigned to the famiy Ornithorhynchidae (Archer et al. 1992, 1993). The genus and species *Ornithorhynchus anatinus* is known from the Pleistocene and Recent of mainland Australia and from the Recent of Tasmania (Archer, Plane, and Pledge 1978).

Order Didelphimorphia

American Opossums

This marsupial order of 4 Recent families, 15 genera, and 66 species is found naturally from southeastern Canada, through the eastern United States and Mexico, into South America to about 47° S in Argentina. Opossums also occur on some of the islands in the Lesser Antilles. The vernacular name "opossum" generally applies to members of this order, which now is found only in the New World. The term "possum" is used mainly for members of certain marsupial families in the Australasian region.

This order has essentially the same living content as the family Didelphidae, as the latter was accepted by nearly all authorities, for example, Archer (1984a) and Kirsch (1977b), until about a decade ago. Subsequent taxonomists, including Aplin and Archer (1987), Marshall, Case, and Woodburne (1990), and Wilson and Reeder (1993), have tended to give ordinal rank to the group. Hershkovitz (1992a, 1992b), who is followed here with respect to names and sequence of families and genera, regarded the group as an order but applied to it the respective terms Didelphoidea and Didelphida and used the name Didelphimorphia for a cohort. Szalay (1993, 1994) also used the name Didelphida for the order but included therein what is here regarded as the order Paucituberculata, as well as the Didelphimorphia, which he reduced to subordinal rank.

The members of this order are small to medium in size. For *Gracilinanus*, which includes the smallest species, adult head and body length is about 70–135 mm and tail length is 100–155 mm. In *Didelphis*, the largest genus, adult head and body length is 325–500 mm and tail length is 255–535 mm. The tail is usually long, scaly, very scantily haired, and prehensile, but in some forms it is short and/or rather hairy. Some genera have long, projecting guard hairs in the pelage. The muzzle is elongate, and the ears have well-developed conches. The limbs are short in many species, with the hind limbs slightly longer. All four feet have five separate digits. The great toe is large, clawless, and opposable. A distinct marsupial pouch is present in *Didelphis*, *Chironectes*, *Philander*, and *Lutreolina*. In the other genera it is absent, or it may consist of two longitudinal folds of skin, separate at both ends, near the median line of the body. Females have 5–27 mammae.

The dental formula for the order is: (i 5/4, c 1/1, pm 3/3, m 4/4) $\times$ 2 = 50. The teeth are rooted and sharp. The upper incisors are conical, small, and unequal; the first is larger than, and separated from, the others. The canines are large. The last premolar is preceded by a multicuspidate, molariform, deciduous tooth, and the molars are tricuspidate. The facial part of the skull is long and pointed, but the cranial part is small. Reig, Kirsch, and Marshall (1987) provided extensive further details on the morphological characters of this group of marsupials.

Opossums are active mainly in the evening and at night. Most forms are arboreal or terrestrial, and one genus (*Chironectes*) is semiaquatic. Opossums are insectivorous, carnivorous, or, more commonly, omnivorous.

As in other marsupials, the gestation period is short and the developmental period is long. The gestation period of *Didelphis*, about 12.5–13 days, is among the shortest found in mammals. The front legs of a newborn opossum are well developed, and the fingers are supplied with sharp, deciduous claws (at least in *Didelphis*); these claws drop off some time after the young reaches the pouch. The hind legs of opossums at birth are much smaller than the forelegs and practically useless. In young opossums the passage from the nasal chamber to the larynx is so separated from the passage to the esophagus that the baby can swallow and breathe at the same time. Contrary to an earlier opinion, the mother does not pump milk into the young; they suckle normally, though continuously, under their own power (Banfield 1974; Lowery 1974). The part of the brain that regulates body temperature is not functional in a young opossum; the baby is kept warm solely by the mother's body heat.

The pouch young of opossums and probably other pouch marsupials breathe and rebreathe air that contains 8–20 times the normal content of carbon dioxide. This may serve some unknown, useful role, or it may merely show unusual tolerance. After leaving the pouch the young of some species travel with the mother for awhile, usually riding by clinging to the fur on her back.

A number of carnivores prey on opossums, and people eat some of the larger species. Didelphid pelts are sometimes used for inexpensive trimmings and garments. Opossums, particularly members of the genus *Didelphis*, are widely used in laboratory research. South American didelphids are susceptible to yellow fever virus.

The statement that opossums copulate through the nose and that sometime later the tiny young are blown into the pouch is entirely false. This idea may have originated with the observation that the pouch is often filled with newborn young soon after the female has investigated the pouch with her nose. Or possibly the tale stems from the possession of a forked penis by the male.

The known geological range of this order is early Cretaceous to early Miocene and Pleistocene to Recent in North America, late Cretaceous to Recent in South America, Eocene to early Miocene in Europe, and Oligocene in North Africa (Hershkovitz 1995; Kirsch 1977a, 1977b; Simons and Bown 1984).

Eight baby murine opossums *(Marmosa murina)* attached to the mother's nipples, photo from New York Zoological Society. B. A female North American opossum *(Didelphis virginiana)* with a well-developed baby trying to get into her pouch, photo by Ernest P. Walker.

DIDELPHIMORPHIA; **Family MARMOSIDAE**

Pouchless Murine, or Mouse, Opossums

This family of 8 Recent genera and 53 species is found from northeastern Mexico, throughout Central and South America, to southern Argentina. Although traditionally considered a part of the Didelphidae, and maintained as such by Gardner (*in* Wilson and Reeder 1993), the Marmosidae were elevated to familial rank by Hershkovitz (1992*a*, 1992*b*). The latter authority, who is followed here with respect to sequence of genera, recognized five subfamilies: the Marmosinae, with *Gracilinanus, Marmosops, Marmosa,* and *Micoureus;* the Thylamyinae, with *Thylamys;* the Lestodelphyinae, with *Lestodelphys;* the Metachirinae, with *Metachirus;* and the Monodelphinae, with *Monodelphis.*

According to Hershkovitz (1992*b*), head and body length is 70–310 mm and tail length is 45–390 mm. There is no pouch, there are 9–27 mammae, the fifth digit on the forefoot has a sharp claw, and the hallux is opposable. In the subfamily Marmosinae the tail is long and prehensile, the claws are short, and the anklebones are specialized. The Thylamyinae have a prehensile but incrassate (seasonally thickened) tail, long and stout claws, parallel-sided nasal bones, and unspecialized anklebones. In the Lestodelphyinae the tail is short, incrassate, and nonprehensile, the claws are stout, the skull has a sagittal crest, and the foot bones are specialized. The relatively large and terrestrial Metachirinae have a long, nonprehensile tail and unspecialized foot bones. The shrewlike Monodelphinae have a short and nonprehensile tail, long claws, a sagittally crested cranium, and unspecialized anklebones.

Recently Hershkovitz (1995) described the genus *Adinolon,* apparently a component of the family Marmosidae from the latter part of the early Cretaceous of Texas. It may be the oldest known member of the Didelphimorphia and is among the earliest of marsupials.

DIDELPHIMORPHIA; MARMOSIDAE; **Genus GRACILINANUS**
Gardner and Creighton, 1989

Gracile Mouse Opossums

There are nine described species (Hershkovitz 1992*b*; Pacheco et al. 1995; Tate 1933):

G. agilis, Colombia, Peru, eastern and south-central Brazil, Bolivia, Paraguay, Uruguay, northeastern Argentina;
G. microtarsus, southeastern Brazil;
G. marica, Colombia, northern Venezuela;
G. emiliae, northern Brazil;
G. perijae, northern Colombia;
G. aceramarcae, Peru, western Bolivia;
G. dryas, Colombia, northwestern Venezuela;
G. longicaudus, central Colombia;
G. kalinowskii, southern Peru.

Hershkovitz (1992*b*) reported the presence of an undescribed species, perhaps with some affinity to *G. kalinowskii,* in northwestern Ecuador. He also indicated that *G. microtarsus* may be only a subspecies of *G. agilis.* The species of *Gracilinanus* long were placed in *Marmosa,* but Creighton (1985) suggested their distinction and most subsequent authorities have agreed. Except as noted, the information for the remainder of this account was taken

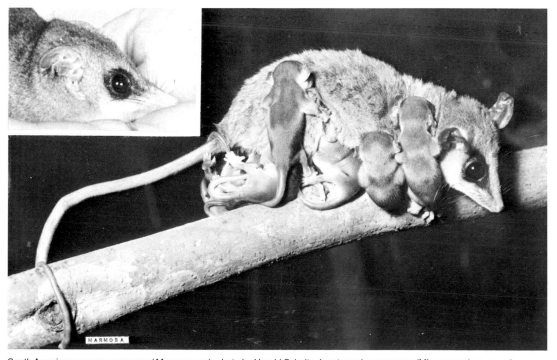

South American mouse opossums (*Marmosa* sp.), photo by Harald Schultz. Inset: murine opossum *(Micoureus demerarae)* showing partially coiled ear, photo by Ernest P. Walker.

from Gardner and Creighton (1989) and Hershkovitz (1992*b*).

Head and body length is 70–135 mm and tail length is 90–155 mm. Weight in *G. agilis* is 23–34 grams (Redford and Eisenberg 1992). The upper parts range from dull brownish gray to bright reddish brown; the underparts are paler, varying from almost white through cream to orange. There is a dark brown or blackish eye ring, the upper surface of the muzzle is pale, and the tail is scaled and weakly bicolored or unicolored fuscous. There is no pouch. Mammae number about 11–15 and are mostly abdominal, but usually a few are pectoral. The closely related genera *Marmosops*, *Marmosa*, and *Micoureus* lack pectoral mammae.

The skull of *Gracilinanus*, unlike those of *Marmosa* and *Micoureus*, lacks postorbital processes, the rostrum is slender, and the hard palate is highly fenestrated. The auditory bullae are tripartite and are relatively large compared with those of *Marmosa*, *Marmosops*, and *Micoureus*. The second upper premolar is always larger than the third, and the lower canine teeth are relatively shorter than those of *Marmosa*, *Micoureus*, and *Thylamys* but not as short or premolariform as in *Marmosops*. The caudal scales of *Gracilinanus* (except *G. kalinowskii*) and *Thylamys* are arranged in an annular pattern, whereas those of *Marmosa*, *Marmosops*, and *Micoureus* have a spiral pattern. Unlike that of *Thylamys*, the tail of *Gracilinanus* does not become seasonally thickened (incrassate) by fat deposition.

These tiny opossums inhabit forests and woodlands from coastal areas to elevations of about 4,500 meters in the Andes. They dwell in trees or shrubs but frequently forage on the ground. Locomotion is by a short, swift quadrupedal gait interchanged with overhand climbing and headfirst descents, assisted by use of the prehensile tail. *G. agilis* has been found in nests made of vegetation 1.6 meters above the ground (Redford and Eisenberg 1992). Individuals have been observed to enter torpor in response to cool weather. The diet consists of fruit, plant exudates, insects, and other small invertebrates. Breeding may continue throughout the year, at least when food is abundant. Litter size in *G. marica* has been reported to be six young (Hayssen, Van Tienhoven, and Van Tienhoven 1993). *G. ac-*

eramarcae, known only from two restricted areas of declining habitat, is classified as critically endangered by the IUCN. *G. emiliae* and *G. dryas*, also losing their limited habitat, are classified as vulnerable, while *G. agilis*, *G. microtarsus*, and *G. marica* are designated as near threatened.

DIDELPHIMORPHIA; MARMOSIDAE; **Genus MARMOSOPS**
Matschie, 1916

There are nine species (Gardner *in* Wilson and Reeder 1993; Gardner and Creighton 1989; E. R. Hall 1981; Handley and Gordon 1979; Pine 1981):

M. noctivagus, Amazonian Brazil, Ecuador, Peru, Bolivia;
M. dorothea, Bolivia;
M. incanus, eastern and southeastern Brazil;
M. invictus, Panama;
M. parvidens, Colombia, Venezuela, Guyana, Surinam, Peru, Brazil;
M. handleyi, known only from the type locality in central Colombia;
M. fuscatus, eastern Colombia, western Venezuela, Trinidad;
M. cracens, northwestern Venezuela;
M. impavidus, western Panama to Peru and southern Venezuela.

Marmosops long was considered a synonym or subgenus of *Marmosa* but now is generally recognized as a full genus (Corbet and Hill 1991; Gardner *in* Wilson and Reeder 1993; Gardner and Creighton 1989; Hershkovitz 1992*a*, 1992*b*). The information for the remainder of this account was taken from Eisenberg (1989) and Hershkovitz (1992*b*).

Head and body length is 90–160 mm, tail length is 105–220 mm, and weight is about 24–85 grams. The upper parts are pale to dark brown or gray, the underparts are pale gray to white, and there is a dark eye ring. The tail is prehensile and, unlike that of *Thylamys*, is never incrassate

South American mouse opossum *(Marmosops incanus)*, photo by Luiz Claudio Marigo.

(seasonally thickened by fat deposition). The caudal scales are arranged in a spiral pattern, not in an annular pattern as in *Gracilinanus*. Unlike those of *Marmosa* and *Micoureus*, the skull lacks strong postorbital processes and a lambdoidal crest. There is no pouch and there are only nine mammae, none of which are pectoral. Pregnant female *M. parvidens* have been found to contain six to seven embryos. *Marmosops* is nocturnal and arboreal, generally inhabits moist forests in either lowlands or mountains, and feeds on insects and fruit. Most species of *Marmosops* are losing their habitat to human activity. *M. handleyi*, known only from a single locality, is classified as critically endangered by the IUCN. *M. cracens*, also known only from one site, is classed as endangered, while *M. dorothea* is classed as vulnerable and *M. incanus*, *M. invictus*, *M. parvidens*, *M. fuscatus*, and *M. impavidus* are designated as near threatened.

DIDELPHIMORPHIA; MARMOSIDAE; **Genus**
MARMOSA
Gray, 1821

Murine, or Mouse, Opossums

There are two subgenera and nine species (Cabrera 1957; Collins 1973; Gardner *in* Wilson and Reeder 1993; E. R. Hall 1981; Handley and Gordon 1979; Husson 1978; Kirsch 1977a; Kirsch and Calaby 1977; Pine 1972b; Reig, Kirsch, and Marshall 1987; Streilen 1982d):

subgenus *Marmosa* Gray, 1821

M. lepida, Surinam to Ecuador and Bolivia;
M. murina, northern and central South America, Tobago;
M. rubra, eastern Ecuador, northeastern Peru;
M. tyleriana, southern Venezuela;

M. robinsoni, Belize, Ruatan Island (Honduras), Panama to Ecuador, Venezuela, Trinidad and Tobago, Grenada;
M. xerophila, northeastern Colombia, northwestern Venezuela;
M. mexicana, northeastern Mexico to western Panama;
M. canescens, western Mexico, Yucatan, Tres Marías Islands;

subgenus *Stegomarmosa* Pine, 1972

M. andersoni, southern Peru.

Until recently *Marmosa* commonly was considered to include the species now assigned to the genera *Gracilinanus*, *Marmosops*, *Micoureus*, and *Thylamys* (see accounts thereof).

Head and body length is 85–190 mm and tail length is 125–230 mm. Weight in *M. robinsoni*, one of the larger, more common species, is 60–130 grams for males and 40–70 grams for females. *M. murina*, a smaller species, weighs only 13–44 grams (Eisenberg 1989). In *M. mexicana* weight is 29–92 grams, but averages are 63.7 grams for males and only 35.2 grams for females (Alonso-Mejía and Medellín 1992). The upper parts range from gray through dark brown to bright reddish brown; the underparts are paler, varying from almost white through buff to yellowish. Almost all forms have dusky brown or black facial markings around the eyes. The fur is short, fine, and velvety in most species. The tail is strongly prehensile, has only a short basal brush of fur, and either is uniformly colored or has the ventral surface slightly paler than the dorsal. There is no pouch. Mammae number 9–19 depending on the species.

According to Hershkovitz (1992b), *Marmosa* differs from *Gracilinanus* in lacking pectoral mammae, in having a pronounced superior postorbital process on the skull, and in having its caudal scales arranged in a spiral, rather than

Mouse opossum *(Marmosa mexicana)*, photo by Barbara L. Clauson.

annular, pattern. *Marmosa* differs from *Marmosops* in having a throat gland present in the mature males and a lambdoidal crest on the skull. *Marmosa* differs from *Micoureus* in having a throat gland, shorter fur, and a ridged and beaded superior border of the frontal bone of the skull. *Marmosa* differs from *Thylamys* in having the tail never incrassate, weaker claws on the forefoot, the nasal bones of the skull flared at the frontomaxillary suture, and the second premolar tooth usually larger than the third (rather than the opposite).

Most species of *Marmosa* are forest dwellers. The genus ranges vertically from sea level to about 3,400 meters. Most specimens of *Marmosa* collected by Handley (1976) in Venezuela were taken near streams or in other moist areas, but *M. xerophila* was almost always found in dry situations. Mouse opossums are nocturnal and usually arboreal, though some species are terrestrial. They are often found on banana plantations and among small trees and vine tangles. They build nests of leaves and twigs in trees or shelter in abandoned birds' nests. They have also been located in ground nests, in holes in hollow logs, and under rocks. *M. mexicana* sometimes digs its own burrow but more commonly nests in trees (Alonso-Mejía and Medellín 1992). Most specimens of *M. canescens* taken in Sinaloa, Mexico, were found by day in their nests in hollows of dead cacti (Armstrong and Jones 1971). The species *M. robinsoni* does not seem to have a fixed abode but apparently spends the day in any suitable shelter when daylight overtakes it. This species reportedly is nomadic in Panama, occupying home ranges of about 0.22 ha. for two or three months and occurring at population densities of 0.31–2.25/ha. (Fleming 1972; Hunsaker 1977b). In northern Venezuela *M. robinsoni* was estimated to occur at densities of 0.25–4.25 adults per hectare, and males were found to be more nomadic than females (O'Connell 1983). Mouse opossums are generally solitary, usually hunting and nesting alone. Fleming (1972) obtained data suggesting that while there was overlap between the home ranges of male *M. robinsoni* and between those of opposite sexes, the females had nonoverlapping home ranges and appeared intolerant of one another.

Murine opossums are courageous fighters for their size. Like some of the other marsupials, they can lower their ears by crinkling them down much as a sail is furled. Their eyes reflect light as brilliant ruby red. When moving along tree limbs and vines, *M. robinsoni* curves its tail loosely around the branch, except when climbing vertically, and sometimes leaps across gaps. None of its motions, however, are particularly rapid. The diet of *Marmosa* consists mainly of insects and fruits but also includes small rodents, lizards, and birds' eggs. Large grasshoppers are killed by a number of bites about the head and thorax; only the harder parts and lower legs are discarded. One mouse opossum in Mexico was noted hanging by its tail and eating a wild fig, which it held in its forefeet. Raw sugar is a particular favorite of *M. mexicana*. Banana and mango crops sometimes are damaged by these marsupials. Occasionally they are found in bunches of bananas in warehouses and stores in the United States, having remained in the bunch when it was shipped from the tropics. Occurrences in the New Orleans area, apparently resulting from such journeys, were documented by Lowery (1974).

Some species of *Marmosa* may breed throughout the year, while others are seasonal. According to Hunsaker (1977b), *M. robinsoni* has breeding peaks in February, June, and July through December, an estrous cycle averaging 23 days, and a gestation period of 14 days. Litter size is seven to nine in the wild and slightly smaller in captivity. Weaning occurs when the young are 60–70 days old, they leave

the mother a few days later, and young females have their first estrus when 265–75 days old. Life expectancy probably is under one year in the wild but is about three years in captivity. Reproductive ability declines in the second year of life.

Some slightly different information compiled by Collins (1973) is that gestation is just under 14 days, average size of 70 laboratory litters was about seven, females reach sexual maturity at approximately six months, and average life span in captivity is one to two years. One male *M. robinsoni* obtained at an unknown age lived for almost five more years, an apparent record longevity.

Godfrey (1975) found that litter size in a laboratory colony of *M. robinsoni* ranged from 1 to 13 and usually was 7–9. Litter sizes of up to 15 have been reported in Venezuela (O'Connell 1983). A study in the Panama Canal Zone found *M. robinsoni* to be a seasonal breeder, with reproduction occurring mainly from late March to September and litter size averaging 10 (Fleming 1973). According to Hunsaker and Shupe (1977), females of *M. murina* produce three litters annually and breed throughout the year. Eisentraut (1970) reported that *M. murina* has a gestation period of 13 days and a maximum litter size of 13. A female *M. canescens* taken on 5 September contained 13 embryos, one found on 4 September was accompanied by 8 nursing young, and females taken in October were lactating (Armstrong and Jones 1971).

As noted by Grimwood (1969), because of their small size and nocturnal habits populations of mouse opossums are not directly threatened by people; however, some of the high-altitude species might be jeopardized by widespread clearing of brush in the Andes region. *M. andersoni*, known only from a single montane locality, is now classified as critically endangered by the IUCN. *M. xerophila*, also restricted to a small area, is classed as endangered, while *M. lepida* is designated as near threatened.

DIDELPHIMORPHIA; MARMOSIDAE; Genus **MICOUREUS** *Lesson, 1842*

There are four species (Cabrera 1957; Gardner *in* Wilson and Reeder 1993; Kirsch 1977a; Kirsch and Calaby 1977):

M. demerarae, Venezuela to Paraguay;
M. alstoni, Belize to Colombia;
M. constantiae, Mato Grosso of Brazil, Bolivia, northern Argentina;
M. regina, Colombia, Ecuador, Peru, Bolivia.

Micoureus long was considered a synonym or subgenus of *Marmosa* but now is generally recognized as a full genus (Corbet and Hill 1991; Gardner *in* Wilson and Reeder 1993; Gardner and Creighton 1989; Hershkovitz 1992a, 1992b). *M. demerarae* formerly was known as *Marmosa cinerea*. The information for the remainder of this account was taken from Eisenberg (1989), Hershkovitz (1992b), Redford and Eisenberg (1992), and Reig, Kirsch, and Marshall (1987).

Head and body length is 120–215 mm, tail length is 170–270 mm, and weight in *M. demerarae* is 50–230 grams. The upper parts are various shades of gray or brown, the underparts are cream or buff to yellow, and there is usually a dark eye ring. The pelage is thick, lax, and crinkly. The prehensile tail may be uniformly colored or bicolored, with the long terminal portion distinctly paler than the proximal portion. The fur extends at least 20 mm onto the tail, whereas in *Marmosa* it usually extends less than 20 mm. *Micoureus* also differs from *Marmosa* in lacking a throat gland and in having the superior border of the frontal bone of the skull projected as a ledge. There are 9–15 mammae, none of which are pectoral.

The only well-known species is *M. demerarae*, which lives mainly in tropical forest, forages in trees and on the ground, is nocturnal, and eats mostly insects and some fruit. The female builds an open nest by transporting leaves with her mouth or prehensile tail. Breeding in Venezuela is tied to rainfall and does not take place during the winter dry season. *M. alstoni* and *M. constantiae* are now designated as near threatened by the IUCN.

DIDELPHIMORPHIA; MARMOSIDAE; **Genus THYLAMYS**
Gray, 1843

Southern Mouse Opossums

There are five species (Cabrera 1957; Collins 1973; Gardner *in* Wilson and Reeder 1993; Gardner and Creighton 1989):

T. elegans, central Peru, Bolivia, Chile, northwestern
  Argentina;
T. macrura, Paraguay, southern Brazil;
T. pallidior, Bolivia, Argentina;
T. pusilla, central and southern Brazil, Argentina,
  Paraguay, southern Bolivia;
T. velutinus, southeastern Brazil.

Thylamys long was considered a synonym or subgenus of *Marmosa* but now is generally recognized as a full genus (Corbet and Hill 1991; Gardner *in* Wilson and Reeder 1993; Gardner and Creighton 1989; Hershkovitz 1992*a*, 1992*b*). *T. macrura* formerly was known as *Marmosa grisea*. Except as noted, the information for the remainder of this account was taken from Eisenberg (1989), Hershkovitz (1992*b*), and Redford and Eisenberg (1992).

Head and body length is 68–150 mm, tail length is 90–160 mm, and weight is about 18–55 grams. The upper parts are various shades of gray or brown, the underparts

are yellowish or white, and there is a dark eye ring. The fur is very dense and soft. The tail is naked or only finely haired, except at the base, and is seasonally incrassate, that is, becoming thickened through storage of fat. In this last regard *Thylamys* differs from *Gracilinanus*, *Marmosops*, *Marmosa*, and *Micoureus*. It also differs in having stout and well-projecting claws on the forefoot, parallel-sided nasal bones, and the third upper premolar larger than the second. There are 15 mammae, including two pectoral pairs, and there is little or no pouch development.

These mouse opossums occur in a variety of habitats from sea level to the mountains and are both arboreal and terrestrial. *T. elegans* has been found in wet forest, brush, and scrub and may nest in trees, rocky embankments, or holes in the ground made by other animals. It stores fat and hibernates during the winter. *T. pusilla* frequently occurs in much drier areas, such as thorn forest, and has been found active even when there is snow on the ground. The diet of *T. elegans* consists primarily of insects but also includes fruit and small vertebrates. *T. elegans* seems to occur at low densities in all habitats; two individuals had an average home range of 289 sq meters. In Chile it reproduces from September to March, during which time a female can have two litters. It has as many as 15 young, usually 8–12. The IUCN now designates *T. macrura* as near threatened.

DIDELPHIMORPHIA; MARMOSIDAE; **Genus LESTODELPHYS**
Tate, 1934

Patagonian Opossum

The single species, *L. halli,* is known only from nine specimens taken at three localities in Patagonia, Argentina (L. G. Marshall 1977), and from a locality in the mountains of west-central Argentina (Redford and Eisenberg 1992). It occurs farther south than any other living marsupial.

Head and body length is 132–44 mm and tail length is 81–99 mm. One specimen weighed 76 grams (Redford and Eisenberg 1992). The fur is not particularly long, but it is dense, fine, and soft. The back is dark gray; the face is somewhat darker with no markings; the sides of the body are

Southern mouse opposum *(Thylamys elegans)*, photo by Milton H. Gallardo.

Patagonian opossum *(Lestodelphys halli)*, photo by Oliver P. Pearson.

clear gray; and the forearms, hands, ankles, feet, and underparts are white. The cheeks, a patch over the eyes, and a patch at the posterior base of the ear are also white. There are dark shoulder and hip patches. The tail, which is furred like the body for about 20 mm at the base, then thickly covered with short, fine hairs, is dark grayish brown above and whitish below and at the tip.

In general appearance individuals of this genus resemble certain species of the subfamily Marmosinae, but they differ in characters of the feet and in the distinctive skull and dental features. The feet are stronger than those of the Marmosinae; the Patagonian opossum probably is more terrestrial than the murine opossums. In *Lestodelphys* the claw of the thumb and other digits extends considerably beyond the soft terminal pad; in the Marmosinae it is markedly shorter than the others and does not extend beyond the pad. The Patagonian opossum has a short, broad skull, small incisors, long, sharp, straight canines, and large molars. In *Lestodelphys* the bullar floor of the skull is complete and does not have a gap between the petrous and alisphenoid components such as are found in all other genera of the Marmosidae (Hershkovitz 1992*b*). As in the Thylamyinae, the tail occasionally becomes thickened near the base due to a seasonal accumulation of fat. The ears are short, rounded, and flesh-colored.

This opossum is said to inhabit pampas areas. One specimen was caught in a steel trap baited with a dead bird. Thomas (1921:138) commented: "This interesting little opossum . . . appears, from the structure of its skull, to be of a more carnivorous and predaceous nature than any of the other small members of the family. Ordinarily *Marmosa* feeds mainly on insects and fruit, and as insects are rare and fruit almost non-existent in its far-southern habi-

tat, this opossum has had to acquire peculiar habits, and no doubt lives largely on mice and small birds." Redford and Eisenberg (1992) cited Oliver Pearson as reporting that captive *Lestodelphys* "killed live mice at lightning speed, eating everything—bones, teeth, and fur. A 70 gram animal will eat an entire 35 gram mouse in one night." Because it apparently occurs in a very restricted area of declining habitat, *L. halli* now is classified as vulnerable by the IUCN.

DIDELPHIMORPHIA; MARMOSIDAE; Genus METACHIRUS
Burmeister, 1854

Brown "Four-eyed" Opossum

The single species, *M. nudicaudatus*, is found from extreme southern Mexico to northeastern Brazil and northeastern Argentina (Husband et al. 1992; Medellín et al. 1992; Redford and Eisenberg 1992). Because of technical problems of nomenclature there has been recent argument regarding which generic name properly applies to this species. Hershkovitz (1976, 1981) supported *Metachirus*, which has been in more general use, while Pine (1973*a*) favored *Philander* Tiedemann, 1808, at least for the time being. E. R. Hall (1981) followed Pine in applying *Philander* to the brown "four-eyed" opossum, but Corbet and Hill (1991), Gardner (1981 and *in* Wilson and Reeder 1993), and Kirsch and Calaby (1977) agreed with Hershkovitz. The generic designation *Philander* is actually applicable to the gray and black "four-eyed" opossums and is so employed here, again in accordance with Hershkovitz (1976, 1981) and his backers,

Brown "four-eyed" opossum *(Metachirus nudicaudatus)*, photo by Cory T. de Carvalho.

rather than the name *Metachirops,* used by Pine (1973*a*) and E. R. Hall (1981).

Head and body length is 190–310 mm and tail length is 195–390 mm (Hershkovitz 1992*b*). Recorded weights in Argentina and Paraguay have ranged from 91 to 480 grams (Redford and Eisenberg 1992). A male and a female from Barro Colorado Island, Panama Canal Zone, each weighed 800 grams. The back and sides are brown, often dark cinnamon brown, and the rump may be washed with black. The face is dusky, almost black in some individuals, with a creamy white spot over each eye. These spots, suggesting eyes, are usually smaller and more widely separated than those of *Philander.* The underparts are buff to gray. The tail is furred for a short distance basally. The pelage is short, dense, and silky.

Although the common names and general appearance of *Metachirus* and *Philander* are similar, the two are not closely related. *Metachirus* may be distinguished externally by its brown coloration and longer tail. Unlike the gray and black "four-eyed" opossums, the females of *Metachirus* lack a pouch, having instead simple lateral folds of skin on the lower abdomen, in which are located the mammae. Females with five, seven, and nine mammae have been recorded (Collins 1973). Kirsch (1977*b*) observed that whereas *Philander* is probably the most aggressive of didelphids, *Metachirus* is almost quiet when held in the hand.

The brown "four-eyed" opossum lives in dense forests or in thickets in open, brushy country. It builds round nests of leaves and twigs in tree branches but occasionally makes its shelter under logs or rocks. Handley (1976) reported that all 18 specimens taken in Venezuela were caught on the ground, mostly near streams or in other moist areas. Members of this genus are completely nocturnal, rarely moving from the nest until dark. The diet includes fruits, insects, mollusks, amphibians, reptiles, birds, eggs, and small mammals. *Metachirus* has been accused of damaging fruit crops in some areas.

Limited data indicate that this opossum is seasonally polyestrous (Fleming 1973). It reportedly breeds in November in Central America, has litters of one to nine young, and probably has a maximum life span of three to four years (Hunsaker 1977*b*). The single, 51-mm young of a female obtained on 18 December was then already able to stand alone. It later rode on its mother's back or hips and was fully independent by early February (Collins 1973).

DIDELPHIMORPHIA; MARMOSIDAE; **Genus**
MONODELPHIS
Burnett, 1830

Short-tailed Opossums

There are 2 subgenera and 15 species (Anderson 1982; Cabrera 1957; Collins 1973; Gardner *in* Wilson and Reeder 1993; Kirsch and Calaby 1977; Massoia 1980; Pine 1975, 1976*a*, 1977, 1979; Pine and Abravaya 1978; Pine, Dalby, and Matson 1985; Pine and Handley 1984; Soriano 1987):

subgenus *Monodelphis* Burnett, 1830

M. brevicaudata, Venezuela and Guianas to northern Argentina;
M. adusta, eastern Panama to western Venezuela and northern Peru;
M. osgoodi, southern Peru, western Bolivia;
M. kunsi, Bolivia, Brazil;
M. domestica, eastern and central Brazil, Bolivia, Paraguay;
M. maraxina, Isla Marajo in northeastern Brazil;
M. americana, northern and central Brazil;
M. sorex, southern Brazil, southeastern Paraguay, northeastern Argentina;
M. emiliae, Amazonian Brazil, northeastern Peru;
M. iheringi, southern Brazil;
M. theresa, eastern Brazil, Andes of Peru;
M. unistriata, southeastern Brazil;

subgenus *Minuania* Cabrera 1919

M. rubida, eastern Brazil;
M. scalops, southeastern Brazil;
M. dimidiata, southeastern Brazil, Uruguay, eastern Argentina.

Head and body length is 110–200 mm and tail length is 45–85 mm. Fadem and Rayve (1985) listed the weight of *M. domestica* as 90–150 grams in males and 80–100 grams in females. Other recorded weights are 24–78 grams for *M. brevicaudata,* 40–84 grams for *M. dimidiata,* 58–95 grams for *M. domestica,* and 741 grams for *M. scalops* (Eisenberg 1989; Redford and Eisenberg 1992). The tail is about half as long as the head and body, always shorter than the body alone. In most forms the tail is sparsely haired, with only a few millimeters at the base well furred. Coloration varies greatly depending upon the species. One group has a bright red-brown back and face, three faint, dusky brown stripes on the back, and buffy to buffy gray underparts. A second

Short-tailed opossum *(Monodelphis brevicaudata)*, photo by Robert S. Voss.

group has chestnut brown back and sides and paler under-parts, with a wash of buffy tips to the hairs. A third group has a well-marked gray dorsal stripe from nose to rump, bright red-brown sides, ashy to buffy gray underparts, and a black tail. A fourth group, which includes *M. adusta* of Panama, has a dark brown back and sides and grayish un-derparts. The fur is short, dense, and rather stiff. The pouch is not developed. There are 11 to 17 or more mammae, de-pending on the species; most are arranged in a circle on the abdomen, but there also are some pectoral mammae (Hershkovitz 1992*b*). The tail is only slightly prehensile but is not incrassate.

The habits of these opossums are not well known. They apparently are among the least well adapted members of the Didelphidae for arboreal life and are usually found on the ground, though they can climb fairly well. Most speci-mens of *M. brevicaudata* collected in Venezuela by Hand-ley (1976) were taken on the ground in moist areas. About half were caught in evergreen forest and about half in open areas. Nests of *Monodelphis* usually are built in hollow logs, in fallen tree trunks that bridge streams, or among rocks. Most species of the subgenus *Monodelphis* are thought to be nocturnal. Their food consists of small ro-dents, insects, carrion, seeds, and fruits. The species *M. do-mestica* received its name because of its habit in Brazil of living in dwellings, where it destroys rodents and insects. It is welcomed by the householders. One opossum was seen running through the double cane walls of an Indian hut carrying a piece of paper by curling its tail downward around the paper. Streilen (1982*b*, 1982*e*) found this species to be an efficient predator and particularly adept at captur-ing scorpions. Individuals were highly intolerant of one an-other, though conflicts rarely resulted in serious injury.

In a study in Argentina, Pine, Dalby, and Matson (1985) found *M. dimidiata* to be diurnal, with most activity con-centrated in the late afternoon. This species was considered to be predominantly insectivorous and not an efficient

predator of rodents. Breeding occurred in the summer months of December and January, and the young dispersed from March to May. The species evidently is semelparous; that is, individuals breed but once in their life, and very few, if any, survive past their second summer. One captive fe-male produced a litter of 16 young.

In the tropical part of their range short-tailed opossums apparently breed throughout the year. The number of young in the subgenus *Monodelphis* varies from 5 to 14 depending on the species, and the newborn cling to the nip-ples of the mother. Later the young ride on the back and flanks of the female. In *M. domestica*, according to Fadem and Rayve (1985), females have as many as four litters an-nually, the gestation period is 14–15 days, there are usual-ly 5–12 offspring, postpartum dependence lasts about 50 days, sexual maturity is attained at 4–5 months, and breed-ing has occurred at up to 39 months of age in males and 28 months in females. The period of estrus in that species was found to last 3–12 days, and the estrous cycle showed a bi-modal distribution, lasting about 2 weeks in one group of captive females and about 1 month in another group. One specimen of *M. domestica* was captured in the wild and subsequently lived four years and one month (Marvin L. Jones, Zoological Society of San Diego, pers. comm., 1995).

Most species of *Monodelphis* evidently are declining be-cause of habitat destruction. *M. kunsi,* which is thought to have been reduced by at least 50 percent in the last decade, is classified as endangered by the IUCN. *M. osgoodi, M. maraxina, M. sorex, M. emiliae, M. theresa, M. unistriata, M. rubida,* and *M. scalops* are classified as vulnerable, and *M. americana, M. iheringi,* and *M. dimidiata* are designat-ed as near threatened.

DIDELPHIMORPHIA; Family CALUROMYIDAE

This family of two living genera and four species is found from southern Mexico, throughout Central and South America, to northern Argentina. This family usually has been considered a subfamily of the Didelphidae that includes *Glironia* and was maintained as such by Gardner (*in* Wilson and Reeder 1993). However, Hershkovitz (1992*a*, 1992*b*) regarded the Caluromyidae as a full family that excludes *Glironia*, which he placed in the separate family Glironiidae.

According to Hershkovitz (1992*a*), a critical distinction expressing the separate evolution of the Caluromyidae and the Didelphidae is found in the anklebones—the astragalus and calcaneus. In the Didelphidae this joint has a primitive pattern in which two separate facets on the calcaneal surface articulate with a corresponding pair of separate facets on the astragalar plantar surface. The Caluromyidae, however, have a partially derived pattern in which there has been coalescence of the once dual facets of the calcaneus into a single continuous facet. The Caluromyidae also can be distinguished by karyotype and retention of a cloaca.

Hershkovitz (1992*b*) pointed out that a major distinction between the Caluromyidae and Glironiidae is found in the auditory or tympanic bullae, the dome-shaped housing for the inner ear at the posterior base of the skull. In the Caluromyidae the bullae are bipartite, with the floor formed by the close junction (but not fusion) of the alisphenoid and petrous bones, thereby enclosing the ectotympanic bone. In the Glironiidae, and also the Didelphidae and Marmosidae (except *Thylamys*), the bullae are tripartite, with the ectotympanic forming a portion of the floor and there being a gap between the alisphenoid and petrous bones. Also, in the Caluromyidae the skull has a sagittal crest and the bony palate is nearly entirely ossified, whereas the Glironiidae lack this crest and have midpalatal and posterolateral vacuities.

DIDELPHIMORPHIA; CALUROMYIDAE; Genus CALUROMYS
J. A. Allen, 1900

Woolly Opossums

There are three species (Cabrera 1957; E. R. Hall 1981; Kirsch and Calaby 1977; Massoia and Foerster 1974):

C. derbianus, southern Mexico to Ecuador;
C. lanatus, Colombia to northern Argentina and southern Brazil;
C. philander, Venezuela to southern Brazil.

Head and body length is 180–290 mm, tail length is 270–490 mm, and weight (Bucher and Fritz 1977) is 200–500 grams. The pelage is long, fine, and woolly and extends along almost all of, or more than the basal half of, the tail. Some forms are pale gray or otherwise not well marked, but woolly opossums usually have an ornate color pattern, including the diagnostic dark median stripe on the face, which extends between the ears and the eyes almost to the nose. Some individuals show a pale stripe extending backward from the shoulder region. The ornate color pattern may also include indistinct dark patches around the eyes, a light creamy white to buffy white face, and a reddish and blackish body.

These opossums can generally be distinguished by the striped face, woolly pelage, and characteristics of the tail. The prehensile tail of *Caluromys* is longer than the head and body; its terminal part is not haired and varies from cream-colored to pinkish.

Woolly opossums generally inhabit forested country and are more arboreal than the other large opossums. Nearly all specimens of *C. lanatus* and *C. philander* collected by Handley (1976) in Venezuela were taken in trees, usually near streams or other moist areas. These animals live in tree hollows or limbs and are active mainly during the evening, night, or early morning. They are quite agile, presenting a sharp contrast to *Didelphis*. A nest of leaves made by *C. derbianus* was found in a vine tangle in a small tree. This species, like *Didelphis*, reportedly coils its tail to carry nesting materials (Hunsaker and Shupe 1977). Woolly opossums are ap-

Ecuadorean woolly opossum *(Caluromys lanatus)*, photo by Ernest P. Walker.

parently fairly common throughout their wide range, but their numbers never seem to reach the population levels of *Philander* and *Didelphis*. Perhaps their arboreal habits make them less conspicuous than terrestrial didelphids.

Woolly opossums are omnivorous. Their diet in the wild consists of a variety of fruits, seeds, leaves, soft vegetables, insects, and small vertebrates. They reportedly feed on carrion as well. One captive group had a decided preference for meat (Collins 1973), while another colony had a special liking for fruit (Bucher and Fritz 1977).

According to Eisenberg (1989), the nightly foraging area of *C. philander* varies from 0.3 to 1.0 ha. depending on food availability. Individuals are solitary and adult interaction tends to be agonistic, but there seems to be no strong territorial defense and home ranges overlap. Females of that species can produce three litters a year. Studies of captive *C. philander* by Atramentowicz (1992) indicate that gestation lasts 25 days, apparently the longest in the Didelphimorphia; litter size is 1–7, and the young leave the pouch at 3 months and are weaned at 4 months.

Captive females of both *C. derbianus* and *C. lanatus* have modal estrous cycle lengths of 27–29 days and are cyclic throughout the year (Bucher and Fritz 1977). In the wild in Nicaragua *C. derbianus* also seems to be reproductively active all year (Phillips and Jones 1968). Litter size there averaged 3.3 (2–4). In Panama the breeding period reportedly begins with the onset of the dry season in February (Enders 1966), and litter size averaged between 3 and 4 (1–6). Sexual maturity is said to be attained by *C. derbianus* at seven to nine months. A number of captive *Caluromys* have survived for more than three years (Collins 1973), and known record longevity is six years and four months for a specimen of *C. philander* (Jones 1982).

All three species are thought to be declining because of habitat destruction. *C. derbianus* is classified as vulnerable by the IUCN, and *C. lanatus* and *C. philander* are designated as near threatened. In the past these animals were trapped

extensively for their pelts, but at present their fur is not popular. They are said to damage fruit crops occasionally in South America, but otherwise they are of little economic importance. Recently they have been shown to have potential use in laboratory research (Bucher and Fritz 1977).

DIDELPHIMORPHIA; CALUROMYIDAE; **Genus**
CALUROMYSIOPS
Sanborn, 1951

Black-shouldered Opossum

According to De Vivo and Gomès (1989), Emmons (1990), and Izor and Pine (1987), the single species, *C. irrupta,* is known with certainty only from three localities in southern Amazonian Peru and one on the upper Jarú River in west-central Brazil; Simonetta's (1979) report of a specimen from extreme southern Colombia is doubtful. The generic distinction of *Caluromysiops* from *Caluromys* has been questioned (Gewalt *in* Grzimek 1990), but Hershkovitz (1992*a*) put the two in completely separate tribes based in part on differences in the molar teeth.

Head and body length is 250–330 mm and tail length is 310–40 mm. The upper parts are gray, with two separate black lines that begin on the forefeet and run onto the back, join on the shoulders, then separate again and run parallel to each other down the back and over the rump to the hind limbs. The face has faint, dusky lines running through the eyes. The underparts are gray with buffy tips to the hairs, and there is a faint dusky line along the middle of the belly. The upper side of the tail for the basal two-thirds to three-quarters of its length is slightly darker than the gray of the body; the remainder of the tail is creamy white. The tail is well furred, except the underside of the last three-fourths, which is naked. The fur is long, dense, and woolly.

Black-shouldered opossum *(Caluromysiops irrupta)*, photo from New York Zoological Society.

The skull is similar to that of *Caluromys*, but the molars are much larger and the rostrum is relatively shorter.

The black-shouldered opossum is thought to inhabit humid forests. According to Hunsaker (1977*b*), it is nocturnal and arboreal in habit and probably has a diet like that of *Caluromys*. Emmons (1990) stated that it uses the upper levels of the forest and rarely descends even to the middle levels; it moves slowly and will spend hours in the same flowering tree, feeding periodically on nectar. Collins (1973) reported that the few specimens maintained in captivity thus far have been hardy in comparison with other didelphids. One survived for 7 years and 10 months. *Caluromysiops* is declining because of habitat destruction and is now classified as vulnerable by the IUCN.

DIDELPHIMORPHIA; **Family GLIRONIIDAE; Genus GLIRONIA**
Thomas, 1912

Bushy-tailed Opossum

The single genus and species, *Glironia venusta*, is known only from eight specimens collected in the Amazonian regions of Ecuador, Peru, Brazil, and Bolivia (Da Silva and Langguth 1989; Emmons 1990; Marshall 1978*c*). Gardner (*in* Wilson and Reeder 1993), along with most other recent authorities, placed the genus in the subfamily Caluromyinae of the family Didelphidae. However, Hershkovitz (1992*a*, 1992*b*) considered *Glironia* and the Caluromyinae to represent distinct families.

Head and body length is 160–205 mm and tail length is 195–225 mm. The upper parts are fawn-colored or cinnamon brown; a dark brown to black stripe extends through each eye and gives the appearance of a mask; the tail is tipped with white or has only a sprinkle of white hairs, and the underparts are gray or buffy white. The texture of the fur varies from soft and velvety to dense and woolly.

This genus is much like *Marmosa* in general appearance, but the tail is well furred and bushy to the tip. The extent of the naked area on the tail of *Glironia* varies, but there is at least a trace of a ventral naked area on the terminal few centimeters. The characters considered by Hershkovitz (1992*b*) to distinguish the Glironiidae as a family distinct from the Caluromyidae are given above in the account of the latter group.

Four of the known specimens of *Glironia* were collected by commercial animal dealers, and all were taken in heavy, humid tropical forests. *Glironia* is presumed to be arboreal because of its large, opposable hallux (Marshall 1978*c*). The diet is unknown but probably is comparable to that of *Marmosa*—insects, eggs, seeds, and fruits. Emmons (1990) reported seeing an individual at night in dense vegetation about 15 meters above the ground. It ran about the vines quickly, often jumping from one to another in a manner unlike that of other opossums, and seemed to be hunting insects. *Glironia* is declining because of habitat destruction and is now classified as vulnerable by the IUCN.

DIDELPHIMORPHIA; **Family DIDELPHIDAE**

Pouched Opossums

This family of four Recent genera and eight species is found from southeastern Canada, through the eastern United States and Mexico, to central Argentina. Corbet and Hill (1991), Gardner (*in* Wilson and Reeder 1993), and most other authorities have considered the Didelphidae to include the Marmosidae, the Caluromyidae, and the Glironiidae. However, the latter three groups were distinguished as full families (see accounts thereof) by Hersh-

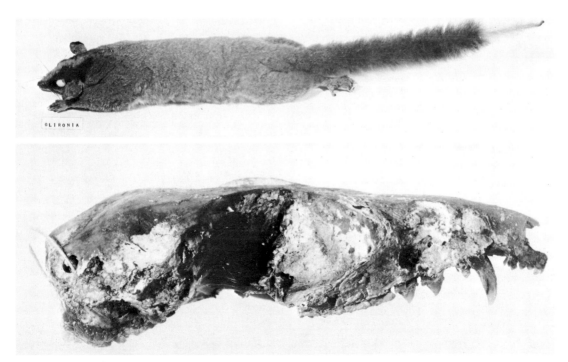

Bushy-tailed opossum *(Glironia venusta)*, photos from British Museum (Natural History).

A North American opossum *(Didelphis virginiana)* feigning death. The opposable great toe and naked, prehensile tail are both plainly shown. Photo from U.S. Fish and Wildlife Service.

kovitz (1992*a*, 1992*b*), who is followed here in that regard and with respect to sequence of genera.

Head and body length is 250–500 mm, tail length is 210–535 mm, and weight is 200–5,500 grams. There is much variation in pelage. The tail is thinly haired or naked for most of its length. Unlike those of the other families in the Didelphimorphia, female didelphids usually have a distinctive pouch, though there is some disagreement regarding its development in *Lutreolina*. According to Hershkovitz (1992*b*), the Didelphidae also generally are distinguished from the Marmosidae in having a more massive skull, with more prominent postorbital processes and sagittal crest, and stouter claws that are recurved and extend well beyond the digital tips (though not on the forefoot of *Lutreolina*). The skull of the Glironiidae lacks a sagittal crest. The skull of the Didelphidae differs from that of the Caluromyidae (see account thereof) in having large maxillary-palatine vacuities and tripartite tympanic bullae.

DIDELPHIMORPHIA; DIDELPHIDAE; **Genus PHILANDER**
Tiedemann, 1808

Gray and Black "Four-eyed" Opossums

There are two species (Cabrera 1957; Gardner *in* Wilson and Reeder 1993; Gardner and Patton 1972; E. R. Hall 1981):

P. opossum (gray "four-eyed" opossum), northeastern Mexico to northeastern Argentina;
P. andersoni (black "four-eyed" opossum), eastern Colombia, Ecuador, southern Venezuela, western Brazil, Peru.

The latter species formerly was known as *P. mcilhennyi*. Pine (1973*a*) argued that the generic name *Metachirops* Matschie, 1916, is the correct designation for both species and that *Philander* actually applies to the brown "four-eyed" opossum (here called *Metachirus*). Both Husson (1978) and E. R. Hall (1981) agreed, but most others have not. For further comment see the account of *Metachirus*.

Head and body length is 200–350 mm and tail length is 253–329 mm. Emmons (1990) reported natural weight to be 200–660 grams, but Collins (1973) stated that healthy, well-fed captive males weighed 800–1,500 grams and females, 600–1,000 grams. The fur is rather straight and short in *P. opossum* but much longer in *P. mcilhennyi*. The upper parts are gray to black, with a white spot above each eye, and the underparts vary from yellowish to buffy white. The white spots above the eyes account for the vernacular name "four-eyed" opossum. The tail, which is furred for about 50–60 mm from the rump and naked toward the tip, is black or grayish black on its basal half and white toward the end, with a pink tip. The body is slim and usually lean, and the head is large, with an elongate, conical muzzle. The ears are naked. The tail is slender, tapering, and prehensile. Females have a distinct pouch. The number of mammae varies from five to nine.

These opossums inhabit forested areas and are often

Gray "four-eyed" opossum *(Philander opossum)*, photo from West Berlin Zoo (Gerhard Budich).

found near swamps and rivers. They are smaller and more agile than *Didelphis* and quick in their actions. Although good climbers and swimmers, they are mainly terrestrial. All 46 specimens taken in Venezuela by Handley (1976) were caught in moist areas, nearly always on the ground. They build globular nests, about 30 cm in diameter, in the lower branches of trees or in bushes, and they may also inhabit ground nests and burrows. They are thought to be mainly nocturnal, though Husson (1978) stated that in Surinam they are as active in the day as at night. Their diet includes small mammals, birds and their eggs, reptiles, amphibians, insects, freshwater crustaceans, snails, earthworms, fruits, and probably carrion. Occasional damage to fruit crops and cornfields has given them a bad reputation in certain areas. In Panama, Fleming (1972) found maximum population densities of 0.55–0.65/ha. Unlike *Didelphis*, these opossums are not known to feign death when danger threatens, but will open their mouths wide, hiss loudly, and fight savagely. When disturbed they may utter a long, chattering cry.

The species *P. opossum* reportedly breeds year-round in some areas, possibly including Veracruz, Mexico, but is seasonal in others (Hunsaker and Shupe 1977). Collins (1973) cited records of females with pouch young being taken in Nicaragua from February to October, in Panama from April to July, and in Colombia in June, September, and October. Jones, Genoways, and Smith (1974) caught a female with six nursing young in March on the Yucatan Peninsula. Phillips and Jones (1969) collected data indicating that the main reproductive season in Nicaragua extends from March through July. Litter size there averaged 6.05 (3–7). According to Fleming (1973), *P. opossum* is seasonally polyestrous in Panama, with two or more litters probably being produced by each female from January to November. Litter size averaged 4.6 (2–7). Husson (1978) found females with young in Surinam during January, March, and April. Litter size there ranged from 1 to 7 but averaged only 3.4. These records support the statement by Phillips and Jones (1969) that litter size tends to be larger in the north. Collins (1973) reported that the small pouch young of a female received on 7 May were weaned on 23 July, that first estrus in females occurred at 15 months of age, and that maximum known longevity for captives was only 2 years and 4 months.

DIDELPHIMORPHIA; DIDELPHIDAE; **Genus**
DIDELPHIS
Linnaeus, 1758

Large American Opossums

There are four species (Cabrera 1957; Gardner 1973 and *in* Wilson and Reeder 1993; E. R. Hall 1981):

D. albiventris, Colombia and Venezuela to central
Argentina;
D. marsupialis, southern Tamaulipas (eastern Mexico) to
Brazil and Bolivia;
D. aurita, eastern Brazil, Paraguay, northeastern
Argentina;
D. virginiana (Virginia opossum), naturally from New
Hampshire to Colorado, and from southern Ontario to
Costa Rica.

Cerqueira's (1985) separation of *D. aurita* from *D. marsupialis* was not accepted by Corbet and Hill (1991).

Head and body length is 325–500 mm and tail length is 255–535 mm. The animals weigh about 0.5–5.5 kg. The pelage, consisting of underfur and white-tipped guard hairs, is unique in the family Didelphidae. Guard hairs are lacking or few in number in the other opossums. Coloration is gray, black, reddish, or, rarely, white. Head markings are sometimes present in the form of three dark streaks, one running through each eye and another running along the midline of the crown of the head. The basal tenth of the tail is furred, and the remainder is almost naked. At the northern limits of their range many of the individuals that survive the winter lose part of the tail and the ears from frostbite.

As in the other didelphids, all four feet have five digits, and all digits have sharp claws, except that the first toe of the hind foot is clawless, thumblike, and opposable to the other digits in grasping. The female has a well-developed pouch in which are commonly located 13 mammae arranged in an open circle with 1 in the center. Number and arrangement of mammae, however, are variable.

Opossums usually inhabit forested or brushy areas but also have been found in open country near wooded watercourses. The species *D. virginiana* generally has been reported to favor moist woodlands or thick brush near streams or swamps (Banfield 1974; Jackson 1961; McManus 1974; Schwartz and Schwartz 1959). According to Cerqueira (1985), *D. albiventris* is found in open and deciduous forests and in mountainous areas, while *D. marsupialis* is restricted to humid, broad-leaved forests. In

Opossums *(Didelphis virginiana)*, photo by Leonard Lee Rue III.

Venezuela, Handley (1976) collected most specimens of *D. marsupialis* on the ground in moist areas but found most *D. albiventris* in dry situations. Opossums are largely nocturnal, spending the day in rocky crevices, in hollow tree trunks, under piles of dead brush, or in burrows (sometimes dug by themselves). They construct rough nests of leaves and grasses, using their curled-up tail and their mouth to transport dry vegetation. They are mainly terrestrial, moving with a slow, ambling gait, and are strong swimmers. They can climb well, with the help of the prehensile tail, and can even hang by the tail (Lowery 1974). In the northern part of its range, *D. virginiana* accumulates fat in the autumn and may become inactive during severe winter weather, remaining in its nest for several days, but it does not hibernate. Females have a greater tendency toward winter inactivity and also are more sedentary in their movements at other times. Opossums have an extremely varied diet that includes small vertebrates, carrion, invertebrates, and many kinds of vegetable matter.

Records compiled by Hunsaker (1977*b*) showed that for *D. virginiana* in the United States, population density averaged 0.26/ha. (0.02–1.16/ha.), home range averaged about 20 ha. (4.7–254.0 ha.), and nightly foraging distance was 1.6–2.4 km; and that for *D. marsupialis* in Panama, density was 0.09–1.32/ha., with the animals being nomadic and remaining in an area for only two or three months. According to Hunsaker and Shupe (1977), *D. virginiana* also is nomadic and stays in a particular area for six months to a year; there is no territoriality, but individuals will defend the space occupied at a given time. In a study in Venezuela, Sunquist, Austad, and Sunquist (1987) found the home range size of *D. marsupialis* to average 11.3 ha. in the dry season and 13.2 ha. in the wet season. Male home ranges overlapped one another extensively, and each overlapped the ranges of several females. The latter occupied exclusive home ranges for at least part of the year.

Opossums usually are thought to be solitary and antisocial, either avoiding one another or acting aggressively.

However, Holmes (1991) found that most interaction among unrelated individuals kept in large enclosures was neutral or affiliative, with females sometimes nesting together. The animals readily formed stable dominance hierarchies, with females usually dominant. There seems agreement that there almost always is extreme agonistic behavior when two males meet, but if opposite sexes meet during the breeding season, initial aggressive displays turn to courtship and the two animals may spend several days together. If a male is placed with a female that is not in estrus, she becomes aggressive, but the male does not return her attacks. The vocal repertoire of *D. virginiana* consists of a hiss, a growl, and a screech, which are used in agonistic and defensive situations, and a metallic lip clicking, which is heard under a variety of conditions, including mating (McManus 1970, 1974).

Death feigning, referred to popularly as "playing possum" and technically known as catatonia, is a passive defensive tactic of *D. virginiana* employed occasionally, but not always, in the face of danger. In this state the opossum becomes immobile, lies with the body and tail curled ventrally, usually opens the mouth, and apparently becomes largely insensitive to tactile stimuli. The condition may last less than a minute or as long as six hours (McManus 1974). Although catatonia seems partly under the conscious control of the animal, physiological changes suggest a state analogous to fainting in humans (Lowery 1974). Death feigning may cause a pursuing predator to lose the visual cue of motion or to become less cautious in its approach, thereby giving the opossum a better chance of escape. McManus (1970) thought that anal secretions, sometimes accompanying catatonia, might contribute to deterring a predator. He also suggested that catatonia may have evolved primarily as a means of reducing intraspecific aggression. If so, the condition may be vaguely comparable to submission in certain other animals. According to Hunsaker and Shupe (1977), death feigning has been reported for, but is rare in, *D. albiventris*.

Except as noted, the following life history data on *D. virginiana* were derived from Collins (1973), Hunsaker (1977*b*), Lowery (1974), Schwartz and Schwartz (1959), and Tyndale-Biscoe and Mackenzie (1976). Females are polyestrous, have an estrous cycle of about 28 days, and are receptive for 1 or 2 days. First matings usually occur in January or February in most of the United States but can be as early as mid-December in Louisiana (Edmunds, Goertz, and Linscombe 1978) or as late as March in Wisconsin (Jackson 1961). There are two litters per year in most areas, though Jackson (1961) thought there usually was only one in Wisconsin, and the possibility of a third occurring in the South has been suggested. The second peak of mating takes place in the spring or early summer, and an average of 110 days separates the two litters. The gestation period is 12.5–13 days. The tiny young are remarkably undeveloped but have pronounced claws, which are used in the scramble from the birth canal to the mother's pouch. The young measure only 10 mm in length and weigh 0.13 grams. Twenty could fit in a teaspoon, and it would take 217 to make one ounce. There usually are about 21 young, but there is a report of 56 being born at once. Since there normally are only 13 mammae, and since the young must continually grasp the mammae to suckle, some newborn often cannot be accommodated and quickly perish. Additional mortality usually occurs later. The number of pouch young generally observed is 5–13 and reportedly averages about 8–9 in northern areas and 6–7 in the South. The young first release their grip on the mammae when they are about 50 days old, begin to leave the pouch temporarily at 70 days, and are completely weaned and independent at 3–4 months. After the pouch becomes too small to hold all the young, but before weaning, some of them ride on the mother's back. Sexual maturity is attained by about 6–8 months, and females apparently have only 2 years of reproductive activity. Very few opossums survive their third year of life, though some laboratories have maintained captives for three to five years.

For *D. marsupialis* in Panama, Fleming (1973) found that females were polyestrous, breeding began in January, and two or possibly three litters were born between then and October. Litter size averaged 6 (2–9). Tyndale-Biscoe and Mackenzie (1976) reported that reproduction in this species also began in January on the llanos of eastern Colombia. A second litter was produced there in April or May, and there was no breeding from September to December. Mean litter size was 6.5 (1–11). In western Colombia the breeding period was the same, but a third litter was known to be produced in August. Mean litter size there was 4.5 (1–7). No reproductive season could be determined for *D. albiventris* in Colombia, but litter size averaged only 4.2 (2–7). In Brazil *Didelphis* reportedly has its first litter in late August or September (Collins 1973).

The Virginia opossum has adapted better to the presence of people than have most mammals. When European colonists first arrived in North America *D. virginiana* apparently did not occur north of Pennsylvania. Subsequently it extended its range, the first being taken in Ontario in 1858 (Hunsaker 1977*b*) and in New England during the early twentieth century (Godin 1977). It also moved westward on the Great Plains, its progress facilitated by human agricultural development (Armstrong 1972; Jones 1964). In 1890 captives were released in California, and boosted by other introductions, the Virginia opossum eventually spread all along the West Coast. At present it is well established from southwestern British Columbia to the vicinity of San Diego. Introduced populations also occupy parts of Arizona, western Colorado, and Idaho (E. R. Hall 1981).

Opossums are sometimes hunted or trapped by people for food and sport and as predators of poultry. Leopold (1959) wrote that in Mexico they are considered chicken and egg thieves, they are hunted for food and local use of their fur, and certain of their parts are believed to have medicinal value. In Peru both *D. marsupialis* and *D. albiventris* have a bad reputation for poultry killing, but populations have not been adversely affected by human settlement (Grimwood 1969). The value of opossum fur and the consequent commercial harvest have varied widely. During a single year in the 1920s, when prices were high, 518,295 opossum skins were sold in Louisiana alone (Lowery 1974) and 350,286 from Kansas were sold (Hall 1955). In the 1970–71 annual season in the United States 101,278 pelts of *D. virginiana* were reported sold at an average price of $0.85. The corresponding figures for the 1976–77 season were 1,069,725 and $2.50 (Deems and Pursley 1978). The take for 1983–84 in Canada and the United States was 515,832 skins, and the average price in Canada was $1.75 (Novak, Obbard, et al. 1987). The number of skins taken in the 1991–92 season in the United States was 145,290, the average price $1.26 (Linscombe 1994). In addition to its other uses, *Didelphis* is finding increasing employment in laboratory research.

DIDELPHIMORPHIA; DIDELPHIDAE; **Genus
CHIRONECTES**
Illiger, 1811

Water Opossum, or Yapok

The single species, *C. minimus*, occurs from southern Mexico and Belize to northeastern Argentina (Marshall 1978*a*; McCarthy 1982*b*). It also may occur on Trinidad.

Head and body length is 270–400 mm, tail length is 310–430 mm, and weight reportedly ranges from 604 to 790 grams (Marshall 1978*a*). Collins (1973), however, stated that the weight of a captive female stabilized at about 1,200 grams. The pelage is relatively short, fine, and dense. The back is marbled gray and black, the rounded black areas coming together along the midline. The muzzle, a band through the eye to below the ear, and the crown are blackish; a prominent grayish white, crescentlike band passes from the front of one ear to the other, just above the eyes. The chin, chest, and belly are white. The striking color pattern is unique among marsupials and may serve as camouflage while the yapok is swimming, the dorsal bands blending with ripples and presenting a disruptive appearance to aerial predators (Brosset 1989). The long, ratlike tail is well furred only at the base and is black near the body and yellowish or whitish toward the end. The facial bristles are stout, long, and placed in tufts. One of the wrist bones is enlarged and simulates, in some respects, a sixth digit on the forefoot. The ears are moderately large, naked, and rounded.

This opossum is the only marsupial well adapted for a semiaquatic life. It has dense, water-repellent pelage, webbed hind feet, a streamlined body, and, in females, a rear-opening, waterproof pouch. Both sexes actually have a pouch. In the male it cannot be fully closed, as in the female, but the scrotum is pulled up into it when the animal is swimming or moving swiftly. The female is able to swim with the young in her pouch. A well-developed sphincter muscle closes the pouch, creating a watertight compartment, and the young can tolerate low oxygen levels for many minutes (Marshall 1978*a*; Rosenthal 1975*b*).

The yapok is confined mainly to tropical and subtropical habitats, where it frequents freshwater streams and lakes.

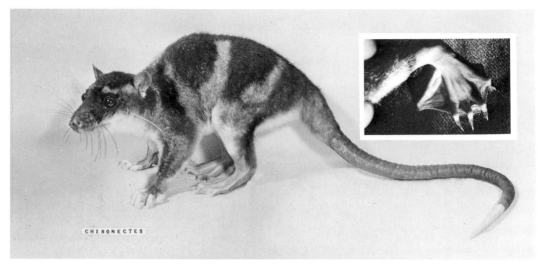

Water opossum, or yapok *(Chironectes minimus)*. Inset: hind foot. Photos by Richard Peacock.

In some areas it is found at considerable elevations along mountain rivers. Grimwood (1969) reported that one was killed in Peru at an altitude of 900 meters, and Handley (1976) collected specimens at up to 1,860 meters in the mountains of Venezuela. This opossum generally is considered rare throughout its range, but this view may stem from the nocturnal habits of the animal and the inaccessibility of its habitat. The main den, a subterranean cavity, usually is reached through a hole in the stream bank just above water level. A ground nest of leaves or grasses in a dimly lit area may be used as a place of rest during the day. Like *Didelphis,* the yapok collects nesting material with the forepaws, pushing it under the body into a bundle that is then held by the tail (Marshall 1978a).

The water opossum is an excellent swimmer and diver, the broadly webbed hind feet serving as effective paddles. Only the rear feet are used in swimming, thereby avoiding the reduced propulsion efficiency that might result from interference between the fore- and hind limbs. The pelage does not soak up water, so the nose can remain above the surface without assistance from the front feet (Fish 1993). *Chironectes* can climb but rarely does so; its tail, though prehensile, is too thick to be of effective use. It is largely carnivorous, feeding on such aquatic life as crayfish, shrimp, and fish. Prey is located in the water by contact with the forefeet, which are not used for propulsion but are held out in front while the animal is swimming (Marshall 1978a). The yapok also may eat some aquatic vegetation and fruits (Hunsaker 1977b).

In Brazil the young are born in December and January; one female was seen in February with 5 young. Litters of 2–5 have been recorded, with the average reportedly being 3.5 (Collins 1973; Hunsaker 1977b). Rosenthal (1975b) stated that three captive-born litters at the Lincoln Park Zoo numbered 4, 5, and 5. Experience with these litters indicated that the young of *Chironectes* develop more rapidly than those of any other didelphid. According to Collins (1973), the greatest known longevity for a captive yapok is 2 years and 11 months.

Grimwood (1969) observed that yapok skins had been considered worthless but were beginning to command a price in Peru. He thought that intensive commercial hunting could readily reduce populations and that protective measures were necessary. Emmons (1990) noted that the

genus seems rare in most regions but is common in some Central American rivers. McLean and Ubico (1993), who reported the first record for Guatemala, suggested that it was very rare in that country. The IUCN designates *Chironectes* as near threatened.

DIDELPHIMORPHIA; DIDELPHIDAE; **Genus LUTREOLINA**
Thomas, 1910

Thick-tailed Opossum

The single species, *L. crassicaudata,* is known to occur east of the Andes in Bolivia, southern Brazil, Paraguay, Uruguay, and northern Argentina. Another population, far to the north, long was known only from two specimens taken in Guyana (Marshall 1978d). Recently, however, Handley (1976) collected four more specimens from eastern Venezuela, and Lemke et al. (1982) reported that nine had been taken in eastern Colombia. It is possible that additional investigation will reveal that this species also occupies the intervening region of central South America.

Head and body length is 210–445 mm and tail length is 210–310 mm. Adults weigh 200–540 grams. The fur is short, dense, and soft but not water-repellent. The upper parts are a rich, soft yellow, buffy brown, or dark brown, and the underparts are reddish ochraceous or pale to dark brown. The pelage of some live individuals has a peculiar purplish tinge. There may be faint markings on the face, but there are no eye spots or other prominent markings.

The body form is long and low, almost weasel-like. The ears are short and rounded and barely project above the fur. The limbs and feet are short and stout, and the pads are small and narrow. The tail is characteristic in its extremely thick, heavily furred base; only about 5 cm of the undersurface of the tip is naked, though in some individuals all the terminal half is thinly haired, showing the scales. The tail is not as prehensile as in other didelphids. The thumbs and great toe are not fully opposable. Although some reports have indicated otherwise, specimens examined by Lemke et al. (1982) did have a well-developed pouch. Reig, Kirsch, and Marshall (1987) also confirmed the presence of

Thick-tailed opossums *(Lutreolina crassicaudata)*, female on left, male on right, photo from New York Zoological Park through Joseph Davis.

a pouch. The mammae, at least in one specimen, numbered nine.

The thick-tailed opossum is restricted mostly to grassland, savannahs, and gallery woodland, often near the shores of streams and lakes. It is considered to be the species of didelphid most adapted to life on the pampas. It climbs well, is agile on the ground, and seems suited for wetland habitat. Although one captive reportedly was clumsy in the water, most information indicates that *Lutreolina* is an excellent swimmer under natural conditions (Grzimek 1975; Hunsaker 1977b; Marshall 1978d). In wooded areas it often shelters in tree holes, but in wetlands it constructs a snug round nest of grasses and rushes among the reeds, and on the pampas it often utilizes abandoned armadillo and viscacha burrows.

This opossum is nocturnal, emerging after dark to prey on small mammals, birds, reptiles, fishes, and insects. It occasionally raids chicken houses and pigeon lofts. It is said to be sometimes savage in temperament but can be tamed. *Lutreolina* actually appears somewhat more sociable than other didelphids, and a group of one male and two females was maintained together successfully (Collins 1973). An average core home range of 0.38 ha. was determined in Argentina (Eisenberg 1989). This opossum breeds in the spring and again later in the year after the young of the first litter have become independent. The young are raised in a nest of dry grass. The gestation period is thought to be about two weeks. One specimen lived for three years in captivity (Collins 1973). Roig (1991) reported that *Lutreolina* had disappeared from a large part of central and northern Argentina because of human habitat disruption.

Order Paucituberculata

"Shrew" Opossums

This order, containing the single Recent family Caenolestidae, with two genera and seven species, is found in western South America (Bublitz 1987). The group sometimes is placed only at the level of a superfamily (Marshall 1987) or infraorder (Szalay 1994), but full ordinal status was supported by Aplin and Archer (1987), Kirsch (1977b), and Marshall, Case, and Woodburne (1990). Some authorities have suggested that there may be only a single valid genus and as few as three species (Marshall 1980). In contrast, Gardner (in Wilson and Reeder 1993) recognized three genera and five species. All species prefer densely vegetated, humid habitat, typically in scrub adjacent to meadows of high, moist Andean paramo (Kirsch 1977a).

These marsupials are somewhat shrewlike in appearance. Head and body length is 90–135 mm and tail length is 65–135 mm. The head is elongate and conical in shape. The eyes are small and vision is poor, but the sense of smell is well developed. Sensory vibrissae are present on the snout and cheeks. The tail is entirely covered with stiff, short hairs. The hind limbs are slightly longer than the forelimbs, and all have five separate, clawed digits. The humerus is large and heavy in comparison with the slender forearm; the hind foot is relatively long and narrow. Females of *Caenolestes* have four mammae, and females of *Rhyncholestes* have five. A marsupium is absent in adult females, though a rudimentary pouch may possibly be present in the young.

The teeth are rooted, sharp, and cutting. The dental formula is: (i 4/3–4, c 1/1, pm 3/3, m 4/4) $\times$ 2 = 46 or 48. According to Marshall (1984), there is one large, laterally compressed, procumbent incisor in each lower jaw, followed by six or seven tiny, spaced, vestigial teeth, among them a vestigial lower canine. The molar teeth are foursided or almost triangular in outline, and there is a sharp reduction in size from the first to the fourth. Szalay (1994) stated that the Paucituberculata are diagnosed by emphasized vertical shearing between the third upper and the first lower molar and by a carpus in which the lunate and magnum are in contact but are indented by a slight lateral wedge of the scaphoid.

Comparatively few specimens have been collected, probably more because of the inhospitable nature of the habitat than because of the rarity of the animals (Kirsch and Waller 1979). Habits are not well known but may be like those of large forest shrews. Caenolestids use runways on the surface of the ground and are mainly terrestrial but climb well. They are active during the evening and night and prey on invertebrates and small vertebrates.

The living Paucituberculata are highly relictual; there are seven extinct families of the order, with a morphological and ecological diversity comparable to that of the Australian order Diprotodontia (Aplin and Archer 1987). The geological range of the family Caenolestidae is early Eocene to Recent in South America (Kirsch 1977a). Fossil specimens from the Tertiary were known before description of the living representatives of the family (Grzimek 1975).

Common "Shrew" Opossums

There are five species (Barkley and Whitaker 1984; Bublitz 1987; Cabrera 1957; Kirsch and Calaby 1977):

C. fuliginosus, Andes of northern and western Colombia, extreme western Venezuela, and Ecuador;
C. caniventer, Andes of southwestern Ecuador and northern Peru;
C. convelatus, Andes of western Colombia and north-central Ecuador;
C. inca, Andes of south-central Peru;
C. gracilis, Andes of southeastern Peru.

Common "shrew" opossum *(Caenolestes obscurus),* photo by John Kirsch through Larry Collins.

Common "shrew" opossum *(Caenolestes inca)*, photo by John Kirsch through Larry Collins.

The last two species are sometimes put in the genus *Lestoros* Oehser, 1934 (synonyms of which are *Orolestes* Thomas, 1917, and *Cryptolestes* Tate, 1934). Various authorities long argued that *Lestoros* would be shown to be invalid once a sufficient number of specimens had been examined (Marshall 1980). Bublitz (1987) did study a large collection and did place *Lestoros* in the synonymy of *Caenolestes*. Gardner (*in* Wilson and Reeder 1993) retained *Lestoros* as a full genus but treated *gracilis* as a synonym of *C. inca*.

Head and body length is 90–135 mm and tail length is 93–139 mm. Tail length is approximately the same as head and body length in all species. Kirsch and Waller (1979) recorded weights of 25.0–40.8 grams for adult males and 16.5–25.4 grams for females. The pelage is soft and thick over the entire body, but it appears loose and uneven because of the different textures of the hairs. Coloration varies somewhat with the species, but all have dark upper parts, generally deep sooty brown, blackish brown, fuscous black, or plumbeous black. Some species are uniformly colored, or nearly so, whereas some are markedly lighter below. The tail is scantily haired and about the same color as the back, except that in some specimens the tip is white.

These marsupials could with equal justification be called shrewlike or ratlike in general form. The head is elongate, the eyes are very small, and the ears are rounded and project above the pelage. The tail tapers gradually and is nonprehensile. In the best-known species, *C. fuliginosus*, the forefoot has five digits: the outer toes are small and bear blunt nails, and the other three digits have sharp, curved claws. On the hind foot the great toe is small and bears a small nail, and the other four digits have well-developed, curved claws. Females have four mammae but no pouch.

Common "shrew" opossums occur in the alpine forest and meadow zone of the Andes at altitudes of about 1,500–4,000 meters. They prefer cool, wet areas covered by thick vegetation. They are nocturnal and terrestrial and move from one favored area to another by means of runways through the surface vegetation. Kirsch and Waller (1979) indicated that the habitat of the southern Peruvian species is considerably drier than that of the more northerly species. These authorities also reported that *C. fuliginosus* and *C. inca* may bound at high speed but are not saltatorial. They are primarily terrestrial, but they are ag-

ile climbers and use their tail as a prop as they move up a vertical surface. *C. fuliginosus* reportedly has well-developed senses of hearing and smell but poor vision. It had long been considered to be mainly insectivorous, but Kirsch and Waller (1979) found that it was readily trapped by baits of meat and could efficiently kill newborn rats. Stomach contents of a series of *C. caniventer* from Peru suggest that the diet consists largely of invertebrate larvae but also includes small vertebrates, fruit, and other vegetation (Barkley and Whitaker 1984).

Of the six female *C. fuliginosus* collected by Kirsch and Waller (1979) in southern Colombia from 25 August to 4 September 1969, one was small, four were lactating, and one had enlarged but empty uteri. None had attached young, so apparently the breeding season must begin several weeks before August. Three of the lactating females had all four mammae enlarged, and one had three of the mammae enlarged, suggesting a litter size as great or greater than the mother's capacity to suckle. Barnett (1991) collected a pregnant female *C. caniventer* with two fetuses in southern Ecuador on 9 September.

PAUCITUBERCULATA; CAENOLESTIDAE; **Genus RHYNCHOLESTES**
Osgood, 1924

Chilean "Shrew" Opossums

There are two species (Bublitz 1987):

R. raphanurus, Chiloe Island in southern Chile;
R. continentalis, mainland Chile just north of Chiloe Island.

Gardner (*in* Wilson and Reeder 1993) considered *continentalis* a synonym of *R. raphanurus*.

Head and body length is 110–28 mm and tail length is 65–87 mm. A specimen taken by Pine, Miller, and Schamberger (1979) weighed 21 grams. The pelage is loose and soft. The coloration is dark brown above and below with no markings, and the tail is blackish. The external appearance is much like that of *Caenolestes* except that the tail is short-

Chilean "shrew" opossum *(Rhyncholestes raphanurus)*, photo by P. L. Meserve.

er and, at least periodically, thickened at the base. The skull is narrow and elongate, especially the facial part. The lateral upper incisors are unique among living marsupials in that they have two cusps. Females have seven mammae, the seventh being in a medial position; there is no pouch (Patterson and Gallardo 1987).

According to Patterson and Gallardo (1987), *Rhyn-* *cholestes* apparently is restricted to temperate rainforests at elevations from sea level to 1,135 meters. Most specimens have been taken on the ground, alongside logs, and in dense cover. Some have been collected near burrow entrances at the bases of trees or under fallen logs. The genus appears to be mainly terrestrial and nocturnal; it is insectivorous but also eats earthworms and plant matter. Patterson, Meserve, and Lang (1990) described it as semifossorial, feeding chiefly on soil-inhabiting invertebrates and fungi. Kelt and Martínez (1989) found that the tails of animals caught during the autumn were thicker than those of specimens taken in the summer, thus suggesting that fat was being deposited in preparation for winter torpor. However, some individuals were taken on packed snow during midwinter. Kelt and Martínez also found lactating females in February, March, May, October, November, and December.

These marsupials appear to be rare; only three specimens were known through the 1970s, but subsequently larger numbers have been collected. Miller et al. (1983) warned that the dense forest habitat is now shrinking because of logging. Such declines have led the IUCN to classify *R. raphanurus* as vulnerable. Patterson, Meserve, and Lang (1990) noted that *Rhyncholestes* tends to occupy tall, wet forest but exhibits broad elevational range, wide habitat tolerance, and local abundance.

Order Microbiotheria

MICROBIOTHERIA; **Family MICROBIOTHERIIDAE;
Genus DROMICIOPS**
Thomas, 1894

Monito del Monte

This order contains one family, the Microbiotheriidae, with the single living genus and species *Dromiciops australis*, which occurs only from the vicinity of Concepción south to Chiloe Island in south-central Chile and east to slightly beyond the Argentine border in the mountains. Except perhaps for the extinct Bibymalagasia, the Microbiotheria have the most restricted Recent distribution of any mammalian order. Until about the 1970s *Dromiciops* generally was placed in the family Didelphidae. Subsequent investigation indicated that it is distinct from that group and more appropriately referred to the Microbiotheriidae, an otherwise extinct family of New World marsupials, which, in turn, was considered part of the group referred to here as the order Didelphimorphia (Archer 1984a; Kirsch 1977b; Kirsch and Calaby 1977; Marshall 1982). Even more recently the characters of *Dromiciops* have suggested that it represents a full order distinct from all other living groups of marsupials (Aplin and Archer 1987; Gardner *in* Wilson and Reeder 1993; Hershkovitz 1992a; Szalay 1982).

The affinities of *Dromiciops* have become a central point of interest and contention with respect to marsupial phylogeny. On the basis of both morphological and biochemical evidence, Reig, Kirsch, and Marshall (1987) regarded the genus as very distinctive but clearly part of the Didelphimorphia and having no special similarity to Australian marsupials. Aplin and Archer (1987) divided the marsupials into two great cohorts, the Ameridelphia, which includes most living New World genera, and the Australidelphia, comprising all living Australasian genera. Considering anatomy, serology, and cytology, they, and also Marshall, Case, and Woodburne (1990), assigned *Dromiciops* and the

Monito del monte *(Dromiciops australis)*, photo by Luis E. Peña

order Microbiotheria to the Australidelphia. Based to a considerable extent on the analysis of limb bones, Szalay (1994) recognized approximately the same cohort division but designated a new order, Gondwanadelphia, with the New World Microbiotheria and the Australasian Dasyuromorphia as suborders. It was suggested that this group arose in South America, spread to Antarctica and then Australia when the southern continents were joined during the late Cretaceous, and became the founding stock of the entire great Australasian marsupial radiation.

Hershkovitz (1992a, 1992b, 1995) disagreed with Szalay's assessment of limb bones and with the assignment of *Dromiciops* and the Microbiotheria to an otherwise Australian cohort. He argued that *Dromiciops* is the sole living member of an entirely separate cohort, the Microbiotheriomorphia, that diverged from the other marsupials before the development of the existing orders. All living marsupials besides *Dromiciops* can be united in the cohort Didelphimorphia; an early migration of the latter group from South America led to the Australasian radiation. The development of the Microbiotheriomorphia may have occurred as early as the Jurassic in either North or South America. The retention today by *Dromiciops* of a basicaudal cloaca such as found in monotremes but in no other marsupials is suggestive of a very primitive and independent origin. Moreover, *Dromiciops* has a number of key characters that are found regularly in placental mammals but have been lost in other marsupials, such as an entotympanic bone at the base of the auditory bullae and a nonstaggered third lower incisor tooth. All other marsupials with comparable dentition are seen to have undergone a reduction in muzzle length, with associated crowding of teeth, so that the third lower incisor is not in line with the other teeth and is usually supported by a bony buttress; this condition is characteristic of the cohort Didelphimorphia.

Head and body length of *Dromiciops* is 83–130 mm, tail length is 90–132 mm, and weight is 16–42 grams (Marshall 1978b; Redford and Eisenberg 1992). The fur is silky, short, and dense. The upper parts are brown, with several ashy white patches on the shoulders and rump arranged in vague whorls in a manner somewhat like the color pattern of *Chironectes*. The underparts are buffy to pale buffy. Other than pronounced black rings around the eyes there are no well-defined markings on the face.

Dromiciops may be recognized by its short, furry ears; thick, hairy tail; and features of the skull and dentition. The dental formula is the same as in the Didelphidae, but the canines, last molars, and first incisors are relatively smaller than those of typical didelphids. The tympanic bullae of *Dromiciops* are greatly inflated and occupy a relatively large area on the base of the skull. The face is quite short. The basal third of the moderately prehensile tail is furred like the body; the terminal two-thirds is also well furred, but the hairs are straight and slightly different in color, being nearly all dark brown. The only naked part of the tail is a narrow strip 25–30 mm long on the underside of the tip. Females have four mammae in a small but distinct pouch.

The monito del monte inhabits dense, humid forests, es-pecially areas with thickets of Chilean bamboo (*Chusquea* sp.). It sometimes is said to be arboreal, but Pearson (1983) considered it scansorial, noting that while it is a good climber, it may take refuge underground when released from a trap. Patterson, Meserve, and Lang (1990) stated that the highly scansorial habits of this genus are underscored by its prehensile tail, opposable toes, and unqualified aversion to entering enclosed live traps. It makes small round nests, about 200 mm in diameter, of sticks and the water-repellent leaves of *Chusquea,* lined with grasses and mosses. The nests may be under rocks or fallen tree trunks, in hollow trees, on branches, or suspended in lianas or patches of *Chusquea* (Marshall 1978b; Rageot 1978). *Dromiciops* is almost entirely nocturnal. In some areas, at least, it hibernates for lengthy periods when cold temperatures prevail and food is scarce. Hibernation is preceded by an accumulation of fat in the basal part of the tail, which may more than double the weight of an individual in one week (Kelt and Martínez 1989; Rageot 1978). *Dromiciops* also has been observed to enter periods of daily torpor even when food was readily available (Grant and Temple-Smith 1987). The natural diet seems to consist mainly of insects and other invertebrates, though Rageot (1978) observed that captives could be maintained on vegetable matter, and Pearson (1983) reported that captives ate large quantities of apple, as well as grubs, flies, and small lizards. The voice has been recorded as a chirring trill with a slight coughing sound at the end.

This marsupial reportedly lives in pairs, at least during the mating season in spring (October–December). Females with young have been recorded from November to May, and litters of one to five have been observed (Collins 1973; Marshall 1978b). Four stages have been described in the development of the young: (1) in the pouch, attached to the mammae; (2) in the nest; (3) nocturnal trips on the mother's back; and (4) loose association with other members of the family. Both males and females become sexually mature during their second year of life. Rageot (1978) reported catching a female that then lived in captivity for two years and two months.

Dromiciops has declined in recent years through loss of its restricted habitat and now is classified as vulnerable by the IUCN, though apparently it once was common in some areas. It is called "colocolo" by the natives of the Lake Region of Chile, who have several superstitions about it. Some believe that it is very bad luck to see a colocolo or to have one living about the house. Some residents of the area have even been known to burn their houses to the ground after seeing one of these inoffensive little animals running about in the house.

Geologically, according to Marshall (1982) and as accepted by Aplin and Archer (1987), the family Microbiotheriidae is known only from the late Oligocene and early Miocene of southern Argentina. However, Reig, Kirsch, and Marshall (1987) considered the family to comprise additional extinct genera from the late Cretaceous of North and South America and from the middle Paleocene to the early Miocene of South America.

Order Dasyuromorphia

Australasian Carnivorous Marsupials

This order of 3 Recent families, 19 genera, and 64 species is found in Australia, Tasmania, and New Guinea and on some nearby islands. Essentially this order has the same living content as the family Dasyuridae as the latter was accepted by Simpson (1945) and most other authorities until about the 1960s. Subsequent investigation led to the distinction of the Myrmecobiidae and Thylacinidae as separate families but eventually to their union with the remaining Dasyuridae at the ordinal level (Aplin and Archer 1987; Groves *in* Wilson and Reeder 1993; Marshall, Case, and Woodburne 1990). Szalay (1994) regarded the Dasyuromorphia as a suborder of his new order Gondwanadelphia, within which he also placed the Microbiotheria as a suborder, but Hershkovitz (1992a, 1995) saw no close relationship between those groups.

The members of the Dasyuromorphia are small to medium in size and vary widely in appearance; head and body length is about 50–1,300 mm. They have a general superficial resemblance to the Didelphimorphia and indeed sometimes were combined with the latter group in an order designated the Marsupicarnivora or Polyprotodontia. However, the Dasyuromorphia have only four upper and three lower incisor teeth on each side of the jaw compared with five upper and four lower in the Didelphimorphia. In both orders the incisors are polyprotodont, being relatively numerous, small, and sharp. In this regard the two orders differ from the diprotodont marsupials, which have only one or two lower incisor teeth on each side of the jaw, including a very strongly developed pair. The two orders also share a didactylous condition–having all their digits separate–whereas in the Australasian orders Diprotodontia and Peramelemorphia the second and third digits of the hind foot are syndactylous, being joined together by integument. Unlike the Didelphimorphia, the Dasyuromorphia lack a caecum and never have a prehensile tail. According to Szalay (1994), the talonids of the molar teeth of the Dasyuromorphia are reduced relative to the size of the trigonids compared with the dentition of the Didelphimorphia, and there is a lack of magnum contact with the lunate in the carpus.

Clemens, Richardson, and Baverstock (1989) indicated that the Didelphimorphia are characterized by retention of many primitive features, including polyprotodonty and didactyly, but also reduction of the number of incisors, presence of a large neocortex, the loss of the calcaneal vibrissae, and a number of adaptations of tarsal morphology for ter-

restrial existence. The known geological range of the order is middle Miocene to Recent in Australasia. However, according to Morton, Dickman, and Fletcher (1989), the group appears to have differentiated from the syndactylous Australian marsupials during the Eocene, at about the time when the last connection between Antarctica and Australia was broken. Subsequently, in the Oligocene the Thylacinidae split off from the basic dasyurid stock, and then in the Miocene the Myrmecobiidae and Dasyuridae also diverged.

DASYUROMORPHIA; Family **DASYURIDAE**

Marsupial "Mice" and "Cats," and Tasmanian Devil

This family of 17 Recent genera and 62 species occurs in Australia, Tasmania, New Guinea, and some adjacent islands. The sequence of genera presented here follows that of Archer (1982b, 1984a), who recognized the following subfamilies: Muricinae, for the genus *Murexia*; Phascolosoricinae, for *Phascolosorex* and *Neophascogale*; Phascogalinae, for *Phascogale* and *Antechinus*; Sminthopsinae, for *Planigale, Ningaui, Sminthopsis,* and *Antechinomys* (which Archer considered part of *Sminthopsis*); and Dasyurinae for the remaining genera. Based on an analysis of mitochondrial DNA, Krajewski et al. (1994) recommended abolishing the subfamilies Muricinae and Phascolosoricinae and transferring their constituent genera to, respectively, the Phascogalinae and Dasyurinae; a close relationship of *Murexia* and the New Guinean species of *Antechinus* was suggested.

Dasyurids are small to medium in size. The genus *Planigale* includes the world's smallest marsupials, some measuring only 95 mm in total length and weighing about 5 grams. The largest living member of the family is the Tasmanian devil *(Sarcophilus),* which may weigh over 10 kg. The tail of a dasyurid is long, hairy, and nonprehensile. The limbs usually are about equal in length, and the digits are separate. The forefoot has five digits, the hind foot four or five. The longest front toe is the third; digits 2 through 5 of the hind foot are well developed, but the great toe, which lacks a claw, is small or absent. Morton, Dickman, and Fletcher (1989) noted that arboreal species have broad hind feet and a mobile hallux, while species that are both arboreal and terrestrial have longer hind feet and a reduced hallux. Some genera walk plantigrade, that is, on the soles of the feet, and others walk digitigrade, on the toes. The mar-

A–C. Narrow-footed marsupial "mice" *(Sminthopsis crassicaudata)*, photos by Shelly Barker. D. Seven baby narrow-footed marsupial "mice" *(Sminthopsis* sp.) attached to their mother's nipples, photo by Vincent Serventy.

supium is absent in some; when present it usually opens posteriorly and often is poorly developed. The pouches of some genera become conspicuous only during the breeding season. Females have 2–12 mammae, the usual number being 6 or 8.

The basic pattern of the dasyurid dentition resembles that of *Didelphis*. The incisors are small, pointed, or blade-like, and the canines are well developed and large with a sharp cutting edge. The molars have three sharp cusps. The teeth in this family are specialized for an insectivorous or carnivorous diet—rooted, sharp, and cutting. The dental formula is: (i 4/3, c 1/1, pm 2–3/2–3, m 4/4) × 2 = 42 or 46.

Dasyurids have acute senses and are considered alert and intelligent. They are active animals and move rapidly. Most of the members of this family are terrestrial, though some marsupial "mice" are primarily arboreal. The usual shelter is a hollow log, a hole in the ground, or a cave. The mouse-like forms are usually silent, but the Tasmanian devil growls and screeches loudly. Dasyurids are active mainly at night and prey upon almost any living thing that they can overpower.

Morton, Dickman, and Fletcher (1989) provided an extensive summary of the physiology and natural history of the Dasyuridae. They identified six different life history strategies based on variation in age of sexual maturity, frequency of estrus, seasonality, and duration of male reproductive effort. As explained by Schmitt et al. (1989), male dasyurids tend to die in large numbers at the end of their very first breeding season, such mortality being nearly complete in certain genera, notably *Antechinus* and *Phascogale*. This phenomenon is thought to result from stress mediated through an increase in the concentration of plasma glucocorticoid and to be exacerbated by a reduction in the concentration of plasma-corticosteroid-binding globulin associated with a progressive rise in plasma androgen. Increased glucocorticoids, which inhibit most stages of the inflammatory and immune responses of animals, results in debilitating effects manifested by anemia, lymphocytopenia and neutrophilia, splenic hypertrophy, gastrointestinal hemorrhage and disease, immune suppression and disease, degeneration of major organs, and negative nitrogen balance.

The geological range of the family is middle Miocene to Recent in Australia and middle Pliocene to Recent in New Guinea (Archer 1982*b*; Marshall 1984).

DASYUROMORPHIA; DASYURIDAE; **Genus MUREXIA**
Tate and Archbold, 1937

Short-haired Marsupial "Mice"

There are two species (Flannery 1995; Kirsch and Calaby 1977; Ziegler 1977):

M. longicaudata, New Guinea, Japen Island (Yapen) to northwest, and Aru Islands to southwest;
M. rothschildi, southeastern Papua New Guinea.

A specimen from Normanby Island, southeast of New Guinea, originally was assigned to *M. longicaudata* but actually seems to represent an unknown species of *Antechinus* (Flannery 1995).

Head and body length is 105–285 mm and tail length is 145–283 mm. Flannery (1990*b*) listed weights of 32–40 grams for male *M. rothschildi*, 114–434 grams for male *M. longicaudata*, and 57–88 grams for the much smaller female *M. longicaudata*. The latter species is dull grayish

Short-haired marsupial "mouse" *(Murexia longicaudata)*, photo by P. A. Woolley and D. Walsh.

brown above with buffy white underparts and has a long, sparsely haired tail with a few longer hairs at its tip. In *M. rothschildi* the upper parts are similarly colored, but a broad black dorsal stripe is present; the underparts in this species are light brown. The fur is short and dense.

The skull is heavy and strong with deep zygomatic arches. The foot pads are striated, indicating that the animals probably are scansorial (Collins 1973). Two pairs of mammae are present.

The species *M. longicaudata* is found in all lowland and midmountain forests of New Guinea, from sea level to 2,200 meters. *M. rothschildi*, which is known from only 10 specimens, has an altitudinal range of 600–2,000 meters. These animals are presumed to be carnivorous and at least partly diurnal and arboreal (Flannery 1990b; Ziegler 1977). Woolley (1989) found *M. longicaudata* to utilize a spherical nest about 20 cm wide composed of leaves and located in a tree about 4 meters above the ground.

DASYUROMORPHIA; DASYURIDAE; **Genus PHASCOLOSOREX**
Matschie, 1916

There are two species (Flannery 1990b):

P. dorsalis, highlands of eastern and extreme western New Guinea;

P. doriae, western New Guinea.

There is some question whether *P. doriae* and *P. dorsalis* are really two separate species, for they are remarkably alike in basic characters and differ most markedly only in coloration and body size. *P. doriae* is generally the larger animal; it ranges in head and body length from 117 to 226 mm and in tail length from 116 to 191 mm. Its general body color is dark grizzled orange brown with bright rufous on its underside, whereas *P. dorsalis* is a grizzled gray brown with chestnut red underneath. The head and body length of *P. dorsalis* is 134–67 mm and the tail length is 110–60 mm.

One of the most distinctive characters of the genus *Phascolosorex* is the presence of a thin black stripe that runs from the head region all the way down the middle of the back to the base of the tail. *Murexia rothschildi*, with

Short-haired marsupial "mouse" *(Murexia rothschildi)*, photo by P. A. Woolley and D. Walsh.

which it may otherwise be confused, has a broad black stripe down its back. The tail is black and short-haired except at the base, where it is covered by thick, short, soft fur like that covering the body. The tail may or may not terminate in white. The ears are relatively short and more sparsely haired than the body. The feet, which are black in *P. doriae* and brown in *P. dorsalis,* possess striated pads and relatively short claws. There are four mammae in the pouch area.

These marsupials occur in mountain forests. *P. dorsalis* is more abundant at high altitudes but has been collected at 1,210–3,100 meters; *P. doriae,* the more geographically restricted species, has been recorded only in the lower to middle elevations, at 900–1,900 meters. The differences in locality of the two species when they both occur in the same general area are more strikingly vertical than horizontal. They usually are thought to be nocturnal and scansorial in habit (Collins 1973). However, extended laboratory observations by Woolley et al. (1991) demonstrate that *P. dorsalis* is predominantly diurnal; in this regard it is unique among the Dasyuridae. Lidicker and Ziegler (1968) referred to *P. dorsalis* as a "relatively rare high mountain species." Dwyer (1983) described it as "a rainforest species

said to be terrestrial and often diurnal in activity. An immature male of 13.3 grams was hand caught in daylight in September while 'sunning' itself on a rock in a small forest clearing. A second immature male of 15 grams was collected in September from beneath the fronds of *Pandanus.*"

DASYUROMORPHIA; DASYURIDAE; **Genus NEOPHASCOGALE**
Stein, 1933

Long-clawed Marsupial "Mouse"

The single species, *N. lorentzii,* is found in the western and central mountains of New Guinea (Flannery 1990b).

This marsupial resembles a tree shrew in general appearance. Head and body length is 170–230 mm and tail length is 185–215 mm. The color ranges from deep rufous to dull, pale cinnamon on the upper parts, and the fur is profusely speckled with white hairs or white-tipped hairs. The underparts are rich rufous with whitish, subterminal bands to the hairs. The head is deep rufous brown, and the

Phascolosorex dorsalis, photo by P. A. Woolley and D. Walsh.

backs of the ears are white. The limbs are rusty in color and blend to brown on the feet. The tail is rufous, except for the last third, which is white. The fur is long, soft, and dense. A few specimens taken have been melanistic.

There are long claws on all the feet, and the pads are striated. Structural differences that support separation from *Phascogale* are mainly in the dentition: there is no gap between incisors 1 and 2, and the last premolar is considerably smaller than the first two. The long-clawed marsupial "mouse" inhabits humid moss forests at altitudes of 1,500–3,400 meters. It is partly diurnal and probably is largely arboreal (Ziegler 1977). Collins (1973) reported that two specimens that survived several months in captivity were fed strips of beef, cicadas, and cockchafer beetles. Live insects were consumed greedily.

DASYUROMORPHIA; DASYURIDAE; **Genus
PHASCOGALE**
Temminck, 1824

Brush-tailed Marsupial "Mice," or Tuans

There are two species (Cuttle *in* Strahan 1983; Kitchener *in* Strahan 1983; Ride 1970; Smith and Medlin 1982):

P. tapoatafa, extreme northern and southwestern Western Australia, northern Northern Territory, northern and southeastern Queensland, eastern New South Wales, southern Victoria, southeastern South Australia;
P. calura, originally known from northeastern and southern Western Australia, southern Northern Territory, southeastern South Australia, southwestern New South Wales, and northwestern Victoria.

In *P. tapoatafa* head and body length is 160–230 mm, tail length is 170–220 mm, and weight is 110–235 grams (Cuttle *in* Strahan 1983). *P. calura* is considerably smaller, with a head and body length of 93–122 mm, a tail length of 119–45 mm, and a weight of 38–68 grams (Kitchener *in* Strahan 1983). Both species are grayish above and whitish below. The tail in *P. tapoatafa* is black except for the gray base; in *P. calura* it is black except for a reddish basal area.

Phascogale is distinguished from the other broad-footed marsupial "mice" by a tail whose terminal half is covered with a silky brush of long black hairs. In *P. tapoatafa* these hairs are capable of erection, producing a striking "bottle brush" effect. This is normal when the animal is active. Soderquist (1994) pointed out that this action is intended to distract predators and deflect their strike away from the body. When the animal is at rest the hairs are pressed along the tail and are not conspicuous. The ears are relatively large, thin, and almost naked. Females have eight mammae (Cuttle 1982). There is no true pouch, but there is a pouch area marked by light-tipped brown hairs coarser in texture than the body fur, which begins to develop protective folds of skin about two months before parturition. Strong, curved claws are present on all digits except the innermost digit of each hind foot, which is clawless.

These marsupial "mice" are usually arboreal, frequenting both the heavy, humid forest and the more sparsely timbered arid regions. Normally they build their nests of leaves and twigs in the forks or holes of trees, but in some localities they build them on the ground. Tuans are nocturnal and active and are as agile as squirrels. They capture and eat small mammals, birds, lizards, and insects. Apparently they do not eat carrion. They sometimes prey on poultry and can be destructive, but the benefits derived by their extermination of mice and insects probably outweigh any harm they may do. Ride (1970) wrote that tuans also feed on nectar. When disturbed, *P. tapoatafa* utters a low, rasping hiss, which apparently is an alarm note. When an-

Long-clawed marsupial "mouse" *(Neophascogale lorentzii)*, photo by P. A. Woolley and D. Walsh.

gered, tuans emit a series of staccato "chit-chit" sounds. Sometimes, when excited, tuans slap the pads of their forefeet down together while holding an alert, rigid pose, thus producing a sharp rapping sound. At times they also make a rapid drumming noise by quick vibrations of the tail.

Radio-tracking studies by Soderquist (1995) of *P. tapoatafa* in Victoria showed that female home ranges averaged 41 ha. and had little or no overlap with one another. Adult females generally were highly agonistic, but a mother sometimes relinquished a portion of her range to a newly independent daughter. Male ranges averaged 106 ha. and overlapped extensively with one another and with those of females. Observations in the wild (Soderquist and Ealey 1994) indicated that females are able to dominate males and to deter mating. Young animals disperse at an average age of 162 days, but females either stay within their mother's range or settle adjacent to it (Soderquist and Lill 1995).

The reproductive biology of *P. tapoatafa* is very much like that of *Antechinus stuartii* (Cuttle 1982 and *in* Strahan 1983; Soderquist 1993). The breeding season is restricted, females are monestrous, and ovulation is synchronized. Studies in Victoria indicate that mating occurs in June (winter), gestation lasts about 30 days, and births take place in late July and early August. Litters usually consist of eight young but sometimes contain as few as one. They remain attached to the nipples for about 40–50 days and are

then left in the nest while the mother forages. They begin to emerge just before weaning at an age of around 5 months and reach adult size at about 8 months. Males disappear entirely from the wild population just after mating and before they are a year old, evidently succumbing to stress-related diseases (see also account of family Dasyuridae). Captive males have lived for more than three years but are not reproductively viable after their first breeding season. Females may survive to breed for a second year in the wild but are known to have survived to a third year only in captivity. The reproductive pattern of *P. calura* is similar (Kitchener 1981 and *in* Strahan 1983). Females give birth to eight young from mid-June to mid-August, and they are weaned before the end of October.

The IUCN designates *P. calura* as endangered. It has disappeared from more than 90 percent of its range and now survives only in isolated reserves of southwestern Western Australia (Kennedy 1992). Kitchener (*in* Strahan 1983) noted that this species depends on large areas of dense, undisturbed vegetation and that its decline was probably associated with destruction of such habitat through grazing by domestic cattle and sheep. Thornback and Jenkins (1982) noted, however, that much suitable habitat remains and that predation by introduced cats and foxes may be a factor. *P. tapoatafa* still is widespread, but it has lost considerable habitat (Kennedy 1992) and is designated as near threatened by the IUCN.

Brush-tailed marsupial "mouse" *(Phascogale calura)*, photo by M. Archer.

DASYUROMORPHIA; DASYURIDAE; Genus
ANTECHINUS
Macleay, 1841

Broad-footed Marsupial "Mice"

There are 10 species (Baverstock et al. 1982; Kirsch and Calaby 1977; Ride 1970; Van Dyck 1980, 1982a, 1982b; Woolley 1982b; Ziegler 1977):

A. melanurus, New Guinea;
A. naso, New Guinea;
A. wilhelmina, central mountains of New Guinea;
A. godmani, northeastern Queensland;
A. stuartii, Victoria, eastern New South Wales,
    Queensland;
A. flavipes, eastern and southern Australia;
A. leo, Cape York Peninsula of northern Queensland;
A. bellus, Northern Territory;
A. swainsonii, southeastern Australia, Tasmania;
A. minimus, coastal southeastern Australia, Tasmania,
    islands of Bass Strait.

The genera *Parantechinus, Dasykaluta,* and *Pseudantechinus* (see accounts thereof) sometimes have been considered part of *Antechinus.* The species once known as *Antechinus maculatus* has been transferred to the genus *Planigale* (Archer 1975, 1976). On the basis of electrophoretic analysis, Dickman et al. (1988) and McNee and

Cockburn (1992) reported that *A. stuartii* actually comprises two morphologically and behaviorally cryptic species, one found in Queensland and northeastern New South Wales, the other in southern New South Wales and Victoria.

Head and body length is about 75–175 mm and tail length is about 65–155 mm. Within any species males generally are larger than females. Green (1972) reported that in *A. swainsonii,* a species of about average size, a series of male specimens weighed 48–90 grams and a series of females weighed 31–55 grams. In *A. minimus,* a smaller species, males weighed 30–57 grams, and females 24–52 grams. Emison et al. (1978) found a sample of *A. stuartii* to average 35.2 grams for males and 26.4 grams for females. The fur of *Antechinus* is short, dense, and rather coarse. Coloration of the upper parts varies from pale pinkish fawn through gray to coppery brown; the underparts are buffy, creamy, or whitish. The tail, usually the same color as the back, is short-haired in most species.

The short, broad feet are distinctive. The great toe is present but is small and clawless. All species have transversely striated pads on the bottom of the feet. In the semiarboreal species, such as *A. flavipes,* the pads are prominent and strongly striated; in *A. swainsonii* and *A. minimus,* which are poor climbers, the pads are small and the striations faint (especially so in *A. minimus,* which occurs in treeless habitat). These two closely related species have long and strong foreclaws modified for digging. *A. flavipes* has short, hooked claws, is much more active, and is a great climber.

A pouch is known to develop during the breeding sea-

Broad-footed marsupial "mouse" *(Antechinus flavipes),* photo by Howard Hughes through Australian Museum, Sydney.

Marsupial "mouse" *(Antechinus stuartii)*, photo by H. J. Aslin.

son in most species but may be absent in others or may consist only of prominent skin folds around the mammae. The number of mammae varies between and within species, having been reported as 10–12 in *A. flavipes,* 6–10 in *A. stuartii,* 6–8 in *A. swainsonii,* 6–8 in *A. minimus* (Collins 1973), 6 in *A. bellus* (Taylor and Horner 1970*a*), 4 in *A. melanurus* (Dwyer 1977), and 10 in *A. leo* (Van Dyck 1980).

The species found in New Guinea, as well as *A. godmani* and *A. swainsonii,* occur mainly in dense, moist forests (although the latter species is a ground dweller); *A. stuartii* and *A. flavipes* can inhabit a wide variety of forest and brushland habitat as long as there is sufficient cover; *A. bellus* dwells on tropical savannahs; and *A. minimus* occurs in grassy areas on the mainland but in Tasmania is restricted to wet sedgelands (Emison et al. 1978; Green 1972; Ride 1970; Robinson et al. 1978; Woolley 1977; Ziegler 1977). Broad-footed marsupial "mice" are secretive, active, nocturnal animals characterized by rapid movements. The claws and ridged foot pads of some species enable them to run upside down on the ceilings of rock caverns. They construct nests in hollow trees, fallen logs, rock crevices, abandoned birds' nests, or pockets in cave ceilings (Collins 1973).

The diet consists mainly of invertebrates. Nagy et al. (1978) found *A. stuartii* in winter to consume about 60 percent of its weight in arthropods each day. Small vertebrates, including house mice that have been introduced in Australia, are also taken. Green (1972) reported that captive *A. swainsonii* accepted items ranging in size from mosquitoes to house mice and domestic sparrows.

Population density of *A. minimus* on Great Glennie Island in Bass Strait was estimated at 80/ha. (Wainer 1976). Density of the best-known species, *A. stuartii,* was determined as 3/ha. and 18/ha. in two sections of wet sclerophyll forest (Nagy et al. 1978). Home ranges of three females near Brisbane were 0.80, 0.98, and 1.15 ha., and males utilized larger but uncalculated areas (Wood 1970). Studies by Cockburn and Lazenby-Cohen (1992), Lazenby-Cohen (1991), and Lazenby-Cohen and Cockburn (1988) indicate that outside of the breeding season both male and female *A. stuartii* forage in clearly defined individual home ranges but that neither sex is territorial. As temperatures drop during the winter individuals may leave their nocturnal foraging range to spend the day in a communal nest that may hold 18 animals simultaneously. The total number of individuals visiting a single nest during the winter may be more than 50. When the mating season begins males abandon their foraging range and aggregate in a few of the nest trees, where they spend most of the night. Males always nest communally and always in groups that include females. Such mating aggregations are comparable to the "leks" that have been described for various large hoofed

mammals. Prior to the mating season females nest alone or in small unisexual groups. Even during the breeding season females continue to use their foraging range and may nest alone, but they make excursions to the aggregations, thus determining the time and place of mating.

Each investigated Australian species (*A. flavipes, A. stuartii, A. bellus, A. swainsonii,* and *A. minimus*) has been shown to be monestrous, to have a breeding season restricted to about three months, and to produce only one litter per year. This reproductive period often occurs in winter (July–September) but may be in autumn or spring in some areas. Births usually are highly synchronized; in any one population of *A. stuartii* they all take place within two weeks (Collins 1973; Lee, Bradley, and Braithwaite 1977). Wood (1970) reported that in the population of *A. stuartii* near Brisbane mating occurred in late September, births occurred in October, and lactation lasted until February. Wainer (1976) found that in *A. minimus* on Great Glennie Island mating occurred in late May and most litters were born in July. The young remained in the pouch until late August or September and then suckled in the nest until November. Available data indicate that in New Guinea *A. melanurus* and probably *A. naso* breed throughout the year (Dwyer 1977). Gestation has been reported as 23–27 days in *A. flavipes* and 26–35 days in *A. stuartii.* Litter size is 3–12 (Collins 1973). The young of *A. stuartii* measure 4–5 mm in length at birth and weigh 0.016 gram. They are weaned and independent after about 90 days and attain sexual maturity at about 9–10 months (Collins 1973; Marlow 1961). Young males are forced by their mothers to disperse, and thus the subsequent winter aggregations consist largely of unrelated males, but young females remain in their natal vicinity (Cockburn and Lazenby-Cohen 1992).

A remarkable feature of the biology of *Antechinus* is the abrupt and total mortality of males following mating, when they are 11–12 months old (Arundel, Barker, and Beveridge 1977; Barker et al. 1978; Emison et al. 1978; Lazenby-Cohen and Cockburn 1988; Lee, Bradley, and Braithwaite 1977; Morton, Dickman, and Fletcher 1989; Nagy et al. 1978; M. P. Scott 1987; Wainer 1976; Wilson and Bourne 1984; Wood 1970). Although the period of mortality varies geographically, it occurs at the same time each year in any given population. This phenomenon is known to take place in five species—*A. flavipes, A. stuartii, A. bellus, A. swainsonii,* and *A. minimus*—but is best understood and is especially sudden in *A. stuartii.* In this species the males become more active and aggressive during the breeding season. As the breeding season approaches a climax they move about considerably, even in daylight and mainly from one male aggregation to another. They apparently are subject to intensive physiological stress resulting from continuous competition for the females that visit the male aggregations as well as from gluconeogenic mobilization of body protein, a process that sustains them temporarily. Increasing levels of testosterone and other androgens evidently depress plasma-corticosteroid-binding globulin and result in a rise in corticosteroid concentration. The trauma associated with stress and endocrine changes causes suppression of the immune system, major ulceration of the gastric mucosa, exposure to parasites and pathological conditions, and death. Even males captured during the mating season die in the laboratory at the same time as those remaining in the wild. Males taken before sexual maturity, however, have been maintained until at least two years and eight months (Rigby 1972), and captive females have survived for more than three years (Collins 1973). In the wild, many females also die after rearing their first litter, but some live for at least another year (Lee, Bradley, and Braithwaite 1977). Limited evidence suggests that a synchronized male die-off may not occur in *A. melanurus* of New Guinea (Dwyer 1977; Woolley 1971*b*).

Broad-footed marsupial "mice" generally are not under serious human pressure, though some populations have been reduced by predation from domestic cats. Green (1972) considered *A. minimus* to be "rare and endangered" in Tasmania because of its limited habitat and the threats posed by mineral development, grazing of domestic livestock, and flooding by hydroelectric dams. Aitken (1977*b*) also expressed concern for the habitat of the mainland populations of *A. minimus*. The subspecies *A. minimus maritimus* and also *A. swainsonii insulanus* of coastal Victoria are designated as near threatened by the IUCN. Laurance (1990) determined that deforestation had reduced the historical range of *A. godmani* by about 30 percent, resulting in extensive fragmentation of remaining populations. Both *A. godmani* and *A. leo* are now classed as near threatened by the IUCN. Cockburn and Lazenby-Cohen (1992) cautioned that the cutting of large trees with hollows might adversely affect the breeding requirements of *A. stuartii*.

DASYUROMORPHIA; DASYURIDAE; **Genus**
PLANIGALE
Troughton, 1928

Planigales, or Flat-skulled Marsupial "Mice"

There are five species (Aitken 1971*b*, 1972; Archer 1976; Baverstock et al. 1982; Lumsden, Bennett, and Robertson 1988; Ziegler 1972):

P. maculata, northern and eastern Australia, mainly near the coast;
P. novaeguineae, southern New Guinea;
P. ingrami, northern parts of Western Australia, Northern Territory, and Queensland;
P. tenuirostris, southern Queensland, northern New South Wales, eastern South Australia;
P. gilesi, eastern South Australia, southwestern Queensland, northern and western New South Wales, northwestern Victoria.

Archer (1976) observed that *P. maculata* and *P. novaeguineae* are very similar and might prove conspecific; he observed as well that it is not clear whether *P. ingrami* and *P. tenuirostris* are separate species. The species *P. gilesi* is distinct from all the others in possessing only two, rather than three, premolars in both the upper and the lower jaw. In contrast, analysis of mitochondrial DNA (Painter, Krajewski, and Westerman 1995) indicates that *P. gilesi* is immediately related to *P. novaeguineae* and *P. ingrami* and that *P. maculata* is the most divergent member of the genus, possibly with closer affinity to *Sminthopsis*. Such analysis also suggests that two specimens from the Pilbara region of northwestern Western Australia represent undescribed species, one related to *P. maculata* and one possibly with affinity to *P. ingrami*.

Head and body length for the genus is about 50–100 mm and tail length is about 45–90 mm. The smallest species, *P. ingrami* and *P. tenuirostris,* weigh only about 5 grams. A series of specimens of *P. maculata,* one of the largest species, averaged 15.3 grams for males and 10.9 grams for females (Morton and Lee 1978). The upper parts are pale tawny olive, darker tawny, or brownish gray; the underparts are olive buff, fuscous, or light tan. The feet are light grayish olive or pale brown, and the tail is grayish or brownish. The fur is soft and dense on the body; the tail is short-haired and nontufted. The central areas of the foot pads are usually smooth but sometimes are striated, the latter condition perhaps being most common in *P. novaeguineae*. In appearance and behavior planigales resemble the true shrews, *Sorex*.

These marsupials are remarkable for their extremely flat skull, which has an almost straight upper profile and a depth of as little as 6 mm. The same condition, however, is closely approached in some species of *Antechinus* (Archer 1976; Ride 1970). The pouch becomes fairly well developed during the breeding season and opens to the rear. The known number of mammae is 5–10 (possibly up to 15) in *P. maculata,* 6–12 in *P. ingrami* and *P. tenuirostris,* and 12 in *P. gilesi* (Archer 1976).

Planigales occur mainly in savannah woodland and grassland, though *P. maculata* also has been reported from rainforest. They may shelter in rocky areas, clumps of grass, the bases of trees, or hollow logs, and in captivity they have been observed to build saucer-shaped nests of dry grass (Archer 1976). Although primarily terrestrial, they climb fairly well (Collins 1973). Most seem to be nocturnal, but Aitken (1972) reported that *P. gilesi* was active for short periods throughout the day. Planigales are avid predators, feeding on insects, spiders, small lizards, and small mammals such as *Leggadina* (Collins 1973). They are capable of catching and eating grasshoppers almost as large as themselves. Captive females of *P. ingrami* ate six to eight grasshoppers 50 mm long every day and appeared to thrive on this diet.

Flat-skulled marsupial "mouse" *(Planigale tenuirostris)*, photo by Dick Whitford.

In captivity, females of *P. maculata* are polyestrous, can breed throughout the year, and can produce several litters annually. Gestation is 19–20 days and litter size is 5–11 young. Males are capable of breeding until they are at least 24 months old (Aslin 1975). In a study of wild populations of this species in the Northern Territory, Taylor, Calaby, and Redhead (1982) found breeding to occur in all months; litter size ranged from 4 to 12 young and averaged 8. Read (1984) reported that *P. gilesi* has an estrus of 3 days and an estrous cycle of 21 days, while *P. tenuirostris* has an estrus of 1 day, an estrous cycle of 33 days, and a gestation period of 19 days; the breeding season of each species extends from July–August to mid-January, and some females produce two litters per season. Three reports on *P. ingrami* are that it breeds from February to April and has litters of 4–6, that in northeastern Queensland it breeds during the wet season (December–March) and has litters of 4–12, and that females collected in the northern part of Western Australia in December–January had pouch young (Archer 1976). The young of this species are capable of an independent existence at approximately three months of age and are mature within their first year (Collins 1973).

The subspecies *P. ingrami subtilissima*, found in the Kimberley Division in the northern part of Western Australia, and the species *P. tenuirostris* are classified as endangered by the USDI. These designations seem to have been applied mainly because relatively few specimens of either taxon had been collected. Recently both forms have been found to occur over a more extensive region than previously thought (Thornback and Jenkins 1982). The IUCN does not assign a classification to either form but does now designate the species *P. novaeguineae* as vulnerable. *P. novaeguineae* is restricted to a small area of suitable habitat that could be quickly eliminated by human encroachment.

DASYUROMORPHIA; DASYURIDAE; Genus NINGAUI
Archer, 1975

Ningauis

There are three species (Archer 1975; Baverstock et al. 1983; Kitchener, Stoddart, and Henry 1983):

N. timealeyi, northwestern Western Australia;
N. yvonneae, southern Western Australia, southern South Australia, central New South Wales, western Victoria;

N. ridei, interior Western Australia, southwestern Northern Territory, western South Australia.

Head and body length is 46–57 mm, tail length is 59–79 mm, and weight is 2–13 grams (Dunlop *in* Strahan 1983; McKenzie *in* Strahan 1983). The upper parts are dark brown to black, the underparts are usually yellowish, and the sides of the face are salmon to buffy brown. The tail is thin, without a brush or crest. *N. ridei* and *N. yvonneae* have seven mammae, and *N. timealeyi* has six (Kitchener, Stoddart, and Henry 1983). This genus is considered to be most similar to *Sminthopsis* but to differ in smaller size, longer hair, broader hind feet with enlarged apical granules, and various cranial and dental characters. Archer (1981) stated that *Ningaui* resembles *Sminthopsis* in having wide molars and narrow nasals and in lacking posterior cingula on the upper molars but differs from *Sminthopsis* in lacking squamosal-frontal contact in the skull.

Specimens have been collected in dry grassland and savannah, and the various species appear adapted to life under arid conditions. They are nocturnal, sheltering by day in dense hummocks, small burrows, or hollow logs. They prey on insects, other invertebrates, and possibly small vertebrates. Captives have climbed along spinifex leaves and used their tails in a semiprehensile manner (McKenzie *in* Strahan 1983). Captive *N. yvonneae* were often found to enter daily torpor, especially in response to low temperature and withdrawal of food (Geiser and Baudinette 1988). According to Joan M. Dixon (National Museum of Victoria, pers. comm., c. 1981), a female ningaui captured in Victoria on 28 December 1977 lived until 6 April 1980. It was fed on a wide variety of insects as well as oranges and apples. The animal was vocal throughout its period in captivity and used a high-pitched rasping noise to demand food.

Dunlop (*in* Strahan 1983) wrote that male *N. timealeyi* become aggressive toward each other during the breeding season and that females with pouch young drive other adults away. In years with good rainfall the reproductive period of this species may extend from September to March (spring and summer); in other years breeding is restricted to November–January. Usually five to six young are carried in the simple pouch to the stage of weaning. By March in most years the population of *N. timealeyi* consists predominantly of the now independent young. They attain sexual maturity in late winter.

In a study of a group of *N. ridei* captured in the Northern Territory, Fanning (1982) found females to be polyestrous and the breeding season to extend from early Sep-

Ningaui *(Ningaui ridei)*, photo by N. L. McKenzie.

Narrow-footed marsupial "mouse" *(Sminthopsis longicaudata)*, photo by P. A. Woolley and D. Walsh.

tember to late February (spring and summer). Males uttered a distinctive mating call, "tsitt," and females responded similarly. Gestation lasted from about 13 to 21 days, and most litters contained seven young. They remained attached to the nipples until they reached about 42–44 days, then were left in a nest constructed by the mother, and became independent at 76–81 days. Fleming and Cockburn (1979) reported that a female *N. yvonneae* with five young was taken in western Victoria in November 1978 and that immature animals were captured there in June, August, and December 1977.

DASYUROMORPHIA; DASYURIDAE; Genus SMINTHOPSIS
Thomas, 1887

Dunnarts, or Narrow-footed Marsupial "Mice"

There are 20 species (Archer 1979*b*, 1981; Aslin 1977; Baverstock, Adams, and Archer 1984; Cole and Gibson 1991; Collins 1973; Hart and Kitchener 1986; Kirsch and Calaby 1977; Kitchener, Stoddart, and Henry 1984; Mahoney and Ride *in* Bannister et al. 1988; McKenzie and Archer 1982; Morton 1978*a;* Morton, Wainer, and Thwaites 1980; Pearson and Robinson 1990; Ride 1970; Van Dyck 1985, 1986; Van Dyck, Woinarski, and Press 1994; Waithman 1979):

S. murina, northeastern Queensland, New South Wales, Victoria, southeastern South Australia;

S. dolichura, southern Western Australia, southern South Australia;

S. fuliginosus, southwestern Western Australia;

S. gilberti, southwestern Western Australia;

S. griseoventer, southern Western Australia;

S. aitkeni, Kangaroo Island off southeastern South Australia;

S. leucopus, northeastern Queensland, eastern New South Wales, southern and southeastern Victoria, Tasmania;

S. longicaudata, northwestern Western Australia;

S. ooldea, southeastern Western Australia, southern Northern Territory, western South Australia;

S. psammophila, southwestern Northern Territory, south-central WesternAustralia, southern South Australia;

S. granulipes, inland parts of southwestern Western Australia;

S. macroura, central Western Australia to western Queensland and northern New South Wales;

S. bindi, northern Northern Territory;

S. virginiae, southern New Guinea, Aru Islands, northeastern Western Australia to northeastern Queensland;

S. douglasi, northwestern Queensland;

S. butleri, northeastern Western Australia;

S. archeri, southwestern New Guinea, Cape York Peninsula of northern Queensland;

S. youngsoni, northern Western Australia, Northern Territory;

Narrow-footed marsupial "mouse" *(Sminthopsis crassicaudata)*, photo by Stanley Breeden.

S. hirtipes, west-central Western Australia, southern
Northern Territory;
S. crassicaudata, mostly inland parts of the southern half
of Australia.

Groves (*in* Wilson and Reeder 1993) noted that *S. butleri*
also occurs in New Guinea, but Flannery (1990*b*) indicated
that the pertinent records are referable to *S. archeri.*

Head and body length is about 70–120 mm and tail
length is about 55–130 mm except in *S. longicaudata.* In
this species the length of the tail is approximately 200 mm,
or about twice the length of the head and body. Adult *S.
crassicaudata* (average head and body length about 83 mm)
usually weigh 10–15 grams. An adult male *S. leucopus*
(head and body length 112 mm) weighed 30 grams (Green
1972). The average weight of *S. macroura* is 19 grams in
males and 16 grams in females (Godfrey 1969). The fur is
soft, fine, and dense. The back and sides are buffy to grayish, the underparts are white or grayish white, the feet are
usually white, and the tail is brownish or grayish. Some
species have a median facial stripe.

This genus is differentiated from other marsupial
"mice" largely by features of the skull and dentition. The
feet are slender, and the pads are striated or granulated. The
hind part of the soles lacks pads. Most species have 8–10
mammae, some have 6, and one species apparently has only
2. The pouch is relatively better developed than in most
other marsupial "mice."

In some species the tail accumulates fat and becomes car-
rot-shaped during times of abundant food. This fat reserve
may be utilized when food is scarce, and the tail will then
become thin. In *S. murina, S. leucopus, S. virginiae, S. long-*

icaudata, and *S. psammophila* the tail never becomes fat,
not even under the best of conditions (Ride 1970).

Most of the permanently thin-tailed species live in
moist forest or savannah, but *S. longicaudata* and *S. psammophila* occupy arid grassland and desert. As would be expected, those species capable of storing food in the tail are
mainly inhabitants of dry country (Ride 1970), though *S.
crassicaudata* is sometimes found in moist areas (Morton
1978*a*). Dunnarts dig burrows or construct nests of grasses and leaves, which are placed in hollow logs or under
bushes or stumps. During the summer *S. crassicaudata* reportedly shelters among rocks. This species proceeds by
means of bipedal bounds when traveling at top speed, but
over short distances it has a peculiar quadrupedal ramble
during which it holds its tail above the ground in a stiff upward curve. It and the other species are mainly terrestrial,
but some are agile climbers. Most species are known to be
strictly nocturnal. Ewer (1968) found that *S. crassicaudata*
sheltered by day and then had alternating periods of activity and rest during the night. Individuals of that species,
along with *S. macroura* and *S. murina,* also have been reported occasionally to enter a state of torpor during periods of low food supplies (Geiser et al. 1984; Morton 1978*d;*
Ride 1970). Dunnarts are mainly insectivorous, but small
vertebrates, such as lizards and mice, also are eaten. Most
insects are caught on the ground, but *S. murina* sometimes
leaps high into the air to catch moths in flight.

For *S. crassicaudata* in two parts of Victoria, Morton
(1978*b*) reported population densities of 0.5/ha. and 1.3/ha.
Individuals occupied home ranges, but precise size could
not be determined as the areas utilized shifted over a period of months. Home ranges overlapped among the same

sex and between sexes even during the breeding season, and no territorial behavior by males was observed. Breeding females, however, tended to be sedentary and possibly defended a small territory around the nest. Both sexes usually nested alone during the breeding season, but at other times up to 70 percent of the population shared nests in groups of 2–8 that included members of either or both sexes. Such groups apparently were nonpermanent, random aggregations. If nest sharing did occur during the breeding season, it generally involved pairs of a male and an estrous female. Males tolerated one another to some extent during the breeding season but may have attempted to monopolize estrous females. Collins (1973) reported that captive dunnarts could be maintained in pairs or small groups if sufficient space and multiple nesting facilities were provided but that an adult male might be attacked by a female with a litter. Aslin (1983) found captive adult *S. ooldea* to be generally intolerant of one another, even of the opposite sex, and to fight to the death when caged together. This species attains sexual maturity at 10 months and gives birth to five to eight young from September to January.

The reproductive traits of *S. crassicaudata* are fairly well known (Collins 1973; Ewer 1968; Godfrey and Crowcroft 1971; Morton 1978c). Females are polyestrous and are able to produce litters continuously in captivity. Estrus occurs in repeated cycles of 25–37 days, extending through a season of at least 6 months and perhaps the entire year. In the wild, breeding appears to be restricted to a period of 6–8 months, starting in June or July, during which each female is thought to give birth to two litters. The gestation period is about 13 days. Litter size is 3–10, usually 7–8 in the wild and slightly smaller in captivity. The young are carried in the mother's pouch until they are about 42 days old and then are left in the nest until they are about 63 days old. They are then practically self-sufficient, and the family soon breaks up. Minimum age of sexual maturity is 115 days in females and 159 days in males.

The breeding season of *S. macroura* was found to last from July to February in a captive colony in South Aus-

tralia (Godfrey 1969). Females are polyestrous, with an average cycle length of 26.2 days. Gestation is 12.5 days. Litter size is one to eight, but litters of only one or two are not reared. The young are carried in the pouch for 40 days, then suckled in the nest for 30 more days. For *S. murina* gestation reportedly is 13–16 days and there are up to eight young (Collins 1973).

Morton (1978c) found that in the wild few female *S. crassicaudata* lived past 18 months and few males lived past 16 months. In captivity dunnarts seem to live and reproduce over a longer period (Collins 1973). The longevity record is held by a specimen of *S. macroura* that survived for 4 years and 11 months (Jones 1982).

The IUCN classifies the species *S. psammophila, S. douglasi,* and *S. aitkeni* as endangered and *S. butleri* as vulnerable. All are highly restricted in distribution and/or population size, and further declines are expected. *S. longicaudata* and *S. psammophila* are listed as endangered by the USDI and are on appendix 1 of the CITES. *S. longicaudata* had been known by only five specimens, the last found in 1975 (Thornback and Jenkins 1982), but was recently rediscovered alive in the Northern Territory (*Oryx* 28 [1994]: 97). Aitken (1971c) reported that *S. psammophila* had been known from only a single specimen collected in 1894 in the Northern Territory but that a colony was discovered on the Eyre Peninsula of South Australia in 1969. This colony is near a wildlife reserve with comparable habitat, so the species may have a good chance for survival in the area. Additional specimens have since been found in both South and Western Australia (Pearson and Robinson 1990). The habitat of *S. douglasi* has been almost entirely cleared for grazing (Thornback and Jenkins 1982). It was known by only four specimens, the last collected in 1972, but was rediscovered alive by Woolley (1992). Archer (1979a) stated that two other species, *S. granulipes* and *S. hirtipes,* should be considered vulnerable or of uncertain status. Both have restricted ranges and may be subject to adverse habitat modification. In addition to the above full species, the IUCN now classifies the subspecies *S. murina tatei,* of northeast-

Julia Creek dunnarts *(Sminthopsis douglasi)*, female with 60-day-old young, photo by P. A. Woolley and D. Walsh.

ern Queensland, as near threatened and an unnamed sub-species of *S. griseoventer* on Boullanger Island as critically endangered. The latter comprises a single population of fewer than 250 individuals and is continuing to decline.

DASYUROMORPHIA; DASYURIDAE; Genus
ANTECHINOMYS
Krefft, 1866

Kultarr

The single species, *A. laniger*, is distributed over much of the inland region from Western Australia to Queensland and New South Wales. Archer (1977) recognized only this one species but observed that there were two distinctive allopatric forms. Lidicker and Marlow (1970) considered these two forms to be full species: *A. laniger* in New South Wales and Queensland and *A. spenceri* in more westerly areas. On the basis of skeletal and dental characters, Archer (1981) treated *Antechinomys* only as a subgenus of *Sminthopsis*, but Woolley (1984) pointed out that biochemical analysis and phallic morphology indicate that the two genera are distinct. The former arrangement was followed by Groves (*in* Wilson and Reeder 1993), the latter by Mahoney and Ride (*in* Bannister et al. 1988).

Head and body length is 80–110 mm and tail length is about 100–145 mm. Males are usually larger than females. The upper parts are grayish; the underparts usually are whitish, with gray hairs at the base. There is a dark ring around the eye and a dark patch in the middle of the forehead. The ears are relatively large. The fur is long, soft, and fine with few guard hairs except on the back and rump. More than the basal half of the very long tail is fawn-colored, then the hairs increase in length, and the terminal third is covered with long brown or black hairs. The upper third of the limbs is furred like the body; the remaining two-thirds is covered by short, fine, white fur. The face is

well provided with vibrissae, and usually there are long vibrissae arising from the carpal pads on the wrists.

The body form is well adapted for a bounding locomotion. The feet are narrow and have granular pads. Each hind foot is elongated and has one large, well-haired, cushion-like pad on the sole and only four toes, the first being absent. The forelimbs also are graceful and elongated. It was once thought that the long hind feet betokened a bipedal, hopping means of progression comparable to that of such rodents as *Notomys, Dipodomys,* and *Jaculus;* however, recent studies have demonstrated that the kultarr actually moves in a graceful gallop, springing rapidly from its hind feet and landing on its forefeet (Ride 1970).

During the breeding season the pouch becomes fairly well developed on females. It consists of folds of skin that enlarge from the sides to partially cover the nipples. The fold is least developed on the posterior side, but unlike in most dasyurids, the pouch does not open to the rear (Lidicker and Marlow 1970). The number of mammae has been recorded at 4, 6, 8, and 10, there usually being 8 in the form once called *A. laniger* and 6 in the form once called *A. spenceri* (Archer 1977).

The kultarr occurs in a wide variety of mostly dry habitats—savannah, grassland, and desert associations. It apparently does not dig its own burrow, but nests in logs, stumps, or vegetation (Collins 1973). Perhaps it also uses the burrows of other animals or nests in deep cracks in the ground. It appears to be strictly terrestrial and nocturnal. The natural diet consists mainly of insects and other small invertebrates. It is not definitely known whether vertebrates, such as lizards and mice, are taken under normal conditions (Archer 1977). Torpor has been induced experimentally through the withholding of food and may be an adaptive mechanism to deal with deteriorating environmental conditions in the wild (Geiser 1986).

Field and laboratory investigations by Woolley (1984) show that there is a long breeding season. Animals in southwestern Queensland are in reproductive condition from mid-winter to mid-summer (July–January). Females

Kultarr *(Antechinomys laniger)*, photo from Australian Museum, Sydney, through Basil Marlow. Inset photo by W. D. L. Ride.

(Parantechinus apicalis), photo by P. A. Woolley and D. Walsh.

are polyestrous, being able to enter estrus up to six times during the season. The estrous cycle lasts about 35 days, gestation 12 days or less. A female may rear up to six young per litter. Weaning occurs after about 3 months and sexual maturity at 11.5 months. Both sexes may live to breed in more than one season. Collins (1973) referred to two adults each of which had lived for nearly three years in captivity.

Although the kultarr has a wide range, it once seemed to be rare. The USDI designates *A. laniger* as endangered, but this classification is intended to apply only to the population of the species that inhabits New South Wales and Queensland. Thornback and Jenkins (1982) noted that the species as a whole is relatively common and widespread.

DASYUROMORPHIA; DASYURIDAE; Genus PARANTECHINUS
Tate, 1947

There are two species (Archer 1982b; Begg *in* Strahan 1983; Woolley *in* Strahan 1983):

P. bilarni, Arnhem Land in northern Northern Territory;
P. apicalis (dibbler), southwestern Western Australia.

Neither Kirsch and Calaby (1977) nor Ride (1970) recognized *Parantechinus* as generically distinct from *Antechinus,* but recent systematic work involving cranial and dental features, biochemical analysis, and phallic morphology suggests that the two genera are not closely related (Archer 1982b; Baverstock et al. 1982; Woolley 1982b).

In *P. bilarni* head and body length is 57–100 mm, tail length is 82–115 mm, and weight is 12–44 grams. The upper parts are grizzled brown, the underparts are pale gray, and there are cinnamon patches behind the large ears. The tail is long and never fat (Begg *in* Strahan 1983). In *P. api-*

calis head and body length is about 140–45 mm, tail length is 95–115 mm, and weight is 30–100 grams. The upper parts are brownish gray speckled with white, and the underparts are grayish white tinged with yellow. This species is distinguished from the species of *Antechinus* by its tapering and hairy tail, a white ring around the eye, and the freckled appearance of its rather coarse fur (Woolley 1991c and *in* Strahan 1983). The pouch of females consists merely of folds of skin on the lower abdomen enclosing the mammae, which number six in *P. bilarni* and eight in *P. apicalis.*

P. bilarni occurs in rugged, rocky country commonly covered with open eucalyptus forest and perennial grasses. Its diet consists mainly of insects. Mating occurs from late June to early July, litters contain four to five young, weaning occurs late in the year, and sexual maturity is attained by the following June (Begg *in* Strahan 1983).

P. apicalis was found in dense heathland and apparently made nests in dead logs or stumps. Captives tend to be nocturnal but may emerge from cover during the day to bask in the warmth of the sun. They burrow through leaf litter, which suggests that they do the same in the wild to search for insects. They also eat chopped meat, honey, and nectar. The latter food, plus the climbing ability of captives, indicates that the nectar of flowers is sought in the wild (Woolley *in* Strahan 1983). Mating in *P. apicalis* occurs early in the year, in March or April, females are monestrous, and gestation lasts 44–53 days (Woolley 1971b, 1991c). There are up to eight young per litter; they remain dependent for 3–4 months and reach sexual maturity at 10–11 months. Recently, Dickman and Braithwaite (1992) showed that some populations of *P. apicalis* experience a synchronized male die-off following the mating season, as occurs in most species of *Antechinus* (see account thereof). However, observations of other populations in the wild and captivity indicate that such is not inevitable (Woolley 1991c).

P. bilarni was not discovered until 1948, and *P. apicalis* is one of the rarest of mammalian species. Geologically Recent fossil remains indicate that it once occurred as far north as Shark Bay on the west-central coast of Western Australia, but it had not been collected for 83 years when, in 1967, two specimens were taken alive in the extreme southwestern corner of that state. Despite intensive efforts, only seven more individuals were captured until 1984, and two were found dead. Clearing of habitat for agricultural purposes has been responsible for the decline within historical time, but a 75-ha. reserve now protects the place where the dibbler was rediscovered (Morcombe 1967; Woolley 1977, 1980; Woolley and Valente 1982). In 1984 another dead specimen was found in Fitzgerald River National Park on the south coast of Western Australia, and subsequently another 17 individuals were trapped in and near the park. Meanwhile, in 1985 the species was discovered in abundance on Boullanger and Whitlock islands off the west coast (Fuller and Burbidge 1987; Muir 1985). *P. apicalis* is classified as endangered by both the IUCN and USDI.

DASYUROMORPHIA; DASYURIDAE; **Genus** **DASYKALUTA**
Archer, 1982

The single species, *D. rosamondae,* is found in the Pilbara region of northwestern Western Australia (Woolley *in* Strahan 1983). Since its description in 1964 this species had been placed in the genus *Antechinus,* but Archer (1982b) considered it to represent a separate and not closely related genus. The latter arrangement was accepted by Groves (*in* Wilson and Reeder 1993), though Mahoney and Ride (*in* Bannister et al. 1988) referred the species to *Parantechinus.*

According to Woolley (*in* Strahan 1983), head and body length is 90–110 mm, tail length is 55–70 mm, and weight is 20–40 grams. The overall coloration is russet brown to coppery. The fur is rather rough, the head and ears are

short, and the forepaws are strong and well haired on the back. *Dasykaluta* has a general form similar to that of *Dasycercus* but is distinguished by its small size, coloration, and lack of black hair on the tail. Females have eight mammae.

Dasykaluta dwells in spinifex grassland and feeds on insects and small vertebrates. Observations on both wild-caught and laboratory-maintained animals (Woolley 1991a) show that females are monestrous and there is a short annual breeding season. Mating occurs in September and the young are born in November. The total period of pregnancy averages 50 days but ranges from 38 to 62 days, the great variation perhaps being associated with arrested development in some cases. Normally there are eight young, but there may be as few as one. They open their eyes after 58–60 days and are weaned at 90–120 days. Both sexes attain sexual maturity at about 10 months and thus are able to participate in the first breeding season following their birth. However, as in *Antechinus* (see account thereof), all mature males perish shortly thereafter. Females produce only one litter annually, but some live to breed for a second and possibly a third season.

DASYUROMORPHIA; DASYURIDAE; **Genus** **PSEUDANTECHINUS**
Tate, 1947

There are three species (Kitchener 1988; Kitchener and Caputi 1988; Woolley *in* Strahan 1983):

P. macdonnellensis, Western Australia, Northern Territory;
P. ningbing, northeastern Western Australia;
P. woolleyae, northwestern Western Australia.

Neither Kirsch and Calaby (1977) nor Ride (1970) recognized *Pseudantechinus* as being generically distinct from

Dasykaluta rosamondae, photo by P. A. Woolley and D. Walsh.

Pseudantechinus macdonnellensis, photo by C. W. Turner through Basil Marlow.

Antechinus, but recent systematic work involving cranial and dental features, biochemical analysis, and phallic morphology suggests that the two genera are not closely related (Archer 1982b; Baverstock et al. 1982; Woolley 1982b). The generic distinction of *Pseudantechinus* was accepted by Groves (*in* Wilson and Reeder 1993) and Morton, Dickman, and Fletcher (1989), but Corbet and Hill (1991) and Mahoney and Ride (*in* Bannister et al. 1988) considered it a synonym of *Parantechinus.* Electrophoretic studies by Cooper and Woolley (1983) suggested that the closest relative of *P. ningbing* is actually *Dasycercus cristicauda.* The IUCN recognizes *P. mimulus,* restricted to a small area of northeastern Northern Territory, as a species separate from *P. macdonnellensis* and classifies it as vulnerable.

According to Woolley (*in* Strahan 1983), head and body length in *P. macdonnellensis* is 95–105 mm, tail length is 75–85 mm, and weight is 20–45 grams. The upper parts are grayish brown, there are chestnut patches behind the ears, and the underparts are grayish white. The tail tapers and becomes very thick at the base for purposes of fat storage. *P. ningbing* is similar in general coloration and form but is smaller, and its tail is longer and has long hairs covering the base. Females have six mammae.

P. macdonnellensis is found mainly on rocky hills and breakaways but also lives in termite mounds in some areas. It is predominantly nocturnal but may emerge from shelter among the rocks to sunbathe. The diet consists mainly of insects. Observations by Woolley (1991b) show that there is a short annual breeding season during the austral winter. Mating occurs in June and early July in the eastern part of the range and in August and early September farther west. The gestation period is 45–55 days. There commonly are five or six young, they open their eyes after 60–65 days, they are weaned at about 14 weeks, and they

are able to breed in the first season following their birth. Unlike the usual situation in *Antechinus,* there is no mass die-off of males after reproduction; indeed, males are potentially capable of breeding for at least three years. Females are monestrous and may breed in at least four seasons. *P. ningbing* also apparently lives into at least a second breeding season. It mates in June and gives birth during a period of just two to three weeks in late July and early August after a gestation of 45–52 days. The young are weaned at about 16 months and reach sexual maturity at 10–11 months (Woolley 1988).

DASYUROMORPHIA; DASYURIDAE; **Genus MYOICTIS** *Gray, 1858*

Three-striped Marsupial "Mouse"

The single species, *M. melas,* is found in New Guinea and on Japen, Waigeo, and Salawatti islands to the northwest and the Aru Islands to the southwest (Flannery 1990b, 1995).

Head and body length is about 170–250 mm and tail length is approximately 150–230 mm. A single adult weighed 200 grams (Flannery 1990b). This animal was named *melas,* meaning "black," because the first specimen known to science was a melanistic individual. Normal individuals are among the most colorful of all marsupials: the upper parts are a richly variegated chestnut mixed with black and yellow. Except in occasional melanistic individuals there are three dark longitudinal stripes. The head is dark rusty red, often with a black stripe on the nose, and the crown may be tawny brown. The chin and chest are pale

Pseudantechinus ningbing, a newly discovered species without tail thickening, photo by P. A. Woolley and D. Walsh.

Three-striped marsupial "mouse" *(Myoictis melas),* photo by P. A. Woolley and D. Walsh.

rufous, and the remainder of the underparts is yellowish gray or whitish. The tail is evenly tapered and covered with long reddish hairs for most of its length; the tip is black above and rufous brown below.

The general body form is similar to that of a small mongoose. The pads are striated. The pouch is only slightly developed, and there are six mammae.

Ziegler (1977) wrote that this marsupial occurred in most rainforests of the lowlands and midmountains of New Guinea but was relatively uncommon. He observed that its external similarities to the largely terrestrial, open forest chipmunks and other sciurids suggested comparable habits and parallel activity patterns. Collins (1973), however, thought that this genus was nocturnal and scansorial in habit. An earlier report stated that one individual, a male, was shot on a recumbent, decayed log in undergrowth, while an accompanying female escaped. A. R. Wallace stated in 1858 that in the Aru Islands these animals "were as destructive as rats to everything eatable in houses." Flannery (1990b) questioned that statement but noted that native hunters had indicated that *Myoictis* enters villages at night in pursuit of murid rodents.

DASYUROMORPHIA; DASYURIDAE; Genus
DASYUROIDES
Spencer, 1896

Kowari

The single species, *D. byrnei*, is known to occur in the south of the Northern Territory of Australia, the southwest of Queensland, and the northeast of South Australia (Aslin 1974). *Dasyuroides* was considered a synonym of *Dasycercus* by Corbet and Hill (1991), Groves (*in* Wilson and Reeder 1993), and Mahoney and Ride (*in* Bannister et al. 1988) but was maintained by Morton, Dickman, and Fletcher (1989).

Head and body length is 135–82 mm, tail length is 110–40 mm, and weight is 70–140 grams, with males being about 30 grams heavier than females (Aslin 1974; Aslin *in* Strahan 1983). The back and sides are grayish with a faint rufous tinge. The underparts are pure creamy white, and the feet are white. Less than the basal half of the tail is rufous; the remainder is densely covered on both the upper and lower surfaces with long black hairs that form distinct dorsal and ventral crests. The soft and dense pelage is composed mainly of underfur with few guard hairs. The body form is strong and stout. The hind feet are very narrow and lack a first toe; the bottoms of the feet are hairy. The tail is not thickened. Pouch development is sufficient to conceal the young completely, and there usually are six (five to seven) mammae (Aslin 1974).

The kowari inhabits desert associations and dry grassland. It may shelter in the hole of another mammal or dig its own burrow, and both sexes construct therein a nest of soft materials (Aslin 1974; Ride 1970). It is primarily terrestrial but climbs well and is capable of vertical leaps of at least 45.7 cm (Collins 1973). In the field, Aslin (1974) observed *Dasyuroides* only after dark; in captivity one individual was largely nocturnal, while another was sporadically active both day and night. Collins (1973) stated that the kowari enjoys sun- and sandbathing. He also reported that the diet consists of insects, arachnids, and probably small vertebrates such as birds, rodents, and lizards. Aslin (1974) reported that one wild individual had fed on *Rattus villosissimus* and some insects. Torpor has been induced experimentally in *Dasyuroides* through a moderate reduction in diet, and it is likely that individuals also enter torpor in the wild in response to declining food supplies (Geiser, Matwiejczyk, and Baudinette 1986).

Aslin (1974) estimated that a population of at least 14 adults occupied an area of less than 750 ha. and suggested that some kind of social aggregation might occur in the wild. Collins (1973) observed that a captive pair or small group could be maintained together but that serious fighting might result if the enclosure were too small. Aslin (1974) noted, however, that aggressive behavior is stylized in such a way as to prevent serious injury in intraspecific conflict. She also reported that each sound made by adults—hissing, chattering, and snorting—is restricted to threat situations and that both sexes use scent produced by sternal and cloacal glands to mark parts of their home range.

The following reproductive information is based on studies of both wild and captive individuals (Aslin 1974, 1980; Collins 1973; Woolley 1971a). Each female may have up to four estrous periods a year, which recur at two-month intervals if young are not being suckled. Mating takes place from April to December, mostly from May to July, and females may produce two litters in a season. Gestation is 30–36 days. Litter size is 3–7, having averaged 5.1 in captivity and 5.8 in the wild. The young are 4 mm long at birth, first detach from the nipples at 56 days, are weaned and practically independent at about 100 days, reach sexual maturity at about 235 days, and attain adult weight at 1 year. The young may ride on the mother's back or sides when

Kowari *(Dasyuroides byrnei)*, photo by Constance P. Warner.

they are 2–3 months old, and she remains tolerant of them for a while after weaning. Both sexes are capable of breeding through their fourth year of life. One captive kowari lived for 7 years and 9 months (Marvin L. Jones, Zoological Society of San Diego, pers. comm., 1995).

The IUCN now classifies *D. byrnei* as vulnerable but recognizes it as part of the genus *Dasycercus*. Kennedy (1992) reported that it has declined by 50 to 90 percent, apparently because of fox predation and habitat destruction through grazing of domestic livestock.

DASYUROMORPHIA; DASYURIDAE; Genus
DASYCERCUS
Peters, 1875

Mulgara

The single species, *D. cristicauda*, inhabits the arid region from the Pilbara in northwestern Western Australia to southwestern Queensland (Ride 1970). Remains found in owl pellets indicate that its range may have extended as far as western New South Wales in the nineteenth or early twentieth century (Ellis 1992). Some authorities include *Dasyuroides* (see account thereof) within *Dasycercus*.

Head and body length is 125–220 mm, tail length is 75–125 mm, and weight is 60–170 grams (Woolley *in* Strahan 1983). Individuals of this genus sometimes exceed *Dasyuroides* in size but usually are smaller. The upper parts vary from buffy to bright red brown, and the underparts are white or creamy. The close and soft pelage consists principally of underfur with few guard hairs. The tail is usually thickened for about two-thirds of its length and is densely covered near the body with coarse, chestnut-colored hairs. Beginning at about the middle of its length the tail is covered with coarse, black hairs that increase in length toward the tip to form a distinct dorsal crest.

This desert dweller is compactly built, with short limbs, ears, and muzzle. The pouch area consists of only slightly developed lateral skin folds. The mammae usually number six or eight, or rarely four.

Dasycercus occupies sandridge or stony desert and spinifex grassland. Woolley (1990) reported two kinds of burrows excavated in the sandhills. One had a grass-lined nest area at a depth of about 0.5 meter that was connected to the surface by one large tunnel and several near vertical popholes. This kind of burrow was found to be occupied by a female and pouch young. The other kind, occupied only by a single male, had up to five entrance holes, a complex system of interconnecting tunnels, and one grass-lined nest at a depth of up to 1 meter. The mulgara is terrestrial but is capable of climbing (Collins 1973). It seems to be both diurnal and nocturnal. Ride (1970) noted that it avoided exposure to heat by remaining in its burrow during the hot part of the day. Collins (1973), however, observed that it was partially diurnal, would bask in the sun whenever the opportunity arose, and would even make use of sun lamps. When sunning itself, it flattens its body against the substrate and the tail twitches sporadically. The mulgara can withstand considerable exposure to both heat and cold. However, captives were found to enter torpor for up to 12 hours, during which metabolic rate and body temperature were greatly reduced (Geiser and Masters 1994). Studies of captives indicate that the physiological adaptations of *Dasycercus* are such that it is able to subsist without drinking water or even eating succulent plants because it is able to extract sufficient water from even a diet of lean meat or mice (Ride 1970).

This genus is relatively uncommon but reportedly increases in numbers when a house mouse plague occurs within its range. *Dasycercus* attacks a mouse with lightning action and then devours it methodically from head to tail, inverting the skin of the mouse in a remarkably neat fashion as it does so. The mulgara also preys on other small vertebrates, arachnids, and insects. It reportedly can skillfully dislodge insects from crevices by means of its tiny forepaws.

Dasycercus can be maintained in captivity in pairs except when the females have young (Collins 1973). They generally do not fight among themselves and appear quite solicitous of each other. According to Woolley (*in* Strahan 1983), however, it is usual to find only a single wild individual in a burrow. Females with up to eight young have been captured between June and December. Mating in captivity has been observed from mid-May to mid-June, and young have been born in late June, July, and August after a gestation period of five to six weeks. The young suckle for 3–4 months and become sexually mature at 10–11 months. Individuals of both sexes have been known to come into

Mulgara *(Dasycercus cristicauda),* photo by Shelley Barker. Insets: undersides of hand and foot, photos from *Report on the Work of the Horn Scientific Expedition to Central Australia, Zoology,* B. Spencer, ed.

breeding condition each year for 6 years, suggesting a fairly long life span.

Archer (1979a) considered the mulgara to be a vulnerable or possibly endangered species. It apparently has disappeared or become very rare in Queensland and South Australia. Its decline may be associated with habitat disturbances and predation caused by the European introduction of livestock, cats, foxes, and rabbits (Kennedy 1992). It now is classified as vulnerable by the IUCN. *D. hillieri*, restricted to the vicinity of Lake Eyre and subject to extreme population fluctuations, is recognized as a separate species by the IUCN and is classified as endangered.

DASYUROMORPHIA; DASYURIDAE; Genus DASYURUS
E. Geoffroy St.-Hilaire, 1796

Native "Cats," Tiger "Cats," or Quolls

There are six species (Archer 1979a; Kirsch and Calaby 1977; Ride 1970; Van Dyck 1987; Ziegler 1977):

D. hallucatus, northern parts of Western Australia and Northern Territory, northern and eastern Queensland;

D. viverrinus, southeastern Australia, Tasmania, Kangaroo Island, King Island;

D. geoffroii, Australia except extreme north;

D. spartacus, southwestern Papua New Guinea;

D. albopunctatus, New Guinea;

D. maculatus, eastern Australia, Tasmania.

These species sometimes were placed in separate genera or subgenera as follows: *D. maculatus* in *Dasyurops* Matschie, 1916; *D. viverrinus* in *Dasyurus* E. Geoffroy St.-Hilaire, 1796; *D. geoffroii* in *Dasyurinus* Matschie, 1916; and *D. hallucatus* and *D. albopunctatus* in *Satanellus* Pocock, 1926. Based on paleontological data, Archer (1982b) again recognized *Satanellus* as a full genus. This view was not accepted by Van Dyck (1987), who considered *D. albopunctatus* to be closely related not to *D. hallucatus* but to his newly described *D. spartacus.* The assignment of all of the above species to *Dasyurus* was accepted by Corbet and Hill (1991), Groves (*in* Wilson and Reeder 1993), Mahoney and Ride (*in* Bannister et al. 1988), and Morton, Dickman, and Fletcher (1989).

Measurements are: *D. hallucatus* and *D. albopunctatus,* the two smallest species, head and body length approximately 240–350 mm, tail length 210–310 mm; *D. spartacus,* head and body 305–80 mm, tail 240–85 mm; *D. viverrinus,* head and body 350–450 mm, tail 210–300 mm; *D. geoffroii,* head and body 290–650 mm, tail 270–350 mm; and *D. maculatus,* the largest species, head and body about 400–760 mm, tail 350–560 mm. Weights in a series of specimens of *D. viverrinus* were 850–1,550 grams for males and 600–1,030 grams for females (Green 1967). An adult female *D. geoffroii* weighed 550 grams. Weight in *D. maculatus* usually is about 2–3 kg.

The upper parts are mostly grayish, or olive brown to dark rufous brown. In *D. viverrinus* there is a less common black phase that appears to be unrelated to sex, often in the same litter with individuals of the more common color. Regardless of the basic coloration, all individuals of the genus have prominent white spots or blotches on the back and sides. Only in *D. maculatus* do the spots usually extend well onto the tail. In *D. geoffroii* the face is paler and grayer than the upper parts, the spots extend onto the head, and the terminal half of the tail is black. In *D. hallucatus* and *D. albopunctatus* the tip and entire ventral surface of the tail are dark brown or black. In *D. spartacus* the body color is deep bronze to tan brown with small white spots, and the tail is black with no spots. In all species the underparts are paler than the back, usually yellowish or white. The coat generally is short, being soft and thick in *D. viverrinus* and *D. geoffroii* and coarse with little underfur in *D. hallucatus* and *D. albopunctatus.*

The species *D. viverrinus* differs from all the others in lacking a first toe on the hind foot. The pads of the feet are granulated in *D. viverrinus* and *D. geoffroii* and striated in the other species. In *D. hallucatus* and *D. albopunctatus* pouch development consists only of lateral folds of skin around the 8 mammae. In the other species there is a shallow pouch formed by a flap of skin and usually 6 mammae (sometimes 8 in *D. viverrinus*).

Habitats include dense, moist forest for *D. maculatus,* drier forest and open country for *D. viverrinus,* savanna for *D. geoffroii,* woodland and rocky areas for *D. hallucatus,* low savannah for *D. spartacus,* and a variety of conditions up to an altitude of 3,500 meters for *D. albopunctatus* in New Guinea. All species are primarily terrestrial but able to climb well (Collins 1973). All are nocturnal, but some occasionally are seen by day. *D. maculatus* may follow its prey for over 4 km in one night. By day *D. viverrinus* shelters in rock piles or hollow logs and *D. hallucatus* reportedly nests in hollow trees or even abandoned buildings. Female *D. geoffroii* excavate burrows in which to give birth, the main tunnel descending to a nursery chamber nearly 1 meter below the ground (Serena and Soderquist 1989a). All species are predators but also will eat vegetable matter (Ride 1970). The diet of the smaller species contains a high proportion of insects, while *D. maculatus* may take mammals as large as wallabies *(Thylogale).* Native "cats" occasionally raid poultry yards and are therefore disliked by farmers, but they probably also benefit human interests by destroying many mice and insect pests (Ride 1970).

Radio-tracking studies of *D. geoffroii* in Western Australia (Serena and Soderquist 1989b) showed females to use home ranges averaging about 337 ha., including a core area of about 90 ha. with numerous dens, where most activity is concentrated. There was usually little or no overlap of these core areas, suggesting that females are intrasexually territorial, but sometimes a daughter shared her mother's core area. Males used much larger ranges, including core areas of at least 400 ha., and there was broad overlap of their core areas with those of both other males and females. Both sexes were found to be essentially solitary and to scent-mark with feces.

In captivity all 6 species can be maintained as pairs except when the female has young (Collins 1973). Most native "cats" appear to be monestrous, winter breeders, producing a single litter between May and August (Ride 1970). There is a lengthy courtship in *D. maculatus,* during which the female may be bitten severely about the head and shoulders. In Victoria the pouch of females of this species enlarges in June and July whether young are born or not. The estrous cycle is 21 days long (Hayssen, Van Tienhoven, and Van Tienhoven 1993), gestation also is 21 days, and litter size is four to six. The young first detach from the nipples at 7 weeks, are one-third grown and independent at 18 weeks, attain sexual maturity at 1 year, and reach full size by 2 years.

Green (1967) collected pouch young of *D. viverrinus* in Tasmania during June and July. A few litters contained 5 young, but most had 6, the same as the number of mammae. Females of this species, however, reportedly some-

Australian native cats: Top, *Dasyurus viverrinus,* photo from U.S. National Zoological Park; Middle, *D. geoffroii;* Bottom, *D. hallucatus,* photos by Shelley Barker.

Australian native cat *(Dasyurus (Satanellus) albopunctatus)*, photo by P. A. Woolley and D. Walsh.

times have 8 mammae and 8 young and occasionally produce up to 24 embryos, far more than can be nurtured. Fletcher (1985) reported that mating activity in captive *D. viverrinus* reaches a peak in late May and early June, that females of this species are polyestrous, and that one female had an estrous cycle of 34 days. There are old and highly doubtful reports that gestation in *D. viverrinus* is as short as 8 days, but Fletcher found the period to vary from 20 to 24 days. The young first detach from the nipples at 8 weeks and become independent by 18 weeks (Collins 1973). There also is evidence of polyestry in *D. geoffroii* (Serena and Soderquist 1990). In that species gestation is 16–23 days, up to 6 young are born, they separate from the mother's nipples by 62 days, and females reach sexual maturity at 1 year (Collins 1973; Ride 1970; Serena and Soderquist 1988). The 6–8 young of *D. hallucatus* measure only 3 mm in length, are weaned after 3 months, and reach full maturity by 10–11 months (Collins 1973).

In a study of *D. hallucatus* in Western Australia, Schmitt et al. (1989) found females to produce a single litter in July or August; lactation continued through November. Most males died during the post-mating period from July to September, evidently because of high testosterone levels and resulting trauma, as has been reported for *Antechinus* (see account thereof). However, males with lower testosterone levels, apparently the less dominant individuals, had increased prospects of survival. Dickman and Braithwaite (1992) determined that a given population of *D. hallucatus* may experience a complete male die-off in some years but not in others. The following longevity data are available for native "cats": individuals of *D. maculatus* have been maintained for 3–4 years in zoos (Collins 1973); maximum known life span of *D. viverrinus* is 6 years and 10 months (Jones 1982); a specimen of *D. geoffroii* was kept in captivity more than 3 years (Archer 1974b); and *D. albopunctatus* has been maintained for up to 3 years (Collins 1973).

All species of native "cats" have suffered since the settlement of Australia by Europeans and evidently are continuing to decline. The IUCN classifies *D. geoffroii, D. spartacus, D. albopunctatus,* and *D. maculatus* generally as vulnerable and *D. hallucatus* and *D. viverrinus* as near threatened. The subspecies *D. maculatus gracilis,* of northeastern Queensland, is singled out as endangered. *D. viverrinus* also is listed as endangered by the USDI. *D. geoffroii* survives only in remote parts of southwestern Western Australia, central Australia, and Queensland. During the 1930s *D. viverrinus* was common on the southeastern mainland, being found even in the suburbs of large cities such as Adelaide and Melbourne. It subsequently became very rare in this region (Emison et al. 1975; Grzimek 1975). In Tasmania, however, *D. viverrinus* remains common and even has increased in numbers in some areas (Green 1967). In addition to direct human persecution and clearing of cover, native "cats" have been adversely affected by introduced placental predators and competitors, such as domestic cats and dogs and foxes. During the first decade of the twentieth century a severe epidemic seems to have greatly reduced the populations of native "cats," as well as certain other marsupials, in southeastern Australia and Tasmania, and in some areas there apparently never was a substantial recovery. *D. maculatus* is now uncommon to rare on the mainland, having declined through such factors as destruc

Large spotted native "cat" *(Dasyurus maculatus)*, photo by C. W. Turner through Basil Marlow.

tion of forests and widespread trapping and poisoning (Mansergh 1984*c*).

DASYUROMORPHIA; DASYURIDAE; **Genus**
SARCOPHILUS
E. Geoffroy St.-Hilaire and F. Cuvier, 1837

Tasmanian Devil

The single living species, *S. harrisii,* is now found only in Tasmania. Once it also occupied much of the Australian mainland (Calaby and White 1967), but it probably disappeared before European settlement, through competition with the introduced dingo *(Canis familiaris dingo).* Subfossil remains indicate that it or a closely related species still was present as recently as 3,120 years ago in the Northern Territory (Dawson 1982*a*), 400 years ago in southwestern Australia (Baynes 1982), and 600 years ago in Victoria (Guiler *in* Strahan 1983). Living specimens collected in Victoria in 1912 and 1971 are generally thought to have escaped from captivity (Guiler 1982; Ride 1970).

Head and body length is about 525–800 mm and tail length is about 230–300 mm. A series of specimens of males weighed 5.5–11.8 kg and a series of females weighed 4.1–8.1 kg (Green 1967). The coloration is blackish brown or black except for a white throat patch, usually one or two white patches on the rump and sides, and the pinkish white snout.

The general form, except for the tail, resembles that of a small bear. The head is short, broad, and covered with mass-

es of muscle; the skull and teeth are extremely massive and rugged. The molar teeth are developed into heavy bone crushers in a manner reminiscent of the hyenas. The first toe is not present, and the granular pads of the feet lack striations. The pouch during the breeding season is a completely closed receptacle, unlike that of many dasyurids. There are 4 mammae.

Sarcophilus is found throughout Tasmania, except in areas where cover has been extensively cleared, and is most numerous in coastal heath and sclerophyll forest (Guiler 1970*a*). It is nocturnal, sheltering by day in any available cover, such as caves, hollow logs, wombat holes, or dense bushes. Both sexes make nests of bark, button grass, or leaves. The Tasmanian devil is terrestrial but capable of climbing and occasionally emerges from its lair to bask in the sun (Collins 1973). Its movements are slow and clumsy, and its habit of continually nosing the ground suggests a well-developed sense of smell (Ride 1970). The diet consists of a wide variety of invertebrates and vertebrates, including poisonous snakes, and a small amount of plant material. Guiler (1970*a*) found the main foods to be wallabies (*Macropus* and *Thylogale*), wombats, sheep, and rabbits, most of which were taken as carrion. Both laboratory and field observations indicate that *Sarcophilus* is an inefficient killer that does not usually hunt large, active prey (Buchmann and Guiler 1977). It is, however, an efficient scavenger of carrion, consuming all parts of a carcass, including the fur and bones. It is said formerly to have fed on the remains of animals killed by the larger Tasmanian wolf *(Thylacinus).*

Population densities in two parts of Tasmania were about 3/sq km and 25/sq km, the latter figure being considered abnormally high (Buchmann and Guiler 1977; Guiler 1970*a*). In an area of abundant food home ranges were found to be small, with individuals traveling about 3.2 km during a night. In an area of less food home ranges were larger and nocturnal movement covered about 16 km. There was considerable overlap in home ranges, both between and among sexes, and no evidence of territorial behavior (Guiler 1970*a*).

As in so many animals, disposition of *Sarcophilus* is individually variable; some seem quite vicious, others more tractable. Eric R. Guiler, of the University of Tasmania, reported (pers. comm., 1972) that most of the more than 7,000 Tasmanian devils he handled were docile to the point of being lethargic and could be handled with ease; he said that ferocity has been greatly exaggerated. Captives, however, generally are highly aggressive toward their own kind, and there is severe fighting over food (Buchmann and Guiler 1977). Agonistic encounters are accompanied by much vocalization, starting with low growls and proceeding to a rising and falling vibrato or even loud screeching. Eventually two individuals may form a stable dominant-subordinant relationship with reduced aggression. Observations of wild animals congregated to feed on carcasses indicated considerable agonistic interaction but few serious physical clashes (Pemberton and Renouf 1993). Collins (1973) reported that pairs could be maintained in captivity if the animals were gradually introduced to each other. Turner (1970) observed that adult males would assist females in cleaning the offspring and that after the young were too big to fit in the mother's pouch they might cling to the back of either parent. Guiler (1971*c*) cautioned, however, that at a later stage the father might devour the young.

Sarcophilus is monestrous. Most mating occurs in March, but some may take place a little later. The gestation period is about 31 days and litter size is two to four. The young weigh 0.18–0.29 grams at birth, can first release the

Tasmanian devil *(Sarcophilus harrisii)*, photo from U.S. National Zoological Park.

nipples at 90 days, leave the pouch at about 105 days, and are weaned by 8 months (Collins 1973; Guiler 1970b, 1971c). Females do not become sexually mature until they are two years old. Maximum known longevity is eight years and two months (Jones 1982).

The Tasmanian devil had become rare by the first decade of the twentieth century, apparently because of persecution by European settlers, destruction of forest habitat, and a severe epidemic in the early twentieth century. There was concern that it, along with the Tasmanian wolf, might be facing extinction. More recently, however, *Sarcophilus* has recovered and even become abundant in many areas. Its increase in numbers seems to have resulted partly from newly available food supplies in the form of carrion from livestock and commercial trapping operations (Green 1967; Guiler 1982; Ride 1970). The Tasmanian devil is sometimes considered a nuisance and has been subject to official control measures. Buchmann and Guiler (1977), however, stated that its reputation as a killer of livestock is unmerited, that sheep probably are eaten only as carrion, and that money spent on control is wasted.

DASYUROMORPHIA; Family MYRMECOBIIDAE; Genus MYRMECOBIUS
Waterhouse, 1836

Numbat, or Banded Anteater

The single known genus and species, *Myrmecobius fasciatus*, occurred in suitable areas from southwestern Western Australia, through northern South Australia and extreme southwestern Northern Territory, to southwestern New South Wales (Friend 1989; Ride 1970). This genus sometimes has been placed in the family Dasyuridae, but the sum of available information indicates that it represents a distinct family (Archer and Kirsch 1977). There are Pleistocene records of *M. fasciatus* from Australia (Archer 1984a); otherwise the Myrmecobiidae are known only from the Recent.

Head and body length is 175–275 mm, tail length is 130–200 mm, and usual weight is 300–600 grams. In *M. fasciatus fasciatus*, the subspecies that inhabits southwestern Western Australia, the foreparts of the body are grayish brown with some white hairs. In *M. fasciatus rufus*, which occurred farther to the east, the main body color is rich brick red. Both subspecies have six or seven white bars between the midback and the base of the tail, producing an effect of transverse light and dark stripes. There is a dark cheek stripe running through the eye, with a white line above and below. The body pelage is short and coarse, and both the head and hindquarters are remarkably flattened above. The body is larger posteriorly than anteriorly. The tail is long, somewhat bushy, and nonprehensile. It is covered with long, stiff hairs that are sometimes bristly and form a brush. The snout is long and tapering, the mouth small; the slender tongue can be extended at least 100 mm. The tip of the nose is naked. The front legs are comparatively thick and widely spaced; the forefoot has five toes and the hind foot has four, all bearing strong claws. A complex gland on the chest opens onto the skin through a number of conspicuous pores. The female has four mammae but lacks any suggestion of a pouch. When the young are attached to her nipples they are protected only by the long hair on her underparts. She supplies them with milk and warmth, but they are dragged beneath her as she travels.

There may be 4 or 5 upper molar teeth and 5 or 6 lower molars (Archer and Kirsch 1977). Therefore, the total number of teeth may be as high as 52, more than that of any other land mammal. The overall dental formula is: (i 4/3, c 1/1, pm 3/3, m 4–5/5–6) × 2 = 48–52. Most of the teeth are small, delicate, and separated from one another. The size is not constant; the molars on the right and left sides of the skull often vary in length and width. The numbat's bony palate extends farther back than in many mammals and may be associated with the long, extensible tongue. This type of palate is also present in the armadillos (Xenarthra,

Numbat *(Myrmecobius fasciatus)*, photo by L. F. Schick.

Dasypodidae) and the pangolins, or scaly anteaters (Pholidota, Manidae). According to Friend (1989), the palate has no vacuities, the alisphenoid tympanic wing forms virtually the entire floor of the middle ear, processes from the frontal and jugal bones form an incomplete bony bar behind the orbit of the skull, and the lachrymal bone is very large and extends a long way out onto the face.

The numbat inhabits open scrub woodland, generally where eucalyptus trees predominate, and desert associations. There had been some question whether this animal constructs its own burrow. Available evidence now suggests that the female may dig a short tunnel with a chamber at the end where she builds a nest and gives birth to her young (Christensen 1975; Ride 1970). The numbat also is known to shelter by night in a bed of leaves or grass placed in a hollow log or the burrow of some other animal.

Radio-tracking studies by Christensen, Maisey, and Perry (1984) show that the numbat uses logs for overnight shelter during warm weather and changes to burrows when the weather becomes cold and that an individual may utilize many such shelters. One burrow consisted of a tunnel about 1 meter long that opened to a nest chamber about 200 mm across and 100 mm below ground level. These studies indicate that home ranges are large, with some individuals using over 100 ha. during a two-month period.

Unlike most marsupials, *Myrmecobius* is active during the day. It is nimble, climbs readily, and spends much of its time in search of food. It trots and leaps about with jerky movements and carries its tail in line with its body, but with a slight upward curve. When the animal is startled or frightened it may sit bolt upright, flatten its entire body and fluff its tail, or run into a hollow log. When caught, it reportedly emits snuffling and hissing sounds but does not attempt to bite. During cooler months it often basks in the sun.

Termites and ants form the regular diet, and other invertebrates are eaten only occasionally. Termites actually are the preferred food; most of the ants consumed are of small predatory species that rush in when the numbat uncovers a termite nest and are lapped up along with the termites. A captive numbat consumed from 10,000 to 20,000 termites of the smaller species daily. These are swallowed whole, whereas those of the larger species are masticated. Termites are obtained from rotten logs, dead trees, and subsurface soil. The strong foreclaws are used in scratching into soil and decayed wood, and the long snout sometimes functions as a lever. The extensible, cylindrical tongue extracts the termites from crevices in the wood and is manipulated with great speed and dexterity. In southwestern Australia, wandoos *(Eucalyptus redunca elata)* are often attacked by a species of termite, so that the woodland floor is littered with hollow branches from the infested trees. These logs provide both food and shelter for the numbat.

Except during the breeding season the numbat is usually solitary, though Collins (1973) observed that captives could be maintained in pairs. The gestation period is 14 days (Friend 1989). In southwestern Australia the young are born from January to April or May. Litters of two to four have been recorded, but four seems to be the usual number. The young are carried by the female, attached to her nipples, for about four months and then are suckled in a nest for about two more months. Juveniles have been seen foraging for themselves by October. A male in the Taronga Zoo in Australia survived for five years, and its mate was still living after six years in captivity (Strahan 1975).

The numbat is classified as vulnerable by the IUCN and as endangered by the USDI. According to J. A. Friend (1990*a*), the decline of the species began in the east shortly after European settlement around 1800 and moved progressively westward. It now has disappeared entirely from New South Wales, South Australia, and the Northern Territory and survives only in a few suitable areas in extreme southwestern Western Australia. The eastern subspecies, *M. f. rufus*, is apparently extinct. There had been speculation that the main cause of the decline was clearing of land for agriculture, which eliminates the dead and fallen trees from which termites can be obtained; however, the continued abundance of the echidna *(Tachyglossus)* in the same region suggests no shortage of food for a mammal that depends on termites and ants. Probably an overriding factor in the decline of the numbat has been predation by introduced carnivores, especially the red fox *(Vulpes vulpes)*, exacerbated by the loss of vegetative cover to fire and clearing. Unlike the echidna, which can dig into the ground and is protected by its spiny pelage, the numbat depends on cover to escape predators. The numbat actually remained common in many of the arid parts of South Australia and Western Australia until about 1930 but then underwent a drastic population crash in the 1940s and 1950s. This decline coincided both with the establishment of the red fox in the region and with the departure of the aboriginal people from the deserts. These people traditionally had burned off small patches of vegetation throughout the year, but the cessation of this practice led to a buildup of scrub and eventually to huge summer wildfires that destroyed all cover. In addition, introduced rabbits *(Oryctolagus)* increased in the same region and further encouraged numbers of fox and domestic cat *(Felis catus)*. Even in the 1970s the numbat was locally common on reserves in southwestern Western Australia, but populations there underwent another crash at that time and disappeared entirely from some reserves, again apparently because of fox predation. Subsequent control of the fox has allowed a modest increase and successful reintroduction of the numbat in some areas.

DASYUROMORPHIA; Family THYLACINIDAE; Genus THYLACINUS
Temminck, 1824

Thylacine, or Tasmanian Wolf or Tiger

The single Recent genus and species, *Thylacinus cynocephalus,* apparently was found only in Tasmania within historical time. There were many reports of animals bearing some resemblance to the thylacine from the Australian mainland during the eighteenth, nineteenth, and twentieth centuries (Heuvelmans 1958; Ride 1970); none have been confirmed, though Archer (1984a) noted that there are some hints that a population may have survived in the north until the early twentieth century. During the late Pleistocene and early Recent periods the genus was widespread in Australia and New Guinea, with the latest definite subfossil mainland records dating from just over 3,000 years B.P. The most likely cause for the disappearance of the thylacine from the mainland and New Guinea was competition with domestic dogs introduced by aboriginal human populations starting several thousand years ago (Archer 1974a; Dawson 1982b; Partridge 1967). These dogs, which in Australia have the name "dingo" *(Canis familiaris dingo),* became feral and spread over a large region but did not enter Tasmania.

The thylacine often was placed in the family Dasyuridae but subsequently came to be considered distinct from that group and to have closer affinity with three extinct South American families of large predatory marsupials (Kirsch 1977b). Those three families, plus the Thylacinidae, were then grouped in the superfamily Borhyaenoidea, which was thought to have originated before the process of continental drift had separated South America, Antarctica, and Australia. More recent studies of morphology, serology, and mitochondrial DNA (Archer 1982c, 1984a; Krajewski et al. 1992; Sarich, Lowenstein, and Richardson 1982; Szalay 1982) have indicated once again that the Thylacinidae are most closely related to the Dasyuridae, though the two groups are considered very distinct families within the order Dasyuromorphia.

Head and body length of *Thylacinus* is 850–1,300 mm, tail length is 380–650 mm, shoulder height is 350–600 mm, and weight is 15–30 kg (Moeller *in* Grzimek 1990). The up-

Tasmanian wolf *(Thylacinus cynocephalus)*, photo from U.S. National Zoological Park.

per parts are tawny gray or yellowish brown with 13–19 blackish brown transverse bands across the back, rump, and base of the tail; the underparts are paler. The face is gray with some indistinct white markings around the eyes and ears. The fur is short, dense, and coarse. The dental formula is: (i 4/3, c 1/1, pm 3/3, m 4/4) $\times$ 2 = 46. The unusual dental features, as noted below, distinguish *Thylacinus* from the Dasyuridae. In addition, the epipubic bones of *Thylacinus* are vestigial and cartilaginous, the clavicles are reduced, and the foramen pseudovale is lacking. Unlike that of the Myrmecobiidae, the palate of *Thylacinus* has large vacuities (Marshall 1984).

With its general external resemblance to a canid, the thylacine represents one of the most remarkable examples of convergent evolution found in mammals. Its scientific name means "pouched dog with a wolf head." The jaws have a remarkably wide gape. Long canine teeth, shearing premolars, and grinding molars show further similarity with the dog family. The overall build is doglike, but the hind end tapers gradually, rather than abruptly, to the rigid tail. The feet also resemble those of dogs, and like the canids, the thylacine is digitigrade, walking on its toes. The pads of the feet are granulated, not striated, and the forefoot leaves a five-toed print. In females there is a rear-opening pouch consisting of a crescent-shaped flap of skin enclosing the four mammae. Despite the superficial appearance of the thylacine, an analysis of its skeletal proportions by Keast (1982) indicated that it is essentially like a large dasyurid and does not have the specialized pursuit adaptations of the placental wolves *(Canis)*.

The preferred habitat of the thylacine probably was open forest or grassland, but its last populations may have occupied dense rainforest in southwestern Tasmania. Its lair reportedly was in a rocky outcrop or a hollow log. Although nocturnal, it was sometimes seen basking in the sun. The thylacine seems to have been mostly solitary but sometimes reportedly hunted in pairs or small family groups. It was said to trot relentlessly after its prey until the victim was exhausted and then to close in with a rush. It fed mainly on kangaroos, wallabies, smaller mammals, and birds. Various calls were reported, including a whine (which may have been for communication), a low growl expressing irritation, and a coughing bark when hunting.

Based on a study of bounty records, Guiler (1961) found that births occurred throughout the year, with a pronounced peak in the summer (December–March). The two to four young were thought to leave the pouch after three months but to remain with the mother until they were about nine months old. Although many individuals were successfully maintained in zoos, the thylacine was never bred in captivity. Record longevity was nearly 13 years (Moeller *in* Grzimek 1990).

As in the case of most large predatory mammals, there is controversy regarding the extent of damage the thylacine caused to domestic livestock. Nonetheless, it was generally considered to be a sheep killer in Tasmania, and intensive private bounty hunting began about 1840. By 1863 the thylacine already had been restricted to the mountains and more inaccessible parts of the island. From 1888 to 1909, 2,184 thylacines were destroyed for payment of a government bounty then in effect, and many more were killed for the private bounties. A rapid decline in the population was noticed about 1905, and by 1914 the thylacine was a rare species. In addition to human hunting, trapping, and poisoning, the demise of the thylacine has been blamed on disease, habitat modification, and increased competition from the settlers' domestic dogs.

The last definite record of a wild individual being killed was in 1930 (Ride 1970). The last captive animal died in 1936 (Paddle 1993). Several organized searches made over the next 40 years failed to find conclusive evidence that the species still existed. A number of recent reports, including the supposed killing of a thylacine in western Tasmania in 1961, were doubted by Brown (1973). Many alleged sightings both in Tasmania and on the mainland of Australia were discussed in Rounsevell and Smith (1982) and Smith (1982). Some persons still have hope that the Tasmanian wolf survives, and a 647,000-ha. game reserve was established in the southwestern part of the island in 1966, partly to protect the species. The thylacine has received complete legal protection since 1938, is designated as endangered by the USDI, and is on appendix 1 of the CITES. It now is classified as extinct by the IUCN.

The geological range of the family Thylacinidae is middle Miocene to Recent in Australasia (Archer 1984*a*).

Order Peramelemorphia

Bandicoots

This order of 2 families with 8 Recent genera and 22 species occurs in Australia, Tasmania, and New Guinea and on certain nearby islands. Until recently the group generally was considered a suborder or superfamily within a larger assemblage of syndactylous marsupials that also included the Diprotodontia. However, Aplin and Archer (1987) explained that there evidently is closer affinity to the Dasyuromorphia and that in any case full ordinal status is warranted. The latter procedure was followed by Groves (in Wilson and Reeder 1993) though not by Szalay (1994). Marshall, Case, and Woodburne (1990) accepted this order but used the name Peramelina for it. There also have been suggestions of a direct relationship between the Peramelemorphia and the New World Didelphimorphia (Gordon and Hulbert 1989). A history of the varying systematic outlook on the group was provided by Strahan (in Seebeck et al. 1990).

The term "bandicoot" is a corruption of a word in the Telugu language of the people of the eastern Deccan Plateau of India meaning "pig rat," which originally was applied to a large species of rodent (Bandicota indica) from India and Sri Lanka. It seems likely that the name was first applied to the Australian marsupials by the explorer Bass in 1799, and the use of the name is of questionable desirability because of confusion with Bandicota. It has been suggested that the semantic association of marsupial bandicoots with placental rats has reduced the popularity of the former as subjects of natural history investigation. This situation now seems to have changed; detailed summaries of recent studies of the ecology, behavior, physiology, morphology, and systematics of the Peramelemorphia have been provided by Gordon and Hulbert (1989) and Lyne (1990).

The genera expressing both extremes of size in this order are endemic to New Guinea. The smallest is Microperoryctes, with a head and body length of 150–75 mm; the largest is Peroryctes, with a head and body length of up to 558 mm and a weight sometimes over 4.7 kg. The muzzle is elongate and pointed. The hindquarters are not as elongated as in the kangaroos (Macropodidae), but the hind limbs are larger than the forelimbs. The tail is hairy, of variable length, and nonprehensile.

The second and third digits of the hind foot are syndactylous, that is, bound together by integument; only the tops of the joints and the nails are separate. The combined toes function somewhat like a single digit and are useful in grooming. The main digit of the hind foot is the fourth; the fifth is well developed in most species but is shorter than the fourth; the first toe of the hind foot lacks a nail and is generally poorly developed, and in the genera Macrotis and Chaeropus it is absent. Except in Chaeropus the forefoot has five digits: the third is the longest, the second is slightly shorter, the fourth is somewhat shorter than the second, and the first and fifth are vestigial. In Chaeropus the second and third digits of the forefoot are about the same size, the fourth is vestigial, and the first and fifth are absent. Those digits of the forefoot that are well developed have sharp nails for digging.

Although the Peramelemorphia are syndactylous, like the Diprotodontia, their dentition is polyprotodont, as in the Dasyuromorphia. The incisors are numerous and small; the uppers are flattened and unequal, and the lowers slope forward; the crown of the last lower incisor has two lobes. The canines are slender and pointed. The premolars are narrow and pointed, and the molars are four-sided or triangular in outline with four external cusps. The dental formula is: (i 4–5/3, c 1/1, pm 3/3, m 4/4) $\times$ 2 = 46 or 48.

The members of this order are alert and sprightly, generally traveling on all four feet with a galloping action. They are terrestrial and mainly nocturnal, sheltering by day in grassy nests on the surface of the ground. Most species have mixed feeding habits. They drink water in captivity, but dew seems sufficient for their needs in the wild when water is not present.

The Peramelemorphia are the only marsupials in which a chorioallantoic placenta develops after the transient yolk-sac placenta. This chorioallantoic placenta, however, unlike that of placental mammals, lacks villi. It evolved independently from the placenta of the latter group and is probably functionally correlated with a high rate of reproduction; the Peramelemorphia are characterized by the shortest gestation period known in mammals, rapid litter succession in polyestrous females, rapid development of pouch young, minimal parental care, and early sexual maturity (Gordon and Hulbert 1989). Females have 6, 8, or 10 mammae (usually 8) and generally have two to five young per litter. The pouch opens downward and backward. Newly born young of some bandicoots have deciduous claws that are shed soon after the young reach the pouch.

Bandicoots have suffered to varying degrees since the settlement of Australia by Europeans. Several kinds have been adversely affected by such factors as habitat modification and introduced predators. In general, those that live in the coastal forests and woodlands of Australia and in Tasmania have survived well. In contrast, those that live on the inland plains and deserts have declined severely (Aitken 1979).

The geological range of the Peramelemorphia is middle Miocene to Recent in Australasia (Archer 1984a).

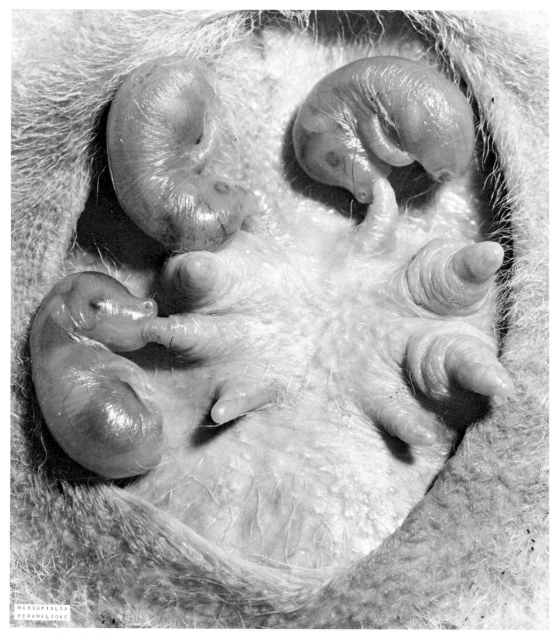

Long-nosed bandicoot *(Perameles nasuta)*. Litter of three pouch young at four days (estimated) after birth. Three of the unoccupied teats are still enlarged and were suckled by a previous litter. Photo by A. Gordon Lyne through Australian Commonwealth Scientific and Industrial Research Organization.

PERAMELEMORPHIA; **Family PERAMELIDAE**

Dry-Country Bandicoots

This family of 4 Recent genera and 11 species is found in Australia and southern New Guinea. Until recently this family generally was considered to include the Peroryctidae, but Groves and Flannery (1990) regarded the latter group as a separate family. In addition, Archer and Kirsch (1977) had placed the genus *Macrotis* in still another family, the Thylacomyidae, but such arrangement was not con-

sidered appropriate by Groves and Flannery (1990) or Baverstock et al. (1990). Szalay (1994) did not accept the Peroryctidae as a full family and treated the Thylacomyidae as a subfamily of the Peramelidae. Kirsch et al. (1990) supported recognition of the members of Peroryctidae as a monophyletic group but also indicated that the Thylacomyidae are even more distinct from the Peramelidae.

Head and body length is 200–550 mm, tail length is 75–275 mm, and weight is around 200–1,600 grams. Coloration varies considerably. Depending on the genus, the pelage may be soft or coarse, the skull may be narrow or relatively broad, the tail may be relatively long or short,

A young long nosed bandicoot *(Perameles nasuta)* about 54 days old entering the rear-opening pouch of its mother. The tail of a young already in the pouch shows beneath the young that is entering. Photo by A. Gordon Lyne.

and the ears may be long or short. The dental formula is consistently: (i 5/3, c 1/1, pm 3/3, m 4/4) × 2 = 48.

According to Groves and Flannery (1990), the members of this family occur mainly in dry country, while those of the Peroryctidae are found in rainforest. The morphological distinctions between the two families, however, are not thought to express this difference in habitat. The Peramelidae are characterized as being flat-skulled, while the Peroryctidae are cylindrical-skulled. In the Peramelidae the anteromedial orbital margin slopes diagonally outward, forming a remarkable ridge, or crest, along its edge; the facial extension of the lacrimal bone is very narrow, its su-

ture with the maxilla parallel to the antorbital ridge; the lower margin of the mandible is convex; the auditory bullae (bones housing the inner ear) are complete; and the molar teeth are squared.

The geological range of the Peramelidae is Miocene to Recent in Australasia (Gordon and Hulbert 1989).

PERAMELEMORPHIA; PERAMELIDAE; Genus MACROTIS
Reid, 1837

Bilbies, or Rabbit-eared Bandicoots

There are two species (Johnson 1989):

M. lagotis (greater bilby), formerly from Western Australia to southwestern Queensland and New South Wales;
M. leucura (lesser bilby), formerly east-central Western Australia, southern Northern Territory, and northern South Australia.

The generic name *Thylacomys* Owen, 1840, sometimes has been applied to these species, but for technical reasons it is no longer considered valid. Based on cranial and dental morphology, serology, and karyology, Archer and Kirsch (1977) had considered *Macrotis* to represent a distinct family. In accordance with the rules of zoological nomenclature, that family is properly called Thylacomyidae even though the name is based on an incorrect generic name. In any event, Aplin and Archer (1987) suggested that the assignment of *Macrotis* to a separate family may have been premature, and Groves and Flannery (1990) restored the genus to the Peramelidae, noting a particularly close relationship to *Chaeropus*. However, new analyses of DNA by Kirsch et al. (1990) support maintenance of the Thylacomyidae.

According to Johnson (1989), head and body length of *M. lagotis* is 290–550 mm, tail length is 200–290, and weight is 600–2,500 grams; respective figures for *M. leucura* are 200–270 mm, 120–70 mm, and 311–435 grams.

Rabbit-eared bandicoot *(Macrotis lagotis)*, photo by Basil Marlow.

Males are much heavier than females (Johnson and Johnson 1983). The pelage is long, silky, and soft. The upper parts are fawn gray, blue gray, or ash gray, often pale vinaceous along the sides, and the underparts are white. The color of the body fur extends onto the base of the tail. In *M. lagotis* the remainder of the proximal half of the tail is black, and the terminal half is white, with a conspicuous dorsal crest of long hairs. The tail of *M. leucura* is entirely white except for a gray line extending from the body onto the upper surface of the base. The end of the tail bears a prominent dorsal crest of hairs, as in *M. lagotis*, and in both species the extreme tip of the tail is naked.

The general form is light and delicate with a long, tapered muzzle. The ears are very long, pointed, and finely furred. As in all the Peramelemorphia except *Chaeropus*, the forefeet have three functional toes that bear stout, curved claws; the remaining toes are small. On the hind foot the first toe is not present, and in this characteristic both *Macrotis* and *Chaeropus* differ from the other genera of the Peramelemorphia. In *Macrotis*, as in the rest of the order, the second and third hind toes are partially united, the fourth toe is the largest, and the fifth is of moderate size. The pouch opens downward and slightly backward, enclosing eight mammae.

Groves and Flannery (1990) pointed out that *Macrotis* shares many characters with *Chaeropus*. However, the skull of *Macrotis* has an extremely broadened braincase and narrowed snout, the most excessively flattened cranium in the Peramelemorphia, and enormous, pear-shaped auditory bullae. The dental formula is identical to that of the other peramelids. Archer and Kirsch (1977) thought that the teeth of *Macrotis* differed from those of the other peramelids in several ways, particularly with regard to the structures involved in the squaring of the molars. Groves and Flannery (1990), however, noted that while the molars of *Macrotis* have undergone an enormous expansion of the metacone, those of *Chaeropus* show a less modified version of the same condition.

Bilbies live in a variety of mainly dry habitats, including woodland, savannah, shrub grassland, and sparsely vegetated desert. Their distribution seems to have been associated with the presence of suitable soils for burrowing. Unlike most peramelids, bilbies are powerful diggers and live mainly in burrows of their own making. These burrows are characteristic in that they usually descend from a single opening as a fairly steep, ever-widening spiral to a depth of 1–2 meters. According to Ride (1970), *M. leucura* burrows only in sandhills, never on flats, and the burrows of *M. lagotis* in Western Australia were found to be most common in shrub grassland. An area occupied by the latter species had 58 burrows, none more than 168 meters from any other. It seems likely that each animal has a number of burrows within its range.

Bilbies are nocturnal and terrestrial. They progress on all four feet by means of a shuffling gait; hopping is facilitated by the relatively long hind feet. These animals usually do not lie down to sleep; instead they squat on their hind legs and tuck their muzzle between their forelegs; the long ears are laid back and then folded forward over the eyes and along the side of the face. They are largely carnivorous, taking mainly insects but also small vertebrates and some vegetable matter. Conical scratching in the earth around trees and bushes gives evidence of their foraging for subterranean insect larvae. Raw or cooked meat, insects, snails, birds, mice, bread, and cake are readily accepted by captives, but roots and fruit usually are rejected.

These bandicoots are found alone or in small colonies usually consisting of a single adult male, a female, and an independent young (Johnson 1989). Burrows generally seem to have only one occupant (Ride 1970) but reportedly sometimes contain a pair or a female with young. Groups with more than one male or with several females can be maintained in captivity if the enclosure is sufficiently large (Collins 1973). In this situation the males form a rigid dominance hierarchy without severe fighting; the dominant male chases its subordinates away from the burrows but freely shares these places with females (Johnson and Johnson 1983).

According to McCracken (1990), *M. lagotis* is physiologically capable of breeding throughout the year and captives have been observed to do so. The former wild populations in the temperate zone of South Australia had a distinct breeding season from about March to May, but the existing populations in arid regions may breed at any time of year, perhaps depending on rainfall and food availability. Ovulation occurs toward the end of lactation; thus births occur as soon as the young of the previous litter are weaned. The average estrous cycle is 20.6 days and the average gestation is 14 days. Litter size is usually one or two young; triplets are rare. Pouch young develop rapidly and have a pouch life averaging about 80 days. The young are then left in a burrow for about 2 more weeks and suckled at regular intervals. Johnson (1989) added that sexual maturity occurs at about 175–220 days in females and 270–420 days in males. The longevity record for a captive *M. lagotis* is 7 years and 2 months (Collins 1973).

Both *M. leucura* and *M. lagotis* are listed as endangered by the USDI and are on appendix 1 of the CITES. The IUCN now designates *M. leucura* as extinct and *M. lagotis* as vulnerable. *M. leucura* has not definitely been collected since 1931. This species formerly was common but was reduced drastically in the early twentieth century through such factors as trapping for its pelt, predation by introduced foxes, and competition with introduced rabbits for burrows. The same problems beset the once even more abundant *M. lagotis*. This species may even have occupied Victoria (Scarlett 1969) as well as suitable habitat all across the southern half of Australia. However, Southgate (1990) reported that it had disappeared from New South Wales by 1912 and from South Australia by about 1970 and is now restricted to a few isolated colonies in Western Australia, the Northern Territory, and southwestern Queensland; the progression of the decline seems to have coincided with the spread of foxes and rabbits. Johnson (1989) suggested that its decline in arid areas is associated with the disappearance of the traditional aboriginal practice of burning off vegetation gradually, thereby producing a patchwork of habitat types in different stages of regeneration. The current regime of large wildfires results in a homogeneity of habitats that is less favorable to the species. Thornback and Jenkins (1982) noted that its numbers may be kept at very low densities by the depletion of food supplies by grazing cattle.

PERAMELEMORPHIA; PERAMELIDAE; **Genus CHAEROPUS**
Ogilby, 1838

Pig-footed Bandicoot

The single species, *C. ecaudatus*, formerly occurred in suitable localities from southwestern and west-central Western Australia, through South Australia and the southern part of the Northern Territory, to southwestern New South Wales and western Victoria (Baynes 1984; Ride 1970). The name *ecaudatus* is an unfortunate misnomer as the original description was based on a mutilated specimen from which the tail was missing.

Pig-footed bandicoot *(Chaeropus ecaudatus)*, photo of mounted specimen by E. F. Aitken.

The head and body length is 230–60 mm and the tail length is 100–140 mm. The hair, though coarse, is not spiny. The upper parts are grizzled gray tinged with fawn, or almost orange brown, and the underparts are white or pale fawn. The tail is gray or fawn below and on the sides and black above. There is an inconspicuous crest above, with a few white hairs toward the tip.

The body form is light and slender. The head is broad with a long, sharply pointed snout. The ears are long and narrow. The limbs are long, slender, and peculiarly developed. The first and fifth digits of the forefeet are absent, the fourth is minute, and the second and third are well developed with long, sharp nails and conspicuous pads under their tips. On the hind feet the first digit is absent, the fifth is barely evident, the fourth is large and long and bears a short, stout nail and a well-developed pad, and the second and third are small and united. The forefeet thus resemble those of Artiodactyla, though in pigs the functional digits are the third and fourth rather than the second and third. The hind feet show some resemblance to those of Perissodactyla, except that in the horses the functional digit is the third, not the fourth.

Ride (1970) described the habitat as sclerophyll woodland, mallee, heath, and grassland. A nest of grass, sticks, and leaves is constructed in a hollow in the ground and lined with soft grasses. *Chaeropus* is cursorial; the gait was described by G. Krefft as resembling that of a "broken-down hack in a canter, apparently dragging the hind quarters after it." It is said to squat in the open with its ears laid back and to run for shelter in hollow logs and trees when chased. *Chaeropus* may not be strictly nocturnal, as on several occasions individuals were seen out in daylight. The genus is omnivorous. Breeding occurs in May or June. The maximum recorded litter size is only two, but females possess eight mammae (Collins 1973).

This genus may well be extinct. The last specimen collected was taken at Lake Eyre in South Australia in 1907. There were unconfirmed reports from central Australia in the 1920s. The decline is thought to have been caused by introduced predators, such as foxes and house cats, destruction of habitat by domestic livestock, and competition with introduced rabbits (Thornback and Jenkins 1982). According to Friend (1990b), there is extensive evidence that *Chaeropus* actually persisted in the interior of Western Australia well into the twentieth century. Its final disappearance may not have been until the 1930s or 1940s and may have been associated with the same factors that affected *Isoodon* (see account thereof). The pig-footed bandicoot is designated as extinct by the IUCN and endangered by the USDI and is on appendix 1 of the CITES.

PERAMELEMORPHIA; PERAMELIDAE; **Genus ISOODON**
Desmarest, 1817

Short-nosed Bandicoots

There are four species (Close, Murray, and Briscoe 1990; Ellis, Wilson, and Hamilton 1991; Groves *in* Wilson and Reeder 1993; Lyne and Mort 1981; Ride 1970; Seebeck et al. 1990):

I. macrourus, southern New Guinea, northeastern Western Australia, northern Northern Territory, eastern New South Wales;

I. obesulus, southeastern New South Wales, southern Victoria, southeastern South Australia and nearby Kangaroo Island and Nuyts Archipelago, southwestern Western Australia and nearby Recherche Archipelago, Tasmania, West Sister Island in Bass Strait;

I. auratus, northern and eastern Western Australia and nearby Barrow Island, Northern Territory, inland South Australia, western New South Wales;

I. peninsulae, Cape York Peninsula of northern Queensland.

The name *Thylacis* Illiger, 1811, often used instead of *Isoodon* for these species, is considered here to be a synonym of *Perameles. I. peninsulae* was treated as an extremely disjunct population of *I. obesulus* by Groves (*in* Wilson and Reeder 1993) and Seebeck et al. (1990), but Close, Murray, and Briscoe (1990), using chromosome and electrophoretic analysis, found it to be the most distinctive member of *Isoodon.*

In *I. macrourus* head and body length is 300–470 mm, tail length is 80–215 mm, and weight is 500–3,100 grams.

Short-nosed bandicoot *(Isoodon macrourus)*, photo by Stanley Breeden.

In *I. obesulus* head and body length is 280–360 mm, tail length is 90–140 mm, and weight is 400–1,600 grams. In *I. auratus* head and body length is 210–95 mm, tail length is 88–121 mm, and weight is 267–670 grams. The pelage in this genus is coarse and glossy with distinct hairs. The upper parts of *Isoodon* generally are a fine mixture of blackish brown and orange or yellow, and the underparts are yellowish gray, yellowish brown, or white.

An elongated snout is characteristic of all bandicoots, but in this genus the head is relatively broader than in *Perameles* and the jaws are slightly shorter and stouter. The ears are short and rounded. The well-developed pouch opens backward, as in other bandicoots, and the pouch contains eight mammae.

Short-nosed bandicoots inhabit open woodland, thick grass along the edges of swamps and rivers, and thick scrub on dry ridges. They are active and inoffensive animals and are often mistaken for large rats or rabbits from a distance. When walking they move the forelegs and hind legs separately, and they do not bound when running. They often make long tunnels through the grass. Their nests of sticks, leaves, and grass, sometimes mixed with earth, are located on the ground or in a hollow log. These bandicoots level off the surrounding vegetation in making their nests. They are terrestrial and generally nocturnal, but captive *I. macrourus* have been found to be active during part of the day (Collins 1973). *I. obesulus* seems to detect the approach of bad weather, as it will enlarge its nest before heavy rains. After entering or leaving its nest, it closes the opening behind it. This species also has been known to construct a short burrow (Kirsch 1968).

Isoodon feeds in definite areas, the location changing from time to time, as evidenced by scratch marks in the ground. *I. obesulus* uses a rapid movement of its forefeet to crush its living prey. This species feeds on a variety of items, at least in captivity, but prefers insects and worms. Individuals that were not fed for several days refused potatoes, carrots, and turnips despite their hunger.

Population densities of *I. obesulus* have been reported as 0.75–2.0 individuals per 10 ha. in Tasmania (Ride 1970) and about 1.5 per ha. in the Franklin Islands (Copley et al. 1990). Home ranges in this species were found to be 4.05–6.48 ha. for four males and 2.31 ha. for one female. Home ranges of *I. macrourus* in Queensland were 1.7–5.2 ha. for males and 0.9–2.1 ha. for females (Stodart 1977).

The species *I. obesulus* has been reported to be highly aggressive and territorial, with probably very little overlap in home ranges of individuals of the same sex. Territories of males have been reported to be larger than those of females, with each male's territory probably overlapping those of several females (Ride 1970). However, recent field studies suggest that the degree of range exclusion in this species may be a function of food availability and population density (Broughton and Dickman 1991). *Isoodon* is solitary in the wild, apparently coming together only to mate (Stodart 1977). Two individuals can rarely be kept together in captivity as they are intolerant of each other and pugnacious. Two males should never be placed in the same enclosure (Collins 1973). They fight with open mouths, but their long claws usually inflict greater injury than their teeth.

Short-nosed bandicoots are polyestrous, with the estrous cycle averaging 22 days in *I. macrourus* (Gemmell 1988; Lyne 1976). Breeding reportedly can take place year-round for *I. macrourus* in Queensland and New South Wales and in May–February for *I. obesulus* in Tasmania (Ride 1970). Several litters can be produced by each female during one season, there usually being two or three in *I. obesulus* in Tasmania (Ride 1970) and four in *I. macrourus* (Friend 1990). In a population of *I. obesulus* in the Franklin Islands, off South Australia, breeding occurred throughout the year and females produced up to five litters during the period (Copley et al. 1990). A captive *I. macrourus* bore eight litters totaling at least 32 young in 17 months (Collins 1973). The gestation period in *I. macrourus* has been calculated as from 12 days and 8 hours to 12 days and 11 hours, the shortest recorded for any mammal (Lyne 1974). In *I. macrourus* there are up to 7 young per litter, with reported averages being 2.7 (Friend 1990) and 3.4 (Stodart 1977). In *I. obesulus* there are up to 5 young (Ride 1970), though Copley et al. (1990) reported an average ranging from 1.6 in the summer to 2.4 in the spring. According to Collins (1973), probably no more than 4 young of any one litter survive, so that four small, unused mammae are immediately available for the next litter and reproduction can proceed at a rapid rate, just as in *Perameles*. The young of *I. macrourus* leave the pouch after 7–8 weeks, are weaned at 8–10 weeks, become sexually mature at only about 4 months, and reach adult size at about 1 year (Collins 1973; Friend 1990). Gemmell (1990) reported longevity of *I. macrourus* to be up to 28 months in captivity, but Lobert and Lee (1990) found wild *I. obesulus* to live up to 4 years.

I. auratus now is classified as vulnerable by the IUCN. It formerly occupied much of northern and interior Australia but now is known to survive only in remote parts of the Kimberley area of northern Western Australia, on nearby Augustus Island, and on Barrow Island farther to the west. Its decline may be associated with habitat destruction by introduced grazing animals and predation by introduced cats, foxes, and pigs. However, the main period of decline took place in the arid interior from about the 1930s to the 1950s, when there was a large-scale movement of aboriginal people from their native lands to permanent settlements and cattle stations. In their former nomadic movements these people had burned off vegetation gradually, thereby producing a mosaic of habitats that provided abundant food and cover to small mammals. Disappearance of these conditions led to massive destruction of habitat by wildfires started by lightning and the subsequent loss of habitat diversity. In contrast, *I. macrourus*, which occurs predominantly in zones of higher rainfall and greater vegetative cover, has not suffered such a substantial decline. *I. obesulus* has lost 50–90 percent of its original habitat; the subspecies *I. o. nauticus*, of South Australia, is classified as vulnerable by the IUCN, and the subspecies *I. o. obesulus*, of eastern mainland Australia, and *I. o. fusciventer*, of Western Australia, are designated as near threatened. *I. peninsulae*, also designated as near threatened, appears to be uncommon, but its status is not well understood (Ashby et al. 1990; Friend 1990b; Gordon, Hall, and Atherton 1990; Johnson and Southgate 1990; Kemper 1990; Kennedy 1992).

PERAMELEMORPHIA; PERAMELIDAE; Genus **PERAMELES**
E. Geoffroy St.-Hilaire, 1803

Long-nosed Bandicoots

There are four species (Kirsch and Calaby 1977; Ride 1970; Seebeck et al. 1990):

P. nasuta, eastern Queensland, New South Wales, and Victoria;

P. gunnii, southern Victoria, Tasmania;

P. bougainville, from islands in Shark Bay at west edge of Western Australia across arid parts of southern Australia to western New South Wales and Victoria;

P. eremiana, southern Northern Territory, northern South Australia, east-central Western Australia.

Head and body length is 200–425 mm and tail length is 75 to about 170 mm. Weights are 500–1,900 grams in *P. nasuta*, 450–900 grams in *P. gunnii*, and 190–250 grams in *P. bougainville* (Burbidge *in* Strahan 1983; Seebeck *in* Strahan 1983; Seebeck et al. 1990; Stodart *in* Strahan 1983). The sleek-looking pelage is composed chiefly of coarse, distinct hairs. The upper parts may be drab, light brown with a slight pinkish tinge, dull orange, yellowish brown, grayish brown, or gray. Black or black-tipped hairs are often interspersed with lighter ones. In all species except *P. nasuta* there are transverse or diagonal dark and light bars on the back and rump, forming in some cases an elaborate pattern. The underparts are white or whitish. These bandicoots have long, tapered snouts and conspicuous pointed ears.

Habitats include rainforest and sclerophyll forest for *P. nasuta*, woodland and open country with good ground cov-

er for *P. gunnii*, heaths and dune vegetation for *P. bougainville*, and spinifex grassland for *P. eremiana* (Ride 1970). Collins (1973) wrote that in heavy vegetation long-nosed bandicoots construct a nest consisting of an oval mound of twigs, leaves, and humus on the surface of the ground. In more open areas they excavate a nest chamber, line it with plant fibers, and then cover it with a mound of twigs and leaves. Abandoned rabbit burrows, rock piles, and hollow logs are also used as nest sites. Long-nosed bandicoots are nocturnal, terrestrial, and highly active. Their rapid running has been described as a kind of gallop, and they have been seen to jump straight up into the air and then to take off immediately in another direction.

The members of this genus are largely insectivorous, though they also feed on worms, snails, lizards, mice, and plant material. *P. nasuta* scratches and digs in the ground for insects and grubs, including weevils found on fruit tree roots and the larvae of the scarabaeid beetle, which feeds on the grass roots of lawns and pastures. Stodart (1977) observed that the claws are used to make a hole big enough for the snout to enter.

An investigated population of *P. gunnii* seemed variable in density, declining in 13 months from about 15/ha. to 4.5/ha. (Ride 1970). Home range in this species has been reported as about 1–4 ha. for females and 18–40 ha. for males. There may be considerable overlap in female ranges, and presumably several are included within the home range of each male. The species *P. nasuta* has been reported to be extremely solitary, with males acting aggressively when coming into contact with one another (Ride 1970). Stodart (1977) found that in enclosures this species was not gregarious, individuals tended to ignore each other, and there was no territorial defense. Collins (1973) wrote that pairs or family groups could be kept together if the enclosure were sufficiently large but that two or more males should not be placed in the same enclosure.

Long-nosed bandicoots are polyestrous, with the average estrous cycle in *P. nasuta* lasting 21 days (Lyne 1976). In some areas reproduction is said to occur year-round. Stodart (1977), however, reported that while males can breed throughout the year, females are inactive during the autumn. She also observed that there was a 62- to 63-day interval from one period of female receptivity through pregnancy to the next period of receptivity. In Tasmania *P. gunnii* breeds from May to February, each female producing three to four litters per season at intervals of 60 days (Ride 1970; Stodart 1977). Robinson et al. (1978) caught a female *P. bougainville* with two pouch young on Bernier Island, Western Australia, in April. Gestation in *P. nasuta* has been timed at about 12.5 days. Litter size in both *P. nasuta* and *P. gunnii* is usually two or three but can be as small as one or as large as five. Since females have eight mammae, the number of young seems rather low, but actually there is an advantage, considering the short interval between births. A new litter is born at the same time that the previous one is weaned. The mammae used by the previous litter have greatly expanded and are too large for the newborn to grasp. Because usually at least four other mammae have not been used, however, the newborn are assured nourishment (Collins 1973). Young of *P. gunnii* leave the pouch at 48–53 days and are weaned at 59–61 days. Females are sexually mature by 3 months and can bear one or two litters in the same breeding season in which they were born. Adult size and male sexual maturity are attained at 4–6 months (Collins 1973; Ride 1970). In *P. nasuta* the young are carried in the pouch 50–54 days, then remain briefly in the nest, and begin to forage with the female at 62–63 days (Stodart 1977). In this species sexual maturity comes at 4 months for females and 5 months for males

Long-nosed bandicoot *(Perameles nasuta)*, photo by Stanley Breeden.

(Ride 1970). Captive specimens of *P. gunnii* have lived for at least 3 years (Collins 1973).

The species *P. eremiana* and *P. bougainville* are listed as endangered by the USDI, and *P. bougainville* also is on appendix 1 of the CITES. The IUCN designates *P. bougainville* generally as endangered, the subspecies *P. b. bougainville*, of eastern Australia, as extinct, *P. gunnii* generally as vulnerable, the mainland population of *P. gunnii* as critically endangered, and *P. eremiana* as extinct. These species declined severely following European settlement, apparently because of clearing of natural vegetation and other habitat modification, inadvertent destruction in poisoning and trapping efforts to control introduced rabbits, and the spread of introduced predators, such as the fox, domestic dog, and domestic cat (Aitken 1979; Menkhorst and Seebeck 1990). Apparently the most drastic losses in the inland arid country were coincident with the large-scale departure of aboriginal people from the 1930s to the 1950s. The gradual, patchwork burning practiced by these people has been replaced by intensive, lightning-caused wildfires that destroy habitat diversity (Johnson and Southgate 1990).

P. eremiana has not been collected since a specimen was taken in Western Australia in 1943. *P. bougainville*, formerly occurring all across southern Australia, had disappeared from New South Wales by the 1860s and from South Australia and the mainland of Western Australia by about 1930. At present it is known to survive only on Bernier and Dorre islands in Shark Bay, Western Australia, where it is common but highly vulnerable to the potential introduction of predators (Friend 1990*b*; Kemper 1990). *P. nasuta* may have fared better, but *P. gunnii* apparently is declining in Tasmania (Robinson, Sherwin, and Brown 1991) and is nearly extinct on the mainland. It formerly occupied a large part of Victoria, but since the 1960s it has been essentially restricted to suburban gardens and lightly farmed areas around the city of Hamilton in the southwestern part of the state (Emison et al. 1978; Minta, Clark, and Goldstraw 1989). Survival in this area may depend on the maintenance of gardens and the willingness of the gardeners to accept the digging activity of the bandicoot. However, the population seems to be in an uncontrollable decline, perhaps associated with loss of genetic viability. Numbers were calculated at 633 individuals in 1985 (Sherwin and Brown 1990), 246 in 1988 (Dufty 1991), and fewer than 100 in 1992 (Robinson, Murray, and Sherwin 1993). Recent reintroduction projects have established small populations at two other sites (Reading et al. 1992).

PERAMELEMORPHIA; Family PERORYCTIDAE

Rainforest Bandicoots

This family of 4 Recent genera and 11 species is found in New Guinea, on a number of nearby islands, and in extreme northeastern Australia. Until recently this group generally was considered a part of the Peramelidae (see account thereof), but Groves and Flannery (1990) designated it a separate family.

Head and body length is 150–558 mm, tail length is 50–335 mm, and weight is around 200–4,700 grams. The pelage may be soft or spiny depending on the genus, and coloration varies considerably. The tail may be relatively long or short, but the ears are short in all genera. The dental formula is mostly the same as in the Peramelidae, but in the genera *Echymipera* and *Rhynchomeles* there are only four upper incisors on each side.

Groves and Flannery (1990) distinguished this family on the basis of characters of the skull and teeth. Whereas the Peramelidae are flat-skulled, the Peroryctidae are "cylindrical skulled," the crania being much higher and subcylindrical in cross section in both the rostral and the neurocranial portion. In the Peroryctidae the foramen rotundum is prolonged into a tube; the alveolar plate in the molar region is reduced, so that the zygomatic arch swings forward low over the posterior molars; the infraorbital fossa is low and compressed dorsoventrally; the main palatal vacuities are long and narrow and do not tend to fuse across the midline; the mesopterygoid fossa is broad and parallel-sided; the antorbital surface is flattened; the gonial hooks are directed posteriad; the superior temporal lines are nearly parallel throughout, not converging further back; and the auditory bullae are incomplete.

Gordon and Hulbert (1989) noted that fossils of *Echymipera* and *Peroryctes* have been recovered in Australia from sites as old as middle and late Miocene. Flannery (1990*b*) indicated that there are no paleontological

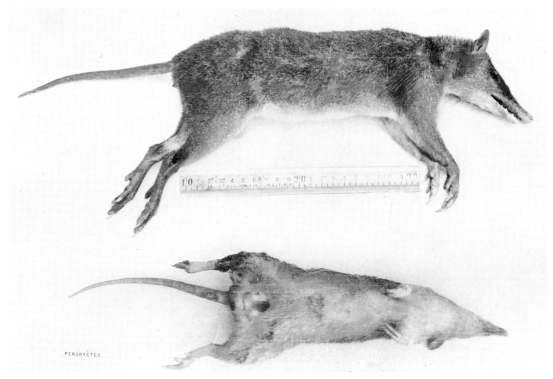

New Guinean bandicoot (*Peroryctes* sp.), ventral view of pouch of the female and the long claws, photo from American Museum of Natural History, Archbold Expedition Collection.

data for New Guinea but that the Peroryctidae appear to be a derived group that originated and radiated almost exclusively there. All species can be found in rainforest and seem to be insectivorous and/or omnivorous.

PERAMELEMORPHIA; PEHORYCTIDAE; **Genus ECHYMIPERA**
Lesson, 1842

New Guinean Spiny Bandicoots

There are five species (Collins 1973; Flannery 1990*a*, 1995; George and Maynes 1990; Menzies 1990*a*; Ziegler 1977):

E. kalubu, New Guinea and many nearby islands, Bismarck Archipelago;
E. rufescens, New Guinea and Misool Island (Mysol) to west, Kei and Aru islands, D'Entrecasteaux Archipelago, Cape York Peninsula of Queensland;
E. clara, northern New Guinea, Japen Island;
E. echinista, known only from Western Province of Papua New Guinea;
E. davidi, known only from Kiriwina Island in the Trobriand Islands off southeastern Papua New Guinea.

The species *E. rufescens* had long been known in Australia only by a single specimen collected in 1932, but in 1970 five more individuals were captured (Hulbert, Gordon, and Dawson 1971), and subsequently the species was found over a large part of the Cape York Peninsula (Gordon and Lawrie 1977).

Head and body length is about 200–500 mm and tail length is 50–125 mm. Often the tail is missing, perhaps bitten off, or it may detach easily. Weight is 500–2,000 grams in *E. rufescens* (Gordon *in* Strahan 1983), 450–1,500 grams in *E. kalubu,* and 825–1,700 grams in *E. clara* (Flannery 1990*b*). The upper parts are bright reddish brown, dark coppery brown, black mixed with yellow, or black interspersed with tawny; the underparts usually are buffy or brownish. The entire pelage is stiff and spiny. The snout is comparatively long and sharp. There are only four upper incisor teeth on each side of the jaw. There are three pairs of mammae in *E. kalubu* and four pairs in *E. rufescens* (Lidicker and Ziegler 1968).

These bandicoots generally inhabit rainforest. Flannery (1990*b*) listed the following altitudinal ranges: *E. kalubu,* 0–2,000 meters; *E. rufescens,* 0–1,200 meters; and *E. clara,* 300–1,200 meters. They are apparently terrestrial, nocturnal, and omnivorous (Collins 1973). Van Deusen and Keith (1966) reported that *E. clara* feeds in part on the fruit of *Ficus* and *Pandanus* and in turn forms an important food staple for the native people within its range.

In a spool-and-line tracking study Anderson et al. (1988) found *E. kalubu* to forage for fallen fruit and to root in the forest floor for earthworms and grubs. Average nightly movement was 344 meters. Individuals maintained regular home ranges; two males utilized 1.0 ha. and 2.1 ha. over a period of three nights. There were several nests within each range, located in hollow logs, leaf piles, and shallow burrows. Ranges overlapped, though there were probably exclusive core areas.

Available evidence indicates that these bandicoots are solitary and highly intolerant of their own kind. According to Flannery (1990*b*), breeding of *E. kalubu* occurs throughout the year, average birth interval is probably about 120 days, one to three pouch young are usually present, and fe-

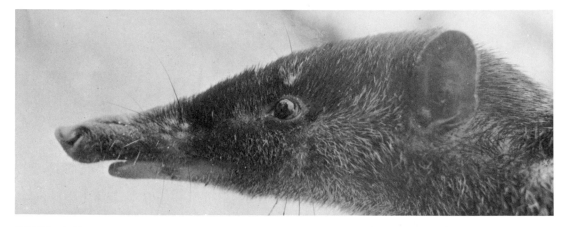

New Guinean spiny bandicoot *(Echymipera rufescens):* Top, photo by G. Gordon; Bottom, photo by Sten Bergman.

males begin breeding at a very early age; pouch young of *E. rufescens* were recorded in New Guinea in May, October, March, and August. On the basis of negative evidence, Gordon and Lawrie (1977) suggested that *E. rufescens* in Australia might be a seasonal breeder, with anestrus during the dry season. A specimen of *E. rufescens* lived in captivity for two years and nine months (Jones 1982).

G. G. George (1979) considered *E. clara* to be threatened in Papua New Guinea because of constant hunting and expanded agricultural activity. Kennedy (1992) designated *E. clara, E. rufescens,* and *E. echinista* as potentially vulnerable because of their restricted ranges.

PERAMELEMORPHIA; PERORYCTIDAE; **Genus**
RHYNCHOMELES
Thomas, 1920

Seram Island Long-nosed Bandicoot

The single species, *R. prattorum,* is known only by seven specimens from Seram Island, located in the South Moluc-cas between New Guinea and Sulawesi. These specimens are in the British Museum of Natural History. The species is named after the collectors, Felix Pratt and his sons Charles and Joseph. Holocene fossils recently discovered on Halmahera Island in the North Moluccas share some similarities with *Rhynchomeles* and suggest that the genus may once have been more widespread (Flannery 1995).

Head and body length is 245–330 mm and tail length is 105–30 mm (Flannery 1995). The fur is crisp, glossy, and completely nonspinous. There is little underfur. This bandicoot is dark chocolate brown above and below with a patch of white on the chest. The head is somewhat lighter than the back, and there may be a whitish area on the forelimbs. The tail is blackish brown and almost naked.

As the generic name suggests, the muzzle is long because of the elongation of the nasal bones of the skull. The ears are small and oval in shape. As in *Echymipera,* there is no fifth upper incisor. Groves and Flannery (1990) noted that *Rhynchomeles* differs from *Echymipera* in only a few features but is characterized by the extreme length and slenderness of its muzzle, its very small last molar, and its generally well-spaced cheek teeth.

All known specimens were obtained during February 1920 in heavy jungle in very precipitous limestone country on Mount Manusela, at an elevation of 1,800 meters

Ceram Island long-nosed bandicoot *(Rhynchomeles prattorum)*, photos from British Museum (Natural History)

(Flannery 1995). Kennedy (1992) referred to *Rhynchomeles* as very rare and recommended an IUCN classification of vulnerable. However, the current IUCN designation is "data deficient." Archer (1984*a*) stated that the genus is probably now extinct, but Flannery (1995) doubted that and even suggested that it may be locally common. Kitchener et al. (1993) were unable to locate it during an expedition in 1991, but conversations with local villagers indicated that it might survive in certain areas of undisturbed montane forest.

PERAMELEMORPHIA; PERORYCTIDAE; Genus MICROPERORYCTES
Stein, 1932

New Guinean Mouse Bandicoots

There are three species (Groves and Flannery 1990; Ziegler 1977):

M. murina, known only by three specimens taken in the Weyland Mountains of western New Guinea and one collected on the Vogelkop Peninsula at the extreme western tip of the island;
M. papuensis, southeastern Papua New Guinea;
M. longicauda, Central Range of New Guinea.

Microperoryctes long was considered to comprise only the species *M. murina,* but recent investigations prompted the transfer of *M. papuensis* and *M. longicauda* to this genus from *Peroryctes* (Flannery 1990*b*; Groves *in* Wilson and Reeder 1993; Groves and Flannery 1990).

Head and body length of *M. murina* is 150–75 mm and tail length is 105–10 mm. The three specimens from the Weyland Mountains have a dark gray, unpatterned coloration above and below and a grayish white scrotum. The feet have scattered white hairs. The short-haired tail is dark

fuscous above and below. The specimen from the Vogelkop differs in having light grayish buff fur, a dark gray middorsal stripe on the back, and a white tip on the tail. In *M. longicauda* head and body length is 239–303 mm, tail length is 141–258 mm, and weight is 350–670 grams; in *M. papuensis* head and body length is 175–200 mm, tail length is 135–55 mm, and weight is 137–84 grams (Aplin and Woolley 1993; Flannery 1990*b*; Lidicker and Ziegler 1968). *M. longicauda* is reddish brown or pale brown speckled with black above, and some subspecies have a dark middorsal line, paired lateral rump stripes, and/or dark eye stripes; the underparts are rufous or buff. In *M. papuensis* the upper parts are dark with a prominent black middorsal line; there also are dark lateral rump stripes and eye stripes as in some subspecies of *M. longicauda,* and the underparts are rich orange buff.

Microperoryctes is characterized by a broadened braincase but a very narrow snout, a highly fenestrated palate with two pairs of vacuities that are shifted forward, straight posterior palatal margins, a deeply wavy coronal suture, small and incomplete auditory bullae lacking anterior spurs, trituberculate molars equal in size to or slightly larger than premolars, granular soles of feet, small ears, a soft and dense pelage, and only three pairs of mammae (Gordon and Hulbert 1989; Groves and Flannery 1990).

Ziegler (1977) stated that *M. murina* lives in moss forest at an altitude of 1,900–2,500 meters. He observed that its short, soft fur suggests a semifossorial existence. He also listed an altitudinal range of 1,000–4,500 meters for *M. longicauda.* The hind foot seems to be adapted for running in *M. longicauda.* Dwyer (1983) collected a 170-gram immature male *M. longicauda* in June. According to Flannery (1990*b*), *M. longicauda* inhabits primary forests and has been taken in nests among roots and at the base of a tree; a female taken in March had four furred young and a female taken in October had three unfurred young. Aplin and Woolley (1993) reported that most specimens of *M. papuensis* have been taken in lower montane forest or associated secondary growth at 1,200–1,450 meters but that

New Guinean mouse bandicoot *(Microperoryctes longicauda)*, photo by J. I. Menzies.

New Guinean bandicoot *(Peroryctes raffrayana)*, photo by Pavel German.

the species also has been found at elevations of up to 2,650 meters. Limited data suggest that *M. papuensis* breeds year-round and litter size is one. Kennedy (1992) designated *M. murina* as potentially vulnerable because of its apparent rarity and restricted range.

PERAMELEMORPHIA; PERORYCTIDAE; **Genus PERORYCTES**
Thomas, 1906

New Guinean Bandicoots

There are two species (Flannery 1995; Ziegler 1977):

P. raffrayana, New Guinea and Japen Island to northwest;
P. broadbenti (giant bandicoot), southeastern New Guinea.

Kirsch and Calaby (1977) did not separate *P. broadbenti* from *P. raffrayana*, but Ziegler (1977) argued that the two are distinct and seem to occur together in southeastern Papua New Guinea. The latter arrangement was accepted by Groves and Flannery (1990), who also transferred the species *P. longicauda* and *P. papuensis* to the genus *Microperoryctes*. Kirsch et al. (1990) agreed that *P. longicauda* is not immediately related to the other species of *Peroryctes* but suggested that it has closer affinity to *Echymipera*.

Measurements (based in part on Lidicker and Ziegler 1968) are: *P. raffrayana*, head and body length 175–384 mm, tail length 110–230 mm, and *P. broadbenti*, head and body 394–558 mm, tail 118–335 mm. Flannery (1990*b*) listed weights of 650–1,000 grams for *P. raffrayana*, and George and Maynes (1990) referred to a specimen of *P.*

broadbenti that weighed 4.9 kg. In *P. raffrayana* the upper parts are dark brown with a slight mixture of black, and the underparts are white, brownish yellow, or buff. *P. broadbenti* also is dark brown dorsally but has reddish buff on the flanks and is near white ventrally. The fur of *Peroryctes* is comparatively soft and long, not spinous as in *Echymipera*. Flannery (1990*b*) stated that *Peroryctes* has much longer and narrower hind feet than do other peroryctids and can be further differentiated from *Microperoryctes* by having a single pair of posterior palatal vacuities rather than two.

New Guinean bandicoots dwell mainly in upland forests. Ziegler (1977) gave altitudinal ranges of 60–3,900 meters for *P. raffrayana* and 0–2,700 meters for *P. broadbenti*. He also noted that *P. raffrayana* is rare below 500 meters, occurs primarily in dense forest, and avoids grassland. The hind foot seems to be adapted for leaping in *P. raffrayana*. Collins (1973) wrote that New Guinean bandicoots are terrestrial, nocturnal, and apparently solitary. Flannery (1990*b*) reported that *P. raffrayana* apparently is a rare inhabitant of undisturbed forest, that it may feed on fruit, and that some specimens have been taken from nests; a female with a pouch young was found in August or September. Dwyer (1983) collected a 600-gram female *P. raffrayana* with a 13-gram pouch young in December. A specimen of *P. raffrayana* lived in captivity for three years and three months (Jones 1982).

According to George and Maynes (1990), the great size of *P. broadbenti* renders it vulnerable to hunting pressure. It occupies a very restricted range, is presumably rare, and may already be extinct, though there remain extensive tracts of suitable habitat with a sparse human population. *P. raffrayana* is more widespread and common but also is subject to hunting.

Order Notoryctemorphia

NOTORYCTEMORPHIA; **Family NOTORYCTIDAE;**
Genus NOTORYCTES
Stirling, 1891

Marsupial "Mole"

This order contains one family, the Notoryctidae, with the single known genus and species *Notoryctes typhlops,* which occupies parts of northern and east-central Western Australia, southern Northern Territory, and western South Australia (Corbett 1975). A second species, *N. caurinus,* was described from the coast of northern Western Australia in 1920. Since then, most authorities have considered it part of *N. typhlops,* but Johnson and Walton (1989) stated that there is no substantive basis for synonymy. *N. caurinus* was accepted by Walton (*in* Bannister et al. 1988) and Groves (*in* Wilson and Reeder 1993) but not by Corbet and Hill (1991). In any case, over the years *Notoryctes* has been placed in various orders of other Australian mammals either at the familial or the subordinal level. However, studies of morphology, serology, and karyology (Archer 1984*a;* Calaby et al. 1974; Kirsch 1977*b*) did not provide a clear idea of the affinities of this unusual mammal with the other marsupial groups. Finally, Aplin and Archer (1987) proposed that it be placed in a full order. The same order was accepted by Marshall, Case, and Woodburne (1990). Based on an analysis of DNA, Westerman (1991) concluded that *Notoryctes* is not closely related to other marsupial groups and belongs in its own order.

Notoryctes has about the size and general proportions of placental moles (Talpidae), with a thick, powerful, and somewhat elongate body. Head and body length is 90–180 mm and tail length is 12–26 mm. A female (head and body 114 mm) weighed approximately 66 grams. Another specimen weighed 40.5 grams (Howe 1975). The coloration has been reported to vary from almost white through pinkish cinnamon to rich golden red. However, Howe (1975) observed that the reddish color probably results from iron staining. A live specimen lost this coloration after a week in captivity, then again became red after its terrarium was filled with red sand. In life the pelage, which consists almost entirely of underfur, is remarkably iridescent, fine, and silky. The short, cylindrical, stumpy tail is hard and leathery, marked by a series of distinct rings, and terminates in a horny knob. There is a horny shield on the nose. The vertebrae in the neck are fused, enhancing the rigidity of the body as the animal digs. The third and fourth digits of the

Marsupial "mole" *(Notoryctes typhlops),* photo by D. Roff.

Marsupial "mole" *(Notoryctes typhlops).* Inset shows broad flat surface of claws. Photos of mounted specimen from Royal Scottish Museum.

forefoot are greatly enlarged and bear large, triangular claws that form a cleft spade or scoop. The remaining three digits of the forefoot are small, but the first and second bear claws and are opposed to the third and fourth. The middle three digits of the hind foot also possess enlarged claws. The only external indication of ears is a small opening beneath the fur on either side of the head. The eyes of *Notoryctes* are vestigial, measure 1 mm in diameter, and are hidden under the skin. There is no lens or pupil, and the optic nerve to the brain is reduced. The female has a distinct pouch that opens posteriorly and contains two mammae. The testes of the male are situated between the skin and the abdominal wall (K. A. Johnson *in* Strahan 1983), and there is no scrotum.

The skull is conical in shape with an expanded occipital region and overhanging nasal bones. The sutures between the bones of the posterior half of the braincase are obliterated. The variable dental formula is: (i 3–4/3, c 1/1, pm 2/2–3, m 4/4) × 2 − 40–44. The incisors, canines, and premolars are simple and often blunt, though the last upper premolar is bicuspidate; the upper molars are tritubercular. The teeth are well separated from each other. The epipubic (marsupium) bones are reduced.

The discovery of this molelike marsupial in 1888 created a stir among mammalogists comparable to the sensation that accompanied the discovery of the duck-billed platypus. The marsupial "mole" affords an interesting example of evolutionary parallel between marsupial and placental mammals. *Notoryctes* resembles the golden moles (order Insectivora, family Chrysochloridae) in general body form, burrowing habits, texture of fur, and even external features of the brain (Kirsch 1977*b*).

The habitat of *Notoryctes* is sandridge desert with acacias and shrubs, especially along river flats (Corbett 1975; Ride 1970). The marsupial "mole" apparently is not as subterranean as the true moles (Talpidae). It may burrow a relatively short distance and then emerge and shuffle along the surface, propelling itself mainly by the hind feet (Ride 1970). It leaves a peculiar triple track, produced by the two hind feet and the tail. When burrowing, it often moves along at about 8 cm below the surface but then may dip vertically to depths exceeding 2.5 meters (K. A. Johnson *in* Strahan 1983). The shielded snout acts as a bore, the foreclaws function as scoops, and the hind feet throw up the sand, which then falls in behind. Thus, *Notoryctes* practically "swims" through the ground and generally does not leave a long-lasting tunnel as the true moles do. Females,

however, probably construct a deep, permanent burrow in which to bear young.

Notoryctes is reported to be active, timorous, apparently solitary, and both diurnal and nocturnal (Collins 1973). One captive proved to be extremely active and moved continuously about its enclosure in search of food. Its nose was always held downward. It fell asleep suddenly on several occasions and awoke just as suddenly to resume its feverish activity. Despite the appearance of being highly nervous, it did not seem to resent handling; it would even consume milk rapidly while being held and then would suddenly fall asleep again. This individual fed ravenously on earthworms, but some other captives have refused them (Collins 1973). Actually, it is unlikely that they are part of the natural diet since only one rare species of earthworm occurs in the range of *Notoryctes.* Available information indicates that the preferred diet consists of the larvae of certain beetles and moths (Corbett 1975; Howe 1975). Examination of the digestive tracts of 10 specimens revealed mostly ants, some other insect remains, and some seeds (Winkel and Humphrey-Smith 1987). One captive made low-intensity, sharp squeaking noises when restrained (Howe 1975). The only information on reproduction is that in the Northern Territory females give birth about November, that the pouch of one contained a single young, and that there is a museum specimen with two sucklings (Collins 1973; Johnson and Walton 1989; Ride 1970).

The IUCN recognizes *N. typhlops* and *N. caurinus* as separate species and now classifies both as endangered. *Notoryctes* is thought to have declined by at least 50 percent in the last decade because of loss of habitat, and a similar rate of decline is projected over the next 10 years. Surprisingly, there was once an active trade in the skin of this small and poorly known creature. Several thousand pelts are said to have been brought by Aborigines to trading posts in the Northern Territory from about 1900 to 1920 and later sold on the market for about two British pounds each (then about U.S. $9). Subsequently, the collection of specimens has remained relatively steady at about 5–15 every 10 years (Johnson and Walton 1989).

Geologically this order is known only from the Recent in Australia, and no fossils can be assigned directly to it (Kirsch 1977*a*). However, Aplin and Archer (1987) noted the recent discovery of fossils from the middle Miocene of Queensland, which suggest a very distant relationship between *Notoryctes* and the Peramelemorphia.

Order Diprotodontia

Koala, Wombats, Possums, Wallabies, and Kangaroos

This order of 10 Recent families, 40 genera, and 131 species is found throughout Australia, Tasmania, and New Guinea and on many islands of the East Indies from Sulawesi to the Solomons. Although this group once was generally considered to form only a marsupial suborder or superfamily, there long has been recognition of its unity and there has been increasing acceptance of its ordinal status. Aplin and Archer (1987), whose sequence is followed here, designated two suborders: Vombatiformes, with the families Phascolarctidae and Vombatidae; and Phalangerida, with the superfamily Phalangeroidea for the family Phalangeridae, the superfamily Macropodoidea for the Potoroidae and Macropodidae, the superfamily Burramyoidea for the Burramyidae, the superfamily Petauroidea for the Pseudocheiridae and Petauridae, and the superfamily Tarsipedoidea for the Tarsipedidae and Acrobatidae. The position of the Macropodoidea in this sequence is questionable; although there is evidence of common ancestry of that group and the Phalangeroidea (Hume et al. 1989), Flannery (1989b) suggested that the Macropodoidea may represent an evolutionary branch separate from all other components of the Phalangerida. Marshall, Case, and Woodburne (1990) accepted the same order and subordinal division but indicated certain differences with respect to superfamilies and families. Szalay (1994) considered the Diprotodontia a semiorder of the Syndactyla, which he regarded as also including what are here designated the Peramelemorphia and Notoryctemorphia.

The Diprotodontia are the largest and most diverse marsupial order, with extensive variation in size and shape, but there are several critical characters common to the group. As the name of the order implies, the dentition is diprotodont; the two middle incisor teeth of the lower jaw are greatly enlarged and project forward. Usually there are no other remaining lower incisors or canine, but if any are present they are very small, and there is a gap between the incisors and the lower cheek teeth. A second major unifying character is syndactyly; the second and third digits of the hind foot are joined together by integument, are relatively small, and retain claws (see Hall 1987b for a detailed discussion of the condition and its phylogenetic significance). No other marsupial order is both diprotodont and syndactylous. Other diagnostic characters of the Diprotodontia include reduction of the upper incisors to three or fewer, selenodont (having crescent-shaped cusps) upper molar teeth, a fasciculus aberrans (connection between the cerebral hemispheres) and large neocortex in the brain, a superficial thymus gland, an expanded squamosal epitympanic wing in the roof of the tympanic bullae, and an unusually complex morphology of the glenoid fossa in the basicranial region of the skull (Aplin 1987; Aplin and Archer 1987; Archer 1984a). Serological investigations also support the content of the Diprotodontia as set forth herein (Baverstock 1984; Baverstock, Birrell, and Krieg 1987).

The order is predominantly herbivorous, though some members consume invertebrates and even small vertebrates. The smallest species, *Acrobates pygmaeus*, may weigh less than 15 grams, is adapted for gliding like a flying squirrel, and eats mainly insects. The largest living species, *Macropus rufus*, weighs up to 100 kg, is structured for a leaping mode of progression, and like many placental bovids is a grazer. *Diprotodon optatum*, which became extinct about 10,000 years ago, had the size and general appearance of a modern hippopotamus. Other fossils extend the geological range of the Diprotodontia back to the late Oligocene and show that the order was highly diverse even then.

DIPROTODONTIA; **Family PHASCOLARCTIDAE;**
Genus PHASCOLARCTOS
De Blainville, 1816

Koala

The single living genus and species, *Phascolarctos cinereus*, has a modern range extending from southeastern Queensland through eastern New South Wales and Victoria to southeastern South Australia (Ride 1970). During the late Pleistocene the koala also occurred in southwestern Western Australia, where suitable habitat still seems to exist. *Phascolarctos* often has been placed in the family Phalangeridae, but it is now considered to represent a distinct and very primitive family with some affinity to the Vombatidae (Archer 1984a; Harding, Carrick, and Shorey 1987; Kirsch 1977b).

Head and body length is 600–850 mm, the tail is vestigial, and weight is 4–15 kg. Ride (1970) stated that average weight in Victoria was about 10.4 kg for males and 8.2 kg for females. The dense, woolly fur is grayish above and whitish below. The rump often is dappled, and the ears are fringed with white.

The compact body, large head and nose, and big, hairy ears produce a comical, appealing appearance. Cheek pouches and a 1.8- to 2.5-meter caecum aid in digesting the bulky, fibrous eucalyptus leaves that form the main diet.

Koalas *(Phascolarctos cinereus)*, photo by Garth Grant-Thomson through Frank W. Lane. Insets: undersurfaces—A. Left foot; B. Left hand, photos from *The Mammals of South Australia*, F. Wood Jones.

Both the forefoot and the hind foot have five digits, all strongly clawed except the first digit of the hind foot, which is short and greatly broadened. The second and third digits of the hind foot are relatively small and partly syndactylous but have separate claws. The first and second digits of the forefoot are opposable to the other three. The palms and soles are granular. Females have two mammae; their marsupium opens in the rear and extends upward and forward.

The skull is massive and flattened on the sides. The tympanic bullae are elongated and flattened from side to side. The dental formula is: (i 3/1, c 1/0, pm 1/1, m 4/4) × 2 = 30. The lower incisor and first upper incisor are large, the canine is small, and the molars are blunt and tubercular.

The koala is confined to eucalyptus forests. It is largely nocturnal and arboreal, coming to the ground only to move between food trees or to lick up soil or gravel, which serves as a digestive aid. However, although it is often viewed as a sluggish animal and usually does move at a sedate pace, its relatively long legs can propel it rapidly over the ground or up the trunk of a tree (Lee and Carrick 1989). In addition to the leaves and young bark of about 12 species of eucalyptus, or "gum," tree, the koala may also eat mistletoe and box *(Tristania)* leaves. The animal has a characteristic eucalyptuslike odor.

In a study in Victoria, Mitchell (1990*a*) found koalas to be inactive for more than 20 hours a day, to be usually solitary, and to occupy restricted home ranges centered on a few large food trees for periods of years. Home range size averaged 1.70 ha. for adult males and 1.18 ha. for females. There was extensive overlap between male and female ranges, with pairs tending to use the same trees in their common areas. In a study of the introduced koala population on Kangaroo Island, South Australia, Eberhard (1978) found each individual to remain in an area of about 1.0–2.5 ha. containing a few favorite food trees. Home ranges of adults were largely separate, but there was some overlap between sexes. Of 943 sightings only 11 percent were of pairs and only 1 percent of three animals. Trespass into an occupied tree led to a savage attack by the resident. Males uttered a loud call during the breeding season and apparently demarcated a territory through this vocalization and the production of a powerful scent from a sternal gland. Mitchell (1990*b*) distinguished a series of sounds, including bellows, most frequently emitted at night and probably helping to space the dominant males, and various wails and screams associated with both sexual and antagonistic encounters; he also observed urine marking as well as rubbing of trees with the sternal glands. Gordon, McGreevy, and Lawrie (1990) suggested that the males in any given established population tend to be mostly more than five years or less than two years of age, with the older group comprising dominant residents and the younger group made up of immature animals that will soon disperse to seek openings in other populations. In contrast, Mitchell and Martin (1990) indicated that young females tend to remain in the vicinity of their mothers.

During the breeding season males may attempt to defend a territory containing several females. The season reportedly extends from September to January in New South Wales and from November to February in Victoria. Eberhard (1978) reported that on Kangaroo Island births occurred from late December to early April, with a peak in February. The following natural history data were summarized by Eberhard (1978), Handasyde et al. (1990), Martin and Handasyde (1990), and Martin and Lee (1984). Females are seasonally polyestrous, with an estrous cycle of about 35 days, and usually breed once every year. Loss of a pouch young during the breeding season is followed by further ovulations and sometimes another birth. The gestation period is variable but averages about 35 days, and the usual litter size is 1, though twins have been recorded regularly. The young weigh as little as 0.36 gram at birth, have a pouch life of 5–7 months, and are weaned at 6–12 months. Toward the end of their pouch life for a period of up to six weeks the young feed regularly on material passed through the mother's digestive tract. Both sexes may attain sexual maturity at 2 years but may not begin to breed that soon. Full physical maturity is reached during or after the fourth year in females and the fifth year in males, and the latter probably do little mating before that time. Reproductively active females as old as 12 years have been observed in the wild. One female aged 17–18 years was captured, and there have been reports that captive koalas can live 20 years.

Until the early twentieth century the koala seems to have numbered in the millions in southeastern Australia despite grievous losses from forest fires and epidemics. In the early twentieth century, clearing of woodland habitat and increased access to its range, combined with a demand for its beautiful, warm, durable fur, signaled the end of the vast koala populations (Eberhard 1978; Grzimek 1975; Ride 1970). The commercial kill increased until in 1924 more than 2 million koala skins were exported. By the end of that year the species appeared to have been exterminated in South Australia and nearly wiped out in Victoria and New South Wales. There still was a huge population in Queensland, but in 1927 the state government, yielding to pressure from economic interests, allowed an open season, and nearly 600,000 more pelts were exported. Subsequent public outcry both in Australia and abroad resulted in legal protection, but only a few thousand scattered animals were left.

Since the 1920s, intensive conservation efforts, including the breeding and transplantation of thousands of individuals by the Victoria Fisheries and Wildlife Division, have allowed partial recovery of the species in some parts of southeastern Australia. However, even though the koala is now legally protected from direct killing by people, its habitat is being reduced and fragmented by settlement, road construction, fires, logging, and massive cutting of forests for production of woodchips (mostly for export to Japanese paper mills). More than half of the tall to medium-sized trees, on which the koala depends for survival, have already been destroyed (Phillips 1990). The IUCN classifies the koala as near threatened. The Australasian Marsupial and Monotreme Specialist Group of the IUCN Species Survival Commission has designated the koala as "potentially vulnerable" and indicated that the species has lost 50–90 percent of its habitat (Kennedy 1992). Another problem is the widespread prevalence of the bacterial pathogen *Chlamydia*, which infects the reproductive tract of females, thereby reducing fertility and thus restricting the size of some populations but is not considered an overriding threat to the survival of the species as a whole (Martin and Handasyde 1990). There is much controversy regarding the precise status of this highly popular mammal and what conservation measures should be undertaken. Overall estimates of numbers remaining in the wild range from about 40,000 to more than 400,000 (Payne 1995; Thompson 1995).

The geological range of the family Phascolarctidae is middle Miocene to Recent in Australia (Archer 1984*a*; Kirsch 1977*a*).

DIPROTODONTIA; Family VOMBATIDAE

Wombats

This family of two Recent genera and three species is known from eastern and southern Australia, Tasmania, and islands of Bass Strait. The sequence of genera presented here follows that of Kirsch and Calaby (1977). The name Phascolomyidae has sometimes been used for this family.

Wombats resemble small bears in general appearance. The body is thick and heavy; the head and body length of adults is about 700–1,200 mm, and the weight is 15–35 kg. The muzzle is naked and the pelage is coarse in *Vombatus*, but in *Lasiorhinus* the muzzle is haired and the pelage is soft. Underfur is almost lacking. The eyes are small. The limbs are short, equal in length or nearly so, and extremely strong. There are five digits on all the limbs, and all have claws except the first toe, which is vestigial. The second and third digits of the hind foot are partly united by skin. The marsupium opens posteriorly and encloses a single pair of mammae. Wombats have traces of cheek pouches. They also have a group of glands of unusual structure inside the stomach that may be associated with digestion of a special type of plant food. The skull is massive and flattened.

The dentition is remarkably similar to that of rodents. Wombats also resemble rodents in their manner of feeding and in the rapid side-to-side movements of their jaws. All the teeth of wombats are rootless and ever-growing, which compensates for wear. The two rootless incisors in each jaw are large and strong, have enamel on their front and lateral surfaces only, and are separated by a wide space from the cheek teeth. The premolars are small, single-lobed, and close to the molars; the molar teeth have two lobes and are rather high-crowned. The dental formula is: (i 1/1, c 0/0, pm 1/1, m 4/4) $\times$ 2 = 24.

Wombats are shy, timid, and difficult to observe in the wild; they are sometimes active during the day but are considered nocturnal. They live in burrows and are rapid, powerful diggers. Under certain conditions they may construct burrow systems more than 30 meters in length. Wombats dig with their front feet, thrusting the soil out with the hind feet, and use their strong incisors to cut such obstructions as roots. The burrow entrance normally is a low arch that fits the body. A grass or bark nest is located near the end of the burrow. A shallow resting place is usually excavated against a tree or log near the mouth of the burrow as a site for sunbathing. Paths often lead from burrows to feeding areas. Wombats eat mainly grasses, roots, bark, and fungi.

Wombats do well in captivity and often become interesting and affectionate pets. They have suffered serious reduction in numbers and range. People are their chief enemy. Colonies have been exterminated near settled areas because of damage to crops and because domestic livestock may injure a leg by breaking through into the burrows. Wombats also have been destroyed in the campaign against rabbits, as these introduced pests often shelter in wombat burrows. There are many regions in Australia where these unusual mammals could reside without disturbing human developments. Adequate protection would probably enable wombats to maintain a stable population. It is surprising how little detailed information is available regarding them.

The geological range of this family is Miocene to Recent in Australia (Kirsch 1977a).

DIPROTODONTIA; VOMBATIDAE; **Genus VOMBATUS**
E. Geoffroy St.-Hilaire, 1803

Common, or Coarse-haired, Wombat

The single species, *V. ursinus,* formerly occupied the coastal region from southeastern Queensland to southeastern South Australia, as well as Tasmania and the islands of Bass Strait (Ride 1970). The mainland populations sometimes have been designated as a separate species with the name *V. hirsutus.*

The head and body length is 700–1,200 mm and the tail is a mere stub. Adults generally weigh from 15 kg to about 35 kg. The general coloration is yellowish buff, silver gray, light gray, dark brown, or black. The fur in this genus is coarse and harsh, not soft and silky as in *Lasiorhinus.*

Vombatus also differs from *Lasiorhinus* in skull and dental features and in its hairless nose and somewhat rounded ears. In the fine-haired wombats the nose is haired and the ears are relatively pointed. The teeth are rootless, that is, they grow continuously from pulpy bases. Both

genera of wombats are the only living marsupials that possess two rootless incisors in each jaw, an arrangement also seen in rodents. In the lesser curvature of the stomach is a peculiar gland patch that is also found in the koala *(Phascolarctos)* and the beaver *(Castor).* The pouch of the female opens posteriorly and contains two mammae.

The common wombat occurs in upland forests, especially in rocky areas. It constructs holes that have a single entrance but may branch into a complex network of tunnels under the surface (Ride 1970). The nesting chamber, lined with vegetation, usually is placed about 2–4 meters from the entrance. This wombat is nocturnal except for occasional bouts of basking in the sun (Collins 1973). It feeds mainly on grass, roots, and fungi, apparently preferring fresh seed stems; it uses its forefeet to tear and grasp pieces of vegetation. It also has been seen foraging among sea refuse along the shore.

This wombat is quick in its movements and can run rapidly, at least for short distances. When touched, especially near the hindquarters, it kicks backward with both hind feet, and when annoyed it may emit a hissing growl. Often, however, it becomes a playful and affectionate pet. Individuals formerly occurring on King Island in Bass Strait were reportedly domesticated by fishermen. These animals, in a reversal of normal routine, would feed in forests during the day and return in the evening to the cabins that served as their retreat.

According to McIlroy (*in* Strahan 1983), population densities of 0.3/ha. occur in areas of favorable habitat. Home ranges measure 5–23 ha., contain a number of burrow systems, and overlap extensively. Collins (1973) wrote that the common wombat is solitary except during the breeding season and that keeping more than one individual in an enclosure often results in fighting and injury. He added, however, that pairs and groups of compatible individuals have been maintained together successfully. Ride (1970) cited a study of a wild population in Victoria in

Coarse-haired wombat *(Vombatus ursinus),* photo from New York Zoological Society.

which usually only a single wombat occupied each burrow, but individuals were sociable and would visit one another's burrows. Wells (1989) indicated that territories are maintained by rubbing scent on logs and branches and by depositing feces along trails. There are a number of vocalizations, the most common being a harsh cough.

Ride (1970) also stated that births probably occur in late autumn (about April–June in Australia) and that the young are independent by the following summer (December–March). According to a report from the Perth Zoo, however, an adult pair were together in late November 1968, the female was found to have a pouch young on 30 July 1969, the young left the pouch first on 12 October 1969 and permanently by 29 October 1969, and the mother and offspring had to be separated in May 1970 (Collins 1973). The estrous cycle is 32–34 days (Hayssen, Van Tienhoven, and Van Tienhoven 1993). Wells (1989) reported that gestation lasts approximately 21 days and that the young are fully weaned at 12 months. In Tasmania, births were found to occur throughout the year, though about 48 percent took place from October to January (Green and Rainbird 1987). A single young is the usual case, but twins are known to occur. Sexual maturity is attained after 2 years (McIlroy *in* Strahan 1983). This genus appears to be capable of long life in captivity; the record longevity is 26 years and 1 month (Collins 1973).

Although the common wombat has declined through human persecution, some of its mainland and Tasmanian populations probably are secure in their mountain forest habitat (Ride 1970). By the late nineteenth century, however, the species had been exterminated from all islands of Bass Strait except Flinders Island, where it still occurs. The subspecies there, *V. ursinus ursinus*, is classified as vulnerable by the IUCN.

DIPROTODONTIA; VOMBATIDAE; Genus LASIORHINUS
Gray, 1863

Hairy-nosed, or Soft-furred, Wombats

There are two species (Kirsch and Calaby 1977; Ride 1970):

L. latifrons, southeastern Western Australia to southeastern South Australia;
L. krefftii, formerly east-central and southeastern Queensland, south-central New South Wales.

The population of *L. krefftii* in east-central Queensland sometimes is considered a distinct species with the name *L. barnardi*. The population of *L. krefftii* in southeastern Queensland, along the Moonie River, has been called *L. gillespiei*, but that name is a synonym of *L. krefftii*.

Head and body length is 772 mm to about 1,000 mm, tail length is 25–60 mm, and weight is 19–32 kg (Gordon *in* Strahan 1983; Wells *in* Strahan 1983). The upper parts are usually dappled with gray and black or brown; the cheeks, neck, and chest are often white, and the remainder of the underparts is gray.

This genus is distinguished from *Vombatus* by skull, dental, and external characters. The specific name *latifrons* refers to the great width of the anterior part of the skull. Externally the fine-haired wombat can be differentiated from *Vombatus* by its haired nose and relatively longer,

more pointed ears. The fur in *Lasiorhinus* is soft and silky, not coarse and harsh as in *Vombatus*. The marsupial pouch is well developed and opens to the rear; there are two mammae.

These wombats occupy relatively dry country—savannah woodland, grassland, and steppe with low shrubs. They construct complex tunnel systems comprising a large number of separate burrows that join together to form a warren; the entrances are connected above ground by a network of trails, and additional trails radiate out to feeding areas and other warrens (Ride 1970). The burrows of one colony in the early twentieth century covered an area 800 meters long and about 80 meters wide. Wells (*in* Strahan 1983) pointed out that while these wombats appear to be slow and bumbling, they are exceedingly alert and can run as fast as 40 km/hr for short distances. They are said to be more docile in captivity than *Vombatus*. They are nocturnal and feed mainly on grass.

Wells (1978) found the density of a population of *L. latifrons* in South Australia to be 1/4.8 ha. Home ranges of individuals varied from 2.51 ha. to 4.18 ha. and seemed centered on a warren. There was some evidence of territorial marking and defense. Ride (1970) observed that despite the gregarious appearance suggested by the warrens, hairy-nosed wombats seem basically solitary, with each individual possessing its own burrow and feeding area. Wells (*in* Strahan 1983), however, stated that each warren system is inhabited by 5–10 wombats, with about equal numbers of each sex. Johnson and Crossman (1991) confirmed the latter position but noted that individuals of the same sex kept to separate ranges when they left the burrow system to feed. They also found that, in contrast to the situation in *Phascolarctos* (see account thereof) and most other mammals that have been studied, dispersal from the natal colony was common for young females but rare for young males.

The young of *L. latifrons* are born in the spring (starting in October) in South Australia (Wells 1978). Crowcroft (1977) found gestation for two births of this species to be 20–21 and 21–22 days. There usually is one young, but twins have been reported. The young first leave the pouch at approximately 8 months, vacate it permanently at 9 months, are weaned at 12 months, may reach full size by 2 years, and are usually sexually mature at 3 years (Wells 1989). A captive specimen of *L. latifrons* lived for 24 years and 6 months (Jones 1982).

Hairy-nosed wombats declined in numbers and distribution following persecution by European settlers. Ride (1970) expressed serious concern about the steadily contracting range of *L. latifrons* in South Australia. Aitken (1971a), however, thought that while the distribution of this species had become somewhat fragmented, it still was abundant over large areas. He stated that its density had been reduced in the late nineteenth century, probably because of competition from introduced rabbits. He added that the recent human take has been negligible but there was some killing based on the unfounded belief that the wombat damaged fences, competed with livestock for grazing, and dug burrows that allowed rabbits (considered pests) to find shelter.

The status of *L. krefftii* is worse. The once plentiful population in the Riverina area of south-central New South Wales disappeared long ago (Ride 1970). The population in southeastern Queensland, formerly called *L. gillespiei*, is thought to have been extinct since about 1900 (Goodwin and Goodwin 1973). The only remaining population of *L. krefftii*, in east-central Queensland, seems to have been long restricted to a small area of about 15.5 sq km. In 1971 most of this area was protected by being declared a nation-

Soft-furred wombat, or hairy-nosed wombat *(Lasiorhinus latifrons)*, photo by Ernest P. Walker.

al park, but the habitat may be jeopardized by cattle grazing in the park (Thornback and Jenkins 1982). The population fell to only about 25 individuals in 1981 but since has increase to about 70 and has been found to remain genetically viable (Taylor, Sherwin, and Wayne 1994). *L. krefftii* is classified as critically endangered by the IUCN and as endangered by the USDI and is on appendix 1 of the CITES.

DIPROTODONTIA; Family PHALANGERIDAE

Possums and Cuscuses

This family of 6 living genera and 22 species occurs in Australia, Tasmania, and New Guinea and on islands from Sulawesi to the Solomons. The family sometimes has been considered to be much larger and to include all the genera that herein, in accordance with Aplin and Archer (1987) and Kirsch and Calaby (1977), are allocated to the families Phascolarctidae, Burramyidae, Pseudocheiridae, Petauridae, Tarsipedidae, and Acrobatidae. The sequence of genera presented here basically follows that of Flannery, Archer, and Maynes (1987), who recognized two subfamilies: Ailuropinae, with the single genus *Ailurops;* and Phalangerinae, with *Strigocuscus, Trichosurus, Spilocuscus,* and *Phalanger. Wyulda,* considered only a subgenus of *Trichosurus* by those authorities, is here given generic rank. Kirsch (1977b) used an arrangement by which the species here assigned to *Trichosurus* and *Wyulda* were placed in the subfamily Trichosurinae and those assigned to all other genera were placed in the subfamily Phalangerinae.

Head and body length is 320–650 mm and tail length is 240–610 mm. The soft, dense pelage is woolly in all genera except *Wyulda.* All four limbs have five digits, and all the digits except the first toe of the hind foot have strong claws; this first toe is clawless but opposable and provides for a firm grip on branches. The second and third digits of the hind foot are partly syndactylous, being united by skin at the top joint, but the nails are divided and serve as hair combs. The largest digits of the hind foot are the fourth and fifth. In the genera other than *Trichosurus* and *Wyulda* the first and second digits of the forefoot are opposable to the other three. The well-developed marsupium opens to the front, and there are two to four mammae.

Flannery, Archer, and Maynes (1987) listed the following diagnostic characters of the Phalangeridae: the mastoid wing on the rear face of the cranium is reduced in size and the rear face of the cranium above the mastoid is composed of an extension of the squamosal; almost the entire plantar surface of the hind foot is covered by a large striated pad; the distal portion of the tail is nearly naked and at least part is covered in scales or tubercles; and the upper second premolar is greatly reduced or lost but the upper first premolar is large. The dental formula of the family is: (i 3/1–2, c 1/0–1, pm 2–3/3, m 5/5) × 2 = 40–46 (Archer 1984a). The first incisors are long and stout. The second and third incisors, if present, are vestigial, as are the lower canine and some of the premolars. The molars have sharp cutting edges. The skull is broad and flattened.

The brush-tailed possum *(Trichosurus vulpecula)* has an unusual set of glands: both sexes possess two pairs of anal glands, one pair producing oil and the other pair producing cells. The former is a scent gland, but the function of the latter is not known. The cell-producing glands do not liquefy their secretion, which eventually appears in the urine as cells. This phenomenon has not been reported for placental mammals.

The geological range of this family is late Oligocene to Recent in Australasia (Archer 1984a).

Cuscus (*Phalanger* sp.), photo by Bernhard Grzimek.

·*Ailurops ursinus*, photo by Colin P. Groves.

DIPROTODONTIA; PHALANGERIDAE; **Genus**
AILUROPS
Wagler, 1830

Large Celebes Cuscus

The single species, *A. ursinus*, occurs on Sulawesi and the
nearby Togian, Peleng, and Talaud islands. This species usually had been assigned to the genus *Phalanger*, but Flannery, Archer, and Maynes (1987) and George (1987) considered it to represent its own distinct genus and subfamily.

According to Tate (1945), one specimen had a head and
body length of 564 mm and a tail length of 542 mm. Miller
and Hollister (1922) indicated that weight is about 7 kg. For
another specimen, a male, Flannery and Schouten (1994)
listed a head and body length of 610 mm, a tail length of
580 mm, and a weight of 10 kg. The upper parts vary greatly in color, being black, grayish, or brown, and the under-

parts are usually whitish. The limbs are relatively long, the
feet and claws very large, the rostrum unusually short and
broad, the rhinarium large and naked, the canine tooth
short, and the third upper molar large. George (1987) noted
that the pelage consists of fine but wiry underfur and
coarse, bristly guard hairs and that the ears are short and
well furred internally.

Flannery, Archer, and Maynes (1987) listed 10 cranial
characters that distinguish *Ailurops* from all other phalangerids. In the others, the mastoid and ectotympanic are
usually broadly continuous, but in *Ailurops* these bones
are separated by a deep and continuous groove. In the others the entire basicranial region, particularly the squamosal and mastoid, are much more pneumatized than in *Ailurops*. In the others the mastoid is restricted to a thin
ventral band on the posterior face of the cranium, but in
Ailurops the mastoid has an extensive wing on the rear face
of the cranium. The third upper incisor of the others is reduced in size, but in *Ailurops* this tooth is larger than the

second upper incisor. The first upper premolar is double-rooted in the others but single-rooted in *Ailurops*.

Although *Ailurops* is a distinctive and relatively large animal, up to 1,200 mm in total length, and has the most northwesterly distribution of any marsupial, almost nothing is known of its status and natural history. Kennedy (1992) stated that it occurs in rainforests and designated it as "potentially vulnerable" because it is presumed to be hunted as game. Flannery and Schouten (1994) cited information that it is most abundant at around 400 meters in elevation, is largely diurnal and folivorous, is generally found in pairs, and occurs at an estimated density of one pair per 400 ha.

DIPROTODONTIA; PHALANGERIDAE; **Genus STRIGOCUSCUS**
Gray, 1861

Ground Cuscuses

There are two species (Corbet and Hill 1992; Flannery and Schouten 1994):

S. celebensis, Sulawesi and the islands of Sangihe and Siau to the north and Muna to the southeast;
S. pelengensis, Peleng and Taliabu islands to the east of Sulawesi.

Strigocuscus usually has been placed in *Phalanger,* but Flannery, Archer, and Maynes (1987) regarded the two as generically distinct. Those authorities also tentatively considered *Strigocuscus* to include the species that here is des-

ignated *Phalanger ornatus* in accordance with Corbet and Hill (1992), Flannery and Schouten (1994), George (1987), and Groves (1987*a* and *in* Wilson and Reeder 1993) as well as the species that here is designated *Phalanger gymnotis* in accordance with Flannery and Schouten (1994) and Springer et al. (1990). Flannery, Archer, and Maynes (1987) tentatively included still another species, *S. mimicus* from southern New Guinea and the eastern Cape York Peninsula of Queensland, in *Strigocuscus,* but that species was regarded as part of *Phalanger orientalis* by George (1987), Groves (1987*a* and *in* Wilson and Reeder 1993), and Menzies and Pernetta (1986), and it was included in *Phalanger intercastellanus* by Flannery and Schouten (1994). Finally, Menzies and Pernetta (1986) regarded *S. pelengensis* as a subspecies of *S. celebensis,* and Flannery, Archer, and Maynes (1987) included it in the genus *Phalanger,* but the species was reassigned to *Strigocuscus* by Flannery and Schouten (1994).

In *S. celebensis* head and body length is 294–380 mm and tail length is 270–373 mm (Groves 1987*a*). In *S. pelengensis* head and body length is 350–70 mm, tail length is 245–300 mm, and weight is 1,070–1,150 grams (Flannery and Schouten 1994). *S. celebensis* is pale buff in color, and *S. pelengensis* is clear tawny brown to reddish above and paler brown or yellowish below.

According to Flannery, Archer, and Maynes (1987), characters of the skull indicate that *Strigocuscus* is more closely related to *Trichosurus* and *Wyulda* than to *Phalanger.* In *Strigocuscus, Trichosurus,* and *Wyulda* the rostrum is relatively narrower than in other phalangerids, the lachrymal is retracted from the face, the ectotympanic is almost totally excluded from the anterior face of the postglenoid process, and the third upper premolar tooth is set at a more oblique angle relative to the molar row than it is

Ground cuscus *(Strigocuscus pelengensis),* photo by Tim Flannery.

in other phalangerids. George (1987) indicated that *Strigocuscus* is characterized by the very large size of the third upper premolar, a widening of the zygomatic arches at the orbits, and short paroccipital processes. Flannery (1995) noted that *Strigocuscus* also shows affinity to *Trichosurus* by the lack of a dorsal stripe and the presence of hairs on the portion of the tail that is usually naked in *Phalanger.*

According to Flannery (1995), *Strigocuscus* is arboreal and sometimes is found in secondary forests and gardens. Individuals have been noted sleeping in the crowns of coconut palms. *S. celebensis* has been reported to be nocturnal and frugivorous and to usually occur in pairs. *S. pelengensis* evidently breeds throughout the year and normally rears a single young.

DIPROTODONTIA; PHALANGERIDAE; **Genus TRICHOSURUS**
Lesson, 1828

Brush-tailed Possums

There are three species (Flannery and Schouten 1994; Kerle, McKay, and Sharman 1991; McKay *in* Bannister et al. 1988):

T. vulpecula, throughout suitable areas of Australia, including Tasmania and Kangaroo and Barrow islands;
T. johnstonii, between Koomboolomba and Kuranda in northeastern Queensland;
T. caninus, mountainous districts of southeastern Queensland, eastern New South Wales, and eastern Victoria.

Flannery, Archer, and Maynes (1987) considered *Wyulda* (see account thereof) a subgenus of *Trichosurus. T. arn-*

hemensis, found in extreme northern parts of Western Australia and Northern Territory and on Barrow Island, was long considered a separate species and was maintained as such by Flannery and Schouten (1994) and Groves (*in* Wilson and Reeder 1993). However, morphological, karyological, electrophoretic, and ecological studies by Kerle, McKay, and Sharman (1991) indicate that it is best treated as a subspecies of *T. vulpecula.*

Head and body length is 320–580 mm, tail length is 240–350 mm, and weight is 1.3–5.0 kg. Crawley (1973) found the average adult weight of an introduced population of *T. vulpecula* in New Zealand to be 2.46 kg for males and 2.33 kg for females. The coloration of *T. vulpecula* is extremely variable, with gradations within the main color phases—gray, brown, black and white, and cream. Ride (1970) noted that a black phase was particularly common in Tasmania. There also is sexual dimorphism in color; adult males usually blend to reddish across the shoulders. In *T. caninus* the general color is dark gray or black with a glossy sheen (Ride 1970). The coat is thick, woolly, and soft. The tail is well furred and prehensile, and at least the terminal portion is naked on the lower side. In the subspecies *arnhemensis* the tail is less well haired than in other populations of *T. vulpecula* and appears thinner (Ride 1970). Females of the genus have a well-developed pouch that opens forward and encloses two mammae.

Brush-tailed possums have a well-developed glandular area on the chest. This brown-pigmented, sternal patch is more prominent in the male than in the female. Observations of animals in the field and in captivity have revealed that these areas are rubbed against stumps or fallen logs to form "scent posts." Copious amounts of an anal gland secretion with a strong musky odor are also produced and may have a function in demarcating territory, as well as being a defense mechanism.

All species are nocturnal and arboreal and usually nest in tree hollows. They generally occupy forest habitat, but

Brush-tailed possum *(Trichosurus vulpecula)* male about 159 days old, photo by A. Gordon Lyne.

T. vulpecula also may be found in areas devoid of trees, where it shelters by day in caves or burrows of other animals. It is even found in semideserts of central Australia, where it takes shelter in eucalyptus trees along watercourses. Resident populations occur in most cities in parks and suburban gardens, where the animals shelter on the roofs of houses. The diet consists mostly of young shoots, leaves, flowers, fruits, and seeds. Insects also are eaten, and there are well-documented records of *T. vulpecula* killing young birds. Harvie (1973) found that pasture species of grass and clover formed about 30 percent of the diet of an introduced population of *T. vulpecula* in New Zealand.

Because *T. vulpecula* has long been established in New Zealand and is considered to be of much economic importance, it has been studied there at least as much as in Australia. Some reported population densities for both countries are: vicinity of Canberra, 1/0.67 ha.; dry woodland and sclerophyll forest of southeastern Australia, 1/2.2–2.7 ha.; gardens in Australia, as high as 1/0.72 ha.; North Island of New Zealand, 1/0.16 ha. in August and 1/0.09 ha. in November; and certain parts of New Zealand, 1/0.124 ha. and 1/0.206 ha. (Crawley 1973; How 1978; Ward 1978). Home range in this species was found to average 3 ha. for males and 1 ha. for females in the Canberra vicinity and 0.81 ha. for males and 0.46 ha. for females on the North Island of New Zealand (Crawley 1973). Ward (1978) found nightly ranges of four individuals in New Zealand usually to be under 1,000 sq meters and annual home ranges to be 0.28–3.21 ha. For the species *T. caninus* in southeastern Australia, density was 1/3.3 ha. and home range was 7.67 ha. for males and 4.85 ha. for females (How 1978).

Information on social structure in *Trichosurus* is still limited. According to Collins (1973), two captive males cannot be kept in the same enclosure, but one male and several females can coexist. There is strong evidence that wild *T. caninus* are paired, with an adult male and female sharing a home range (How 1978). The species *T. vulpecula* seems basically solitary. In its low-density populations a territorial system may function, with individuals of the same sex maintaining discrete ranges. In its high-density populations, however, as in parts of New Zealand, there is considerable overlap of home range both among the same sex and between sexes, and territorial behavior is not evident. Spacing between individuals is maintained through agonistic encounters and olfactory and vocal communication (Crawley 1973; How 1978).

The mating call of *T. vulpecula* is a loud, rolling, guttural sound terminating in a series of staccato "ka-ka-ka" syllables. When mildly disturbed, all species make a metallic clicking sound and then, if further aroused, a series of harsh exhalations with the mouth open. Further disturbance causes the animal to rise on its hind limbs with the forelimbs raised and outstretched. The possum then extends itself to full height and screams. All species are powerful fighters.

Females are polyestrous; the estrous cycle averages about 25 days in *T. vulpecula* (Van Deusen and Jones 1967) and about 26.4 days in *T. caninus* (Smith and How 1973). Breeding may extend throughout the year, but generally for *T. vulpecula* in Australia there are two distinct seasons or peaks, one in spring and one in autumn. In some areas a female may produce a litter in both seasons, but usually most females give birth during the autumn (How 1978). The subspecies *T. vulpecula arnhemensis* in the tropics of extreme northern Australia has been found to breed continuously throughout the year, with females evidently conceiving before weaning the pouch young or returning to estrus in about 10 days if the pouch young is lost (Kerle and Howe 1992). In New Zealand it is extremely rare for a female to have two litters in one year (Crawley 1973). For

T. vulpecula in New South Wales, Smith, Brown, and Frith (1969) found 89.6 percent of births to occur in autumn (late March–June) and 9.1 percent to occur in spring (September–November). For *T. caninus* in New South Wales, How (1976) found 87.5 percent of the births to take place from February to May, mostly in March and April, with a few scattered over the rest of the year. For an introduced population of *T. vulpecula* in New Zealand, Crawley (1973) reported a single well-defined breeding season from April to July, with a few births also occurring later.

Gestation averages about 17.5 days in *T. vulpecula* (Van Deusen and Jones 1967) and 16.2 days in *T. caninus* (Smith and How 1973). Usually a single young is produced; of 60 births of *T. vulpecula* recorded at the London Zoo, 6 resulted in twins and 1 in triplets (Collins 1973). Development appears to take place more rapidly in *T. vulpecula* (How 1976, 1978). The young leave the pouch at about 4–5 months, are weaned by 6–7 months, and separate from the mother at 8–18 months. Females generally attain sexual maturity and begin to breed at 9–12 months, though in some populations this phase is delayed 1 or even 2 years. In *T. caninus* the young emerge from the pouch at 6–7 months, weaning occurs at 8–11 months, dispersion occurs at 18–36 months, and females do not reach sexual maturity until 24–36 months. Males of *T. vulpecula* are sexually mature by the end of their second year of life (Crawley 1973; Smith, Brown, and Frith 1969). In *T. vulpecula* mortality is high among dispersing young; only 25 percent reach the age of 1 year (How 1978). Adult mortality has been found to be relatively low—20 percent in one study—and life expectancy is 6–7 years. In wild populations individuals of *T. vulpecula* over 13 years old and of *T. caninus* up to 17 years have been found (How *in* Strahan 1983). *Trichosurus* also does well in captivity; one specimen of *T. vulpecula* survived for 14 years and 8 months (Jones 1982).

Because of its fecundity and adaptability to a variety of conditions, *T. vulpecula* has been compared to the Virginia opossum *(Didelphis virginiana)* in North America. Both species seem able to live near people and even to expand their range in the face of human development. However, the subspecies *T. vulpecula hypoleucus,* restricted to extreme southwestern Western Australia, is designated as near threatened by the IUCN. Ride (1970) wrote that Australians had more contact with *T. vulpecula* than with any other native mammal. This species can find shelter in artificial structures and can subsist on garden vegetation, but unlike the Virginia opossum, it now is considered by some to be a major pest. In Australia it is said not only to damage flowers, fruit trees, and buildings but also to adversely affect regenerating eucalyptus forests and introduced pine plantations and to carry diseases that are potentially harmful to humans and livestock (How 1976; McKay and Winter 1989; Ride 1970). However, this possum has been taken extensively in Australia for the fur trade, with more than a million pelts going on the market in some years. At present, commercial hunting is restricted to Tasmania and serves to suppress populations that are alleged to be damaging forestry and agriculture. As recently as 1980 more than 250,000 skins were exported with a value of about U.S. $6 each, but subsequently prices fell and the market declined. In 1990 and 1991 exports totaled less than 15,000 skins at a value of about $2 each (Callister 1991).

A desire to share in the fur market led to the introduction of *T. vulpecula* in New Zealand, first in 1840 and more extensively in the 1890s. It adapted well, greatly increased in range and numbers, and now often is treated more as a pest than as a fur bearer (Crawley 1973; Fitzgerald 1976, 1978; Gilmore 1977; Harvie 1973). It reportedly damages gardens, orchards, crops, pasture, exotic tree plantations, and remaining native forests. All legal restrictions on killing it in New Zealand were removed in 1947, a bounty was in effect from 1951 to 1960, and the species has been subject to intensive programs of poisoning, trapping, and shooting. Its overall numbers, however, do not seem to have been greatly affected. From 1962 to 1974 the annual number of skins exported from New Zealand ranged from 346,000 to 1,605,000.

DIPROTODONTIA; PHALANGERIDAE; **Genus WYULDA**
Alexander, 1919

Scaly-tailed Possum

The single species, *W. squamicaudata,* is restricted to the Kimberley Division in northern Western Australia (Ride 1970). It was known only from four specimens until 1965, when eight more were collected (Fry 1971). It subsequently has been observed with more regularity. Flannery, Archer, and Maynes (1987) considered *Wyulda* a subgenus of *Trichosurus.* That position was followed by Corbet and Hill (1991) but not Groves (*in* Wilson and Reeder 1993) or McKay (*in* Bannister et al. 1988), and *Wyulda* was restored to generic rank by Flannery and Schouten (1994).

Head and body length is 290–395 mm, tail length is about 250–325 mm, and adult weight range is 1.4–2.0 kg (Humphreys et al. 1984). The pelage is short, soft, fine, and dense. The general dorsal color is pale or dark ashy gray. A dark stripe, obscure or distinct, runs along the middorsal line from the shoulders to the rump. The nape, shoulders, and rump of one specimen were mottled with buff, and its

throat and chest were gray. The sides and back are usually the same color, though the sides are somewhat paler. The underparts are creamy white.

The prehensile tail is densely furred at the base and has nonoverlapping, thick scales for the remainder of its length. Short, bristly hairs are present around the scales. This is the only member of the family Phalangeridae with a tail of this kind. The head is short and wide. The claws are short and not strongly curved.

This possum inhabits areas with trees and rocks in broken sandstone country (Ride 1970). It is nocturnal and scansorial, apparently sheltering by day among the rocks and emerging at night to feed on leaves, blossoms, fruit, insects, and possibly small vertebrates (Collins 1973). After leaving the rocks, it reportedly climbs the nearest tree and then crosses from tree to tree without coming down. It appears to be solitary, and three female specimens each carried a single young (Ride 1970). One population occurred at a density of about 1/ha. The young are born from March to August and are weaned sometime after 8 months; males do not reach sexual maturity until past 18 months, and females not until their third year (Humphreys et al. 1984). A captive individual was said to be gentle and affectionate and to make a chittering noise like a bird (Fry 1971). It was still alive after being maintained in a private home for 4 years and 4 months (Collins 1973).

The genus has been designated as endangered by the USDI since 1970, though Winter (1979) stated that it is probably more plentiful than previously thought, and Thornback and Jenkins (1982) reported no evidence of any major threat. More recently, Flannery and Schouten (1994) reaffirmed that *Wyulda* should be regarded as endangered. While it reportedly is abundant at one site, it has a very

Scaly-tailed possum *(Wyulda squamicaudata),* photo by A. G. Wells / National Photographic Index of Australian Wildlife.

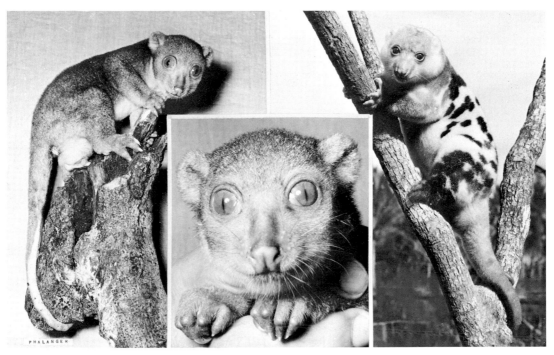

Spotted Cuscus *(Spilocuscus maculatus)*, photos by Sten Bergman.

limited and patchy distribution and may be declining as its habitat is disrupted by the expanding pastoral industry. The IUCN currently designates it as near threatened.

DIPROTODONTIA; PHALANGERIDAE; **Genus
SPILOCUSCUS**
Gray, 1861

Spotted Cuscuses

There are four species (Feiler 1978*a*, 1978*b*; Flannery 1990*b*; Flannery, Archer, and Maynes 1987; Flannery and Calaby 1987; Flannery and Schouten 1994; G. G. George 1979, 1987):

S. rufoniger, northern New Guinea;
S. maculatus, New Guinea and nearby islands, Bismarck Archipelago, Seram and nearby islands, Cape York Peninsula of Queensland;
S. kraemeri, Manus and some small nearby islands in the Admiralty Group northeast of New Guinea;
S. papuensis, Waigeo Island off the western tip of New Guinea.

Spilocuscus usually has been placed in *Phalanger,* but Flannery, Archer, and Maynes (1987) and George (1987) regarded the two as generically distinct. Those authorities also suggested that *S. papuensis* is a species distinct from *S. maculatus;* that view was followed by Corbet and Hill (1991) but not Groves (*in* Wilson and Reeder 1993). Head and body length is 338–640 mm, tail length is 315–590 mm, and weight is 2.0–7.0 kg (Flannery 1995; Flannery and Calaby 1987; Flannery and Schouten 1994; Winter *in* Strahan 1983). On average females are larger than males. Except for *S. papuensis,* the sexes also are col-

ored differently. In *S. maculatus* the adult males, unless completely white, are gray spotted with white above and white below. The females are uniformly gray and usually unspotted. In *S. rufoniger* and the much smaller *S. kraemeri* there is a dark saddle on the back of females but only an area of mottling or spots on the males. In *S. papuensis* both sexes are marked with very small spots. The young go through a sequence of color changes. The fur is dense and woolly, the snout is short, and the ears are almost invisible. Flannery, Archer, and Maynes (1987) listed the following unique characters that distinguish *Spilocuscus* from related genera: there is sexual dichromatism as noted above; in both sexes the frontal bones of the skull are markedly convex and accommodate a large sinus that does not open into the nasal cavity; there is a well-developed protocone on the first upper molar; and the alisphenoid and basoccipital consistently form a more extensive suture that is developed earlier in life than it is in other phalangerids. Females have four mammae.

Winter (*in* Strahan 1983) wrote that *S. maculatus* occurs mainly in rainforest from sea level to an elevation of 820 meters but also has been seen in mangroves and open forests. It does not use a den or nest but is reputed to build a small sleeping platform of leaves. It is mainly nocturnal and arboreal, climbs slowly and deliberately, maintains a strong grip with its feet, and bounds at the speed of a fast human walk when on the ground. Under natural conditions it has been observed to eat fruit, flowers, and leaves, but captives have readily taken dog food and chickens. Males are aggressive and cannot be housed together in captivity. There apparently is an extended breeding season in both New Guinea and Australia. Although three young have been recorded from a pouch, it is likely that only one is reared. A specimen of *S. maculatus* lived for more than 11 years in captivity.

All species of *Spilocuscus* are on appendix 2 of the CITES (though under the single name *S. maculatus*). The IUCN designates *S. rufoniger* as endangered. According to Thornback and Jenkins (1982), that species is known only by 18 specimens, is subject to intensive hunting, and may be losing its rainforest habitat to logging and agricultural expansion. Flannery and Schouten (1994) reported that a few more specimens had been collected recently but considered the species to be seriously endangered by hunting for use as food and disruption of habitat. There is a disjunct population of *S. maculatus* on Salayar Island, south of Sulawesi, where it may have been introduced by human agency (George 1987). *S. kraemeri* apparently was brought to the area from which it is currently known only 1,000–2,000 years ago. Since it probably could not have evolved its distinctive characters over so short a period, there may be an as yet undiscovered parent population (Flannery 1995; Flannery and Schouten 1994).

DIPROTODONTIA; PHALANGERIDAE; **Genus**
PHALANGER
Storr, 1780

Cuscuses

There are 11 named species (Colgan et al. 1993; Feiler 1978*a*, 1978*b*; Flannery 1987*b*, 1990*b*; Flannery, Archer, and Maynes 1987; Flannery and Boeadi 1995; Flannery and Schouten 1994; G. G. George 1979, 1987; Groves 1976, 1987*a*, 1987*b*; Kirsch and Calaby 1977; Laurie and Hill 1954; Menzies and Pernetta 1986; Ride 1970; Springer et al. 1990; Ziegler 1977):

P. rothschildi, Obi and Bisa islands in the northern Moluccas;

P. ornatus, Halmahera, Bacan (Batjan), Ternate, Tidore, and possibly Morotai islands in the Northern Moluccas;

P. alexandrae, Gebe Island in the Northern Moluccas;

P. matanim, known only by five specimens from the mountains of western Papua New Guinea;

P. orientalis, Molucca Islands, Buru, Timor, Seram, northern New Guinea and many nearby islands, Bismarck Archipelago, Solomon Islands;

P. intercastellanus, southern New Guinea, Aru Islands, D'Entrecasteaux Islands, Louisiade Archipelago, eastern Cape York Peninsula of Queensland;

P. vestitus, highlands of western, central, and eastern New Guinea;

P. gymnotis, New Guinea and Aru Islands to southwest;

P. carmelitae, highlands of central and eastern New Guinea;

P. sericeus, highlands of central and eastern New Guinea;

P. lullulae, Woodlark Island east of New Guinea and (possibly through introduction) nearby Alcester and Madau islands.

Flannery (1995) and Flannery and Schouten (1994) indicated the existence of two undescribed species, one on Mount Karimui in east-central New Guinea, which may be closely related to *P. intercastellanus*, and one on Gebe and perhaps throughout the northern Molucca Islands, which has been referred mistakenly to *P. ornatus*. *Phalanger* often has been considered to include the genera *Ailurops*, *Strigocuscus*, and *Spilocuscus* (see accounts thereof). Flannery, Archer, and Maynes (1987) referred *P. ornatus* and *P. gymnotis* to *Strigocuscus*, but Flannery and Schouten (1994) reassigned both to *Phalanger* based in part on the molecular analyses of Springer et al. (1990). *P. vestitus* sometimes has been referred to as *P. interpositus* and in-

Gray cuscus *(Phalanger mimicus)*, photo by Queensland National Parks and Wildlife Service / National Photographic Index of Australian Wildlife.

cludes *P. permixtio* of east-central New Guinea as a synonym. George (1987) regarded *P. leucippus* of mainland New Guinea as a species distinct from *P. gymnotis*.

Head and body length is 325–600 mm, tail length is 240–610 mm, and weight is 1,045–4,850 grams (Flannery 1990*b*; Flannery and Schouten 1994; Lidicker and Ziegler 1968). The fur in most forms is thick and woolly. Coloration in the genus ranges from white, reds, and buffs through various shades of brown to light grays and different intensities of black. Some members of this genus have suffusions of yellow or tawny over the shoulder region, and others have dark dorsal stripes that extend from the head to the rump. In *P. vestitus* some individuals are pale silvery brown and short-haired, others are dark brown and long-haired, and still others represent intermediate forms.

These are heavy and rather powerfully built animals. Their yellow-rimmed protruding eyes, bright yellow noses, inconspicuous ears, and prehensile tails give them a somewhat monkeylike appearance. The terminal portion of the tail is covered with scales and lacks hair. The fingers are not of equal length; the claws are long, stout, and curved; and the soles are naked and striated. Females have four mammae in a well-developed pouch.

Cuscuses inhabit mainly tropical forests and thick scrub. They are arboreal animals with strongly prehensile tails but sometimes descend to the ground. They are nocturnal, resting by day curled up in the thick foliage of a vine tangle, in a tree or tree hollow, under tree roots, or among rocks. Cuscuses are slow-moving and somewhat sluggish, resembling the slow loris *(Nycticebus)* in their movements.

Cuscus *(Phalanger gymnotis)*, photo by Pavel German.

Even when handled gently cuscuses usually emit a penetrating musk odor. Snarls and barks are the sounds they make. Their diet consists mainly of fruits and leaves but also includes insects, small vertebrates, and birds' eggs. Cuscuses are solitary but can be maintained in pairs in roomy enclosures (Collins 1973). The breeding season appears to be extensive and may last throughout the year in some species; females have up to three young, though they may rear only one (Flannery 1995; Winter *in* Strahan 1983).

According to Flannery (1990*b*), *P. gymnotis* is common in many parts of New Guinea from sea level to 2,700 meters but avoids swampy areas and floodplains. It usually rests during the day in burrows under tree roots, in caves, or in human-made tunnels. It is capable of climbing and tends to feed in trees at night. It reportedly eats a wide variety of fruit in the wild, though captives have killed small vertebrates. Native people say that females carry fruit back to the den in their pouch. Severe fighting evidently is frequent, even between opposite sexes when females were in an anestrous condition. Breeding apparently is continuous throughout the year, and there is usually a single pouch young. A captive was known to be at least 11 years old.

Recent archeological research indicates that many of the island populations of *P. orientalis* were introduced through human agency in prehistoric time. Based on electrophoretic and morphological analyses, Colgan et al. (1993) suggested that the species was introduced to New Ireland from New Britain between 10,000 and 19,000 years ago and then spread to the Solomon Islands between 2,000 and 6,000 years ago. Flannery, Archer, and Maynes (1987) noted that *P. orientalis* probably was introduced on Timor about 4,000–5,000 years ago.

The species *P. orientalis* is on appendix 2 of the CITES. The IUCN classifies *P. matanim* as endangered and *P. rothschildi* and *P. vestitus* as vulnerable. All are declining because of loss of forest habitat and hunting by people for use as food. The restricted population of *P. intercastellanus* in

Australia is designated as near threatened by the IUCN. Thornback and Jenkins (1982) wrote that *P. lullulae* had not been collected since 1953 and that its limited habitat is unprotected and suitable for logging or agriculture. However, Flannery and Schouten (1994) reported that it recently has been rediscovered in abundance on Woodlark Island.

DIPROTODONTIA; **Family POTOROIDAE**

"Rat"-Kangaroos

This family of 5 Recent genera and 10 species occurs in Australia, especially the southern and eastern parts, and on a few nearby islands and in Tasmania. In the past the Potoroidae often were regarded only as a subfamily of the Macropodidae, but there now is general recognition that the two are distinct families (Calaby and Richardson *in* Bannister et al. 1988; Corbet and Hill 1991; Groves *in* Wilson and Reeder 1993; Strahan 1983). The sequence of genera presented herein follows that of Flannery (1989*b*), who, like most other authorities, recognized two living subfamilies: Hypsiprymnodontinae, with the single genus *Hypsiprymnodon;* and Potoroinae with *Potorous, Bettongia, Caloprymnus,* and *Aepyprymnus.* Szalay (1994) continued to include the Potoroinae as a subfamily of the Macropodidae but considered the Hypsiprymnodontinae a completely separate family.

Seebeck and Rose (1989) provided a detailed description of the Potoroidae. Head and body length is 153–415 mm, tail length is 123–387 mm, and weight is 360–3,500 grams. The body is covered with dense fur. The upper parts range in color from dark chocolate brown, through browns and grays, to rufous and sandy; the underparts are usually pale. Unlike in the Macropodidae, coat color is usually uniform with no stripes or other markings. All living genera are compact animals with variously elongate muzzles, short and rounded ears, and short but muscular forelegs bearing small paws with short, forward-pointing spatulate claws. The hind limbs are well developed and heavily muscled, and the hind feet are elongate. In *Hypsiprymnodon* the proportional difference between the forelimbs and the hind limbs is less than in the other genera. Also, whereas all the other genera have lost the first digit of the hind foot, it is retained by *Hypsiprymnodon*. The second and third digits of the hind foot are syndactylous (united by integument). The fourth toe is the largest. The long tail is furred in most genera but naked and scaly in *Hypsiprymnodon*; it is prehensile in *Bettongia*. The stomach is simple in *Hypsiprymnodon* and complex in all other genera that have been studied. The female reproductive tract of potoroids differs from that of most macropodids in the presence of an anterior vaginal expansion.

The skull is short and broad in most genera but elongate and narrow in some species of *Potorous*. In all genera except *Hypsiprymnodon* the squamosal bone has wide contact with the frontal bone, thereby separating the parietal and alisphenoid bones. In the Macropodidae parietal-alisphenoid contact is usual. In the Potoroidae the masseteric canal of the mandible is confluent with the inferior dental canal and extends farther anteriorly than it does in the Macropodidae. The dental formula in most genera is: (i 3/1, c 1/0, pm 1/1, m 4/4) × 2 = 30; in *Hypsiprymnodon* the incisors are 3/2. The dentition is typically diprotodont, the pair of lower incisors being enlarged and projecting forward. There are two deciduous premolars on each side of both jaws, but these are shed near maturity and replaced by a single large, permanent tooth. This premolar is bladelike and has vertical corrugations. Unlike in the Macropodidae, there is no forward progression of molar teeth. The molars are mostly quadrituberculate, and they decrease in size from the anterior to the posterior.

Habitats range from hummock grassland to tropical rainforest, but most species are found in forest or woodland in southern Australia. Most are mainly nocturnal. Although omnivorous, many species now are known to feed primarily on fungi. Reproduction generally is continuous, mating is promiscuous, and there is little evidence of territoriality. The reproductive biology of the Potoroidae is much like that of the Macropodidae, but the "rat"-kangaroos generally have more than one young per year and have a relatively shorter pouch life. Despite their fecundity, potoroids are reduced in numbers over most of their range because of the effects of European settlement. Only 3 of the 10 modern species remain relatively common (Rose 1989; Seebeck, Bennett, and Scotts 1989).

The known geological range of the Potoroidae is middle Miocene to Recent in Australia.

DIPROTODONTIA; POTOROIDAE Genus
HYPSIPRYMNODON
Ramsay, 1876

Musky "Rat"-kangaroo

The single species, *H. moschatus*, occupies approximately 320 km of the coast of northeastern Queensland (Calaby 1971).

This is the smallest of the "rat"-kangaroos: head and body length is 208–341 mm, tail length is 123–65 mm, and weight is 337–680 grams (Johnson and Strahan 1982). The pelage is close, crisp, and velvety and consists mainly of underfur. The general coloration is rich brown or rusty gray, brightest on the back and palest on the underparts. Some

Musky "rat"-kangaroo *(Hypsiprymnodon moschatus)*, photo by P. M. Johnson.

individuals have a white area on the throat that extends as a narrow line to the chest.

This genus is unique among the Potoroidae in that the hind foot has a well-developed first digit that is movable and clawless but nonopposable to the other digits. The tail, which is used to gather nesting material, also differs from that of all other members of the family in being almost completely naked and scaly; only the extreme base is hairy. The muzzle is naked, and the ears are rounded, thin, and naked except at their posterior base. The limbs are more equally proportioned than those of other "rat"-kangaroos. The claws are quite small, weak, and unequal in length. The females have four mammae and a well-developed pouch. The specific name refers to the musky scent emitted by both sexes.

This "rat"-kangaroo occurs in rainforests, often in dense vegetation bordering rivers and lakes. It seems to be fairly common but is shy, quick in its movements, and difficult to observe in its dense habitat. It differs from other members of the Potoroidae in being completely diurnal (Seebeck and Rose 1989). One individual was seen sunbathing, lying spread-eagle on a fallen tree trunk. Most observers state that *Hypsiprymnodon* runs on all four limbs instead of hopping on its hind feet. According to Johnson and Strahan (1982), adults have been seen to climb on fallen branches and horizontal trees, and juveniles have ascended a thin branch inclined at about 45°. At night and in the middle of the day the musky "rat"-kangaroo sleeps in a nest in a clump of vines or between the plank buttresses of a large tree.

The diet is different from that of other "rat"-kangaroos. It consists to a great extent of insects and worms; the berries of a palm *(Ptychosperma)* and tuberous roots also are eaten. This animal sits on its haunches while eating. Food is obtained by turning over debris and by digging.

Johnson and Strahan (1982) stated that *Hypsiprymnodon* appears to be solitary, but feeding aggregations of up to three individuals have been observed. Breeding occurs from February to July (the rainy season), and usually two young are born. They leave the pouch after about 21 weeks, and then for several more weeks they spend a considerable part of the day in the nest. Females attain sexual maturity at just over a year.

The range of the musky "rat"-kangaroo has been affected by clearing of rainforest for agricultural development. Where it remains, however, it seems relatively common (Seebeck, Bennett, and Scotts 1989). Kennedy (1992) considered it "potentially vulnerable" because of its restricted rainforest habitat.

DIPROTODONTIA; POTOROIDAE Genus **POTOROUS**
Desmarest, 1804

Potoroos

Three species now are recognized (Johnston and Sharman 1976, 1977; Kirsch and Calaby 1977; Seebeck and Johnston 1980):

P. tridactylus (long-nosed potoroo), southeastern Queensland, coastal New South Wales, Victoria, southeastern South Australia, southwestern Western Australia, islands of Bass Strait, Tasmania;
P. longipes, eastern Victoria;
P. platyops (broad-faced potoroo), southwestern Western Australia.

The subspecies *P. tridactylus apicalis* of Tasmania and the islands in Bass Strait was formerly considered a distinct species. Subfossil specimens indicate that the range of *P. platyops* formerly extended to southeastern South Australia and that *P. longipes* also occurred in southeastern New South Wales, but it is not clear whether these records date from historical time (Seebeck 1992*b*; Seebeck and Rose 1989).

Head and body length is 243–415 mm, tail length is 198–325 mm, and weight is 660–2,200 grams (Johnston *in* Strahan 1983; Kitchener *in* Strahan 1983; Seebeck *in* Strahan 1983). The pelage, at least in *P. tridactylus*, is straight, soft, and loose. The upper parts are grayish or brownish, and the underparts are grayish or whitish. The tail of *P. tridactylus* is often tipped with white. The muzzle is elongated in *P. tridactylus* and shortened in *P. platyops*. Females have four mammae located in a well-developed pouch that opens forward. The hind feet of *Potorous* are shorter than the head.

Following three years of field and laboratory observations, Buchmann and Guiler (1974) were able to clear up the uncertainty regarding potoroos' means of locomotion. There are three principal methods: (1) the "quadrupedal crawl," a slow (20–23 meters per minute) plantigrade movement in which the hind feet are employed for the main forward propulsion and then the weight is shifted to the forefeet, which is used in leisurely feeding and foraging; (2) the "bipedal hop," a series of synchronous digitigrade thrusts by the hind limbs, moving the animal 30–56 meters per minute, used for escape and chasing in aggression; and (3) "jumping," a single bound 2.5 meters long and 1.5 meters high resulting from a powerful thrust of the hind legs, used in initial escape or aggression.

Potoroos inhabit dense grassland or low, thick scrub especially in damp places (Ride 1970). They are nocturnal but occasionally may engage in early morning basking in the sun (Collins 1973). During the day potoroos shelter in shallow "squats," usually excavated at the base of a tussock or under dense shrubs, and do not construct complex nests (Seebeck, Bennett, and Scotts 1989). *P. tridactylus* digs small holes in the ground when feeding. These holes are not round like those made by bandicoots. Well-defined trails lead from one feeding site to another. The overall home range of *P. tridactylus* is relatively large, perhaps because fungi form a major food source. One study in Tasmania found that fungi were of substantial importance from May to December and constituted more than 70 percent of the food in May and June (Guiler 1971*a*). *P. tridactylus* also depends on insects more than do most potoroids, especially in summer (Guiler 1971*b*). In addition, the diet includes grass, roots, and other forms of vegetation.

P. tridactylus has been found at population densities of 0.2–2.55/ha. (Seebeck, Bennett, and Scotts 1989). Its home range in Victoria averages only about 2.0 ha. for males and 1.5 ha. for females, but those of *P. longipes* may exceed 10.0 ha. Male ranges may overlap those of several females, but female ranges are often exclusive (Seebeck and Rose 1989). Observations of *P. tridactylus* indicate that males are territorial and defend a small part of their home range but tend to avoid conflict under natural conditions. They may be kept in captivity together with other males and several females, but when a female is in estrus the males will fight fiercely, with one eventually establishing dominance (Collins 1973; Ride 1970; Russell 1974*b*).

Potoroos (*Potorous* sp.), photo by Ernest P. Walker.

Potoroos are polyestrous, with an estrous cycle of about 42 days, and apparently lack a well-defined breeding season. In Tasmania *P. tridactylus* bears young throughout most of the year, with peaks from July (winter) to January (summer). Each female is able to breed twice a year. Nondelayed gestation is 38 days and normal litter size is 1. Four days after the season's first young is born the female mates again, but because of embryonic diapause development is arrested and parturition does not occur for about 4.5 months unless the first young is lost. The newborn measures 14.7–16.1 mm in length, is able to detach from the nipple at 55 days, and leaves the pouch at about 130 days. Females are able to breed at 1 year (Bryant 1989; Collins 1973; Ride 1970). Reproduction in *P. longipes* is similar, but pouch life is 140–50 days and sexual maturity may not be attained until 2 years (Seebeck 1992*a*). A wild specimen of *P. tridactylus* is known to have lived at least 7 years and 4 months (Guiler and Kitchener 1967), and captives have reached an age of at least 12 years (Johnston *in* Strahan 1983).

The species *P. platyops* of southwestern Australia apparently has been extinct since about 1875 and may have declined considerably prior to the coming of European settlers; it is designated extinct by the IUCN. The subspecies of *P. tridactylus* in southwestern Australia, *P. t. gilbertii*, also may be extinct, none having been collected in more than 80 years (Calaby 1971). Recently, however, there have been unconfirmed sightings in southwestern Australia of a mammal that might represent *Potorous* (Thornback and Jenkins 1982). The IUCN now treats *gilbertii* as a separate and living species and classifies it as critically endangered; it is thought to number fewer than 50 individuals and to be declining. On the mainland of southeastern Australia *P. tridactylus* apparently has declined because of its dependence on dense vegetation and its resultant vulnerability to clearing operations and brush fires, but it is not as rare as once thought. It no longer occurs in South Australia but is widespread from Victoria to southern Queensland and remains common in much of Tasmania. In Bass Strait it still is present, though rare, on Flinders and King islands, but it seems to have disappeared from Clarke Island, which is heavily grazed by livestock and introduced rabbits (Calaby 1971; Poole 1979; Seebeck 1981). It was considered "potentially vulnerable" by Kennedy (1992). The IUCN designates the eastern mainland subspecies, *P. tridactylus tridactylus,* as vulnerable, noting that its total number is less than 10,000. *P. longipes*, with fewer than 2,500, is classified as endangered by the IUCN. It apparently is very rare, and its restricted habitat is threatened by logging (Thornback and Jenkins 1982).

DIPROTODONTIA; POTOROIDAE **Genus BETTONGIA**
Gray, 1837

Short-nosed "Rat"-kangaroos

There are four species (Kirsch and Calaby 1977; Ride 1970; Sharman et al. 1980):

B. penicillata (woylie), formerly from southwestern
    Western Australia to central New South Wales;
B. tropica, eastern Queensland;
B. gaimardi, southeastern Queensland, coastal New South
    Wales, southern Victoria, Tasmania;
B. lesueur (boodie), Western Australia, including certain
    coastal islands, to southwestern New South Wales.

Another species, *B. cuniculus,* formerly was recognized from Tasmania but is now included in *B. gaimardi*. Sharman et al. (1980) concluded that there is no chromosomal basis for the specific distinction of *B. tropica* from *B. penicillata*. This position was accepted by Groves (*in* Wilson and Reeder 1993) but not by Calaby and Richardson (*in* Bannister et al. 1988), Corbet and Hill (1991), or Flannery (1989*a*).

Head and body length is 280–450 mm and tail length is 250–330 mm. Weight is 1.1–1.6 kg in *B. penicillata* and 1.2–2.25 kg in *B. gaimardi* (Christensen *in* Strahan 1983; K. A. Johnson *in* Strahan 1983; Rose *in* Strahan 1983). The upper parts are buffy gray to grayish brown; the underparts are paler. The tail is crested in all species except *B. lesueur* and usually is white-tipped in *B. gaimardi* and *B. lesueur*. The unworn adult pelage in *B. penicillata* is often crisp or even harsh; in *B. lesueur* it is soft and dense. The tip of the muzzle is naked and flesh-colored and the ears are short and rounded. The hind feet are longer than the head. Females have four mammae and a well-developed pouch.

Habitats include grassland, heath, and sclerophyll woodland. These "rat"-kangaroos are nocturnal. They construct nests, usually of grass but sometimes using sticks or bark, and they carry nesting material in the curled-up tip of their tail. The nests generally are located at the base of a grass tussock or overhanging bush. There sometimes is an earth excavation at the base, but *B. lesueur* constructs a large burrow for community dwelling and may modify a rabbit warren for its own use. This species is entirely bipedal in locomotion, never using its forefeet, even when moving slowly (Ride 1970). The claws of all species are used for digging. When fleeing, *B. penicillata* travels with head held low, back arched, and the tail brush displayed conspicuously. Most species seem primarily herbivorous, feeding on roots, tubers, seeds, and legume pods, but there are some reports of feeding on marine refuse, carrion, and meat (Collins 1973; Ride 1970). There is increasing evidence that *Bettongia* and some other potoroids depend to a large extent on fungi (Seebeck, Bennett, and Scotts 1989). In Tasmania, sporocarps of mycorrhizal fungi make up the bulk of the diet of *B. gaimardi* year-round, though other foods also are thought to be necessary for nutritional balance (R. J. Taylor 1992, 1993*a*).

B. penicillata has been found at population densities of 0.07–0.45/ha. and in home ranges averaging 35 ha. for males and 20 ha. for females (Seebeck, Bennett, and Scotts 1989), an unusually large area for an animal of its size. Taylor (1993*b*) found an average of 61 ha. for individuals of *B. gaimardi* tracked longer than two months, with the ranges of males being generally larger than those of females.

In *B. lesueur* a male and several females form a social group and occupy a burrow system. Large warrens may have 40–50 occupants (Seebeck, Bennett, and Scott 1989). Males are aggressive toward one another and seem to defend groups of females but not a particular territory. Females generally are amicable but sometimes will establish a territory and exclude other females. They are polyestrous, with the modal length of the estrous cycle being 23 days, and can produce up to three litters per year. Breeding may go on throughout the year in some areas, but in the Bernier Island population of *B. lesueur* most births occur between February and September. The modal length of undelayed gestation is 21 days and the usual litter size is one, though twins occasionally occur. Just after one young is born the female mates again, but because of embryonic diapause development is delayed, and parturition of the second young does not take place for about four months un-

Short-nosed "rat"-kangaroo *(Bettongia lesueur)*, photo from New York Zoological Society. Inset: *B. penicillata*, photo from *A Natural History of the Mammalia*, G. R. Waterhouse.

less the first young is lost. The young weigh about 0.317 grams at birth, leave the pouch permanently at about 115 days, and attain adult size at 280 days. Females apparently are capable of giving birth at about 200 days (Ride 1970; Tyndale-Biscoe 1968). Reproductive information on other species is more limited, but in *B. gaimardi* and *B. penicillata* the estrous cycle, gestation period, embryonic diapause, and pouch life are nearly the same as in *B. lesueur* (Rose 1978, 1987). Record longevity for the genus is held by a specimen of *B. gaimardi* that was still living after 11 years and 10 months in captivity (Jones 1982).

Modern human agency seems to have harmed *Bettongia* more than any other polytypic genus of marsupials. All four species are listed as endangered by the USDI and are on appendix 1 of the CITES. The IUCN designates *B. tropica* as endangered, *B. lesueur* generally as vulnerable, the subspecies *B. l. graii* of mainland Australia as extinct, *B. penicillata* generally as conservation dependent, the subspecies *B. p. penicillata* of eastern Australia as extinct, the mainland subspecies *B. gaimardi gaimardi* as extinct, and the Tasmanian subspecies *B. g. cuniculus* as near threatened. *B. tropica* had been known from only six specimens, the latest collected in 1932, and was thought to be possibly extinct, but a population recently was discovered in the Davies Creek National Park in northeastern Queensland (Poole 1979). The species *B. gaimardi* has not been recorded from the mainland since 1910; it still is found in reasonable numbers in many localities of Tasmania but may be jeopardized there by logging and poisoning (Calaby 1971; Ride 1970; Rose 1986).

B. lesueur once was among the most widely distributed of native Australian mammals, occurring in all mainland states except perhaps Queensland. Although the last specimen in New South Wales was collected in 1892, *B. lesueur* remained common in parts of central and southwestern

Australia until the 1930s. By the early 1960s it had disappeared completely from the mainland and become restricted to Barrow, Boodie, Bernier, and Dorre islands off the west coast. Its drastic decline apparently resulted from competition with introduced rabbits for burrows and food, habitat disruption and disturbance by domestic livestock, predation by introduced foxes, and direct killing by people. Until recently the remnant island populations were thought to be viable and well protected (Australian National Parks and Wildlife Service 1978; Calaby 1971; Poole 1979). However, surveys by Short and Turner (1993) revealed that the species had been wiped out on Boodie Island, apparently as an inadvertent consequence of a recent poisoning campaign against *Rattus rattus*. There were about 5,000 individuals left in the other three island populations, but they were considered seriously endangered, especially because of the potential introduction of cats or other predators.

Another severe decline has been suffered by *B. penicillata*, which formerly occurred all across the southern part of the continent, including certain islands and the northwestern corner of Victoria. Its range may even have extended well into the Northern Territory and Queensland (Seebeck, Bennett, and Scotts 1989). It was described as "very abundant" in New South Wales in 1839–40 but disappeared from that state shortly thereafter. It was common in South Australia at the end of the nineteenth century but was gone from there by 1923. Today it is restricted to a few tracts of woodland at the extreme southwestern tip of the continent. The primary cause of its decline seems to have been clearing of brush for agricultural development (Australian National Parks and Wildlife Service 1978; Calaby 1971; Poole 1979; Ride 1970). Recently, *B. penicillata* was reintroduced on several islands off the coast of South Australia (Delroy et al. 1986).

DIPROTODONTIA; POTOROIDAE **Genus**
CALOPRYMNUS
Thomas, 1888

Desert "Rat"-kangaroo

The single species, *C. campestris*, was described in 1843 on the basis of three specimens from an unknown locality in South Australia. There were no further records until 1931, when a specimen was collected in the northeastern corner of South Australia. A subsequent survey indicated that in the early 1930s the species occurred in an area about 650 km north to south and 250 km wide in the Lake Eyre Basin in northeastern South Australia and southwestern Queensland. In addition, apparently Recent cave remains were found in southeastern Western Australia, and late Pleistocene fossils were located in New South Wales (Australian National Parks and Wildlife Service 1978; Calaby 1971; Ride 1970).

Head and body length is 254–82 mm, tail length is 297–377 mm, and weight is 637–1,060 grams (Smith *in* Strahan 1983). The pelage is soft and dense. The coloration of the upper parts is clear pale yellowish ochre, which matches that of clay pans and plains; the underparts are whitish. The ears are longer and narrower than those of any other "rat"-kangaroo, and the muzzle is naked. The long, cylindrical tail is evenly short-haired without a trace of a crest. There is a well-defined neck gland, at least in the skin of the type specimen. The most conspicuous feature, however, is the relatively enormous hind foot. The forelimb is small and delicate; the bones of its three segments weigh only 1 gram, whereas the bones of the hind limb weigh 12 grams.

A peculiar feature of the hopping gait of *Caloprymnus* is that the feet are not brought down in line with one another; rather, the right toe mark registers well in front of the left toe mark. This "rat"-kangaroo seldom dodges or doubles back when moving rapidly and is noted for endurance rather than speed. A young adult male exhausted two galloping horses in a 20-km run. The gait, when the animal moves on all four limbs and uses the tail as a support, is normal for the Potoroidae.

The area known to have been inhabited consists mainly of gibber plains, clay pans, and sandridges. There is a sparse cover of saltbush and other shrubs. *Caloprymnus* is a nest builder, not a burrower, and shelters in simple leaf and grass nests in scratched-out excavations despite the glaring heat of its habitat. It has the unusual habit of protruding its head through a gap in the roof of the nest to observe its surroundings. It apparently is nocturnal (Collins 1973). It has been reported to feed mainly on the foliage and stems of plants and to feed less on roots than do the other "rat"-kangaroos. However, recent examination of the stomach and colon contents of a preserved specimen revealed extensive remains of beetles (Dixon 1988).

Females with a single pouch young have been taken in June, August, and December. The breeding season is apparently irregular: one female had a small, naked young in her pouch at the same time that two other females were carrying well-furred and almost independent "joeys." According to Jones (1982), long ago a specimen lived in captivity for 13 years.

The desert "rat"-kangaroo apparently was rare for many years but then became fairly common in its restricted range when severely dry conditions abated in the early 1930s. By 1935, however, it again was rare, and there have been no reliable records since that year. Possibly it is now extinct, or perhaps a small population still survives, awaiting the time when it again may increase in response to proper conditions. The species is designated as extinct by the IUCN and as endangered by the USDI and is on appendix 1 of the CITES.

DIPROTODONTIA; POTOROIDAE **Genus**
AEPYPRYMNUS
Garrod, 1875

Rufous "Rat"-kangaroo

The single species, *A. rufescens*, formerly occurred from northeastern Queensland to northeastern Victoria (Ride 1970). Remains have been found in cave deposits in southwestern Victoria and on Flinders Island, near Tasmania (Calaby 1971).

Desert "rat"-kangaroo *(Caloprymnus campestris)*, from *The Red Centre*, Hedley H. Finlayson (Angus & Robertson, publishers).

Rufous "rat"-kangaroo *(Aepyprymnus rufescens)*, photo by John Warham.

This is the largest of the "rat"-kangaroos. Head and body length is 380–520 mm and tail length is 350–400 mm. According to Johnson (1978), adults stand 350 mm tall, and weight in a series of specimens ranged from 2.27 to 2.72 kg for males and 1.36 to 3.60 kg for females. The pelage is crisp and often harsh. The upper parts, grizzled in appearance, are rufescent gray, and the underparts are whitish. The tail is thick, evenly haired, and not crested. This genus can be distinguished by its ruddy color, black-backed ears, whitish hip stripe (usually not distinct), and hairy muzzle. Females have four mammae in a well-developed pouch.

At present this "rat"-kangaroo is found from sea level to the tops of plateaus, in open forest and woodland with a dense grass floor. It is nocturnal and constructs nests in which it shelters during the day. These nests consist of a shallow excavation lined and covered with grass or bark (Seebeck, Bennett, and Scotts 1989). An individual constructs one or two clusters of nests that provide protection from the sun and concealment from predators (Wallis et al. 1989). Although not particularly fast, *Aepyprymnus* is wonderfully agile, adept at dodging, and difficult to approach on foot. On horseback, however, a person may get quite close. When startled it usually seeks shelter in a hollow log, if available. It has little fear of people at night, is often attracted by a bush camp, and, if not molested by dogs, can be enticed up to a tent door to receive scraps of food. When taken young and treated kindly, it becomes tame and responsive. In the wild its diet consists mainly of grasses (Johnson 1978), but in captivity it accepts a variety of foods and apparently thrives on them. Like most of the coastal species, it has little resistance to drought; during dry periods it excavates holes in creek beds to reach the water level.

The rufous "rat"-kangaroo is generally solitary, though a female often is accompanied by a nearly full-grown offspring and loose feeding aggregations sometimes form. Captive males are extremely aggressive toward one another. The species is polyestrous, with an estrous cycle of about 21–25 days. Breeding apparently occurs over most of the year. In the Dawson Valley of central coastal Queensland females with pouch young were taken from January to March. Near Killarney in southeastern Queensland three females with pouch young were caught in June, and local residents reported finding pouch young all year. Females in a captive colony at Canberra had large pouch young in February, July, and August and smaller ones in July, August, and December. The gestation period is 22–24 days. Litters usually consist of a single offspring, but twins occasionally occur. The pouch young releases the nipple at 7–8 weeks, vacates the pouch permanently at about 16 weeks, and then remains with the mother for another 7 weeks. Sexual maturity is attained at about 11 months in females and 12–13 months in males (P. M. Johnson 1980; P. M. Johnson *in* Strahan 1983; Moors 1975; Rose 1989). One specimen reportedly lived more than eight years in captivity (Collins 1973).

The rufous "rat"-kangaroo still is widely distributed along the coast from northeastern Queensland to central New South Wales. It is fairly common in much of this region and seems able to coexist with grazing beef cattle (Calaby 1971). Farther south the species has disappeared, the last record in Victoria dating 1905. Its decline may have resulted from predation by introduced foxes. Because of the continuing pressures on its habitat it was designated as "potentially vulnerable" by Kennedy (1992).

DIPROTODONTIA; Family **MACROPODIDAE**

Wallabies and Kangaroos

This family of 12 Recent genera and 61 species is native to Australia, Tasmania, New Guinea, some nearby islands,

Western gray kangaroo *(Macropus fuliginosus)* and her young, about 12 months old, photo by Mary Eleanor Browning.

and the Bismarck Archipelago. The Potoroidae, or "rat"-kangaroos (see account thereof), sometimes have been considered a subfamily of the Macropodidae but now are generally considered a separate family. Flannery (1983, 1989b), whose phylogenetic sequence is followed here, recognized two living macropodid subfamilies: Sthenurinae, with the single living genus *Lagostrophus*, and Macropodinae for the other 10 genera. Analysis of albumin immunologic relationships (Baverstock et al. 1989) supports the same basic division but suggests certain sequential alternatives, especially an association of *Petrogale* with *Dorcopsulus*. A somewhat different arrangement, suggested by the work of Kirsch (1977b), would put the genus *Onychogalea* in a position intermediate to the Potoroidae and Macropodidae, while the genera *Macropus* and *Wallabia* would form one main line at the opposite end of the family from the Potoroidae; the genus *Lagorchestes* would form another main line, and all the other genera would compose a fourth branch of the Macropodidae.

The adults in this family vary in head and body length from less than 300 mm to as much as 1,600 mm in *Macropus*. Full-grown individuals of the latter genus weigh as much as 100 kg (Hume et al. 1989). The head is rather small in relation to the body, and the ears are relatively large. The tail is usually long, thick at the base, hairy, and nonprehensile. It is used as a prop or additional leg, as a balancing organ when leaping, and sometimes for thrust. In all Recent members of this family except *Dendrolagus* the hind limbs are markedly larger and stronger than the forelimbs. The forelimbs are small with five unequal digits. The hind foot is lengthened and narrowed in all genera, hence the family name Macropodidae, meaning "large foot." Digit 1 of the hind foot is lacking in all genera, the small digits 2 and 3 are united by skin (the syndactylous condition), digit 4 is long and strong, and digit 5 is moderately long. The well-developed marsupial pouch opens forward and encloses four mammae.

In kangaroos and wallabies the dentition is suitable for a grazing or browsing diet. The first upper incisors are prominent and with the other incisors form a U- or V-shaped arcade; the two remaining lower incisors are very large and forward-projecting (the diprotodont condition), and except in the Sthenurinae their tips fit within the upper incisor arcade and press on a pad on the front of the palate; there are no lower canines, and the upper canines are small or absent, thus leaving a space between the incisors and cheek teeth; the premolars are narrow and bladelike; and the molars are broad, generally high-crowned, rectangular teeth with two transverse ridges separated by a deep trough crossed by a longitudinal ridge that is weakly formed in the browsing species but strong in grazers (Hume et al. 1989). In many macropodid genera, as in elephants, manatees, and certain pigs, an anterior migration of the molariform teeth occurs throughout life, making room for the late-erupting rear molars. The fourth molar may not erupt until well after adulthood is attained. The dental formula in this family is: (i 3/1, c 0–1/0, pm 2/2, m 4/4) × 2 = 32 or 34.

Most species are nocturnal, but they may sunbathe on warm afternoons and some are active at intervals during

Tree kangaroo (*Dendrolagus goodfellowi*), photo from Zoological Society of London.

the day. All the members of this family except *Dendrolagus* progress rapidly by leaps and bounds, using only the hind limbs. Representatives of the genus *Macropus* illustrate the peak of development of the jumping mode of progression. Macropodids commonly move on all four limbs and often use the tail as a support when feeding, progressing slowly, or standing erect.

The members of this family are mainly grazers or browsers, feeding on many kinds of plant material. The occurrence of ruminantlike bacterial digestion in the kangaroos and wallabies enables them to colonize areas that would be nutritionally unfavorable to most other large mammals. In this kind of digestion the food is fermented by a dense bacterial population in the esophagus, stomach, and upper portion of the small intestine, thus providing the available energy for chemical breakdown of foods over a longer period of time and enhancing the uptake of nitrogen and other nutrients.

Females of most macropodid genera usually give birth to one young at a time. A phenomenon known as embryonic diapause, or delayed birth, approximately equivalent to delayed implantation in placental mammals, occurs in some, perhaps most, members of this family. Following the birth of one young the female often mates again. Usually this mating occurs only a day or two after the birth, but in some species it is late in the pouch life of the young. In one species, *Wallabia bicolor*, the mating takes place just before the young is born (Kaufmann 1974). The development of the embryo resulting from this mating is arrested at about the 100-cell stage. The embryo remains in this state of diapause until the first young nears the end of its pouch life, dies, or is abandoned. At such a time the embryo resumes development, and a second young is soon produced. This reproductive approach seems well adapted to the variable, often severe climate of inland Australia.

Should adverse conditions lead to the death of a pouch young or force the female to discard it, a successor is in reserve for another try at rearing during the season. If conditions are consistently favorable, both offspring can be raised, the second being born shortly after the first permanently vacates the pouch. Other views regarding embryonic diapause are that it functions simply to prevent two young from crowding the pouch at the same time or that it helps to ensure synchrony between the embryo and the uterus (Russell 1974*a*). A feature unique to the macropodids and potoroids is the capability of females to produce milk in one mammary gland that has a very different nutritional composition from that of the other gland when young at different stages of development are being reared (Merchant 1989).

Modern humans have greatly affected many species of this family. Perhaps the single most important instance was the introduction of vast herds of domestic livestock, which cropped grasslands, thereby eliminating the cover needed by smaller macropodids. Some of the larger species of *Macropus,* however, appear to have at least temporarily benefited from the elimination of tall, dry grass by livestock and the resultant production of more favorable vegetation (Newsome 1975).

The geological range of the family Macropodidae is Miocene to Recent in Australasia (Kirsch 1977*a*).

DIPROTODONTIA; MACROPODIDAE; **Genus LAGOSTROPHUS**
Thomas, 1887

Banded Hare Wallaby, or Munning

The single species, *L. fasciatus,* occupied Western Australia, including islands in Shark Bay. It may also have been present in historical time in South Australia, where its remains have been found at an archeological site on the lower Murray River (Calaby 1971).

Head and body length is 400–460 mm and tail length is 320–400 mm. Prince (*in* Strahan 1983) listed weight as 1.3–2.1 kg, occasionally up to 3.0 kg. The fur is thick, soft, and long. *Lagostrophus* can be readily distinguished by its banded color pattern. The dark transverse bands on the posterior half of the body contrast sharply with the general grayish coloration. The underparts are buffy white with gray hair bases, and the hands, feet, and tail are gray. This species, like those in the genus *Lagorchestes,* is referred to as a hare wallaby because of its harelike speed, its jumping ability, and its habit of crouching in a "form." The muzzle in *Lagostrophus* is rather long and pointed, and the nose is naked rather than hairy. The tail is evenly haired throughout, except for an inconspicuous pencil of longer hairs at the tip. The claws of the hind feet are hidden by the fur. Unlike the condition in all other genera of the Macropodidae, the lower incisors of *Lagostrophus* occlude with the upper incisors, not with a pad between the upper incisors (Hume et al. 1989). Additional technical characters of the dentition that distinguish *Lagostrophus* were listed by Flannery (1983, 1989*b*).

On the mainland this marsupial inhabits prickly thickets on the flats and the edges of swamps. On the islands it lives in thickets of a thorny species of *Acacia.* Runs and "forms" are made in these tangled masses. The species is nocturnal, emerging at night from retreats in the scrub to feed on various plants and fruits.

Banded hare wallaby *(Lagostrophus fasciatus)*, photo by A. G. Wells / National Photographic Index of Australian Wildlife.

The banded hare wallaby is gregarious, congregating in spaces under the low-hanging limbs of bushes in dense thickets (Ride 1970). According to Collins (1973), available data indicate that *Lagostrophus* is a seasonal breeder, with a reproductive peak in the first half of the year. A postpartum estrus may occur, followed by embryonic diapause while the first young still is suckling. Normal litter size is one, but indications of twin offspring were present in one specimen. The Australian National Parks and Wildlife Service (1978) stated that in the island populations young animals of all ages can be observed at any one time, thus suggesting that the breeding season is an extended one, occurring at least from February to August. According to Prince (*in* Strahan 1983), young spend about six months in the pouch and are weaned after another three months; both sexes are capable of breeding in the first year but usually do not do so until the second year.

The banded hare wallaby has not been recorded on the mainland since 1906. Its disappearance may have been associated with clearing of vegetation for agriculture, competition for food with rabbits and livestock, and predation by introduced foxes. It still is found on Bernier and Dorre islands, where it is now well protected, and its populations fluctuate between high numbers and relative scarcity (Ride 1970; Thornback and Jenkins 1982). Surveys by Short and Turner (1992) indicated minimum numbers of about 3,900 on Bernier and 3,800 on Dorre Island. Earlier, when sheep were brought temporarily to Bernier Island, the number of banded hare wallabies was reduced severely. A population on nearby Dirk Hartog Island disappeared entirely in the 1920s following the establishment of sheep. In the 1970s an unsuccessful attempt was made to reintroduce *L. fasciatus* on that island (Poole 1979); further efforts are being planned (Kennedy 1992; Short et al. 1992). The species is designated as endangered by the USDI and is on appendix 1 of the CITES. The IUCN classifies the mainland subspecies *L. f. albipilis* as extinct and the island subspecies *L. f. fasciatus* as vulnerable.

DIPROTODONTIA; MACROPODIDAE; Genus DORCOPSIS
Schlegel and Müller, 1842

New Guinean Forest Wallabies

Four species now are recognized (Groves and Flannery 1989):

D. luctuosa, coastal lowlands of southeastern New Guinea;
D. muelleri, lowlands of western New Guinea, including Misool, Salawatti, and Japen islands;
D. hageni, lowlands of northern New Guinea;
D. atrata, Goodenough Island off southeastern New Guinea.

The genus *Dorcopsulus* (see account thereof) sometimes has been considered a subgenus or synonym of *Dorcopsis.* The name *D. veterum* sometimes has been used in place of *D. muelleri,* but the latter designation was recommended by George and Schürer (1978) and accepted by Groves and

New Guinean forest wallaby (*Dorcopsis* sp.), photo from Zoological Society of London.

Flannery (1989). Groves (*in* Wilson and Reeder 1993) reported the range of *D. muelleri* to extend to the Aru Islands. Skeletal remains indicate that *D. muelleri* or a closely related species was present on Halmahera Island in the North Moluccas until at least 1,870 years B.P. (Flannery, Bellwood, et al. 1995).

Head and body length is 340–970 mm and tail length is 270–550 mm. Flannery (1990*b*) listed weights of 3.6–11.6 kg for *D. luctuosa*, 5.0 kg for *D. muelleri*, and 5.0–6.0 kg for *D. hageni*. Adults of *D. atrata*, a medium-sized species, weigh about 3.9–7.5 kg. The pelage is short and sparse in *D. hageni* but long and thick in the other species. In *D. muelleri* the coloration of the upper parts is dull brown or blackish brown tipped with light buff and the underparts are gray or white. *D. hageni* is light brown or fuscous above with a narrow white dorsal stripe and grayish white below. *D. luctuosa* is dark gray above and drab gray to creamy orange below. In *D. atrata* the upper parts are black or blackish brown and the underparts are also blackish brown. In all species the nose is large, broad, and naked; the ears are small and rounded; and the hairs are reversed on the nape. The tail is evenly haired except for a fifth or less of the terminal half, which is naked. Females have four mammae and a well-developed pouch that opens forward.

On New Guinea these wallabies generally inhabit lowland rainforests up to 400 meters in elevation. *D. atrata* of Goodenough Island lives in oak forest at elevations of 1,000–1,800 meters in the forested mountains (Flannery 1995). These animals do not appear to be as adapted for hopping as most other wallabies. They are presumed to be mainly nocturnal, but there is some evidence that they move about in daytime in dense forest. Observations of a captive colony revealed a crepuscular pattern of activity (Bourke 1989). The diet consists of roots, leaves, grasses, and fruit. Ganslosser (*in* Grzimek 1990) pointed out a number of unusual behaviors of *Dorcopsis*. When it is sitting or hopping slowly only the end of the tail is in contact with the ground. During mating the male bites the female's neck, a primitive habit also seen in the Didelphidae and Dasyuridae.

Observations of *D. luctuosa* in captivity indicated little agonistic behavior and a tendency to form social groups (Bourke 1989). Females usually give birth to one young at a time (Collins 1973). *D. muelleri* and *D. atrata* may breed throughout the year (Flannery 1995). According to Flannery (1990*b*), naked pouch young of *D. hageni* were found in January and April, a captive male *D. luctuosa* first emerged from the pouch on 22 May 1983 and reached sexual maturity by June 1985, and a specimen of *D. muelleri* was born in captivity on 16 October, first left the pouch on 13 April, and permanently left on 17 May. Another captive *D. muelleri* lived for 7 years and 7 months (Jones 1982).

Natives on Goodenough Island regard *D. atrata* as a valuable food animal. G. G. George (1979) considered it a threatened species, and the IUCN now classifies it as endangered. Thornback and Jenkins (1982) indicated that it is susceptible to hunting and possibly habitat destruction. Flannery (1995) noted that it occupies less than 100 sq km of habitat and is heavily hunted but may still be common.

DIPROTODONTIA; MACROPODIDAE; **Genus**
DORCOPSULUS
Matschie, 1916

New Guinean Forest Mountain Wallabies

There are two species (Flannery 1989*b*, 1990*b*):

D. vanheurni, mountains of New Guinea;
D. macleayi, mountains of southeastern Papua New
  Guinea.

Kirsch and Calaby (1977) suggested that it would be reasonable to put these two species in the genus *Dorcopsis*, and such was done by Ziegler (1977) and Corbet and Hill (1991). Kirsch and Calaby (1977) also stated that *D. van-*

heurni probably is conspecific with *D. macleayi*. Nonetheless, *Dorcopsulus* was accepted as a distinct genus, and *D. vanheurni* as a separate species, by Flannery (1989*b*, 1990*b*) and Groves (*in* Wilson and Reeder 1993).

Head and body length is 315–460 mm, tail length is 225–402 mm, and weight is 1,500–3,400 grams (Flannery 1990*b*). Both species are deep gray-brown above with a darker mark above the hips and light brownish gray below. The chin, lips, and throat are whitish. The pelage is long, thick, soft, and fine. The tail is evenly haired, the terminal quarter to half being bare with a small white tip. Females have four mammae and a well-developed pouch that opens forward. From *Dorcopsis, Dorcopsulus* is distinguished by being smaller and more densely furred and in having more of the tail naked. Flannery (1990*b*) noted that with its small size, low-crowned molar teeth, and elongate premolars *D. vanheurni* offers a striking parallel with several members of the Potoroidae in Australia and may fill the same ecological niche.

These wallabies inhabit mountain forests, the known altitudinal ranges being 800–3,100 meters for *D. vanheurni* and 1,000–1,800 meters for *D. macleayi*. Although they are presumed to be nocturnal, Lidicker and Ziegler (1968) observed that *D. vanheurni* apparently is active by day. *D. macleayi* is reported to be fond of the fruit and leaves of *Ficus* and various other trees (Flannery 1990*b*). Two adult female *D. vanheurni* collected by Lidicker and Ziegler (1968) in October were each carrying a single pouch young and did not have visible embryos. A female *D. macleayi* taken in January had two pouch young, and one taken in March had a single pouch young (Flannery 1990*b*).

Lidicker and Ziegler (1968) expressed concern about the ease with which *D. vanheurni* could he caught by small domestic dogs. Flannery (1990*b*) indicated that it was hunted by large-scale drives and burning of the forest. G. G. George (1979) considered *D. macleayi* to be a threatened species, and Thornback and Jenkins (1982) indicated that it is susceptible to hunting and possibly habitat destruction; the IUCN now classifies it as vulnerable.

DIPROTODONTIA; MACROPODIDAE; **Genus**
DENDROLAGUS
Schlegel and Miller, 1839

Tree Kangaroos

There are 10 species (Flannery 1989*b*, 1990*b*, 1993*b*, 1995; Flannery, Boeadi, and Szalay 1995; Flannery and Seri 1990; Groves 1982*c*):

D. inustus, northern and western New Guinea, Japen and possibly Waigeo and Salawatti islands;

D. lumholtzi, northeastern Queensland;

D. bennettianus, northeastern Queensland;

D. ursinus, far western New Guinea;

D. matschiei, Huon Peninsula of eastern Papua New Guinea and (perhaps through human introduction) nearby Umboi Island;

D. spadix, south-central Papua New Guinea;

D. goodfellowi, eastern New Guinea;

D. mbaiso, Maokop (Sudirman Range) of west-central New Guinea;

D. dorianus, far western, central, and southeastern New Guinea;

D. scottae, Toricelli Mountains of north-central New Guinea.

DENDROLAGUS

Dusky tree kangaroo *(Dendrolagus ursinus)*, photo from New York Zoological Society. Inset: Lumholtz's kangaroo *(D. lumholtzi)*, photo from Denver Museum of Natural History.

Groves (1982*c*) originally considered *D. spadix* and *D. goodfellowi* subspecies of *D. matschiei*, and this arrangement was used by Corbet and Hill (1991). Later, however, Groves (*in* Wilson and Reeder 1993) followed Flannery and Szalay (1982) and Flannery (1989*b*, 1990*b*) in accepting both as full species.

Head and body length is 520–810 mm and tail length is 408–935 mm. Flannery (1990*b*) listed weights of 6.5–14.5 kg, and Flannery and Seri (1990) noted that the maximum recorded weight for a tree kangaroo is 20.0 kg in a wild-caught *D. dorianus*. The pelage usually is fairly long; in some forms it is soft and silky, in others coarse and harsh. Coloration in *D. ursinus* is blackish, brown, or gray above and white or buff below. *D. scottae* is uniformly blackish (Flannery and Seri 1990), *D. spadix* uniformly brownish, and *D. inustus* grizzled in color (Flannery 1990*b*). Both *D. dorianus* and *D. bennettianus* are some shade of brown over most of the body. In *D. lumholtzi* the upper parts are grayish or olive buff, the underparts are white, and the feet are blackish. *D. matschiei* is among the most brilliantly colored of marsupials: its back is red or mahogany brown, its face, belly, and feet are bright yellow, and its tail is mostly yellow. *D. goodfellowi* is similar but also has yellow lines along each side of the spine, and the tail is mostly the same color as the dorsum with some yellow markings (Lidicker and Ziegler 1968). *D. mbaiso* is generally dark above but has distinctly white facial markings and underparts (Flannery, Boeadi, and Szalay 1995).

The forelimbs and hind limbs are of nearly equal proportions. The cushionlike pads on the large feet are covered with roughened skin, and some of the nails are curved. The long, well-furred tail is of nearly uniform thickness and acts as a balancing organ; it is not prehensile but is often used to brace the animal when climbing. The thick fur on the nape and sometimes on the back grows in a reverse direction and apparently acts as a natural water-shedding device as the animals sit with the head lower than the shoulders. Females have a well-developed pouch and four mammae.

Tree kangaroos dwell mainly in mountainous rainforests. Reported altitudinal ranges (in meters) are: *D. inustus*, 100–1,400; *D. bennettianus*, 450–760, *D. ursinus*, 50–2,000; *D. matschiei*, 1,000–3,300; *D. spadix*, 0–800; *D. goodfellowi*, 1,200–2,900; *D. mbaiso*, 3,250–4,200; *D. dorianus*, 600–4,000; and *D. scottae*, 1,200–1,400 (Flannery 1990*b*; Flannery, Boeadi, and Szalay 1995; Flannery and Seri 1990; Ride 1970; Ziegler 1977). They are very agile in trees and reportedly are active both day and night. Often they travel rapidly from tree to tree, leaping as much as 9 meters downward to an adjoining tree. They also jump to the ground from remarkable heights, up to 18 meters or perhaps even more, without injury. They shelter in small groups in trees during the day and spend much of their life there but also frequently descend to the ground. When descending trees they usually back down, unlike the possums (Phalangeridae). On the ground they progress by means of relatively small leaps, leaning well forward to counterbalance the long tail, which is arched upward. Recent studies suggest that *D. dorianus*, the heaviest arboreal marsupial, may actually have become mainly a ground-dweller (Moeller *in* Grzimek 1990) but that *D. lumholtzi* spends only 2 percent of its time on the ground, the remainder being spent in the middle and upper layers of the canopy (Hutchins and Smith 1990). The diet consists mainly of leaves and fruit, obtained either in trees or on the ground. Captives have been observed to eat chickens (Flannery 1990*b*).

There is evidence that related female *D. dorianus* form coalitions that interact in a friendly manner and cooperate in agonistic displays toward unfamiliar males (Ganslosser 1984). Recent field studies of *D. lumholtzi* indicate that it is relatively solitary and nongregarious; females seem to occupy nonoverlapping home ranges of about 1.8 ha., several of which are overlapped by the range of a territorial male (Hutchins and Smith 1990). Groups of female *D. lumholtzi* with a single male can be maintained amicably in captivity, but two males in the presence of females will fight savagely. The species appears to have no definite mating season (P. M. Johnson *in* Strahan 1983). Based on births in captivity, there seems to be no well-defined breeding season in *D. matschiei*; the gestation period is approximately 32 days (Olds and Collins 1973). New Guinea females of various species have been found with pouch young in April, May, September, October, and December (Flannery 1990*b*).

Tree kangaroo *(Dendrolagus mbaiso)*, photo by Tim Flannery.

The average litter size is one. A young captive individual was observed first to emerge completely from the pouch at about 305 days and to suckle with only the head in the pouch at about 408 days (Collins 1973). Tree kangaroos seem capable of long life in captivity; the record for *D. ursinus* is 20 years and 2 months (Collins 1973), and specimens of *D. goodfellowi* and *D. matschiei* were still living at 21 years and 23 years and 10 months, respectively (Marvin L. Jones, Zoological Society of San Diego, pers. comm., 1995).

Clearing of rainforest in northeastern Queensland has considerably reduced the range of *D. bennettianus* and *D. lumholtzi*, especially that of the latter species. Although both still appear to be common in protected reserves, there is concern that habitat fragmentation may prevent dispersal and increase inbreeding (Calaby 1971; Hutchins and Smith 1990; Poole 1979). In New Guinea there have been widespread declines of most species mainly because of hunting by people for use as food, and such problems are intensifying with the spread of modern weapons and other technologies into remote areas (Flannery 1990*b*; G. G. George 1979; Thornback and Jenkins 1982). The recently discovered *D. scottae* may number in the low hundreds in a restricted habitat of only 25–40 sq km and is gravely endangered by hunting and disturbance (Flannery and Seri 1990). The IUCN now classifies *D. matschiei*, *D. goodfellowi*, and *D. scottae* as endangered, *D. mbaiso* and *D. dorianus* as vulnerable, and *D. lumholtzi* and *D. bennetianus* as near threatened. *D. inustus* and *D. ursinus* are on appendix 2 of the CITES.

DIPROTODONTIA; MACROPODIDAE; **Genus SETONIX**
Lesson, 1842

Quokka

The single species, *S. brachyurus*, inhabits southwestern Australia, including Rottnest Island, near Perth, and Bald Island, near Albany (Ride 1970).

Head and body length is about 475–600 mm, tail length is 250–350 mm, and adult weight is 2–5 kg. The hair is short and fairly coarse, and the general coloration is brownish gray, sometimes tinged with rufous. There are no definite markings on the face. The ears are short and rounded. The tail, which is only about twice as long as the head, is sparsely furred and short. Females have four mammae and a well-developed pouch.

On islands the quokka occurs in a variety of habitats with sufficient cover, but on the mainland it seems to be restricted to dense vegetation in swamps amidst dry sclerophyll forest (Ride 1970). An important factor in the quokka's ecology, at least on Rottnest Island, is the diurnal shelter in a thicket or some other shady location where the animal can avoid the heat (Nicholls 1971). An individual returns to the same shelter every day through most of the year but may change sites in May or June. At night the quokka emerges from its shelter to feed. It makes runways and tunnels through dense grass and undergrowth. When moving quickly it hops on its hind legs; when moving slowly it does not use its tail as a third prop for the rear end of the body as do kangaroos and the larger wallabies. Like most macropodids, it is terrestrial, but it can climb to reach twigs up to about 1.5 meters above the ground (Ride 1970). The quokka is herbivorous, feeding on a variety of plants.

Ruminantlike digestion in the Macropodidae was first demonstrated in *Setonix*. The pregastric bacterial digestion in the quokka is much like that in sheep, for most of the 15 or so morphological types of bacteria present in the large stomach region of the quokka are comparable to those in the rumen of sheep. Most of the bacterial fermentation in the stomach of the quokka takes place in the sacculated part. This wallaby seems to occupy a position intermediate to the ruminant and the nonruminant herbivores in its efficient digestion of fiber and the rate of food passage.

During the wet season the feeding area or home range used by a quokka on Rottnest Island covers about 10,000–125,000 sq meters. In the dry season (November–April) this area increases to 20,000–170,000 sq meters (Nicholls 1971). In addition, some individuals move up to 1,800 meters to soaks or fresh-water seepages during the summer since there is almost no free surface water on Rottnest Island at that time. Other individuals do without water (Ride 1970). Population density is 1/0.4 ha. to 1/1.2 ha. (Main and Yadav 1971).

Kitchener (1972) found that on Rottnest Island the quokka population is organized into family groups. Adult males dominate the other members of the family and also form a linear hierarchy among themselves that is usually stable. On hot summer days, however, the adult males may fight intensively for possession of the best shelter sites. Apparently the availability of such sites, rather than food, is

Quokka *(Setonix brachyurus)*, photo from New York Zoological Society.

the main factor in limiting the population. Other than the conflict for shelters there is little evidence of territoriality, and groups of 25–150 individuals may have overlapping home ranges (Nicholls 1971). During the summer some parts of the population concentrate around available fresh water (Ride 1970).

Reproduction in the quokka has been studied in some detail (Collins 1973; Ride 1970; Rose 1978; Shield 1968). Females are polyestrous, with an average estrous cycle of 28 days. They are capable of breeding throughout the year in captivity, but in the wild anestrus occurs from about August through January. On Bald Island most births take place in March or April. Nondelayed gestation is 26–28 days. Litter size is normally one, and in the wild usually only one young can be successfully reared each year. One day after the young is born, however, the female mates again. Embryonic diapause, which was first demonstrated in *Setonix*, then occurs. If the young already in the pouch should die within a period of about 5 months, the embryo resumes development and is born 24–27 days later. If the first young lives, the embryo degenerates when the female enters anestrus. In captivity or under unusually good conditions in the wild the second embryo can resume development even if the first young is successfully raised. The young initially leaves the pouch at about 175–95 days but will return if alarmed or cold and will continue to suckle for three to four more months. Earliest recorded sexual maturity is 389 days for males and 252 days for females. In the wild on Rottnest Island, however, it is unlikely that females give birth until well into their second year. According to Shield (1968), a number of captive females lived for more than 7 years, a marked 10-year-old wild female carried a young, and a wild male lived more than 10 years.

Until the 1930s the quokka was very common in coastal parts of the mainland of southwestern Australia. Subsequently it disappeared, except for a few small colonies on the mainland and the two relatively numerous island populations (Calaby 1971). The Bald Island population is thought to number 200–600 individuals (Main and Yadav 1971). The total number on Rottnest is larger, but Ride (1970) expressed concern for the future of the species there

because of the development of the island for recreational purposes. There recently has been some recovery of the mainland population (Kitchener *in* Strahan 1983), but an attempt to reintroduce it to a reserve near Perth was unsuccessful (Short et al. 1992). The IUCN now classifies the quokka as vulnerable, noting that it has declined by at least 50 percent over the last decade and now numbers fewer than 2,500 mature individuals. The species is listed as endangered by the USDI.

DIPROTODONTIA; MACROPODIDAE; **Genus THYLOGALE**
Gray, 1837

Pademelons

Six species are reported to exist (Flannery 1992, 1995; Ride 1970; Ziegler 1977):

T. billardierii (red-bellied pademelon), southeastern South Australia, Victoria, Tasmania, islands of Bass Strait;

T. thetis (red-necked pademelon), eastern Queensland, eastern New South Wales;

T. stigmatica (red-legged pademelon), south-central New Guinea, eastern Queensland, eastern New South Wales;

T. brunii (dusky pademelon), southern New Guinea, Aru and Kei Islands;

T. browni, northern and eastern New Guinea, Bismarck Archipelago;

T. calabyi, southeastern Papua New Guinea.

Hope (1981) described an additional species, *T. christenseni* from remains at an archeological deposit in western New Guinea and stated that both that species and *T. brunii* survived in that area until less than 5,000 years ago. Flannery (1990b) suggested the possibility that both might yet persist there. Maynes (1989) indicated that the presence of *T. brunii* on the Kei Islands and of *T. browni* on Umboi and New Britain in the Bismarck Archipelago might be attrib-

Red-legged pademelons *(Thylogale stigmatica)*, photo by Ernest P. Walker.

utable to introduction by the Melanesians. Flannery (1995) stated that *T. browni* was brought by people to New Ireland in the Bismarck Archipelago about 7,000 years ago. Van Gelder (1977b), on the basis of captive hybridization between *T. thetis* and *Macropus rufogriseus*, recommended that *Thylogale* be considered a synonym of *Macropus*.

Head and body length is about 290–670 mm, tail length is 246–510 mm, and weight is 1.8–12.0 kg (Flannery 1990b; K. A. Johnson *in* Strahan 1983; Johnson and Rose *in* Strahan 1983; P. M. Johnson *in* Strahan 1983). The fur is soft and thick. *T. thetis* is grizzled gray above, becoming rufous on the shoulders and neck, and often has a light hip stripe. In *T. stigmatica* the upper parts are mixed gray and russet and the yellowish hip stripe is conspicuous. *T. brunii* is gray brown to chocolate brown above with a dark cheek stripe from behind the eye to the corner of the mouth, above which is a white area; it also has a prominent hip stripe. The underparts in these three species are considerably paler than the dorsal parts. The species *T. billardierii* is grayish brown above tinged with olive; a yellowish hip stripe often is evident, and the undersurface is rufous or orange.

This genus is distinguished by features of the incisor teeth; a groove on the third upper incisor is positioned almost at the rear of the crown (Flannery 1990b). The comparatively short tail is sparsely haired and thickly rounded. Females have four mammae and a well-developed pouch.

Habitats include rainforest, sclerophyll forest, savannah, thick scrub, and grassland. In New Guinea *T. brunii* has an altitudinal range of 0–4,200 meters (Ziegler 1977), but *T. stigmatica* is found only from sea level to 60 meters (Flannery 1990b). Some species are nocturnal or crepuscular, sleeping by day in dense thickets and under bushes. However, K. A. Johnson (1980) found *T. thetis* generally to spend the night grazing in pasture and much of the day traveling widely through the rainforest to seek food and sites for basking.

Pademelons may move about in undergrowth through tunnel-like runways. They are inoffensive and apparently quite curious, as they sometimes will allow a close approach before bounding away. They have a habit, comparable to that of rabbits, of thumping on the ground with their hind feet, presumably as a signal or warning device. In the absence of predators or other disturbing factors they graze in the dusk and early morning on grass near thickets, into which they hop quickly when alarmed. The diet also includes leaves and shoots.

Captive pademelons may be maintained in groups (Collins 1973). *T. billardierii* reportedly is gregarious in the wild. In captivity males of this species were observed to make numerous aggressive displays but not actually to fight. Females also were aggressive toward one another, sometimes fought, and seemed to form a linear dominance hierarchy (Morton and Burton 1973). This species is a seasonal breeder in Tasmania, with most births in April, May, and June (Rose and McCartney 1982). *T. brunii* apparently breeds throughout the year in New Guinea (Flannery 1990b). Births in captivity have been recorded in January, February, and July for *T. billardierii* and in September for *T. thetis*. Embryonic diapause is known to occur in these two species. In *T. billardierii* the estrous cycle averages 30 days, nondelayed gestation is 29.6 days, and the period from the end of embryonic diapause to birth is 28.5 days (Rose 1978). Litter size usually is one, but twins have been recorded (Collins 1973). Pouch life of *T. billardierii* is 202 days, weaning occurs about 4 months after the young permanently leave the pouch, and sexual maturity comes at about 14–15 months (Morton and Burton 1973; Rose and

McCartney 1982). A specimen of *T. billardierii* lived in captivity for 8 years and 10 months (Collins 1973).

Johnson and Vernes (1994) maintained captive colonies of one adult male and eight adult female *T. stigmatica* and found births to occur throughout the year in captivity. Supplementary studies indicated that young were produced in the wild in Queensland from October to June. A postpartum estrus and mating generally followed birth. The estrous cycle was 29–32 days, gestation was 28–30 days, the pouch was permanently vacated at 184 days, weaning occurred an average of 66 days later, and mean age at sexual maturity was 341 days for females and 466 days for males.

The species *T. billardierii* disappeared from the mainland of Australia during the nineteenth century and also has been exterminated on some of the islands of Bass Strait. It still survives on other of these islands and is common in many parts of Tasmania (Calaby 1971), where it is sometimes considered a pest to agriculture and silviculture and is taken in large numbers for its skin (Ride 1970). Population levels now are thought to be high enough to sustain the regulated hunting and control operations that are carried out (Poole 1978). The distribution of *T. thetis* and *T. stigmatica* has been reduced in Australia by land clearance for agriculture, dairying, and forestry, but these two species still are common in some areas (Calaby 1971; Poole 1979). The status of the New Guinea populations seems to be worse, with widespread disappearances and range fragmentation, apparently caused by excessive hunting by people and their dogs (Flannery 1992). The IUCN now classifies *T. brunii* and *T. browni* as vulnerable and *T. calabyi* as endangered.

DIPROTODONTIA; MACROPODIDAE; Genus PETROGALE
Gray, 1837

Rock Wallabies

There are 14 species (Eldridge and Close 1992, 1993; Eldridge, Johnston, and Lowry 1992; Flannery 1989b; Kitchener and Sanson 1978; Maynes 1982):

P. xanthopus, eastern South Australia, northwestern New South Wales, southwestern Queensland;

P. persephone, east-central coast of Queensland;

P. rothschildi, Pilbara District and Dampier Archipelago of northwestern Western Australia;

P. lateralis, suitable areas of Western Australia, Northern Territory, South Australia, and western Queensland, including many islands off the western and southern coasts but not Tasmania;

P. penicillata, extreme southeastern Queensland, eastern New South Wales, Victoria;

P. herberti, southeastern Queensland;

P. sharmani, east-central coast of Queensland;

P. mareeba, northeastern Queensland;

P. coenensis, northeastern Cape York Peninsula of Queensland;

P. inornata, east-central Queensland;

P. assimilis, northern Queensland;

P. godmani, eastern Cape York Peninsula of Queensland;

P. brachyotis, northeastern Western Australia, northern Northern Territory, including Groote Eylandt;

P. burbidgei, Kimberley area of northeastern Western Australia.

Petrogale often is considered to include the genus and species *Peradorcas* (see account thereof) *concinna*. *P. pur-*

A. Rock wallaby *(Petrogale penicillata)*, photo by Bob McIntyre through Cheyenne Mountain Zoo. B. Ring-tailed rock wallabies *(P. xanthopus)*, photo from Australian News and Information Bureau.

pureicollis of western Queensland sometimes has been treated as a species separate from *P. lateralis* (Kirsch and Calaby 1977). Much of the recent taxonomic work on *Petrogale* has been based on evaluation of chromosomes, with certain populations that appeared morphologically separable being found to have identical karyotypes and with populations that appeared morphologically indistinguishable being designated as distinct though cryptic species based on karyological differences. Some of these designated species have been reported to form hybrid zones, and other data suggest six various alternatives to the above phylogenetic arrangement (Eldridge and Close 1993; Sharman, Close, and Maynes 1990).

Excluding the very small *P. burbidgei*, the genus is characterized as follows: head and body length about 500–800 mm; tail length, 400–700 mm; adult weight, 3–9 kg; and thick, long, dense fur. In *P. burbidgei*, according to Kitchener and Sanson (1978), head and body length is 290–353 mm, tail length is 252–322 mm, and weight is 0.96–1.43 kg.

The general coloration of the upper parts varies from pale sandy or drab gray to rich dark vinaceous brown; the underparts are paler, usually buffy, yellowish gray, yellowish brown, or white. Several species have stripes, patches, or other striking markings. *P. xanthopus*, the ring-tailed or yellow-footed rock wallaby, is one of the most brightly colored members of the kangaroo family. It is grayish with a white cheek stripe, yellow on the back of the ears, a dark streak from between the ears to the middle of the back, brown patches on the yellow limbs, white underparts, and a tail ringed with brown and pale yellow. In *P. rothschildi* and the subspecies *P. lateralis purpureicollis* the upper back becomes a brilliant purple at certain times of the year. In some other subspecies of *P. lateralis* the tail and armpits are black, there is a white cheek stripe, and the ears have a

prominent black patch and whitish margins (Ride 1970). In *P. burbidgei*, according to Kitchener and Sanson (1978), the upper parts are mostly ochraceous tawny and the underparts are ivory yellow.

The tails of rock wallabies are long, cylindrical, bushy, and thickly haired at the tip. They are less thickened at the base than in *Thylogale*, *Macropus*, and *Wallabia* and are used primarily for balancing rather than as props for sitting. The hind foot of *Petrogale* is well padded; the sole is roughly granulated, permitting a secure grip on rock, and edged with a fringe of stiff hair that extends onto the digits. The central hind claws are short, only exceeding the toe pads by 2–3 mm. Females have a well-developed, forward-opening pouch and four mammae.

These wallabies usually inhabit rocky ranges and boulder-strewn outcrops with an associated cover of forest, woodland, heath, or grassland. They are nocturnal, spending their days in rock crevices and caves; occasionally they emerge on warm afternoons for a sunbath. Their agility among the rocks is astonishing; some of their leaps measure up to 4 meters horizontally. They also can scramble up cliff faces and leaning tree trunks with relative ease. The friction of their fur and feet on regular paths, over many generations, imparts a glasslike sheen to limestone. Rock wallabies usually progress in a series of short or long leaps. They travel awkwardly in open country. Their diet consists mainly of grasses, though in dry seasons they can exist for long periods without water by eating the juicy bark and roots of various trees.

The species *P. penicillata* reportedly lives in small groups that have a defended territory, and there is a linear dominance hierarchy among males established through fighting (Russell 1974b). Observations of wild *P. assimilis* indicate that a male and a female may form a stable rela-

Rock wallaby (*Petrogale* sp.), photo by Lothar Schlawe.

tionship involving regular social grooming and the sharing of resting sites and other parts of an exclusive home range (Barker 1990). Females are polyestrous (Rose 1978). In a study of captive *P. penicillata* Johnson (1979) found an estrous cycle of 30.2–32.0 days and a gestation period of 30–32 days. There was usually a postpartum mating, and the resultant embryo developed and was born 28–30 days after the premature removal of the original young. Normal pouch life was determined to be 189–227 days, and sexual maturity was attained at 590 days by males and 540 days by females. The following additional life history data were summarized by Collins (1973). Births in captivity have been recorded for *P. xanthopus* during every month and for *P. penicillata* during March, June, August, and September. Gestation in *P. xanthopus* is 31–32 days. Litter size usually is one, but twins occur on occasion. A specimen of *P. penicillata* lived for 14 years and 5 months in captivity.

While rock wallabies have disappeared from some areas and their numbers have declined considerably in the sheep country of southern Australia, they seem to have maintained themselves better than other small macropodids, and several species are common (Calaby 1971; Poole 1979). Portions of their rocky habitat are inaccessible to sheep and rabbits, which compete for forage, though feral goats are a problem in some areas. Perhaps the species most severely affected by European settlement is *P. xanthopus,* classified as endangered by the USDI and generally as near threatened by the IUCN, which not only suffered through habitat alteration but was heavily hunted for its beautiful pelt. It has disappeared in some areas but still occurs in the Flinders Range of South Australia and is locally common in other areas (Calaby 1971; Copley 1983; Gordon, McGreevy, and Lawrie 1978). The subspecies in the Flinders Range, *P. x. xanthopus,* is classified as vulnerable by the IUCN. The population in Queensland is estimated to contain 5,000–10,000 individuals and to be vulnerable to land development (Gordon et al. 1993). The closely related species *P. persephone* also appears to be rare and endangered (Maynes 1982) and is termed endangered by the IUCN. The species *P. sharmani, P. coenensis,* and *P. burbidgei* are designated near threatened by the IUCN.

The most widely distributed species, *P. lateralis* and *P. penicillata,* have been greatly reduced in numbers and range in both southeastern and southwestern parts of the continent. Despised as agricultural pests and valued for their skins, hundreds of thousands were killed in the nineteenth and early twentieth centuries. Remaining populations are continuing to decline because of various factors, including destruction of vegetative cover by introduced rabbits, competition with introduced goats for food and rock shelters, and predation by introduced foxes (D. J. Pearson 1992; Short and Milkovits 1990); both are designated vulnerable by the IUCN. *P. penicillata* was thought to have disappeared in Victoria by 1905 (Harper 1945), but two small colonies are now known to remain (Emison et al. 1978; Wakefield 1971). In the northern half of Australia most species of *Petrogale* seem to be common; the subspecies *P. lateralis purpureicollis* reportedly is thriving in northwestern Queensland (Poole 1979). The conservation of *P. lateralis, P. rothschildi,* and *P. brachyotis* may be assisted by the presence of numerous natural island populations (Calaby 1971; Main and Yadav 1971; Poole 1979). A population of *Petrogale penicillata* became established on the island of Oahu, Hawaii, in 1916 (Lazell, Sutterfield, and Giezentanner 1984; Maynes 1989).

DIPROTODONTIA; MACROPODIDAE; **Genus**
PERADORCAS
Thomas, 1904

Little Rock Wallaby

The single species, *P. concinna,* is found in the Kimberley area of northeastern Western Australia and in Arnhem Land in the northern Northern Territory. *Peradorcas* has a history of being switched back and forth by various taxonomists, some regarding it as a distinct genus and others placing it in *Petrogale.* Although the latter procedure was suggested by Poole (1979) based on serological and karyological study and has been followed in most recent systematic treatments (Calaby and Richardson *in* Bannister et al. 1988; Corbet and Hill 1991; Flannery 1989*b;* Groves *in* Wilson and Reeder 1993), *Peradorcas* was retained as a full genus by Ganslosser (*in* Grzimek 1990), Nelson and Goldstone (1986), and Sanson (*in* Strahan 1983) based on morphological and behavioral data. Recent

Little rock wallaby *(Peradorcas concinna)*, photo by G. D. Sanson / National Photographic Index of Australian Wildlife.

chromosomal analyses have indicated that if *Peradorcas* is treated as a separate genus, it should include the evidently closely related *Petrogale brachyotis* and *P. burbidgei* even though the latter two species do not possess continually erupting molar teeth, the character on which *Peradorcas* was originally founded (Eldridge, Johnston, and Lowry 1992).

Head and body length is 290–350 mm, tail length is 220–310 mm, and weight is 1.05–1.70 kg (Sanson *in* Strahan 1983). The fur is short, soft, and silky. The upper back is rusty red to grayish, the rump is reddish to orange, the underparts are white or grayish white, and the tail becomes darker toward the tip. The hind foot is well padded and the sole is roughly granulated. Females have four mammae in the pouch.

Peradorcas resembles *Petrogale burbidgei* but is distinguished by its longer ears and dental characters. Its molar dentition is unique among marsupials in that there are supplementary replacement molars behind the last regular molar. The actual number of molar teeth is not known, but study suggests that as many as nine molars may erupt successively and that there are seldom more than five molars in place at any one time.

According to Sanson (*in* Strahan 1983), the little rock wallaby shelters in sandstone crevices and caves during the day in the dry season and emerges at night to forage for ferns that grow on adjacent blacksoil plains. These ferns are extremely abrasive, and this factor may be associated with the development of the continual molar replacement system found in *Peradorcas*. In the wet season, activity is partly diurnal and the diet includes grasses that grow on the margins of the plains that are not inundated.

Breeding probably occurs throughout the year. In studies of a captive colony Nelson and Goldstone (1986) found *Peradorcas* to have a postpartum estrus and embryonic diapause. The estrous cycle lasted 31–36 days, and gestation about 30 days. Dominant females had shorter cycles than did subordinates. The single young opens its eyes at about 110 days, leaves the pouch at about 160 days, and is independent at about 175 days. Weaning is much more sudden

than in *Petrogale*. Sexual maturity is attained in the second year.

Peradorcas inhabits a relatively small area that is subject to some environmental stress. The IUCN includes *P. concinna* in the genus *Petrogale* and designates the species as near threatened.

DIPROTODONTIA; MACROPODIDAE; **Genus LAGORCHESTES**
Gould, 1841

Hare Wallabies

There are four species (Flannery 1989*b;* Ride 1970):

L. conspicillatus, northern Western Australia, including Barrow Island, to northern Queensland;

L. leporides, formerly eastern South Australia, western New South Wales, northwestern Victoria;

L. hirsutus, formerly Western Australia, including islands of Shark Bay, Northern Territory, western South Australia;

L. asomatus, Lake McKay on Western Australia–Northern Territory border.

Head and body length of *L. conspicillatus* is 400–470 mm, tail length is 370–490 mm, and weight is 1.6–4.5 kg; respective figures for *L. hirsutus* are 310–90 mm, 245–380 mm, and 780–1,740 grams (Burbidge and Johnson *in* Strahan 1983). *L. leporides* is usually gray brown above with reddish sides and grayish white underparts; the coloration of *L. hirsutus* is much the same except that long reddish hairs are present on the lower back, imparting a shaggy appearance. *L. conspicillatus* is gray brown or yellowish gray

Hare wallaby *(Lagorchestes* sp.), photo from West Berlin Zoo through Heinz-Georg Klös.

above and whitish or reddish below with reddish patches encircling the eyes and white hip marks. The guard hairs are long and coarse, and all the species have long, soft, thick underfur.

The generic name means "dancing hare." These wallabies are not much larger than hares and resemble them in their movements and, to some extent, in their habits. The nose in *Lagorchestes* is wholly or partially covered by hair; the central hind claw is long and strong and is not hidden by the fur of the foot, and the tail is evenly haired throughout. The members of this genus have 34 teeth, 2 more than other wallabies.

Hare wallabies live in open grassy or spinifex plains with or without shrubs or trees (Ride 1970). They generally are solitary and nocturnal, resting by day in a "hide" or "form" lightly scratched in the ground in the shade of a bush or by a tuft of grass. *L. hirsutus* emits a whistling call when pursued. Jumps of 2–3 meters have been credited to *L. leporides* when pressed, and one individual chased by dogs for 0.4 km doubled back on its track and leaped over the head of a man standing in its path. Hare wallabies are herbivorous, feeding on various grasses and herbs (Collins 1973).

Burbidge and Johnson (*in* Strahan 1983) reported that on Barrow Island *L. conspicillatus* is a selective feeder, browsing mainly on colonizing shrubs and also eating the tips of spinifex leaves; it does not drink even when water is available. It constructs several hides within a home range of about 8–10 ha. Breeding occurs throughout the year, but on Barrow Island there are birth peaks in March and September. Young vacate the pouch at about 150 days, and females become sexually mature at 12 months. Johnson (1993) noted an estrous cycle of 30 days and a gestation period of 29–31 days in this species.

The species *L. asomatus* is known only from a single skull taken in 1932. *L. leporides,* common in South Australia and New South Wales until the mid–nineteenth century, has not been recorded since 1890 (Calaby 1971). Both species are designated extinct by the IUCN. *L. conspicillatus* has declined over much of its range, especially in Western Australia, but still occurs regularly in parts of the Northern Territory and northern Queensland (Calaby 1971; Poole 1979). Its decline seems to be associated with the destruction of vegetative cover by sheep, competition for habitat with introduced rabbits, and predation by introduced cats (Ingleby 1991*b*). A population of about 10,000 individuals is present on Barrow Island off the west coast of Western Australia (Short and Turner 1991). *L. conspicillatus* is designated generally as near threatened by the IUCN, and the subspecies *L. c. conspicillatus,* of Barrow Island, is classified as vulnerable. *L. hirsutus* disappeared from most of its mainland range long ago, but about 20 years ago the presence of two small colonies each containing only 6–10 animals was confirmed in the Tanami Desert Sanctuary in the Northern Territory. One of these colonies has since increased to about 20 individuals and has been studied in some detail (Lundie-Jenkins, Corbett, and Phillips 1993); the other was destroyed, apparently by introduced foxes. *L. hirsutus* is moderately abundant and well protected on Bernier and Dorre islands in Shark Bay. Populations there are known to fluctuate, but recent surveys carried out by Short and Turner (1992) during a drought and thus probably when numbers were minimal indicated the presence of 2,600 individuals on Bernier Island and 1,700 on Dorre. *L. hirsutus* is listed as endangered by the USDI and is on appendix 1 of the CITES. The IUCN classifies the island subspecies *L. hirsutus bernieri* and *L. h. dorreae* as vulnerable, the formerly widespread mainland subspecies *L. h. hirsutus* as extinct, and the surviving unnamed subspecies in the central desert as critically endangered.

Ride (1970) suggested that an important factor in the decline of hare wallabies on the mainland may have been alteration of grassland habitat through trampling and grazing by sheep and cattle. The aborigines of Australia avidly hunted hare wallabies for food but actually may have benefited the animals. These people regularly set winter fires in order to clear areas for easier hunting and thereby produced a mosaic of different regenerating vegetative stages. This process not only provided food for *Lagorchestes* but also prevented the buildup of brush and the devastation of the habitat by lightning-caused fires during the summer. The decline of hare wallabies coincided with the removal of the aborigines from large areas and the reduction of winter fires. The two recently discovered mainland colonies of *L. hirsutus* were in localities where regular winter burning is still practiced (Australian National Parks and Wildlife Service 1978; Bolton and Latz 1978; Ingleby 1991*b*).

DIPROTODONTIA; MACROPODIDAE; **Genus ONYCHOGALEA**
Gray, 1841

Nail-tailed Wallabies

There are three species (Calaby 1971; Gordon and Lawrie 1980; Ride 1970):

O. unguifera, northern Western Australia to northeastern Queensland;
O. lunata, formerly southern Western Australia, southwestern Northern Territory, South Australia, probably southwestern New South Wales;
O. fraenata, formerly inland parts of eastern Queensland, New South Wales, and northern Victoria.

In the two living species, *O. unguifera* and *O. fraenata,* head and body length is 430–700 mm, tail length is 360–730 mm, and weight is 4–9 kg; *O. lunata* was smaller (Burbidge *in* Strahan 1983; Gordon *in* Strahan 1983). The fur is soft, thick, and silky. In *O. unguifera* the upper parts are fawn-colored. The general color in the other two species is gray. In *O. fraenata,* the bridled wallaby, the white shoulder stripes run from beneath the ears along the back of the neck and down around the posterior part of each shoulder to the white undersurface of the body; the center of the neck is black or gray. In *O. lunata* the white shoulder stripes form a crescent around the posterior region of each shoulder and do not extend onto the neck, which is dark rufous. In all three species the underparts are white and white hip stripes, often indistinct, are present.

Nail-tailed wallabies inhabit woodland, savannah, brushland, or steppe. They are mainly nocturnal, and although they occasionally move about in daylight, they spend most of the day in a shallow nest scratched out beneath a tussock of grass or a bush. They seem to depend on thick vegetation for cover. When disturbed, *O. unguifera* emits a quickly repeated "u-u-u" and then flees. All species are remarkably swift when startled from their shelter and often escape by running into thick brush or a tree hollow. Unlike other wallabies, they carry their arms outward almost at a right angle to the body. When hopping they move their arms in a rotary motion, which has led to the common name "organ grinders." The diet is somewhat different from that of other wallabies, for it seems to consist mainly of the roots of various species of coarse grass and other herbaceous vegetation.

Nail-tailed wallaby *(Onychogalea fraenata)*, photo by S. Carruthers / National Photographic Index of Australian Wildlife.

Nail-tailed wallabies are shy and usually solitary. Females are said to give birth to a single young, usually in early May. A specimen of *O. fraenata* was maintained in captivity for 7 years and 4 months (Jones 1982).

Both *O. fraenata* and *O. lunata* are listed as endangered by the USDI and are on appendix 1 of the CITES. The IUCN designates the former as endangered but the latter as extinct. Both have suffered drastic declines largely through habitat alteration by farming and grazing and perhaps because of predation by introduced foxes and dogs. Newsome (1971a) suggested that removal of the thickets in which these wallabies sheltered through livestock grazing and deliberate clearing left them homeless, insecure, and greatly vulnerable to predation. Still reasonably common in some parts of its range until the 1930s, *O. lunata* now is very rare, if not extinct (Thornback and Jenkins 1982). The latest reliable records are from central Australia in the early 1960s (Poole 1979; Ride 1970). *O. fraenata* was common in parts of Queensland and New South Wales in the mid–nineteenth century but subsequently became rare. There were no confirmed sightings from 1937 to 1973, and the species was considered possibly extinct. In the latter year, however, a small population was discovered in an area of about 100 sq km in central Queensland. About half of the area has been purchased by the state government to form a reservation for the species (Gordon and Lawrie 1980; Poole 1979). The third species of the genus, *O. unguifera*, is still widespread in northern Australia and reportedly is common in several localities (Calaby 1971). It may have survived through an association with wetter habitats than those used by the other species of *Onychogalea* and also by many other now endangered or extinct Australian marsupials. It thus would be less dependent on the core areas of rich vegetation that were needed by the others during droughts but were degraded by livestock, introduced animals, and other problems caused by humans (Ingleby 1991a).

DIPROTODONTIA; MACROPODIDAE; **Genus WALLABIA**
Trouessart, 1905

Swamp Wallaby

The single species, *W. bicolor*, occurs in eastern Queensland, eastern New South Wales, Victoria, and southeastern South Australia (Ride 1970). Some of the species here assigned to the genus *Macropus* often have been put in *Wallabia*. Calaby (1966), however, argued that because of its karyology, reproductive physiology, behavior, and dental morphology *W. bicolor* should be placed in a monotypic genus. Calaby and Richardson (*in* Bannister et al. 1988), Flannery (1989b), and most other recent systematic accounts have followed this arrangement, though Kirsch (1977b), on the basis of serological investigation, thought that *W. bicolor* probably could be included in *Macropus*. Van Gelder (1977b) recommended that *Wallabia* be considered a synonym of *Macropus* since hybridization in captivity had occurred between *W. bicolor* and *M. agilis*.

Head and body length is 665–847 mm, tail length is 640–862 mm, and weight is 10.3–15.4 kg in females and 12.3–20.5 kg in males (Merchant *in* Strahan 1983). The fur is long, thick, and coarse. In the north the back and head are reddish brown and the belly is orange; in the south the general coloration is brownish or grayish black with grayish sides and sometimes brownish red underparts. The paws, toes, and terminal part of the tail are usually black. There may be a distinct light-colored stripe extending from the upper lip to the ear. Females have 4 mammae and a well-developed, forward-opening pouch.

Swamp wallaby *(Wallabia bicolor)*, photo from Zoological Society of San Diego.

Swamp wallaby *(Wallabia bicolor)*, photo by Joan M. Dixon.

Despite its common name, *W. bicolor* is not restricted to swamps. It does inhabit moist thickets in gullies and even mangroves but also is found in open forest in upland areas as long as there are patches of dense cover. It is nocturnal and hops heavily with the body well bent over and the head held low (Ride 1970). A study in Victoria found this species to occupy elongated home ranges, with long axes measuring up to 600 meters, and to be mainly a browser (Edwards and Ealey 1975). In Queensland the diet includes pasture, brush, and agricultural crops (Kirkpatrick 1970*b*).

The swamp wallaby usually is solitary, but unrelated animals may gather at an attractive food source (Kirkpatrick 1970*b*). There does not appear to be any territorial defense (Edwards and Ealey 1975). Individuals of both sexes maintain small, overlapping home ranges of about 16 ha. (Troy and Coulson 1993). Females are polyestrous, with an estrous cycle averaging 31 (29–34) days (Kaufmann 1974). Births in captivity have been recorded from January to May and from October to December (Collins 1973), but there appears to be no sharply demarcated breeding season in the wild. Under good conditions females can breed continuously and give birth about every eight months (Edwards and Ealey 1975; Kirkpatrick 1970*b*). Average undelayed gestation is 36.8 (35–38) days, and litter size usually is one, but twins have been reported (Collins 1973; Kaufmann 1974). This is the only marsupial with a gestation period longer than its estrous cycle. Thus, there is a prepartum estrus and mating during the last 3–7 days of pregnancy. Presumably, after this mating there is a near-term embryo in one uterus and a segmenting egg in the other. After the first embryo is born, its suckling induces diapause in the other embryo. Should the first young die or be removed from the pouch, the second embryo will resume development and another birth will occur in about 30 days. Otherwise the second embryo will remain dormant until late in the pouch life of the first young and then will resume development in time to be born when the first young vacates the pouch at an age of about 250 days (Collins 1973; Kaufmann 1974; Russell 1974*a*). Sexual maturity is attained by both sexes at about 15 months, and maximum life span in the wild may be 15 years (Kirkpatrick 1970*b*). A captive individual lived 12 years and 5 months (Collins 1973).

Although the swamp wallaby has declined in numbers and distribution because of clearing of its habitat, it still is widespread and common (Calaby 1971). Approximately 1,500 skins are marketed each year in Queensland, and a larger number of animals are shot as agricultural pests, but such killing is not considered a threat to the survival of the species (Kirkpatrick 1970*b*). The swamp wallaby was introduced on Kawau Island, New Zealand, about 1870 and still occurs there (Maynes 1977*a*).

DIPROTODONTIA; MACROPODIDAE; **Genus MACROPUS**
Shaw, 1790

Wallabies, Wallaroos, and Kangaroos

There are 3 subgenera and 14 species (Caughley 1984; Collins 1973; Dawson and Flannery 1985; Flannery 1989*b*, 1995; Kirsch and Calaby 1977; Kirsch and Poole 1972; Richardson and Sharman 1976; Ride 1970):

subgenus *Notamacropus* Dawson and Flannery, 1985

M. irma (western brush wallaby), southwestern Western Australia;

M. greyi (toolache wallaby), formerly southeastern South Australia, Victoria;

M. parma (parma wallaby), eastern New South Wales;

M. dorsalis (black-striped wallaby), southeastern Queensland, eastern New South Wales;

M. agilis (agile wallaby), southern New Guinea and (possibly through introduction) several islands to the east, northeastern Western Australia, northern Northern Territory, northern and eastern Queensland;

M. rufogriseus (red-necked wallaby), coastal areas from eastern Queensland to southeastern South Australia, Tasmania, islands in Bass Strait;

M. eugenii (tammar wallaby), southwestern Western Australia, southern South Australia, several offshore islands;

M. parryi (whiptail wallaby), eastern Queensland, northeastern New South Wales;

subgenus *Macropus* Shaw, 1790

M. giganteus (eastern gray kangaroo), eastern and central Queensland, New South Wales, Victoria, extreme southeastern South Australia, Tasmania;

M. fuliginosus (western gray kangaroo), southern Western Australia, southern South Australia, western and central New South Wales, southern Queensland, western Victoria, Kangaroo Island;

subgenus *Osphranter* Gould, 1842

M. bernardus (black wallaroo), Arnhem Land of northern Northern Territory;

M. robustus (wallaroo, euro), Australia except Tasmania;

M. antilopinus (antilopine wallaroo), northeastern Western Australia to northern Queensland;

M. rufus (red kangaroo), Australia except extreme north, east coast, and extreme southwest.

Until recently the genus *Macropus* usually was restricted to the last six species listed above, with the remaining eight species being placed in the genus *Wallabia*. The latter genus is now considered to include only *W. bicolor*, though Kirsch (1977*b*) and Van Gelder (1977*b*) suggested that even that species might belong in *Macropus*. The red kangaroo (*M. rufus*) often has been put in the separate genus or subgenus *Megaleia* Gistel, 1848.

The kangaroos and wallaroos (the last six species listed above) include the largest living marsupials. Measurements (from Grzimek 1975) for *M. giganteus* and *M. fuliginosus* are: head and body length for males, 1,050–1,400 mm, and for females, 850–1,200 mm; tail length for males, 950–1,000 mm, and for females, approximately 750 mm. For *M. robustus* and *M. antilopinus* measurements are: head and body length for males, 1,000–1,400 mm, and for females, 750–1,000 mm; tail length for males, 800–900 mm, and for females, 600–700 mm. Measurements for *M. rufus* are: head and body length for males, 1,300–1,600 mm, and for females, 850–1,050 mm; tail length for males, 1,000–1,200 mm, and for females, 650–850 mm. Weight is 20–90 kg but seldom exceeds about 55 kg in males and 30 kg in females. When standing in a normal, plantigrade position, male gray and red kangaroos usually are about 1.5 meters tall and sometimes reach nearly 1.8 meters. Reports of kangaroos 2.1 meters (7 feet) tall seem unfounded (W. E. Poole, Commonwealth Scientific and Industrial Research Organization, Division of Wildlife Research, pers. comm., 1977). When standing on their hind toes, however, as when in an aggressive position, males may reach or slightly exceed 2.1 meters. Wallaroos are shorter on average than kangaroos, but they are more heavily built. Male *M. antilopinus*, the largest wallaroo, may weigh just as much as male gray and red kangaroos (Russell 1974*a*). *M. bernardus*, the smallest wallaroo, is only two-thirds as large as *M. robustus* and *M. antilopinus* (Richardson and Sharman 1976). The fur of kangaroos and wallaroos is generally thick and coarse, being especially long and shaggy in the wallaroos.

Red-necked wallabies *(Macropus rufogriseus)*, photo by Bernhard Grzimek.

Eastern gray kangaroos *(Macropus giganteus)*, photo by Lothar Schlawe.

The muzzle is completely hairless in wallaroos, partly haired in the red kangaroo, and fully haired in the gray kangaroos. General coloration is as follows (Ride 1970; Russell 1974*a*): *M. giganteus,* silvery gray; *M. fuliginosus,* light gray brown to chocolate; *M. robustus,* red brown to very dark blue gray (looking black); *M. antilopinus,* reddish tan or bluish gray; *M. bernardus,* males dark sooty brown (looking black), females paler; *M. rufus,* males usually rich reddish brown, females usually bluish gray, but in a few areas males are blue and females red.

Wallabies of the genus *Macropus* (the first eight species listed above) are generally smaller than kangaroos and wallaroos. Overall, head and body length is about 400–1,050 mm, tail length is 330–750 mm, and weight is 2.5–24 kg. Most species average about 700–800 mm in head and body length, 600–700 mm in tail length, 20 kg in weight for males, and 12 kg in weight for females. There are, however, two especially small species, *M. eugenii* and *M. parma.* A series of mature specimens of *M. parma* from New South Wales measured as follows: head and body length, 424–527 mm; tail length, 416–544 mm; and weight, 2.6–5.9 kg (Maynes 1977*b*). The sexes were approximately the same size. In coloration *M. parma* is rich brown with a dark dorsal stripe extending from neck to shoulders and has white underparts and throat; *M. eugenii* is grayish brown, often with reddish shoulders; *M. agilis* has a yellowish brown color, a white cheek stripe, and a fairly distinct whitish hip stripe; *M. rufogriseus* has a fawny gray body with red nape and shoulders; *M. dorsalis* has a dark brown spinal stripe and a distinct white hip stripe; *M. parryi* is pale gray with a marked white face stripe and a very long, slender tail; *M. irma* also has a distinct white face stripe, black and white ears, and a crest of black hair on the tail; *M. greyi* is colored like *M. irma,* but its tail is crested with pale hair.

The enlarged hindquarters in *Macropus* are powerfully muscled, and the tapered tail acts as a balance and rudder when the animal is leaping and as a third leg when it is sitting. The tail is strong enough to support the weight of the entire animal. Females have a well-developed, forward-opening pouch and four mammae.

Habitat varies widely in this genus. Most wallabies rely on dense vegetation for cover but usually move into open forest or savannah to feed. *M. parma* and *M. dorsalis* generally live in moist forest. The gray kangaroos, *M. giganteus* and *M. fuliginosus,* are found in forests and woodland and also seem to require heavy cover. *M. robustus* and *M. bernardus* occur mainly in mountains or rough country. *M. antilopinus* and *M. rufus* dwell on grass-covered plains or savannah but, again, depend on places with denser vegetation for shelter. *M. robustus* sometimes shelters in caves. All species are primarily crepuscular or nocturnal, feeding from late afternoon to early morning and resting during the day, but they sometimes move about in daylight. The most diurnal species is *M. parryi* and the most nocturnal in habit are *M. parma* and *M. rufogriseus* (Kaufmann 1974; Maynes 1977*b*; Russell 1974*a*).

When moving slowly, as in grazing or browsing, these animals exhibit an unusual, "five-footed" gait, balancing on their tail and forearms while swinging their hind legs forward, then bringing their arms and tail upward. They are remarkably developed in the leaping mode of progression. At a slow pace the leaps of wallaroos and kangaroos usually measure 1.2–1.9 meters, and at increased speeds they may leap 9 or more meters. One gray kangaroo jumped nearly 13.5 meters on a flat. Normally they do not jump higher than 1.5 meters. Speeds of about 48 km/hr are probably attained for short distances when these animals are pressed in relatively open country.

All species of *Macropus* are herbivorous, and most are mainly grazers. In some areas well over 90 percent of the

Tammar wallaby *(Macropus eugenii)*, photo by L. F. Schick / National Photographic Index of Australian Wildlife.

food eaten by kangaroos and wallaroos is grass (Russell 1974*a*). Some species can go for long periods without water—two to three months have been recorded for *M. robustus*. Where they shelter in cool caves they can exist indefinitely without water, even when outside temperatures exceed 45° C. Some species dig in the ground for water or eat succulent roots. In certain areas where there is almost no fresh water *M. eugenii* obtains what it needs from salty plant juices and even is able to drink sea water (Ride 1970).

Caughley, Sinclair, and Wilson (1977) estimated that in 496,000 sq km of New South Wales, overall population densities were 3.18/sq km for the combined populations of *M. giganteus* and *M. fuliginosus* and 4.18/sq km for *M. rufus*. In the Pilbara District of northwestern Western Australia the density of *M. robustus* averaged 9.3/sq km but in some favorable areas was as high as 40/sq km. Kaufmann (1974) found a density of 1/2 ha. to 1/4 ha. for *M. parryi*. Other densities, as summarized by Main and Yadav (1971),

Parma wallaby *(Macropus parma)*, photo by Wolfgang Dressen.

Agile wallaby *(Macropus agilis)*, photo from San Diego Zoological Society.

are: 1/1.6 ha. for *M. eugenii*, 1/10.3 ha. for *M. robustus*, and 1/24 ha. to 1/89 ha. for *M. rufus*. The red kangaroo sometimes becomes relatively mobile, covering considerable distances to seek food and readily shifting its range in response to environmental conditions. One marked individual moved 338 km from the point of release (Bailey 1992). In contrast, *M. giganteus* and *M. robustus* are sedentary, seemingly preferring to remain in a restricted home range even when food and water become scarce. In such a way of life the euro is assisted by its ability to withstand dehydration better than the kangaroos, to survive on food containing less nitrogen, and to minimize water loss by sheltering in caves during the daytime (Russell 1974a). In a semiarid part of New South Wales, Croft (1991a, 1991b) found *M. robustus* to utilize weekly home ranges of 116–283 ha. in summer and winter, and over a period of years no individual moved more than 7 km. In the same locality, *M. rufus* used weekly ranges of 259–516 ha., but during a prolonged dry spell there was a general movement to a more favorable area 20–30 km away.

Social structure is variable in *Macropus* and does not seem fully related to the size, taxonomy, or ecology of the animals. The small *M. parma*, the medium-sized *M. rufogriseus* and *M. agilis*, and the large *M. robustus* are often solitary but may aggregate temporarily in the vicinity of favored resources (food, water, shelter); the medium- and large-sized species *M. dorsalis*, *M. parryi*, *M. giganteus*, *M. fuliginosus*, *M. antilopinus*, and *M. rufus* occur in organized groups (often called "mobs"); and the small *M. eugenii* is intermediate in social activity (Croft 1989; Russell 1974a). With respect to *M. parma*, Maynes (1977b) made sightings of 52 solitary animals, 14 pairs, and 5 groups of 3. McEvoy (1970) reported that the only associations of *M.*

rufogriseus are of females and offspring, which separate not more than a month after the young leave the pouch, and mated pairs, which stay together for only 24 hours. In *M. robustus* the young may remain with its mother for several months after pouch life ends, and a few males may follow a near-estrous female, but other observed groups represent only chance aggregations at some favorable site (Kirkpatrick 1968). *M. agilis* sometimes has been referred to as gregarious, but several studies have indicated that it is most frequently seen alone, in temporary aggregations in the vicinity of resources, or in consorting units (Dressen 1993). It also may be found in small groups made up mainly of females sharing the same resting and feeding areas. Such groups are not necessarily related, however, as it is thought that young animals quickly separate from the females (Kirkpatrick 1970a).

The gray and red kangaroos and the antilopine wallaroo usually occur in small groups of 2 to at least 10 individuals that appear to be more than just random, temporary assemblies (Russell 1974a, 1974b). Such groups are in a different category from the large but unorganized aggregations of kangaroos that sometimes collect at a favorable feeding site. Newsome (1971b), for example, observed a gathering of 1,500 *M. rufus* in central Australia and stated that aggregations of 50–200 were common during periods of drought.

A male *M. rufus* may gain temporary control of several females, along with their young, but there usually is no permanent association between adults of opposite sexes (Russell 1974b, 1979). As in other species of *Macropus*, there appears to be no territorial defense, but the male may fight fiercely with other males that challenge his possession of the females, and there may also be agonistic behavior relative to competition for food and resting sites. Croft (1980) stated that in such situations male *M. rufus* might suddenly rear up and begin hitting each other from an upright position (boxing) and also kick with the hind feet. Collins (1973) reported that captive *M. rufus*, as well as other species of *Macropus*, could be kept in groups but that the presence of more than one adult male would lead to trouble.

The most social of the larger species of *Macropus* may be *M. giganteus*. Mobs of more than 10 animals comprising mainly adult females and young are often observed. Large males are thought to join the groups only when females are in estrus (Kirkpatrick 1967). Males form dominance hierarchies through aggressive interaction to determine access to females, and females also form them (Grant 1973; Russell 1974b). Kaufmann (1975) studied two mobs each comprising 20–25 animals, which were divided into subgroups averaging 3.7 individuals. He found a persistent group structure but no territorial defense or permanent association between males and females.

The most gregarious species in the genus are *M. dorsalis* and *M. parryi*. The latter, known as the "whiptail wallaby," was studied intensively in northeastern Queensland by Kaufmann (1974). It is perhaps the most social of all marsupials. In the study area were three loosely organized but discrete mobs each having a year-round membership of 30–50 individuals. The animals in any one mob did not keep together in a single group but usually split into varying subgroups of fewer than 10 individuals. All the animals in a particular mob, however, used the same home range, measuring 71–110 ha., which had little overlap with the home ranges of other mobs. Members of adjacent mobs mingled peaceably in the zones of overlap but kept mostly to their own undefended ranges. Within each mob, males established a dominance hierarchy through ritualized bouts of pawing, which did not cause injury. Larger males were dominant over smaller ones. The hierarchy functioned only to determine access to estrous females and ensured that fathering of offspring was limited to higher-ranking adult males. Courtship involved wild chasing of females by males. As they approached maturity, some subadult males left their natal mobs and joined other mobs, but no females were observed to do this.

Extensive information now is available on reproduction in *Macropus*; however, only a brief compilation can be provided here. A detailed summary also was given by Hume et al. (1989). The females of all species that have been investigated are polyestrous, with lengthy reproductive seasons but relatively brief intervals of receptivity lasting perhaps only a few hours (Kaufmann 1974). Calculated average lengths of estrous cycles (in days) are: *M. eugenii*, 28.4; *M. parma*, 41.8; *M. agilis*, 30.6; *M. rufogriseus*, 32.4; *M. parryi*, 42.2; *M. giganteus*, 45.6; *M. fuliginosus*, 34.9; *M. robustus*, 32.8; *M. rufus*, 34.3 (Kaufmann 1974; Poole and Catling 1974; Rose 1978). A capability to breed throughout the year has been demonstrated in *M. parma*, *M. agilis*, *M. rufogriseus*, *M. parryi*, *M. giganteus*, *M. fuliginosus*, *M. robustus*, and *M. rufus* (Collins 1973; Kaufmann 1974; Maynes 1977*b*; Merchant 1976; Poole 1975; Russell 1974*a*). Births in captive *M. dorsalis* have been recorded in March, September, and October (Collins 1973). Most other species are not well known in this regard. *M. eugenii*, however, is an exception to the general pattern. This species apparently has a rigid breeding season extending from January to June or August, with most births occurring between mid-January and mid-February; females can produce only one young per year. The season is the same in a natural population on Kangaroo Island off South Australia, in an introduced population on Kawau Island off northern New Zealand, and in captivity in Australia. The season is reversed when animals are taken to the Northern Hemisphere (Maynes 1977*a*). A restricted breeding season also is found in populations of *M. rufogriseus* on Tasmania and islands in Bass Strait (Poole 1979).

Most of those species of *Macropus* capable of continuous breeding in captivity or under favorable natural conditions may nonetheless be dependent on environmental factors. Hence, the introduced population of *M. parma* on Kawau Island bred throughout 1970 and 1971 but only from March to July in 1972, when less food was available (Maynes 1977*a*). The red kangaroo *(M. rufus)* and the euro *(M. robustus)* are opportunistic breeders, with both sexes remaining potentially fertile throughout the year. Young are produced continuously when there is adequate rainfall and forage, but breeding may cease altogether during prolonged drought (Newsome 1975; Tyndale-Biscoe 1973). In the wild the gray kangaroos *(M. giganteus* and *M. fuliginosus)* breed mainly from September to March, a period that follows winter rainfall and coincides with the time of maximum growth of vegetation. In New South Wales most births occur from November to January (Poole 1973, 1975, 1976).

Nondelayed, average gestation periods (in days) are: *M. eugenii*, 28.3; *M. parma*, 34.5; *M. agilis*, 29.2; *M. rufogriseus*, 29.6; *M. parryi*, 36.3; *M. giganteus*, 36.4; *M. fuliginosus*, 30.6; *M. robustus*, 32.3; *M. rufus*, 33.0 (Kaufmann 1974; Maynes 1973; Poole 1975; Rose 1978). Litter size usually is one, but twins have been recorded in *M. eugenii*, *M. rufogriseus* (also one set of triplets), *M. rufus* (Collins 1973), and *M. giganteus* (Poole 1975).

Embryonic diapause occurs in most species of *Macropus*. In these species, unlike in most mammals, conception alone does not affect the estrous cycle, and the next period of receptivity and mating come at the same time as they would have if the female had not become pregnant (Russell

Whiptail wallaby *(Macropus parryi)*, photo by I. R. McCann / National Photographic Index of Australian Wildlife.

1974*a*). After the second mating, estrus finally is suppressed through the suckling stimulus of the newly born first young. The embryo resulting from the second mating develops only to the blastocyst stage and then becomes quiescent until the first young is approaching the end of pouch life or perishes or is experimentally removed. Subsequently the second embryo resumes development, and birth occurs at a time when the pouch is vacant.

Embryonic diapause is perhaps best known in *M. rufus* (Russell 1974*a*; Tyndale-Biscoe 1973). A female of this species mates within two days after giving birth. The resultant embryo develops into a blastocyst of approximately 85 cells and then becomes dormant, provided the original young still is suckling in the pouch. When the pouch young is about 204 days old, or at any time sooner if the pouch young dies or is removed, the embryo resumes development. In 31 days, and within a day after the first young permanently leaves the pouch, a second birth occurs. Immediately thereafter the female again becomes receptive, and another mating can take place. If the pouch young is lost before it is 204 days old the female becomes receptive 33 days after such loss. As a result of this process, a female red kangaroo can produce one young approximately every 240 days as long as favorable conditions hold. Under such conditions most adult females examined in the field are found to have one quiescent embryo, one pouch young, and one accompanying young outside of the pouch. When food supplies are low a pouch young may die from inadequate lactation, but the diapausing embryo will then resume growth and may reach term by the time conditions are better. If there has been no environmental improve-

Western gray kangaroo *(Macropus fuliginosus)* with young, photo by Mary Eleanor Browning.

ment, the second young also may die, but another fertilization would have taken place and a third embryo would be on its way. In severe droughts, which are common in much of Australia, females become anestrous and breeding ceases.

The reproductive patterns of three other species are known to resemble closely those of *M. rufus* in that there is continuous breeding under suitable conditions, a postpartum estrus before the pouch young is two days old, and embryonic diapause. These species, the average period from premature loss of pouch young to next birth, and the age of pouch young when the second young normally is born are: *M. agilis,* 26.5 days, 7 months; *M. rufogriseus,* 26.7 days, 9 months; and *M. robustus,* 30.8 days, 8–9 months (Kaufmann 1974; Kirkpatrick 1968, 1970a; McEvoy 1970).

In still another three species the pattern is somewhat different. *M. eugenii* is not a continuous breeder but does have a postpartum estrus. If the pouch young is lost while the breeding season is still in progress, the next birth occurs in 27 days (Rose 1978). If the first pouch young survives, the diapausing embryo is retained through the anestrous period of the female and resumes development at the beginning of the next breeding season, about 11 months after fertilization (Russell 1974a). In *M. parma* some females have an estrus 4–13 days after giving birth, but most do not become receptive until the pouch young is 45–105 days old, and a few do not enter estrus at all while carrying a pouch young. If fertilization does occur when a pouch young is present, embryonic diapause occurs and removal of the pouch young is followed by birth in 30.5–32.0 days. If the pouch young survives, it vacates the pouch when it is about 212 days old, and the next young is born 6–11 days later (Maynes 1973). In *M. parryi* there is no postpartum estrus, and mating usually does not occur again until the pouch young is 150–210 days old. Embryonic diapause then may follow. If the pouch young dies, the next young is born

about a month later; otherwise the next young is born just after the first young vacates the pouch at 9 months (Kaufmann 1974).

Embryonic diapause also is known to occur in *M. irma* (Collins 1973), *M. dorsalis* (Poole and Catling 1974:300), and *M. giganteus.* In the latter species, at least, this phenomenon appears rare, having been found in only about 5 percent of wild females with pouch young that were shot for study (Clark and Poole 1967; Poole 1973), and in only seven captive females during 10 years of observation (Poole and Catling 1974). In the latter animals diapause followed matings that took place 160–209 days after the first young was born. In both *M. giganteus* and *M. fuliginosus* (in which embryonic diapause has not been confirmed) females with pouch young may become receptive as early as 150 days after the first young is born. Conception, however, does not generally occur until there will be time before birth for the first young to vacate the pouch, which it does at an average age of about 320 days. The interval between successive births is approximately one year in both species (Poole 1977). If the pouch young dies, the female returns to estrus within an average of 10.92 days in *M. giganteus* and 8.25 days in *M. fuliginosus* (Poole and Catling 1974).

The red kangaroo, largest of marsupials, weighs only 0.75 grams at birth (Sharman and Pilton 1964). At that time it has a large tongue and well-developed nostrils, forelimbs, and digits but is embryonic in other external features. Like other newborn marsupials, it scrambles from the birth canal to the pouch without the assistance of the mother and grasps one of the mammae. It first releases the nipple at about 70 days, first protrudes its head from the pouch at 150 days, temporarily emerges at 190 days, and permanently vacates the pouch at 235 days. It then continues to suckle by placing its head in the pouch, and it is finally weaned at about 1 year.

Despite differences in size, most other species of *Macropus* do not take much less time to develop than *M. rufus,* and some take longer. For example, age (in days) at first departure from the pouch, permanent emergence, and weaning, respectively, for three species are: *M. parma,* 175, 207–18, 290–320 (Maynes 1973); *M. agilis,* 176–211, 207–37, 365 (Merchant 1976); and *M. parryi,* 240, 275, 450 (Kaufmann 1974).

Another large species, *M. robustus,* has a pouch life of about the same length as that of *M. rufus* (Collins 1973), but in the gray kangaroos development is longer, average age at first emergence from the pouch being 283.9 days in *M. giganteus* and 298.4 days in *M. fuliginosus.* Respective figures for final departure are 319.2 days and 323.1 days. Lactation in both species usually exceeds 18 months, longer than in any other marsupials (Poole 1975).

Sexual maturity also takes fairly long in gray kangaroos (Poole 1973; Poole and Catling 1974). Average age for captive males was 29.0 months in *M. fuliginosus* and 42.5 months in *M. giganteus.* A few wild males mated as early as 20 months, whereas some did not begin until 72 months. Average age for first mating in captive females was about 22 months in *M. fuliginosus* and 21 months in *M. giganteus.* Wild females generally became sexually mature between 20 and 36 months. In *M. rufus* females usually mature at 15–20 months, and males at about 20–24 months, but the period is lengthened under adverse environmental conditions. In the rigorous habitat of northwestern Australia *M. robustus* has not been observed to breed before 25 months, but captive males may breed at 22 months and females may breed at about 18 months. Sexual maturity in *M. parma* was attained at 11.5–16.0 months in captivity, as early as 12 months in the wild in Queensland, and as early as 19 months, but usually not until 2 years, on Kawau

Island, New Zealand (Maynes 1973, 1977*a,* 1977*b*). In *M. agilis* males mature at 14 months and females at 12 months (Merchant 1976). In *M. parryi* maturity occurs at 18–29 months, but because of social factors males are prevented from mating until they are 2–3 years old (Kaufmann 1974).

Although most individuals probably do not even live to maturity, wallabies, kangaroos, and wallaroos have a potentially great life span. Some notable longevities in captivity are: *M. eugenii,* 9 years and 10 months; *M. agilis,* 10 years and 2 months; *M. rufogriseus,* 15 years and 2 months; *M. dorsalis,* 12 years and 5 months; *M. parryi,* 9 years and 8 months; *M. giganteus,* 24 years; *M. robustus,* 19 years and 7 months; *M. antilopinus,* 15 years and 11 months; and *M. rufus,* 16 years and 4 months (Collins 1973; Jones 1982; Kirkpatrick 1965). Some outstanding estimated ages for wild individuals are: *M. parma,* 7 years; *M. rufogriseus,* 18.6 years; *M. giganteus,* 19.9 years; *M. robustus,* 18.6 years; and *M. rufus,* 27 years (Bailey 1992; Kirkpatrick 1965; Maynes 1977*b*).

The effect of people on *Macropus* has varied considerably. One species, *M. greyi,* is designated as extinct by the IUCN. It was still common in southeastern Australia in the first years of the twentieth century but subsequently declined rapidly through competition from livestock, bounty and sport hunting, and killing for its beautiful pelt. The last wild individuals were recorded in 1924, though a few captives survived until about 1937 (Calaby 1971; Harper 1945). Two additional species were once thought to be extinct. *M. bernardus,* which occupies a restricted area in the Northern Territory, had not been collected since 1914 but was rediscovered in 1969 (Goodwin and Goodwin 1973; Parker 1971). It now is considered to be maintaining its original numbers and distribution (Calaby 1971) but is designated as near threatened by the IUCN. *M. parma,* of the

Wallaroo *(Macropus robustus)*, photo from San Diego Zoological Society.

Antilopine wallaroo *(Macropus antilopinus)*, photo by I. R. McCann / National Photographic Index of Australian Wildlife.

coastal forests of New South Wales, seemed to have disappeared through clearing of its habitat, with the last known specimens taken in 1932. During the 1870s, however, the species, along with several other kinds of wallabies, had been introduced on Kawau Island, New Zealand. There a relatively dense population built up that was unknown to science until 1965. Then in 1967 *M. parma* again was found in its original range in New South Wales. Later investigation showed that the species still occurred along several hundred kilometers of the coast but at low densities (Maynes 1974, 1977*b;* Poole 1979; Ride 1970). *M. parma* now is classified as near threatened by the IUCN and as endangered by the USDI. A closely related species, *M. eugenii,* also was introduced to Kawau Island in the 1870s and still occurs there. It was common on the mainland of Australia until about 1920 but subsequently has nearly vanished in South Australia and become restricted to a few isolated colonies in Western Australia because of habitat loss and competition with domestic sheep. A number of natural island populations also were extirpated or reduced, but the species still occurs in satisfactory numbers on Kangaroo Island, South Australia, and on several islands off the coast of Western Australia (Calaby 1971; Harper 1945; Poole 1979; Poole, Wood, and Simms 1991). The IUCN designates *M. eugenii* generally as near threatened but notes that the subspecies *M. e. eugenii,* of Nuyt's Archipelago off South Australia, is extinct in the wild.

The five other species of wallaby in the genus *Macropus* have fared better in modern times. Although Harper (1945) expressed concern about heavy sport and skin hunting of *M. parryi* and *M. irma,* and even though these and other species have lost habitat in recent years, all five still are widely distributed and relatively common (Calaby 1971; Poole 1979). In certain areas suitable habitat actually has increased through scattered opening of the forest for agriculture (Kaufmann 1974; Ride 1970). *M. irma* is designated as near threatened by the IUCN.

In Tasmania *M. rufogriseus* probably is more common now than when settlement began (Calaby 1971). It is considered a pest because of its damage to regenerating eucalyptus forests (Ride 1970). It and other species are sometimes shot in large numbers for alleged destruction of crops and pasture. In addition, several species are still important in the skin trade, though hunting is now regulated in order to ensure that overall populations are not adversely affected. In 1976, 45,259 *M. parryi* were killed for commercial purposes in Queensland (Poole 1978). *M. dorsalis* was introduced on Kawau Island, New Zealand, in the 1870s and still occurs there. *M. rufogriseus* was brought to the South Island of New Zealand at about the same time, eventually increased to an estimated 750,000 individuals, and then was reduced by control in the 1960s to only about 3,500 (Wodzicki and Flux 1971).

Surprisingly, perhaps, the effects of European settlement thus far have seemed less detrimental to the larger wallaroos and kangaroos than to many smaller species of marsupials. Gray kangaroos, however, have disappeared from many densely settled localities, and there is concern for some remaining populations, such as that of *M. giganteus* in southeastern South Australia (Poole 1977). The subspecies *M. giganteus tasmaniensis* of Tasmania was greatly reduced in numbers and distribution by the early twentieth century through excessive sport and commercial hunting and today occurs in just a few small parts of its original range (Barker and Caughley 1990; Harper 1945). It is now listed as endangered by the USDI and as near threatened by the IUCN. *M. fuliginosus fuliginosus* of Kangaroo Island off South Australia, recently confirmed as a highly distinctive subspecies (Poole, Carpenter, and Simms 1990) and also designated as near threatened by the IUCN, may be subject to loss of habitat as human activity increases in its limited range (Poole 1976, 1979). Otherwise the gray kangaroos, as well as the red kangaroo *(M. rufus)* and the two larger wallaroos *(M. robustus* and *M. antilop-*

Red kangaroos *(Macropus rufus),* photos by Lothar Schlawe.

inus), still occur over most of their original ranges and generally are not thought to be in any immediate jeopardy (Calaby 1971; Poole 1979). Indeed, there is evidence that at least some of these species increased in numbers following settlement. They seem to have been aided by establishment of artificial water holes for livestock, as well as by the grazing of stock, which crops the long, dry grass avoided by the kangaroos and causes the sprouting of soft, green shoots. Favorable grazing conditions for the kangaroos thus were created over vast areas (Newsome 1971*a;* Poole 1979; Russell 1974*a*). The subspecies *M. robustus isabellinus,* restricted to Barrow Island off Western Australia, is classified as vulnerable by the IUCN.

According to Newsome (1975), livestock and introduced rabbits may yet cause the collapse of the red kangaroo. He postulated that simultaneous competition with other herbivores (sheep, cattle, or rabbits), especially during drought, could eventually bring about conditions under which *M. rufus* would become rare. He also observed that in northwestern Australia *M. rufus* initially increased in numbers following introduction of sheep but then lost in competition with the sheep and *M. robustus.* The latter species had invaded the plains because overgrazing by sheep had eliminated the luxuriant grass cover and allowed the spread of highland vegetation favored by *M. robustus.* Edwards (1989) indicated that the ecological relationship between kangaroos and sheep is complex and not fully understood but that competition does occur under certain circumstances.

For the present, at least, kangaroos are relatively common in Australia. According to Grigg et al. (1985), surveys from 1980 to 1982 indicated that there were then about 8,351,000 red, 1,774,000 western gray, and 8,978,000 eastern gray kangaroos in Australia. Mainly because of a severe drought, by 1984 numbers had fallen to 6,330,000 red, 1,162,000 western gray, and 5,791,000 eastern gray kangaroos. New surveys reported by Fletcher et al. (1990) suggested that populations had recovered to approximately their 1980 levels, but no overall estimate was calculated. The USDI (1995) used a total estimate of 26,200,000 red and gray kangaroos for 1992 but cited other estimates of barely half that number and noted that there subsequently had been a renewed decline in association with another drought. In any case, the total number of these three species plus that of the wallaroos probably approximates or exceeds the number of humans (about 18 million) in Australia. No other terrestrial region in the world of comparative size still has more large wild mammals than people.

The major question, however, is not whether there are plenty of kangaroos but whether populations can sustain present hunting levels, especially considering the additional factors of human habitat modification and drought. Kangaroos have long been heavily hunted, first by the aborigines and then by European settlers for meat and skins, and later by stockmen because of alleged damage to fences and pasture. During the late nineteenth and early twentieth centuries millions of kangaroos were killed for bounty payments alone.

Subsequently an industry developed based on the harvesting and processing of about 300,000–400,000 skins per year (Poole 1978). In the late 1950s, improved refrigerating equipment allowed a great increase in the take of kangaroos for use as pet food. Such hunting, along with continued killing for skins and pest control, was not well regulated and coincided with severe droughts in parts of Australia. Kangaroo numbers dropped alarmingly over large areas (Ride 1970; Russell 1974*a*).

The total recorded harvest of *M. giganteus, M. robustus,* and *M. rufus* in Queensland amounted to at least 200,000 animals annually from 1954 to 1983 and was over 1 million in several of those years. In 1983 the kill was 502,161 gray kangaroos, 62,100 wallaroos, 263,346 red kangaroos, and about 21,000 of the smaller wallabies (Poole 1984). In New South Wales the legal take in 1975 was 60,300 gray kangaroos and 48,100 red kangaroos, or 3.8 percent and 2.3 percent, respectively, of the estimated populations in the state (Caughley, Sinclair, and Wilson 1977). In 1992 the total commercial kill of the three species was 2,676,000, or about 10 percent of the total population estimated by the USDI (1995).

Much controversy centers on how many kangaroos can be safely harvested and whether commercial utilization jeopardizes the species or helps by giving them a value that discourages their mass destruction as pests (Australian National Parks and Wildlife Service 1988; Poole 1984). During the 1970s the Australian state governments increased their control of kangaroo hunting. From 1973 to 1975 the Australian federal government prohibited the general export of kangaroo products but subsequently ended the ban as the states established acceptable management programs (Poole 1978). Most exports of kangaroo skins had been going to the United States, but in 1974 the USDI listed *M. giganteus, M. fuliginosus,* and *M. rufus* as threatened. Accompanying regulations prohibited the importation of these species and products thereof until there was satisfactory certification that wild populations would not be adversely affected. Such certification subsequently was received, and importation was opened in 1981. The USDI proposed removing the three species from threatened status in 1983, but in 1984, following drastic population declines attributed to a severe drought in Australia, the proposal was withdrawn. Gaski (1988) reported that the number of skins imported to the United States averaged more than 1 million annually in 1963–66, was 158,254 in 1981, and was about 50,000 in 1986. Many conservationists continue to oppose commercial importation and do not consider kangaroo populations to be safely managed in Australia. Over such opposition, and despite the role of its regulations in improving such management, the USDI did proceed to cancel the regulations and remove the three species from threatened status in 1995.

DIPROTODONTIA; Family BURRAMYIDAE

Pygmy Possums

This family of two Recent genera and five species is found in Australia, Tasmania, and New Guinea. The members of the Burramyidae sometimes have been placed in the family Phalangeridae but now are considered to represent a distinct group (Kirsch 1977*b*). The sequence of genera presented here follows that of Kirsch and Calaby (1977). The genera *Distoechurus* and *Acrobates* also were placed in the Burramyidae when that family was first resurrected in the 1970s, but recently they were moved to a new family, the Acrobatidae (Aplin and Archer 1987; Archer 1984*a*). Nonetheless, the karyotype of *Distoechurus* has been found to be very similar to that of the burramyids (Westerman, Sinclair, and Woolley 1984).

Head and body length is 60–120 mm and tail length is 70–175 mm. The pelage is soft. Digital structure is much like that of the Phalangeridae, but neither genus has opposable digits on the forefoot. The dental formula is: (i 3/2, c 1/0, pm 2–3/3, m 3–4/3–4) × 2 = 34–40. According to Archer (1984*a*), the Burramyidae are characterized by reduction of the first upper and second lower premolars, a bicuspid tip on

"Dormouse" possum *(Cercartetus lepidus)*, photo by Peter Crowcroft.

the third lower premolar, and a well-developed posteromesial expansion of the tympanic wing of the alisphenoid. Both genera have prehensile tails and are good climbers.

The known geological range of this family is middle Miocene to Recent in Australasia, though a possible burramyid was reported from beds of late Oligocene age in Tasmania (Marshall 1984).

DIPROTODONTIA; BURRAMYIDAE; **Genus**
CERCARTETUS
Gloger, 1841

"Dormouse" Possums

There are four species (Aitken 1977a; Barritt 1978; Dixon 1978; Kirsch and Calaby 1977; Ride 1970; Wakefield 1970a):

C. concinnus, southwestern Western Australia, southern South Australia, western Victoria, southwestern New South Wales;

C. nanus, southeastern South Australia to southeastern Queensland, Tasmania;

C. lepidus, southeastern South Australia, western Victoria, Kangaroo Island, Tasmania;

C. caudatus, New Guinea, northeastern Queensland.

C. lepidus long was known only from Tasmania, except for fossil and subfossil remains on the mainland. In 1964, however, it was discovered alive on Kangaroo Island, and in the 1970s specimens were collected in southeastern South Australia and western Victoria. The species *C. lepidus* and *C. caudatus* sometimes have been placed in a separate genus, *Eudromicia* Mjöberg, 1916.

Head and body length is 70–120 mm, tail length is 70–175 mm, and adult weight is about 15–40 grams. In *C. concinnus* the upper parts are reddish brown and the underparts and feet are white; *C. nanus* is grayish or fawn-colored above and slaty below; and in *C. lepidus* and *C. caudatus* the upper parts vary from dull rufous and rich brown to bright pale fawn and the underparts are white to yellowish. *C. caudatus* has well-marked, broad black bands passing from the nose through the eyes, not quite reaching the ears. The pelage is dense and soft. The prehensile, cylindrical tail is well furred at the base and scantily haired for the remainder of its length. With the approach of winter *C. nanus* becomes very fat and its tail becomes greatly enlarged, especially at the base, so that stored food will be available during periods of torpor.

The common name, "dormouse" possum, reflects the superficial resemblance of this marsupial to the dormouse *(Glis glis)* so well known in Europe. In addition, both genera are nocturnal and undergo hibernation in cold weather. "Dormouse" possums have large, thin, almost naked ears. The forefoot looks somewhat like a human hand, and the great toe of the hind foot is thumblike and widely opposable. The claws are small, but the pad of each digit is expanded into two lobes. The pouch is well defined in females. The normal number of mammae is four except in *C. concinnus,* which has six (Collins 1973). *C. nanus* usually has four functional and two nonfunctional teats (Ward 1990a).

These tiny, arboreal possums dwell in forest, heath, or shrubland. They construct a small, dome-shaped nest of grass or bark in a hollow limb, hollow stump, crevice in a tree trunk, or thick clump of vegetation. They sometimes reside in abandoned birds' nests. *C. nanus* usually makes a nest of soft bark and may travel as much as 500 meters to

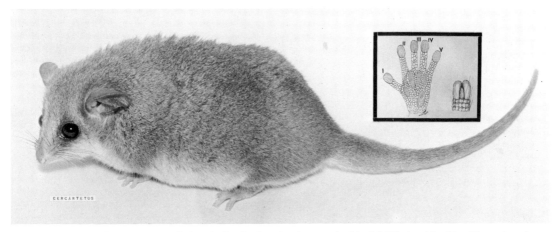

"Dormouse" possum *(Cercartetus nanus),* photo by Stanley Breeden. Inset: underside of right foot and tip of toe *(C. concinnus),* photo from *The Mammals of South Australia,* F. Wood Jones.

secure the desired kind of bark. Possums of this genus hang by the tail when reaching from one branch to another and also skillfully use the tail when climbing or descending. They are nocturnal, leaping and running freely at night. They generally sleep soundly during the day but sometimes become active in cloudy weather. The species *C. nanus* and *C. lepidus* have been found to undergo alternate periods of activity and dormancy throughout the year in Tasmania. Neither species experienced prolonged hibernation; the longest period of torpor was 6 days for *C. lepidus* and 12 days for *C. nanus*. Body temperature during dormancy was about equal to air temperature (Hickman and Hickman 1960). Experimental inducement of deep hibernation in *Cercartetus nanus* for periods of up to 35 days has been achieved by Geiser (1993). All species seem to be omnivorous, feeding on leaves, nectar, fruit, nuts, insects and their larvae, spiders, scorpions, and small lizards. One captive specimen of *C. nanus*, however, was a strict vegetarian, refusing to eat meat or insects, and lived a record 8 years (Perrers 1965).

These possums have been maintained in pairs in captivity, and a group of three males and one female was found hibernating together, but generally *Cercartetus* seems to be solitary in the wild (Collins 1973). *C. concinnus* and *C. nanus* may breed throughout the year in Victoria, but most births of the latter species occur from November to March. After the birth of a litter females return to estrus and mate again; there probably is a subsequent period of embryonic diapause, with the young being born soon after the previous litter is weaned. Most females produce two litters per year, but some produce three (Ward 1990*a*, 1990*c*). Minimum gestation in *C. concinnus* is 51 days (Hayssen, Van Tienhoven, and Van Tienhoven 1993). In New Guinea a wide range of birth dates has been recorded for *C. caudatus*, but there is a breeding recession during the dry season from mid-May to early August (Dwyer 1977). Captive *C. caudatus* have bred twice yearly, in January and February and from late August to early November; litters have contained one to four young, which emerged from the pouch at 34 days of age, became independent at 92 days, and first bred at 15 months (Atherton and Haffenden 1982). Average litter size is five in *C. concinnus*, with a maximum of six. Ward (1990*a*) recorded litter sizes of four to six in *C. nanus*, though modal size was four; pouch life is 30 days, the young become independent immediately after weaning and reach sexual maturity at as early as 4.5–5.0 months. Available records from Tasmania indicate that litters of about four *C. lepidus* are born between September and January and that they approach adult size and become independent at 3 months (Green *in* Strahan 1983). *C. nanus* appears to do well in captivity; in addition to the specimen mentioned above that lived 8 years, a number of individuals have survived for 4–6 years (Collins 1973); maximum longevity in the wild is at least 4 years (Ward 1990*a*).

The subspecies *C. caudatus macrurus*, of northeastern Queensland, is designated as near threatened by the IUCN.

DIPROTODONTIA; BURRAMYIDAE; **Genus**
BURRAMYS
Broom, 1895

Mountain Pygmy Possum

The single species, *B. parvus*, now inhabits the mountains of eastern Victoria and extreme southeastern New South Wales. Until recently it was known only from fossilized cranial and dental material found at the Wombeyan Caves, near Goulburn in New South Wales. The remains are from the late Pleistocene and are estimated to be about 15,000 years old. There was controversy regarding the specimens, with some scientists thinking that the fossils represented a kind of miniature kangaroo. In the early 1950s, however, the material was examined by W. D. L. Ride, who showed that *Burramys* actually was a small possum related to *Cercartetus*. Additional fossils were located in the late 1950s. Then in August 1966 a live possum of an unknown kind was found in a ski hut on Mount Hotham, Victoria. Upon examination it was identified as *B. parvus*. In the words of Ride (1970:16): "The dream dreamed by every paleontologist had come true. The dry bones of the fossil had come together and were covered with sinews, flesh and skin." Subsequently, from 1970 to 1972, 3 more specimens were taken at Mount Hotham; 1 was trapped in the Falls Creek area of the Bogong High Plains, Victoria; and 19 were collected in the Kosciusko National Park, New South Wales (Dimpel and Calaby 1972; Dixon 1971). More recently many hundreds have been live-trapped in the course of field investigations (Mansergh and Scotts 1990).

Head and body length is 101–30 mm, tail length is 131–60 mm, and weight is 30–60 grams (Calaby *in* Strahan 1983). The upper parts are brownish gray, and the underparts are paler. The body fur continues onto the tail for about 1 cm, and for the rest of its length the prehensile tail is almost naked. The premolar teeth are unusually large. There is a distinct pouch with two pairs of mammae (Dixon 1971).

All specimens have been taken in mountainous areas at altitudes of about 1,500–1,800 meters. The species apparently is restricted to dense patches of shrubs associated with snow gum trees and large boulders in the subalpine to alpine zones (Dimpel and Calaby 1972). The climate in the Kosciusko National Park is severe, with an average annual precipitation of about 125–200 cm, much of it winter snow. Calaby, Dimpel, and Cowan (1971) suggested that *Burramys*, unlike most possums, is not arboreal but mainly terrestrial and adapted to climbing shrubs. They also thought that the animal probably lives in holes among or under rocks. When released under controlled conditions *Burramys* ran rapidly on the ground and over rocks in a ratlike manner and did not hesitate to dive into holes. It climbed shrubs rapidly by grasping the stems with its forefeet and hind feet. Dixon (1971) reported that a captive was well adapted for climbing and jumping but made only occasional use of its prehensile tail.

A number of recent investigations have substantially increased our knowledge of this newly discovered genus (Fleming 1985; Kerle 1984*a*, 1984*b*; Mansergh 1984*a*, 1984*b*; Mansergh and Scotts 1990). *Burramys* is strictly nocturnal. In captivity it has entered prolonged bouts of deep torpor lasting up to 20 days (Geiser and Broome 1991), and it apparently undergoes winter hibernation in the wild. The diet consists mainly of seeds, fruits, worms, and arthropods. *Burramys* is the only marsupial known to establish caches of durable foods; such storage would facilitate winter survival. Nonbreeding captives are tolerant of one another, thus suggesting that huddling, another winter survival strategy, occurs in the wild. Breeding females, however, defend their nests. The pygmy possum population in 6 ha. of suitable habitat on Mount Higginbotham, Victoria, has been estimated to contain 84–122 breeding females. Densities in other areas are much lower.

Females are polyestrous, with a mean estrous cycle of 20.3 days. The short alpine summer requires a brief gestation period and rapid development of young. In the wild,

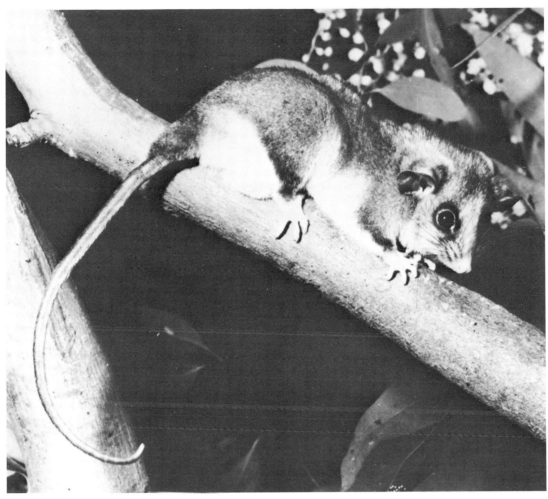

Mountain pygmy possum *(Burramys parvus)*, photo from Australian News and Information Bureau.

births occur in mid- and late November (spring), though in captivity they have been recorded in several other months. Limited evidence indicates that gestation lasts 13–16 days. Litters contain up to eight young, though only four actually survive to be carried in the pouch. They leave the pouch after 3 weeks, open their eyes at 5–6 weeks, are weaned at 8–9 weeks, and attain adult weight at 3–4 months. Females invariably breed in the first spring after they are born, but some males do not. Females have been known to live for at least 11 years in the wild, but males only up to 4 years.

The mountain pygmy possum is listed as endangered by the IUCN and USDI. Its restricted range may be jeopardized by habitat destruction and other environmental modifications associated with the development of ski resorts. The largest amount of remaining habitat, about 8 sq km in Mount Kosciusko National Park, New South Wales, is estimated to contain 500 adult *Burramys*. Another population, of much greater density but limited to a 6-ha. area on Mount Higginbotham, Victoria, contains 400 individuals. The remaining tracts of good habitat are all small and are separated from one another by large, unsuitable areas (Caughley 1986; Mansergh 1984*a*).

DIPROTODONTIA; **Family PSEUDOCHEIRIDAE**

Ring-tailed and Greater Gliding Possums

This family of 6 Recent genera and 16 species inhabits Australia, Tasmania, New Guinea, and certain nearby islands. The members of the Pseudocheiridae long were regarded as components of the family Phalangeridae. Kirsch (1977*b*), however, explained that they differ in serology, karyology, and other characteristics, and he placed them in the subfamily Pseudocheirinae of the family Petauridae. On the basis of additional serological studies, Baverstock (1984) elevated the Pseudocheirinae to familial rank but indicated that this group is closely related to the Petauridae. This view has been supported by Aplin and Archer (1987), Baverstock, Birrell, and Krieg (1987), and most other recent authorities. Szalay (1994), however, regarded the group only as a tribe within the family Petauridae. The sequence of genera presented herein attempts to express the affinities discussed by Flannery and Schouten (1994) but does not precisely follow their order of presentation.

New Guinean ring-tailed possum *(Pseudochirulus forbesi)*, photo by Pavel German.

Head and body length is 163–480 mm and tail length is 170–550 mm. The pelage may be woolly or silky. According to McKay (1989), the dentition of the Pseudocheiridae is characterized as follows: three upper incisors on each side of the jaw, only the first pair prominent, and also a reduced upper canine; three upper premolars, the third elongated and bearing two or three cusps; four strongly selenodont upper molars, with crescent-shaped ridges connecting the cusps; the large, procumbent first lower incisor compressed and bladelike; a second, vestigial lower incisor may be present in some species but lost in others; three large lower premolars; and four lower molars, all approximately equal in size. The skull is characterized by a robust zygomatic arch and posterior palatal vacuities. The forefoot has the first two digits at least partly opposable to the other three, an adaptation to grasping small branches. The tail is prehensile and may have a ventral friction pad of naked calloused skin or only a small naked area at the tip. In the digestive

tract the caecum is greatly enlarged to act as a fermentation chamber.

All members of the Pseudocheiridae are arboreal. One genus has attained a gliding ability through development of a membrane uniting the front and hind limbs. This genus, *Petauroides,* has an analogous phylogenetic position to that of *Petaurus* in the family Petauridae (Kirsch 1977*b*).

The geological range of the Pseudocheiridae is Miocene to Recent in Australasia (McKay 1989). Details on Miocene pseudocheirid fossils and their significance to the phylogeny of the family were presented by Woodburne, Tedford, and Archer (1987).

DIPROTODONTIA; PSEUDOCHEIRIDAE; **Genus PSEUDOCHIRULUS**
Matschie, 1915

New Guinean and Queensland Ring-tailed Possums

There are seven species (Flannery 1995; Flannery and Schouten 1994; Kirsch and Calaby 1977; Musser and Sommer 1992; Ride 1970; Ziegler 1977):

P. canescens, New Guinea and Japen and Salawatti islands to the west;
P. mayeri, Central Range of New Guinea;
P. caroli, west-central New Guinea;
P. schlegeli, known only by the type from the Vogelkop Peninsula of western New Guinea and by one other specimen probably from the same area;
P. forbesi, New Guinea;
P. herbertensis, from Kuranda to Ingham in northeastern Queensland;
P. cinereus, from Mount Lewis to Thornton Peak in northeastern Queensland.

These species long were placed in the genus *Pseudocheirus* together with *P. peregrinus.* However, Flannery and Schouten (1994), interpreting the molecular analyses of Baverstock (1984) and Baverstock et al. (1990), treated *Pseudochirulus* as a separate genus with a content as given above. Further studies suggest that *P. forbesi* might comprise several species (Musser and Sommer 1992).

Head and body length is about 167–368 mm, tail length is 151–395 mm, and weight is 105–1,450 grams (Flannery and Schouten 1994; Winter *in* Strahan 1983). In most species the fur is dense, soft, and woolly. The upper parts of the body are gray or brown, often very dark, and the underparts are white, yellowish, or almost as dark as the back. Some species have dark and light markings on the head, a median stripe on the back, or stripes around the thighs. The distal end of the tail is usually bare for some length on the undersurface and only sparsely haired on the upper surface. In most species the prehensile end of the tapered tail is usually curled into a ring, hence the common name of these possums. The first two digits of the forefoot are opposable to the other three digits. Females have four mammae, but normally only two are functional.

All species are arboreal forest dwellers. Their habits are not well known but are thought to be comparable to those of *Pseudocheirus.* Some species are known to build nests like those of *Pseudocheirus* or to shelter in hollow trees. The diet probably is largely folivorous (leaf-eating) and also includes fruit. *P. herbertensis* appears to be solitary except when males are attracted to estrous females. Mating

New Guinean ring-tailed possum *(Pseudochirulus mayeri)*, photo by Tim Flannery.

may occur throughout the year but seems to peak in April and May. Usually two young are reared. They emerge from the pouch after four to five months and may then be carried on the mother's back for a short period; thereafter they are left in the nest and make increasingly longer forays alone (Winter *in* Strahan 1983). *P. forbesi* also appears to breed year-round but usually has only a single young (Flannery 1990*b*). *P. herbertensis* may be in jeopardy because of logging of its tropical forest habitat (Winter 1984*a*). That species, as well as *P. cinereus* of the same region, is designated as near threatened by the IUCN.

DIPROTODONTIA; PSEUDOCHEIRIDAE; **Genus PSEUDOCHEIRUS**
Ogilby, 1837

Common Ring-tailed Possum

Flannery and Schouten (1994) recognized a single species, *P. peregrinus,* occurring along the eastern coast of Australia from the Cape York Peninsula of northeastern Queensland to southeastern South Australia, in southwestern Western Australia, and on Kangaroo Island and Tasmania. Kennedy (1992) regarded the isolated subspecies of *P. peregrinus* in Western Australia as a separate species, *P. occidentalis,* and it is possible that other populations also warrant specific status. The genera *Pseudochirulus, Pseudochirops, Petropseudes,* and *Hemibelideus* (see accounts thereof) sometimes have been included in *Pseudocheirus.*

Head and body length is 287–353 mm, tail length is 287–360 mm, and weight is 700–1,110 grams (Flannery

and Schouten 1994; McKay *in* Strahan 1983). The fur is short, dense, and soft. The coloration is highly variable, ranging from predominantly gray to rich red, and the underparts are paler. There are always distinctive white ear tufts and a white tail tip. The tail is tapered, with the fur progressively shorter distally and a long, bare friction pad on the undersurface of the distal portion. This prehensile end of the tail is commonly curled into a ring. The first two digits of the forefoot are opposable to the other three digits. Females have four functional mammae.

The common ring-tailed possum is scansorial and nocturnal. It may be found in rainforest, sclerophyll forest, woodland, or brush. It generally shelters by day in a large, dome-shaped nest of interwoven leaves, bark, and ferns located in the branches of a shrub or tree, or it may rest in a tree hollow, which it lines with leafy material. The height of the nest varies from only a few centimeters above the ground in heavy cover to about 25 meters in mistletoe clusters (Collins 1973). In the wild this possum seems to be strictly herbivorous, feeding on a variety of leaves, fruits, flowers, bark, and sap.

The movements of this possum are generally slow, and its quiet, retiring manner may make it seem uninteresting to human observers. When it moves from one limb to another it usually keeps hold of the old resting place with its tail until it has grasped the new branch with its forefeet. It sleeps with its head beneath its body and between its hind feet. Although it is agile and graceful, its usual reaction to a sudden encounter is to remain motionless with a vacant stare. If the intruder remains, the possum may slowly creep away. On the ground it travels at fair speed with a waddling gait.

The scarce western subspecies, *P. p. occidentalis,* has been

Ring-tailed possum *(Pseudocheirus peregrinus)*, photo by G. Weber / National Photographic Index of Australian Wildlife.

reported to occur at densities of about 0.1–4.5/ha. and to have an average home range of 1.0–2.5 ha. (Jones, How, and Kitchener 1994). Density of the eastern populations commonly is 12–16/ha., up to 19/ha. in favorable habitat, and home range averages 0.37 ha. In areas where density is high individuals are very aggressive toward one another. Otherwise this genus is semisocial and even gregarious, with reports in the east of up to three adults per nest, up to eight individuals constructing their nests in close proximity, and groups of males and females remaining together all year (Collins 1973; How 1978; How et al. 1984; McKay *in* Strahan 1983; Ride 1970). *Pseudocheirus* is polyestrous, with an estrous cycle of about 28 days. Breeding season varies: late April–December in Victoria, early winter and early summer near Sydney, and late summer and autumn in northern Queensland. Females sometimes rear two litters in a year. In the east litter size is one to three, with a mode of two. The young leave the pouch at about 18 weeks, are weaned at 6–7 months, and disperse at 8–12 months. Females reach sexual maturity at about 1 year (Collins 1973; How 1978; How et al. 1984; Ride 1970; Tyndale-Biscoe 1973). The western subspecies has birth peaks in April–June and September–November, there usually is only a single young, and pouch life is about 104 days (Jones, How, and Kitchener 1994). Life span in the wild is four to five years, and the record captive longevity is eight years (Collins 1973; How 1978).

In Tasmania, where presumably the cold winters result in desirable pelts, *P. peregrinus* is an important fur bearer; approximately 7.5 million were taken for this purpose from 1923 to 1955 (Ride 1970). In parts of Western Australia and Victoria *P. peregrinus* has become uncommon or restricted in distribution because of loss of habitat to agricultural development (Emison et al. 1978; Winter 1979). The subspecies *P. peregrinus occidentalis* of Western Australia has lost 50–90 percent of its habitat (Kennedy 1992) and is now classified as vulnerable by the IUCN.

DIPROTODONTIA; PSEUDOCHEIRIDAE; **Genus PSEUDOCHIROPS**
Matschie, 1915

Coppery and Silvery Ring-tailed Possums

There are five species (Flannery 1990*b*; Flannery and Schouten 1994; Kirsch and Calaby 1977; McKay *in* Bannister et al. 1988; Ride 1970; Ziegler 1977):

P. cupreus, Central Range of New Guinea;
P. albertisii, northern and western New Guinea and nearby Japen Island;
P. coronatus, Arfak Mountains of far western New Guinea;
P. corinnae, Central Range of New Guinea;
P. archeri, northeastern Queensland.

Coppery ring-tailed possum *(Pseudochirops cupreus)*, photo by Pavel German.

Kirsch and Calaby (1977) and most other authorities once considered *Pseudochirops* to be a subgenus of *Pseudocheirus* and also to include the species *P. dahli*. More recent serological and morphological studies have suggested that *Pseudochirops* should be elevated to generic rank (Baverstock 1984) and that *P. dahli* should be placed in the separate genus *Petropseudes* (see account thereof). Corbet and Hill (1991) restricted *Pseudochirops* to the single species *P. archeri*, but Groves (*in* Wilson and Reeder 1993) accepted the general arrangement given above. *P. coronatus* usually was considered a synonym of *P. albertisii* until it was restored to specific rank by Flannery and Schouten (1994).

Head and body length is 289–410 mm, tail length is 258–371 mm, and weight is 640–2,250 grams (Flannery 1990*b*; Flannery and Schouten 1994; Winter *in* Strahan 1983). The upper parts are coppery or silvery green, the underparts are paler, and there are dark dorsal stripes and pale facial markings on some of the species. The tail is prehensile and has distally bare areas similar to those of *Pseudocheirus*. McKay (1989) characterized *Pseudochirops* as follows: fur short, dense, and fine; tail shorter than head and body, tapering rapidly from a thickly furred base to sparsely furred tip, friction pad long; pupil a vertical slit. Flannery (1990*b*) added that the molar teeth of *Pseudochirops* are larger and more complex than those of *Pseudocheirus*.

The following natural history information was taken from Flannery (1990*b*) and Winter (*in* Strahan 1983). The New Guinea species are found mainly in undisturbed montane forests at elevations of 1,000–4,000 meters and may be declining in response to human habitat disruption. *P. archeri*, of Australia, lives in upland rainforest and seems to be able to survive in areas that have been partially logged over. Both *P. archeri* and *P. corinnae* are unusual in that they sleep on exposed branches rather than in a nest. In contrast, *P. cupreus* nests both in tree hollows and in burrows under tree roots. All species presumably are nocturnal but sometimes move about by day. All are primarily folivorous but may also eat fruit. *P. archeri* is solitary and breeds mainly in the latter half of the year. *P. cupreus* breeds throughout the year. A single young is normally produced.

The IUCN classifies *P. albertisii* and *P. corinnae* of New Guinea as vulnerable; both are thought to be undergoing a substantial decline. *P. archeri*, of Australia, is designated near threatened.

DIPROTODONTIA; PSEUDOCHEIRIDAE; **Genus**
PETROPSEUDES
Thomas, 1923

Rock Possum

The single species, *P. dahli*, is found in northeastern Western Australia, northwestern Northern Territory, and northwestern Queensland (Flannery and Schouten 1994). Although *Petropseudes* long was considered a full genus or subgenus, Kirsch and Calaby (1977) treated it as a synonym of *Pseudochirops*. More recent systematic accounts have restored it to generic rank (Groves *in* Wilson and Reeder 1993; McKay *in* Bannister et al. 1988). Part of the following information was taken from Nelson and Kerle (*in* Strahan 1983).

Head and body length is 325–450 mm, tail length is 200–275 mm, and weight is 1,280–2,000 grams. The fur is long and woolly. The upper parts are grayish, more or less suffused with rufous, the rump is brighter rufous, and the

Rock possum *(Petropseudes dahli)*, photo by Pavel German.

underparts are lighter gray. The short but prehensile tail is rufous gray and lacks a white tip; it is thickly furred at the base, but the tip and the terminal two-thirds of the undersurface are almost naked. Unlike in the other pseudocheirids, the first two digits of the forefoot are not opposable to the other three digits. The claws are reduced and blunt. Females have a well-developed, anteriorly opening pouch and two mammae.

This possum inhabits rocky outcrops on savannahs; it is nocturnal, spending the day in caves or crevices among the rocks and usually selecting the darkest recesses. It is not known to make a nest, and when at rest it often lies squeezed flat in a crevice or sits in a bent-over position in a corner. When the animal is in the latter position the tail is curved forward and upward, but when the animal is walking the tail is held straight out. *Petropseudes* is a good climber and enters trees at night to feed on fruits, flowers, and leaves. It generally associates in pairs, a male and a female, sometimes accompanied by a young, but aggregations of up to nine have been observed. Breeding apparently occurs throughout the year, and there normally is a single offspring.

Rock possum *(Petropseudes dahli)*, photo by Pavel German.

Brush-tipped ring-tailed possum *(Hemibelideus lemuroides)*, photo from Queensland Museum.

DIPROTODONTIA; PSEUDOCHEIRIDAE; **Genus**
HEMIBELIDEUS
Collett, 1884

Brush-tipped Ring-tailed Possum

The single species, *H. lemuroides,* is found only in a small highland area in northeastern Queensland (Winter *in* Strahan 1983). *Hemibelideus* sometimes is considered a subgenus of *Pseudocheirus,* but new serological and morphological studies suggest that the two are generically distinct (Baverstock 1984; Johnson-Murray 1987). Chromosomal studies point to possible close affinity between *Hemibelideus* and *Petauroides* (McQuade 1984).

Head and body length is 313–52 mm, tail length is 335–73 mm, and weight is 810–1,270 grams (Winter *in* Strahan 1983). The upper parts vary in color from light fawn to dark blackish gray, and the underparts are yellowish gray. The head is dark brown with a tinge of red, and the limbs are dark brown with black near the ends. The ears project only a short distance beyond the thick fur of the head. The pelage is soft and woolly, even on the feet. The prehensile tail is black, bushy, thickly covered with fur for the whole length on the upper surface, and slightly tapering; the naked underside of the tip is very short.

Hemibelideus possesses suggestions of gliding membranes, for it has small folds of skin (less than 25 mm in width) along the side of the body. It has been observed making long jumps from tree to tree and from tree to ground. Some zoologists think that this genus bears characters transitional between the other ring-tailed possums and the greater gliding possum *(Petauroides).* It is said to be very agile and active in the trees—some of its jumps almost have the appearance of true glides. It seems to use its furry tail as a rudder during these jumps.

According to Winter (*in* Strahan 1983), this possum is found in rainforest at elevations above about 450 meters. It is strictly arboreal and nocturnal, spends the day in a tree hollow, and emerges just after dark to forage. Leaps from branch to branch often cover 2–3 meters. The diet consists mostly of leaves. *Hemibelideus* is frequently seen in groups of two individuals (mother and young or male and female) and in family groups of three. Up to three may share a den, and feeding aggregations of as many as eight have been found in a single tree. Females have two mammae in the pouch, but usually only one young is reared. Young have been recorded in the pouch from August to November and riding on the mother's back from October to April. Winter (1984a) cautioned that *Hemibelideus* may be in jeopardy because of logging of its tropical forest habitat. Laurance (1990) found that the genus could be relatively abundant on large tracts of primary forest but that it declined by more than 97 percent when such areas were fragmented into small areas, thereby preventing normal movement through trees. The IUCN now designates *Hemibelideus* as near threatened.

DIPROTODONTIA; PSEUDOCHEIRIDAE; **Genus**
PETAUROIDES
Thomas, 1888

Greater Gliding Possum

The single species, *P. volans,* occurs in the coastal region from eastern Queensland south to southern Victoria (Ride 1970). The use of the name *Petauroides* in place of the former *Schoinobates* Lesson, 1842, was explained by McKay (1982).

Head and body length is 300–480 mm and tail length is 450–550 mm. Adult weight is 0.9–1.5 kg (Flannery and Schouten 1994; How 1978). The fur is soft and silky. Coloration is variable, ranging from black to smoky gray or creamy white on the upper parts and sooty gray, grayish white, or pure white on the underparts. All-white and white-headed individuals are common. The long tail is prehensile and evenly furred except the underpart of the tip, which is naked. Females have a well-developed pouch and two mammae.

This attractive marsupial glides with the use of a patagium, consisting of a fold of skin extending from the elbow to the leg. The position of the arms of *Petauroides* when gliding is entirely different from that of gliding squirrels and of the lesser gliding possums *(Petaurus)*: they are bent at the elbows, so that the forearms are directed in toward the head and the hands almost meet on the front part of the chest. Grzimek and Ganslosser (*in* Grzimek 1990) noted that the gliding membrane apparently has a secondary use as a blanket: the animal wraps it around its body as a protection against loss of heat.

According to Wakefield (1970*b*), the attributes of the greater gliding possum have been confused with those of the largest of the lesser gliding possums, *Petaurus australis*. Whereas the latter species can glide up to about 114 meters, is highly maneuverable in the air, and calls loudly while gliding, *Petauroides* is a sedentary, slow-moving, silent animal of "minor gliding ability." McKay (*in* Strahan 1983), however, wrote that the glides of *Petauroides* may cover a horizontal distance of up to 100 meters and involve changes in direction of as much as 90 degrees.

The habitat is sclerophyll forest and tall woodland (Ride 1970). Although this possum apparently travels across open ground at times, it spends nearly its entire life in trees. It is nocturnal, sheltering by day in hollows high up in trees. Some individuals make a nest of stripped bark or leaves in their den, but often no material is added. There is some indication that this possum transports nesting material in its rolled-up tail. The musty eucalyptus smell of this animal permeates the shelter. The diet is very specialized, consisting mainly of the leaves, bark, and bud debris of certain species of *Eucalyptus* (Marples 1973).

Population densities of 0.24–0.83/ha. have been reported. Captive pairs or groups can be maintained in a sufficiently large enclosure. Wild adult males utilize home ranges that are separate from those of other males but overlap those of females. The ranges of females overlap to some extent, but the females tend to avoid one another. Home range size averaged about 1.5 ha. in one area and 2.6 ha. in another. Some males are monogamous and others are bigamous. There is some interaction between the sexes all year, but contact peaks just prior to and during the mating season. Females are polyestrous and have one litter per season. In New South Wales mating occurs from March to May and births occur from April to June. In Victoria the young usually are born in July and August. Litters contain a single young. It first releases the nipple at about 6 weeks but spends up to 6 months in the pouch and then another 4 months as a dependent nestling. It may sometimes be carried on the back of the mother. There evidently is no paternal care. Full independence comes at about 10–13 months, and sexual maturity is attained in the second year of life. Longevity may be 15 years in the wild (Collins 1973; Henry 1984; How 1978; Kerle and Borsboom 1984; Smith 1969; Tyndale-Biscoe and Smith 1969).

Kennedy (1992) regarded *Petauroides* as "potentially vulnerable" because of the continued fragmentation and disturbance of the coastal forests on which it depends. McKay (*in* Strahan 1983) observed that the conservation of

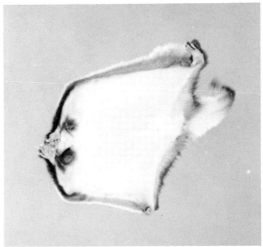

Greater gliding possum *(Petauroides volans)*, photos by Alan Root through Bernhard Grzimek.

Petauroides is utterly dependent on the maintenance of sufficient amounts of old-growth forest and that its abundance in undisturbed areas contrasts strongly with its absence from pine plantations and its scarcity in regenerated forest that lacks old trees with suitable hollows for nesting.

DIPROTODONTIA; **Family PETAURIDAE**

Gliding and Striped Possums

This family of 3 Recent genera and 10 species inhabits Australia, Tasmania, New Guinea, and certain nearby islands. The members of the Petauridae sometimes have been placed in the family Phalangeridae. Kirsch (1977*b*), however, explained that the Petauridae differ from the Phalangeridae in serology, karyology, and other characteristics. He regarded the Pseudocheiridae (see account thereof) as one subfamily of the Petauridae and also recognized two other subfamilies: Petaurinae, for the genera *Gymnobelideus* and *Petaurus,* and Dactylopsilinae, for the genus *Dactylopsila.* Flannery and Schouten (1994) maintained the latter two subfamilies, though Edwards and Westerman (1992) found DNA analysis to indicate that *Gymnobelideus* actually is more closely related to *Dactylopsila* than to *Petaurus.* On the basis of serological studies, Baverstock (1984) and Baverstock, Birrell, and Krieg (1987) suggested that the genera *Distoechurus* and *Acrobates* have affinity to the Petauridae, but these genera now are placed in a new family, the Acrobatidae (see account thereof).

Head and body length is 120–320 mm and tail length is 150–480 mm. The pelage may be woolly or silky. All members of the Petauridae are arboreal. One genus, *Petaurus,* has attained a gliding ability through development of a membrane uniting the front and hind limbs. According to McKay (1989), the dentition of the Petauridae is characterized as follows: three upper incisors on each side of the jaw, the first pair normally longer than the succeeding pairs and projecting anteriorly; the upper canine small in the Dactylopsilinae but larger and laterally compressed in the Petaurinae; three single-cusped upper premolars, the first and second laterally compressed and the third conical; four bunodont upper molars, decreasing in size from first to last, each having four low pointed cusps; the procumbent first lower incisor slightly curved in the Petaurinae and strongly curved and greatly enlarged in the Dactylopsilinae; a second, vestigial lower incisor; three small lower premolars; and four lower molars, decreasing in size from first to fourth. The skull is characterized by a slender and rounded zygomatic arch and a palate without posterior vacuities. The forefoot shows no specialization for climbing other than enlarged claws; in the Petaurinae the second to fifth digits are all subequal in length, but in the Dactylopsilinae the fourth digit is considerably elongated as an adaptation for feeding. The tail is semiprehensile; in the Petaurinae it is entirely furred, but in the Dactylopsilinae there is a small ventral naked patch at the tip. The caecum of the digestive tract is large in the Petaurinae but very small in the Dactylopsilinae.

The geological range of the Petauridae is Miocene to Recent in Australasia (Kirsch 1977*a*).

Striped possum: Left, *Dactylopsila trivirgata* young, photo by John Warham; Right, *D. palpator,* photo by Howard Hughes through Australian Museum, Sydney.

DIPROTODONTIA; PETAURIDAE; **Genus PETAURUS**
Shaw and Nodder, 1791

Lesser Gliding Possums

There apparently are five species (Colgan and Flannery 1992; Flannery 1990*b*; Flannery and Schouten 1994; Mc-Kay *in* Bannister et al. 1988; Ride 1970; Smith 1973; Suckling *in* Strahan 1983; Ziegler 1981):

P. breviceps (sugar glider), Halmahera Islands in the North Moluccas, New Guinea and many nearby islands, Bismarck Archipelago, Louisiade Archipelago, northern and eastern Australia;
P. norfolcensis (squirrel glider), eastern Queensland, eastern New South Wales, Victoria, southeastern South Australia;
P. gracilis, northeastern Queensland;
P. abidi, northwestern coastal Papua New Guinea;
P. australis (fluffy glider), coastal parts of Queensland, New South Wales, and Victoria.

Flannery (1990*b*) indicated that the forms *biacensis* and *tafa* of New Guinea may be species distinct from *P. breviceps*. Flannery and Schouten (1994) treated *P. biacensis*, which is found on Biak and Supiori islands off northwestern New Guinea, as a separate species but treated *tafa* as a subspecies of *P. breviceps*. Those authorities also indicated that a population in the D'Entrecasteaux Islands off southeastern New Guinea may be an undescribed species related to *P. breviceps*.

Head and body length is 120–320 mm and tail length is 150–480 mm. Weights are 79–160 grams in *P. breviceps*, 200–260 grams in *P. norfolcensis* (Suckling *in* Strahan 1983), and 228–332 grams in *P. abidi* (Flannery 1990*b*). The larger *P. australis* weighs 435–710 grams (Henry and Craig 1984). The fur is fine and silky. In *P. breviceps*, *P. norfolcensis*, *P. gracilis*, and *P. abidi* the upper parts are generally grayish and the underparts are paler. A dark dorsal stripe runs from the nose to the rump, and there are stripes on each side of the face from the nose through the eye to the ear. In *P. australis* the upper parts usually are dusky brown, markings are less conspicuous, the feet are black, and the underparts are orange yellow.

Lesser gliding possums resemble flying squirrels *(Glaucomys)* in form and in having a large gliding membrane, but the tail of the former is furred all around and not as flattened as that of *Glaucomys*. As in *Glaucomys*, but not the greater gliding possum *(Petauroides)*, the gliding membrane extends all the way from the outer side of the forefoot to the ankle and is opened by spreading the limbs straight out. Females have a well-developed pouch during the breeding season, and the number of mammae usually is four but occasionally is two. The pouch morphology of *P. australis* is unique among marsupials in having two compartments separated by a well-furred septum (Craig 1986).

All species inhabit wooded areas, preferably open forest, and are arboreal and largely nocturnal. They generally shelter by day in a leaf nest in a tree hollow. *P. breviceps* usually collects leaves by hanging by its hind feet, passing the leaves via the forefeet to the hind feet and then to the tail, which then coils around the nest material. The tail cannot be used for gliding when it is employed in this manner, so the possum transports its load of leaves by running along branches to its hollow.

These possums are extremely active. *P. breviceps* can glide up to about 45 meters and has been observed to leap at and catch moths in flight, and *P. australis* can glide at

A. Lesser gliding possum *(Petaurus norfolcensis)*, photo by Ernest P. Walker. B. *Petaurus* sp., photo by Bernhard Grzimek.

least 114 meters; vertical and lateral angle can be changed in the air (Ride 1970; Smith 1973). All species are omnivorous, feeding on sap, blossoms, nectar, insects and their larvae, arachnids, and small vertebrates (Collins 1973; Smith 1973; Wakefield 1970*b*). *P. australis* probably eats mostly nectar and arboreal arthropods, but it also removes bark from various eucalyptus trees to get the sugary sap. It makes characteristic V-shaped cuts in the bark that channel the sap to its mouth, placed at the bottom of the V. In a study of *P. australis* in coastal New South Wales, Goldingay (1990) found that individuals devoted 90 percent of the time outside of their dens to foraging, 70 percent of that time feeding on *Eucalyptus* nectar, and probably assisted in the cross-pollination of the eucalyptus trees.

In a study in remnant natural vegetation in Victoria, Suckling (1984) found the average population density of *P. breviceps* to vary from 2.9/ha. in summer to 6.1/ha. in autumn. Average home range was about 0.5 ha., though males extended their range slightly during the breeding season. In a study of the same species in New South Wales, Quin et al. (1992) found home ranges nearly 10 times as large. The species of *Petaurus* are social to some extent and produce a variety of sounds. *P. breviceps* has an alarm call resembling the yapping of a small dog and a high-pitched cry of anger (Smith 1973). While gliding, pairs of *P. australis* communicate by loud calls that are audible to humans at a distance of several hundred meters (Wakefield 1970b). The loud vocalizations of that species may also serve to define territories of a group (Goldingay 1994). Collins (1973) reported that captive *P. australis* can be maintained in pairs and that *P. breviceps* and *P. norfolcensis* can be kept in groups but that established groups might attack newly introduced individuals. Fleming (1980) reported that individuals of *P. breviceps* huddle together to conserve energy during cold weather and that they may simultaneously enter daily torpor when winter food supplies are low.

In the wild *P. breviceps* nests in groups of up to seven adult males and females and their young (Ride 1970; Smith 1973; Suckling 1984). All members probably are related and descended from an original colonizing pair. Groups appear mutually exclusive, territorial, and agonistic toward one another. One or two dominant, usually older males are responsible for most territorial maintenance, aggression against intruders, and fathering of young. There are a few lone adults, and the young generally leave their natal group at 10–12 months. This species has a complex chemical communication system based on scents produced by frontal, sternal, and urogenital glands of males and by pouch and urogenital glands of females. Each animal of a group has its own characteristic smell, which identifies it to other individuals and which is passively spread around the group's territory. In addition, a dominant male actively marks the other members of the group with his scent.

Studies of *P. australis* (Craig 1985, 1986; Goldingay 1992; Goldingay and Kavanagh 1990, 1993; Henry and Craig 1984) show that the species is thinly distributed. Most estimates of population density range from about 1/25 ha. to 1/6 ha. A population is divided into pairs or small groups of adults, with or without dependent young, that occupy largely separate home ranges of 30–100 ha. These areas include a variety of habitats, especially eucalyptus forest, favored for denning and feeding. The pairs are stable, monogamous units; one unit existed throughout a 41-month study period. Polygynous units consisting of one adult male and several adult females also may be found, especially where food resources are abundant. Breeding occurs mainly from about June to December in Victoria and New South Wales, but at one site in the latter state females were observed to give birth predominantly from February to April, presumably in association with availability of food resources. Breeding is year-round in Queensland. Almost always a single young per pair is raised in a year, though there is a record of twins. Both parents provide care for the young. Dispersal and independence has been observed at various ages from 9 to 24 months.

The species *P. breviceps* is polyestrous, with an estrous cycle averaging about 29 days. A female may produce a second litter during a breeding season if the first is lost or weaned (Smith 1971, 1973). In captivity there seems to be no definite breeding season for *P. breviceps* and *P. norfol-*

censis (Collins 1973). The same is true for *P. breviceps* in the wild in New Guinea and Arnhem Land in the Northern Territory, but in southeastern Australia the young of this species are born only from June to November (Flannery 1990b; Smith 1973). Gestation is about 16 days in *P. breviceps* and slightly less than 3 weeks in *P. norfolcensis*; litter size usually is one or two in these species but may be as large as three in *P. breviceps* (Collins 1973; Smith 1973). The young of *P. breviceps* weigh about 0.19 grams at birth, first release the nipple at about 40 days, first leave the pouch at about 70 days, first leave the nest at 111 days, and are independent shortly thereafter (Collins 1973; Smith 1973). Sexual maturity in this species appears to come late in the first year for females and early in the second year for males. Lesser gliding possums may have a long life span in captivity, the recorded maximums being 14 years in *P. breviceps*, 11 years and 11 months in *P. norfolcensis*, and 10 years in *P. australis* (Collins 1973; Jones 1982). Wild *P. australis* are known to have lived at least 6 years (Goldingay and Kavanagh 1990).

The species *P. breviceps* is comparatively common in Australia. It was introduced into Tasmania in 1835 and has since spread over the island (Smith 1973). The considerably larger *P. australis*, however, appears to have become rare in some areas. It is a wide-ranging species and depends on maintenance of substantial tracts of woodland (Emison et al. 1975, 1978; Mackowski 1986). Both *P. australis* and *P. norfolcensis*, the latter also being rare and confronted with habitat fragmentation (Kennedy 1992), are designated as near threatened by the IUCN.

P. abidi and *P. gracilis*, which occur in very restricted areas of threatened habitat, are classified by the IUCN as vulnerable and endangered, respectively. *P. gracilis* apparently numbers fewer than 2,500 mature individuals and is continuing to decline.

DIPROTODONTIA; PETAURIDAE; **Genus GYMNOBELIDEUS**
McCoy, 1867

Leadbeater's Possum

The single species, *G. leadbeateri*, occurs in eastern Victoria and possibly in southeastern New South Wales (Ride 1970). It had been known only from six specimens collected between 1867 and 1909 but was rediscovered in 1961 (Lindenmayer and Dixon 1992).

Head and body length in four specimens is 152–68 mm and tail length is 190–99 mm. Weight is 120–65 grams (Smith 1984a). The sexes are equal in size. The fur is soft but neither as long nor as silky as in *Petaurus*. The upper parts are gray or brownish gray with a dark middorsal stripe extending from the forehead to the base of the tail. The fur of the ventral surface and the inner surface of the limbs is a dull creamy yellow at the tips and light gray beneath. The markings about the ears and the eyes are very similar to those of the sugar glider, *Petaurus breviceps*.

In many respects this possum closely resembles the sugar glider, but it lacks the gliding membrane. Unlike in other possums, the tail is flattened laterally, narrow at the base, and bushes out evenly to the tip. It is not prehensile to any marked degree and appears to be used for balance when the animal is climbing or jumping. The digits are very wide or spatulate at the tip and bear short, strong claws. *Gymnobelideus* is very active and appears to rely on the grip of the toe pads rather than the claws when climbing. The well-developed pouch of the female contains four mammae.

Leadbeater's possum *(Gymnobelideus leadbeateri)*, photos by Norman A. Wakefield, of Fisheries and Wildlife, Victoria, through J. McNally.

This possum is found in dense, wet sclerophyll forest at altitudes of up to about 1,200 meters. It is arboreal and nocturnal and constructs a nest of loosely matted bark about 10–30 meters up in a large hollow tree. The diet consists mainly of plant and insect exudates and also includes a variety of arthropods. A notch may be gnawed into a tree in order to obtain gums (Smith 1984*b*).

According to Smith (1984*a*), populations have a density of 1.6–2.9/ha. and are grouped into colonies of one to eight individuals. Each colony occupies a den tree centered in a defended territory of 1–2 ha. There are prolonged territorial disputes that involve chasing and grappling. Colonies usually consist of a single adult female, her mate, one to two unrelated adult males, and one or more generations of offspring. Aggression between adult females is prevalent, but males appear to move freely between colonies. Births occur in all months except January and February, with peaks from April to June (autumn) and October to December (spring). Females are polyestrous and can give birth within 30 days of losing a litter. Gestation probably lasts 15–17 days. Litters usually contain one to two young. They remain in the pouch for about 90 days and then spend another 5–40 days in the nest. Female offspring disperse at 7–14 months, and males at 11–26 months. Since the young

females are excluded from established colonies, they suffer a high rate of mortality. A captive female, however, was reproductively active until the age of 9 years. A wild male was still alive at 7.5 years.

The first four known specimens of this species apparently came from the Bass River Valley and Koo-wee-rup Swamp, which are lowland areas southeast of Melbourne. Each of these was taken long before it was actually described, and little or no information was available regarding the living population that it represented. This lack of information, plus the clearing of the involved habitat, led to the assumption that the species was extinct. In 1931, however, it was learned that another specimen had been taken in 1909 in a highland area farther inland. Initial searches for a population in this area failed, but in 1961 individuals were observed in the Cumberland Valley of eastern Victoria. Subsequent investigation has shown that the species occurs in numerous localities over a large area of mountain forest and that it apparently increased in numbers and range following severe brush fires in 1939 (Ride 1970).

Recently, however, forest clearing has destroyed a substantial area of habitat. If additional timbering is carried out as planned, the old trees that the species depends upon for nesting could be eliminated over almost its entire known range. There is some question about the extent of future logging and the degree to which the needs of the species will be considered relative to such activity (Thornback and Jenkins 1982). The species and its remnant habitat also are closely tied to a narrow set of climatic conditions that could be severely affected by the "greenhouse effect" (Lindenmayer et al. 1991). The population is estimated to number 5,000 individuals and is expected to undergo further decline and fragmentation (Kennedy 1992). Leadbeater's possum now is classified as endangered by the IUCN and USDI.

DIPROTODONTIA; PETAURIDAE; Genus **DACTYLOPSILA**
Gray, 1858

Striped Possums

There are two subgenera and four species (Colins 1973; Flannery 1990*b;* Kirsch and Calaby 1977; Ride 1970):

subgenus *Dactylopsila* Gray, 1858

D. trivirgata, New Guinea and certain nearby islands, northeastern Queensland;
D. tatei, Fergusson Island off southeastern Papua New Guinea;
D. megalura, central and western New Guinea;

subgenus *Dactylonax* Thomas, 1910

D. palpator, central and eastern New Guinea.

Dactylonax was considered a full genus by McKay (*in* Bannister et al. 1988). Ziegler (1977) did not consider *D. tatei* and *D. megalura* to be specifically distinct from *D. trivirgata.* However, the above arrangement was followed by Corbet and Hill (1991) and Groves (*in* Wilson and Reeder 1993).

Head and body length is 170–320 mm and tail length is 165–400 mm. *D. trivirgata* weighs 246–470 grams; a male *D. palpator* weighed 550 grams, a female 320 grams (Flannery 1990*b;* Van Dyck *in* Strahan 1983). In the subgenus *Dactylopsila* the fur is thick, close, woolly, and rather harsh; in *Dactylonax* it is silky and dense. All species have three parallel, dark stripes on the back; in the subgenus *Dactylopsila* these stripes are black on a basal color of white or gray, and in *Dactylonax* they are smoky brown on a basal color of grayish tawny. In *D. trivirgata* and *D. megalura* there is a black chin spot. The bushy tail is well haired except at the undersurface of the tip and is mostly dark-colored except that the tip is usually whitish. Females have two mammae and a well-developed pouch. Flannery (1990*b*) noted that the pouch of *D. palpator* is divided into two subunits, each with a single nipple.

This genus is characterized by large first incisor teeth and a slender, elongated fourth digit with a hooked nail on the forefoot. Both features are more pronounced in the subgenus *Dactylonax* and are shared by the aye aye *(Daubentonia),* a primate found in Madagascar. The claw of the fourth front digit of *Dactylonax* is much smaller than those of the other digits.

Striped possums live in rainforest or sclerophyll forest and are nocturnal and arboreal. The members of the subgenus *Dactylopsila,* at least, are superb climbers that seldom if ever descend to the ground. They shelter by day in

Striped possum *(Dactylopsila tatei),* photo by Pavel German.

dry leaf nests in hollows, where they lie curled into a ball, flat and not rolled up in a sitting position like most possums. The subgenus *Dactylonax,* however, seems less adept at climbing, may spend considerable time hunting insects in rotting logs on the forest floor, and nests on the ground among tree roots as well as in tree hollows (Flannery 1990*b*). The extremely unpleasant and penetrating odor of striped possums is of glandular origin, but it cannot be ejected in "skunklike" manner. Nonetheless, the odor, together with the black and white markings, provide a striking case of convergence with North American skunks (Flannery and Schouten 1994). When angered, these marsupials utter a loud and prolonged "throaty gurgling shriek."

A captive striped possum cleaned itself elaborately upon emerging at night and then began to hunt for food with long-legged striding movements, so that it seemed to flow rather than jump from branch to branch. Striped possums sniff loudly around likely food sources, use their incisors to tear wood and gnaw out wood-boring grubs, and use their long finger to hook grubs from deep and narrow holes. They also tap their forefeet rapidly on loose bark, presumably to disturb insects beneath it. Their natural diet consists mainly of insects but also includes fruits and leaves. A captive occasionally attacked and consumed mice. Smith (1982) reported that *D. trivirgata* evidently feeds mainly on ants and other social insects, which it obtains by breaking into nests with its incisors.

Flannery (1990*b*) reported that *D. trivirgata* and *D. palpator* have been found in small female groups but that males of the latter species have always been found alone. Females with single young have been collected in New Guinea throughout much of the year. According to Van Dyck (*in* Strahan 1983), mating in *D. trivirgata* may occur in Australia from February to August, and up to two young are born. One specimen of *D. trivirgata* lived in captivity for 9 years and 7 months (Marvin L. Jones, Zoological Society of San Diego, pers. comm., 1995).

Both *D. tatei* and *D. megalura* are found in restricted areas of vulnerable habitat (Kennedy 1992). They are classified by the IUCN as endangered and vulnerable, respectively. *D. tatei* long was known only by nine specimens collected in 1935 and could not be located by several expeditions from the 1950s to the 1980s, but another specimen was taken in 1992 (Flannery and Schouten 1994).

DIPROTODONTIA; **Family TARSIPEDIDAE; Genus TARSIPES**
Gray, 1842

Honey Possum

The single known genus and species, *Tarsipes rostratus* is found in the southwestern part of Western Australia (Ride 1970). This genus once was usually placed in the family Phalangeridae. More recently, it came to be recognized morphologically and serologically as one of the most divergent of Australian marsupials and was put in its own superfamily (Kirsch 1977*b*; Kirsch and Calaby 1977). Later serological and morphological studies suggested that this superfamily also should include the genera *Acrobates* and *Distoechurus* (Aplin and Archer 1987; Baverstock et al. 1987). The use of the name *T. rostratus* in place of the previously used *T. spenserae* was explained by Mahoney (1981).

Head and body length is 65–90 mm, tail length is 70–105 mm, and weight is 7–11 grams for males and 8–16 grams for females (Russell and Renfree 1989). The head is pale brown, and the body is grayish brown with three dark stripes along the back. The central stripe, which is almost black, extends from the head to the base of the tail; the two outer stripes are fainter and do not reach the tail. The underparts are pale yellowish or white, the limbs are pale rufous, and the feet are white. Some individuals are more grayish, but the stripes are always present. The fur is rather coarse, short, and close. The long whiplike tail is almost hairless and has a naked prehensile undertip.

This possum can be distinguished from the other small Australian marsupials by its coloration and its extremely long snout, which is about two-thirds of the length of the rest of the head. The long tongue, bristled at the tip, can be extended about 25 mm beyond the nose. Ridges on the hard palate scrape honey and pollen off the tongue when it is withdrawn through a channel formed by flanges on the upper and lower lips, assisted by the long, slender lower incisors. The pair of procumbent lower incisors are the only well-developed teeth; the at most 20 other teeth are reduced to small pegs, reflecting the soft diet of the animal. All the digits except the united second and third of the hind feet, which are equipped with functional claws, are expand-

Honey possum *(Tarsipes rostratus),* photo by P. A. Woolley and D. Walsh.

ed at the tip and bear a short nail much like those of the tarsier *(Tarsius),* hence the generic name *Tarsipes* for the honey possum. Females have a well-developed pouch and four mammae.

The honey possum dwells on tree and shrub heaths and is an active, nimble climber. It moves short distances to trees and shrubs that are producing its favorite flowers, and several animals may assemble in one desirable place. It often hangs upside down, especially when feeding on flowers. Sometimes it shelters in deserted birds' nests, though it often constructs its own refuge. It has three peaks of activity in each 24-hour period: between 0600 and 0800, 1700 and 1900, and 2330 and 0130 (Vose 1973). Nocturnal movements are concerned primarily with feeding. The diet consists almost exclusively of nectar and pollen, there being little evidence that insects are a significant component (Russell and Renfree 1989). Like some bats and hummingbirds, the honey possum is well adapted for probing into flowers and licking up its food. Nearly every floret in a favorite flower is explored.

Individuals usually occupy permanent, overlapping home ranges of about 1 ha. *Tarsipes* is very gregarious in the laboratory, huddling together in groups of two or more. Animals are often caught in a torpid state during cold or wet weather, and presumably huddling helps to minimize body heat loss. Large adult females interact very little with one another and are dominant to males; when accompanied by young, they may exclude other individuals from their range. Females with pouch young have been found throughout the year but predominantly in early autumn

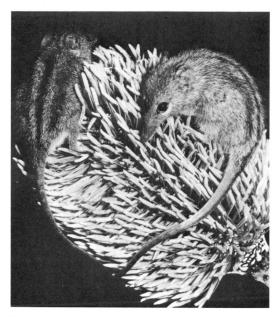

Honey possum *(Tarsipes rostratus)*, photo by M. B. Renfree.

(February–March), winter (May–July), and spring (September–October), when pollen and nectar are most abundant. Few young seem to be born in the summer months of December and January. Females are polyestrous, and some breed at least twice a year. Embryonic diapause such as in the Macropodidae also occurs in *Tarsipes*. Dormant embryos, carried in the uterus while the female is lactating, resume development after the young have left the pouch. The active gestation period appears to be 21–28 days. Litter size is usually two or three, but from one to four young have been recorded. With a weight of under 5 milligrams each, the young are the smallest known mammalian neonates. Lactation lasts about 10 weeks; for the the first 8 weeks the young remain in the pouch and for the last 2 weeks they are left in a shelter while the mother forages. They may then ride about on her back for another 1–2 weeks. Adult size is attained at about 8 months, and females may breed in their first year of life. Following the spring breeding peak there is a drastic population decline and most individuals apparently do not survive more than a year (Renfree 1980; Renfree, Russell, and Wooller 1984; Renfree and Wooller *in* Strahan 1983; Russell and Renfree 1989).

Vose (1973) cautioned that the existence of the honey possum in the wild is directly dependent on a continued supply of blossoms from eucalyptus, *Banksia* and *Callistemon*. Large-scale urbanization and habitat destruction in southwestern Australia is affecting the feeding of this species. Kennedy (1992) indicated that it already had lost up to 50 percent of its habitat and designated it "potentially vulnerable."

The geological range of the family Tarsipedidae is late Pleistocene to Recent in Australia (Marshall 1984).

DIPROTODONTIA; Family ACROBATIDAE

Feather-tailed Possums

This family of two genera and two species occurs only in New Guinea and eastern Australia. The two genera, *Dis-*

toechurus and *Acrobates,* once were usually placed in the family Phalangeridae, but serological data led to their transfer to the Burramyidae (Kirsch 1977*b*). For a time they were considered to have affinity to the Petauridae, but new serological and morphological studies indicated a closer relationship to the Tarsipedidae and, in any case, that they warranted inclusion in an entirely separate family (Aplin and Archer 1987; Archer 1984*a*; Baverstock, Birrell, and Krieg 1987). The latter arrangement was followed by Groves (*in* Wilson and Reeder 1993), though Corbet and Hill (1991) continued to include the two genera in the Burramyidae, and Szalay (1994) considered them to constitute only a tribe of the Petauridae.

Head and body length is 60–120 mm and tail length is 65–155 mm. The distichous or "pen" tail is distinctive in having paired lateral fringes of long stiff hairs that give a featherlike appearance. The genus *Acrobates* has a narrow patagium, or gliding membrane, that extends from the forelimbs to the hind limbs. The dental formula is: (i 3/2, c 1/0, pm 3/3, m 4/4) $\times$ 2 = 40. Aplin and Archer (1987) provided a detailed technical diagnosis of the family based in part on the structures of both the external and the internal ear. The external ear is uniquely complex, with a prominent anterior helical process, paired antitragal processes, and a well-defined bursa. In the internal ear the auditory bulla lacks an alisphenoid component and instead is formed from the petrosal, squamosal, and ectotympanic bones; the bulla is underrun medially and posteriorly by secondary tympanic processes from basi- and exoccipital bones; the primary tympanic cavity is complexly compartmentalized by numerous septa. Other diagnostic characters include a distinct lingual eminence at the rear of the tongue, a long upper canine that projects well below the level of the upper incisor teeth, greatly enlarged posterior palatal vacuities, and an intestinal caecum moderately to very elongate. The Burramyidae and Petauridae lack these and other key features, but the Tarsipedidae show some similarity in the structure of the internal ear and associated basicranial components.

In a study of the organogenesis and fetal membranes of *Distoechurus,* Hughes et al. (1987) found that the bilaminar yolk sac is totally invasive and that the maternal endometrial glands exhibit total degeneration. These two features are without parallel in marsupials so far studied and may provide further evidence of the distinctiveness of the Acrobatidae.

DIPROTODONTIA; ACROBATIDAE; Genus DISTOECHURUS
Peters, 1874

Feather-tailed Possum

The single species, *D. pennatus,* is found in suitable habitat throughout New Guinea (Ziegler 1977).

Head and body length is 103–32 mm, tail length is 126–55 mm, and weight is 38–62 grams (Flannery 1990*b*). The general body coloration is dull buff, light brown, slightly darker and more olivaceous, or slightly darker and grayer. In contrast to the dull, plain body, the head is strikingly ornamented. The face is white, with two broad well-defined dark brown or black bands that pass from the sides of the muzzle through the eyes to the top of the head just between the ears, and there is a conspicuous black patch just below each ear. The basal part of the tail is well furred, and the remainder is nearly naked but fringed laterally with long, relatively stiff hairs. The pelage is soft, thick, and woolly.

Feather-tailed possum *(Distoechurus pennatus)*, photo by P. A. Woolley and D. Walsh.

Gliding membranes are not present. The claws are sharp and curved; the terminal pads of the digits are not expanded. The eyes are large, and the ears are small and naked. The tip of the tail is prehensile. Females have one medially placed teat in a well-developed pouch, which opens anteriorly. Two lateral teats have also been reported.

This possum is primarily an inhabitant of young regrowth and disturbed forest but also is found in gardens and may occur in primary rainforest. The elevational range is sea level to 1,900 meters. It sometimes nests in tree hollows but probably also nests in leafy vegetation. It is active, arboreal, and nocturnal and has been reported to eat blossoms, fruit, insects, and other invertebrates. Most individuals seem to nest alone, but an adult male and an adult female with a single pouch young were collected together in a leafy nest in a small tree in January. Other females that were lactating or that had pouch young have been collected from June to October. Individuals have been kept in captivity for as long as 20 months (Collins 1973; Flannery 1990*b*; Woolley 1982*a*; Ziegler 1977).

DIPROTODONTIA; ACROBATIDAE; **Genus**
ACROBATES
Desmarest, 1817

Pygmy Gliding Possum, or Feathertail Glider

The single species, *A. pygmaeus*, occurs along the eastern coast of Australia, from the extreme northern tip of the Cape York Peninsula to southeastern South Australia (Ride 1970). Another named species, *A. pulchellus*, was based on a single specimen from a small island to the north of New Guinea and is usually considered to represent an introduced pet *A. pygmaeus* (Groves *in* Wilson and Reeder 1993). However, Flannery (1990*b*) suggested that *Acrobates* probably does occur in New Guinea but has not yet

been discovered because the most likely habitat has not been well explored.

Head and body length is 65–80 mm, tail length is 70–80 mm, and weight is 10–14 grams (Russell *in* Strahan 1983). The pelage is soft and silky. The upper parts are grayish brown, and the underparts and inner sides of the limbs are white. The ears are sparsely haired and bear tufts of hair at their base. The tail is light brown throughout, the ventral surface being somewhat paler than the dorsal surface.

This beautiful, delicate creature is the smallest marsupial capable of gliding flight and is indeed, as the name *Acrobates pygmaeus* indicates, a pygmy acrobat. The gliding membrane is a very narrow fold of skin, fringed with long hairs along its margin, that extends from each limb along the sides of the body. The structure of the tail is such that it provides an additional plane: it is flattened by a fringe of hairs along the sides, a characteristic that has given *Acrobates* the common name "feathertail glider." The fringe of hairs spreads from each side to a total width of 8 mm. The tip of the tail is without hair below, and the tail is somewhat prehensile. The tips of the fingers and toes have expanded, deeply scored or striated pads, which assist the animal in clinging to surfaces. The fourth digit on all four limbs is the longest, and the claws of the digits are sharp. The well-developed pouch encloses four mammae and is lined with yellow hair.

This active, agile marsupial inhabits sclerophyll forest and woodland and is much like a flying squirrel *(Glaucomys)* in its actions. It can glide from one tree to another as well as travel short distances, from limb to limb, by air. Glides of up to 20 meters can be achieved (Flannery and Schouten 1994). Because of its small size and nocturnal habits, *Acrobates* often is overlooked. Some individuals are caught by domestic cats or are discovered when trees are felled. This animal makes small, spherical nests of dry leaves in hollow branches and knotholes of trees at a height of 15 meters or more. During cold days it may become torpid in the nest. Jones and Geiser (1992) experimentally induced deep torpor, during which body temperature fell to 2° C, for periods of up to 5.5 days. Insects, including larvae,

Pygmy gliding possum *(Acrobates pygmaeus)*, photo by Stanley Breeden.

probably form the main part of the diet. Nectar and other plant products also are eaten.

The feathertail glider does not appear to be territorial, shows remarkable conspecific tolerance, and has an extended breeding season (Collins 1973; Fleming and Frey 1984; Ward 1990*b*; Ward and Renfree 1988*a*, 1988*b*). Family units consisting of one or both parents plus the offspring of one or two litters are present for much of the year, and such units freely share their nests with other individuals. The resulting aggregations may be the basis of reports that families contain up to 16 individuals. Artificial nest boxes have yielded up to 14 animals, though groups more commonly have 2–5 individuals. In Victoria adult males are usually solitary from January to April and then form pairs with the females from May to August. Births occur there from July to February, with a peak from August to November. Females undergo a postpartum estrus and a period of embryonic diapause and normally produce two litters during the season. A female may carry as many as 4 young in her pouch, but litter size averages 2.5 at weaning. When they are well furred, the young ride on the mother's back. The young are left in the nest at 60 days and weaned at 95–100 days. A second litter may then be born immediately. The young from the first litter remain with the mother while she raises the second litter. Sexual maturity is attained at 8 months in females and by 12–18 months in males. According to Jones (1982), a captive specimen lived 7 years and 2 months.

Order Xenarthra

Xenarthrans

This order, the living components of which are 4 families, 13 genera, and 29 species, inhabits the south-central and southeastern United States, Mexico, Central America, and South America. At least 2 additional genera and 2 species are thought to have occurred in the West Indies until about 500 years ago. The name Xenarthra usually has been applied at the subordinal level, with the name Edentata being given to an order that included the Xenarthra and the supposedly ancestral suborder Palaeanodonta. However, Glass (1985) explained that the Palaeanodonta are ancestral to the Pholidota, not to the Xenarthra, and that the name Edentata should be considered a synonym of the name Pholidota. In a recent review of fossil and morphological evidence, Rose and Emry (1993) also indicated a connection between the Palaeanodonta and the Pholidota, found no association between the Pholidota and the Xenarthra, and concluded that the Xenarthra form a separate order that long has been isolated from all other placental mammals. Although Novacek (1992) has continued to suggest a possible affinity between the Pholidota and the Xenarthra, there now does seem to be a consensus in using the name Xenarthra at the ordinal level (Barlow 1984; Corbet and Hill 1991; Engelmann 1985; Gardner *in* Wilson and Reeder 1993; Wetzel 1985a). The living Xenarthra are sometimes divided into the following groups (Barlow 1984; E. R. Hall 1981; Romer 1968):

Infraorder Pilosa (sloths)
    Superfamily Megalonychoidea
        Family Megalonychidae
        Family Bradypodidae
Infraorder Vermilingua (anteaters)
    Superfamily Myrmecophagoidea
        Family Myrmecophagidae
Infraorder Cingulata (armadillos)
    Superfamily Dasypodoidea
        Family Dasypodidae.

Engelmann (1985) indicated that the Vermilingua are part of the Pilosa. Based on albumin systematics, however, Sarich (1985) indicated that the sloths, anteaters, and armadillos diverged at least 75–80 million years ago and that they are at least as distinct from one another as are carnivores, bats, and primates. That study supports other work suggesting that the Xenarthra separated from other placental mammals prior to the evolution of the latter into modern orders.

The three xenarthran infraorders are very different from one another and highly specialized. Living members vary in head and body length from about 125 mm for *Chlamyphorus* to 1,200 mm for the giant anteater, *Myrmecophaga*. The hand usually has two or three digits that are much larger than the others, when present, and all the fingers have long, sharp, strong claws. The foot usually has five toes. The bones of the lower arm are separate, whereas the bones of the lower leg are either separate, united at the ankle, or joined at both ends. The mammae are located near the armpit, on the chest, or on the abdomen. Females have a common urinary and genital duct. The testes are located in the abdominal cavity between the rectum and the urinary bladder. Xenarthrans have a double posterior vena cava that returns blood from the posterior part of the body to the heart, whereas most other mammals have only a single vena cava. The skull is elongated in anteaters but short and rounded in sloths. Clavicles are present in living xenarthrans, but the fossil glyptodonts (large, armored relatives of the armadillos) lacked these bones. The neck vertebrae vary in number from six to nine, and the pelvis is narrow and elongate.

Although the former technical designation of this order, Edentata, signifies toothlessness, only the anteaters actually have no teeth. In fact the giant armadillo *(Priodontes)* may have as many as 100 teeth (one of the largest numbers of teeth in mammals), but these are usually small and vestigial. None of the living xenarthrans has incisors or canines. The cheek teeth of the tree sloths and armadillos lack enamel. They are open at the root, a condition incident to continuous growth throughout life. The premolars usually resemble the molars, so the cheek teeth present a uniform series, except in *Choloepus*, in which caninelike teeth are present. Some fossil forms had canines, and the glyptodonts had lobed teeth.

The order is distinguished from all others by what are known as xenarthrous vertebrae. There are secondary and sometimes even further articulations between the vertebrae of the lumbar series. They lend support, particularly to the hips, which is especially valuable to the armadillos for digging. There are also ischial articulations with the spinal column, thus incorporating into the sacrum vertebrae what in other mammals would be caudal elements (Glass 1985; E. R. Hall 1981).

Armadillos and the giant anteater live on the ground and are active by day or night. Other anteaters dwell mainly in trees and are usually nocturnal. Sloths are restricted to trees and are active mainly at night. Xenarthrans are chiefly solitary but may form small, loose associations. All have a good sense of smell, anteaters have poor eyesight and hearing, sloths have good vision but poor hearing, and armadillos have relatively good vision and hearing. Anteaters and some armadillos are basically insectivorous.

Xenarthrans: Top left, two-toed sloth *(Choloepus didactylus)*; Top right, lesser anteater *(Tamandua tetradactyla)*; Bottom, giant armadillo *(Pridontes maximus)*, photos by Bernhard Grzimek.

Other armadillos are omnivorous, feeding on invertebrates, small vertebrates, some plant material, and carrion. Sloths are herbivorous.

The Xenarthra now form an exclusively New World order, though there are some fossil records from the Eocene of Europe and possibly Asia. South American fossils are known from as far back as the Paleocene. According to Thenius (*in* Grzimek 1990), the Xenarthra may have been distributed worldwide in the Cretaceous but evidently became restricted to South America and remained there for most of their history, evolving into numerous groups. Not until the middle Pliocene, a few million years ago, did some xenarthrans begin to invade North America. These were giant sloths, and they were followed in the Pleistocene (Ice Age) by glyptodonts and armadillos. Anteaters and tree sloths moved no farther north than southern Mexico. The Xenarthra once were far more diverse than today; there are known to be 10 times as many fossil as living genera.

The Pilosa sometimes were said to include an additional late Recent superfamily, the Mylodontoidea, and family, the Mylodontidae. This view was based on the supposed survival of the genus *Mylodon* Owen, 1840, one of the giant ground sloths, into modern times. In April 1888 specimens assigned to the species *M. listai* were found in a cave

named Cueva Eberhardt by Dr. Nordenskjöld. The cave is situated near Bahia Ultima Esperanza (51°35′ S, 72°38′ W), Territorio de Magallanes, Argentina. The specimens supposedly were found in direct association with human remains, tools, and bones and pieces of fur of the guanaco (*Lama guanicoe*). The sloth remains consisted of several pieces of hide with long, reddish hair and studded with oblong dermal ossicles similar to those previously known from other species of *Mylodon*. The find was reported by Einar Lonnberg in 1899 in a paper entitled "On some remains of '*Neomylodon listai*' Ameghino." There long was speculation that these specimens represented a population that may still have been living at the time of discovery (Allen 1942; Heuvelmans 1958), but they now have been radiocarbon dated to about 13,500 B.P. Ground sloth remains from other sites in South America have been dated as late as 8639 B.P., and it is not likely that the Mylodontidae survived much longer than that (E. Anderson 1984; P. S. Martin 1984). Nonetheless, Oren (1993) noted that native peoples throughout Brazilian Amazonia report the current or very recent existence of a creature the descriptions of which are surprisingly consistent with the characters expected of a relatively small (approximately 1.8 meters long), forest-dwelling, mylodontid ground sloth.

XENARTHRA; **Family MEGALONYCHIDAE**

West Indian Sloths and Two-toed Tree Sloths

This family includes a single living genus, *Choloepus*, with the two species of two-toed tree sloths of Central and South America and at least two genera and two species of sloths that evidently survived in the West Indies until around the time of the European invasion of the Americas, 500 years ago.

Another 11 genera of sloths are known from the West Indies, including one from Curacao (Woods 1990; Woods and Eisenberg 1989), and there is a possibility that some of them lived into historical times. Humans began moving into the West Indies about 7,000 years ago, though the initial invasion of advanced peoples, who introduced agriculture and pottery, occurred around 2,000 years ago (Rouse 1989). P. S. Martin (1984) suggested that most of the West Indian sloths did not disappear until after this human invasion. In his compilation of the living and recently extinct mammals of North America, E. R. Hall (1981) included the megalonychid genus *Acratocnus* Anthony, 1916, with the single species *A. odontrigonus* known from skeletal material collected in caves on Puerto Rico. However, Morgan and Woods (1986) reported that the youngest radiocarbon date associated with a sloth in Puerto Rico is 13,180 B.P. and that apparently none of the megalonychid genera of Puerto Rico and Cuba survived past the late Pleistocene or were associated with people. Accordingly, the only West Indian sloths included herein are two described genera from Hispaniola.

Choloepus traditionally was placed together with *Bradypus* in the family Bradypodidae, but available evidence now suggests that these two tree sloth genera represent convergent surviving lines from separate ancestral stocks that also gave rise to two different groups of ground sloths. *Choloepus*, which sometimes is put in its own family, the Choloepidae, is associated with the Megalonychidae, whereas the Bradypodidae show affinity to the long-extinct family Megatheriidae. Webb (1985) suggested that the early Miocene ancestors of both the Megalonychidae and the Megatheriidae were arboreal and that a third family of sloths, the Mylodontidae, was always terrestrial.

The three families, Megalonychidae, Mylodontidae, and Megatheriidae, are distinguished mainly by body size, number of digits, skull shape, and form and number of teeth. Sloths varied in size from about that of a medium dog to that of an elephant. All three families included some members that attained great size and were fully terrestrial. Those genera of the Megalonychidae that may have been present in the West Indies in late Recent times were not among the larger kinds. In appearance they seemed intermediate to living anteaters and tree sloths. They originally were thought to have been ground dwellers, but some may have been partly or largely arboreal. For the most part, they were probably inoffensive plant eaters. Additional characters of the Megalonychidae, as expressed in the surviving *Choloepus*, are given below in the account of that genus.

XENARTHRA; MEGALONYCHIDAE; **Genus SYNOCNUS**
Paula Couto, 1967

Lesser Haitian Ground Sloth

The single species, *S. comes*, is known only by skeletal remains from Haiti (E. R. Hall 1981). According to Hall, this genus is characterized by trigonal canine teeth, a sagittal crest on the skull, a large preorbital fossa, slender limbs, and wide caudal vertebrae. There is a conspicuous tubercle at the middle of the shaft of the femur and a shorter and less outwardly bowed neck on the femur. The animal probably weighed about 23 kg. Webb (1985) stated that in size and many anatomical features this genus is closely comparable to *Choloepus*; both genera have enlarged caniniform teeth and short crania strongly flexed downward. *Synocnus* may have been semiarboreal. Its remains were associated with fragments of pottery and with the bones of humans and domestic pigs. Morgan and Woods (1986) indicated that some of the sites where *Synocnus* has been found date from 3715 B.P. The genus seems to have been exterminated by people in Recent times.

XENARTHRA; MEGALONYCHIDAE; **Genus PAROCNUS**
Miller, 1929

Greater Haitian Ground Sloth

This genus of a single species, *P. serus*, was founded on a thigh bone collected in a cave on Haiti (E. R. Hall 1981). Larger and heavier than *Synocnus*, the animal probably weighed about 70 kg. Its remains were found in the same limestone cave as those of *Synocnus* and were associated with fragments of pottery and with the bones of humans and domestic pigs.

XENARTHRA; MEGALONYCHIDAE; **Genus CHOLOEPUS**
Illiger, 1811

Two-toed Tree Sloths

The single known genus, *Choloepus*, contains two species (Cabrera 1957; E. R. Hall 1981; Wetzel and Avila-Pires 1980):

C. didactylus, east of the Andes in Colombia, Venezuela, Guianas, Ecuador, Peru, and northern Brazil;
C. hoffmanni, Nicaragua to Peru and central Brazil.

Head and body length is 540–740 mm, the tail is absent or vestigial, and weight is 4.0–8.5 kg. The pelage consists of long guard hairs and short underfur. The coloration is gray-

Greater Haitian ground sloth *(Parocnus serus)* (the scale is in inches), photo by P. F. Wright of bone material in U.S. National Museum of Natural History.

Two-toed sloth *(Choloepus didactylus)*, photo by Bernhard Grzimek.

ish brown with a paler face. The shoulders and top of the head are the darkest. The coat often has a greenish cast, produced by algal growth on the skin during wet seasons.

According to Aiello (1985), the hairs of both *Choloepus* and *Bradypus* differ in form and structure from those of all other mammals, apparently being specialized to encourage colonization by algae. Although the sloths usually are said to benefit from this arrangement by being camouflaged, they actually may gain nutrients either by absorption through the skin or by licking the algae.

The limbs are long, but unlike in *Bradypus,* the forelegs are only slightly longer than the hind legs. All limbs terminate in narrow, curved feet. In *Choloepus* there are only two digits on each forefoot, and these are closely bound together with skin for their entire length; each is armed with a large hooklike claw about 75 mm in length. The hind feet have three toes with hooklike claws. The claws and limbs are used for hanging from tree branches.

The skull is rounded and short in the facial region. The teeth usually number 5/4 on a side ($\times$ 2 = a total of 18 teeth). The teeth grow throughout life and have cupped grinding surfaces. Nearly all mammals have seven neck vertebrae, but two-toed sloths usually have six or seven, occasionally eight. The ears of *Choloepus* are inconspicuous, and the eyes are directed forward. The stomach is complex for the digestion of vegetation. Females have two mammae located in the chest region.

Two-toed tree sloths, like the related three-toed sloths, spend nearly their entire life upside down, even eating, sleeping, mating, and giving birth in that position. They sleep with the head raised and placed between the forelegs on the chest. Often all four feet are placed so close together that the animal has the appearance of a bunch of dried leaves. Frequently they choose a position that permits them

to rest their back on a lower limb or sit in a fork; the feet, however, are always hooked to a branch. Almost all their movements are extremely slow and by means of a hand-over-hand motion. They swim voluntarily, using a breast stroke and with the body right side up. When on the ground, where they do come about once a week to urinate and defecate, they have some difficulty standing, but they can rise on their palms and soles and crawl for short distances. Climbing is accomplished with seemingly little effort and some speed even when supports are vertical (Mendel 1981).

Unlike that of most mammals, the body temperature of sloths varies considerably depending on the temperature of their surroundings. As a result, they are physiologically restricted to a limited equatorial habitat of constant temperature. *Choloepus* has the lowest and most variable body temperature of any mammal, ranging from a low of 24° C to a high of 33° C.

It may seem strange that such slow-moving creatures have been able to survive in regions where there are large birds of prey and carnivorous mammals that can climb well. When sloths descend to the ground, they slowly drag their bodies along the surface, which makes them easy prey. Often they are killed by jaguars, ocelots, and other cats. They do defend themselves, however, and can strike fairly quickly with their long forefeet, which have large, hooked, sharp claws. They also use their teeth effectively, and both means of defense can inflict severe wounds. If left alone, however, sloths are inoffensive. They have color vision, a good sense of smell, and a poorly developed sense of hearing. They probably owe their survival to such factors as protective coloration, remaining motionless during the day, nocturnal habits, heavy fur, thick skin, and an extreme tenacity to life.

Two-toed sloth *(Choloepus didactylus)*, photo by Carl B. Koford. Inset: newborn young, photo by Ernest P. Walker.

Two-toed sloths inhabit tropical forests and are almost entirely arboreal and nocturnal. According to Sunquist and Montgomery (1973), *C. hoffmanni* in the Panama Canal Zone shows almost no crepuscular activity; most individuals remain inactive until the hour before sunset. In the study area *Choloepus* was much more active than *Bradypus* and was seldom found occupying the same tree on successive days. It has a larger home range, often 2–3 ha. (Eisenberg 1989). Because of its greater mobility, *Choloepus* is exposed to a wider variety of foods and tends to eat a wider variety than *Bradypus*. Its diet consists of leaves, tender twigs, and fruits. The arms are often used to pull food within reach of the mouth.

According to Meritt and Meritt (1976), in Panama *C. hoffmanni* is found at a ratio of 11 females to 1 male. They speculated that such a proportion would be extremely advantageous to a species that has an unusually long gestation period. On the other hand, they noted that in *C. hoffmanni* the females tend to gather in groups but males are loners, so the sampling techniques may have missed the latter and not adequately expressed the true sex ratio. Merritt (1983) indicated that several individuals could be kept together in captivity.

There is a single young; it hooks its tiny claws into the long hair on the mother's breast and abdomen, where it clings while she climbs or rests. Eisenberg and Maliniak

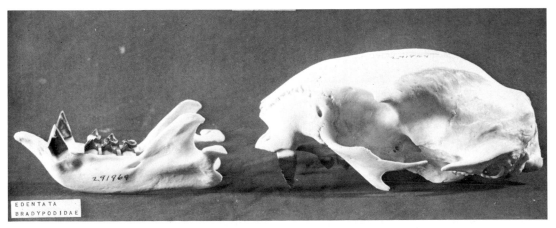

Skull of two-toed sloth *(Choloepus hoffmanni)*, photo by P. F. Wright of skull in U.S. National Museum of Natural History.

(1978), using a controlled breeding program and subjecting females to x-ray examination, estimated the true gestation period of *C. hoffmanni* at around 11.5 months. They found that when lactation and successful rearing take place, interbirth intervals may average 14–16 months. Sexual maturity may not occur until 3.5 years in females and occurs at 4–5 years in males. Females may live for 20 years with no decline in fecundity. Meritt (1985b) reported that the young weigh 350–454 grams at birth, first hang upside down at 20–25 days, and regularly feed away from the mother at 5 months; they may, however, continue a regular association with the latter for at least 2 years. Veselovsky (1966) reported that a female *C. didactylus* at the Prague Zoo gave birth to a single young 5 months and 20 days after mating had been observed. Therefore, there seemingly may be a considerable difference in gestation period for the two species of the genus. *Choloepus* readily adapts to and breeds in captivity, and there are numerous records of individuals living more than 10 years (McCrane 1966; Snyder and Moore 1968; Veselovsky 1966). According to Jones (1982 and pers. comm., 1995), a captive specimen of *C. didactylus* was still living after 27 years and 9 months, and a captive *C. hoffmanni* was still living after 32 years and 1 month.

XENARTHRA; Family BRADYPODIDAE; Genus BRADYPUS
Linnaeus, 1758

Three-toed Tree Sloths

The single known genus, *Bradypus*, contains two subgenera and three species (Wetzel 1985a; Wetzel and Kock 1973):

subgenus *Bradypus* Linnaeus, 1758

B. variegatus, Honduras to northern Argentina;
B. tridactylus, southern Venezuela, Guianas, northern Brazil;

subgenus *Scaeopus* Peters, 1864

B. torquatus, coastal forests of eastern Brazil.

Paula Couto (1979) considered *Scaeopus* to be a full genus, but such a view was not followed by Corbet and Hill (1991) or Gardner (*in* Wilson and Reeder 1993).

Head and body length is 413–700 mm, tail length is 20–90 mm, and adult weight is 2.25–6.20 kg. The pelage consists of long, thick overhairs that are grooved longitudinally and short underfur of fine texture. Most hairs point downward when the animal is hanging beneath a branch. Thus, most of the hairs are directed opposite to the hairs of most other mammals. The general coloration is grayish brown, slightly darker on the head and face, and on the shoulders there is usually a light area with brown markings. Frequently the coloration appears to be greenish because of algal growth on the coat. *B. torquatus*, the maned sloth, has long, dark hairs on the head and neck.

These sloths have some resemblance to *Choloepus*, but the forelegs are substantially longer than the hind legs and there are three digits on both the forefeet and the hind feet. The digits are closely united, and each terminates in a long, hooklike claw. There is a tail in *Bradypus*, though it is short, stout, and blunt, having the appearance of the stump of an amputated limb. Whereas most mammals have seven neck

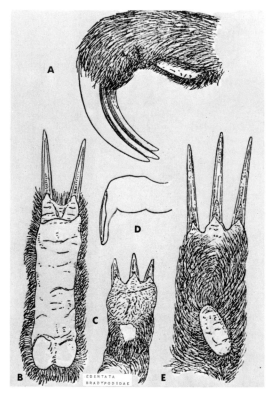

A. Right forefoot of *Bradypus tridactylus* from inner side with hair parted to reveal pad. B. The same of *Choloepus*, showing the suppression of the fourth digit and the entirely naked sole when the hairs are turned aside. C. The same of a young specimen. D. The same of a fetal young from the side showing the extension of the digital pad to guard the tip of the claw. E. The same as above with digits abnormally spread. Photos from *Proc. Zool. Soc. London*, 1924.

vertebrae, *Bradypus* has eight or nine, thus allowing greater flexibility, a desirable feature in an animal that is so limited in its other movements. In this genus the head can be turned through an arc of 270°. The eighth and ninth neck vertebrae sometimes bear a pair of short, movable ribs. The head is small and round, and the eyes and ears are small. As in the Megalonychidae, there are 5 upper and 4 lower teeth on each side, thus a total of 18 teeth. Females have two mammae in the chest region.

Three-toed sloths inhabit forests and spend practically their entire life in trees, where they hang beneath the limbs or sit in a fork. Arboreal locomotion is usually by using the claws as hooks to move along beneath a branch or to climb a tree. In the Panama Canal Zone, Sunquist and Montgomery (1973) found *B. variegatus* to be active both by night and by day. Sloths apparently descend to the ground about once or twice a week to urinate and defecate and on occasion to move from one tree to another. Normal terrestrial locomotion is by a slow crawl on the soles and forearms (Mendel 1985). They swim readily. If attacked on the ground, they endeavor to catch their adversary with their claws, with which they can inflict severe wounds. Although normally the movement of sloths in trees is slow and methodical, they can progress rapidly when pressed.

Sunquist and Montgomery (1973) reported that in *B. variegatus*, at least, individuals changed trees about four times less often from day to day than did individuals of

Three-toed sloth *(Bradypus torquatus)*. Young is clinging to its mother with its right arm and leg visible on the side of her body. Photo by Joao Moojen. B. Three-toed sloth *(B. tridactylus)*, photo from New York Zoological Society.

Choloepus hoffmanni. Moreover, only 11 percent of the individuals of *B. variegatus* showed movements of 38 meters or more per day, whereas 54 percent of the individuals of *C. hoffmanni* had a daily movement of 38 meters or more. Since *Bradypus* tends to stay in one tree for a long period, feeding is restricted to that tree and to other plants supported by the tree. *Bradypus* therefore tends to be a more specialized feeder than *Choloepus,* the members of which are exposed to many more species of plants from which to choose. Young leaves, tender twigs, and buds, especially those of *Cecropia,* form almost the entire diet of *Bradypus.* These items are pulled to the mouth with slow movements of the forelegs. Crandall (1964) stated that three-toed sloths are more difficult to breed in captivity than are two-toed sloths, possibly because of their more specialized diet in the wild.

Montgomery and Sunquist (1978) reported that the population density of *B. variegatus* on Barro Colorado Island was 6–7/ha. and that individual home range usually was less than 2 ha. Members of the genus *Bradypus* are solitary. In contrast to *Choloepus hoffmanni,* in which 90 percent or more of the adults taken in the wild have proved to be females, there seems to be a one-to-one sex ratio in *B. variegatus.* In some populations of *Bradypus* breeding appears to take place in March–April, while in others, including *B. torquatus* (Pinder 1993), it occurs throughout the year. There is usually a single young. The gestation period of *B. tridactylus* reportedly is 106 days (Hayssen, Van Tienhoven, and Van Tienhoven 1993). Studies of *B. variegatus* on Barro Colorado Island (Montgomery and Sunquist 1978) indicate that females give birth in successive years, the gestation period is 5–6 months, and the young cease nursing at 3–4 weeks but depend on the female for mobility for another 5 months. Merrett (1983) stated that the single young weighs 200–250 grams at birth and is carried on the mother's abdomen.

All species of *Bradypus* may be jeopardized by habitat

Three-toed sloth *(Bradypus tridactylus)*, photo by Luiz Claudio Marigo.

Silky anteater *(Cyclopes didactylus)*, photo by Bernhard Grzimek.

The hands of the giant anteater *(Myrmecophaga tridactyla)* as it walks on its knuckles to protect its long nails, photo by P. F. Wright of mounted specimen in U.S. National Museum of Natural History.

destruction and excessive hunting. One species in particular, *B. torquatus,* is classified as endangered by the IUCN and the USDI. The Atlantic coastal forest of eastern Brazil, which this sloth inhabits, is being rapidly cut over for lumber extraction and charcoal production and to make way for plantations and cattle pasture (Thornback and Jenkins 1982). It recently has been found to have an unusual distribution, with large gaps between major populations, one result being that it is absent from most of the protected forest reserves within its overall range. The subspecies *B. variegatus brasiliensis* occurs in the same region of eastern Brazil and also is threatened by habitat destruction and hunting pressure but has a more continuous distribution (Oliver and Santos 1991). Another subspecies, *B. variegatus boliviensis,* is on appendix 2 of the CITES.

XENARTHRA; Family MYRMECOPHAGIDAE

Anteaters

This family of three genera and four species occurs in southern Mexico, Central America, and South America as far south as Paraguay and northern Argentina.

Head and body length ranges from 153 mm in *Cyclopes* to about 1,200 mm in *Myrmecophaga.* Males are usually larger than females. The hair is coarse in *Myrmecophaga* and *Tamandua* but fine and silky in *Cyclopes.* Anteaters have elongated, tapered snouts and tubular mouths. The long tongue has minute posteriorly directed spines (filiform papillae) and is covered with a sticky secretion from the large salivary glands when the animal is feeding. The tail is prehensile in *Cyclopes* and *Tamandua* but not in *Myrmecophaga.* There are two to four fingers and four or five toes. The claws on the fingers are long and sharp and can be used as powerful weapons of defense, though their usual function is the opening of ant and termite nests. The ears are short and rounded and the eyes are small. Anteaters usually have no teeth. The skull is elongate and would appear to be fragile, but the bony walls are thick and unusually hard. The mammae of females are located in the chest and abdominal regions.

Anteaters inhabit tropical forests and savannahs, sheltering in trees, hollow logs, or burrows constructed by other animals. They feed on ants, termites, and other insects. They use their strong foreclaws to rip open ant and termite nests, and they capture the insects with their long, sticky tongue. They have a good sense of smell, but sight and hearing are not well developed. Anteaters occur singly or in pairs, usually a female and her young.

The geological range of this family is middle Eocene in Europe, early Miocene to Recent in South America, and Recent in Central America and Mexico (Barlow 1984). Shaw and McDonald (1987) also identified a fossil of *Myrmecophaga tridactyla* from an early Pleistocene site in northwestern Sonora, Mexico, more than 3,000 km north of the present range of the species.

Giant anteaters *(Myrmecophaga tridactyla):* young riding on back of parent, and head with tongue extended, photos from New York Zoological Society; skull, photo by P. F. Wright of specimen in U.S. National Museum of Natural History.

XENARTHRA; MYRMECOPHAGIDAE; **Genus**
MYRMECOPHAGA
Linnaeus, 1758

Giant Anteater

The single species, M. *tridactyla,* is found from southern Belize to northern Argentina (Cabrera 1957; E. R. Hall 1981).

Head and body length is usually 1,000–1,200 mm, tail length is 650–900 mm, and weight is usually 18–39 kg. Moeller (*in* Grzimek 1990) indicated that individuals in zoos have weighed as much as 60 kg. The body is narrow. The color is gray, and the diagonal stripe is black with white borders. The hair is coarse, stiff, and longest on the tail. This animal is easily recognized by its large size, cylindrical snout, diagonal stripe, and bushy tail.

The powerful claws on the hands and the long, extensible tongue are the instruments for food gathering. Termite and ant mounds are ripped apart with the claws, and the eggs, cocoons, and adult ants are picked up with the saliva-coated tongue. Moeller (*in* Grzimek 1990) noted that *Myrmecophaga* may consume as many as 35,000 ants or termites in a single day. Beetle larvae are also eaten in the wild, and fruit is taken on occasion, at least in captivity. The usual diet in zoos is eggs and milk beaten together, mealworms, or ground beef. There are three large claws and one small claw on each hand and five relatively small claws on each foot. The salivary glands appear to secrete only when the animal is feeding. The tongue can be extended as much as 610 mm, but its diameter is only 10–15 mm at the widest point, just outside the mouth.

The giant anteater is found in savannahs, grasslands, swampy areas, and humid forests. It seems to be active mainly during daylight in areas uninhabited by people and during the night in densely populated areas. In a Brazilian national park, Shaw, Machado-Neto, and Carter (1987) found activity to begin at 1300–1400 hours, peak at 1800–1900 hours, and gradually diminish until 0200 hours.

Montgomery and Lubin (1977) found that a radio-tracked individual was active from 1900 to 0800 hours. *Myrmecophaga* walks with its nose close to the ground and with the side and knuckles of the hands on the ground. It takes to water readily and can swim across wide rivers. Unlike *Tamandua* and *Cyclopes,* the adult giant anteater does not climb trees, but Widholzer and Voss (1978) indicated that it is adept at climbing out of zoo enclosures. Although it is a powerful digger, it does not construct burrows but merely seeks secluded spots in which to curl up, tucking its head between its forelegs and covering its head and body with the fanlike tail. A slow, clumsy gallop is the usual reaction to danger. It does not fight unless forced, and then it uses its hands to grasp and claw its adversary.

Home ranges of at least 9 sq km (Pinto da Silveira 1969) and 25 sq km (Montgomery and Lubin 1977) have been reported, but extensive studies in a Brazilian national park (Shaw, Carter, and Machado-Neto 1985; Shaw, Machado-Neto, and Carter 1987) suggest considerably smaller sizes. Ranges there averaged 3.67 sq km for adult females and 2.74 sq km for adult males. The female ranges overlapped one another by an average of about 29 percent, and the male ranges by about 4 percent. Minimum population density in the area was calculated to be 1.3/sq km. Animals usually were seen alone. Observations of agonistic interaction, generally between males, suggest that occupied space is defended and may contradict earlier reports that the genus is not territorial. Such interaction varied from slow circling to chases, serious fighting, and injuries. Other observations also suggest that the genus is usually solitary in the wild except for females with young, but Widholzer and Voss (1978) stated that a captive group of three adult males and two adult females did well together.

According to Merrett (1983), breeding occurs throughout the year in captivity and probably also in the wild. The gestation period is about 190 days, but there are records of periods as short as 142 days. The single young weighs 1–2 kg at birth, opens its eyes after 6 days, and is weaned at 4–6 weeks. It is carried on the mother's back for up to a year, by which time it is nearly full grown. Sexual maturity is attained at 2.5–4 years. The young remains with the mother until she again becomes pregnant. According to Jones (1982), a captive specimen lived 25 years and 10 months.

The giant anteater is classified as vulnerable by the IUCN and is on appendix 2 of the CITES. The genus reportedly is now uncommon and localized in most regions. It has disappeared from much of Central America as a result of habitat loss and expansion of the human population. In South America it is jeopardized by hunting, not for use as food, but simply as a trophy or curiosity (Thornback and Jenkins 1982).

XENARTHRA; MYRMECOPHAGIDAE; Genus **TAMANDUA**
Gray, 1825

Lesser Anteaters, or Tamanduas

There are two species (Wetzel 1975):

T. mexicana, southern Mexico to northwestern Venezuela and northwestern Peru west of the Cordillera Oriental;
T. tetradactyla, South America east of the Andes, from Venezuela and Trinidad to northern Argentina and southern Brazil.

Head and body length is 470–770 mm and tail length is 402–672 mm. Meritt (1975) reported that captive specimens weighed 2–7 kg. The body is covered with short, dense, coarse hair, and coloration varies considerably. Wetzel (1975) observed that *T. mexicana* always has a vivid black area on the trunk, continuous from the shoulders to the rump and widening behind the shoulders to encircle the

body. In *T. tetradactyla* this black "vest" is present only in specimens from the southeastern portion of the range of the species, this being the area most removed from the range of *T. mexicana.* The general background coloration in both species varies from blonde to tan or brown. The underside of the tail and its entire terminal portion are naked with irregular black markings. There are four clawed digits on the hand. The claw of the third digit is curved and is the largest; the claw of the first digit is the smallest. Each foot has five clawed digits. The snout is elongated. The mouth opening is about the diameter of a lead pencil. To compensate for the lack of teeth, a portion of the stomach is a muscular gizzard comparable to that of gallinaceous birds.

Tamanduas inhabit tropical forests and savannahs. They walk on the outside of the hand to avoid forcing the tips of the large claws into their palms. An individual may have a daily period of activity that begins at any time of day or night and continues for about eight hours (Montgomery 1985c). Tamanduas commonly shelter in hollow trees. Their movements on the ground appear rather clumsy, and unlike *Myrmecophaga,* they do not seem to be able to gallop. Nonetheless, Montgomery (1985c) reported *T. mexicana* to be scansorial, moving, feeding, and resting both on the ground and in trees. If attacked while in a tree, tamanduas defend themselves by assuming a tripod position formed by the hind feet and the tail, which leaves the arms free. They outstretch the arms and bare their claws until the enemy comes into reach. If, however, tamanduas are attacked while on the ground, they protect the back by leaning against a tree or rock, and they seize the opponent with their strong forearms. In both cases their protection lies in the great strength of their arms and the tearing power of their foreclaws. They feed on ants, tree and ground termites, and bees. Montgomery (1985b) found *T. mexicana* on Barro Colorado Island to consume about 9,000 ants per

Lesser anteater *(Tamandua tetradactyla),* photo by Ernest P. Walker. Inset photo by P. F. Wright of specimen in U.S. National Museum of Natural History.

day. Population density in that area was 0.05/ha. and home range was about 25 ha. Montgomery and Lubin (1977) reported home ranges of 350 ha. and 400 ha. for two *T. tetradactyla*.

Limited reproductive data indicate that usually a single young is born in the spring. Pinto da Silveira (1968) suggested that females are polyestrous, with a gestation period of 130–50 days. Merrett (1983), however, cited reports of 160 and 190 days. He noted also that twins have been recorded, that the young are carried on the back or flanks of the mother, and that the two separate after about a year. A captive specimen lived 9 years and 6 months (Jones 1982).

XENARTHRA; MYRMECOPHAGIDAE; **Genus**
CYCLOPES
Gray, 1821

Silky Anteater

The single species, *C. didactylus,* is found from southern Mexico to Bolivia and Brazil (Cabrera 1957; E. R. Hall 1981).

Head and body length is 153–230 mm and tail length is 165–295 mm. Wetzel (1985a) listed weight as 175–357 grams. The relatively long, prehensile tail is naked on the underside. The pelage is soft and silky with a buffy gray to golden yellow color; it is darker above with a dark line along the top of the head, neck, and back. The tip of the nose is pink, the soles of the feet are reddish, and the eyes are black.

The claws on the second and third fingers of the hand are large, curved, and sharp. The long hind feet have a heel pad and a peculiar "joint" in the sole that permits the claws to be turned back under the foot as grasping organs.

The silky anteater inhabits tropical forests and is active only at night. Sunquist and Montgomery (1973) reported that an individual followed by radiotelemetry was active almost continuously at night from within 15 minutes of sunset until 1.5–3 hours before dawn. Bouts of activity averaged about 4 hours in length. During the day this individual rested on or among shaded vines, in or below the crowns of trees; no two days were spent in the same tree.

The silky anteater is reported to climb about treetops in search of insects and termites (white ants), which it secures with its long, wormlike, sticky tongue. Apparently it almost never descends to the ground, though it can walk well on flat surfaces by placing the sides of the forefeet on the level surface and turning the claws inward. W. J. Schaldach, Jr., obtained a specimen in the city of Matias Romero, Oaxaca, that had walked into the kitchen of his house. In Brazil, C. T. Carvalho observed a *Cyclopes* crossing from one wooded area to another on the ground in the park at the Museo Goeldi.

It is said that the silky anteater frequents the silk-cotton tree *(Ceiba),* which has seed pods that are a massive ball of soft, silverish fibers. The sheen of this silky mass and that of the little anteater are so strikingly alike that when the animal is placed next to a freshly opened pod a person can scarcely tell the difference. This protective coloration helps the silky anteater escape the keen eyes of its chief predators, the harpy eagle, various eagle-hawks *(Spizaetus, Spizastur),* and the spectacled owl *(Pulsatrix).*

When alarmed and taking a defensive attitude, the silky anteater raises up on its hind legs, grasps opposite sides of a limb with its feet, and wraps its tail securely around a twig. The forefeet, with their powerful sharp claws, are held

Silky anteater *(Cyclopes didactylus).* The naked underside of the prehensile tail and the strong claws on the front feet are clearly shown. Photo from New York Zoological Society.

close to the face to strike swiftly and forcibly if the enemy comes within reach. This animal is, however, slow-moving and inoffensive and merely tries to defend itself in the only way possible.

Montgomery (1985*b*, 1985*c*) reported a population density of 0.77/ha. on Barro Colorado Island. An adult male had a home range of about 11 ha. that overlapped the ranges of at least two adult females but not the range of an adjacent male. Female home ranges averaged 2.8 ha. and did not overlap one another. Most of the females were assumed to be carrying or accompanied by young. Depending on their age and sex, individuals ate 700–5,000 ants per day.

Being small, arboreal, and nocturnal, the silky anteater is naturally difficult to locate, and even the natives seldom see it. Little is known of its habits, and it seldom has been kept in captivity for more than short periods, though Merrett (1983) noted that one specimen was maintained for 2 years and 4 months. He suggested also that the gestation period is 120–50 days. One observation indicates that the single young is placed by the mother in a nest of dry leaves in a hole in a tree trunk. Moeller (*in* Grzimek 1975) reported that the young are raised by both parents, which regurgitate semidigested insects to feed it. Sometimes the male anteater carries the baby on his back.

XENARTHRA; Family DASYPODIDAE

Armadillos

This family of 8 living genera and 20 species is found from the southern United States to the Strait of Magellan. The sequence of genera presented here follows that of Cabrera (1957), who recognized two subfamilies: the Dasypodinae for most genera and the Chlamyphorinae for *Chlamyphorus*. A species in this family, *Dasypus novemcinctus*, is the only xenarthran that occurs in the United States.

The word *armadillo* is of Spanish origin and refers to the armorlike covering of these animals. The skin is remarkably modified to provide a double-layered covering of horn and bone over most of the upper surface and sides of the animals and some protection to the underparts and limbs. The covering consists of bands or plates connected or surrounded by flexible skin. The scutes on the back and sides, which are quadrate or polygonal in shape, are usually arranged into shoulder and hip shields with movable rings between them. This body armor is flexible in most kinds of armadillos when they are alive but becomes rigid after they die; it often is sold to tourists as baskets. The skin between the bands and plates provides further flexibility. The limbs, which are covered on their outer or exposed margins with irregular horny shields, can be withdrawn into the space between the body and the shoulder and hip shields. The top of the head has a shield, and the tail is usually encased by bony rings or plates. The scutes are covered by a layer of horny epidermis. The undersurface of the body and the inner surfaces of the limbs are covered with soft, haired skin. In many species hairs also project through openings between the scutes on the back.

As a result of the armorlike covering, the body is heavy; the giant armadillo *(Priodontes)* weighs as much as 60 kg. Head and body length in the family is 125–1,000 mm and tail length is 25–500 mm. The scutes are brown to pinkish in color, and the pelage is grayish brown to white. Most species of armadillo have moderate-sized ears. The snout varies considerably in length and has a long, protrusible tongue. The hands have three, four, or five digits with powerful, curved claws, and the feet have five digits with claws.

The head and left hand of a three-banded armadillo (*Tolypeutes* sp.), with the very large claw bearing much of the weight carried by the left arm, photo by Ernest P. Walker.

The lower bones of the legs are united proximally and distally. In females the mammae are usually located in the chest region, but an abdominal pair is occasionally present.

The skull is dorsoventrally flattened, and the lower jaw is elongate. The teeth are small, peglike, ever-growing, and numerous. There are usually 7—9 teeth in each half of the jaw, but *Priodontes* may have more than 40 small teeth in each jaw.

Armadillos generally inhabit open areas such as savannahs and pampas but also occur in forests. They travel singly, in pairs, or occasionally in small bands. They are terrestrial in habit, powerful diggers and scratchers, and either diurnal or nocturnal. When not active they usually live in underground burrows. They generally walk on the tips of the claws of their fingers and the soles of their feet with the heels of their feet touching the ground. They can run fairly rapidly. Although they can defend themselves with their claws and sometimes by biting, the usual reaction to danger is to run or burrow rapidly into the ground and then anchor themselves in the burrow. If overtaken while running, or if they do not have the opportunity to burrow, some species draw in their feet so that the edges of the armor are in contact with the ground, and a few species roll themselves into a ball. Armadillos have relatively good senses of sight, smell, and hearing. They feed on insects and other invertebrates, some small vertebrates, plant material, and carrion.

The gestation period is prolonged by delayed implantation. In *Dasypus* birth occurs about 120 days after implantation. Armadillos generally give birth to several identical young produced from a single ovum (up to 12 but usually 2–4 and often only 1). The young are covered with a soft, leathery skin that gradually hardens with age.

The geological range of this family is late Paleocene to

Recent in South America and Pliocene to Recent in North America (Barlow 1984).

XENARTHRA; DASYPODIDAE; Genus
CHAETOPHRACTUS
Fitzinger, 1871

Hairy Armadillos, or Peludos

There are three species (Cabrera 1957; Grimwood 1969; Myers and Wetzel 1979; Pefaur et al. 1968; Wetzel 1985*b*; Ximenez, Langguth, and Praderi 1972):

C. vellerosus, western Bolivia to Paraguay and central Argentina;
C. nationi, Bolivia, northern Chile;
C. villosus, northern Paraguay and probably southern Bolivia to central Argentina.

Moeller (*in* Grzimek 1975) considered *Chaetophractus,* as well as *Zaedyus,* to be congeneric with *Euphractus,* and *C. vellerosus* and *C. nationi* to be subspecies of *C. villosus.*

Head and body length is 220–400 mm and tail length is 90–175 mm. Wetzel (1985*b*) listed average weights of about 0.84 kg for *C. vellerosus* and 2.02 kg for *C. villosus.* The head shield of *C. nationi* is about 60 mm long and 60 mm wide. The armor consists of the shield on the head (which has a granular surface), a small shield between the ears on the back of the neck, and the carapace (which protects the shoulders, back, sides, and rump). The banded portion of the carapace has about 18 bands, of which usually 7–8 are movable. These animals have more hairs than do most armadillos. Hairs project from the scales of the body armor, and the limbs and belly are covered with whitish or light brown hairs.

These armadillos usually inhabit open areas and seem best adapted to semidesert conditions. *C. nationi* is found only in grasslands at high altitudes. They are powerful diggers and live in burrows. Greegor (1985) reported that the burrows of *C. vellerosus* are usually on sloping sand dunes

and several meters long and more than a meter deep. Activity is largely nocturnal in summer, to avoid the desert heat, and diurnal in winter. The species *C. villosus,* when pursued, at first attempts to run away, often emitting a characteristic snarling sound. If unable to find a hole, it tries to burrow into the ground. If overtaken while running or if it does not have chance to burrow, this species draws in its feet so that the edges of its armor are in contact with the ground and thus effectively protects itself against canid and avian predators. *C. villosus* anchors itself in its burrow by spreading its feet sideward and bending its body so that the free hind edges of the bands grasp the walls of the burrow. Greegor (1980*b*) found that during a three-day period one individual *C. vellerosus* moved an average of 1,032 meters per night within an overall area of 3.4 ha.

The members of this genus regularly burrow under animal carcasses to obtain maggots and other insects and are said sometimes to burrow into carcasses. Under certain conditions they obtain grubs and insects from a few centimeters below the surface of the ground by the unusual method of forcing a hole in the ground with the head and then turning the body in a circle so that a conical hole is formed without any digging. They have been observed to kill small snakes by throwing themselves upon the snakes and cutting them with the edges of the shell. According to Greegor (1980*a*, 1985), *C. vellerosus* feeds mostly on insects during the summer and takes a substantial number of rodents, lizards, and other small vertebrates. It also relies heavily on plant material, especially in the winter, when over half of its diet consists of vegetation.

Mating in *C. villosus* takes place in September in the Argentine province of Santa Fe. Merrett (1983) stated that in this species the gestation period is 60–75 days and that there is said to be more than one litter annually. Litters usually consist of two young, often one male and one female. The young weigh about 155 grams at birth, open their eyes after 16–30 days, are weaned at 50–60 days, and reach sexual maturity at 9 months. One specimen of *C. villosus* lived in captivity for 23 years and 6 months and another was still living after 22 years and 7 months (Marvin L. Jones, Zoological Society of San Diego, pers. comm., 1995).

Hairy armadillo *(Chaetophractus vellerosus)*, photo by Ernest P. Walker.

Hairy armadillos may be systematically hunted in areas where they burrow extensively in loose farm soil. Their flesh is thought to be good and is frequently eaten by people. Miller et al. (1983) regarded *C. nationi* as scarce in Chile, where it is hunted for its meat and shell. *C. nationi* now is classified as vulnerable by the IUCN and is on appendix 2 of the CITES.

Six-banded Armadillo

The single species, *E. sexcinctus*, occurs in Surinam and east of the Andes from the Amazon Basin of Brazil to northern Argentina (Wetzel 1985*b*).

According to Redford and Wetzel (1985), head and body length is 401–95 mm, tail length is 119–241 mm, and weight is 3.2–6.5 kg. Most individuals have a moderately hairy covering. The prevailing color is yellowish to reddish brown. Although it bears a general resemblance to certain other genera of armadillos, *Euphractus* is distinguished by its pointed and flattened head and its six to eight movable bands. The head shield is composed of rather large plates arranged in a fairly definite pattern. The plates of the well-armored tail are arranged in two to four distinctive bands at the base. Holes in a few plates above the base of the tail seem to be openings for scent glands, since the characteristic odor of the animal has been traced to this region. All five toes on the forefoot have claws, of which the second is the longest.

The habitat is usually dry savannah or the drier parts of wet savannah. Moeller (*in* Grzimek 1975) stated that *Euphractus* is among the most prevalent of armadillos in Argentina. Its den is found in dry areas, and it moves abroad even in bright daylight. It continually digs new passageways in search of food. It generally burrows just one or two meters into the earth and then widens the underground area enough to turn around. Defecation always takes place outside the den. When digging, this animal does not throw the dirt to the side (as do moles) but scratches it up with the forefeet and then throws it behind with the hind feet. The diet consists of plant material and insects. Moeller observed that the only time large numbers of *Euphractus* are seen together is when they gather around the carcass of a dead animal, where they feed on maggots and bits of meat; otherwise *Euphractus* is essentially a solitary creature. Sometimes this armadillo becomes abundant around plantations, where it causes damage by eating sprouting corn and other crops. Its burrows can cause dangerous accidents when horses stumble into them on the pampas.

In Brazil, Carter (1983) found minimum home range to average 93.3 ha. At the Wroclaw Zoo in Poland young may be born at any time of the year after a gestation period of 60–65 days; litters contain one to three young and may include one or both sexes (Gucwinska 1971). A female taken in July in the Mato Grosso of Brazil was pregnant with two embryos, a male and a female, almost ready to be born. Meritt (1976) stated that captive female *Euphractus* have given birth to twins, usually one of each sex. The young weigh 95–115 grams at birth, open their eyes after 22–25 days, take solid food after a month, and mature by 9 months (Redford and Wetzel 1985). One specimen of this genus lived 18 years and 10 months in captivity (Jones 1982).

Pichi

The single species, *Z. pichiy*, occurs in central and southern Argentina and in the Andean grasslands of Chile, south to the Strait of Magellan (Wetzel 1985*b*). Moeller (*in* Grzimek 1975) considered *Zaedyus* to be congeneric with *Euphractus*.

Head and body length is 260–335 mm, tail length is 100–140 mm, and weight is about 1–2 kg. The head shield and body carapace are dark brown with yellow or whitish lateral edges, and the tail shield is yellowish. The posterior edges of the dorsal plates are thickly set with fine blackish

Six-banded armadillo *(Euphractus sexcinctus)*, photo from New York Zoological Society. Inset: sleeping pose showing nonarmored underparts, photo by Ernest P. Walker.

Pichi *(Zaedyus pichiy)*, photo from East Berlin Zoo.

hairs interspersed with longer yellowish brown and whitish bristles. The underparts of the body are covered with coarse, yellowish white hairs. The digits on all limbs are distally separate and have well-developed claws. The ears are very small. This armadillo resembles *Euphractus* externally, but Jorge, Meritt, and Benirschke (1977) found that karyologically *Zaedyus* appears to be more closely related to *Chaetophractus* than to *Euphractus*. It differs from *Chaetophractus vellerosus*, which is about the same size, in having much shorter ears, a narrower head shield, and pointed marginal scutes.

The usual defense reaction of the pichi is to draw in its feet so that the edges of its armor are in contact with the ground. This armadillo shelters in shallow holes, often in sandy soil, and can anchor itself in its burrow by wedging the serrated edges of the carapace into the surrounding dirt. According to Hatcher (*in* Allen 1905), the pichi hibernates in winter, at least in some localities. It eats insects, worms, and any other small animal food it can find, including dead creatures. Redford and Eisenberg (1992) wrote that it also eats plant material, especially pods of the *Prosopis* tree.

Merrett (1983) stated that *Zaedyus* is solitary, probably breeds throughout the year, and has a gestation period of about 60 days. One to three young are born, usually two. They weigh 95–115 grams at birth, are fully weaned at six weeks, and become sexually mature after nine months to a year. According to Jones (1982), a captive pichi lived for nine years.

This genus is fairly common in South America. Its flesh is said to have an excellent flavor and is highly prized by the natives as food. In some areas the pichi is a house pet.

XENARTHRA; DASYPODIDAE; **Genus PRIODONTES** *F. Cuvier, 1827*

Giant Armadillo

The single species, *P. maximus*, is found throughout most of South America east of the Andes, from northwestern Venezuela to northeastern Argentina (Cabrera 1957; Handley 1976; Wetzel 1985*b*).

Head and body length is 75–100 cm and tail length is about 50 cm. Adults weigh as much as 60 kg, but maximum recorded weights may apply to overstuffed zoo residents. Redford and Eisenberg (1992) listed weights of 18.7–32.3 kg. The hair covering is scant; only a few hairs are scattered between the plates. The coloration is dark brown except on the head and tail and a band around the lower edge of the shell, which is whitish. There are 11–13 movable bands on the back and 3–4 bands on the back of the neck. The carapace is very flexible. The head shield is oval and is not expanded between the eyes. The plates on the tail are closely set and are not arranged in rows. The claws of the forefeet are powerful, and the claw on the third finger measures about 203 mm along the curve. The giant armadillo is so much larger than other armadillos that size alone serves to distinguish adults. It occasionally possesses as many as 100 small teeth, but these are shed as the animal ages. Females have two mammae.

The giant armadillo is a powerful and rapid digger and shelters in burrows of its own construction. It has been reported to occur primarily in unbroken, relatively undisturbed forest and usually near water. In a study in Brazil, however, Carter (1983) found 68 percent of its burrows in grassland, 28 percent in brushland, and only 3 percent in woodland. Nearly half of these burrows were located in active termite mounds. The average distance to the woods for burrows located in grassland or brushland was 192 meters.

Giant armadillo *(Priodontes maximus)*, photo by Ernest P. Walker.

Burrows averaged 41 cm in width and 31 cm in height. Burrows of different ages were found in close proximity, indicating that *Priodontes* reuses certain areas. Carter and Encarnaçao (1983) found this genus also to remain longer in a burrow than did other armadillo genera investigated; one female stayed in one hole for 17 days. Activity is strictly nocturnal. Carter (1985) reported mean nightly movement to be 2,765 meters and minimum home range to average 452.5 ha. Despite its rigid appearance, the giant armadillo is fairly agile and often balances itself on its hind legs and tail with its forefeet off the ground. Unlike certain other armadillos, it cannot completely enclose itself in the carapace; if closely pursued, it may try to dig itself in. It also digs extensively to obtain food. The diet consists primarily of termites, but ants, other insects, spiders, worms, larvae, snakes, and carrion are also consumed. *Priodontes* frequently is accused of eating garden vegetables, but it probably digs in gardens in search of insects, not vegetables.

Merrett (1983) wrote that the gestation period is said to be 4 months and that there are one or two young. They weigh up to 113 grams each and have tough, leathery skin. They are weaned at 4–6 weeks and reach sexual maturity at 9–12 months. Life span is said to be 12–15 years.

The IUCN now classifies the giant armadillo as endangered, noting that it has declined by at least 50 percent in the last decade; the species also is listed as endangered by the USDI and is on appendix 1 of the CITES. It has a wide range in South America, but populations have been exterminated or reduced by overhunting, settlement, and agricultural development. This armadillo is valued for use as food, and it is avidly hunted whenever encountered by people. The species is still common in the Guianas and southwestern Brazil, but it has disappeared from the vicinity of all human settlements in Peru. Only six specimens were in captivity in 1979 (Thornback and Jenkins 1982).

XENARTHRA; DASYPODIDAE; **Genus CABASSOUS**
McMurtie, 1831

Naked-tailed Armadillos

There are four species (Cuarón, March, and Rockstroh 1989; McCarthy 1982*b*; Wetzel 1980, 1985*b*):

C. unicinctus, east of the Andes from Venezuela to southern Brazil;

C. centralis, extreme southern Mexico and Belize to western Colombia and northwestern Venezuela;

C. chacoensis, the Gran Chaco of southeastern Bolivia, western Paraguay, and northern Argentina and probably the adjacent part of Brazil;

C. tatouay, southern Brazil, eastern Paraguay, Uruguay, northeastern Argentina.

Naked-tailed armadillos are closely related to *Priodontes* and resemble that genus closely except in size. Head and body length is 300–490 mm and tail length is 90–200 mm. Meritt (1985*a*) reported weights of 2.0–3.5 kg for *C. centralis* and 2.2–4.8 kg for *C. unicinctus.* Redford and Eisenberg (1992) listed weights of 3.4–6.4 kg for *C. tatouay,* the largest species. The coloration above is dark brownish to almost black, the lateral edges of the carapace are yellowish, and the underparts are dull yellowish gray. There are five large claws on the forefeet; the middle claw is especially large and sickle-shaped. The snout is short and broad, the head is broad, and the ears are widely separated. The movable transverse bands across the middle of the back are the most numerous in the family, varying in number from 10 to 13. The tail is slender and shorter than the head and body and only slightly armored with small, thin, widely spaced plates. The scapular and pelvic shields are attached to the body almost to the base of the limbs. The lack of complete armor on the tail is unique among armadillos.

The species *C. centralis* walks on the tips of the claws of its forefeet and on the soles of its hind feet. When danger threatens, it runs quite rapidly for short distances, burrows into the ground, or goes into water. It feeds mainly on ter-

Naked-tailed armadillo *(Cabassous centralis)*, photo by Lloyd G. Ingles.

mites and ants dug out of litter and soil, which it apparently locates by scent. The sicklelike claw is used to cut small roots as the animal digs after insect colonies in dead roots or stumps; the prey is extracted from the tunnels by the long, extensible tongue. *C. centralis* sometimes completely buries itself while digging for insects. It is slower than *Dasypus novemcinctus.*

According to Meritt (1985*a*), naked-tailed armadillos live in a variety of habitats, including grasslands, semiarid and moist lowlands, upland areas, and riversides. They live in burrows, the entrances of which usually are in open ground or at the base of an embankment. Captives do not construct a nest even when provided with material. Activity is nocturnal and begins shortly after sunset. The diet probably consists largely of ants and termites. Animals are usually found alone in the wild, but breeding groups of up to two males and four females have been established in captivity, with the individuals generally ignoring one another. The single young weighs about 100 grams at birth. *C. tatouay,* found mainly in a region undergoing rapid development, is designated as near threatened by the IUCN.

XENARTHRA; DASYPODIDAE; Genus **TOLYPEUTES** *Illiger, 1811*

Three-banded Armadillos

There are two species (Wetzel 1985*b*):

T. tricinctus, east-central Brazil;
T. matacus, central and eastern Bolivia, the Mato Grosso of central Brazil, the Chaco region of Paraguay, northern and central Argentina.

Head and body length is 218–73 mm, tail length is 60–80 mm, and weight is 1.00–1.59 kg (Redford and Eisenberg

1992; Wetzel 1985*b*). The overall coloration is blackish brown. Most individuals have three movable bands, but some have only two bands and others have four. These are the only armadillos that can completely enclose themselves by rolling into a sphere. Unlike most genera in the family, in *Tolypeutes* the sides of the two large shells are free from the skin; there is considerable space in the shell into which the head, legs, and tail can be fitted when the animal "rolls up." The second, third, and fourth toes of the hind foot are grown together, with nails almost like hoofs; the first and fifth toes are slightly separated from the others and have normal claws. The short, thick tail, with prominent tubercles, is almost inflexible. *T. tricinctus* has five claws on the forefoot, but *T. matacus* has only four.

Jorge, Meritt, and Benirschke (1977) found a remarkable difference in karyotype between *T. matacus* and other armadillos. In this species there are no truly acrocentric autosomes, and the diploid number (2n = 38) is the lowest found among the xenarthrans they examined. All other armadillos have diploid numbers from 50 to 64.

In the Mato Grosso *T. matacus* has been found in grassy or marshy areas between scattered forests. This species apparently does not dig holes but utilizes the abandoned burrows of anteaters. It runs relatively rapidly with a peculiar gait, only the tips of the foreclaws touching the ground. When danger threatens, it leaves a small opening between the edges of the shell and the extremities; then, if it is touched on the chest or abdomen, it snaps the shells together like a steel trap. This behavior seems to be quite an effective defense against natural enemies.

In a study of the food habits of *T. matacus* in the Chaco of Argentina, Bolkovic, Caziani, and Protomastro (1995) found beetle larvae to be the most frequently consumed item throughout the year, ants and termites to be important in the dry season from July to November, and fruits to be especially significant during the summer rains. Ants and termites are obtained by probing into the ground, under bark, and into nests with the powerful forelegs and claws.

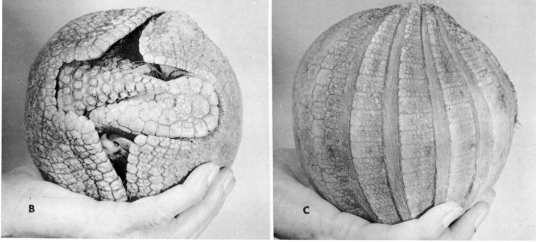

Three-banded armadillo *(Tolypeutes tricinctus):* A. Unrolled, ready for walking. Much of the time it touches only the tips of the very strong claws of the forefeet to the ground. B. Almost completely closed for protection against enemies. The large, somewhat triangular shield at the right is the top of the head; the slightly smaller triangular member studded with tubercles is the tail. C. Back of the animal showing bands and flexible skin that connects them. Photos by Ernest P. Walker.

In captivity, fruits, leaves, boiled rice, soft bread in milk or tea, ants, ant eggs or termite larvae, and mealworms are eaten.

Sources cited by Wetzel (1982) indicate that population densities of *T. matacus* reach 7/sq km and that this species is primarily solitary, though groups of up to 12 individuals may be found together in a single shallow nest during the cold season. According to Meritt (1976), *Tolypeutes* is probably uniparous. From 1969 to 1974 at the Lincoln Park Zoo there were 18 live births of *T. matacus,* in January, February, March, May, June, August, October, November, and December. In each case there was only a single young. Meritt stated that in the Paraguayan Chaco most births take place from November to January but that in captivity

the season seems to shift a little; the majority of births at Lincoln Park occurred from October to December. The only information on gestation is that an adult male and female produced a young 120 days after they were placed together. Meritt observed that a newborn *Tolypeutes* is a miniature adult. The claws are fully developed and hardened, and the eyes and ear pinnae are closed, but they open in the third or fourth week of life. The flexible carapace has a leathery texture, and the individual scute markings are already apparent. Right from birth the baby is capable of coordinated movements, including walking and rolling itself into a sphere. Birth weight is about 113 grams (Hayssen, Van Tienhoven, and Van Tienhoven 1993). According to Merrett (1983), weaning is completed by 72 days and sexual maturity comes at 9–12 months. One individual lived for over 17 years after arrival at the San Antonio Zoo (Marvin L. Jones, Zoological Society of San Diego, pers. comm., 1995).

The IUCN classifies *T. tricinctus* as vulnerable. Thornback and Jenkins (1982) reported it to be rare, subject to exploitation for use as food, and easily captured. Wetzel (1982) indicated that it was known by not more than six museum specimens, was not located during several intensive field surveys, and probably had disappeared over much of the southeastern highlands of Brazil. Cardoso da Silva and Oren (1993) noted that the difficulty in locating *T. tricinctus* may be associated with its restricted habitat within patches of tropical deciduous forest on the more elevated sectors of the Brazilian plateau. Such habitat is naturally fragmented and highly threatened by agricultural development, cutting for charcoal, and mining for the underlying calcareous deposits. Roig (1991) reported that *T. matacus* also has been eliminated over a large part of central Argentina. It is designated as near threatened by the IUCN.

XENARTHRA; DASYPODIDAE; **Genus DASYPUS**
Linnaeus, 1758

Long-nosed Armadillos

There are three subgenera and six species (Redford and Eisenberg 1992; Wetzel 1982, 1985*b*; Wetzel and Mondolfi 1979):

subgenus *Dasypus* Linnaeus, 1758

D. novemcinctus, south-central and southeastern United States to Peru and Uruguay, Grenada in the Lesser Antilles, Trinidad and Tobago;
D. septemcinctus, eastern and southern Brazil, eastern Bolivia, Paraguay, extreme northern Argentina;
D. hybridus, Paraguay, northern and central Argentina, Uruguay, southern Brazil;
D. sabanicola, llanos of Colombia and Venezuela;

subgenus *Hyperoambon* Peters, 1864

D. kappleri, east of the Andes from Colombia and northern Bolivia to the Guianas and northeastern Brazil;

subgenus *Cryptophractus* Fitzinger, 1856

D. pilosus, known only from the mountains of southwestern Peru.

Head and body length is 240–573 mm, tail length is 125–483 mm, and weight is 1–10 kg. Hair is almost lacking on the upper parts and is sparsely scattered on the underparts. The hairs are pale yellowish, and the remainder of the body is mottled brownish and yellowish white. Coloring is the same in both sexes, and there is no seasonal variation. The body of these armadillos is broad and depressed, the muzzle is obtusely pointed, and the legs are short. The forefoot has four toes, of which the middle two are the largest; the hind foot has five toes. The number of movable bands in this genus varies from 6 to 11 depending on the species and, within any given species, on geographic location. *Dasypus novemcinctus*, often called the "nine-banded armadillo," actually usually has eight bands in the northern and southern parts of its range and nine in the central part (northern South America).

Long-nosed armadillos are partial to dense shady cover and limestone formations from sea level to 3,000 meters. They construct burrows at a depth of 0.5–3.5 meters and up to 7.5 meters long in which they assemble large nests of leaves and grass. Nests of *D. novemcinctus*, resembling miniature haystacks, also have been found above the ground in clumps of saw palmetto (Layne and Waggener 1984). These armadillos are mainly nocturnal but frequently are seen foraging during daylight. Layne and Glover (1985) reported that activity in south-central Florida peaked when temperatures were 20°–25° C and thus concluded that animals were more nocturnal in summer. When rooting about they grunt almost constantly and do not appear to be alarmed by humans. Hunters report that if they stand still, armadillos will occasionally bump against their feet without noticing. When alarmed, howev-

Long-nosed armadillo *(Dasypus novemcinctus)*, photo by Ernest P. Walker.

er, they hurry away toward a burrow. If overtaken, they characteristically curl up as much as possible so that the armor protects their soft underparts. They forage in a jerky, nervous fashion, poking into many holes, crevices, and leaf piles in search of arthropods, small reptiles, and amphibians. According to studies cited by Lowery (1974) and Wetzel and Mondolfi (1979), the diet of *D. novemcinctus* consists predominantly of animal matter, especially insects such as beetles and ants.

Population densities of up to 13/sq km for *D. novemcinctus* and of about 280/sq km for *D. sabanicola* have been reported in South America (Wetzel 1982). Layne and Glover (1977) found the average home range of 12 *D. novemcinctus* in Florida to be 5.7 (1.1–13.8) ha. The home ranges of different individuals overlapped, and no antagonism was observed. Several individuals may frequent a common burrow, though usually these animals are all of the same sex. In a study of a dense population of *D. novemcinctus* in southern Texas, McDonough (1994) observed considerable evidence of aggression, including chases and "boxing," in which animals balanced on their hind feet and tails and clawed at each other with their front feet. Fully adult males commonly were aggressive toward younger males, especially during the mating season, and near-term and lactating females were aggressive toward young born the previous year.

In North America, *D. novemcinctus* commonly mates in July and August, but implantation is delayed until November. Data compiled by Hayssen, Van Tienhoven, and Van Tienhoven (1993) indicate that mating may occur over a much longer period in Texas, from around June to November, and that implantation may be delayed up to 5 months. Births have been recorded in Texas during March and April and in Mexico from February to April. The period of actual developmental gestation is about 120 days. The normal litter consists of 4 young, all of the same sex, from a single fertilized egg. Each weighs about 30–50 grams. They are weaned after 4–5 months and attain sexual maturity at about 1 year of age. In *D. hybridus*, of South America, implantation occurs about 1 June, and births take place in October (Barlow 1967). Litter size has been reported to be 4–12 young in *D. hybridus*, 2–12 in *D. kappleri*, 4 in *D. sabanicola*, and usually 4–8 in *D. septemcinctus* (Hayssen, Van Tienhoven, and Van Tienhoven 1993). A captive *D. novemcinctus* was still living after 22 years (McDonough 1994).

D. pilosus, highly restricted in range and declining, is classified as vulnerable by the IUCN, but some of the other species of *Dasypus* may be increasing. *D. novemcinctus* is the only xenarthran that occurs as far north as the United States. Apparently this species extended its range into Texas around 1880 and since then has gradually moved to the north and east (Fitch, Goodrum, and Newman 1952). Lowery (1974) stated that it first appeared in numbers in Louisiana not long before 1925 and subsequently spread to every parish in the state. It also moved into Oklahoma, Missouri, Arkansas, Mississippi, and Alabama. The species was deliberately introduced in Florida in the 1920s and, mostly on its own, proceeded to occupy much of that state as well as Georgia and South Carolina (Cleveland 1970; Fitch, Goodrum, and Newman 1952). To the northwest, the range now has reached Colorado (Meaney, Bissell, and Slater 1987). The movement of *D. novemcinctus* into the United States during the last century is probably only the most recent of several northward fluctuations of the range of *Dasypus*, resulting from varying climatic conditions. It has been present intermittently in Florida since Blancan times (about a million years ago), and during part of the late Pleistocene a large, extinct species, *D. bellus*, had much the same range in the United States as that now occupied by *D. novemcinctus* (Klippel and Parmalee 1984).

Although armadillos are sometimes accused of stealing eggs, starting erosion, and undermining buildings, they are of economic importance through their destruction of noxious insects. They also construct burrows in which other animals seek shelter. Some persons value armadillos as a source of food, and their armored hides are made into baskets that are sold to tourists. They are easily tamed but should not be confined to small quarters. *D. novemcinctus* does not thrive under the conditions usually provided in zoos. This species recently has proved valuable to medical researchers engaged in studies of multiple births, organ transplants, birth defects, and such diseases as leprosy, typhus, and trichinosis (Yager and Frank 1972). It is particularly noted for its susceptibility to leprosy both in the laboratory and under natural conditions (Storrs and Burchfield 1985).

XENARTHRA; DASYPODIDAE; **Genus CHLAMYPHORUS**
Harlan, 1825

Pichiciegos

There are two species (Wetzel 1985*b*):

C. truncatus (pink fairy armadillo), central Argentina;
C. retusus, Chaco region of western and central Bolivia, Paraguay, and extreme northern Argentina.

The latter species sometimes is placed in a separate genus usually called *Burmeisteria* Gray, 1865. According to Wetzel, however, if *C. retusus* is retained in its own genus, the correct name is *Calyptophractus* Fitzinger, 1871.

C. truncatus is the smallest armadillo. Redford and Eisenberg (1992) listed head and body lengths of 84–117 mm and tail lengths of 27–35 mm. Carter and Encarnação (1983) indicated that weight is 85 grams. The armor is pale pink in color. The body under the shell, the underparts, and the legs are covered with fine, soft, white hairs. The head and body armor is anchored to two large, rough prominences on the bones above the eyes and by a narrow ridge of flesh along the spine. There is free movement between all the bands across the neck and back. *C. truncatus* is the only armadillo whose dorsal shell is almost separated from the body. The separate plate at the rear is securely attached to the pelvic bones. The spatula-shaped tail projects from a notch at the lower edge of the rear plate. The tail cannot be raised and is dragged on the ground. The lips are horny and stiff, the nose is pointed, and the nostrils are ducted downward. Females have two mammae.

In *C. retusus* head and body length is 140–75 mm and tail length is about 35 mm. The plates are whitish and yellowish brown in color and the woolly hair is whitish. The head is covered by a shield, the dorsal carapace is attached to the skin of the back for its entire extent, and the tail is partly covered with plates. The underparts are covered with woolly hair. *C. retusus* resembles *C. truncatus* but is larger, has the entire carapace attached to the body, and has a less defined head shield that lacks a posterior row of large scutes.

Pichiciegos inhabit dry grasslands as well as sandy plains with thornbushes and cacti. *C. truncatus* is nocturnal and tends to be sluggish except when digging. One observer re-

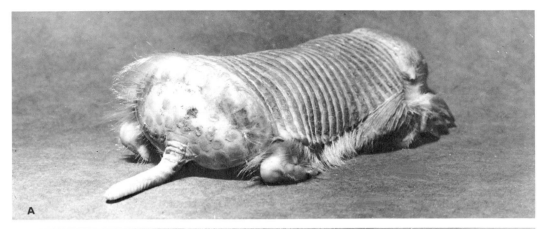

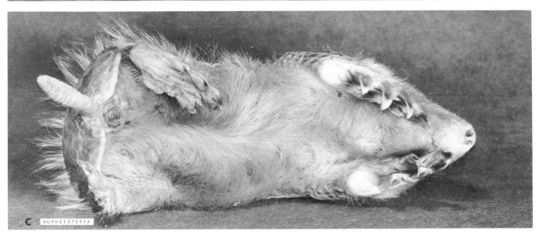

Chlamyphorus retusus: A. Rear view; B. Side view; C. Hairy underside. Photos by Ernest P. Walker.

ported seeing an individual of this species while he was riding a horse, but before he could dismount the pichiciego had burrowed out of sight. To accomplish such rapid digging, the animal supports the rear end of the body with the rigid tail and uses the hind feet to kick away the ground loosened by the forefeet. After entering the burrow, it may use the armor plate at the rear end of the body effectively to close the burrow opening, much like corking a bottle. This small armadillo often burrows near an anthill. It seems to prefer dry soil that feels uncomfortably warm to the human hand. When infrequent rains moisten the soil, the animal may leave its burrow. *C. retusus* is not as good

a digger as *C. truncatus;* it probably presses itself against the ground when surprised in the open. Cries almost like those of human infants have been noted in *C. retusus.*

The diet apparently consists mostly of insects and their larvae. Several hundred ants were found in the digestive tract of one specimen of *C. truncatus,* and this species also is known to feed on worms, snails, roots, and other plant material. *Chlamyphorus* is difficult to keep in captivity, but Rood (1970*b*) maintained a specimen of *C. truncatus* on a diet of bread, milk, whole oats, and occasional beetles and beetle larvae. This captive animal slept in an underground burrow and came aboveground only once each day to eat.

CHLAMYPHORUS

A captive *C. retusus* did well on boiled rice and grapefruit. An individual of this species that had been in captivity in Bolivia for two months in a yard where it could burrow appeared to be in good health when received at the U.S. National Zoo. It was given soil in which to burrow and a wide assortment of foods, including mealworms, raw eggs, and meat, but it died after a few days. The immediate cause of death seems to have been a respiratory infection.

The pink fairy armadillo *(C. truncatus)* is classified as endangered by the IUCN and the USDI. Numbers are declining, and habitat has been greatly reduced. Plowing of land for agriculture is apparently the main threat, but predation by domestic dogs is also a serious problem. The range of this species is now restricted to the remote desert areas of western and central Argentina. *C. retusus* also may be in jeopardy because of loss of habitat to agriculture and excessive collection (Thornback and Jenkins 1982); it is classified as vulnerable by the IUCN.

Lesser pichiciego *(Chlamyphorus truncatus)*, photos by J. P. Rood.

Order Insectivora

Insectivores

This order does not comprise a particularly large number of members, but it is difficult to classify. Various kinds of small fossil and living mammals with primitive characters have been placed in the order Insectivora, especially groups that could not readily be assigned to some other major order. The family Cynocephalidae (herein put in the separate order Dermoptera) sometimes has been included in the Insectivora. A common arrangement was to divide the Insectivora into two suborders: the Lipotyphla, with the living families Erinaceidae, Chrysochloridae, Tenrecidae, Solenodontidae, Nesophontidae, Soricidae, and Talpidae; and the Menotyphla, with the families Macroscelididae and Tupaiidae.

In contrast, Van Valen (1967) considered the order Insectivora to be restricted to the families Tupaiidae, Macroscelididae, Cynocephalidae (put in the suborder Dermoptera), Erinaceidae, Talpidae, Nesophontidae, and Soricidae. He placed the families Tenrecidae, Solenodontidae, and Chrysochloridae in a separate order, the Deltatheridia. Eisenberg (1981) also united the Tenrecidae and Chrysochloridae in a separate order, the Tenrecomorpha.

McKenna (1975) elevated the Insectivora to the rank of a grandorder containing two orders: the Erinaceomorpha, with the family Erinaceidae; and the Soricomorpha, with the living families Chrysochloridae, Tenrecidae, Solenodontidae, Nesophontidae, Soricidae, and Talpidae. The Macroscelididae were placed in a separate order, the Macroscelidea, which together with the order Lagomorpha was put in the grandorder Anagalida. Still another grandorder, the Archonta, was considered to include the orders Scandentia (with the family Tupaiidae), Dermoptera (with the family Cynocephalidae), Chiroptera, and Primates. A modified version of this arrangement was suggested by Novacek (1986), who ranked the Insectivora as a superorder, with the Lipotyphla being the only living order and containing two suborders: the Erinaceomorpha for the Erinaceidae, and the Soricomorpha for the Chrysochloridae, Tenrecidae, Solenodontidae, Nesophontidae, Soricidae, and Talpidae. Subsequently, MacPhee and Novacek (1993) designated a third suborder, the Chrysochloromorpha, for the family Chrysochloridae.

Notwithstanding the above, most recent systematic accounts continue to recognize the Insectivora as an order, but one that comprises only the Lipotyphlan families (Corbet and Hill 1991; Hutterer *in* Wilson and Reeder 1993; Yates 1984). The Tupaiidae, Cynocephalidae, and Macroscelididae are not thought to be immediately related to this group and are placed in separate orders. This arrangement is generally accepted here, but with consideration of the three suborders recognized by MacPhee and Novacek (1993). The following sequence covers 7 insectivore families with 68 genera and 440 species.

Insectivores are usually small and have long, narrow snouts and five-clawed digits on each limb. The body covering is usually short, close-set fur with only one kind of hair or sometimes spines. The ears are small; the eyes are almost always minute and sometimes have no external opening. These animals generally move about on the sole of the foot with the heel touching the ground. Neither the thumb nor the big toe is opposable. The radius and ulna are separate, but the tibia and fibula are often fused near the ankle. Clavicles are present in all genera except *Potamogale*. In the males the testes are abdominal, inguinal, or borne in a sac in front of the penis; a baculum is present in some forms. In many members of this order the genital and urinary systems have a common exit.

The teeth of insectivores are generally primitive in structure. The deciduous teeth are usually shed early and are seldom functional. All the teeth are rooted. The incisors are enlarged or reduced, and the canines may be shaped like incisors or premolars. The lower molars usually have five pointed cusps, but the uppers have three or four tubercles. The dental formula is often: (i 3/3, c 1/1, pm 4/4, m 3–4/3–4) $\times$ 2 = 44, 46, or 48, but there are many variations. The skull is low and flat in outline and often long and slender. The braincase is small and is little elevated above the facial line. The orbits generally open posteriorly, and the zygomatic arches are reduced or absent. The cerebral hemispheres of the brain are smooth, lack fissures, and do not extend backward over the cerebellum.

The members of this order are terrestrial, fossorial, or semiaquatic, and with the exception of some shrews and aquatic forms they are almost completely nocturnal. As their name indicates, they are generally insectivorous, though some are carnivorous. Many forms of this order have the reputation of consuming great quantities of food, but some of their food contains very few nutrients. Insectivores have various means of protection, including nocturnal, subterranean, or aquatic habits: they seek cover in deep forests, holes deserted by other animals, stems and roots of plants, or the boughs of trees, and sometimes they are covered with a shield of spines.

Some members of this order seem to have departed less from the form of the generalized, primitive mammalian type than have any other Recent placentals. For this and other reasons certain insectivores are believed to resemble the basic stock of certain placental lines of descent. The geological range of the Insectivora is late Cretaceous to Recent.

INSECTIVORA; **Family ERINACEIDAE**

Gymnures and Hedgehogs

This family of 10 genera and 20 species is found in Africa, Europe, and Asia as far north as the limit of deciduous forest and on Tioman Island, Sumatra, Java, Borneo, and Mindanao Island in the Philippines. The sequence of genera presented here basically follows that of Corbet (1988a), who recognized two distinctive subfamilies: the Galericinae, comprising the gymnures or moonrats, and the Erinaceinae, containing the hedgehogs. Consideration also has been given to the work of Frost, Wozencraft, and Hoffmann (1991), who indicated a number of modifications and pointed out that the appropriate name for the first subfamily is Hylomyinae rather than Galericinae (or Echinosoricinae, which also is sometimes used).

Some members of the Hylomyinae resemble certain microtine rodents in general appearance but have longer muzzles. All members of this subfamily lack spines. Hedgehogs have barbless spines on the back and sides and hair on the face, limbs, and underparts. In the family, head and body length is 105–460 mm and tail length is 10 to approximately 325 mm. In total length *Echinosorex* is the largest genus and *Neotetracus* is the smallest. The snout of gymnures and hedgehogs is elongate and blunt, and the eyes and ears are well developed. The tail is hairy. The members of this family have five digits on each foot, except for some forms of *Atelerix*, which have only four digits on the hind foot. Gymnures and hedgehogs walk on the soles of their feet with their heels touching the ground. The two lower bones of the hind limb, the tibia and fibula, are united. The urogenital opening in the females is well separated from the anus. Females have two to five pairs of mammae. The penis is directed anteriorly and is partly pendulous from the abdomen.

The dental formula is: (i 2–3/3, c 1/1, pm 3–4/2–4, m 3/3) $\times$ 2 = 36–44. The first incisor is usually larger than the others. The upper molars have four main cusps and a small, slender median cusp. The skull varies from narrow and elongate to short and broad, and the braincase is small.

These animals shelter in and under logs, among rocks, under the roots of trees and brush piles, in termite mounds, and in burrows. Hedgehogs generally dig their own burrows. The hedgehogs in the Rajasthan Desert of India *(Hemiechinus* and *Paraechinus)* loosen the sand by lateral strokes of their forelimbs with force sufficient to throw the earth behind them. When an appreciable amount of sand has accumulated behind the animal, it backs up in the burrow and kicks the soil out with a rapid alternate action of the hind limbs. The burrows of hedgehogs are generally located under a hedge or dense bush. Hedgehogs enter and leave their burrows headfirst, turning inside the tunnel. Some of the gymnures are diurnal and some are nocturnal; all the hedgehogs, on the other hand, are active at night, spending the day in a leaf nest or dry cavity, rolled up in a ball or fully stretched out. *Erinaceus europaeus* hibernates in the colder parts of its range. The normal heartbeat in this species is about 188 beats per minute; during hibernation it drops to 21 beats per minute. Hedgehogs of the genera *Hemiechinus* and *Paraechinus* in the Rajasthan Desert of India spend a "passive period" during the winter.

Hearing seems to be the keenest sense in hedgehogs. When threatened, they often roll themselves into a ball, bring the snout and limbs close under the body to hide them almost completely, and erect the spines, so that the animal is practically a ball of spines. The ability to roll into a ball is aided by the contraction of a longitudinal muscle on either side of the body that acts as a "drawstring." Hedgehogs sometimes emit a hissing, snakelike noise when frightened, and a loud, screaming cry is a call of distress; they utter grunting and snuffling noises when they are feeding. Hedgehogs progress by means of a waddling walk or trot with the spines directed backward. Although terrestrial, they can climb and swim well. The lesser gymnure, *Hylomys*, sometimes climbs in trees, and the moon rat, *Echinosorex*, may be aquatic to some extent.

Gymnures feed on insects, other invertebrates, and some plant matter. *Echinosorex* probably includes fish and frogs in its diet, and hedgehogs feed on a wide range of animal matter, including carrion and probably fruits and roots. Most members of this family seem to have one or two breeding seasons a year and a litter size of one to seven. At

Lesser gymnure *(Hylomys suillus)*, photo by Lim Boo Liat.

birth the young of *Hemiechinus auritus collaris* have soft, 2-mm spines with no fixed coloration—some are dirty white and others black. Five hours after birth the spines are 8 mm long. In adults of this subspecies the spines measure 30 mm and are largely dirty white with black tips. In *Erinaceus europaeus* the young are born with soft whitish spines, which harden and assume the adult appearance in about three weeks. At about four weeks the young of this species are following the mother, and at about one year they are fully grown.

The geological range of this family is middle Paleocene to early Pliocene in North America, early Miocene to Recent in Africa, late Paleocene to Recent in Europe, and Eocene to Recent in Asia (Yates 1984).

INSECTIVORA; ERINACEIDAE; Genus HYLOMYS
Müller, 1839

Lesser Gymnure

The single species, *H. suillus,* is found in Yunnan (Burmese border area of China), Burma, Indochina, Thailand, the Malay Peninsula and nearby Tioman Island, Sumatra, Java, and Borneo (Corbet 1988a; Ellerman and Morrison-Scott 1966). *Hylomys* also sometimes is considered to include *Neohylomys* and *Neotetracus* (see accounts thereof).

Head and body length is 90–147 mm and tail length is about 10–30 mm. Reported weights are 45–80 grams (Medway 1978) and 15–20 grams (Lekagul and McNeely 1977). The pelage is soft and dense. The upper parts are rusty brown, sometimes with a faint black nape stripe or dorsal stripe, and the underparts are grayish, buffy, or yellowish. This genus resembles certain microtine rodents in appearance and has a longer muzzle. Females have four mammae. As in *Echinosorex,* there is a strong characteristic odor.

The lesser gymnure usually inhabits humid forests with thick undergrowth in hilly or mountainous areas. It is terrestrial and normally confined to the forest floor, but it

sometimes has been seen climbing in low bushes (Lekagul and McNeely 1977). A series was trapped under logs in a wild banana grove in Thailand. *Hylomys* moves on the ground with short leaps, and when threatened it travels with considerable speed. It has been reported to be active both by day and by night and to use definite paths or runways, often passing under logs slightly off the ground. It feeds by searching through litter with its long, mobile snout. The natural diet seems to consist mainly of invertebrates, such as insects and earthworms, but some captives have also eaten fruit. Rudd (1980) found individuals to have a home range about 40 meters in diameter.

Medway (1978) wrote that *Hylomys* occurs singly or in groups of two or three. He recorded that a pregnant female with two embryos was taken in March. Lekagul and McNeely (1977) stated that breeding probably occurs throughout the year, with two or three young born after a gestation period of 30–35 days.

The IUCN recognizes *H. parvus,* restricted to a small area of declining habitat on the slopes of Korinchi Peak in central Sumatra, as a separate species and classifies it as critically endangered. It was not treated as a species or subspecies by Corbet and Hill (1992).

INSECTIVORA; ERINACEIDAE; Genus NEOTETRACUS
Trouessart, 1909

Chinese Gymnure

The single species, *N. sinensis,* occurs in the south-central Chinese provinces of Sichuan, Yunnan, and Guizhou and also in the northern parts of Burma and Viet Nam (Corbet 1988a; Frost, Wozencraft, and Hoffmann 1991). *Neotetracus* was synonymized with *Hylomys* by Van Valen (1967), and that arrangement was supported by Frost, Wozencraft, and Hoffmann (1991), though they indicated that recognition of *Neotetracus* as a separate genus also was an acceptable course. The latter procedure was followed by Corbet (1988a), Corbet and Hill (1991), Heaney and Morgan

Chinese gymnure *(Neotetracus sinensis),* photo from *Traité de Zoologie . . .* , Pierre-P. Grasse. Inset photo from *Genera Mammalium Insectivora,* Angel Cabrera.

Hainan moon rat *(Neohylomys hainanensis),* photo by Wang Sung.

(1982), and Yates (1984) but not by Corbet and Hill (1992) or Hutterer (*in* Wilson and Reeder 1993).

Head and body length is 101–25 mm and tail length is 43–70 mm (Allen 1938; Heaney and Morgan 1982). The coat is soft, dense, and quite long. The upper parts are olive-brown, cinnamon brown, or mixed cream color and black, and the sides of the head and neck sometimes are tinged with rufous. An indistinct, blackish dorsal stripe may be present. The underparts are reddish, buffy gray, or cream-colored over a blackish base. The tail is thinly covered with minute hairs. Females have eight mammae.

Like *Hylomys,* this genus resembles certain microtine rodents in general appearance. *Neotetracus* is closely related to *Hylomys,* and the two genera share many unique characters, such as the presence of posterior processes of the maxillae and anterior processes of the parietals, which extend across the frontals and meet or nearly meet dorsal to the orbits (Heaney and Morgan 1982). From *Hylomys,* *Neotetracus* is distinguished by its longer tail, shorter snout, and fewer teeth. Corbet (1988*a*) stated that in *Hylomys* the upper canine is larger than the adjacent teeth and there are four lower premolars on each side, but in *Neotetracus* the upper canine is no larger than the adjacent teeth and there are three lower premolars on each side.

Neotetracus inhabits cool, damp forests from about 1,000 to 2,800 meters in elevation. It is apparently strictly terrestrial and nocturnal. It is common in parts of its range and is found in runways and burrows with moss and fern cover and under logs and rocks. It is not known whether *Neotetracus* makes the runways, for they also are used by other small mammals. The stomach of one individual contained earthworms, but that of another contained only vegetable matter. A female with four embryos was taken in April, and two females, one with four embryos and the other with five, were obtained in August. These records may indicate either a long breeding season or two litters per year. *Neotetracus* occupies a relatively restricted habitat zone subject to increasing human encroachment and now is designated as near threatened by the IUCN (under the generic name *Hylomys*).

Hainan Moonrat

The single species, *N. hainanensis,* is known only from Hainan Island off southern China. *Neohylomys* was synonymized with *Hylomys* by Van Valen (1967), and that arrangement was supported by Frost, Wozencraft, and Hoffmann (1991), though they indicated that recognition of *Neohylomys* as a separate genus also was an acceptable course. The latter procedure was followed by Corbet (1988*a*), Corbet and Hill (1991), and Yates (1984). Synonymization was accepted by Corbet and Hill (1992) and Hutterer (*in* Wilson and Reeder 1993) and also, with question, by Heaney and Morgan (1982).

Head and body length in seven specimens was 120–47 mm, tail length was 36–43 mm, and weight was 50–69 grams. The back is rusty brown with gray, and there is a long, black middorsal stripe. The sides are washed with olive-yellow, and the underparts are pale gray or yellowish white. The ears, feet, and tail are nearly naked, having only minute, scattered hairs.

Females have two pectoral and four abdominal mammae. From *Hylomys,* *Neohylomys* is distinguished by larger size, a relatively much longer tail, and a more pronounced dorsal stripe. Corbet (1988*a*) stated that in contrast to *Hylomys,* which has four lower premolar teeth on each side, *Neohylomys* has only three.

According to Corbet (1988*a*), *Neohylomys* has been reported to occur in tropical rainforest and subtropical evergreen forest. It originally was described as subterranean, but it is not clear whether it forages underground or (more likely) merely uses burrows as refuges. The IUCN classifies it as endangered (under the generic name *Hylomys*), noting that its restricted habitat is declining.

Malayan gymnure, or moon rat *(Echinosorex gymnurus)*, photos by Lim Boo Liat.

INSECTIVORA; ERINACEIDAE; **Genus ECHINOSOREX**
De Blainville, 1838

Moon Rat, or Gymnure

The single species, *E. gymnura,* is found on the Malay Peninsula (including parts of southern Burma and Thailand), Sumatra, and Borneo (Frost, Wozencraft, and Hoffmann 1991). A generic name that often has been used for the moon rat is *Gymnura* Lesson, 1827, which is preoccupied by *Gymnura* Kuhl, 1824, a genus of fishes.

Head and body length is 260–460 mm, tail length is about 165–300 mm, and adult weight is 0.5–2.0 kg; apparently females are slightly larger than males (Davis 1962; Lekagul and McNeely 1977; Medway 1978; Ralls 1976). The scantily haired tail reveals that the scales are arranged in rows around the tail except near the body, where they are arranged diagonally. The tail is compressed for the terminal third of its length. The underside of the long, mobile nose is grooved from its tip to a point between the upper incisor teeth. Body form is exceedingly narrow, perhaps an adaptation for seeking food in narrow crevices. The rough and harsh pelage consists of a short, thick underfur covered by a dense layer of long, coarse hair. The scaly tail is almost naked. The color is usually black, but the head and shoulders and the distal half of the tail are whitish. The face is generally marked with black spots near the eyes. White

forms also occur. There are two pectoral and two inguinal mammae.

The moon rat inhabits lowlands, including primary and secondary forest, mangroves, rubber plantations, and cultivated areas. It is terrestrial and nocturnal, resting during the day in hollow logs, under the roots of trees, in empty holes, in crevices, or (in mangrove swamps) in nipa palms. It prefers wet areas and is usually found near streams. A nest is constructed of leaves or other available material (Gould 1978a). There is some controversy regarding food habits. Davis (1962) discounted reports that the moon rat was mainly a hunter of aquatic vertebrates and indicated that it fed almost exclusively on earthworms and arthropods found on the forest floor. Lim (1967) also found the latter two items to predominate in the diet and reported in addition the taking of crabs, land mollusks, and other invertebrates. Lekagul and McNeely (1977) stated that the moon rat often enters water to hunt for frogs, fish, crustaceans, mollusks, and insects. Gould (1978a) found captives to catch and eat goldfish from an aquarium and to cache pieces of fish in nest boxes. Whittow, Gould, and Rand (1977) suggested that wild specimens ate fruits of the cultivated oil palm and that captives fed on fruit, fish, and clams.

The moon rat is generally solitary and intolerant of conspecifics. Captives were found to lack loud vocalizations but to emit hiss-puffs and low roars during encounters. They also marked the entrances of their nest boxes with power-

Philippine gymnure *(Podogymnura truei)*, dead specimens, photo by Lawrence R. Heaney.

ful scent from a pair of small glands located near the anus (Gould 1978a). This scent has been described as resembling that of rotten onions or ammonia.

Lekagul and McNeely (1977) stated that breeding occurs throughout the year, with usually two litters per year, an average litter size of 2, and a gestation of about 35–40 days. The young weigh about 14–15 grams at birth (Hayssen, Van Tienhoven, and Van Tienhoven 1993). Medway (1978) recorded pregnancies in May, June, September, and November; an average litter size of 1.9 (1–2); and a record life span in captivity of 55 months.

INSECTIVORA; ERINACEIDAE; Genus **PODOGYMNURA**
Mearns, 1905

Philippine Gymnures, or Philippine Wood-shrews

There are two species (Heaney and Morgan 1982):

P. truei, Mindanao;
P. aureospinula, Dinagat Island (Philippines).

In *P. truei* head and body length is approximately 130–50 mm and tail length is about 40–70 mm. The pelage is long, soft, and full. The color of the upper parts is gray mixed with reddish brown hairs, and the underparts are hoary, slightly mixed with brown hairs. The moderately haired tail is buffy to purplish flesh color. In two specimens of *P. aureospinula* head and body length was 190 and 211 mm and tail length was 59 and 73 mm. The latter species is distinguished by its spiny dorsal pelage, which is generally golden brown with black speckling. The underparts lack spines and are mostly brownish gray. One specimen was a female and had two pairs of mammae. *Podogymnura* is considered closely related to *Echinosorex*, and the two genera share a number of cranial and dental characters, such as a long rostrum and long, well-developed canine teeth. *Podogymnura*, however, is smaller and has a much shorter tail and ess prominent temporal, sagittal, and nuchal crests (Heaney and Morgan 1982).

P. truei has been collected on Mount Apo at elevations of 1,700–2,100 meters, on the east slope of Mount McKinley at 1,800–2,300 meters, and on Mount Katanglad at 1,600 meters. Specimens have been taken near a hole at the base of a large tree, among tangled roots of trees, among roots in thick moss, near logs along creeks in dense forest with fern undergrowth, under grass on the edge of a lake, and near boulders in a fern-covered valley. Perhaps habits are comparable to those of true shrews. Many specimens have been caught in traps baited with bird flesh, and one stomach examined contained insects and worms. The native Bagobo name means "ground pig." *P. aureospinula* was taken in an area of rolling country and low hills that had been logged but still contained many patches of remnant dipterocarp forests.

The IUCN now classifies both *P. truei* and *P. aureospinula* as endangered. Both occur in restricted patches of forest habitat that are being destroyed by logging and the practice of slash-and-burn agriculture.

INSECTIVORA; ERINACEIDAE; Genus **ERINACEUS**
Linnaeus, 1758

Eurasian Hedgehogs

Three species now are recognized (Corbet 1988a; Zaitsev 1984):

E. europaeus, western and central Europe, northern European Russia, Britain, Ireland, Sicily, Sardinia, Corsica;
E. concolor, eastern Europe, western Siberia, Asia Minor to northern Iran and Palestine;
E. amurensis, southeastern Siberia, Manchuria, Korea, northeastern China.

Erinaceus sometimes is considered to include *Atelerix* (see account thereof). The range of *E. europaeus* has expanded northward in Scandinavia during the twentieth century (Kristiansson 1981).

Head and body length is 135–300 mm, tail length is 10–50 mm, and adult weight is usually 400–1,100 grams. The body, except for the face, legs, and underparts, is cov-

Eurasian hedgehog *(Erinaceus europaeus)*, photos by Ernest P. Walker.

ered with dense spines. The upper parts are often chocolate-colored with pale yellow tips on the spines, but some forms are almost black and others are nearly white. The underparts are usually brownish or grayish.

Erinaceus is distinguished from other genera of hedgehogs chiefly by cranial features. The hallux is well developed, whereas it is small or absent in the closely related genus *Atelerix* (Corbet 1988*a*). The spines are smooth, not rugose as in *Paraechinus*. The ears are relatively short. Females have five pairs of mammae.

Hedgehogs occur in forest, grassland, scrub, or cultivated areas, at either high or low altitudes, wherever they can find adequate food and cover. They generally do not enter extensive coniferous forests because of the lack of ground cover. They are mainly terrestrial but are good swimmers and climbers. Their spines, besides serving as protection against enemies, act as a cushion when they fall or drop deliberately from a height. The usual gait is a slow, rolling walk, but hedgehogs can run fairly rapidly. They are active mostly at night, resting by day under brush piles or in leaf nests in rock crevices and burrows. *E. europaeus* hibernates from about October to early April in the colder parts of its range, but individuals often emerge for brief periods in the winter.

Hedgehogs feed mainly on invertebrates but also consume frogs, snakes, lizards, and young birds and mice—in fact, nearly anything, including carrion. *E. europaeus* has a hunting range that extends about 200–300 meters from its shelter. An individual may occupy the same home range for years (Fons *in* Grzimek 1990). A particularly active male was found to travel over 2 km per night (Morris 1988).

Hedgehogs perform an unusual "self-anointing" comparable in certain respects to "anting" in birds. Almost any substance is a potential stimulus for this activity. The "self-anointing" consists in licking the substance until frothy saliva accumulates in the mouth, then raising the forelegs, turning the head, and, with the tongue, placing the froth on the spines, first on one side of the body and then on the other. This performance may go on almost continuously for as long as 20 minutes. Brockie (1974) suggested that this activity is a form of sexual signal since in the wild it is confined to the breeding season. Brodie (1977), however, reported that the substances that release "self-anointing" behavior are either novel or irritating and hence the process serves as protection against predators. He also observed hedgehogs to rub toads against their spines and suggested that the hedgehogs might be making defensive use of the toxins produced by the skins of the toads.

The typical defensive posture of these hedgehogs is to roll themselves into a ball, covering the vulnerable belly, face, and limbs with the bristling spines. Hedgehogs adjust readily to captivity and become quite docile. Captives can be maintained on chopped meat and a porridge of bread soaked in milk.

In a radio-tracking study of *E. europaeus* in suburban London, Reeve (1982) found that at least 33 individuals inhabited an area of 40 ha. Home range averaged 32 ha. for males and 10 ha. for females. There was considerable overlap in the ranges of both sexes and no evidence of territoriality. Nonetheless, hedgehogs are generally solitary and intolerant of conspecifics. In another radio-tracking study in Italy, Boitani and Reggiani (1984) also found no territoriality, but they did determine that adults of the same sex never approach to within 20 meters of one another. Fons (*in* Grzimek 1990) did describe vigorously defended territories of 1.8–2.5 ha. but noted that in the spring, after initial emergence from hibernation, individuals do not stay in these prescribed areas. At this time they wander over a large area looking for food and mates, and there may be intensive fighting between males. *E. europaeus* usually has two definite breeding seasons per year. In England pregnancies occur from May to October, with peaks in May–July and September. Gestation is 31–35 days, and the recorded number of embryos is two to seven. The young weigh about 15 grams at birth, are weaned after 38–45 days, and reach sexual maturity at about 10 months. *E. europaeus* has been known to live 7 years both in the wild and in captivity (Corbet and Southern 1977; Hayssen, Van Tienhoven, and Van Tienhoven 1993).

INSECTIVORA; ERINACEIDAE; **Genus ATELERIX**
Pomel, 1848

African Hedgehogs

There are four species (Corbet *in* Meester and Setzer 1977; Meester et al. 1986):

A. albiventris, Senegal to Sudan and Zambia;
A. sclateri, northern Somalia;
A. frontalis, southwestern Angola, Namibia, eastern Botswana, western Zimbabwe, South Africa;
A. algirus, Morocco to Libya, Canary Islands, Balearic Islands, Spain, southeastern France (probably introduced in Europe and on islands).

Atelerix often is considered a subgenus of *Erinaceus* (Yates 1984). Robbins and Setzer (1985) showed that *Atelerix* warrants generic distinction, and this view was followed by Corbet (1988a), Corbet and Hill (1991), Frost, Wozencraft, and Hoffmann (1991), and Meester et al. (1986). Robbins and Setzer argued further that the species *A. frontalis* and *A. algirus* should be placed in still another genus, *Aethechinus* Thomas, 1918, but Corbet (1988a), Frost, Wozencraft, and Hoffmann (1991), and Meester et al. (1986) disagreed.

Except as noted, the remainder of this account refers to the species *A. albiventris* and *A. frontalis,* with information taken from Kingdon (1974a) and Smithers (1983). Head and body length is 170–235 mm, tail length is 17–50 mm, and weight is 236–700 grams. The upper parts, except for the front of the head and ears, are covered with short spines. The spines are mainly white at the base and at the tips and have a central band that is dark brown or black. This band varies in width geographically and imparts a darker or lighter appearance to the animals in different areas. The face, limbs, and tail are covered with dark brown or grayish brown hair, and the underparts vary in color from white to black. Females have three pairs of mammae.

Robbins and Setzer (1985) reported that from the genus *Erinaceus* the species *A. albiventris* is distinguished by being smaller and having a smaller skull (condylobasal length usually less than 45 mm), having a broad (rather than narrow) posterior palatal shelf in the skull, having the upper third incisor and canine with two (rather than one) roots, having a reduced or absent third upper premolar, and generally lacking a hallux. The species *A. frontalis* and *A. al-*

African hedgehog *(Atelerix albiventris),* photos by Ernest P. Walker.

girus approach *Erinaceus* in size (condylobasal length 45–55 mm) and have a normal third upper premolar and a hallux but otherwise resemble *A. albiventris*.

African hedgehogs have a wide variety of habitats, including grassland, scrub, savannah, and suburban gardens; they appear to be absent from deserts, marshes, and dense forests. A general requirement is the presence of dry shelter where the animals can rest and bear their young. They are predominantly nocturnal and rest by day, curled up in a ball, under matted grass or in leaf litter, a rocky crevice, or a hole in the ground.

Such refuges are changed on a daily basis unless they are being used to rear young or as a hibernation site. During the warm rainy season, when food is abundant, the hedgehogs gain weight, and they enter torpor when the weather is cool and dry. In southern Africa such hibernation takes place from about June to September, with the animals remaining torpid for periods of up to six weeks but emerging during warm intervals. They generally move along at a slow toddle but are capable of surprising bursts of speed. Their vision may be poor, and they seem to locate food primarily by scent. The omnivorous diet includes a variety of invertebrates, frogs, lizards, snakes, mice, eggs, fruit, and fungi. A self-anointing procedure like the one described in the account of *Erinaceus* also occurs in *Atelerix*.

These hedgehogs are generally solitary; when two individuals meet there is much growling, snorting, and butting of heads. The main defense mechanism is to roll into a ball, with the head, limbs, and underparts protected by the sharp spines of the upper parts. Home range may be somewhat smaller than in *Erinaceus*. The young are born during the warm and wet months, mainly from October to March in southern Africa. Gestation lasts approximately 35 days. Litters contain 1–10 young, usually 4–5. They weigh about 10 grams at birth, open their eyes after about 14 days, and are weaned and begin to accompany the mother at around 6 weeks. Based on studies in captivity, Brodie, Brodie, and Johnson (1982) reported that sexual maturity is attained at 61–68 days and that females can give birth several times per year.

A. frontalis is on appendix 2 of the CITES and was designated as rare in South Africa by Smithers (1986). This species may have declined in numbers because of its popularity as a pet and as food and through loss of habitat to agriculture.

INSECTIVORA; ERINACEIDAE; **Genus PARAECHINUS**
Trouessart, 1879

Desert Hedgehogs

There are three species (Corbet 1978, 1988*a*; Corbet *in* Meester and Setzer 1977; Frost, Wozencraft, and Hoffmann 1991; Nader 1991; Roberts 1977):

P. hypomelas, Aral Sea area, southeast to Pakistan and northern India, and southwest to southern Iran and the southern Arabian Peninsula;

P. micropus, Pakistan, western and southern India;

P. aethiopicus, Morocco and Mauritania, east through Egypt and Sudan to northern Somalia, the Arabian Peninsula, and Iraq.

Frost, Wozencraft, and Hoffmann (1991) regarded *Paraechinus* only as a subgenus of *Hemiechinus*. Corbet and Hill (1992) suggested that the isolated population of *P. micropus* in southern India may have been introduced through human agency.

Head and body length is 140–272 mm and tail length is about 10–40 mm. Walton and Walton (1973) reported that the heaviest specimen in a series of *P. micropus*, a male, weighed 435 grams and that a lactating female weighed 312 grams. The coloration is variable in this genus; there is a tendency to melanism and also to albinism. The spines may be banded with dark brown, with black and white, or with yellow, but often one of these colors predominates. Some forms have a brown muzzle and a white forehead and sides. The underparts may be blotched dark brown and white, the

Desert hedgehog *(Paraechinus aethiopicus)*, photo from Department of Zoology, Hebrew University, Jerusalem, through J. Wahrman.

variation ranging from entirely brown to entirely white. *Paraechinus* has a wide and prominent naked area on the scalp, whereas this area is very narrow in *Erinaceus,* moderately wide in *Atelerix,* and lacking altogether in *Hemiechinus* (Corbet 1988a).

These hedgehogs inhabit deserts and other arid areas. They are mainly terrestrial and nocturnal. The habits of *P. micropus* in the Rajasthan Desert of India are much like those of *H. auritus* in the same area. When frightened or chasing a toad the Rajasthan Desert hedgehogs travel at a speed of 635 mm/sec, whereas they normally trot at a speed of 305 mm/sec. *P. micropus* digs its own burrow, which is about 457 mm long and has a single opening. It is a rather sedentary species, with an individual keeping to the same burrow and range all year. It does not hibernate but may remain torpid in its burrow if food or water is scarce. *P. hypomelas* appears to be more nomadic than other hedgehogs and is not an active burrower (Roberts 1977).

In the Salloum village area of Egypt *P. aethiopicus* is reported to rest in cliffs during the day and to forage on the coastal plain at night. The diet consists mainly of insects but includes such items as small vertebrates, the eggs of ground-nesting birds, and scorpions. *P. micropus* takes food to its burrow for future use and apparently does not eat plant material. *P. hypomelas* eats fruit as well as small vertebrates and insects (Roberts 1977). Captive individuals of *P. micropus* have survived from four to six weeks without food and water. Like *Erinaceus, Paraechinus* has been reported to "self-anoint" (Brodie 1977; Walton and Walton 1973).

These hedgehogs are solitary. In captivity they generally do not molest one another if fed well, but sometimes they eat their young. *P. micropus* reportedly breeds from April to September (Roberts 1977; Walton and Walton 1973). Female *P. hypomelas* are known to have given birth in April and May (Nader 1991). The litter size in the genus ranges from one to six but normally is one to two in *P. micropus* and three to four in *P. hypomelas* (Roberts 1977). Young *P. aethiopicus* weigh about 8–9 grams at birth and are weaned after 40 days (Hayssen, Van Tienhoven, and Van Tienhoven 1993). A captive *P. hypomelas* lived to

about 7 years and 2 months (Marvin L. Jones, Zoological Society of San Diego, pers. comm., 1995).

Corbet (1988a) indicated that increasing desertification of the range of *Paraechinus* is resulting in fragmentation of populations. Harry Hoogstraal (pers. comm.) reported the subspecies *P. aethiopicus wassifi* of Egypt to be extremely scarce, if not extinct.

INSECTIVORA; ERINACEIDAE; **Genus HEMIECHINUS** *Fitzinger, 1866*

Long-eared Desert Hedgehogs

There are two species (Corbet 1988a):

H. auritus, steppe zone from eastern Ukraine to Mongolia in the north and from Libya to western Pakistan in the south, Cyprus;
H. collaris, Pakistan, northwestern India.

Mesechinus (see account thereof) has usually been considered part of *Hemiechinus*. Although Roberts (1977) and some other authorities have referred to *H. megalotis* of Afghanistan and adjacent areas as a full species, it was listed only as a subspecies of *H. auritus* by Corbet (1978) and Frost, Wozencraft, and Hoffmann (1991). The population of *H. auritus* on Cyprus seems to have been introduced by people, perhaps very recently (Boye 1991).

Head and body length is approximately 150–280 mm and tail length is 10–55 mm. Maheshwari (1982) reported the weight of *H. auritus* to be 248–72 grams. The spines are usually banded with dark brown and white, and the underparts are generally whitish. Unlike *Erinaceus* and *Paraechinus, Hemiechinus* lacks a median spineless tract on the top of the head. Also, in *Hemiechinus* the ears are longer and more prominent than in the other two genera, apparently being desert adaptations for heat radiation.

These hedgehogs inhabit subdesert country and steppes over most of their range. In Egypt *H. auritus* is rare in

Long-eared hedgehog *(Hemiechinus auritus aegyptiacus)*, photo by Ernest P. Walker.

poorly vegetated areas and is commonly found in gardens in association with people (Hoogstraal 1962). In Egypt this species utilizes existing shelter sites and seldom digs its own burrow. Reports from other areas, however, indicate that it does regularly excavate burrows up to 150 cm in length (Atallah 1977; Harrison 1964; Roberts 1977). In the Rajasthan Desert of India the burrows usually are under small bushes, have a single opening, and are occupied by one individual throughout the year. During the breeding season the females widen the ends of the burrows to accommodate the young. These hedgehogs are strictly nocturnal and mainly terrestrial. In the Punjab of northern India *H. auritus* may hibernate up to 3.5 months in winter, and in the mountains of Pakistan hibernation lasts from October to March. In warmer areas there is no prolonged winter hibernation, but during periods of food scarcity there may be summer estivation (Roberts 1977). The diet includes insects and other invertebrates, small vertebrates, eggs, carrion, fruit, and seeds (Atallah 1977; Roberts 1977). *H. auritus* is remarkably resistant to hunger and thirst; two individuals survived for 10 weeks in a laboratory without food and water. As in *Erinaceus*, "self-anointing" has been reported in *Hemiechinus* (Brodie 1977).

In a study of *H. auritus* in Israel, Schoenfeld and Yom-tov (1985) found home range to average 2.8 ha. for females and 1.9 ha. for males. *Hemiechinus* is solitary, and captives are sometimes cannibalistic but generally do not molest one another if properly fed. In the Rajasthan Desert the breeding season is May–October, mainly July–September, and litter size is one to six. In Pakistan breeding occurs in spring and summer, gestation reportedly is 35–42 days, and litter size is usually two to three in the east and five to six in the western mountains (Roberts 1977). The young have been reported to weigh around 8–13 grams at birth, to be weaned after about a month, and to be sexually mature by only 6 weeks of age (Hayssen, Van Tienhoven, and Van Tienhoven 1993). According to Jones (1982), a captive lived 6 years and 9 months.

INSECTIVORA; ERINACEIDAE; **Genus MESECHINUS**
Ognev, 1951

Daurian Hedgehogs

There are two species (Corbet 1988a; Frost, Wozencraft, and Hoffmann 1991):

M. dauuricus, eastern Gobi Desert region from Lake
   Baikal to northern China;
M. hughi, Shaanxi and Shanxi provinces in central China.

Mesechinus usually has been treated as a synonym of *Hemiechinus* but was distinguished at the generic level by Frost, Wozencraft, and Hoffmann (1991). This arrangement was followed by Hutterer (*in* Wilson and Reeder 1993). *M. hughi* was not considered a species distinct from *M. dauuricus* by Corbet and Hill (1991, 1992).

According to Allen (1938), one specimen of *M. dauuricus* had a head and body length of 244 mm and a tail length of 28 mm. The spines, which cover the back, are dark brown at the base, then dull whitish for nearly half their length; there is then a brownish black band and finally a white tip. The underparts are whitish and the feet and tail are chestnut brown. Frost, Wozencraft, and Hoffmann (1991) distinguished *Mesechinus* as follows: rostrum broad, anterior incisors not closely approximating (rostrum narrow in *Hemiechinus,* anterior incisors closely approximated);

lacrimal/maxilla suture not fused in young adults (fused and indistinct in *Hemiechinus*); suprameatal fossa shallow, its anterior and posterior borders narrowly separated (fossa well developed in *Hemiechinus,* its borders widely separated); basisphenoid not inflated (inflated in *Hemiechinus*); squamosal forming a major part of bullar roof (does not participate in bullar roof of *Hemiechinus*); pelage spines not grooved (grooved in *Hemiechinus*); and ventral pelage coarse (soft and densely furred in *Hemiechinus*).

M. hughi is classified as vulnerable by the IUCN. The species occupies a relatively small area of suitable habitat that is shrinking because of human encroachment.

INSECTIVORA; **Family CHRYSOCHLORIDAE**

Golden Moles

This family of 7 genera and 19 species inhabits Africa from Cameroon and Uganda southward to the Cape of Good Hope. The sequence of genera presented here follows that suggested by Bronner (1995) and Meester et al. (1986). The same genera were recognized by Corbet and Hill (1991) and Hutterer (*in* Wilson and Reeder 1993), but other classifications of the family have been proposed. Both Petter (1981a) and Yates (1984) accepted only the following as full genera: *Chrysospalax, Eremitalpa, Amblysomus* (including *Calcochloris* and *Chlorotalpa*), *Chrysochloris,* and *Cryptochloris.* Simonetta (1968) divided the Chrysochloridae into subfamilies (not recognized by the other authorities cited here), considered *Cryptochloris* to be a subgenus of *Chrysochloris,* and regarded *Carpitalpa* (here included in *Chlorotalpa*) as a full genus.

Golden moles somewhat resemble the true moles (Insectivora, Talpidae) in appearance. Head and body length is 76–235 mm, and the tail is not visible externally. The skin is tough and loosely attached to the body; the hair is thick with dense woolly underfur. The pelage has a metallic luster or iridescence of red, yellow, green, bronze, or violet. The muzzle terminates in a smooth, leathery pad used in working in the soil, and the nostrils are located under a fold of skin at the front of the long snout. The eyes are vestigial and covered with hairy skin; in the genus *Eremitalpa* the eyelids fuse at an early age and the skin covering the eye increases in thickness. The ears are small and concealed in the pelage. Golden moles have short forelimbs with four-clawed digits; the third digit is the longest and has a powerful claw, and the other three vary in relative length according to the species. The anterior walls of the chest cavity are deeply hollowed to provide space for the thick, muscular arms. The hind limbs are also short, the two lower bones (the tibia and fibula) being fused near the ankle. The hind feet have naked soles and five sharp-clawed toes connected by membranous skin. Females have one pair of abdominal mammae and one pair of inguinal mammae. The urogenital system has a single external opening.

Except in the genus *Amblysomus* the dental formula is: (i 3/3, c 1/1, pm 3/3, m 3/3) $\times$ 2 = 40. *Amblysomus* has the same number of incisors, canines, and premolars but only 2 molars above and 2 below, making a total of 36 teeth. The teeth are slightly separated from each other. The incisors are enlarged; the first premolar is much like the canines and the 2 posterior incisors, and the last 2 premolars are like the molars: narrow and wedge-shaped with a blade-like medial half. The molars have two lateral cusps and one medial cusp, with an enamel ridge on each side of the medial cusp. Golden moles have high-crowned teeth, the crowns of the lower teeth being twice as high as those of

Large golden mole *(Chrysospalax trevelyani)*, photo of mounted specimen from National Museum of Wales.

the upper teeth. The skull is conical in shape, broad at the braincase, and not constricted at the orbits. The olfactory region and the premaxillary bones form a narrow, elongate snout that widens at the end into two processes on which the nose pad is attached.

Golden moles inhabit sandy areas, plains, forests, and cultivated areas. They burrow by means of powerful thrusts, as do the true moles (Talpidae), but use the armored nose more than do the true moles. Some forms burrow just below the surface of the ground; others generally burrow deeper. Local soil conditions probably determine the depth of the tunnels. Members of the genus *Chrysospalax* (and probably some others) rest in chambers and passages in mounds of earth reached by a system of tunnels made in part by them and in part by certain rodents. Golden moles are active day or night. They are believed to be inactive for varying periods during the colder months. The majority of forms seem to find food underground and come to the surface only after a rain brings worms and other invertebrates to the surface. *Chrysospalax*, however, may seek food on the surface at night. The golden moles of this genus have an extraordinary sense of orientation, for they can return directly to the burrow entrance with rapid speed. Most forms feed on invertebrates; *Eremitalpa* includes species of sand-burrowing skinks in its diet, and *Cryptochloris* feeds on legless lizards in addition to various invertebrates. At least one genus *(Cryptochloris)* has the unusual habit of feigning death when picked up or even when turned over with a spade. This may be similar to the involuntary "fainting" of *Didelphis*. Golden moles seldom bite when handled.

In South Africa they breed during the winter (the rainy months). The young, usually two per litter, are born and raised in a grass nest deep in the burrow.

The geological range of this family is Miocene, Pleistocene, and Recent in Africa. The Miocene records are from Kenya. *Proamblysomus*, an extinct genus from the Pleistocene in South Africa known from skull fragments found in the Sterkfontein caves is a small golden mole closely resembling *Amblysomus*.

INSECTIVORA; CHRYSOCHLORIDAE; **Genus CHRYSOSPALAX**
Gill, 1883

Large Golden Moles

There are two species (Meester *in* Meester and Setzer 1977):

C. villosus, Transvaal, Natal, and eastern Cape Province in South Africa;
C. trevelyani, eastern Cape Province in South Africa.

Head and body length is 125–75 mm in *C. villosus* and 198–235 mm in *C. trevelyani*. Weight is 108–42 grams in *C. villosus* (Smithers 1983). The fur is harsher in texture than that of the other genera of golden moles. *C. villosus* is rufous, yellowish brown, dark brown, or dark slaty. *C. trevelyani* is usually reddish brown (Poduschka 1980). The pelage in both species has a metallic sheen or luster.

Distinguishing features of this genus are the large body size and the shape of the zygomatic arch of the skull, which is produced upward posteriorly and meets the lambdoid crest at the back. The claw of the fourth finger is usually much more developed in *C. trevelyani* than in *C. villosus*. Females of *C. villosus*, at least, have four mammae.

C. villosus has been found in grassy areas, especially in meadowlike ground bordering marshes. *C. trevelyani* is a forest inhabitant. These golden moles live in chambers and passages in mounds reached by a system of tunnels made in part by the golden moles and in part by blesmols *(Cryptomys)* or mole-rats *(Bathyergus)*. Poduschka (1980) reported that *C. trevelyani* makes numerous hills 40–60 cm

Large golden mole *(Chrysospalax villosus)*, photo by Graham C. Hickman.

in diameter and about 25 cm high in which there are plugged openings to the tunnels below. He questioned earlier reports that this species actively seeks prey above ground, that it hibernates in winter, and that it can swim across a stream. Hickman (1986) confirmed that *C. trevelyani* is a poor swimmer. However, Maddock (1986) determined that *C. trevelyani* does emerge nightly to forage or to move up to 95 meters across the surface to another tunnel and that it does occasionally enter nonforested areas. He reported also that its tunnels are up to 13.6 meters long and that the entrances are left open. He found surface runways through the leaf litter but no molehills. Whereas Poduschka indicated that *C. trevelyani* feeds mainly on giant worms *(Microchaetus)*, Maddock found a wider dietary range.

Both species are often washed out of their shallow burrows after heavy rains. When *C. villosus*, at least, does feed above ground, it roots about like a little pig in search of worms and insects. It has a remarkable sense of direction in regard to the exact location of its burrow; when threatened, it dashes rapidly and surely for the entrance. These and other golden moles have been blamed for damage to vegetables and crops that was done by rodents, particularly the blesmol *(Cryptomys)*. Reproductive information does not seem to have been recorded, except that two embryos have been found in female *C. villosus*.

The species *C. trevelyani* was designated as vulnerable by Smithers (1986) and is now classified as endangered by the IUCN. Its forest habitat is being reduced and fragmented through cutting and overgrazing, and a local increase in the human population is leading to predation by domestic dogs. The six subspecies of *C. villosus* also were considered vulnerable by Smithers (1986), all evidently being extremely rare; the entire species is now classified as vulnerable by the IUCN.

INSECTIVORA; CHRYSOCHLORIDAE; **Genus**
CRYPTOCHLORIS
Shortridge and Carter, 1938

De Winton's and Van Zyl's Golden Moles

There are two species (Meester *in* Meester and Setzer 1977):

C. wintoni, Port Nolloth in coastal Little Namaqualand (South Africa);
C. zyli, near Lambert's Bay in southwestern Cape Province (South Africa).

Head and body length is 80–90 mm. The fur is short, soft, and dense. The upper parts are a drab lead color with a violet iridescence, and the underparts are lead gray, not much paler than the back. Whitish buff facial markings are present in *C. zyli*. The colors of both are remarkably like those of *Eremitalpa granti*.

Distinguishing features are the temporal bullae, the three well-developed foreclaws, and the large inner digit pad on the forefoot. The flattened and roundly oval body shape is most pronounced in *C. zyli*. As in *Eremitalpa*, this shape may be an adaptation for burrowing through loose sand.

This genus inhabits the same range as *Eremitalpa*; in the Lambert's Bay area it occupies the white coastal sand dunes-salongside Grant's desert golden mole. Its habits also appear to resemble those of *Eremitalpa*. De Winton's golden mole generally burrows just below the surface. Occasionally it tunnels deeper, often to the base of a bush, where presumably it secures shelter and rears its young. In those areas where the sand is extremely powdery the roofs of its surface tunnels collapse almost immediately and form shallow furrows. After a rain the fresh workings of this golden mole are indicated on the surface by minute cracks that disappear as soon as the ground dries. Unsuccessful attempts to dig these animals out usually result in a desertion of the disturbed ground the following night, when the moles travel overland to a new area, often a distance of several hundred meters. The Cape Province form of *Cryptochloris* utters a fairly sharp squeak when handled; when first picked up or when turned over with a spade it feigns death.

The diet consists of legless lizards *(Typhlosaurus)* and various invertebrates. These lizards, which grow to a length of nearly 200 mm, shelter in the sand under bushes. It has been conjectured that *Cryptochloris* uses its long foreclaws to kill and devour them.

C. zyli, known only by a single specimen from a small area of declining habitat, is classified as critically endangered by the IUCN. *C. wintoni*, known by two specimens and also thought to be losing its habitat, is classed as vulnerable.

INSECTIVORA; CHRYSOCHLORIDAE; **Genus**
CHRYSOCHLORIS
Lacépède, 1799

Cape Golden Moles

There are three species (Aggundey and Schlitter 1986; Lamotte and Petter 1981; Meester *in* Meester and Setzer 1977):

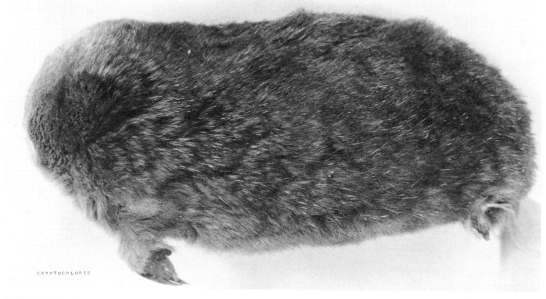

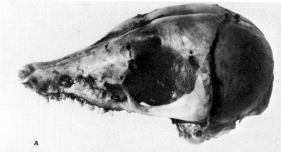

Cryptochloris wintoni, Top photo from British Museum (Natural History); A. Photo from British Museum (Natural History); B. Photo by Don Davis.

C. asiatica, western Cape Province in South Africa;
C. visagiei, known only from the type specimen collected at Gouna in southwestern South Africa;
C. stuhlmanni, western Cameroon, northern and northeastern Zaire, Uganda, Kenya, Tanzania.

Head and body length is 90–140 mm. The ground color is tawny olive, brownish, or grayish. Depending on the angle from which the animal is viewed, the fur has an iridescent luster of greenish, violet, or purplish in many of the forms. The face and nose pad are often paler in color than the back. The underparts are light, almost white or creamy in some individuals. The fur is soft, fine, and dense.

Distinguishing features of this genus are the temporal bullae and the two well-developed foreclaws. The second foreclaw is shorter than the third, the first claw is still shorter, and the fourth digit is represented by a small tubercle. Females have four mammae.

In Uganda these golden moles live at elevations as high as 2,800 meters. They usually burrow just below the surface of the ground, so that fresh workings may be traced by the cracked and ridged ground surfaces above the shallow tunnels. These tunnels sometimes radiate from the base of a bush and sometimes penetrate the large mounds of mole-rats *(Bathyergus).* Occasionally, however, golden moles burrow downward to such depths that they must remove soil from the burrow, which they push to the surface of the ground as mounds of fresh soil. They are common in certain areas and are often found in gardens, where they can burrow in the loose, cultivated soil. Sometimes they make short overland journeys at night. The diet consists of invertebrates, such as beetles, grubs, and worms. These golden moles become active on the surface in rainy weather, when they root like little pigs for earthworms in the damp earth.

The breeding season in the Cape Province seems to be during the rainy months, April–July. In this area the animals make a round nest of grass in which the young are born. At birth they are about 47 mm in length when stretched straight. The young suckle until they are nearly full grown, that is, for two to three months, as the teeth do not cut through the gums until the young are almost mature.

The Bakiga natives in Kigezi use the skins of *C. stuhlmanni* as charms. Meester (1976) designated *C. visagiei* as rare but noted that the single specimen thereof might be nothing more than an aberrant *C. asiatica.* The IUCN now classifies *C. visagiei* as critically endangered, noting that its restricted habitat is declining.

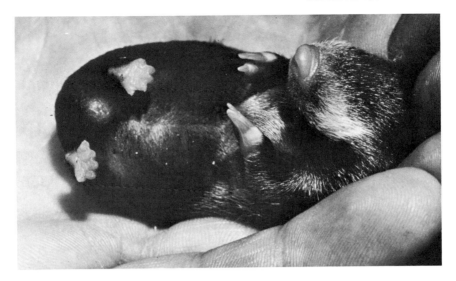

African golden mole *(Chrysochloris asiatica)*, photos by John Visser.

INSECTIVORA; CHRYSOCHLORIDAE; **Genus**
EREMITALPA
Roberts, 1924

Grant's Desert Golden Mole

The single species, *E. granti,* occurs in southwestern Cape Province and Little Namaqualand in South Africa and in the Namib Desert of Namibia (Meester et al. 1986).

Head and body length is 76–88 mm. This animal is the smallest of the golden moles. One individual weighed 15 grams. The fur is long and silky. In subadults the upper parts are a beautiful aluminum gray, changing with age to sandy buff or gray. The face is whitish or buffy in the adults, with indications of pale cheek markings in the immature. The underparts vary from buffy white to pale rufous. The tips of the hairs, especially on the lower back, shine like spun glass, but there is no iridescence. Superficially the general appearance is similar to that of *Chrysochloris; Eremitalpa,* however, lacks temporal bullae.

Distinguishing features are the absence of temporal bullae, the three long foreclaws, and a prominent thickened pad on the hind foot that probably takes the place of a heel. The claws are broad, flat, and leaflike. The body shape of *Eremitalpa,* like that of *Cryptochloris,* is flattened and roundly oval, like a small tortoise, and may be an adaptation for tunneling through loose sand. *Cryptochloris* also occurs in the sand dunes of Little Namaqualand. Fons (*in* Grzimek 1990) stated that *Eremitalpa* moves by "literally swimming" in the sand.

The habitat in Little Namaqualand is the coastal strip of white sand dunes. *Eremitalpa* probably does not range inland because of the relatively firm and level sandveld. It makes shallow, winding tunnels in the sand dunes. These tunnels often can be traced for 45 meters or more and occasionally connect with deeper excavations that descend several meters below the ground. It has been speculated that the young are borne in these shafts. Mounds of sand are not thrown up in the manner of some moles that burrow in harder soil. Surface openings occur irregularly wherever excavations are present.

Fielden (1991) and Fielden, Perrin, and Hickman (1990) indicate that unlike most other subterranean mammals, *Eremitalpa* lacks a permanent burrow system and is a nocturnal surface forager. Individuals cover different parts of their range each night, usually forage along a path of several hundred meters, and seldom return to the rest site occupied the previous day. Termites are the major prey, but the diet includes a variety of other invertebrates and also small lizards. Most rest sites are shallow excavations under vegetative cover. No deep underground nesting chambers have been discovered, though they presumably would be needed for the rearing of young. Home ranges are stable and average 4.63 ha., those of males usually being larger than those of females. Ranges overlap, but individuals are solitary and avoid one another while foraging. Two pregnant females, each with a single fetus near full-term, were taken in October.

Smithers (1986) designated *E. granti* as rare, noting that its limited habitat has been fragmented by dune removal associated with diamond mining. The IUCN classifies this species as vulnerable.

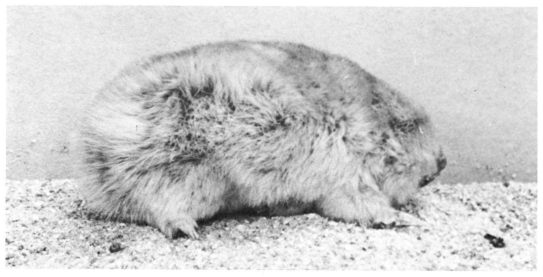

Grant's desert golden mole *(Eremitalpa granti)*, photo by James A. Batemen of mounted specimen.

INSECTIVORA; CHRYSOCHLORIDAE; **Genus**
CALCOCHLORIS
Mivart, 1867

Yellow Golden Mole

The single species, *C. obtusirostris*, occurs in southeastern Zimbabwe, southern Mozambique, and eastern South Africa (Meester et al. 1986). *Calcochloris* was considered a synonym of *Amblysomus* by Petter (1981*a*), Simonetta (1968), and Yates (1984). Morphological and karyological studies by Bronner (1991, 1995) support the generic distinction of *Calcochloris.*

Head and body length is 97–108 mm and weight is about 20–30 grams. The coloration of the upper parts varies geographically from glossy brown (with yellow underfur) to bright golden yellow. The chin, throat, limbs, and belly are yellow. The sides of the face are yellow, and there is a broad yellow or buffy band across the top of the snout. The claw on the third digit of the front foot is about 10 mm long, and the fourth digit has a claw about 2 mm long (Smithers 1983). From *Amblysomus, Calcochloris* is distinguished by its striking coloration, broader skull, and more molariform first upper and lower premolar teeth (Meester et al. 1986).

According to Smithers (1983), the yellow golden mole is confined to light sandy soils, sandy alluvium, and coastal sand dunes. In captivity it makes no attempt to burrow into heavy clay or schist soils but tunnels to a depth of 20 cm in loose, sandy soil. In the wild it moves in subsurface runs leading from chambers around the bases of trees, and if disturbed it may burrow deeply into the substrate. It is highly sensitive to the movements of prey dropped onto the surface of the soil and quickly burrows to the vicinity. At times it may emerge and move porpoiselike across the surface at surprising speed. The diet consists of earthworms and insects. Smithers (1986) designated the species as rare.

INSECTIVORA; CHRYSOCHLORIDAE; **Genus**
CHLOROTALPA
Roberts, 1924

African Golden Moles

There are five species (Meester *in* Meester and Setzer 1977; Meester et al. 1986):

C. arendsi, eastern Zimbabwe and adjacent parts of Mozambique;

C. duthieae, southern Cape Province in South Africa;

C. sclateri, Cape Province, eastern Orange Free State, southeastern Transvaal, Lesotho;

C. leucorhina, Cameroon, southern Zaire, northern Angola;

C. tytonis, known only from the type specimen collected at Giohar in Somalia.

Chlorotalpa was considered part of *Amblysomus* by both Petter (1981*a*) and Yates (1984). Simonetta (1968) placed *C. arendsi* in the separate genus *Carpitalpa* Lundholm, 1955, included *C. leucorhina* and *C. tytonis* in *Amblysomus,* and retained *Chlorotalpa* only for the species *C. duthieae* and *C. sclateri.* Morphological and karyological studies by Bronner (1991, 1995) support the generic distinction of *Chlorotalpa.*

In *C. arendsi* head and body length is 109–39 mm and weight is 38–76 grams; the upper parts are glossy black or very dark brown with a sheen of green and purple, while the sides of the face and the underparts are paler; and the claw on the third digit of the front foot is 10 mm long. In *C. duthieae* head and body length is 95–111 mm; the upper parts are very dark brown with a distinct green sheen, while the underparts are a shade lighter and the sides of the face are yellowish; and the first digit of the front foot is reduced. And in *C. sclateri* head and body length is about 100 mm; the upper parts are glossy brown, the flanks have a reddish tinge, and the underparts are a dull gray; and the third digit of the front foot has a claw about 9 mm long (Smithers 1983).

Chlorotalpa resembles *Amblysomus* in having short and dense fur, two well-developed foreclaws, and two pairs of mammae in the female and in not having temporal bullae. *Chlorotalpa* is distinguished by having 10 upper and lower teeth in each side of the jaw. *Amblysomus* usually has only 9 such teeth; although the species *A. gunningi* has a tenth, this tooth differs in appearance from the homologous tooth in *Chlorotalpa* (Smithers 1983).

Almost nothing is known about the natural history of these golden moles. *C. arendsi* is found in montane grassland and makes subsurface runs between clumps of tussock grass; *C. duthieae* occurs in alluvial sand and sandy loam in coastal areas; and *C. sclateri* appears to be associated with rocky hillsides. The diet probably consists principally of earthworms and insects (Smithers 1983). The IUCN classifies *C. tytonis* as critically endangered, noting that its restricted habitat is in decline, and classifies *C. duthieae* and *C. sclateri* as vulnerable.

INSECTIVORA; CHRYSOCHLORIDAE; **Genus**
AMBLYSOMUS
Pomel, 1848

South African Golden Moles

There are two subgenera and five species (Bronner 1991, 1995; Meester 1972; Meester et al. 1986):

subgenus *Neamblysomus* Roberts, 1924

A. gunningi, eastern Transvaal;
A. julianae, southern and eastern Transvaal;

subgenus *Amblysomus* Pomel, 1848

A. septentrionalis, southern Transvaal;

A. iris, coastal areas of southern and eastern Cape Province, Natal;
A. hottentotus, coastal areas of southern and eastern Cape Province, Natal, eastern Transvaal, northeastern Orange Free State, Lesotho, Swaziland, southern Mozambique.

Bronner (1995) stated that an undescribed sixth species related to *A. septentrionalis* is found in the eastern Transvaal. *Amblysomus* sometimes is considered to include *Chlorotalpa* and *Calcochloris* (see accounts thereof).

Head and body length is approximately 95–145 mm. The weight of *A. hottentotus* is 40–101 grams (Kuyper 1985). Two specimens of *A. julianae* weighed 21 and 23 grams (Smithers 1983). The fur is short, fine, and dense. The upper parts are golden brown with a bronze sheen in *A. gunningi,* very dark smoky brown with a distinct greenish sheen in *A. iris,* dark reddish brown with a bronze sheen in *A. hottentotus,* and cinnamon brown in *A. julianae;* the underparts and sides of the head are generally paler (Smithers 1983).

Distinguishing features of this genus are the absence of temporal bullae and the presence of two well-developed foreclaws. The four mammae are in two pairs, one inguinal and one on the flanks.

These golden moles vary geographically in South Africa because of local isolation in much the same manner as the American pocket gophers. Some species inhabit areas of peaty soil in the sheltered ravines of mountains or in forests; others occur in escarpment forests, and certain forms are present at times in forests but are usually more common in open areas where there is a good grass cover. Some forms reportedly usually burrow just below the surface of the ground, so that raised ridges mark the course of the tunnel. Others are said to burrow deeper and to throw up mounds of fresh earth. Local soil conditions may be the major factor in determining the depth of the tunnel. These animals are often active in the daytime and during the

African golden mole *(Amblysomus hottentotus)*, photo by Graham C. Hickman.

rainy season. They make special, grass-lined nesting chambers in which they bear and nurse their young. Tunnels are often located near the nest chamber so that if an enemy, such as a mole snake, appears the golden mole can burrow rapidly through the walls and escape. Hickman (1986) found *A. hottentotus* to be a competent swimmer. The diet consists of worms, larvae, pupae, and insects.

According to Kuyper (1985), *A. hottentotus* occupies a wide range of habitats from sea level to an elevation of 3,300 meters. It is active at intervals throughout the day and night. When the temperature is above about 30° C or below 15° C, it enters daily torpor to conserve energy. Excavated burrows have varied from 9.5 to 240 meters in length. The more permanent burrows are 29–94 cm below the surface and have several vertical passages that extend to a greater depth where the animal can sleep or escape danger. It burrows through about 4–12 meters of soil per day; during the winter, when rainfall is low, its movements are more restricted and at greater depths. It is solitary, fighting sometimes occurs, and a dominant individual may take over a neighbor's burrow. There is no distinct breeding season; births have occurred throughout the year. There are up to three young per litter; they weigh 4.5 grams at birth and are evicted from the maternal burrow when they reach 35–45 grams.

Meester (1976) designated *A. julianae* as rare because of its limited distribution and impending habitat destruction in the Pretoria area. This species is known only by three small and isolated populations (Bronner 1990). The IUCN now classifies *A. julianae* as critically endangered because of the continuing decline of its restricted habitat; *A. gunningi* is classed as vulnerable.

INSECTIVORA; Family TENRECIDAE

Tenrecs, or Madagascar "Hedgehogs"

This family of 10 Recent genera and 25 species is found in western and central equatorial Africa, Madagascar, and the Comoro Islands. The sequence of genera presented here follows that of Van Valen (1967), who recognized three subfamilies: the Potamogalinae, with the genera *Potamogale* and *Micropotamogale*, the Oryzorictinae, with the genera *Geogale, Oryzorictes, Microgale,* and *Limnogale,* and the Tenrecinae, with the genera *Tenrec, Setifer, Hemicentetes,* and *Echinops.* Hutterer (*in* Wilson and Reeder 1993) placed *Geogale* in a separate subfamily, the Geogalinae. The Potamogalinae have sometimes been accorded familial rank, but most recent authorities have treated them

as a subfamily, especially since the discovery of *Micropotamogale,* a genus that seems intermediate to *Potamogale* and the other tenrecids (Corbet *in* Meester and Setzer 1977; Yates 1984).

Head and body length is 40 to approximately 400 mm. The Tenrecinae have only a rudimentary tail; in the other subfamilies tail length is 30–290 mm. *Microgale longicaudata* has 47 tail vertebrae, more than any other mammal except some of the tree pangolins *(Manis).* In *Microgale* the tip of the tail is modified for prehension. The members of the subfamily Tenrecinae have a body covering of spines or bristly hairs. Other genera have soft fur. The two lower bones of the hind leg can be united or free. The mammae are often numerous: *Tenrec* has 12 pairs. Males have a retractile penis; the testes are located in a scrotum in the Potamogalinae, in the pelvic region in *Microgale,* and against the kidneys in the other genera.

The Potamogalinae of Africa resemble otters in general appearance. The other branches of the family probably became established in Madagascar in the Cretaceous and subsequently developed a number of specialized body forms. *Tenrec* resembles *Didelphis* (an American genus of opossums) in certain dental, cranial, and skeletal features; *Setifer* and *Echinops,* the Madagascar "hedgehogs," resemble the true hedgehogs (Erinaceidae) in several dental features and in the spiny pelage. *Oryzorictes* has acquired definite fossorial characters; *Microgale* has some resemblances to the terrestrial shrews; and *Limnogale* resembles the semiaquatic *Desmana* of Eurasia.

The dental formula in the Potamogalinae is: (i 3/3, c 1/1, pm 3/3, m 3/3) × 2 = 40. The first upper incisor and the second lower incisor are caninelike, and the first upper incisors are separated by a greater space than are any other teeth. The canines are low and resemble the premolars. The upper molars of the otter shrews have V-shaped or weak W-shaped cusps. In the other genera of the family the dental formula is: (i 2/3, c 1/1, pm 3/3, m 4/3) × 2 = 40. The last molar, however, does not erupt until after the first molar has been shed, so that the complete dentition is not present at any one time. The incisors and canines in these genera are variable; in *Tenrec* the canines are enlarged. The first premolar is absent. The crowns of the molars are V-shaped, occasionally W-shaped.

The skull has long, narrow nasal bones and a small braincase that is not constricted between the orbits. Those lobes of the brain associated with the sense of smell are well developed. Clavicles are present except in the Potamogalinae.

The known geological range of the family is middle Eocene to middle Oligocene in North America, Miocene to Recent in Africa, and Pleistocene to Recent in Madagascar (Van Valen 1967).

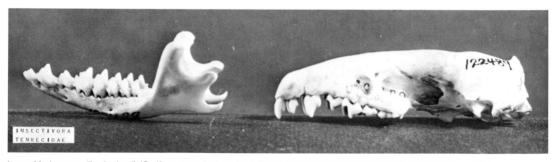

Large Madagascar "hedgehog" *(Setifer setosus),* photo by P. F. Wright of skull in U.S. National Museum of Natural History.

Giant African water shrew *(Potamogale velox)*, photo of mounted specimen from Bristol City Museum, England.

INSECTIVORA; TENRECIDAE; **Genus POTAMOGALE**
Du Chaillu, 1860

Giant African Water Shrew, or Giant Otter Shrew

The single species, *P. velox*, is found in the rainforest zone of central Africa, from Nigeria to western Kenya, northern Zambia, and central Angola (Aggundey 1977; Corbet *in* Meester and Setzer 1977).

Head and body length is 290–350 mm and tail length is 245–90 mm. Nicoll (1985) reported weight to be 340–97 grams. The fur is short, soft, and dense with a protective coat of coarse guard hairs. The upper parts are dark brown or blackish, and the underparts are whitish or yellowish.

This animal resembles an otter in appearance, hence one of its vernacular names. The flattened muzzle has stiff white whiskers. External ears are present; the eyes are minute; and the nostrils are covered by flaps that act as valves when the animal is submerged. The body is cylindrical, and the thick, powerful tail is strongly compressed laterally. The short, rather weak limbs have five nonwebbed digits. A longitudinal flange of skin is present along the inner border of the hind foot, so that it may be pressed smoothly against the body and tail in swimming. The tail seems to be the only means of propulsion in the water. On land this animal walks on the soles of its feet with the heels touching the ground. The two lower bones of the hind leg are united distally. Females have two mammae located in the groin region.

Habitats include both sluggish lowland streams and cold, clear mountain torrents from sea level to about 1,800 meters. In some areas the otter shrew frequents small forest pools during the rainy season and migrates overland to streams at the beginning of the dry season. It is an extremely agile and rapid swimmer. It becomes active in the late afternoon, after sheltering during the day in holes and tunnels in stream banks. The entrance to its burrow is below the water level. On land its movements are rather clumsy, but it can move at considerable speed.

This species, one of the largest of the insectivores, has been reported to feed on crabs, fish, and amphibians. In some areas freshwater crabs are eaten almost exclusively. *Potamogale* turns a crab over on its back and tears the flesh out of its body and two claws. A captive adult consumed 15–20 crabs per night and left those he could not eat. This animal seems to hunt mainly by touch and scent and to come out on land to eat and apparently to void excrement.

According to Happold (1987), *Potamogale* is solitary, each adult utilizes 500–1,000 meters of a stream, females probably have two litters annually, and litter size is one or two young. Births may occur over a considerable part of the year. Kuhn (1971) reported that all pregnant females he examined contained two embryos.

Potamogale is widely hunted for its skin and also is trapped accidentally, but if its forest habitat remains intact, even as a narrow strip along a riverbank, a viable population apparently can be maintained. Unfortunately, logging and subsequent soil erosion, with increased opaqueness of watercourses, is leading to local disappearances (Nicoll and Rathbun 1990). *Potamogale* now is classified as endangered by the IUCN.

INSECTIVORA; TENRECIDAE; **Genus MICROPOTAMOGALE**
Heim de Balsac, 1954

Dwarf African Water Shrews, or Dwarf Otter Shrews

There are two species (Corbet *in* Meester and Setzer 1977; Kingdon 1974a):

M. lamottei, southeastern Guinea and adjacent parts of Liberia and probably of the Ivory Coast;
M. ruwenzorii, northeastern Zaire, southwestern Uganda.

The latter species is sometimes placed in the genus or subgenus *Mesopotamogale* Heim de Balsac, 1956.

Head and body length is 120–200 mm, tail length is 100–150 mm, and weight is about 135 grams (Kingdon 1974a; Kuhn 1971; Rahm 1966). In *M. ruwenzorii* the coloration is brownish gray above and gray below, the feet are webbed between the digits, and the tail is short-haired throughout and roughly oval in cross section with a slight keel along the upper and lower surfaces. *M. lamottei* has unwebbed feet and a round tail. From *Potamogale*, *Micropotamogale* is distinguished by its smaller size, zalambdodont rather than dilambdodont dentition, its fleshy

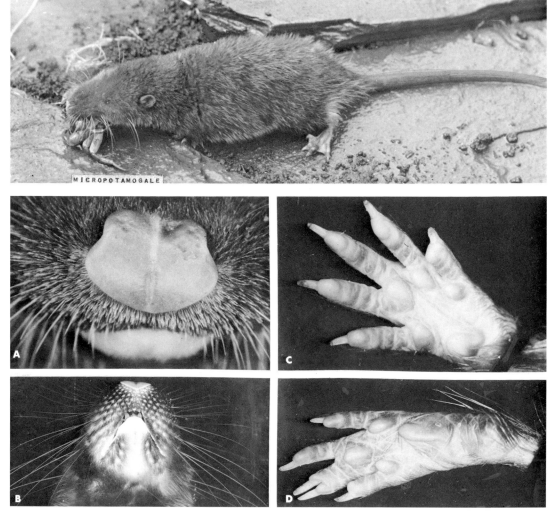

Small African otter shrew *(Micropotamogale ruwenzorii)*, photo by U. Rahm. Detail photos by Hans-Jurg Kuhn of *M. lamottei:* A. Nose shield; B. Underside of head; C. Right hand; D. Left foot.

rather than hornlike or leathery rhinarium, and the remarkable evolution of the middle ear, which is transformed into an almost perfect bulla with a carotid canal rather than only a groove. A female *M. lamottei* had four pairs of mammae (Kuhn 1971).

These mammals live along upland forest streams. They are nocturnal and semiaquatic. *M. ruwenzorii,* found at an altitude of about 1,220 meters, is known to dig tunnels and make sleeping chambers with hay or grass. Its dives are brief, and it frequently comes out of the water. It brings larger prey out of the water to eat, and it sometimes hunts on dry land (Kingdon 1974a). Foods recorded for the genus include worms, insects and their larvae, crabs, fish, and small frogs. A female *M. ruwenzorii* collected in late September contained two embryos (Kingdon 1974a). A female *M. lamottei* taken on 5 December in Liberia had four embryos, each 18.5 mm long (Kuhn 1971).

These animals reportedly have considerable strength. They do damage in fish traps, generally killing all the fish in the trap and then, unable to escape, drowning. Many of them have been found in the holes of the diamond diggers along small creeks in the Saniquelli District. Two specimens of *M. ruwenzorii* were caught in fishing nets. Such accidental killing is one problem, but a more serious concern is the loss of the very restricted habitat in which the two species occur. The future of *M. lamottei,* in particular, seems bleak (Nicoll and Rathbun 1990). It is known almost entirely from an area of less than 1,500 sq km around Mount Nimba at the junction of Guinea, Liberia, and Ivory Coast. Mining activities have devastated the Liberian sector and may be spreading into Guinea. Both species now are classified as endangered by the IUCN.

INSECTIVORA; TENRECIDAE; Genus GEOGALE
Milne-Edwards and Grandidier, 1872

The single species, *G. aurita,* is found on both the west and the east coast of Madagascar (Heim de Balsac 1972). The

Geogale aurita, photo by Martin E. Nicoll.

genus and species *Cryptogale australis* Grandidier, 1928, described from cranial remains found in the Grotte d'Androhomana, south of Fort Dauphin, Madagascar, is included in *Geogale aurita* (Genest and Petter *in* Meester and Setzer 1977).

Head and body length of the type specimen is 71 mm and tail length is 32 mm. Weight is 5.0–8.5 grams (Stephenson 1993). The pelage is soft. The coloration of the type is reddish brown above and soiled yellowish white below. The muzzle projects beyond the lower jaw. The tail is cylindrical, scaly, and covered with fine hairs. All the limbs have five digits.

Heim de Balsac (1972) stated that *Geogale* inhabits the semiarid zone of western Madagascar but has also been found in a very humid area on the east coast. Gould and Eisenberg (1966) reported that a single specimen was taken from beneath a fallen tree, where it was sleeping in a cavity of sand. It was found in June in dry open woods and apparently was in torpor. Nicoll and Rathbun (1990) reported finding numerous individuals within fallen logs, all torpid by day but active by night. They noted that *Geogale* takes a wide variety of invertebrate prey but shows a marked preference for termites.

Stephenson (1993) established a captive colony with 21 individuals found in crevices within decomposing logs on the forest floor. Mating occurred from September to February, births from November to March during the austral summer. Females exhibited a postpartum estrus. Average gestation was 57 days, though in some cases it lasted as long as 69 days, possibly in association with the females entering torpor. Litter size averaged 3.9 and ranged from 2 to 5 young. They weighed about 0.7 grams at birth, opened their eyes at about 24 days, and were weaned after about 5 weeks.

According to Nicoll and Rathbun (1990), a male caught as an adult lived 25 months in captivity. The subspecies *G. a. aurita* is widespread in southern and western Madagascar but may be threatened by forest clearing. The subspecies *G. a. orientalis* is known only by a single specimen taken on the northeastern coast.

INSECTIVORA; TENRECIDAE; Genus ORYZORICTES
Grandidier, 1870

Rice Tenrecs

There are three species (Genest and Petter *in* Meester and Setzer 1977; Stephenson 1994*a*):

O. hova, central and northeastern Madagascar;
O. talpoides, northwestern Madagascar;
O. tetradactylus, northern Madagascar.

The last species is sometimes placed in the separate genus or subgenus *Nesoryctes* Thomas, 1918.

Head and body length is 85 to about 130 mm and tail length is 30–50 mm. Stephenson (1994*a*) reported weights of 29.5 grams for a male and 41.5 grams for a female *O. hova.* Fons (*in* Grzimek 1990) listed a weight range of 42–48 grams for *O. talpoides.* The fur is velvety in *O. hova* and *O. talpoides;* the hair texture of *O. tetradactylus* is coarser than that of the other two species. The upper parts are gray brown or dark brown, and the underparts are grayish or buffy brown. The tail is bicolored like the body.

These animals are somewhat modified for a burrowing life and thus resemble moles rather than shrews. The forefeet are stout, larger than the hind feet, and have strong digging claws. The species *O. tetradactylus* has four digits on the forepaws, whereas the other two species have five. The body is more robust than in the shrewlike species of the genus *Microgale.*

Rice tenrecs are inhabitants of rainforests and marshy areas, especially the moist banks of rice fields. They are said to be mainly nocturnal, but because of their burrowing habits they may move about at all hours without being observed. These insectivores have been collected in the banks of rice paddies in tunnels that resembled those of moles (Talpidae). In some parts of Madagascar their burrowing activities in the water-retaining walls of the fields has resulted in damage to the dikes. The diet is assumed to consist mainly of insects and other invertebrates, such as mollusks. Observations by Stephenson (1994*a*) indicate that *O. hova* sometimes forages above ground on the forest floor,

Rice tenrec *(Oryzorictes)*, photos by Edwin Gould.

using its muzzle to probe beneath the leaf litter and humus, and may feed on earthworms. Fons (*in* Grzimek 1990) indicated that *O. hova* gives birth to three young at a time.

INSECTIVORA; TENRECIDAE; **Genus MICROGALE**
Thomas, 1882

Long-tailed Tenrecs, or Shrewlike Tenrecs

There are 13 species (Jenkins 1988, 1992, 1993; MacPhee 1987*a*):

M. cowani, northern, central, and eastern Madagascar;
M. thomasi, southeastern Madagascar;
M. soricoides, east-central Madagascar;
M. parvula, northern Madagascar;
M. pulla, northeastern Madagascar;
M. dryas, northeastern Madagascar;
M. gracilis, southeastern Madagascar;
M. longicaudata, northern and eastern Madagascar;
M. principula, southeastern Madagascar;
M. pusilla, eastern and southwestern Madagascar;
M. brevicaudata, western and northeastern Madagascar;
M. dobsoni, eastern Madagascar;
M. talazaci, northern and eastern Madagascar.

Until MacPhee's revision, far more species were generally recognized, and some were put in the separate genera or subgenera *Leptogale* Thomas, 1918; *Nesogale* Thomas, 1918; and *Paramicrogale* Grandidier and Petit, 1931. MacPhee's views were followed by Corbet and Hill (1991) and Hutterer (*in* Wilson and Reeder 1993), but Nicoll and Rathbun (1990) suggested that there are a number of additional valid species. Based on observations of live specimens, Stephenson (1995) recommended that *M. melanorrhachis, M. longirostris,* and *M. taiva* of east-central Madagascar should again be distinguished from *M. cowani.*

Head and body length is 40–130 mm, and tail length is 43–160 mm except in the species *M. brevicaudata,* in which the tail is only about 35 mm long. Individuals of some smaller species of *Microgale* weigh as little as 5 grams. The following weights for certain medium to large species were given by Eisenberg and Gould (1970): *M. cowani,* 10.9–12.1 grams; *M. talazaci,* 39–61 grams; and *M. dobsoni,* 34–45 grams. With the onset of winter *M. dobsoni* accumulates fat reserves in the body and tail, and weight may then reach 84.7 grams in captivity. The tail of *M. thomasi* is also known to accumulate fat. The fur of *Microgale* is soft. The upper parts are buffy, dark rufous brown, olive brown, or nearly black; the underparts are usually buffy, buff gray, or lead-colored. In some specimens of *M. cowani* there is a black dorsal stripe.

These animals are shrewlike and apparently have become adapted to the same sort of habitat that on the African mainland would be occupied by various shrews (Soricidae). The tip of the tail in the long-tailed forms is usually modified for prehension. The forelimbs are not adapted for digging, and all the limbs have five digits. The ears are prominent, projecting well above the fur.

Long-tailed tenrecs live in a variety of habitats but usually where there is dense vegetation near the surface. They seem to be both diurnal and nocturnal and apparently resemble shrews in their way of life. Eisenberg and Gould (1970) proposed the following functional classification: (1) a semifossorial form *(M. brevicaudata),* in which the tail is relatively short; (2) surface-foraging forms *(M. cowani, M. gracilis, M. thomasi, M. dobsoni),* in which the tail length approaches the head and body length; (3) surface foragers and climbers *(M. talazaci, M. pusilla, M. parvula),* in which the tail is up to 1.5 times as long as the head and body; and (4) possible climbers and ricochetors among branches *(M. longicaudata),* in which the tail is 1.5–2.6 times as long as the body.

According to Eisenberg and Gould (1970), a captive *M.*

Long-tailed tenrec, or shrewlike tenrec *(Microgale gracilis)*, photo by Martin E. Nicoll.

dobsoni did not enter deep torpor, but when it had accumulated fat reserves it showed a tendency to become inactive, sleep a great deal, eat less, and decline in body temperature. It constructed a nest when a suitable box was provided. *M. talazaci* did not accumulate extensive amounts of fat and showed no tendency toward inactivity. In the wild, *M. talazaci* lives in a more stable habitat (rainforest) than *M. dobsoni* and has not evolved the latter's adaptations for an extended dry season. *M. talazaci* ap-

peared to make use of an extensive tunnel system, as well as regular runways on the surface. Both species, and probably all other species of *Microgale,* feed primarily on insects. *M. dobsoni* was observed to move about sniffing the substrate, occasionally to insert its nose under litter, and to rush at insects when it detected them by hearing. Gould (1965) presented evidence that *M. dobsoni* employs echolocation to assist it in its movements.

Observations by Eisenberg and Gould (1970) indicate

Long-tailed tenrec, or shrewlike tenrec *(Microgale talazaci)*, photo by Martin E. Nicoll.

Long-tailed tenrec, or shrewlike tenrec *(Microgale dobsoni)*, photo by John Eisenberg.

that *M. dobsoni* is basically solitary and well spaced in the wild. Strange individuals are antagonistic toward one another and sometimes fight, but a male and a female may establish a stable relationship. Several sounds are produced during agonistic encounters. The behavior of *M. talazaci* is much the same, but opposite sexes may maintain an association throughout the year. Field studies indicate that *M. dobsoni* breeds in the spring and summer and that *M. talazaci* may have a more extended season. In *M. dobsoni* gestation is 62–64 days, litter size is 1–2, minimum age of conception is 22 months, and maximum known longevity in captivity is 5 years and 7 months. In *M. talazaci* gestation is 58–63 days, litter size averages 2 (1–3), birth weight is 3.6 grams, weaning takes place at 28–30 days, sexual maturity is reached at about 21 months, and maximum longevity is 5 years and 10 months (Eisenberg 1975; Eisenberg and Maliniak 1974; Hayssen, Van Tienhoven, and Van Tienhoven 1993).

Nicoll and Rathbun (1990) expressed concern for a number of species of *Microgale*. They noted that *M. longicaudata*, *M. parvula*, and *M. principula*, as well as many of the forms that were recognized as species prior to MacPhee's (1987a) revision, had not been collected for more than 50 years. The IUCN now classifies *M. dryas* as critically endangered, *M. parvula* and *M. principula* as endangered, and *M. gracilis*, *M. pulla*, and *M. thomasi* as vulnerable; in all cases the main problem is restricted habitat that is declining because of human activity.

INSECTIVORA; TENRECIDAE; **Genus LIMNOGALE**
Forsyth Major, 1896

Web-footed Tenrec

The single species, *L. mergulus*, is known from a few scattered localities in eastern Madagascar (Eisenberg and Gould 1970; Nicoll and Rathbun 1990).

Head and body length is 122–70 mm and tail length is 119–61 mm (Eisenberg and Gould 1970). Stephenson (1994b) indicated a weight range of 62–90 grams. The fur is close, dense, and soft and resembles that of an otter in texture. The upper parts are brownish with an admixture of reddish and blackish hairs. The underparts are pale yellowish gray.

The head is short, broad, and flattened. The eyes are small, and the short ears are almost hidden in the fur. This aquatic mammal has fringed forefeet and webbed toes. The powerful tail is thick and almost square in the proximal half and laterally compressed in the distal half. Presumably both the hind feet and the tail are used in swimming, but the tail is probably the main organ of propulsion. Females have six mammae.

This tenrec inhabits the banks of streams and the shores of marshes and lakes at altitudes of 600–2,000 meters. It constructs burrows at the water's edge. Apparently it is completely carnivorous, foraging in the water for small frogs and fish, freshwater shrimp, crayfish, and aquatic insect larvae. Its habitat is frequently associated with the aquatic lace plant, *Aponogeton*, the bases of which harbor abundant aquatic invertebrates. Breeding is estimated to take place in December and January, and the average litter size is probably about three (Eisenberg and Gould 1970).

The web-footed tenrec is now classified as endangered by the IUCN, and Nicoll and Rathbun (1990) regarded it as the most severely threatened species of the Tenrecidae. It is found mainly along fast-flowing streams, which are becoming increasingly isolated as agricultural expansion fragments the remnant upland forests. In addition, it is being adversely affected by the intensive collection of the plant *Aponogeton* for ornamental use, by sedimentation of watercourses, and by accidental capture in fish traps. It is known only from a few sites, at some of which it apparently has become extremely rare.

INSECTIVORA; TENRECIDAE; **Genus TENREC**
Lacépède, 1799

Tenrec

The single species, *T. ecaudatus*, inhabits Madagascar and has been introduced on the Comoro Islands in the channel between Madagascar and Africa as well as on Reunion and Mauritius in the Indian Ocean (Eisenberg and Gould 1984). In the U.S. National Museum of Natural History, in Washington, D.C., there is a male tenrec that was captured in July 1892 by Dr. W. L. Abbott on the Island of Mahe in the Seychelles, 1,500 km north of Mauritius. The label reads: "Introduced from Madagascar via Bourbon (Mauritius). Introduced about 10 years ago and now very abundant."

Head and body length is approximately 265–390 mm and tail length is 10–16 mm. Weight in captivity is usually 1.6–2.4 kg (Eisenberg and Gould 1970). The general coloration is grayish brown or reddish brown, but some individuals are dark brown on the back and rump. The pelage, which is not dense, consists of hairs and spines. Young animals have strong white spines arranged in longitudinal rows along the back; in adults these are replaced by a crest of long, rigid hairs. The adult pelage is coarse and spiny but not sharp-tipped. The skull is cylindrical and elongated. The forelimbs are shorter than the hind limbs, and the digits are all well developed. The tenrec is strong, muscular, and powerful. It does not curl up in a protective ball as *Setifer* does. When angry, alarmed, or excited the tenrec is capable of erecting the mane of the upper back, which has longer hairs than the rest of the body. This, combined with the stout body form and elongated nose, gives the animal the appearance of a little pig. The tenrec emits a low hissing noise when annoyed and also squeaks or squeals. Females have 12 pairs of mammae.

This hedgehoglike mammal seems to be common in most parts of Madagascar, from sea level up to about 900 meters, except in the arid southwestern districts. Its habitat generally has brush or undergrowth for cover and a source of free water. The animal appears equally well adapted to inland plateaus and coastal rainforests. Observations of wild and captive individuals demonstrate a bimodal pat-

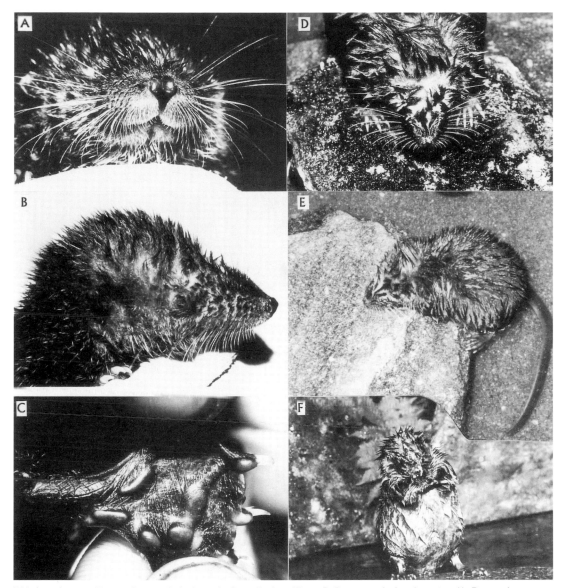

Web-footed tenrec *(Limnogale mergulus)*, photos by Steve Zack and Bruce D. Patterson of a female captured in Ranomafana National Park.

tern of activity with peaks between the hours of 1800 and 2100 and between 0100 and 0500 (Eisenberg and Gould 1970). This mammal shelters in a nest in a hollow log or under a rock but hibernates in a ground burrow one to two meters long. The burrow is plugged with soil before the animal goes into its long sleep. The tenrec hibernates during the dry season, or southern winter (May–October), possibly for only part of this period. Torpid individuals are cold to the touch and have a breathing rate of about 30 respirations per minute. Apparently the sleep is continuous, as all the animals examined in the dormant state have been fat and have not had any trace of food or feces in the intestine. The average body temperature of *Tenrec* is 28°–29° C, with a variation of about 11°, from 24° C to 35° C. The tenrec digs with its snout for a variety of invertebrates. It also eats some vegetable matter and may occasionally take small vertebrates.

Eisenberg and Gould (1970) concluded that except for the mother-young group, individual adult tenrecs forage and hibernate alone. They tend to avoid one another and may fight if they cannot. Association between sexes apparently is brief and probably occurs in the austral spring (October–November). The young are born in December and January following a gestation period of 56–64 days. It is unlikely that there is more than one litter per year, though it is possible that a second litter could be produced if the first is lost immediately after being born in December. Litter size averages about 15 young and ranges from 1 to 32, making *Tenrec* one of the most prolific of mammals (Eisenberg 1975; Louwman 1973). The young weigh 10–18 grams at birth (Fons *in* Grzimek 1990). They forage with the mother when they are about 3–6 weeks old and may remain together briefly after separation from the mother. They are full grown by March or April (Eisenberg and Gould 1970).

Web-footed tenrec *(Limnogale mergulus)*, photos from . . . *Zoologie de Madagascar*, G. Grandidier and G. Petit.

Females can first conceive at 6 months (Hayssen, Van Tienhoven, and Van Tienhoven 1993). One captive individual survived 6 years and 4 months (Jones 1982).

INSECTIVORA; TENRECIDAE; Genus SETIFER
Froriep, 1806

Large Madagascar "Hedgehog"

The single species, *S. setosus,* inhabits northern and eastern Madagascar (Eisenberg and Gould 1970). *Dasogale* Grandidier, 1928, with the species *D. fontoynonti,* is now considered to be part of *Setifer* and probably to represent an immature *S. setosus* (MacPhee 1987b). *Ericulus* I. Geoffroy St. Hilaire, 1837, also is a synonym of *Setifer.*

Head and body length is 150–220 mm, tail length is 15–16 mm, and weight in the wild is 180–270 grams (Eisenberg and Gould 1970). The upper parts are covered with close-set, sharp, and bristly white-tipped spines that extend from the forehead along the back and sides and even cover the short tail projection. The muzzle, limbs, and underparts are covered with soft, scanty hair. As in *Tenrec*, there are a number of vibrissae, about 50–75 mm long, growing from the face. A finely mottled color effect results from the white-tipped spines; some individuals are comparatively pale with broadly white-tipped spines, whereas others are blackish with narrowly white-tipped spines. The underparts are buffy, dark brown, or whitish.

Tenrec *(Tenrec ecaudatus)*, immature, photo by Howard E. Uible.

Tenrec *(Tenrec ecaudatus)*, photo by Martin E. Nicoll.

The general appearance is that of a small species of hedgehog *(Erinaceus),* but the body is relatively longer and the muzzle is less pointed. The tip of the muzzle projects beyond the lower lip. The ear is not longer than the muzzle and is much like the ear of *Tenrec* in general form. The hind limbs are only slightly longer than the forelimbs, and all the digits are present and well developed. Females have five pairs of mammae.

This animal dwells in dry forest and agricultural areas and is abundant in the eastern highlands of Madagascar. It constructs short tunnels leading to a leaf-lined nest. It appears to be strictly nocturnal. Although an overall population may be active throughout the year, individuals have the capacity to become torpid and may spend most of the winter within a restricted foraging radius, passing much time in a torpid state. For defense the large Madagascar

Large Madagascar "hedgehog" *(Setifer setosus)*, photo by Martin E. Nicoll.

"hedgehog" rolls itself into a protective ball of spines that point in every direction. For general locomotion it employs a crossed extension coordination pattern. When running it rises up on its toes, keeping its heels off the ground. It climbs very well but slowly. Captives have eaten earthworms, grasshoppers, raw ground meat, and carcasses of mice but have seemed incapable of killing larger prey. Wild individuals have been observed to take carrion and to scavenge around garbage dumps (Eisenberg and Gould 1970).

Adults appear to be generally solitary and to avoid one another, but there may be temporary associations between two adults of the opposite or the same sex. Several sounds have been recorded during agonistic or sexual encounters. In one investigated population mating took place from late September to mid-October (Eisenberg and Gould 1970). Gestation is variable, being shorter when ambient temperature is higher. Gestation in captivity was reported as 65–69 days by Eisenberg (1975) and 51–61 days by Mallinson (1974). Average litter size is three young, with a range of one to five (Eisenberg 1975). The young weigh about 25 grams at birth, they are weaned after just 15–30 days, and females can conceive by the time they are 6 months old (Hayssen, Van Tienhoven, and Van Tienhoven 1993). A captive lived 10 years and 6 months (Jones 1982).

INSECTIVORA; TENRECIDAE; Genus HEMICENTETES
Mivart, 1871

Streaked Tenrec

The single species, *H. semispinosus,* occurs in Madagascar (Genest and Petter *in* Meester and Setzer 1977). The two subspecies, *H. s. semispinosus,* of the eastern rainforests, and *H. s. nigriceps,* of higher elevations, have sometimes been considered distinct species.

Head and body length is 160–90 mm and the tail is vestigial. Field weight for adults ranges from about 80 to 280 grams (Eisenberg and Gould 1970). The pelage is spiny and sharply pointed. The ground color is black, with chestnut brown markings in *H. s. semispinosus* and whitish markings in *H. s. nigriceps.* The former subspecies has a central stripe from the muzzle to behind the ears, a prominent crest of erectile spines on the nape, and a number of longitudinal stripes on the back and sides. The stripes in *H. s. nigriceps* are not as distinct as those in *H. s. semispinosus,* though the crest is equally pronounced. Patches of chestnut brown or white are also present on the upper parts. The underparts are chestnut brown in *H. s. semispinosus* and whitish or buffy in *H. s. nigriceps.* The hairs covering the underparts are quite spiny in *H. s. semispinosus* and soft in *H. s. nigriceps.*

The genus is easily recognized by its streaked color pattern. The skull of *Hemicentetes,* in comparison with that of *Tenrec,* is exceedingly elongated and tapered.

The streaked tenrec inhabits rainforest and brushland and has extended its range into cultivated rice fields. An individual may construct a burrow consisting of a nest chamber at the end of a tunnel about 200–500 mm long and 50–150 mm deep; or there may be an extensive burrow system for a group of animals, with tunnels several meters in length. In locomotion *Hemicentetes* uses a diagonal coordination pattern of limb movement. The heel may be lifted off the ground in running, but generally the animal is plantigrade. It is capable of climbing but does not often do so. Available evidence indicates that *H. s. nigriceps* is strictly nocturnal but *H. s. semispinosus* may be active at any time of the day. A state of torpor, in which body temperature is less than 1° C above ambient temperature, may be attained during the winter. Stephenson and Racey (1994) found that *H. s. semispinosus,* which lives at lower, warmer elevations, may be able to avoid torpor when favorable environmental conditions prevail but that the upland *H. s. nigriceps,* which they considered to be a separate species, must hibernate throughout the winter. The diet consists largely of earthworms. *Hemicentetes* forages by inserting its nose at the roots of grasses or under leaves (Eisenberg and Gould 1970). When disturbed or touched it spreads the crest of spines on its nape, lowers its head, and then quick-

Streaked tenrec *(Hemicentetes semispinosus nigriceps)*, photo by Martin E. Nicoll.

Streaked tenrec *(Hemicentetes semispinosus semispinosus)*, photo by Eugene Maliniak.

ly thrusts its head upward. A small area of heavy spines in the middle of the back is often vibrated rapidly, independently of the rest of the spines on the back. Gould (1965) presented evidence that *Hemicentetes* uses echolocation.

The streaked tenrec produces a number of sounds that are audible to humans. During agonistic behavior there may be a noisy "crunch" and a "putt-putt." Sound also results from the vibration of the specialized quills in the mid-dorsal region. The latter process, known as stridulation,

seems to be a means of communication and apparently is especially important in helping the mother and young locate one another (Eisenberg and Gould 1970).

Hemicentetes may occur in groups of various composition. One observed group consisted of an extended colony of 18 individuals, including 2 adult males, 2 adult females, and 14 juveniles. Most other groups are smaller, usually having 1 or 2 adults and several young. Apparently there is a lengthy period of association between the sexes. Mat-

Streaked tenrecs *(Hemicentetes semispinosus semispinosus)*, mother and young, photo by Martin E. Nicoll.

ing begins in September or October and continues through December. The young are born from November to March, following a gestation period of about 55–63 days. Litter size reportedly averages 2.8 (2–4) in *H. s. nigriceps* and 6.6 (2–11) in *H. s. semispinosus*. Development is extraordinarily fast. The young are weaned after 17–20 days. Females attain sexual maturity within only 5 weeks of their birth. Maximum captive longevity is about 2 years and 7 months (Eisenberg 1975; Fons *in* Grzimek 1990; Eisenberg and Gould 1970).

INSECTIVORA; TENRECIDAE; Genus ECHINOPS
Martin, 1838

Small Madagascar "Hedgehog"

The single species, *E. telfairi*, inhabits southern and south-western Madagascar (Genest and Petter *in* Meester and Setzer 1977).

Head and body length is 140–80 mm, there is no external tail, and weight in captivity is 110–250 grams. Coloration is variable in a given population, ranging from a pale, almost white color to a very dark color resulting from more intense melanin deposition in the annular bands of the quills (Eisenberg and Gould 1970). *Echinops* closely resembles *Setifer* but is slightly smaller, has shorter claws, and has only 32 teeth (*Setifer* has 36).

The small Madagascar "hedgehog" is confined to more arid parts of Madagascar. It dens in tree cavities, and the location of its nests indicates considerable arboreal ability. It is nocturnal and hibernates for three to five months during the cold season (Godfrey and Oliver 1978). It forages both in trees and on the ground, and captives take a variety of animal food, including insects, baby mice, and chopped meat (Eisenberg and Gould 1970).

Echinops exhibits considerable social tolerance: captive mated pairs or several adult females have shared a small cage even when a litter was being raised, and occasionally two adults have been found hibernating together. Males may be highly aggressive toward one another. Communication is by tactile, chemical, and a few auditory signals. Mating probably begins in October, during the austral spring, shortly after the adults emerge from torpor, and births take place in December or January. Females appear to be polyestrous and to exhibit several cycles of 6 days each during the mating season. Gestation evidently lasts 42–49 days. Litters contain 1–10 young, usually 5–7. The young weigh about 8 grams at birth, open their eyes by 9 days, and are weaned and functionally independent at 30–35 days. They are sexually mature after they complete their first winter's hibernation

Small Madagascar "hedgehog" *(Echinops telfairi)*, photo by Martin E. Nicoll.

Small Madagascar "hedgehog" *(Echinops telfairi)*, photo by Howard E. Uible.

(Eisenberg 1975; Eisenberg and Gould 1967, 1970; Godfrey and Oliver 1978; Mallinson 1974). Maximum known longevity in captivity is 13 years (Jones 1982).

INSECTIVORA; Family SOLENODONTIDAE; Genus SOLENODON
Brandt, 1833

Solenodons

The single known genus, *Solenodon*, contains four species (E. R. Hall 1981; Morgan and Ottenwalder 1993; Morgan and Woods 1986; Woods and Eisenberg 1989):

S. cubanus, Cuba;
S. arredondoi, known only by skeletal remains from western Cuba;
S. marcanoi, known with certainty only by skeletal remains from Hispaniola;
S. paradoxus, Hispaniola.

Some authorities, such as Yates (1984), also have placed the extinct West Indian genus *Nesophontes* in the Solenodontidae, but most, including Hutterer (*in* Wilson and Reeder 1993) and Woods and Eisenberg (1989) now regard *Nesophontes* as the only genus of another family, the Nesophontidae. *S. cubanus* sometimes has been placed in the separate genus or subgenus *Atopogale* Cabrera, 1925, and *S. marcanoi* sometimes has been placed in the separate genus or subgenus *Antillogale* Patterson, 1962, but neither designation was accepted by Hutterer (*in* Wilson and Reeder 1993) or Morgan and Ottenwalder (1993). Only *S. cubanus* and *S. paradoxus* are definitely known as living species in modern times. *S. marcanoi* originally was described from a late Pleistocene site in the Dominican Re-

public, but its remains subsequently were found in association with *Rattus* in southwestern Haiti (Morgan and Woods 1986). It thus apparently survived past the time of the initial European colonization, and Woods and Eisenberg (1989) suggested that it may still be extant. *S. arredondoi* originally was described, without being named, from what was thought to be a late Pleistocene site (Morgan, Ray, and Arredondo 1980). However, when Morgan and Ottenwalder (1993) named the species and reported additional material, they indicated that the age was uncertain but that extinction likely had resulted from human habitat destruction and predation by domestic dogs.

Solenodons resemble large, stoutly built shrews. In the two species known to exist, *S. paradoxus* and *S. cubanus*, head and body length is 280–390 mm, tail length is 175–255 mm, and weight is about 1 kg. Morgan and Ottenwalder (1993) indicated that *S. marcanoi* is smaller than those two species but that *S. arredondoi*, originally described as a "giant extinct insectivore," probably had a head and body length of 450–550 mm and a weight of 1.5–2.0 kg. The pelage is coarse in *S. paradoxus* and longer and finer in *S. cubanus*. The tail and feet are nearly naked. *S. paradoxus* varies in color from mixed blackish and buff to a deep reddish brown, usually with a small, square spot of white in the middle of the nape; *S. cubanus* is blackish brown with white or buff.

The elongate, bare-tipped cartilaginous snout is supported at the base by a small round bone. The nostrils open to the side. The eyes are small, and the partly naked ears extend beyond the pelage. There are five claws on each foot; those of the forefoot are longer and more curved than those of the hind foot. The tibia and fibula may be fused near the ankle in old age. The secretion from glands in the armpit and the groin has a goatlike odor. The urogenital and anal openings are separate; there is no cloaca. The two mammae are located in the inguinal region.

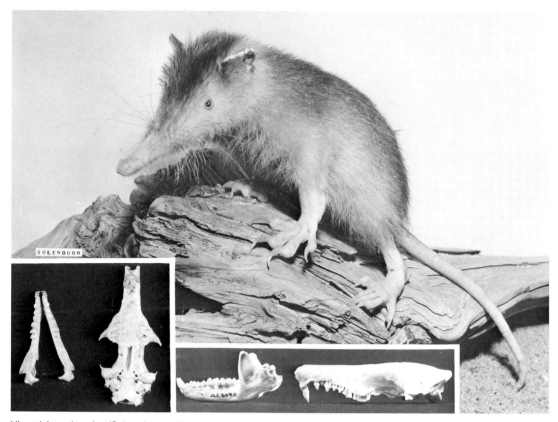

Hispaniolan solenodon *(Solenodon paradoxus)*, photo from New York Zoological Society. Insets: *S. paradoxus*, photos by P. F. Wright of specimen in U.S. National Museum of Natural History.

Solenodon *(Solenodon paradoxus)*, photo from Antwerp Zoo.

The dental formula is: (i 3/3, c 1/1, pm 3/3, m 3/3) × 2 = 40. The first upper incisor is larger than the other, and there is a diastema between the first and second upper incisors. The upper molars have three cusps arranged in a V. The skull is long, slender, and constricted between the orbits. The lambdoidal crest, on the dorsal surface of the cranium, projects eavelike over the occipital region. The sub-

maxillary glands of *S. paradoxus* and presumably those of *S. cubanus* produce a toxic saliva. The duct of the gland ends at the base of the deeply grooved second lower incisor. *S. paradoxus* does not seem to be immune to its own venom, as there have been cases of death after fights among cage mates even though the wounds were slight.

Solenodons inhabit forests and brushy areas, often

around plantations. They are mainly nocturnal, sheltering by day in caves, rocky crevices, hollow trees, logs, or burrows they dig themselves. They generally do not build nests except during the breeding season (Eisenberg and Gould 1966). They have been found to construct extensive burrow systems through deep layers of humus; here they forage and nest beneath the surface (Eisenberg and Gonzalez Gotera 1985). In moving about they pursue an irregular, somewhat zigzag course with a stiff, waddling gait. Only their toes come into contact with the ground. When disturbed they use a quadrupedal ricochet, with forelimbs and hind limbs alternately striking the ground. They can run surprisingly fast but can easily be caught by people. They seem incapable of jumping but can climb (Eisenberg and Gould 1966). Solenodons obtain food by rooting in the ground with their snouts and by tearing into rotten logs and trees with their foreclaws. They eat various invertebrates, reptiles, fruits, vegetables, and occasionally poultry.

The vocalizations resemble those of the Tupaiidae and Soricidae. A high-frequency "click" is produced ranging from 9,000 to 31,000 cps with a duration of 0.1–3.6 msec. Perhaps this pulse functions in echolocation. Solenodons may fight each other on first meeting, but eventually they establish a dominance relationship and live together in captivity in relative harmony.

According to Fons (*in* Grzimek 1990), *S. paradoxus* is relatively social. Up to eight individuals may live in the same burrow, and almost always several are found together in a shelter. Older offspring stay near their parents. Very little is known about reproduction, but mating is believed to be irregular and aseasonal. There usually is just a single young, occasionally two, with a birth weight of 40–55 grams. The offspring take their first solid food at about 13 weeks. A captive specimen of *S. cubanus* lived 6 years and 6 months and a captive *S. paradoxus* lived 11 years and 4 months.

S. marcanoi is regarded as extinct by the IUCN. *S. cubanus* and *S. paradoxus* are listed as endangered by the IUCN and the USDI. Despite legal protection, these animals have declined because of deforestation, increasing human activity, and predation by introduced dogs and cats (Thornback and Jenkins 1982). Earlier in this century it appeared that *S. cubanus* might be extinct but that *S. paradoxus* was not in immediate danger (Allen 1942). Recently *S. cubanus* has been discovered in many parts of eastern Cuba, though it is rare. *S. paradoxus* is still found in both the Dominican Republic and Haiti, but it seems in imminent danger of extirpation in the latter country.

Except as indicated above, the geological range of the family Solenodontidae is Recent in Cuba and Hispaniola. Reports of Oligocene genera in the family are now considered invalid. MacFadden (1980), however, concluded that *Solenodon* and *Nesophontes* are island relics of a group of soricomorph insectivores that were widespread in North America and the Caribbean region during the late Mesozoic and early Cenozoic.

INSECTIVORA; Family NESOPHONTIDAE; Genus NESOPHONTES
Anthony, 1916

Extinct West Indian Shrews

The single known genus, *Nesophontes,* contains eight described species (E. R. Hall 1981; Morgan and Woods 1986):

N. micrus, Cuba, Isle of Pines;
N. longirostris, Cuba;
N. major, Cuba;
N. submicrus, Cuba;
N. paramicrus, Haiti;
N. hypomicrus, Haiti;
N. zamicrus, Haiti;
N. edithae, Puerto Rico and Vieques Island to southeast.

This genus is known only from skulls and fragmentary skeletal material found in owl pellets. It is likely that not all of the above species are valid, but all were accepted by Hutterer (*in* Wilson and Reeder 1993). Morgan and Woods (1986) did not list *N. longirostris* as a species but did indicate that an additional, undescribed species has been found in the Cayman Islands. All of the above species evidently survived past the end of the Pleistocene, but *N. major, N. submicrus,* and *N. edithae,* unlike the other five, are not known to have lived into post-Colombian times.

McDowell (1958) included *Nesophontes* in the family Solenodontidae because of similarities in the bones and the outer portions of the molars of *Nesophontes* and *Solenodon* and considered both genera closely related to the Soricidae. Yates (1984) followed this arrangement, but Van Valen (1967) maintained the Nesophontidae as a family separate from the Solenodontidae and even placed each in a separate order. Most recent authorities, including Hutterer (*in* Wilson and Reeder 1993) and Morgan and Woods (1986), have recognized both Nesophontidae and Solenodontidae as separate families within the order Insectivora.

The Nesophontidae probably ranged in size from about that of a mouse to that of a chipmunk. *N. zamicrus* was the smallest species and *N. edithae* appears to have been the

Extinct West Indian shrew *(Nesophontes edithae),* reconstruction from C. A. Woods.

largest, its skull length being as much as 52 mm in males and 40.8 mm in females. Although that species sometimes has been regarded as the most primitive member of the genus (and, if so, should appear first in the above list), Woods and Eisenberg (1989) suggested that the opposite may be true, which would be more in keeping with current biogeographic hypotheses; *N. edithae* may have become very large because of a lack of competition from *Solenodon*, which did not occur in Puerto Rico.

Nesophontes lacked a jugal bone and zygomatic arch and seems to have lacked auditory bullae. The braincase was low, the rostrum elongate and tubular. The dental formula is: (i 3/3, c 1/1, pm 3/3, m 3/3) × 2 = 40. The genus probably had an elongated and flexible snout; a long, slender head; small eyes; and a tail about as long as the body. The limbs were moderately long, and there were probably five fingers and five toes, each bearing a claw. The thorax in *Nesophontes* appears to have been narrow instead of widened as in *Solenodon*.

As indicated above, most species of *Nesophontes* probably became extinct after the arrival of the Spanish in the West Indies, as their bones have been found together with those of *Rattus* and of *Mus* in caves. The last two genera probably arrived with the Spanish. The factors leading to the probable extinction of the West Indian shrews are not definitely known, but the introduction of rats from Europe and the burning of forests for cultivation are thought to have contributed to it. Nonetheless, remains of *Nesophontes* found in 1930 in pellets from a barn owl under an overhanging ledge in Haiti were so recent and fresh in appearance as to suggest the possibility that some of these insectivores were still living. Some of the bones retained bits of dried tissue (Allen 1942). MacFadden (1980) wrote that there is a possibility that *Nesophontes* might yet be extant, though a survey of Hispaniola and Puerto Rico by Woods, Ottenwalder, and Oliver (1985) found no conclusive evidence of its survival. The IUCN now designates *N. micrus, N. paramicrus, N. hypomicrus, N. zamicrus,* and the undescribed species in the Cayman Islands as extinct.

INSECTIVORA; Family SORICIDAE

Shrews

This family of 22 Recent genera and 322 species is found on all major land areas except the arctic islands, Ungava, Greenland, Iceland, the West Indies, Australia, Tasmania, New Zealand, and some of the Pacific islands. However, it

is found only in the northernmost part of South America. The sequence of genera presented here follows basically that suggested by George (1986), Repenning (1967), and Yates (1984), who recognized two living subfamilies: the Soricinae, with the genera *Sorex, Blarinella, Cryptotis, Blarina, Notiosorex, Megasorex, Neomys, Nesiotites, Soriculus, Nectogale, Chimarrogale,* and *Anourosorex;* and the Crocidurinae, with *Myosorex, Crocidura, Paracrocidura, Suncus, Sylvisorex, Ruwenzorisorex, Feroculus, Solisorex, Diplomesodon,* and *Scutisorex.* Recent immunological studies by George and Sarich (1994) suggest that *Notiosorex* is the most divergent member of the Soricinae and thus perhaps should be placed at the end of the subfamilial sequence. *Myosorex* has here been placed prior to *Crocidura* in accordance with Maddalena and Bronner (1992), who considered that it holds a central phylogenetic position in the Soricidae and that it may even be a surviving component of the subfamily Crocidosoricinae, which otherwise became extinct by the Pliocene and which Reumer (1987, 1994) thought gave rise to the two living subfamilies.

Shrews are small, short-legged, mouselike animals with long, pointed noses. One of the two smallest mammals in the world, *Suncus etruscus,* which weighs approximately 2 grams as an adult, is a member of this family. Head and body length of shrews ranges from 35 to 180 mm, and tail length is 9–120 mm. Adults weigh from 2 to about 35 grams, usually 3–18 grams. The fur is short, dense, and usually some shade of brown or gray. Odoriferous glands are present on the flanks of some, and the odor is most noticeable during the breeding season. The eyes are small and sometimes hidden in the fur, and vision is poorly developed. Hearing and smell, however, are acute. The pinna of the ear is present, but in some forms it is reduced. The foot is not specialized except in the aquatic species. *Nectogale elegans* is the only member of the family with webbed feet; in the other aquatic forms the feet, toes or fingers, and tail are fringed with stiff hairs. The fringe of hairs increases the surface area of the foot, aids in swimming, and traps air bubbles, so that the shrew can actually run on the surface of water. The two bones of the lower leg are fused. In most genera the genital and urinary systems have a common opening through the skin; in *Sorex* the openings are separate, and in *Myosorex* an intermediate condition exists.

The first set of teeth is shed in the embryonic stage, so that the teeth at birth are the permanent set. There are 26–32 teeth, normally 6 on each side of the lower jaw; *Myosorex* and, rarely, *Surdisorex norae* have 7 lower teeth. The first upper incisor is enlarged and has a hook at the end. This tooth also has a cusp projecting ventrally at the base. The first lower incisor is enlarged and projects forward and

Masked shrew *(Sorex cinereus),* photo by Leonard Lee Rue III.

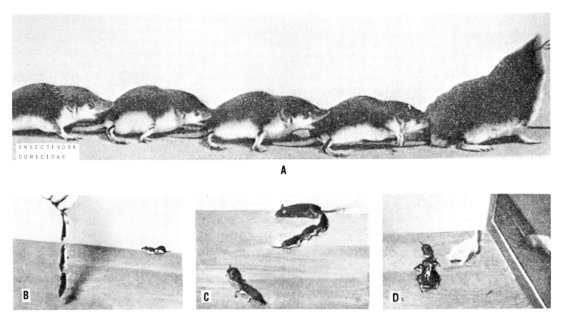

Young shrews *(Crocidura leucodon):* A. Holding on to their mother and each other in their early explorations; B. Holding on to each other even when lifted; C. Attempting to hold on to a mouse; D. Effect of introducing a white mouse. Photos from *Handbuch der Zoologie,* "Das Verhalten der Insektivoren," K. Herter.

slightly upward. The teeth anterior to the last premolar have only one cusp; the upper molars have their cusps arranged in a W-shaped pattern. The skull is long and narrow, and the sutures fuse early in life. The brain is small with smooth cerebral hemispheres.

Shrews usually inhabit moist areas. *Notiosorex* and *Diplomesodon,* however, are found in arid regions, though their exact habitats are not definitely known. Because of the small size of shrews, temperature and evaporation may be of more importance in governing the distribution of this group than that of most other mammals. Shrews frequently have markedly restricted habitats. They are mainly terrestrial, but some take to water freely and others burrow a little. There are records of tree climbing by *Blarina* and *Sorex.* Most forms travel at a swift run. The members of this family are active throughout the year; some forms are active day and night, others only at night. Shrews are extremely nervous animals; when they are frightened, their heart may beat 1,200 times per minute, and they often die of fright from loud noises, even from thunder. Some experiments with the oxygen consumption of shrews have demonstrated a higher metabolic rate at rest than for various wild mice and voles of the same weight. Other experiments, however, indicate that shrews have a metabolic rate comparable to that of other mammals of the same size. It has been suggested that the high rates of oxygen consumption exhibited by the shrews in the former group of experiments are due to external factors such as nervous stimulation rather than to higher inherent rates of tissue metabolism.

Shrews are generally solitary except when they pair off during the breeding season. One exception is *Cryptotis parva,* a species that is somewhat gregarious and colonial. Shrew populations vary greatly with the seasons of the year. The members of this family live under litter or beneath the surface of the ground in runways they construct or in those dug by mice and other small burrowing animals. Shrews may employ echolocation.

Shrews are mainly insectivorous and carnivorous, but some eat seeds, nutmeats, and probably other plant material. The salivary glands of some forms secrete a poisonous substance that usually quickly immobilizes small prey, and there are records that shrew bites have poisoned humans, causing great pain.

It has been stated that certain species of *Sorex, Notiosorex,* and *Crocidura* eat their own feces and perhaps those of other creatures. By doing this they obtain vitamins B, K, and others that are present in such material. The same habit is observed in captive rabbits and in some rodents.

The breeding season is usually March–November for species in the northern temperate zone, but these animals apparently breed throughout the year in the tropics. The known gestation periods are 17–28 days. There are one or several litters a year with 2–10 young in each litter. The young are born naked and blind in a nest of dried grass or leaves that is placed under a shelter or in a ground cavity; weaning appears to occur at 2–4 weeks in most forms. Some zoologists believe the life span in the wild to be approximately 12–18 months, but it may be longer.

Their heavy consumption of insects and insect larvae probably makes shrews valuable in agricultural areas. Because of their aggressive and voracious behavior and their unusual appearance they are the object of many superstitions.

Probably the most common cause of death shortly after capture is shock. Therefore, immediately after capture shrews should have the opportunity to hide in soil, leaves, moss, or underground shelters so that they are shielded to some degree from the new and strange sounds, smells, and sights that probably terrify them.

The geological range of this family is late Eocene to Recent in North America, Pleistocene to Recent in northwestern South America, early Oligocene to Recent in Europe and Asia, and Miocene to Recent in Africa (Yates 1984).

Long-tailed shrew *(Sorex fontinalis)*, photo by Ernest P. Walker.

INSECTIVORA; SORICIDAE; **Genus SOREX**
Linnaeus, 1758

Long-tailed Shrews

There are 4 named subgenera and 68 species (Caire, Vaughan, and Diersing 1978; Carraway 1990; Conway and Schmitt 1978; Corbet 1978; Corbet and Hill 1992; Dannelid 1990, 1991; Diersing and Hoffmeister 1977; Dolgov and Hoffmann 1977; Ellerman and Morrison-Scott 1966; Frey and Moore 1990; S. B. George 1988; George and Sarich 1994; E. R. Hall 1981; Hauser et al. 1985; Hennings and Hoffmann 1977; Hoffmann 1985, 1987; Hoffmann and Fisher 1978; Junge and Hoffmann 1981; Junge, Hoffmann, and De Bry 1983; Kirkland and Levengood 1987; Kirkland, Schmidt, and Kirkland 1979; Kirkland and Van Deusen 1979; Long and Hoffmann 1992; Mitchell 1975; Nagorsen and Jones 1981; F. W. Scott 1987; Tomasi and Hoffmann

1984; Van Zyll de Jong 1982, 1983, 1991; Yoshiyuki and Imaizumi 1986; Zaytsev 1988):

subgenus *Homalurus* Schultze, 1890

S. alpinus, mountains of Europe from Pyrenees to Carpathians;

subgenus *Sorex* Linnaeus, 1758

S. samniticus, central Italy;
S. minutus, Europe, western Siberia, mountains of Central Asia;
S. thibetanus, mountains from Tajikistan to southeastern Tibet;
S. buchariensis, Pamir Mountains of Tajikistan;
S. volnuchini, Caucasus and adjacent areas;
S. caecutiens, taiga and tundra zones from eastern Europe to Korea, Sakhalin and Hokkaido islands;
S. shinto, Japan except Hokkaido;
S. sadonis, Sado Island off western Honshu;
S. isodon, southeastern Norway to eastern Siberia, Sakhalin;
S. roboratus, Siberia, northern Mongolia;
S. sinalis, central China;
S. cansulus, known only by three specimens from Gansu (central China);
S. unguiculatus, southern Pacific coast of Siberia, Sakhalin, Hokkaido (Japan);
S. hosonoi, central Honshu (Japan);
S. cylindricauda, Sichuan (south-central China);
S. bedfordiae, central and south-central China, Nepal, Assam, northern Burma;
S. minutissimus, taiga zone from Norway to Japan and central China;
S. raddei, Caucasus, northeastern Turkey;
S. mirabilis, North Korea, Ussuri region of southeastern Siberia;
S. gracillimus, southeastern Siberia, Manchuria, Sakhalin, Hokkaido and Kurils (Japan);
S. araneus, Europe, western Siberia;
S. coronatus, France and adjoining areas;
S. granarius, Spain, Portugal;
S. daphaenodon, Siberia, Mongolia, Manchuria, Sakhalin;
S. satunini, Caucasus, northern Turkey;

European shrew *(Sorex araneus)*, photo by P. Morris.

Long-tailed shrew *(Sorex minutus)*, photo by P. Rödl.

S. asper, Kyrgyzstan and adjacent Sinkiang;

S. excelsus, central China;

S. tundrensis, Siberia and some adjacent parts of Europe and Asia, Alaska, northwestern Canada;

S. arcticus, Yukon to Wisconsin and Nova Scotia;

subgenus *Otisorex* De Kay, 1842

S. camtschatica, Kamchatka Peninsula of eastern Siberia;

S. leucogaster, Paramushir Island in the northern Kurils;

S. portenkoi, northeastern Siberia;

S. ugyunak, tundra region from Alaska to Hudson Bay;

S. jacksoni, St. Lawrence Island (Alaska);

S. hydrodromus, St. Paul Island (Alaska);

S. haydeni, southern Alberta and Manitoba to Iowa and northern Kansas;

S. cinereus, Alaska, Canada, northern United States;

S. fontinalis, Pennsylvania, West Virginia, Maryland, Delaware;

S. lyelli, east-central California;

S. preblei, central Oregon to eastern Montana and western Colorado;

S. milleri, northeastern Mexico;

S. longirostris, southeastern United States, Illinois;

S. vagrans, southwestern Canada, western conterminous United States;

S. oreopolus, central Mexico;

S. sonomae, southwestern Oregon, northwestern California;

S. bairdii, northwestern Oregon;

S. monticolus, Alaska to northwestern Mexico;

S. pacificus, western Oregon;

S. veraepacis, southern Mexico, Guatemala;

S. macrodon, eastern Mexico;

S. ornatus, California, Baja California;

S. tenellus, southwestern Nevada, east-central California;

S. nanus, central Montana to eastern Arizona and New Mexico;

S. palustris, southeastern Alaska south to New Mexico and east to Nova Scotia, Appalachian Mountains;

S. bendirii, southwestern British Columbia to northwestern California;

S. fumeus, southeastern Canada, northeastern United States;

S. gaspensis, southeastern Quebec, New Brunswick, Cape Breton Island;

S. dispar, Nova Scotia to eastern Tennessee;

S. trowbridgii, southwestern British Columbia to central California;

S. merriami, western United States;

S. arizonae, southeastern Arizona, southwestern New Mexico, northwestern Mexico;

S. saussurei, Mexico, Guatemala;

S. ventralis, southern Mexico;

S. emarginatus, west-central Mexico;

S. sclateri, southern Mexico;

S. stizodon, southern Mexico;

subgenus *Microsorex* Coues, 1877

S. hoyi (pygmy shrew), most of Alaska and Canada south of the tundra, parts of the Rocky Mountain region south to central Colorado, parts of the north-central and northeastern United States.

Microsorex long was regarded as a full genus, but Diersing (1980) concluded that it is only a subgenus of *Sorex*. That view received general acceptance, but subsequently S. B. George (1988) concluded that *Microsorex* is merely a part of the subgenus *Otisorex*. Studies of mitochondrial DNA by Stewart and Baker (1994) associated *Microsorex* with the species *Sorex vagrans,* *S. monticolus,* and *S. palustris.* Long (1972*b,* 1974) thought that a separate species of *Microsorex,* *S. thompsoni,* occupied southeastern Canada and the northeastern United States, but Van Zyll de Jong (1976) considered *S. thompsoni* only a subspecies of *S. hoyi.* S. B. George (1988) also indicated that another, still unnamed subgenus comprises the species *S. trowbridgii,* *S. merriami,* and *S. arizonae* and tentatively *S. saussurei,* *S. ventralis,* *S. emarginatus,* *S. sclateri,* and *S. stizodon.* Van Zyll de Jong (1982) used the name *S. pribilofensis* in place of *S. hydrodromus* for the species on St. Paul Island, but Hutterer (*in* Wilson and Reeder 1993) and Jones et al. (1992) did not. In contrast to Jones et al. (1992), Hutterer (*in* Wilson and Reeder 1993) did not distinguish *S. fontinalis* from *S. cinereus* but did distinguish *S. alaskanus* of the Glacier Bay area of Alaska from *S. palustris.* Van Zyll de Jong and Kirkland (1989) had argued that *S. fontinalis* and possibly *S. milleri* are not specifically distinct from *S. cinereus.* Hutterer (*in* Wilson and Reeder 1993) also listed *S. kozlovi* of eastern Tibet and *S. planiceps* of Kashmir as species distinct from *S. thibetanus,* but Hoffmann (1987) regarded both as subspecies. Dannelid (1991) restricted the subgenus *Homalurus* to the species *S. alpinus,* as shown above, but Ivanitskaya (1994) considered *Homalurus* to also comprise

S. minutus, S. buchariensis, S. volnuchini, S. caecutiens, S. isodon, S. roboratus, and *S. unguiculatus* (presumably the related species *S. thibetanus, S. shinto, S. sadonis, S. sinalis,* and *S. cansulus* also would be placed in *Homalurus*). The affinities of *S. samniticus* are uncertain, but it appears to be a primitive and phylogenetically isolated species (Dannelid 1991; Ivanitskaya 1994).

Head and body length is 46–100 mm, tail length is 25–82 mm, and adult weight is 2.1–18.0 grams. *S. hoyi* averages less in weight than any other American mammal. The tail is a third to more than a half of the total length. It is hairy in the young and naked in old adults. The fur is sleek. The pelage may be uni-, bi-, or tricolored, with the colors varying from tan to black.

The body is slender and the snout is long, slender, highly movable, and incessantly rotating, with conspicuous vibrissae. The eyes are minute but visible, and the ears usually project slightly above the pelage. Females have three or four pairs of mammae. The tips of the five upper unicuspid teeth are colored brownish to purplish, but the coloration varies according to the species.

Long-tailed shrews inhabit most of the temperate and arctic parts of the Northern Hemisphere. They occur over a great altitudinal range and in many kinds of vegetation but generally prefer moist areas. A recent study in Manitoba found the best habitat for *S. cinereus, S. arcticus, S. monticolus,* and *S. palustris* to be grass-sedge marsh and willow-alder fen (Wrigley, Dubois, and Copland 1979).

These shrews may be active by day or night and do not become seasonally inactive. During the summer *S. vagrans* has two nocturnal peaks of activity and a morning peak but is relatively inactive during the afternoon. *S. palustris* is active all day but has peaks before sunrise and after sunset (Banfield 1974). *S. araneus* and *S. minutus* are most active in daylight (Pernetta 1977). Long-tailed shrews have periods of deep slumber, but during their waking hours they are extremely active, darting rapidly over the ground, traveling through subsurface tunnels, or burrowing through snow. They make their own runways but also use those of mice and moles. Most species are ineffective burrowers, though *S. araneus* is an exception. Some species climb readily, and *S. araneus* sometimes occupies nests of *Micromys* in bushes. At least in certain species movement is assisted by echolocation (Buchler 1976; Gould, Negus, and Novick 1964). In the American water shrew, *S. palustris,* the hind foot has a fringe of stiff hairs that aids in swimming and diving. *S. palustris,* the world's smallest homeothermic diver, is capable of surviving forced dives of 30–47.7 seconds (Calder 1969). This species has been observed "running" on water, evidently deriving support from surface tension and buoyancy from air trapped by the hairs of the feet. The Pacific water shrew, *S. bendirii,* also can "run" on top of the

A

B

Pygmy shrew *(Sorex hoyi):* A. Drawing by P. F. Wright; B. Photo by Roger Rageot.

water for 3–5 seconds (Pattie 1973). The diet of long-tailed shrews consists largely of a variety of insects, worms, and other small invertebrates. They also feed on vertebrates, usually in the form of carrion, and occasionally eat plant material. Several species store food.

The pygmy shrew *(S. hoyi)* is usually found in grassy areas within the forest. Most areas it occupies contain water, and there are numerous records from bogs, marshes, and other wet habitat (Banfield 1974; Long 1972*a*, 1974). It is so small that its insectlike holes in leaf mold are not quite large enough to admit a pencil. It can travel in the tunnels of large beetles. It is active on the ground surface or in the runways and tunnels of mice and other shrews. It is a capable climber and jumper and is very quick in its movements. A captive, when running, held the tail straight out from the body with a slight upward curve and constantly twitched its snout. This individual reacted to any noise or disturbance with a violent quiver and a quick dash to shelter. It made a nest in a ball of cotton, with openings at either end, repeating this on every occasion when the cage was cleaned and supplied with fresh cotton. High-pitched thin squeaks are uttered on occasion. The bites of this shrew feel like pin pricks, but they do not pierce the skin of human hands. The pygmy shrew usually forages for food among dead plant material, where it finds insects, other invertebrates, and the carcasses of dead animals. It kills grasshoppers by biting and tearing at the head and abdominal regions. This species apparently occurs at lower densities than most other shrews.

In a forest in the Czech Republic, Nosek, Kozuch, and Chmela (1972) found the following population densities: *S. araneus,* 17.5/ha.; *S. minutus,* 4.6/ha.; and *S. alpinus,* 1.1/ha. In England and Sweden the density of *S. araneus* is about 2–10/ha. (Buckner 1969; Hansson 1968). In British Columbia the density of *S. vagrans* was found to be about 12/ha. (Hawes 1977). The average density of *S. cinereus* is about 7–10/ha. but sometimes is higher than 100/ha., and *S. fumeus* has attained even higher densities (Jackson 1961). Siberian populations of *S. caecutiens* have been found to undergo regular cycles, increasing to high densities and then crashing every fourth year (Hanski 1989).

Average home range was calculated at about 372 sq meters for *S. vagrans* in California (Ingles 1961) and 532 sq meters for *S. araneus* in the Czech Republic (Nosek, Kozuch, and Chmela 1972). For British Columbia, Hawes (1977) reported the following home range sizes with no significant difference between sexes: nonbreeding *S. vagrans,* 338–1,986 sq meters; and breeding *S. vagrans,* 605–5,261 sq meters. Buckner (1969) found that in England *S. araneus* occupied a well-defined home range in autumn and winter of about 2,800 sq meters and that females' home ranges were completely separate from one another. In spring the females continued to maintain territories, though of a reduced size, whereas males abandoned their regular home ranges and became nomadic. Pernetta (1977) reported that in England both *S. araneus* and *S. minutus* marked their home ranges with lateral scent glands. Home range sizes in that study were: *S. araneus,* 800 sq meters in November–December and 1,100 sq meters in January–March; and *S. minutus,* 1,700 sq meters in November–December and 1,400 sq meters in January–March. Hawes (1977) found that territories of *S. vagrans* were maintained only by young of the year, in order to assure themselves enough food over winter, and that such territoriality broke down with the onset of reproductive maturity. Cantoni (1993) reported that *S. coronatus* was strictly territorial during the autumn and winter, with little overlap between

the individual home ranges of around 200 sq meters. In the spring breeding season females remained territorial and showed little increase in home range size, but the ranges of males became two to three times as large and overlapped extensively with the ranges of females and other males.

Long-tailed shrews are basically solitary and highly aggressive toward one another. If several are kept together in close quarters, only one individual may survive. Ognev (1962) stated that males of *S. araneus* came together in large groups during the breeding season and fought over females. Gashwiler (1976) found the breeding season of *S. trowbridgii* in western Oregon to extend from February to October, a period that encompasses nearly all reported times of breeding for the genus. In Canada and the northern United States most species breed from spring to late summer, and there may be two litters per year (Banfield 1974; Godin 1977). In England *S. araneus* and *S. minutus* have breeding peaks in May–June and August–September (Pernetta 1977). Limited data indicate that in some areas breeding of *S. hoyi* takes place from late spring to late summer and that there is only one litter per year; embryo counts have ranged from 3 to 8 (Long 1972*a*, 1974). Recently, however, Feldhamer et al. (1993) reported an interesting case of apparent temporal resource partitioning in western Kentucky. Most of the births of *S. hoyi* there occurred from January to March and there was a lesser peak from August to December. In contrast, births of the other small shrew in the area, *S. longirostris,* were concentrated in spring and summer, none at all occurring from November to March.

Gestation in *Sorex* is not well known but is thought to range from 18 to 28 days. Godfrey (1979) determined gestation in captive *S. araneus* to be 24–25 days. Searle and Stockley (1994) added that gestation in that species may increase to 27 days when there is a postpartum estrus, that females are polyestrous and usually produce three litters per season, and that a given litter may be fathered by several different males. Litter size in the genus varies from 2 to 12, with most species averaging about 4–7. The young are born in a ball nest of vegetation, weigh about 0.5 grams each, and are weaned and independent after about 3–5 weeks. After leaving the nest, young long-tailed shrews may exhibit a behavior known as caravaning, in which they follow each other or the mother, one after the other, each holding on to the hind end of the one in front with the teeth (M. K. Goodwin 1979; Harper 1977). In times of low population density some young females breed in their first year of life; otherwise they do not breed before the spring of the following year. Life span is usually 1–2 years.

The IUCN recognizes *S. kozlovi,* known only from a single locality in eastern Tibet, as a species separate from *S. thibetanus* and considers it critically endangered. Also in Eurasia, *S. cansulus* is classified as critically endangered, *S. sadonis* and *S. cylindricauda* as endangered, and *S. sinalis, S. hosonoi,* and *S. leucogaster* as vulnerable. All are restricted to relatively small patches of suitable habitat that are declining because of human encroachment. In North America *S. jacksoni, S. hydrodromus, S. sclateri,* and *S. stizodon* are classified as endangered by the IUCN and *S. milleri* and *S. arizonae* are classed as vulnerable, each because its very small population may be subject to rapid decline. *S. oreopolus* and *S. macrodon* are designated as near threatened. Olterman and Verts (1972) listed two species, *S. preblei* and *S. merriami,* and one subspecies, *S. vagrans trigonirostris,* as rare in Oregon.

Several other North American subspecies are of concern from a conservation standpoint. *S. longirostris fischeri,* of the Dismal Swamp in Virginia and North Carolina, is designated as threatened by the USDI because human modification of its habitat and resultant drying may lead to its being genetically swamped by the more numerous *S. longirostris longirostris* of neighboring uplands (Rose, Everton, and Padgett 1987). A third subspecies, *S. longirostris eionis*, recently has been found to have a wide distribution in Florida, though upland portions of its habitat are being converted to human use (C. A. Jones et al. 1991). *S. palustris punctulatus* of the southern Appalachians is endangered because of siltation and pollution of the streams on which it depends and numerous other adverse human impacts on its habitat (Handley 1980*b*, 1991*b*).

At least four California subspecies are critically endangered (D. F. Williams 1986). *S. ornatus relictus*, which once may have occupied marshlands throughout most of the Tulare Basin, has nearly disappeared because of drainage and cultivation of its habitat. *S. ornatus sinuosus*, found along San Pablo and Suisun bays, has been greatly reduced through destruction and modification of the tidal marshes on which it depends. *S. ornatus willetti*, known only by two specimens collected on Santa Catalina Island, was reduced to the point of extinction as its natural woodland and wetland habitat was overrun by introduced livestock. *S. vagrans halicoetes*, restricted to the salt marshes of the south arm of San Francisco Bay, is now in jeopardy because of loss of most of its habitat to human development.

INSECTIVORA; SORICIDAE; **Genus BLARINELLA**
Thomas, 1911

Asiatic Short-tailed Shrews

There are two species (Hoffmann 1987; Hutterer *in* Wilson and Reeder 1993):

B. quadraticauda, montane taiga forests of Gansu, Shaanxi, Sichuan, and Yunnan provinces in central and southern China;
B. wardi, Yunnan, northern Burma.

Corbet and Hill (1992) did not accept *B. wardi* as a species distinct from *B. quadraticauda*.

Head and body length is 60–82 mm and tail length is 30–60 mm. Weight is about 10–15 grams (Fons *in* Grzimek 1990). The upper parts are brownish gray with a silvery or smoky gray reflection, and the underparts are paler. The hands, feet, and tail are dull grayish or dark brown.

The structure of these animals is somewhat modified for burrowing. The body is rather stout, and the tail is slender and short, about half the length of the head and body. The small ears are concealed by the pelage. The claws on both hands and feet are large. There are normally five upper unicuspids—the second and third incisors, the canine, and two premolars—but these are of such proportions and are so slanted that only three are generally visible from the side. Oldfield Thomas considered this genus to be more allied to the North American *Blarina* than to any of the Old World genera of shrews, whereas Allen (1940) regarded it as being like *Sorex*, with the beginnings of fossorial modifications in its exterior form.

The habits of this shrew do not seem to be recorded; presumably it tunnels in loose soil and plant debris. One specimen was found dead with a beetle in its jaws, and one was obtained on a mossy bank in birchwood. *B. wardi* is designated as near threatened by the IUCN.

Asiatic short-tailed shrew *(Blarinella quadraticauda)*, photo by Howard E. Uible of specimen in U.S. National Museum of Natural History.

INSECTIVORA; SORICIDAE; **Genus CRYPTOTIS**
Pomel, 1848

Small-eared Shrews

There are 18 species (Cabrera 1957; Choate 1970; Choate and Fleharty 1974; E. R. Hall 1981; Hutterer 1980, 1986*f*, and *in* Wilson and Reeder 1993; Owen 1986; Pitts and Smolen 1988; Woodman and Timm 1992, 1993):

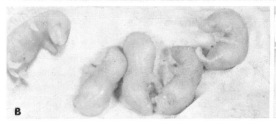

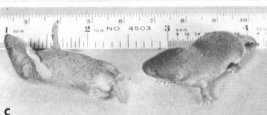

C. *mexicana*, eastern and southern Mexico;

C. *goldmani*, central and southern Mexico, Guatemala;

C. *goodwini*, extreme southern Mexico, Guatemala, western El Salvador;

C. *parva* (least shrew), eastern United States, southeastern Ontario, Texas, eastern New Mexico, northeastern and central Mexico to Panama;

C. *magna*, southern Mexico;

C. *hondurensis*, highlands of Honduras and possibly adjoining countries;

C. *mayensis*, lowlands of Yucatan, Belize, and northern Guatemala (also known from remains found in owl pellets at an isolated montane site in Guerrero, Mexico);

C. *merriami*, highlands from extreme southern Mexico to northern Nicaragua;

C. *nigrescens*, highlands of Costa Rica and western Panama;

C. *mera*, mountains of extreme eastern Panama and probably adjacent Colombia;

C. *colombiana*, known only from the type locality in northwestern Colombia;

C. *gracilis*, highlands of Costa Rica and extreme western Panama;

C. *endersi*, highlands of western Panama;

C. *meridensis*, Andes of northwestern Venezuela;

C. *thomasi*, Andes of Colombia, Ecuador, and northern Peru;

C. *montivaga*, Andes of south-central Ecuador;

C. *squamipes*, Andes of southwestern Colombia;

C. *avia*, Andes of western Colombia.

Many more species once were recognized. The number was reduced largely by Choate's (1970) study of populations in Mexico and Central America, though recently it was increased again by Woodman and Timm's (1993) investigation of the same group. The name *Cryptotis surinamensis*, once believed to refer to a shrew from Surinam, is now thought to apply to a mislabeled specimen of *Sorex araneus* from Europe (Cabrera 1957; Husson 1978; Tate 1932).

Head and body length is approximately 55–100 mm, tail length is 12–42 mm, and adult weight is about 4–7 grams. The upper parts are brownish or blackish, and the underparts are the same color or paler. The snout is pointed, the

Least shrew *(Cryptotis parva):* A. Adult; B. Babies 2 days old; C. Babies 7 days old; photos by Ernest P. Walker. D. Photo by Karl H. Maslowski.

eyes are minute, and the ears are inconspicuous. There are four upper unicuspids: the second and third incisors, the canine, and a premolar; the premolar is smaller than the canine and is usually minute.

Cryptotis has only 30 teeth, whereas *Sorex* and *Blarina* have 32. Where *Blarina* and *Cryptotis parva* occupy the same range they can be readily distinguished by the lead-colored fur and greater size of *Blarina* and the brownish gray fur, lighter underparts, and smaller size of *Cryptotis*.

These shrews generally frequent forests, but *C. parva*, the only species found north of Mexico, usually inhabits open, grassy fields. This species uses the runways and burrows of moles, voles, or other small mammals but also makes its own tunnels in loose, soft soil. A nest is constructed in the burrow or under a log, rock, or other object. The nest is roughly spherical, about 50–125 mm in diameter, and composed of dry grass and leaves (Barbour and Davis 1974; Godin 1977). *C. parva* and probably all other members of the genus are active throughout the year and at all hours of the day. They move about with the snout quivering, apparently trying to detect odors. These shrews feed on invertebrates, small lizards and frogs, and carrion.

C. parva sometimes kills and stores insects in a tunnel. A population density of about 31/ha. was reported for *C. parva* by Kale (1972). Home ranges for two individuals of this species were about 4,400 and 12,000 sq meters (Choate and Fleharty 1973).

The best-known species, *C. parva*, is somewhat gregarious and colonial. Several adults of this species sometimes occupy a nest. Two captive individuals seemed to cooperate in digging, one doing most of the tunneling while the other pushed away the loosened soil and packed it aside. *C. parva* emits a variety of sounds, including a high proportion of clicking noises that seem to correspond to friendly social interaction (Gould 1969).

Cryptotis probably breeds throughout the year in the southern part of its range and from about March to November in the north. The following life history data were obtained in Mock and Conaway's (1975) study of captive *C. parva*. There was no evidence of a true estrous cycle; females became sexually receptive 24–48 hours after being placed with males. Females usually bred again 1–4 days after giving birth, and the gestation period was 21–22 days. The mean size of 327 litters was 4.56 (1–9). The young were weaned at 18–19 days, by which time they had attained adult weight. Sexual maturity occurred as early as 31 days in females and 36 days in males and usually had been attained by 50 days in both sexes. Average life expectancy in the laboratory was 8 months, with a maximum longevity of 31 months. The IUCN now classifies *C. endersi* as endangered and *C. hondurensis* and *C. gracilis* as vulnerable. Each is restricted to a small area of suitable habitat that could be quickly eliminated by human activity.

INSECTIVORA; SORICIDAE; Genus **BLARINA**
Gray, 1838

Short-tailed Shrews

There are three species (Ellis, Diersing, and Hoffmeister 1978; French 1981; Genoways and Choate 1972; George, Choate, and Genoways 1981, 1986; George et al. 1982; Jones, Choate, and Genoways 1984; Moncrief, Choate, and Genoways 1982; Schmidly and Brown 1979; Tate, Pagels, and Handley 1980):

B. brevicauda, southern Saskatchewan and central Nebraska to Nova Scotia and central Georgia;
B. carolinensis, southeastern United States;
B. hylophaga, southern Nebraska and Iowa to eastern Texas.

The subspecies *B. carolinensis peninsulae* and *B. c. shermani* of Florida may actually represent isolated populations of *B. brevicauda,* or they may even be separate species (George, Choate, and Genoways 1986; Jones, Choate, and Genoways 1984).

Head and body length is about 75–105 mm, tail length is 17–30 mm, and weight is 15–30 grams. The upper parts are slate-colored and the underparts are only slightly paler. The body is robust, the snout is pointed, the eyes are small, and the ears are hidden by the fur. The only mammalian genus in its range that resembles *Blarina* is *Cryptotis,* which is smaller, browner in color, and lighter underneath. *Blarina* has five unicuspid teeth in the upper jaw: the second and third incisors, the canine, a normal premolar, and a minute premolar. Females have six mammae.

The submaxillary glands secrete a poison that acts on the nerves of creatures bitten by these shrews. It is quickly effective on small animals and can cause pain for several days in humans. Martin (1981) reported that the main function of this venom evidently is the immobilization of insects for consumption by *Blarina* at a later time.

Short-tailed shrews are found in nearly all land habitats. They are the most fossorial of American shrews (Banfield 1974). Runways in leaves, plant debris, or snow and subterranean burrows are constructed with the strong paws and stout cartilaginous nose. Surface and subsurface runways of moles and rodents are also used. In addition, these shrews may shelter in logs, stumps, or crevices of building foundations (Godin 1977). They are effective climbers: one *B. brevicauda* climbed 1.9 meters up a tree 203 mm in diameter to obtain suet. Two kinds of nests are made: a small resting nest about 160 mm long by 80 mm wide and a much larger, more elaborate mating nest 150–250 mm long by 100–150 mm wide. The nests are constructed of shredded grass or leaves and placed in tunnels or under logs or rocks (Godin 1977). These shrews are active all year and are seen by day or night. The diet consists of invertebrates, small vertebrates, and plant material. *B. brevicauda* stores snails and beetles and in captivity puts nutmeats, sunflower seeds, and other edibles where they can be obtained later. It feeds frequently, usually in the early morning and late afternoon. *Blarina* often serves as an important check on larch sawflies and other destructive insects. Population density usually is about 3–30/ha. depending on the quality of the habitat, and home range is about 0.2–0.8 ha. (Banfield 1974; Schwartz and Schwartz 1959).

In captivity individual *Blarina* often seem to live together peacefully if provided with enough room. Data collected by Platt (1976), however, indicate that in the wild *B. brevicauda* is a solitary territorial species. Investigated populations contained both resident and nomadic compo-

Greater North American short-tailed shrew *(Blarina brevicauda)*, photo by Ernest P. Walker.

nents. Residents occupied small, stationary areas when prey (mice) densities were high and larger, shifting areas when prey densities were low. Individual ranges overlapped slightly during the nonbreeding season. In the breeding season there was no overlap of ranges of animals of the same sex but some overlap of ranges of animals of opposite sexes. Residents marked their ranges with scent and threatened and fought intruders. Nomadic shrews, either young of the year or adults emigrating from unfavorable areas, moved through resident populations and briefly occupied undefended areas. A variety of sounds (chirps, buzzes, twitters) and postures are employed by *Blarina* during intraspecific threat situations; a clicking sound accompanies courtship behavior (Gould 1969).

The breeding season generally extends from early spring to early autumn, though Dapson (1968) reported scattered reproductive activity throughout the year. The estrous cycle is reported to be 2–4 days (Hayssen, Van Tienhoven, and Van Tienhoven 1993). There may be up to three litters per year (Godin 1977). Gestation averages 21 (17–22) days (Gould 1969). Litter size is 3–10, usually about 5–7. The young leave the nest at 18–20 days and are weaned a few days later (Blus 1971; Gould 1969). Sexual maturity is attained at about 6 weeks in females and 12 weeks in males (Banfield 1974). Although few wild individuals survive for more than a year, captives have lived up to 33 months (George, Choate, and Genoways 1986).

The subspecies *B. carolinensis shermani*, found only at one locality in southwestern Florida, may already have been wiped out by human development of its habitat and predation by domestic cats (Layne 1978 and *in* Humphrey 1992). The subspecies *B. hylophaga plumbea* is known only by seven specimens from Aransas National Wildlife Refuge on the Texas coast. It is isolated by about 500 km from all other localities known to have populations of *Bla-*

rina except another isolated site about 217 km to the north, where small groups of both *B. hylophaga* and *B. carolinensis* were recently discovered (Baumgardner, Dronen, and Schmidly 1992).

INSECTIVORA; SORICIDAE; **Genus NOTIOSOREX**
Coues, 1877

Gray Shrew, or Desert Shrew

The single species, *N. crawfordi*, is found in the southwestern United States east to central Texas and western Arkansas, Baja California, and northern and central Mexico (E. R. Hall 1981; Yensen and Clark 1986).

Head and body length is 48–69 mm, tail length is 22–31 mm, and adult weight is about 3–5 grams. General coloration is grayish. The ears are conspicuous. Females have six mammae. In the upper jaw, immediately behind the first incisor, are located three small teeth, each with one cusp; these are the unicuspids, and in this genus they represent the second and third incisors and the single canine.

Notiosorex is found in a wide variety of xeric habitats. The most commonly occupied community is a semidesert scrub association with such vegetation as mesquite, agave, and scrub oak (Armstrong and Jones 1972b). Nests are usually constructed on the surface of the ground beneath dead or dying agaves, among human litter, beneath piles of firewood and building materials, or in the houses of wood rats. The nests are small, birdlike, and built of available materials such as bark, leaves, and hair (Hoffmeister and Goodpaster 1962). This species has also been found in beehives; it is so small that it can enter and leave a beehive through the regular entrances made for the bees. It does not make

Gray shrew *(Notiosorex crawfordi)*, from *Journal of Mammalogy*, photos by Joseph Dixon.

regular runways. A captive individual was active at night, twitching its snout and vibrissae while foraging about. Lindstedt (1980) determined that *Notiosorex* has a greater ability to cool itself by evaporation than do other shrews and that energy metabolism, water loss, and daytime activity are all lower than in other genera. Hoffmeister (1986) suggested that this shrew is able to reduce its metabolic rate when food is scarce and to enter deep torpor. It also apparently can live without free water. The diet includes a wide variety of insects and other invertebrates, and captives have eaten dead vertebrates.

A population of about 24 individuals was found on approximately 1 ha. in Arizona (Hoffmeister and Goodpaster 1962). Captives seem to be able to live together with little antagonism. The known vocalization is a high-pitched squeak emitted during occasional bouts of fighting or disturbance. Parturition probably occurs throughout the year, it is believed that there are at least two litters annually, and as many as six young have been recorded per birth (Hoffmeister 1986). The young are nearly adult-sized at about 40 days (Armstrong and Jones 1972b). A captive female and her young stayed together for 40 weeks (Hoffmeister and Goodpaster 1962).

INSECTIVORA; SORICIDAE; **Genus MEGASOREX**
Hibbard, 1950

Giant Mexican Shrew

The single species, *M. gigas*, is found in southwestern Mexico in the states of Nayarit, Jalisco, Colima, Michoacán, Guerrero, and Oaxaca (E. R. Hall 1981). Some authorities, including Hall, have referred this species to *Notiosorex*, but Armstrong and Jones (1972a) recognized *Megasorex* as a full genus. The latter arrangement has now been generally accepted (Corbet and Hill 1991; Hutterer *in* Wilson and Reeder 1993). In an electrophoretic analysis George (1986) found *Megasorex* to be closer to *Neomys* than to *Notiosorex*.

Head and body length is 83–90 mm and tail length is 39–50 mm. According to Armstrong and Jones (1972a), weight is 9.5–11.7 grams, the upper parts are dark brown to grayish brown, the underparts are slightly paler, the skull is large, and the braincase rises slightly above the plane of the rostrum.

Comparatively little is known about this genus. Only about 20 specimens have been reported. These have been taken in moist areas near streams, in tropical forests, in deciduous forests, and in semiarid areas. The known altitudinal range is from near sea level to about 1,700 meters. The only reproductive information available is that a female was lactating when caught on 14 June in Guerrero, Mexico (Armstrong and Jones 1972a).

INSECTIVORA; SORICIDAE; **Genus NEOMYS**
Kaup, 1829

Old World Water Shrews

There are three species (Corbet 1978; Nadachowski 1984):

N. fodiens, Europe (including British Isles), western
     Siberia, northern Asia Minor, Pacific coast of Siberia,
     North Korea, Sakhalin;
N. anomalus, mainland of western and southeastern
     Europe;

N. schelkovnikovi, Armenia and Georgia (Caucasus of
     Eurasia).

Head and body length is 67–96 mm; tail length is 45–77 mm; and weight is about 12–18 grams. The upper parts are cinnamon brown to black; the underparts are usually whitish, gray, or dark brown. *N. fodiens* has a keel of stiff hairs on the underside of the tail that is lacking in *N. anomalus*. Both species have fringes of stiff hairs on the hands and feet that aid in swimming. Females of *N. fodiens* have 10 mammae.

Although these shrews often forage on land and may even shelter far from water, they are generally near a stream, lake, or marsh. They move quickly in swimming, diving, and searching for food. *N. fodiens* makes a few consecutive dives in one place, then moves a meter or two along the bank and dives again. Köhler (1991) reported details on diving behavior and the electrostatic water-repellent properties of the pelage. These shrews can remain submerged from 5 to 20 seconds. Upon landing, they shake themselves dry or enter their burrows and emerge a moment later dry and glossy; apparently the tunnels squeeze out the water and the soil absorbs it. They are active all year, day or night. A mass movement of water shrews was noted in England; the observers reported that hundreds, possibly thousands, of these animals were swimming upstream close together in a narrow creek. *N. fodiens* produces a poisonous secretion in the submaxillary glands to weaken its prey. Churchfield (1985) reported that it feeds on a variety of invertebrate and vertebrate prey, from tiny freshwater dipteran larvae to sizable fish and frogs.

N. fodiens is usually solitary, but home ranges overlap. It is territorial in captivity, and individuals are aggressive toward one another (Corbet and Southern 1977). In a field study in Switzerland, Cantoni (1993) found this species to be territorial during the winter, each individual defending most of a river area of about 450 sq meters. During the breeding season females remained territorial, but males tended to wander across the ranges of several females. The breeding season in England is from April to September, and there are at least two, perhaps several, litters per season. According to Michalak (1983), litters in captivity are produced about 2 months apart, the gestation period is 19–21 days, litter size averages 5.76 (3–8), and lactation lasts 38–40 days. Sexual maturity is reached at 3–4 months (Hayssen, Van Tienhoven, and Van Tienhoven 1993). A captive lived to an age of 3 years and 1 month (Marvin L. Jones, Zoological Society of San Diego, pers. comm., 1995).

INSECTIVORA; SORICIDAE; **Genus NESIOTITES**
Bate, 1945

Mediterranean Shrews

Three species are thought to have lived in historical time (Vigne 1987):

N. hidalgo, Balearic Islands of the western Mediterranean
     Sea;
N. corsicanus, Corsica;
N. similis, Sardinia.

Each species is known only by fossil remains. Reumer and Sanders (1984) stated that *N. hidalgo* was present on Mallorca and Menorca when humans first arrived there, about 7,000 years ago, and that it disappeared, along with several other now extinct species, over the next few millennia.

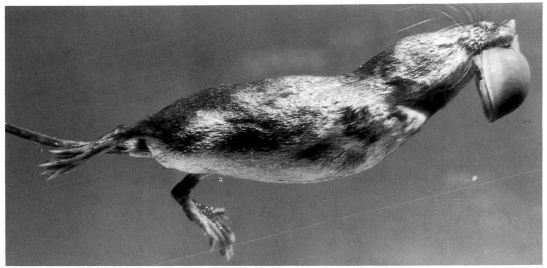

Old World water shrew *(Neomys fodiens)*: Top photo by Dieter Köhler; bottom photo by Liselotte Dorfmüller.

Vigne (1987) estimated times of extinction as about 5000 B.P. for *N. hidalgo,* 2500 B.P. for *N. corsicanus,* and 4000 B.P. for *N. similis.* Reumer (1989) treated *Nesiotites* as a subgenus of *Episoriculus,* which here, as in Corbet and Hill (1991) and Hutterer (*in* Wilson and Reeder 1993), is considered a subgenus of *Soriculus.*

These shrews were medium in size, with some resemblance to *Sorex* in dentition and with the articular structure of the mandible as in *Neomys* (Repenning 1967). They are thought to be most closely related to some of the living shrew genera of Central and Southeast Asia. Those genera are amphibious, but the relatively deep skull and specialized humerus of *Nesiotites* suggest terrestrial habits (Kurten 1968).

The reasons for the extinction of the genus are not definitely known, though Kurten (1968) noted that the shrew genus *Crocidura* was absent in the Pleistocene of the islands where *Nesiotites* lived but is found there now. Possibly *Crocidura* was accidentally introduced through human agency and proved to be more adaptable than *Nesiotites.* Reumer (1989) pointed out that both *Crocidura* and *Suncus* appeared on Corsica during the Neolithic and that subfossil remains indicate that a subsequent increase in those two genera corresponded with a decline in *Nesiotites.* It also is tempting to speculate that the fate of *Nesiotites* paralleled that of *Sorex ornatus willetti* of Santa Catalina Island off the coast of southern California, which seems to have been brought to the verge of extinction through degradation of its restricted habitat by the sheep, cattle, goats, and pigs introduced by people in modern times (D. F. Williams 1986). Such a process may have begun thousands of years earlier in the Mediterranean.

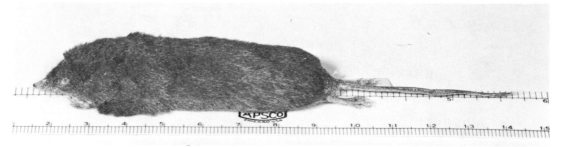

Asiatic shrew *(Soriculus caudatus)*, photo by Gwilym S. Jones.

INSECTIVORA; SORICIDAE; Genus SORICULUS
Blyth, 1854

Asiatic Shrews

There are 3 subgenera and 10 species (Corbet 1978; Ellerman and Morrison-Scott 1966; Hoffmann 1986*b*; Jameson and Jones 1978; Lekagul and McNeely 1977; Mitchell 1977):

subgenus *Soriculus* Blyth, 1854

S. nigrescens, Kumaon (northern India), Nepal, Sikkim, Bhutan, Mishmi (southeastern Tibet), northern Burma;

subgenus *Episoriculus* Ellerman and Morrison-Scott, 1951

S. leucops, Nepal, Sikkim, northeastern India, southern China, northern Burma, northern Viet Nam;
S. macrurus, Nepal, southern China, northern Burma, northern Viet Nam;
S. caudatus, Kashmir to south-central China and northern Burma;
S. fumidus, central Taiwan;

subgenus *Chodsigoa* Kastschenko, 1907

S. hypsibius, central and southern China;
S. lamula, central and southern China;
S. salenskii, known only by the holotype from Sichuan (central China);
S. smithii, central China;
S. parca, southern China, northern Burma, northern Thailand, northern Viet Nam.

Both Van Valen (1967) and Repenning (1967) regarded all three subgenera as full genera, but Hoffmann (1986*b*) considered the differences between them to be small. Corbet and Hill (1991, 1992) and Hutterer (*in* Wilson and Reeder 1993) included *Episoriculus* and *Chodsigoa* in *Soriculus.* Hutterer also indicated that *S. sodalis* of Taiwan might be a species distinct from *S. fumidus.* Yu (1993) collected two specimens on Taiwan that appeared referable to the subgenus *Chodsigoa.*

Head and body length ranges from about 44 to 99 mm and tail length is about 38–120 mm. Fons (*in* Grzimek 1990) listed weights of 5–6 grams for *S. salenskii* and 12–16 grams for *S. nigrescens.* Alexander, Lin, and Huang (1987) reported a weight of 4–7 grams for *S. fumidus.* The coloration is reddish brown, dark brown, grayish, or blackish; the underparts are usually somewhat paler. The fur is dense and soft.

The subgenus *Chodsigoa* has only 28 teeth, but the subgenera *Soriculus* and *Episoriculus* have 30 teeth; the last upper unicuspid is not present in *Chodsigoa.* There are 32 teeth in *Sorex.* The single species of the subgenus *Soriculus* is a stocky shrew with enlarged foreclaws and a short tail that is usually equal to less than 70 percent of the head and body length. This species is rather large, the head and body length usually being more than 80 mm. The members of the subgenus *Episoriculus* are rather slender with small foreclaws and an extremely long tail that is nearly as long as the head and body. These shrews are rather small, the head and body length usually measuring less than 70 mm.

Asiatic shrews inhabit mainly damp areas in forests, thickets, and cultivated fields. On Taiwan they have been taken at altitudes of 1,000–3,200 meters, while in Nepal they occur between 2,200 and 4,200 meters (Jameson and Jones 1978; Mitchell 1977). *S. fumidus* evidently mates during the dry season (October–April) on Taiwan and produces young in the spring (Yu 1993). *S. nigrescens* may nest in a stone fence; the nest consists of a compact ball of dry grasses and fibrous material about 12–15 cm in diameter, lined with dry grasses. *S. caudatus* feeds on earthworms, insects, and other invertebrates and has also been known to eat small mammals caught in traps. In Nepal *S. nigrescens* breeds from about May to October with peaks in May and August. There are two or three litters per year with an average size of five (three to seven). *S. caudatus* has breeding peaks in April–May and August, with one or two litters per year and an average litter size of five (three to six) (Mitchell 1977). *S. salenskii,* known only by a single specimen from a small area of declining habitat, is classified as critically endangered by the IUCN.

INSECTIVORA; SORICIDAE; Genus NECTOGALE
Milne-Edwards, 1870

Web-footed Water Shrew

The single species, *N. elegans,* occurs in Tibet, south-central China (Shaanxi, Sichuan, Yunnan), Nepal, Sikkim, Bhutan, and northern Burma (Ellerman and Morrison-Scott 1966; Mitchell 1977).

Head and body length is 90–128 mm and tail length is 89–110 mm. Weight is 25–45 grams (Fons *in* Grzimek 1990). The upper parts are slaty gray with an abundance of long white-tipped hairs, which impart a silvery appearance to the fur; the underparts are silvery gray, sometimes washed with buff. The fur is soft.

This shrew has a long snout, and the ear is so reduced that it can scarcely be detected. The feet are covered dorsally by small scales that are transverse on the digits. The

Tibetan water shrews *(Nectogale elegans)*, photo from *Recherches pour servir à l'histoire naturelle des mammifères*, M. H. Milne-Edwards and M. Alphonse Milne-Edwards.

digits have a fringe of short, stiff white hairs along their edges. Disklike pads on the feet may assist the shrews in traversing wet stones and perhaps in holding prey. The tail is peculiar: the general color is black, but two lateral fringes of white hairs come together and continue on the undersurface to the tip, two other lateral fringes begin at the first third of the tail length and fade out at the terminal third, and a dorsal fringe of stiff white hairs begins at about the beginning of the terminal third and continues to the tip. This gives the tail the appearance of being four-sided at the base, triangular in the middle third, and laterally compressed in its terminal part. The teeth seem to be specialized for eating fish. Milne-Edwards stated that the hair of this shrew shows minute, rainbowlike iridescence when wet.

The web-footed water shrew is found along streams in mountain forests. It swims and dives remarkably well and shelters in burrows in stream banks. Its diet is composed of small fish and probably other aquatic life.

INSECTIVORA; SORICIDAE; **Genus CHIMARROGALE** *Anderson, 1877*

Asiatic Water Shrews

Corbet and Hill (1992) and Hoffmann (1987) recognized four species:

C. himalayica, Kashmir, northern India, Himalayan region, central and southern China, Taiwan, northern Burma and Thailand, Laos, Viet Nam;
C. platycephala, Japan;

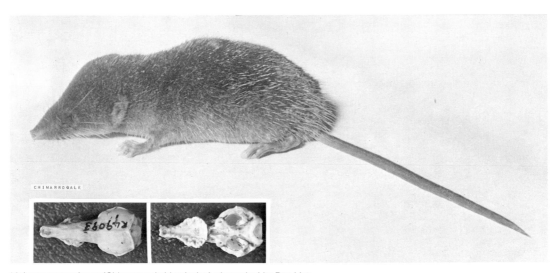

Malayan water shrew *(Chimarrogale himalayica)*, photos by Lim Boo Liat.

C. phaeura, southern Malay Peninsula, Sumatra, Borneo; *C. styani,* south-central China, northern Burma.

Jones and Mumford (1971) did not consider *C. platycephala* and *C. phaeura* to be specifically distinct from *C. himalayica.* In contrast, Hutterer (*in* Wilson and Reeder 1993) restricted *C. phaeura* to Borneo and recognized the populations of the Malay Peninsula and Sumatra as the separate species *C. hantu* and *C. sumatrana.*

These are relatively large shrews. Head and body length is 80–135 mm and tail length is 60–126 mm. Eleven specimens from Japan weighed about 27–43 grams (Arai et al. 1985). The coloration above is slaty gray, sometimes with a slight brown tinge, or blackish; it is paler gray or white below. Silver-tipped hairs are usually scattered throughout the pelage. The tail is either sharply bicolored (dark above and pale below) or uniformly dark. The fur is fine and soft.

This genus is modified for aquatic life. The eyes are small, and the reduced ears have a valvular flap that seals the ear opening when submerged. There is a swollen area at the bases of the vibrissae. As in most other fur-clad mammals that frequent the water, the pelage is somewhat water-repellent. The feet are fringed with stiff hairs on both lateral edges of the digits. The tail is relatively long.

These shrews usually inhabit streams in mountain forests at altitudes of up to 3,300 meters. They reportedly are able to swim well under water and occasionally are caught in fish traps. The diet consists of insects, aquatic larvae, small crustaceans, and fish. Arai et al. (1985) stated that the main breeding season of *C. platycephala* in Japan is spring and early summer, though it is possible that females give birth more than once a year. They noted also that this species depends on clear streams in the mountains and may be declining because of pollution, deforestation, and river diversion. The IUCN recognizes *C. hantu* of the Malay Peninsula and *C. sumatrana* of Sumatra as species distinct from *C. phaeura* of Borneo (see above) and classifies the first two as critically endangered, the third as endangered.

Each evidently is restricted to a very limited habitat that is declining because of human encroachment.

INSECTIVORA; SORICIDAE; Genus **ANOUROSOREX**
Milne-Edwards, 1870

Mole Shrew, or Sichuan Burrowing Shrew

The single species, *A. squamipes,* occurs in central and southern China, Assam, northern Burma, northern Thailand, probably Laos, northern Viet Nam, and Taiwan (Corbet and Hill 1992; Jameson and Jones 1978; Lekagul and McNeely 1977).

Head and body length is 85–113 mm and tail length is 7–17 mm. Alexander, Lin, and Huang (1987) recorded weights of 14–25 grams. The fur is soft, velvety, and dense. The hairs are longest on the rump, forming an elevated tuft, almost a brush. There is often a mucilaginous exudate on these elongated hairs. The upper parts are dark olive gray to fuscous, and the underparts are somewhat paler. The eyes are minute and the ears are concealed in the fur. The feet are short, broad, naked, and scaled and have comparatively long claws. The tail is scaly and slightly shorter than the hind foot. The nose is long and rather sharp, ideally shaped for searching for food in the leaf mold and loose humus of the forest floor.

This shrew lives in mountain forests, mainly at altitudes of 1,500–3,100 meters. On Taiwan it has also been taken in a cornfield and in streamside secondary growth (Jameson and Jones 1978). It reportedly lives under the ground, digging burrows among the roots of plants and feeding on insects, their larvae, and earthworms. Alexander, Lin, and Huang (1987) collected five pregnant females on Taiwan in April, August, and December with 2–4 embryos each. They

Mole shrew *(Anourosorex squamipes),* photo from *Recherches pour servir à l'histoire naturelle des mammifères,* M. H. Milne-Edwards and M. Alphonse Milne-Edwards.

indicated that breeding there probably occurs year-round, as also has been reported in mainland China. However, Yu (1993) collected pregnant females on Taiwan, again with litters of 2–4, only in the wet season (May–September).

INSECTIVORA; SORICIDAE; Genus **MYOSOREX**
Gray, 1838

Mouse Shrews, or Forest Shrews

There are 3 subgenera and 15 species (Heim de Balsac and Meester *in* Meester and Setzer 1977; Hutterer *in* Wilson and Reeder 1993; Maddalena and Bronner 1992; Meester and Dippenaar 1978):

subgenus *Congosorex* Heim de Balsac and Lamotte, 1956

M. polli, known only by the type specimen from south-central Zaire;

subgenus *Surdisorex* Thomas, 1906

M. norae, east side of Aberdare Range in central Kenya;
M. polulus, west side of Mount Kenya;

subgenus *Myosorex* Gray, 1838

M. schalleri, eastern Zaire;
M. cafer, South Africa, eastern Zimbabwe, western Mozambique;
M. tenuis, Transvaal;
M. sclateri, Natal;
M. longicaudatus, near Knysna in south-central South Africa;
M. geata, southwestern Tanzania;

M. varius, South Africa, Lesotho;
M. blarina, Ruwenzori Mountains of eastern Zaire and Uganda;
M. babaulti, mountains west and east of Lake Kivu in eastern Zaire, Rwanda, and Burundi;
M. okuensis, Cameroon;
M. rumpii, Cameroon;
M. eisentrauti, Bioko Island (Fernando Poo).

Based mainly on sources published prior to 1970, Hutterer (*in* Wilson and Reeder 1993) listed *Congosorex* and *Surdisorex* as full genera, but most recent systematic accounts (Corbet and Hill 1991; Heim de Balsac and Meester *in* Meester and Setzer 1977; Meester et al. 1986; Yates 1984) have included those two taxa within *Myosorex.*

Head and body length is 60–110 mm and tail length is 24–67 mm. In *M. longicaudatus* the tail is especially long and prehensile. Weight is 16–22.5 grams in *M. polulus* and 22.5–27.5 grams in *M. norae* (Duncan and Wrangham 1970).

In the subgenera *Congosorex* and *Myosorex* the coloration is buffy, fawn, or shades of brown, gray, or black, often with a speckled effect caused by light hairs interspersed with the dark hairs. Some species have pale gray feet. There are no long hairs on the tail. The fur is soft, fine, and velvety. A large gland, present in males but vestigial in females, is located behind the forelimb in the species *M. varius* and *M. cafer* and probably in others.

In the subgenus *Surdisorex* the color is a deep lustrous brown with a pale frosting; the underparts are paler and buffier. When the pelage is worn the sides and flanks become buffy, leaving a dark dorsal stripe. In the original description of *S. norae* the tail was said to be dark brown above and below, but other authors have stated that the tails of both species are faintly bicolor, brownish above and buffy below. The fur is rather long, dense, and coarser than in *Myosorex.* The ears are small and almost completely

Mouse shrew *(Myosorex varius)*, photo by John Visser.

hidden in the fur. The tail in this subgenus is less than twice the length of the hind foot, but in *Myosorex* and *Congosorex* the tail is about three times the length of the hind foot. The foreclaws of *Surdisorex* are 5–6 mm in length, longer and sturdier than the foreclaws of the other two subgenera.

The subgenera *Congosorex* and *Myosorex* usually have a third premolar behind the second premolar in the lower jaw, making a total of seven lower teeth. This tooth is present in most of the living forms; in those specimens in which the tooth is absent its alveolus is present. In the subgenus *Surdisorex* an extra tooth is present on rare occasions in the lower jaw of *M. norae*, usually in the inner angle between the second and third teeth of the mandible. It apparently never occurs in *M. polulus*.

Mouse shrews generally inhabit moist areas, especially forests or scrub and the dense vegetation lining the banks of mountain streams, but they are not restricted to such habitats. *M. norae* and *M. polulus* are inhabitants of the unique moss and heather cloud forests of the Kenyan mountains at altitudes of 2,800–3,300 meters. These shrews are said to be active both day and night, though Baxter, Goulden, and Meester (1979) found captive *M. varius* and *M. cafer* to be predominantly nocturnal. In captivity *M. varius* constructs small blind burrows and makes almost spherical nests of grass with one or two openings, often near rocks (Goulden and Meester 1978). *M. norae* and *M. polulus* are also thought to do some burrowing and to live in tunnels and burrows in the moss and leaf mold. The diet of *M. varius* consists mainly of insects, but small birds and mammals will be eaten when available. Duncan and Wrangham (1970) found *M. polulus* and *M. norae* to feed primarily on earthworms.

These shrews are basically solitary except during the breeding season. They emit a short, sharp squeak when alarmed or fighting. Goulden and Meester (1978) found that when a male and a female *M. varius* were placed together, they would first avoid one another and squeak aggressively or briefly fight, but after one to five hours they would begin to sleep and eat together.

According to Baxter and Lloyd (1980), the breeding season of this species lasts from September to March; females exhibit a postpartum estrus and may have two or more litters of two to five young each during the period. The young weigh about one gram each at birth, open their eyes at 15–18 days, and are weaned by 24 days. They may move about by caravaning (described below for *Suncus*). For Kenya, Kingdon (1974a) reported pregnant *M. polulus* in September and October, pregnant *M. norae* in October, and a litter size of one to two.

The IUCN (which accepts *Congosorex* and *Surdisorex* as full genera) classifies *M. polli*, *M. schalleri*, and *M. rumpii* as critically endangered, *M. geata* and *M. eisentrauti* as endangered, and *M. norae*, *M. polulus*, *M. tenuis*, *M. sclateri*, *M. longicaudatus*, *M. blarina*, and *M. okuensis* as vulnerable. All inhabit relatively small areas and are thought to be declining because of human habitat destriction.

INSECTIVORA; SORICIDAE; Genus CROCIDURA
Wagler, 1832

White-toothed Shrews

There are 4 named subgenera and 158 recognized species, more than in any other mammalian genus (Aggundey and Schlitter 1986; Catzeflis 1983; Catzeflis et al. 1985; Chasen 1940; Contoli 1992; Corbet 1978, 1984; Corbet and Hill 1992; Crawford-Cabral 1987; Davison 1984; Dippenaar

1980; Dippenaar and Meester 1989; Ellerman and Morrison-Scott 1966; Goodman 1989; Harrison and Bates 1986; Heaney and Ruedi 1994; Heaney and Timm 1983b; Heaney et al. 1991; Heim de Balsac and Hutterer 1982; Heim de Balsac and Meester *in* Meester and Setzer 1977; Howell and Jenkins 1984; Hutterer 1981a, 1981b, 1981c, 1981d, 1983a, 1983b, 1983c, 1986a, 1986b, 1986d, 1991, 1994b, *in* Wilson and Reeder 1993, and pers. comm., 1988; Hutterer and Dippenaar 1987a, 1987b; Hutterer and Happold 1983; Hutterer and Harrison 1988; Hutterer and Jenkins 1980, 1983; Hutterer, Jenkins, and Verheyen 1991; Hutterer and Kock 1983; Hutterer, Lopez-Jurado, and Vogel 1987; Hutterer, Maddalena, and Molina 1992; Hutterer, Sidiyene, and Tranier 1991; Hutterer and Yalden 1990; Jameson and Jones 1978; Jenkins 1976, 1982, and pers. comm., 1988; Kitchener et al. 1994; Laurie and Hill 1954; Lekagul and McNeely 1977; Maddalena 1990; Maddalena and Ruedi 1994; Martín, Hutterer, and Corbet 1984; Medway 1977, 1978; Mitchell 1975; Molina and Hutterer 1989; Pieper 1978; Rabor 1952; Redding and Lay 1978; Reumer and Payne 1986; Rickart et al. 1993; Ruedi et al. 1990, 1993; Sara, Lo Valvo, and Zanca 1990; Stogov 1985; Taylor 1934; Vesmanis 1976, 1977, 1986; Vogel 1988; Vogel, Hutterer, and Sara 1989; Vogel, Maddalena, and Catzeflis 1986; Vogel, Maddalena, and Schembri 1990; Yalden, Largen, and Kock 1976):

subgenus *Heliosorex* Heller, 1910

C. dolichura, Nigeria to Uganda;
C. roosevelti, Central Africa;
C. crenata, southern Cameroon, Gabon, Zaire;
C. muricauda, Guinea to Ghana;
C. ludia, northern Zaire;
C. latona, northeastern Zaire;
C. polia, northeastern Zaire;
C. grassei, Cameroon, Central African Republic, Gabon;
C. kivuana, Kivu between Zaire and Rwanda;
C. niobe, eastern Zaire, western Uganda, possibly southwestern Ethiopia;
C. balsamifera, known only by mummified remains found near Thebes (Egypt);
C. lanosa, eastern Zaire, Rwanda;
C. congobelgica, northeastern Zaire;
C. desperata, southern Tanzania;
C. telfordi, east-central Tanzania;
C. maurisca, Uganda, Kenya;
C. monax, southern Kenya, northern Tanzania;
C. ultima, western Kenya;
C. littoralis, northeastern Zaire;
C. stenocephala, east-central Zaire;
C. manengubae, Cameroon;
C. usambarae, northern Tanzania;
C. tansaniana, northern Tanzania;

subgenus *Praesorex* Thomas, 1913

C. goliath, southern Cameroon, Gabon, Zaire;

subgenus *Afrosorex* Hutterer, 1986

C. fischeri, Kenya, Tanzania;
C. macarthuri, Kenya, Somalia;
C. lamottei, savannah zone from Senegal to Cameroon;
C. voi, Mali to Kenya and Somalia;
C. parvipes, Cameroon and southern Sudan to Angola and Zambia;
C. lusitania, Sahelian zone from Morocco and Senegal to Ethiopia;
C. smithii, Senegal, Ethiopia, Somalia;

White-toothed shrew *(Crocidura suaveolens)*, photo by Liselotte Dorfmüller.

subgenus *Crocidura* Wagler, 1832

C. luna, northern Zaire and Kenya to eastern Angola and Zimbabwe;

C. fumosa, Kenya;

C. montis, Uganda, Kenya, northern Tanzania;

C. selina, Uganda;

C. raineyi, central Kenya;

C. baileyi, highlands of Ethiopia;

C. glassi, highlands of Ethiopia;

C. lucina, highlands of Ethiopia;

C. thalia, highlands of Ethiopia;

C. macmillani, highlands of Ethiopia;

C. bottegi, Guinea east to southern Ethiopia and northern Kenya;

C. obscurior, Sierra Leone to Ivory Coast;

C. pasha, Sudan, Ethiopia;

C. allex, Kenya, northern Tanzania;

C. nanilla, Mauritania to northern Kenya and southeastern Zaire;

C. nana, Ethiopia, Somalia;

C. crossei, Sierra Leone to Cameroon;

C. ebriensis, Ivory Coast;

C. fuscomurina, Senegal to Ethiopia and south to South Africa, Zanzibar and Pemba islands;

C. bovei, southern Zaire;

C. planiceps, Nigeria, Sudan, Zaire, Uganda, Ethiopia;

C. douceti, Guinea, Ivory Coast, Nigeria;

C. elgonius, western Kenya, northeastern Tanzania;

C. maquassiensis, Zimbabwe, Transvaal;

C. pitmani, Zambia;

C. eisentrauti, western Cameroon;

C. vulcani, western Cameroon;

C. denti, Cameroon, Gabon, Zaire;

C. jacksoni, eastern Zaire, Uganda, southern Kenya, northern Tanzania;

C. hirta, southern Somalia to Zaire and South Africa;

C. erica, western Angola;

C. nigricans, Angola, southern Zambia;

C. xantippe, Kenya, Tanzania;

C. greenwoodi, southern Somalia;

C. fulvastra, Mali, Nigeria, Sudan, Ethiopia, Kenya;

C. mariquensis, southern Africa,

C. mutesae, Uganda;

C. suahelae, Central Africa;

C. flavescens, South Africa, southern Mozambique;

C. olivieri, Egypt, Senegal to Ethiopia and south to South Africa;

C. zaphiri, Ethiopia, Kenya;

C. viaria, savannah zone from southern Morocco to Tanzania;

C. caliginea, northeastern Zaire;

C. buettikoferi, Guinea-Bissau to Nigeria;

C. attila, southeastern Nigeria, southern Cameroon, Zaire;

C. hildegarde, sub-Saharan Africa;

C. phaeura, southwestern Ethiopia;

C. harenna, Bale Mountains of Ethiopia;

C. bottegoides, Bale Mountains of Ethiopia;

C. gracilipes, northern Tanzania;

C. cyanea, southern Africa;

C. silacea, southern Africa;

C. theresae, Guinea to Ghana;

C. foxi, Nigeria;

C. longipes, western Nigeria;

C. nigrofusca, southern Sudan and Ethiopia to Angola and Zambia;

C. macowi, northern Kenya;

C. poensis, Guinea east to Cameroon and south to northern Angola, islands of Bioko (Fernando Poo) and Principe;

C. batesi, southern Cameroon, Gabon;

C. nigeriae, Nigeria, Cameroon, Bioko;

C. thomensis, Sao Tomé Island off West Africa;

C. nimbae, Guinea, Sierra Leone, Liberia;

C. wimmeri, southern Ivory Coast;

C. picea, western Cameroon;

C. zimmeri, southeastern Zaire;

C. turba, Cameroon to Kenya and Zambia;

C. tarella, Uganda;

C. grandiceps, Guinea, Ivory Coast, Ghana, Nigeria;

C. ansellorum, northern Zambia;

C. yankariensis, savannah zone of Nigeria, Sudan, Ethiopia, Somalia, and Kenya;

C. cinderella, Gambia, Mali;

C. somalica, Mali, Sudan, Ethiopia, Somalia;

C. dhofarensis, Oman;

C. arabica, Oman, South Yemen;

C. russula, western Europe, northern Africa, and (probably through human introduction) Sardinia and Ibiza (Balearic Islands);

C. cossyrensis, Pantelleria Island (between Sicily and Tunisia);

C. osorio, Gran Canaria and possibly Tenerife in the Canary Islands;

C. pullata, Afghanistan to southern China and northern Thailand, Taiwan;

C. suaveolens, temperate woodland and steppe zones of Eurasia, Arabian Peninsula, northern Africa, and (probably through human introduction) Minorca, Corsica, Crete, Lesbos, and Cyprus;

C. gueldenstaedti, Georgia (Caucasus) and possibly some adjacent areas;

C. horsfieldi, Kashmir east to southern China and Malay Peninsula, Sri Lanka, Ryukyu Islands, Taiwan, Hainan;

C. zarudnyi, southeastern Iran, western Pakistan, Afghanistan;

C. susiana, southwestern Iran;

C. pergrisea, western Himalayas;

C. armenica, Armenia;

C. serezkyensis, mountains from southwestern Asia Minor to Kazakhstan;

C. aleksandrisi, northeastern Libya;

C. floweri, Egypt;

C. religiosa, Nile Valley of Egypt;

C. leucodon, France to the Caucasus and Palestine, Lesbos Island (Aegean Sea);

C. zimmermanni, highlands of Crete;

C. sicula, Sicily and nearby Egadi Islands, Gozo Island and formerly nearby Malta;

C. canariensis, Fuerteventura and Lanzarote in the Canary Islands;

C. whitakeri, Morocco and Western Sahara to Tunisia, coastal Egypt;

C. tarfayaensis, Atlantic coast from Morocco to Mauritania;

C. sibirica, south-central Siberia, Mongolia;

C. lasiura, southeastern Siberia, Korea, northeastern China;

C. attenuata, southern China, northern India, Nepal, Bhutan, northern Burma, Viet Nam, Malay Peninsula, Taiwan, Sumatra, Java, Batanes Islands off northern Philippines, Christmas Island (Indian Ocean);

C. hispida, Middle Andaman Island (Bay of Bengal);

C. andamanensis, South Andaman Island (Bay of Bengal);

C. jenkinsi, South Andaman Island (Bay of Bengal);

C. nicobarica, Great Nicobar Island (Bay of Bengal);

C. dsinezumi, Japan, Quelpart Island (South Korea);

C. orii, Ryukyu Islands;

C. miya, Sri Lanka;

C. fuliginosa, northeastern India to southern China and Indochina, Malay Peninsula and nearby islands, Sumatra, Bangka Island, Java, Bali, Borneo;

C. malayana, peninsular Malaysia and some nearby islands;

C. palawanensis, Palawan Island (Philippines);

C. tenuis, Timor;

C. orientalis, western Java;

C. baluensis, Sumatra, northern Borneo;

C. paradoxura, Sumatra;

C. beccarii, Sumatra;

White-toothed shrews *(Crocidura flavescens)*, photo by John Visser.

C. monticola, Malay Peninsula, Java, Borneo;

C. minuta, Java;

C. neglecta, Sumatra;

C. maxi, Java, Flores, Sumbawa, Sumba, Timor, Amboina, Aru and some nearby small Islands in the East Indies;

C. elongata, Sulawesi;

C. rhoditis, Sulawesi;

C. lea, northeastern Sulawesi;

C. levicula, central Sulawesi;

C. nigripes, Sulawesi;

C. grandis, Mindanao;

C. negrina, Negros Island (Philippines);

C. mindorus, Mindoro Island (Philippines);

C. beatus, Mindanao, Leyte, Biliran, and Maripipi islands (Philippines);

C. grayi, Luzon, Mindoro, and Catanduanes islands (Philippines).

The above list includes all species accepted by Hutterer (*in* Wilson and Reeder 1993). The following species also are included: *C. bovei,* which Hutterer (1983c) considered a possible subspecies of *C. fuscomurina* but did not subsequently discuss, and which was accepted as a species by Corbet and Hill (1991); *C. balsamifera,* which was described by Hutterer (1994b); *C. ebriensis,* which was designated a species distinct from *C. crossei* by Maddalena and Ruedi (1994); *C. vulcani,* which was not placed within any other species by Hutterer but which was accepted as a full species by Corbet and Hill (1991); and *C. baluensis* and *C. orientalis,* which were regarded as species distinct from *C. fuliginosa* by Corbet and Hill (1992).

The general arrangement of this list attempts to express the subgeneric allocations suggested by Hutterer (1986a, 1994b) as well as the overall view that *Crocidura* originated in Africa, that the species *C. luna* and *C. bottegi* are among the most primitive, and that there is a basic phylogenetic division between a tropical African and a Palaearctic-Oriental group, with the species *C. russula* intermediate to the two groups (Maddalena 1990; Maddalena and Ruedi 1994). The detailed sequence of the list attempts to follow the interrelationships indicated in the various sources cited above. There is, however, much disagreement. For example, Maddalena's electrophoretic studies aligned the species *C. lamottei* and *C. lusitania,* which are in Hutterer's subgenus *Afrosorex,* with species outside of that subgenus, *C. flavescens* and *C. crossei,* respectively. The single species of Hutterer's subgenus *Praesorex, C. goliath,* was listed as a subspecies of the species *C. odorata* by Heim de Balsac and Meester (*in* Meester and Setzer 1977). However, Hutterer (*in* Wilson and Reeder 1993) indicated that *odorata* is a subspecies of *C. olivieri* and reiterated that *C. goliath* is a full species. Detailed morphological studies by Butler, Thorpe, and Greenwood (1989) and McLellan (1994) would call for some rearrangement of the sequence of African species in the above list and also suggest that *Crocidura* may not be a monophyletic genus. McLellan's work indicated a closer affinity between the subgenus *Heliosorex* and the genus *Sylvisorex* and between the subgenus *Afrosorex* and the genus *Suncus* than either subgenus has with the rest of *Crocidura.*

Head and body length is 40–180 mm and tail length is 40–110 mm. Fons (*in* Grzimek 1990) listed the following weights: *C. suaveolens,* 3–13 grams; *C. russula,* 6–12 grams; *C. leucodon,* 6–13 grams; *C. fuliginosa,* 10–18 grams; *C. longipes,* 12–24 grams; *C. foxi,* 14–20 grams; *C. grandiceps,* 19–27 grams; *C. nigeriae,* 22–25 grams; and *C. olivieri,* 37–65 grams. The general coloration is some shade of fawn, brown, gray, or black, and the fur is close and velvety. The tail is long, though rarely longer than the head

and body. In most members of this genus the tail is covered with both long white and short bristly hairs. The foreclaws are not enlarged. Females in many, if not all, species have six mammae. Scent glands are present but are better developed in males. The dental formula is: (i 3/1, c 1/0, pm 1/2, m 3/3) × 2 = 28. In the upper jaw the three teeth immediately behind the first incisor are small and bear only a single cusp (unicuspids). The teeth are white; they never have brownish or purplish tips.

White-toothed shrews inhabit damp and dry forests, grassland, cultivated areas, and occasionally human settlements and buildings. They can dig their own burrows but often use those of other animals. *C. suaveolens,* for example, has been taken in Egypt in the burrows of the fat sand rat *(Psammomys).* A nest is constructed of grass or twigs. White-toothed shrews tunnel through loose humus and leaf mold and often are active under fallen trees and heaps of brushwood or stone. Captive *C. russula* have been found to enter periods of daily torpor (Fons *in* Grzimek 1990). The diet of *Crocidura* consists of invertebrates and the bodies of freshly killed animals. In captivity many species will consume all but the skin, tail, and parts of the limbs of a small mammal; the brain is always eaten first. They also feed on frogs, toads, and lizards.

According to Fons (*in* Grzimek 1990), the population density of *C. suaveolens* on Corsica is about 2/ha. in October, increases to 3.8/ha. in April, and reaches a maximum of 5.3/ha. in July. Home range in that species is 56–395 sq meters. White-toothed shrews are aggressive and voracious. If disturbed, they crouch on the ground with head raised and teeth bared and emit a single sharp, metallic squeak. Some species show social tolerance, and the central European species are known to form winter communities of as many as eight individuals in one nest. In a study of *C. russula* in Switzerland, Cantoni and Vogel (1989) found that during the winter all individuals in an area shared a common nest and there was also considerable overlap in home range. With the onset of the breeding season in the spring, however, each mature female became territorial and shared a nest with only a single male.

Breeding apparently occurs throughout the year in *C. fuliginosa* of southeastern Asia (Medway 1978) and in some Zambian species (Sheppe 1972). In *C. russula* of the Mediterranean region the breeding season extends from March to September (Atallah 1977). Captive females of *C. russula* had postpartum mating within a few hours and gave birth about once a month; gestation averaged 28.5 (24–32) days. The average litter size was 3 (1–7) in 937 litters (Hellwing 1970, 1973). Litter size averages 1–2 in *C. fuliginosa* (Lekagul and McNeely 1977; Medway 1978) and 2–5 in some Zambian species (Sheppe 1972). For the genus as a whole, known litter size is 1–10. The young weigh about 1 gram at birth and are weaned at around 20 days (Innes 1994); they are hairless for the first week and fully haired at 16 days, open their eyes at 13 days, and are sexually mature by 2–3 months. In the central European species (and probably the whole genus) the young follow the mother, one after another, each holding on to the hind end of the one in front with its teeth, thus forming a caravan. The maximum recorded age for captive individuals is 4 years.

There is a surprising degree of conservation concern for these small insectivores. The IUCN now formally classifies 67 of the species as follows: *C. polia, C. desperata, C. telfordi, C. ultima, C. raineyi, C. macmillani, C. eisentrauti, C. caliginea, C. phaeura, C. harenna, C. gracilipes, C. macowi, C. picea, C. ansellorum, C. dhofarensis, C. jenkinsii,* and *C. negrina,* critically endangered; *C. selina, C. longipes, C. wimmeri, C. susiana, C. floweri, C. hispida, C. andaman-*

ensis, C. nicobarica, C. orii, C. miya, C. malayana, C. paradoxura, C. beccarii, C. grandis, and C. mindorus, endangered; C. ludia, C. latona, C. kivuana, C. congobelgica, C. monax, C. stenocephala, C. usambarae, C. tansaniana, C. fischeri, C. fumosa, C. baileyi, C. glassi, C. lucina, C. allex, C. elgonius, C. pitmani, C. erica, C. xantippe, C. greenwoodi, C. flavescens, C. attila, C. bottegoides, C. foxi, C. thomensis, C. zimmeri, C. osorio, C. pergrisea, C. zimmermanni, C. canariensis, C. palawanensis, C. tenuis, C. orientalis, C. beatus, and C. grayi, vulnerable; and C. thalia, near threatened. Each of these species occurs in very restricted habitat that is thought to be declining because of human activity.

In describing C. bottegoides and C. harenna Hutterer and Yalden (1990) noted that their remnant forest habitat in the Bale Mountains of southern Ethiopia is very vulnerable. Another new species was considered to be so threatened by the destruction of the isolated forests in the Uzungwe Mountains of southern Tanzania that it was given the name C. desperata (Hutterer, Jenkins, and Verheyen 1991). C. maquassiensis of South Africa and Zimbabwe is known only from six specimens and was listed as rare by Smithers (1986). C. goliath of Cameroon had been known by only three specimens taken early in the twentieth century and was thought extinct (Goodwin and Goodwin 1973), but Hutterer (in Wilson and Reeder 1993) indicated that it still survives. The subspecies C. attenuata trichura did apparently disappear following human settlement of Christmas Island at the end of the nineteenth century (Harper 1945). C. zimmermanni, of Crete, evidently became greatly restricted in distribution following the usurpation of its habitat by the human-introduced C. suaveolens (Vogel, Maddalena, and Catzeflis 1986). Another native Mediterranean species, C. sicula, disappeared from Malta after human colonization but has survived on the small nearby island of Gozo as well as on Sicily (Hut-

terer 1991; Vogel, Maddalena, and Schembri 1990; Vogel et al. 1990). The two endemic species of the Canary Islands, C. canariensis and C. osorio, have restricted habitats and may be jeopardized by rapid urbanization and increasing desiccation (Hutterer, Maddalena, and Molina 1992; Molina and Hutterer 1989).

INSECTIVORA; SORICIDAE; Genus PARACROCIDURA
Heim de Balsac, 1956

There are three species (Hutterer 1986e and in Wilson and Reeder 1993):

P. schoutedeni, Cameroon, Central African Republic, Gabon, Congo, south-central Zaire;
P. maxima, eastern Zaire, Rwanda, Uganda;
P. graueri, known only by a single specimen from the Itombwe Mountains of eastern Zaire.

Head and body length is 65–96 mm, tail length is 33–46 mm, and weight in two specimens was 13 and 16 grams. The snout is short but relatively large. The ears are very visible and short, and the hair on the ears is so thin that the gray color of the underlying skin can be clearly seen. The limbs and claws are short. The short tail is covered with short black hair, and its diameter rarely exceeds 3.5 mm. In the thickness of both the tail and the short dark whiskers on the face Paracrocidura is similar to Crocidura.

The body generally is covered with thin, short, ashy black or dark brown hair. A large glandular axillary spot covered with black hair is usually present. The feet and hands are covered with thin hair, through which the gray skin is seen, except at the carpal and tarsal tubercles. The

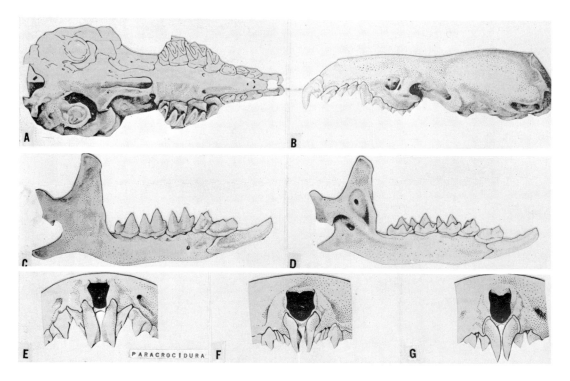

Paracrocidura schoutedeni, skull: A. Ventral view; B. Lateral view, mandible; C. Lateral view; D. Medial view, incisors; E. P. schoutedeni; F. Chimarrogale himalayica styani; G. Crocidura monax (F and G included for comparison). Photos from Revue de Zoologie et de Botanique Africains.

skin is generally dark superficially and somewhat lighter in the deeper layers.

The distinct dental picture has been the primary cause for the establishment of *Paracrocidura* as a separate genus. In two dental characteristics *Paracrocidura* is not comparable to any other of the Soricidae: the straightness of the posterior borders of the fourth premolar and first and second molars, and the development of the first premolar. Other dental characteristics are the aliform processes at the inner edge of the upper incisors and the presence of a tricuspid second lower premolar. *Paracrocidura* has 28 white teeth, of which 3 are maxillary unicuspids.

Specimens have been taken in rainforest at elevations of about 200–2,350 meters. Some have been collected in thickly vegetated areas along streams. The morphology of *Paracrocidura* suggests that it is terrestrial and forages in the leaf litter of the forest floor. *P. graueri*, which is losing its highly restricted habitat to human encroachment, is classified as critically endangered by the IUCN.

INSECTIVORA; SORICIDAE; **Genus SUNCUS**
Ehrenberg, 1833

Musk Shrews, or Pygmy Shrews

There are 15 species (Chasen 1940; Corbet 1978; Corbet and Hill 1992; Ellerman and Morrison-Scott 1966; Heim de Balsac and Meester *in* Meester and Setzer 1977; Hutterer *in* Wilson and Reeder 1993; Hutterer and Happold 1983; Kitchener, Schmitt, and Maharadatunkamsi 1994; Kock 1974b; Laurie and Hill 1954; Medway 1977, 1978; Mitchell 1977; Roberts 1977; Taylor 1934):

S. murinus, apparently native to southern Asia from Afghanistan to southern China and the Malay Peninsula and perhaps also to Taiwan, Sri Lanka, and some islands of the East Indies but occurs also (probably through human introduction in historical time and partly as a human commensal) in coastal Africa from Egypt to Tanzania, in coastal parts of the Arabian Peninsula and Iran, in Japan, in Madagascar, and on Zanzibar, the Comoro Islands, the Maldives and some other islands of the Indian Ocean, the Philippines, most islands of East Indies as far east as Seram and Kei Besar, and Guam;

S. montanus, Sri Lanka;

S. zeylanicus, Sri Lanka;

S. stoliczkanus, Pakistan, India, Nepal;

S. dayi, southern India;

S. ater, Borneo;

S. mertensi, Flores Island (Indonesia);

S. etruscus, southern Europe, northern Africa, southern Asia, Sri Lanka;

S. malayanus, Malay Peninsula;

S. hosei, Borneo;

S. madagascariensis, Madagascar;

S. lixus, Zaire and Kenya to northeastern South Africa;

S. varilla, Nigeria and Tanzania to South Africa;

S. infinitesimus, central and southern Africa;

S. remyi, Gabon.

Ellerman and Morrison-Scott (1966) suggested that *Suncus* was not more than a subgenus of *Crocidura*. Lekagul and McNeely (1977) did not distinguish the two, and this view was supported by Harrison and Bates (1991). There also have been recent suggestions that some species of *Crocidura* (see account thereof) may belong in *Suncus*, though such has not yet received general acceptance. The genus *Sylvisorex* sometimes is considered a subgenus of *Suncus*, but most recent authors have arranged the two as distinct genera. The genus *Podihik* Deraniyagala, 1958, with the single species *P. kura* of Sri Lanka, apparently represents *Suncus etruscus* (J. E. Hill, British Museum of Natural History, pers. comm.). The name *Podihik* was not accepted by Eisenberg and McKay (1970) in their list of the mammals of Sri Lanka or by most subsequent authorities. Corbet and Hill (1992) listed *S. malayanus* of the Malay Peninsula and *S. hosei* of Borneo as subspecies of *S. etruscus*, but Corbet and Hill (1991) and Hutterer (*in* Wilson and Reeder 1993) treated them as separate species. Hutterer also regarded *S. fellowesgordoni* of Sri Lanka as a species distinct from *S. etruscus*, but such was not done by Corbet and Hill (1991, 1992). *S. etruscus* has been reported from Ethiopia and some other parts of both East and West Africa, but Hutterer considered such records doubtful. Three named species of the Philippines, *S. occultidens*, *S. palawanensis*, and *S. lu-*

Musk shrew *(Suncus murinus)*, photo by Lim Boo Liat.

Musk shrew *(Suncus etruscus)*, photo by Jan M. Haft.

zoniensis, were placed in the synonymy of *S. murinus* by Corbet and Hill (1992) and Hutterer (*in* Wilson and Reeder 1993).

Suncus includes some of the largest of the true shrews but also one of the two smallest mammals in the world, *S. etruscus*. This species has a head and body length of about 35–48 mm, a tail length of 25–30 mm, and an adult weight of about 2.5 grams. The genus as a whole ranges up to 150 mm in head and body length and 100 mm in total length. Average weight of an introduced population of *S. murinus* on Guam is 30 grams for males and 20 grams for females; average weight in Calcutta is 105.6 grams for males and 67.7 grams for females (Hasler, Hasler, and Nalbandov 1977). The color varies from gray to shades of brown. The fur is short, soft, and in some species velvety. The members of this genus have long hairs on the tail that project beyond the short bristly hairs. The males' strong musky odor derives from a large, well-developed scent gland on the flanks.

These shrews may be found in forests, cultivated fields, or human habitations. *S. murinus*, the best-known species, is nocturnal, sleeping by day in nests in burrows or inside buildings. In contrast, *S. varilla* depends on termite mounds for cover and breeding sites; it burrows into the center and usually below ground level and constructs a ball-shaped grass nest about 100 mm in diameter (Lynch 1986). The diet of *S. murinus* and other commensal forms of *Suncus* consists mainly of insects but also includes meat, bread, other human food items, refuse, and small animals that can be overpowered.

These shrews usually are thought to be solitary and intolerant of one another; their vocal repertoire includes a high proportion of chirps and buzzes, sounds associated with aggressive behavior (Gould 1969; Roberts 1977). In a study of a laboratory colony of *S. etruscus*, however, Fons (1974) found that pairs and young lived peacefully together during the breeding season. Lynch (1990, 1991) reported natural populations of *S. varilla* in South Africa to be relatively stable, with densities of 0.25–8.00/ha. The species apparently is monogamous, pairs persist in harmony throughout the year, and young accompany their mothers long after weaning (up to 9 months). Breeding in South Africa occurs from August to March, each female produces only one litter during that period, most young females do not give birth until their second year of life, and longevity is 24–30 months.

Pregnant females and fertile males of *S. murinus* have been collected at all times of the year in the Malay Peninsula, with most pregnancies occurring from October to December (Medway 1978). On Guam *S. murinus* also seems to breed throughout the year. The population on Salsette Island, India, seems to breed mainly during the monsoon months, when food is plentiful. In Nepal *S. murinus* breeds from April to September (Mitchell 1977). Both parents of this species collect nesting material, and the young do not leave the nest until they are about three-fourths grown. Laboratory studies have shown gestation to last about 27.5 days in *S. etruscus* (Vogel 1970) and 30 days in *S. murinus* (Hasler, Hasler, and Nalbandov 1977). Litter size in *S. etruscus* was reported to be 4–6 in Pakistan (Roberts 1977) and 2–5 in captivity (Vogel 1970). Litter size in *S. murinus* averaged 4 (1–8) in Nepal (Mitchell 1977) and 2.7 (1–5) in the Malay Peninsula (Medway 1978). In a laboratory investigation of this species litter size averaged 2.1 (1–4) in a population originating in Guam and 2.8 (1–6) in a population originating in Madagascar (Hasler, Hasler, and Nalbandov 1977). Weaning usually occurs at 17–20 days (Gould 1969), but Medway (1978) reported that one litter of *S. murinus* survived forcible weaning at 12 days and reached adult weight and apparent sexual maturity at 36 days. Young musk shrews sometimes travel by caravaning, with the first holding onto the mother's tail and each of the others holding the tail of the one ahead. Life span in captivity is 1.5–2.5 years (Medway 1978).

S. murinus has become accustomed to living in and about human habitations. This species has been repeatedly introduced by people to localities from East Africa and Madagascar to as far east as Guam, in much the same accidental manner as house rats and mice. Hutterer and Tranier (1990) found that the first reliable records of the species in the western Indian Ocean region date from the early nineteenth century and that occurrences correspond largely to the trade routes of Arabian dhows. *S. murinus* damages stored goods and is a nuisance around homes because of its noisy habits and offensive odor, but in return it destroys many insects and perhaps rodent pests. In China *S. murinus* is known as the "money shrew" because of a fancied resemblance between its rather constant, small chatter and the sound of jingling coins. In contrast to the widespread commensals, several species of *Suncus* occur in highly restricted natural habitat. *S. remyi* is known only from two areas of undisturbed forest in Gabon and has not been found since its original description in 1965 (Hutterer *in* Wilson and Reeder 1993; Nicoll and Rathbun 1990). The IUCN classifies *S. remyi,* as well as *S. ater* and *S. mertensis,* as critically endangered, noting that each occupies a very restricted zone of declining habitat. For the same reason, *S. zeylanicus, S. dayi,* and *S. fellowesgordoni* (recognized as a species distinct from *S. etruscus*) are classified as endangered by the IUCN, and *S. montanus* and *S. hosei* are designated as vulnerable.

INSECTIVORA; SORICIDAE; Genus SYLVISOREX
Thomas, 1904

Forest Musk Shrews

There are 11 species (Elbl, Rahm, and Mathys 1966; Heim de Balsac and Meester *in* Meester and Setzer 1977; Hutterer 1986*b* and *in* Wilson and Reeder 1993; Hutterer, Dieterlen, and Nikolaus 1992; Hutterer and Verheyen 1985; Jenkins 1984):

S. ollula, eastern Nigeria to Zaire;

S. oriundus, northeastern Zaire;

S. lunaris, mountains of eastern Zaire, Rwanda, Burundi, and Uganda;

S. morio, mountains of southwestern Cameroon;

S. isabellae, Bioko Island (Fernando Poo);

S. howelli, known only by two specimens from northeastern and east-central Tanzania;

S. megalura, Guinea to Ethiopia, and south to Angola and Mozambique;

S. camerunensis, mountains of southeastern Nigeria and adjacent Cameroon;

S. granti, montane forests of Zaire, Rwanda, Uganda, Kenya, and Tanzania;

S. vulcanorum, highlands of eastern Zaire, Rwanda, Burundi, and Uganda;

S. johnstoni, Cameroon, Congo, Gabon, Bioko, Zaire, Burundi, Uganda, Tanzania.

Sylvisorex sometimes is considered a subgenus of *Suncus,* but most recent authors (Corbet and Hill 1991; Heim de Balsac and Meester *in* Meester and Setzer 1977; Yates 1984) have given it full generic rank. Recently it has been suggested that some species of *Crocidura* (see account thereof) may belong in *Sylvisorex.*

Head and body length is about 45–100 mm and tail length is about 40–90 mm. Kingdon (1974*a*) listed weights of 3–12 grams. In some species the tail is barely half the length of the head and body; in others the two portions are about the same length. The fur is soft and velvety and usually longer than that of *Suncus.* Also unlike *Suncus, Sylvisorex* has no long hairs on the tail. The upper parts are generally slaty gray and the underparts are somewhat paler.

These shrews are generally thought to inhabit deep forest. Fons (*in* Grzimek 1990) wrote that *S. megalura* is de-

Forest musk shrew *(Sylvisorex megalura),* photos by P. Vogel.

cidedly arboreal, with captives spending much of their time climbing in trees. Ansell (1960), however, reported that a specimen of *S. megalura* was taken in a grassy area well away from trees. He also stated that this species is considered both arboreal and terrestrial and both diurnal and nocturnal, that it is solitary, and that a female with two near-term embryos was taken in Zambia in October. Nicoll and Rathbun (1990) expressed concern about three apparently rare and narrowly restricted species, *S. howelli, S. ollula,* and *S. vulcanorum,* noting that the last is almost certainly threatened with extinction. The IUCN now classifies *S. howelli, S. oriundus,* and *S. isabella* as vulnerable and *S. morio* as endangered.

INSECTIVORA; SORICIDAE; Genus RUWENZORISOREX
Hutterer, 1986

The single species, *R. suncoides,* is known from the Ruwenzori region of east-central Zaire, Rwanda, Burundi, and Uganda. This species formerly was assigned to *Sylvisorex,* but Hutterer (1986*c* and *in* Wilson and Reeder 1993) considered it to represent a distinct genus.

In three adult specimens head and body length is 92–95 mm and tail length is 55–62 mm. In one of these specimens the weight is 18.2 grams. The overall coloration is shiny black, though the upper sides of the feet are paler. The body is stocky, the head relatively short, the muzzle weak, the hind foot relatively long, and the tail relatively short. The tail is almost naked and has no long hairs. The external ears are small and round.

Ruwenzorisorex is distinguished mainly by cranial and dental characters. The skull is relatively flat and rectangular in profile. The first three upper unicuspid cheek teeth are nearly equal in size and the fourth is much smaller. The fourth upper premolar is massive and has one long cutting edge. The first and second upper molars are strongly heteromorph. The first lower incisor is massive and has a smoothly sloping profile. The first and second lower molars are extremely high-crowned but have the paraconid much reduced. The mental foramen of the mandible is located below the first lower molar. In *Sylvisorex* the first upper unicuspid is distinctly larger than the next three, the fourth upper premolar is more pointed, the first and second upper molars are nearly monomorph, the first lower incisor is relatively smaller and has a distinctly undulating profile, the first and second lower molars have relatively well-developed paraconids, and the mental foramen is located below the fourth lower premolar.

All specimens have been taken in primary rainforest on the slopes of mountains. The elevational range is about 800–2,100 meters. Localities of collection have been damp, thickly vegetated, and mossy. The body structure of *Ruwenzorisorex* suggests a terrestrial form of life, and its massive teeth indicate that its diet includes mollusks. Until its recent discovery in Burundi the genus was known only by five specimens. Nicoll and Rathbun (1990) stated that it is apparently rare and of great conservation concern. It is classified as vulnerable by the IUCN.

INSECTIVORA; SORICIDAE; Genus FEROCULUS
Kelaart, 1852

Kelaart's Long-clawed Shrew

The single species, *F. feroculus,* is restricted to the island of Sri Lanka, where it inhabits the hillsides of the central mountain mass at elevations of approximately 1,850–2,150 meters. This interesting shrew was discovered on the Nuwara Eliya plateau by Kelaart in 1850. It appears to be restricted to the highest plateaus and hillsides of the central mountain cluster. Specimens have been trapped among weeds and dense undergrowth in a wet ravine. This shrew is poorly represented in study collections—fewer than 10 specimens have come to the attention of the scientific world since the first one was taken in 1850. Most of the specimens are in the Colombo Museum in Sri Lanka, and the remainder are in the British Museum of Natural History. Corbet and Hill (1992) indicated that there is an old, unconfirmed report that *Feroculus* also occurs in the Palni Hills of southern India and that this genus is most closely related to *Sylvisorex* and *Suncus.*

Head and body length is 106–18 mm and tail length is 56–73 mm. One individual weighed 35 grams. The upper parts are uniformly slaty or ashy black and the underparts are paler slate. The forefeet are nearly white and the hind feet are fleshy gray. The tail is dusky with a few whitish hairs at the tip. The fur is close, soft, and short; the tail is scantily covered by fine hairs, with a few long, bristlelike hairs scattered along it. The forefeet are large with long claws, and the hind feet are smaller with small claws. The dental formula is: (i 3/2, c 1/0, pm 2/1, m 3/3) × 2 = 30, which differs from *Solisorex,* also of Sri Lanka.

Very little is known about the habits of this shrew. Its discoverer, Kelaart, thought that it was a water shrew, but later investigators, including Hutterer (1985), believe it to be a semiburrowing form. Specimens have been taken in traps baited with meat and coconut flesh. One specimen's stomach was filled with vegetable matter. Another specimen was kept alive for several days on earthworms. According to Kelaart, the creature was not timid and exhibited spirited behavior on all occasions. The IUCN classifies *Feroculus* as endangered, noting that its restricted habitat is declining due to human encroachment.

Kelaart's long-clawed shrew *(Feroculus feroculus),* photo from *Manual of the Mammals of Ceylon,* W. W. A. Phillips.

Pearson's long-clawed shrew *(Solisorex pearsoni)*, photo from *Manual of the Mammals of Ceylon*, W. W. A. Phillips.

INSECTIVORA; SORICIDAE; **Genus SOLISOREX**
Thomas, 1924

Pearson's Long-clawed Shrew, or Mole Shrew

The single species, *S. pearsoni*, apparently is restricted to the central highlands of Sri Lanka at elevations of approximately 1,100–1,850 meters. This shrew was discovered in the Hakgala Gardens, near Nuwara Eliya, at an elevation of 1,600 meters, by Dr. J. Pearson of the Colombo Museum. Specimens of this little-known genus are in the Colombo Museum in Sri Lanka and the British Museum of Natural History. Corbet and Hill (1992) stated that *Solisorex* is most closely related to *Crocidura*.

Head and body length is 125–50 mm and tail length is 59–70 mm. The fur is close, soft, and fine, though not as velvety as the fur of most shrews. The general color above is dark grayish brown, with light-tipped hairs, which give the coat a glossy sheen; the underparts are about the same color, perhaps somewhat paler. The feet are brown. An immature specimen caught at Gammaduwa was darker in coloration than adult animals.

The forefeet are large, with long foreclaws. The slender tail is closely haired and lacks the scattered long hairs that are present in the tail of the other peculiar Sri Lankan genus of Soricidae, *Feroculus*. The ears are small and fully furred. The teeth are large and heavy, and the anterior incisors are well developed. The dental formula of *Solisorex* is: (i 3/2, c 1/0, pm 1/1, m 3/3) × 2 = 28, which differs from *Feroculus*. This species is easily distinguishable by its exceptionally long foreclaws and by its short tail with no long hairs scattered along its length.

The habits of this shrew are almost unknown, or at least they are not recorded in the literature. Four individuals have been taken in long grass, in traps baited with scraps of meat. One animal caught alive at Gammaduwa was kept alive for several days; it uttered an occasional low, squeaking cry. Dr. Pearson surmised that this animal is essentially a jungle dweller and probably more carnivorous than most shrews. Hutterer (1985) indicated that *Solisorex* is semifossorial. The genus, restricted to a small area of deteriorating habitat, is classified as endangered by the IUCN.

INSECTIVORA; SORICIDAE; **Genus DIPLOMESODON**
Brandt, 1852

Piebald Shrew, or Turkestan Desert Shrew

The single species, *D. pulchellum*, inhabits Kazakhstan from the Volga River to Lake Balkhash, as well as parts of Uzbekistan and Turkmenistan (Corbet 1978).

Head and body length is 54–76 mm, tail length is 21–31 mm, and adult weight is generally 7–13 grams. The color pattern is unusual and striking: the upper parts are grayish, with an elongated oval patch of white in the middle of the back; the underparts, feet, and tail are white. All hairs are gray at the base. The fur is soft and fine, and the vibrissae are long and numerous, resembling the whiskers of desert rodents more than the vibrissae of other shrews. The palm and digits of the foreclaw are fringed on both sides with rather long, stiff, elastic hairs; the foot is also fringed, though to a lesser extent. These elastic hairs, by increasing the surface area of the paws, give the animal support on loose sand. The ears are rather large. There are only two unicuspid teeth on each side.

This shrew inhabits dry steppe and subdesert. It is active throughout the year, sometimes by day but mainly at night. It is usually solitary and frequently changes shelters, using crevices, holes, threshed stacks of the buckwheatlike forage crop "soulkhir" *(Agriophyllum arenarium)*, and human dwellings. The piebald shrew digs swiftly in the sand, though it probably does not dig its own burrow. In the Volga-Ural sandy areas this shrew occurs with *Hemiechinus auritus* and *Crocidura suaveolens;* these two insectivores seem to be keen competitors. In the years when *Diplomesodon* is most numerous in the Volga-Ural area, its numbers are not fewer than 1–2/ha. of hilly tract, with the greatest population in the spring.

The diet consists mostly of lizards and insects. One

Piebald shrews *(Diplomesodon pulchellum)*, Top, mother with two young in caravan position, photo by Izmaeel Moshin through Sergei Popov, Moscow Zoo; bottom, adult, photo by Olga Ilchenko through Sergei Popov, Moscow Zoo.

piebald shrew kept in captivity consumed large numbers of lizards: in one 24-hour period this individual killed 11 lizards and fed on 5. Lizards are killed by biting the head, and the parts not touched are usually the feet and tail. The bones of the skull and skeleton are consumed. Generally only the tough chitin of insects is left untouched.

In the Volga-Ural sandy areas the breeding season is April–August, with an average of five young per litter. Females probably give birth more than once in a season.

INSECTIVORA; SORICIDAE; Genus SCUTISOREX
Thomas, 1913

Armored Shrew, or Hero Shrew

The single species, *S. somereni,* is found in the tropical rainforests of northern Zaire, Rwanda, Burundi, and Uganda (Heim de Balsac and Meester *in* Meester and Setzer 1977; Hutterer *in* Wilson and Reeder 1993).

Head and body length is approximately 120–50 mm and tail length is 68 to about 95 mm. Weight is 30–90 grams (Fons *in* Grzimek 1990). The pelage is long, thick, and coarse. There are no long hairs on the tail. The general col-

oration is grayish, sometimes with a slight buffy wash.

This genus is distinguished by extremely strong vertebrae that, because of their shape and size and the large number of facets for articulation, create a sturdy spinal column. The vertebrae in *Scutisorex* have a unique feature: they possess not only lateral interlocking spines but also dorsal and ventral spines, a condition unrecorded for any other mammal. Despite this feature, the members of this genus can bend their backs considerably, both dorsoventrally and laterally.

The extraordinary strength of armored shrews is illustrated by the following quotation from Herbert Lang, as cited by Allen (1917):

> Whenever [the natives] have a chance they take great delight in showing its resistance to weight and pressure. After the usual hubbub of various invocations, a full-grown man weighing some 160 pounds steps barefooted upon the shrew. Steadily trying to balance himself upon one leg, he continues to vociferate several minutes. The poor creature seems certainly to be doomed. But as soon as his tormentor jumps off, the shrew after a few shivering movements tries to escape, none the worse for this mad experience. . . . During this demonstration the head is always left free. The strength of the vertebral column, to-

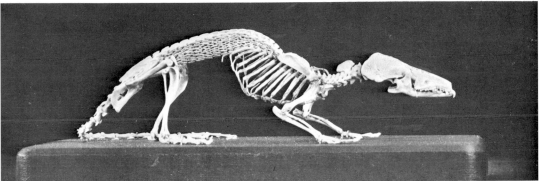

Hero shrew *(Scutisorex somereni)*, photo from *Bull. Amer. Mus. Nat. Hist.* Skeleton photo from American Museum of Natural History.

gether with the strong convex curve behind the shoulder . . . evidently protects the heart and other viscera from being crushed.

The natives believe that the shrews can confer magical powers of bravery and immunity from wounds if parts of them are eaten or worn as talismans.

Scutisorex is active both day and night. Its movements are more deliberate than those of many small mammals. Also, its attacks on prey are not distinguished by the rapid, jerky movements of many insectivores. This genus searches for food under leaves, decayed wood, and stones. Its diet seems to consist of small animals and perhaps some plant food. Kingdon (1974a) reported the collection of two lactating females in May. Fons (*in* Grzimek 1990) noted that litter size is one to three young.

INSECTIVORA; **Family TALPIDAE**

Moles, Shrew Moles, and Desmans

This family of 17 genera and 42 species is found in Eurasia north to about 63° N and south to the Mediterranean, the Himalayas, parts of Indochina and the Malay Peninsula, and in North America from southern Canada to northern Mexico. Van Valen (1967) as well as Hutterer (*in* Wilson

and Reeder 1993) recognized three subfamilies: Uropsilinae, with the genus *Uropsilus;* Desmaninae, with the genera *Desmana* and *Galemys;* and Talpinae, with all other genera in the family. Biochemical systematic studies by Yates and Greenbaum (1982), however, support earlier suggestions that the Talpinae are restricted to the genus *Talpa* and some closely related Old World genera and that the North American genera should be placed in two distinct subfamilies: Scalopinae, with *Neurotrichus, Scalopus, Parascalops,* and *Scapanus;* and Condylurinae, with *Condylura.* Subsequently, Yates and Moore (1990) indicated that *Neurotrichus* has a closer affinity to the Japanese *Urotrichus* than to the other North American genera. Yates and Moore also suggested other phylogenetic associations, which are considered in the following sequence of genera.

Head and body length is 63–215 mm and tail length is 15–215 mm, usually 15–85 mm. *Desmana,* the Russian desman, is the largest member of the family, and the several genera of shrew moles and the long-tailed mole, *Scaptonyx,* are the smallest in head and body length. Moles and desmans have elongated and cylindrical bodies. The long, tubular, naked, tough muzzle extends beyond the margin of the lower lip. In the star-nosed mole of North America, *Condylura,* the nose is divided at the end into a naked fringe of 22 fleshy "tentacles." The eyes are minute, hidden or nearly hidden in the pelage and in some cases covered with skin. Despite some reports to the contrary, there is no external ear, only a tiny hole opening to the surface and obscured by dense fur. The neck is short. The limbs are short and have five digits; in most genera the terminal bones of the digits have a deep median groove or two lobes. The hand is permanently turned outward because the radius articulates with the humerus in an S-shaped cavity. The humerus is massive and articulates with the short, thick clavicle. The scapula is long and narrow, and the sternum is large, projected anteriorly, and, in the burrowing forms, deeply keeled. The tibia and the fibula are joined along their distal half. Females have three or four pairs of mammae. The penis is directed toward the rear of the body, and the scrotum is represented by only a slight bulge in the skin.

In most forms of moles almost all the hairs are about the same length, soft, flexible, and of small diameter near the body, so that the fur is much like velvet and will lie in any direction, enabling a mole to go forward or backward in small burrows. Desmans have long, oily guard hairs interspersed with the shorter hairs.

The dental formula is: (i 2–3/1–3, c 1/0–1, pm 3–4/3–4, m 3/3) × 2 = 32–44. The first upper incisor is flat without an elongate crown. The upper molars have three cusps, and the crowns are W-shaped. The skull is flattened, long, and narrow, with an elongated snout; the sutures disappear at an early age.

This family includes forms that burrow extensively, spending most of their lives underground, and some aquatic or semiaquatic forms that occasionally burrow. Moles make tunnels of two types: shallow subsurface tunnels, usually marked by surface ridges of soil that the moles have pushed up with their backs, and deeper tunnels, generally marked by cone-shaped surface mounds of earth. These are the familiar "molehills," composed of earth that has been pushed out through the tunnel. Surface mounds are not formed if the mole compresses the soil around the deep tunnel sufficiently to provide the required space; molehills are usually 15–25 cm in height and 18–36 cm in diameter. The more permanent deep tunnels are used for shelter and often for rearing the young; the shallow tunnels are used for feeding and resting. *Urotrichus* and *Neurotrichus* are often active on the ground and occasionally climb low bushes. Moles traverse their burrows with considerable speed, and on top of the ground they can dig into hard soil in a matter of seconds. The snout is used to push and shove the earth. A keen ability to sense ground vibrations and direction compensates for a virtual lack of sight. *Desmana, Galemys,* and *Condylura* use all four feet and the tail when swimming. They shelter in burrows often located in stream banks.

Moles and desmans are often solitary, but several individuals of the same species sometimes occupy an extensive tunnel system together, and as many as 8 adult Russian desmans have been found in a den. In areas of abundant food supply the Old World genus *Talpa* digs community tunnels, which are used by as many as 40 moles. A nest for rearing the young is located in a deep tunnel or at ground level under some object on the surface.

These animals are active day and night and do not estivate or hibernate. Most members of this family are insectivorous, feeding mainly on worms, insect larvae, and other small invertebrates encountered while digging or traveling through burrows. Some forms, such as *Scapanus townsendii,* consume some plant material, and the semiaquatic genera feed on aquatic insects, crustaceans, mollusks, and fish. Many moles have a strong odor, which perhaps prevents attacks on them. Cats sometimes catch and kill moles but rarely eat them.

The tunneling activities of moles and desmans are of value in maintaining and developing the soil. The trade in moleskins is now insignificant, but when this fur was fash-

ionable America imported as many as 4 million moleskins a year from England.

To keep moles out of a garden, empty bottles can be placed at an angle to the surface of the ground, with the bottoms in the mole burrows and the necks sticking out; this causes a piping noise to be made when the wind blows. Moles are said to disappear almost overnight when this is done.

The geological range of this family is early Oligocene to Recent in North America, early Eocene to Recent in Europe, and late Miocene to Recent in Asia (Yates 1984).

INSECTIVORA; TALPIDAE; **Genus UROPSILUS**
Milne-Edwards, 1871

Asiatic Shrew Moles

There are four species (Hoffmann 1984; Hutterer *in* Wilson and Reeder 1993):

U. soricipes, Sichuan (central China);
U. gracilis, Sichuan and Yunnan (central and southern China), northern Burma;
U. investigator, Yunnan (southern China);
U. andersoni, Sichuan (central China).

The species *U. gracilis* and *U. investigator* once were put in the separate genus *Nasillus* Thomas, 1911, and *U. andersoni* was put in *Rhynchonax* Thomas, 1911.

Head and body length is 63–88 mm and tail length is 50–78 mm. Weight of *U. soricipes* is about 12–20 grams (Fons *in* Grzimek 1990). The upper parts vary from dark brown to slaty, and the underparts are slaty or grayish. These animals are shrewlike in appearance and have a long tail. The tail is sometimes as long as the head and body and usually equal to more than 80 percent of this length. The long, scaly snout is formed of two tubular nostrils close together with a groove along the top. The external ears extend beyond the fur of the head and are conspicuous. The hands are small, like those of shrews, and not adapted for burrowing. The claws are curved and weak. The upper surface of the hands and feet and the lower third of the arms and legs are very scantily haired, but they are covered with small, rounded scales. The tail of *Uropsilus* is thinly covered with tiny dark hairs and encircled with about 80 rows of scales. *Nasillus* and *Rhynchonax* were based on skulls having different dental formulas, but a variable number of teeth is now recognized as of little value in the classification of moles (Talpidae).

Asiatic shrew moles live in forest and alpine habitats at elevations of 1,250–4,500 meters. They probably frequent leaf mold and feed on invertebrates. They have been taken in Burma under logs and rocks. The species *U. soricipes* and *U. investigator* are classified as endangered by the IUCN. Both are restricted to small areas of suitable habitat and are declining because of human activity.

INSECTIVORA; TALPIDAE; **Genus DESMANA**
Güldenstaedt, 1777

Russian Desman

The single species, *D. moschata,* occurs in the Don, Volga, and Ural river drainages of southwestern Russia and adja-

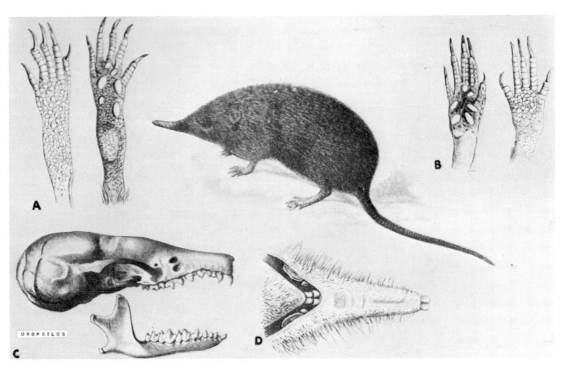

Asiatic shrew mole *(Uropsilus soricipes)*, photo from *Nouv. Arch. Mus. H. N., Paris, Bull.* "Recherches d'histoire naturelle des mammales," Moupin. A. Foot, upper and lower surfaces; B. Hand, lower and upper surfaces; C. Skull, lateral view; D. Undersurface of head; photos from *Recherches pour servir à l'histoire naturelle des mammifères,* M. H. Milne-Edwards and M. Alphonse Milne-Edwards.

Russian desmans *(Desmana moschata)*, photo from *The Royal Natural History,* Richard Lydekker, ed. Inset: cutaway of glands at base of tail, photo from *Genera Mammalium, Insectivora,* Angel Cabrera.

Russian desman *(Desmana moschata)*, photo by H. Chachin through Dmitry Bibikov.

levels. The diet consists of insects, crustaceans, mollusks, fish, and amphibians that are captured in the water.

This insectivore is a comparatively social species. As many as eight adults have been found in one den. Limited evidence indicates that there may be two birth seasons, in early June and early November. The gestation period is 40–50 days, litter size is 2–5 young, and weaning occurs at about 30 days (Fons *in* Grzimek 1990; Ognev 1962).

The Russian desman is listed as vulnerable by the IUCN and is termed endangered by Russia. The species was relatively abundant until the late nineteenth century, when there was an intensification of the demand for its fur. By the turn of the century about 20,000 skins were being processed annually (Fons *in* Grzimek 1990), and there was a subsequent drastic reduction in numbers. The species also declined because of water pollution, drainage of wetlands, impoundments, clearing of the banks on which it depends for denning, and competition from introduced nutria and muskrats. The desman is now protected by law and small introduced populations have been established in the Dnepr and upper Ob river basins. However, estimated total numbers have fallen from around 70,000 in 1973 to about 40,000 today (Khakhin 1994).

INSECTIVORA; TALPIDAE; **Genus GALEMYS**
Kaup, 1829

Pyrenean Desman

The single species, *G. pyrenaicus,* is found in the mountains of southern France, northern Spain, and northern Portugal (Niethammer 1970).

Head and body length is 110–56 mm, tail length is 123–56 mm, and weight is 35–80 grams. The long tail is compressed near the end and has a slight ridge of stiff hairs on the bottom. The snout is approximately 20 mm long. The eyes are very small, and the hind legs are very large. The color above is deep brown, varying from grayish brown to clear dark brown. The fur gives off brilliant metallic glints in somewhat the same manner as the fur of the golden moles. The fur becomes more lustrous when the animal is underwater. The underparts are pale, usually a clear gray; the long nose and the feet are dark, almost black; and the tail is whitish. Both the forefeet and the hind feet are webbed and fringed with stiff hairs. The soft and dense fur resembles otter fur in its water-repellent quality. *Galemys* differs from *Desmana* in being smaller and in having a rounded tail and a proportionally longer snout.

Unlike *Desmana,* the Pyrenean desman is found along swift-flowing streams (Richard and Viallard 1969). It is mainly nocturnal, spending the day in crevices in the riverbanks or in burrows made by water rats *(Arvicola).* A captive constructed a sleeping nest of grass and dead leaves (Niethammer 1970). *Galemys* can swim very rapidly, mainly by using its webbed hind feet for propulsion. On land it has an ungainly walk and rarely runs, but it climbs nimbly, using its strong claws (Richard and Viallard 1969). Vision is poor, but the animal is assisted in moving by echolocation and the sense of touch is well developed in the long, prehensile snout (Richard 1973). Evidently, however, the snout is not electrosensitive (Schlegel and Richard 1992). According to Richard and Viallard (1969), this desman brings its prey to the land; feeds on small fish, crustaceans, insects, and worms in the wild; and eats the equivalent of at least two-thirds of its body weight daily in captivity. Niethammer (1970) reported that the natural diet is principally rheophilous insect larvae but that captives

cent parts of Ukraine and Kazakhstan (Corbet 1978). Fossil remains indicate that the species also once inhabited western Europe and the British Isles.

Head and body length is 180–220 mm and tail length is 170–215 mm. Weight is 100–220 grams (Fons *in* Grzimek 1990). The long snout, much wider than it is deep and grooved above and below, is flexible and used for exploration. The tail is compressed or flattened from side to side; it is large at the base, where special scent glands give the entire animal, including the flesh, a musky odor. Rings of scales encircle the tail, with a few hairs growing between them. The hind feet are webbed to the tips of the toes, and along the edges of the feet fringes of stiff hairs further increase the surface for swimming. The forefeet are partially webbed and also have hair-fringed edges. The furry coat is made up of a short, dense, plushlike underfur and longer, coarser guard hairs. The color above is a rich reddish brown, shading to ashy gray beneath with a silvery sheen in certain lights. It is somewhat like the fur of otters *(Lutra).* Females have four pairs of mammae. The Russian desman is often called a muskrat because of its musky odor and because the general form is very much like that of the American muskrat *(Ondatra).*

The Russian desman is aquatic, preferring quiet freshwater streams, lakes, and ponds. Both the tail and the feet are used to propel the animal through the water. The den consists of a chamber in the stream bank above the high water level and near the surface of the ground. It has no entrances above the ground but is located below bushes or a stump so that air is provided through the root system. A nest of leaves or moss is constructed in the chamber. A tunnel up to six meters long leads from the chamber to an entrance below the level of the water. This animal appears to be mainly nocturnal but occasionally is seen by day. It is somewhat nomadic, probably because of fluctuating water

Pyrenean desman *(Galemys pyrenaicus):* Top right, foot; Bottom right, nose. Photos by J. Niethammer.

readily accepted mice, fish, other vertebrates, mussels, earthworms, and mealworms.

A banding study showed that the desman spent the entire year within a home range and that males covered a larger area than females (Richard and Viallard 1969). A radio-tracking study (Stone 1987) determined the most common social arrangement to be an adult pair inhabiting a length of stream, with the male's territory extending an average of 429 meters and completely enclosing the female's territory, which averaged 301 meters. The shared area was defended from other desmans through vigilance and scent marking. There was little or no overlap between the ranges of adults of the same sex. In addition, there were solitary adults of both sexes, and these had somewhat larger home ranges than did the pairs. The solitary animals moved through their entire ranges in a 48-hour period; the paired individuals covered their territories on a daily basis. In the wild, mutual avoidance results in few agonistic encounters, but captives confined together show extreme aggressiveness, usually resulting in the death of one individual.

According to Palmeirim and Hoffmann (1983), estrus in females commences in January, pregnant individuals have been found from the beginning of February through the end of June, and lactation continues until late August. The existence of three annual peaks in pregnancies—in February, March, and May—suggests that females are polyestrous. There evidently is a postpartum estrus, and the gestation period is about 30 days. Among 53 pregnant females the number of embryos averaged 3.6 (1–5). The first weaned juveniles have been taken in late March, and maturity may be attained in the second year of life. Richard (1976) reported that a marked specimen lived to at least 3.5 years in the wild.

The Pyrenean desman is listed as vulnerable by the IUCN and was termed endangered by Richard (1973) and Queiroz (1994). Its range has declined steadily because of pollution of the swift mountain streams upon which it depends. In addition, its habitat is being fragmented through human activity, it is persecuted by fishermen who consider it a competitor, it may be subject to excessive scientific collection, and it evidently is being displaced by mink that have escaped from fur farms and become established in the wild (Poduschka and Richard 1986).

INSECTIVORA; TALPIDAE; **Genus SCAPTONYX**
Milne-Edwards, 1871

Long-tailed Mole

The single species, *S. fusicaudus,* has been recorded from Sichuan and Yunnan in China and from northern Burma (Ellerman and Morrison-Scott 1966). This genus is represented in study collections by fewer than a dozen specimens, most of which are in the British Museum of Natural History; others are in the Paris Museum of Natural History and the American Museum of Natural History.

Head and body length is approximately 65–90 mm and the tail is about one-third this length. The fur is soft, short, and velvety; the hairs are dark slate in color, with brown tips.

The general body form is molelike; the slightly broadened hands have stout digging claws. There is practically no external ear, and the tail is only thinly covered with short, stiff hairs. Milne-Edwards remarked that this animal looks like a mole with the feet of *Urotrichus,* or like *Urotrichus* with the head of a mole.

The habits of *Scaptonyx* apparently have not been recorded. Specimens have been taken at elevations of approximately 2,150–4,500 meters. One specimen was collected during the daytime in a mousetrap set across an open tunnel. Another was taken on a mossy bank in a fir forest on the Mekong-Salween divide (Allen 1938).

Scaptonyx fusicaudus, photo from *Recherches pour servir à l'histoire naturelle des mammifères*, M. H. Milne-Edwards and M. Alphonse Milne-Edwards.

INSECTIVORA; TALPIDAE; **Genus TALPA**
Linnaeus, 1758

Old World Moles

There are nine species (Abe 1985; Corbet 1978; Ellerman and Morrison-Scott 1966; Filippucci et al. 1987; Hutterer *in* Wilson and Reeder 1993; Krystufek 1994; Vohralik 1991):

T. europaea, Europe except Ireland and parts of
   Mediterranean region, east through Russia as far as the
   rivers Ob and Irtysh;
T. romana, extreme southeastern France, Italy, a historical
   record from Sicily;
T. stankovici, Montenegro, Macedonia, Greece, Corfu,
   probably Albania;
T. levantis, southeastern Bulgaria, Thrace, northern
   Turkey;
T. caucasica, northern Caucasus from Sea of Azov to
   Caspian Sea;
T. caeca, southern Europe, possibly Asia Minor;
T. occidentalis, Portugal, southern Spain;
T. streeti, Kurdistan in northwestern Iran;
T. altaica, central Siberia, northern Mongolia.

Until recently the genera *Scaptochirus, Euroscaptor, Parascaptor,* and *Mogera* usually were regarded as subgenera or synonyms of *Talpa.* However, Hutterer (*in* Wilson and Reeder 1993) and Yates and Moore (1990) recognized all four as distinct genera. The following additional names, based mainly on the variable dental formulas of moles, are still regarded as synonyms of *Talpa: Chiroscaptor* Heude, 1898; *Eoscalops* Stroganov, 1941; *Asioscalops* Stroganov, 1941; and *Asioscaptor* Schwarz, 1948.

Head and body length is 95–180 mm, tail length is about 15–34 mm, and weight is 65–120 grams. The normal gray color in *T. europaea* varies from almost black to whitish and may also be cinnamon, cream, or golden. Occasionally there are irregular blotches of buff or white. The fur is dense, short, and velvety. These moles molt four times a year.

The forefeet, wider than they are long, are permanently turned outward with heavy straight claws. The hand is larger than in the other European and Asiatic moles. There is no external ear, and the long snout and the tail are scantily haired. The limbs are very short: the skin of the body is attached around the wrists and ankles, so that the legs can scarcely be seen. The feet are almost naked and flesh-colored. The eyes are minute, completely hidden in the fur, and probably useless except to distinguish between light and darkness. Moles' bodies are strong and highly specialized in structure, particularly in the forepart. They run backward almost as freely as forward and can turn about easily in the narrow burrows. The peculiar fur will lie in any direction in which it is brushed and is of very beautiful, soft texture.

Talpa and *Euroscaptor* have 11 teeth in each jaw, and *Mogera, Parascaptor,* and *Scaptochirus* have fewer than 11 (Allen 1938). The dental formulas are as follows: *Talpa* and *Euroscaptor,* (i 3/3, c 1/1, pm 4/4, m 3/3) $\times$ 2 = 44; *Mogera,* (i 3/3, c 1/0, pm 4/4, m 3/3) $\times$ 2 = 42; *Parascaptor,* (i 3/3, c 1/1, pm 3/4, m 3/3) $\times$ 2 = 42; and *Scaptochirus,* (i 3/3, c 1/1, pm 3/3, m 3/3) $\times$ 2 = 40 (Lekagul and McNeely 1977).

Old World moles spend practically all their time in tunnels both close to the surface of the ground and to depths of nearly a meter. The burrows usually have two circular tunnels concentric at different levels with connections between them, a central nest, and lateral tunnels into adjacent areas. These burrows are planned more systematically than those of the American mole. In lowlands, where the ground water level is high, the nest chamber of dry grass and leaves is built on top of the ground in a heap of soil about a meter high. *T. europaea* is active both day and night. Worms and insects compose the diet, but these moles have been known to attack, kill, and devour snakes, lizards, mice, and small birds. Excess worms are often stored; the earthworms cannot burrow out because their anterior segments are nipped off or mutilated. Apparently, moles must eat at frequent intervals, as they often die if deprived of food for 10–12

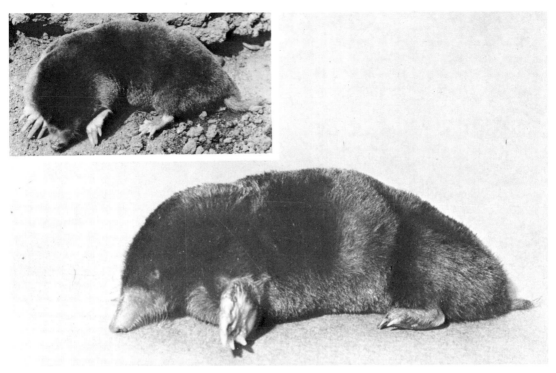

East Asian mole *(Mogera latouchei)*, photo by Robert E. Kuntz. Inset: Old World mole *(Talpa europaea)*, photo by Douglas English.

hours. In a study in Scotland the density of *T. europaea* was found to be 11.1/ha. in an area of abundant earthworms and 6.7/ha. in an area with fewer earthworms (Funmilayo 1977). Individuals of this species were found to occupy largely exclusive home ranges of about 2,000 sq meters during most of the year, but males covered areas three times as great during the spring breeding season, at which time their ranges overlapped those of neighboring females (Stone and Gorman 1985). Closely comparable data were obtained in studies of *T. romana*, though small core areas, intensively exploited for food for brief periods, also were reported (Loy, Dupré, and Capanna 1994).

In *T. europaea* males generally outnumber females by four or five to one. One or, rarely, two litters are born per year, generally in the spring. Litter size is two to seven, usually three or four. The young weigh about 3 grams at birth (Hayssen, Van Tienhoven, and Van Tienhoven 1993). Corbet and Southern (1977) stated that in England the breeding season of *T. europaea* is short (March–May) and that estrus may last only 20–30 hours. The gestation period in this species is about 28 days; delayed gestation has been reported for another species. The young leave the nest 33–34 days after birth but remain near the nest for some time. They are not sexually mature until 6 months old.

The IUCN classifies *T. streeti* as critically endangered. This species is found only in small patches of suitable habitat that have been subject to intensive environmental disruption and military activity.

INSECTIVORA; TALPIDAE; Genus **SCAPTOCHIRUS**
Milne-Edwards, 1867

Short-faced Mole

The single species, *S. moschatus,* is found in northeastern China (Corbet 1978). Until recently *Scaptochirus* usually was considered a subgenus or synonym of *Talpa,* but Hutterer (*in* Wilson and Reeder 1993) and Yates and Moore (1990) recognized each as a distinct genus. Corbet and Hill (1992) did not place *S. moschatus* in a separate genus but acknowledged its distinction and also noted a record from a Holocene cave site in Jiangxi Province in southeastern China.

According to Allen (1938), head and body length is about 140 mm, tail length is about 10–16 mm, and coloration is a nearly uniform clear, grayish brown. From *Talpa, Scaptochirus* differs in having a more reduced tail, a less delicate skull, a shorter and broader rostrum, larger molar teeth with higher crowns, and one less upper and lower premolar. Localities of collection include dry, sandy areas, where there probably are few earthworms and where *Scaptochirus* may instead prey on beetle larvae.

INSECTIVORA; TALPIDAE; Genus **EUROSCAPTOR**
Miller, 1940

Southeast Asian Moles

There are six species (Corbet 1978; Corbet and Hill 1992; Ellerman and Morrison-Scott 1966; Hutterer *in* Wilson and Reeder 1993; Lekagul and McNeely 1977):

E. mizura, Honshu (Japan);

E. micrura, mountains of Nepal, Sikkim, Assam, northern Burma, and southern China;

E. klossi, highlands of Thailand, Laos, southern Burma, and Malay Peninsula;

E. grandis, southern China, Viet Nam;

E. parvidens, Viet Nam;

E. longirostris, southern China.

Until recently *Euroscaptor* generally was considered a subgenus or synonym of *Talpa,* but Hutterer (*in* Wilson and Reeder 1993) and Yates and Moore (1990) recognized the two as generically distinct. Corbet and Hill (1992) did not list *Euroscaptor* as a separate genus but did indicate that the above species form a closely related group that is distinct from the species of moles in China, Japan, and Southeast Asia that are here placed in *Mogera, Scaptochirus,* and *Parascaptor.* Based on molecular analysis, Yates and Moore (1990) reported that *Euroscaptor* and *Mogera* are phylogenetically closer to each other than either is to *Talpa.* Such an alignment would seem plausible geographically, considering the restriction of *Talpa* to Europe and southwestern and northern Asia, even though morphological criteria would suggest affinity between *Euroscaptor* and *Talpa.*

Four individuals of *E. mizura* averaged 104 mm in head and body length, 23 mm in tail length, and 29 grams in weight (Abe, Shiraishi, and Arai 1991). According to Lekagul and McNeely (1977), head and body length of the species here referred to as *E. klossi* is 130–60 mm and tail length is 7–10 mm. The thick, dark, brown fur of *E. klossi* has a silvery gloss, but molting occurs four times a year, at which time the gloss is lost. The tail is much reduced and concealed by fur. The hands are pinkish and relative to body size are the largest found in moles. The incisor teeth are rather small, but the canines are conspicuously large and have two roots. The three following premolars are small, the fourth larger. The last molar tooth is the largest. Abe, Shiraishi, and Arai (1991) indicated that the dental formula of *Euroscaptor* is the same as that of *Talpa* but the third upper molars of the former are relatively larger and have a relatively large metacone. Like that of *Mogera,* but unlike that of *Talpa,* the pelvic girdle of *Euroscaptor* has two pairs of dorsal foramina.

Little natural history information has been obtained on this group of moles specifically, but Lekagul and McNeely (1977) stated that *E. klossi* is usually found in forests, especially in hilly or mountainous areas where the forest floor has deep peat held together by roots. It tunnels through this material, just under the surface, hunting for insects and other small prey. It also is found in the clearings made by slash-and-burn agriculture, where it forages around fallen trees, apparently searching for grubs and termites. Each mole has a territory of about 100–200 sq meters. The IUCN classifies *E. parvidens* as critically endangered amd *E. mizura* as vulnerable; each is restricted to a small area of deteriorating habitat.

INSECTIVORA; TALPIDAE; **Genus MOGERA**
Pomel, 1848

East Asian Moles

There are seven species (Corbet 1978; Corbet and Hill 1992; Ellerman and Morrison-Scott 1966; Hutterer *in* Wilson and Reeder 1993; Yoshiyuki and Imaizumi 1991):

M. wogura, Honshu, Kyushu, and several smaller islands of Japan;

M. robusta, southeastern Siberia, Manchuria, Korea;

M. kobeae, Kyushu, Shikoku, southern Honshu;

M. minor, Honshu;

M. tokudae, Sado Island off western Honshu;

M. etigo, central Honshu;

M. insularis, southeastern China, Taiwan, Hainan.

Until recently *Mogera* generally was treated as a subgenus or synonym of *Talpa,* but Hutterer (*in* Wilson and Reeder 1993) and Yates and Moore (1990) recognized the two as generically distinct. Corbet and Hill (1992) did not distinguish *Mogera* as a full genus but did consider the species therein to form a related group.

In *M. robusta* head and body length is 140–200 mm and tail length is about 20 mm. The upper parts are pale brownish gray, the underparts are pale gray, and the feet are pale yellow (Ognev 1962). In *M. insularis* from the mainland of China head and body length is 87–115 mm and tail length is 14–20 mm; the pelage is a nearly uniform slate color. And in *M. i. hainana,* from Hainan Island, head and body length is 115–34 mm, tail length is 9–14 mm, and the pelage is brownish above and slightly more smoky below (Allen 1938). From *Talpa, Mogera* differs in lacking canine teeth in the lower jaw and in the strong development of the last lower premolars, which function in place of the canines. Very little information is available on these moles. The IUCN classifies *M. tokudae* and *M. etigo* as endangered; both occur only in small areas of declining habitat.

INSECTIVORA; TALPIDAE; **Genus NESOSCAPTOR**
Abe, Shiraishi, and Arai, 1991

Shenkaku Mole

The single species, *N. uchidai,* is known only by the type specimen from Uotsuri-jima in the Japanese Shenkaku Islands, northeast of Taiwan (Abe, Shiraishi, and Arai 1991). This island group is thought to have been isolated from both the mainland and the Ryukyu Islands for more than 6 million years. *Nesoscaptor* was accepted as a genus by Hutterer (*in* Wilson and Reeder 1993), but Corbet and Hill (1992) suggested that it likely should be included in *Mogera.*

Head and body length of the type specimen is 130 mm, tail length is 12 mm, and weight is 43 grams. The upper parts are dark grayish brown and the underparts are slightly paler. The genus resembles *Euroscaptor* and *Mogera,* but unlike the former, it has only two lower incisor teeth on each side, and unlike the latter, it has a lower canine tooth and only 3 upper and 3 lower premolars. There thus is a total of only 38 teeth. The last lower premolar retains two distinctive cusps, a character not shared with any other genus of moles. The molar teeth are relatively large; the rostrum of the skull is very short and broad and markedly bent downward; and the posterior tip of the large tympanic bullae extends back beyond the line connecting the mastoid processes. Like that of *Euroscaptor* and *Mogera,* but unlike that of *Talpa,* the pelvic girdle of *Nesoscaptor* has two pairs of dorsal foramina. Whereas *Euroscaptor* usually has only five sacral vertebrae and *Mogera* has six, there are seven in *Nesoscaptor.*

The IUCN already classifies this newly described genus as endangered. It has a very restricted range and is thought to be declining.

INSECTIVORA; TALPIDAE; **Genus PARASCAPTOR**
Gill, 1875

Indian Mole

The single species, *P. leucura,* is found in Assam (northeastern India), Burma, and adjacent parts of southern China (Allen 1938). Until recently *Parascaptor* generally was considered a subgenus or synonym of *Talpa,* but Hutterer (*in* Wilson and Reeder 1993) and Yates and Moore (1990) recognized the two as generically distinct. Ellerman and Morrison-Scott (1966) and Lekagul and McNeely (1977) listed *leucura* only as a subspecies of *Euroscaptor micrura* and indicated that the range included Thailand, Indochina, and the Malay Peninsula, but such information may have been based primarily on acceptance of the form *klossi* as part of *P. leucura.* Currently, *klossi* is regarded as a species of *Euroscaptor* (see account thereof). Corbet and Hill (1992) regarded *P. leucura* as a species distinct from *E. micrura* but doubted that it warranted generic separation.

According to Allen (1938), head and body length of *Parascaptor* is 110–15 mm, tail length is 10–15 mm, and coloration is a uniform brown. The skull and teeth resemble those of *Talpa,* but there are only three upper premolars on each side. Specimens have been collected at elevations of up to 3,000 meters. One was taken in a place where there were numerous burrows in a riverbed, amidst grass and shrubs.

INSECTIVORA; TALPIDAE; **Genus UROTRICHUS**
Temminck, 1841

Japanese Shrew Moles

There are two species, both restricted to Japan (Corbet 1978):

U. talpoides, Honshu, Shikoku, Kyushu, Dogo, and
    Tsushima islands;
U. pilirostris, Honshu, Shikoku, and Kyushu islands.

Yates (1984) and Yates and Moore (1990) placed *U. pilirostris* in a separate genus, *Dymecodon* True, 1886, but that genus was not recognized by Corbet and Hill (1991) or Hutterer (*in* Wilson and Reeder 1993).

Head and body length is 64–102 mm and tail length is 24–41 mm. Ishii (1982) reported adult weight of *U. talpoides* to be 14–20 grams. The fur is soft and dense but not quite as velvety as that of some members of the family Talpidae. The color is dark brown to black, with a metallic sheen in reflected light. These moles are somewhat shrewlike in general appearance. The forefeet are only slightly broadened. Ears are present, but they are small and hidden in the fur. The tail is densely haired and often enlarged with fat.

U. talpoides occurs in forest and grassland, but *U. pilirostris* is restricted to montane coniferous forests (Cor-

Japanese shrew mole *(Urotrichus talpoides):* A. Photo by Yoshinori Imaizumi; B. Photo from *Proc. Zool. Soc. London.*

bet 1978). Both species are reportedly common in suitable habitat. They burrow just under the surface of the ground, but *U. talpoides* is also active on the surface and occasionally climbs low bushes and trees. In winter this species is often found dead in bird nesting boxes in trees at heights of two to four meters above the ground. This species of *Urotrichus* is said to be rather curious and less cautious than the common mole *(Talpa)*. The diet consists of insects, spiders, worms, and other invertebrates. *U. talpoides* is reported to be easily caught in traps baited with flour paste or soybeans.

In studies near Tokyo, Ishii (1993) found the minimum average home range size to be about 1,300 sq meters outside of the breeding season, with no significant difference between males and females. At that time, ranges were mutually exclusive for individuals of the same sex, but there was extensive overlap between the ranges of animals of opposite sexes. During the breeding season the minimum average home range size of males increased to about 2,800 sq meters with considerable overlap. Ishii (1982) reported males in reproductive condition from mid-February to May, pregnant females in March and April, and young in April and May. Farther south, mating has been reported to start as early as January, with births as early as March. In another area a second and minor peak in births occurred from June to September, and some females then gave birth again. Gestation was estimated to last 4 weeks, as was lactation. Pregnant females contained three or four embryos.

INSECTIVORA; TALPIDAE; Genus NEUROTRICHUS
Günther, 1880

American Shrew Mole

The single species, *N. gibbsii*, ranges from southwestern British Columbia to central California (E. R. Hall 1981). It occurs mainly on the Pacific slope of the Cascade Mountains and in the Coastal Range of California but has also been found on the east slope of the Cascades in suitable areas (Williams 1975). Yates and Greenbaum (1982) suggested that the morphological similarity between *Neurotrichus* and *Urotrichus*, found on the opposite side of the Pacific, in Japan, may be the result of convergence. Subsequently, however, Yates and Moore (1990) indicated that chromosomal data support an earlier view that the two are more closely related to one another than either is to any other living genus.

Head and body length is 69–84 mm and tail length is 31–42 mm. Adults generally weigh 7–11 grams. The common name, "shrew mole," is apt, as this genus combines a pelage composed of both guard hairs and underfur directed posteriorly, as in a shrew, with the large head and heavy dentition of a mole. The fur is soft and dense, with a slight metallic gloss in certain lights. The color varies from dark gray to sooty blue black.

Neurotrichus is the smallest of the American moles, with a tail about half as long as the head and body. The eyes are nearly concealed, and the nostrils open on the sides of the tapering snout. Despite some reports to the contrary, there are no pinnae, the only external evidence of ears being tiny holes opening to the surface. The forefeet are only slightly broadened, and the three middle claws on all the feet are elongated. The tail is rather thick, constricted at the base, encircled with rows of scales, and covered with coarse hairs.

The elevational range of the shrew mole is sea level to 2,500 meters. It prefers a deep, soft soil under a cover of trees or shrubs; infrequently it inhabits dry hillsides. It travels and forages on the surface in a complex system of runways just beneath the surface cover of dead vegetation. It also makes true burrows, which are 25.4 mm or less in diameter and located at a maximum depth of 305 mm. Molehills are not formed, the burrow being made by pressing aside and packing the earth. A small hole leading from a burrow to the surface functions as a ventilation duct for an animal sleeping or resting in the burrow. The shrew

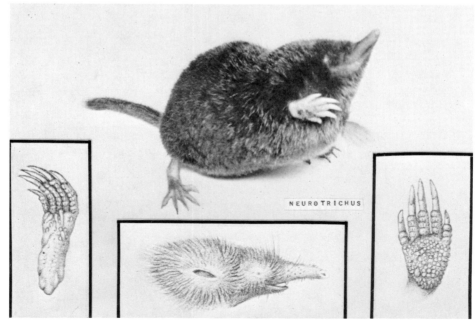

American shrew mole *(Neurotrichus gibbsii)*, photo by Victor B. Scheffer. Inset: photos from *Proc. Zool. Soc. London.*

mole is quite agile; it climbs low bushes, apparently in search of insects, and can swim rapidly. The ability of *Neurotrichus* to bring its forelegs so far under its body as to place the palms flat on the ground gives this genus greater agility than the true moles. The shrew mole is both diurnal and nocturnal. It appears to tap the surface with its flexible nose when hunting. The diet consists of various invertebrates, such as earthworms, isopods, and insects and their larvae, and some vegetable matter. Salamanders are also known to have been eaten.

Population density in favorable habitats is usually about 12–15/ha. (Carraway and Verts 1991). The shrew mole is somewhat gregarious, apparently traveling in loose groups. The only recorded vocalization is a faint twittering audible for several feet. There may be several litters annually, as young have been noted in all months except December and January. Litter size is one to four. The newborn measure about 30 mm in total length and weigh less than 1 gram.

INSECTIVORA; TALPIDAE; Genus SCAPANULUS
Thomas, 1912

Gansu Mole

The single species, *S. oweni,* has been recorded only from a limited area on the borders of Gansu, Shaanxi, and Sichuan in central China (Corbet 1978).

Head and body length is 98–108 mm and tail length is approximately 35–38 mm. The general color is drab gray, with a silvery appearance in certain lights. The individual hairs are actually slaty with brown tips.

This mole resembles the North American moles of the genera *Scapanus* and *Scalopus* in such structural features as the reduction of the upper canines and the enlargement of the anterior incisors. The external appearance of *Scapanulus* is molelike, but the hands, which are broader than in *Scaptonyx,* are not as wide proportionally as in *Talpa.* The claws of the Gansu mole are long and flattened but slender. The snout is rather long, tapering, and grooved on the underside. The tail is relatively thick and well haired.

Two specimens collected near Taochow, Gansu, were taken in mossy undergrowth in a fir forest. No other natural history data are available. Only about six specimens of this genus are known to be in study collections. These are in the British Museum of Natural History, the American Museum of Natural History, the Museum of Comparative Zoology at Harvard, and the United States National Museum of Natural History.

INSECTIVORA; TALPIDAE; Genus PARASCALOPS
True, 1894

Hairy-tailed Mole

The single species, *P. breweri,* ranges in North America from southern Quebec and Ontario to central Ohio, and south as far as western North Carolina in the Appalachian Mountains (E. R. Hall 1981).

Head and body length is 116–40 mm and tail length is 23–36 mm. Adults weigh from 40 to about 85 grams. The fur is thick and soft but slightly coarser than that of the eastern American mole, *Scalopus.* The color is blackish. White spots are often present on the breast or abdomen; the snout, tail, and feet may become almost pure white with age.

The snout is shorter than in *Scalopus* or *Scapanus* and has a median longitudinal groove on its anterior half. The nostrils are lateral and directed upward. There are no external ears, and the eyes are nearly hidden in the fur. The palms of the hands are as broad as they are long, and the digits are not webbed. The tail is thick, fleshy, constricted at the base, annulated with scales, and densely covered with long hairs. Females have four pairs of mammae.

The hairy-tailed mole has an elevational range from about sea level to 900 meters. It frequents light, well-drained soils in forests and open areas. *Parascalops,* like other moles, pushes up surface ridges of soil from shallow subsurface tunnels and mounds of earth through a vertical passage from deep tunnels. The ridges formed by *Parascalops* are not as pronounced as those of *Scalopus,* and

Gansu mole *(Scapanulus oweni)*, photos by P. F. Wright of specimen in U.S. National Museum of Natural History.

Brewer's hairy-tailed mole *(Parascalops breweri)*. The white mark on the face occurs more frequently among older animals. Photos by Neil D. Richmond from Carnegie Museum through J. Kenneth Doutt.

its mounds (about 15 cm in diameter) are smaller than those of *Scalopus* and *Condylura*. The shallow tunnels are much more extensive than the deep tunnels, which are used mainly in the winter. Whenever a warm period occurs in winter the moles appear in the surface tunnels. The tunnel systems of this mole are sometimes used for a number of years and become quite extensive, different individuals occasionally constructing an interlocking network of passages that are also used by mice, shrews, and other moles. The hairy-tailed mole is active by day and night. It may travel on the surface of the ground at night, so it is sometimes captured by owls or other animals. The maximum diameter of the home range seems to be about 30 meters. The diet of *Parascalops* consists of earthworms, insects, and other invertebrates. It is capable of eating several times its own weight in one day (Hallett 1978), though Jensen (1983) found food consumption of captives to be much lower than previously reported.

The following life history information is based mainly on Hallett (1978). Population density reportedly averages about 3/ha. but occasionally reaches 25–30/ha. Individuals spend the winter separately in deep tunnels, using nests about 400 mm below the surface. During the spring mating season several males may appear in the range of a female. Prior to giving birth the female again becomes solitary. The nest for raising the young is a ball of dry vegetation about 150 mm in diameter located about 250 mm below the surface of the ground. After mating the males associate freely, and by late summer adults and young of both sexes utilize the same tunnel systems. No evidence of agonistic behavior has been reported. In New Hampshire mating occurs in late March or early April. Females produce one litter per year. Estimated gestation is four to six weeks, and litter size usually is four or five. The young remain in the nest for about 1 month, by which time they are eating solid food. Females attain sexual maturity

at 10 months. Individuals estimated to be in their fourth year of life have been collected.

INSECTIVORA; TALPIDAE; **Genus SCALOPUS**
E. Geoffroy St.-Hilaire, 1803

Eastern American Mole

The single species, *S. aquaticus,* occurs in extreme southern Ontario and in the United States from southeastern Wyoming and central Texas to Massachusetts and Florida (E. R. Hall 1981). There are also small, isolated colonies in southwestern Texas and in Coahuila and Tamaulipas in northwestern Mexico.

Head and body length is 110–70 mm, tail length is 18–38 mm, and weight is about 40–140 grams. Males are usually larger than females. The color is blackish, grayish, brownish, or coppery. The fur is soft, dense, and velvety.

The nose is a distinct snout with nostrils opening upward, and the eyes and ears are not evident externally. The palms of the hands are broader than they are long, and the fingers and toes have small areas of skin-covered tissue connecting them. The name *aquaticus* refers to this suggestion of webbing rather than to any aquatic habits; like other moles, however, *Scalopus* can swim if necessary. The tail is less than one-fourth of the total length, tapering, indistinctly annulated with scales, and nearly naked. Females have six mammae.

This mole inhabits well-drained soil in fields, meadows, pastures, and open woodlands. It forms mounds of earth, but much less frequently than other species of American moles. In its pursuit of food it makes shallow subsurface tunnels, leaving raised ridges of soil. It also constructs deep, more permanent passages that are used as living quarters

Eastern American mole *(Scalopus aquaticus)*, photo from U.S. Fish and Wildlife Service.

and as thoroughfares to feeding grounds (Yates and Schmidly 1978). In the northern parts of its range this mole tunnels deeper in the winter than it does in the summer. It may dig up to 4.5 meters in an hour; one individual dug more than 31 meters of shallow tunnels in one day. The nest of dry vegetation is located about 30 cm below the ground in a chamber 10–15 cm in diameter or is constructed under a boulder, a stump, or the roots of a plant.

Burrowing requires great strength. The mole forces its hands forward into the earth, then outward and back, while at the same time forcing its head and body forward and raising its body to force the earth upward. This produces the well-known ridges of earth on the surface, indicating the course of the burrows that are near the surface. It is difficult to realize the great power in the arms and body unless one holds a live mole in one's hands.

Scalopus is active throughout the year and at all hours of the day, but Harvey (1976) found that in Kentucky there were two main periods of activity daily: 0800–1600 and 2300–0400 hours. He also found home range to average 1.09 ha. for males and 0.28 ha. for females.

The diet of *Scalopus* consists mostly of earthworms but also includes insects and their larvae and some vegetable matter. A captive on exhibition in the National Zoological Park for about 17 months worked actively in the soil. Twice a day it was fed a total of about 25 grams of mealworms and horse meat. At its death it was quite fat and weighed 88 grams. Its death was probably from shock after it caught one of its claws in a crack in the cage. Captives lap water from a dish like a dog.

Harvey (1976) found a high degree of overlap in home range. Two or three individuals sometimes occupy the same tunnel system. The breeding season may be restricted to the early spring in the north but probably starts as early as January in Texas and Louisiana (Yates and Schmidly 1977, 1978). There is only one litter per year. Estimates of gestation vary from less than 4 weeks to 45 days. Litter size is two to five. The young are independent at about 1 month and sexually mature by the following breeding season. About 70 percent of the young survive for more than a year, females generally outlive males, and maximum longevity is at least 3 and possibly more than 6 years (Davis and Choate 1993). Harvey (1976) tracked one wild individual for 36 months.

INSECTIVORA; TALPIDAE; **Genus SCAPANUS**
Pomel, 1848

Western American Moles

There are three species (E. R. Hall 1981):

S. townsendii, extreme southwestern British Columbia to northwestern California;
S. orarius, southwestern British Columbia to west-central Idaho and northwestern California;
S. latimanus, south-central Oregon to northern Baja California.

Head and body length is 111–86 mm, tail length is 21–55 mm, and adult weight is about 50–170 grams. The fur is soft and silky. The color is generally blackish brown to almost black, but some individuals are light gray.

The snout is shorter and less truncate than in *Scalopus,* and the nostrils open upward. The eyes are visible, but the ears are not externally evident except as tiny openings on the surface. The palms of the forefeet are as broad as they are long, and the nonwebbed fingers possess large digging claws. The tail is thick, tapered toward the hip, slightly constricted near the base, indistinctly encircled with rows of scales, and covered with coarse hairs. Females have four pairs of mammae.

S. orarius occurs with *S. townsendii* in some areas but prefers better-drained soil and enters dense deciduous forests, which *S. townsendii* avoids. *S. latimanus* prefers moist soils. The elevational range exceeds 2,700 meters. These moles are active either day or night, but they are so hidden in their burrows that they are seldom seen. *S.*

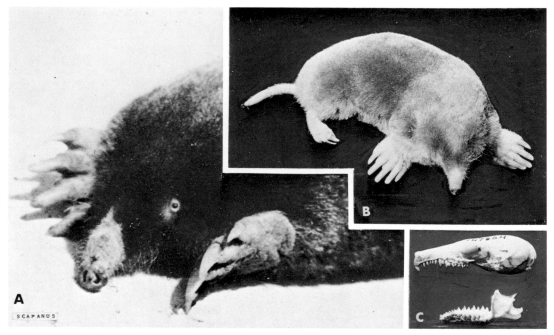

A. Townsend mole *(Scapanus townsendii)*, photo by Theo H. Scheffer from U.S. Fish and Wildlife Service. B. Western American mole *(S. orarius)*, photo by Victor B. Scheffer. C. *S. latimanus dilatus*, photo by P. F. Wright of specimen in U.S. National Museum of Natural History.

townsendii seems to be mainly nocturnal. The extensive tunnel systems of western American moles are indicated by surface ridges and mounds. During periods of dry weather they dig and seek food, and perhaps water, at deeper levels. A Scheffer mole *(S. orarius schefferi)* in the lower Fraser Valley of British Columbia dug a tunnel to a depth of 2.1 meters during a dry period, apparently in search of water. McCully (1967) reported finding burrows of *S. latimanus* in the intertidal zone of an exposed ocean beach in California. An individual *S. townsendii,* working a field of about 1,000 sq meters, formed 302 mounds, or "molehills," in 77 days. Such mounds of earth are mainly from deep burrows, as the moles raise the surface of the ground when tunneling near the surface and do not push the earth of the burrow. The tunnels of these moles are used by mice and shrews. During the breeding season *S. townsendii* constructs nest chambers about 1,600 cubic cm in size and about 15–20 cm deep, with 3–11 tunnel entrances, and lined internally with fine, dry grass and externally with grass, moss, or leaves (Carraway, Alexander, and Verts 1993). The diet of *Scapanus* consists of earthworms, insect larvae, and other small invertebrates. *S. townsendii* eats some vegetable food, such as the bulbs of plants.

Population densities of 0.42–12.40/ha. have been reported for *S. townsendii* in Oregon (Carraway, Alexander, and Verts 1993). The members of this genus tend to be solitary or slightly colonial except during the breeding season. Giger (1973) found that there was no overlap in the areas used by adults of *S. townsendii* and that captives were aggressive toward one another, sometimes fighting violently. During the breeding season male *S. orarius* enlarge their home ranges by constructing long runways to communicate with the tunnel systems of neighboring females; the normal home range of males has been estimated to average 0.12 ha., and that of females varies from 0.15 ha. to 0.35 ha. (Hartman and Yates 1985). The two to five young are born

in March or April after a gestation period of about one month. The young weigh about five grams at birth and are independent by June. Hartman and Yates (1985) stated that females become reproductively active at approximately 9–10 months.

In areas where bulbs are grown *S. townsendii* is considered a pest. This species also is reported to cause serious damage to forage crops. However, the mole's aeration of the soil and destruction of insects and other invertebrate pests is at least partial compensation. The skins of *S. townsendii* were once actively marketed and used for caps, purses, and garment trimmings. Around 1920 each skin was worth $0.50–$0.60 (Carraway, Alexander, and Verts 1993).

INSECTIVORA; TALPIDAE; Genus CONDYLURA
Illiger, 1811

Star-nosed Mole

The single species, *C. cristata,* occurs from Manitoba to Labrador in Canada and southward to Minnesota, Wisconsin, Indiana, Ohio, North Carolina, and Georgia (E. R. Hall 1981).

Head and body length is 100–127 mm, tail length is 56–84 mm, and adult weight is 40–85 grams. The dense, rather coarse fur is water-repellent. The color is blackish brown to nearly black. The eyes of *Condylura* are small but visible, and the ears are barely evident externally. The palms of the forefeet are as broad as they are long. The tail is constricted at the base, scaled, annulated, and sparsely haired. It increases in diameter to about the size of a lead pencil in both sexes during winter and early spring. This is mainly fat, probably a reserve supply of energy for the breeding season. Females have eight mammae.

Star-nosed mole *(Condylura cristata)*, photo by Karl H. Maslowski. Inset: close-up of nose, photo by Vernon Bailey.

The distinctive feature of this genus, from which the common name is derived, is the peculiar muzzle, which is ringed in front with 22 pink fleshy appendages (known as "rays" or "tentacles"). The rays contain sensory receptors, and a corresponding pattern has been found in the organization of the primary somatosensory cortex of the brain (Catania et al. 1993). Experiments indicate that the rays are electrosensitive and have a variety of functions in navigation, environmental testing, and the location and manipulation of prey and nesting materials (Gould, McShea, and Grand 1993).

The following ecological information is based in part on Banfield (1974) and Godin (1977). Star-nosed mole distribution is largely dependent on the presence of damp or muddy soil. The animal constructs a network of shallow tunnels but also spends considerable time foraging on the surface at night. Where the water table permits, tunnels as deep as 60 cm may be excavated. The wet soil is compacted and pushed to the surface to form molehills, which are about 30–60 cm wide and 15 cm high. Some of the tunnels may lead directly into the water. A nest of dry vegetation about 15 cm in diameter is placed well above the water level beneath a stump or log. This mole is active throughout the year both by day and by night. In winter it burrows through the snow or travels on the surface of the snow. The home range probably extends over about 0.4 ha. Population density averages about 2/ha. but may be five times as great in swampland.

The star-nosed mole is an expert swimmer and diver. It uses all four feet in swimming and traveling under the ice in winter. Considerable food is obtained from the bottoms of ponds and streams. When the animal is searching for food the fleshy rays are in constant motion, except for the two median upper tentacles, which are held rigidly forward; these appendages are drawn together when the mole eats. The diet consists of aquatic insects, crustaceans, small fish, and earthworms.

The star-nosed mole is often found in small colonies. Unlike the case with most moles, a male and a female may live together during the winter. The single litter of two to seven young is born in a nest of dry vegetation usually between mid-April and mid-June. The nose "star" is conspicuous at birth. The young become independent at about 3 weeks, when they are approximately two-thirds grown. Sexual maturity is attained at about 10 months.

Handley (1991b) considered the subspecies *C. c. parva* to be of special concern. It is found in the southern Appalachians and along the east coast from Virginia and West Virginia to southeastern Georgia. Although there is a surprisingly robust population in the vicinity of Richmond, Virginia, the subspecies is absent from many apparently suitable areas, and much of its known habitat is fragmented.

Order Scandentia

Tree Shrews

This order, containing the single family Tupaiidae, with 5 Recent genera and 16 species, is found in forested areas of eastern Asia from India and southwestern China eastward through the Malay Peninsula to Borneo and the Philippines. Two subfamilies usually are recognized: Tupaiinae, with the genera *Tupaia, Anathana, Dendrogale,* and *Urogale;* and Ptilocercinae, with *Ptilocercus* (Luckett 1980*b*).

Few other mammalian families are so difficult to classify. Frequently the Tupaiidae have been grouped with the Macroscelididae in either the separate order or the insectivore suborder Menotyphla. Simpson (1945) recognized the Tupaiidae as belonging to the superfamily Tupaioidea, within the infraorder Lemuriformes of the primate suborder Prosimii. This arrangement was widely accepted and is still used by some authorities (Mahe 1976). Recently, however, there has been an increasing tendency to omit the tree shrews from classifications of the Primates (Hershkovitz 1977; Petter and Petter-Rousseaux 1979). Studies of behavior and reproduction led Martin (1968) to consider the Tupaiidae advanced insectivores intermediate to Lipotyphla and Primates but led Sorenson (1970) to conclude that "the tree shrews are indeed the bottom rung of the primate ladder." Following a review of the available evidence, Campbell (1974) decided that the tree shrews should be classified with the Insectivora. Nonetheless, several recent studies have suggested that the Tupaiidae have no immediate relationship with the Macroscelididae and should be placed in a separate order closely related to the Primates, Chiroptera, and Dermoptera (Dene, Goodman, and Prychodko 1978; Goodman 1975; Luckett 1980*a*; McKenna 1975). Such an order, with the name Scandentia, now has received general recognition (Corbet and Hill 1991; Napier and Napier 1985; Novacek 1992; Wilson *in* Wilson and Reeder 1993; Yates 1984).

In external appearance tree shrews resemble long-snouted squirrels, but with the exception of *Ptilocercus* tupaiids can readily be distinguished from squirrels by the absence of long whiskers. *Ptilocercus,* the pen-tailed tree shrew, is easily identified by its tail, which is naked except for a whitish feather-shaped arrangement of the hairs near the end. Tree shrews have a slender body, most adults generally weighing less than 400 grams. Head and body length is about 100–220 mm and tail length is approximately 90–225 mm. The pelage is of the usual mammalian kind, consisting of long, straight guard hairs and shorter, softer, more woolly underfur. Some forms have pale shoulder stripes, and others have facial markings. The ears are squirrel-like, that is, comparatively small and cartilaginous, except in the pen-tailed tree shrew, in which they are larger and more membranous. The feet of tree shrews are naked beneath; the soles are supplied with tuberclelike pads. The long and supple digits bear sharp, moderately curved claws.

Some primatelike characters of tree shrews are the relatively large braincase, the remarkable resemblance of the carotid and subclavian arteries to those of humans, and the permanent sac, or scrotum, for the testes in males. The orbits are completely encircled by bone. The upper incisors are large, caninelike, and separated, and the small canines resemble the premolars. The upper molars are broad, with a W-shaped pattern of cusps. The dental formula is: (i 2/3, c 1/1, pm 3/3, m 3/3) $\times$ 2 = 38.

In addition to their external resemblance to squirrels, tree shrews are generally diurnal, and some of their actions and movements are comparable. They are swift runners. *Ptilocercus* differs in being mainly nocturnal; when on the ground it progresses in a series of hops. Tupaiids are good climbers, seeking their food in trees as well as on the ground. They have keen senses of smell and hearing and good vision. Although they are often short-tempered in the presence of others of their own kind, usually they are easily tamed by people. Their diet consists primarily of insects and fruits but occasionally includes other animal food and various kinds of plant material. Tree shrews are generally fond of water for both drinking and bathing.

There long were doubts concerning the identification of fossil specimens assigned to this family, but Jacobs (1980) described tupaiids from the middle Miocene of Pakistan and discussed other fossil material that seems to belong to this or to closely related groups. Butler (1980) stated that the Tupaiidae appear to have been distinct from other mammals at least as far back as the beginning of the Tertiary and possibly in the Cretaceous. Yates (1984) gave the geological range of the family, and the order Scandentia, as Paleocene and Eocene in North America, middle Miocene in Pakistan, possibly Paleocene in Europe, and Recent in the current range.

Tree Shrews

Corbet and Hill (1992) recognized the following 11 species:

T. glis, Malay Peninsula below the Isthmus of Kra, Singapore, Sumatra, Bangka, Java, Borneo, and many

244

Tree shrew *(Tupaia glis)*, photo by Ernest P. Walker.

small nearby islands, Philippines (Palawan, Balabac, Culion, Busuanga, and Cuyo islands);

T. belangeri, eastern Nepal and Bangladesh to southeastern China, Indochina, and the Malay Peninsula above the Isthmus of Kra, Hainan;

T. nicobarica, Nicobar Islands (Bay of Bengal);

T. splendidula, southern Borneo, Natuna Islands, Karimata Island;

T. montana, northwestern Borneo;

T. javanica, Sumatra, Nias Island, Java, Bali;

T. minor, Malay Peninsula, Sumatra, Borneo, and some nearby islands;

T. gracilis, Borneo and some nearby islands, Bangka Island;

T. picta, Borneo;

T. tana, Sumatra, Borneo, and some nearby islands;

T. dorsalis, Borneo.

Wilson (*in* Wilson and Reeder 1993) also recognized *T. chrysogaster* of the Mentawai Islands off southwestern Sumatra, *T. longipes* of Borneo, and *T. palawanensis* of the Philippines as species distinct from *T. glis.* Certain of the above species are sometimes placed in the separate genus or subgenus *Lyonogale* Conisbee, 1953, but there is disagreement about its status. *Lyonogale* was considered a full genus by Butler (1980), Luckett (1980*b*), and R. D. Martin (1984) but was included within *Tupaia* by Corbet and Hill (1991, 1992), Wilson (*in* Wilson and Reeder 1993), and Yates (1984). Moreover, R. D. Martin (1984) considered *Lyonogale* to comprise the species *T. tana* and *T. dorsalis,* but Dene, Goodman, and Prychodko (1978) recognized it as a subgenus comprising *T. montana, T. minor, T. tana,* and *T. palawanensis.* These last authorities also regarded *T. chinensis* of Thailand and southern China to be a species distinct from *T. belangeri.*

Head and body length is about 140–230 mm and tail length varies from slightly less to a little more than head and body length. The weight is about 100–300 grams. Compared with other tree shrew genera, *Tupaia* is scantily haired. The upper parts are ochraceous, reddish, olive, or shades of brown and gray to almost black. The underparts are whitish or buff to dark brown. A light shoulder stripe may be present. According to R. D. Martin (1984), the species *T. tana* and *T. dorsalis* are distinguished by a conspicuous black dorsal stripe, well-developed canine teeth, robust claws, and large size.

The generic name is derived from *tupai,* a Malay word for squirrels. *Tupaia* resembles a squirrel externally but can easily be distinguished by the longer nose and the absence of long black whiskers. *Tupaia* is distinguished from the other genera of tree shrews by the following features: the lower lobe of the ear is smaller than the upper part; the naked area on top of the nose is cut squarely across instead of being slightly prolonged backward in the midline; and the tail is covered by long hairs. Studies by Dr. Heinrich Spranckel, of Frankfurt, Germany, on the skin glands of various mammals have demonstrated the presence of a throat gland in *T. glis* comparable to the glands in certain insectivores and different from the glands in other primates.

These tree shrews are diurnal. A few forms seem to be mainly arboreal and shelter in nests some distance above the ground, but most members of this genus actually spend most of their time on the ground and in low bushes, nesting in tree roots and fallen timber. They seem to be constantly on the move, searching for food in all possible crevices. The diet consists mainly of insects but also includes other animal food, fruits, seeds, and leaves. Observations by Emmons (1991*a*) indicate that frugivory may be more significant than usually thought. Tree shrews may hold food between their forepaws and sit on their haunches while eating. They are fond of water for both drinking and bathing.

The natural population density of *T. glis* has been reported to be about 6–12/ha. in Thailand and 2–5/ha. in peninsular Malaysia. Territoriality has been demonstrated in some species and is probable in others (Langham 1982; Lekagul and McNeely 1977; Sorenson 1974). There is no convincing evidence that under natural conditions groups contain more than one pair of sexually mature individuals, but the pair bond between male and female seems pronounced (Martin 1968). In a study of *T. glis* on Singapore, Kawamichi and Kawamichi (1979, 1982) found males to pair with 1–3 females. In this area there was little overlap between the home ranges of adult residents of the same sex, but those of opposite sexes overlapped completely, and a male's range sometimes included the ranges of more than one female. Aggressive chases were directed only against individuals of the same sex. Home range size in the study area averaged 10,174 sq meters for males and 8,809 sq meters for females. Groups changed composition mainly through loss of juveniles, with young males departing sooner than females.

Captive studies have shown that the males of most species, as well as the females of some, establish linear dominance hierarchies based mainly on aggressive interaction (Hasler and Sorenson 1974; Sorenson 1970, 1974). Among captive groups of *T. glis*, however, one male despot harasses and dominates all other males. Only this dominant male mates with the females present. *T. montana* is a more sociable species, there being two top-ranking, mutually tolerant males in a captive group. Eight distinct vocalizations have been identified in *Tupaia* (Binz and Zimmermann 1989). These include loud "squeals" of aggression, modulating "screams" to indicate immediate danger, "chatters" in response to disturbance, and rhythmic "clucking" and "whistles" associated with courtship and mating.

In some wild populations breeding apparently takes place year-round; nests are built in holes in fallen trees, hollow bamboos, or other suitable sites (Lekagul and McNeely 1977). In peninsular Malaysia, Langham (1982) found the main reproductive season to extend from February to June. In Singapore, Kawamichi and Kawamichi (1982) reported that *T. glis* probably is reproductively inactive from August through November and then has estrus in December, births in February followed by a postpartum estrus, and more births in April. Tree shrews are relatively easy to keep in captivity, and extensive observations have been made on reproductive activity therein. Females of *T. montana* exhibit a 9- to 12-day estrous cycle, a 23- to 29-day pseudopregnancy cycle, and a 49- to 51-day gestation period (Sorenson and Conway 1968). In *T. glis* the estrus cycle is 8–39 days, reported gestation is 40–52 days, and litter size is one to three (Gensch 1963a; Hasler and Sorenson 1974; Mallinson 1974). Birth weight is about 10–12 grams (Hayssen, Van Tienhoven, and Van Tienhoven 1993).

In a laboratory investigation of *T. glis* Martin (1968) found that there was usually a fertile postpartum estrus and that there was strong evidence for delayed implantation, with the blastocyst implanting about halfway through the interbirth interval. He observed the young to be born and reared in a separate "juvenile nest," while the parents slept in a "parental nest." The young apparently were visited and suckled by the female only once every 48 hours. The young emerged and moved to the parental nest to sleep at about 36 days, by which time they were weaned. Sexual maturity was attained at about 3 months in both sexes, and females produced their first young at 4.5 months. Emmons and Biun (1991) demonstrated basically the same "absentee" system of maternal care in wild *T. tana* on Borneo and suggested that it serves to protect the young from predation. The mother gave birth to a litter in a nest within a tree hole and then usually visited the young every other morning. The young emerged from the nest at about 34 days and the mother subsequently spent much more time with them; the male had no contact with the young. Record longevity for *Tupaia* is held by a captive *T. glis* that lived 12 years and 5 months (Jones 1982).

All species of *Tupaia* are on appendix 2 of the CITES. This designation may stem in part from the original general addition of the entire order Primates to the appendixes of the CITES in 1975, when that order commonly was thought to include the Tupaiidae. No detailed information is available on the status of any of the species, but it is likely that at least some are declining in association with destruction of tropical forest habitat. The IUCN recognizes *T. longipes*, *T. chrysogaster*, and *T. palawanensis* as distinct species (see above) and classifies the first as endangered, the other two as vulnerable; *T. niobarica* also is designated as endangered.

SCANDENTIA; TUPAIIDAE; **Genus ANATHANA**
Lyon, 1913

Indian Tree Shrew

The single species, *A. ellioti*, is found in India south of the Ganges River (Roonwal and Mohnot 1977).

Head and body length is 175–200 mm and tail length is 160–90 mm. R. D. Martin (1984) listed weight as 160 grams. The upper parts are usually speckled yellow and brown, with the middle of the back, the rump, and sometimes the upper tail tinged with reddish, producing a yellowish to reddish brown effect. Some individuals are blackish and orangish above. The underparts are whitish or buffy. A whitish or cream-colored shoulder stripe is pres-

Anathana ellioti, photo from Zoological Society of London.

ent. This genus differs from *Tupaia* in its larger and thickly haired ears and in cranial and dental features, but the general appearance is close to that of *Tupaia*.

This tree shrew is little known compared with some species of *Tupaia*, but it probably lives in much the same way. It is sometimes found in forests, and unlike *Tupaia*, it does not frequent human dwellings (Roonwal and Mohnot 1977). Chorazyna and Kurup (1975) observed several individuals on a slope covered with shrubs and stones at an elevation of 1,400 meters. At night the animals sheltered in holes among the rocks, sometimes using a system of corridors with two or three entrances. They usually left the shelters at dawn and returned two hours before sunset. Most of their active time was spent foraging—seeking insects on the ground, digging out worms, and sometimes leaping after flying insects. They also ate fruits. They climbed skillfully, even on vertical rocks, but rarely entered trees. They appeared to be largely solitary, always foraging alone, and there was usually one individual per hole, but occasionally groups of three or four were seen playing together for up to 20 minutes. A call described as a staccato squeak was heard twice. Roonwal and Mohnot (1977) reported that a pregnant female was taken in June. Hayssen, Van Tienhoven, and Van Tienhoven (1993) listed a record of five embryos being found in a female. *Anathana* is on appendix 2 of the CITES.

SCANDENTIA; TUPAIIDAE; **Genus DENDROGALE**
Gray, 1848

Small Smooth-tailed Tree Shrews

There are two species (Corbet and Hill 1992; Lekagul and McNeely 1977; Medway 1977):

D. murina, eastern Thailand, Cambodia, southern Viet Nam;
D. melanura, northwestern Borneo.

Head and body length is 100–150 mm and tail length is 90–145 mm. Lekagul and McNeely (1977) gave the weight of *D. murina* as 35–55 grams. *D. murina* is light in color and has facial markings, whereas *D. melanura* is dark and lacks facial markings. In *D. murina* the upper parts are brownish or blackish, and buff, ochraceous, or tawny. The underparts and the inner sides of the legs are buffy. The side of the head has a blackish line from the base of the whiskers through the eye to the ear and light, usually buffy, lines above and below the black band. The tail is dark above and darker distally; it is ochraceous buff below with a dark line down the center. The claws in this species are small. In *D. melanura* the upper parts are mixed blackish and ochraceous buff or cinnamon rufous, the darker color predominating; the underparts and the inner sides of the legs are ochraceous. Short ochraceous lines are sometimes present immediate-

Small smooth-tailed shrew *(Dendrogale murina)*, photo from *Natuurkundie Commissie in Indei. . . , Verhandlingen over de Natuurlijke Geschiedenis der Nederlandsche Overzeesche Besittingen door de Leden, Zoologie,* C. J. Temminck.

ly above and below the eye. The claws in this species are long.

These are the only small members of the family Tupaiidae with round, uniformly even-haired tails. Shoulder stripes are not present.

D. melanura is found in mountains from about 900 to 1,500 meters; one specimen was taken on the top of Mount Dulit, where it was living amid the moss-covered, stunted jungle. During the day these tree shrews are quite active, running about on the lower branches of trees and shrubs looking for insects. *Dendrogale* is more arboreal than either *Tupaia* or *Anathana*. It emits a shrill call when climbing. In captivity it will accept fruits and chopped meat. It is on appendix 2 of the CITES. *D. melanura,* restricted to a small zone of declining habitat, is classified as vulnerable by the IUCN.

SCANDENTIA; TUPAIIDAE; **Genus UROGALE**
Mearns, 1905

Philippine Tree Shrew

The single species, *U. everetti,* is widely distributed on Mindanao and also has been recently collected on Dinagat and Siargao islands in the Philippines (Heaney and Rabor 1982).

Head and body length is 170–220 mm and tail length is 115–75 mm. A full-grown male weighs about 350 grams. A blackish and tawny mixture on the upper parts produces a brownish color; the underparts are orangish to orangish red, being brightest on the chest region. There often is an indistinct orangish shoulder stripe. The specimens from Dinagat are relatively light in color with a metallic golden sheen dorsally; those from Siargao are much darker, with the dorsum nearly black. This tree shrew is distinguished externally by the elongated snout and the even-haired, rounded tail. The second pair of upper incisors is enlarged and caninelike.

Several calls are uttered: a whimpering, a snorting, chirping squeals, and a protracted distress call. Males tend to be more inquisitive than females in captivity. Natives say that *Urogale* nests in the ground and in cliffs. It usually sleeps curled up in a tight ball, sometimes with the top of its head under its body. The Philippine tree shrew is a good climber and swift runner.

Urogale is omnivorous. In captivity it feeds on such items as meat, insects, fruits and vegetables, mice, lizards, and earthworms. Eggs are opened with a skill suggesting that it eats them in the wild. The Mindanao tree shrew has a large appetite; it will consume several bananas or several two-ounce pieces of raw beef daily. In the wild it feeds mainly in the morning. Water is readily lapped up.

Urogale has bred in several zoos. The gestation period is probably 54–56 days, the number of young is one or two, and the female is receptive to the male soon after giving birth. On one occasion the female would not allow the male to come near the nest for several days before a birth. One young male had a complete body covering of hair and opened its eyes 19 days after it was born. It climbed about the cage with considerable agility and shared its mother's meals in the fifth week. It weighed 123 grams at 6 weeks of age, 145 grams at 7 weeks, and 185 grams at 12 weeks. A captive specimen lived 11 years and 6 months (Jones 1982).

Urogale is on appendix 2 of the CITES and is classified as vulnerable by the IUCN. Its habitat, like that of many other mammals of the Philippines, is rapidly being destroyed by human activity.

SCANDENTIA; TUPAIIDAE; **Genus PTILOCERCUS**
Gray, 1848

Pen-tailed Tree Shrew, or Feather-tailed Tree Shrew

The single species, *P. lowii,* inhabits the Malay Peninsula (including extreme southern Thailand), Sumatra, Bangka Island, Siberut Island, northern and western Borneo, and several other small nearby islands (Corbet and Hill 1992).

Head and body length is 100–140 mm and tail length is

Mindanao tree shrew *(Urogale everetti),* photo by Ernest P. Walker.

Pen-tailed tree shrew *(Ptilocercus lowii)*, photo by Lim Boo Liat.

130–90 mm. Lekagul and McNeely (1977) reported the weight to be 25–60 grams. The fur is soft in texture, usually dark grayish brown above and yellowish gray beneath. The tail is naked and dark, except for the terminal part, which has whitish hairs on opposite sides, producing a featherlike form as all the hairs are in the same plane. The entire tail resembles an old-fashioned quill pen, hence the common names.

The head is moderately tapering, and the whiskers are elongated and rather rigid. In contrast with the small thick ears of *Tupaia*, the ears of *Ptilocercus* are rather large, thin, and membranous. The limbs are nearly equal in length. The hands, feet, and pads are relatively larger than in the other genera of tree shrews. The five fingers and five toes bear short, sharp claws.

The feather-tailed tree shrew lives in primary or secondary forests from sea level to about 1,000 meters; it is nocturnal and arboreal (Lekagul and McNeely 1977). It is an expert climber, using the tail for support and balance.

Perhaps the tail also serves as an organ of touch, and it may assist in leaps. On the ground this mammal proceeds in a series of hops, with the tip of the tail inclined upward. It nests in holes in tree trunks or branches or among epiphytes about 12–20 meters above the ground. Nests consist of dried leaves, twigs, and fibers of soft wood. The diet consists of insects, small vertebrates such as geckos, and fruits (Gould 1978a; Lekagul and McNeely 1977).

Captives may be fierce at first but usually soon become tame. When annoyed *Ptilocercus* emits a hoarse, snarling hiss with the mouth open. A pair observed in captivity carried their tails downward and outstretched, at the same time moving them back and forth like a pendulum. *Ptilocercus* sleeps with its body rolled up in a tight ball, and the tail may be curled so that the plume covers the face.

Although *Ptilocercus* usually occurs in pairs, as many as five individuals have been found in a single nest. One specimen lived 2 years and 8 months in captivity (Lekagul and McNeely 1977). The genus is on appendix 2 of the CITES.

Order Dermoptera

DERMOPTERA; Family CYNOCEPHALIDAE; Genus
CYNOCEPHALUS
Boddaert, 1768

Colugos, or Flying Lemurs

The single Recent family, Cynocephalidae, contains one known genus, *Cynocephalus*, and two species (Chasen 1940; Corbet and Hill 1992; Lekagul and McNeely 1977):

C. variegatus, southern parts of Thailand and Indochina, Malay Peninsula, Sumatra, Java, Borneo, and many nearby islands;
C. volans, southern Philippines.

The systematic position of this order is questionable, but it commonly has been placed near the Insectivora and the Chiroptera. Van Valen (1967) considered the Dermoptera to be only a suborder of the Insectivora. Like some other authors, he employed the familial name Galeopithecidae instead of Cynocephalidae and used the generic name *Galeopithecus* Pallas, 1783, for the species *C. volans*, while placing *C. variegatus* in a separate genus, *Galeopterus* Thomas, 1908. More recently it has been widely recognized that the Dermoptera are not immediately related to the Insectivora but form part of a supraordinal group, the Archonta, together with the Chiroptera, the Scandentia, and the Primates. Within that assemblage some authorities see closest affinity between the Dermoptera and Chiroptera (Simmons 1995), but there is some evidence that the Chiroptera are distantly related to the other three orders and that the Dermoptera are close to the Primates (Novacek 1992, 1993). Based on analysis of fossil and morphological data, Beard (1993) considered the Dermoptera to be immediately related to the Primates and even to include several fossil families (e.g., Plesiadapidae) that usually are assigned to the latter order. While accepting *Cynocephalus* as the appropriate generic name for the living flying lemurs, Beard used Galeopithecidae as the familial name.

Head and body length is 340–420 mm, tail length is 175–270 mm, and weight is usually 1.0–1.75 kg. There is great variation in color and pattern, but apparently the upper parts of males tend to be some shade of brown and those of females are grayish. Scattered white spots are present on the upper parts of *C. variegatus*. The underparts of both species are paler and without spots. The shaded and mottled color pattern of *Cynocephalus* blends well with the bark of trees.

Colugos have a large gliding membrane attached to the neck and the sides of the body. The development of this membrane is greater than that found in other volant mammals, such as flying squirrels (Rodentia) and gliding possums (Diprotodontia), whose gliding surface is stretched only between the limbs, with the fingers, toes, and tail left free. In *Cynocephalus* the membrane extends along the limbs to the tips of the fingers, toes, and tail.

The arms, legs, and tail are long and slender. The feet are broad, and all digits are tipped with sharp, recurved claws. The head is broad, the ears are short, and the eyes are large. The skull is unique among mammals, showing many primitive characters. Females have a single pair of mammae located at the sides of the body, almost in the armpits.

The dental formula is: (i 2/3, c 1/1, pm 2/2, m 3/3) × 2 = 34. The lower incisors are developed into peculiar "comb teeth" that vaguely resemble the teeth of true lemurs (Primates). The difference is that each lower incisor of *Cynocephalus* may have as many as 20 prongs radiating from one root, whereas in true lemurs each prong of the comb is a single tooth. The comb teeth of *Cynocephalus* may act as food strainers or as scrapers, or they may be used to groom the fur. Aimi and Inagaki (1988) thought both these functions were probable, but Wischusen (1990) observed no specialized use of the lower incisors during foraging or grooming.

According to Lekagul and McNeely (1977), colugos inhabit both lowland and mountainous areas and may be found in primary or secondary forests, coconut groves, and rubber plantations. They are totally arboreal, seldom if ever descending to the ground, where they are nearly helpless, cannot stand erect, and try to climb any object encountered. They are skillful though slow climbers, ascending unbranched vertical trunks in a series of lurches with head up and limbs spread to grasp the tree. When moving about on branches or while feeding, colugos are in an upside-down position. At such times the gliding membrane is drawn down under the forelegs so that it will not catch on branches. There is no indication that the tail is prehensile. Lekagul and McNeely (1977) wrote that in primary and secondary forest colugos spend the day in holes or hollows of trees about 25–50 meters above the ground but that in coconut plantations they curl up in a ball or hang from a palm frond, with all four feet close together. In the Philippines, Wischusen (1990) found most day roosts in thick foliage. He also observed that females carrying young usually transfered from tree to tree by moving through the branches rather than by gliding. Total nightly movements covered 1,011–1,764 meters. Although occasionally seen by day, colugos are generally nocturnal. They come out of their shelters around dusk, climb a short distance up a tree, and glide off to seek food. They usually glide to the same spots on the same food trees night after night. Each glide may

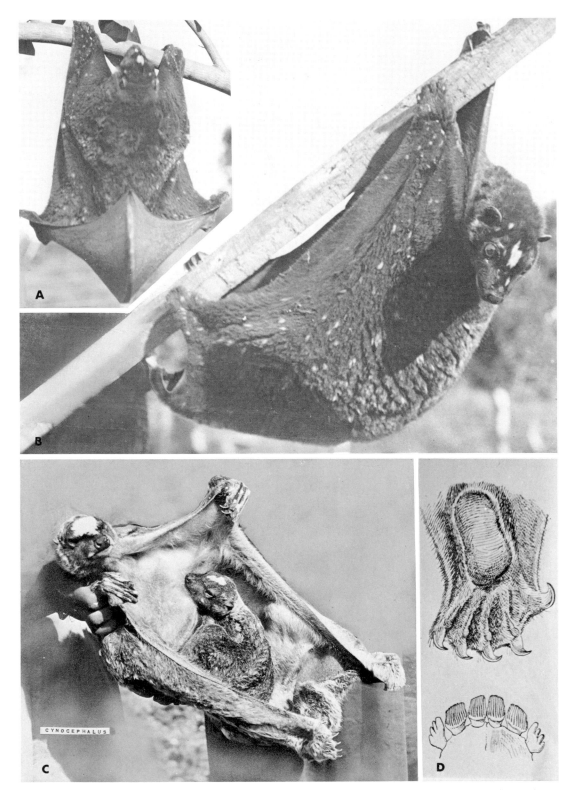

Philippine gliding lemurs *(Cynocephalus volans):* A. Lemur with tail turned back to prevent soiling of interfemoral membrane, photo by John N. Hamlet; B. Lemur clinging to branch, photo by John N. Hamlet; C. Mother and nursing baby, photo by Charles H. Wharton; D. Plantar surface of hind foot, and teeth, photo from *Genera Mammalium, Insectivora and Galeopitheca,* Angel Cabrera.

cover 100 meters or more. One individual lost only 10.5–12 meters of elevation during a measured glide of 136 meters between two trees.

Wischusen (1990) found *C. volans* to feed exclusively on young foliage. Although the diet sometimes has been reported to include fruits, buds, and flowers, as well as leaves, the stomach contents of about eight specimens revealed only green vegetable matter. Captive specimens were forced to feed on fruit, but they took it most reluctantly. Colugos pull food within reach of the mouth and then bite off leaves or parts of fruit. In the wild they probably obtain water by licking wet leaves.

Several individuals may share the same shelter, but little is known about social life. MacKinnon (1984) noted that apart from mothers with young, animals move about singly, but several cover the same area and use the same feeding trees. Six independent animals were found in an area of less that 0.5 ha. on a coconut plantation in Java. In a radio-tracking study of *C. volans,* Wischusen (1990) determined monthly home ranges to measure 6.4–13.4 ha. and to overlap extensively. Allogrooming and other friendly interaction was observed between adults of opposite sexes and between adults and young, but adult males sometimes displayed hostility toward one another. Colugos utter a rasping cry that probably is an alarm call.

Wischusen (1990) observed female *C. volans* carrying young from December to June and suggested that breeding may continue throughout the year. Indeed, female *C. volans*, with young 223–54 mm long, also have been noted in March, April, and May. Gestation has been reported to last about 60 days, though observations of *C. volans* by Wischusen (1990) suggest a period of approximately 150 days, perhaps involving delayed implantation. Litter size is one, rarely two. Birth weight is about 36 grams (Hayssen, Van Tienhoven, and Van Tienhoven 1993). Medway (1978) stated that lactating females with unweaned young have proved to be pregnant, and evidently births may follow in rapid succession. Lekagul and McNeely (1977) wrote that the young is born in a marsupial-like undeveloped state and that the gliding membrane of the mother can be folded into a soft, warm pouch to hold the young. The mother may leave the young in a nest tree or carry it with her while foraging. In the latter case the young clings with its claws and milk teeth to the belly fur and nipples of the mother. Young *C. volans* are not weaned until about 6 months and may not attain full adult size until 3 years (Wischusen 1990; Wischusen, Ingle, and Richmond 1992). In captivity colugos are difficult to keep and usually die quickly because of improper diet or dampness of the underparts, probably resulting from unsuitable cage facilities. One individual, however, was kept as a pet for 17 years and 6 months, after which it escaped.

Wischusen (1990) expressed concern about the very rapid loss of forested habitat within the ranges of both species of *Cynocephalus* and said that they may be threatened with extinction in the very near future. *C. volans,* which is limited to a small and rapidly developing region, may be in particular danger. It may originally have numbered 9,450,000 individuals, but the current estimate is 1,000,000, many of which are in isolated forest fragments and have little chance of genetic exchange. The very slow rates of maturation and reproduction would compound the difficulty of recovery. *C. volans* now is classified as vulnerable by the IUCN.

The family Cynocephalidae long was known only from the Recent of Asia, but a fossil from the late Eocene of southern Thailand was recently reported (Beard 1993). Another family, the Plagiomenidae, occurred in North America from the middle Paleocene to the early Eocene and has been considered part of or closely related to the Dermoptera (Rose 1975). However, detailed morphological studies by MacPhee, Cartmill, and Rose (1989) have not confirmed such affinity. There also has been disagreement about whether still another early Tertiary family, the Paromomyidae, had a gliding ability (Krause 1991).

Order Chiroptera

Bats

Bats inhabit most of the temperate and tropical regions of both hemispheres but are absent from certain remote oceanic islands. They are not found in the colder parts of either hemisphere beyond the limit of tree growth. Among mammals only the rodents exceed bats in number of species.

The order Chiroptera is divided into two suborders: the Megachiroptera, which contains the single family Pteropodidae, and the Microchiroptera, which includes all the other families. Unfortunately, these terms are somewhat misleading as some of the Megachiroptera are smaller than some of the Microchiroptera. Studies of brain morphology and other aspects of the nervous system led Pettigrew (1986, 1987) and Pettigrew et al. (1989) to suggest that the Megachiroptera are actually descendants of an early branch of the order Primates and thus bats and the power of flight developed twice in the course of mammalian evolution. This hypothesis was not supported by analysis of mitochondrial DNA (Bennett et al. 1988). Intensive controversy subsequently developed relative to this question of chiropteran monophyly. Simmons (1994) summarized the results of no less than 30 relevant studies, utilizing many different biochemical, molecular, and morphological approaches, that have been published since 1988. She also reviewed the morphology of the dentition, skull, cranial vascular system, postcranial musculoskeletal system, fetal membranes, and nervous system of bats and related mammals. The evidence overwhelmingly supported the view that bats are monophyletic, all having arisen from a single ancestor, and that powered flight evolved only once in mammals.

There are still questions regarding the closest relative of the Chiroptera, though most of the work reviewed by Simmons suggests affinity to the Dermoptera (see account thereof) and also tends to uphold the validity of the Archonta, a supraordinal grouping that includes the Chiroptera, the Dermoptera, the Primates, and the Scandentia. Within that grouping, the Chiroptera and Dermoptera may form a monophyletic clade, the Volitantia, primitive members of which were specialized for finger gliding and hanging from branches; from such ancestry bats would have evolved powered flight and hanging by their hind limbs (Simmons 1995). Examination of the specific muscles involved in flight provides additional support for a common origin of bats and for an immediate relationship of bats and dermopterans (Thewissen and Babcock 1992). Notwithstanding the above, Pettigrew (1994) has continued to advocate a biphyletic origin of bats, noting that such is supported by certain DNA sequence data.

The order Chiroptera (including the Megachiroptera) is further divided here into 18 families, 192 genera, and 977 species. The sequence of families and genera presented largely follows the arrangement of Koopman (1984b), who recognized 4 superfamilies within the Microchiroptera: Emballonuroidea, with the families Emballonuridae, Craseonycteridae, and Rhinopomatidae; Rhinolophoidea, with Nycteridae, Megadermatidae, and Rhinolophidae; Phyllostomoidea, with Mormoopidae, Noctilionidae, and Phyllostomidae; and Vespertilionoidea, with Thyropteridae, Myzopodidae, Furipteridae, Natalidae, Mystacinidae, Molossidae, and Vespertilionidae. Some modifications in the sequence of vespertilionoid families are given here, the Hipposideridae are here regarded as a family distinct from the Rhinolophidae, and the Mystacinidae have been transferred to the Phyllostomoidea.

Head and body length varies from about 25 to 406 mm. The pelage is usually long and silky with a dense underfur. Some species have color phases. Scent glands, which produce a strong musky odor and are variable in position, are usually present. The testes of male bats descend into a temporary sac during the breeding season; a baculum is usually present. Female bats usually have one pair of functional mammae located in the chest region, but females of *Lasiurus* have four functional mammae. Simmons (1993a) reported the presence of an additional pair of nipples located in the pubic (lower abdominal) region apparently having a lacteal function in the families Rhinopomatidae, Craseonycteridae, Megadermatidae, and Rhinolophidae (including the Hipposideridae). Depending on its phylogenetic significance, this character could indicate the need for a change in the superfamilial groupings given above.

Bats are the only mammals that fly, though several gliding mammals are referred to as "flying." The membranes that extend from the sides of the body, legs, and tail are actually extensions of the skin of the back and belly. They are elastic and thin and consist of two layers of skin with no flesh between them and only a small amount of connective tissue, in which the blood vessels and nerves are located. The wing membrane is supported by the elongated fingers of the forelimbs. The length of the third finger is usually equal to that of the head and body plus the legs. The knee is directed outward and backward as a result of the rotation of the leg for the support of the wing membrane. A cartilaginous spur, the calcar or calcaneum, on the inner side of the ankle joint, rarely absent, helps to spread the tail membrane. The thumb is short and has a sharp-hooked claw except in the smoky bats (Furipteridae). The thumb is often used for clinging to surfaces when the bat alights. The second finger also has a claw in most of the genera of Old World fruit bats (Pteropodidae). With the exception of the thumb, all the fingers help to extend the wing, like the ribs

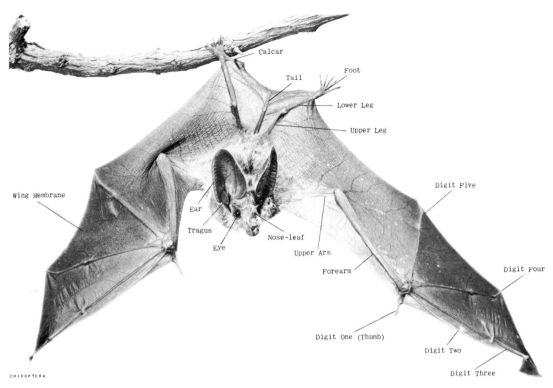

A Mexican big-eared bat *(Macrotus waterhousii mexicanus)* about to take flight. This photo, by Ernest P. Walker, illustrates many of the parts of its external form.

of an umbrella. The toes have sharp, curved claws. The long bones are slender, light, and tubular and contain marrow. Various structural features impart a rigid support to the arm, so that it can be used as a wing. For example, some of the joints in the vertebral column are fused together and the ribs are flattened. The shoulder girdle is better developed than the pelvic girdle and is firmly anchored by a well-developed collarbone (clavicle), which reaches to the breastbone (sternum). The sternum is usually centrally keeled, or ridged, for the attachment of enlarged muscles used in the downward stroke of the wings.

The maximum number of milk teeth is 22; these may be used by a young bat to cling to the mother's fur as she flies. The permanent teeth number 20–38. The cheek teeth are usually smooth, with a longitudinal groove in the Pteropodidae and generally three cusps in the other forms. The sutures of the skull tend to disappear with age, becoming fused. The cerebral hemispheres are smooth and do not extend backward over the cerebellum.

The heads of bats vary greatly; some resemble those of familiar animals, such as dogs or mice, while others have elaborate structures that give them a unique, otherworldly appearance. Many species have a nose leaf or other facial ornamentation through which echolocation sounds are emitted and augmented (Kunz and Pierson 1994). The external ears of many bats also are remarkably modified with folds and crenulations that may aid in sound reception. Microchiropterans have a tragus and/or an antitragus, fleshy projections at the front of the ear orifice. The tragus is thought to aid echolocating bats in finding prey (Kunz and Pierson 1994). Although bats are sometimes said to have poor vision, most have well-developed eyes that are comparable in sensitivity to those of many other mammals. For

many kinds of bats, however, vision may be less significant than hearing and olfaction. Like those of many nocturnal animals, the retinae of bats consist almost entirely of rod cells and lack the cones that are associated with color vision. However, most of the megachiropterans have large eyes and are capable of color vision, perhaps in association with their largely diurnal, tree-roosting habits (Kunz and Pierson 1994).

Echolocation is known to be used by more than half of the species of bats for nocturnal orientation and prey capture. This phenomenon involves the emission of vocal sounds through the mouth or nose as a bat flies. These sounds are usually above the limit of human hearing and are reflected back to the bat in flight as echoes. They enable the bat to avoid hitting obstacles when flying in darkness and to locate the position of flying insects. Bats are unable to guide themselves by their own voice when their ears are plugged. The New World bats that feed on fruit or the blood of large sleeping animals send out pulses having only about one-thousandth the sound energy of those used by bats that feed on flying insects and fish. Echolocation has been found in all the families investigated thus far, though it apparently is absent in all but one genus *(Rousettus)* of the Megachiroptera (Kunz and Pierson 1994).

In addition to the high-frequency vocal sounds emitted in echolocation, bats produce sounds to express emotion or for communication. Ernest P. Walker detected a vibration of the entire body, legs, and wings at 52 cps in the common big brown bat of North America *(Eptesicus)* and similar vibrations in other North American insectivorous bats. Apparently this vibration is under completely controlled by the bat and seems to occur only when the bat is resting and contented. It ceases when the bat goes to sleep. This vibra-

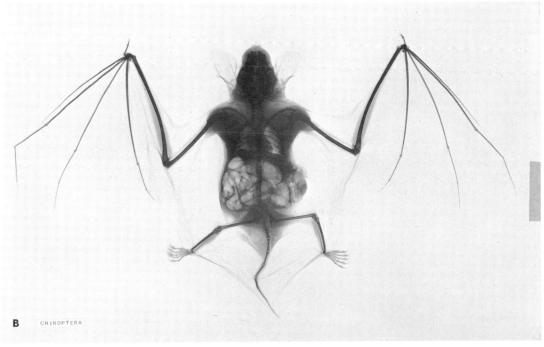

A. A big brown bat *(Eptesicus fuscus)* almost torpid, resting on glass illuminated from beneath to show the blood vessels, some filled and some empty. The crepelike or wrinkled appearance normal to bat wings that are not fully stretched is also shown. Photo by Ernest P. Walker. B. X-ray of a bat showing the remarkable development of the arms into wings and the peculiar modification in the position of the legs. Photo by Lew Gust.

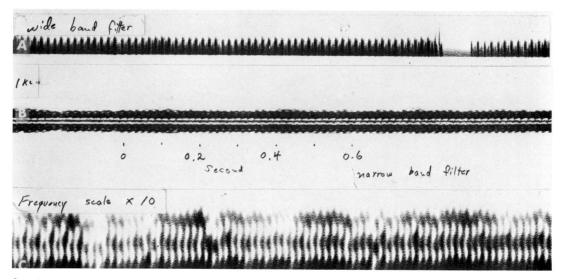

Sound spectrogram made from sound tape recording of the low-frequency "body buzz" of a big brown bat *(Eptesicus fuscus)* at rest and contented. Pictures made with a Sona-Graph: A. With wide-band filter; B. With the narrow-band filter; C. With a scale magnifier using a factor of 10. The frequency is below 60 cps and is probably about 52 if these bats are uniform in their rate of vibration. Pictures made by Dr. Wesley E. Lanyon, American Museum of Natural History, from sound tape recorded by Ernest P. Walker.

tion probably exists in many if not all bats, but it apparently has been overlooked by all but the few persons who have kept bats as pets. Bats also utter many other vocal sounds that humans cannot hear.

Bats shelter in caves, crevices, tree cavities, and buildings, and some sleep in exposed locations on trees. In the colder parts of their ranges in the temperate zone bats either hibernate during winter or migrate to areas where food is available. There is a great reduction in metabolism during hibernation, so that the rate of oxygen consumption is only about one-hundredth of the normal active rate. When sleeping during the day many bats become semitorpid and there is a great reduction in body temperature, but as they awaken their temperature rises. The rate of oxygen consumption during sleep may be only one-tenth of the active rate. Some bats thus spend much of their life in hibernation or a condition that approaches hibernation or estivation. Resting bats thoroughly groom themselves using the tongue and toes. Back scratching is no problem for a bat: it merely reaches up with either hind foot and scratches any portion of its back, the top of its head, its face, or its mouth. Bats do not carry bed bugs, as has often been alleged; some of the parasites occasionally found on bats do very closely resemble bed bugs.

When they are at rest, bats normally hang head downward, though a few rest on horizontal surfaces in caves and elsewhere. Occasionally fruit bats, while scrambling about in trees, get into a head-upward position. When a bat is hanging head downward in an elevated position it is very easy for it to take flight by merely letting go, dropping, and spreading its wings. Bats have no difficulty in taking to flight from a level surface: they leap into the air, using both arms and legs to launch themselves. (The illustrations of some of the genera here are from old publications in which the animal was drawn with the head upward.)

Considering their structure, it is not surprising that bats fly with their legs as well as their wings; one could say that they "swim" through the air. Photographs of bats in flight reveal that their legs work in unison with their wings, just as swimmers use their legs and arms at the same time.

Some bats pick insects off foliage or even off the ground. Probably all bats require more water or moist food than do other mammals of comparable weight because their wings have such great evaporation surfaces in comparison with their weight.

When catching insects in flight a bat uses its tail and wing membranes to prevent the escape of its prey. When it has an insect in its mouth on which it does not have a firm grip the bat spreads its legs out and forward, so that the membrane is spread into an excellent lap, and bends its head forward into this apron. Thus, it keeps the insect under restraint while it manipulates it about in its mouth and renders it helpless. Bats do this in flight, when hanging up, or on a level surface. Often in this struggle they fall onto their backs, but only for an instant. They quickly right themselves, for they are superb acrobats.

Bats may be divided into six groups according to the main kinds of feeding habits. The following list is not a systematic classification, however, as some families of bats may include several of the kinds listed below.

Insectivorous bats, which obtain most of their insect food while flying. Many also eat some fruit. They are of relatively small size and comprise the majority of the bat population.

Fruit-eating bats, which feed almost exclusively on fruit and some green vegetation. No doubt they also eat some insects or insect larvae found in or on fruits. They often work together in groups, sometimes traveling long distances to fruit-bearing trees or shrubs. In some areas these bats do considerable damage to fruit grown for human consumption; however, most of them depend upon wild fruit. They are of necessity tropical bats, as they can survive only where fruit is constantly ripening. They range in size from the large, so-called flying foxes, with a wingspread of up to 1.7 meters, to those with a wingspread of about 250–300 mm.

Flower-feeding bats, which eat mainly pollen and nectar and possibly some of the insects found in flowers. They are almost exclusively small and usually have a long, pointed head and a long tongue with a brushlike tip to help them

A. Four female Mexican fruit bats *(Artibeus jamaicensis)* with their young, suspended from the underside of a leaf frond, photo by Cory T. de Carvalho. B. Rough-legged bats *(Myotis dasycneme)* clustered in a church attic—the way bats frequently assemble in caves and buildings for daily sleeping or for hibernation, photo by P. F. Van Heerdt.

obtain their food. They are inhabitants of the tropics or subtropics.

True vampire bats, which eat blood obtained by making a small incision in the skin of an animal while it is asleep. There are only three small species in this group. The danger to the bitten animal is subsequent contraction of virus-caused diseases, such as rabies, and secondary infection of the wound.

Carnivorous bats, which prey on other small mammals, birds, lizards, and frogs. They have a widely varied diet, however, and do not eat only animals. They are of moderate size.

Fish-eating bats, which catch fish at or near the water's surface. These bats obtain the fish with their feet, which are large, powerful, and equipped with hooked claws.

Bats destroy many harmful insects and pollinate many flowers. The guano deposits found in "bat caves" are valu-

able as fertilizer. People eat some of the larger bats. Some animals that feed on bats are the birds of prey, snakes, and other mammals, including other bats.

Most bats have only one young per year. This low reproductive rate is offset by the fact that bats live longer then most mammals of their size. Ernest P. Walker had a pet *Eptesicus* that attained an age of at least 13 years. The Zoological Society of London had a fruit bat for 17 years, and there are records of bats marked in the wild that lived at least 30 years. In the hibernating forms ovulation occurs in the spring, breeding generally having taken place the preceding autumn (delayed implantation). The sperm is retained in the reproductive tract of the female during the winter. In the nonhibernating forms ovulation and breeding occur in the spring or in a favorable season.

The congregation of bats in caves facilitates the preservation of their bodies after death. Some of the water that drips into caves is heavily impregnated with minerals; this slows the rate of decay and, in time, helps to form a deposit over the dead bats. Ernest P. Walker encountered such a condition in a lead mine in Mexico. The mine was merely a fissure two to eight feet wide that had been worked out. On the floor of the fissure were thousands of dead bats that had recently succumbed to disease. There was little decomposition, so that a deposit of soil and suitable chemical conditions or the dripping of water onto the dead bats would eventually make a fossil deposit.

Most authorities believe bats and gliding lemurs evolved from insectivorous, probably arboreal ancestors. Fossil remains show that these mammals were plentiful and well developed at least as long ago as the Eocene. These fossils also indicate that bats of that time were very similar to bats of the present. Indeed, Novacek (1985) found that the oldest known bats, from the early Eocene of Wyoming and the middle Eocene of Germany, show certain basicranial features that suggest refinement of ultrasonic echolocation comparable to that of modern microchiropterans.

Due to the abundance of forms and the myriad legends and superstitions concerning them, bats have been given vernacular names in practically all languages. Most of these names are all-inclusive terms meaning simply "bat," but many are derived by combining words used for other animals. Thus, the Aztec term *Quimichpapalotl* means "butterfly mouse," the German *Fledermaus* means "flying mouse," and the French *chauve-souris* means "bald mice."

A "bare-backed" fruit bat or flying fox (*Pteropus neohibernicus* [?]) from New Guinea walking along a branch, photo by Sten Bergman.

CHIROPTERA; **Family PTEROPODIDAE**

Old World Fruit Bats

This family of 42 living genera and 169 species is found in the tropical and subtropical regions of the Old World, east to Australia and the Caroline and Cook islands. Koopman (*in* Wilson and Reeder 1993) and Koopman and Jones (1970) recognized two living subfamilies: Pteropodinae, for most of the genera; and Macroglossinae, for the genera *Eonycteris, Megaloglossus, Macroglossus, Syconycteris, Melonycteris,* and *Notopteris.* Corbet and Hill (1992) recognized two additional subfamilies: Harpyionycterinae, for the single genus *Harpyionycteris;* and Nyctimeninae, for *Nyctimene* and *Paranyctimene.* Based on chromosomal studies, Rickart, Heaney, and Rosenfeld (1989) included *Harpyionycteris, Nyctimene,* and *Paranyctimene* in the Pteropodinae, but they also reported a number of findings that might suggest some modification in the following sequence of genera.

Head and body length is 50–400 mm depending on the species. The tail is short, rudimentary, or absent, except in *Notopteris,* in which it has 10 vertebrae. The tail membrane is only a narrow border. Adults range in weight from approximately 15 grams for the smallest nectar- and pollen-feeding members of the family to over 1,500 grams for the largest fruit eaters. This family contains the largest living bats, some species of *Pteropus* and *Acerodon* having a wingspread of 1.7 meters. The fur is as much as 3 cm in length, dense, and variable in color but with prevailing tinges of dark brown. The genera *Epomophorus, Micropteropus, Epomops, Scotonycteris, Nanonycteris, Hypsignathus, Casinycteris,* and *Plerotes,* among others, exhibit secondary sexual characters, including, in males, much greater size, glandular pouches with tufts of hair on the shoulders, and large pharyngeal sacs in the chest region. The males of *Hypsignathus* have greatly folded lips, and the males of some genera have erectile hairs on the nape. All the members of the family Pteropodidae except the genera *Dobsonia, Eonycteris, Notopteris,* and *Neopteryx* have a claw on the second finger in addition to the claw on the thumb; all the members of the suborder Microchiroptera lack a claw on the second finger. The external ear of Old World fruit bats is elongate, oval, and rather simple; the margin of the ear forms a complete ring, and there is no tragus. In the Microchiroptera the external ear is often complex; the margin does not form a closed tube, and a tragus is usually present. Two genera of Pteropodidae, *Nyctimene* and *Paranyctimene,* have tubular nostrils that open laterally. The members of this family have large, well-developed eyes. The surface area for the rods, the receptors in the eye for black-and-white vision, is greatly increased by small fingerlike projections, enclosing blood vessels, of the inner coat of the eyeball that penetrate the outer layer of the retina; these projections apparently are lacking in the Microchiroptera. Only the genus *Rousettus* is known to emit ultrasonic sounds during flight, in addition to utilizing its eyes for vision. The remaining genera of Pteropodidae are believed to guide themselves visually, unless perhaps the tube-nosed fruit bats *Nyctimene* and *Paranyctimene* orient by means of echolocation.

The penis resembles that of some Primates. Females have one pair of mammae in the chest region.

The dental formula varies from (i 2/2, c 1/1, pm 3/3, m 2/3) × 2 = 34 in *Pteropus* and *Rousettus* to (i 1/0, c 1/1, pm 3/3, m 1/2) × 2 = 24 in *Nyctimene* and *Paranyctimene.* The incisors are small, and the canines are always present, even in those forms in which the dentition tends to be reduced. The back teeth are low, elongate, and widely separated. The crowns of the molars are smooth and marked with a longitudinal groove except in *Pteralopex* and *Harpyionycteris,* which have cuspidate molars. The dentition is adapted for a soft diet. The palate of the bats of this family generally has eight transverse ridges against which the tongue crushes the food. The bony palate gradually narrows behind the last molars, whereas in the Microchiroptera the bony palate is not continued behind the last

A. Fruit bats *(Pteropus tonganus)* roosting near the village of Tonga, Fiji Islands. This picture illustrates the characteristic manner in which the fruit bats sleep during the day. Photo from Public Relations Office, Fiji, through R. A. Derrick. B. A group of African fruit bats *(Rousettus aegyptiacus)* hanging in a cave in the same manner as many of the insectivorous bats, which sometimes congregate by the thousands. Photo by Don Davis.

molar. The tongue is covered with well-developed papillae in some forms, and it can be protruded far beyond the end of the mouth in the nectar and pollen feeders of the subfamily Macroglossinae.

These bats are active mostly in the evening and at night but have been observed flying in the daytime. All fruit bats are somewhat irregular in their presence in a region because they often leave areas where fruit is not available. They often make long flights between their roosting and feeding areas. The large forms are slow but powerful fliers; some of the large fruit bats may fly as far as 15 km to their feeding areas. *Eidolon,* which has long narrow wings, is adapted to flying long distances. The bats of this family climb actively in trees for fruit and in their roosts in trees and caves and under the eaves of buildings. When they are at rest, hanging by their feet, their head is at a right angle to the axis of the body. The larger forms are often gregarious, roosting in large groups, whereas the smaller forms are generally solitary.

Most species of this family locate food by smell. Many if not all of these bats crush ripe fruit pulp in the mouth, swallow the juice, and spit out most of the pulp and seeds, often pressed into almost uniformly shaped pieces; some of the softer pulp is swallowed. Material that is swallowed goes through the very simple digestive tract in a very short time, perhaps half an hour on the average. In *Epomophorus* and related genera the structure of the lips, windpipe, and gullet forms a type of suction apparatus that probably aids in feeding on the softer parts of fruits. Bats can bite into fruit while hovering, or they may hold onto a branch with one foot and press the fruit to the chest with the other foot in order to bite into it; sometimes they carry small fruit to a branch and hang head downward while eating. The members of this family also chew flowers to obtain nectar and juices. The members of the subfamily Macroglossinae seem to feed mainly on pollen and nectar. *Nyctimene* sometimes includes insects in its diet.

Some forms appear to breed throughout the year, but most members of this family probably have well-defined breeding seasons. Pregnant females of several species shelter apart from the males. There is usually only one young in a birth. Individuals of this family have lived in captivity for more than 30 years.

Old World fruit bats as a group are threatened by human destruction of their tropical forest habitat, disruption of the fragile environment on the isolated islands to which many of the species are endemic, direct killing because they are considered agricultural pests, and intensive hunting for local use as food and for a lucrative commercial trade. Detailed summaries of the conservation problems of many species, together with information on systematics and natural history, were compiled recently by Mickleburgh, Hutson, and Racey (1992) and Wilson and Graham (1992).

The geological range of this family is early Oligocene in Europe, early Miocene to Recent in Africa, Pleistocene to Recent in Madagascar and the East Indies, and Recent in other parts of its range (Koopman 1984*a*).

CHIROPTERA; PTEROPODIDAE; Genus EIDOLON
Rafinesque, 1815

Straw-colored Fruit Bats

There are two species (Bergmans 1990):

E. helvum, southwestern Arabian Peninsula, most of the forest and some of the savannah zones of Africa from

Straw-colored fruit bat *(Eidolon helvum),* photo by John Visser.

Senegal to central Ethiopia and south to South Africa, islands of Gulf of Guinea and off East Africa;

E. dupreanum, Madagascar.

Head and body length is about 143–215 mm, tail length is 4–20 mm, forearm length is 109–32 mm, and wingspan can reach 762 mm. Fayenuwo and Halstead (1974) reported that during the breeding season in Nigeria, adult weight increased from about 230 to 330 grams in males and from 240 to 350 grams in females. The hairs on the neck are longer and more woolly than those on the body, and the interfemoral membrane is hairy in the middle of its upper surface. Coloration is yellowish brown or brownish above and tawny olive or brownish below. The cinnamon hairs on the glandular skin on the foreneck and sides of the neck are most conspicuous in adult males; otherwise the sexes are much the same in color and size.

In contrast to *Epomophorus* and similar genera, *Eidolon* has a more pointed head and lacks the white patches at the base of the ear. The wings are long and narrow, adapted to flying long distances. The wings are also used in climbing about branches in the roosts. When hanging by the hind feet this genus turns the second phalanges of the third and fourth digits inward, so that they fold against the lower surface of the wing, as in the vespertilionid genus *Miniopterus.*

This bat inhabits forest and savannah country and is found at elevations of up to 2,000 meters in the Ruwenzori Mountains (Kingdon 1974a). It is gregarious and prefers to roost in tall trees by day but has also been found in lofts and in caves in rocks. During the daytime it is often noisy and restless and even flies about from place to place. At night groups fly out of the roosts in search of ripe fruit. The large roosts are usually 60 km or more apart, suggesting a foraging range of at least 30 km for a colony. Thomas (1983) determined that *Eidolon* also makes extensive seasonal migrations: a colony of about 500,000 individuals left its roost in the southern forest of the Ivory Coast following the birth of young in February, moved northward into the savannah zone, and had migrated at least as far as the Niger River Basin by the middle of the wet season in July. It is probable that even greater distances (a round trip of over 2,500 km) are covered by some colonies in East Africa.

The roosts of *Eidolon* are within easy reach of forests or fruit plantations. Juices of various fruits are the preferred food, though this bat also feeds on the blossoms and perhaps young shoots of the silk-cotton tree *(Ceiba)*. *Eidolon* will eat directly into the fruit of the palm *Borassus* and has the unusual habit of chewing into soft wood, apparently to obtain moisture.

The straw-colored fruit bat sometimes occurs in enormous colonies of 100,000 to 1,000,000 individuals of both sexes. Studies in Nigeria and Uganda (Fayenuwo and Halstead 1974; Kingdon 1974a; Mutere 1967) indicate that mating occurs in the colonies from April to June and that there is delayed implantation. Most of the major roosts are abandoned from June to September, apparently because social bonds then break down. There is a reconvergence of the colonies in September–October, which is followed in the females by synchronous implantation of the ova in the uteri. Births occur in February and March in the midst of clusters formed by hundreds of females and some males. There is a single young per female, and one newborn weighed 50 grams. It appears that the reproductive cycle is geared to the rainfall pattern so as to ensure that the young are weaned when conditions are optimal. Implantation coincides exactly with the start of the dry season, and births with the beginning of the wet season. According to DeFrees and Wilson (1988), the total minimum gestation period is 9 months, though the actual period of embryonic development is 4 months. However, Mickleburgh, Hutson, and Racey (1992) indicated that some colonies in Central Africa give birth only 4 months after conception, there being no delayed implantation. The known longevity record for *Eidolon* is 21 years and 10 months (Jones 1982).

In some areas *Eidolon* is hunted and eaten by humans, but in others it is traditionally protected (Happold and Happold 1978b). In the Ivory Coast *Eidolon* is considered a threat to introduced pine plantations because the bats gnaw the bark, wood, and leaves, causing death to the trees (Malagnoux and Gautun 1976). *Eidolon* may also eat and destroy dates to such a degree that protective measures are required. According to Mickleburgh, Hutson, and Racey (1992), the genus is generally still abundant, even roosting in the center of some large cities. However, certain populations, particularly in West Africa, could be seriously threatened by killing because they are considered pests in fruit orchards and because they are subject to both local and commercial hunting for use as food.

CHIROPTERA; PTEROPODIDAE; **Genus ROUSETTUS**
Gray, 1821

Rousette Fruit Bats

There are 9 species (Aggundey and Schlitter 1984; Baeten, Van Cakenberghe, and De Vree 1984; Bergmans 1977; Bergmans and Hill 1980; Chasen 1940; Corbet 1978; Corbet and Hill 1992; Ellerman and Morrison-Scott 1966; Flannery 1995; Ghose and Ghosal 1984; Hayman and Hill *in* Meester and Setzer 1977; Juste and Ibáñez 1993b; Kock 1978b; Koopman 1979, 1986, 1989a, and *in* Wilson and Reeder 1993; Laurie and Hill 1954; Lekagul and McNeely 1977; McKean 1972; Nader 1975; Rookmaaker and Bergmans 1981; Sung 1976):

R. aegyptiacus, Turkey and Cyprus to Pakistan, Arabian Peninsula, Egypt, most of Africa south of the Sahara, islands in Gulf of Guinea;

R. obliviosus, Grand Comoro and Anjouan islands between Africa and Madagascar;

R. madagascariensis, lowland forests throughout Madagascar;

R. amplexicaudatus, southern Burma, Thailand, Cambodia, Viet Nam, Malay Peninsula, Sumatra and nearby Mentawai and Enggano islands, Krakatau, Java, Bali, Borneo, throughout the Philippines, Sulawesi, Togian Islands, Lombok, Sumba, Flores, Timor, most other islands in the Lesser Sundas, Molucca Islands, New Guinea, Bismarck Archipelago, Solomon Islands;

R. spinalatus, northern Sumatra, northern Borneo;

R. leschenaulti, Pakistan, India, southern Tibet, Nepal, Sikkim, Bhutan, Burma, Thailand, Indochina, extreme southeastern China, Sri Lanka, Sumatra and nearby Simeulue Island, Java, Bali, Lombok;

R. lanosus, Sudan, northeastern Zaire, Kenya, Uganda, Rwanda, Tanzania;

R. celebensis, Sulawesi and the nearby Talaud, Togian, Sangihe, and Sulu Islands;

R. angolensis, Senegal to Kenya, and south to Angola and Zimbabwe.

Boneia (see account thereof) sometimes is considered a subgenus of *Rousettus.* The species *R. angolensis* often is placed in another separate genus or subgenus, *Lissonycteris* Andersen, 1912 (see Bergmans 1979c). The species *R. lanosus* often is placed in the subgenus *Stenonycteris* Andersen, 1912. Hill (1983a) suggested that *R. celebensis* also might belong in the latter subgenus. Koopman (*in* Wilson and Reeder 1993) accepted both *Lissonycteris* and *Stenonycteris* as subgenera. Corbet and Hill (1991, 1992) accepted the former but not the latter; they also noted that the name *aegyptiacus* should be spelled *egyptiacus*.

Head and body length is about 95–177 mm, tail length is 10–22 mm, and forearm length is 65–103 mm. Reported weights are about 81–171 grams for *R. aegyptiacus* (Jones 1971a), 54–75 grams for *R. amplexicaudatus,* and 45–106 grams for *R. leschenaulti* (Lekagul and McNeely 1977). Males are substantially larger than females. Certain members of this genus have glandular hairs on the foreneck and the sides of the neck; these are bright tawny olive in the adult males of some species. The upper parts are usually brownish, the underparts somewhat lighter. At least two species, *R. amplexicaudatus* and *R. leschenaulti,* have a pale collar of short hairs. Otherwise *Rousettus* tends to have a drab appearance.

Rousette bats occur in a variety of habitats from lowlands to mountains. They roost in ancient tombs and tem-

Rousette fruit bat (*Rousettus* sp.): A. Photo from New York Zoological Society; B. Photo from Lim Boo Liat.

ples, rock crevices, garden trees, and date plantations but are most common in caves. According to Lekagul and McNeely (1977), colonies of *R. leschenaulti* shift in response to the supply of fruit, and *R. amplexicaudatus* may make a foraging round trip of 40–50 km in a night. While flying in darkness the latter species utters a high-pitched buzzing call that is the basis of a simple echolocation mechanism (Medway 1978); *R. aegyptiacus* and probably *R. spinalatus* also use echolocation (Francis 1989; Herbert 1985). Some segments of colonies of *R. aegyptiacus* may make limited seasonal migrations (Jacobsen and Du Plessis 1976). The

diet consists of fruit juices and flower nectar. In the course of obtaining nectar from different flowers these bats carry pollen from one place to another. They have been noted at dusk flying around flowering acacia trees.

Lekagul and McNeely (1977) stated that roosting group size in *R. leschenaulti* varies from 2 or 3 to as many as 2,000 and that there is no sexual segregation. Jacobsen and Du Plessis (1976) described two roosting colonies of *R. aegyptiacus* in the eastern Transvaal each consisting of 7,000–9,000 individuals. The bats crowded closely together in caves, maintaining bodily contact. Fighting was common and was accompanied by loud screams and hacking coughs. The genus in general tends to be noisy and restless in the daytime.

Extensive data summarized by Mickleburgh, Hutson, and Racey (1992) indicate that rousette fruit bats have extended breeding seasons, sometimes with two annual peaks, which follow the rains. In Cyprus *R. aegyptiacus* has two cycles per year, though most females breed only in one. In Israel there are two main peaks, in early April and from late August to early September, but females may experience a postpartum estrus and thus give birth in both seasons; some first conceive at 7–8 months and others at 15–16 months. Most births of this species in Pakistan occur in early spring, the breeding peak is November–December in Liberia, and births occur from October to June in the Cape Province of South Africa. *R. aegyptiacus* appears to be a seasonal breeder in the Transvaal (Jacobsen and Du Plessis 1976). Mating there occurs primarily from June to September, gestation lasts about 4 months, and births take place from October to December. There is usually a single young, but occasionally twins are born. The females carry the young at first, then leave them at the roosts. By early March the young are weaned and able to fly on their own. According to Bergmans (1979c), pregnant females were taken in Congo in November and December; lactating females were collected there in February and March; and females give birth twice a year in Uganda, in March and September. Most births in the London Zoo have been in March, April, and May, with one young per birth the usual number and twins occurring about every fourth birth. Weight at birth is about 20 grams (Hayssen, Van Tienhoven, and Van Tienhoven 1993).

In *R. leschenaulti* in India as well there are two periods of birth, occurring about March and July–August. In an intensive study of a colony in Maharashtra State, Gopalakrishna and Choudhari (1977) found that pregnancies lasted from November to March and from March to July, with females undergoing a postpartum estrus shortly after the first birth. The gestation period was about 125 days, and there was a single young per female. Sexual maturity was attained at 5 months by females but not until 15 months by males. For *R. amplexicaudatus* in the Solomon Islands, Phillips (1968) reported that adult females were lactating in December–January but not in March–June. On Timor, however, R. E. Goodwin (1979) found most adult *R. amplexicaudatus* to be in breeding condition in March, April, and May. For the same species on Bougainville Island, McKean (1972) reported pregnant females in July and September and a lactating female in September. In the Philippines, females of *R. amplexicaudatus* probably give birth to one young in each of two periods separated by 4–5 months; gestation lasts 3.5–4.5 months, lactation 2.5–3.0 months, and females reach sexual maturity at 6–9 months (Mickleburgh, Hutson, and Racey 1992). *R. angolensis* in Congo may breed either continuously throughout the year or biannually in April–June and October–November (Bergmans 1979a). The same may be true for that species in Nigeria and Cameroon, while farther west in Africa there is an extended season starting in July, peaking in September, and tailing off by the dry season in March (Mickleburgh,

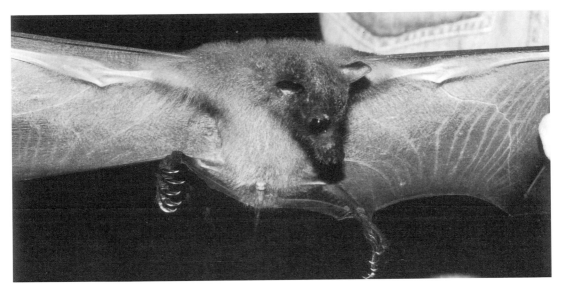

Boneia bidens, photo by Jan M. Haft.

Hutson, and Racey 1992). The known longevity record for the genus is 22 years and 11 months (Jones 1982).

Most species and populations of *Rousettus* are not thought to be threatened, though some are hunted for use as food or because they are thought to damage fruit crops. The subspecies *R. aegyptiacus arabicus* of southwestern Asia apparently has become rare (Mickleburgh, Hutson, and Racey 1992). The IUCN classifies *R. madagascariensis* and *R. spinalatus* as vulnerable and *R. obliviosus* as near threatened; each occupies a small area and is subject to adverse human activity.

CHIROPTERA; PTEROPODIDAE; Genus BONEIA
Jentink, 1879

The single species, *B. bidens*, is known only from northern Sulawesi between the towns of Gorontalo and Manado. The type locality is Boné (hence the generic name), a mountain range and river near Gorontalo. In their review of this species Bergmans and Rozendaal (1988) considered *Boneia* to be a synonym of *Rousettus*, but they noted that further study might result in retention of *Boneia* as a valid subgenus. It was indeed regarded as a subgenus by Corbet and Hill (1992), but it was maintained as a full genus by Koopman (*in* Wilson and Reeder 1993).

Head and body length is about 138 mm and tail length is about 25 mm. According to Bergmans and Rozendaal (1988), forearm length is 94.3–103.5 mm and weight is 150–94 grams. General coloration is dark brown on the back, russet on the rump, and dull brownish on the undersides. Males have a mantle, lateral tufts on the neck, and long hairs on the throat. The color of these features, which are absent in females, ranges from light golden buffy to much darker brown. *Boneia* closely resembles *Rousettus* but is distinguished by having a skull with a longer occipital region, a relatively broader palate, and premaxillae separated in front. The dental characters that originally were said to distinguish *Boneia* are now known to be present in some species of *Rousettus*.

Bergmans and Rozendaal (1988) reported that *B. bidens* is found in forests and areas where forest, bush, and farm habitats intermingle. Specimens have been taken at elevations of 200–1,060 meters. *Boneia*, like *Rousettus*, probably is capable of echolocation and, also like the latter, lives gregariously in caves. Two females obtained on 17 June and 17 July each contained a single small embryo. The IUCN classifies *B. bidens* (which it places in the genus *Rousettus*) as near threatened.

CHIROPTERA; PTEROPODIDAE; Genus MYONYCTERIS
Matschie, 1899

Little Collared Fruit Bats

There are three species (Bergmans 1976, 1980; Juste and Ibáñez 1994; Schlitter and McLaren 1981):

M. torquata, Sierra Leone to western Uganda, and south to northern Angola and northern Zambia, Bioko;
M. relicta, southeastern Kenya to central Tanzania;
M. brachycephala, Sao Tomé Island (Gulf of Guinea).

Koopman (1975) once thought that *Myonycteris* could best be treated as a subgenus of *Rousettus*, but subsequently he (1984a and *in* Wilson and Reeder 1993), as well as Corbet and Hill (1991), retained it as a full genus. Head and body length is about 85–165 mm, tail length is 4–13 mm, and forearm length is 55–70 mm. Weight is 27–54 grams (Bergmans 1976). Coloration of the upper parts of *Myonycteris* is light brown, tawny, or dark brown, sometimes with a reddish or yellowish appearance. Males have a tawny olive or dull yellow orange ruff of hairs on the foreneck. This patch of coarse hairs, which extends onto the chest, is associated with a cutaneous glandular formation. Such ruffs and patches of hair are found in many genera of the family Pteropodidae and are always associated with skin glands.

Juste and Ibáñez (1993a) reported that the species *M. brachycephala* consistently has a total of only three lower incisor teeth, whereas the other two species have a total of four, two on each side of the jaw. The asymmetric arrange-

Little collared fruit bat *(Myonycteris torquata)*, photo by Merlin D. Tuttle, Bat Conservation International.

ment in *M. brachycephala* is unique among those mammals that have a constant, heterodont dental formula.

These fruit bats occur in forest, woodland, and savannah. They apparently roost in bushes and low trees, often in direct sunlight. Nothing is known about the natural diet, but these bats have been caught near mangoes, guavas, and bananas, and captive specimens have taken soft fruits, honey, and butter (Bergmans 1976).

Studies by Thomas (1983) in the Ivory Coast indicate that *M. torquata* migrates from forest to savannah areas as the rainy season progresses in April. At this time the females are pregnant, and they return to the forests later in the wet season to give birth (about August–September). A second peak in births may occur in February–March. Most males do not return to the forests but follow the rains into the northern part of the country and do not move south again until about October. In Liberia, Wolton et al. (1982) found the same species to have an extended breeding season, beginning in July and continuing well into the dry season, at least until March, with most females giving birth in September, at the peak of the rains.

A study of captives from Gabon found males to be sexually active all year and females to be polyestrous; births occurred twice a year, in December–January and June. Wild females taken in Gabon were pregnant or lactating from November to March and sexually inactive in June and July. Lactating females also were collected in Cameroon in February and in the Central African Republic in May

(Bergmans 1976). Three females taken in Equatorial Guinea in October had one embryo each (Jones 1971*a*). A female taken in eastern Congo in September had a half-grown young (Kingdon 1974*a*). Those and other records indicate that there usually is a single young, but twins have been reported (Hayssen, Van Tienhoven, and Van Tienhoven 1993).

The IUCN now classifies *M. relicta* and *M. brachycephala* as vulnerable. Both have very restricted ranges in areas that are experiencing rapid destruction of forest habitat. *M. torquata* is confronted by the same problem and has become rare in some areas but is still widely distributed (Mickleburgh, Hutson, and Racey 1992).

CHIROPTERA; PTEROPODIDAE; **Genus PTEROPUS**
Erxleben, 1777

Flying Foxes

There are 17 species groups and 60 currently recognized species (Andersen 1912; Bergmans 1990; Bergmans and Rozendaal 1988; Chasen 1940; Cheke and Dahl 1981; Corbet and Hill 1992; Daniel 1975; Ellerman and Morrison-Scott 1966; Felten 1964; Felten and Kock 1972; Flannery 1995; Hall 1987*a*; Hayman and Hill *in* Meester and Setzer 1977; Heaney and Rabor 1982; Hill 1958, 1971*b*, 1979, 1983*a*; Hill and Beckon 1978; Kitchener and Maryanto 1995; Kitchener, Packer, and Maharadatunkamsi 1995; Klingener and Creighton 1984; Koopman 1984*c* and *in* Wilson and Reeder 1993; Koopman and Steadman 1995; Laurie and Hill 1954; Lawrence 1939; Lekagul and McNeely 1977; Musser, Koopman, and Califia 1982; Phillips 1968; Ride 1970; Roberts 1977; Sanborn 1931, 1950; Sanborn and Nicholson 1950; Tate 1934; Taylor 1934; Van Deusen 1969; Wodzicki and Felten 1975, 1981; Yoshiyuki 1989):

hypomelanus group

P. hypomelanus, Maldive Islands (Indian Ocean), coastal areas and numerous small islands from southern Burma and southern Viet Nam through Malaysia and Indonesia to New Guinea and the Solomons, and throughout the Philippines;
P. brunneus, Percy Island off northeastern Queensland;
P. faunulus, Nicobar Islands (Bay of Bengal);
P. speciosus, Mindanao and Basilan islands (Philippines), Sulu Archipelago and Talaud islands to south, Besar and Mata Siri (Java Sea);
P. griseus, Sulawesi and many nearby islands, Banda Islands in the southern Moluccas, Timor and nearby islands;
P. pumilus, Philippines and Talaud Islands to south;
P. admiralitatum, Admiralty Islands (north of New Guinea), Solomon Islands, Bismarck Archipelago;
P. howensis, known only from Ontong Java Atoll in the northern Solomons;
P. ornatus, New Caledonia, Loyalty Islands (southwestern Pacific);
P. dasymallus, southern Kyushu, Ryukyu Islands, Daito Islands, Taiwan;
P. subniger, Reunion and Mauritius islands (Indian Ocean);

mariannus group

P. mariannus, Okinawa, Ryukyu Islands, Guam, Northern Mariana Islands, Palau Islands, Yap, Ulithi Islands, Kosrae Island (western Pacific);

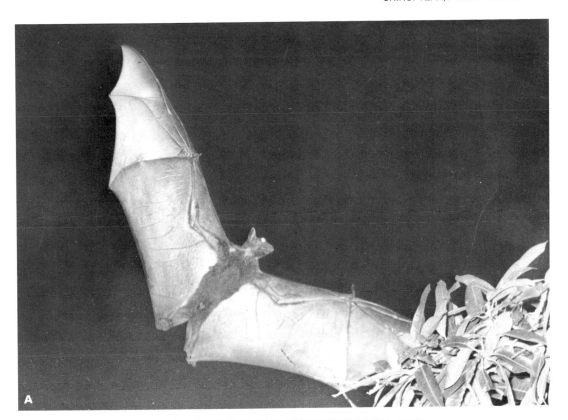

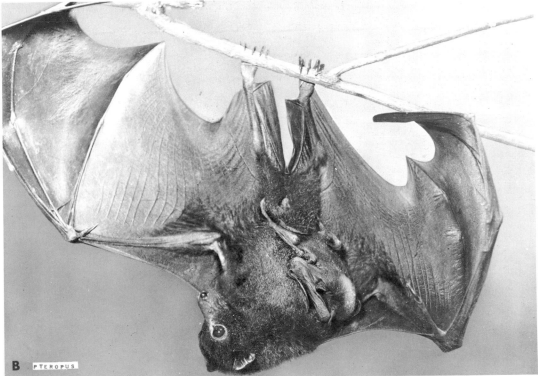

A. *Pteropus alecto*, photo by John Warham. B. *P. vampyrus*, photo by Bernhard Grzimek.

P. tonganus, Karkar Island off northeastern New Guinea, Rennell Island in the Solomons, Vanuatu (New Hebrides), New Caledonia, Fiji Islands, Uvea Island, Tonga Islands, Samoa, Niue Island, Cook Islands, many other islands of the South Pacific;

caniceps group

P. caniceps, Halmahera and Sula islands in the Moluccas and doubtfully nearby Sulawesi;
P. argentatus, known only by a single specimen supposedly from Amboina Island in the Moluccas;

rufus group

P. rufus, Madagascar;
P. seychellensis, Comoro Islands between Africa and Madagascar, Mafia Island off east coast of Tanzania, Seychelles Islands (Indian Ocean);
P. aldabrensis, Aldabra Islands (Indian Ocean);
P. voeltzkowi, Pemba Island off northeast coast of Tanzania;
P. niger, Reunion and Mauritius islands (Indian Ocean), known from subfossil remains on nearby Rodriguez Island;

melanotus group

P. melanotus, Andaman Islands, Nicobar Islands, Nias and Enggano islands off western Sumatra, Christmas Island (south of Java);

melanopogon group

P. melanopogon, Molucca and Aru islands;
P. livingstonii, Johanna and Moheli islands in the Comoros;

rayneri group

P. rayneri, Solomon Islands;
P. cognatus, San Cristobal Island in the southern Solomons;
P. rennelli, Rennell Island in the southern Solomons;
P. chrysoproctus, Amboina, Buru, Seram, and several other islands between Sulawesi and New Guinea;
P. fundatus, Banks Islands in northern Vanuatu (New Hebrides);

lombocensis group

P. lombocensis, Lombok, Sumbawa, Komodo, Flores, Lembata, Pantar, Alor, and Timor islands in the Lesser Sundas;
P. rodricensis, Rodriguez Island and formerly Mauritius and nearby Round Island (Indian Ocean);
P. molossinus, Caroline Islands (western Pacific);

samoensis group

P. samoensis, Samoa and Fiji islands (South Pacific), also skeletal remains from 'Eua in The Tonga Islands dating from less than 3,000 years ago;
P. anetianus, Vanuatu (New Hebrides) Islands (southwestern Pacific);

pselaphon group

P. vetulus, New Caledonia (southwestern Pacific);
P. insularis, Chuuk (Truk) Islands in the Carolines (western Pacific);

P. phaeocephalus, Mortlock Island in the Carolines (western Pacific);
P. tokudae, Guam (western Pacific);
P. pselaphon, Bonin and Volcano islands south of Japan;
P. pilosus, Palau Islands (western Pacific);
P. tuberculatus, Vanikoro Island (South Pacific);
P. nitendiensis, Santa Cruz Islands in the eastern Solomons;
P. leucopterus, Luzon, Dinagat, and Catanduanes islands (Philippines);

temmincki group

P. temmincki, Buru, Amboina, and Seram islands in the Moluccas, doubtfully Timor;
P. capistratus, Bismarck Archipelago;
P. personatus, Halmahera Islands in the northern Moluccas;

vampyrus group

P. giganteus, Pakistan, India, Nepal, Sikkim, Bhutan, Burma, Sri Lanka, Maldive Islands (Indian Ocean), Andaman Islands (Bay of Bengal), one questionable record from Qinghai Province in western China;
P. intermedius, southeastern Burma, western Thailand;
P. vampyrus, southern Thailand and Indochina, Tenasserim, Malaysia, Indonesia, Philippines, Timor, possibly Andaman and Nicobar islands in the Bay of Bengal;
P. lylei, southern Thailand, Cambodia, Viet Nam;

alecto group

P. alecto, Sulawesi, Salayer Island, Bawean and Kangean islands in the Java Sea, Lombok, Sumba and Savu islands, southern New Guinea, northern and eastern Australia;

conspicillatus group

P. conspicillatus, Halmahera Islands, New Guinea and nearby islands, northeastern Queensland;
P. ocularis, Seram and Buru islands in the southern Moluccas;

neohibernicus group

P. neohibernicus, New Guinea and some nearby islands, Admiralty Islands, Bismarck Archipelago;

macrotis group

P. macrotis, New Guinea, Aru Islands;
P. poliocephalus, southeastern Queensland, New South Wales, Victoria;
P. pohlei, Japen Island off northwestern New Guinea;

scapulatus group

P. scapulatus, extreme southern New Guinea, northern and eastern Australia, rarely on Tasmania, one record from New Zealand;
P. woodfordi, Solomon Islands;
P. mahaganus, Santa Ysabel and Bougainville (Solomons);
P. gilliardorum (formerly *gilliardi*), New Britain Island (Bismarck Archipelago).

Ride (1970) considered *P. brunneus* a vagrant *P. hypomelanus,* but Koopman (1984c) reluctantly continued to treat it as a full species, and most subsequent authorities (e.g., Corbet and Hill 1991; Mahoney and Walton *in* Bannister et al. 1988; Mickleburgh, Hutson, and Racey 1992) have followed suit. Koopman (*in* Wilson and Reeder 1993) listed *P. mearnsi* of Mindanao and Basilan islands as a separate species but noted that it probably is a synonym of *P. speciosus.* Corbet and Hill (1992), however, treated it as a synonym of *P. hypomelanus* and considered *P. speciosus* a synonym of *P. griseus. P. speciosus* was regarded as a full species by Flannery (1995), Koopman (*in* Wilson and Reeder 1993), and Mickleburgh, Hutson, and Racey (1992). *P. pelewensis* of the Palau Islands (western Carolines), *P. yapensis* of Yap (western Carolines), and *P. ualanus* of Kosrae Island (far eastern Carolines) were maintained as distinct species by Flannery (1995), and *P. loochoensis* of Okinawa was treated as a distinct species by Yoshiyuki (1989), but all four were included in *P. mariannus* by Corbet and Hill (1991), Koopman (*in* Wilson and Reeder 1993), Mickleburgh, Hutson, and Racey (1992), and most other recent authorities. *P. vanikorensis,* supposedly from Vanikoro Island in the South Pacific, also sometimes has been considered distinct from *P. mariannus* but probably is based on an erroneous record. Both *P. cognatus* and *P. rennelli* usually were included within *P. rayneri,* and *P. capistratus* within *P. temminicki,* until all were distinguished as full species by Flannery (1995). *P. sanctacrucis* of the Santa Cruz Islands was listed as a species distinct from *P. nitendiensis* by Koopman (*in* Wilson and Reeder 1993) and Mickleburgh, Hutson, and Racey (1992) but not by Corbet and Hill (1991) or Flannery (1995). Corbet and Hill's (1992) recognition of *P. intermedius* as a species distinct from *P. vampyrus* was not followed by Koopman (*in* Wilson and Reeder 1993).

Head and body length is 170–406 mm, forearm length is 85–228 mm, and wingspan is 610–1,700 mm; there is no tail. The largest species is *P. vampyrus.* Some reported weights are: *P. personatus,* 104–53 grams; *P. hypomelanus,* 300–455 grams; *P. rayneri,* 800–870 grams (Flannery 1995); *P. mahaganus,* 1,220–50 grams (McKean 1972); male *P. giganteus,* 1,300–1,600 grams; and female *P. giganteus,* 900 grams (Roberts 1977). The coloration is grayish brown or black with the area between the shoulders often yellow or grayish yellow. These mammals have a very noticeable characteristic odor.

Flying foxes inhabit forests and swamps, often on small islands near coasts. Indra Kumar Sharma (pers. comm.) stated that having a large body of water nearby is essential for *P. giganteus.* Significant numbers of *P. mariannus* have been found to move between islands in the southern Marianas on an irregular basis (Wiles and Glass 1990). Data compiled by Pierson and Rainey (1992) indicate that most species roost in emergent trees that rise above the forest canopy and that foraging areas are always well removed from the roosts. Colonies may use the same roosting site year after year, sometimes for many decades. Populations on large land masses may travel 40–60 km to reach a feeding area, and colonies on offshore islands often travel across water to forage on the mainland. Several species and subspecies of oceanic islands have become partially or entirely diurnal, but most of these bats are nocturnal. During daylight there is much noise and motion in the roosts, and individuals sometimes fly from one place to another. At dusk the bats fly to fruit trees to feed. These bats, particularly the larger species, are strong fliers. Sterndale (1884) mentioned a *P. giganteus* that landed on a boat more than 200 miles from land, exhausted but alive, probably having been blown offshore by high winds. Flying

Fruit bat *(Pteropus dasymallus),* photo by Gwilym S. Jones.

foxes eat, rest, and digest their food for several hours while at their feeding trees; then they return to the roosting site. In a radio-tracking study of *P. giganteus* Walton and Trowbridge (1983) found a female to leave her roost at 1800 hours, arrive at her main feeding area after 2000 hours, having stopped several times along the 12- to 14-km route, and then return directly to the roost at about 0400 hours, usually taking less than an hour. The principal food of *Pteropus* is fruit juices, which the bats obtain by squeezing pieces of the fruit pulp in their mouths. They swallow the juice and spit out the pulp and seeds. If the pulp is very soft, like banana, they swallow some of it. They also chew eucalyptus and probably other flowers to obtain the juices and pollen. They drink while flying to and from their feeding and sleeping locations. Some drink sea water, apparently to obtain mineral salts lacking in the plant food. *Pteropus* maintains a body temperature of 33°–37° C; during cool weather this temperature range is maintained by constant activity. Population density of *P. mariannus* on various islands in the Marianas was found to vary from less than 1 to nearly 100 bats per ha. (Wiles, Lemke, and Payne 1989).

Rainey and Pierson (1992) reported that most species that have been studied are moderately to strongly colonial, roosting in groups of a few dozen to a few hundred thousand individuals; such groups once may have numbered in the millions of animals. A few species, such as *P. samoensis,* commonly roost in male-female pairs or small groups. In some populations, such as that of *P. tonganus* in Vanuatu, large mixed-sex colonies are present for part of the year, but pregnant females depart to form their own nursery groups. Within the colonies of some species each dominant male may attempt to control a "harem" of about 10 females for purposes of mating, and subordinate males roost together separately. According to Ride (1970), Australian flying foxes congregate in great "camps" during the summer, when blossoms are abundant. Some of these

Flying fox *(Pteropus personatus)*, photo by F. G. Rozendaal.

roosts contain nearly 250,000 individuals at densities of 4,000–8,000/ha. The young are born shortly after the colonies form and remain with the females for three to four months. Mating takes place in the camps, and soon thereafter a nongregarious phase, with segregation of the sexes, seems to start. This phase culminates in a dispersed population that persists throughout the winter. Prociv (1983) reported that a camp of 100,000 *P. scapulatus* was found near Brisbane in December, with most females associated with a male in harem groups and some males forming their own groups. Mating occurred at this time, and births took place in April. Data compiled by Mickleburgh, Hutson, and Racey (1992) indicate that *P. scapulatus* forms camps of up to 1 million individuals in November or December; pairs establish territories within the roosting areas. After mating, females leave and form smaller groups of their own, within which the young are born. When the large camps again come together in the following year the juveniles congregate in groups separate from the adults.

Lekagul and McNeely (1977) reported that *P. lylei* also formed very large colonies and that groups of *P. vampyrus* often exceeded 100 individuals. Camps of *P. vampyrus* in the Philippines once contained up to 100,000 bats (Mickleburgh, Hutson, and Racey 1992). On Niue Island, in the South Pacific, *P. tonganus* has been reported to roost singly, in pairs, or in larger groups, rarely up to 100 individuals in size (Wodzicki and Felten 1975). Neuweiler (1969) studied a colony of 800–1,000 *P. giganteus* in India and found a vertical rank order among males, each having a particular roosting place in the tree. Mating in the colony occurred from July to October and births took place mostly in March. The young were carried by the females for the first few weeks of life but subsequently were left in the tree. When the young males became independent, they separated from the colony and gathered on a neighboring tree. Gould (1977) observed individual *P. vampyrus* to defend an entire tree while feeding.

Data compiled by Pierson and Rainey (1992) indicate little variation in the reproductive pattern among the many species of *Pteropus*. Births seem to be highly synchronized and seasonal for most species, generally occurring during the first half of the year at northern latitudes and during the second half of the year at southern latitudes, though there are exceptions. A few species may have more than one birth peak per year, but only one, *P. mariannus,* seems to be truly aseasonal, with young being found in every month. Captive female *P. rodricensis* are capable of producing one young every nine months, but it is likely that wild females

of most species give birth only once annually. There usually is a single young; that of *P. poliocephalus* weighs 46–92 grams at birth (Hayssen, Van Tienhoven, and Van Tienhoven 1993).

Data for many species indicate that, except in *P. mariannus* and perhaps certain populations of *P. pumilus* and *P. tonganus,* there usually is a regular period of about 1–3 months in any given area during which all or most births take place (Mickleburgh, Hutson, and Racey 1992; O'Brien 1993). This period commonly occurs around February to May in *P. giganteus* and *P. vampyrus* on the mainland of southern Asia, on the Maldive Islands in the northern Indian Ocean, and in the Philippines. It occurs from around August to November in *P. niger* and *P. seychellensis* of the southern Indian Ocean, *P. alecto* of Australia, and *P. ornatus* of New Caledonia. *P. poliocephalus* of Australia has a mating peak in March and April and a birth peak from September to October. In contrast, *P. scapulatus* of the same region gives birth mainly in April or May, during the austral autumn, when this nomadic species must be highly mobile in conjunction with the flowering of its food plants. Generally in *Pteropus,* gestation lasts about 140–92 days depending on species, and lactation lasts 3–6 months. The young may be carried by the mother for about the first 3–6 weeks of life. Sexual maturity usually is attained at 18–24 months, but at only 6 months in *P. melanotus.* One captive female *P. rodricensis* was observed to physically assist another during parturition by such means as supporting her in a suitable posture, grooming her and the newborn, and maneuvering the pup into a suckling position (Kunz, Allgaier, et al. 1994). The known longevity record for the genus is held by a *P. giganteus* that lived for 31 years and 5 months in captivity (Jones 1982).

Flying foxes have been found to play an important role, especially on Pacific islands, in pollination and seed dispersal for a great variety of plants that are useful to people for lumber, food, medicine, and other products (Cox et al. 1991, 1992; Fujita and Tuttle 1991; Wiles and Fujita 1992). In many areas, however, *Pteropus* is considered a serious pest by fruit growers. Colonies of *P. lylei* in Thailand are said sometimes to devastate a crop overnight (Lekagul and McNeely 1977). In Australia large numbers of flying foxes have been poisoned in an effort to protect orchards and gardens (Ride 1970). Raids on bananas there are thought to have increased in recent years, as natural food supplies have disappeared through land clearing (Tidemann 1987a). Information summarized by Mickleburgh, Hutson, and Racey (1992) indicates that all or some of the four wide-ranging Australian species—*P. alecto, P. conspicillatus, P. poliocephalus,* and *P. scapulatus*—are legally unprotected in the various states and subject to large-scale shooting and poisoning. *P. hypomelanus* and *P. giganteus* also are regarded as pests by fruit growers in the Maldive Islands and undergo intensive and officially sanctioned killing. *P. giganteus* is hunted as a pest in some parts of India but in certain areas is considered sacred and is protected. Rainey and Pierson (1992) discussed a number of other threats to *Pteropus,* including habitat destruction, hunting for use as food or medicines, commercial trade, and predation by snakes introduced on islands. As a consequence of numerical reduction by such human-caused factors, remnant flying fox populations become increasingly exposed to natural dangers, particularly cyclones and epidemics. Certain species, such as *P. melanotus* and *P. samoensis,* are especially vulnerable to hunting because they seem not to have developed a fear of humans and tend to be diurnal (Tidemann 1987b). The slow reproductive rate of *Pteropus* reduces prospects for recovery, especially for small populations on isolated islands (Robertson 1992).

All of these problems have made flying foxes the object of unusual conservation concern (see extensive reports in Mickleburgh, Hutson, and Racey 1992; and Wilson and Graham 1992). Indeed, the IUCN now classifies 33 species of *Pteropus*, more than any other mammalian genus except *Crocidura, Hipposideros,* and *Myotis*. The IUCN classifications of *Pteropus* are: *P. brunneus, P. subniger, P. loochoensis* (recognized as a species distinct from *P. mariannus;* see above), *P. tokudae,* and *P. pilosus,* extinct; *P. voeltzkowi, P. livingstonii, P. rodricensis, P. molossinus, P. insularis,* and *P. phaeocephalus,* critically endangered; *P. dasymallus, P. leucopterus,* and *P. mariannus* (including *P. pelewensis, P. yapensis,* and *P. ualanus;* see above), endangered; *P. speciosus, P. pumilus, P. howensis, P. ornatus, P. niger, P. fundatus, P. samoensis, P. faunulus, P. pselaphon, P. tuberculatus,* and *P. nitendiensis, P. sanctacrucis* (recognized as a species distinct from *P. nitendiensis;* see above), *P. ocularis, P. pohlei, P. mahaganus,* and *P. gilliardorum* (under the name *P. gilliardi;* see above), vulnerable; and *P. chrysoproctus, P. vetulus,* and *P. temmincki,* near threatened. The following species are on appendix 1 of the CITES: *P. insularis, P. mariannus* (including *P. pelewensis, P. yapensis,* and *P. ualanus*), *P. molossinus, P. phaeocephalus, P. pilosus, P. samoensis,* and *P. tonganus;* all other species of *Pteropus* are on appendix 2. The USDI lists only *P. rodricensis, P. tokudae,* and the Guam population of *P. mariannus* as endangered. Ironically, the USDI does list all species of *Pteropus* as "injurious," thereby prohibiting the importation of live specimens.

The IUCN classifications tell only part of the story, as many additional subspecies and populations have suffered. Roberts (1977) noted that *P. giganteus giganteus* was much rarer in Pakistan than it had been 10–20 years earlier because orchard owners had become less tolerant and there was increasing disruption of colonies as people sought to kill bats for their fat, thought to be a cure for rheumatism. Mickleburgh, Hutson, and Racey (1992) indicated that the subspecies *P. giganteus ariel* and *P. hypomelanus maris* of the Maldive Islands are both endangered by deliberate and officially sanctioned killing, supposedly to protect fruit crops. *P. melanotus natalis* of Christmas Island in the Indian Ocean may still number up to 10,000 individuals, but accidental introduction of the rat snake *(Lycodon aulicus)* in 1987 could eventually result in problems as devastating as those caused by the brown tree snake *(Boiga)* in the Pacific (see below). *P. poliocephalus* of Australia has declined both because of clearing of native habitat for agricultural purposes and control to protect the crops that are subsequently planted. Heaney and Heideman (1987) noted that aggregations of about 150,000 *P. vampyrus lanensis* were common in the Philippines in the 1920s but that the largest groups now seen contain no more than a few hundred individuals and that the subspecies could be extinct in 20 years; the apparent reasons for the decline are excessive hunting for meat and deforestation.

According to Mickleburgh, Hutson, and Racey (1992), of the species that do have IUCN classifications those found on island groups around the far western rim of the Pacific—*P. dasymallus, P. gilliardorum, P. leucopterus, P. mahaganus, P. nitendiensis, P. pselaphon, P. pumilus, P. speciosus, P. tuberculatus,* and *P. vetulus*—are generally threatened by a combination of human destruction of restricted forest habitat and local hunting for food. These factors recently led to the total disappearance from the wild of the subspecies *P. dasymallus formosus* of Taiwan; a captive group of 8–10 individuals may not be viable *(Bats* 38 [July 1995]: 5). The same problems also confront the species inhabiting more remote islands to the east, though a more notable threat across much of the Pacific is a vast commercial trade that has developed in the last few decades. On Guam and some of the islands of the Northern Marianas fruit bats are considered delicacies, especially *P. mariannus* and related species; this market has led to the importation of great numbers of bats killed in the Marianas and on many other islands. Data compiled by Wiles (1992) show that 220,899 fruit bats are known to have been imported to Guam from 1975 to 1989. Commerce peaked in the late 1970s, when approximately 20,000–29,000 bats were imported annually. At that time most of the bats came from the Mariana and Caroline islands. In the mid-1980s many of the imports were coming from as as far away as New Guinea, the Philippines, and Samoa, but since 1987 the Carolines have again been the main source. In 1989 the retail price of a fruit bat on Guam was U.S. $12–$18.

The severity of this trade led to the placing of nine species of *Pteropus* on appendix 2 of the CITES in 1987. Nonetheless, Bräutigam and Elmqvist (1990) pointed out that there was a failure to implement the mandated controls, and no fewer than 30,000 frozen flying fox carcasses were imported into Guam over the next 30 months. Finally, in 1989, the seven Pacific species thought to be most threatened *(P. mariannus, P. tonganus, P. molossinus, P. samoensis, P. insularis, P. phaeocephalus,* and *P. pilosus)* were placed on appendix 1 of the CITES, and all other species of *Pteropus* were put on appendix 2. According to Sheeline (1993), however, large-scale trade has continued either illegally or because of inadequate restrictions. There was a declared importation of nearly 21,000 fruit bats from 1990 to 1992, mostly from Palau and only into the Commonwealth of the Northern Marianas, indicating a market at least as great as on Guam. Prices recently have reached U.S. $25–$40 per bat.

Flying fox *(Pteropus woodfordi)*, photo by Jan M. Haft.

Flying fox *(Pteropus tonganus)*, photo by Jan M. Haft.

Until quite recently this commerce could not be regulated by CITES provisions as it was occurring technically within areas entirely under the jurisdiction of the United States; imports from Palau averaged 13,000 bats annually from 1990 to 1993. In October 1994 Palau became independent, and now there are questions both about its ability to control exploitation of fruit bats and about the ability of the USDI to enforce CITES restrictions on import to Guam and the Northern Marianas. The market demand is so great in the latter two areas that there is potential for significant illegal activity and for disagreements between federal authorities and local interests (Wiles 1994). The political ramifications could be a factor in the apparent reluctance of the USDI to apply formal endangered or threatened classifications to fruit bat populations outside of Guam.

The problems confronting *P. mariannus* extend to a number of other Pacific flying foxes. *P. tokudae,* a small species of Guam and the Northern Marianas, has not been definitely recorded since 1968, when an individual was killed by hunters. *P. pilosus,* a large species not seen since the 1870s, apparently was wiped out in an earlier phase of native hunting in Palau. *P. insularis, P. molossinus,* and *P. phaeocephalus,* all found in the Caroline Islands within the Federated States of Micronesia, have become endangered because of commercial hunting for export to Guam and conversion of natural forest habitat for agriculture. The subspecies *P. tonganus tonganus,* though not classified by the IUCN or the USDI, was regarded as indeterminate by Mickleburgh, Hutson, and Racey (1992). It still occurs over a vast stretch of the South Pacific but is declining through destruction of forests, cyclones, and hunting, especially since shotguns have become available. Numbers in American Samoa were estimated at 140,000 in 1975 but now are thought to be barely one-tenth of that. Similar drastic reductions have occurred in Western Samoa, Tonga, Fiji, Niue, the Cook Islands, and other parts of the range.

Another South Pacific species, *P. samoensis,* has a much more restricted distribution. The subspecies *P. s. nawaiensis* of the Fiji Islands was designated indeterminate in Mickleburgh, Hutson, and Racey (1992), but *P. s. samoensis* of American and Western Samoa was regarded as endangered. The latter is jeopardized by both local exploitation for food and commercial hunting for export to Guam and the Northern Marianas, being especially vulnerable because it is a relatively large, slow animal and primarily diurnal. It also is losing forest habitat to logging and agriculture, the surviving bats being subject to devastation by periodic typhoons. In 1984 the USDI was petitioned to officially classify *P. samoensis* as endangered but eventually announced that such a measure was unwarranted. This finding seems surprising in light of the many problems confronting the species and recent surveys (Craig, Trail, and Morrell 1994) indicating that declines are continuing, with only 200–400 individuals left on Tutuila, the largest island in American Samoa and the one that formerly supported the highest bat population.

There also are serious conservation problems involving the species restricted to small islands in the southern Indian Ocean. *P. voeltzkowi,* which is known to occur only in a few colonies on Pemba Island off East Africa, has declined severely because of hunting and deforestation (Mickleburgh, Hutson, and Racey 1992). *P. niger* and *P. rodricensis* have been seriously affected by the clearing of forests and two resultant factors: easier access by hunters and loss of buffering protection against cyclonic winds (Cheke and Dahl 1981; Hamilton-Smith 1979). *P. niger* has disappeared from Reunion Island, and its population on Mauritius has been reduced to not more than 3,000 individuals. *P. rodricensis* may once have occurred on Mauritius and nearby

The population of *P. mariannus* on Guam declined as a result not only of market hunting but also of habitat modification by people and predation by the brown tree snake *(Boiga irregularis).* This large reptile, accidentally introduced just after World War II, is now so prevalent that practically no young survive to enter the population. Overall numbers on Guam fell from around 3,000 in 1957 to 1,000 by 1972 and to fewer than 50 in 1978 but subsequently increased, probably through migration from the nearby island of Rota. Numbers have since fluctuated at around 500–800. The population on Rota had about 2,000–2,500 bats in 1988 but was since reduced by poaching to 1,000–1,400. Numbers also fell drastically on Saipan, from thousands in the early 1970s to only 75–100 in the late 1980s, and on most other islands in the Northern Marianas and Carolines. Although the IUCN classifies *P. mariannus* only as vulnerable, except for the probably extinct subspecies *loochoensis* of the Ryukyus, Mickleburgh, Hutson, and Racey (1992) referred to all the other subspecies—*mariannus* (Guam, Rota, Tinian, Saipan, Aguijan), *paganensis* (Almagan, Pagan), *pelewensis* (Palau Islands), *ualanus* (Kosrae Island), *ulthiensis* (Ulithi Islands), and *yapensis* (Yap), together with unnamed populations in the Northern Marianas—as endangered because of excessive hunting.

The populations in the Palau Islands are of particular concern as they have been the source of about half of all fruit bats imported to Guam and the Northern Marianas.

Round Island but disappeared there shortly after settlement. There may have been over 1,000 on Rodriguez Island in 1955, but by 1974, because of continued deforestation and cyclones, only about 75 were left. Subsequent conservation efforts allowed the population to again reach 1,000 in 1990, but a cyclone led to another decline; in 1991 there were about 350 in the wild and 250 in nine captive groups (Caroll 1984; Mickleburgh and Carroll 1994; Mickleburgh, Hutson, and Racey 1992). *P. subniger*, also a resident of Reunion and Mauritius, became extinct in the mid–nineteenth century, when its last habitat, in the high forests, was destroyed; *P. livingstonii*, once abundant in the Comoros, has been eliminated by deforestation, except, perhaps, for 120 individuals on Johanna (Anjouan) Island, about 20 on Moheli, and a dozen in captivity (Cheke and Dahl 1981; Mickleburgh and Carroll 1994; Mickleburgh, Hutson, and Racey 1992; Reason and Trewhella 1994). Yet another Indian Ocean species, *P. seychellensis*, still occurs in several island populations that contain thousands of individuals each, but they are jeopardized by agricultural usurpation of their habitat, market hunting for tourist restaurants, and collision with and electrocution by electrical and telephone wires (Cheke and Dahl 1981; Nicoll and Racey 1981; Verschuren 1985). The subspecies *P. seychellensis aldabrensis*, of Aldabra Atoll, has been reduced to only a few hundred individuals.

CHIROPTERA; PTEROPODIDAE; Genus ACERODON
Jourdan, 1837

There are six species (Corbet and Hill 1992; Flannery 1995; Laurie and Hill 1954; Musser, Koopman, and Califia 1982; Taylor 1934):

A. celebensis, Sulawesi, Salayer Islands to northeast, Mangoli and Sanana islands to east;

A. mackloti, Lombok, Sumbawa, Flores, Sumba, Timor, and Alor islands in the Lesser Sundas;

A. leucotis, Palawan, Busuanga, and Balabac islands (Philippines);

A. humilis, Talaud Islands between Sulawesi and Philippines;

A. jubatus, Philippines;

A. lucifer, Panay Island (Philippines).

Head and body length is 178–290 mm and forearm length is 125–203 mm; there is no tail. The wingspan of the largest species, *A. jubatus*, is 1.51–1.7 meters. Rickart et al. (1993) listed weights of 720–1,140 grams for that species. R. E. Goodwin (1979) reported that eight specimens of *A. mackloti* weighed 450–565 grams. Coloration is variable. The forehead and sides of the head are often dark brown or black, the nape may be orange or golden yellow, the shoulders are usually reddish brown or chestnut, the lower back is usually darker than the area between the shoulders, and the undersides are usually dark brown or black. This genus is externally indistinguishable from *Pteropus* but differs in dental features.

In the Philippines, *Acerodon* has been found roosting in clumps of bamboo, hardwood trees, and swampy forest, usually on small offshore islands. Taylor (1934) mentioned a colony of *A. jubatus* and *Pteropus vampyrus* that covered an area of 8–10 ha. and contained some 150,000 bats. These bats flew 9–16 km into the mountains in the evening to feed on wild fruit.

On Timor, R. E. Goodwin (1979) located two colonies of *A. mackloti*, each spread out over the crowns of large fig trees on the edge of open forest. The staple food in this area was at least two species of fig. As the sun was setting the bats began to leave their roosts to make rounds of the fruit trees. They usually traveled in small, well-spaced groups of two to six individuals. If more than one bat landed in the same food tree, there would be noisy squabbling until a dominant individual drove the others off. R. E. Goodwin collected pregnant females, each with a single embryo, on Timor on 19 March and 5 May. On Lombok Island Kitchener et al. (1990) collected female *A. mackloti* in October that were lactating and also in an early stage of pregnancy, which suggests that the species is at least seasonally polyestrous. Data gathered by Rickart et al. (1993) indicate that female *A. jubatus* in the Philippines attain sexual maturity at 2 years and give birth to a single young in May and June.

Acerodon jubatus, photo by Paul D. Heideman.

Acerodon celebensis, photo by F. G. Rozendaal.

Neopteryx frosti, photo by F. G. Rozendaal.

The species *A. lucifer* is considered extinct by the IUCN. According to Heaney and Heideman (1987) and Utzurrum (1992), no specimens have been taken since the original series was collected on Panay in 1888 and 1893. The forests on which the species depended for roosting sites and food have been largely destroyed by human activity. *A. jubatus* is classified as endangered by the IUCN; according to Mickleburgh, Hutson, and Racey (1992), it has declined drastically throughout the Philippines and probably faces imminent extinction. Colonies that comprised many thousands of individuals earlier in the century have mostly disappeared or have only a few hundred bats. The main problems are destruction of forest habitat and intensive subsistence and commercial hunting such as described in the above account of *Pteropus*. For similar reasons the IUCN also designates *A. humilus* and *A. leucotis* as vulnerable and *A. celebensis* as near threatened. Both *A. lucifer* and *A. jubatus* are on appendix 1 of the CITES; all other species of *Acerodon* are on appendix 2.

CHIROPTERA; PTEROPODIDAE; Genus NEOPTERYX
Hayman, 1946

The single species, *N. frostii*, now is known by one specimen collected by W. J. C. Frost in 1938 or 1939 at Tamalanti in western Sulawesi, as well as by three others obtained in 1985 in the northeastern part of the island (Bergmans and Rozendaal 1988).

Head and body length of the type specimen is 105 mm, and there is no tail. According to Bergmans and Rozendaal (1988), forearm length was 104.9–110.6 mm in three adult specimens and weight was 190 grams in an adult male and 250 grams in a pregnant female. The thick, short fur is generally tawny or brownish in color. There is a paler, woolly mantle. The muzzle is sepia with creamy white stripes on top and on each side.

The wings are attached near the midline of the back, giving the impression that the back is without fur, as the fur on the back is hidden by the naked wing membranes. The thumb has a well-developed claw, but there is no claw on the index finger. The only other fruit bats of the family Pteropodidae that lack the claw and tail are ced. Hayman (1946) stated that the conspicuous narrowing of the anterior portion of the palate, combined with the weakening of the teeth to such an extent (the canines both above and below are remarkably short and weak), suggests a special adaptation to a particular type of food supply.

Hayman (1946) stated that *Neopteryx* constitutes the fifth known genus in the Megachiroptera lacking a claw on the index finger. The absence of a tail distinguishes it at once from *Dobsonia*, *Eonycteris*, and *Notopteris*, which all lack the index claw, and *Nesonycteris*, the only fruit bat previously known to combine absence of tail with absence of index claw, has distinctive cranial and dental characters. The sharply narrowed rostrum and degenerate dentition together distinguish the skull from that of any other fruit bat. It would appear that this genus and species represents a specialized branch of the pteropine group.

The type specimen was collected at an elevation of 1,000 meters. Two others were taken in lowland primary forest. One, a pregnant female with one embryo, was caught in March. The IUCN classifies *N. frosti* as vulnerable.

Pteralopex acrodonta, photo by Pavel German.

CHIROPTERA; PTEROPODIDAE; **Genus**
PTERALOPEX
Thomas, 1888

There are four named species (Flannery 1991; Hill and Beckon 1978; Phillips 1968):

P. anceps, Bougainville and Choiseul islands in the
  Solomons;
P. atrata, Santa Ysabel and Guadalcanal islands in the
  Solomons;
P. pulchra, Guadalcanal Island in the Solomons;
P. acrodonta, Taveuni Island in Fiji.

An unnamed species inhabits New Georgia Island in the Solomons and also formerly occurred on nearby Kolombangara (Flannery 1995).
  Head and body length is 162–275 mm, there is no tail, forearm length is 111–69 mm, and weight is 241–506 grams. The fur is usually thick, woolly, and brownish or black, but *P. pulchra* has a yellow belly and *P. acrodonta* is entirely olive-green to yellow (Flannery 1995). The wing membranes are sometimes mottled with white beneath. *Pteralopex* differs from *Pteropus* in having cuspidate teeth, and it can be distinguished externally from *Pteropus pselaphon* in that the wing membranes originate in the midline of the back. The original describer, Thomas, regarded *Pteralopex* as an isolated survivor from the time when the modern ancestors of the Old World fruit bats had teeth with cusps. Most Old World fruit bats do not now possess

cuspidate teeth, apparently having lost them in their evolution.
  P. acrodonta reportedly roosts in pairs in the fern clumps growing 6–10 meters from the ground on trunks of the larger trees in open, tall forest (Hill and Beckon 1978). An adult female *P. anceps* taken in July on Bougainville was lactating (Phillips 1968). One individual was shot at night while feeding on green coconuts. The unnamed species on New Georgia is actually the best known (Flannery 1995). It is common in old fields adjacent to primary forest with large fig and and Ngali nut trees; it roosts in hollows within these trees about 10 meters above the ground. Roosts have been observed to contain one or two individuals, though there are reports of up to 10 bats inhabiting a single tree. The diet includes figs, pollen, and nuts. Most births seem to occur from April to June. The species disappeared from Kolombangara Island in the mid-1970s apparently because of the logging of its forest habitat.
  All of the named species of *Pteralopex* are classified as critically endangered by the IUCN. *P. acrodonta* is known only by five specimens from a single mountain on Taveuni Island, *P. anceps* has not been located in recent field searches, *P. atrata* apparently also has declined sharply but was sighted in 1991, and *P. pulchra* is known only by a single specimen (Flannery 1995; Mickleburgh, Hutson, and Racey 1992).

CHIROPTERA; PTEROPODIDAE; **Genus**
STYLOCTENIUM
Matschie, 1899

The single species, *S. wallacei,* is known from Sulawesi and the nearby Togian Islands (Bergmans and Rozendaal 1988; Hill 1983*a*).
  Head and body length is 152–78 mm, there is no tail, forearm length is about 90–103 mm, and weight is 174–218 grams. The fur is soft and silky. The back and rump are usually light gray in color, often with a pale, reddish brown wash. The underparts are a light reddish brown. The head is reddish brown with badgerlike white stripes and spots that enhance the appearance of this bat. There is no appreciable sexual difference in coloration.
  This genus differs from *Pteropus* in dental features, the most conspicuous being the absence of the first lower incisor and the third lower molar. *Styloctenium* has been described as a "rather slightly specialized offshoot of . . . *Pteropus*" (Andersen 1912). It apparently is rare, though some specimens recently have been observed in the wild and maintained in captivity. The IUCN now classifies *S. wallacei* as near threatened.

CHIROPTERA; PTEROPODIDAE; **Genus DOBSONIA**
Palmer, 1898

Bare-backed Fruit Bats

There are 14 species (Bergmans 1975*a*, 1978*a*, 1979*b*; Bergmans and Sarbini 1985; Boeadi and Bergmans 1987; Corbet and Hill 1992; De Jong and Bergmans 1981; Flannery 1995; Hill 1983*a*; Koopman 1979; Laurie and Hill 1954; Lidicker and Ziegler 1968; McKean 1972; Phillips 1968; Rabor 1952; Ride 1970):

D. minor, central Sulawesi, New Guinea and nearby Japen
  Island;

Styloctenium wallacei: Left, male feeding on fruit; Right, female with clinging young. Photos by Jan M. Haft.

D. exoleta, Sulawesi and Togian and Sula islands to east;

D. chapmani, Negros and Cebu islands (Philippines);

D. moluccensis, Buru, Amboina, and Seram islands in the southern Moluccas;

D. magna, New Guinea and Japen and Sulawati islands to west, Halmahera Islands in the northern Moluccas, Cape York Peninsula of northern Queensland;

D. anderseni, Bismarck Archipelago;

D. pannietensis, Trobriand Islands, D'Entrecasteaux Islands, and Louisiade Archipelago southeast of New Guinea;

D. emersa, Biak, Owii, and Numfoor islands off western New Guinea;

D. beauforti, Waigeo, Biak, Owii, and several other islands off western New Guinea;

D. praedatrix, Bismarck Archipelago;

D. inermis, Solomon Islands;

D. peronii, islands from Bali to Timor in the East Indies;

D. viridis, Sulawesi, southern Molucca Islands from Mangole and Buru to Banda and Kai;

D. crenulata, Sangihe Islands northeast and Togian Islands east of Sulawesi, Halmahera Islands in the northern Moluccas.

D. chapmani was listed as a subspecies of *D. exoleta* by Corbet and Hill (1992) but was considered a distinct species by Koopman (*in* Wilson and Reeder 1993) and most other recent authorities. *D. magna* was regarded as a subspecies of *D. moluccensis* by Corbet and Hill (1992), Koopman (*in* Wilson and Reeder 1993), and Mahoney and Walton (*in* Bannister et al. 1988) but was considered a distinct species by Bergmans and Sarbini (1985) and Flannery (1995). *D.*

anderseni was included within *D. moluccensis* by Hill (1983a) and Koopman (*in* Wilson and Reeder 1993) and within *D. pannietensis* by Corbet and Hill (1991) and Mickleburgh, Hutson, and Racey (1992) but was considered a distinct species by Bergmans and Sarbini (1985) and Flannery (1995). Hill (1983a) suggested that *D. pannietensis* should be included in *D. moluccensis,* but Corbet and Hill (1991), Flannery (1995), and Koopman (1982b and *in* Wilson and Reeder 1993) retained it as a full species. *D. remota* of the Trobriand Islands is sometimes designated a species distinct from *D. pannietensis* but was included within the latter, at least tentatively, by all recent authorities. *D. crenulata* was considered a subspecies of *D. viridis* by Corbet and Hill (1991, 1992), Hill (1983a), Koopman (*in* Wilson and Reeder 1993), Koopman and Gordon (1992), and Mickleburgh, Hutson, and Racey (1992) but was recognized as a full species by De Jong and Bergmans (1981) and Flannery (1995). Taylor (1934) stated that *D. peronii* had been reported from Samar Island in the eastern Philippines but thought it probable that the record either was an error or referred to *D. exoleta* or to an unknown species.

Head and body length is 102–242 mm, tail length is about 13–39 mm, forearm length is 74–161 mm, and weight is 68–600 grams; *D. minor* is the smallest species and *D. magna* the largest (Flannery 1990b, 1995). There is considerable variation in color, but most species are dull brown washed with olive, or grayish black.

There is no claw on the index (second) finger. The ears are pointed. The naked wing membranes are attached to the body along the spinal column in the midline of the back, as in *Pteronotus,* thus covering the hair of the body in the region that suggests the vernacular name. This form of at-

Bare-backed fruit bat *(Dobsonia moluccensis)*, photo by P. Morris.

tachment may supply a more effective wing area to aid these bats in hovering than exists for the bats whose wing membranes are attached along the sides of the body.

Unlike most of the pteropodids, *Dobsonia* seems to prefer caves and tunnels rather than trees for roosting. However, it has been taken in tree roosts, and it is known to forage in areas of thick vegetation. Dwyer (1975a) found *D. magna* in New Guinea to be abundant at an altitude of 2,700 meters. It roosted there in the twilight zone of caves and flew out by night to feed on a wide variety of fruits; it could easily be detected by its noisy flight. Bergmans (1978a) observed that *D. peronii* inhabited caves and trees near the coast and inland from sea level to about 1,200 meters. Other populations of *Dobsonia* have been noted in rock shelters on the coast at sea level and in coral limestone caves. In Australia specimens have been collected at night while feeding on flowering bloodwood trees and native figs.

According to Flannery (1990b), *D. magna* sometimes forms immense colonies in New Guinea with many thousands of bats per cave. However individuals of that species also may roost singly and commonly forage alone. Mating, at least in some parts of New Guinea, occurs from April to June at the close of the wet season; births take place from August to November, mostly in September and October, at the close of the dry season; and weaning occurs from February to April, at the height of the wet season. The single young weighs about 50–60 grams and reaches sexual maturity at two years of age. Some additional information compiled by Flannery (1990b, 1995) follows: pregnant fe-

Bare-backed fruit bat *(Dobsonia viridis)*, photo by Pavel German.

male *D. minor*, each with a single embryo, have been collected in New Guinea in January, late May, and early June; *D. anderseni* forms large colonies, lactating females and young have been collected in January and June–August; pregnant and lactating female *D. beauforti* have been taken in October, November, and December; pregnant and lactating female *D. crenulata* have been taken from November to January on Halmahera Island; pregnant and lactating female *D. inermis* have been taken in March, July, and November; and *D. praedatrix* roosts alone or in small groups in foliage and may have two birth peaks, one around the middle of the year and one in December–January. R. E. Goodwin (1979) reported that a colony of 300 *D. peronii*, composed of adults and young adults of both sexes, occupied a cave on Timor; a pregnant female was taken on this island on 20 March. An estimated 300 *D. chapmani* were found roosting in a cave on Negros Island. Births of that species occurred in May or June, and the young flew by August or September (Mickleburgh, Hutson, and Racey 1992).

Flannery (1995) recently has found several species to be common. However, Bergmans (1978a) observed that people had destroyed over 50 percent of the forests on some of the islands inhabited by *Dobsonia* and that the bats probably had suffered accordingly. Heaney and Heideman (1987) reported that such deforestation, with consequent loss of food sources and greater access to human hunters, may have contributed to the extinction of *D. chapmani* of the Philippines. This species also declined as a result of killing and disturbance when people entered its cave roosts to mine guano. It has not been recorded since 1964, though it may possibly survive on northern Negros and the neighboring island of Panay (Mickleburgh, Hutson, and Racey 1992). It now is designated as extinct by the IUCN; *D. beauforti* is classified as endangered, *D. emersa* and *D. peronii* as vulnerable, and *D. minor, D. exoleta,* and *D. praedatrix* as near threatened.

CHIROPTERA; PTEROPODIDAE; Genus APROTELES
Menzies, 1977

The single species, *A. bulmerae,* was first discovered among fossil remains about 9,000–12,000 years old collected in the mountains of Chimbu Province in central Papua New Guinea (Menzies 1977). Shortly thereafter it was realized that the same species was represented by specimens killed in 1975 at Luplupwintem Cave in the Hindenburg Ranges of far western Papua New Guinea (Hyndman and Menzies 1980). A small living population was discovered in the same area in 1992 (Flannery and Seri 1993). In 1993 relatively recent bones of *A. bulmerae,* along with many of *Dobsonia,* were found in a cave about 300 km to the east-southeast, in Eastern Highlands Province of Papua New Guinea;

Aproteles bulmerae, photo by Tim Flannery.

about 700 living fruit bats also were in the cave, but their identity could not be confirmed (Wright et al. 1995).

The mounted skin of one specimen taken in 1975 was prepared but lost, and the genus long was known only from cranial material. Affinity to *Dobsonia* is indicated, but *Aproteles* differs in lacking lower incisors and in having less crowded, simpler teeth and a longer, more tapering rostrum (Menzies 1977). Three complete specimens now are available (Flannery and Seri 1993), though only one, a female, is an adult. It has a head and body length of 242 mm, a tail length of 32 mm, a forearm length of 166 mm, and a weight of 600 grams. The head is dark brown, the upper part of the mantle is paler, the distal part of the mantle is whitish, and the rump and venter are brown. Externally *A. bulmerae* differs from *Dobsonia magna* in being larger and in having a more massive head and finer fur. In both species the wings meet in the midline of the back, giving the back a naked appearance.

The fossil material was found at an altitude of 1,530 meters in the refuse dump of a rock shelter. The bones are those of bats that had been killed and cooked by the human inhabitants of the shelter. Many fragments of *Aproteles* were recovered, and Menzies (1977) suggested that this genus, like *Dobsonia,* was a cave dweller because a tree-roosting bat would hardly have been accessible in such quantity to people of the Stone Age. Most of the specimens obtained subsequently were from a large cave at an altitude of 2,300 meters. The colony apparently contained thousands of bats until the mid-1970s, though only 137 were counted leaving the roost on the evening of 1 May 1992. One specimen, a subadult female, was shot near a village about 32 km away in February 1984 while it fed on the fruit of a fig tree. The adult female was taken on 3 May 1992 and was then carrying a dependent young male that probably had been born a few weeks earlier. Other evidence also indicates that births occur seasonally in April (Flannery and Seri 1993).

Hyndman and Menzies (1980) suggested that intensive human hunting had eliminated *Aproteles* in the central highlands of Papua New Guinea but that the genus managed to survive in sparsely populated areas farther to the west. However, when an effort was made in November 1977 to locate *Aproteles* in the cave where it had been found in 1975, it was learned that local hunters had already killed or driven away nearly the entire colony. Additional searches of the cave in 1985 and 1987 also were unsuccessful and the species was feared extinct. In 1991 it was learned that a specimen had been collected in 1984 but had been mistakenly identified as *Dobsonia.* New searches were stimulated and *Aproteles* was rediscovered at Luplupwintem Cave in May 1992. It is likely that some bats had always been present but had not been detected because of the great difficulty involved in accessing the cave. The colony there had traditionally been protected by the native people of the area, but an inflow of outside cash in the mid-1970s led to the purchase of caving equipment and guns and to the decimation of the bat colony (Flannery and Seri 1993). *A. bulmerae* now is listed as critically endangered by the IUCN and as endangered by the USDI.

CHIROPTERA; PTEROPODIDAE; Genus HARPYIONYCTERIS
Thomas, 1896

Harpy Fruit Bats

There are two species (Corbet and Hill 1992; Peterson and Fenton 1970):

genus to be an isolated group, but he thought it should be placed near *Rousettus* and *Boneia*. Knud Andersen, however, considered it a close relative of *Dobsonia*. Tate (1951*a*) associated it with the tube-nosed fruit bats. Corbet and Hill (1992) placed it in its own subfamily, the Harpyionycterinae, which in turn they placed near the Nyctimeninae, a subfamily for *Nyctimene* and *Paranyctimene*.

The type specimen was shot at dusk as it was flying around some high trees at an elevation of 1,660 meters on the island of Mindoro. Other specimens have been taken at altitudes of about 150–1,500 meters in Sulawesi and the Philippines. Peterson and Fenton (1970) speculated that the multicusp teeth of this genus might be adapted for extracting the juice of a particular type of tough-textured fruit and that the bats might not normally ingest the fruit fibers. Bergmans and Rozendaal (1988) reported the following for *H. celebensis*: pregnant females, each with a single embryo, obtained in January and September; a nursing female observed in January; and immature specimens found in January, May, June, and July.

Rickart et al. (1993) reported the collection of pregnant female *H. whiteheadi whiteheadi*, each with a single embryo, on Leyte, Biliran, and Maripipi islands in the Philippines from 20 March to 8 May. Citing the work of Heideman and Heaney on *H. whiteheadi negrosensis*, the subspecies of Negros Island in the Philippines, Mickleburgh, Hutson, and Racey (1992) recorded that maximum density of about 1/1.4 ha. occurs in lower montane forest. Foraging takes place at the upper levels of trees and the diet includes the fruits of pandans and figs. There appear to be two synchronous birth periods, from January to early February and from July to early August, both during the rainy season. Most females produce one young in each period. Gestation lasts about 4–5 months, lactation 3–4 months. Some females give birth for the first time at one year. Mickleburgh, Hutson, and Racey designated this subspecies as vulnerable. Although it is not threatened by hunting, it requires primary or only lightly disturbed forest and cannot survive in areas cleared for agriculture or urban development.

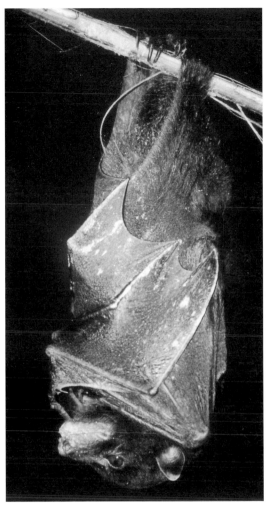

Harpy fruit bat *(Harpyionycteris celebensis)*, photo by Jan M. Haft.

H. whiteheadi, Luzon, Mindoro, Negros, Leyte, Biliran, Maripipi, Mindanao, and Camiguin islands in the Philippines;
H. celebensis, Sulawesi.

H. celebensis was listed as a subspecies of *H. whiteheadi* by Mickleburgh, Hutson, and Racey (1992) but as a full species by Corbet and Hill (1991, 1992) and Koopman (*in* Wilson and Reeder 1993).

Head and body length is 140–53 mm and forearm length is about 82–92 mm; there is no tail. Bergmans and Rozendaal (1988) recorded weights of 83–142 grams. The color is chocolate brown to dark brown above and paler below. The hind feet are very short, and the small interfemoral membrane is hidden in thick fur.

This genus differs from all other fruit bats in that the molar teeth have five or six cusps and the lower canines have three. Also, the premaxillary bones of the skull, the upper incisors, and the upper and lower canines are more strongly inclined forward than in other fruit bats. The canines cross each other almost at right angles when the jaws are closed. Oldfield Thomas, the describer, considered this

CHIROPTERA; PTEROPODIDAE; Genus PLEROTES
Andersen, 1910

The single species, *P. anchietai*, is known from central Angola, southeastern Zaire, and northeastern Zambia (Bergmans 1989).

An adult female had a head and body length of 87 mm and a wing expanse of 343 mm. Head and body length was 95 mm in another adult female and 96 mm in a male (Bergmans 1989). There is no external tail, and the forearm length is about 48–53 mm. The pelage is long and fine, grayish brown above and much paler below. There is a small tuft of white hairs at the base of the ears. A detailed description of the genus and a summary of data on the nine known specimens were provided by Bergmans (1989).

This bat may be distinguished from all other Old World fruit bats (Pteropodidae) by the white patches at the base of the ears and the absence of a "spur," or calcar, a bone that projects from the heel to extend the interfemoral membrane. Males probably possess shoulder pouches and hair tufts. According to the original describer of this species, a cheek pouch on each side of the muzzle surrounds the eyes.

The skull is delicate and the dentition is unusually weak for a bat of this group, in which weak dentition prevails. The teeth are much reduced in size, and the molars have very little structure on the surface. The dentition suggests

Harpy fruit bat *(Harpyionycteris celebensis)*, photo by F. G. Rozendaal.

flower and nectar feeding, though nothing is known specifically about the habits of *Plerotes*. It probably roosts in dense foliage during the day. It appears to be highly specialized and to be threatened by loss of woodland habitat (Mickleburgh, Hutson, and Racey 1992). It now is classified as vulnerable by the IUCN.

CHIROPTERA; PTEROPODIDAE; **Genus HYPSIGNATHUS**
H. Allen, 1861

Hammer-headed Fruit Bat

The single species, *H. monstrosus*, has been reported to occur from Gambia to southwestern Ethiopia and south to northern Angola and Zambia (Hayman and Hill *in* Meester and Setzer 1977; Koopman 1975; Langevin and Barclay 1990; Largen, Kock, and Yalden 1974). However, Koopman (*in* Wilson and Reeder 1993) gave the east-west range as Sierra Leone to Kenya, noting that the records from Gambia and Ethiopia are doubtful. Bergmans (1989) further restricted the range to the lowland moist forest zone from Sierra Leone to extreme western Kenya, with the most southerly records being from central Zaire and northwestern Angola.

Head and body length is about 193–304 mm, there is no tail, and forearm length is 118–37 mm. The wingspan in males is as much as 907 mm. *Hypsignathus* is the largest bat in Africa (Langevin and Barclay 1990), and the genus has the greatest sexual dimorphism in the Chiroptera; Bradbury (1977) found that males, which averaged 420 grams, were nearly twice as heavy as females, which averaged 234 grams. The coloration is grayish brown or slaty brown. The breast is paler, and the lighter color extends up around the neck, forming a sort of collar. A white patch is present at the base of the ear. Shoulder pouches and epauletlike hair tufts are lacking in both sexes.

Male *Hypsignathus* may be recognized in flight by the large, square, truncate head. The muzzle is thick and hammer-shaped, hence the common name. Other distinctive features are enormous and pendulous lips, ruffles around the nose, a warty snout, a hairless, split chin, and highly developed voice organs in adult males. Females have a foxlike muzzle similar to that of *Epomophorus*.

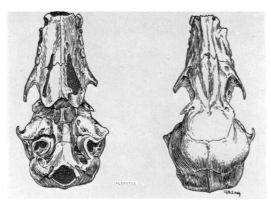

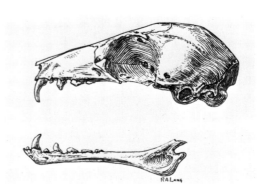

Plerotes anchietai, photos from *Annals of Carnegie Museum*.

A. Hammer-headed bats *(Hypsignathus monstrosus)*, photo by V. Aellen. B. Head; C. Mouth; D. Lips; photos by J. L. Perret.

In referring to this genus, Lang and Chapin (1917) commented: "In no other mammal is everything so entirely subordinated to the organs of voice." The adult male has a pair of air sacs that open into the sides of the nasopharynx and can be inflated at will, as well as a great enlargement of the voice box (larynx) and vocal cords. The larynx "is nearly equal in length to one half of the vertebral column," actually filling most of the chest cavity, pushing the heart and lungs backward and sideward. The voice thus produced, a continuous croaking or quacking, is quite remarkable and probably attracts the females. The gregarious chorus reminded Lang and Chapin of "a pondful of noisy American wood-frogs, greatly magnified and transported to the treetops."

The hammer-headed bat inhabits forests, being most common in swamps, mangroves, and palms along rivers. It usually roosts in foliage but has been found in a cave. Bradbury (1977) stated that *Hypsignathus* roosted at a height of 20–30 meters during the day and would forage up to 10 km from the roost at night. With the ripening of certain fruits, this bat often seeks the high forest or native clearings to feed. It may take the juices of mangoes, soursops, and bananas. Van Deusen (1968) reported that *Hypsignathus* killed and ate tethered chickens.

In a lengthy study in Gabon, Bradbury (1977) determined that *Hypsignathus* had a "lek" mating system comparable to that of certain bovids. Twice a year, from June to August and from December to February, adult males assembled nightly at traditional sites. Each assembly numbered from 25 to 132 bats, and each bat set up a territory roughly 10 meters in diameter. The remarkable call was given from about 1830 to 2300 and from 0300 to 0500 every night, and the males also displayed by flapping their wings. Females visited the assemblies, hovered about, and eventually settled beside one male to mate. Only 6 percent of the males accounted for 79 percent of the matings, and the relatively few successful males tended to be aggregated at certain apparently favorable locations within the overall assemblies. In comparison with the noise and activity at the nocturnal assemblies, day roosts were relatively quiet, with the bats resting alone or in small groups. At night the sexually active males would move to an assembly area, display for a while, and then fly out to forage. Examination of captives indicated that sexual maturity was attained at 6 months by females but not until 18 months by males. Wolton et al. (1982) suggested that in Liberia there is a birth peak during the middle of the rainy season, in August–September, and possibly another peak at the end of the rains, in October–December. Kingdon (1974a) observed that available records suggested breeding peaks around February and July in East Africa and that the usual number of young is one, but twins had been reported. Weight at birth is about 40 grams (Hayssen, Van Tienhoven, and Van Tienhoven 1993). Langevin and Barclay (1990) noted that captive *Hypsignathus* maintain the same breeding seasons (June–August and December–January) as those of wild populations and that females have a postpartum estrus and can give birth twice a year.

Epauleted bat *(Epomops franqueti)*: Top photo from Amsterdam Zoo; Middle photo by Jane Burton; Bottom photo by Gudrun Haft.

CHIROPTERA; PTEROPODIDAE; **Genus EPOMOPS**
Gray, 1870

Epauleted Bats

There are three species (Bergmans 1979*a*, 1982, 1989; Bergmans and Jachmann 1983; Hayman and Hill *in* Meester and Setzer 1977):

E. franqueti, Ivory Coast to southern Sudan, south to Angola and Zambia;

E. buettikoferi, Guinea to Nigeria;

E. dobsoni, Rwanda, Tanzania, Angola, southern Zaire, Zambia, Malawi, northeastern Botswana.

Bergmans (1989) indicated that *E. dobsoni* shows some morphological affinity to *Epomophorus* and *Nanonycteris* and may not warrant retention in *Epomops*.

Head and body length is 135–80 mm, there is no tail, and forearm length is 74–102 mm. Reported weights are: male *E. franqueti,* 59–160 grams; female *E. franqueti,* 56–115 grams; male *E. buettikoferi,* 160–98 grams; and female *E. buettikoferi,* 85–132 grams (Bergmans 1975*b*; Jones 1971*a*, 1972). Coloration is tawny, brownish, or grayish, with much variation. Whitish hair tufts are present at the base of the ears. Males have two pairs of pharyngeal sacs, large shoulder pouches, and epauletlike hair tufts. These tufts, whitish or yellowish in color, are conspicuous in dried specimens, but in the live bats, as in related genera, the shoulder pouch may be so drawn in that the epaulet is almost completely concealed. Also as in the related bats, the shoulder pouches of *Epomops* are lined with a glandular membrane.

The adult males of *E. franqueti* have a bony voice box, as in *Hypsignathus,* and they emit a call that at close range sounds somewhat like "kûrnk!" or "kyûrnk!" but at a distance has a whistled, almost musical, effect. This high-pitched note can be heard throughout the night and may be emitted continuously for several minutes by a given male.

Epauleted bats are usually found in forests but also appear in open country. They roost in trees and bushes and are quite alert by day. Jones (1972) found them mainly in large trees 4–6 meters above the ground. Their diet consists of the juice and softer parts of such fruits as guavas, bananas, and figs. The manner of feeding for *Epomops* and related genera is as follows: the expansible lips encircle the fruit; the pointed canines and premolars pierce the rind; the jaws squeeze the fruit, which is then pressed upward by the tongue against the thick palate ridges; and suction is accomplished by the large pharynx, which communicates anteriorly with the mouth by a small opening and is supported behind by the hyoid apparatus. Suction, rather than mastication, appears to be the major process in feeding.

Epauleted bats are not gregarious, usually resting alone or in groups of two or three. In Uganda *E. franqueti* has two breeding seasons, with implantation occurring in April and late September and births taking place in September and late February (Okia 1974*a*). These sequences are timed so that births occur at the beginning of the two rainy seasons or shortly thereafter. Gestation lasts five to six months. Bergmans (1979*c*) suggested that the reproductive pattern in Congo is not much different from that found in Uganda. Gallagher and Harrison (1977) stated that breeding females of *E. franqueti* were taken in Zaire in December and January. Bergmans, Bellier, and Vissault (1974) reported the collection of possibly lactating females of *E. franqueti* and *E. buettikoferi* in the Ivory Coast in October and November. Weight at birth is about 20 grams (Hayssen, Van Tienhoven, and Van Tienhoven 1993).

Additional studies in the Ivory Coast by Thomas and Marshall (1984) indicate that *E. buettikoferi* has birth peaks from about February to April and from September to November. Both periods seem to be timed so that lactation coincides with the area's two rainy seasons, when fruit availability is at a maximum. Each parturition period is followed by a postpartum estrus and apparently by immediate embryonic development. The gestation period is 5–6 months and lactation lasts 7–13 weeks. Young females can mate at

6 months and give birth at 12 months. Males are sexually mature at 11 months.

The IUCN classifies *E. buettikoferi* as vulnerable. Mickleburgh, Hutson, and Racey (1992) indicated that it may even be endangered because of the severe forest destruction in West Africa.

CHIROPTERA; PTEROPODIDAE; **Genus EPOMOPHORUS**
Bennett, 1836

Epauleted Fruit Bats

There are seven species (Bergmans 1988; Claessen and De Vree 1990, 1991; Gaucher 1992):

E. gambianus, Senegal and southern Mali to southern Sudan and Ethiopia;
E. crypturus, southern Zaire and Tanzania to eastern South Africa;
E. labiatus, northeastern Nigeria and southern Congo to northern Ethiopia and Malawi, one record from extreme southwestern Saudi Arabia;
E. minimus, Ethiopia, southern Somalia, Uganda, Kenya, northern Tanzania;
E. angolensis, western Angola, northwestern Namibia;
E. wahlbergi, Cameroon and Somalia south to South Africa;
E. grandis, southern Congo, northeastern Angola.

Of several other species that often have been recognized (Hayman and Hill *in* Meester and Setzer 1977), Bergmans

Epauleted fruit bat *(Epomophorus wahlbergi)*, photo by John Visser.

(1988) regarded *E. anurus* as a synonym of the largely sympatric *E. labiatus; E. reii,* of Cameroon, as a synonym of *E. gambianus;* and *E. pousarguesi,* of the Central African Republic, as a subspecies of *E. gambianus.* Bergmans also considered *E. crypturus* a subspecies of *E. gambianus,* but Claessen and De Vree (1990) concluded that the two are specifically distinct. The two taxa are largely, though not entirely, distinguishable by morphometrical analysis and are geographically separated by more than 1,000 km. Koopman (*in* Wilson and Reeder 1993) followed Bergmans's interpretation. Bergmans considered *E. minor,* found from southern Sudan and Ethiopia to Malawi, to be a distinct species, but Claessen and De Vree (1991) allotted it in part to *E. labiatus* and in part to a new species, *E. minimus;* Koopman followed the latter interpretation. Bergmans also transferred *E. grandis* to *Epomophorus* from the genus *Micropteropus,* but Boulay and Robbins (1989), citing Claessen, suggested that *grandis* should be restored to *Micropteropus.*

Head and body length is usually 125–250 mm, a vestigial external tail may be present beneath the interfemoral membrane, forearm length is about 60–100 mm, and the wingspan in males is about 508 mm. Weight is about 40–120 grams (Kingdon 1974*a*). These bats are grayish brown, russet, or tawny in color with a white patch at the base of the ear in both sexes. Males have a conspicuous pair of shoulder tufts around large glandular sacs. Air sacs are present on the necks of males. Epauleted fruit bats have expansible, pendulous lips, often with peculiar folds.

These bats occur in woodland and savannah; true forest bats, they prefer the edges of forests (Kingdon 1974*a*). They roost in such places as large hollow trees, thick foliage, accumulated roots along stream banks, and below the thatch of open sheds. They often roost where there is considerable light, and sometimes groups of about half a dozen hang from the midribs of palm fronds in plain sight. They

Epauleted fruit bat *(Epomophorus wahlbergi)*, photo by Erwin Kulzer.

seem to be quite alert when roosting during the day. Fenton et al. (1985) found *E. wahlbergi* to switch day roosts every few days and to fly up to 4 km from these roosts to the nocturnal feeding area, females usually traveling farther than males.

The flight of one species, *E. labiatus*, has been compared to that of a raven. While flying and while feeding, members of this genus emit various squeaks, chucklings, and sharp metallic calls. These bats are likely to appear wherever fig, mango, guava, or banana trees are in fruit; some species, such as *E. wahlbergi*, make local migrations in search of ripe fruit. *E. gambianus* has been noted feeding on the nectar of the flowers of *Parkia clappertoniana* in Ghana. Control measures, in the form of poisoned fruit for bait, are sometimes utilized in areas where they feed extensively on cultivated fruit. Detailed accounts of diet and relationship with various plants were provided in Mickleburgh, Hutson, and Racey (1992).

These fruit bats roost in small groups; there is some indication that in *E. labiatus* the females with young roost apart from the males, but about 20 *E. wahlbergi*, including young and old of both sexes, were once collected from the lower fronds of a single palm tree. According to Wickler and Seibt (1976), the latter species roosts during the day in bisexual groups of 3–100 individuals. One colony of about 100 bats in Kenya was found in the same locality every January for five years. During the breeding season the males would leave the main roost at night and fly to another location, where they would begin to emit courting calls. The calls seemed to space out the males and attract females. While calling, the males would display the normally concealed epaulets. Similar roosting and territorial behavior has been recorded for *E. gambianus* (Boulay and Robbins 1989).

Data from Congo suggest defined seasonal reproduction for *E. wahlbergi*, involving two cycles per year, with births occurring around the end of February and the beginning of September (Bergmans 1979c). A similar pattern has been reported for that species in Zaire (Mickleburgh, Hutson, and Racey 1992). Okia (1974b) found that in Uganda *E. labiatus* was a continuous but strictly cyclic breeder with two distinct breeding seasons separated by about a month. Gestation apparently lasted about five to six months, from April to September and from October to March. Birth weight in *E. labiatus* is 11 grams (Hayssen, Van Tienhoven, and Van Tienhoven 1993). Koopman, Mumford, and Heisterberg (1978) collected a lactating female *E. gambianus* in Burkina Faso in January and pregnant females in March, May, and September. Thomas and Marshall (1984) reported that in the Ivory Coast *E. gambianus* breeds twice annually, with females having a postpartum estrus and births occurring during the rainy seasons, about April–May and October–November. Boulay and Robbins (1989) stated that pregnant or lactating female *E. gambianus* were collected in Niger during the dry season, from February to May. Female *E. crypturus* normally bear one young per year in the period from September to February; twins are known (Mickleburgh, Hutson, and Racey 1992).

The IUCN classifies *E. grandis* as endangered because of the decline of its restricted habitat; *E. angolensis* is designated as near threatened. Mickleburgh, Hutson, and Racey (1992) also designated *E. gambianus pousarguesi* of the Central African Republic as rare. Most other species and subspecies are considered relatively common, and some reportedly cause damage to fruit crops.

CHIROPTERA; PTEROPODIDAE; **Genus** **MICROPTEROPUS** *Matschie, 1899*

Dwarf Epauleted Bats

There are two species (Ansell 1974; Bergmans 1979c, 1989; Hayman and Hill *in* Meester and Setzer 1977; Kock 1987):

M. pusillus, Gambia to Ethiopia, and south to Angola and southern Zaire;
M. intermedius, southwestern Zaire, extreme northeastern Angola.

Bergmans (1988) transferred another species, *M. grandis*, to the genus *Epomophorus* (see account thereof). Bergmans (1989) indicated that the cranial characters of *Micropteropus* do not warrant more than subgeneric distinction from *Epomophorus* but that the two tentatively could be maintained as separate genera on the basis of apparent chromosomal differences.

Head and body length in *M. pusillus* is 67–105 mm, forearm length is about 46–65 mm, and coloration is brownish above and lighter below. Average weight in a series of specimens from Equatorial Guinea was about 20 grams for males and 22 grams for females (Jones 1972). A rudimentary external tail is present in *Micropteropus*. The pelage is moderately long, thick, and very soft.

In general appearance *Micropteropus* resembles *Epomophorus*, but the lips are not as expansible, the muzzle is shorter, and the ears are relatively shorter. Small whitish tufts of hair are present at the base of the ears, at least in *M. pusillus*, and males have shoulder pouches and epauletlike hair tufts. A collector's note on a specimen of *M. pusillus* reads: "When erected the tuft had a vibratory movement." The females of this genus lack the shoulder tufts, but female *M. pusillus* often have distinct though shallow pouches somewhat like those of subadult males.

Most records of *M. pusillus* are from open woodlands, but the species has also been found in high forest. Specimens have been taken from between the leaves of dense bushes, usually close to the ground, and from the lower fronds of palm trees growing among vines and small trees in the bed of a stream bordering on an arid, grassy plain. This species is often exposed to direct sunlight in its roosts and is not wary, seldom flying from its shelter. It may fly if unduly disturbed, but even then it usually settles down within 10 meters of its original resting place. *M. pusillus* is mainly a wanderer in search of ripe fruit. In captivity it feeds slowly and deliberately, consuming its food by a slow sucking action of the lips and mouth. The uneaten fruit pulp is dropped after fluids are removed (Jones 1972). This species also has been reported to visit the sausage tree (*Kigelia africana*) to lap nectar (Rosevear 1965).

Studies in the Ivory Coast by Thomas and Marshall (1984) indicate that *M. pusillus* has birth peaks from about March to May and from September to November. Both periods seem to be timed so that lactation coincides with the area's two rainy seasons, when fruit availability is at a maximum. Each parturition period is followed by a postpartum estrus and apparently by immediate embryonic development. The gestation period is 5–6 months, and lactation lasts 7–13 weeks. Young females can mate at 6 months and give birth at 12 months. Males are sexually mature at 11 months. Available data from Congo also suggest two reproductive cycles per year in *M. pusillus*, with births occurring there mainly in February and September

Dwarf epauleted bat *(Micropteropus pusillus)*, photos by Merlin D. Tuttle.

(Bergmans 1979c). Most births of *M. pusillus* in the Garamba National Park, Zaire, apparently take place about the end of February and in November or December. Hill and Morris (1971) collected a pregnant female in Ethiopia on 30 August. Bergmans (1989) reported that a pregnant female *M. intermedius*, with a single embryo, was taken in June or July. That species is known only by four specimens and is classified as vulnerable by the IUCN.

CHIROPTERA; PTEROPODIDAE; Genus NANONYCTERIS
Matschie, 1899

Little Flying Cow

The single species, *N. veldkampi*, occurs from Guinea to the Central African Republic; earlier reports of its presence in Zaire apparently were in error (Bergmans 1982, 1989; Hayman and Hill *in* Meester and Setzer 1977).

Head and body length is about 54–75 mm, a rudimentary tail may be present, and forearm length is 43–54 mm. A series of specimens from Nigeria weighed 19–33 grams (Happold and Happold 1978a). The back may be light russet and the underparts near cream buff, but much color variation exists. Both sexes have a small white patch of hair at the base of the ears. Adult males have shoulder pouches with epauletlike hair tufts. Cranial and dental features distinguish this genus from related genera (see account of *Scotonycteris*). The head of *Nanonycteris* is said to have a calflike appearance.

This bat seems to favor closed forest or fringed forest within the Guinea type of open woodland (Rosevear 1965). In Ghana it was observed feeding on the nectar of the flowers of *Parkia clappertoniana*. *Nanonycteris* arrived at the *Parkia* trees at dusk, usually just before the departure of *Epomophorus gambianus*, which also was feeding on the flowers. *Nanonycteris* on several occasions harassed individuals of *Epomophorus* that were hanging onto the flowers; this usually resulted in *Epomophorus* releasing its position and flying off. Whereas *Epomophorus* tended to feed only in the upper parts of the trees, *Nanonycteris* tended to feed on flowers all over the tree and continued to feed in the trees in varying numbers throughout most of the night.

Nanonycteris remained on flowers from 1 to 30 seconds,

Little flying cow *(Nanonycteris veldkampi)*, photo by Merlin D. Tuttle.

considerably less than the time spent by *Epomophorus*. Baker and Harris (1957), who observed and photographed this bat, wrote:

> [*Nanonycteris*] clasps the inflorescence with its wings, holding on with its thumbs, and applies its face to the depressed ring containing the nectar. The nectar is than lapped with the narrow, rough tongue. Departure from an inflorescence is rapid and details cannot be observed easily, but from the photographs it would seem that the bat first throws itself backwards. It may then drop below the inflorescence, completing a somersault before opening its wings and flying away . . . or it may do a half-roll. . . . No evidence was obtained of any flower-eating by the bats. . . . There was . . . no sign of pollen in the stomach, . . . only nectar.

Evidently in response to food availability, *Nanonycteris* migrates northward from the forests of the Ivory Coast and onto the savannahs during the rainy season. The annual movement is nearly 750 km (Thomas 1983). In northern Ghana, Marshall and McWilliam (1982) found it only during the wet season. There it roosts alone or in small groups of well-spaced individuals. Apparently, females are polyestrous and have an extended parturition period during the early rains, about May and June. Studies by Wolton et al. (1982) in Liberia also show that *Nanonycteris* makes an annual migration; in that area there apparently is a long breeding season, from October until at least mid-March, with no marked peaks, but the first births do not occur before mid-November.

CHIROPTERA; PTEROPODIDAE; **Genus** SCOTONYCTERIS
Matschie, 1894

There are two species (Bergmans 1973; Bergmans, Bellier, and Vissault 1974; Hayman and Hill *in* Meester and Setzer 1977):

S. zenkeri, Liberia to eastern Zaire, island of Bioko (Fernando Poo);
S. ophiodon, Liberia to southern Congo.

In *S. zenkeri* head and body length is 65–80 mm and forearm length is 47–56 mm. In *S. ophiodon* head and body length is 104–43 mm and forearm length is about 75–88 mm. The tail is rudimentary in this genus. A series of specimens of *S. zenkeri* from Nigeria weighed 20–27 grams (Happold and Happold 1978*a*). Liberian series weighed 18–26 grams for *S. zenkeri* and 65–72 grams for *S. ophiodon.* Coloration is about the same in both species, being russet, rust brown, or dark brown above and paler below. The small white patch at the base of the ear present in other bats that resemble *Epomophorus* is inconspicuous in *S. ophiodon* and absent in *S. zenkeri.* Both species usually have a white patch in front of and between the eyes and a white patch behind the eyes. One specimen of *S. ophiodon* had yellow spots behind the eyes.

As determined by features of the skull and teeth, this genus closely resembles *Epomophorus* and related genera. Externally it is said to be scarcely distinguishable from *Casinycteris* (but see account thereof). *Scotonycteris* is best distinguished from *Casinycteris* by its normal palate, the palate in the latter genus being extremely shortened. From *Epomophorus, Scotonycteris* differs in not having a flattened skull. From *Nanonycteris* and *Cynopterus, Scotonycteris* differs in having premaxillae that do not taper above and have deeper alveolar branches, more strongly diverging upper tooth rows, a longer and converging postdental palate, smaller postorbital processes,

Scotonycteris zenkeri, photo from *Proc. Zool. Soc. London.* Inset: palate, photo from *Zoologische Beitrage,* Bonner.

Scotonycteris sp., photo by Norman J. Scott, Jr.

Casinycteris argynnis, photo by J. L. Perrot through V. Aellen.

more horizontal zygomata owing to broader temporal fossae, and a stronger mandible (Bergmans 1990).

Both species of *Scotonycteris* seem to be solitary tree dwellers, but little was known about them until studies by Wolton et al. (1982) at Mount Nimba, Liberia. *S. zenkeri* was taken only at elevations below 800 meters and seemed to prefer primary closed forest to secondary growth or cultivated land. Its morphology indicated that it feeds on small, thin-skinned fruits. Collection of pregnant females in August and November and (citing Bergmans, Bellier, and Vissault 1974) of a lactating female in December suggested an extended breeding season.

Wolton et al. collected 10 specimens of *S. ophiodon* and considered the species very rare. All but one of these bats were found at elevations from 1,000 to 1,200 meters. The species seemed confined to the immediate vicinity of Mount Nimba, and there was concern that it could be adversely affected by habitat disruption. In contrast to *S. zenkeri*, which took a number of long rests during the night, *S. ophiodon* fed almost continuously. A captive ate bananas as its first choice, then guavas and plantains. Females appeared more active and vocal than did males and emitted a high-pitched whistle through the night. Available data indicated to Wolton et al. that most females are pregnant in August and September and that there may be a second breeding at the end of the year. Indeed, there is an earlier report that a female *S. ophiodon* collected in December in Ghana had a youngster clinging to it. Both species evidently produce a single young at a time (Hayssen, Van Tienhoven, and Van Tienhoven 1993).

CHIROPTERA; PTEROPODIDAE; **Genus**
CASINYCTERIS
Thomas, 1910

The single species, *C. argynnis*, is now known from 12 localities in southern Cameroon and Zaire (Bergmans 1990; Meirte 1983).

Head and body length is about 90–95 mm, the tail is vestigial, and forearm length is about 50–63 mm. Bergmans (1990) listed weights of 26–30 grams. The fur is light brown in color, but the muzzle, eyelids, ears, and wings are said to be yellowish or bright orange when the bat is alive. Tufts of white hairs are present at the base of the ears, and an oblong white patch extends back between the eyes, with another present behind the eyes.

Casinycteris sometimes is said to be remarkably similar to *Scotonycteris* in external appearance, but Bergmans (1990) indicated that such remarks are based on nonliving material and that the upturned rostrum, relatively larger ears, and other features of *Casinycteris* make it look quite different. Internally *Casinycteris* differs from *Scotonycteris* in its extremely shortened palate and dental features. Thomas (1910a) wrote: "This striking bat . . . is remarkable for possessing a palate quite unlike that of other fruit-eating bats, and more recalling that found in some of the Microchiroptera. The astonishing resemblance of the type species to *Scotonycteris* . . . is also noticeable. Probably both bats bear a protective resemblance to the leaves, fresh or dry, of some local tree." Andersen (1912) also discussed the relationship between these two genera: "It might seem a little strange that these two genera, though so closely related as to be evidently modifications of one type of bat, are nevertheless inhabitants of the same faunistic area. It appears reasonable to suppose, however, that the profound difference in the posterior portion of the bony roof of the mouth must be connected either with . . . a difference in diet or a difference in the manner of feeding on the same food, so that . . . in any case there would . . . be but little competition between them."

Casinycteris evidently is restricted to the central forest block of Africa and may be locally abundant (Meirte 1983). However, it has been reported to be threatened by forest destruction (Mickleburgh, Hutson, and Racey 1992). It is now designated as near threatened by the IUCN. One specimen was collected while hanging alone about three meters from the ground in the dense foliage of a bush. Births reportedly occur in May, lactating females have been taken from

May to July, and there is a single young (Hayssen, Van Tienhoven, and Van Tienhoven 1993).

CHIROPTERA; PTEROPODIDAE; **Genus**
CYNOPTERUS
F. Cuvier, 1825

Short-nosed Fruit Bats, or Dog-faced Fruit Bats

There are four species (Corbet and Hill 1992; Flannery 1995; Hill 1983a):

C. brachyotis, Nepal, southern and probably northeastern India, Guangdong Province of southeastern China, southern Burma, Thailand, Indochina, Malay Peninsula, Sri Lanka, Andaman and Nicobar islands, Sumatra and many nearby islands, Java, Kangean Islands, Bali, Borneo and many associated islands, throughout the Philippines, Talaud and Sangihe islands, Sulawesi, Togian Islands, Lesser Sunda Islands from Lombok to Sumba and Lembata, Sula Islands;

C. sphinx, probably southeastern Pakistan, India to Indochina, Malay Peninsula, extreme southern China, Hainan, Sri Lanka, Andaman and Nicobar islands, Sumatra, Krakatau, Mentawai Islands, Java, Bali, Borneo, North Natuna Islands, Serasan Island, Sulawesi, Salayer Island;

C. titthaecheilus, Sumatra, Nias Island, Krakatau, Sebesi, Sertung, Panjang, Java, Bali, Lombok, possibly Timor;

Short-nosed fruit bat *(Cynopterus horsfieldi)*, photo by Lim Boo Liat.

C. horsfieldi, southern Thailand, Malay Peninsula, Sumatra, Simeulue Island, Nias Island, Labuan, Sertung, Java, Borneo, Lombok.

Kitchener and Maharadatunkamsi (1991) described another species, *C. nusatenggara,* from the Lesser Sunda Islands (Lombok, Sumbawa, Moyo, Komodo), also recognized *C. minutus* of the Greater Sundas (Sumatra and Nias Island to the west, Java, Borneo, Sulawesi) and *C. luzoniensis* of Sulawesi and the Philippines as species distinct from *C. brachyotis,* and made some other revisions in the systematics of *Cynopterus.* That revision was not followed by Corbet and Hill (1992), Flannery (1995), and Rickart et al. (1993). Koopman (*in* Wilson and Reeder 1993) accepted *C. nusatenggara* as a species but otherwise maintained the content of *Cynopterus* as given above.

Head and body length is 70–127 mm, tail length is 6–15 mm, forearm length is 55–92 mm, wingspan is 305–457 mm, and adult weight is about 25–100 grams. The fur is dense and variable in color but usually some shade of olive brown. These bats have prominent, almost tubular nostrils, and the upper lip is divided by a deep vertical groove. Their vocalizations in captivity have been described as loud, squeaking cries of two distinct notes.

These bats are found in forests and open country from sea level to 1,850 meters. Caves, deserted mines, and under the eaves of houses are common roosting sites; hollow trees are used only rarely. In Sri Lanka they prefer to roost in the folded leaves of the talipot palm. Groups of 6–12 may roost in one palm or in a group of palms. In flight short-nosed fruit bats are usually seen among bushes and low trees. They may travel 97–113 km in one night to feed on such fruits as palms, figs, guavas, plantains, mangoes, and chinaberries, as well as on such flowers as those of the sausage tree *(Oroxylum indicum).* Local abundance is often associated with the fruiting of such trees, so that they occasionally damage fruit plantations. These bats seem to subsist mainly on the juice of fruits rather than the pulp. By carrying off the fruit of the date palm *(Phoenix sylvestris)* and eating it elsewhere and disseminating the seeds through its droppings, *C. sphinx* becomes an agent of seed dispersal for this plant. Kitchener, Gunnell, and Maharadatunkamsi (1990) reported that pollen is an important component of the diet of *Cynopterus.*

This is the only Old World bat known to make shelters: *Cynopterus* occasionally bites off the "center seed string" in the fruit clusters of the kitul palm, thus leaving a hollow in which to hang. Details of such construction were discussed by Bhat and Kunz (1995). In southern India, Balasingh, Koilraj, and Kunz (1995) also observed *C. sphinx* to sever the stems of the curtain creeper and the stems and leaves of the mast tree, creating partially enclosed cavities ("stem tents") in which to roost. These enclosures may be 5–10 meters above the ground and have an internal volume of about 0.1–0.3 cubic meters. Construction of a tent takes place mostly at night and may require up to 50 days to complete. Evidently the work is accomplished by one male, which usually occupies the tent alone at night. For reproductive purposes a dominant male recruits a group of 2–19 females, which he defends against other males and which share the tent with him and their young during the day. There are two annual periods of tent construction, February–March and September–October, which generally coincide with biannual breeding seasons.

Data compiled by Mickleburgh, Hutson, and Racey (1992) indicate that most populations of *C. brachyotis* in the Philippines give birth twice annually; pregnant females have been found in nearly all months. Gestation there lasts approximately 3.5–4.0 months, lactation about 6–8 weeks.

Most females become pregnant at about 6–8 months, and males become sexually mature at about 1 year. Medway (1978) stated that pregnant females of *C. brachyotis* and *C. horsfieldi* had been collected throughout the year on the Malay Peninsula and that breeding was apparently nonseasonal. A single young was produced and was carried by the female during the early part of its life. According to Lekagul and McNeely (1977), breeding in Thailand was also nonseasonal, but most pregnancies in *C. brachyotis* occurred from March to June, with peaks also in January and September. Most pregnancies in *C. sphinx* reportedly were found in February and June, and gestation in that species was 115–25 days. Observations by Sreenivasan, Bhat, and Geevarghese (1974) suggested a well-defined breeding season for *C. sphinx* in Mysore, India. Two periods of pregnancy occurred between December or January and August, with a three-month gestation period. None of the pregnant females collected in this study had more than one embryo. Sandhu (1984) found births in this species to take place in February–March and again in June–July, with the second gestation period overlapping the first lactation period. Gestation lasted about 120 days. The young weighed about 11 grams at birth, were weaned 40–45 days later, and were carried by the mother for 45–50 days. Sexual maturity was attained by females at 5–6 months and by males at 15–20 months.

In certain areas, such as northern Thailand, dog-faced fruit bats are caught and sold in the markets for medicinal use. Some People in the Orient are said to eat these bats because they believe them to be "strength-giving." Information compiled by Mickleburgh, Hutson, and Racey (1992), however, indicates that hunting is not a major threat, that these bats adapt well to disturbed habitats and agricultural areas, and that most populations remain common, at least on the mainland and larger islands. The IUCN recognizes *C. nusatenggara* as a separate species (see above) and designates it as near threatened.

CHIROPTERA; PTEROPODIDAE; Genus MEGAEROPS
Peters, 1865

There are four species (Corbet and Hill 1992; Ellerman and Morrison-Scott 1966; Francis 1989; Hill 1983a; Hill and Boeadi 1978; Lekagul and McNeely 1977; Taylor 1934; Yenbutra and Felten 1983):

M. ecaudatus, southern Thailand, possibly Viet Nam, Malay Peninsula, Sumatra, Borneo;
M. kusnotoi, Java;
M. niphanae, northeastern India, Thailand, Viet Nam;
M. wetmorei, Malay Peninsula, Borneo, southern Philippines.

Yenbutra and Felten (1983) described *M. niphanae* on the basis of specimens from Thailand. Meanwhile, Hill (1983a) reported, with question, the presence of the previously described *M. ecaudatus* in northeastern India. Both authorities suggested that the specimens from each of the involved areas, as well as from Indochina, showed morphological affinity. Corbet and Hill (1992) subsequently indicated that the range of the two species probably is as shown above.

Head and body length is 76–102 mm and forearm length is about 49–59 mm. *M. wetmorei* has a rudimentary tail, whereas the other three species lack a tail. Lekagul and McNeely (1977) reported weights of 20–38 grams for *M. ecaudatus*. Coloration in *M. ecaudatus* is yellowish brown

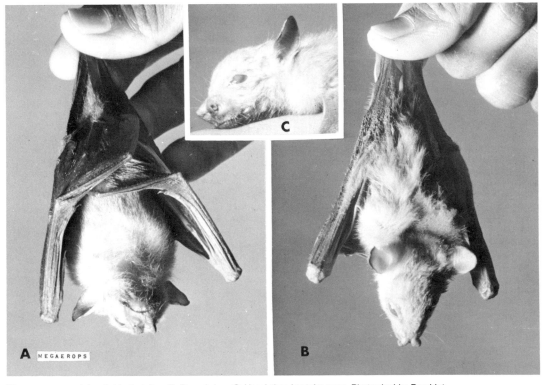

Megaerops ecaudatus: A. Ventral view; B. Dorsal view; C. Head showing tube nose. Photos by Lim Boo Liat.

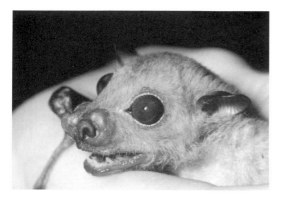

Megaerops ecaudatus, photo by Klaus-Gerhard Heller.

Ptenochirus jagori, photo by Paul D. Heideman.

above; the nape of the neck is pale gray, contrasting with the brown of the back; and the breast and belly are pale silvery gray. *M. wetmorei* has a silvery to ashen gray head, the ashen gray back is slightly lighter than the head, and the individual hairs have pale brown tips. The coloration beneath is slightly lighter than on the back. In *M. kusnotoi* the dorsal pelage is grayish brown and the ventral surface is a slightly paler gray.

The members of this genus are similar to *Cynopterus*, but the braincase is deeper. There is only one pair of incisors in the lower jaw, whereas *Cynopterus* has two pairs; the second upper incisor is reduced in length. The base of the thumb is partially enveloped in the wing membrane, so that it folds inward when the wing is folded.

In Thailand *M. ecaudatus* has been found both in mountains and on plains, in evergreen and dry evergreen forest, ranging from 500 to 3,000 meters (Lekagul and McNeely 1977). On the Malay Peninsula this species occurs in forest and open country and in lowlands and mountainous areas. Observations on the Malay Peninsula suggest an extended breeding season from about February to June (Medway 1978). *M. ecaudatus* may be fairly common, but *M. kusnotoi* is classified as vulnerable by the IUCN, and the subspecies *M. wetmorei albicollis* of Brunei and peninsular Malaysia was designated as indeterminate by Mickleburgh, Hutson, and Racey (1992) as its restricted lowland forest habitat is threatened by human development.

CHIROPTERA; PTEROPODIDAE; **Genus PTENOCHIRUS**
Peters, 1861

There are two species (Corbet and Hill 1992):

P. minor, Biliran, Leyte, Dinagat, Mindanao, and very
    doubtfully Palawan islands (Philippines);
P. jagori, throughout the Philippines.

In *P. minor* head and body length is 86–101 mm, tail length is 9–15 mm, and forearm length is 68–77 mm (Yoshiyuki 1979). In *P. jagori* head and body length is about 90–130 mm, tail length is 4–14 mm, forearm length is 72–90 mm, and weight is 61–102 grams (Rickart et al. 1993). Coloration is dark brown above and paler below. These bats resemble *Cynopterus* but differ from that genus in having only one pair of lower incisors instead of two and much reduced second upper incisors.

Ptenochirus has been taken from sea level to about 1,300

meters mostly in primary rainforest, and both species appear to be abundant. *P. jagori* probably usually roosts in tree hollows and perhaps in foliage but may be found in cave entrances in localities where caves are abundant. It feeds heavily on figs and bananas and also has been reported to eat other fruits and flowers, including those of coffee trees, coconut palms, and kapok *(Ceiba pentandra)*. It has been found at densities of about 1–3/ha. and apparently roosts singly or in groups of up to 10 or more. In the central Philippines *P. jagori* probably has two synchronized birth periods each year that are separated by four months and vary in timing from island to island. Gestation lasts about four months and lactation about three months. There usually is a single young, occasionally twins. Five-year-old individuals were still healthy and reproducing, but signs of age were apparent (Heideman and Heaney 1989 and *in* Mickleburgh, Hutson, and Racey 1992). On Mount Makiling, Luzon, Ingle (1992) found the birth periods to be April–May and September. Mudar and Allen (1986) found *P. jagori* to be the most abundant bat in their study area in northeastern Luzon. Although collected in a variety of habitats, it was most common over rivers, streams, and ponds. Pregnant and lactating females were taken in May, August, and October. Rickart et al. (1993) collected pregnant female *P. minor*, each with a single embryo, from early March to late May.

CHIROPTERA; PTEROPODIDAE; **Genus DYACOPTERUS**
Andersen, 1912

Dyak Fruit Bat

The single species, *D. spadiceus*, is known from a few specimens collected on the Malay Peninsula, Sumatra, Borneo, Luzon, and Mindanao (Heaney 1991; Hill 1961a; Kock 1969b; Peterson 1969; Start 1972a, 1975). Corbet and Hill (1992) considered the specimens from Sumatra and probably Luzon and Mindanao to represent a second species, *D. brooksi*, but such distinction was not accepted by Koopman (*in* Wilson and Reeder 1993).

Head and body length is 107–52 mm, tail length is about 13–18 mm, forearm length is 76–92 mm, and reported weight is 70–100 grams. The coloration is blackish on the face, yellowish on the shoulders, brown on the back and sides, and dull whitish on the chest and belly. There is a

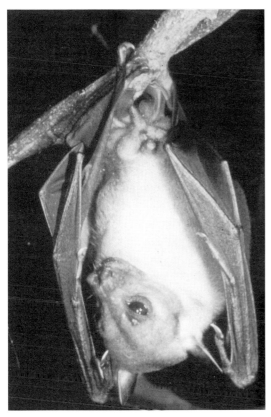

Dyak fruit bat *(Dyacopterus spadiceus)*, photo by Charles Francis through Thomas H. Kunz.

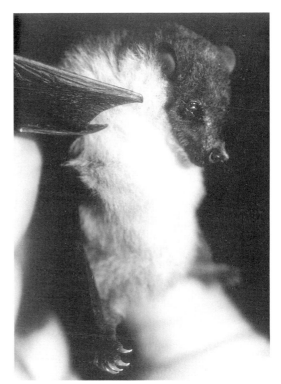

Black-capped fruit bat *(Chironax melanocephalus)*, photo by F. G. Rozendaal.

small opening just behind the orbit of the eye. In this respect, *Dyacopterus* resembles the genera *Cynopterus*, *Ptenochirus*, and *Megaerops*.

Dyacopterus has been found roosting in a tree trunk in a forest and has been netted in a paddy field (Medway 1978). It also has been caught near caves (Payne, Francis, and Phillipps 1985). Although little else has been reported about its social or reproductive habits, Francis et al. (1994) announced that during July and August 1992 in Malaysia a series of 10 mature males was collected; all of these males possessed functional mammary glands and were lactating. Of 6 mature males and 8 mature females taken in September 1992 none was lactating. Of 5 mature females taken in August 1993 one was lactating and 2 were pregnant; one of 4 mature males taken at the same time also was lactating. Previously, male lactation had been reported only in highly inbred domestic animals and under abnormal circumstances in humans. The phenomenon could suggest that *Dyacopterus* is a monogamous genus in which males share in the care of the young. The genus is classified as near threatened by the IUCN and may be jeopardized by deforestation (Mickleburgh, Hutson, and Racey 1992).

CHIROPTERA; PTEROPODIDAE; **Genus CHIRONAX**
Andersen, 1912

Black-capped Fruit Bat

The single species, *C. melanocephalus,* occurs on the Malay Peninsula (including southern Thailand), Sumatra, Nias Is-

land, Java, Borneo, and Sulawesi (Chasen 1940; Hill 1974*b*; Hill and Francis 1984; Lekagul and McNeely 1977).

Head and body length is about 65–70 mm, there is no external tail, forearm length is 40–50 mm, and weight is 12–17 grams (Lekagul and McNeely 1977; Payne, Francis, and Phillipps 1985). The head is black, the back is gray brown to reddish brown, the undersides are pale buff, and the wings are not spotted.

Chironax closely resembles *Balionycteris* but differs in dental characters, the lack of spots on the wings, and its smaller size. *Chironax* differs from *Cynopterus* in several features, the most conspicuous being the absence of a tail and the absence of a small opening just behind the orbit. In the absence of a postorbital foramen *Chironax* resembles the genera *Balionycteris, Thoopterus, Penthetor,* and *Sphaerias.*

Chironax is usually found at elevations above 600 meters. Groups of about two to eight individuals have been found during the daytime, resting on the undersides of tree ferns several meters above the ground, in virgin forests in Java. These individuals seemed to have been feeding mainly on wild figs *(Ficus).* On the Malay Peninsula pregnant females have been recorded only from January to April, suggesting that breeding there is restricted to the early months of the year (Medway 1978; Mickleburgh, Hutson, and Racey 1992). Pregnant females, each with a single embryo, were collected in Sulawesi in March and April (Bergmans and Rozendaal 1988).

Short-nosed fruit bat *(Thoopterus nigrescens)*, photo by Jan M. Haft.

CHIROPTERA; PTEROPODIDAE; **Genus**
THOOPTERUS
Matschie, 1899

Short-nosed Fruit Bat

The single species, *T. nigrescens*, now is known by the type specimen from Morotai Island in the North Moluccas, by many more specimens recently collected on Sulawesi and from Sangihe Island just to the northeast and Mangole Island just to the east of Sulawesi (Bergmans and Rozendaal 1988; Corbet and Hill 1992; Flannery 1995; Hill 1983*a*). It has been reported from Luzon and other islands in the Philippines, but its occurrence there is doubtful.

Head and body length is 94–109 mm (Flannery 1995) and the tail is a mere spicule. Forearm length is about 70–82 mm and weight is 67–99 grams (Bergmans and Rozendaal 1988). The coloration of the head and body is grayish brown. *Thoopterus* lacks postorbital foramina. It possesses grooved upper canines without ridges at the base; the fourth premolar and the first molar are greatly increased in breadth. The membranes attached to the hind feet are inserted on the second toe.

The species *T. nigrescens* was formerly placed in *Cynopterus* but was removed from that genus because of structural peculiarities. Hill (1983*a*) suggested affinity between *Thoopterus*, *Penthetor*, and *Latidens*. *Thoopterus* is almost identical to *Latidens* in external form and color and has a similarly long, strong rostrum. Dentally, however, *Thoopterus* approaches *Penthetor*. The three genera are easily distinguished by the number of incisor teeth, *Thoopterus* having 8; *Penthetor*, 6; and *Latidens*, 4.

According to Bergmans and Rozendaal (1988), *Thoopterus* has been collected in forests at elevations of 50–1,780 meters and evidently uses communal roosts.

Pregnant females, each with a single embryo, were taken in January. Subadults were taken in January, March, October, and December. The genus is designated as near threatened by the IUCN.

CHIROPTERA; PTEROPODIDAE; **Genus SPHAERIAS**
Miller, 1906

Mountain Fruit Bat

The single species, *S. blanfordi*, is now known from southern Tibet, southern China, northern India, Nepal, Bhutan, Burma, and northwestern Thailand (Corbet and Hill 1992).

Head and body length is about 64–80 mm, there is no external tail, and forearm length is about 50–52 mm. The color is dull grayish brown above and paler below. The absence of a tail is a character exhibited by several other supposedly related genera, but *Sphaerias* is unique in having the interfemoral membrane reduced to a narrow rim along the femur and the upper part of the tibia.

This bat seems to be confined to montane forest from 800 to 2,700 meters. Specimens from Thailand were collected in a pine and oak forest (Lekagul and McNeely 1977).

CHIROPTERA; PTEROPODIDAE; **Genus**
BALIONYCTERIS
Matschie, 1899

Spotted-winged Fruit Bat

The single species, *B. maculata*, occurs on the Malay Peninsula, the Riau Archipelago near Singapore, and Borneo (Lekagul and McNeely 1977).

Mountain fruit bat *(Sphaerias blanfordi)*, photo by B. Elizaboth Horner and Mary Taylor of specimen in Harvard Museum of Comparative Zoology.

Head and body length is about 50–66 mm, there is no external tail, and forearm length is 39–43 mm. Reported weight is 9.5–14.5 grams (Lim, Shin, and Muul 1972; Medway 1978). The color is sooty brown above and grayish below; the head is blackish, and the dark brown wings are marked with yellow spots. The only other fruit bats with well-defined yellow spots on the wings are the tube-nosed bats, *Nyctimene* and *Paranyctimene*. *Balionycteris* closely resembles the black-capped fruit bat, *Chironax*, in structural features.

The color pattern is suggestive of foliage and tree-dwelling habits. On the Malay Peninsula this bat is locally common in the forests of the lowlands and foothills. It is not normally a cave dweller but has been recorded from a cave in Sabah. In Selangor small groups of adults and young have been found roosting in the crowns of palms and in clumps of ferns epiphytic on forest trees (Medway 1978). On the Malay Peninsula females produce one young at a time; pregnancies or carried young have been recorded in nearly every month. The subspecies in that area and in the Riau Archipelago, *B. m. seimundi*, may be threatened by development and deforestation and was designated as indeterminate by Mickleburgh, Hutson, and Racey (1992).

CHIROPTERA; PTEROPODIDAE; Genus AETHALOPS
Thomas, 1923

Pygmy Fruit Bat

The single species, *A. alecto*, is known from the Malay Peninsula, Sumatra, western Java, northern Borneo, Bali,

and Lombok (Corbet and Hill 1992; Hill 1983a; Kitchener et al. 1993; Medway 1977, 1978). *A. aequalis* of Borneo was considered a separate species by Kitchener et al. (1990, 1993), but such distinction was not accepted by Corbet and Hill (1992), Koopman (*in* Wilson and Reeder 1993), or Mickleburgh, Hutson, and Racey (1992).

This is about the smallest of the Old World fruit bats. Head and body length is 65–73 mm, there is no external tail, and forearm length is 42–52 mm. One specimen weighed 19.3 grams (Medway 1978). The coloration is often black or one of various shades of dark gray.

This genus is characterized externally by the furred, narrow interfemoral membrane, the absence of a tail, the minute calcar bone on the foot, and the small ears. Boeadi and Hill (1986) stated that the inner upper incisors of *Aethalops* are less robust than the outer pair, whereas the reverse is true in the related genera *Balionycteris* and *Chironax*. *Aethalops* also differs from *Chironax* in having two, not four, lower incisors, and from *Balionycteris* in lacking a small last upper molar.

According to information compiled by Mickleburgh, Hutson, and Racey (1992), *Aethalops* is found only in hill and mountain forests at elevations of about 900–2,700 meters. It is presumed to feed on the soft fruits of trees and vines, though observations by Kitchener, Gunnell, and Maharadatunkamsi (1990) indicated that pollen also is important in the diet. It roosts alone or in groups of two or three. Pregnant females have been recorded on the Malay Peninsula from February to June and on Lombok in October. The genus seems to be uncommon in most areas and may be threatened by deforestation and development. It is classified as near threatened by the IUCN.

Spotted-winged fruit bat *(Balionycteris maculata)*, photo by Klaus-Gerhard Heller.

CHIROPTERA; PTEROPODIDAE; **Genus PENTHETOR**
Andersen, 1912

Dusky Fruit Bat

The single species, *P. lucasi,* inhabits the Malay Peninsula, Singapore and the nearby Riau Archipelago, Sumatra, and Borneo (Corbet and Hill 1992; Mickleburgh, Hutson, and Racey 1992).

Head and body length is about 114 mm, tail length is 8–10 mm, forearm length is 57–67 mm, and weight is 30–55 grams. The fur is coarse and smoky brown in color. This fruit bat may be recognized by the combination of one pair of lower incisors and the extreme thinness of the tail.

Andersen (1912) stated that *Penthetor* is without doubt the Indo-Malayan representative of the Austro-Malayan *Thoopterus*. He considered its principal claim to standing as a genus distinct from *Thoopterus* to be the absence of the first incisor on the lower jaw and the shortening of the second incisor on the upper jaw. Also, the toothrows extend farther backward; the incisors are more sharply needle-pointed; and the premolars and molars, though similar in outline and characters, are very similar to those of *Thoopterus* but lack all trace of the surface cusps so conspicuous in that genus. The tail is not reduced to a mere spicule, the insertion of the membranes on the hind feet is different, the digits are considerably shorter, and the tibia is unusually long.

The following account is based primarily on Medway (1978). On the Malay Peninsula the dusky fruit bat is irregularly distributed in lowland and hill forests. It normally roosts in caves, rock shelters, or crannies between large boulders, a habit that limits its distribution. It emerges immediately after dusk to feed at the nearest fruit plantation. Seeds and husks of many kinds of fruit often litter the cave floor under roosts, indicating that fruits are carried back to the cave to be eaten. This bat appears to be gregarious. Reproduction on the Malay Peninsula apparently is seasonal, as a large proportion of females examined in September were pregnant, a few were pregnant in June, but none were pregnant in January, February, March, or July. Start (1972*b*) reported that a female taken in Sarawak in May was lactating and that one taken in January had an advanced embryo.

CHIROPTERA; PTEROPODIDAE; **Genus HAPLONYCTERIS**
Lawrence, 1939

Fischer's Pygmy Fruit Bat

The single named species, *H. fischeri,* was long known only by a single specimen collected in 1937 on the slopes of Mount Halcon, Mindoro, Philippine Islands. Since the 1970s numerous additional specimens have been taken on Luzon, Mindanao, Negros, Leyte, Catanduanes, Biliran, Bohol, and Dinagat islands (Heaney, Heideman, and Mudar 1981; Heaney and Rabor 1982; Heideman 1988; Mudar and Allen 1986; Peterson and Heaney 1993). A record from Palawan is probably erroneous (Mickleburgh, Hutson, and Racey 1992). Specimens were first collected from Sibuyan Island in 1989 and may represent a second, larger species (Utzurrum 1992). Genetically the Sibuyan population is strikingly different from those on other islands (Peterson and Heaney 1993).

A series of 27 specimens from northeastern Luzon averaged 75.3 mm in total length and 52.1 mm in forearm length (Mudar and Allen 1986). A series of 15 specimens from Negros averaged 17.9 grams in weight (Heaney, Heideman, and Mudar 1981). Series totaling 34 specimens from Leyte and Biliran had an average head and body length of 68–80 mm, an average forearm length of 44–53 mm, and an average weight of 16–21 grams (Rickart et al. 1993). There is no tail. The thumb is long, about 25 mm in length, and the hind foot is short, about 13 mm in length. The coloration is given by the describer as "cinnamon brown" on the back, pale "mummy brown" in the shoulder region, darker in the head region, and "wood brown" beneath, with a slight tinge of silver along the midline of the belly.

None of the other genera similar to *Cynopterus* has such a relatively long thumb. One pair of incisors, one pair of canines, and four pairs of premolars and molars also set *Haplonycteris* apart from related genera. The describer stated: "*Haplonycteris* belongs in the group of small bats that resemble *Cynopterus* in which the postorbital foramen has become obliterated. In the reduction of the tooth formula with the strengthening of the remaining teeth, it appears to be the most highly evolved of this group. The unusually developed cusps and transverse ridges also indicate a greater degree of differentiation" (Lawrence 1939).

Mudar and Allen (1986) found *Haplonycteris* to exhibit a marked preference for forest habitat and often to be the only bat caught in nets set away from water or agricultur-

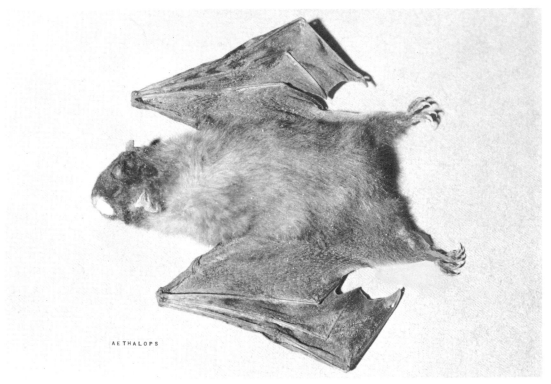

Pygmy fruit bat *(Aethalops alecto),* photo by R. Elizabeth Horner and Mary Taylor of specimen in Harvard Museum of Comparative Zoology.

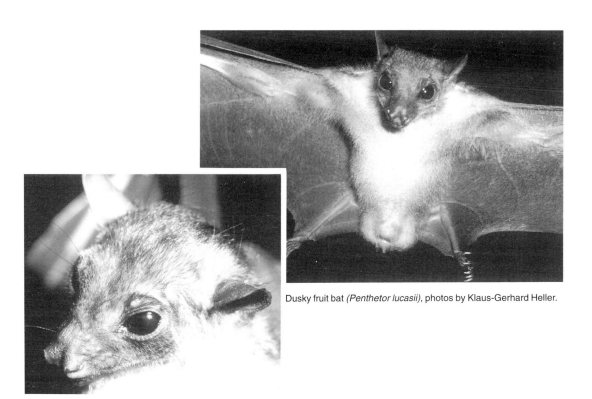

Dusky fruit bat *(Penthetor lucasii),* photos by Klaus-Gerhard Heller.

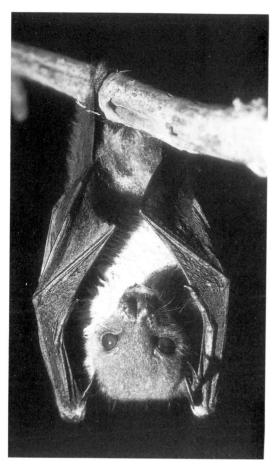

Fischer's pygmy fruit bat *(Haplonycteris fischeri)*, photo by Paul D. Heideman.

al fields. Data compiled by Mickleburgh, Hutson, and Racey (1992) indicate that *Haplonycteris* can be abundant in primary forest but is rare elsewhere, that it has been collected at elevations from below 100 meters to 1,550 meters, and that it feeds on figs and other fruits. Heideman and Heaney (1989) reported a population density of 3.7/ha.

Heideman (1988, 1989) reported that this bat's annual reproductive pattern seems to be associated with seasonal patterns of abundance of the fruit upon which it feeds. Females give birth just once annually. Births are highly synchronous within a given population, occurring in May or June on Negros Island but up to 3 months later in other areas. Although mating may occur throughout the year, synchronization of births is achieved by a postimplantation delay in embryonic development. Females also undergo a postpartum estrus 1–3 weeks after giving birth, at which time most become pregnant. The period of delay lasts up to 8 months and is followed (starting about March on Negros) by 3 months of rapid embryonic growth. The total maximum period of 11.5 months from mating to birth is the longest known in bats. A single young is born, and lactation lasts about 10 weeks. Most females born in June on Negros become pregnant in September–November, at 3–5 months of age. Males become sexually mature at 10–11 months. Heideman and Heaney (1989) reported that some individuals live more than 10 years.

Haplonycteris fischeri is classified as vulnerable by the IUCN. While too small and inconspicuous to be jeopardized by hunting, it is seriously threatened by habitat destruction. The forests on which it depends for survival have been reduced from about 75 percent of the land area of the Philippines to perhaps 10–20 percent in the last 100 years. Populations of the bat have declined correspondingly (Mickleburgh, Hutson, and Racey 1992).

CHIROPTERA; PTEROPODIDAE; Genus OTOPTEROPUS
Kock, 1969

The single species, *O. cartilagonodus,* is now known from much of Luzon Island in the northern Philippines (Heideman, Cummings, and Heaney 1993; Kock 1969a; Mickleburgh, Hutson, and Racey 1992; Mudar and Allen 1986; Ruedas, Demboski, and Sison 1994; Utzurrum 1992).

This is a very small, long-haired, tailless flying fox related to *Cynopterus.* Head and body length is 55–87 mm, forearm length is about 41–52 mm, and weight is 12–21 grams. The coloration of the back is dark blackish brown; the belly is lighter with more gray coloration. The hairs at the middle of the breast and belly are tipped with white. The eyes in *Otopteropus* are very large. The most noteworthy character of the genus is the ears, the edges of which are marked with reddish thickenings. On the front edge of the ear the thickening is less broad than it is thick; on the hind edge is a well-marked lobe. The naked nose has tubular nostrils. Although there is no tail, a cartilaginous calcarium is present. The uropatagium is thickly furred on the upper side and on the distal edge of the underside. The tibia are thickly furred, the feet and toes only thinly so. The wing membranes have a reticulated pattern. The dental formula for this genus is: (i 1/1, c 1/1, pm 3/3, m 1/1) × 2 = 24. Kock (1969a) wrote that the big eyes and small hearing apparatus of this genus, as well as the relatively narrow teeth and weak lower jaw, indicate a very species-specific life. The long, thick hair suggests adaptation for higher altitudes.

According to Mickleburgh, Hutson, and Racey (1992), *Otopteropus* has been taken at elevations of 200–1,750 meters throughout the mountains of Luzon. It is moderately common in primary lowland forest on southern Luzon and is present but uncommon in montane and mossy forest. Fig seeds have been found in the feces of many individuals.

Births apparently occur synchronously within a given population. On Mount Isarog in southern Luzon in March and May of 1988, Heideman, Cummings, and Heaney (1993) collected 22 females, each pregnant with a single embryo, whose development indicated that births would occur in late May, June, or early July. The simultaneous absence of lactating females, reproductively active males, and immature bats indicated that no births had occurred in the previous 6–10 months. That information, together with examination of the embryos and reproductive tracts, suggested that mating had occurred long before and that *Otopteropus* experiences an extended postimplantation delay in embryonic development such as has been described for *Haplonycteris.* Ruedas, Demboski, and Sison (1994) reported that a series from the Zambales Mountains of southern Luzon collected from mid-February to early March included both pregnant and nonpregnant females.

The IUCN designates *Otopteropus* as vulnerable. It has never been recorded outside of primary or good secondary forest and undoubtedly has declined with the progressive destruction of its habitat (Mickleburgh, Hutson, and Racey 1992).

Otopteropus cartilagonodus, photo by Lawrence R. Heaney.

CHIROPTERA; PTEROPODIDAE; **Genus**
ALIONYCTERIS
Kock, 1969

The single species, *A. paucidentata*, long was known only by seven specimens from Mount Katanglad, Bukidnon Province, Mindanao, Philippines (Kock 1969c). In 1992 a few additional specimens were collected at the same site (Utzurrum 1992).

Head and body length is 64–70 mm, forearm length is 45–50 mm, and weight is 14–18 grams (Ingle and Heaney 1992). The fur is brownish black in color, and on the dorsal side there are numerous soft guard hairs that are about double the length of the underfur. The blackish ears are naked, their edges being slightly thickened, as is usual in related genera. The rear legs, feet, and toes are thickly haired, as is the proximal third of the underarms.

Alionycteris is a very small, long-haired flying fox related to *Cynopterus*. It has long thumbs, no external tail, no interfemoral membrane, and no postorbital foramina. The divided naked nose is composed of two cylinders 4 mm in length that stand free on the end. The dental formula, which distinguishes this genus from all other known genera, is: (i 1/1, c 1/1, pm 3/3, m 1/2) × 2 = 26.

Kock (1969c) wrote that the various dental and cranial characters suggest that this bat consumes very soft food. The thick hair covering the body suggests that *Alionycteris* is adapted for life at higher elevations in the mountains. It now is classified as vulnerable by the IUCN. Given its re-

Alionycteris paucidentata, photos by Lawrence R. Heaney.

stricted range and the current rate of deforestation on Mindanao, it seems likely that it is declining and that its existence is threatened (Mickleburgh, Hutson, and Racey 1992).

CHIROPTERA; PTEROPODIDAE; Genus LATIDENS
Thonglongya, 1972

The single species, *L. salimalii*, long was known only by one specimen collected in 1948 from High Wavy Mountains, Madura district, South India, at an altitude of 750 meters. The specimen is in the collections of the Bombay Natural History Society, Bombay, India. In April 1993, during a survey by the Harrison Zoological Museum, six more specimens were captured in the same area (Hutson 1994).

This is a medium-sized bat similar in general appearance to *Cynopterus* but without an external tail. Bates et al. (1994) reported head and body length to be 102–9 mm in six specimens and forearm length to be 66–69 mm in seven specimens. The pelage is short and soft, generally dark brown to black, and lightly grizzled with pale hairs. The fur of the underparts, including chin and throat, is shorter and thinner than that of the upper parts and is light gray brown in color. The ears and wings are black. The ear membrane is thin and oval with no white rim. The wing membrane, brownish and rather thin, starts at the first toe of the foot and has no white along the fingers, unlike in *Cynopterus*. The index claw is present. The baculum is much larger than that of *Cynopterus* and other related genera.

Latidens possesses only one pair of upper and lower incisors, a character known among the Megachiroptera in *Dobsonia, Haplonycteris,* and *Harpyionycteris*. However, *Harpyionycteris* differs from *Latidens* in possessing strongly proclivous upper incisors and upper and lower canines. *Latidens* is readily separated from the other genera mentioned above in having the cheek teeth 4–4/5–5, whereas in *Dobsonia* they are 5–5/6–6 and in *Haplonycteris* 4–4/4–4.

The type specimen was captured on a coffee estate. The six specimens taken in 1993 were caught during the night while entering a small cave that obviously was not their day roost (Hutson 1994). The area is within a broad-leaved, montane forest interspersed with coffee bushes, at an altitude of about 460 meters (Bates et al. 1994). *Latidens* now is classified as critically endangered by the IUCN, which notes that the genus probably has declined to fewer than 50 individuals.

CHIROPTERA; PTEROPODIDAE; Genus NYCTIMENE
Borkhausen, 1797

Tube-nosed Fruit Bats

There are 13 species (Corbet and Hill 1992; Flannery 1990*b*, 1995; Heaney and Peterson 1984; Hill 1983*a*; Kitchener, Packer, and Maryanto 1993; Koopman 1979, 1982*b*, 1984*c*; Laurie and Hill 1954; Peterson 1991; Phillips 1968; Ride 1970; Smith and Hood 1983; Troughton 1931):

N. minutus, Sulawesi, Amboina Islands;

N. rabori, Negros Island (Philippines);

N. cephalotes, Sulawesi, Timor, Talaud Islands, Sula Islands, South Moluccas, south-central coast of New Guinea and Numfoor and Umboi islands off north coast;

N. albiventer, Halmahera Islands, New Guinea and some nearby islands, Banda and Aru islands, Bismarck Archipelago;

N. draconilla, south-central New Guinea;

N. bougainville, Solomon Islands;

Latidens salimalii, photo by N. M. Thomas, Harrison Zoological Museum, from *Bats of the Indian Subcontinent*, by P. J. Bates and D. L. Harrison (1997).

Tube-nosed fruit bat *(Nyctimene rabori)*, photo by Paul D. Heideman.

N. *robinsoni*, eastern Queensland;
N. *major*, several small islands off the northwestern coast of New Guinea, Bismarck Archipelago, Solomon Islands, Trobriand Islands, D'Entrecasteaux Islands, Louisiade Archipelago;
N. *sanctacrucis*, known only by the type specimen from Santa Cruz Island in the southwestern Pacific;
N. *cyclotis*, New Guinea, Biak and Numfoor islands just to the northwest, and New Britain Island just to the east;
N. *certans*, central and eastern New Guinea;
N. *aello*, New Guinea, Mysol and Salawati islands just to the west, and Kairiru Island just off the north-central coast;
N. *celaeno*, western and northwestern New Guinea.

Koopman (*in* Wilson and Reeder 1993) recognized N. *vizcaccia* and N. *masalai* of the Bismarcks and Solomons as distinct species and included *bougainville* within N. *vizcaccia*, but Flannery (1995) treated *vizcaccia* and *masalai* as synonyms of N. *albiventer* and regarded *bougainville* as distinct. Koopman listed N. *malaitensis* of the Solomons as a full species, but Flannery considered it a subspecies of N. *bougainville*. Flannery also suggested that N. *draconilla* is a synonym of N. *bougainville*. Peterson (1991) treated N. *papuanus* of New Guinea as a full species, but both Koopman and Flannery included it within N. *albiventer*. Kitchener, Packer, and Suyanto (1995) elevated N. *albiventer keasti* of the Kai, Banda, and Tanimbar islands, southwest of New Guinea, to specific level. Records of N. *aello* or N. *celaeno* from the Halmahera Islands are actually attributable to N. *albiventer* (Koopman and Gordon 1992).

Head and body length is 68–136 mm, tail length is 15–30 mm, and forearm length is 50–86 mm. Flannery (1990*b*, 1995) listed weights of 22–93 grams. The coloration is usually buffy gray above, but on some species it is buffy or creamy. The underparts are usually paler. A dark brown spinal stripe is usually present, and the wings, forearms, and ear membranes are speckled with yellow. The only other fruit bats with contrasting yellow spots are *Paranyctimene* and *Balionycteris*.

The common name refers to the tubular nostrils that project from the upper surface of the muzzle to a length of about 6 mm. This genus is distinguished from *Paranyctimene*, the only other genus of bats with which it could be confused, by dental features.

Nyctimene has been taken singly in foliage and on tree trunks, but the mottled and broken color pattern increases its chance of concealment. In its resting position the animal hangs freely, wrapped in the wing membranes. When these bats begin calling or when there is a disturbance the nostrils are stretched outward and moved with a slight trembling motion. Possibly the peculiar shape of the nostrils serves some purpose in giving ultrasonic sounds for orientation. While in flight these bats emit a high, whistling note.

McKean (1972) reported the collection of *Nyctimene* in the following habitats: N. *aello*, primary and secondary rainforest, and swamp forest; N. *albiventer*, primary and secondary rainforest, primary montane forest, sago swamp, and the vicinity of native gardens; and N. *cyclotis*, primary rainforest. Altitude ranged from sea level to 1,650 meters. Spencer and Fleming (1989) found N. *robinsoni* to roost alone and to generally forage within 200 meters of its day roost.

Vestjens and Hall (1977) reported that the stomachs of three N. *albiventer* contained insects but that most other stomachs had fruit or vegetable matter. Earlier observations indicated that captive individuals preferred soft, juicy fruit and would not take insects that were offered. To eat, the animal hangs horizontally or obliquely in a fruit bush with its thumbs inserted into the fruit. It turns its lips up on the fruit and bites pieces with its lower jaw, the upper teeth merely aiding the lips in supporting the lower jaw. It

Tube-nosed bat *(Nyctimene major)*, photo by L. J. Brass.

shoves the bitten pieces toward its breast and belly with its muzzle, then masticates them mainly for the juice. A fringe of fleshy lobes on the inner edge of the lips seems to aid the bat. Captives that fed on guavas were not observed to bite into the inside of the fruit, and the nostrils did not come in contact with the fruit at any time. The pulp of young coconuts is also eaten.

McKean and Price (1967) reported that two females of *N. robinsoni* taken in late August in Queensland contained single embryos. McKean (1972) reported the following reproductive information: pregnant females of *N. aello* collected in New Guinea in January and February; a lactating female *N. albiventer* taken on Bougainville in July; pregnant females of *N. albiventer*, each with one embryo, taken in Papua New Guinea in January, July, and August; and lactating females of *N. albiventer* taken in Papua New Guinea in February, May, and August. Flannery (1995) added the following: all female *N. albiventer* taken on New Britain in December, but none taken on Salawati in October, were lactating; pregnant female *N. bougainville* taken in October and May; pregnant female *N. cephalotes* taken in the Sula Islands in November; a lactating female *N. cyclotis* taken on Supiori Island in September; and lactating female *N. major* taken in December and January. According to Mickleburgh, Hutson, and Racey (1992), pregnant *N. cephalotes* were collected in Sulawesi in October and January; female *N. robinsoni* give birth to one young from October to December; and female *N. rabori* produce one young in April or early May, have a gestation of 4.5–5.0 months and a lactation period of about 3–4 months, and first become pregnant at about 7–8 months.

The IUCN classifies *N. cyclotis*, *N. certans*, and *N. aello* as near threatened, *N. minutus*, *N. masalai* (recognized as a species distinct from *N. albiventer*; see above), *N. draconilla*, *N. malaitensis* (recognized as a species distinct from *N. bougainville*; see above), and *N. celaeno* as vulnerable, *N. rabori* as critically endangered, and *N. sanctacrucis*

as extinct. The last has not been seen since 1907 despite a subsequent expedition to locate it, and it probably disappeared because of disturbance. *N. rabori* depends almost entirely on undisturbed, upland forest of Negros Island, where it occurs at low densities; at current rates of forest destruction and fragmentation there will be insufficient habitat remaining in another 10 years (Mickleburgh, Hutson, and Racey 1992). Two specimens, possibly referable to *N. rabori*, were collected recently on Sibuyan Island (Goodman and Ingle 1993).

CHIROPTERA; PTEROPODIDAE; **Genus PARANYCTIMENE**
Tate, 1942

Lesser Tube-nosed Fruit Bat

The single species, *P. raptor*, is now known to occur in Papua New Guinea (Laurie and Hill 1954; McKean 1972) but probably is widespread throughout New Guinea (Flannery 1990*b*; Koopman 1982*b*).

Head and body length is 63–93 mm, tail length is 17–21 mm, forearm length is 48–54 mm, and weight is 22–33 grams (Flannery 1990*b*). The coloration is grayish brown above and dull yellowish buff below. There is a greenish cast to the membranes and the fur around the nostrils. A dorsal spinal stripe is lacking. The spotted color pattern and the tubular nostrils are similar to those of the genus *Nyctimene*.

Tate stated that this bat can be distinguished from *Nyctimene* by the high molar and premolar cusps, the extreme height and slenderness of the upper and lower canines and premolars, the elongation of the postdental palate, and the absence of the dorsal stripe. Both upper and lower canines are daggerlike, apparently having seizing or grappling functions. As in *Nyctimene*, the lower incisors are absent.

McKean (1972) reported the collection of 27 specimens in primary and secondary forest at altitudes ranging from sea level to 260 meters. Of these specimens 7 were pregnant females collected in January, February, and May, each with one embryo. Flannery (1990*b*) reported an altitudinal range of near sea level to 1,000 meters, that he netted two lactating females in May, and that an apparent group consisting of three males and a female carrying one young was taken in July or August. The type specimen of *P. raptor*, an adult female taken in August, was carrying a nursing young. The species appears to be uncommon and is classified as near threatened by the IUCN.

CHIROPTERA; PTEROPODIDAE; **Genus EONYCTERIS**
Dobson, 1873

Dawn Bats

There are two species (Corbet and Hill 1992; Flannery 1995):

E. spelaea, far southwestern and northeastern India, Bangladesh, extreme southern China, Burma, Thailand, Indochina, Malay Peninsula, Andaman Islands, Sumatra, Java, Bali, Borneo, throughout the Philippines, Sulawesi, Muna Island, Lombok, Sumba, Timor, Sula Islands, Halmahera Islands;

E. major, Borneo, throughout the Philippines, doubtfully North Pagai in the Mentawai Islands.

Lesser tube-nosed fruit bat *(Paranyctimene raptor)*, photo by Pavel German.

Dawn bat *(Eonycteris spelaea)*, photo by Klaus-Gerhard Heller.

E. rosenbergi of northern Sulawesi and *E. robusta* of the Philippines sometimes have been regarded as separate species (Rickart et al. 1993) but were included in *E. spelaea* and *E. major,* respectively, by both Corbet and Hill (1992) and Koopman (1989a and *in* Wilson and Reeder 1993).

Head and body length is about 85–125 mm, tail length is about 12–33 mm, and forearm length is 60–85 mm. Weight of *E. spelaea* is 55–82 grams in males and 35–78 grams in females (Beck and Lim 1973; Bhat, Sreenivasan, and Jacob 1980). The upper parts are brownish, usually dark brown, and the underparts are paler. The sides of the neck in females tend to be thinly haired, whereas the same area in males is covered with a "ruff" of long hairs slightly darker in color than the hairs of the head and chest.

In *Eonycteris* and five other genera of fruit bats in the subfamily Macroglossinae the tongue is long, slender, and protrusible with brushlike projections for picking up nectar and pollen, and the snout is long and slender. The cheek teeth are small, barely showing above the gums. The macroglossine type of tongue, in combination with the length of the tail, is characteristic of *Eonycteris.* In all other genera of macroglossine bats except *Notopteris* the tail is reduced to a spicule or is absent; in *Notopteris* it is as long as the forearm. The absence of a claw on the index finger also distinguishes *Eonycteris. E. major* has a kidney-shaped gland on either side of the anus in both sexes.

Dawn bats occur in a variety of habitats, including forests and cultivated areas. They usually roost in caves, but *E. major* also has been taken in hollow trees in Borneo (Medway 1977). The natural food seems to consist mainly of pollen and nectar (Medway 1978).

E. spelaea has been noted darting in and out among agave flowers, which are remarkable for their long and projecting stamens. The bats alighted occasionally for a few minutes. Their preference for this particular type of agave was so marked "that neither shooting nor artificial light would frighten them away." The stomach of one of these bats contained only pollen. D. P. Erdbrink stated that in his

Dawn bat *(Eonycteris spelaea)*, photo by Jan M. Haft.

experience the feeding habits of *Eonycteris* are so similar to those of *Nanonycteris* that the description of the latter given by Baker and Harris (1957) and quoted in the account of *Nanonycteris* could apply to *Eonycteris* verbatim.

These bats are gregarious. Several hundred individuals of *E. major* were found clinging to the ceiling of a cave in Luzon. Bhat, Sreenivasan, and Jacob (1980) found that roosting colonies were divided into clusters that were segregated by sex. Beck and Lim (1973) reported that colonies of *E. spelaea* in Malaysia range from a few dozen individuals in shallow limestone shelters to tens of thousands in the larger massifs. In a study at one of the major roosting caves on the Malay Peninsula Beck and Lim found that at any one time more than 50 percent of the adult females were either pregnant or lactating or both. The females were polyestrous, and successive pregnancies could begin in the late stages of lactation. The gestation period seemed to be slightly longer than 6 months, possibly as long as 200 days. The usual number of young was one, but two fetuses were found in rare instances. Shortly after parturition the young took hold of a nipple and remained firmly attached for 4–6 weeks as the female flew about. After this period the young

could detach readily, and they made short flights on their own, but weaning did not occur until at least 3 months. Sexual maturity was attained sometime after the first year of life, in males possibly not until after the second year.

Although neither species is given an overall classification by the IUCN, Mickleburgh, Hutson, and Racey (1992) designated the subspecies *E. major robusta* of the Philippines as rare and *E. spelaea glandifera* of Java, the Lesser Sunda Islands, Borneo, the Philippines, and Sulawesi as vulnerable. Populations in Java and the Lesser Sundas actually were considered endangered because of habitat destruction, cave disturbance, and intensive hunting for use as food. These same problems apply to *robusta* in the Philippines, which has become very rare in association with elimination of primary forests. Populations of *glandifera* in the Philippines seem more adaptable to habitat modification but are extremely vulnerable to hunting and disturbance by people. Colonies there, which may have a potential size of tens of thousands of individuals, now mostly contain well under 500 bats. Even some caves that harbored up to 500 in the 1970s or early 1980s were found to have none or fewer than 50 in the late 1980s.

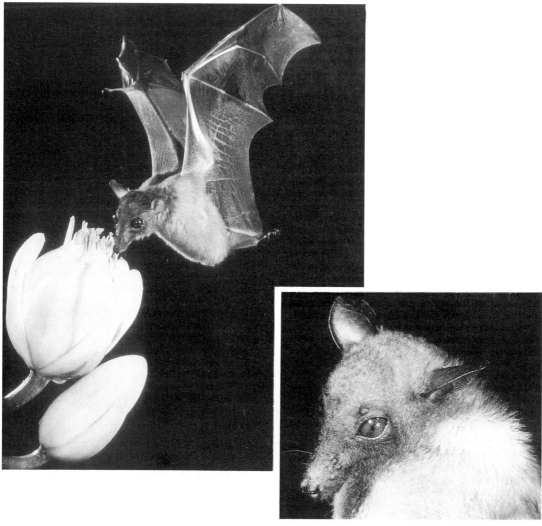

African long-tongued fruit bat *(Megaloglossus woermanni)*, photos by Merlin D. Tuttle, Bat Conservation International.

CHIROPTERA; PTEROPODIDAE; **Genus**
MEGALOGLOSSUS
Pagenstecher, 1885

African Long-tongued Fruit Bat

The single species, *M. woermanni*, occurs from Guinea to Uganda and southern Zaire (Bergmans and Van Bree 1972; Hayman and Hill *in* Meester and Setzer 1977).

Head and body length is 60–82 mm, the tail is a mere spicule or is absent, and forearm length is 37–50 mm. Jones (1971*a*) reported a weight of 8.4–15.6 grams in nine females and 13.2 grams in one male. Coloration is dark brown above and lighter brown below with an indistinct dark longitudinal stripe from the middle of the crown to the nape. Adult males have creamy or buff-colored hairs on the foreneck and sides of the neck.

This is the only genus of the subfamily Macroglossinae in which the tail is reduced or absent and the fifth metacarpal bone of the hand is shorter than the third. Other diagnostic characters are the long, slender, and protrusible tongue with brushlike projections, the clawed index finger, and the fifth hand bone shorter than the third.

Wolton et al. (1982) collected this bat at all altitudes in both primary and secondary vegetation. It has been found roosting in plantain leaves, in the foliage of shrubs, and in a native hut. It feeds on pollen and nectar (Rosevear 1965). Bergmans and Van Bree (1972) reported the apparently unusual event of two individuals of this genus being collected in a cave. Data compiled by Mickleburgh, Hutson, and Racey (1992) indicate that *M. woermanni* is a lowland forest species, that it sometimes is solitary but may congregate in unisexual or bisexual groups depending on the time of year, and that its reproductive pattern is either seasonally polyestrous, with two breeding seasons a year, or aseasonally polyestrous. Kingdon (1974*a*) stated that a lactating female was taken in mid-December and that two lactating females and a flying juvenile were taken in Uganda in late February. Jones (1971*a*) reported the collection of a lactating female in April and a pregnant female in September in Equatorial Guinea and two pregnant females in August in Ghana. In Congo pregnant females were collected in March and nursing young in February and May (Bergmans 1979*c*). In Liberia pregnant or lactating females were taken in July and August (Wolton et al. 1982). Czekala and Benirschke (1974) reported a pregnant female with two fetuses but indicated that normally a single young was produced.

CHIROPTERA; PTEROPODIDAE; **Genus**
MACROGLOSSUS
F. Cuvier, 1824

Long-tongued Fruit Bats

There are two species (Chasen 1940; Corbet and Hill 1992; Flannery 1995; R. E. Goodwin 1979; Hill 1983*a;* Koopman 1979, 1984*c*, 1989*a;* Laurie and Hill 1954; Lekagul and McNeely 1977; Ride 1970, Taylor 1934):

M. minimus, southern Thailand and Indochina, Malay Peninsula, Java and some nearby islands, Kangean Islands, Borneo, Natuna Islands, throughout the Philippines, Sulawesi and nearby islands, Sangihe Islands, Lombok, Timor, Molucca Islands, New Guinea

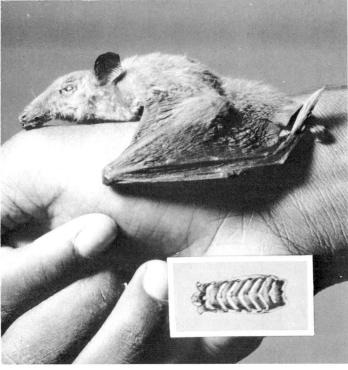

Asiatic long-tongued fruit bat *(Macroglossus minimus)*, photos by Lim Boo Liat. Inset: palate *(Macroglossus* sp.*)*, photo from *Jahrb. Hamburgischen, Wissenschaft. Anstalten.*

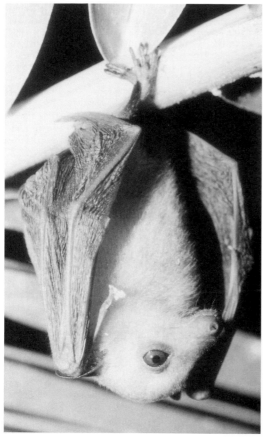

Long-tongued fruit bat *(Macroglossus minimus)*, photo by Jan M. Haft.

and nearby islands, Bismarck Archipelago, Solomon Islands, northern Australia;

M. sobrinus, northeastern India to Indochina, Malay Peninsula, Sumatra, Nias Island, Siberut, Sipora, North Pagai Island, Krakatau, Java, Bali.

Other species sometimes have been distinguished, but Koopman (*in* Wilson and Reeder 1993) accepted only these two. Kitchener and Foley (1985) reported that measurements of specimens from Bali overlap both *M. sobrinus* and *M. minimus.*

Head and body length is about 60–85 mm, the tail is rudimentary, and forearm length is 36–49 mm. Lekagul and McNeely (1977) gave weights of 18.5–23 grams for *M. sobrinus* and 12–18 grams for *M. minimus.* Coloration is russet brown above and paler below with a more or less definite longitudinal stripe of darker brown from the crown to the nape. Males lack the neck tufts present in adult males of *Eonycteris* and *Megaloglossus. Macroglossus* is distinguished from the other genera of the subfamily Macroglossinae by cranial and dental features. Its call has been described as a piercing high shriek.

According to Lekagul and McNeely (1977), in Thailand *M. sobrinus* is found in evergreen forest from sea level to 2,000 meters and *M. minimus* is usually confined to coastal areas, especially in the vicinity of mangroves. However, in the Philippines *M. minimus* occurs from sea level to at least 1,500 meters and uses a wider variety of habitats than does any other fruit bat in the country (Mickleburgh, Hutson,

and Racey 1992). Although long-tongued fruit bats shelter under the branches of trees and under roofs, the preferred daytime retreat seems to be in the rolled leaves of hemp and banana plants. On the island of Bali an individual of *M. sobrinus* was found in a pisang leaf. *Macroglossus* seems to prefer the pollen and nectar of a variety of plants, especially wild and cultivated bananas, coconuts, and mangroves (Mickleburgh, Hutson, and Racey 1992); the juices of soft fruits also are taken.

Lekagul and McNeely (1977) mentioned that *M. sobrinus* has been found roosting in groups of 5–10 individuals. However, like many of the smaller fruit bats, *M. minimus* is usually solitary, which reduces the chance of discovery. In the Philippines, five men looked in hemp plants for one day and found only one *Macroglossus.* Usually only one individual or a mother with a single young is found at a roost site there. Reproduction there is apparently aseasonal (Mickleburgh, Hutson, and Racey 1992). Most females there probably have two young per year, giving birth to the second soon after weaning the first, as a result of a postpartum estrus. Breeding of *M. minimus* is asynchronous and occurs throughout the year in Papua New Guinea, but females there have been found to store sperm in the reproductive tract (Hood and Smith 1989). Two female *M. minimus* collected in August on Buru Island had embryos almost ready to be born. Phillips (1968) reported that lactating female *M. minimus* were collected in the Solomons in January, March, and December. Lim, Shin, and Muul (1972) reported that a female *M. minimus* captured in Sarawak in mid-June had a single embryo and that pregnant females of this species had been found on the Malay Peninsula in April and June. McKean (1972) found pregnant females of *M. minimus,* each with a single embryo, in every month of the year during which any females were collected in New Guinea and the Solomons. Smith and Hood (1981) thought this species bred throughout the year in the Bismarck Archipelago.

CHIROPTERA; PTEROPODIDAE; **Genus SYCONYCTERIS**
Matschie, 1899

Blossom Bats

There are three species (Flannery 1990*b,* 1995; Hill 1983*a;* Koopman 1982*b,* 1984*c;* Laurie and Hill 1954; Lidicker and Ziegler 1968; Maryanto and Boeadi 1994; McKean 1972; Ride 1970; Rozendaal 1984; Ziegler 1982*a*):

S. australis, southern Molucca Islands (Buru, Seram, Amboina, Kei), New Guinea and some nearby islands, Bismarck Archipelago, Trobriand Islands, Louisiade Archipelago, D'Entrecasteaux Islands, eastern Queensland and New South Wales;
S. hobbit, west-central to eastern New Guinea;
S. carolinae, Halmahera and Batjan islands in the northern Moluccas.

Kitchener, Packer, and Maryanto (1994) noted the distinction of the subspecies *S. australis major* on Amboina and Seram and suggested that it may warrant recognition as a full species.

In *S. australis* and *S. hobbit* head and body length is about 50–75 mm, the tail is a mere spicule, and forearm length is 38–50 mm. McKean (1972) gave the weight of *S. australis* as 11.5–25.0 grams, and Ziegler (1982*a*) listed weights of 15.1–15.7 grams for *S. hobbit.* In the much larg-

Syconycteris australis, photo by Stanley Breeden.

Guinea; a single case of twin fetuses was detected. Flannery (1995) reported the collection of pregnant and lactating female *S. carolinae* in early January.

The IUCN now classifies *S. hobbit* and *S. carolinae* as vulnerable. Each is know to occur only within a small area of deteriorating habitat.

CHIROPTERA; PTEROPODIDAE; **Genus**
MELONYCTERIS
Dobson, 1877

There are three species (Flannery 1993*a*):

M. melanops, New Britain, New Ireland, Long, Tolokiwa, and Moiko islands in the Bismarck Archipelago;

M. woodfordi, Buka, Bougainville, Alu, Fauro, Mono, Oblari, Choiseul, Ysabel, Nggela, and Florida islands in the Solomons;

M. fardoulisis, Vella Lavella, Kolombangara, New Georgia, Guadalcanal, Malaita, Makira (San Christobal), and possibly Russell islands in the Solomons.

Koopman (*in* Wilson and Reeder 1993) listed another species, *M. aurantius,* but Flannery (1993*a*) considered it a subspecies of *M. woodfordi. M. woodfordi* and *M. aurantius* sometimes had been put in the separate genus or subgenus *Nesonycteris* Thomas, 1887, but Phillips (1968) placed that name in the synonymy of *Melonycteris.*

Head and body length is about 77–106 mm, there is no tail, and forearm length is about 42–65 mm. Weight is 25–61 grams (Flannery 1995). In *M. melanops* the head is

er *S. carolinae* head and body length is 83–98 mm, forearm length is 56–62 mm, and weight is 35–47 grams (Flannery 1995). The coloration is reddish brown or grayish brown to dark brown above and lighter below. Males do not have neck tufts.

As in the other bats of the subfamily Macroglossinae the tongue is long, slender, and protrusible with brushlike projections that pick up nectar and pollen. This genus is distinguished by dental features, especially the large size of the upper incisors and the great difference in size between the first and second lower incisors.

McKean (1972) collected *S. australis* in a variety of habitats covering almost all forest and woodland formations occurring in Papua New Guinea and at altitudes ranging from sea level to 2,860 meters. Ziegler (1982*a*) found *S. hobbit* only in montane moss forest at elevations above 2,200 meters. One *S. australis* was caught on a fruit that it was devouring. *Syconycteris,* however, is specialized for feeding predominantly on nectar and pollen. Law (1994) reported *S. australis* to occur at population densities of 1–17.5/ha. in New South Wales, abundance being correlated with the availability of nectar and pollen. In a radio-tracking study Law (1993) determined that the roosts of *S. australis* were used for just one day and were located about 1 km from the foraging area. In that investigation all individuals were found to roost alone, but Ride (1970) had indicated that in New South Wales *S. australis* may form "camps" like those of *Pteropus* for two to four weeks in October.

McKean (1972) found pregnant or lactating females of *S. australis* in New Guinea during most months from January to September. Each pregnant female contained only one embryo. Based on the examination of extensive preserved material, Lawrence (1991*a*) concluded that breeding of *S. australis* continues throughout the year in New

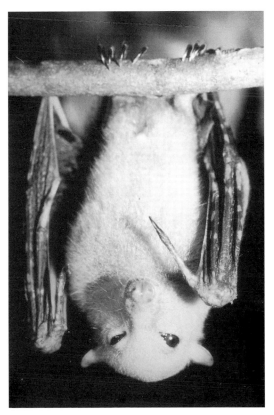

Melonycteris woodfordi, photo by Jan M. Haft.

dark brown, the back is pale to golden cinnamon, and the underparts are dark brown, almost black. In *M. woodfordi* the head and nape are pale russet, the back is orange to russet cinnamon, and the underparts are grayish brown, paler than the back. In *M. fardoulisi* the head and back are generally brown and the underparts are drab or brown; males are darker than females (Flannery 1993*a*). All three species have long, slender, protrusible tongues with bristlelike papillae.

M. melanops may be distinguished from all other fruit bats of the family Pteropodidae by its tongue and by the color of the underparts, which are darker than the back. This species normally has a claw on the index (second) finger and has four lower incisors. *M. woodfordi* and *M. fardoulisi* differ from all other Old World fruit bats by usually lacking a claw on the index finger and by possessing five pairs of upper cheek teeth, six pairs of lower cheek teeth, two pairs of upper incisors, and one pair of lower incisors. None of the species of *Melonycteris* has the neck tufts that are present on adult males of *Eonycteris* and *Megaloglossus*.

These bats are assumed to feed on nectar, pollen, and the juices of soft fruits. According to Flannery (1995), *M. woodfordi* is the most common species, normally being found in disturbed habitats and agricultural areas, but its specific diet is unknown and no roosting site has ever been found. *M. melanops* also is relatively common but has only been collected within dense stands of banana trees or on cocoa plantations. *M. fardoulisi* is considered a vulnerable species but has been found in disturbed areas as well as primary montane forest; it has been observed feeding on the flowers of *Heliconia*. Pregnant or lactating female *M. woodfordi* were collected in November, May, and March, suggesting that reproduction is nonseasonal; the young are never carried by the mother. Pregnant and lactating female *M. melanops* were taken in June and July, and lactating females in December, suggesting possibly that births occur from June to August, followed by a long period of lactation. Pregnant and lactating female *M. fardoulisi* were taken in March. That species and *M. aurantius* (recognized as distinct from *M. woodfordi;* see above) are classified as vulnerable by the IUCN; both are expected to decline because of loss of suitable habitat.

CHIROPTERA; PTEROPODIDAE; **Genus NOTOPTERIS**
Gray, 1859

Long-tailed Fruit Bat

The single species, *N. macdonaldi*, occurs in Vanuatu (New Hebrides Islands), the Fiji Islands, and New Caledonia in the southwestern Pacific (Hill 1983*a*). It also is represented by skeletal remains found at an archeological site, dating from less than 3,000 years ago, on 'Eua in the Tonga Islands (Koopman and Steadman 1995). It has been reported from Ponapé in the Carolines, but Corbet and Hill (1991) listed that record with question. Flannery (1995) treated *N. neocaledonica* of New Caledonia as a separate species, but such was not done by Koopman (*in* Wilson and Reeder 1993) or other recent authorities.

Long-tailed fruit bat *(Notopteris macdonaldi)*, photos by Anthony Healey

Head and body length is 93–110 mm, tail length is 43–62 mm, and forearm length is 59–69 mm; weight was 70–73 grams in 5 males and 56–64 grams in 5 females (Flannery 1995). Nelson and Hamilton-Smith (1982) reported an average weight of 51 grams for 12 females and 61 grams for two males. *Notopteris* is the only member of the family Pteropodidae with a long, free tail. Coloration is olive brown to dark brown, and neck tufts are lacking.

Notopteris resembles *Dobsonia* in that it lacks a claw on the index finger and has the wing membranes attached along the midline of the back, thus covering the fur on the back of the body. The tongue is like those of other bats in the subfamily Macroglossinae, that is, long, slender, and protrusible with bristlelike papillae.

This bat frequently roosts in caves, though natives on New Caledonia have seen it roosting in hollow trees. Flannery (1995) suggested that it travels daily from lowland roosting caves to foraging areas in the montane forests, though it also has been taken while feeding in gardens near sea level. About 300 long-tailed fruit bats have been observed resting in clusters of 5–25 individuals in a cave on New Caledonia. Their roosting habits are similar to those of some of the Microchiroptera, especially *Tadarida brasiliensis*. The diet is assumed to comprise nectar, pollen, and the juices of fruits. A captive female was kept alive for three days on the juices from canned pears, peaches, and sugar water; it refused solid food. Death was attributed partly to a lack of food.

The sexes apparently do not segregate at any time. On New Caledonia pregnant or lactating females have been collected in July and December and young have been noted in August, December, and January; females evidently are polyestrous (Nelson and Hamilton-Smith 1982). In Fiji pregnant and lactating females were collected in May, juveniles and females carrying young were taken in October, and lactating females were taken in November. In Vanuatu lactating females and largely independent young were collected in May. As in some other bats, the single young of *Notopteris* is carried by its mother for a period following birth, clinging to her underparts with mouth, wing hook, and claws.

The IUCN classifies *N. macdonaldi* as vulnerable. Mickleburgh, Hutson, and Racey (1992) designated the subspecies *N. m. macdonaldi* of Vanuatu and Fiji as vulnerable because of human disturbance of roosting caves and hunting for use as food. Flannery (1995), however, regarded that subspecies as secure but *neocaledonica* of New Caledonia as vulnerable.

CHIROPTERA; **Family RHINOPOMATIDAE; Genus RHINOPOMA**
E. Geoffroy St.-Hilaire, 1818

Mouse-tailed Bats, or Long-tailed Bats

The single known genus, *Rhinopoma*, contains four species (Hill 1977b; Nader and Kock 1983b; Van Cakenberghe and De Vree 1994):

R. microphyllum, Morocco, Mauritania, Burkina Faso, Nigeria, Cameroon, Egypt, Sudan, Israel and Saudi Arabia to Afghanistan and southern India, Sumatra;
R. muscatellum, United Arab Emirates, Oman, Iran, southern Afghanistan, Pakistan, India;
R. hardwickei, Morocco and Senegal to Egypt and

Mouse-tailed bat *(Rhinopoma hardwickei)*, photo by P. Morris.

Somalia, Israel and Yemen to Afghanistan and India, Socotra Island;
R. macinnesi, Ethiopia, Kenya, Somalia.

In addition to the above, Van Cakenberghe and De Vree (1994) described the subspecies *R. hardwickei sondaicum* based on two specimens probably collected in 1899 from indefinite localities in the Sunda Archipelago, a region extending approximately from Sumatra to Sulawesi and Timor. Van Cakenberghe and De Vree also referred to early records of *Rhinopoma* from Burma, Thailand, and Malaysia, but any specimens involved apparently have been lost. These records have been variously attributed to *R. microphyllum* and *R. hardwickei* (Corbet and Hill 1992; Koopman *in* Wilson and Reeder 1993; Lekagul and McNeely 1977).

Head and body length is 53–90 mm, tail length is 43–75

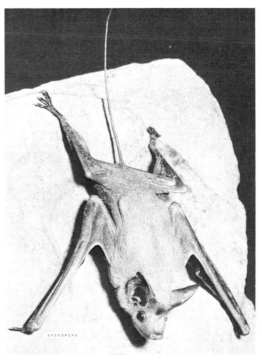

Mouse-tailed bat *(Rhinopoma microphyllum)*, photo by Erwin Kulzer.

mm, forearm length is 45–75 mm, and adult weight is about 6–14 grams. The soft fur is lacking on the face, rump, and posterior portion of the abdomen. The coloration is grayish brown or dark brown above and usually paler below.

The bats of this genus are the only ones in the suborder Microchiroptera with a tail nearly as long as the head and body combined. The tail membrane is extremely short and narrow. The large ears, connected by a band of skin across the forehead, extend beyond the nostrils when the ears are laid forward. A well-developed tragus is present. The snout bears a small, rounded nose leaf. The valvular nostrils appear as slits that open forward and can be closed at will. The nasal chambers are swollen laterally, the muzzle having a distinct ridgelike outgrowth of skin. The nasal bones are expanded laterally and vertically, and the frontal bones are depressed, forming a concavity in the forehead. The teeth are of the normal insectivorous kind. The dental formula is: (i 1/2, c 1/1, pm 1/2, m 3/3) × 2 = 28.

Mouse-tailed bats are usually found in treeless and arid regions. They roost in caves, rock clefts, wells, pyramids, palaces, and houses. *R. microphyllum* has inhabited certain pyramids in Egypt for 3,000 years or more. These bats often hang by the thumbs as well as the feet. *R. muscatellum* has an unusual characteristic flight. It rises and falls, much like some small birds. Usually flying at least six to nine meters above the ground, this species travels by a series of glides, some of great length, and occasionally it flutters.

Fat, sometimes equaling the normal weight of the individual, accumulates in the autumn beneath the naked skin, especially in the abdominal region; this fat is absorbed during the winter. *R. microphyllum*, at least, does not hibernate but does remain in a torpor during the winter and does not move about in search of food. During this period the bat is able to survive for several weeks in captivity without food and water. The resorption of the accumulated fat in the winter suggests that torpidity is probably an adaptation to circumvent the scarcity of insect food during the cold season (Anand Kumar 1965).

These bats sometimes roost in colonies numbering many thousands of individuals. *R. hardwickei*, however, may roost alone or in small groups of 4–10. Several such groups may form a large, loose group of 80–100 individuals. The smaller groups may be sexually segregated for at least part of the year (Lekagul and McNeely 1977). Colonies of 25–200 individual *R. hardwickei* have been observed in the Arabian region, including a small nursery with females and young that was found in August (Harrison and Bates 1991).

The reproductive pattern of *R. microphyllum* in Jodhpur, India, can be related to its seasonal movement, which occurs just before and after its winter torpor. The females are monestrous, mating in March and giving birth to a single young in July–August. Gestation lasts at least 123 days. The young are weaned after 6–8 weeks and attain sexual maturity in their second year of life (Anand Kumar 1965). The following reproductive data for *R. hardwickei* were summarized by Qumsiyeh and Jones (1986): young are born in June and July in India, females pregnant with single embryos have been taken in Egypt and Sudan in late March, and lactating females have been taken in the same countries in August. Data on *R. hardwickei* compiled by Hayssen, Van Tienhoven, and Van Tienhoven (1993) indicate that the gestation period is 95–100 days, birth weight is 2–3 grams, lactation lasts 2 months, the interbirth interval is 1 year, and sexual maturity is attained by females at about 9 months and by males at 16–17 months. According to Harrison and Bates (1991), pregnant female *R. muscatellum* were collected in Oman in June and July and lactating females were found in late August and September.

No fossils referable to the Rhinopomatidae have been found. The family currently is widely distributed, though generally in a region containing limited amounts of habitat that may be subject to human disruption. The IUCN now classifies *R. macinnesi* and *R. hadithaensis*, a newly described species of Iraq, as vulnerable.

CHIROPTERA; **Family EMBALLONURIDAE**

Sac-winged Bats, Sheath-tailed Bats, and Ghost Bats

This family of 13 Recent genera and 49 species is widely distributed in the tropical and subtropical regions of the world. The sequence of genera presented here largely follows that of Robbins and Sarich (1988), who, utilizing immunological and electrophoretic data, recognized two subfamilies, Taphozoinae, with *Taphozous* and *Saccolaimus*, and Emballonurinae, with the remaining genera. Studies of hyoid morphology (Griffiths and Smith 1991) generally supported this arrangement but indicated a close relationship between the genera *Diclidurus* and *Rhynchonycteris*. Morphometric analysis (Freeman and Lemen 1991) also indicated a division of the family into one group with *Taphozous* and *Saccolaimus* and another with most other genera but suggested that *Diclidurus* represents a third unit mainly because of its wing morphology. The latter position seems to correspond to an earlier view, still followed by Corbet and Hill (1991) and Jones and Hood (1993), that *Diclidurus*, together with *Cyttarops*, should be placed in a separate subfamily, the Diclidurinae.

The size ranges from small to relatively large. Head and body length is 37–157 mm, tail length is 6–36 mm, and forearm length is 37–97 mm. Adults weigh 5–105 grams. Most members of this family are gray, brown, or black in color, but *Diclidurus*, the ghost bat, has an unusual color for bats, being white or white mixed with gray. In the subgenus *Peronymus* of *Peropteryx* the wings are white from the forearm outward. In the proboscis bat, *Rhynchonycteris*, the upper surface of the forearm is dotted with tufts of hair.

The face and lips are smooth. There is no nose leaf. The ears are often united across the top of the head, and a tragus is present. Many forms have glandular wing sacs that open on the upper surface of the wing. These glands secrete a red substance of strong odor. The sacs can be seen by holding the bat with its head up and its belly toward the light and gently extending its wings; if present, they appear near the shoulder. The wing sacs are larger and better developed in the males; perhaps the odor of their secretion attracts the opposite sex. In the genus *Taphozous* a glandular mass is located between the lower jaws. The common name "sheath-tailed bats" refers to the nature of the tail attachment. In the members of this family the tail pierces the tail membrane, and its tip appears completely free on the upper surface of the membrane, the base of the tail being loosely enclosed in the membrane. Thus, in flight the tail membrane can be lengthened by stretching the hind limbs, the membrane then slipping quite easily over the tail vertebrae. Thus, by pulling their hind legs in or moving them out these bats can "set sail." Most forms can steer and turn exceptionally well. The hind legs are slender and the foot is normal.

The molar teeth have a **W** pattern of cusps and ridges. The dental formula varies. In *Emballonura* it is (i 2/3, c 1/1, pm 2/2, m 3/3) × 2 = 34; in the American genera and

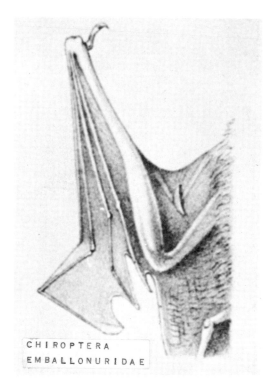

CHIROPTERA; EMBALLONURIDAE; **Genus**
TAPHOZOUS
E. Geoffroy St.-Hilaire, 1818

Tomb Bats

There are 2 subgenera and 13 species (Aggundey and Schlitter 1984; Chasen 1940; Corbet 1978; Corbet and Hill 1992; Ellerman and Morrison-Scott 1966; Flannery 1990*b*, 1995; R. E. Goodwin 1979; Hayman and Hill *in* Meester and Setzer 1977; Hill 1983*a*, 1991; Juste and Ibáñez 1994; Kitchener 1980*b*; Kock 1974*a*; Kitchener et al. 1993; Koopman 1975, 1982*b*, 1984*c*; Laurie and Hill 1954; Lawrence 1939; Lekagul and McNeely 1977; McKean and Friend 1979; Ride 1970; Roberts 1977; Taylor 1934):

subgenus *Taphozous* E. Geoffroy St.-Hilaire, 1813

T. mauritianus, savannah regions throughout Africa south of the Sahara, Bioko, Madagascar, Mauritius, Reunion, Assumption Island, Aldabra Island;

T. hildegardeae, coastal and central Kenya, northeastern Tanzania, Zanzibar;

T. perforatus, Senegal to India, and south to Zimbabwe;

T. longimanus, India to southern Indochina and Malay Peninsula, Sri Lanka, Sumatra, Java, Bali, Borneo, Flores;

T. melanopogon, India to Indochina, Yunnan (southern China), Malay Peninsula, Sri Lanka, Andaman Islands, Hainan, Sumatra, Java, Kangean Islands, Bali, Borneo, Natuna Islands, Talaud and Sangihe islands, Sulawesi, Togian Islands, Lombok, Sumbawa, Savu, Roti, Semau, Timor, Alor, Kei Islands, Molucca Islands;

T. philippinensis, Philippines;

T. theobaldi, central India, southern Burma, Thailand, Viet Nam, Java, Borneo, Sulawesi;

T. georgianus, Western Australia, Northern Territory, Queensland;

T. hilli, Western Australia, Northern Territory;

T. kapalgensis, northern Northern Territory;

T. australis, probably south-central and southeastern New Guinea, northeastern Queensland, islands of Torres Strait;

subgenus *Liponycteris* Thomas, 1922

T. hamiltoni, southern Sudan, Kenya;

T. nudiventris, Mauritania to Egypt and Tanzania, Palestine and Arabian Peninsula to Burma.

Wing of emballonurid bat, showing the glandular wing sac that is present in many members of the family Emballonuridae, photo from *Biologia Centrali-Americana, Mammalia,* E. R. Alston.

Coleura it is (i 1/3, c 1/1, pm 2/2, m 3/3) $\times$ 2 = 32; and in *Taphozous* it is (i 1/2, c 1/1, pm 2/2, m 3/3) $\times$ 2 = 30.

Shelters include rocky crevices, caves, ruins, houses, trees, curled leaves, and hollow logs. These bats are agile in their retreats, scrambling about with considerable dexterity and often clinging to vertical walls or crawling into crevices. At rest they often fold the tips of their wings back on their upper surface. Some forms roost in large colonies; others assemble in groups of about 10–40 individuals. These groups seem to operate as units in most of their activities. Other forms are generally solitary. Some species occasionally shelter with other species. When they do, the species usually remain separate. In *Rhynchonycteris* the females roost apart from the males when the young are born, and different shelters are used by adult male and female *Balantiopteryx plicata* during the summer; most of the other forms seem to remain together all year. Some species of *Taphozous* breed throughout the year, but most members of this family probably have a fairly well-defined breeding season. Sheath-tailed bats find objects in the dark by echolocation, but they may also guide themselves by vision. Some species are rapid fliers. They feed on insects and occasionally on fruit.

The geological range of this family is middle Eocene to early Miocene in Europe, early Miocene to Recent in Africa, Pleistocene to Recent in South America, and Recent in other parts of the range (Koopman 1984*a*).

Saccolaimus (see account thereof) sometimes is considered a subgenus of *Taphozous.* Koopman (*in* Wilson and Reeder 1993) listed *T. philippinensis* as a species but stated that it probably is a subspecies of *T. melanopogon,* Rickart et al. (1993) did treat it as a subspecies, and Corbet and Hill (1991) did not list *T. philippinensis* at all. Kitchener et al. (1993) elevated *T. melanopogon achates,* from Savu, Roti, and Semau, in the southeastern part of the Lesser Sunda Islands, to specific status. Flannery (1995) treated *achates* separately but noted that it is extremely similar to *T. melanopogon,* except that males of the former have a black beard and those of the latter have a brown beard.

Head and body length is about 62–100 mm, tail length is 12–35 mm, forearm length is 50–75 mm, and weight is 10–50 grams. The upper parts are grayish or various shades of brown, sometimes with a reddish or cinnamon cast. Some species have whitish spots on the body. The under-

A. Tomb bat *(Taphozous hildegardeae)*. B. *T. mauritianus*. Photos by David Pye.

parts are creamy, pale brown, or white. Males of the species *T. longimanus* are usually cinnamon brown in color, whereas most females of this species are dark gray.

A wing pocket or pouch is present in all species. Most species also have a glandular sac in the lower throat. This sac is more highly developed in the males than in the females and may be entirely absent in the latter. The species *T. melanopogon* lacks this sac and instead has small pores that open into the area where the sac would be.

Tomb bats are found in a variety of habitats, including rainforest, open woodland, and fairly arid country. They roost in tombs, old buildings, rock crevices, caverns in rocky deserts, shallow caves, cliffs along seashores, and trees. *T. longimanus* is often found in the top of coconut palms. Tomb bats are quite agile and active in crawling around crevices and rock walls and sometimes cling on vertical walls. They are numerous in certain large tombs. Hibernation is suggested by the seasonal accumulation of fat.

An almost inaudible "tic-tic-tic" may be uttered when roosting, whereas the call when wounded and on the ground is shrill and piercing. Some species emit a loud cry in flight. Tomb bats are strong fliers. They may begin feeding before dusk at heights of 60–90 meters, coming down to lower levels as the evening progresses. The diet consists of flying insects. In a study in India, Subbaraj and Chandrashekaran (1977) found that *T. melanopogon* emerged from its roost at about the same time every day, regardless of the time of sunset, so that the bats began flying at sunset in some months and well after dark in others. Kingdon (1974a) wrote that *T. mauritianus* has three hours of in-

tensive activity after sunset and then alternates long rests with short flights. According to Roberts (1977), *T. nudiventris* makes seasonal migrations, with colonies reoccupying a summer roost in a matter of a few days.

Lekagul and McNeely (1977) reported the following roosting group sizes in Thailand: *T. longimanus*, 2–20; *T. melanopogon*, 150–4,000; and *T. theobaldi*, in the thousands. Kingdon (1974a) wrote that *T. nudiventris* shelters in groups of 200–1,000 and forms all-female groups during late pregnancy and lactation, and Roberts (1977) mentioned one colony of more than 2,000. While roosting, each individual *T. melanopogon* occupies a definite vertical territory, and males and females may be in separate areas. Pregnant female *T. melanopogon* have been collected in India from January to May, the gestation period is about 4 months, births occur from May to August, weight of the single young at birth is about 20 grams, lactation lasts 2 months, and females can first conceive at 8 months (Hayssen, Van Tienhoven, and Van Tienhoven 1993).

Tomb bat *(Taphozous philippinensis)*, photo by Paul D. Heideman.

Tomb bat (*Taphozous* sp.), photo by Erwin Kulzer.

In *T. longimanus* there may be continuous breeding, with each female undergoing more than one pregnancy per year (Lekagul and McNeely 1977). In a study in India, Gopalakrishna (1955) found *T. longimanus* to breed all year, with each female having a rapid succession of pregnancies and producing one young at a time. In Pakistan one colony of *T. nudiventris* reportedly arrives at its summer roost every year in the first week of March, and each female gives birth to a single young in mid-April. The young are carried by the females until the age of eight weeks (Roberts 1977). Female *T. mauritianus* are polyestrous and give birth to a single young during November–December and March–April in southern Africa (Happold and Happold 1990). The birth seasons of *T. mauritianus* in northeastern Zaire are April–May and November–December (Kingdon 1974a). In coastal Kenya male *T. hildegarde* hold territories within caves throughout the year and attempt to control groups of females for purposes of mating; although there are two annual periods of increased size in sexual glands, mating was found to occur only around July (McWilliam 1988a). In studies in Western Australia, Kitchener (1973, 1976) found *T. georgianus* to be monestrous, with an anestrous period from mid-autumn to mid-winter. A single young was produced from late November to late April following a gestation period of about four months. In central Queensland, Jolly (1990) found *T. georgianus* to mate in late August and early September and to give birth in late November or early December. The single young was born well furred, its eyes were open, and it weighed about 7–8 grams. The young attained independence at 3–4 weeks and was nearly full grown at 3 months. Females generally mated at about 9 months, but males were not able to mate until approximately 21 months.

The IUCN recognizes *T. troughtoni* of the Cape York Peninsula of northern Queensland as a species distinct from *T. georgianus* and classifies that species as critically endangered. Its numbers are thought to have fallen to fewer than 50 individuals in recent years, and a further decline is expected as its restricted habitat comes under increasing human pressure. *T. hildegardeae*, *T. achates* (recognized as a species distinct from *T. melanopogon*; see above), *T. kapalgensis*, and *T. hamiltoni* also are thought to be declining and are classed as vulnerable; *T. australis* is designated as near threatened.

CHIROPTERA; EMBALLONURIDAE; Genus SACCOLAIMUS
Temminck, 1838

There are 5 species (Chasen 1940; Corbet 1978; Corbet and Hill 1992; Ellerman and Morrison-Scott 1966; Feiler 1980; Flannery 1990b, 1995; R. E. Goodwin 1979; Hayman and Hill *in* Meester and Setzer 1977; Hill 1983a; Koopman 1975, 1982b, 1984c; Laurie and Hill 1954; Lawrence 1939; Lekagul and McNeely 1977; Ride 1970; Taylor 1934):

S. peli, Liberia to western Kenya and northeastern Angola;
S. saccolaimus, India, Bangladesh, Burma, Malay Peninsula, Sri Lanka, Great Nicobar Island, Sumatra, Java, Borneo, Talaud Islands, Sulawesi, Timor, Halmahera Islands, New Guinea, Guadalcanal in the Solomon Islands, northwestern Northern Territory, northeastern Queensland;
S. pluto, Philippines;
S. flaviventris, southeastern Papua New Guinea, throughout Australia except Tasmania;
S. mixtus, southern and eastern New Guinea, Cape York Peninsula of northern Queensland.

Saccolaimus was considered only a subgenus of *Taphozous* by Corbet and Hill (1991, 1992), but the two were regarded as generically distinct by Barghoorn (1977), Freeman and Lemen (1991), Griffiths and Smith (1991), Koopman (*in* Wilson and Reeder 1993), and Robbins and Sarich (1988). Koopman (*in* Wilson and Reeder 1993) listed *pluto* as a species but noted that it almost certainly is a subspecies of *S. saccolaimus*; Corbet and Hill (1992) included *pluto* in *S. saccolaimus*.

In the largest species, *S. peli*, of Africa, head and body length is 110–57 mm, tail length is 27–36 mm, forearm length is 84–97 mm, and weight is 92–105 grams (Kingdon 1974a). In the three species that occur in Australia head and body length is 72–100 mm, tail length is 20–35 mm, forearm length is 62–80 mm, and weight is 30–60 grams (Hall *in* Strahan 1983; Richards *in* Strahan 1983). The upper parts vary in color from reddish brown to jet black; the underparts usually are paler, sometimes entirely white. In *S. saccolaimus* the pelage is irregularly flecked with patches

Papuan sheath-tailed bat *(Saccolaimus mixtus)*, photo by B. G. Thomson / National Photographic Index of Australian Wildlife.

of white and the posterior portion of the back is naked. There is a large pouch under the chin in both sexes. From *Taphozous*, *Saccolaimus* differs in lacking a wing pouch and in having a deep groove on the lower lip (Lekagul and McNeely 1977). *Saccolaimus* also is considered generically distinct on the basis of numerous cranial characters, especially in that the auditory bullae are completely ossified (Barghoorn 1977).

According to Kingdon (1974*a*), *S. peli* shelters by day in deep forests and comes out at night to forage along the margins of the forest and in clearings and river valleys. Its flight is acrobatic and is accompanied by loud calls. It pursues moths and beetles. *S. saccolaimus* is found in a variety of woodland and forest habitats and roosts in hollow trees, caves, old tombs, buildings, and openings between boulders (Hall *in* Strahan 1983). *S. flaviventris* may migrate to warmer areas for the winter (Richards *in* Strahan 1983).

S. flaviventris is usually solitary but occasionally forms colonies of fewer than 10 individuals (Richards *in* Strahan 1983). Chimimba and Kitchener (1987) reported *S. flaviventris* to begin mating in August and to give birth to a single young from December to mid-March. Groups of *S. saccolaimus* in Thailand contain five or six individuals (Lekagul and McNeely 1977). According to Hall (*in* Strahan 1983), this species is gregarious but does not form tight clusters; females are lactating during the wet season in Australia and produce a single young. Pregnant female *S. peli* have been reported in June and December in eastern Zaire (Kingdon 1974*a*). The IUCN classifies *S. mixtus* as vulnerable, noting that available habitat is in decline. *S. peli* and *S. flaviventris* are designated as near threatened.

CHIROPTERA; EMBALLONURIDAE; Genus MOSIA
Gray, 1843

Lesser Sheath-tailed Bat

The single species, *M. nigrescens*, is found on Sulawesi, the Togian Islands, the Moluccas from Halmahera to Buru and the Kei Islands, much of New Guinea and several nearby islands, the Bismarck Archipelago, Woodlark Island, and the Solomon Islands (Corbet and Hill 1992; Flannery 1990*b*, 1995). *Mosia* long was included in the synonymy of *Emballonura* but was restored to generic rank by Griffiths, Koopman, and Starrett (1991). *Mosia* was accepted as a full genus by Koopman (*in* Wilson and Reeder 1993), was thought by Corbet and Hill (1992) to justify subgeneric rank, and was not recognized by Flannery (1995), pending further taxonomic study.

According to Flannery (1990*b*, 1995), head and body length is 32–41 mm, tail length is 7–12 mm, forearm length is 31–36 mm, and weight is 2.5–4.1 grams. It is light brown in color and has a short, blunt, sparsely haired face with large eyes. Males are smaller than females and, indeed, are among the smallest of all mammals. Griffiths, Koopman, and Starrett (1991) gave *Mosia* generic distinction primarily because it does not share an unusual derived character of the hyoid musculature found in related genera. In *Emballonura* and *Coleura* the sternohyoid muscle is deflected dorsally and laterally by a prominent postlaryngeal tracheal expansion. In *Mosia*, as in most other bats, there is no dorsal deflection. In addition, *Mosia* has a long, slender (wormlike) penis, whereas *Emballonura* and *Coleura* have a short, broad penis. Finally, the tragus of *Mosia* differs from that of the other two genera in being longer, narrower, and thinner.

Mosia has been collected in secondary forest, freshwa-

ter mangrove, and village environs; it roosts under the leaves of broad-leaved trees, in the roofs of native huts (McKean 1972), and in the twilight zone of caves (Flannery 1995). It often flies before dusk in the highest levels of the rainforest and also around dwellings and gardens. Vestjens and Hall (1977) found 44 stomachs to contain mostly wingless ants, suggesting that the bats had fed on or near the ground, on trees, or possibly at their roosting site. Available data suggest that *Mosia* may differ from *Emballonura* in roosting more frequently in bisexual groups (Flannery 1995). Pregnant females were collected in New Guinea in February, May, June, and July (McKean 1972). There is a single young (Hayssen, Van Tienhoven, and Van Tienhoven 1993).

CHIROPTERA; EMBALLONURIDAE; Genus EMBALLONURA
Temminck, 1838

Old World Sheath-tailed Bats

There are 9 species (Bruner and Pratt 1979; Corbet and Hill 1992; Ellerman and Morrison-Scott 1966; Flannery 1990*b*, 1994, 1995; Griffiths, Koopman, and Starrett 1991; Hayman and Hill *in* Meester and Setzer 1977; Hill 1956, 1971*a*, 1983*a*, 1985*b*; Hill and Beckon 1978; Laurie and Hill 1954; Lekagul and McNeely 1977; Lemke 1986; McKean 1972; Medway 1977; Smith and Hood 1981; Tate and Archbold 1939; Taylor 1934):

E. atrata, Madagascar;

E. alecto, Borneo, Philippines, Talaud Islands, Sulawesi, Sula Islands, South Molucca Islands (Amboina, Seram, Kei Islands, Tanimbar Islands);

E. monticola, Malay Peninsula (including southern Burma and Thailand), Riau Archipelago, Sumatra, Bangka and Billiton islands, Nias Island, Mentawai Islands, Enggano Island, Krakatau, Java, Borneo, Natuna Islands, Sulawesi;

E. semicaudata, Northern Marianas, Guam, Palau Islands, Chuuk (Truk), Pohnpei, Fiji, Samoa Islands, Tonga (western and South Pacific);

E. raffrayana, Halmahera Islands, Seram, New Guinea and some nearby islands, Solomon Islands;

E. beccarii, New Guinea and some nearby islands, Kei Islands, Bismarck Archipelago, Trobriand Islands;

E. dianae, Papua New Guinea, New Ireland in the Bismarck Archipelago, Solomon Islands (Santa Ysabel, Guadalcanal, Malaita, Rennell);

E. furax, southwestern to southeastern New Guinea;

E. serii, New Ireland in the Bismarck Archipelago.

Another species, *E. nigrescens*, recently was restored to the genus *Mosia* (see account thereof). A record of *E. semicaudata* from Vanuatu is probably incorrect (Flannery 1995).

Head and body length is about 38–62 mm, tail length is 10–20 mm, and forearm length is 30–53 mm. Lekagul and McNeely (1977) gave the weight of *E. monticola* as 30–40 grams. However, Flannery (1990*b*, 1995) listed much lower weights, about 4–8 grams, for populations occurring in New Guinea and the southwestern Pacific. Coloration is rich brown to dark brown above and paler below. *Emballonura* and *Mosia* are the only genera in the family with two pairs of upper incisors. Wing sacs are not present.

These bats occur in a variety of habitats and are frequently observed in settled areas, but relatively little is known about them. *E. monticola* is a forest species; it has

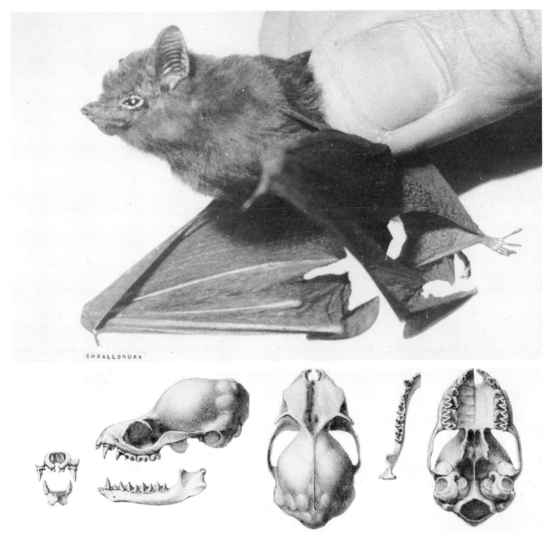

Old World sheath-tailed bat *(Emballonura monticola)*, photo by Lim Boo Liat. Skull photos of *Emballonura* from *Die Fledermäuse des Berliner Museums für Naturkunde*, Wilhelm K. Peters.

been seen flying in dense shade during the day (Lekagul and McNeely 1977). This species roosts in a variety of rather exposed sites, including inside hollow logs, under overhanging earth banks, between boulders, and in rock shelters; in caves it rarely penetrates beyond the twilight zone, and it may occupy brightly lit sectors (Medway 1978). The usual diet of *Emballonura* is insects, but fruit is eaten occasionally.

According to Flannery (1995), the sexes of *Emballonura* usually roost separately, though a bisexual colony was found in the Halmahera Islands in December. *E. monticola* is moderately gregarious, usually associating in groups of 2–20 (Medway 1978), but one colony of 100–150 was found in a cave (Lekagul and McNeely 1977). Small colonies of up to 20 *E. alecto* were found roosting among rocks in the Philippines; pregnant females were caught there in April and June (Ingle 1992). Pregnant female *E. monticola*, bearing one embryo each, were found on the Malay Peninsula in February, March, October, and November (Medway 1978). Adult female *E. raffrayana* with dependent young

were taken in New Guinea in May (Flannery 1990*b*).

The IUCN classifies *E. atrata, E. dianae,* and *E. furax* as vulnerable; each is in decline through loss of habitat. *E. raffrayana* is designated as conservation dependent, and *E. semicaudata* is classed as endangered. This last species once was widespread and common in Polynesia and Micronesia but has disappeared or become rare in most parts of its range, apparently because of human disturbance of roosting caves and associated habitat (Flannery 1995; Grant, Banack, and Trail 1994). Bruner and Pratt (1979) stated that the subspecies *E. semicaudata rotensis,* of Rota Island in the Marianas, had not been seen for many years and might be extinct. Perez (1972) suggested that this same bat may once have been abundant on Guam but had become extremely rare. Lemke (1986) obtained reports that *E. semicaudata rotensis* was present in the Marianas in some numbers through the 1960s but subsequently had declined, perhaps in association with the swiftlets that occupied the same caves. During intensive surveys of the Marianas in 1983–84 he located only two groups of about four bats

Old World sheath-tailed bat *(Emballonura monticola)*, photo by Klaus-Gerhard Heller.

each, both on Aguijan Island. This subspecies may thus rank as one of the world's most critically endangered mammals. Surprisingly, although it is found in territory under the jurisdiction of the United States, the USDI failed to act on a petition to formally classify it as endangered.

CHIROPTERA; EMBALLONURIDAE; **Genus COLEURA**
Peters, 1867

African and Arabian Sheath-tailed Bats

There are two species (Hayman and Hill *in* Meester and Setzer 1977):

C. afra, southwestern Arabian Peninsula, Guinea-Bissau to Somalia, and south to Angola and Mozambique;

C. seychellensis, Seychelles Islands.

Head and body length is 55–65 mm, tail length is 12–20 mm, and forearm length is 45–56 mm. Nicoll and Suttie (1982) reported weight of *C. seychellensis* to average 10.2 grams in adult males and 11.1 grams in parous females. Coloration is reddish brown, dark brown, or sooty brown above and somewhat paler below. This genus resembles *Emballonura* but has only one pair of upper incisors instead of two, and there are differences in the skulls. *Coleura* does not have wing sacs.

Coleura roosts in caves and houses, generally in crevices and cracks. Bats of this genus usually do not roost upside down; instead they crawl into a cranny or press their underside flat against a stone wall. *C. afra* has been found in the same cave with *Triaenops persicus* and *Asellia tridens.* Thousands of bats have been found in a cave partially filled with water during high tide on the rocky coast of the Indian Ocean. Large colonies of *C. afra* and *Taphozous hildegardeae* were in the front of this cave, and smaller colonies of *Triaenops afer* and *Hipposideros caffer* were in the deeper passages. The species were segregated. The behavior of *Coleura* and *Taphozous* was much the same in this cave; they rested flat against the walls, individually or close together, and only when approached within about 50 cm did they fly away. They would not leave the cave during the day, even when noise was used to try to drive them out. Members of the genus *Coleura* sometimes become pests when large numbers successfully colonize a house. Overlapping tiles or corrugated iron sheets are suitable resting places. The diet is mainly insects.

McWilliam (1987a) analyzed the population structure of *C. afra* roosting in coral caves along the coast of Kenya. Colonies consisted of up to 50,000 individuals or more, but each bat had a precise roosting place to which it returned. Colonies were divided into clusters, each with about 20 individuals, comprising mainly a harem of adult females and their young, plus a single adult male. Solitary bats, mainly bachelor males, roosted on the periphery of the clusters. Females were found to be polyestrous and to give birth twice a year, primarily in April and November. These periods evidently were timed so that lactation would continue for about three months during the rainy seasons, when availability of insects was at a maximum. Many of the young females remained in their natal clusters for at least a year, and since they attained sexual maturity during this period

African sheath-tailed bats *(Coleura afra)*, photo by Bruce J. Hayward.

each cluster tended to be a permanently related unit. Pregnant female *C. afra* also have been collected in December in Tanzania and in April in Yemen. In Sudan breeding has been found to be markedly seasonal, with births occurring in October, at the end of the rains. Only one young is produced at a time (Kingdon 1974*a*).

Little was known of *C. seychellensis* until studies by Nicoll and Suttie (1982). Colonies roosted in caves and evidently were divided into harem groups like those of *C. afra*. Births occurred during the November–December rainy season over three years, though the presence of young in one year suggested that females are polyestrous. This species once was abundant but has declined drastically, perhaps because of environmental disruption and predation by introduced barn owls. Only two occupied caves were located during searches on three islands. The IUCN now classifies *C. seychellensis* as critically endangered, noting that fewer than 50 individuals are thought to survive.

CHIROPTERA; EMBALLONURIDAE; **Genus**
RHYNCHONYCTERIS
Peters, 1867

Proboscis Bat, or Sharp-nosed Bat

The single species, *R. naso,* occurs from southern Mexico to eastern Peru, northern Bolivia, and southeastern Brazil, as well as on the island of Trinidad (Cabrera 1957; E. R. Hall 1981; Jones and Hood 1993; Plumpton and Jones 1992).

Head and body length is 37–43 mm, tail length is about 12 mm, and forearm length is 35–41 mm. Four males and three females from Trinidad weighed 2.1–4.3 grams. The whitish-tipped hairs and chocolate brown underfur impart an unusual grizzled yellowish gray color. Two curved lines, white or gray in color, are usually present on the lower back and rump. The skull is very small, with the rostrum elongate and the frontal area sharply depressed. This genus may be identified by the small tufts of grayish hair on the forearms, the elongate and pointed muzzle, and the absence of wing sacs.

This bat is generally found in forests, usually along waterways. During the day it roosts adjacent to the water on the branches or boles of trees, on rocks, or on the sides of cliffs (Bradbury and Emmons 1974; Husson 1978). It has also been discovered in the curled leaves of the heliconia, or false banana plant. Because of the color pattern, groups of roosting *Rhynchonycteris* resemble patches of lichens, and their habit of slightly curving their head and nose backward increases the resemblance to the curled edge of a lichen. The latter position, however, may not be assumed until the bat is threatened. Proboscis bats usually reassemble at the home perch within 20 minutes after being scattered by a disturbance. Groups of roosting *Rhynchonycteris* have also been reported to bear a remarkable resemblance to cockroaches (Dalquest 1957). There is usually a definite space between roosting bats, and in common with *Saccopteryx* and *Peropteryx, Rhynchonycteris* folds its wings at about a 45° angle to the body when inactive (Goodwin and Greenhall 1961). Husson (1978) stated that groups of 8–10 hang downward in a vertical row on tree branches, with each bat directly above another, about 10 cm apart.

In studies in Trinidad and Costa Rica, Bradbury and Vehrencamp (1977) found *Rhynchonycteris* to forage almost entirely over water, up to a height of 3 meters. The

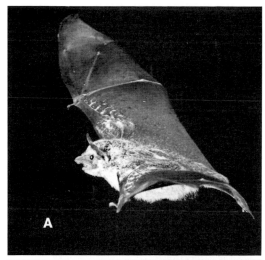

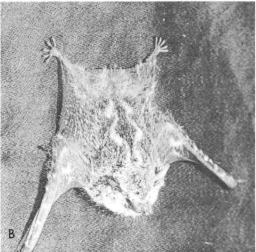

Proboscis bat *(Rhynchonycteris naso):* A. Photo by J. Scott Altenbach; B & C. Photos by David Pye.

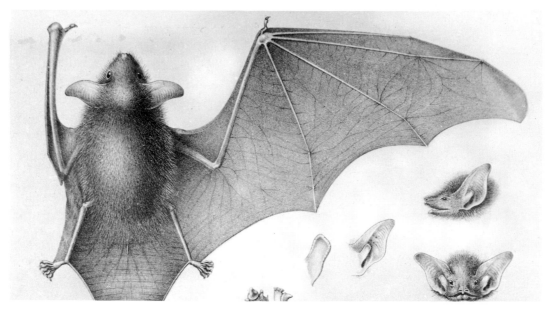

foraging range averaged 1.1 ha., and population density in the study area was 7.6/ha. The diet of the proboscis bat consists of small insects.

Colonies of *Rhynchonycteris* roost together and seem to associate as a unit in most of their activities. Bradbury and Vehrencamp (1977) found colony size to range from 3 to 45 individuals and to average 5–11 depending on area. Each colony used three to six roosting sites, among which it moved at intervals. In some colonies there appeared to be a dominant male, which resided within the colony more constantly than other adult males. The dominant male typically foraged alone at the boundaries of the colony's foraging range and chased intruding conspecifics. The reproductive females of the colony, as well as their young when capable, foraged together in a small area. Other males and nonreproductive females foraged in solitary "beats." At two to four months the young of both sexes dispersed to adjacent colonies. In Costa Rica births did not occur during the early dry season (January–March) but were recorded in April, June, July, and October. Bradbury and Emmons (1974) recorded births in Trinidad in July and August and found that, contrary to earlier reports, females with young did not roost away from the main colony. Graham (1987) found pregnant females in Peru in May, July, August, and September, and Webster and Jones (1984c) reported one taken in Ecuador on 12 June. Plumpton and Jones (1992) noted that breeding evidently occurs throughout the year in at least some parts of the range, with females giving birth up to twice annually.

Shaggy-haired bat *(Centronycteris maximiliani):* Top, photo from *Die Fledermäuse des Berliner Museums für Naturkunde*, Wilhelm K. Peters; Bottom, photo by Richard K. LaVal.

CHIROPTERA; EMBALLONURIDAE; **Genus CENTRONYCTERIS**
Gray, 1838

Shaggy-haired Bat

The single species, *C. maximiliani*, occurs from southern Mexico to Peru and Brazil (Cabrera 1957; E. R. Hall 1981; Tuttle 1970).

Head and body length is about 50–62 mm, tail length is 18–23 mm, and forearm length is 43–48 mm. Coloration of the long, soft hair is raw umber or tawny. The hairs in front of the eyes and those on the interfemoral membrane are sometimes reddish in color. There are no stripes.

This genus resembles *Saccopteryx* but is more slender in body form, and as far as is known, there is no wing sac. Also, the lower border of the orbit of the skull projects only slightly, so that the toothrow is visible from above and is not hidden as in *Saccopteryx*.

Starrett and Casebeer (1968) observed that the rarity of specimens of this genus suggests that it requires relatively heavy forest in which to live. Its very slow and highly maneuverable, floppy flight makes it well adapted to hunting insects among trees and in natural and artificial pathways and clearings. Baker and Jones (1975) observed several individuals flying extremely slowly and in a straight path along the right of way of a telegraph line in a second-growth forest. *Centronycteris* has also been collected from holes in the boles of trees. A pregnant female with a single fetus was taken in Nicaragua on 26 May (Greenbaum and Jones 1978). In Costa Rica two pregnant females were collected on 15 May and a subadult was taken on 15 September (LaVal 1977).

CHIROPTERA; EMBALLONURIDAE; **Genus**
BALANTIOPTERYX
Peters, 1867

Least Sac-winged Bats

There are three species (Cabrera 1957; E. R. Hall 1981; Jones and Hood 1993):

B. plicata, western and southern Mexico, including southern Baja California, to Costa Rica;
B. io, southern Mexico to Belize and Guatemala;

B. infusca, northwestern Ecuador and probably the adjacent part of Colombia.

Head and body length is about 48–55 mm, tail length is about 12–21 mm, forearm length is 35–49 mm, and adult weight is about 4–9 grams. The upper parts are dark chestnut brown, dark brown, or dark gray, and the underparts are usually paler. There are no dorsal stripes. The wing sac is in the center of the membrane between the upper arm and the forearm. This genus can be distinguished from the other bats of the family Emballonuridae that have wing sacs by the greatly inflated rostrum.

In Jalisco, Mexico, *B. plicata* is common in many areas and has a known altitudinal range of sea level to about 1,300 meters (Watkins, Jones, and Genoways 1972). *Balantiopteryx* roosts in caves and culverts and under boulders, often near lakes and rivers. Individuals have been found in limestone caves in the crevices at the tops of stalactites, but these bats usually hang in exposed places in fairly well illuminated spots with appreciable space between each other. Starrett and Casebeer (1968) found one colony of *B. plicata* in a series of fissures in the face of a sea cliff that was accessible only at low tide and another in the shadow of a huge rafter near the mouth of a partially collapsed manganese mine. Least sac-winged bats are among the first bats to appear in the evening, often before

Least sac-winged bat *(Balantiopteryx plicata)*, photo by Lloyd G. Ingles. Inset showing wing sac, photo from . . . *American Bats of the Subfamily Emballonurinae,* Colin C. Sanborn.

Least sac-winged bat *(Balantiopteryx plicata)*, photo by M. Brock Fenton.

sunset. Their flight is erratic but relatively slow, and the diet consists of insects.

These bats are gregarious. Approximately 300 *B. io* were found clinging in a random manner to the ceiling of a cave in late July and early August in Oaxaca, Mexico. Pregnant females of this species, each with a single embryo, have been taken from March to July (Arroyo-Cabrales and Jones 1988*a*). Bradbury and Vehrencamp (1977) investigated one cave in Costa Rica that contained 1,500–2,000 *B. plicata*. These bats were divided into separate colonies of 50–200 each. The bats foraged alone or in groups, and no evidence of territoriality was observed. There were usually more males than females in a colony, but some males moved away during the rainy season. There was a single highly synchronized birth season starting in late June, and females were capable of reproduction in their first year of life. In Mexico pregnant female *B. plicata* have been reported from May to August, and lactating females have been found in September (Jones, Choate, and Cadena 1972; Watkins, Jones, and Genoways 1972). The gestation period is about 4.5 months; the single young weighs about 2 grams at birth and is precocious. It is carried by the mother for its first week of life, can fly alone at 2 weeks, and is fully weaned at about 9 weeks (Arroyo-Cabrales and Jones 1988*b*).

B. infusca occupies a restricted area that is suffering intensive human habitat disruption. The IUCN classifies it as endangered and *B. io* as near threatened.

CHIROPTERA; EMBALLONURIDAE; **Genus**
SACCOPTERYX
Illiger, 1811

White-lined Bats

There are four species (Anderson, Koopman, and Creighton 1982; Cabrera 1957; E. R. Hall 1981; Jones and Hood 1993; Koopman 1982*b*):

S. bilineata, southern Mexico to Bolivia and southeastern Brazil, Trinidad;

S. leptura, southern Mexico to Bolivia and southeastern Brazil, Trinidad;

S. canescens, northern Colombia to eastern Peru and northeastern Brazil;

S. gymnura, northeastern Brazil.

Head and body length is 37–55 mm, tail length is 12–20 mm, forearm length is 35–52 mm, and adult weight is 3–11 grams. According to Bradbury and Emmons (1974), female *S. bilineata* are larger; average weight on Trinidad was 7.5 grams in males and 8.5 grams in nonpregnant females. Coloration is dark brown above and paler brown below. As in *Rhynchonycteris*, two whitish longitudinal lines are present on the back, but *Saccopteryx* does not have tufts of hair on the forearm. A pouchlike gland is present in the membrane that extends from the shoulder to the wrist. This is a scent gland that can be opened by muscular action. It is well developed in males but vestigial or absent in females and may serve to attract the opposite sex.

In Venezuela, Handley (1976) found *S. bilineata* and *S. leptura* mainly in forests but *S. canescens* mainly in open areas. Most specimens of each species were collected near streams or in other moist places. These bats roost in less shaded locations than do many other bats. In southern Mexico they have been found under concrete bridges, in culverts under highways, on the trunks of tall trees such as wild figs *(Ficus)*, inside hollow trees, and at or near the entrances of limestone caves. Tuttle (1970) found *S. bilineata* to be one of the first bats to fly after sundown. In studies in Costa Rica and Trinidad, Bradbury and Vehrencamp (1977) found roosting *S. leptura* to prefer the exposed boles of large trees in riparian forest and *S. bilineata* to roost mostly in hollow trees or the buttress cavities of large trees. Each colony of *S. bilineata* had a series of sites where it moved seasonally to forage for insects. Population density for *S. leptura* was 2.5/ha. in Costa Rica and 17.6/ha. in Trinidad, and respective foraging ranges averaged 0.14 ha. and 1.4 ha. Population density of *S. bilineata* averaged about 0.7/ha., and each colony used an annual foraging area of about 6–18 ha.

Detailed investigations in Costa Rica and Trinidad have revealed marked differences in the social structure of *S. bilineata* and *S. leptura* (Bradbury and Emmons 1974; Bradbury and Vehrencamp 1977). Social groups of *S. bilineata* are composed of a single adult male and a harem of 1–8 females. A number of such groups may be found in a single tree, and together they form a colony of 40–50 individuals, but each male actively defends an area of about 1–3 sq meters of vertical roost and performs elaborate visual and vocal displays to attract a harem of females. There may also be lone adult males in a colony that seek to form their own harems. When a male with a harem is approached by another male, there is high-frequency barking and sometimes striking with wings but no noticeable injuries. Such territorial defense and competition for females goes on throughout the year. There is much movement of females between harems, though the females of an established harem appear aggressive toward newcomers. Individual bats forage alone but adjacent to other colony members. The foraging sites of a colony are divided into exclusive territories defended by harem males against other males. Each of these group territories is in turn divided into separate territories for females, from which they chase outside females. When a female changes harems at a roost site she also changes her foraging site.

In *S. leptura* group size is about 1–5 in Trinidad and 2–9 in Costa Rica. This species appears to have monogamous bonds, and the most common kind of group is simply an adult male and female, but other combinations also occur. Adult males do not defend roosting territories, are not aggressive toward one another, and do not perform rituals to

A. White-lined bat *(Saccopteryx bilineata)*, photo by Rexford Lord. B & C. Two-lined bats *(S. leptura)*, photos by Francis X. Williams, California Academy of Sciences, through Robert T. Orr.

attract females. Unlike *S. bilineata*, in which a harem occupies a single roost for a lengthy period, *S. leptura* roams among a number of roosts. Group composition changes frequently, and both sexes may shift location. *S. leptura* forages either in groups or alone, but even when individuals forage alone the foraging areas are adjacent and form a group territory that is actively defended. In Costa Rica the young of *S. leptura* are born at the beginning of the rainy season in May and again in November. In Trinidad births occur only in May. The females carry their young for 10–15 days, after which the young can fly by themselves. The young are weaned at about 2.5 months.

In Trinidad the reproductive season of *S. bilineata* is remarkably synchronized, with each female producing a single young from late May to mid-June, just after the rainy season. The females never leave the young at the roosts; they carry them during nightly foraging or possibly drop them off in another tree for the night. The young can fly by themselves at about 2 weeks. By 10–12 weeks weaning has occurred, at which time nearly all young females disperse to other sites. The young males, however, tend to remain near the parents and appear to await the opportunity to take over a harem or split off some females for themselves. Female *S. bilineata* first give birth at 12 months (Hayssen, Van Tienhoven, and Van Tienhoven 1993).

In addition to the above information provided by Bradbury and Emmons (1974) and Bradbury and Vehrencamp (1977), records of pregnant female *S. bilineata* have originated in Jalisco, Mexico, in April (Watkins, Jones, and Genoways 1972); the Yucatan Peninsula in February (Jones, Smith, and Genoways 1973); Guatemala and Belize in March, April, and May (Rick 1968); El Salvador in March and Costa Rica in March, April, May, September, and De-

cember (LaVal and Fitch 1977); Panama in January, February, March, and April but not in July (Fleming, Hooper, and Wilson 1972); and Peru in June, July, and August (Tuttle 1970). Pregnant *S. leptura* have been collected in northeastern Peru during October and December (Ascorra, Gorchov, and Cornejo 1993).

The IUCN classifies *S. gymnura* as near threatened.

CHIROPTERA; EMBALLONURIDAE; **Genus CORMURA**
Peters, 1867

Wagner's Sac-winged Bat

The single species, *C. brevirostris*, is found from Nicaragua to Peru, the Guianas, and central Brazil (Cabrera 1957; E. R. Hall 1981; Koopman 1982a).

Head and body length is 50–60 mm, tail length is 6–12 mm, and forearm length is 43–50 mm. Eisenberg (1989) listed average weights of 8.6 grams for males and 10.2 grams for females. The upper parts are either deep blackish brown or reddish brown and the underparts are paler. A long sac-shaped pouch containing a scent gland is located in the small triangular membrane in front of the upper arm and forearm. The open end of the sac is at the end nearest the wrist.

Cormura resembles *Peropteryx* in the absence of striping, in coloration, and in size, but it can be distinguished from that genus by its shorter feet and more robust structure and by the differently placed attachments of the wing membrane to the feet, that is, from the ankle in *Cormura* and from the femur in *Peropteryx*.

Wagner's sac-winged bat *(Cormura brevirostris)*, photos by Merlin D. Tuttle.

Eisenberg (1989) indicated that *Cormura* is rare and is associated with streams and moist areas, preferably in multistratal tropical evergreen forests. One specimen was collected in a hollow fallen tree, and another was taken under the projecting end of a fallen tree. Fleming, Hooper, and Wilson (1972) reported that the stomachs of nine specimens contained 100 percent insect food. These authors also indicated that pregnant females had been found in Panama in April and May but not in June, July, September, and October. Pregnant females have been taken in French Guiana in August and September (Hayssen, Van Tienhoven, and Van Tienhoven 1993).

CHIROPTERA; EMBALLONURIDAE; **Genus**
PEROPTERYX
Peters, 1867

Doglike Bats

There are two subgenera and four species (Anderson, Koopman, and Creighton 1982; Brosset and Charles-Dominique 1990; Cabrera 1957; Goodwin and Greenhall 1961; E. R. Hall 1981; Jones and Hood 1993; Koopman 1978*b*; Lemke et al. 1982; Myers, White, and Stallings 1983; Tuttle 1970):

subgenus *Peropteryx* Peters, 1867

P. macrotis, southern Mexico to Paraguay and southern Brazil, Grenada, Tobago;
P. trinitatis, Venezuela to French Guiana, Trinidad;
P. kappleri, southern Mexico to Peru and southern Brazil;

subgenus *Peronymus* Peters, 1868

P. leucopterus, eastern Colombia, Venezuela, Guianas, eastern Peru, northern Brazil.

Peronymus was recognized as a full genus by Baker, Genoways, and Seyfarth (1981) and Corbet and Hill (1991) but not by Cabrera (1957), Eisenberg (1989), Koopman (1978*b*, 1984*b*, and *in* Wilson and Reeder 1993), Jones and Hood (1993), or Robbins and Sarich (1988). *P. trinitatis* was included within *P. macrotis* by Koopman (*in* Wilson and Reeder 1993) and tentatively by Jones and Hood (1993) but was recognized as a separate species by Brosset and Charles-Dominique (1990) and Handley (1976).

Head and body length is about 45–55 mm, tail length is about 12–18 mm, and forearm length is 38–54 mm. Bradbury and Vehrencamp (1977) reported that the weight of *P. kappleri* averaged 9 grams for females and 11 grams for males. Eisenberg (1989) listed the following average weights: *P. macrotis,* 5.1 grams in males, 6.1 grams in females; *P. trinitatis,* 3.6 grams in males, 4.4 grams in females. Coloration of the upper parts is usually some shade of brown and the underparts are paler. In *P. leucopterus* the wings are whitish or pale gray brown from the forearm outward.

The ears of *P. macrotis* and *P. kappleri* are separate at the base, but in *P. leucopterus* they are joined by a low membrane. *Peropteryx* differs from *Saccopteryx* in the absence of dorsal stripes and in the position and size of the wing sacs. In *Peropteryx* the wing sacs are near the upper edge of the antebrachial membrane and open outward.

Doglike bats occur in forests, swamps, savannahs, and cultivated areas (Handley 1976). They usually roost in shallow caves or rock crevices where light can enter, but they have been taken from the dark recesses of ruins in the Yucatan and from underground passages of ruins in Chiapas, Mexico. A few have been found in dead trees, and several have been found hanging from palm thatching at the entrance to a ruin. *Peropteryx* has been discovered roosting with *Saccopteryx* under the arch of a natural bridge. As might be expected of bats that roost in exposed locations, they are alert in the daytime. They sometimes hang from a horizontal surface but often suspend themselves from a vertical surface by spreading out the forearms and feet. These bats, in common with *Saccopteryx* and *Rhynchonycteris,* fold their wings at about a 45° angle to the body when roosting.

In a study in Costa Rica, Bradbury and Vehrencamp (1977) found colonies of *P. kappleri* roosting in fallen logs and the cavities of large trees, usually about one meter above the ground. Population density was about 0.6/ha. Colonies averaged 4.3 (1–6) individuals, and usually several adults of each sex were present. There was no harem formation and no territoriality. Births began in May and possibly continued until October. LaVal (1977) reported a pregnant female *P. kappleri* in April in Costa Rica. Pregnant females of *P. leucopterus* have been found in March, April, May, and June.

In Colombia, Giral, Alberico, and Alvaré (1991) found organized colonies of 5–47 individual *P. kappleri* in abandoned coal mines. There was only one group per mine, and each demonstrated high fidelity to particular roost sites, though there was some visitation between groups. Males were not territorial and females seemed to lead in move-

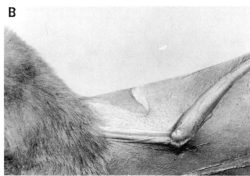

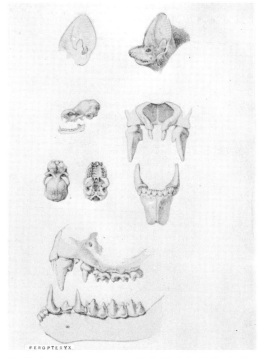

Doglike bat *(Peropteryx macrotis):* A. Photo by P. Morris; B. Photo by Merlin D. Tuttle. Drawings of ear, head, and cranial details from *Die Fledermäuse des Berliner Museums für Naturkunde,* Wilhelm K. Peters.

ments and roost selection. Births occurred during much of the year, but there was a marked peak in April and a lesser one in October–November, coinciding with the rainy seasons. Most females gave birth once a year, always to a single young, but a few had two litters. Females established small, separate roosting territories where they left their young while nursing. The young weighed about 1.5 grams at birth. They reached adult size at 55 days and shortly thereafter permanently dispersed from the natal colony.

In east-central Brazil, Willig (1983) found *P. macrotis* roosting in small aggregations of up to 10 individuals. Each group contained only 1 adult male, thus suggesting the maintenance of harems. Reproductive data for *P. macrotis* include: in the Yucatan 14 of 16 females taken on 18 April were pregnant, each with a single embryo, and a lactating female was taken in August (Jones, Smith, and Genoways 1973); in Guatemala pregnant females with one embryo were taken from February to April, and a lactating female was collected in July (Jones 1966; Rick 1968); in southwestern Colombia 15 of 33 adult females collected in July were pregnant (Arata and Vaughan 1970); and in Peru pregnant females were taken in August (Tuttle 1970).

CHIROPTERA; EMBALLONURIDAE; **Genus CYTTAROPS**
Thomas, 1913

The single species, *C. alecto,* occurs from Nicaragua to northeastern Brazil and the Guianas (Cabrera 1957; E. R. Hall 1981).

Head and body length is about 50–55 mm, tail length is about 20–25 mm, and forearm length is about 45–47 mm. The coloration is a uniform dull smoky gray. The most distinctive feature of this bat is the tragus of the ear; the lower half of the outer margin consists of a large, angular structure that is unique among bats. There is no wing sac, and apparently no glands are associated with the tail.

The type specimen was collected in a garden. A series of specimens from Costa Rica was collected during the day as the bats hung in groups of one to four under the fronds of

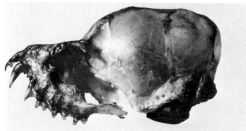

Cyttarops alecto, photos from British Museum (Natural History).

coconut palms located near buildings. The groups were composed of mixed sexes and ages. Insect remains were found in the digestive tracts of some of these specimens (Starrett 1972). The IUCN classifies *C. alecto* as near threatened.

CHIROPTERA; EMBALLONURIDAE; **Genus**
DICLIDURUS
Wied-Neuwied, 1820

Ghost Bats, or White Bats

There are four species (Cabrera 1957; Ceballos and Medellín 1988; Goodwin and Greenhall 1961; E. R. Hall 1981; Handley 1976; Husson 1978; Jones and Hood 1993; Jones, Swanepoel, and Carter 1977; Koopman 1982a; Ojasti and Linares 1971; Tuttle 1970):

D. albus, southern Mexico to Panama, Colombia, Ecuador, eastern Peru, Venezuela, Surinam, northern and eastern Brazil, Trinidad;

D. ingens, southeastern Colombia, Venezuela, Guyana, Amazonian Brazil;

D. scutatus, Venezuela, Guianas, eastern Peru, northern Brazil;

D. isabellus, southern Venezuela, Amazonian Brazil.

The last of the above species long was known from only a single specimen and was placed in the separate genus *Depanycteris* Thomas, 1920. Handley (1976), however, who collected 28 specimens of this species in Venezuela, placed it in the genus *Diclidurus*. *D. virgo* of South America sometimes has been treated as a species distinct from *D. albus*. Corbet and Hill (1991) listed *Depanycteris* as a full genus and *Diclidurus virgo* as a separate species, but Jones and

White bat *(Diclidurus albus):* Top, photo by Cornelio Sanchez Hernandez; Bottom, photo by Merlin D. Tuttle.

Hood (1993) and Koopman (*in* Wilson and Reeder 1993) did not.

Head and body length is 50–80 mm, tail length is 12–25 mm, and forearm length is 45–73 mm. Eisenberg (1989) listed the following average weights: *D. albus*, 14.9 grams in males, 16.2 grams in females; *D. scutatus*, 13.6 grams in males, 12.9 grams in females; and *D. isabellus*, 13.4 grams in males, 15.6 grams in females. *D. ingens* was reported to be larger than the others, but no specific weights were given. The coloration is unusual for bats, generally being white or white mixed with gray. The hairs of the upper parts are slate-colored at the base and white or gray for most of their length. The hairs of the underparts are slate-colored on the basal half and white on the outer half. The wings are also white. There is no wing sac. These are the only white bats except for *Ectophylla alba* and occasional albinos. In the type specimen of *D. isabellus* the ventral surface of the body is dull buffy white, but the upper parts are pale brown.

The eyes are large and the ears are short and rounded. The tail is shorter than the tail membrane and protrudes upward through it. Near the place where the tail penetrates the membrane there appears to be either a glandular structure or a small pouch. The short thumb has only a rudimentary claw and is partly concealed in the wing membrane.

Handley (1976) collected most specimens of *D. ingens*, *D. albus*, and *D. scutatus* in open yards, but he obtained *D. isabellus* only in evergreen forest; all species were found predominantly near streams or in other moist places. According to Ceballos and Medellín (1988), *D. albus* has been taken in tropical rainforests, tropical dry-deciduous and semideciduous forests, coconut plantations, and disturbed vegetation; it roosts by day under the leaves of palms and may undertake local seasonal or migratory movements. Starrett and Casebeer (1968) reported the collection of *D. albus* in Costa Rica, both in the lowlands and at altitudes of 1,100–1,500 meters. Some specimens were taken as they pursued moths and other insects. The call of *D. albus* was reported to be a "unique musical twittering not heard in other Costa Rican bats." Jones (1966) observed about 100 *D. albus* in the floodlights of an oil rig in Guatemala; the bats foraged at heights of about 3–135 meters.

Ceballos and Medellín (1988) stated that *D. albus* is solitary for most of the year but that early in the breeding season up to four individuals, usually a male and several females, may be found roosting within 5–10 cm of each other. Mating probably occurs in January and February, and pregnant females, each with a single embryo, have been taken from January to June.

The IUCN classifies *D. ingens* as vulnerable, noting that it is declining in response to habitat disruption. *D. isabellus* is designated as near threatened.

CHIROPTERA; **Family CRASEONYCTERIDAE; Genus**
CRASEONYCTERIS
Hill, 1974

Kitti's Hog-nosed Bat

The single known genus and species, *Craseonycteris thonglongyai*, has thus far been found only in a small area along the Khwae Noi River ("River Kwai"), Kanchanaburi Province, western Thailand (Duangkhae 1991; Hill 1974a; Lekagul and McNeely 1977).

About the size of a large bumblebee, this bat may be the world's smallest mammal. Head and body length is 29–33

Kitti's hog-nosed bat *(Craseonycteris thonglongyai)*, photo by Jeffrey A. McNeeley.

mm, there is no tail, forearm length is 22–26 mm, and weight is about 2 grams. There seem to be two color phases, one in which the upper parts are brown to reddish and one in which the upper parts are distinctly gray. The underparts are paler; the wings and interfemoral membrane are dark. The ears are relatively large, and the eyes are small and largely concealed by fur.

Distinctive external characters include the very small size, lack of a tail (though there are two caudal vertebrae), lack of a calcar, presence of a large interfemoral membrane, and a prominent glandular swelling at the base of the underside of the throat in males. The muzzle is rather piglike, slightly swollen around the nostrils and chin. The wide, slightly crescent-shaped nostrils open directly in the face of the nose pad and are separated by a relatively wide septum, which broadens to form a small nose pad. The large ears reach beyond the tip of the muzzle when laid forward. The tragus is a little less than half the length of the ear, basally narrow, then broadly expanding to reach its greatest width about halfway along its length. The wing is relatively long and wide with a long tip adapted for hovering flight. The membrane in front of the forearm and upper arm (propatagium) is broad, the thumb is short with a well-developed claw, and the hind foot is long, narrow, and slender. The penis is relatively large. Females have one pair of pectoral and one pair of pubic nipples, the latter probably being vestigial.

The skull is very small, with a slightly inflated globose braincase. There are no lambdoidal crests, postorbital processes, or supraoccipital ridges, but a rather prominent sagittal crest occurs in both sexes. The zygomata are slender but complete; the bullae are relatively large and flattened on the inner face, and the palate is short and wide. The dental formula is: (i 1/2, c 1/1, pm 1/2, m 3/3) $\times$ 2 = 28. The upper incisors are relatively large and have strong cingula at the base. The lower incisors are long and narrow,

about equal in size, and tricuspid, the central cusp being the largest.

The Craseonycteridae are considered to be related to the Rhinopomatidae and the Emballonuridae but to differ sufficiently to warrant familial status. *Craseonycteris* most nearly resembles *Rhinopoma*, the sole genus of the family Rhinopomatidae, but differs in having nostrils that are not valvular or slitlike, large ears not joined anteriorly by a band of integument, a tragus that is widest in the middle, no tail, a more inflated braincase, and relatively larger incisors.

This bat was discovered in October 1973 at Sai Yok, an area of forest intermixed with agricultural fields and riddled with limestone caves. Partly in order to protect the bat, about 500 sq km of the area was designated a national park in 1981. Duangkhae (1991) reported that *Craseonycteris* had been located in 21 caves, some in or near the park but some farther south along the River Kwai. The genus is confined to the limestone region of western Thailand and evidently only to a small portion thereof, as it has not been located in surveys north of 14°45' N or south of 13°45' N. It seems to prefer complex caves within areas of dry evergreen or deciduous forest near the river and within a short flight of fields of cassava or kapok, where prey insects can be found. It roosts mostly at the far end of caves or at the top of the dome chamber, where it seeks small holes or crevices formed by stalactites.

Duangkhae (1990) found that there were two short foraging periods, one for about 18 minutes around dawn and one for 30 minutes at dusk, with torpor likely during the interim. Specific flyways were followed to foraging areas. Flight activity began much earlier in summer than in winter. Surlykke et al. (1993) observed *C. thonglongyai* flying out of many small limestone caves along the River Kwai and hunting in open areas. Flight was fast and fluttering with many sharp turns. The bat usually stayed about 2–5 meters above the ground and several meters from the nearest vegetation. It did not appear to glean from foliage as earlier observations had suggested. Echolocation was utilized intensively and insects probably were seized directly with the mouth rather than with the wings or interfemoral membrane. According to Hill and Smith (1981), the stomach contents of one specimen consisted of the fragments of small insects and a spider.

Duangkhae (1990, 1991) reported that colony size in the 21 caves where *Craseonycteris* was located averaged about 100 individuals and ranged from 1 to 500. Individuals roosted alone, well separated from one another. Each individual also had its own foraging area and would chase intruders away. Breeding began in the dry season, in late April, and continued through May, with each female giving birth to a single young. Mothers always carried their babies while in the roost but left them when going out to forage.

Craseonycteris is classified as endangered by the IUCN and the USDI. Surlykke et al. (1993) reported that ultrasonic detection equipment had located it in many more caves than had been reported previously and in some caves where it was thought to have disappeared. Nonetheless, there is concern for the genus because of its restricted habitat. Fewer than 2,500 individuals are estimated to survive, and a decline is continuing. Duangkhae (1990) noted that it is undergoing considerable disturbance, mostly by tourists and scientists wanting to observe or collect the world's smallest mammal. The vegetation in the vicinity of its caves, flyways, and foraging areas is subject to annual burning, which could be especially harmful to the young.

CHIROPTERA; **Family NYCTERIDAE; Genus NYCTERIS**
G. Cuvier and E. Geoffroy St.-Hilaire, 1795

Slit-faced Bats, or Hollow-faced Bats

The single known genus, *Nycteris,* contains 13 species (Bergmans and Van Bree 1986; Corbet and Hill 1992; Crawford-Cabral 1989a; Ellerman and Morrison-Scott 1966; Griffiths 1994; Hayman and Hill *in* Meester and Setzer 1977; Juste and Ibáñez 1994; Koch-Weser 1984; Koopman 1975, 1989c, 1992; Koopman, Mumford, and Heisterberg 1978; Medway 1977; Nader and Kock 1983a; Rautenbach, Fenton, and Braack 1985; Van Cakenberghe and De Vree 1985, 1993a, 1993b):

N. arge, Sierra Leone to northeastern Angola and western Kenya, Bioko;

N. nana, Ivory Coast to northeastern Angola and western Kenya;

N. major, Liberia to eastern Zaire;

N. intermedia, Ivory Coast to western Tanzania and Angola;

N. tragata, Malay Peninsula (including Tenasserim), Sumatra, Borneo;

N. javanica, Java, Kangean Islands, Bali;

N. grandis, Liberia to Kenya and Zimbabwe, Zanzibar;

N. hispida, Senegal to Somalia and south to Angola and Mozambique, two specimens also taken at the Cape of Good Hope in South Africa;

N. aurita, Ethiopia, Somalia, Kenya, Tanzania;

N. macrotis, Gambia to Sudan and Somalia and south to

Slit-faced bat *(Nycteris thebaica),* photo by Merlin D. Tuttle, Bat Conservation International.

Angola and Botswana, Mozambique, Zanzibar, Madagascar;

N. woodi, Cameroon, Somalia to Transvaal;

N. thebaica, open country throughout most of Africa, Palestine, Arabian Peninsula, island of Corfu (Greece);

N. gambiensis, Senegal to Burkina Faso and Benin.

The above sequence reflects the rather unusual case of complete agreement between several systematic approaches about which species of a polytypic genus are the most primitive and which are phylogenetically related (Griffiths 1994; Van Cakenberghe and De Vree 1985, 1993a, 1993b). An old report of *N. javanica* from Timor was not accepted by Bergmans and Van Bree (1986) or Goodwin (1979). *N. madagascariensis* of Madagascar is sometimes considered a full species and sometimes a synonym of *N. thebaica,* but Van Cakenberghe and De Vree (1985) indicated that it is a synonym of *N. macrotis. N. vinsoni* of Mozambique is sometimes considered a separate species, related to *N. thebaica,* but Koopman (1992) regarded it as a subspecies of *N. macrotis.*

Head and body length is 40–93 mm, tail length is about 43–75 mm, forearm length is 32–64 mm, and adult weight is 10–43 grams. The pelage is long and rather loose. The coloration is rich brown or russet to pale brown or grayish. *N. macrotis* exhibits considerable orange color variation in Zaire, and an albino specimen of *N. nana* has been found.

The muzzle is divided by a longitudinal furrow; the nostrils are located in the anterior end of this groove. This furrow is margined and concealed by nose leaves and expands posteriorly into a deep pit on the forehead. The interorbital region of the skull is deeply concave. The lower lip has a granular surface at the tip. The large ears are longer than

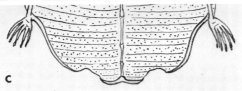

A. Slit-faced bat (*Nycteris* sp.), photo by V. Aellen. B. *N. thebaica*, photo by Erwin Kulzer. C. *N. javanica* showing terminal cartilage, drawing by William J. Schaldach, Jr., of specimen in U.S. National Museum of Natural History.

the head and united at their bases by a low membrane. A small tragus is present. The wings are broad. The long tail is completely enclosed within the large interfemoral membrane and terminates in a T-shaped tip, which serves as a support for the free edge of the tail membrane. This is a distinctive character that is unique among mammals.

The upper incisors closely resemble the lower incisors in size and form. The dental formula is: (i 2/3, c 1/1, pm 1/2, m 3/3) × 2 = 32.

Most species inhabit woodland savannah or dry country, but some, including *N. grandis* and *N. javanica*, occupy dense forest (Hayman and Hill *in* Meester and Setzer 1977; Lekagul and McNeely 1977; Rosevear 1965). Roosting sites include hollow trees, dense foliage, rocky outcrops, caves, buildings, ruins, culverts, abandoned wells, and even porcupine and aardvark burrows. These bats feed on a variety of arthropods, including moths, butterflies, other insects, spiders, and sun spiders. Scorpions form a staple part of the diet of *N. thebaica*. Most species flutter around trees and bushes, but they also come into lighted verandas and rooms in search of insects. Captive *N. grandis* have captured and consumed other bats and frogs (Fenton, Gaudet, and Leonard 1983), and this species is now known to regularly feed on fish, frogs, birds, and bats in the wild (Fenton, Thomas, and Sasseen 1981; Fenton et al. 1993).

Most species shelter alone, in pairs, or in small family groups, but *N. hispida* occasionally gathers in groups of about 20 and *N. thebaica* and *N. gambiensis* are even more gregarious. Smithers (1971) reported a colony of 500–600 *N. thebaica* in one cavern. He also observed that the young of this species were born in Botswana during the warm, wet summer months from about September to February. Bernard (1982) found this same species to mate in early June in South Africa; gestation then lasted five months and lactation another two months. Based on Bernard's studies and additional observations from various areas, Happold and Happold (1990) concluded that female *N. thebaica* give birth synchronously during a restricted season once a year; that species evidently is not polyestrous, but *N. macrotis* apparently is. According to Hayssen, Van Tienhoven, and Van Tienhoven (1993), *N. macrotis* produces two or three litters annually and has a gestation period of 2.5 months and a lactation period of 2 months. Kingdon (1974*a*) wrote that thousands of *N. thebaica* had been reported roosting together in caves in South Africa, but Skinner and Smithers (1990) reported colony size there to be "up to hundreds." Other data summarized by Kingdon include a pregnant female *N. macrotis* found in Uganda in late December; a lactating female *N. macrotis* found in Tanzania in November; and embryos or young of *N. grandis* found in Gabon in April, August, and November. Pregnant female *N. thebaica* with one embryo have been recorded from Egypt in April (Gaisler, Madkour, and Pelikan 1972), the southwestern Arabian Peninsula in April and October (Harrison and Bates 1991), and Zambia in August (Ansell 1960). In South Africa *N. thebaica* gives birth in November and December (Herselman and Norton 1985). *N. javanica* breeds throughout the year, and females mate again shortly after giving birth (Lekagul and McNeely 1977). Young *N. nana* nurse from 45 to 60 days.

No fossils referable to this family have been found. The group is now widely distributed, though some species occur in areas confronted with rapid loss of habitat to human encroachment. The IUCN recognizes *N. vinsoni*, restricted to a small part of Mozambique and continuing to decline, as critically endangered. *N. javanica* and *N. major* are classified as vulnerable, and *N. intermedia*, *N. aurita*, and *N. woodi* are designated as near threatened.

 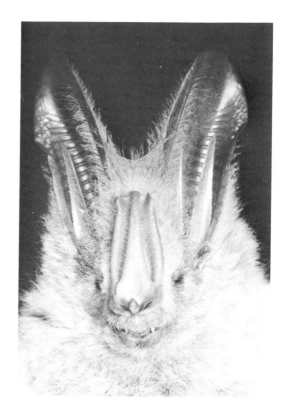

African yellow-winged bat (Lavia frons), photos by David Pye.

CHIROPTERA; Family MEGADERMATIDAE

False Vampire Bats and Yellow-winged Bats

This family of four Recent genera and five species occurs in Africa, south-central and southeastern Asia, the East Indies, and Australia. The sequence of genera presented here follows that of Griffiths, Truckenbrod, and Sponholtz (1992), whose investigation of hyoid morphology indicated that *Macroderma* is phylogenetically primitive and somewhat removed from a related group formed by *Cardioderma, Megaderma,* and *Lavia*. Based mainly on studies of dentition, Hand (1985) had suggested that *Lavia* was the most primitive living genus and that *Macroderma* was an advanced form closely related to *Lyroderma,* which in turn was regarded as generically distinct from *Megaderma*.

Head and body length is 65–140 mm, an external tail is lacking, and forearm length is 50–115 mm. The Australian member of this family, *Macroderma gigas,* is the largest of the bats other than some of those in the family Pteropodidae and the New World species *Vampyrum spectrum;* it is known as the "Australian giant false vampire."

These bats are noted for their large, basally united ears (largest in the genus *Lavia*), divided tragus, and long, erect nose leaf. The genera differ from each other in size and in dental and skull features.

Upper incisors are lacking. The upper canines project noticeably forward, the shaft having a large secondary cusp. The dental formula in the subgenera *Megaderma* and *Lyroderma* of the genus *Megaderma* is (i 0/2, c 1/1, pm 2/2, m 3/3) $\times$ 2 = 28, but in the genera *Cardioderma,* *Macroderma,* and *Lavia* it is (i 0/2, c 1/1, pm 1/2, m 3/3) $\times$ 2 = 26.

Megaderma, Cardioderma, and *Macroderma* roost in caves, rock crevices, buildings, and hollow trees and are often the sole occupant of their retreat, but *Lavia* generally shelters in trees and bushes. These bats shelter singly or in groups of as many as 50–100 individuals. Although not particularly rapid fliers, they are skillful and dexterous in flight. *Lavia* is often active during the day, but the other members of this family hunt only at night. These bats, as well as others that have a nose leaf, possibly fly with their mouths closed, emitting high-frequency sounds through their nostrils.

The common name "false vampire bat" reflects the old but erroneous belief that these bats feed on blood, as do the true vampire bats (Desmodontinae) of the New World. The diet of *Megaderma* and *Macroderma* consists of insects and a variety of small vertebrates, and *Lavia* feeds mainly, if not entirely, on insects. Although *Megaderma lyra* may first consume the blood and then eat the flesh of an animal it has captured, no member of this family preys on another animal solely for its blood.

The geological range of this family is early Oligocene to early Pliocene in Europe, early Oligocene to Recent in Africa, Pleistocene to Recent in Asia, and middle Miocene to Recent in Australia (Hand 1985; Koopman 1984a; Koopman and Jones 1970).

Australian giant false vampire bat *(Macroderma gigas)*, photo by Stanley Breeden.

CHIROPTERA; MEGADERMATIDAE; **Genus**
MACRODERMA
Miller, 1906

Australian Giant False Vampire Bat, or Ghost Bat

The single species, *M. gigas,* currently has a scattered distribution in the Pilbara and Kimberley regions of northern Western Australia, the northern part of the Northern Territory, and coastal Queensland as far south as Rockhampton. Skeletal remains, guano deposits, and early sightings and collections indicate that the same species occurred throughout that part of Australia north of 27° S until the nineteenth century and that it was present even farther south several hundred years ago (Churchill and Helman 1990; Molnar, Hall, and Mahoney 1984). Based on habitat suitability and reported observations, Filewood (1983) stated that there is little doubt that *M. gigas* also occurs in New Guinea.

This is among the largest of the bats of the suborder Microchiroptera. Head and body length is 100–140 mm, there is no external tail, forearm length is 105–15 mm, and wingspan is about 600 mm. Hudson and Wilson (1986) listed weight as 130–70 grams, but Nelson (1989) referred to a maximum of 216 grams for a nonpregnant female. The hairs of the back are usually white with pale grayish brown tips that impart an ashy gray effect, and the head, ears, nose leaf, membranes, and underparts are generally whitish. Several specimens from Queensland were deep brownish drab above and plumbeous below, with darkened skin. The vernacular name "ghost bat" refers to the usual pale coloration.

The ghost bat occurs in tropical areas, roosting in caves, rock crevices, and old mine shafts. It emerges from its shelter after sunset and flies a straight, smooth course. Tidemann et al. (1985) found it to fly an average of 1.9 km from the roost to the foraging area. There it would hang from a tree at heights of up to 3 meters and watch for prey. *Macroderma* is carnivorous; it drops on small mammals from above, envelops them with its wings, and kills by biting the head and neck. The most common prey in the wild is house mice, but native rodents, small marsupials, other bats, birds, reptiles, and insects are also taken (Douglas 1967). The prey is carried to a high point or back to the main roost to be consumed, and uneaten remains often accumulate under such sites. Individuals of other species of bats have been found in a cave occupied by the ghost bat; these smaller bats, *Taphozous georgianus* and *Pipistrellus pumilus,* were wedged deeply into crevices, possibly to avoid attacks by the ghost bat.

Macroderma roosts alone or in small groups. Ride (1970) stated that field observations of this genus in Western Australia showed that females give birth to a single young about November in southern areas and about September in the far north. It seems likely that at these times the females congregate without the males in maternity colonies, while the males gather in all-male colonies. In a study of a cave colony of about 150 individuals in central Queensland, however, Toop (1985) found some males to always be present with the females. Mating occurred in April; the group then dispersed for the cooler months and came together again prior to the birth season, which occurred from mid-October to late November. The young were initially carried by the mothers and then left at the roost. They began flying on their own at 7 weeks, weaning was complete by March, and both sexes reached reproductive maturity in their second year of life. Nelson (1989) reported that two individuals had been kept in captivity for just over 16 years.

Based on numerous remains and guano deposits, it can be supposed that *Macroderma* formerly was common in the southern part of Australia, where it does not now occur. It has been suggested that the absence of *Macroderma* from this region may be related to increasingly arid conditions. Such aridity would deplete the insect life and so result in a reduction in the numbers of the insectivorous bats upon which the ghost bat feeds. However, *Macroderma* still was present in parts of southern Australia until the time of European settlement and persisted at about the central latitudes of the continent through the early twentieth century. Its disappearance from these areas may be associated with changing fire regimes, the disappearance of small native prey mammals, and other environmental disruption associated with settlement (Churchill and Helman 1990).

The current status of the ghost bat is questionable, partly because of its secretive habits. Although it was once thought to be extremely rare even where it did survive, Ride (1970) indicated that field studies had shown it still to have a wide distribution and to be relatively common in some areas. Subsequently there has been increasing concern that quarrying operations and other human activity are jeopardizing the remaining colonies. In 1989 conservationists lost a long battle when commercial interests destroyed roosting caves critical to much of the population of *Macroderma* in Queensland. About 1,500 more bats, perhaps 25–30 percent of the entire species, roost at the Kohinoor Gold Mine, south of Darwin in the Northern Territory; they could be jeopardized if mining activity intensifies (Hutson 1991). The IUCN classifies the ghost bat as vulnerable.

African false vampire bat *(Cardioderma cor)*, photo by D. W. Yalden.

African false vampire bat *(Cardioderma cor)*, photo by Erwin Kulzer.

CHIROPTERA; MEGADERMATIDAE; **Genus**
CARDIODERMA
Peters, 1873

African False Vampire Bat

The single species, *C. cor*, is found in East Africa from northeastern Ethiopia and southern Sudan to northern Zambia. Although it is sometimes considered a subgenus of *Megaderma*, most recent authors have given *Cardioderma* full generic rank (Corbet and Hill 1991; Griffiths, Truckenbrod, and Sponholtz 1992; Hand 1985; Koopman *in* Wilson and Reeder 1993).

Head and body length is 70–77 mm, there is no tail, forearm length is 54–59 mm, and adult weight is 21–35 grams. This bat is uniformly blue gray in color. From *Megaderma*, *Cardioderma* differs in having a narrower frontal shield, only one upper premolar, and no reduction of the third upper molar (Lekagul and McNeely 1977).

Most of what is known about this genus stems from a detailed study by Vaughan (1976) in bushland in southern Kenya. The bat was found to roost in hollow baobab trees in groups of up to 80 individuals. At night each bat utilized an exclusive foraging area; two such areas measured 0.10 ha. and 0.11 ha. in mid-April, during the wet season, and two others measured 0.55 ha. and 1.01 ha. in late August, during the dry season, when food was less abundant. During the night individuals spent considerable time perching in low vegetation, listening for terrestrial prey. Flights to capture prey and to move between perches were brief, usually lasting less than five seconds and covering less than 25 meters. Food was generally captured on the ground and consisted mainly of large, ground-dwelling beetles, but centipedes, scorpions, and, rarely, small bats were also taken. During part of the rainy season, leaf gleaning and aerial capture of insects were important. During the March–April wet season individuals spent considerable time "singing" and establishing their exclusive foraging areas. The "song" consisted of four to nine high-intensity pulses, each with a sharp chip like that given by some passerine birds. There was also a less frequent, more variable and louder flight call. Females gave birth twice each year, during the rainy periods, in March or early April and in November. More than one young per female was never observed. Kingdon (1974*a*) indicated that the breeding season may be more extended in some areas and stated that gestation lasts about three months. Additional studies in Kenya have suggested the existence of long-lasting male-female pairs and territories (McWilliam 1987*b*). In Tanzania mating has been reported during September–October, births in early January, and weaning at the end of March (Hayssen, Van Tienhoven, and Van Tienhoven 1993). The IUCN regards *Cardioderma* as near threatened.

CHIROPTERA; MEGADERMATIDAE; **Genus**
MEGADERMA
E. Geoffroy St.-Hilaire, 1810

Asian False Vampire Bats

There are two subgenera and two species (Corbet and Hill 1992; Ellerman and Morrison-Scott 1966; Roberts 1977):

subgenus *Megaderma* E. Geoffroy St.-Hilaire, 1810

M. spasma, India to Indochina and Malay Peninsula, Sri Lanka, Andaman Islands, Sumatra and several nearby

Asian false vampire bat *(Megaderma spasma)*, photo by Paul D. Heideman.

islands to west, Krakatau, Java, Kangean Islands, Borneo, Natuna Islands, Philippines, Sulawesi, Sangihe Islands, Togian Islands, Lombok, Molucca Islands;

subgenus *Lyroderma* Peters, 1872

M. lyra, eastern Afghanistan and Sri Lanka to southeastern China and northern Malay Peninsula.

Head and body length is about 65–95 mm, there is no external tail, and forearm length is 50–75 mm. Lekagul and McNeely (1977) listed weights of 23–28 grams for *M. spasma* and 40–60 grams for *M. lyra*. *M. spasma* is smoky bluish gray above and brownish gray below. *M. lyra* is grayish brown above and whitish gray below.

These bats roost in groups in caves, pits, buildings, and hollow trees and are usually the sole occupant of their retreat. *M. spasma* generally inhabits wetter areas than does *M. lyra* (Lekagul and McNeely 1977). In a radio-tracking study of *M. lyra* in southern India, Audet et al. (1991) found individuals to return nightly to nonexclusive foraging areas about 0.1 sq km in size and located 0.5–4.0 km from the day roosts. *M. lyra* is carnivorous, feeding on insects, spiders, and small vertebrates such as other bats, rodents, birds, frogs, and fish. Captured prey is carried to the roost to be eaten, and vertebrate remains may be found beneath the roost. *M. spasma* is less carnivorous, apparently favoring grasshoppers and moths, but it also carries prey to a feeding roost (Lekagul and McNeely 1977). Both species often feed among trees and undergrowth, usually flying within three meters of the ground, and they will enter houses to pick lizards and insects off the wall.

Both species usually roost in groups of about 3–30, but a seasonal colony of 1,500–2,000 *M. lyra* has been reported in India. The sexes of *M. spasma* live together year-round, whereas the adult males and females of *M. lyra* appear to segregate when birth is imminent. Mating in India takes place from about November to January, gestation lasts 150–60 days, and the young are born from April to June, before the onset of the rains. There is usually one young, occasionally two. The young grow rapidly, are initially carried by the females, and are suckled for 2–3 months. In *M. lyra* the males are sexually mature at 15

months and the females at 19 months. In the Philippines, Ingle (1992) found several groups of 1–7 *M. spasma* roosting in cavities within large trees; there apparently was a seasonal birth period in April–May.

CHIROPTERA; MEGADERMATIDAE; **Genus LAVIA** Gray, 1838

African Yellow-winged Bat

The single species, *L. frons*, occurs from Gambia to Ethiopia and south to northern Zambia (Hayman and Hill *in* Meester and Setzer 1977).

Head and body length is 58–80 mm, there is no external tail, and forearm length is 49–63 mm. Kingdon (1974*a*) gave a weight of 28–36 grams. The color of the body is blue gray, pearl gray, or slaty gray; the lower back is sometimes brownish or olive green; and the underparts occasionally are yellow. The ears and wings are reddish yellow.

This bat inhabits forests and open country, generally being found around swamps, lakes, and rivers where trees and bushes fringe the water and afford roosting sites. It prefers to roost where it can observe its surroundings, hence usually in an environment where the herbaceous growth is not too dense. On the savannahs of Kenya, Vaughan and Vaughan (1986) found *Lavia* to roost in acacia trees at 5–10 meters above the ground. It also has been found in the cavity of a tree and in buildings. This bat is alert and often active during the day. It sometimes betrays its presence by the motion of its long, sensitive ears. Vaughan and Vaughan reported that the head is held up almost constantly in daylight, the body revolves, and the eyes were never seen to be closed. When disturbed it generally flies a considerable distance away, even on a bright day.

Unlike other members of this family, *Lavia* is not known to feed on small vertebrates. The diet is mainly, if not entirely, insects. Vaughan and Vaughan (1986) suggested that *Lavia* is closely associated with acacia trees, which attract many insects during periods of flowering. The method of feeding has been compared to that of a flycatcher: the bat often hangs from a branch at night, constantly scanning the

African yellow-winged bat *(Lavia frons)*, photo by Bruce J. Hayward. Inset: skull, photo by P. F. Wright of specimen in U.S. National Museum of Natural History.

area until an insect is detected; it then swoops down to the insect and returns to the same or another perch to consume the catch. This bat has been flushed from grass, where it may have been feeding. It can dodge about between bushes and trees with considerable skill.

Vaughan and Vaughan (1986) found *Lavia* to live in permanent, territorial pairs. This lifestyle may be an adaptation to environmentally produced stress resulting from the recurring periods of low food availability when acacias are not in flower. Monogamy is necessary to allow the female to concentrate her energy on raising young. The members of a pair roost together, less than a meter apart, and interact at night through flying rituals. They maintain an exclusive foraging territory of 0.60–0.95 ha. The male regularly patrols this area and drives away intruding conspecifics. The roosting site is usually near the edge of the

territory, more than 20 meters from the closest neighboring pair. A single young is born to the pair in early April, at the start of the long rainy season. Vaughan and Vaughan (1987) added that the young clings tenaciously to its mother for several weeks, even when she forages. The young then is left at the roost for about a week and begins flying alone in early May. Weaning occurs 20 days after the first solo flight, when the young is about 55 days old, but it continues to share its parents' territory and roost for at least another 30 days.

Kingdon (1974*a*) provided the following additional reproductive data: in eastern Zaire pregnancies are most numerous from January to April, but there are scattered records at other times; in Garamba National Park, in extreme northeastern Zaire, there appears to be a defined breeding season, with births starting in April; and gestation lasts about three months. Koopman, Mumford, and Heisterberg (1978) reported the collection of pregnant females in Burkina Faso, each with a single embryo, in March, May, and September. Births have been reported in Tanzania during January, April, August, and November (Hayssen, Van Tienhoven, and Van Tienhoven 1993).

CHIROPTERA; **Family RHINOLOPHIDAE; Genus RHINOLOPHUS**
Lacépède, 1799

Horseshoe Bats

The single Recent genus, *Rhinolophus,* contains 12 species groups and 62 species (Aellen 1973; Aggundey and Schlitter 1984; Allen 1939*b;* Baeten, Van Cakenberghe, and De Vree 1984; Bergmans and Van Bree 1986; Bogdanowicz 1992; Bogdanowicz and Owen 1992; Brosset 1984; Chasen 1940; Corbet 1978; Corbet and Hill 1992; DeBlase 1980; Ellerman and Morrison-Scott 1966; Flannery 1990*b*, 1995; Francis 1997; R. E. Goodwin 1979; Hayman and Hill *in* Meester and Setzer 1977; Heaney et al. 1991; Herselman and Norton 1985; Hill 1962, 1982*a*, 1983*a*, 1986; Hill, Harrison, and Jones 1988; Hill and Schlitter 1982; Hill and Thonglongya 1972; Hill and Yoshiyuki 1980; Koopman 1975, 1979, 1982*b*, 1989*c*, and *in* Wilson and Reeder 1993; Laurie and Hill 1954; Lawrence 1939; Lekagul and Mc-Neely 1977; McFarlane and Blood 1986; McKean 1972; Medway 1977, 1978; Qumsiyeh 1985; Rickart et al. 1993; Ride 1970; Taylor 1934; Topal and Csorba 1992):

megaphyllus group

R. megaphyllus, Thailand, Malay Peninsula and islands along coast, Lombok, Sumbawa, Komodo, Molucca Islands, Papua New Guinea, Bismarck Archipelago, Woodlark Island, St. Aignan's Island in Louisiade Archipelago, eastern Australia;
R. celebensis, Java, Madura, Kangean Islands, Bali, Talaud and Sangihe islands, Sulawesi, Togian Islands, Timor;
R. virgo, Philippines;
R. borneensis, Cambodia, possibly Viet Nam, Borneo and several nearby islands, Java;
R. nereis, Siantan and Bunguran islands between Malay Peninsula and Borneo;
R. anderseni, probably Luzon and Palawan Island in the Philippines;
R. malayanus, Thailand, Indochina, Malay Peninsula;

A. Lesser horseshoe bat *(Rhinolophus hipposideros)* carrying young in flight, photo by Ronald Thompson. B. Greater horseshoe bat *(R. ferrumequinum)*, photo by Liselotte Dorfmüller. C. Lesser horseshoe bat *(R. hipposideros)* in sleeping position, photo by Eric J. Hosking.

rouxi group

R. stheno, southern Thailand, Malay Peninsula, Sumatra, Java;
R. thomasi, Yunnan (southern China), Burma, Thailand, Indochina;
R. rouxi, India and Sri Lanka to southeastern China;
R. affinis, northern India and Nepal to southern China and Indochina, Malay Peninsula, Andaman Islands, Sumatra, Mentawai Islands, Java, Kangean Islands, Lombok, Sumbawa, Sumba, Flores, Solor, Borneo, Siantan and Bunguran islands;
R. acuminatus, southern Thailand, Laos, Cambodia, Malay Peninsula, Sumatra, Nias and Enggano islands, Java, Bali, Borneo, Palawan, Lombok;

pusillus group

R. lepidus, Afghanistan and India to southern China and Burma, Malay Peninsula and nearby islands, Sumatra;
R. pusillus, northern India and Nepal to southern China and Indochina, Malay Peninsula and nearby islands, Hainan, Sumatra, Mentawai Islands, Java, Madura, Borneo, Siantan Island, Lombok;
R. cornutus, Japan, Ryukyu Islands, possibly eastern China;

R. subbadius, Nepal, northeastern India, Burma, northern Viet Nam;
R. monoceros, Taiwan;
R. cognatus, Andaman Islands (Bay of Bengal);
R. imaizumii, Iriomote Island (southern Ryukyu Islands);

euryotis group

R. euryotis, Sulawesi, Togian Islands, Molucca Islands, Aru Islands, New Guinea, Bismarck Archipelago;
R. creaghi, Borneo, Madura Island northeast of Java;
R. canuti, Java, Timor;
R. arcuatus, Sumatra, Borneo, Philippines, Sulawesi, Buru and Amboina islands, New Guinea;
R. coelophyllus, Burma, Thailand, Malay Peninsula;
R. shameli, Burma, Thailand, Cambodia;
R. inops, Philippines;
R. rufus, Philippines;
R. subrufus, Luzon, Mindoro, Tablas, Negros, Camiguin, and Mindanao (Philippines);

pearsonii group

R. pearsonii, northern India and Nepal to southern China and Viet Nam, possibly Malay Peninsula;
R. yunanensis, Yunnan (southern China), northeastern India, northern Burma, Thailand;

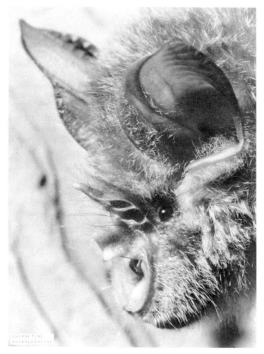 

Horseshoe bat *(Rhinolophus ferrumequinum):* Left photo, showing the complex nose leaf of the Rhinolophidae, by Walter Wissenbach; Right photo by P. Rödl.

philippinensis group

R. *philippinensis,* Borneo, Philippines, Sulawesi, Timor, Kei Islands, east-central New Guinea, northeastern Queensland;
R. *marshalli,* Thailand, Viet Nam;
R. *macrotis,* northern Pakistan, northern India, Nepal, southern China, Thailand, Indochina, Malay Peninsula, Sumatra, Guimaras and Negros islands (Philippines);
R. *mitratus,* known only by a single specimen from the state of Orissa in eastern India;

trifoliatus group

R. *trifoliatus,* northeastern India, Tenasserim, southern Thailand, Malay Peninsula, Sumatra, Nias and Bangka islands, Java, Borneo;
R. *luctus,* northern India to southeastern China and Viet Nam, Malay Peninsula, Taiwan, Hainan, Sumatra, Java, Bali, Borneo;
R. *beddomei,* peninsular India, Sri Lanka;
R. *rex,* Sichuan and Guizhou (south-central China);
R. *paradoxolophus,* known only by one specimen from northern Viet Nam, one from northeastern Thailand, and one from central Laos;
R. *sedulus,* Malay Peninsula, Borneo;
R. *maclaudi,* Guinea and Liberia in West Africa, Ruwenzori region of eastern Zaire, western Rwanda, and western Uganda;

ferrumequinum group

R. *ferrumequinum,* entire southern Palaearctic region from Great Britain and Morocco to Afghanistan and Japan;

R. *clivosus,* central and southwestern Asia, most of the open country of Africa from Algeria and Cameroon eastward and south to South Africa, Liberia;
R. *bocharicus,* Azerbaijan, northeastern Iran, Turkmenistan, Uzbekistan, northern Afghanistan, possibly northern Pakistan;
R. *darlingi,* Nigeria and Namibia to Tanzania and Transvaal;

fumigatus group

R. *fumigatus,* Africa south of the Sahara;
R. *eloquens,* southern Sudan, Somalia, northeastern Zaire, Rwanda, Uganda, Kenya, northern Tanzania, Zanzibar, Pemba Island;
R. *hildebrandti,* East Africa from southern Sudan and Somalia to Transvaal;
R. *silvestris,* Gabon, Congo;
R. *deckenii,* Uganda, eastern Kenya and Tanzania, Zanzibar and Pemba islands;

capensis group

R. *capensis,* southern Africa;
R. *simulator,* Guinea, northern Nigeria, Cameroon, Sudan to South Africa;
R. *denti,* Guinea to Ghana, southern Africa;
R. *swinnyi,* southern Zaire and Tanzania to South Africa;
R. *blasii,* Italy to Afghanistan, Morocco to Ethiopia and Transvaal;
R. *landeri,* Senegal to Somalia, and south to Namibia and eastern Transvaal;
R. *guineensis,* Guinea, Sierra Leone, Liberia;
R. *alcyone,* forest zone from Senegal to Uganda;
R. *adami,* Congo;

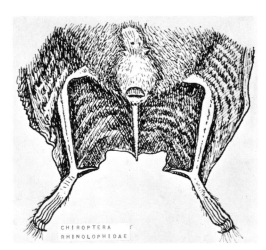

Tail membrane of a bat of the family Rhinolophidae, photo from *Exploration du Parc Albert National, Mamm.*

euryale group

R. *euryale,* Mediterranean region of southern Europe and northern Africa, east to Iran and Turkmenistan;

R. *mehelyi,* southern Europe, Morocco to Afghanistan;

hipposideros group

R. *hipposideros,* British Isles to Arabian Peninsula and Central Asia, Morocco to Ethiopia.

The Hipposideridae (see account thereof) are sometimes considered a subfamily of the Rhinolophidae. The above arrangement of species groups, the assignment of species thereto, and the sequence of species therein are based in part on the work of Bogdanowicz (1992) and Bogdanowicz and Owen (1992). Their conclusions indicate that the Rhinolophidae originated in Southeast Asia, that the *megaphyllus* group contains the most primitive species, and that the Palaearctic and Ethiopian species are generally the most advanced. R. *maclaudi* of Africa seems to be phylogenetically closest to Asian species. Some of the other sources cited above, particularly Corbet and Hill (1992), also have been utilized in organizing the list of species. Koopman's (1975) arrangement of African species and his (1989a) discussion of Indo-Malayan species of *Rhinolophus* would indicate some further modifications of the above sequence. R. *simplex* of Lombok and Sumbawa, R. *keyensis* of the Molucca Islands, and R. *robinsoni* of Thailand and the Malay Peninsula were listed as separate species by Koopman (*in* Wilson and Reeder 1993) but were included in R. *megaphyllus* by Corbet and Hill (1992). R. *osgoodi* of Yunnan was listed as a separate species by Koopman but was included in R. *lepidus* by Corbet and Hill. R. *bocharicus* was included in R. *clivosus* by Koopman but considered a distinct species by Corbet and Hill. R. *pumilus* and R. *perditus* of Okinawa and the Ryukyu Islands were treated as separate species by Yoshiyuki (1989) but included in R. *cornutus* by Corbet and Hill and by Koopman. Recognition here of R. *beddomei* as a separate species follows Topal and Csorba (1992), though it was included in R. *luctus* by Corbet and Hill and by Koopman. Bergmans and Rozendaal (1982) described R. *tatar* from northeastern Sulawesi, but Hill (1983a) considered it a subspecies of R. *euryotis.* The species R. *paradoxolophus* was formerly placed in the separate genus *Rhinomegalophus* Bourret, 1951, but is now

considered a species of *Rhinolophus* (Hill 1972b; Thonglongya 1973) and may be conspecific with R. *rex* (Corbet and Hill 1992).

Head and body length is about 35–110 mm, tail length is 15–56 mm, and forearm length is 30–75 mm. The weight of R. *hipposideros,* one of the smaller species, is 4–10 grams; R. *ferrumequinum,* a larger species, weighs 16.5–28 grams. Color varies greatly, ranging from reddish brown to deep black above and paler below.

These bats have a peculiar, complex, nose-leaf expansion of the skin surrounding the nostrils. It consists of three parts. The lower part, which is horseshoe-shaped, covers the upper lip, surrounds the nostrils, and has a central notch in the lower edge. Above the nostrils, the appendage is a pointed, erect structure, the lancet, attached only by its base. Both the horseshoe and the lancet are flattened from front to back. The sella, located between the horseshoe and the lancet, is flattened from side to side; it is connected at its base by means of folds and ridges. The shape and arrangement of the nose leaf varies from species to species. These bats generally fly with their mouth closed and emit ultrasonic sounds through the nostrils. The sounds thus emitted may be oriented with the aid of the nose leaf.

The ears are large and lack a tragus. Two teatlike processes not connected with a mammary gland, known as "dummy teats," are found on the abdomens of females in addition to the two functional mammae on the chest. An infant horseshoe bat may grasp the dummy teats of its mother while she carries it during flight.

Young horseshoe bats shed milk teeth before birth. The teeth of adults exhibit the normal cuspidate pattern found in insectivorous bats. The dental formula is: (i 1/2, c 1/1, pm 2/3, m 3/3) × 2 = 32. The nasal region of the skull is considerably expanded. The first toe has two bones, and all the other toes have three; the Hipposideridae, in contrast, have only two bones in each toe. The eyes of horseshoe bats are quite small, and the field of vision seems to be partly obstructed by the large nose leaf, so sight is probably of little importance.

These bats, roosting where they can hang freely, do not close their wings alongside their body as most bats do but wrap them around the body. The small bare patch on the back at the base of the tail is covered by the upturned tail and membrane; the bat is thus completely enclosed in its flight membranes and resembles the pod of a fruit or the cocoon of an enormous insect. When the bat is at rest the basal axis of the head makes a right angle with the vertebral column, so that it looks in the direction of its ventral surfaces.

The wings are broad with rounded ends. Horseshoe bats generally have a fluttering, butterflylike, or hovering flight. Their relatively short tails and small tail membranes are not large enough to form a pouch for holding insects. When a large insect is caught in flight, it may be tucked into the wing membrane under the arm while the bat manipulates it with its mouth. When they catch large prey, horseshoe bats sometimes alight in order to eat more easily.

Horseshoe bats occur throughout the temperate and tropical zones of the Old World, being found in a great variety of forested and nonforested habitats at both high and low altitudes. They roost in caves, buildings, foliage, and hollow trees. The species living in temperate regions hibernate during the winter in retreats other than their summer roosts, but they awaken readily and change their hibernating sites occasionally, sometimes flying 1,500 meters or more to a new place. The body temperature of R. *ferrumequinum* has been recorded in Berlin as from 8° C in hibernation to 40° C in periods of normal activity. These bats begin feeding on insects and spiders later in the evening than most bats and often return to the roost to eat their catch. They usually hunt within six meters of the ground and will also feed on the ground. Like many bats, they generally have regular feeding territories or hunting areas.

Most species roost in moderate-sized groups, but some are solitary. In some species the sexes live together all year, whereas in others the females form maternity colonies. Lekagul and McNeely (1977) wrote that *R. lepidus* occurs in groups of 3–4 to 400 and, unlike most species of *Rhinolophus*, forms roosting clusters. The clusters are segregated by sex during the birth season. Individuals of this species may establish well-defined foraging territories near their roosts. Bhat and Sreenivasan (1990) located numerous colonies of *R. rouxi* with up to 4,000 adults each; nearly all males departed as the time for parturition approached. In general, individuals of *Rhinolophus* are solitary hunters, while those of *Hipposideros* forage in small groups.

In some species, including those that hibernate, mating occurs during the autumn, but there is a delay in ovulation, so that actual fertilization does not occur until the spring. In other species mating and fertilization occur in the spring. Menzies (1973) stated that both *R. hipposideros* in Europe and *R. megaphyllus* in Australia undergo delayed fertilization. In *R. landeri* of Nigeria mating takes place in November, no development occurs in December–January, actual gestation takes place in February–April, births occur in late April and early May, and lactation goes on through June. Female *R. blasii* and *R. fumigatus* in Zambia and Malawi evidently give birth once annually during a restricted period during the early wet season (November to early January) and then lactate for about a month (Happold and Happold 1990). In general for *Rhinolophus* gestation takes about 7 weeks, a single young is produced in late spring, and sexual maturity is attained by 2 years (Lekagul and McNeely 1977). Birth weight is around 2–6 grams (Hayssen, Van Tienhoven, and Van Tienhoven 1993). Medway (1978) reported pregnant females of several species of *Rhinolophus* on the Malay Peninsula from February to April but also stated that pregnancies in *R. trifoliatus* had been recorded in March, May, June, September, October, and November. Sreenivasan, Bhat, and Geevarghese (1973) studied a colony of 2,000 *R. rouxi* in India and found a short, well-defined breeding season. Pregnant females were present only from January to April, lactation lasted from April to June, and there was arrested reproductive activity from June to November. Additional data on *R. rouxi* compiled by Bhat and Sreenivasan (1990) indicate that some females attain sexual maturity in their first year of life. Lekagul and McNeely (1977) stated that longevity in *Rhinolophus* seldom exceeds 6–7 years, but there are several cases of remarkably long life spans. The known record is that of an individual *R. ferrumequinum* that was captured in France in 1982, 29 years after initial banding, and then seen again in 1983 (Caubere, Gaucher, and Julien 1984).

There is considerable conservation concern for these bats. The IUCN recognizes *R. keyensis* of the Molucca Islands as a species distinct from *R. megaphyllus* (see above) and classifies it as endangered; fewer than 2,500 mature individuals are thought to survive. *R. imaizumii*, of Iriomote Island, declining rapidly because of human habitat disruption, also is designated as endangered. The IUCN assigns the following additional classifications: *R. cognatus*, *R. subrufus*, *R. rex*, *R. paradoxolophus*, *R. capensis*, *R. euryale*, *R. mehelyi*, and *R. hipposideros*, vulnerable; *R. ferrumequinum*, conservation dependent; and *R. celebensis*, *R. virgo*, *R. nereis*, *R. thomasi*, *R. cornutus*, *R. monoceros*,

R. creaghi, *R. canuti*, *R. shameli*, *R. rufus*, *R. yunanensis*, *R. philippinensis*, *R. maclaudi*, *R. silvestris*, *R. blasti*, *R. guineensis*, and *R. alcyone*, near threatened.

R. ferrumequinum was considered vulnerable by Smit and Van Wijngaarden (1981), endangered in Europe by Stebbings (1982), and endangered throughout the world by Stebbings and Griffith (1986). This species is declining rapidly because of disturbance of its roosts in caves and buildings, vandalism, habitat modifications resulting in loss of large insect prey, and increasing use of insecticides that may be absorbed by the bats. In England during the last century numbers have fallen by more than 98 percent, to about 2,200 individuals. Similar problems have befallen *R. hipposideros*, *R. blasii*, *R. euryale*, and *R. mehelyi*, which also were listed as endangered in Europe by Stebbings and Griffith (1986). A slow recovery of *R. euryale* has been noticed since the most dangerous pesticides were banned in the 1980s (Brosset et al. 1988).

The geological range of this family is middle Eocene to Recent in Europe, Miocene to Recent in Africa, Miocene to Recent in Australia, and Recent in other regions now occupied (Corbet and Hill 1992).

CHIROPTERA; Family HIPPOSIDERIDAE

Old World Leaf-nosed Bats

This family of 9 Recent genera and 69 species inhabits tropical and subtropical regions in Africa and southern Asia, east to the Philippine Islands, the Solomon Islands, and Australia. Some authorities (e.g., Ellerman and Morrison-Scott 1966; Koopman 1984a and in Wilson and Reeder 1993) consider the Hipposideridae to be only a subfamily of the Rhinolophidae. Others (e.g., Corbet and Hill 1991, 1992; Hayman and Hill in Meester and Setzer 1977; Lekagul and McNeely 1977) maintain both groups as distinct families, and this procedure is followed here. The Hipposideridae differ from the Rhinolophidae in the form of the nose leaf, the foot structure, the absence of the lower small premolar, and the structure of the shoulder and hip girdles.

The muzzle has an elaborate leaflike outgrowth of skin. This nose leaf consists of an anterior horseshoe-shaped part, sometimes with smaller accessory leaflets, and an erect transverse leaf, corresponding to the lancet in the nose leaf of the Rhinolophidae, usually divided into three cell-like parts, the apices of which may be produced into points. It lacks a sella, which is the median projection of the nose leaf in the Rhinolophidae.

Head and body length is 28–110 mm. An external tail is absent in some forms but may be as long as 60 mm in others; when present the tail is enclosed in the tail membrane. Forearm length is 30–110 mm. *H. commersoni gigas*, with a forearm length of 110 mm, is one of the largest bats in the suborder Microchiroptera. In several species the sexes differ in body size, size of nose leaves, and color of fur. The ears are well developed or short and united across the forehead in some genera. A tragus is lacking. In these bats each toe has only two bones, whereas in the Rhinolophidae each toe consists of three bones, except the first toe, which has two bones. The dental formula is: (i 1/2, c 1/1, pm 1–2/2, m 3/3) × 2 = 28 or 30. The nasal region of the skull is only slightly inflated.

There is a high degree of fusion in the elements of the shoulder girdle in both the Rhinolophidae and Hipposideridae, but in the latter family the fusion of the first and second ribs involves the entire bone, to and including the corresponding thoracic vertebrae, which produces a solid ring of bone consisting of the seventh cervical and first and second thoracic vertebrae, the first and second ribs, and the entire presternum. The pelvic girdle of the Hipposideridae is similar to that of the Rhinolophidae posteriorly, but anteriorly there is an additional bridge of bone connecting the acicular process with the front of the ilium, which produces a preacetabular foramen slightly exceeding the thyroid foramen in size (Hall 1989*a*, 1989*b*).

These bats utilize caves, underground chambers made by people, buildings, and hollow trees as retreats. *Hipposideros fulvus* has been found in burrows of the large porcupine *(Hystrix)* in Africa. Some species associate in groups of hundreds, others in small groups, and some forms roost singly. Sometimes they associate with other species of bats. Some forms of *Hipposideros* hibernate; *Asellia* may hibernate; and there is a record of apparent hibernation for *Coelops*.

Horseshoe bats catch most of their insect prey in flight. As in other insectivorous bats, the jaws are worked with a slightly side-to-side movement as well as up and down, so that the cusps of the upper and lower teeth sweep past each other in a shearing action that cuts up the hard parts of insects. These bats possibly fly with the mouth closed, emitting ultrasonic sounds through the nostrils. *Asellia tridens* emits far-reaching, beamed pulses, like the members of the family Rhinolophidae, as well as shorter pulses, like the members of the family Vespertilionidae.

The geological range of this family is Eocene to Recent in Eurasia and Miocene to Recent in Africa and Australia (Corbet and Hill 1992).

CHIROPTERA; HIPPOSIDERIDAE; **Genus**
HIPPOSIDEROS
Gray, 1831

Old World Leaf-nosed Bats

There are 7 species groups and 57 species (Aggundey and Schlitter 1984; Balete, Heaney, and Crombie 1995; Bergmans and Van Bree 1986; Brosset 1984; Chasen 1940; Corbet and Hill 1992; Cranbrook 1984; Ellerman and Morrison-Scott 1966; Flannery 1990*b*, 1995; Flannery and Colgan 1993; Gaucher and Brosset 1990; R. E. Goodwin 1979; Gould 1978*b*; Happold 1987; Harrison and Bates 1991; Hayman and Hill *in* Meester and Setzer 1977; Heaney et al. 1991; Hill 1963*b*, 1983*a*, 1985*a*; Hill and Francis 1984; Hill and Morris 1971; Hill and Rozendaal 1989; Hill and Yenbutra 1984; Hill, Zubaid, and Davison 1985, 1986; Jenkins and Hill 1981; Juste and Ibáñez 1994; Kitchener and Maryanto 1993*b*; Kock and Bhat 1994; Koopman, 1989*c* and *in* Wilson and Reeder 1993; Largen, Kock, and Yalden 1974; Laurie and Hill 1954; Lekagul and McNeely 1977; Mahoney and Walton *in* Bannister et al. 1988; Nader 1982; Roberts 1977; Rautenbach, Schlitter, and Braack 1984; Rickart et al. 1993; Schlitter et al. 1986; Smith and Hill 1981; Topal 1993):

megalotis group

H. megalotis, Djibouti, Somalia, Ethiopia, Kenya, one record from Jeddah in western Saudi Arabia;

bicolor group, *bicolor* subgroup

H. bicolor, southern Thailand, Malay Peninsula, Teratau and Tioman islands, Sumatra, Bangka Island, Nias Island, Enggano Island, Java, Borneo, Philippines, Flores, possibly Timor;

H. pomona, southwestern and northeastern India to southern China and Indochina, northern Malay Peninsula;

H. macrobullatus, Kangean Islands, southwestern Sulawesi, Seram;

H. ater, India to Thailand, Malay Peninsula, Sri Lanka, Nicobar Islands, Sumatra, Java, Bali, Philippines, Sulawesi, possibly Sangihe and Talaud islands, Togian Islands, Lombok, Molucca Islands, Aru Islands, New Guinea, Bismarck Archipelago, northern Australia;

H. fulvus, Afghanistan, Pakistan, India, Sri Lanka;

H. cineraceus, northeastern Pakistan, northern India, Burma to northern Viet Nam and Malay Peninsula, Riau Archipelago, Sumatra, Krakatau, Borneo, Kangean Islands northeast of Java, possibly Luzon;

H. durgadasi, central India;

H. halophyllus, southern Thailand;

H. nequam, known only by a single specimen from the Malay Peninsula;

H. calcaratus, New Guinea, Japen Island, Bismarck Archipelago, Solomon Islands;

H. maggietaylorae, New Guinea, Bismarck Archipelago;

H. coronatus, Polillo and Mindanao (Philippines);

H. ridleyi, Malay Peninsula, Singapore, northern Borneo;

H. jonesi, Guinea to Nigeria;

H. dyacorum, Malay Peninsula, Borneo;

H. sabanus, Malay Peninsula, Sumatra, Borneo;

H. doriae, Sarawak (Borneo);

H. obscurus, Luzon, Catanduanes, Dinagat, Maripipi, and Mindanao (Philippines);

H. marisae, Guinea, Liberia, Ivory Coast;

bicolor group, *galeritus* subgroup

H. pygmaeus, Luzon, Negros, and Bohol (Philippines);

H. galeritus, India, Malay Peninsula (including southern Thailand and Penang Island), Sri Lanka, Java, Borneo, Sanana (Sula Islands east of Sulawesi);

H. hypophyllus, Bangalore region of southern India;

H. cervinus, Malay Peninsula and many nearby islands, Sumatra, Bangka Island, Sipura and North Pagai islands, Enggano Island, Borneo, Natuna Islands, Kangean Islands, Mindanao, Talaud Islands, Sulawesi, Togian Islands, Molucca Islands, Aru Islands, New Guinea, Bismarck Archipelago, Trobriand Islands, D'Entrecasteaux Islands, Louisiade Archipelago, Solomon Islands, Vanuatu (New Hebrides), Cape York Peninsula of northern Queensland;

H. crumeniferus, Timor;

H. breviceps, North Pagai Island west of Sumatra;

H. curtus, Cameroon, island of Bioko (Fernando Poo), possibly Nigeria;

H. fuliginosus, Liberia to Cameroon, eastern Zaire, Ethiopia;

H. caffer, southwestern Arabian Peninsula, Morocco, most of Africa south of Sahara except central forested region;

H. ruber, Senegal to Ethiopia, and south to Angola and Zambia;

H. lamottei, known only from Mount Nimba area on Guinea-Liberia border but probably more widespread;

H. beatus, Guinea-Bissau, Sierra Leone, Liberia, Ghana, Ivory Coast, Nigeria, Cameroon, Rio Muni (Equatorial Guinea), Gabon, northern Zaire;

H. coxi, known only by a single specimen from Sarawak (Borneo);

H. papua, Halmahera and Bacan islands in the northern Moluccas, northwestern New Guinea and nearby Biak and Sipori islands;

cyclops group

H. cyclops, Guinea-Bissau to southern Sudan and Kenya,
Bioko;
H. camerunensis, Cameroon, eastern Zaire, western
Kenya;
H. muscinus, southern New Guinea;
H. wollastoni, western and central New Guinea;
H. semoni, eastern New Guinea, northern Queensland;
H. corynophyllus, central and possibly eastern New
Guinea;
H. edwardshilli, north-central New Guinea;
H. stenotis, northern Australia;

pratti group

H. pratti, southern China, northern Viet Nam;
H. lylei, Yunnan (southern China), eastern Burma,
western Thailand, Malay Peninsula;

armiger group

H. armiger, Nepal and Assam to southern China and
Indochina, Malay Peninsula, Taiwan, Hainan;
H. turpis, southern Ryukyu Islands, Viet Nam, extreme
southern Thailand;

speoris group

H. abae, Guinea-Bissau to southern Sudan and Uganda;
H. larvatus, Assam to extreme southeastern China and
Viet Nam, Malay Peninsula, Hainan, Sumatra,
Simeulue and Nias islands, North Pagai Island,

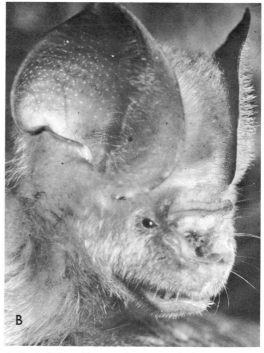

A. Head of *Hipposideros armiger terasensis* showing
construction of nose leaf in the family Hipposideridae, photo by
Robert E. Kuntz. B. Head of *H. ruber*, photo by P. Morris.

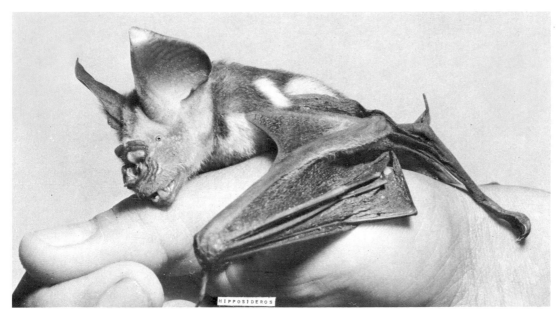

Large Malayan leaf-nosed bat *(Hipposideros diadema)*, photo by Lim Boo Liat.

Krakatau, Java, Madura, Kangean Islands, Bali, Borneo, South Natuna and Serasan islands, Sumbawa, Flores, Sumba, Savu, Roti, Semau, possibly Timor; *H. speoris*, southern India, Sri Lanka;

diadema group

H. lankadiva, peninsular India, Sri Lanka;
H. schistaceus, Bellary (southern India);
H. diadema, southern Burma and Thailand, Indochina, Malay Peninsula, Nicobar Islands, Sumatra, Nias and Enggano islands, Krakatau, Java, Kangean Islands, Bali, Borneo, Natuna Islands, Philippines, Talaud and Sangihe islands, Sulawesi, Togian Islands, Lombok, Sumbawa, Sumba, Flores, Savu, Timor, Tanimbar Islands, Molucca Islands, New Guinea, Bismarck Archipelago, Solomon Islands, coastal Northern Territory, northeastern Queensland;
H. demissus, San Cristobal Island in the Solomons;
H. lekaguli, southern Thailand, Malay Peninsula, Luzon and Mindoro (Philippines);
H. dinops, Sulawesi, Peleng Island, Solomon Islands;
H. inexpectatus, northern Sulawesi;
H. commersoni, Gambia to Somalia, south to Namibia and Transvaal, Madagascar, Bioko.

Both Corbet and Hill (1992) and Koopman (*in* Wilson and Reeder 1993) expressed doubt about the specific validity of *H. crumeniferus* and *H. doriae*. Koopman also included *H. durgadasi* in *H. cineraceus*, but Corbet and Hill considered it distinct. Kock and Bhat (1994) cast some doubt on the validity of *H. pomona*. Kitchener and Maryanto (1993) distinguished three new species from eastern populations of *H. larvatus*: *H. sorenseni* on central and western Java, *H. madurae* on central Java and nearby Madura, and *H. sumbae* on Sumbawa, Flores, Sumba, Savu, Roti, and Semau islands. Yoshiyuki (1991*b*) concluded that *H. terasensis* of Taiwan is a distinct species, but it was included in *H. armiger* by both Corbet and Hill (1992) and Koopman (*in* Wilson and Reeder 1993).

Head and body length is about 35–110 mm, tail length is 18–70 mm, and forearm length is about 33–105 mm. Medway (1978) listed the following weights: *H. bicolor*, 8–10 grams; *H. cineraceus*, 7–8 grams; *H. galeritus*, 6–7 grams; *H. lylei*, 33–17 grams; *H. armiger*, 40–60 grams; *H. larvatus*, 14–20 grams; and *H. diadema*, 34–50 grams. Kingdon (1974*a*) gave weights of 30–39 grams for *H. cyclops* and 74–180 grams for *H. commersoni*. Coloration of the upper parts varies from reddish to some shade of brown and the underparts are paler. Some species have two color phases, reddish and gray.

These bats are distinguished from those of the genus *Rhinolophus* by features of the nose leaf and ear, characters of the teeth, the greater posterior width of the skull, and the characters separating the families Hipposideridae and Rhinolophidae, as explained in the accounts thereof. Many species of *Hipposideros* have a sac behind the nose leaf that can be everted at will; this sac secretes a waxy substance and is found chiefly in the males.

Members of this genus usually roost in hollow trees, caves, and buildings. *H. fulvus* has been found in burrows of the large crested porcupine *(Hystrix)* in Africa. Some species are gregarious; others associate in small family groups or at times may be alone. The nose leaf and ears often twitch while these bats are hanging. Some species hibernate.

These bats fly lower than most bats and catch insects such as beetles, termites, and cockroaches; cicadas apparently are an important food source for some species. *H. armiger* appears to locate these insects by their calls. *H. commersoni* in Africa eats the beetle larvae that are inside wild fig fruits and incidentally consumes some fig pulp. *Hipposideros* often returns to its roost to eat its catch and seems to have certain feeding territories that it regularly patrols.

Hipposideros has been found to use echolocation, emitting calls via the nostrils, and appears to be specialized for short-range hunting (Hall 1989*b*). The species *H. caffer* may emit a long, drawn-out whistle when handled. A female of another species returned several times to its captured young in response to its sharp twitter.

sisting of about 500 bats, was in a dark room in a ruined temple. They also stated that 13 females taken in Egypt in April–May were pregnant with one embryo each. In Yemen pregnant females have been found in April. Yom-Tov, Makin, and Shalmon (1992) caught lactating female *A. tridens* in Israel during July. Qumsiyeh (1985) reported colonies of 300–1,000 or more in Egypt and stated that in Iraq births begin in early June, gestation is assumed to be 9–10 weeks, and lactation lasts 40 days.

The IUCN classifies *A. patrizii* as vulnerable. It is restricted to a few areas that are subject to increasing environmental disturbance.

CHIROPTERA; HIPPOSIDERIDAE; Genus ANTHOPS
Thomas, 1888

Flower-faced Bat

The single species, *A. ornatus,* is known only from Bougainville, Choiseul, Santa Ysabel, and Guadalcanal in the Solomon Islands (Flannery 1995). Oldfield Thomas had six specimens when he named and described this genus and species. These specimens are in the British Museum of Natural History. The U.S. National Museum of Natural Histo-

ry and the Field Museum of Natural History also have specimens. The species is restricted to a region subject to increasing habitat disruption and now is classified as vulnerable by the IUCN.

Head and body length is about 47–50 mm, the tail is very short, and forearm length is about 48–51 mm. The weight of a single male was 8 grams (Flannery 1995). The pelage is long, soft, and silky, and the coloration is grizzled grayish buff.

This genus is similar to *Hipposideros,* differing from it in the greatly reduced tail and in skull and dental features. The tail in *Anthops* is less than half as long as the femur. The horseshoe-shaped part of the nose leaf has two lateral leaflets, the inner one being quite small and the outer one large. Traces of a small frontal sac are present.

CHIROPTERA; HIPPOSIDERIDAE; Genus ASELLISCUS
Tate, 1941

Tate's Trident-nosed Bats

There are two species (Corbet and Hill 1992; Flannery 1995; Laurie and Hill 1954; Lekagul and McNeely 1977;

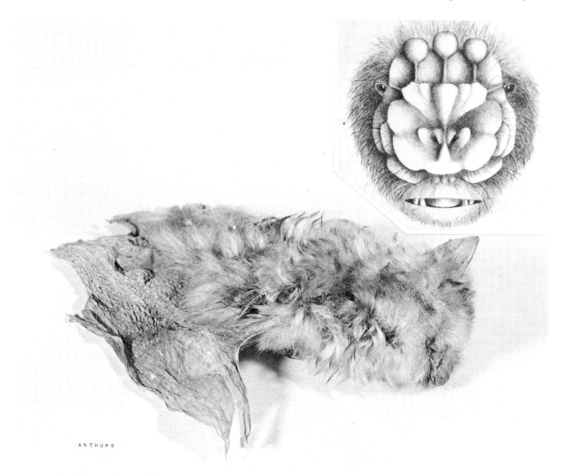

Flower-faced bat *(Anthops ornatus),* photo by Howard E. Uible of specimen in U.S. National Museum of Natural History. Inset photo from *Proc. Zool. Soc. London.*

Tate's trident-nosed bat (*Aselliscus* sp.), photo by Boonsong Lekagul.

McKean 1972; Schlitter, Williams, and Hill 1983; Zubaid 1988):

A. stoliczkanus, extreme southeastern China, northern Indochina, eastern Burma and Tenasserim, western Thailand, Penang and Tioman islands off the coast of the Malay Peninsula;
A. tricuspidatus, Molucca Islands, New Guinea and some nearby islands, Bismarck Archipelago, Trobriand and D'Entrecasteaux islands, Louisiade Archipelago, Solomon and Santa Cruz islands, Vanuatu (New Hebrides).

Head and body length is about 38–45 mm, tail length is 20–40 mm, and forearm length is 35–45 mm. McKean (1972) reported a weight of 3.5–4.0 grams for *A. tricuspidatus* and a weight of 6–8 grams for *A. stoliczkanus*. The coloration of *A. tricuspidatus* is bright brown above and buffy brown below. *A. stoliczkanus* has been described as both brownish and sooty, so perhaps there are two color phases in this species.

The upper margin of the transverse nose leaf is divided into three points, and two lateral leaflets margin the horseshoe. No frontal sac is present in either sex. The tail extends beyond the membrane as in *Asellia*. *Aselliscus* is best distinguished by features of the skull and teeth.

Available records indicate that these bats are cave dwellers. Smith and Hood (1981) encountered *A. tricuspidatus* in great abundance in caves and tunnels on New Britain and New Ireland islands. Individuals there were always found hanging singly and evenly spaced (30–40 cm apart) within discrete groups that usually contained 40–50 bats. An especially large colony of several hundred individuals was seen in a limestone cave; the bats there hung from the tips or sides of stalactites. Females were seen carrying a single young. McKean (1972) reported that 62 specimens of *A. tricuspidatus* were taken while roosting in limestone caves in New Guinea at altitudes of 98–260 meters. Of 26 females collected there on 1 July, 12 were pregnant with one embryo each; the other 14 apparently were immature. According to Flannery (1990b, 1995), lactating female *A. tricuspidatus* were taken in Papua New Guinea in March and on Espiritu Santo Island in May; pregnant females were collected in Vanuatu only in October, and females with young were found there in December. Phillips (1967) reported that 4 female *A. stoliczkanus* taken in early June in Laos were carrying young. Topal (1974) collected pregnant or lactating females of the same species in Viet Nam during May.

CHIROPTERA; HIPPOSIDERIDAE; **Genus**
RHINONYCTERIS
Gray, 1847

Golden Horseshoe Bat

The single species, *R. aurantius*, occurs from the Pilbara and Kimberley regions of northern Western Australia, through the northern part of the Northern Territory, and along the Gulf of Carpentaria into northwestern Queensland (Churchill, Helman, and Hall 1987). The generic name sometimes is spelled *Rhinonicteris* (as by Koopman *in* Wilson and Reeder 1993).

Jolly (*in* Strahan 1983) indicated that head and body length is 45–53 mm, tail length is 24–28 mm, forearm length is 47–50 mm, and weight is 8–10 grams. Subsequently, Jolly (1988) reported an average weight of 8.4 grams for males and 7.6 grams for females. The most common color pattern is bright orange upper parts and paler underparts, but there is much variation. Specimens have ranged from dark rufous brown, through orange, dark lemon, and pale lemon, to white (Churchill, Helman, and Hall 1987). The fur is fine and silky.

According to Hill (1982b), *Rhinonycteris*, *Triaenops*, and *Cloeotis* form a group of genera characterized principally by a number of common features of the nose leaf. All have a straplike projection extending forward from the internarial region over the anterior leaf to its edge, and all have a strongly cellular posterior leaf. None of the remaining six genera of the Hipposideridae have the anterior median straplike process.

According to F. W. Jones (1923–25),

the crown of the head is well raised above the face, the face itself being almost entirely occupied by the nose-leaf. The ears are short; laid forward they cover the eye; they are sharply pointed at the tip. The inner margin of the auricle sweeps forward with a fairly uniform convexity to the tip; the outer margin starts straight, or slightly concave, from the tip and then sweeps round, boldly convex, joining the side of the head behind and slightly below the eye. The inner margins are widely separated by the crown of the head. The nose-leaf is large and complex; it consists of two parts, a lower part which forms a horse-shoe ring below the nostrils, and an upper part which surmounts the nostrils. The lower part consists of two bilateral leaves, notched in the middle line below, and there joined by a projecting shelf, which springs from the interspace between the nostrils. From the inner margin of each of the leaves a small curved projection passes along the outer margin of each nostril. The complex upper portion of the nose-leaf surmounts the nostrils like a crown, a curious middle-line process jutting out just above the nostrils. Above this process the crown-like portion of the leaf is extensively honeycombed and sculptured.

According to Churchill, Helman, and Hall (1987), the golden horseshoe bat is found in a variety of habitats, including grassland, open forest, dense palm forest, and mangroves. It usually roosts in caves and mines, often together with *Macroderma*. Hot and humid sites are preferred. *Rhinonycteris* evidently is unable to lower its body temperature and does not enter torpor but becomes lethargic if exposed to ambient temperatures under 20° C. It emerges about 0.5–1.5 hours after sunset and has been observed to forage for insects one to three meters above the vegetation. Its flight is relatively faster than that of other hipposiderids. Specimens also have been collected in and

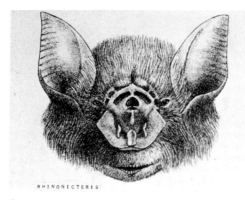
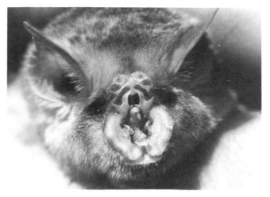

Golden horseshoe bat *(Rhinonycteris aurantius):* Left, photo from *Catalogue of the Chiroptera in the Collection of the British Museum*, G. E. Dobson; Right, photo by Simon Jolly.

around buildings. Vestjens and Hall (1977) reported that three stomachs of *Rhinonycteris* collected in the Northern Territory contained only moths, and five stomachs contained moths and traces of other insects. Churchill, Helman, and Hall (1987) noted that roost sites commonly contain several hundred bats; spacing between roosting individuals is at least 15 cm. Breeding is thought to occur during the wet season, from October to April.

Churchill (1991) stated that *Rhinonycteris* is known only from 10 caves and mines throughout its range and considered it to be rare and vulnerable. The total population was estimated at 35,570 bats, of which nearly 25,000 were at a single site, Tolmer Falls Cave in the Northern Territory. Other sites contained from a few to a few thousand individuals. The roosts were characterized by temperatures of 28°–32° C and 85–100 percent relative humidity. Colony size varied seasonally, with almost all bats abandoning their cave roosts during the wet season. Possibly they move into forests at that time. *Rhinonycteris* apparently has highly specialized habitat requirements within a climatically lim-

ited geographic range. It requires high humidity and during the dry season depends on the relatively few roosting sites that provide adequate conditions. Jolly (1988) indicated that such sites are generally jeopardized by current and potential human activity. Cutta Cutta Cave in the Northern Territory evidently had about 50,000 golden horseshoe bats in 1970, but after development of the cave as a tourist attraction there was a drastic decline, to fewer than 3,000. *Rhinonycteris* now is classified as vulnerable by the IUCN.

CHIROPTERA; HIPPOSIDERIDAE; **Genus TRIAENOPS** *Dobson, 1871*

Triple Nose-leaf Bats

There are two species (Corbet 1978; Hayman and Hill *in* Meester and Setzer 1977; Hill 1982*b;* Kock and Felten 1980):

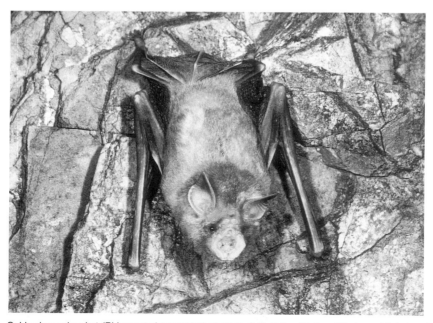

Golden horseshoe bat *(Rhinonycteris aurantius)*, photo by G. B. Baker / National Photographic Index of Australian Wildlife.

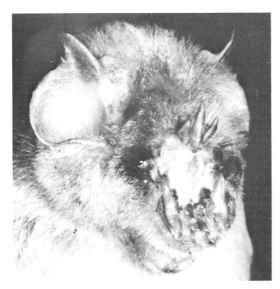

Triple nose-leaf bat *(Triaenops persicus)*, photo by David Pye.

T. furculus, western and northern Madagascar, Aldabra
and Cosmoledo islands;
T. persicus, southwestern Iran, southern Pakistan, Oman,
Yemen, possibly Egypt, East Africa from Ethiopia to
Mozambique, Congo, eastern Madagascar.

T. rufus and *T. humbloti* of eastern Madagascar sometimes
have been treated as distinct species, but they were included within *T. persicus* by Corbet and Hill (1991) and Koopman (*in* Wilson and Reeder 1993).

Head and body length is 35–62 mm, tail length is 20–34
mm, forearm length is 45–55 mm, and adult weight is usually 8–15 grams. *T. persicus* exhibits great diversity in coloration—grays, browns, and reds; these may be correlated
with age. Individuals in some areas are pale buff, almost
white. *Triaenops* resembles *Hipposideros* to some extent

but differs from that genus in features of the nose leaf, as
described above in the account of *Rhinonycteris. Triaenops*
evidently is closely related to *Rhinonycteris* but differs in
having a tridentate posterior section of the nose leaf such
as is also found in *Cloeotis.* Cranially, *Triaenops* differs
from both genera in its raised and inflated rostrum and relatively much larger cochleae but resembles *Rhinonycteris*
in the structure of the anteorbital region and zygoma and
in the curiously thickened premaxillae (Hill 1982*b*).

T. persicus has been found in a cave by the sea in southern Yemen, associated with *Coleura afra* and *Asellia tridens.* This species was collected in Oman while flying
around the opening of an underground water tunnel. It apparently roosted in the tunnel in small numbers along with
hundreds of *Asellia tridens.* The *Triaenops* emerged quite
early in the evening, while there was still light, and flew
low over the ground. Thousands of *T. persicus* and *Coleura afra* were found in a cave with many passages and chambers at different levels. The *T. persicus* in this cave assembled in groups just behind the entrance before flying out to
feed on insects. At least some of the young of *T. persicus* are
born in January; 24 males and 6 females, 2 pregnant, were
taken from the Amboina Caves in Tanzania, only a few meters above sea level, in December. *T. furculus* is thought to
be declining because of habitat loss and now is designated
as vulnerable by the IUCN.

CHIROPTERA; HIPPOSIDERIDAE; **Genus CLOEOTIS**
Thomas, 1901

African Trident-nosed Bat

The single species, *C. percivali,* occurs from southern Zaire
and Kenya to the Transvaal in South Africa (Hayman and
Hill *in* Meester and Setzer 1977).

Head and body length is 33–50 mm, tail length is 22–33
mm, and forearm length is 30–36 mm. Ansell (1986) listed
weights of 3.8–5.9 grams. Two subspecies are recognized: *C.
p. percivali* and *C. p. australis.* The type specimen of *C. p.
percivali* is colored as follows: bright buffy face, grayish

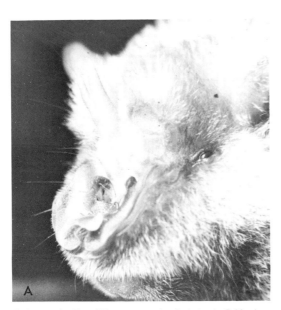

Triple nose-leaf bat *(Triaenops persicus)*, photos by P. Morris.

African trident-nosed bat *(Cloeotis percivali)*, photo by Herbert Lang through J. Meester.

crown, grayish brown back, and smoky brown wings and membranes, with the hairs on the abdomen having slaty gray bases and yellowish white tips. *C. p. australis* is usually buffy brown or dark brown above and yellowish below.

In this genus the well-developed tail is longer than the femur. Three pointed processes, tridentlike, are present at the back of the nose leaf. The ears are rounded and short, scarcely projecting above the level of the hair. *Cloeotis* resembles *Triaenops* in that its nose leaf has a tridentate posterior portion but differs in such cranial characters as its larger braincase and lower rostrum (Hill 1982b).

In South Africa this bat has been found in large numbers in caves with narrow entrances. Records from Zimbabwe and Botswana also indicate that *Cloeotis* is a cave dweller and that the diet is insectivorous (Smithers 1971). In Zambia, Whitaker and Black (1976) found this bat to feed almost entirely on adult Lepidoptera. Colonies may contain hundreds of individuals roosting in tight clusters.

Pregnant females, each with a single embryo, were collected in Zambia and Zimbabwe in October (Ansell 1986; Smithers 1971, 1983). *G. percivali* is designated as near threatened by the IUCN.

CHIROPTERA; HIPPOSIDERIDAE; **Genus COELOPS**
Blyth, 1848

There are two species (Corbet and Hill 1992; Cranbrook 1984; Hill 1972a, 1983a; Lekagul and McNeely 1977; Taylor 1934):

C. frithii, eastern India, Burma, southeastern China, northern Indochina, Thailand, Malay Peninsula, Taiwan, Sumatra, Java, Bali;
C. robinsoni, known by two specimens from the Malay Peninsula, one from Teratau Island off southern Thailand, one from Borneo, and one from Mindoro Island in the Philippines.

A third specific name, *C. hirsuta,* has sometimes been applied to the specimen from the Philippines (as by Koopman *in* Wilson and Reeder 1993) and to one other from the Malay Peninsula. Hill (1972a), however, suggested that *C. hirsuta* is conspecific with *C. robinsoni,* and neither Corbet and Hill (1992) nor Medway (1978) treated *hirsuta* as a species.

Head and body length is 28–50 mm, the tail is extremely short or absent, and forearm length is 33–47 mm. Lekagul and McNeely (1977) listed weights of 7–9 grams for *C. frithii* and 6–7 grams for *C. robinsoni.* The coloration above is bright brown, ashy brown, dusky, or blackish, and the underparts are brownish or ashy. Lateral leaflets are lacking on the nose leaf, or they are obscured by dense, stiff hairs. A frontal sac is present in some individuals. The ears are short and rounded.

The type specimen of *C. frithii sinicus* was taken in Sichuan from a "warm-air cave in which it was evidently hibernating." On Taiwan, Gwilym S. Jones (pers. comm.) found that *C. frithii* was caught in old Japanese pillboxes as well as in caves; it was "not abundant, nor was it particularly rare." In Java *C. frithii* inhabits forests, sheltering during the day in hollow trees in groups of 16 individuals or

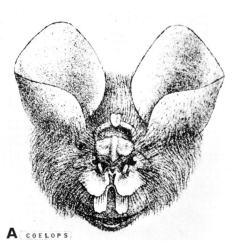

A. *Coelops frithi,* photo from *Catalogue of the Chiroptera in the Collection of the British Museum,* G. E. Dobson. B & C. *C. robinsoni,* photos by P. F. Wright of specimen in U.S. National Museum of Natural History.

fewer. The following are Javan records: 2 females, each pregnant with 1 embryo, taken in January and a female and single young obtained in March. *C. robinsoni* is designated as near threatened by the IUCN.

CHIROPTERA; HIPPOSIDERIDAE; **Genus PARACOELOPS**
Dorst, 1947

The single species, *P. megalotis*, is known only from the holotype, a badly damaged specimen collected at Vinh, Annam, Viet Nam (Corbet and Hill 1992). This specimen, an adult male, was obtained by M. David Beaulieu in 1945 and is now in the Paris Museum. The generic name refers to the funnel-shaped ears.

Measurements of the type are as follows: head and body length, 45 mm; forearm length, 42 mm; and ear length, 30 mm. There is no tail. This bat weighed 7 grams. The hairs are quite long, particularly on the upper parts. The interfemoral membrane is dark brown and lacks hair. The top of the head is bright golden yellow, in sharp contrast with the brownish back. The underparts are light beige, the bases of the hairs are yellow, and the ears are pale brown.

According to Corbet and Hill (1992), *Paracoelops* can be distinguished from the other genera of the *Hipposideridae* by its exceptionally long and broad ears, which extend far beyond the tip of the muzzle when laid forward. The ears are separate and rounded at the top. In addition to the large ears, this medium-sized bat is characterized by the presence of a nose leaf and the absence of a tail. The nose leaf is horseshoe-shaped and surmounted by a rounded leaf with radial striations. The tail membrane, 30 mm long, is not indented by attachment to the end of the tail and is supported by the long heel bones, the calcanea.

Paracoelops has not been reported since its description more than 50 years ago. The IUCN considers it to occur in a very small area of declining habitat and classifies it as critically endangered. Notwithstanding this ominous context, the recent discovery or rediscovery of various mammalian genera and species in Indochina subsequent to the cessation of military hostilities there suggests some hope of finding more information on *Paracoelops*.

CHIROPTERA; **Family MORMOOPIDAE**

Moustached Bats, Naked-backed Bats, or Ghost-faced Bats

This family of two Recent genera and eight species is found in the New World, from southern Arizona and southern Texas south through Mexico, Central America, and South America to the Mato Grosso of Brazil. These bats generally are restricted to tropical habitats below 3,000 meters. Until recently this group usually was considered a subfamily of the Phyllostomidae with the name Chilonycterinae. Some authorities, such as E. R. Hall (1981) and Husson (1978), still give the mormoopids only subfamilial rank, but Jones and Carter (1976), in their comprehensive review of the Phyllostomidae, excluded the Mormoopidae. The increasing acceptance of the latter group as a distinct family is based largely on the revisionary work of Smith (1972), and much of the following familial account is abstracted from his report.

Bats of this family range in size from the small *Pteronotus quadridens*, with a forearm length of 35–40 mm, to the large *P. parnellii*, with a forearm measuring 54–65 mm. Mormoopids differ from the phyllostomid bats in that they do not possess a nose leaf. Instead the lips have been expanded and ornamented with various flaps and folds that form a "funnel" into the oral cavity when the mouth is opened. Short, bristlelike hairs surround this funnel and may act to direct airflow toward the scooplike mouth. The nostrils have been incorporated into the expanded upper lip; above and between them are various bumps and ridges that form a sort of nasal plate. The eyes in the Mormoopidae are small and inconspicuous, in contrast to the large and prominent eyes of the Phyllostomidae.

The wing membrane is variously attached to the side of the body. In several species the membranes meet and fuse on the dorsal midline, giving these species a naked-backed appearance. The attachment is somewhat lower in the oth-

Paracoelops megalotis, photo by F. Petter.

Pteronotus gymnonotus, photo by John P. O'Neill.

er species. A tail, the latter half of which protrudes from the tail membrane, is always present in this family.

The pelage in this family is short, fine, and densely distributed over the body. It was formerly thought that these bats occurred in two distinct color phases, but Smith (1972) has shown that these phases are actually seasonal variations. It is noteworthy that naked-backed bats possess a thick, long pelage beneath the dorsally fused wing membranes.

The tragus of the Mormoopidae is different from that of any other group of bats. It varies in the different genera from a seemingly simple lanceolate structure to one with a secondary fold of skin at a right angle to the main longitudinal axis of the structure. This secondary fold is barely more than a pocketlike structure in the cranial edge of the tragus in *Pteronotus parnellii* and is best developed in the genus *Mormoops*. In addition to the above external characters, the family differs from the Phyllostomidae in a number of cranial and postcranial skeletal characters.

In a study of the forelimbs of the Mormoopidae and the Phyllostomidae, Vaughan and Bateman (1970) found the mormoopids to be characterized by specializations furthering reduction of the weight of the wing. Such adaptations were considered to be associated with maneuverable, rapid flight and the ability to remain continuously in the air for long periods. In contrast, the wings of phyllostomids were found to be less well adapted for efficient flight and to retain muscular patterns allowing food handling and clambering in vegetation.

Mormoopids are gregarious cave dwellers, sometimes roosting in very large colonies. They are exclusively insect eaters. The geological range is Pleistocene to Recent in North America and the West Indies and Recent in South America (Koopman 1984a).

CHIROPTERA; MORMOOPIDAE; **Genus PTERONOTUS** *Gray, 1838*

Naked-backed Bats, Moustached Bats, or Leaf-lipped Bats

There are three subgenera and six species (Eisenberg 1989; E. R. Hall 1981; Ibáñez and Ochoa G. 1989; Koopman 1982a; Smith 1972):

subgenus *Phyllodia* Gray, 1843

P. parnellii, Sonora and Tamaulipas (Mexico) to northern Bolivia and Mato Grosso region of central Brazil, Cuba, Jamaica, Hispaniola, Puerto Rico, Trinidad;

Mustache bat *(Pteronotus parnellii).* Insets: head, photo by Ernest P. Walker; tail, photo by Harold Drysdale, Trinidad Regional Virus Laboratory, through A. M. Greenhall.

PTERONOTUS

Naked-backed bats *(Pteronotus davyi)*, photo by Ernest P. Walker.

subgenus *Chilonycteris* Gray, 1839

P. macleayii, Cuba, Isle of Pines, Jamaica;
P. quadridens, Cuba, Jamaica, Hispaniola, Puerto Rico;
P. personatus, Sonora and Tamaulipas (Mexico) to
    northern Bolivia and central Brazil, Trinidad;

subgenus *Pteronotus* Gray, 1838

P. davyi, Sonora and Nuevo Leon (Mexico) to northern
    Peru and Venezuela, Lesser Antilles, Trinidad;
P. gymnonotus, southern Mexico to Peru, northern
    Bolivia, and Mato Grosso region of central Brazil.

Chilonycteris sometimes is recognized as a full genus that
includes *P. parnellii* as well as the three species indicated
above (see, e.g., Husson 1978). *P. parnellii* also is known
from remains about 4,300 years old found in a fissure deposit on the island of Antigua in the Lesser Antilles and
from late Pleistocene remains found in the Bahamas
(Pregill et al. 1988). Adams (1989) showed the range of *P.
davyi* extending from Venezuela well into eastern Brazil

but noted that it had not been recorded from the Guianas.
Subsequently Koopman (*in* Wilson and Reeder 1993) indicated that the Brazilian record was erroneous. *P. gymnonotus* formerly was known as *P. suapurensis* (see J. D. Smith
1977).

Head and body length is 40–77 mm, tail length is 15–30
mm, and forearm length is 35–65 mm. Eisenberg (1989)
listed the following average weights: *P. parnellii*, 20.4
grams in males, 19.6 grams in females; *P. personatus*, 8.0
grams in males, 6.9 grams in females; *P. davyi*, 9.4 grams
in males, 9.1 grams in females; and *P. gymnonotus*, 12.6
grams in males, 14.0 grams in females. Coloration is variable, often being light or dark brown, grayish brown, or
ochraceous orange, with the underparts usually paler.

The lower lip of *Pteronotus* has platelike outgrowths
and small papillae. A nose leaf is not present, and the tail is
well developed. The wings in *P. davyi* and *P. gymnonotus*
attach along the midback and cover the fur of the back, giving the appearance that the back is naked. These two species
are known as "naked-backed bats." In the other species the
wings are attached along the sides of the body and the back
is furred.

These bats occupy a variety of habitats. In Jalisco, Mexico, three species of *Pteronotus* were collected along waterways lined with dense vegetation at altitudes from sea level to about 1,700 meters (Watkins, Jones, and Genoways 1972). In Venezuela *P. davyi* and *P. gymnonotus* were most often taken in dry, open areas, whereas *P. parnellii* was found mainly in moist places within forests (Handley 1976). Although bats of this genus have been found in houses, they roost primarily in caves and tunnels. Some species may also shelter in hollows of plants, such as thorny bamboo. When hanging in caves these bats seek the darker recesses, seldom near the entrances. They generally hang singly rather than in compact masses. *P. personatus* has been taken from a cave, where it preferred to lie flat on horizontal and nearly horizontal surfaces. The voice of *Pteronotus* has been described as a sibilant, birdlike chirp. Its diet is thought to consist mostly of Lepidoptera and Coleoptera, and it employs a unique echolocation system convergent with that of the Old World families Rhinolophidae and Hipposideridae (Eisenberg 1989).

Bateman and Vaughan (1974) studied a combined group of *P. parnellii*, *P. personatus*, *P. davyi*, and *Mormoops megalophylla* that inhabited a cavern system in Sinaloa, Mexico. The bats began flying shortly after sunset; some returned as early as 1.5 hours after departure, but most appeared to remain away from the roost for 5–7 hours. The flyways to foraging grounds were at least 3.5 km and probably several times that length for some individuals. It was estimated that these bats consumed 1,902–3,805 kg of insects each night. The total number of bats in the cavern system was estimated as 400,000–800,000. Additional investigation indicated that male *P. parnellii* roosted separately from the females for much of the year.

LaVal and Fitch (1977) determined that in Costa Rica *P. parnellii* was seasonally monestrous, with pregnant females being found from January to May but not from July to December. Adams (1989) stated that *P. davyi* exhibits seasonal monestry, with mating probably occurring in January or February. The following data also indicate seasonal reproduction in this genus: in Sinaloa, Mexico, pregnancies in *P. parnellii*, *P. davyi*, and *P. personatus* in May and June and lactating *P. parnellii* in July (Jones, Choate, and Cadena 1972); in Jalisco, Mexico, pregnant female *P. parnellii* and *P. davyi* in May and June but none from July to October (Watkins, Jones, and Genoways 1972); in Michoacán, female *P. parnellii* pregnant in March but not in February, September, or December (Hernandez et al. 1985); in the Yucatan, most female *P. parnellii* pregnant in February but none pregnant in July, August, or January (Jones, Smith, and Genoways 1973); in Guatemala, most female *P. parnellii*, *P. davyi*, and *P. personatus* pregnant in March but none pregnant in January (Jones 1966) and a lactating female *P. parnellii* taken in late July (Rick 1968); in Nicaragua, pregnant female *P. parnellii* in February and March, pregnant *P. davyi* and *P. gymnonotus* in April and May, no pregnancies from June to August, and a lactating female *P. parnellii* in April (Jones, Smith, and Turner 1971); in Haiti, a pregnant female *P. quadridens* in February (Klingener, Genoways, and Baker 1978); on Margarita Island, Venezuela, a pregnant female *P. parnellii* in July but two not pregnant in November (Smith and Genoways 1974); and in Surinam, pregnant females in July (Genoways and Williams 1979b).

All pregnancy records refer to a single embryo per female. The IUCN classifies *P. macleayii* as vulnerable and *P. quadridens* as near threatened. The former, at least, is expected to decline in response to habitat disruption.

Leaf-chinned Bats

There are two species (Graham and Barkley 1984; E. R. Hall 1981; Smith 1972):

M. blainvillii, Cuba, Jamaica, Hispaniola, Puerto Rico, Mona Island;
M. megalophylla, southern Arizona and Texas to Honduras, southern Baja California, Caribbean coast of South America and adjacent islands, northern Ecuador, northwestern Peru.

E. R. Hall (1981), for technical reasons of nomenclature, used the name *Aello* Leach, 1821, for this genus and the name *Aello cuvieri* for the species *Mormoops blainvillii*. That species also is known by a single pre-Columbian specimen from Exuma Island (Bahamas) and by remains about 4,300 years old found in a fissure deposit on the island of Antigua in the Lesser Antilles (Pregill et al. 1988).

Head and body length is about 50–73 mm, tail length is 18–31 mm, forearm length is 45–61 mm, and adult weight is usually 12–18 grams. Coloration in *M. megalophylla* is reddish brown or other shades of brown, buff, and cinnamon. In the pale phase *M. blainvillii* is light brown above and buffy below, whereas in the dark phase the upper parts are dark brown and the underparts are ochraceous tawny.

The lower lip has fleshy peglike projections, the chin has leaflike projections, and there are long stiff hairs at the sides of the mouth, so that it is almost hidden. The upturned nose is short and has grooves, ridges, and pits. The tail is well developed. The braincase is greatly deepened, its floor being so elevated that the lower border of the foramen magnum is above the level of the rostrum. The tongue is protruded while drinking, but the water taken up by the foliation of the lips is sucked in, with the head raised, an action resembling chewing.

M. megalophylla occupies diverse habitats, from desert scrub to tropical forest, and roosts in caves, mines, tunnels, and, rarely, buildings (Barbour and Davis 1969). In southwestern Texas, Easterla (1973) found this species to prefer the hot lowlands. Handley (1976) collected most specimens in Venezuela in moist, forested areas. In Jalisco, Mexico, this species ranges in altitude from sea level to about 2,000 meters (Watkins, Jones, and Genoways 1972). One observer noted that at Alamos, Sonora, Mexico, *M. megalophylla* emerged from retreats and began flight about 10 minutes before full dark. The bats seemed most active at about 2300 hours and again at about midnight. Leaf-chinned bats feed later in the evening than do most bats that hunt insects. They seem to become less active in the winter, but they do not hibernate. *M. megalophylla* appears to hunt just above the ground. It searches for insects over land and water.

Goodwin (1970) stated that *M. blainvillii* has an extraordinarily swift flight even in the narrow passageways of caves. Its wings often produce a characteristic humming sound. It penetrates farther into the depths of caves than does any other Jamaican bat. The species is designated as near threatened by the IUCN.

M. megalophylla is colonial, 500,000 having been found in one cave in Nuevo Leon, but it does not roost in clusters (Barbour and Davis 1969). Easterla (1973) reported up to 4,000 per cave in southwestern Texas. In the latter area pregnant females, with one embryo each, were taken in June and lactating females were found from June to Au-

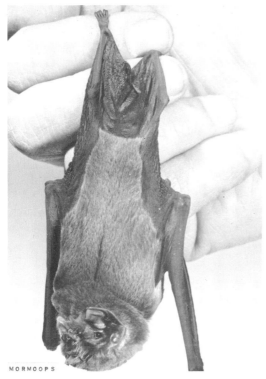

MORMOOPS

Leaf-chinned bat *(Mormoops megalophylla),* showing the remarkable folds of skin on the chin and the peculiar folding of the ear and the shape of the nose, photos by Ernest P. Walker.

gust. A pregnant female with a single embryo was collected in the Yucatan in February (Jones, Smith, and Genoways 1973). Pregnant females also have been taken in Coahuila in March, in Veracruz in April, in Sonora in May, and in Arizona in June.

Leaf-chinned bat *(Mormoops megalophylla)* in flight, photo by Ernest P. Walker.

CHIROPTERA; Family **NOCTILIONIDAE;** Genus
NOCTILIO
Linnaeus, 1766

Bulldog Bats, or Fisherman Bats

The single known genus, *Noctilio,* contains two species (Baker, Genoways, and Patton 1978; Buden 1985; Cabrera 1957; Davis 1976*a;* Dolan and Carter 1979; E. R. Hall 1981; Hershkovitz 1975*b;* Hood and Jones 1984; Polaco 1987):

N. leporinus, Sinaloa and southern Veracruz (Mexico) to northern Argentina and southeastern Brazil, Greater and Lesser Antilles, Bahamas;
N. albiventris, southern Mexico to northern Argentina.

N. albiventris often has been referred to incorrectly as *N. labialis.*

In the larger species, *N. leporinus,* head and body length is 98–132 mm and forearm length is 70–92 mm. Brooke (1994) indicated that weight is 50–90 grams. In *N. albiventris* head and body length is 57–85 mm and forearm length is 54–70 mm; and Hood and Pitocchelli (1983) noted adult weights of 18–44 grams. The upper parts of *N. leporinus* are bright orange rufous in the males and gray or dull brown in the females. In *N. albiventris* the upper parts are grayish brown to yellowish or bright rufous, many males being bright rufous and many females dull brown to drab. Both species usually have a paler middorsal line and paler underparts.

The pointed muzzle and the nose lack excrescences. The lips are full and appear swollen. The upper lips are smooth but are divided by a vertical fold of skin under the nostrils, forming a "harelip" or hood over the mouth; the edges of

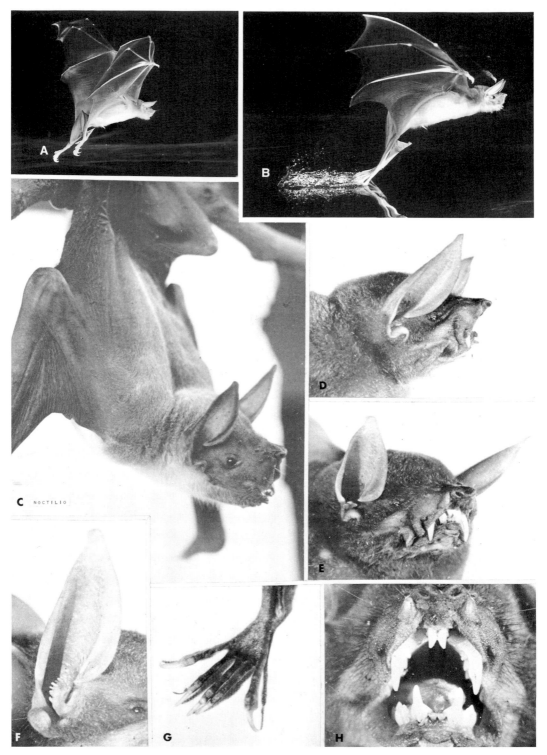

Fisherman bats *(Noctilio)*: A & B *N. leporinus,* photos by J. Scott Altenbach. C. *N. leporinus mexicanus,* photo by Bernardo Villa. D–H, *N. leporinus leporinus,* photos by Harold Drysdale, Trinidad Regional Virus Laboratory, through A. M. Greenhall.

the lower lips are smooth, but there is a lip pad or wart at the middle and a fold of skin under the pad in addition to other semicircular folds of skin on the chin. The cheeks are elastic and can be greatly extended. The nostrils open forward and down, with the somewhat tubular nose projecting slightly beyond the lips. There is no nose leaf. The large, slender, pointed ears are separate, and a tragus with a serrated outer margin is present. The tail is well developed, being more than half as long as the thigh bone, and extends to about the middle of the well-developed tail membrane. It appears close to the upper surface of the membrane and perforates the surface, its tip being free. The calcar, or heel extension, is large and bony, particularly in *N. leporinus*. This species has unusually long hind legs and large hind feet with well-developed claws. The limbs of *N. albiventris* are not so modified for fishing. The hind legs are almost completely free from the wing membrane and are relatively short, particularly the femur. Both species have quite narrow and long wings in comparison with other families of the Microchiroptera.

The dental formula is: (i 2/1, c 1/1, pm 1/2, m 3/3) × 2 = 28. Of the two pairs of upper incisors the first, or middle, pair is larger, nearly concealing the smaller second pair, which is located directly behind the first incisors. The molars exhibit a W pattern of cusps and ridges.

Handley (1976) found both species in a variety of vegetative conditions in Venezuela but always near streams or in other moist places. *N. leporinus* usually roosts in rock clefts and fissures, dark caves, and hollow trees but has also been taken in buildings; *N. albiventris* generally shelters in hollow trees, foliage, and buildings. The former species has been found in sea caves with *Mormoops*, *Carollia*, and *Glossophaga*, and *N. albiventris* has been found in the same hollow trees as *Molossus major*. A roost of *Noctilio* is often detected by the unusually strong and penetrating musky odor of these bats. The flight of *N. leporinus* is rather stiff-winged and not particularly rapid, but powerful. When knocked into water it swims well, using its wings as oars.

Noctilio takes a variety of prey and is one of only three genera of bats now known to catch and eat fish; the others are *Myotis (M. vivesi, M. adversus)* and *Megaderma*. *N. leporinus* fishes over the sea, at the edge of the surf, in large rivers, and in freshwater ponds. It generally feeds at dusk and during the night but has been seen in late afternoon flight over water in the company of pelicans. Presumably the bat catches small fish disturbed by the pelicans. Groups of *N. leporinus* move in zigzag flight, chirping and skimming the surface of the water. It catches small, surface-swimming fish (25–76 mm long) with the sharp, long claws of its feet and lifts them quickly to its mouth. A series of high-speed photographs by Altenbach (1989) demonstrated that the flying bat brings its tail membrane up under its body to help prevent loss of prey during transfer from the feet to the mouth. The fish are either eaten in flight or carried to a roost to be eaten while the bat is resting. The bat does not scoop fish into its tail membrane, as once was claimed, but may transfer fish to the membrane after capture. Each of two captive individuals of this species was estimated to consume 30–40 fish each night. It evidently uses echolocation to find fish and determine their swimming velocity (Wenstrup and Suthers 1984). Fish apparently cannot be found while they are underwater but are echolocated when they jump above the surface. The bat then descends to just above the water and rakes with its claws through areas where it has detected a high level of fish activity (Schnitzler et al. 1994).

There has been controversy over the relative importance of insects and fish in the diet of *N. leporinus*. By collecting and analyzing guano and discarded material beneath roosts over a nine-month period, Brooke (1994) determined that insects, mostly moths and beetles, predominated in the diet during the wet season, from July to December. In the dry season pelagic and freshwater fish represented a greater proportion of the diet than did insects. Small quantities of crabs, scorpions, shrimp, and terrestrial insects also were taken. Whereas fish and terrestrial invertebrates were captured with the feet, aerial insects were caught in the wing and tail membranes.

Although *N. albiventris* had not previously been reported to feed on fish, Suthers and Fattu (1973) induced captive specimens to catch floating insects and pieces of fish on the surface of a pool by dipping their feet into the water in a manner like that of *N. leporinus;* these specimens thrived on a diet of fish. Fleming, Hooper, and Wilson (1972) reported that 28 stomachs of *N. albiventris* from Central America contained 100 percent insect food. Brown, Brown, and Grinnell (1983) found *N. albiventris* to employ echolocation to capture insects from the surface of the water.

According to Eisenberg (1989), *N. leporinus* roosts in colonies of up to 75 bats, females apparently form nursery colonies, and breeding may be highly synchronous in strongly seasonal habitats. In a semiarid region of northeastern Brazil, Willig (1983, 1985b) found *N. leporinus* to roost by day in groups of up to 30 individuals and to forage by night in groups of 5–15. The species exhibited clear seasonal monestry in this region, the later stages of gestation and the period of lactation corresponding with the wet season and greatest availability of fish and insects. Pregnancies occurred from September until January, and lactation was first seen in November and continued until April.

Additional data on social structure and reproduction include the following: in Sinaloa, Mexico, 3 of 7 female *N. leporinus* collected in mid-June were lactating (Jones, Choate, and Cadena 1972); in Jalisco, Mexico, a lactating female *N. leporinus* was taken in January and a female pregnant with a single embryo was taken in April (Watkins, Jones, and Genoways 1972); in the Yucatan a colony of *N. leporinus* collected on 7 July included 3 adult males, 7 lactating females, and 6 young (Jones, Smith, and Genoways 1973); in Nicaragua most female *N. albiventris* collected in April were pregnant with a single embryo, as were 2 female *N. leporinus* taken in March, and 1 female *N. leporinus* collected in June was lactating (Jones, Smith, and Turner 1971); in Costa Rica pregnant female *N. leporinus* were taken in February, April, and August (LaVal and Fitch 1977); in Panama *N. albiventris* was reported to mate in late November or December and to give birth in late April or early May (Anderson and Wimsatt 1963); on Montserrat in the Lesser Antilles 2 of 4 females collected in late July were lactating (Jones and Baker 1979); in Peru pregnant females of both species were collected in July, and *N. albiventris* was reported to forage in groups of 8–15 (Tuttle 1970); and captive infant *N. albiventris* were nursed for almost three months (Brown, Brown, and Grinnell 1983). According to Jones (1982), a captive *N. leporinus* lived 11 years and 6 months.

Martin (1972) stated that specimens of *N. leporinus* had been found in possibly late Pleistocene sites in Puerto Rico and Cuba. Otherwise this family lacks a fossil history.

Saussure's long-nosed bat (*Leptonycteris* sp.) hovering at a spike of *Agave* flowers while obtaining nectar and pollen, photo by Bruce Hayward.

CHIROPTERA; **Family PHYLLOSTOMIDAE**

American Leaf-nosed Bats

This family of 52 Recent genera and 154 species is found in the tropical and subtropical regions of the New World, from the southwestern United States and the West Indies to northern Argentina. For years there has been confusion about the spelling of the name of this family, with some authors using Phyllostomatidae and others writing Phyllostomidae. In a detailed analysis of the Greek and Latin origins of the name Handley (1980*a*) determined that the proper spelling is Phyllostomidae. Although its name is now settled, no other mammalian family, except perhaps the Muridae (Rodentia), has been subject to such intensive analysis and debate with respect to phylogenetic alignment of component genera. The arrangement of subfamilies, tribes, and genera adopted here is based primarily on the classification of Baker, Hood, and Honeycutt (1989), which represented a synthesis of morphological, chromosomal, and biochemical data. Those authorities, however, did not assign *Macrotus* and *Micronycteris* to any subfamily, and the designations utilized here for those genera follow the conclusions of Van Den Bussche (1992), which arc based on analysis of ribosomal DNA. The resulting overall sequence is:

Subfamily Desmodontinae, genera *Desmodus, Diaemus, Diphylla;*

Subfamily Macrotinae, genus *Macrotus;*

Subfamily Micronycterinae, genus *Micronycteris;*

Subfamily Vampyrinae, genera *Vampyrum, Trachops, Chrotopterus;*

Subfamily Phyllostominae, Tribe Phyllostomini, genera *Phylloderma, Phyllostomus, Tonatia, Mimon, Lonchorhina, Macrophyllum;*

Tribe Glossophagini, genera *Glossophaga, Monophyllus, Leptonycteris, Lonchophylla, Lionycteris, Anoura, Scleronycteris, Lichonycteris, Hylonycteris, Choeroniscus, Choeronycteris, Musonycteris, Platalina, Brachyphylla, Erophylla, Phyllonycteris;*

Tribe Stenodermatini, genera *Carollia, Rhinophylla, Sturnira, Uroderma, Platyrrhinus, Vampyrodes, Vampyressa, Mesophylla, Chiroderma, Ectophylla, Enchisthenes, Dermanura, Koopmania, Artibeus, Ardops, Phyllops, Ariteus, Stenoderma, Pygoderma, Ametrida, Sphaeronycteris, Centurio.*

The recognition of the Desmodontinae (vampire bats) as a phyllostomid subfamily rather than a full family now has received general acceptance (Corbet and Hill 1991; Jones and Carter 1976, 1979; Koopman 1984*a* and *in* Wilson and Reeder 1993). Their location at the beginning of the above sequence reflects accumulating evidence of their basal phylogenetic position within the Phyllostomidae (Baker, Hood, and Honeycutt 1989; Van Den Bussche 1992). E. R. Hall (1981) did state that "the sanguinivorous habits and corresponding morphological adaptations . . . as well as other obvious differences, contrast so sharply with those of other bats that the vampires are here accorded family rank." However, Baker, Honeycutt, and Bass (1988) concluded that

Long-tongued bat *(Glossophaga soricina)* hovering at a cluster of flowers of a banana while obtaining nectar and pollen, photo by Bernardo Villa.

"the affiliation of vampire bats within the Phyllostomidae is clear from immunological, morphological, and chromosomal data. To consider the vampires as a separate family obscures the evolution of sanguivory and constructs an unnatural classification." Pine and Ruschi (1976) listed a number of unconfirmed reports suggesting that certain phyllostomids other than the Desmodontinae occasionally feed on blood.

Notwithstanding the above, a more traditional classification of the Phyllostomidae, as set forth by Jones and Carter (1976, 1979), is still largely followed by some authorities, such as Corbet and Hill (1991). That arrangement recognizes six or seven subfamilies: Desmodontinae, Phyllostominae (for the genera placed above in the Macrotinae, Micronycterinae, Vampyrinae, and the tribe Phyllostomini), Glossophaginae (for the genera placed above in the tribe Glossophagini, except for *Brachyphylla*, *Erophylla*, and *Phyllonycteris*), Carolliinae (for the genera *Carollia* and *Rhinophylla*), Sturnirinae (sometimes used for the genus *Sturnira*), Stenodermatinae (for the other genera placed above in the tribe Stenodermatini), and Phyllonycterinae (for the genera *Brachyphylla*, *Erophylla*, and *Phyllonycteris*).

Based primarily on morphological features of those phyllostomids specialized for feeding on nectar, Griffiths (1982) proposed (1) moving the genera *Lonchophylla*, *Lionycteris*, and *Platalina* from the Glossophaginae (as designated in the above traditional view) to a new subfamily, the Lonchophyllinae; (2) recognizing close affinity between the remaining glossophagines and the subfamily Phyllonycterinae; (3) placing *Brachyphylla* in its own subfamily, the Brachyphyllinae; and (4) recognizing that the genera *Glossophaga*, *Monophyllus*, and *Lichonycteris* form a systematic division of the Glossophaginae. These recommendations generally were accepted by Koopman (1984*b* and *in* Wilson and Reeder 1993), but they also came under criticism (Haiduk and Baker 1982; Smith and Hood 1984). Two recent systematic checklists (Corbet and Hill 1991; Jones, Arroyo-Cabrales, and Owen 1988) did not follow Griffiths's proposals, and Griffiths (1985) found that molar

morphology actually indicated that *Brachyphylla* is in the same systematic group as *Phyllonycteris* and *Erophylla*. On the basis of studies using albumin immunology, Honeycutt and Sarich (1987) did suggest that certain species of the Glossophaginae are closely related to the Phyllonycterinae (which they called the Brachyphyllinae). This last point was expressed in the above new classification of Baker, Hood, and Honeycutt (1989), in which the Phyllonycterinae were combined with the Glossophaginae in the tribe Glossophagini.

Phyllostomids range in size from small to the largest of the American bats, *Vampyrum spectrum*. Head and body length is 40 to approximately 135 mm; the tail is absent or 4–55 mm in length; and forearm length is 31 to about 105 mm. When an external tail is lacking, the free edge of the tail membrane is deeply and broadly notched. The fur is variable in color. One species *(Ectophylla alba)* is whitish, and a number of species have two color phases. Several species in the subfamily Stenodermatinae have white spinal stripes and/or white facial stripes.

A nose leaf is usually present, but it is reduced or absent in a few genera. When it is present it is not as complexly developed as in the families of Old World leaf-nosed bats. Those genera in which the nose leaf is reduced or absent often have platelike outgrowths on the lower lip. According to Arita (1990), the nose leaf functions in the echolocation system and its morphology apparently is involved with feeding habits. The lower portion of the leaf, which is associated with the nostrils, is called the "horseshoe"; the upper portion is called the "spear." The position of the nostrils and the structure of the horseshoe are associated with modulation of the frequency and intensity of the echolocation sound. The spear seems to be responsible for vertical direction of the signals, and its development is most pronounced in species that pursue insects in flight. Species that depend on olfaction or sight to locate food may have well-developed horseshoes but relatively small spears. The latter condition is found in the Desmodontinae and in the genera *Brachyphylla*, *Phyllonycteris*, and *Erophylla*, as well as in some large carnivorous species, such as *Vampyrum spectrum* and *Chrotopterus auritus*. In the nectarivorous Glossophagini both the spear and the horseshoe are reduced.

The ears are connected by a band of tissue across the top of the head in some species and are variable in form, but usually they are rather narrow and tend to be pointed. In some genera they are greatly elongated. The tragus is variously thickened or notched. The males of *Phyllostomus* have a glandular throat sac. In the subfamilies Glossophaginae and Phyllonycterinae the snout is elongate and the tongue is long, highly extensible, and covered with bristlelike papillae, much as in the Old World fruit bats, Pteropodidae (Macroglossinae). In the subfamilies Macrotinae and Micronycterinae, in most members of the tribe Phyllostomini, and in some of the Glossophagini the dental formula is: (i 2/2, c 1/1, pm 2/3, m 3/3) $\times$ 2 = 34. Most other genera have 26–32 teeth.

In the subfamily Desmodontinae the short, conical muzzle lacks a true nose leaf, having instead naked pads bearing U-shaped grooves at the tip that have been likened to a nose leaf. The ears are rather small and the tail membrane is short. All the long bones of the wing and leg are deeply grooved for the accommodation of muscles. The dental formulas for the three genera of vampire bats are: *Desmodus*, (i 1/2, c 1/1, pm 2/3, m 0/0) $\times$ 2 = 20; *Diaemus*, (i 1/2, c 1/1, pm 1/2, m 2/1) $\times$ 2 = 22; and *Diphylla*, (i 2/2, c 1/1, pm 1/2, m 2/2) $\times$ 2 = 26. The incisors and the canines are specialized for cutting, being sicklelike or shearlike, with their cutting edges forming a V. The cheek teeth are greatly reduced, and all traces of crushing surfaces are absent.

A. Vampire bat *(Desmodus rotundus)* lapping blood from a saucer. (The bat's left forearm is broken near the elbow.) B. The face of a vampire bat *(D. rotundus),* showing the extremely sharp incisor teeth, which are used to make the small incisions for obtaining blood. Photos by Ernest P. Walker.

Other anatomical indications of the liquid diet are the short esophagus and the slender caecumlike stomach.

Vampire bats seem to prefer retreats of almost complete darkness, sheltering mostly in caves but also in old wells, mine shafts and tunnels, hollow trees, and buildings. Single individuals, small groups, or colonies of thousands utilize a specific roost. About 20 other species of bats have been found in the same retreats as *Desmodus* but not closely associated with it. Vampire bats are shy and agile in their roosts; using their feet and thumbs, they can walk rapidly on horizontal or vertical surfaces. These bats do not leave their shelters until after dark and are most active before midnight. They forage low over the ground, flying fairly straight courses. Vampire bats, like those New World bats known to feed on fruit, emit pulses having only about one-thousandth of the sound energy of those used by bats that feed on flying insects or fish.

Some other species of the Phyllostomidae also roost in dark areas, and still others shelter in fairly well-illuminated places. Some members of this family are gregarious, some associate in small groups, and others roost singly. American leaf-nosed bats are often associated with other species of bats in their retreats, which may be caves, culverts, trees, buildings, or animal burrows. Several species form their own shelters by modifying leaves. It had been thought that the bats of this family were true homeotherms, never exhibiting torpor; however, laboratory studies have demonstrated the ability of three phyllostomids to estivate: *Glossophaga soricina, Platyrrhinus helleri,* and *Carollia perspicillata* (Rasweiler 1973).

Indirect evidence indicates that those species found in the southwestern United States migrate to a warmer climate for the winter. American leaf-nosed bats fly with their mouth closed, transmitting sounds through their nostrils. Except in the insectivorous forms the soft impulses emitted by the members of this family comprise a complex and changing mixture of high frequencies of short duration.

Most forms feed on insects, fruit, nectar, and pollen; *Phyllostomus hastatus, Chrotopterus,* and *Vampyrum* are carnivorous. Some species aid in seed dissemination, and those that seek nectar, pollen, or insects from flowers pollinate the flowers. The scientific names of certain genera suggest that they drink blood, but their food habits are actually much like those of most other phyllostomids.

The Desmodontinae are the true vampires, feeding, as far as is known, only on fresh blood. Their method is to alight on or near the prospective victim and walk or climb onto it. Vampires attack in areas without hair or feathers or where the hair is scant. These include the naked skin around the anus and vagina, the ears and neck of cattle, and the wattles and combs of chickens. When a suitable area is found, the bat makes a quick shallow bite with its very sharp teeth that cuts away a small thin piece of skin. This operation is practically painless and usually does not disturb the sleeping or quiet animal or human being. The bats do not bite deeply or struggle with a victim.

The wound thus produced is 3–6 mm wide, 5–10 mm long, and 1–5 mm deep. The tongue of the bat is then applied to the wound. The lower and lateral surfaces of the tongue have grooves that function like capillary tubes, through which the blood flows. The upper surface of the tongue remains free of blood. The tongue is protruded and retracted slowly and may produce a partial vacuum in the mouth cavity, helping to move the flow of blood into the mouth. The blood sometimes flows from the wound for as long as eight hours as the saliva of the bat may delay coagulation of the blood on the wound. *Desmodus* sometimes consumes so much blood at a meal that it is barely able to fly. The feeding period may not be more than 30 minutes in length.

Practically any warm-blooded animal that is quiet may be attacked by a vampire bat. Domestic animals often are victims because of their accessibility, but vampires seldom bite dogs since dogs can hear sounds of higher frequency than some of the larger mammals are able to hear and so may detect a bat's approach. Stock raising in some tropical areas is uneconomical because of attacks by vampires. These bats bite sleeping humans, but not often.

The quantity of blood lost is usually not great, though it may be smeared extensively and present a startling appearance. The real danger of vampire bites lies in the diseases and infections that may result: these animals can transmit rabies and the livestock disease murrina, and the open wounds may become infected with bacteria and parasitic insect larvae, such as screw worms. The danger of infection is probably greatest when the flow of blood from the wound is not particularly rapid and the vampire, as a result, actually licks the wound. The bats may die of the diseases they transmit.

Fenton (1992) noted that two theories had been advanced on the origin of blood feeding in bats: (1) that vampires evolved from frugivorous bats that had developed incisor teeth capable of cutting into fruits with hard rinds; and (2) that the vampire line began with bats that specialized in feeding on ectoparasites of large mammals, such as ticks. He proposed a third theory, that saguinivory resulted when insectivorous bats, with appropriate teeth, were attracted to the insects that had swarmed about the bleeding wounds of large animals.

Some species of the Phyllostomidae form maternity colonies; in others the sexes remain together throughout the year. There is one, rarely two, young per birth. Breeding usually is soon followed by fertilization and development of the young.

The geological range of this family is Miocene and Pleistocene to Recent in South America and Pleistocene to Recent in the West Indies and North America.

CHIROPTERA; PHYLLOSTOMIDAE; **Genus**
DESMODUS
Wied-Neuwied, 1826

Common Vampire Bat

The single species now known to be extant, *D. rotundus,* is found from northern Mexico to central Chile, Argentina, and Uruguay, as well as on the islands of Margarita and Trinidad off northern Venezuela (Koopman 1988). A closely related and larger species, *D. stocki,* occurred across much of the southern United States and Mexico during the late Pleistocene and evidently survived on San Miguel Island, off southern California, until less than 3,000 years ago (Ray, Linares, and Morgan 1988). The skeletal remains of a giant species, *D. draculae,* recently were discovered in caves in northern Venezuela and southeastern Brazil, and late Pleistocene fossils of comparable size are known from West Virginia and the Yucatan. As the remains from Venezuela and Brazil are unmineralized and were found on the surface in apparent association with living species, there is a "faint possibility" that *D. draculae* still exists (Morgan, Linares, and Ray 1988; Ray, Linares, and Morgan 1988; Trajano and De Vivo 1991).

Head and body length of *D. rotundus* is 70–90 mm, there is no tail, forearm length is 50–63 mm, and adult weight is

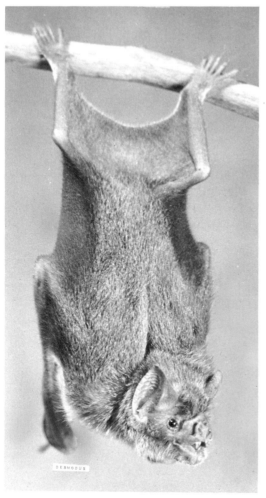

Vampire bat *(Desmodus rotundus)*, photo by Ernest P. Walker. The skull photo, by P. F. Wright of a specimen in U.S. National Museum of Natural History, shows large, sharp, middle incisors used for making incisions for taking blood. This photo, of a very young animal, is of unusual interest in that it also shows unshed milk teeth next to the permanent middle incisors.

15–50 grams. The upper parts are dark grayish brown and the underparts are paler, sometimes with a faint buffy wash.

Desmodus is distinguished from the other true vampire bats by its pointed ears, longer thumb with a distinct basal

pad, naked interfemoral membrane, and dental features. This genus has only 20 teeth, the largest being the 2 chisel-like upper incisors and the 2 upper canines.

This adaptable species inhabits both arid and humid regions of the tropics and subtropics. In Venezuela, Handley (1976) collected 57 percent of his specimens in forests of all types and 43 percent in yards, pastures, and other open habitats. Vampire bats usually reside in caves but also occupy hollow trees, old wells, mine shafts, and abandoned buildings. These retreats usually have a strong odor of ammonia from the pools of digested blood that collect in the crevices. If disturbed, the bats quickly run into more protected crevices. When moving about they somewhat resemble large spiders. They commonly forage in an area within 5–8 km of the diurnal roost, and in some areas this distance may extend to 15–20 km (Greenhall et al. 1983). At a study site in Costa Rica, Turner (1975) estimated that a population of 100–150 vampires occupied 1,300 ha. and utilized 1,200 head of livestock as prey.

The feeding habits of the vampire have been so exaggerated and confused with Old World legends that the animal is of particular interest. Shortly after dark it leaves its roost with a silent, low flight, usually only one meter above the ground. Its usual food is the blood of horses, burros, cattle, and occasionally human beings. Data summarized by Gardner (1977) indicate that in some areas *Desmodus* also preys extensively on domestic turkeys and chickens. It has been conservatively estimated that each individual bat consumes 20 ml of blood per day. Greenhall (1972) listed four phases in the feeding cycle of *Desmodus,* the whole process lasting as long as two hours: (1) selection of a suitable site on a prey animal; (2) preparation, during which the bat licks the site with its tongue; (3) shearing or shaving, in which the bat cuts the hair or feathers; (4) biting, typically involving the neat removal of a circular piece of skin with the upper and lower incisors. After feeding, *Desmodus* retires to its daytime retreat to digest its meal (for additional details see account of the family Phyllostomidae).

The common vampire bat may roost alone, in small groups, or in colonies of more than 2,000 individuals. Most colonies contain 20–100 bats; if more than 50 are present, there are stable, identifiable social units consisting of 8–20 adult females and their young (Schmidt et al. 1978; Wilkinson 1985a, 1985b, 1988). Females within a unit regurgitate blood for their young and for one another. A single adult male resides in the vicinity of each female group and attempts to maintain his position and thus the ability to mate with the females. If the roost is within a hollow tree, this male is positioned near the top. Other males roost in groups nearby and attempt to displace the top male. They often are successful, and thus paternity varies within a social unit. Fighting between males is vicious. Sailler and Schmidt (1978) described six different vocalizations in captive *Desmodus,* mostly associated with aggressive interaction during feeding.

Wilson (1979) described the reproductive pattern of *Desmodus* as continuous polyestry, there being records from many areas suggesting year-round breeding. Some females have a postpartum estrus and produce more than one litter per year. Estrus lasts 2–3 days and birth weight is about 7 grams (Hayssen, Van Tienhoven, and Van Tienhoven 1993). According to Schmidt (1988), there are peak periods of birth in some areas, such as, for example, April–May and October–November in Trinidad. There normally is a single young; a case of twins has been reported, but one of the neonates was not viable. Recorded gestation periods of captive females are 205, 213, and 214 days. At birth the young are well developed and their eyes are open. Although they first feed on regurgitated blood in their second month of life and can forage for blood themselves at 4 months, they are not completely weaned before 9–10 months. Sexual maturity seems to come at about the same time. Lopez-Forment (1980) reported that wild individuals have been banded and recaptured at the same roosts nearly 9 years later and that a captive female lived for at least 19.5 years.

The vampire bat long has been considered a threat both to people and to their domestic animals in Latin America. The main problem at present is the transmission of rabies from *Desmodus* to the cattle on which it feeds. It is estimated that more than 100,000 cattle die in this manner each year and that annual economic losses amount to over U.S. $40 million (Acha and Alba 1988).

CHIROPTERA; PHYLLOSTOMIDAE; Genus DIAEMUS
Miller, 1906

White-winged Vampire Bat

The single species, *D. youngi,* has been recorded at scattered localities from Tamaulipas in northeastern Mexico to northern Argentina and southeastern Brazil, as well as on Margarita Island and Trinidad (Koopman 1988). *Diaemus* was included within *Desmodus* by Handley (1976) and Koopman (1978b), but later Koopman (1988 and *in* Wilson and Reeder 1993) concluded that the two are generically distinct.

Head and body length is about 85 mm, there is no external tail, and forearm length is approximately 50–56 mm. Adults weigh approximately 30–45 grams. The pelage is usually glossy clay color, light brown, or dark cinnamon brown. The edges of the wings are white, and the membrane between the second and third fingers is largely white.

According to Miller (1907), "The peculiar short thumb with single pad under its metacarpal and the slightly recurved lower incisors with their different system of cusps are the principal characters which distinguish this genus from *Desmodus.*" The thumb in *Diaemus* is about one-eighth as long as the third finger; in *Desmodus* it is about one-fifth as long. The thumb of *Diaemus* is about the same length as the thumb of *Diphylla,* but it differs from that of the hairy-legged vampire bat in the presence of the large pad at its base. *Diaemus* is the only bat known to have 22 permanent teeth. Old individuals occasionally lack the second upper molars and so have only 20 teeth.

In the lowlands of Venezuela, Handley (1976) collected specimens mainly in moist, open areas. One individual was found in a cave. As far as is known, *Diaemus* feeds only on fresh blood, apparently preferring the blood of birds and goats. Captives would not feed on defibrinated cattle blood (even when it was mixed with chicken blood) but did feed heavily on chicken blood. A live guinea pig was attacked ravenously by a captive.

Colonies of up to 30 individuals of both sexes have been found in hollow trees on Trinidad, and a single individual was obtained in a well-illuminated cave there. In August, 2 lactating females were taken on Trinidad, and in October an immature male, 4 pregnant females, and 1 lactating female were collected (Wilson 1979).

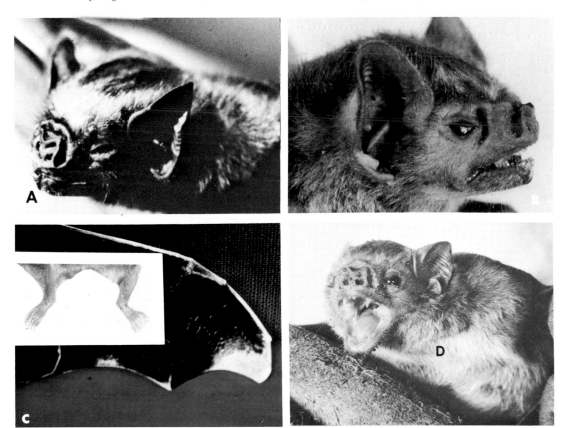

White-winged vampire bat *(Diaemus youngi):* A. Photo by Richard K. LaVal. Inset and B. Photos by Harold Drysdale, Trinidad Regional Virus Laboratory, through A. M. Greenhall. C. Photo from Pan-American Sanitary Bureau. D. Photo by David Pye.

Hairy-legged vampire bat *(Diphylla ecaudata)*, photos by Ernest P. Walker.

CHIROPTERA; PHYLLOSTOMIDAE; Genus DIPHYLLA
Spix, 1823

Hairy-legged Vampire Bat

The single species, *D. ecaudata,* occurs from southern Texas and eastern Mexico to northern Bolivia and south-central Brazil (Koopman 1988). The only record for Texas (and for any vampire bat anywhere within the United States in modern time) is of a single individual found in 1967 (Greenhall, Schmidt, and Joermann 1984).

Head and body length is 65–93 mm, there is no tail, and forearm length is 50–56 mm. Two females from Brazil weighed 25.5 and 26.7 grams, and two males weighed 24.0 and 24.4 grams (Russell E. Mumford, Purdue University, pers. comm.), but Greenhall, Schmidt, and Joermann (1984) stated that weight ranges up to 43 grams. The coloration is dark brown or reddish brown above and somewhat paler below.

Diphylla somewhat resembles *Desmodus* externally but is usually smaller and has shorter and rounder ears, a shorter thumb without the basal pad, a calcar, and longer, softer fur. Another distinctive character of *Diphylla* is the well-haired interfemoral membrane. This genus has 26 teeth, compared with 22 in *Diaemus* and 20 in *Desmodus*. It also differs from the other true vampire bats in its fan-shaped, multilobed outer lower incisor, unique among bats. This tooth resembles the lower incisor in gliding lemurs (Dermoptera). Although *Diphylla* usually is said to have seven lobes, examination of 85 specimens by Bhatnagar, Fentie, and Wible (1992) showed that 67 percent had six lobes, 27 percent had seven lobes, and the rest had either five or eight lobes.

In northern Venezuela, Handley (1976) collected specimens in both moist and dry, forested and open areas. Individuals roosted in caves and houses. Like *Desmodus,* the hairy-legged vampire is shy and agile, but it usually flies to an adjacent perch when disturbed in its roost instead of scrambling into a crevice. This bat has also been found in a mine tunnel and in hollow trees. It appears to prey mainly on birds, such as chickens. Attacks on horses, burros, and cattle also have been reported, but Greenhall, Schmidt, and Joermann (1984) stated that reliable records indicate that avian blood is the only source of nourishment. *Diphylla* bites the legs and cloacal region of chickens; secondary infections may result in the death of the bird.

This genus is not as gregarious as *Desmodus* and does not cluster in its roosts. Pools of digested blood thus do not form, and *Diphylla* may escape notice. Although as many as 35 have been found in a cave, the usual number is 12 or fewer, and often there are only 1–3. Data summarized by Wilson (1979) indicate that there may be two litters per year. Pregnant or lactating females have been reported in Mexico and Central America in March, May, July, August, October, and November. *Diphylla* is designated as near threatened by the IUCN.

CHIROPTERA; PHYLLOSTOMIDAE; Genus MACROTUS
Gray, 1843

Big-eared Bats

There are two species (Davis and Baker 1974; Jones and Carter 1976; Koopman 1989*b*; Wilson 1991):

M. waterhousii, western and central Mexico to the Yucatan and Guatemala, Tres Marías Islands off western Mexico, Cuba, Jamaica, Cayman Islands, Hispaniola, Bahamas;

M. californicus, southwestern United States, northwestern Mexico, including Baja California.

E. R. Hall (1981) did not consider *M. californicus* to be specifically distinct from *M. waterhousii.*

Head and body length is 50–69 mm, tail length is 35–41 mm, forearm length is 45–58 mm, and adult weight is usually 12–20 grams. The upper parts are brownish or grayish, and the underparts are brown or buff, generally with a silvery or whitish wash.

This genus can be distinguished externally from *Mi-*

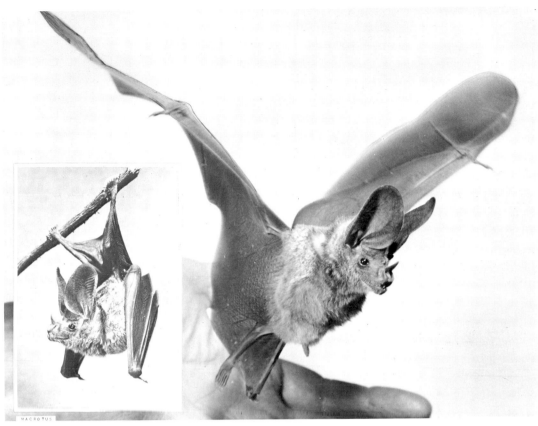

Mexican big-eared bat *(Macrotus waterhousii)*, photos by Ernest P. Walker.

cronycteris by the larger ears and the longer tail. The ears of *Macrotus* are united, and the tail extends slightly beyond the interfemoral membrane. The nose leaf is erect and lanceolate, shaped somewhat like an arrowhead.

Big-eared bats usually frequent arid lowlands, but *M. waterhousii* also occurs in more humid locations in the West Indies. In Jalisco, Mexico, Watkins, Jones, and Genoways (1972) found *M. waterhousii* mainly in relatively arid areas where subtropical vegetation predominated. Caves are a favorite roosting site, but this genus also occupies mine tunnels and buildings. Complete darkness is not required, and the bats are often found within 10–30 meters of the entrance of a cave or in partially lighted buildings. *Macrotus* emerges later than most bats, usually 90–120 minutes after sunset (Anderson 1969). Unlike most bats native to the United States, *Macrotus* cannot crawl, but it is among the most agile and alert in flight (Barbour and Davis 1969). Some individuals of *M. californicus* winter north of the Mexican border, though most probably migrate to warmer regions. Hoffmeister (1986) reported that colonies in most caverns in southern Arizona are about the same size in summer and winter. Three individuals of *M. californicus* were found in a semidormant condition in March in northern Arizona, but the species does not experience deep hibernation. Although they can feed on the ground, the members of this genus feed mainly in flight and then hang to digest their catch. The diet consists of insects and fruits, including those of cacti.

These bats are gregarious, colonies of dozens or hundreds being common in some areas. Studies of a colony of *M. californicus* in southern Arizona showed both sexes to be present in March and April, but during the summer the females segregated in maternity colonies and the males dispersed in small groups. From August to October the sexes reassociated, but during the winter only males were present. Ovulation, insemination, and fertilization occurred in September and October. The embryo then grew slowly until March in a process known as delayed development. Subsequent development was more rapid, and births took place from May to early July after a total pregnancy of about 8 months. Normally there was a single young, rarely two. Weaning took place at about 1 month. Young females could breed during the first autumn after their birth, but males were not sexually mature until the following year. Maximum life expectancy was estimated to be more than 10 years (Anderson 1969; Barbour and Davis 1969). Other reproductive studies, as summarized by Wilson (1979), have found pregnant or lactating *M. californicus* from March to July in California and northwestern Mexico. Pregnant or lactating *M. waterhousii* have been taken in Mexico, Cuba, and the Bahamas from February to July, and lactating females have been found in Jamaica in December. Klingener, Genoways, and Baker (1978) reported that five of six adult females collected in southern Haiti on 16 May were pregnant and that no pregnancies were found in January and August.

Brown and Berry (1991) suggested that prior to the advent of mining, *M. californicus* may have been present in the southwestern United States only in the warmer months as it requires a warm roost of 27° C or more. Large winter colonies now form in just a few geothermally heated mines in the region. These roosts are subject to vandalism, disturbance by explorers, and destruction in the course of new mining operations. The IUCN now classifies *M. californicus* as vulnerable.

CHIROPTERA; PHYLLOSTOMIDAE; Genus
MICRONYCTERIS
Gray, 1866

Little Big-eared Bats

There are 7 subgenera and 11 species (Anderson, Koopman, and Creighton 1982; Ascorra, Gorchov, and Cornejo 1993; Ascorra, Wilson, and Gardner 1991; Brosset and Charles-Dominique 1990; Carter et al. 1981; Davis 1976*b*; Genoways and Williams 1986; Jones and Carter 1976, 1979; McCarthy 1987; McCarthy and Blake 1987; McCarthy and Ochoa G. 1991; McCarthy et al. 1993; Trajano 1982; Williams and Genoways 1980*b*):

subgenus *Micronycteris* Gray, 1866

M. megalotis, Jalisco and Tamaulipas (Mexico) to Bolivia and southern Brazil, Grenada, Trinidad;
M. microtis, Costa Rica to French Guiana;
M. schmidtorum, Yucatan Peninsula of Mexico to northern Peru and extreme eastern Brazil;
M. minuta, Honduras to Bolivia and Brazil, Trinidad;

subgenus *Xenoctenes* Miller, 1907

M. hirsuta, Honduras to Peru and French Guiana, Trinidad;

subgenus *Lampronycteris* Sanborn, 1949

M. brachyotis, southern Mexico to Amazonian Brazil, Trinidad;

subgenus *Neonycteris* Sanborn, 1949

M. pusilla, eastern Colombia, northern Brazil, and possibly adjacent regions of South America;

subgenus *Trinycteris* Sanborn, 1949

M. nicefori, Belize to northeastern Peru and northern Brazil, Trinidad;

subgenus *Glyphonycteris* Thomas, 1896

M. sylvestris, Nayarit and Veracruz (Mexico) to eastern Peru and southeastern Brazil, Trinidad;
M. behnii, Peru, central Brazil;

subgenus *Barticonycteris* Hill, 1964

M. daviesi, known only from Honduras, Costa Rica, Panama, Venezuela, Guyana, Surinam, French Guiana, Amazonian Peru, and Brazil.

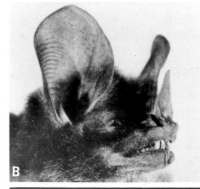

Little big-eared bat *(Micronycteris megalotis):* A. Photo by Bruce J. Hayward; B. Photo by Harold Drysdale, Trinidad Regional Virus Laboratory, through A. M. Greenhall; C. Photo by David Pye.

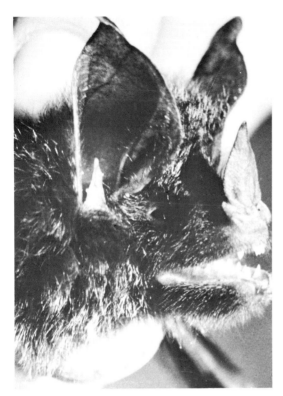

Micronycteris daviesi, photo by Richard K. LaVal.

Barticonycteris was treated as a full genus by Corbet and Hill (1991). In contrast, Genoways and Williams (1986) considered *Barticonycteris* a synonym of *Micronycteris* and put the species *M. daviesi* in the subgenus *Glyphonycteris*. The distinction of *M. microtis* from *M. megalotis* by Brosset and Charles-Dominique (1990) was not followed by Koopman (*in* Wilson and Reeder 1993).

Head and body length is 42–69 mm, tail length is about 10–17 mm, forearm length is 31–57 mm, and adult weight is usually 4–16 grams. Coloration is usually some shade of brown above and brownish or buffy below. The upper parts in *M. megalotis* are dark brown tinged with russet or dark brown without the russet tinge, some specimens exhibiting intermediate colors. *M. minuta* is orange rufous or brown. In *M. brachyotis* the upper parts are olive brown, the throat is yellowish or orangish, and the chest and belly are tawny olive.

A prominent nose leaf is present. In some species the ears are connected by a high notched band; in others they are not. In the bats of this genus the tail extends to the middle of the interfemoral membrane and the middle lower premolar is not reduced. *Barticonycteris* is distinguished from the other subgenera by large size and the presence of a single pair of upper incisors so expanded basally as to fill the intercanine space completely and to extend posteriorly to a line joining the upper canines.

Little big-eared bats occupy a variety of habitats, including desert scrub and tropical forest. Handley (1976) found most species in Venezuela in moist places within forests, but many specimens were also taken in dry, open areas, and *M. nicefori* was found predominantly in dry forest. In Venezuela the most common roosting sites were holes in trees and hollow trees and logs. These bats are also known to roost in caves and crevices, animal burrows, cul-

verts, buildings, and ancient ruins. They often roost where there is some light. Available data, summarized by Gardner (1977), indicate that *Micronycteris* is primarily insectivorous but that a variety of fruits are also consumed, their importance probably varying seasonally.

These bats roost alone or in small groups, usually not more than 20 being found together. Rick (1968) reported the discovery of a colony of *M. brachyotis* in a small chamber in Mayan ruins in Guatemala on 10 July. Present were a single adult male, 9 adult females, and 5 nursing young. Of the 7 adult females that were examined, 6 were lactating and 1 was pregnant with a single embryo. LaVal and Fitch (1977) found pregnant female *M. megalotis* in Costa Rica in April, a lactating female in May, subadults in August and September, and only nonpregnant females from October to February. Wilson (1979) summarized reproductive information for *Micronycteris*, showing that overall for the genus pregnant and lactating females have been found from February to August. These data suggest the following possible patterns: *M. megalotis*, seasonal breeding, with females pregnant in the northern part of the range during the beginning of the rainy season and there being perhaps two annual breeding cycles per female in the southern part of the range; *M. minuta*, breeding initiated at the beginning of the rainy season; *M. hirsuta*, a bimodal reproductive pattern in Trinidad; and *M. sylvestris*, breeding late in the rainy season in the north and early in the rainy season in the south.

Medellín et al. (1983) discovered what was perhaps the largest colony of *Micronycteris* on record. In 1978 they found a group of more than 300 *M. brachyotis* in a coastal cave in southern Veracruz, Mexico. Observations over the next few years suggested that females are polyestrous. Unfortunately, the colony declined in numbers during the period and had disappeared entirely by July 1981. This loss was attributed to human habitat disturbances of the surrounding forest and to deliberate killing by persons who mistakenly believed the bats belonged to a vampire species. Such problems are leading to declines in various other species of *Micronycteris*, with the IUCN now classifying *M. pusilla* and *M. behnii* as vulnerable and *M. sylvestris* and *M. daviesi* as near threatened.

CHIROPTERA; PHYLLOSTOMIDAE; **Genus VAMPYRUM**
Rafinesque, 1815

Linnaeus's False Vampire Bat, or Spectral Vampire

The single species, *V. spectrum*, is found from southern Mexico to Peru, northern Bolivia, and central Brazil and on Trinidad; a single specimen from Jamaica probably represents an accidental occurrence (Anderson 1991; Carter et al. 1981; E. R. Hall 1981; Jones and Carter 1976).

This is the largest New World bat; head and body length is about 125–35 mm, there is no tail, forearm length is about 100–108 mm, and the wingspan is usually 762–914 mm, but some measure as much as 1,016 mm. Adults weigh approximately 145–90 grams. The coloration is reddish brown above and slightly paler below. This bat is readily distinguished from *Chrotopterus* by its larger size, absence of a tail, and four upper and four lower incisors instead of four upper and two lower incisors. As in *Chrotopterus*, the chin is smooth.

In Venezuela this bat occurs in lowlands and foothills;

Tropical American false vampire bats *(Vampyrum spectrum):* A. Photo from New York Zoological Society; inset photo from *Zool. Verhandel.,* "The Bats of Suriname," A. M. Husson; B. Photo by Richard K. LaVal.

Handley (1976) collected all specimens in moist areas, with 40 percent being taken in evergreen forest, 40 percent in yards, and 20 percent in swamps. On Trinidad groups of not more than five individuals have been found roosting in hollow tree trunks, and on the upper Amazon scores were observed flying out of a church in the twilight. In Costa Rica a radio-tracked individual hunted over an area of 3.2 ha. and spent most of its time in deciduous woodland, secondary growth, and forest edge (Vehrencamp, Stiles, and Bradbury 1977). *Vampyrum* once was thought to be a true vampire, but actually it is carnivorous and does not drink blood. According to Gardner (1977), the diet consists of birds, bats, rodents, and possibly some fruit and insects. A

lactating female was taken on Trinidad in May, and reproductively inactive individuals were found in Costa Rica in February, July, August, and October (LaVal and Fitch 1977; Wilson 1979).

Greenhall (1968), reporting on a breeding pair that he successfully kept in captivity for five years, said that the bats readily adapted to captivity and became tame and gentle. They fed on white mice, birds, and raw meat and drank water. The standard daily meal consisted of two adult white mice per bat. Young chicks and small pigeons were also relished. Raw fruits, such as bananas, mangoes, papayas, and citrus, were offered but were never eaten. When living mice were presented, the bats would crawl stealthily down the wall to within a few centimeters of the floor and wait for a mouse to pass underneath. Then a bat would drop on the mouse and kill it. With the mouse clamped tightly between its jaws, the bat would hitch itself backward up the wall to its roost to eat. The bat would hold and steady the mouse with the claws of its thumbs and thoroughly masticate it from head to heel, excluding the tail, which it discarded. Greenhall stated that the bats were careful, slow stalkers and rarely missed their aim. The head of a bird or rodent was carefully seized near the snout or beak, and the large canine teeth were quickly sunk into the skull. The captive *Vampyrum* mated and gave birth to a single young in late June, but Greenhall was not able to ascertain the gestation period. The female took good care of the offspring, and the male was also solicitous of his family. He sometimes wrapped his huge wings around both mother and young, and frequently around the mother alone.

Although *Vampyrum* still has a wide distribution, the region is subject to increasing human encroachment and habitat loss. The great size of the genus, with a wingspan approaching 1 meter, makes it particularly susceptible. It now is designated as near threatened by the IUCN.

CHIROPTERA; PHYLLOSTOMIDAE; **Genus TRACHOPS**
Gray, 1847

Frog-eating Bat

The single species, *T. cirrhosus,* occurs from southern Mexico to Bolivia and southern Brazil (Jones and Carter 1976) and also on Trinidad (Carter et al. 1981).

Head and body length is 76–88 mm, tail length is 12–21 mm, and forearm length is 57–64 mm. LaVal and Fitch (1977) reported an average weight of 32.3 grams for a series of specimens from Costa Rica. The upper parts are dark reddish brown, cinnamon brown, or somewhat darker; the underparts are dull brownish washed with gray.

This genus can be distinguished by the wart-studded lips and the small size and peculiar position of the second lower premolar. The ear is longer than the head. The tail is much shorter than the femur and projects from the upper surface of the interfemoral membrane.

In Venezuela this bat was collected mostly in humid lowland forest and was found to roost in hollow trees (Handley 1976). Other reported roosting sites include caves, a Mayan building (Rick 1968), culverts, and an abandoned railroad tunnel (Starrett and Casebeer 1968). Data summarized by Gardner (1977) indicate that the diet of this bat consists of insects, small vertebrates such as lizards, and possibly some fruit. Ryan, Tuttle, and Barclay (1983) showed that *Trachops* locates frogs and distinguishes frog species by listening for and analyzing the frogs' calls. Ryan and Tuttle (1983) added that such an ability allows the bat

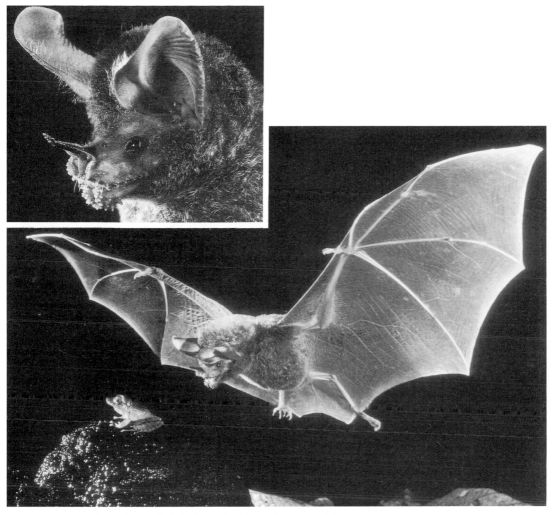

Frog-eating bat *(Trachops cirrhosus)*, photos by Merlin D. Tuttle, Bat Conservation International.

to safely discriminate between poisonous and palatable prey.

Colonies of six individuals of both sexes have been discovered in hollow trees on Trinidad. Pregnant or lactating females have been collected in southern Mexico and Central America in February, March, April, May, August, and December; on Trinidad in March; and in Peru in July. These data suggest an extended breeding season or a geographically variable one (Wilson 1979). There is a single young (Hayssen, Van Tienhoven, and Van Tienhoven 1993).

CHIROPTERA; PHYLLOSTOMIDAE; **Genus
CHROTOPTERUS**
Peters, 1865

Peters's Woolly False Vampire Bat

The single species, *C. auritus*, ranges from southern Mexico to Paraguay and northern Argentina (Jones and Carter 1976).

Head and body length is about 100–112 mm, tail length is 7–17 mm, and forearm length is 75–87 mm. Rick (1968) listed weights of 72.7 grams for a male and 90.5 grams for a female. The long, soft hair is dark brown on the upper parts and grayish brown on the underparts.

On the front of the neck there is an opening into a glandular pocket, apparently somewhat like that on some other bats. The lips and chin are smooth except for a small wart in the center of the lower lip and a narrow elevation on either side. The ears are large, ovate, and separate. The tail is practically absent. In *Chrotopterus* there are four upper and two lower incisors, whereas *Vampyrum* has four upper and four lower incisors.

In Venezuela this bat occurs in forested lowlands, usually near streams or in other moist places (Handley 1976). Reported roosting sites include: caves in the Yucatan (Jones, Smith, and Genoways 1973), a Mayan ruin in Guatemala (Rick 1968), and a hollow tree in Costa Rica (Starrett and Casebeer 1968). The diet includes insects, fruit, and apparently a substantial proportion of small vertebrates such as other bats, opossums *(Marmosa)*, mice, birds, lizards, and frogs (Gardner 1977; Sazima 1978).

Colonies contain 1–7, usually 3–5, individuals (Medel-

Peters's woolly false vampire bat *(Chrotopterus auritus)*, photo by Louise Emmons.

lín 1989). Pregnant females have been taken in southern Mexico in April and July and in Argentina in July; a lactating female was found in the Yucatan in July (Wilson 1979). Taddei (1976) found reproductive activity in southeastern Brazil only in the second half of the year; one female, captured pregnant and maintained in isolation, gave birth to a single young after 99 days.

CHIROPTERA; PHYLLOSTOMIDAE; **Genus**
PHYLLODERMA
Peters, 1865

Peters's Spear-nosed Bat

The single species, *P. stenops*, occurs from southern Mexico to Bolivia and southeastern Brazil (Barquez and Ojeda 1979; E. R. Hall 1981; Jones and Carter 1976, 1979; Trajano 1982). Based on an analysis of genetic data from protein variation, Baker, Dunn, and Nelson (1988) proposed that this species be transferred to the genus *Phyllostomus*. Such was done by Baker, Hood, and Honeycutt (1989) and Van Den Bussche and Baker (1993) but not by Corbet and Hill (1991) or Koopman (*in* Wilson and Reeder 1993).

Head and body length is 85–120 mm, tail length is about 16–22 mm, and forearm length is 67–80 mm. Eisenberg (1989) listed an average weight of about 47 grams. The coloration is dark brown above and buffy to pale brown below. The wing membranes are described as blackish brown with

whitish tips in the subspecies *P. s. septentrionalis* and dark brown in *P. s. stenops*.

This bat generally resembles *Phyllostomus* but may be readily distinguished by the two-lobed middle upper incisors, the narrow lower molars, and the presence of a small third lower premolar. A male specimen in the British Museum of Natural History appears to have a glandular throat sac.

In Venezuela, Handley (1976) collected all specimens in lowlands near streams and in other moist places; about half were taken in forests and about half in more open areas, such as pastures, orchards, and marshes. In Peru, Gardner (1976) collected a specimen in a small clearing in a cloud forest at 2,600 meters. In Costa Rica, LaVal (1977) captured a female adjacent to a small forest stream. The feces of this specimen consisted mostly of the large seeds of one species of the family Annonaceae; however, in captivity this specimen eagerly ate bananas and drank sugar water with a long, extensible tongue. Another individual was caught in Brazil while eating the larvae and pupae from an active nest of a social wasp (Jeanne 1970). LaVal's (1977) specimen, taken on 9 February, was pregnant with one large embryo.

CHIROPTERA; PHYLLOSTOMIDAE; **Genus**
PHYLLOSTOMUS
Lacépède, 1799

Spear-nosed Bats

There are four species (Anderson, Koopman, and Creighton 1982; Baker, Dunn, and Nelson 1988; Brosset and Charles-Dominique 1990; Carter et al. 1981; Jones and Carter 1976; Koopman 1982a; Marshall, Monzón S., and Jones 1991; Myers and Wetzel 1983; Van Den Bussche and Baker 1993):

P. latifolius, southeastern Colombia, Guyana, Surinam, French Guiana, northern Brazil;
P. discolor, southern Mexico to northern Argentina, Trinidad;
P. hastatus, Guatemala to Paraguay and southeastern Brazil, Trinidad;
P. elongatus, east of the Andes in South America from Colombia to Bolivia and southeastern Brazil.

Phylloderma (see account thereof) sometimes is included in *Phyllostomus.*

P. hastatus is one of the larger American bats, with a head and body length of about 100–130 mm, forearm length of about 83–97 mm, wingspread of about 457 mm, and adult weight of about 50–142 grams. In the smallest species, *P. discolor*, head and body length is about 75 mm, forearm length is 55–65 mm, and weight is about 20–40 grams. Tail length in *Phyllostomus* ranges from 10 mm to about 25 mm. The coloration is dark brown or blackish brown, grayish, reddish brown, or chestnut brown above and somewhat paler below.

Distinguishing features include the robust form, the simple, well-developed nose leaf, the widely separated ears, the short tail, and the heavy skull. The glandular throat sac, well developed in the males, is rudimentary in the females. The lower lip has a V-shaped groove edged with wartlike protuberances.

In Venezuela, Handley (1976) found *P. discolor, P. hastatus,* and *P. elongatus* to occur mostly near streams and in other moist places, but many specimens were taken in dry areas. About half of the specimens were collected in forests

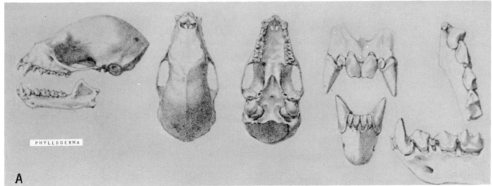

Peters's spear-nosed bat *(Phylloderma stenops):* A. Photo from *Die Fledermäuse des Berliner Museums für Naturkunde*, Wilhelm K. Peters; B. Photo by David Pye.

and about half in more open areas. Roosts included caves, culverts, hollow trees, holes in trees, and buildings. In Peru, Tuttle (1970) found roosting colonies of *P. elongatus* in large hollow trees and those of *P. hastatus* in hollow trees, termite nests, caves, and thatched roofs. Both species were netted most often near gardens where bananas were grown. A radio-tracking study in Trinidad determined that after *P. hastatus* emerged from its roosting caves in the evening, it flew to feeding sites some 1–5 km away (Fenton and Kunz 1977). Data summarized by Gardner (1977) indicated that bats of the genus *Phyllostomus* are omnivorous. *P. hastatus* preys on small vertebrates as well as insects, whereas the diet of *P. discolor* consists mainly of fruit, pollen, nectar, and insects caught in flowers. The diets of *P. elongatus* and *P. latifolius* probably include flower parts, fruits, insects, and small vertebrates such as anoles and geckos gleaned from vegetation. Fleming, Hooper, and Wilson (1972) found stomachs of *P. discolor* and *P. hastatus* taken wild in Central America to contain almost entirely insect food. Captive *P. discolor* are said to feed on fruit and

to refuse meat. Captive *P. hastatus* have eaten fruit and small vertebrates; one such captive fell on a bat of the genus *Molossus,* crushed it with bites to the back and head, and later ate it.

Spear-nosed bats are gregarious; several thousand *P. hastatus* were found in a cave in Panama, and in Trinidad numerous individuals of this species fly together as a group from roost to feeding grounds. In Peru, Tuttle (1970) found colonies of *P. elongatus* to number 7–15 and those of *P. hastatus* to number 10–100 or more. More detailed studies (cited by Fenton and Kunz 1977) have determined that colonies of *P. hastatus* are divided into harems with as many as 30 adult females per dominant male, along with juveniles and nonharem males. Dominant males defend their harems against other males. *P. discolor* also occurs in harems, but the females number only 1–12 per dominant male and tend to be more nomadic.

In a radio-tracking study of cave-dwelling colonies of *P. hastatus* in Trinidad, McCracken and Bradbury (1981) found harem clusters to contain an average of 18 females

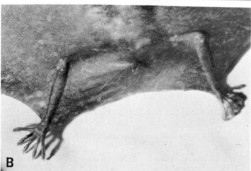

A. Spear-nosed bat *(Phyllostomus hastatus)*, photo by P. Morris. B. *P. d. discolor*, photo by Harold Drysdale, Trinidad Regional Virus Laboratory, through A. M. Greenhall. C. *P. hastatus*, photo by David Pye.

each and to be highly stable; some individuals roosted together for years. The single harem male also sometimes remained for several years, but turnover of male residency was frequent and did not disrupt female social structure. Bachelor male groups, averaging about 17 bats each, were less stable. Resident males repelled intruding males in

quick, violent fights but generally ignored female movements. Each individual in the colony had a separate foraging area; those belonging to the females of a particular cluster adjoined and were sometimes shared, but females of different harems foraged well apart from one another. Harem males also hunted separately. Mating in the colonies took place from about October to February, and there was a synchronized birth period in April and May. The young weighed about 13 grams at birth, were carried for several days, and were then left at the roost while the mothers foraged. They could fly within the cave at about 6 weeks and went out on their own at 2 months. Juveniles of both sexes dispersed after several months and were not recruited into their parental social units. Young females from different clusters formed new stable harems.

Additional reproductive data summarized by Wilson (1979) indicate that *P. discolor* may be an acyclic or continuous breeder in some areas, though possibly it is monestrous in Costa Rica. For *P. hastatus* the data could support either a monestrous pattern (in Nicaragua, Panama, and Trinidad) or a polyestrous one (Colombia); the reproductive strategy possibly varies geographically. In both species pregnant or lactating females have been recorded for most months of the year. One record not listed by Wilson is of a pregnant *P. discolor* collected in Peru in the period 31 October to 8 November (Davis and Dixon 1976). Collections of *P. hastatus* in Peru by Ascorra, Gorchov, and Cornejo (1993) indicated peak breeding during the first half of the wet season, in October–January. The smaller amount of data for *P. elongatus* (pregnancies in Colombia in June and in Peru in March, July, August, October, and November) indicate that breeding occurs mostly in the rainy season.

These bats are widely distributed and not immediately jeopardized with extinction, but their relatively large size and intensifying habitat disruption within their range give some cause for concern. *P. latifolius* is designated as near threatened by the IUCN.

CHIROPTERA; PHYLLOSTOMIDAE; **Genus TONATIA** *Gray, 1827*

Round-eared Bats

There are seven currently recognized species (Anderson 1991; Barquez and Ojeda 1992; Brosset and Charles-Dominique 1990; Davis and Carter 1978; Gardner 1976; Genoways and Williams 1979*b*, 1980, 1984; E. R. Hall 1981; Jones and Carter 1976, 1979; McCarthy 1982*a*, 1987; McCarthy, Cadena G., and Lemke 1983; McCarthy and Handley 1987; McCarthy, Robertson, and Mitchell 1988; McCarthy et al. 1993; Marques and Oren 1987; Medellín 1983; Myers, White, and Stallings 1983; Ojasti and Naranjo 1974; Powell, Owen, and Bradley 1993; Williams, Willig, and Reid 1995):

T. saurophila, southern Mexico and Belize to southern Peru and eastern Brazil, Trinidad;

T. bidens, eastern and southern Brazil, Paraguay, northern Argentina;

T. carrikeri, Colombia, Venezuela, Surinam, northern Brazil, Peru, Bolivia;

T. brasiliense, southern Mexico to northeastern Bolivia and eastern Brazil, Trinidad;

T. sylvicola, eastern Honduras to Bolivia and northern Argentina;

T. evotis, southeastern Mexico to Caribbean versant of Honduras;

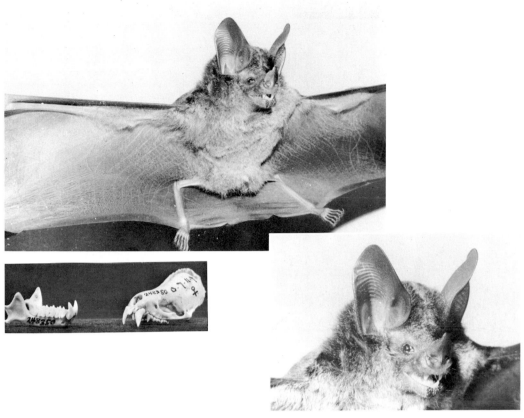

Round-eared bat *(Tonatia bidens)*, photos by Merlin D. Tuttle. Inset: skull *(T. sylvicola)*, photo by P. F. Wright of specimen in U.S. National Museum of Natural History.

T. schulzi, Guyana, central Surinam, French Guiana, northern Brazil.

T. saurophila also is known from late Pleistocene fossil material found on Jamaica (Morgan and Woods 1986). The use of the spelling *sylvicola* rather than *silvicola* follows Patterson (1992*a*).

Head and body length is 42–82 mm, tail length is 12–21 mm, forearm length is 32–60 mm, and adult weight is approximately 9–35 grams. The coloration is ochraceous to dark brown above and paler below. The underparts of *T. carrikeri* are white. The ears are rounded and large, as long as the head or longer. Those of some species are separate, while those of others are united at the base. The tail extends to about the middle of the interfemoral membrane. Medellín and Arita (1989) provided a key showing the distinguishing features of each species.

In Venezuela, Handley (1976) collected four species of *Tonatia,* mostly in moist areas within forests, and reported *T. sylvicola* to roost in hollow trees and termite nests. The species *T. carrikeri* and *T. brasiliense* also have been reported to roost in hollowed-out termite nests, both abandoned ones and those still in use by the insects. The diet of *Tonatia* consists mainly of insects, taken in flight and gleaned, and may also include a variety of fruits (Gardner 1977).

Fenton and Kunz (1977) stated that *T. bidens* and *T. sylvicola* apparently roost alone or in small groups. Tuttle (1970) collected four colonies of *T. sylvicola,* each comprising 6–10 individuals, from hollow termite nests. The type series of *T. carrikeri,* comprising 2 males and 5 females,

some mature and some not yet fully grown, was collected in December from a hollowed-out termite nest hanging from a vine about 5 meters above the ground (Allen 1911). Wilson (1979) suggested that *T. bidens* and *T. brasiliense* might breed twice a year and that *T. sylvicola* apparently gives birth early in the rainy season. Data summarized by him include: pregnant *T. bidens* in January, February, May, July, and August; pregnant *T. brasiliense* in February, April, and July and lactation in August; and pregnant *T. sylvicola* in January, March, July, and August. There is a single young (Hayssen, Van Tienhoven, and Van Tienhoven 1993). Medellín and Arita (1989) indicated that *T. sylvicola* may breed twice a year where resources allow but that *T. evotis* probably is monoestrous, with parturition occurring at the beginning of the rainy season, in April.

The IUCN classifies *T. carrikeri* and *T. schultzi* as vulnerable and *T. evotis* as near threatened. Each is expected to decline in response to habitat disruption.

CHIROPTERA; PHYLLOSTOMIDAE; **Genus MIMON**
Gray, 1847

Gray's Spear-nosed Bats

There are two species (Anderson, Koopman, and Creighton 1982; Brosset and Charles-Dominique 1990; Carter et al. 1981; E. R. Hall 1981; Jones and Carter 1976, 1979; Koopman 1978*b*):

Gray's spear-nosed bat *(Mimon crenulatum)*, photo by Merlin D. Tuttle, Bat Conservation International.

M. bennettii, southern Mexico to northern Colombia, eastern South America from the Guianas to southeastern Brazil;

M. crenulatum, southern Mexico to Bolivia and the Mato Grosso region of central Brazil, Trinidad.

Anthorhina Lydekker, 1891, sometimes is used as a genus or subgenus to include *M. crenulatum* (Husson 1978). However, Gardner and Ferrell (1990) showed that *Anthorhina* actually is a synonym of *Tonatia. Mimon koepckeae,* from Huanhuachayo, Peru, was named as a full species by Gardner and Patton (1972) but was reduced to subspecific rank by Koopman (1978b). *Mimon cozumelae,* from southern Mexico to northern Colombia, has sometimes been considered a species distinct from *M. bennettii.*

Head and body length is approximately 50–75 mm, tail length is about 10–30 mm, and forearm length is 48 to about 57 mm. A female *M. bennettii* weighed 22.9 grams and two males averaged 21.5 grams (Valdez and LaVal 1971). Average weight of *M. crenulatum* is about 12 grams (Eisenberg 1989). The coloration in *M. bennettii* is uniformly pale brownish except for whitish patches behind the ears. The fur in this species is long and woolly. In *M. crenulatum* the upper parts are bright mahogany brown or blackish brown in the fresh pelage; with age these colors become obscured with yellow, orange, and red. This species has a whitish or yellow orange patch behind the ear and a pale-colored spinal line; both the patch and the line are sometimes indistinct or even absent. The underparts are whitish to rusty, occasionally grayish. The fur in *M. crenulatum,* at least in the typical subspecies, is medium long and lax.

The ears are separate. In *M. bennettii* the chin has a broad naked area divided by a longitudinal groove; in *M. crenulatum* the lower lip has a V-shaped notch in front

bordered by wartlike protuberances. This genus may be distinguished from *Chrotopterus* by its smaller size and from *Phyllostomus* by its pointed, rather than rounded, ears.

In Venezuela, Handley (1976) found most specimens of *M. crenulatum* near streams or in other moist places within forests, and some were roosting in hollow trees. This species has also been obtained from hollow, decayed tree stumps in wooded areas of Panama and Ecuador. *M. bennettii* seems to prefer to roost in dark, damp caves below the level of the ground but has also been taken in highway culverts in Oaxaca in the company of *Carollia, Glossophaga, Trachops,* and *Desmodus.* This species and *Phyllostomus discolor* have been noted side by side in the same cave. The diet of *Mimon* probably consists of a variety of arthropods and fruits (Gardner 1977).

M. bennettii does not roost in large groups, the usual number in a given retreat being two to four. Pregnant and lactating females of this species have been taken in southern Mexico and Central America from March to August, and apparently a single young is produced at the beginning of the rainy season in that region (LaVal and Fitch 1977; Wilson 1979). Pregnant female *M. crenulatum* have been taken in southern Mexico in February, in Costa Rica in April, in Venezuela in March, in Surinam in July (Genoways and Williams 1979b), in central Brazil in October (Mares, Braun, and Gettinger 1989), and in Peru in January, July, August, and September (Ascorra, Gorchov, and Cornejo 1993).

CHIROPTERA; PHYLLOSTOMIDAE; Genus LONCHORHINA
Tomes, 1863

Sword-nosed Bats

There are four species (Anderson, Koopman, and Creighton 1982; Brosset and Charles-Dominique 1990; Eisenberg 1989; Hernandez-Camacho and Cadena 1978; Jones and Carter 1976; Ochoa G. and Ibáñez 1982; Ochoa G. and Sanchez H. 1988; Polaco, Arroyo-Cabrales, and Jones 1992; Trajano 1982):

L. orinocensis, southern Venezuela, southeastern Colombia;

L. fernandezi, southern Venezuela;

L. aurita, southern Mexico to Bolivia and southern Brazil, Trinidad, doubtfully New Providence in the Bahamas;

L. marinkellei, known from southern Colombia and central French Guiana.

Head and body length is 51–74 mm, tail length is 32–69 mm, and forearm length is 41–59 mm. Eisenberg (1989) listed average weights of about 14.5 grams for *L. aurita* and 8.7 grams for *L. orinocensis.* The fur is usually light reddish brown. The long, sharply pointed nose leaf is about as long as the large, separate ears. The lower lip is grooved in front and has a raised cushion on either side. The tail extends to the edge of the interfemoral membrane.

Handley (1976) found most *L. aurita* in Venezuela in moist places within forests, whereas nearly all *L. orinocensis* were caught emerging from hot dry roosts in large rocks on prairies. *L. aurita* seems to roost mainly in caves and tunnels. Some 500 were found in a tunnel in a dense Panama forest. They were hanging in clusters toward the back of the tunnel, while bats of another genus, *Carollia,* occupied the front of the tunnel in about equal numbers. *L. au-*

Sword-nosed bat *(Lonchorhina aurita)*, photos by Merlin D. Tuttle, Bat Conservation International.

rita has also been found hanging singly and concealed in large colonies of other species in a water canal. Another large colony was found in a highway culvert in Oaxaca, Mexico, again associated with large numbers of *Carollia*. Brosset and Charles-Dominique (1990) discovered a colony of *L. marinkellei* at the ceiling of a cave above deep pools of water; the bats apparently had been feeding on spiders and insects. In November 1989 the colony consisted of about 300 tightly packed adult females and fully grown but still lactating young.

Wilson (1979) stated that the breeding season in *L. aurita* evidently is correlated with the beginning of the rainy season. Pregnant females of this species have been found in southern Mexico, Central America, and Trinidad from February to April. Hernandez-Camacho and Cadena (1978) reported that the holotype of *L. marinkellei*, an adult female, was pregnant with one embryo when collected on 8 August. Ochoa G. and Sanchez H. (1988) indicated that *L. fernandezi* and *L. orinocensis* breed during the late rainy season, from September to December.

The IUCN classifies *L. fernandezi* and *L. marinkellei* as vulnerable and *L. orinocensis* as near threatened. Each is expected to decline in response to habitat disruption.

CHIROPTERA; PHYLLOSTOMIDAE; **Genus**
MACROPHYLLUM
Gray, 1838

Long-legged Bat

The single species, *M. macrophyllum*, occurs from southern Mexico to Peru, northern Argentina, and southeastern Brazil (Harrison 1975a; Jones and Carter 1976).

Head and body length is 43–62 mm, tail length is about 37 mm, forearm length is about 34–45 mm, and adult weight is usually 6–9 grams. The coloration is sooty brown above and somewhat paler below. This bat can be distinguished externally by the long hind limbs, the long tail enclosed within the broad interfemoral membrane, the large and broad nose leaf, and the slender body form.

Handley (1976) collected most specimens in Venezuela in moist areas within forests; all roosting individuals were found in culverts. Other reports also indicate roosting in culverts, caves, sea caves, irrigation tunnels, and abandoned buildings (Harrison 1975a). The diet is mainly insectivorous and may include some aquatic life (Gardner 1977). Seymour and Dickerman (1982) reported groups of up to 59 individuals in culverts in a coastal marsh of Guatemala.

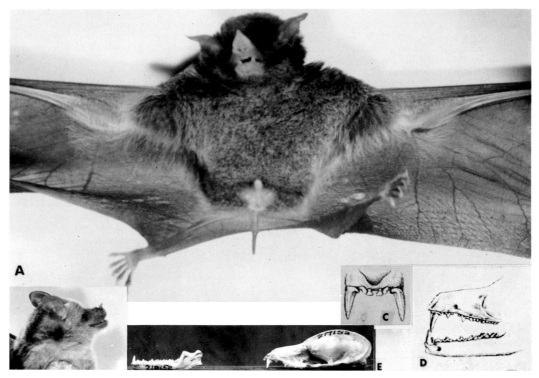

Monophyllus redmani: A & B. Photos by J. R. Tamsitt; C & D. Photos from *Proc. Zool. Soc. London;* E. Photo by P. F. Wright of specimen in U.S. National Museum of Natural History.

CHIROPTERA; PHYLLOSTOMIDAE; Genus
MONOPHYLLUS
Leach, 1821

There are two species (E. R. Hall 1981; Jones and Carter 1976):

M. redmani, Cuba, Jamaica, Hispaniola, Puerto Rico, southern Bahamas;
M. plethodon, Lesser Antilles from Anguilla to Barbados, known from subfossils in Puerto Rico.

Head and body length is about 50–80 mm, tail length is about 4–16 mm, and forearm length is about 35–45 mm. Klingener, Genoways, and Baker (1978) gave a weight range of 8–13 grams for *M. redmani* in Haiti. Homan and Jones (1975a) reported weights of 12.5–17.2 grams for *M. plethodon* on Dominica. Coloration is various shades of brown, sometimes washed with gray underneath.

This genus is distinguished from *Glossophaga* and related bats mainly by dental features. The muzzle is elongate; the long tongue is supplied with papillae; the cheek teeth are narrow and elongate; and the tail usually projects for about half its length beyond the edge of the narrow interfemoral membrane.

M. redmani evidently roosts mainly in caves, and *M. plethodon* has also been collected in caves (Homan and Jones 1975a, 1975b). Goodwin (1970) observed that *M. redmani* prefers large, deep caves with high humidity. H. E. Anthony mentioned a unique metallic buzzing sound suggestive of the droning flight of a huge beetle that is probably made by *M. redmani* when disturbed and flying about. The flight of this species has been described as strong but not particularly erratic or angular. Various authors have in-

dicated that *Monophyllus* feeds on soft fruit or nectar and possibly also insects, but there are no firm data on food habits (Homan and Jones 1975b).

M. redmani has been found roosting alone and in groups; clusters may be formed and apparently are sometimes segregated by sex. Pregnancies in this species have been recorded in December, January, February, and May. Pregnant female *M. plethodon* have been found on Do-

Monophyllus plethodon, photo from J. Knox Jones, Jr.

minica in March and April, and a lactating female was taken on Guadeloupe in late July. Both species are known to produce only one young at a time (Baker, Genoways, and Patton 1978; Klingener, Genoways, and Baker 1978; Homan and Jones 1975a, 1975b; Wilson 1979). *M. plethodon* is designated as near threatened by the IUCN.

CHIROPTERA; PHYLLOSTOMIDAE; Genus
LEPTONYCTERIS
Lydekker, 1891

Saussure's Long-nosed Bats

Two species now are recognized (Arita and Humphrey 1988; McCarthy et al. 1993):

L. nivalis, southeastern Arizona and southern Texas to southern Mexico and Guatemala;

L. curasoae, southern Arizona and southwestern New Mexico to El Salvador, Caribbean coast of northeastern Colombia and northern Venezuela and the nearby islands of Aruba, Bonaire, Curacao, and Margarita.

The subspecies *L. curasoae yerbabuenae,* found from Arizona and New Mexico to El Salvador, sometimes has been treated as a separate species, sometimes with the name *L. sanborni.*

Head and body length is 70–95 mm, the tail is minute and appears to be lacking but actually consists of three vertebrae, and forearm length is 46–57 mm. The weight is usually 18–30 grams. These bats are usually reddish brown or sooty brown above and cinnamon or brown below. This genus is characterized by dental features: the third molar is absent, and the lower incisors are usually present. In the only other genus of bats lacking the third molar *(Lichonycteris)* the lower incisors are absent. The muzzle is elongate; the long tongue is supplied with papillae; and the cheek teeth are elongate and slender. The interfemoral membrane is very narrow.

According to Barbour and Davis (1969), *L. nivalis* favors high pine-oak country and *L. curasoae* occurs in lowland desert scrub. Arita (1991) found the distribution of *L. curasoae* in Mexico to correspond closely with that of the mezcal plant *(Agave angustifolia).* Handley (1976) collected *L. curasoae* mainly in dry thorn forest in Venezuela. Both species roost in caves, tunnels, and buildings. They emerge relatively late in the evening to feed. Populations of *L. nivalis* and *L. curasoae* in the southwestern United States apparently migrate to Mexico for the winter. They may return year after year to the same locality for the summer (Barbour and Davis 1969; Fenton and Kunz 1977; Hayward and Cockrum 1971). Data compiled by Cockrum (1991) indicate that most *L. curasoae* in that region stay south of southern Sonora from November to March. Gravid adult females then enter the northern part of the range in late April and early May, congregating in traditional maternity roosts at lower elevations in areas with growths of saguaro and organ pipe cacti. Adult males and apparently some nongravid females remain in the south until June. The maternity roosts largely disperse in July, at which time at least some of the females and newborn young move to areas of higher elevation where agave nectar is available from late-blooming plants. Also by July the bats that had remained in the south spread northward and into the higher elevations. By late September or early October all bats of the population migrate south.

The diet consists of nectar, pollen, fruit, and insects. The insects may be ingested accidentally while feeding on

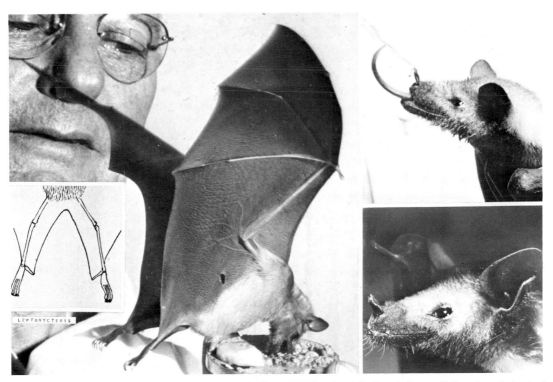

Saussure's long-nosed bat *(Leptonycteris nivalis)*, photos by Ernest P. Walker. Inset: *Leptonycteris* sp., with the calcars that extend the membrane at the heel clearly shown, photo from *Bats,* G. M. Allen.

pollen and nectar. As noted above, *L. curasoae* seems to depend heavily on the flowers of cacti and agaves. Recent investigations of that species by Fleming (1991) indicate that its spring migration corresponds with the south-to-north progression of the blooming of giant cacti. In the autumn agaves provide a southward trail of blooms. *L. nivalis* is known to visit the flowers of *Malvaviscus,* flowers and fruits of cacti, and perhaps the flowers of jimson weed *(Datura).* The bats probably use their long muzzle to reach the spineless parts of the cactus fruits, their canine teeth to tear the skin, and their tongue to lap up the juices. The tongue of *L. nivalis* can be extended to 76 mm. This manner of feeding on fruits may be utilized by other bats as well. The diet of *Leptonycteris* often results in the accumulation of yellow or red droppings on the floors of roosts.

These bats may occur in groups of 10,000 or more. During the spring females of *L. curasoae* in the northern part of their range form maternity colonies numbering into the thousands (Barbour and Davis 1969). Apparently, adult male *L. nivalis* also separate from the females in the summer and do not occupy the northern part of the range of the species (Hensley and Wilkins 1988). Smith and Genoways (1974) reported that in July a colony of *L. curasoae* on Margarita Island, Venezuela, comprised approximately 4,000 females nursing nearly full-grown young. In November the colony contained no young, but pregnant females and adult males in breeding condition were then present. The young of *L. nivalis* apparently are born during the summer in Texas and Mexico, and pregnant female *L. curasoae* have been found in Mexico in February, March, April, July, September, and November (Wilson 1979). There is usually a single young, birth weight is about 7 grams, and lactation has been reported to last 4–8 weeks (Hayssen, Van Tienhoven, and Van Tienhoven 1993).

According to Arita and Wilson (1987), *L. curasoae* and *L. nivalis* are vital in the pollination of certain kinds of cacti and agaves. As the bats seek the nectar from the flowers of these plants their fur becomes coated with pollen, which then is transferred to other flowers that they visit. The clearing of natural agaves may have resulted in greatly reduced bat populations, a situation that in turn jeopardizes the survival of some of the agaves. Colonies of *Leptonycteris* also have been destroyed deliberately by people who mistakenly believe the bats belong to a vampire species. Such problems reportedly were responsible for the elimination of most of the major known colonies in Mexico and for an apparent drastic reduction in the number of these bats that migrate into the southwestern United States. There recently have been encouraging reports that populations of *Leptonycteris* in this region still comprise thousands of bats and may even have recovered to some extent (Cockrum and Petryszyn 1991; Hoyt, Altenbach, and Hafner 1994). Nonetheless, the IUCN classifies *L. nivalis* as endangered, noting a decline by at least 50 percent over the last decade, and designates *L. curasoae* as vulnerable, reflecting a 20 percent decline in the period. The USDI lists both species as endangered.

CHIROPTERA; PHYLLOSTOMIDAE; Genus
LONCHOPHYLLA
Thomas, 1903

There are eight species (Alberico 1987; Brosset and Charles-Dominique 1990; Eisenberg 1989; Gardner 1976; E. R. Hall 1981; Hill 1980*a*; Jones and Carter 1976; Sazima, Vizotto, and Taddei 1978; Taddei, Vizotto, and Sazima 1983; Webster and Jones 1984*c*):

L. hesperia, Ecuador, Peru;
L. mordax, Ecuador, Bolivia, Brazil;
L. concava, Costa Rica, Panama, Colombia;
L. dekeyseri, east-central Brazil;
L. robusta, Nicaragua, Costa Rica, Panama, Colombia, Ecuador, Venezuela, Peru;
L. handleyi, Colombia, Ecuador, Peru;
L. thomasi, Panama, Colombia, Venezuela, Guyana, Surinam, French Guiana, Brazil, Ecuador, Peru, Bolivia;
L. bokermanni, southeastern Brazil.

Koopman (*in* Wilson and Reeder 1993), as well as E. R. Hall (1981), considered *L. concava* a subspecies of *L. mordax,* though it tentatively was treated as a full species by Corbet and Hill (1991) and Jones and Carter (1976).

Head and body length is about 45–65 mm, tail length is 7–10 mm, and forearm length is 30–48 mm. LaVal and Fitch (1977) reported an average weight of 16.3 grams for *L. robusta,* and Willig (1983) listed an average weight of about 8.5 grams for *L. mordax.* These bats are rusty or dark brown above and somewhat paler below.

The muzzle is elongate, the long tongue is equipped with papillae, the cheek teeth are narrow and elongate, and the short tail barely reaches to the middle of the interfemoral membrane. This genus is similar to *Glossophaga* but differs from it in the incompleteness of the zygomatic arch of the skull and in features of the incisor teeth. The nose leaf in *Lonchophylla* is high and narrow, not low and broad as in *Lionycteris.*

In Venezuela, Handley (1976) collected *L. robusta* and *L. thomasi* mostly in moist areas within forests; *L. thomasi* was found to roost in hollow trees. Bats of this genus also roost in caves. The diet includes insects, fruit, pollen, and nectar (Gardner 1977). Pregnant female *L. concava* have been recorded in March and August at Costa Rica (Wilson 1979). LaVal and Fitch (1977) found pregnant female *L. robusta* in Costa Rica in February, May, August, and October and a lactating female in January. In northern Peru, Ascorra, Gorchov, and Cornejo (1993) collected pregnant female *L. mordax* in December and *L. thomasi* in April.

The IUCN classifies *L. hesperia, L. dekeyseri, L. handleyi,* and *L. bokermanni* as vulnerable. All are expected to decline in response to increasing human habitat destruction.

CHIROPTERA; PHYLLOSTOMIDAE; Genus
LIONYCTERIS
Thomas, 1913

The single species, *L. spurrelli,* has been reported from eastern Panama, Colombia, Venezuela, Guyana, Surinam, French Guiana, northern Brazil, and Amazonian Peru (Jones and Carter 1976; Webster and McGillivray 1984; Williams and Genoways 1980*b*). Head and body length is about 50 mm, tail length is about 10 mm, and forearm length is 34–36 mm. The color is reddish brown to dull brown. The muzzle is elongate, the extensible tongue is supplied with bristlelike papillae, and the cheek teeth are narrow but lack the horizontal lengthening of the cheek teeth of other bats of the subfamily Glossophaginae. The interfemoral membrane is well developed, and the tail extends about half its length.

In Venezuela, Handley (1976) collected most specimens in moist, forested areas and found this bat to roost in caves and crevices. In Peru, Tuttle (1970) caught two individuals on the edges of native villages and one among blooming cashew trees. Tuttle also collected a pregnant female with

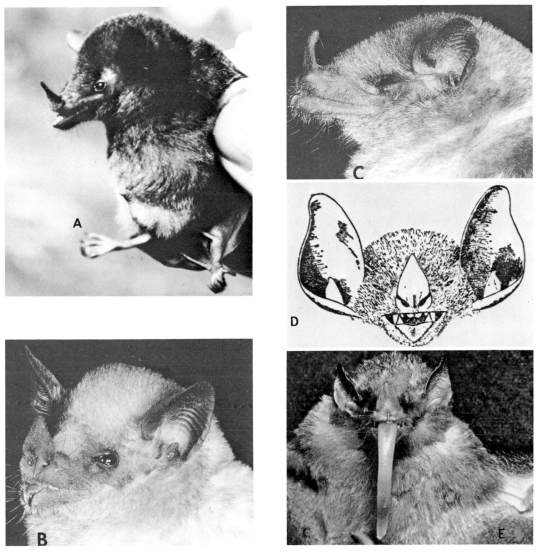

Lonchophylla robusta: A. Photo by Richard K. LaVal; B & C. Photos by David Pye; D. *Lonchophylla* sp., photo from *Bull. Amer. Mus. Nat. Hist.*, "Mammals of Costa Rica," G. G. Goodwin; E. Tongue extended *(L. mordax)*, photo by Stefan Vogel.

one embryo on 5 August. Graham (1987) caught two pregnant females in Peru in August. Brosset and Charles-Dominique (1990) located five cave colonies in French Guiana, four with about 20 individuals each and one with more than 1,000; lactating females and young were observed there during November.

CHIROPTERA; PHYLLOSTOMIDAE; **Genus ANOURA**
Gray, 1838

Geoffroy's Long-nosed Bats

There are five species (Anderson, Koopman, and Creighton 1982; Eisenberg 1989; Handley 1984; Jones and Carter 1976; Lemke and Tamsitt 1979; Molinari 1994; Nagorsen and Tamsitt 1981; Redford and Eisenberg 1992; Tamsitt and Nagorsen 1982):

A. geoffroyi, Mexico to Bolivia and southeastern Brazil, Trinidad;

A. latidens, Venezuela to central Peru;

A. caudifera, Colombia and Guianas to extreme northern Argentina and Brazil;

A. luismanueli, Andes of northwestern Venezuela;

A. cultrata, Costa Rica, Panama, Colombia, Venezuela, probably Ecuador, Peru, Bolivia.

Head and body length is about 50–90 mm, the tail is absent or only 3–7 mm long, and forearm length is 34–48 mm. LaVal and Fitch (1977) reported the average weight of a series of specimens of *A. geoffroyi* to be 15.2 grams. *A. cultrata* weighs 14–23 grams (Tamsitt and Nagorsen 1982), and *A. luismanueli* weighs 7.5–10.0 grams (Molinari 1994). Average weights of about 11 grams for 36 males and 10 grams for 31 females of *A. caudifera* were listed by Eisenberg (1989). Coloration in *A. geoffroyi* is dull brown above, usually silvery gray over the shoulders and the sides

Lionycteris spurrelli, photo by Merlin D. Tuttle.

of the neck, and grayish brown below. *A. caudifera* has dark brown upper parts and paler brown underparts. *A. cultrata* is shiny blackish, gray, or rich orange brown and may be paler ventrally; the pelage is short and crisp.

Members of this genus resemble *Glossophaga* in that the muzzle is elongate, the long tongue is supplied with papillae, and the cheek teeth are narrow and elongate. *A. geoffroyi* lacks a tail and has a rudimentary calcar, whereas *A. caudifera* has a rudimentary tail and a well-developed calcar.

In Venezuela, Handley (1976) found *Anoura* to occur predominantly in moist, forested areas, often at high altitudes. Roosting was in caves and rocks. *A. cultrata* is found in forests from about 220 to 2,600 meters, roosting exclusively in caves or tunnels (Tamsitt and Nagorsen 1982). According to Gardner (1977), the diet of this genus includes fruit, pollen, nectar, and insects, and *A. geoffroyi* is considered a highly insectivorous glossophagine. Sazima (1976) found that all stomachs of *A. geoffroyi* and *A. caudifera* collected in Brazil contained nectar, pollen, and insects, including beetles and moths that were too large to have been absorbed accidentally with nectar.

In Costa Rica, Lemke and Tamsitt (1979) found that *A. geoffroyi* and *A. cultrata* roosted singly or in groups of up to 20. In Peru, Tuttle (1970) reported a colony of about 75 *A. geoffroyi* of both sexes. Data summarized by Wilson (1979) suggest that *A. geoffroyi* forms sexually segregated colonies at certain times of the year in the same caves on Trinidad. In one there were 20 males and 25 females in June, 29 males and 1 female in October, and 32 males and 56 females in November. In that area *A. geoffroyi* apparently breeds late in the rainy season; studies there by Heideman, Deoraj, and Bronson (1992) showed that females

become pregnant in July or August and young are born in late November or early December. Pregnancies in this species have also been reported in July in Nicaragua, in March in Costa Rica, and in June in Peru. Lactating female *A. geoffroyi* have been taken in Mexico in July, November, and December (Carter and Jones 1978; Wilson 1979). Pregnant or lactating female *A. caudifera* have been collected in January, February, May, June, and November (Wilson 1979). Pregnant or lactating female *A. cultrata* have been taken in Costa Rica, Colombia, and Peru in July and August; although this species usually bears a single young, one female gave birth to twin males (Tamsitt and Nagorsen 1982). Birth weight is about 3–5 grams (Hayssen, Van Tienhoven, and Van Tienhoven 1993). One specimen of *A. geoffroyi* was still living after 10 years in captivity (Jones 1982).

Although these bats are widely distributed and not uncommon, their dependence on forests and caves makes them especially susceptible to the intensifying habitat disruption that is occurring in much of Latin America. The IUCN classifies one species, *A. latidens*, as near threatened.

CHIROPTERA; PHYLLOSTOMIDAE; **Genus SCLERONYCTERIS**
Thomas, 1912

The single species, *S. ega*, apparently is known only by two specimens collected in the upper Orinoco drainage in the state of Amazonas, southern Venezuela, and by three specimens from the adjacent state of Amazonas in northwestern Brazil (Ochoa G. et al. 1993).

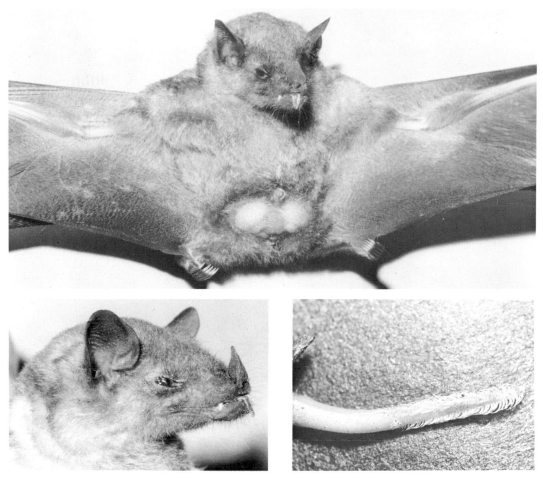

Geoffroy's long-nosed bat (*Anoura* sp.), photos by Merlin D. Tuttle. Tongue photo by P. Morris.

Measurements of the type specimen, an adult female, are as follow: head and body length, 57 mm; tail length, 6 mm; and forearm length, 35 mm. The coloration of the upper parts is dark brown; the underparts are light brown. This bat is similar to *Choeronycteris* and *Choeroniscus*, but

Scleronycteris ega, photo by Merlin D. Tuttle, Bat Conservation International.

the molar and premolar teeth are more normal in structure. The chin is unusually prominent, projecting both forward and downward. Lower incisors are not present.

One of the specimens from Venezuela was netted in a yard near a stream in an evergreen forest at 135 meters (Handley 1976); the other was caught in a primary forest at 190 meters (Ochoa G. et al. 1993). According to Gardner (1977), the diet probably consists of fruit, pollen, nectar, and insects. The IUCN classifies *S. ega* as vulnerable, noting that its restricted known habitat is expected to decline.

CHIROPTERA; PHYLLOSTOMIDAE; **Genus LICHONYCTERIS**
Thomas, 1895

The single species, *L. obscura,* occurs from Guatemala and Belize to French Guiana, eastern Brazil, and central Bolivia (Anderson, Koopman, and Creighton 1982; Brosset and Charles-Dominique 1990; Gardner 1976; Hill 1985*b;* Jones and Carter 1976; Taddei and Pedro 1993).

Head and body length is about 50–55 mm, tail length is about 8–10 mm, and forearm length is about 33 mm. Hayssen, Van Tienhoven, and Van Tienhoven (1993) listed weight as 5–8 grams. Coloration is brownish with a yel-

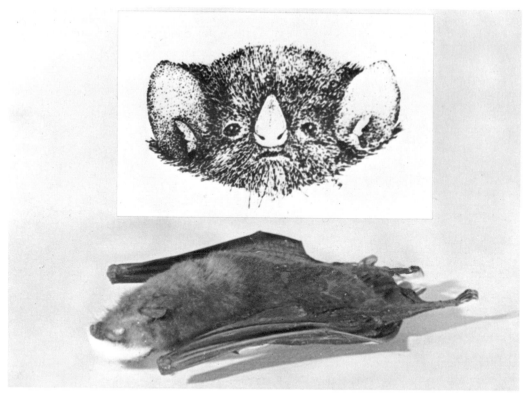

Lichonycteris sp., photo by Howard E. Uible of specimen in U.S. National Museum of Natural History. Inset: face *(L. obscura)*, photo from *Bull. Amer. Mus. Nat. Hist.*, "Mammals of Costa Rica," G. G. Goodwin.

lowish suffusion, or dark brown. The muzzle is elongate, the long tongue is supplied with papillae, and the cheek teeth are elongate and slender. The interfemoral membrane is well developed. This genus differs from *Leptonycteris* in not having lower incisor teeth and in features of the upper incisors.

Specimens have been collected in tropical evergreen forest (Handley 1976; Tuttle 1970). The features of the muzzle, tongue, and teeth suggest that this genus feeds on nectar and fruit. This bat is assumed to pollinate some of the night-blooming plants. Gardner, LaVal, and Wilson (1970) reported that a lactating female *L. obscura* was collected simultaneously with a flying juvenile male on 9 January in Costa Rica. Another female contained a 14-mm embryo on 27 March. Two pregnant females were taken in Guatemala in February (Wilson 1979). A pregnant female with a 13-mm embryo was taken in eastern Brazil on 8 August (Taddei and Pedro 1993).

CHIROPTERA; PHYLLOSTOMIDAE; **Genus**
HYLONYCTERIS
Thomas, 1903

Underwood's Long-tongued Bat

The single species, *H. underwoodi*, occurs from Nayarit in western Mexico to Panama (Jones and Carter 1976; Ramirez-Pulido and Lopez-Forment 1979).

Head and body length is 49–72 mm, tail length is 6–11 mm, and forearm length is 31–36 mm (Eisenberg 1989). Jones and Homan (1974) reported that six males weighed

6–7 grams and two females weighed 8 and 9 grams. The color is uniformly dark gray above and below; or dark brown above, with the crown of the head darker, and slightly paler below.

The muzzle is elongate, the long tongue is supplied with papillae, the cheek teeth are elongate and slender, and the lower incisors are lacking. *Hylonycteris* has three upper and three lower molars on each side, whereas *Lichonycteris* has two upper and two lower molars on each side. The interfemoral membrane is well developed. This bat resembles *Choeronycteris* and *Choeroniscus* but differs from those genera in features of the skull.

Phillips and Jones (1971) collected specimens in dense

Underwood's long-tongued bat *(Hylonycteris underwoodi)*, photo by Merlin D. Tuttle, Bat Conservation International.

forest in Jalisco, Mexico. LaVal (1977) found groups of two and eight individuals roosting, respectively, under a log bridge and in a hollow log. This bat also roosts in caves and tunnels, apparently in small numbers. According to Gardner (1977), the diet consists of pollen, nectar, fruit, and occasionally insects. Data summarized by Wilson (1979) indicate that there is a bimodal reproductive pattern in Costa Rica, with pregnancies reported there from January to April and from August to November. Pregnant or lactating females have been found in Mexico and Guatemala in March, May, September, and November. A single young is produced at a time (Jones and Homan 1974).

CHIROPTERA; PHYLLOSTOMIDAE; Genus CHOERONISCUS
Thomas, 1928

There are four species (Anderson, Koopman, and Creighton 1982; Jones and Carter 1976, 1979; Koopman 1978*b*, 1982*a*, and *in* Wilson and Reeder 1993; Webster and McGillivray 1984; Williams and Genoways 1980*b*):

C. godmani, Sinaloa (western Mexico) to Colombia and Surinam;
C. minor, Colombia, Venezuela, Guyana, Surinam, French Guiana, Ecuador, Peru, Bolivia, Brazil;

C. intermedius, eastern Peru, northern Brazil, Guianas, Trinidad;
C. periosus, Pacific coast of Colombia, western Ecuador.

Head and body length is about 50–55 mm, tail length is about 12 mm, and forearm length is 32–38 mm. Hayssen, Van Tienhoven, and Van Tienhoven (1993) listed weight as 5–9 grams. LaVal and Fitch (1977) reported an average weight of 7.9 grams for a series of specimens of *C. godmani* from Costa Rica. Coloration is usually uniformly dark brown above and somewhat paler below, but the fur of the back may be bicolored.

This genus is characterized by its small size, uniformly dark color, exceptionally long muzzle, small triangular nose leaf, small ears, and short tail. The rostrum is much elongated but less than half the length of the skull. The long tongue is supplied with papillae at its tip, and the cheek teeth are narrow and elongate. There are no lower incisors.

In Venezuela, Handley (1976) collected specimens of *C. godmani* and *C. minor* mostly in moist parts of tropical evergreen forest. The only record of roosting habits appears to be the finding of eight individuals, two males and six females, on the underside of a fallen log over a stream. The diet probably consists of pollen, nectar, fruit, and insects (Gardner 1977). Pregnant female *C. godmani* have been found in Sinaloa in July, in Nicaragua in March, and in Costa Rica in December, January, February, and March (LaVal and Fitch 1977; Wilson 1979). A lactating female *C. god-*

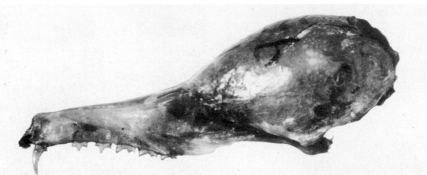

Godman's long-nosed bat (*Choeroniscus* sp.), photo by Merlin D. Tuttle. Skull of Godman's long-nosed bat *(C. minor)*, photo from British Museum (Natural History).

mani was taken in Oaxaca, Mexico, in May, and a pregnant *C. intermedius* was taken in Trinidad in August. Ascorra, Gorchov, and Cornejo (1993) collected lactating *C. intermedius* in northeastern Peru during April and June. There is a single young (Hayssen, Van Tienhoven, and Van Tienhoven 1993). *C. periosus* is known to occur only in a small area that is experiencing intensive habitat destruction; the IUCN classifies it as vulnerable and *C. godmani* and *C. intermedius* as near threatened.

CHIROPTERA; PHYLLOSTOMIDAE; **Genus**
CHOERONYCTERIS
Tschudi, 1844

Mexican Long-nosed Bat, or Hog-nosed Bat

The single species, *C. mexicana*, ranges from the southern part of the southwestern United States to Honduras and Guatemala (Jones and Carter 1976).

Head and body length is about 55–80 mm, tail length is 10–16 mm, forearm length is about 43–47 mm, and weight is usually 10–20 grams. The coloration has been described as dark brown above and paler brown below. In some individuals the males are reddish brown and the females are sooty, both being lighter underneath.

This genus resembles *Choeroniscus*, but the rostrum is longer and more than half the length of the skull. As in some other bats of the subfamily Glossophaginae, the lower incisors are absent. The muzzle is elongate, the long tongue has papillae at its tip, forming a brush, and the cheek teeth are elongate and slender.

The hog-nosed bat has been recorded mainly from arid habitats at elevations of 600–2,400 meters. It roosts in caves, tunnels, and buildings, usually in fairly well-illuminated locations, and has been found under the exposed roots of trees in a shaded area. Indirect evidence indicates

that colonies in southern Arizona migrate south into Mexico for the winter months. The diet consists of fruit, pollen, nectar, and probably insects (Gardner 1977). This bat is thought to pollinate certain plants. Captives lapped up the juices of grapes and plums but rejected the solid parts. No attempt at chewing was observed.

Colonies usually contain several dozen bats, though solitary individuals and groups of 2–12 have also been noted. The sexes usually roost together, though the females may separate from the males when the young are born and being raised (Hoffmeister 1986). In the southwestern United States and in northern and central Mexico pregnant females have been collected in February, March, June, and September, and lactating females have been taken in June, July, and August (Wilson 1979). A single young is born, and the female can carry it in flight (Barbour and Davis 1969). Birth weight is about 4.4 grams (Hayssen, Van Tienhoven, and Van Tienhoven 1993). The young grow rapidly and probably can fly by themselves within a month of birth.

There has been some concern for the survival of *Choeronycteris* in the small part of the United States within its range. The genus in general has a limited habitat in a region undergoing increased environmental disturbance. It is designated as near threatened by the IUCN.

CHIROPTERA; PHYLLOSTOMIDAE; **Genus**
MUSONYCTERIS
Schaldach and McLaughlin, 1960

Banana Bat, or Colima Long-nosed Bat

The single species, *M. harrisoni*, is known only from the states of Jalisco, Colima, Guerrero, and Michoacán in southwestern Mexico (Jones and Carter 1976, 1979; Webster et al. 1982). E. R. Hall (1981) and some other recent authors have treated this genus as a synonym of *Choeronyc-*

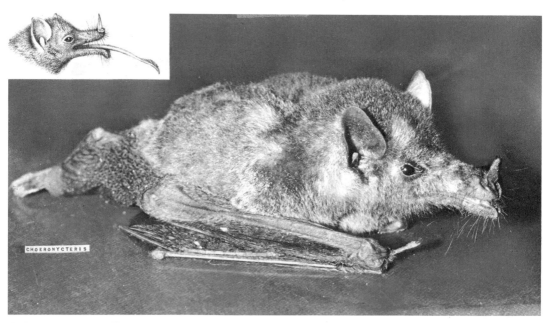

Mexican long-nosed bat *(Choeronycteris mexicana)*, photo by Lloyd G. Ingles. Inset photo from *Catalogue of the Chiroptera in the Collection of the British Museum*, G. E. Dobson.

Banana bat *(Musonycteris harrisoni)*, photo by John Bickman.

teris. On the basis of morphological and karyological features, Webster et al. (1982) considered the two genera to be closely related but distinct.

Head and body length is 70–79 mm, tail length is 8–12 mm, and forearm length is approximately 41–43 mm. The shoulder region and the middle back are brownish light drab, the lower back is dark brown, and the underparts are colored like the shoulders.

As in other bats of the subfamily Glossophaginae, the snout is long, the tongue is extensible, and the cheek teeth are narrow and elongate. This genus is distinguished by skull and dental features, particularly by the extremely long rostrum, which is more than one-half the total length of the skull and longer than the rostrum of any other related genus except the bats of the South American genus *Platalina,* from which *Musonycteris* differs in its smaller size and dental formula. *Musonycteris* seems to be most closely related to *Choeronycteris.*

Specimens of *Musonycteris* have been taken in nylon mist nets set across a small irrigation ditch in a banana grove in arid thorn forest. An extensive tropical deciduous forest was only a short distance from the thorn forest. The banana trees in the grove were blooming when the first three specimens were collected in 1958; presumably these bats were feeding on the flowers. The generic name is derived from the generic name of the banana *(Musa)* and a bat *(Nycteris).* Gardner (1977) stated that the diet probably consists of pollen, nectar, and insects. Wilson (1979) reported that two pregnant females were taken in September in Colima. *Musonycteris* is dependent on a limited habitat within a small area that is experiencing intensive environmental disturbance. It is expected to decline and is classed as vulnerable by the IUCN.

CHIROPTERA; PHYLLOSTOMIDAE; **Genus PLATALINA**
Thomas, 1928

The single species, *P. genovensium,* is now known from five localities in Peru and is represented in eight museum collections (Jiménez and Pefaur 1982).

The approximate measurements of the type specimen are: head and body length, 72 mm; tail length, 9 mm; and forearm length, 46 mm. In addition, Jiménez and Pefaur (1982) referred to two other specimens each with a forearm length of about 48 mm and to one that weighed 47 grams. In coloration *Platalina* is pale brownish throughout.

The muzzle is elongate, the extensible tongue is supplied

Platalina genovensium, photo by Jaime E. Péfaur.

with bristlelike papillae, and the cheek teeth are narrow and elongate. The short tail barely reaches the middle of the interfemoral membrane. This bat differs from the other glossophagine bats in dental and skull features. The generic name refers to the broad and spatulate nature of the inner upper incisor teeth. The nose leaf is somewhat diamond-shaped.

According to A. L. Gardner (U.S. National Museum of Natural History, pers. comm., 1988), *Platalina* has been collected in relatively dry regions and in mines and grottoes at elevations ranging from near sea level to about 2,300 meters. It has a highly derived feeding apparatus, which evolved independently of that of the specialized glossophagine genera. Although little is known of its feeding ecology, the hyoid/lingual morphology suggests that it is well adapted for the nectarivorous niche. Jiménez and Pefaur (1982) noted that it is thought to feed on the pollen and nectar of cacti. Graham (1987) collected two females in September, one of which was pregnant with a single embryo.. The IUCN, noting the highly restricted habitat of *Platalina*, now classifies it as vulnerable.

CHIROPTERA; PHYLLOSTOMIDAE; Genus BRACHYPHYLLA
Gray, 1834

There are two species (Swanepoel and Genoways 1978):

B. cavernarum, Puerto Rico, Virgin Islands, Lesser Antilles south to St. Vincent and Barbados;

B. nana, Cuba, Isle of Pines, Grand Cayman, Hispaniola, Grand Caicos.

B. nana also is known from fossils found on the islands of Jamaica, Cayman Brac, Andros, and New Providence, though apparently these remains represent populations that disappeared prior to historical time (Morgan 1989). Buden (1977) considered *B. nana* a subspecies of *B. cavernarum,* but Swanepoel and Genoways (1983a, 1983b) continued to treat the two as distinct species, noting that *B. nana* is consistently smaller than *B. cavernarum.*

Head and body length is 65–118 mm, the tail is vestigial and concealed in the base of the interfemoral membrane, and forearm length is 51–69 mm. According to Hayssen, Van Tienhoven, and Van Tienhoven (1993), weight is 45–67 grams. The upper parts are ivory yellow, and the hairs are tipped with sepia, except for patches on the neck, shoulders, and sides, which are paler. The underparts are usually brown.

The muzzle is somewhat conical in shape, and the lower lip has a V-shaped groove margined by tubercles. The nose leaf is vestigial, and the ears are small and separate. The interfemoral membrane is well developed. The molar teeth are broad and well ridged. The call of *B. cavernarum* is a strident, rasping squeak.

These bats are primarily cave dwellers, but they also have been found roosting in buildings and in a well. The roosts may be in either dark or well-illuminated locations, but emergence is relatively late in the evening. The diet is opportunistic, including many kinds of fruit as well as flowers, pollen, nectar, and insects (Gardner 1977; Nellis and Ehle 1977; Silva Taboada and Pine 1969; Swanepoel and Genoways 1978).

Nellis and Ehle (1977) reported a colony of about 5,000 *B. cavernarum* on St. Croix Island. In Cuba colonies of 2,000–3,000 *B. nana* are common, and five colonies have been conservatively estimated to contain 10,000 bats each

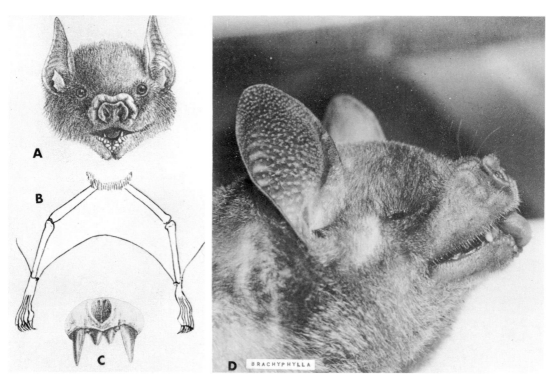

A & C. *Brachyphylla cavernarum,* photos from *Catalogue of the Chiroptera in the Collection of the British Museum.* B. *Brachyphylla sp.,* photo from *Bats,* G. M. Allen. D. *B. nana,* photo by Albert Schwartz.

Brachyphylla cavernarum, photo by Robert J. Baker through J. Knox Jones, Jr.

(Silva Taboada and Pine 1969). About 2,000 *B. cavernarum* were killed by gassing in the ruins of an old sugar factory on St. Croix in 1956 in the mistaken belief that they were *Artibeus jamaicensis*, which did considerable damage to fruit in the area. Sexing of 330 *Brachyphylla* from this colony revealed 276 females and 63 males, all adults or subadults. Adult *B. cavernarum* generally are aggressive, hitting and biting one another, especially when feeding. On St. Croix births during the study of Nellis and Ehle (1977) occurred in a three-week period in late May or early June;

colonies observed at this time comprised females with a single young each and relatively few males and barren females. The young first flew at about two months. Other reproductive data include the taking of pregnant females in Puerto Rico in February, on St. Croix in March, and on Caicos Island in March; and lactating females in Puerto Rico in April and July, on St. Croix in April, on Montserrat in July, and on Guadeloupe in July (Baker, Genoways, and Patton 1978; Jones and Baker 1979; Wilson 1979). Klingener, Genoways, and Baker (1978) reported a lactating female *B. nana* collected in southern Haiti in August. Birth weight is around 10–15 grams (Hayssen, Van Tienhoven, and Van Tienhoven 1993). *B. nana* is designated as near threatened by the IUCN.

CHIROPTERA; PHYLLOSTOMIDAE; **Genus EROPHYLLA**
Miller, 1906

Brown Flower Bats

There are two species (E. R. Hall 1981; Jones and Carter 1976; Morgan 1989):

E. bombifrons, Hispaniola, Puerto Rico;
E. sezekorni, Bahamas, Cuba, Jamaica, Cayman Islands.

Buden (1976) considered *E. bombifrons* a subspecies of *E. sezekorni*, and this procedure was followed, with question, by Baker, August, and Steuter (1978). Koopman (*in* Wilson and Reeder 1993) stated that *E. bombifrons* probably is a distinct species, though he did not list it as such.

Head and body length is about 65–75 mm, tail length is

Brown flower bat *(Erophylla sezekorni)*, photo by Merlin D. Tuttle, Bat Conservation International.

Brown flower bat *(Erophylla bombifrons)*, photo by Robert J. Baker through J. Knox Jones, Jr.

about 12–17 mm, and forearm length is approximately 42–55 mm. In *E. sezekorni* the upper parts are pale yellowish brown or buffy and the underparts are paler; in *E. bombifrons* the upper parts are dark brown and the underparts are slightly paler.

This genus and *Phyllonycteris* externally resemble the bats related to *Glossophaga* in that the skull is long and narrow and the tongue is long, protrusible, and armed with bristlelike papillae. In *Erophylla* the nose leaf is notched or forked at the tip, the ears are separate and about as high as they are broad, the tail projects beyond the narrow interfemoral membrane, and a short but distinct calcar is present.

Erophylla has been reported to roost in caves, both in dark interior portions and on exposed surfaces where much light penetrates (Baker, August, and Steuter 1978). The diet consists of fruit, pollen, nectar, and insects (Gardner 1977). In Cuba colonies of *E. sezekorni* may range in size from a few hundred to several thousand (Silva Taboada and Pine 1969). A colony of several hundred *E. sezekorni* has been reported in a cave in the Bahamas. In this colony the bats hung, singly or in clusters, from the walls and ceiling. When observed in July this colony consisted of adult males, adult females, and young. A colony of many hundreds of *E. sezekorni* in a Cuban cave in February consisted of adult males and females. Some of the females were pregnant with small embryos. A colony possibly totaling 40 *E. bombifrons* was found in a cave on Puerto Rico in July. Apparently the adults of both sexes were present, along with young from two-thirds grown to nearly full grown. Data summarized by Wilson (1979) suggest a restricted breeding season for both *E. bombifrons* and *E. sezekorni*, with pregnant females found from February to June and lactating females from May to July. There normally appears to be a single young, but one case of twins has been reported (J. R. Tamsitt, Royal Ontario Museum, pers. comm.).

CHIROPTERA; PHYLLOSTOMIDAE; Genus PHYLLONYCTERIS
Gundlach, 1861

There are two subgenera and four species (E. R. Hall 1981; Jones and Carter 1976; Morgan and Woods 1986):

subgenus *Phyllonycteris* Gundlach, 1861

P. major, known in historical time only by remains from a fissure deposit on Antigua Island in the Lesser Antilles;
P. obtusa, Hispaniola;
P. poeyi, Cuba, Isle of Pines;

subgenus *Reithronycteris* Miller, 1898

P. aphylla, Jamaica.

P. major was included by E. R. Hall (1981) in his account of post-Columbian mammals even though the species then was known only by remains from cave deposits on Puerto Rico that now are thought to date from the late Pleistocene; *P. major* subsequently was discovered in deposits on Antigua that probably date from less than 3,000 years ago (Morgan and Woods 1986). *P. obtusa* was considered a separate species by E. R. Hall (1981) and Morgan (1989) but was treated as a subspecies of *P. poeyi* by Corbet and Hill (1991), Jones and Carter (1976), and Koopman (*in* Wilson and Reeder 1993). *P. poeyi* also is known from New Providence Island in the Bahamas, but only by remains that apparently date from about 8,000 years ago (Morgan 1989). The subgenus *Reithronycteris* was not used by E. R. Hall (1981) but was accepted by Corbet and Hill (1991) and Koopman (*in* Wilson and Reeder 1993).

In species known from the living state, head and body

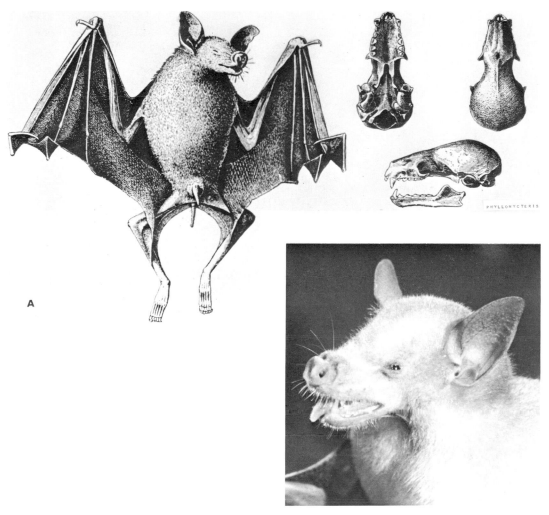

Phyllonycteris poeyi: A. Photo from *Proc. L S. Natl. Mus.*; B. Photo by Robert J. Baker through J. Knox Jones, Jr.

length is about 64–83 mm, tail length is about 7–12 mm, and forearm length is about 43–50 mm. H. F. Howe (1974) reported that three female *P. aphylla* weighed 14.0–14.8 grams. For *P. obtusa*, Klingener, Genoways, and Baker (1978) reported that two females and a male weighed 20.0–21.1 grams. *P. poeyi* is grayish white. The pelage of this species has a silky texture that produces silvery reflections under certain light. A specimen in alcohol of *P. aphylla* had light yellowish brown upper parts and underparts.

Both *Phyllonycteris* and *Erophylla* resemble other bats of the tribe Glossophagini in that the skull is long and narrow and the tongue is long, protrusible, and armed with bristlelike papillae. In *Phyllonycteris* the nose leaf is rudimentary, the ears are moderately large and separate, the tail projects beyond the narrow interfemoral membrane, and a calcar is not present. A groove on the ventral surface of the braincase in *P. aphylla* seems to be unique among mammals.

These bats roost in caves. The diet of *P. poeyi* consists of fruit, pollen, nectar, and insects (Gardner 1977). Captive *P. aphylla* thrived on a diet of bananas, mangoes, papayas, and various kinds of canned fruit nectars (Henson and Novick 1966). *P. poeyi* is gregarious, roosting by the hundreds and

apparently sometimes by the thousands. Most of its young seem to be born in June. *P. aphylla* is also colonial; a female taken in January was pregnant with one embryo (Goodwin 1970). Klingener, Genoways, and Baker (1978) reported that three female *P. obtusa* taken in southern Haiti on 17 December were pregnant with one embryo each. Both *P. aphylla* and *P. obtusa* long were known only from skeletal remains and one specimen in alcohol, but recently they were rediscovered in the living state. The only major known colony of *P. aphylla* contains a few hundred individuals and is considered vulnerable to human disturbance (McFarlane 1986). The IUCN now classifies *P. aphylla* as endangered and *P. poeyi* (including *P. obtusa*) as near threatened. *P. major* is classified as extinct by the IUCN. It apparently disappeared as a consequence of habitat destruction following the human invasion of Antigua about 3,000 years ago (Morgan and Woods 1986).

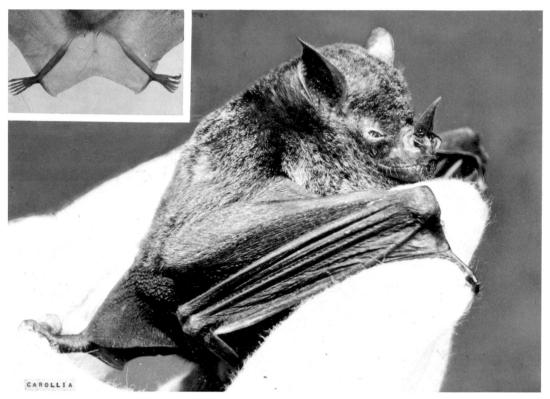

Short-tailed leaf-nosed bat *(Carollia perspicillata)*, photo by Ernest P. Walker. Inset photo by Harold Drysdale, Trinidad Regional Virus Laboratory, through A. M. Greenhall.

CHIROPTERA; PHYLLOSTOMIDAE; **Genus CAROLLIA** *Gray, 1838*

Short-tailed Leaf-nosed Bats

There are four species (Barquez and Ojeda 1992; E. R. Hall 1981; Jones and Carter 1976; Koopman *in* Wilson and Reeder 1993):

C. castanea, Honduras to Bolivia;
C. subrufa, southwestern Mexico to Nicaragua, Guyana;
C. brevicauda, San Luis Potosi (eastern Mexico) to northeastern Brazil and Bolivia;
C. perspicillata, southern Mexico to northern Argentina and southern Brazil, Trinidad and Tobago, Grenada in the Lesser Antilles.

Recorded occurrences of *C. perspicillata* on Jamaica and Redonda Island are questionable.

Head and body length is 48–65 mm, tail length is 3–14 mm, forearm length is 34–45 mm, and weight is usually 10–20 grams. Coloration is generally dark brown to rusty, but bright or pale orange specimens of *C. perspicillata* are common in some areas (Cloutier and Thomas 1992). This genus differs from *Rhinophylla* in that the lower molars are of a different form from the lower premolars and in the presence of a tail.

The species *C. subrufa* is found in relatively dry tropical deciduous forest; the other three species occur primarily in humid tropical evergreen forest (Handley 1976; Pine 1972a). *C. perspicillata* is most common in second growth and frequently forages at ground level (Cloutier and Thomas 1992). Reported roosting sites include caves, mines, rocks, culverts, hollow trees, logs, and buildings. In a radio-tracking study in Costa Rica, Heithaus and Fleming (1978) found *C. perspicillata* to disperse nightly, with each individual going to two to six feeding areas and flying an average of 4.7 km per night. The diet seems to consist mainly of fruits, including guavas, bananas, wild figs, and plantains. Sazima (1976) observed *C. perspicillata* taking nectar from passionflowers. In this species, and probably others as well, insects also seem to be an important food (Ayala and D'Alessandro 1973; Gardner 1977).

These bats roost singly, in small groups, and in colonies of several hundred to several thousand individuals. Males and females usually live together throughout the year. In studies of both wild and captive *C. perspicillata,* Porter (1978, 1979a) found that colonies are divided into heterosexual groups, or harems, comprising a single adult male and from one to eight females and their infants and that other groups are composed only of males or juveniles. Harem males appear to recruit females, vigorously defend them from other males, and exhibit territorial behavior and spatial fidelity. C. F. Williams (1986), however, collected evidence suggesting that harems are actually passive aggregations of females that form at the limited suitable roosts. The presence of harem males apparently is a result rather than a cause of female grouping. Porter (1979b) reported that *C. perspicillata* communicates through a rich variety of vocalizations, including warbles of greeting between male and female and screeches by which a harem male threatens other males and controls his females. Porter and McCracken (1983) added that harem males also guard the offspring of their harem females while the mothers are out foraging and apparently help to reunite separated offspring and mothers.

Data summarized by Wilson (1979) indicate that reproduction in *Carollia* generally fits a bimodal polyestrous pattern. Pregnant female *C. castanea* have been found in Central America in January–May and July–August and in South America in January–April and September–November. Pregnancies in *C. subrufa* have been recorded in Central America in December–May and July–October. Although pregnant female *C. perspicillata* have been collected in all months of the year, there are definite peak periods, occurring in February–May and June–August in Panama and somewhat earlier in other areas, depending on seasonal rainfall pattern. Pregnant female *C. brevicauda* have been found in every month from December to August in Mexico and Central America. Ascorra, Gorchov, and Cornejo (1993) also collected *C. brevicauda* in every month of the year in northeastern Peru but noted a parturition peak in the wet season. In a study in Costa Rica, LaVal and Fitch (1977) found all three resident species of *Carollia* to be seasonally polyestrous, with an apparent minimum in reproductive activity occurring late in the wet season, from October to early January. Usually there is a single young. Birth weight is about 5 grams (Hayssen, Van Tienhoven, and Van Tienhoven 1993). In a study of captive *C. perspicillata*, Laska (1990) found postpartum sexual receptivity to last 28–32 days, gestation 105–25 days, and interbirth interval 120–50 days; young females first gave birth at about 1 year. Fleming (1988) determined that male *C. perspicillata* attain sexual maturity at between 1 and 2 years. Nearly two-thirds of births are of males, though higher male mortality evidently leads to a 1:1 sexual ratio among adults. Average life expectancy in this species is 2.6 years, though a few individuals survive for more than 10 years. According to Jones (1982), a specimen of *C. perspicillata* was still living after 12 years and 5 months in captivity.

These bats, especially *C. perspicillata*, have the reputation of being pests (Pine 1972a). Apparently they have increased in numbers through the availability of crops as a source of food and buildings as roosting sites. Considerable damage has been reported to mangoes, coffee beans, guavas, pawpaws, almonds, and other cultivated products.

CHIROPTERA; PHYLLOSTOMIDAE; **Genus RHINOPHYLLA**
Peters, 1865

There are three species (Alberico 1987; Anderson, Koopman, and Creighton 1982; Baud 1982; Jones and Carter 1976; Webster and Jones 1984c; Webster and McGillivray 1984):

R. pumilio, Colombia, Venezuela, Guyana, Surinam, French Guiana, northern Brazil, eastern Ecuador and Peru, northeastern Bolivia;
R. alethina, western Colombia, northwestern Ecuador;
R. fischerae, Amazonian region of Peru and Brazil, and adjacent parts of Colombia, Ecuador, and Venezuela.

Head and body length is about 43–58 mm, there is no tail, and forearm length is about 29–37 mm. Eisenberg (1989) indicated that weight is about 9–16 grams. The most common coloration is grayish brown. There is a nose leaf. The members of this genus differ from *Carollia* in that the lower molars are not different in form from the lower premolars and in the absence of a tail.

Rhinophylla sp., photo by P. Morris.

In Venezuela, Handley (1976) collected *R. pumilio* mainly in moist parts of tropical evergreen forests. In Peru, Tuttle (1970) took both *R. pumilio* and *R. fischerae* most often in or near gardens where bananas and papayas were grown. Gardner (1977) stated that bats of this genus all are probably frugivores but may consume insects as well. Pregnant or lactating *R. pumilio* have been found in April, May, June, July, and December (Wilson 1979). Pregnant female *R. fischerae*, each with a single embryo, have been collected in Peru in June and July (Graham 1987). *R. alethina* and *R. fischerae* are designated as near threatened by the IUCN.

CHIROPTERA; PHYLLOSTOMIDAE; **Genus STURNIRA**
Gray, 1842

Yellow-shouldered Bats, or American Epauleted Bats

There are 2 subgenera and 13 named species (Alberico 1987; Anderson, Koopman, and Creighton 1982; Barquez and Ojeda 1992; Davis 1980; Gardner and O'Neill 1969, 1971; Handley 1976; Jones and Carter 1976; Koopman *in* Wilson and Reeder 1993; Lemke et al. 1982; Marques and Oren 1987; McCarthy, Barkley, and Albuja V. 1991; Molinari and Soriano 1987; Pacheco and Patterson 1991, 1992; Soriano and Molinari 1987; Tamsitt, Cadena, and Villarraga 1986):

subgenus *Sturnira* Gray, 1842

S. lilium, Sonora and Tamaulipas (Mexico) to northern Argentina and possibly Chile, southern Lesser Antilles, Trinidad, possibly Jamaica;
S. luisi, Costa Rica to Peru;
S. thomasi, Guadeloupe Island in the Lesser Antilles;
S. tildae, northern and central South America, Trinidad;
S. mordax, Costa Rica, Panama;

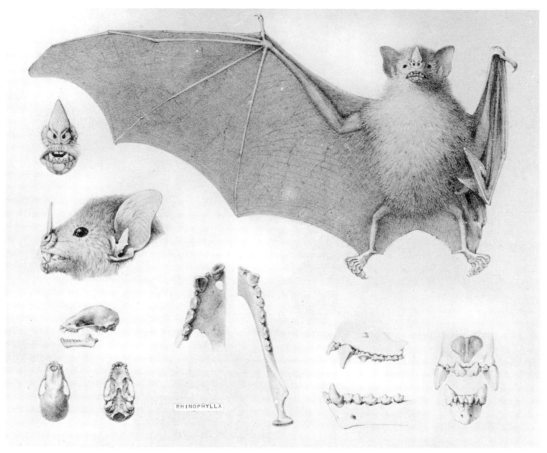

Rhinophylla pumilio, photo from *Die Fledermäuse des Berliner Museums für Naturkunde*, Wilhelm K. Peters.

S. ludovici, Sinaloa and Tamaulipas (Mexico) to Guyana and Ecuador;

S. oporaphilum, Ecuador, Peru, Bolivia, northwestern Argentina;

S. bogotensis, Colombia, Ecuador, Peru;

S. erythromos, Venezuela, Colombia, Ecuador, Peru, Bolivia, northern Argentina;

S. magna, Amazonian region of Colombia, Ecuador, Peru, and Bolivia;

S. aratathomasi, western Venezuela, western Colombia, western Ecuador, northern Peru;

subgenus *Corvira* Thomas, 1915

S. bidens, western Venezuela, Colombia, Ecuador, Peru, northern Brazil;

S. nana, central Peru.

Pacheco and Patterson (1991) indicated that there is an additional, undescribed species in Ecuador. Koopman (*in* Wilson and Reeder 1993) suggested that *S. oporaphilum* includes *S. bogotensis,* but the two were found to be distinct by Pacheco and Patterson (1991, 1992).

Head and body length is 51–101 mm, there is no external tail, and forearm length is 34–61 mm. Eisenberg (1989) listed the following approximate average weights: *S. lilium,* 19 grams, *S. ludovici,* 25 grams, *S. tildae,* 26 grams, *S. bogotensis,* 20 grams, *S. erythromos,* 15 grams, and *S. bidens,*

17 grams. Gardner (1976) reported weight in four specimens of *S. magna* as 41.0–44.5 grams. Thomas and McMurray (1974) gave the weight of *S. aratathomasi* as 46.8–67.1 grams. Except in *S. bidens* and *S. nana* conspicuous tufts of stiff yellowish or reddish hairs are present near the front of the shoulders in males. The general coloration is pinkish buff with a brown tinge, dark ochraceous brown, dark grayish brown, or dark brown; the underparts are usually paler. Some forms have two color phases: bright cinnamon brown and dull pale gray.

The ears are short, the nose leaf is normal, the interfemoral membrane is narrow and furred, and the hind limbs and feet are haired to the claws. The molar teeth are longitudinally grooved with lateral cusps. In most species there are four lower incisors, but in *S. bidens* and *S. nana* there are only two.

In Venezuela, Handley (1976) found the following habitat conditions to prevail: *S. bidens,* high cloud forest; *S. erythromos,* moist parts of mountain forests; *S. lilium,* moist parts of forests and open areas; *S. ludovici,* moist places in forested mountains and foothills; and *S. tildae,* humid lowland forests. *S. magna* is found in forests at elevations of 200–2,300 meters (Tamsitt and Hauser 1985). Roosting sites of the genus include hollow trees and buildings. The diet consists mostly of fruit.

LaVal and Fitch (1977) reported *S. ludovici* to be seasonally polyestrous in Costa Rica, with several peaks of pregnancy and lactation from April to August. Data summa-

Yellow-shouldered bat *(Sturnira lilium)*, photo by Merlin D. Tuttle, Bat Conservation International.

rized by Wilson (1979) indicate that reproduction in this species, as well as in *S. lilium,* follows a pattern of bimodal polyestry. In Colombia pregnant female *S. ludovici* have been reported from February to May and from August to December, and lactating females have been found from May to September. In Jalisco, Mexico, pregnant or lactating female *S. ludovici* have been taken in April, May, July, August, and November. Pregnant and lactating female *S. lilium* have been collected in all months of the year, but in any one region there apparently are peaks of reproductive activity. Other data listed by Wilson (1979) are: pregnant *S. tildae* in March on Trinidad and in July in Brazil; pregnant or lactating *S. mordax* in Costa Rica in February, May, and August; and pregnant *S. erythromos* in Colombia in December and in Peru in August. According to Baker, Genoways, and Patton (1978), two lactating female *S. thomasi* were collected on Guadeloupe in July. Ascorra, Gorchov, and Cornejo (1993) took pregnant *S. tildae* in

northeastern Peru in October and December. Data compiled by Hayssen, Van Tienhoven, and Van Tienhoven (1993) indicate that there is a single young, its birth weight being about 2.2 grams in *S. tildae* and 13.6 grams in *S. aratathomasi.*

Several species of *Sturnira* occur in limited areas undergoing severe environmental disturbance. The IUCN now classifies *S. thomasi* as endangered, *S. nana* as vulnerable, and *S. mordax, S. magna. S. aratathomasi,* and *S. bidens* as near threatened.

Tent-building bat *(Uroderma bilobatum)*, photo by Cory T. de Carvalho. Inset: *Uroderma* sp., photo by P. Morris.

CHIROPTERA; PHYLLOSTOMIDAE; Genus **URODERMA**
Peters, 1865

Tent-building Bats

There are two species (Anderson, Koopman, and Creighton 1982; Jones and Carter 1976, 1979; Ramirez-Pulido and Lopez-Forment 1979):

U. bilobatum, southern Mexico to Bolivia and
  southeastern Brazil, Trinidad;
U. magnirostrum, southern Mexico to Brazil and northern
  Bolivia.

Head and body length is about 54–74 mm, there is no external tail, forearm length is 39–45 mm, and adult weight is usually 13–21 grams. The coloration of the head and body is grayish brown. The ear margins are yellowish white, there are four white facial stripes, and there is a narrow white line down the middle of the lower back.

These bats may be recognized by "the naked or finely haired posterior border of the interfemoral membrane, in combination with the length of the forearm." The nose leaf consists of two parts: a horseshoe-shaped basal portion with unique rounded lobes on either side and an erect, lancet-shaped, finely toothed portion.

In Venezuela, Handley (1976) collected *U. bilobatum* in a variety of forested and nonforested habitats and *U. magnirostrum* mostly in moist, open areas. *U. bilobatum* roosts in the leaves of trees, such as in the fronds of palms and the leaves of bananas. Both species of *Uroderma* probably are mainly frugivorous, but they may also consume pollen, nectar, and insects found in flowers and fruit (Gardner 1977).

Uroderma, especially *U. bilobatum,* has become well known for the construction and utilization of "tents" within living vegetation. Thomas Barbour, who first recorded tent-making in *U. bilobatum,* thought that this habit was rare and had developed originally in response to "the superior shade producing quality" of the leaves of introduced palms and then spread to include the native palms. The introduced palms in Panama, where Barbour made his observations, have large, stiff, palmate fronds, whereas the native palms have pinnate leaves. More recent studies indicate that tent-making is common and widespread, involving both palmate and pinnate palm leaves, many other kinds of plants, and a variety of construction techniques. Kunz, Fujita, et al. (1994) recorded the utilization by *U. bilobatum* of the plants of five plant families, in addition to the Palmae, and of eight different styles of tents: conical, formed when a cluster of leaflets collapses downward after the bat chews through the petioles; palmate umbrella, a rooflike shelter formed when the terminal portions of the leaflets of a fan-shaped palm frond are collapsed by chewing along a semicircular line through the more centrally located veins of the leaflets; pinnate, when an elongated enclosure is formed by biting through the leaflets along both sides of a pinnate palm frond; apical, a small enclosure formed by causing the outer portions of a simple leaf to droop downwards; bifid, formed by cutting across the middle of a broad leaf so that the distal portion drops down; paradox, formed by a V-shaped cut near the base of a broad leaf, which causes the distal half to fold down; boat, formed by chewing through the veins all along both sides of the midrib of a large *Heliconia* leaf, thereby causing the sides of the leaf to collapse downward; stem, a bell-shaped cavity made by severing the stems of vines and small branches. Choe (1994) reported an additional type of tent constructed by *U. bilobatum* that involved multiple leaves: the bat severs the midribs of the leaves of *Coccoloba* so that they collapse and form a conical enclosure around the stem.

Tent construction can be a long and arduous process involving the severance of many large plant veins, but there are several benefits. In addition to providing shelter against sun, wind, and rain, tents may have an antipredator function, since they provide camouflage but offer a view through the open bottom, are commonly composed of tough and spiny materials, and allow the occupants to detect movements in surrounding vegetation (Timm and Clauson 1990). The bitten spots afford good footholds, but *U. bilobatum* also hangs from above and below the cut zone. The distal parts of palm fronds eventually dry up and drop off, and a new leaf has to be cut. Timm (1987), however, stated that since the bats bite the tissue between veins along the midrib and leave the midrib and most veins intact, they do not kill the leaves. The resulting tent is thus available for an extended period; one was observed in use for more than 60 days.

Timm and Lewis (1991) and Lewis (1992) studied a population of *U. bilobatum* roosting in a grove of coconut palms *(Cocos nucifera)* in Costa Rica. Of 56 trees in the grove, 23 contained tents constructed by the bats. These tents were of the pinnate type—elongate and rooflike—and were located at an average height of about 4 meters. The bats arrived at the grove in June, during the early rainy season, apparently to form maternity colonies. They roosted alone or in groups of up to 17 individuals. Most of the females usually roosted together in just one or two of the tents. Most of the males roosted alone, but two were found together with the large groups of females. All of the females were pregnant, and they gave birth fairly synchronously in early July. Previous observations indicate that nursing females roost in groups of up to 40 individuals and do not carry their young with them during their feeding flights. The males at this time roost singly or in groups smaller than those formed by the females. Data summarized by Wilson (1979) suggest that reproduction in this genus follows a pattern of bimodal polyestry, though Ascorra, Gorchov, and Cornejo (1993) noted breeding only in the wet season. Pregnant female *U. bilobatum* have been reported from December to July in Central America and in January, July, August, September, October, November, and December in South America. Pregnant *U. magnirostrum* have been taken in El Salvador in June, in Nicaragua in March and July, in Bolivia in September, and in Brazil in June. Data compiled by Hayssen, Van Tienhoven, and Van Tienhoven (1993) indicate that estimated gestation is 4–5 months, litter size is 1 young, and independence is attained at 1 month.

CHIROPTERA; PHYLLOSTOMIDAE; **Genus**
PLATYRRHINUS
Saussure, 1860

White-lined Bats

There are 10 species (Alberico 1990*b*; Alberico and Velasco 1991; Anderson 1991; Anderson, Koopman, and Creighton 1982; Barquez and Olrog 1980; Brosset and Charles-Dominique 1990; Eisenberg 1989; Handley 1976; Jones and Carter 1976, 1979; Koopman 1982*a* and *in* Wilson and Reeder 1993; Lemke et al. 1982; Myers and Wetzel 1983; Williams and Genoways 1980*b*; Williams, Genoways, and Groen 1983):

P. vittatus, Costa Rica to Bolivia and Venezuela;
P. infuscus, Colombia to western Brazil and Bolivia;
P. aurarius, mountains of southwestern Colombia, Guiana Highlands of Venezuela, central Surinam;

P. chocoensis, western Colombia;
P. dorsalis, Panama to Bolivia;
P. umbratus, Panama, western Colombia, northern Venezuela;
P. brachycephalus, Colombia, Venezuela, Guyana, Surinam, French Guiana, Amazonian Brazil, Ecuador, Peru, northern Bolivia;
P. helleri, southern Mexico to Bolivia and Brazil, Trinidad;
P. lineatus, Colombia and French Guiana to eastern Brazil and northern Argentina;
P. recifinus, eastern Brazil.

Although E. R. Hall (1981) used the name *Platyrrhinus* for this genus, most other authorities long have used *Vampyrops* Peters, 1865. Recently, however, the nomenclatural validity of *Platyrrhinus* has been upheld (Alberico and Velasco 1991; Gardner and Ferrell 1990; Koopman *in* Wilson and Reeder 1993). *P. umbratus* was recognized as a species distinct from *P. dorsalis* by Corbet and Hill (1991) and Koopman (*in* Wilson and Reeder 1993) but not by Jones, Arroyo-Cabrales, and Owen (1988). *P. nigellus* of Peru was listed as a separate species by Owen (1987) but was treated as a subspecies of *P. lineatus* by Willig and Hollander (1987) and most other recent authorities. Alberico (1990*b*) indicated that two unnamed species related to *P. infuscus* are found in western Colombia.

Head and body length is 48–98 mm, there is no external tail, and forearm length is 36–64 mm. Eisenberg (1989) listed the following approximate average weights: *P. aurarius,* 34 grams; *P. umbratus,* 24 grams; *P. brachycephalus,* 15 grams; and *P. helleri,* 14 grams. LaVal and Fitch (1977) listed an average weight of 54.7 grams for *P. vittatus* in Costa Rica. Coloration in the genus is usually dark brown or black. Pale facial and dorsal stripes are usually present. The narrow facial stripes extend from the side of the nose to the ears, and the white or gray dorsal stripe extends from between the ears to the tail membrane. The nose leaf is well developed, and the ears are rounded and of medium size. The interfemoral membrane is narrow and fringed with hair on its posterior edge.

In Venezuela, Handley (1976) collected several species of *Platyrrhinus* mainly in moist, forested areas. These bats have been found under roots on canyon walls and stream banks, in moss on a tree, in clusters of leaves in the tops of trees and on the branches of trees, in caves, and in buildings. *P. lineatus* has been observed in Brazil in November and December roosting in groups of about 3–10 individuals in leaves in the tops of mango trees and on the branches of these trees about halfway up on windy or rainy days. The females with young often roost on thick branches of the trees 3–5 meters above the ground, apart from the males; the female *P. lineatus* without young were generally associated with males, in separate pairs. The diet consists mainly of fruit but includes some insects (Gardner 1977). In Brazil, Sazima (1976) observed *P. lineatus* taking nectar from flowers.

Willig and Hollander (1987) suggested that individual adult male *P. lineatus* maintain harems of about 6–20 females. This species evidently breeds throughout the year in southeastern Brazil; females are polyestrous and may be simultaneously pregnant and lactating. In northeastern Brazil, however, pregnancies were found only from the early dry season (July) to the end of the rainy season (February–March), and there was a bimodal distribution of breeding in this period.

Wilson (1979) listed the following additional reproductive data: *P. vittatus,* pregnant or lactating females taken in Costa Rica from March to July, pregnant females taken in Colombia in May and October; *P. dorsalis,* pregnant or lac-

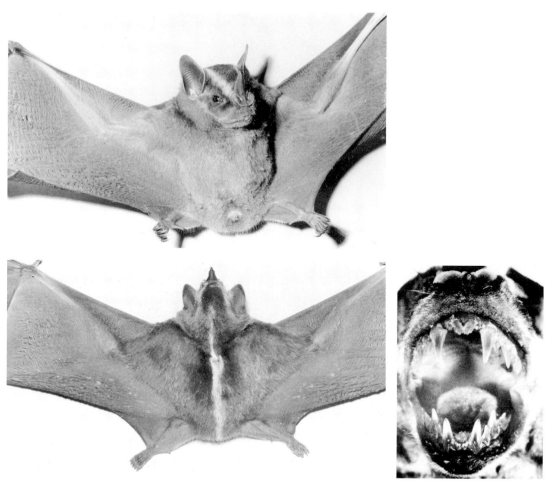

White-lined bat (*Platyrrhinus* sp.), photos by Merlin D. Tuttle. Inset: teeth of white-lined bat *(P. helleri)*, photo by Harold Drysdale, Trinidad Regional Virus Laboratory, through A. M. Greenhall.

tating females found in Colombia from November to August; *P. brachycephalus,* pregnant females taken in February in Venezuela and in August in Peru, a lactating female taken in October in Venezuela; *P. helleri,* pregnant females taken in southern Mexico and Central America from March to August and in January, pregnant or lactating females taken in Colombia from January to August, pregnant females found in French Guiana in August and September, and pregnant females found in Peru in July and August; and *P. lineatus,* pregnant females found in Brazil in January, March, and December. Davis and Dixon (1976) collected pregnant female *P. helleri* and *P. brachycephalus,* each with a single large embryo, in Peru from 31 October to 8 November. Ferrell and Wilson (1991) noted that available data suggest that *P. helleri* is bimodally polyestrous, with a first birth period beginning during the second half of the dry season (March or April) and a second beginning in the middle of the wet season (July or August). The gestation period of *P. lineatus* has been estimated at 3.5 months and its birth weight has been given as 8.3 grams (Hayssen, Van Tienhoven, and Van Tienhoven 1993). Jones (1982) reported that a specimen of *P. lineatus* was still living after 10 years and 2 months in captivity.

The IUCN classifies *P. chocoensis* as endangered, noting that it has declined by at least 50 percent in the past decade through destruction of its limited habitat and that fewer than 2,500 mature individuals may survive. *P. recifinus,* also susceptible to a severe decline, is classed as vulnerable, and *P. infuscus, P. aurarius,* and *P. umbratus* are designated as near threatened.

CHIROPTERA; PHYLLOSTOMIDAE; **Genus VAMPYRODES** *Thomas, 1900*

Great Stripe-faced Bat

The single species, *V. caraccioli,* occurs from southern Mexico to northern Brazil and Bolivia and on Trinidad and Tobago (Anderson, Koopman, and Creighton 1982; Carter et al. 1981; E. R. Hall 1981; Jones and Carter 1976; Willis, Willig, and Jones 1990). The name *V. major* sometimes is used for this species.

Head and body length is about 65–77 mm, there is no external tail, and forearm length is 45–57 mm. Weight is usually 30–40 grams (Willis, Willig, and Jones 1990). Coloration is uniformly grayish brown above and below, or cinnamon brown above and grayish brown below. There

Great stripe-faced bat *(Vampyrodes caraccioli)*, photo by Merlin D. Tuttle.

are four white facial stripes, and a white line extends from the top of the head down the midline of the back. This genus is similar to *Platyrrhinus* but has only two molar teeth on each side instead of three; these differ conspicuously from each other in form owing to the reduction of the metaconid in the second molar to a mere trace.

In Venezuela, Handley (1976) collected this bat mainly in moist areas of tropical evergreen forest. Individuals have been found hanging from the underside of palm fronds on Trinidad. Four males were obtained on Tobago while hanging in a cluster in a shrub in virgin forest; another individual was roosting a short distance from the cluster. Several specimens have been taken in Oaxaca, Mexico, by the use of mist nets placed across small watered arroyos in dense rainforest. This bat is considered frugivorous in diet (Gardner 1977).

On forested Barro Colorado Island in Panama, Morrison (1980) found *Vampyrodes* to roost by day in dense foliage and to emerge about 45 minutes after sunset. Its initial flight to a fruiting tree covered an average of 850 meters, and it subsequently visited several more feeding areas in the course of the night. Roosts were changed almost daily, and foraging was curtailed during periods of bright moonlight, apparently as a means of avoiding predation. Individuals roosted and foraged in relatively stable groups comprising two or three adult females, their young, and probably a single adult male.

Wilson (1979) listed records of pregnant or lactating females for southern Mexico and Central America in January, June, July, and August; for Colombia from January to August and in October and November; and for Peru in July. Davis and Dixon (1976) collected pregnant females, each with a single large embryo, in Peru from 31 October to 8 November. Willis, Willig, and Jones (1990) noted that available data suggest a pattern of two sequential breeding periods followed by a quiescent period from July to September and that females have a postpartum estrus.

CHIROPTERA; PHYLLOSTOMIDAE; **Genus VAMPYRESSA**
Thomas, 1900

Yellow-eared Bats

There are three subgenera and five species (Anderson 1991; Anderson and Webster 1983; Ascorra, Gorchov, and Cornejo 1993; Brosset and Charles-Dominique 1990; Genoways and Williams 1979*b*; Jones and Carter 1976; Koopman *in* Wilson and Reeder 1993; Lemke et al. 1982; Myers, White, and Stallings 1983; Williams and Genoways 1980*b*):

subgenus *Vampyressa* Thomas, 1900

V. pusilla, southern Mexico to Peru, Paraguay, southeastern Brazil, and French Guiana;
V. melissa, eastern slopes of the Andes in southwestern Colombia and Peru, French Guiana;

subgenus *Metavampyressa* Peterson, 1968

V. nymphaea, Nicaragua to western Colombia;
V. brocki, Colombia, northeastern Peru, Guyana, Surinam;

subgenus *Vampyriscus* Thomas, 1900

V. bidens, Colombia, Venezuela, Guyana, Surinam, French Guiana, northern Brazil, Ecuador, Peru, Bolivia.

Detailed morphometric analyses by Owen (1987, 1988) suggest a strikingly different arrangement, with *V. pusilla*, *V. brocki*, and *V. bidens* being more closely related to one another than to the other species. In contrast, Davis (1975) concluded that the species *V. nymphaea* and *V. brocki* are sufficiently similar to *V. bidens* to warrant inclusion in

Vampyressa bidens, photo by Merlin D. Tuttle.

Vampyriscus and that the subgenus *Metavampyressa* is therefore unwarranted; such an arrangement was followed by Corbet and Hill (1991), though Koopman (*in* Wilson and Reeder 1993) continued to recognize *Metavampyressa*. The genus *Mesophylla* (see account thereof) often is included in *Vampyriscus*.

Head and body length is 43 to about 65 mm, there is no external tail, and forearm length is 30–38 mm. Eisenberg (1989) listed average weights of about 9 grams for *V. pusilla* and 12 grams for *V. bidens*. The coloration is smoky gray, whitish brown to pale brown, or dark brown. White facial stripes are present in some species. A dorsal stripe is present in *V. nymphaea* and *V. bidens*. The short, rounded ears have yellow margins. There are two upper and two lower molar teeth on each side. Each of the two middle upper incisors has two lobes at the end. In the subgenus *Vampyriscus* there usually are only two lower incisors.

In Venezuela, Handley (1976) collected *V. pusilla* and *V. bidens* mostly near streams and in other moist parts of evergreen forest. These bats probably roost in trees and shrubs. The diet consists largely of fruit (Gardner 1977). Timm (1984) reported that *V. pusilla* constructs tents from the leaves of *Philodendron* in much the same manner as described above in the account of *Uroderma*. Brooke (1987) described tent-making in *V. nymphaea* and found that the number of individuals per tent varied from two to seven. These groups evidently represented harems, as they consisted of a single adult male, one to three pregnant or lactating females, and their young.

Wilson (1979) listed the following reproductive data: *V. pusilla*, pregnant or lactating females collected in Mexico and Central America from January to March and in July and August and in Colombia in March, April, May, July, August, and November; *V. nymphaea*, pregnant females taken in Central America in February and April, pregnant or lactating females taken in Colombia from October to August; *V. brocki*, one lactating and two pregnant females taken in Colombia in June and July; and *V. bidens*, pregnant females taken in Peru in December. Davis and Dixon (1976)

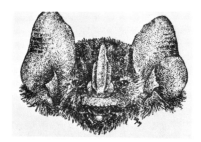

Yellow-eared bat *(Vampyressa pusilla):* Top, drawing of face from *Bull. Amer. Mus. Nat. Hist.*, "The Mammals of Costa Rica," G. G. Goodwin; Bottom, photo by Merlin D. Tuttle.

reported numerous pregnant female *V. pusilla* and *V. bidens,* each with a single large embryo, taken in Peru from 31 October to 8 November. Ascorra, Gorchov, and Cornejo (1993) collected a lactating female *V. pusilla* in northeastern Peru in April. Genoways and Williams (1979b) collected a pregnant female *V. bidens* in Surinam on 11 August.

The IUCN classifies *V. brocki* as vulnerable, noting that its restricted population is subject to rapid decline through human disturbance. *V. melissa* and *V. bidens* are designated as near threatened.

CHIROPTERA; PHYLLOSTOMIDAE; **Genus**
MESOPHYLLA
Thomas, 1901

Little Yellow-faced Bat

The single species, *M. macconnelli,* is found from Nicaragua to French Guiana, Amazonian Brazil, and Bolivia and on Trinidad (Anderson and Webster 1983; Brosset and Charles-Dominique 1990; E. R. Hall 1981; Jones and Carter 1976; Lemke et al. 1982; McCarthy et al. 1993). This species sometimes has been referred to the genus *Ectophylla* (Jones and Carter 1976) and sometimes retained in *Mesophylla* (E. R. Hall 1981). However, there also long were suggestions, on both karyological and morphological grounds, that the species is more closely related to *Vampyressa pusilla* than to *Ectophylla alba* (Greenbaum, Baker, and Wilson 1975; Starrett and Casebeer 1968). Morphometric analyses by Owen (1988) indicated affinity between *Vampyressa nymphaea* and *M. macconnelli* but also considerable similarity between the latter and *Ectophylla*

alba. More recently, *Mesophylla* has been accepted as a full genus by Corbet and Hill (1991), Koopman (*in* Wilson and Reeder 1993), Kunz and Pena (1992), and Van Den Bussche (1992) though not by Baker, Hood, and Honeycutt (1989).

Head and body length is about 40–50 mm, there is no tail, forearm length is 29–34 mm, and weight is about 5–9 grams. The head and anterior back are dull brownish white, darkening posteriorly to brown, the underparts are pale whitish beige, and the wings are gray brown. The ears, nose leaf, wrists, distal forearms, and thumbs are yellow. The fur is close and thick and the tail membrane is naked. Externally *M. macconnelli* resembles *Vampyressa pusilla,* but the latter species can be distinguished by its facial stripes, gray wrists and thumbs, and slight fringe of hair in the center of the tail membrane. Also, whereas the inner cusps of the second lower molar of *Vampyressa* are large, they are reduced or absent in *Mesophylla.* From *Ectophylla, Mesophylla* differs in having a leaflet behind the nose leaf, a basal lappet on the ear, a third lower molar tooth, and no longitudinal ridge on the occlusal surfaces of the second upper and lower molars (Eisenberg 1989; Emmons 1990; E. R. Hall 1981).

In Venezuela, Handley (1976) collected *Mesophylla* mostly near streams and in other moist parts of evergreen forest. Other observations suggest that such habitat is favored throughout the range of the genus (Kunz and Pena 1992). It probably roosts in trees and shrubs and feeds largely on fruit. Kunz, Fujita, et al. (1994) tabulated records of *Mesophylla* utilizing three types of tents—apical, bifid, and paradox—such as described above in the account of *Uroderma.* Koepcke (1984) reported that the tents constructed by *Mesophylla* can be used for five to six months and that groups of up to eight individuals roost therein. Kunz and Pena (1992) stated that female *Mesophylla* ap-

Yellow-eared bat *(Mesophylla macconnelli)*, photo by Merlin D. Tuttle.

pear to be seasonally polyestrous and give birth to a single young at a time. Nursing females use single shelters until their young approach maturity; at other times females change shelters every few days. Pregnant females have been reported from Panama in March, Colombia in January, Venezuela in January, February, and March, Peru in May, July, and August, Brazil in April and August, Bolivia in July, and Trinidad in August. Lactating females are known from Ecuador in June, Peru in June, July, August, and October, Peru and Surinam in September, and Trinidad in August.

CHIROPTERA; PHYLLOSTOMIDAE; **Genus**
CHIRODERMA
Peters, 1860

Big-eyed Bats, or White-lined Bats

There are five species (Anderson, Koopman, and Creighton 1982; Baker et al. 1994; Carter et al. 1981; Jones and Baker 1979; Jones and Carter 1976; Taddei 1979):

C. salvini, Chihuahua (northwestern Mexico) to Bolivia;
C. villosum, southern Mexico to Bolivia and Brazil, Trinidad;
C. improvisum, Guadeloupe and Montserrat in the Lesser Antilles;
C. trinitatum, Panama to Bolivia, Trinidad;
C. doriae, Minas Gerais and Sao Paulo (eastern Brazil).

Head and body length is 55–87 mm, there is no external tail, and forearm length is about 37–58 mm. Eisenberg (1989) listed average weights of about 22 grams for *C. salvini*, 14 grams for *C. trinitatum*, and 22 grams for *C. villosum*. The upper parts are brownish and the underparts are usually paler. Facial and dorsal stripes are absent or faint in *C. villosum*. *C. salvini* and *C. trinitatum gorgasi* usually have conspicuous whitish facial and back stripes. Apparently stripes are not present in *C. doriae*. The coloration of *C. trinitatum gorgasi* is yellowish brown, brown, or grayish dorsally and paler on the abdomen; a distinct white stripe extends from the upper back to the base of the tail.

This genus may be distinguished by the absence of the nasal bones of the skull. *Chiroderma* resembles *Platyrrhinus* in external appearance, but the nose leaf is broader and the forearm and interfemoral membrane are more heavily furred.

Handley (1976) listed the following habitat conditions in Venezuela: *C. villosum*, humid lowlands, mostly in moist, open areas; *C. salvini*, mountains, usually in moist places within evergreen forest or forest openings; and *C. trinitatum*, usually in moist parts of evergreen forest. Watkins, Jones, and Genoways (1972) collected specimens of *C. salvini* mostly over streams or rivers in forested areas. In Peru *C. trinitatum* was collected by Tuttle (1970) in secondary vegetation at the edge of a small garden where bananas and papayas were grown and by Gardner (1976) in disturbed habitats such as clearings around houses and fruit groves. The diet of *Chiroderma* apparently consists largely of fruit (Gardner 1977).

Wilson (1979) listed the following reproductive data: *C. villosum*, pregnant or lactating females collected in southern Mexico and Central America in March, April, May,

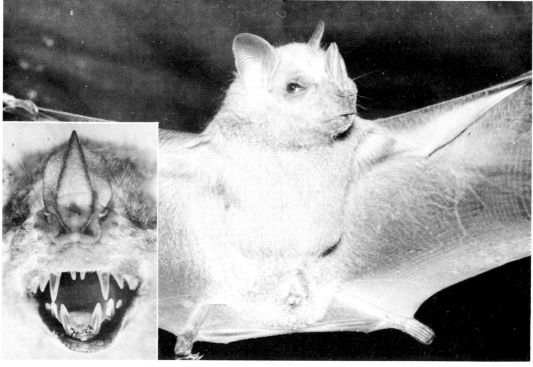

White-lined bat *(Chiroderma salvini)*, photo by Merlin D. Tuttle. Inset: *Chiroderma* sp., photo by Harold Drysdale, Trinidad Regional Virus Laboratory, through A. M. Greenhall.

July, and December, pregnant females collected on Trinidad in August and September and in Colombia in January; *C. salvini*, pregnant or lactating females collected in Mexico in January and February, in Honduras in July and August, and in Colombia from January to July and in October and November; and *C. trinitatum*, pregnant or lactating females taken in Panama in February, May, and September, pregnant females taken on Trinidad in March, in Colombia in July, and in Brazil in June and July. Davis and Dixon (1976) reported pregnant female *C. villosum* and *C. trinitatum*, each with a single large embryo, in Peru from 31 October to 8 November. The gestation period of *C. villosum* is about 4 months, that of *C. doriae* about 3.5 months (Hayssen, Van Tienhoven, and Van Tienhoven 1993).

The IUCN classifies *C. improvisum* as endangered, noting that its restricted population is expected to decline by at least 50 percent over the next decade in response to human habitat destruction. *C. doriae*, also in decline, is designated as vulnerable.

CHIROPTERA; PHYLLOSTOMIDAE; **Genus**
ECTOPHYLLA
H. Allen, 1892

White Bat

The single species, *E. alba*, is found from eastern Honduras to western Panama (Timm 1982), and there also is a record from western Colombia (Koopman *in* Wilson and Reeder 1993). The genus *Mesophylla* (see account thereof) sometimes has been included in *Ectophylla*.

Head and body length is 35–48 mm, there is no external tail, forearm length is 25–35 mm, and average weight is 7.5 grams (Eisenberg 1989). *Ectophylla* resembles a small whitish *Platyrrhinus* in external appearance. The hair on the dorsal and ventral surfaces is lightly tipped with pale gray. The eyes are ringed with dark gray hair. The tragus, chin, ear margins, skin of the wing bones, and nose leaf are bright yellow but fade to a cream color upon death.

In Costa Rica, Gardner, LaVal, and Wilson (1970) reported catching *Ectophylla* in second-growth thickets interspersed with dense stands of wild plantain *(Heliconia)*. According to Gardner (1977), the diet of *Ectophylla* probably consists mainly of fruit. Timm and Mortimer (1976) found *Ectophylla* to alter the large leaves of five species of

White bat *(Ectophylla alba)*, photo by Richard S. Casebeer.

Heliconia for use as diurnal roosts. The bats cut the side veins extending out from the midrib so that the two sides of the leaf fold downward to form a "tent," which the bats then get under. The bats were found roosting singly and in groups of two, four, and six. Brooke (1990) found groups of four to eight individuals roosting in tents of *Heliconia;* she suggested that the risk of predation may be reduced when the white pelage of the bats seems to become greenish by reflecting the color of the surrounding foliage. Tents were used for up to 45 days and groups remained together when they moved to new tents. After parturition tenting groups divided into all-male aggregations and maternity colonies. The latter contained females, nursing young, and a single adult male. Possibly the females experienced a postpartum estrus and the male was a dominant individual that kept other males away. According to Timm (1982), pregnant females have been collected in February, March, June, July, and August in Costa Rica, and lactating females have been found there in March and April. As in most phyllostomids, females apparently bear only a single young.

CHIROPTERA; PHYLLOSTOMIDAE; **Genus**
ENCHISTHENES
Andersen, 1906

Hart's Little Fruit Bat

The single species, *E. hartii*, is found from the states of Jalisco (west-central Mexico) and Tamaulipas (northeastern Mexico), through Central America, to northwestern Peru and central Bolivia and on the island of Trinidad; there also is a single record from the Tucson Mountains of southern Arizona (Anderson, Koopman, and Creighton 1982; Graham and Barkley 1984; E. R. Hall 1981; Jones and Carter 1976). *Enchisthenes* has variously been recognized as a full genus (Corbet and Hill 1991; E. R. Hall 1981) or as a subgenus of *Artibeus* (Jones and Carter 1979; Koopman *in* Wilson and Reeder 1993). Owen (1987, 1988, 1991) provisionally included *Enchisthenes* in *Dermanura* (see account thereof), and affinity between the two also was suggested by Baker, Hood, and Honeycutt (1989) and Van Den Bussche (1992). More recently, Van Den Bussche et al. (1993) concluded that *Enchisthenes* is a unique genus, not closely related to the group comprising *Artibeus*, *Dermanura*, and *Koopmania*.

Head and body length averages about 60 mm, there is no tail, forearm length averages about 39 mm, and weight averages about 17.3 grams (Eisenberg 1989). The upper parts are dark brown, almost black on the head and shoulders, and the underparts are paler. Four narrow, buffy facial stripes are present. Unlike *Artibeus* and *Dermanura*, *Enchisthenes* has a prominent white dorsal stripe (Eisenberg 1989). The tail membrane is short and hairy (Emmons 1990).

Enchisthenes generally has the appearance of a small *Artibeus* but can be distinguished by an upward-pointing projection located on the inner margin of the tragus near the tip. E. R. Hall (1981) noted also that *Enchisthenes* differs in that its inner upper incisor tooth is not bifid and its third molar, both above and below, is well developed and affects the form of the surrounding bone.

Hart's little fruit bat is strongly associated with moist habitats and multistratal tropical evergreen forest but does penetrate cloud forest and sometimes dry deciduous forest; specimens in Venezuela have been collected at elevations of 2–2,250 meters, mostly above 1,000 meters (Eisenberg 1989). On the Amazonian slope of the Andes in southeast-

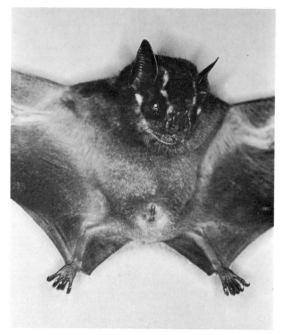

Hart's little fruit bat *(Enchisthenes hartii)*, photo by Merlin D. Tuttle.

ern Peru specimens have been taken from 1,660 to 3,540 meters (Koopman 1978*b*). In Jalisco *Enchisthenes* has been found feeding on small figs *(Ficus)* about a centimeter in diameter. The fruits were removed from the tree while the bat was in flight and then carried to a branch and eaten. One individual was taken in Jalisco when the temperature was only about 3° C. Gardner, LaVal, and Wilson (1970) suggested that *Enchisthenes* is reproductively active throughout the year in Costa Rica; pregnant females, with one embryo each were collected there in January, May, and June, and lactating or postlactating females in May, June, and July. Pregnancies also have been reported in Colombia in April and May (Wilson 1979), and a lactating female was taken in Michoacán in April (Sanchez Hernandez et al. 1985).

CHIROPTERA; PHYLLOSTOMIDAE; Genus **DERMANURA**
Gervais, 1855

Dwarf Fruit Bats

There are nine species (Anderson, Koopman, and Creighton 1982; Baud 1986; Brosset and Charles-Dominique 1990; Eisenberg 1989; Handley 1987; Jones and Carter 1976, 1979; Kalko and Handley 1994; Koopman *in* Wilson and Reeder 1993; Owen 1987, 1988, 1991; Timm 1985; Webster and McGillivray 1984):

D. gnoma, Venezuela, Guyana, French Guiana, Ecuador, eastern Peru, Amazonian Brazil, Trinidad and Tobago, Grenada (Lesser Antilles);
D. glauca, Colombia, Venezuela, Guyana, Ecuador, Peru, Bolivia;
D. watsoni, southern Mexico to Colombia;

D. incomitata, Isla Escudo de Varaguas off Caribbean coast of Panama;
D. azteca, central Mexico to western Panama;
D. tolteca, northern Mexico to western Colombia;
D. phaeotis, Sinaloa and Veracruz (Mexico) to northeastern Peru and Guyana;
D. anderseni, Ecuador, French Guiana, Brazil, Peru, Bolivia;
D. cinerea, Venezuela, Guianas, lower Amazon Basin of Brazil, southeastern Peru.

These species, together with those of *Enchisthenes* and *Koopmania,* long were included in the genus *Artibeus.* Based on detailed statistical analyses of morphological characters, Owen (1987, 1988) concluded that *Artibeus* as then understood was a polyphyletic assemblage, the different components being more closely related to other genera than to one another. He recommended that the smaller species in the group—the above, except *D. gnoma* and *D. incomitata,* which had not yet been described, plus what are now *Enchisthenes hartii* and *Koopmania concolor*—be placed in the separate genus *Dermanura* and that the remaining, larger species be left in *Artibeus* (see account thereof). While noting that *Enchisthenes* sometimes is regarded as a distinct genus, he placed it within *Dermanura.* Jones, Arroyo-Cabrales, and Owen (1988) accepted *Dermanura* as a genus and also stated that *Enchisthenes* (see account thereof) probably deserved generic status. Subsequently, Owen (1991) proposed moving *D. concolor* to the new genus *Koopmania* (see account thereof). Neither Corbet and Hill (1991) nor Koopman (*in* Wilson and Reeder 1993) used *Dermanura* at the generic level, though Koopman did accept it as a subgenus. Lim and Wilson (1993) also recognized *Dermanura* as a subgenus. However, in describing *D. incomitata,* a close relative of *D. watsoni,* Kalko and Handley (1994) placed the new species in *Artibeus* and did not comment on *Dermanura.* In their phylogenetic analyses of the Phyllostomidae both Baker, Hood, and Honeycutt (1989) and Van Den Bussche (1992) treated *Dermanura* as a full genus. More recently, assessment of mitochondrial and nuclear DNA by Van Den Bussche et al. (1993) indicated that *Dermanura, Koopmania,* and *Artibeus* do form a monophyletic assemblage, being more closely related to one another than to any other genus, and that there is thus no phylogenetic necessity to consider them as separate genera. The question whether such status is still warranted based on other factors was left open. The treatment of *Dermanura* herein is based, first, on the striking morphological distinction between that genus and *Artibeus* revealed by Owen's (1987, 1988) multistatistical analyses, indicating even that the two resemble other genera more than they do each other and, second, on the findings of Van Den Bussche et al. (1993) showing that the two genera are nonetheless closely related. *D. gnoma* and *D. watsoni* were included within *D. glauca* by Koopman (*in* Wilson and Reeder 1993) but were recognized as distinct species by Corbet and Hill (1991), Handley (1987), and Owen (1991).

In *Dermanura,* according to Emmons (1990), head and body length is 47–69 mm, there is no tail, forearm length is 34–52 mm, and weight is 10–22 grams. The upper parts are usually gray or gray brown, sometimes pale brown or dark chocolate brown, and the wings and tail membrane are blackish. Facial stripes, when present, usually end at the upper edge of the ear. The nose leaf is spear-shaped and often pale on the sides of the base. The tail membrane may have a fringe of hairs on the edge.

According to Owen (1991), the hypocone of the first upper molar tooth of *Dermanura* is only slightly developed in most species, moderately developed in a few species, but strongly developed in *Artibeus* and *Koopmania*. In *Dermanura*, as in *Artibeus*, there are only two lower premolars, whereas there are three in *Enchisthenes* and *Koopmania*. *Dermanura* also is characterized by a lengthening of the nose leaf, a reduction or loss of the third upper molar, and a reduction of the third lower molar.

Emmons (1990) noted further that dwarf fruit bats are found in mature and disturbed lowland and montane rainforest, plantations and gardens, and cloud forest and deciduous forest; the elevational range is up to 2,400 meters. In Central America, Davis (1970a) found *D. phaeotis* and *D. tolteca* to occur in lowlands but *D. azteca* to be restricted to high elevations. In Jalisco, Mexico, *D. phaeotis* is confined to tropical deciduous forest and thorn forest (Watkins, Jones, and Genoways 1972). *D. cinerea* has been found roosting in the apical, bifid, paradox, and boat types of "tents" described in the account of *Uroderma*, and *D. watsoni* in the palmate umbrella, pinnate, apical, bifid, and boat types; *D. anderseni*, *D. glauca*, *D. gnoma*, *D. phaeotis*, and *D. tolteca* also have been reported to use tents (Choe and Timm 1985; Kunz, Fujita, et al. 1994). These bats feed on a variety of fruit and may also take pollen and insects (Eisenberg 1989; Timm et al. 1989).

Wilson (1979) compiled the following reproductive data: *D. cinerea*, bimodal polyestry suggested in Colombia, Panama, and Nicaragua but pregnant or lactating females reported for nearly every month of the year; *D. phaeotis*, bimodal polyestry, with pregnant or lactating females reported from January to August; *D. tolteca*, apparently an extended breeding season with two births per year, pregnant or lactating females reported from January to October; and *D. azteca*, pregnancies reported during the summer in Mexico. Numerous specimens collected by Sanchez Hernandez et al. (1985) indicate that *D. phaeotis* and *D. tolteca* breed throughout the year in Michoacán, Mexico. There is a single young (Hayssen, Van Tienhoven, and Van Tienhoven 1993).

CHIROPTERA; PHYLLOSTOMIDAE; **Genus KOOPMANIA**
Owen, 1991

The single species, *K. concolor*, is found in southeastern Colombia, southern Venezuela, the Guianas, northern Brazil, and Amazonian Peru (Owen 1991). This species long was assigned to the genus *Artibeus*, but Owen (1987, 1988) provisionally included it in *Dermanura* (see account thereof) and subsequently described for it the new genus *Koopmania*. Koopman (*in* Wilson and Reeder 1993) recognized *Koopmania* as a subgenus of *Artibeus*. Van Den Bussche et al. (1993) considered *Koopmania*, *Dermanura*, and *Artibeus* to be immediately related but left open the question of generic distinction.

Eisenberg (1989) listed the following average measurements: head and body length, 59.6 mm for males, 64.1 mm for females (there is no tail); forearm length, 46.9 mm for males, 48.3 mm for females; and weight, 18.2 grams for males, 20.0 grams for females. Both the upper parts and underparts are medium brown in color, the throat may be pale brown, and there are indistinct facial stripes.

Owen (1991) characterized *Koopmania* as being intermediate in size to *Dermanura* and *Artibeus*, with females averaging larger than males and forearm length ranging from 43 mm to 52 mm. *Koopmania* is distinguished from all related genera in having the plagiopatagium attaching distally to the metatarsal-phalangeal joint (rather than to

the tarsus or metatarsus) and in having no paraoccipital process. Like *Artibeus*, *Koopmania* has a strongly developed hypocone on the first upper molar, whereas this hypocone is only slightly or moderately developed in *Dermanura*. There are three upper and three lower molar teeth in *Koopmania*, the third lower molar being much smaller than that of *Enchisthenes*. The rostrum of *Koopmania* is broad and shorter than that of any species of *Dermanura* except *D. gnoma*. The postpalatal shelf of the skull of *Koopmania* is shorter than in *Dermanura*.

In Venezuela, Handley (1976) found *Koopmania* at elevations of 100–1,000 meters, mostly below 500 meters, and mainly in moist, open areas. The genus sometimes is considered rare, but Brosset and Charles-Dominique (1990) captured it regularly in the forests of French Guiana, always at 15–25 meters above the ground. The diet is presumed to be frugivorous (Acosta and Owen 1993). Wilson (1979) recorded collection of a pregnant female in Colombia in February. The IUCN includes *K. concolor* in the genus *Artibeus* and designates it as near threatened.

CHIROPTERA; PHYLLOSTOMIDAE; **Genus ARTIBEUS**
Leach, 1821

Neotropical Fruit Bats

There are 10 species (Anderson, Koopman, and Creighton 1982; Barquez and Ojeda 1992; Buden 1985; Davis 1984; Eisenberg 1989; Graham and Barkley 1984; Handley 1987, 1989, 1991a; Jones and Carter 1976, 1979; Koepcke and Kraft 1984; Koop and Baker 1983; Koopman 1978b and *in* Wilson and Reeder 1993; Lazell and Koopman 1985; Lim and Wilson 1993; Myers and Wetzel 1983; Patterson, Pacheco, and Ashley 1992; Redford and Eisenberg 1992; Timm 1985; Webster and Jones 1984c; Webster and McGillivray 1984; Wilson 1991):

A. fimbriatus, southern Brazil, Paraguay, northern Argentina;

A. fraterculus, Ecuador, Peru;

A. lituratus, southern Mexico to northern Argentina and Paraguay, Trinidad and Tobago, Lesser Antilles;

A. intermedius, Sinaloa and Tamaulipas (Mexico) to northern South America, Tres Marías Islands off western Mexico;

A. obscurus, Amazon Basin and adjacent parts of Colombia, Venezuela, Guianas, Ecuador, Peru, Bolivia, and Brazil;

A. jamaicensis, Sinaloa and Tamaulipas (Mexico) to Ecuador and Venezuela, throughout Greater and Lesser Antilles, Trinidad and Tobago, Bahamas, lower Florida Keys;

A. planirostris, southern Colombia and Venezuela, through Amazonian Brazil and adjacent regions, to northern Argentina and Paraguay;

A. amplus, Colombia, Venezuela, Guyana;

A. inopinatus, El Salvador, Honduras, Nicaragua;

A. hirsutus, western Mexico.

The genera *Enchisthenes*, *Dermanura*, and *Koopmania* (see accounts thereof) are often included within *Artibeus*. *A. planirostris* was regarded as a subspecies of *A. jamaicensis* by Handley (1991a), but was considered a full species by Corbet and Hill (1991), Koopman (*in* Wilson and Reeder 1993), and Patterson, Pacheco, and Ashley (1992). There is some question about the southerly distribution of *A. jamaicensis*; Koopman (*in* Wilson and Reeder 1993)

Mexican fruit bat *(Artibeus jamaicensis)*. A. Bat hanging where it has been eating. The light-colored, irregular masses are pieces of food that it has chewed, extracted the liquid from, and rejected. The rounded portions are feces that have passed through the bat in 15–20 minutes. B. In flight, seen from below, showing greatly reduced membrane between hind legs. C. Shows rubberlike nose leaf and tiny tubercles in lower lip. Photos by Ernest P. Walker.

noted that its range extends only as far as Ecuador and Venezuela if *planirostris* and *obscurus* are considered distinct species, but other authorities (e.g., Corbet and Hill 1991; Patterson, Pacheco, and Ashley 1992; Redford and Eisenberg 1992), even while recognizing *planirostris* as distinct, have indicated that *A. jamaicensis* occurs as far south as Amazonian Brazil, Paraguay, and northern Argentina. Humphrey and Brown (1986) questioned Lazell and Koopman's (1985) report of a resident population of *A. jamaicensis* in the Florida Keys, but additional sightings of the species there were discussed by Lazell (1989).

Head and body length is 70–103 mm, forearm length is 48–75 mm, and weight is 25–86 grams (Emmons 1990). *A. lituratus* is the largest species. *Artibeus* lacks an external tail and has a narrow interfemoral membrane. The upper parts are dull brownish, grayish, or black, with a silvery tinge, and the underparts are usually paler. Four whitish facial stripes are usually present. There is no light dorsal line. The fur is short, very soft, and velvety in texture. *Artibeus* differs from similar genera in its more pointed ears and in cranial and dental features (see accounts of *Dermanura* and *Koopmania*). The total number of teeth, depending on the number of molars present, is 28, 30, or 32; the number of molars varies with the species and sometimes individually.

In Venezuela, Handley (1976) found these bats under the following conditions: *A. jamaicensis*, mostly moist, open areas and roosting mostly in houses; *A. obscurus*, moist open and forested areas; and *A. lituratus*, moist or dry areas in both forests and openings. *A. jamaicensis* occurs throughout the Yucatan Peninsula and roosts in caves, cenotes, and buildings (Jones, Smith, and Genoways 1973). On Barro Colorado Island, Panama, it roosts mostly in dense foliage and tree hollows (Morrison and Handley 1991). It also has been reported to modify palm fronds for roosting by biting around the midrib so that the leaflets fold downward to make a "tent"; it has been found in the palmate umbrella, pinnate, and apical types of tents, as described in the above account of *Uroderma* (Foster and Timm 1976; Kunz, Fujita, et al. 1994; Timm 1987). In a study in Jalisco, Mexico, Morrison (1978) found that four female *A. jamaicensis* flew an average of 8 km to forage. On Barro Colorado Island individuals of the same species showed average movements of 1–4 km between day roost and feeding site (Handley, Gardner, and Wilson 1991).

Artibeus is primarily frugivorous but also consumes pollen, nectar, flower parts, and insects (Gardner 1977). Certain kinds of leaves may also be significant, at least in the diet of *A. jamaicensis* (Kunz and Diaz 1995). Figs *(Ficus)* are the most important food of *A. jamaicensis* on Barro Colorado Island; a bat there may eat more than its weight in figs during a night (Handley and Leigh 1991). Other fruits known to be taken by *Artibeus* include mangoes, avocados, bananas, espave nuts, and the pulpy layer surrounding the seeds of *Acrocomia* palms. The smaller fruits are carried to feeding sites during the night, but toward morning these bats carry fruit to their regular roosts. Food passes through the digestive tract so rapidly (15–20 minutes) that there is probably little or no bacterial action. Perhaps there is some chemical or enzyme action. The newly passed fecal material often has the odor of the fruit on which the bat fed. Nuts, seeds, and fruit cores accumulate beneath roosting areas; *Artibeus* thus aids in the dissemination of seeds of tropical fruits. Control measures sometimes have been taken to protect cultivated fruit from these bats.

A density of about 200 individual *A. jamaicensis* per sq km was found on Barro Colorado Island, Panama (Leigh and Handley 1991). The population there apparently consists of aggregations that roost and forage within definite areas (Handley, Gardner, and Wilson 1991); night aggregations at feeding sites may consist of hundreds of bats (Handley and Morrison 1991). The population is further divided into social units, including groups of 1–3 bachelor males or juvenile females that roost primarily in foliage and "harems" of 3–14 adult females that share a tree hollow with their young and a single adult male; the latter spends much of his flying time in the vicinity of the roost, apparently defending the site from rival males (Morrison 1979; Morrison and Handley 1991). Kunz, August, and Burnett (1983) also reported evidence of harems in a cave colony of *A. jamaicensis*. Most of the bats in this group were in clusters comprising 2–14 pregnant or lactating females, their young, and a single adult male. Older and heavier males had the largest harems, and there also were groups of bachelor males and nonreproductive females. Klingener, Genoways, and Baker (1978) observed a cluster of 6 *A. jamaicensis* roosting in a shallow embrasure. Females of this species also have been reported to segregate into discrete groups prior to giving birth and until the young are weaned (Fenton and Kunz 1977). Some species, such as *A. lituratus*, are thought to be more solitary than others. However, Husson (1978) reported a cluster of 16 *A. lituratus* on the underside of a leaf of a coconut palm.

There is usually a single young per birth in this genus, but several cases of twins have been reported for *A. jamaicensis*. Hayssen, Van Tienhoven, and Van Tienhoven (1993) listed birth weights of 5.9–15.5 grams. Davis and Dixon (1976) reported pregnant female *A. planirostris*, *A. obscurus*, and *A. lituratus*, each with a single large embryo, in Peru from 31 October to 8 November. The extensive reproductive data compiled by Wilson (1979) can be summarized briefly as follows: *A. hirsutus*, no restricted breeding season, pregnancies reported in Mexico in February and from April to September; *A. inopinatus*, approximately one-month-old individuals taken in Honduras in August; and *A. lituratus*, in Mexico one breeding season from about February to August, in southern Central America probably a pattern of bimodal polyestry with a quiescent period from about September to December, on Trinidad pregnancies reported from February to July and lactation from April to October, in Colombia year-round breeding with pregnancy peaks in December and May. *A. jamaicensis* has a unique seasonally polyestrous cycle in Panama. A birth peak in March–April is followed by a postpartum estrus and a second birth peak in July–August. Then there is another postpartum estrus and implantation, but the blastocysts are dormant from September to November. Subsequently there is normal development and then birth in March or April. In Colombia and the Yucatan Peninsula there may be continuous or acyclic breeding by this species. Pregnant or lactating female *A. jamaicensis* have been found in various areas during all months of the year.

Studies of *A. jamaicensis* on Barro Colorado Island, Panama, suggest that the unusual reproductive pattern—alternating annual episodes of normal embryonic development and delayed development—allows the young to be weaned during periods of maximum food availability (Wilson, Handley, and Gardner 1991). Observations of a captive colony of *A. jamaicensis* (Taft and Handley 1991) indicate that normal, undelayed gestation is 112–20 days but that the mean interbirth interval with delayed embryonic development is 177 days. The single young averages about 14 grams at birth, the lactation period averages 66 days, initial flight occurs at 31–51 days, and adult size is attained at around 80 days. Males attain sexual maturity at 8–12 months and females usually first give birth at 1 year. In the wild both males and females disperse from their natal roosting groups (Morrison and Handley 1991). A female *A.*

jamaicensis on Barro Colorado Island was recaptured 7 years after it was banded (Wilson and Tyson 1970). One specimen of *A. jamaicensis* was still living after 10 years in captivity (Jones 1982).

The IUCN classifies *A. hirsutus* as vulnerable, noting that it is expected to decline by at least 20 percent over the next decade because of both habitat loss and hunting by people. *A. fraterculus* also is classed as vulnerable, and *A. fimbriatus, A. obscurus, A. amplus,* and *A. inopinatus* are designated as near threatened.

CHIROPTERA; PHYLLOSTOMIDAE; **Genus ARDOPS**
Miller, 1906

Tree Bat

The single species, *A. nichollsi,* occurs in the Lesser Antilles from St. Eustatius to St. Vincent (Jones and Carter 1976).

Head and body length is 50–73 mm, there is no external tail, and forearm length is 42–54 mm. Weight is 15–19 grams (Hayssen, Van Tienhoven, and Van Tienhoven 1993). The skull resembles that of *Artibeus* but is broader. The upper incisors are short and thick. The color is dark brown or grayish. Apparently this genus lacks facial and dorsal lines.

Ardops evidently roosts exclusively in trees and other kinds of arborescent vegetation (Jones and Genoways 1973). The diet presumably consists of fruit (Gardner 1977). On Dominica 5 pregnant females, each with a single embryo, were collected from 27 March to 14 April and a lactating female was taken on 19 April. A pregnant female was collected on St. Eustatius on 9 March (Jones and Genoways 1973). Of 14 adult females taken on Guadeloupe in July, 2 were lactating and 6 were pregnant (Baker, Genoways, and Patton 1978). McCarthy and Henderson (1992) collected pregnant and lactating females on Guadeloupe and Marie-Galante in April and stated that *Ardops* has a polyestrous reproductive cycle that appears to be loosely synchronized during two breeding periods. The IUCN classifies the genus as near threatened.

Falcate-winged bat *(Phyllops haitiensis)*, photo by Charles A. Woods.

CHIROPTERA; PHYLLOSTOMIDAE; **Genus PHYLLOPS**
Peters, 1865

Falcate-winged Bats

There are two living species (Jones and Carter 1976):

P. falcatus, Cuba;
P. haitiensis, Hispaniola.

The above two species are closely related and may represent a single species; they were listed as such by Koopman (*in* Wilson and Reeder 1993) though not by Corbet and Hill (1991). A third species, *P. vetus,* is known only by fossils

Tree bat *(Ardops nichollsi):* Left, photo by Robert J. Baker through J. Knox Jones, Jr.; Right, photo from *The Land and Sea Mammals of Middle America and the West Indies,* D. G. Elliot.

Red fruit bat *(Stenoderma rufum)*, photo by J. R. Tamsitt. Inset: skull *(S. rufum)*, photo by P. F. Wright of specimen from Museum of Natural History, University of Kansas.

from Cuba, including remains from a supposedly post-Columbian cave deposit on the Isle of Pines (Morgan and Woods 1986).

Head and body length is about 48 mm, there is no external tail, and forearm length is about 39–45 mm. Coloration is dark brown or grayish. Externally this genus resembles *Artibeus*, but it differs from it and other related genera in cranial and dental features.

Klingener, Genoways, and Baker (1978) reported that *P. haitiensis* was collected more frequently in thickly vegetated ravines than in drier scrub thorn habitats. *P. falcatus* occasionally enters houses, and *P. haitiensis* has been obtained in a house in a town as well as while sleeping in mango trees. The diet is thought to consist mainly of fruit. Klingener, Genoways, and Baker (1978) recorded the following reproductive data for *P. haitiensis* in southern Haiti: all 6 females taken between 4 and 9 January were pregnant; 5 of 8 females caught on 27 May were pregnant; and of 54 females taken between 14 and 27 August, 27 were not pregnant, 8 had an enlarged uterus, 14 had a single embryo, and 5 had an enlarged postpartum uterus. The IUCN classifies *P. falcatus* (including *P. haitiensis*) as near threatened.

CHIROPTERA; PHYLLOSTOMIDAE; **Genus ARITEUS**
Gray, 1838

Jamaican Fig-eating Bat

The single species, *A. flavescens,* occurs on Jamaica (Jones and Carter 1976).

Head and body length is about 50–67 mm, there is no

tail, and forearm length is about 40–44 mm. H. F. Howe (1974) reported that four specimens weighed from 9.2 to 13.1 grams. Coloration is light reddish brown above and paler below. A small white patch is present on each shoulder. Facial stripes and dorsal lines are not present. This bat resembles *Ardops* but differs from it in dental features.

H. F. Howe (1974) collected four specimens in heavily disturbed habitat and observed that *Ariteus* probably is not a cave bat. It feeds on fruits, such as the naseberry *(Achras sapota)* and the rose apple *(Eugenia jambos),* as well as insects. It begins to fly and feed shortly after sunset. Allen (1942) wrote that the status of this bat was not then known. However, evidently it was rare in collections and was thought likely to become scarce because of intensive agricultural development on Jamaica. The IUCN now classifies it as vulnerable, noting that it is subject to a precipitous decline.

CHIROPTERA; PHYLLOSTOMIDAE; **Genus STENODERMA**
E. Geoffroy St.-Hilaire, 1818

Red Fruit Bat

The single species, *S. rufum,* was for nearly a century known only from a single skin and skull collected at an unknown locality but believed to possibly have come from Egypt. In 1916 the species was found as a subfossil on Puerto Rico and so was accepted as a new world taxon, but it was thought to be extinct. In 1957 three living individuals were collected on St. John Island in the U.S. Virgin Islands, and since then others have been found on St. Thomas and Puerto Rico (Gannon, Willig, and Jones 1992).

Head and body length is about 53–73 mm, there is no external tail, and forearm length is about 46–51 mm. Weight is about 25 grams (Hayssen, Van Tienhoven, and Van Tienhoven 1993). The upper parts are reddish brown, or nearly so, and the underparts are less reddish. Morphological comparisons indicate that this bat is most closely related to *Phyllops, Ardops,* and *Ariteus.* Like these other genera, *Stenoderma* has a white spot on both shoulders.

Jones, Genoways, and Baker (1971) collected specimens in nets set in tropical broad-leaved forest and above the forest canopy. According to Gardner (1977), the diet is composed of fruit. Three captive specimens were maintained for three weeks on a diet of mangoes, bananas, and various fruit nectars (Genoways and Baker 1972). Data summarized by Wilson (1979) suggest that *Stenoderma* is polyestrous in Puerto Rico, with pregnancies recorded in February, March, May, July, and August and lactation in February, July, and November. This suggestion was supported by Gannon and Willig (1992), who collected pregnant and lactating females throughout the year on Puerto Rico, during both dry and rainy seasons. Gannon, Willig, and Jones (1992) reported that *Stenoderma* occurs in very restricted habitat and may be threatened by environmental disturbance. The IUCN now classifies it as vulnerable.

CHIROPTERA; PHYLLOSTOMIDAE; **Genus PYGODERMA**
Peters, 1863

Ipanema Bat

The single species, *P. bilabiatum,* has been reported from Surinam, southeastern Brazil, Bolivia, Paraguay, and

Jamaican fig-eating bat *(Ariteus flavescens):* Top, photo from *M. ber. Kö. Preuss. Akad. Wis. Berlin;* Bottom, photo by Robert J. Baker through J. Knox Jones, Jr.

adjacent Argentina; a reported occurrence in Mexico has been shown to be erroneous (Anderson, Koopman, and Creighton 1982; Jones and Carter 1976; Owen and Webster 1983). The vernacular name refers to the type locality, Ipanema, in Sao Paulo, Brazil.

In the nominate subspecies head and body length is 53–65 mm, there is no external tail, forearm length is 36–46 mm, and weight is 19–26 grams (Redford and Eisenberg 1992). In a newly described and larger subspecies from Bolivia head and body length is 84 mm, forearm length is 43 mm, and weight is 27.5 grams (Owen and Webster 1983). The color is dark brown, almost black, above and grayish brown below. There is a white spot on each shoulder near the wing. *Pygoderma* may be distinguished by the unequal size of the upper incisors, the inner pair being the larger, and the greatly shortened and depressed rostrum of the skull. The palate is short and rounded. The lower lip has a central tubercle bordered by smaller protuberances.

According to Webster and Owen (1984), some specimens have been taken at night above trails and streams in mature tropical forests and secondary growth bordering forest. Others have been taken around fruit trees, and stomach contents suggest that *Pygoderma* feeds on pulpy or overripe fruit that has few fibers or seeds. Pregnant females, each with a single fetus, have been collected in March, July, and August in Paraguay and in August in Brazil.

CHIROPTERA; PHYLLOSTOMIDAE; **Genus AMETRIDA**
Gray, 1847

The single species, *A. centurio,* occurs in Venezuela, the Guianas, Brazil, and Trinidad and on Bonaire Island (Jones and Carter 1976); one specimen also was collected in the Canal Zone of Panama (Emmons 1990). For many years a second species, *A. minor,* was thought to exist, but Peterson (1965a) showed that the small *A. minor* actually comprised the male members of *A. centurio.*

Head and body length is 35–47 mm, there is no external tail, and forearm length is 25–33 mm. Males, with an average weight of 8 grams, are smaller than females, which av-

erage 12.6 grams (Eisenberg 1989; Peterson 1965a). The fur is sooty brown to dark brown in color, and there is a white spot on each shoulder near the wing. *Ametrida* differs from *Sphaeronycteris* in the presence of a nose leaf, a shorter facial portion of the skull, and other cranial and dental characters. The small third lower molar is present.

In Venezuela, Handley (1976) collected *Ametrida* mostly near streams and in other moist parts of evergreen forest. Patterson (1992a) reported finding one inside of a large leaf. According to Gardner (1977), the diet is unknown but probably consists of fruit. Pregnant females, each with a single embryo, were collected on Trinidad in July and August (Carter et al. 1981).

CHIROPTERA; PHYLLOSTOMIDAE; **Genus SPHAERONYCTERIS**
Peters, 1882

The single species, *S. toxophyllum,* occurs from Colombia and Venezuela to Amazonian Peru and Bolivia (Jones and Carter 1976).

Average head and body length is about 57 mm, there is no external tail, average forearm length is about 39 mm, and average weight is about 14.8 grams (Eisenberg 1989). The upper parts are cinnamon brown, the individual hairs in the middle of the back being whitish; the underparts are brownish white. There is a horn-shaped growth on the nose that appears to be larger on males than on females. It is probably soft flesh, like the nose-leaf structures on many other bats, but no definite statement has been found regarding this structure on live or recently killed bats of this genus. This bat is similar to *Centurio,* but the facial portion of the skull is shorter and the outgrowths on the face are less extreme. A third lower molar, absent in *Centurio,* is present in *Sphaeronycteris.* As in *Centurio,* there is no true nose leaf. Skull and dental features differentiate *Sphaeronycteris* from another similar genus, *Ametrida.*

In Venezuela, Handley (1976) collected numerous specimens in many kinds of habitat but mostly in moist, open areas. According to Gardner (1977), the diet is unknown but probably consists of fruit. Anderson and Webster (1983) collected a pregnant female in Bolivia on 3 October.

Ipanema bat *(Pygoderma bilabiatum),* photos by R. E. Mumford.

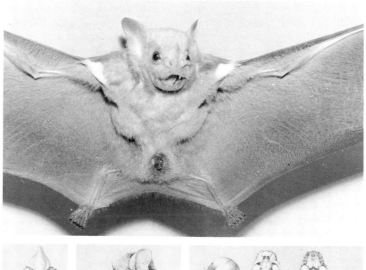

Ametrida centurio, photos by Merlin D. Tuttle. Drawings of face and skull from *Die Fledermäuse des Berliner Museums für Naturkunde*, Wilhelm K. Peters.

CHIROPTERA; PHYLLOSTOMIDAE; **Genus CENTURIO**
Gray, 1842

Wrinkle-faced Bat, or Lattice-winged Bat

The single species, *C. senex*, occurs from Sinaloa and Tamaulipas in northern coastal Mexico to Venezuela, as well as on Trinidad (Jones and Carter 1976, 1979).

Head and body length is 55–70 mm, there is no external tail, forearm length is approximately 41–47 mm, and adult weight is 17–28 grams. The upper parts are medium brown, dark brown, or yellowish brown, with a white spot on each shoulder; the underparts are paler.

Externally this genus may be recognized by the grotesque facial features: the face is short, broad, naked, and covered with wrinkled outgrowths of the skin. There is no true nose leaf. Glands present in the neck probably secrete an odoriferous substance. The skull is characteristic in its high, rounded braincase, extremely short rostrum, and short palate and in the position of the external nares, which are directly over the roots of the upper incisors.

All the reported roosts of this genus have been in trees, such as mango, rayo (*Dracaena* sp.), and *Putranjiva*. The wrinkle-faced bat has always been found roosting singly or in twos or threes, usually under leaves, with a maximum of approximately a dozen individuals in a given tree. In captivity, this bat covers its face with the chin fold when roosting or sleeping. This fold extends over the ears, which lie flat over the top of the head when the bat is at rest. A small

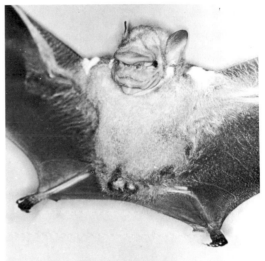

Sphaeronycteris toxophyllum: Top, male with chin flap up; Middle, male with chin flap down; Bottom, female. Photos by Merlin D. Tuttle.

Sphaeronycteris toxophyllum: Top, photo by Thomas O. Lemke; Bottom, photo by Merlin D. Tuttle.

projection on the top of the head acts as a sort of "doorstop," allowing the wrinkled fold of skin to be stretched taut at this point. At least in some individuals, two areas in this facial mask are devoid of hair; these translucent areas, or "windows," cover the bat's eyes when the mask is stretched tight, presumably allowing it to perceive light and perhaps objects even when the face is covered. When *Centurio* leaves its roost, it unmasks itself and departs in a jerky flight resembling that of a large butterfly. The facial mask then appears beneath the chin as a series of wrinkled folds of skin. The males of *Centurio* also have large lappets of loose skin on the chin that are not part of the skin fold; these lappets probably contain scent glands.

In the lowlands of western Venezuela, Handley (1976) collected specimens at a variety of moist and dry sites in both forested and open areas. The diet is assumed to consist mainly of fruit. Captives in Trinidad preferred the soft, mushy parts of ripe bananas and pawpaws, which they appeared to suck up. The many small papillae on the skin between the lips and the gum line may act as strainers when these bats feed on soft fruit.

Pregnant females apparently shelter with males in the same trees, though they may roost by themselves. Pregnant females have been recorded in Mexico and Central America from February to August, in Colombia in April, and on Trinidad in January; lactating females have been found in Mexico in March, April, July, and August (Carter and Jones 1978; Lemke et al. 1982; Wilson 1979). There is a single young (Hayssen, Van Tienhoven, and Van Tienhoven 1993).

CHIROPTERA; **Family MYSTACINIDAE; Genus MYSTACINA**
Gray, 1843

New Zealand Short-tailed Bats

The single known genus, *Mystacina,* contains two species (Flannery 1987a; Hill and Daniel 1985):

Lattice-winged bat *(Centurio senex),* photos by Harold Drysdale, Trinidad Regional Virus Laboratory, through A. M. Greenhall. Skull *(C. senex)* photo by P. F. Wright of specimen in U.S. National Museum of Natural History.

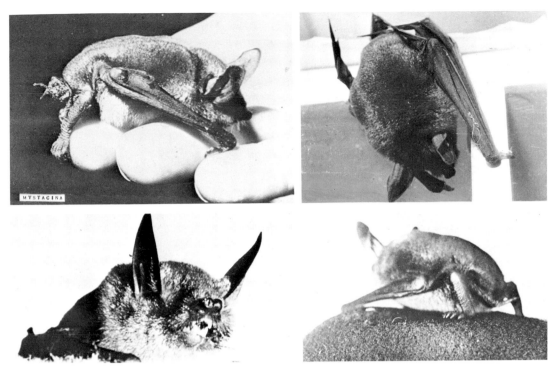

New Zealand short-tailed bat *(Mystacina tuberculata)*, photos by H. P. Collins.

M. tuberculata, probably once found throughout New
    Zealand;
M. robusta, probably once found throughout New
    Zealand.

The systematic position of the Mystacinidae long has been
questionable. In recent years this family usually has been
placed in the superfamily Vespertilionoidea, together with
the Vespertilionidae, Molossidae, and related families.
However, on the basis of biochemical analysis of serum
proteins Pierson et al. (1986) showed the Mystacinidae to
belong in the superfamily Phyllostomoidea and thus to
have affinity to the New World families Mormoopidae,
Noctilionidae, and Phyllostomidae. The familial arrangements of Corbet and Hill (1991) and Koopman (*in* Wilson
and Reeder 1993) continued to indicate an association between the Mystacinidae and the Vespertilionoidea.

Until Hill and Daniel's (1985) revision *M. robusta* usu-

Thumb showing denticle at base of claw, and foot and part of
inner surface of membrane of the New Zealand short-tailed bat
(Mystacina tuberculata), photo from *Proc. Zool. Soc. London.*

New Zealand short-tailed bat *(Mystacina tuberculata),* photo
by Entomology Division, D.S.I.R. (New Zealand).

ally was treated as a subspecies of *M. tuberculata*. Daniel (1979), however, had suggested that the two might warrant elevation to specific status because of the marked difference in their size. Adult *M. tuberculata* have a forearm length of about 40–45 mm and weigh about 12–15 grams, whereas adult *M. robusta* have a forearm length of 45–49 mm and weigh 25–35 grams. Blanchard (1992) noted that head and body length of *M. tuberculata* is 60–68 mm, tail length is 10–12 mm, and wing span is about 300 mm. Flannery (1995) reported that a single male *M. robusta* had a head and body length of 66 mm and a tail length of 18.6 mm; respective measurements for a female were 61 mm and 16.6 mm. The pelage is said to be thicker than in any other species of bats. The upper parts are grayish brown to brown and the underparts are paler.

Structurally these bats are unique: the claws are needle sharp and the wings are remarkably transformed. The thumb has a large claw with a small talon projecting from it, and the claws of the feet also have talons. The membrane is thick and leathery along the sides of the body, forearm, and lower leg. The wings can be rolled up beneath this leathery membrane when the bat is not flying. The base of the tail membrane is also thick and wrinkled. The oblique-ly truncate muzzle has a rudimentary nostril pad and a scattering of stiff bristles with spoon-shaped tips. The nostrils are oblong and vertical. The ears are separate, and the tragus is long and pointed. The short, broad feet, the grooved covering of the sole of the foot, and the short, thick legs suggest climbing habits. As in the sac-winged bats (Emballonuridae), the tail perforates the tail membrane and appears on the upper surface.

The teeth are of the normal cuspidate insectivorous type. The upper premolars are well developed. The dental formula is: (i 1/1, c 1/1, pm 2/2, m 3/3) × 2 = 28.

These bats are remarkably agile, running freely on the ground and even running or climbing rapidly up sloping, smooth surfaces. This agility results from the constriction of the wing membrane; peculiar folding processes along the forearm allow the wing to be used as a normal forelimb in walking and climbing. The basal talons on the claws of the thumbs and toes may be further adaptations for terrestrial, arboreal, and burrowing behavior (Daniel 1979).

Mystacina is found in forests and roosts in hollow trees, caves, crevices, and burrows. Groups of bats apparently use their teeth to burrow roosting cavities and tunnels through the wood of trees (Daniel 1979). They emerge relatively late in the evening. Available evidence indicates that unlike *Chalinolobus*, the other New Zealand bat, *Mystacina* does not hibernate for a prolonged period but awakes spontaneously to feed when winter weather becomes relatively mild (Daniel 1979). The diet of *Mystacina* is now known to be surprisingly broad, consisting of flying and resting arthropods, fruit, nectar, and pollen (Daniel 1976, 1979).

The first colony of *Mystacina* to be studied in detail comprised about 500 individuals of *M. tuberculata* roosting in a giant hollow kauri tree (Daniel 1979). Other observations suggest that some groups are considerably smaller. The investigated colony included about 100–150 newborn young on 16 December 1973. Apparently, *M. tuberculata* is monestrous, with mating occurring in autumn and there being some kind of delay of development, fertilization, or implantation. The single young is born in summer (December–January). A population on Codfish Island, at the southern tip of New Zealand, breeds from February to April; at that time a nursery colony is established and males roost separately in other trees (Flannery 1995). Limited observations of *M. robusta* suggest that it may be polyestrous, with births occurring from spring to autumn.

Although both species are thought once to have oc-curred throughout much of New Zealand, there are relatively few precise records (Daniel and Williams 1984; Flannery 1987a; Hill and Daniel 1985). *M. tuberculata* still is known to be present on North Island, South Island, Little Barrier Island off the north coast of North Island, and several small islands off of Stewart Island. It has been reduced, however, to perhaps 10 sporadic populations that together probably contain only a few thousand individuals and are probably in danger of extermination. From actual collection and observation of living specimens since European settlement, *M. robusta* is known only from one locality near the northern tip of South Island and from Big South Cape and Solomon islands, which are off Stewart Island. However, subfossil remains, which may represent populations from well within historical time, show that *M. robusta* was present at least as far north as central North Island. In any case, modern observations of this species have been restricted to Big South Cape and Solomon islands, where it has not been seen since 1965. It already may be extinct. Both species are thought to have declined through destruction of forest habitat, predation by introduced rats and other exotic mammals, accidental poisoning, and human disturbance of roosts. The IUCN regards *M. tuberculata* as vulnerable and *M. robusta* as extinct.

CHIROPTERA; **Family NATALIDAE; Genus NATALUS** *Gray, 1838*

Funnel-eared Bats

The single living genus, *Natalus*, contains three subgenera and five species (Cabrera 1957; E. R. Hall 1981; Koopman *in* Wilson and Reeder 1993; Morgan 1989; Ottenwalder and Genoways 1982; Williams, Genoways, and Groen 1983; Wilson 1991):

subgenus *Natalus* Gray, 1838

N. stramineus, southern Baja California, northern Mexico to Brazil, Tres Marías Islands off western Mexico, Cuba, Jamaica, Hispaniola, Lesser Antilles;
N. tumidirostris, Colombia, Venezuela, Surinam, Curacao, Bonaire, Trinidad;

subgenus *Chilonatalus* Miller, 1898

N. tumidifrons, Great Abaco and San Salvador (Watling) islands in the Bahamas;
N. micropus, Cuba and nearby Isle of Pines, Jamaica, Hispaniola, Old Providence Island off east coast of Nicaragua;

subgenus *Nyctiellus* Gervais, 1855

N. lepidus, Cuba and nearby Isle of Pines, and Eleuthera, Cat, Great Exuma, and Long islands in the Bahamas.

Chilonatalus was used only as a synonym of *Natalus* by E. R. Hall (1981) but was considered a subgenus by Corbet and Hill (1991) and Koopman (*in* Wilson and Reeder 1993). Morgan (1989) regarded *Nyctiellus* as a distinct genus and *Natalus major* of the Greater Antilles as a species distinct from *N. stramineus*. The subspecies of *N. stramineus* on Cuba, *primus*, is known only from skeletal remains and is thought to be extinct (E. R. Hall 1981). *N. stramineus* also is known by remains found in deposits on the Isle of Pines and Grand Cayman Island south of Cuba and on Andros,

Funnel-eared bat (Natalus stramineus), photo by Harold Drysdale, Trinidad Regional Virus Laboratory, through A. M. Greenhall.

Grand Caicos, and New Providence islands in the Bahamas; Morgan (1989) indicated that the New Providence deposit is about 8,000 years old but that the remains from Grand Caicos are associated with human artifacts and probably date from less than 4,500 years ago. Fossil remains of N. tumidifrons are known from Andros, New Providence, Cat, and Great Exuma in the Bahamas, and those of N. lepidus are known from Andros and Great Exuma (Morgan 1989).

Head and body length is 35–55 mm, tail length is approximately 50–60 mm, and forearm length is 27–41 mm. Adults usually weigh 4–10 grams. The fur is soft and long. The coloration is gray, buff, yellowish, reddish, or deep chestnut. N. stramineus mexicanus has two color phases—a pale phase, in which the upper parts are buffy to pinkish cinnamon, and a dark phase, in which the upper parts are rich yellowish or reddish brown; the underparts are paler in both phases.

These are slim-bodied bats, with long and slender wings, legs, and tail. The ears are large, separate, and funnel-shaped, the surface of the external ear being studded with glandular papillae as in the Old World vespertilionid genus Kerivoula. The tragus is short, variously thickened, and more or less triangular in shape. The small eyes are not prominent. The top of the head is considerably elevated above the concave forehead. Adult males have a large structure on the face or muzzle, the "natalid organ," which is composed of cells that show resemblance to sensory cells but may also have a glandular function. As far as is known, this structure is confined to the Natalidae; it is not always noticeable externally. The muzzle is elongate and lacks a nose leaf. The nostrils are oval, set close together, and open near the margin of the lip. The lower lip is broad, reflected outward in front, and often has transverse grooves. The short thumb, bound to the wing by a membrane at its base, bears a well-developed claw. The very slender tail is completely enclosed within the large tail membrane. The membrane is pointed at its free edge at the tip of the tail. The normal-sized feet bear long, slightly recurved claws.

The upper incisors are separated from one another and from the canines. The first and second premolars are well developed in both jaws. The dental formula is: (i 2/3, c 1/1, pm 3/3, m 3/3) × 2 = 38.

In Venezuela, Handley (1976) collected N. stramineus primarily in lowland forest. In Jalisco, Mexico, Watkins, Jones, and Genoways (1972) found this species from sea level to about 2,500 meters. In general, bats of this genus roost in the darkest recesses of caves and mine tunnels, often with other kinds of bats. N. tumidirostris has been found in a hollow rubber tree. Funnel-eared bats were discovered in a dormant condition in a cave in oak forest near Ciudad Victoria, Tamaulipas, Mexico, by William J. Schaldach, Jr. They were hanging singly. The temperature on this date in January was about 12° C outside the cave and had been fairly low for about three days prior to this date. The fluttering, almost mothlike flight of funnel-eared bats is distinctive. They feed on insects.

These bats may be found in large groups, but sometimes fewer than a dozen individuals are present. Kerridge and Baker (1978) reported finding as many as several hundred N. micropus hanging in loose clusters from ledges in a deep cave on Jamaica. The sexes of N. stramineus appear to segregate when the young are born. In N. lepidus from Cuba and some smaller adjacent islands there seems to be only partial segregation of the sexes, at least in July. Baker, Genoways, and Patton (1978) reported a pregnant female N. stramineus on Guadeloupe in August. This species breeds during the dry season in El Salvador and southern Mexico. Data cited by Hayssen, Van Tienhoven, and Van Tienhoven (1993) indicate that females are monestrous and that embryonic development is very slow, the total gestation period being 8–10 months. In Sonora, northwestern Mexico, implantation occurs in December and early January, births in late July and August. Pregnant female N. stramineus have been found from January through July in Mexico and Central America. There is one young per female; its birth weight is about 2.1 grams, which is more than 50 percent of adult mass.

This family is known from both the Pleistocene and the Recent of South America. It also is known by many fossils from the West Indies, but it is often difficult to determine

Funnel-eared bat *(Natalus stramineus)*, photo by Merlin D. Tuttle, Bat Conservation International.

precisely whether they date from the late Pleistocene (prior to about 10,000 years ago), early Recent, or even historical time (less than 5,000 years ago). It is apparent that *Natalus* disappeared from much of its range because of early human exploitation (see above). Some remaining populations are thought to be susceptible to further declines. The IUCN now classifies *N. tumidifrons* as vulnerable and *N. lepidus* as near threatened.

CHIROPTERA; **Family FURIPTERIDAE**

Smoky Bats, or Thumbless Bats

This family of two genera and two species inhabits southern Central America and tropical South America. There has been controversy regarding the systematic position of this family, though generally the Furipteridae are considered to be closely related to the Natalidae, the Thyropteridae, and the Myzopodidae and to belong in the superfamily Vespertilionoidea (Ibáñez 1985; Koopman 1984b). Fossils referable to this family have not been found.

Smoky bats are small: head and body length is 33–58 mm, tail length is 24–36 mm, and forearm length is 30–40 mm. They are delicate and in general appearance resemble the funnel-eared bats (Natalidae) and the disk-winged bats (Thyropteridae). The thumb is present but is so small that it appears to be absent; it is included within the wing membrane to the base of the small, functionless claw. The wings

are relatively long. The crown of the head is greatly elevated above the face line, and the snout is cut off or truncated, with the end in the form of a disk or pad. The ears are separate and funnel-shaped, as in the Natalidae and Thyropteridae. Their broad bases cover the eyes. The tragus is small with a broad base and somewhat triangular in shape. The oval or triangular nostrils are set close together and open downward. There is no nose leaf. The tail is relatively long, but it does not extend to the free edge of the tail membrane and does not perforate it or appear on its upper surface. The legs are long and the feet are short, the latter ending in long, recurved claws. The fur tends to be coarse. Females have one pair of abdominal mammae (Ibáñez 1985). The dental formula is: (i 2/3, c 1/1, pm 2/3, m 3/3) $\times$ 2 = 36.

CHIROPTERA; FURIPTERIDAE; **Genus FURIPTERUS**
Bonaparte, 1837

The single species, *F. horrens,* occurs from Costa Rica to Peru and Brazil and on Trinidad (Cabrera 1957; E. R. Hall 1981; Tuttle 1970).

Head and body length is about 33–40 mm, tail length is 24–36 mm, forearm length is 30–40 mm, and weight is about 3 grams. Uieda, Sazima, and Storti Filho (1980) found females to be significantly larger than males. The coloration is brownish gray, dark gray, or slaty blue above and somewhat paler below. The fur on the head is long and thick and covers all the head as far as the snout, almost concealing the mouth. Both surfaces of the tail membrane are haired, and the tail is short.

This genus and *Amorphochilus* are distinguished by the reduced thumb, which is included in the wing membrane to the base of the small, functionless claw. Nasal appendages are lacking; the muzzle and lips do not have conspicuous warty outgrowths; and the tail terminates short of the posterior border of the interfemoral membrane, not perforating it or appearing on its upper surface. The snout is piglike in form and turned upward slightly. The tragus resembles a barbed arrowhead in form.

Smoky bat *(Furipterus horrens)*, photo by Merlin D. Tuttle, Bat Conservation International.

Smoky bat *(Furipterus horrens):* Top, photo from *Proc. Zool. Soc. London;* Bottom, photo by Richard K. LaVal.

In Panama several specimens were recently collected in a cave; they were hanging in a large, domed, well-lit chamber in company with several *Mimon bennettii* (Charles O. Handley, Jr., U.S. National Museum of Natural History, pers. comm.). In Venezuela, Handley (1976) collected 6 specimens near streams or in other moist areas, 2 of them in evergreen forest and 4 in yards. In Costa Rica, LaVal (1977) found a colony of more than 59 individuals, all males, on 12 May. The bats were hanging in clusters from the top of a hollow log in primary forest. Some were also observed to forage at heights of 1–5 meters above the forest floor; flight was slow and mothlike. Whenever roosting bats were approached they were alert and took flight immediately.

Uieda, Sazima, and Storti Filho (1980) observed two colonies in caves in northeastern Brazil during January and February. The bats left their caves to forage when darkness was complete, and examination of digestive tracts showed the main prey to be Lepidoptera. One colony contained about 250 individuals divided into groups of 4–30 roosting in holes in the walls. In the other cave there were 150 bats roosting separately from each other. The latter colony was composed of adult males, adult females, and young. Although the females hung in the usual head-down position, the young positioned themselves head-up on the mother's body, perhaps in response to the unusual abdominal location of the mammae in the Furipteridae.

CHIROPTERA; FURIPTERIDAE; Genus
AMORPHOCHILUS
Peters, 1877

The single species, *A. schnablii,* is known from Puna Island off the coast of Ecuador, several other points along the coast of Ecuador and Peru, an inland area in northern Peru, and

northern Chile (Ibáñez 1985; Koopman 1978*b*). The type specimen is in the Berlin Museum. These bats are also represented in the collections of the Field Museum of Natural History, the American Museum of Natural History, the U.S. National Museum of Natural History, the Museum of Comparative Zoology at Harvard University, and the Lima Museum in Peru.

Head and body length is 39–67 mm, tail length is 23–31 mm, and forearm length is 36–39 mm. The long fur is pale gray lightly washed with brown. The tail membrane is pale brown and the wings and ears are brownish gray (Redford and Eisenberg 1992). The reduced thumb is included within the wing membrane to the base of the small, functionless claw. Nasal appendages are lacking, and the tail terminates short of the posterior border of the interfemoral membrane, not perforating it or appearing on its upper surface. This genus differs from *Furipterus* in the height of the braincase and in the presence of conspicuous wartlike outgrowths on the muzzle and lips.

Amorphochilus seems to occur mainly in cultivated valleys and arid regions within its range. It has been collected in an unused sugar mill, dark wine storehouses, and an irrigation tunnel; two individuals have been taken on separate occasions at 0300 hours. Redford and Eisenberg (1992) stated that it is found along the coast near the mouths of rivers, roosts in cracks in rocks, and feeds on small dipterans and lepidopterans. Ibáñez (1985) reported the discovery of a colony in a culvert at a road crossing within an area of tropical desert bush at the mouth of the Javita River, Ecuador. Analysis of the stomach contents of 5 specimens showed Lepidoptera to be the only prey. When the colony was found, on 19 November 1981, it contained 300 individuals and included both sexes but no juveniles. None of the females were lactating, but 8 of the 10 examined contained a single fetus. On the Pacific slope of Peru, Graham (1987) collected juveniles in April and pregnant females in June, Au-

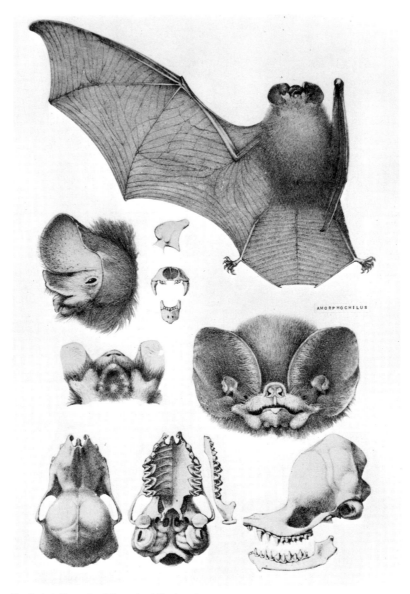

Smoky bat *(Amorphochilus schnablii)*, photo from *Die Fledermäuse des Berliner Museums für Naturkunde*, Wilhelm K. Peters.

gust, and October. The IUCN classifies *Amorphochilus* as vulnerable. It is expected to decline by at least 20 percent over the next decade because of human habitat disturbance.

CHIROPTERA; Family THYROPTERIDAE, Genus THYROPTERA
Spix, 1823

Disk-winged Bats, or New World Sucker-footed Bats

The single known genus, *Thyroptera*, contains three species (Anderson and Webster 1983; Cabrera 1957; E. R.

Hall 1981; Jones, Arroyo-Cabrales, and Owen 1988; Pine 1993; Torres, Rosas, and Tiranti 1988; Wilson 1976):

T. discifera, southern Nicaragua to Guianas, Amazonian Brazil, and Bolivia;
T. lavali, Departamento de Loreto in extreme northeastern Peru;
T. tricolor, southern Mexico to Bolivia and southern Brazil, Trinidad.

In *T. discifera* and *T. tricolor* head and body length is 34–52 mm, tail length is 25–33 mm, and forearm length is 27–38 mm. Pine (1993) indicated that *T. lavali* is larger, with a forearm length of about 40 mm. Findley and Wilson (1974) reported an average weight of 4.2 grams for *T. tricolor*, and Eisenberg (1989) noted that one *T. discifera* weighed 6

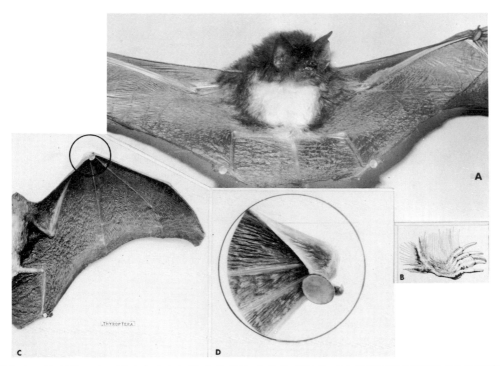

A. Disk-winged bat *(Thyroptera tricolor)*, photo by Merlin D. Tuttle. B. *T. discifera,* drawing showing syndactylism of third and fourth digits of foot, from *Proc. Biol. Soc. Washington.* C. Wing and foot showing position of thumb and disk, photo from Field Museum of Natural History. D. Enlargement of thumb disk.

Disk-winged bat *(Thyroptera tricolor),* photo by Merlin D. Tuttle, Bat Conservation International.

grams. *T. tricolor* is reddish brown or somewhat darker above and white below, the lateral extent of the white area being variable. *T. discifera* is reddish brown above and brown below. The ears in *T. tricolor* are blackish, whereas in *T. discifera* they are yellowish and larger. The pelage of *T. lavali* is appreciably darker above and below than that of the other two species (Pine 1993).

These bats have circular suction disks or cups with short stalks on the soles of the feet and at the base of the well-developed claw of the thumb. In their Old World counterpart, *Myzopoda*, of Madagascar, the suction disks are sessile and the claw on the thumb is vestigial. These concave disks are larger on the thumbs than on the feet. American sucker-footed bats are small and delicately formed. The crown of the head is considerably elevated above the concave forehead. The ears are separate and funnel-shaped, and a tragus is present. The muzzle is elongate and slender. There is a small wartlike projection above the nostrils but no nose leaf. The nostrils are circular and set rather far apart. The tip of the tail extends slightly beyond the free, or rear, edge of the broad tail membrane. As in the smoky bats (Furipteridae), the third and fourth toes are joined.

The upper incisors on each side are widely spaced from each other and from the canines. The premolars are well developed in both jaws, and the molars exhibit a W pattern of cusps and ridges. The dental formula is: (i 2/3, c 1/1, pm 3/3, m 3/3) × 2 = 38.

In Venezuela, Handley (1976) collected most specimens in moist parts of evergreen forest. Unlike most other bats, the members of this family usually hang head upward. Disk-winged bats cling to smooth surfaces by means of the suction disks, the suction of a single disk being capable of supporting the entire weight of the bat. The usual daytime retreat is in a rolled frond or a curled leaf of some plant, such as the heliconia tree or the banana. As many as eight individuals have been found inside a curled leaf, but usually only one or two disk-winged bats rest in a given shelter. When several utilize a trumpet-shaped leaf as a roost, they space themselves evenly one above the other. Two *Thyroptera* have been found with *Rhynchonycteris*, the proboscis bat, in a curled heliconia leaf. The diet consists of insects.

In a study in Costa Rica, Findley and Wilson (1974) found the home range of *T. tricolor* to average about 3,028 sq meters and density to average 3.3 colonies, or 19.8 bats, per hectare. Group size ranged from 1 to 9 and averaged 6. Each roosting aggregation was found to have social cohesion and would rejoin. In mid-August 1963 it was estimated that 60–80 percent of the adult females were pregnant. A lactating female *T. tricolor* carrying a young male was observed in May on Barro Colorado Island, Panama. The young, which weighed approximately 46 percent as much as its mother, remained attached to her when she flew around a room. LaVal and Fitch (1977) reported a juvenile in Costa Rica in June. In eastern Peru, Graham (1987) collected juveniles in March and October, pregnant females in July, and lactating females in December. Of three adult females collected in Bolivia in November two were each nursing a single young and a third was pregnant with one embryo (Anderson and Webster 1983).

No fossils referable to this family have been found. Two of the living species are widely distributed, but *T. lavali* is restricted to a very small area that is experiencing intensive environmental disturbance. It is classified as vulnerable by the IUCN.

CHIROPTERA; Family MYZOPODIDAE; Genus MYZOPODA
Milne-Edwards and Grandidier, 1878

Old World Sucker-footed Bat

The single known genus and species, *Myzopoda aurita*, now is restricted to Madagascar. Known localities recorded are Majunga on the west coast and Maroantsera, Mananara, Mahambo, Tamatave, Mananjary, and the northern vicinity of Fort Dauphin on the east coast (Schliemann and Maas 1978). *Myzopoda* is the only genus of bats now restricted to Madagascar, though in the Pleistocene it also occurred in East Africa (Koopman 1984a). Although both *Myzopoda* and *Thyroptera* possess suction disks, these organs are completely different in the two genera with regard to histological and anatomical details and probably evolved independently (Schliemann and Maas 1978).

Approximate measurements are as follow: head and body length, 57 mm; tail length, 48 mm; and forearm length, 46–50 mm. This bat may be recognized by the sessile adhesive pads or disks on the wrists and ankles, whereas the suction disks in *Thyroptera* are provided with a short stem or stalk. In addition, *Myzopoda* has a large ear with a tragus and a unique mushroom-shaped process. The lips are wide, and the upper lip extends beyond the lower. The thumb has a vestigial claw, and the tail extends beyond the tail membrane.

The skull is short, broad, and rounded. The teeth are of the normal cuspidate insectivorous type. The dental formula is: (i 2/3, c 1/1, pm 3/3, m 3/3) × 2 = 38.

The suction disks do not appear to be as efficient as those of *Thyroptera*, but the bats probably use the pads to hold on to the smooth hard stems and leaves of palms and other smooth surfaces. Thewissen and Etnier (1995), noting that there are many glands on the surface of the pads, suggested that the secretions thereof might provide a kind of glue that functions in adhesion.

The IUCN now classifies *Myzopoda* as vulnerable. It is expected to decline by at least 20 percent over the next decade because of human habitat disturbance.

CHIROPTERA; Family VESPERTILIONIDAE

Vespertilionid Bats

This family of 43 genera and 342 species has a worldwide distribution in the temperate and tropical regions. *Myotis*, a member of this family, has the widest distribution of any genus of bats. The sequence of genera presented herein basically follows that of Corbet and Hill (1991), who recognized six subfamilies: Vespertilioninae, with most genera; Miniopterinae, with the genus *Miniopterus*; Murininae, with the genera *Murina* and *Harpiocephalus*; Kerivoulinae, with *Kerivoula*; Nyctophilinae, with *Nyctophilus* and *Pharotis*; and Tomopeatinae, with *Tomopeas*. Exceptions to Corbet and Hill's arrangement include maintaining a close association of *Glauconycteris* and *Chalinolobus* in keeping with Koopman (*in* Wilson and Reeder 1993) and disassociating *Nycticeius*, *Rhogeessa*, and *Otonycteris* from *Plecotus* and related genera in accordance with Frost and Timm (1992). However, the latter authorities suggested that the appropriate sequence for the plecotine genera would be *Euderma, Idionycteris, Barbastella, Plecotus, Corynorhinus*, which is somewhat different from that used here and from

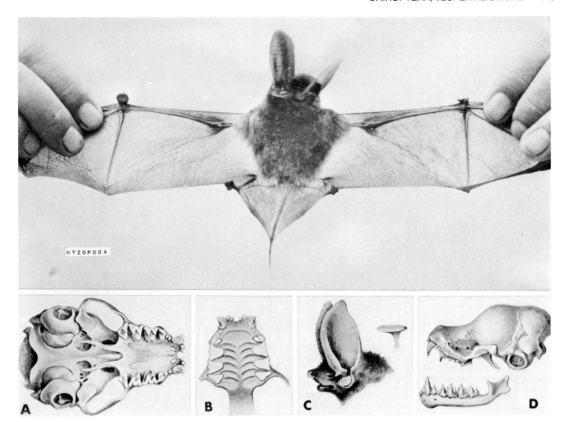

Old World sucker-footed bat *(Myzopoda aurita)*, photo by Harry Hoogstraal. A. Ventral view of skull; B. Palate; C. Head and thumb disk; D. Lateral view of skull, photos from *Proc. Zool. Soc. London.*

that suggested by Tumlison and Douglas (1992). Koopman (1984*b* and *in* Wilson and Reeder 1993) placed the members of the Nyctophilinae in the Vespertilioninae and presented the accepted subfamilies in the following order: Kerivoulinae, Vespertilioninae, Murininae, Miniopterinae, and Tomopeatinae. Based on karyological analysis of 23 genera, Volleth and Heller (1994*b*) proposed that *Myotis* be placed in a separate subfamily (Myotinae), treated the Nyctophilinae as part of the Vespertilioninae, and suggested the following evolutionary sequence: Miniopterinae, Kerivoulinae, Murininae, Myotinae, and Vespertilioninae (the Tomopeatinae were not included in the study). Volleth and Heller (1994*b*) also rejected a recommendation by Mein and Tupinier (1977) that the Miniopterinae be elevated to familial rank. A recent study of protein electrophoresis and mitochondrial DNA indicated that the Tomopeatinae actually are referable to the family Molossidae (Sudman, Barkley, and Hafner 1994).

Head and body length is 32–105 mm, tail length is 25–75 mm, and forearm length is 22–75 mm. The well-developed tail extends to the rear edge of the tail membrane or slightly beyond. Adults weigh from about 4 to 50 grams. The coloration is generally blackish, gray, or various shades of brown, but a number of species are red, yellow, or orange. An ornate color pattern is present in *Scotomanes ornatus*, and the spotted bat, *Euderma*, has a characteristically arranged pattern of white spots. A nose leaf is lacking except in the genera *Nyctophilus* and *Pharotis*. Tubular nostrils are present in two genera, *Murina* and *Harpiocephalus*. In some members of this family large glands are

Spotted bat *(Euderma maculatum)*, with 24-hour-old baby, photo by David A. Easterla.

Myotis lucifuga catching a moth. Inset shows bat with the moth caught in its interfemoral membrane. Photos by Frederick A. Webster and D. A. Cahlander.

Long-eared bats of eastern North America *(Corynorhinus rafinesquii)* hanging in their normal position for resting: A. With ears partially uncoiled from the tight coil when sleeping; B. With ears fully extended when awake. Photos by Ernest P. Walker.

present underneath the skin on the snout, producing large bulges or folds of the skin covering them. Fleshy lobes are associated with the mouth in *Chalinolobus* and *Glauconycteris*. The eyes are minute. The ears are as long as 40 mm, usually separate, and have a tragus and an anterior basal lobe (except in *Tomopeas*). Small suction disks or pads are present on the soles and/or wrists in *Glischropus, Eudiscopus,* and *Tylonycteris,* as well as in some species of *Pipistrellus* and *Hesperoptenus*. Wing glands are present in *Myotis vivesi* and *Cistugo.* Most females have two mammae, except for those of the genus *Lasiurus,* which have four.

The incisors are small and separated medially. The molars have a well-developed W-shaped pattern of cusps and ridges, a few genera showing a tendency toward the reduction of the cusps. The dental formula varies from (i 1/2, c 1/1, pm 1/2, m 3/3) × 2 = 28 to (i 2/3, c 1/1, pm 3/3, m 3/3) × 2 = 38.

These bats range in altitude to the limits of tree growth and from tropical forests to arid areas. Most vespertilionids are cave dwellers, but they also shelter in mine shafts, tunnels, old wells, rock crevices, buildings, tree hollows, the foliage of trees and bushes, hollow joints of bamboo, large tropical flowers, tall grass, abandoned bird nests, storm sewers, and culverts, as well as under rocks and loose bark on tree trunks. The members of this family generally pitch head upward when alighting, hanging by their thumbs and toes, then quickly shuffle around until they are hanging head downward by their toes, their wings folded alongside the body. Often they hang on a vertical surface rather than suspend themselves freely. Some members of this family are solitary, others roost in pairs or in small groups, and still

others generally shelter in colonies. The colonial forms usually return to the same roosting site year after year. Some species remain in colonies throughout the year, whereas others congregate only in winter. A number of forms migrate between summer and winter quarters. Some of the temperate-region forms with tree-dwelling habits migrate to warmer climates for the winter, whereas those having cave-dwelling habits hibernate, though they often change hibernating places during the winter. Homing ability has been demonstrated in several forms. Vespertilionids send out ultrasonic sounds through their mouths.

Nearly all members of this family feed mainly on insects. *Myotis vivesi* is definitely known to feed on fish, and two other species of *Myotis* are suspected to do so. *Antrozous pallidus* in captivity has captured and eaten lizards. Vespertilionids generally capture insects in flight, often using the well-developed tail membrane as a pouch in which to manipulate the larger kinds of prey. A fairly definite feeding territory is often established. The feeding flights generally alternate with periods of rest, when the bat hangs to digest its catch.

Females of many colonial species segregate into maternity colonies to bear and raise the young. Males do not exhibit an active interest in the young. In those forms from temperate climates that hibernate, breeding occurs from August through October and often again in the spring, the two periods producing only a single litter since the sperm from the former period are stored in the reproductive tract of the female over the winter and ovulation and fertilization occur only in the spring. *Lasiurus,* a genus that occurs in temperate regions and is thought to be migratory, breeds mainly in late summer and early autumn, the births in northern latitudes taking place from late May to early July. In those forms that occur in tropical regions breeding is immediately followed by fertilization and the development of the embryos. The gestation period in most forms is 40–70 days, but in some species it is 100 or more days. The number of young is 1–4. The life span in the wild may be no more than 4–8 years for most individuals, but there are records of banded individuals more than 21 years old, and captives are known to have lived more than 20 years.

The geological range of this family is middle Eocene to Recent in Europe, late Oligocene to Recent in North America, middle Miocene to Recent in Africa and Asia, Pleistocene to Recent in the West Indies, South America, and Australasia, and Recent over the remainder of the present range (Koopman 1984a).

CHIROPTERA; VESPERTILIONIDAE; Genus CISTUGO
Thomas, 1912

Wing-gland Bats

There are two species (Hayman and Hill *in* Meester and Setzer 1977; Herselman and Norton 1985):

C. seabrai, Angola, Namibia, northwestern Cape Province in South Africa;
C. lesueuri, known only by eight specimens from the Cape Province in South Africa.

The authorities cited above, as well as Corbet and Hill (1991) and Koopman (*in* Wilson and Reeder 1993), treated *Cistugo* as a subgenus of *Myotis*. However, a recent karyological analysis, together with assessment of morphological characters, led Rautenbach, Bronner, and Schlitter (1993) to conclude that *Cistugo* is a distinct genus, perhaps the most primitive in the family Vespertilionidae.

Head and body length is 40–47 mm, tail length is 40–43 mm, and forearm length is 32–35 mm. The coloration in *C. seabrai* is drab brown to slaty; *C. lesueuri* has yellowish brown upper parts and yellowish white underparts.

Certain dental features and the presence of wing glands distinguish *Cistugo* from *Myotis*. In *C. seabrai* the glands are thicker, broader, and in a different position from those in *C. lesueuri;* in some specimens of the former species two glands set close together are present in each wing. The wing glands of *C. lesueuri* are reported to be evident only in the living animal and in a moistened museum specimen, not in a dry skin.

These bats frequent groves of trees in arid regions. Both species begin flying soon after sunset with a fairly steady and direct flight, but they have been observed to descend later and flutter around orange trees and bushes. Perhaps this indicates that they regularly begin feeding on flying insects and later in the evening hunt for insects on foliage. The type specimen of *C. lesueuri* was taken from a cat. Although *C. seabrai* appears to be locally common, both species occupy very restricted ranges and were designated as indeterminate by Smithers (1986). The IUCN classifies both as vulnerable, placing them in the genus *Myotis* and noting that they are likely to decline by at least 20 percent over the next decade because of habitat loss.

CHIROPTERA; VESPERTILIONIDAE; Genus MYOTIS
Kaup, 1829

Little Brown Bats

There are 6 subgenera and 87 species (Albayrak 1993; Anderson, Koopman, and Creighton 1982; Anderson and Webster 1983; Barquez and Ojeda 1992; Baud 1979; Baud and Menu 1993; Blood and McFarlane 1988; Bogan 1978; Bogdanowicz 1990; Cabrera 1957; Carter et al. 1981; Chasen 1940; Corbet 1978; Corbet and Hill 1991, 1992; DeBlase 1980; Dolan and Carter 1979; Ellerman and Morrison-Scott 1966; Findley 1972; Flannery 1990b, 1995; Gaisler 1983; Gauckler and Kraus 1970; Genoways and Williams 1979a; E. R. Hall 1981; Hayman and Hill *in* Meester and Setzer 1977; Hill 1962, 1972a, 1974b, 1983a; Hill and Beckon 1978; Hill and Francis 1984; Hill, Harrison, and Jones 1988; Hill and Morris 1971; Hill and Thonglongya 1972; Hill and Topal 1973; Hoffmann, Jones, and Campbell 1987; Horacek and Hanák 1984; Kitchener, Cooper, and Maryanto 1995; Koopman 1982a, 1982b, 1984c, 1989c, and *in* Wilson and Reeder 1993); Laurie and Hill 1954; LaVal 1973b; Lekagul and McNeely 1977; Manning 1993; Masson and Breuil 1992; McCarthy and Bitar 1983; Myers and Wetzel 1983; Neuhauser and DeBlase 1974; Polaco, Arroyo-Cabrales, and Jones 1992; Qumsiyeh 1980, 1983; Redford and Eisenberg 1992; Reduker, Yates, and Greenbaum 1983; Rickart et al. 1993; Ride 1970; Roberts 1977; Rzebik-Kowalska, Woloszyn, and Nadachowski 1978; Taylor 1934; Tupinier 1977; Van Zyll de Jong 1984; Yoshiyuki 1984):

subgenus *Myotis* Kaup, 1829

M. myotis, England, Azores, central and southern Europe, Ukraine, most Mediterranean islands, Asia Minor, Lebanon, Palestine;
M. blythii, southern Europe, northern Africa, southwestern Asia, parts of Central Asia, Himalayas, Mongolia, northeastern China;
M. chinensis, eastern and southern China, northern Thailand;
M. sicarius, Nepal, Sikkim;
M. pequinia, eastern China;
M. altarium, southern China, northern Thailand;
M. emarginata, southern Europe to Uzbekistan and Pakistan, northern Africa;
M. tricolor, Liberia, Zaire and Ethiopia to South Africa;
M. morrisi, Nigeria, Ethiopia;
M. goudoti, Madagascar, Anjouan (Comoro Islands);
M. bechsteini, Europe, Iran;
M. keenii, southeastern Alaska, Mackenzie, western British Columbia and Washington, Saskatchewan and Newfoundland to northwestern Florida;
M. evotis, southwestern Canada, western conterminous United States, Baja California;
M. auriculacea, southwestern United States to central Mexico, Guatemala;
M. thysanodes, western North America from southern British Columbia to southern Mexico;

subgenus *Chrysopteron* Jentink, 1910

M. formosa, eastern Afghanistan, northern India to eastern China, Korea, Tsushima Island (off southwestern Japan), Taiwan, Java, Bali, Luzon (Philippines), Sulawesi;

Little brown bat *(Myotis)*: A. *M. nattereri*, photo by Frank W. Lane; B. *M. evotis*, photo by Ernest P. Walker; C. *M. lucifuga*, photo by Ernest P. Walker.

M. hermani, northwestern Sumatra;
M. welwitschii, Zaire and Ethiopia to South Africa;

subgenus *Selysius* Bonaparte, 1841

M. mystacina, entire Palaearctic region from Ireland and Morocco to Japan and the Himalayas, southeastern China, possibly Hainan;
M. brandti, Britain and Spain to Kamchatka Peninsula and Japan;
M. abei, Sakhalin Island;
M. frater, Uzbekistan to Manchuria and southeastern China, Japan;
M. insularis, supposedly Samoa;
M. muricola, Afghanistan and northern India to southern China and Malay Peninsula, Sumatra, Nias Island, Mentawai Islands, Java, Kangean Islands, Bali, Borneo, Natuna Islands, throughout the Philippines, Sulawesi, Lombok, Sumbawa, Sumba, Flores, Amboina Island;
M. patriciae, Mindanao;
M. atra, Mentawai and Siberut islands off western Sumatra, Borneo, Culion Islands (Philippines), Sulawesi, Togian Islands, Molucca Islands, possibly New Guinea and Australia;
M. australis, New South Wales, possibly northern Western Australia;
M. ikonnikovi, eastern Siberia, Mongolia, Manchuria, Korea, Sakhalin, Hokkaido;
M. hosonoi, Honshu;
M. yesoensis, Hokkaido;

M. ozensis, Honshu;
M. siligorensis, northern India to southern China and Malay Peninsula, Borneo;
M. annectans, northeastern India and Bangladesh to Thailand;
M. oreias, Singapore;
M. rosseti, Thailand, Cambodia;
M. ridleyi, Malay Peninsula, probably Sumatra, Borneo;
M. ciliolabrum, southwestern Canada to western Oklahoma and Puebla (south-central Mexico);
M. californica, western North America from extreme southeastern Alaska to Guatemala;
M. leibii, southeastern Canada to eastern Oklahoma and Georgia;
M. scotti, highlands of Ethiopia;
M. sodalis (Indiana bat), eastern United States;
M. nigricans, east and west coasts of central Mexico to northern Argentina, Trinidad and Tobago, Grenada (Lesser Antilles);
M. atacamensis, coastal desert of southern Peru and northern Chile;
M. keaysi, eastern and southern Mexico to Venezuela and northern Argentina, Trinidad;
M. martiniquensis, Martinique and Barbados in Lesser Antilles;
M. carteri, western Mexico;
M. planiceps, northeastern Mexico;
M. findleyi, Tres Marías Islands off Nayarit (western Mexico);
M. elegans, southern Mexico to Costa Rica;

M. dominicensis, Dominica and perhaps Guadeloupe in the Lesser Antilles;

M. nesopolus, northern Venezuela and nearby islands of Bonaire and Curacao;

subgenus *Isotus* Kolenati, 1856

M. nattereri, Europe, Morocco, Algeria, Caucasus, Turkey, Kurdistan (northern Iraq), Palestine, Jordan, Iran, Turkmenistan;

M. schaubi, Armenia, western Iran;

M. bombina, southeastern Siberia, Manchuria, Korea, Japan;

subgenus *Leuconoe* Boie, 1830

M. horsfieldii, India, South Andaman Island, Thailand, Malay Peninsula, Hainan, Sumatra, Java, Bali, Borneo, Mindanao (Philippines), Sulawesi;

M. adversa, Malay Peninsula, Taiwan, Java, Kangean Islands, Borneo, Talaud Islands, Sulawesi, Togian Islands, Lesser Sunda Islands, Molucca Islands, New Guinea, Bismarck Archipelago, D'Entrecasteaux Islands, Solomon Islands, Vanuatu (New Hebrides), northern and eastern Australia;

M. hasseltii, northeastern India, Burma, Thailand, Cambodia, Viet Nam, Malay Peninsula, Sri Lanka, Riau Archipelago, Sumatra, Mentawai Islands, Java, Borneo, possibly Sulawesi;

M. bocagei, Senegal to Malawi and Transvaal, Yemen;

M. montivaga, India, Burma, southeastern China, Malay Peninsula, Borneo;

M. riparia, western Honduras to Uruguay, Trinidad;

M. fortidens, western and southern Mexico, Guatemala;

M. lucifuga, Alaska, Canada, conterminous United States, northern Mexico (a single record from Iceland probably represents accidental introduction by human agency [Koopman and Gudmundsson 1966]);

M. velifer, southwestern United States to Honduras;

M. grisescens (gray bat), eastern Kansas and Oklahoma to western Virginia and northwestern Florida;

M. chiloensis, Chile and adjacent parts of southern Argentina;

M. aelleni, northern Patagonia (Argentina);

M. ruber, southeastern Brazil, Paraguay, northern Argentina;

M. peninsularis, southern Baja California;

M. cobanensis, known only from Coban in Guatemala;

M. daubentoni, Ireland and Portugal to Kamchatka Peninsula and southeastern China, Japan, Hainan;

M. capaccinii, southern Europe, Turkey and Palestine to Uzbekistan, northwestern Africa;

M. longipes, Afghanistan, Kashmir, possibly Viet Nam;

M. macrodactyla, Maritime Provinces in southeastern Siberia, Kuril Islands, Japan;

M. fimbriata, southeastern China;

M. yumanensis, western North America from British Columbia to central Mexico;

M. austroriparia, southeastern United States;

M. albescens, southern Mexico to Uruguay;

M. sima, Amazonian region of Colombia, Ecuador, Peru, Bolivia, and Brazil;

M. levis, southern Brazil, Bolivia, Paraguay, Uruguay, Argentina;

M. oxyota, Costa Rica to Bolivia;

M. volans, western North America from southeastern Alaska to central Mexico;

M. pruinosa, Honshu and Shikoku;

M. macrotarsa, Borneo, Philippines;

M. stalkeri, Gebe and Kei islands in the Moluccas;

M. dasycneme, eastern France to western Siberia, Manchuria;

M. ricketti, eastern China;

subgenus *Pizonyx* Miller, 1906

M. vivesi (fishing bat), coastal parts of western Mexico and Baja California.

In accordance with Woodman (1993), the spelling of specific names is in the feminine form (i.e., *atra* rather than *ater, austroriparia* rather than *austroriparius, macrodactyla* rather than *macrodactylus,* etc.). E. R. Hall (1981) accepted the validity of the above subgenera and also *Hesperomyotis* Cabrera, 1958, for the species *M. sima.* Baud and Menu (1993) concluded that *M. sima* did not warrant subgeneric distinction (and also that this species did not occur beyond the Amazon Basin). Findley (1972) and Koopman (*in* Wilson and Reeder 1993) accepted only *Myotis* (including *Chrysopteron* and *Isotus*), *Selysius,* and *Leuconoe* (including *Hesperomyotis*). Corbet and Hill (1991, 1992) used the above subgenera and *Hesperomyotis,* as well as *Paramyotis* Bianchi, 1916, for the species *M. bechsteini, M. evotis, M. milleri, M. auriculacea,* and *M. keenii,* and *Rickettia* Bianchi, 1916, for *M. ricketti.* In addition, *Cistugo* (see account thereof) often is considered a subgenus of *Myotis.* Koopman's list (*in* Wilson and Reeder 1993) also differed from the above and from the lists of Corbet and Hill (1991, 1992) as follows: *M. altarium* was put in the subgenus *Selysius, M. hermani* was included in *M. formosa, M. abei* was put in the subgenus *Leuconoe, M. patriciae* and *M. atra* were included in *M. muricola, M. annectans* was put in the subgenus *Myotis, M. ciliolabrum* was included in *M. leibii, M. carteri* was included in *M. nigricans,* and *M. peninsularis* was put in the subgenus *Myotis.* Koopman, like most other authorities, listed *M. milleri* of the Sierra San Pedro Martir in northern Baja California as a separate species, but Manning (1993) designated it a subspecies of *M. evotis.* Manning, like certain other authorities, also treated *M. septentrionalis,* found from Saskatchewan and Newfoundland to northwestern Florida, as a species distinct from *M. keenii.* Kitchener, Cooper, and Maryanto (1995) distinguished *M. moluccarum* of Seram, New Guinea, and northern Australia and *M. macropa* of southeastern Australia as species separate from *M. adversa.* Lekagul and McNeely (1977) considered *M. muricola* to be a subspecies of *M. mystacinus.* Yoshiyuki (1989) treated *M. fujiensis* of Honshu and *M. gracilis* of southeastern Siberia and Hokkaido as species distinct from *M. brandti.* E. R. Hall (1981) used the name *M. subulatus* in place of *M. leibii. M. australis* was considered probably conspecific with *M. ater* by Hill (1983*a*). The name *Anamygdon solomonis,* which applies to a bat from the Solomon Islands, was considered by Phillips and Birney (1968) to be a synonym of *Myotis moluccarum,* which now is regarded as a subspecies of *M. adversa.*

Head and body length is about 35–100 mm, tail length is 28–65 mm, and forearm length is 28–70 mm. *M. lucifuga,* a small North American species, usually weighs 5–14 grams. *M. siligorensis,* the smallest Old World species, weighs 2.3–2.6 grams (Lekagul and McNeely 1977). *M. myotis,* the largest species, weighs 18–45 grams. The upper parts are generally some shade of brown, and the underparts are paler. Some forms have light and dark color phases. *M. formosa* is remarkable for its orange coloration. *M. vivesi* is dark buff or pale tan above and whitish below.

The face of a mouse-eared bat (Myotis myotis), considerably enlarged. Note the sharp teeth, which are well adapted for piercing and chewing insects. Photo by Walter Wissenbach.

It differs from most *Myotis* in its greatly elongated feet and large, laterally compressed toes and hind claws. *M. macrotarsa* and *M. stalkeri* of the East Indies are also big-footed and may be closely related to *M. vivesi* (Findley 1972). The tragus of *Myotis* is erect and tapering.

This genus is the most widely distributed group of bats. It is absent only from arctic, subarctic, and antarctic regions and many oceanic islands. It probably has the widest natural distribution of any genus of terrestrial mammals except *Homo*. *M. chiloensis* is the southernmost occurring of any bat, specimens having been reported from Navarino and Wollaston islands, just south of Tierra del Fuego (see Pine, Miller, and Schamberger 1979).

There is considerable variation in habitat. In western North America, for example, *M. evotis* occupies coniferous forest, whereas *M. auricula* is found in arid woodland and desert scrub (Barbour and Davis 1969). In Venezuela, Handley (1976) collected *M. albescens* mostly in moist areas, both forested and open, whereas *M. nesopolus* was taken mainly in dry places. A number of species, including *M. grisescens*, *M. austroriparia*, and *M. yumanensis*, seem closely associated with water, roosting near and foraging over streams, ponds, or reservoirs (Barbour and Davis 1969; Tuttle 1976a). *M. vivesi* shelters under piles of stones along seashores or in sea caves and fissures and forages over the open sea. Patten and Findley (1970) observed a group of about 400 *M. vivesi* around a boat at least 7 km from the shore. *M. hasseltii* also is reported to forage over the ocean and sometimes to roost in adjacent mangroves. Caves are a more common roosting site for *Myotis*, but some species often use hollow trees, rock crevices, and structures erected by people as well. *M. formosa* has been found hanging in trees, bushes, and tall jungle grass.

All species roost by day and forage at night. The feeding flights usually alternate with periods of rest, during which the bats hang to digest their catch. The flight is described as slow and straight in *M. myotis* and erratic and faster in some other species. *M. lucifuga* attains flight speeds of up to 35 km/hr and averages about 20 km/hr (Godin 1977). Roosting sites sometimes change by season, and it is common in North American species for maternity colonies to break up in the autumn and shift to hibernacula. *M. lucifuga* may migrate up to 275 km from their summer to their winter roosts (Godin 1977). This species, like many others found in temperate regions, is known to hibernate during the winter. Probably the greatest range of body temperature in any vertebrate occurs in *M. lucifuga*; it has been cooled to 6.5° C without apparent harm and has also

been found at 54° C (Barbour and Davis 1969). Colonies of *M. grisescens* travel up to 6.6 km from roost to foraging areas; during the summer a colony occupies a definite home range and may move among six or more roosting caves. Large-scale movement by this species to hibernation sites begins in September, and this migration covers 17–437 km (Tuttle 1976a, 1976b). *M. sodalis* concentrates in caves during the autumn, preliminary to winter hibernation, but disperses to other habitats for the summer. A nursery colony was found in Indiana under the loose bark of a tree; the 50 bats present foraged over about 0.82 km of riparian habitat (Humphrey, Richter, and Cope 1977). Summer nursery colonies of *M. daubentoni*, a common Eurasian species, are usually found in hollow trees, whereas winter hibernation sites are caves, mines, cellars, and old underground military fortifications. The distance between summer and winter roosts of *M. daubentoni* is usually small, not over 19 km in England and generally 0.5–88.0 km in Germany. That species arrives at its winter roosts around October or November; hibernation lasts 175–90 days in central Europe (Bogdanowicz 1994).

The diet of *Myotis* usually consists predominantly of insects. *M. daubentoni* feeds largely on Lepidoptera, Diptera, and Hemiptera. It forages mostly less than 2 meters above ground or water level and often catches small insects with the aid of its tail membrane, which is formed into a pouch, or with its wing. There is some evidence that it also captures small fish at the water's surface, perhaps using its relatively large feet (Bogdanowicz 1994). *M. vivesi* definitely catches fish and small aquatic crustaceans. Exactly how this is accomplished is unknown, but the long feet and enlarged hind claws presumably aid in the capture. Using high-speed photography, Altenbach (1989) found that the uropatagium of *M. vivesi*, as well as its feet, is lowered into the water and evidently is involved in taking the prey and transferring it to the bat's mouth. Several other species, including *M. hasseltii* and *M. macrotarsa*, may also catch fish. Recently, Robson (1984) showed that fish are eaten by *M. adversa* of Australia, though its diet consists mainly of insects captured above the surface of the water.

In favorable New England habitat, *M. lucifuga* has a population density of about 10/sq km (Barbour and Davis 1969). *M. daubentoni* has been found at densities of about 1/sq km in the Czech Republic and 2.4/sq km in Scotland. Both summer and winter colonies of that species generally comprise about 100 individuals, though some aggregations are much larger (Bogdanowicz 1994). Most other species also are gregarious to some extent. *M. keenii* often roosts alone but forms small nursery colonies of 1–30 bats, and its hibernating groups may consist of up to 350 individuals (Barbour and Davis 1969; Fitch and Shump 1979). In Thailand, *M. horsfieldii* roosts in caves in groups of 2–3 to more than 100 and *M. hasseltii* is found in groups of up to 25 individuals of both sexes (Lekagul and McNeely 1977).

Usually the females of *Myotis* form maternity colonies to bear young, with few or no males present until after the young are born. Nursery colonies of *M. lucifuga* reportedly numbered 15–1,100 in Alberta (Schowalter, Gunson, and Harder 1979) and 12–1,200 in Vermont (Godin 1977). Colonies of several other species are in the same size range, but those of *M. austroriparia* usually number many thousands, a range of 2,000–90,000 having been reported in Florida. These great aggregations disperse in October. Hibernating groups are often considerably larger than the summer colonies; a single cave in Vermont contained 300,000 hibernating *M. lucifuga*, probably the total population of the species in an area of 22,300 sq km (Barbour and Davis 1969). About 90 percent of all *M. grisescens* east of the Mississippi River and south of Kentucky are thought to hibernate in just three caves, with the populations there being 125,000, 250,000, and 1,500,000 (Tuttle 1976b). In the summer the females collect in smaller groups at other caves to bear young, while the males assemble at sites nearby. After late July the males move into the maternity colonies. In a study of *M. bocagei* in Gabon, Brosset (1976) found the population divided largely into harems, each comprising a single adult male, 2–7 females, and recent young. Such groups roosted in the leaves of banana plants. Solitary males were also present in the population, and there was a rapid turnover of harem males, but breeding females tended to remain in the group for a considerable period. The young departed at 4–5 months, but young males sometimes returned at 12 months to try to take over the harem.

The usual reproductive pattern in temperate regions is: mating during the autumn; storage of sperm in the uterus of the female through winter hibernation; ovulation and fertilization in the early spring; and birth in the late spring or early summer. Considerable variation in the period of birth in *M. volans* was indicated by the discovery of a month-old young in Washington in March and a pregnant female in Colorado in August (Barbour and Davis 1969). The gestation period is 50–60 days in *M. lucifuga* and up to 70 days in *M. myotis*. In a study of *M. velifer* in Kansas, Kunz (1973a) found most matings to take place in October, ovulation in April, births in June, flight capability at 3 weeks, and sexual maturity in the first year of life. O'Farrell and Studier (1973) determined that in a colony of *M. thysanodes* in New Mexico gestation was 50–60 days, births occurred synchronously in a 2-week period in late June and early July, the young could make limited flights at 16.5 days, and during the night a few females always remained in the nursery to guard the young. According to Bogdanowicz (1994), the gestation period of *M. daubentoni* usually is 53–55 days, the young weigh about 2 grams at birth, their eyes open at 8–10 days, they begin to fly in the third week, lactation lasts 35–45 days, and adult size is attained at 9–10 weeks.

Apparently there is more variation in the reproductive pattern in tropical species. On the Malay Peninsula pregnant female *M. mystacina* have been found in all months of the year, most frequently in April and May (Medway 1978). In Queensland *M. adversa* has been found to be polyestrous, with births recorded in March, September, and December (Dwyer 1970). Pregnant females have been

Fishing bat *(Myotis vivesi):* A. Photo by Bruce Hayward; B. Photo by Lloyd G. Ingles; C. Photo by Robert T. Orr.

recorded for *M. yumanensis* in Jalisco in August (Watkins, Jones, and Genoways 1972), for *M. elegans* in the Yucatan in February (Jones, Smith, and Genoways 1973), and for *M. albescens* in Costa Rica in January (LaVal and Fitch 1977). Myers (1977*b*) found *M. albescens* and *M. nigricans* to give birth during the spring (September–November) in Paraguay. *M. albescens* then had a postpartum estrus and probably bred one or two more times during the year. *M. nigricans* had no postpartum estrus but probably also bred once or twice more. Both species were found to reach sexual maturity at less than 1 year. Wilson (1971) determined that on Barro Colorado Island *M. nigricans* was a polyestrous, continuous breeder with a postpartum estrus but that mating ceased in October–December, so that no young were weaned during the dry season, January–March. The gestation period of *M. nigricans* was 50–60 days, weaning took place at 5–6 weeks, and sexual maturity was attained at 2.5–3 months in males and somewhat later in females.

Most species of *Myotis* normally have a single young, but in *M. austroriparia* 90 percent of the births are twins (Barbour and Davis 1969). Life span in *Myotis* is generally about 6 or 7 years, though some individuals survive much longer in the wild. Keen and Hitchcock (1980) reported that two specimens of *M. lucifuga* were recaptured in southeastern Ontario 29 and 30 years, respectively, after banding.

As noted above, *Myotis* has a vast distribution and a great diversity of habitat. Nonetheless, many species are declining, and nearly half (39) are now classified by the IUCN. *M. milleri* (treated as distinct from *M. evotis*; see above) and *M. planiceps* of Mexico are considered extinct. Two other Mexican species, *M. cobanensis* and *M. findleyi*, classified as critically endangered and endangered, respectively, occupy very restricted habitat that is confronted with imminent destruction by people. The same problems confront *M. pruinosa*, of Japan, and *M. stalkeri*, of the Molucca Islands, which also are considered endangered. The remarkable *M. stalkeri* had not been recorded since the collection of the holotype in the Kei Islands in 1907, but a colony of about 100 was found in a cave farther north, on Gebe Island, in 1991, and a single specimen was taken in the Kei Islands in 1993 (Flannery 1995). Other IUCN classifications are: *M. sodalis* and *M. grisescens*, endangered; *M. sicarius*, *M. emarginata*, *M. morrisi*, *M. bechsteini*, *M. yesoensis*, *M. scotti*, *M. atacamensis*, *M. dominicensis*, *M. aelleni*, *M. ruber*, *M. peninsularis*, *M. capaccinii*, *M. dasycneme*, and *M. vivesi*, vulnerable; and *M. myotis*, *M. pequinia*, *M. goudoti*, *M. frater*, *M. hosonoi*, *M. annectans*, *M. rosseti*, *M. ridleyi*, *M. martiniquensis*, *M. nesopolus*, *M. bombina*, *M. montivaga*, *M. fortidens*, *M. chiloensis*, *M. macrodactyla*, *M. macrotarsa*, and *M. ricketti*, near threatened.

There is concern for the future of several North American species of *Myotis*, especially because of the destruction or modification of roosting caves by people (Mohr 1972). *M. grisescens* and *M. sodalis*, listed as endangered by the USDI, concentrate in large numbers in relatively few caves and have declined through such factors as deliberate killing by people, disturbance by spelunkers, and commercialization of caves for tourism. Tuttle (1979) reported that the total number of *M. grisescens* in 22 major summer colonies had declined from 1,199,000 before 1968 to 293,600 in 1976. Subsequent protection allowed recovery of some of these colonies (Tuttle 1987), and overall the species numbers about 1.5 million. Based on surveys of winter colonies, Humphrey (1978) determined that the entire population of *M. sodalis* had fallen from 640,361 in 1960 to 459,876 in 1975. Although additional colonies since have been discovered (Thornback and Jenkins 1982), by 1993 numbers had

fallen to 347,890 (Drobney and Clawson 1995). Another eastern species, *M. austroriparia*, also has disappeared from many traditional roosting caves, and numbers have fallen in those where it does survive (Gore and Hovis 1992). The subspecies *M. lucifuga occulta* and *M. velifer velifer* of the southwestern United States and adjacent Mexico have declined drastically through disturbance of colonies and loss of riparian habitat along the Colorado River (D. F. Williams 1986).

The situation for some Old World species is even worse. *M. dasycneme*, for example, may now number only about 3,000 individuals in western Europe and fewer than 7,000 in its entire range. *M. myotis* has been nearly exterminated in Great Britain, the Low Countries, and Israel, and colonies have been drastically reduced elsewhere. Stebbings and Griffith (1986) listed those species, along with *M. emarginata* and *M. blythii*, as endangered, *M. bechsteini* as rare, and *M. nattereri*, *M. capaccinii*, *M. mystacina*, and *M. brandti* as vulnerable. Problems include loss of natural roosts as forests are cleared, disturbance of hibernating colonies in caves and mines, deliberate exclusion from nursery sites in castles and cathedrals, and pollution. However, the greatest immediate threat in western Europe is the remedial chemical treatment of wood in the buildings on which the bats have come to depend for roosting. The chemicals remain on the surface of treated timber for years, are absorbed through the skin and mouth of the bats, and eventually cause death or reproductive failure. Probably more than 100,000 buildings are treated annually in Britain (Mitchell-Jones, Hutson, and Racey 1993).

CHIROPTERA; VESPERTILIONIDAE; Genus
LASIONYCTERIS
Peters, 1865

Silver-haired Bat

The single species, *L. noctivagans*, occurs in southeastern Alaska, southern Canada, the conterminous United States, Tamaulipas (northeastern Mexico), and Bermuda (E. R. Hall 1981; Yates, Schmidly, and Culbertson 1976). A specimen also was collected in the Caicos Islands, Bahamas (Buden 1985).

Head and body length is 55–65 mm, tail length is 38–50 mm, forearm length is 37–44 mm, and adult weight is 6–14 grams. The individual hairs are dark brown to black with silvery white tips; the coloration of the tips of the hairs imparts the frosted or silvery appearance to the pelage. The underparts are somewhat paler, the silvery wash being less pronounced than on the upper parts. The ears are short and about as broad as they are long, and the tail membrane is furred above on the basal half.

The silver-haired bat is found along streams and rivers in wooded areas and in montane coniferous forests. It is mainly a tree dweller but sometimes hibernates in caves. During the spring and summer it shelters in tree hollows, under loose bark, among leaves, in birds' nests, in the cracks of sandstone ledges, in buildings, under loose boards of buildings, and infrequently in caves. Brack and Carter (1985) reported discovery of an individual in an apparent ground squirrel burrow 30–45 cm below the surface of the ground. According to Banfield (1974), *Lasionycteris* is the first bat to appear in the evening and is often seen in broad daylight. At first it generally flies low over water to drink, but later it climbs to treetop level. In a study in Iowa, Kunz (1973*b*) determined that there was an initial foraging period within three to four hours after sunset and a distinct sec-

Silver-haired bat *(Lasionycteris noctivagans)*, photos by Ernest P. Walker.

ond period peaking six to eight hours after sunset. Barbour and Davis (1969) stated that with the possible exception of *Pipistrellus hesperus, Lasionycteris* is the slowest-flying bat in North America. Its diet consists of insects.

The silver-haired bat is a year-round resident in some parts of its range, but populations in the north and at high altitudes generally migrate to warmer areas for the winter. Hibernation occurs during the winter, at least in some places. Izor (1979) reported that *Lasionycteris* may be present during the winter in suitable caves anywhere in the conterminous United States except perhaps the extreme northern Midwest and Great Plains and that it will also use trees and buildings for hibernation north to approximately the limit of the 6.7° C mean daily minimum isotherm for January. Banfield (1974) stated that the migration to wintering sites begins in late August and September and the northward movement occurs in late May. During these periods some flocks may wander far out to sea, and such activity accounts for the records from Bermuda. *Lasionycteris* apparently has a well-developed homing instinct; Davis and Hardin (1967) reported that one specimen traveled at least 175 km to return to its home roost.

Some old reports indicate that large nursery colonies are formed by this bat, but Barbour and Davis (1969) could find no firm evidence of colonial activity. Parsons, Smith, and Whittam (1986) agreed that the early accounts of large reproductive groups are dubious but also reported the discovery of a maternity colony of about 12 adult females and their young in Ontario on 26 June 1979. Banfield (1974) noted that although *Lasionycteris* lives solitarily in the summer, it congregates in flocks for migration.

During the birth period in June and July adult males appear to reside apart from adult females. Kunz (1982) wrote that mating probably occurs in autumn and is followed by a period of sperm storage in the females during the winter. Ovulation peaks in late April and early May, and gestation lasts 50–60 days. Easterla and Watkins (1970) collected 18 adult female *Lasionycteris* at a site in southwestern Iowa between 9 June and 2 July; 5 were lactating, 7 contained 2 embryos, 2 contained a single embryo, and the reproductive state of 4 was undetermined. The most common litter size in the genus appears to be 2. The young weigh around 2 grams at birth, lactation lasts 36 days, and sexual maturity is attained after 5 months (Hayssen, Van Tienhoven, and Van Tienhoven 1993). At three weeks the young are able to follow the mother in weak flight. Schowalter, Harder, and Treichel (1978) found a 12-year-old individual in Alberta.

According to Krebs, Wilson, and Childs (1995), *Lasionycteris* is an important vector of rabies in the northwestern United States. Of the 14 domestically acquired cases of human rabies reported in the country from 1980 to 1994, 11 were associated with bats, 8 of those with *Lasionycteris*. However, in only 1 case was the exposure attributable to a direct bite from a bat.

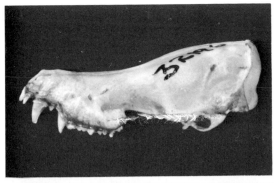

Disk-footed bat *(Eudiscopus denticulus)*, photos from Field Museum of Natural History.

CHIROPTERA; VESPERTILIONIDAE; **Genus**
EUDISCOPUS
Conisbee, 1953

Disk-footed Bat

The single species, *E. denticulus,* is known from the type locality, Phong Saly in Laos, and from Pegu Yoma in south-central Burma (Koopman 1972). This species, originally described as *Discopus denticulus,* is known only from six specimens in the Field Museum of Natural History and two in the American Museum of Natural History.

Head and body length is approximately 40–45 mm, tail length is 39–42 mm, and forearm length is 34–38 mm. The upper parts are cinnamon brown and the underparts are brighter.

Osgood (1932) described disk-footed bats as "externally similar to *Pipistrellus,* but tragus longer and more slender although not pointed at the apex; ears longer and narrowed at the tip; hind feet with highly developed disklike pads even more extreme than in *Tylonycteris* and *Glischropus.* Skull with a broad, greatly flattened braincase, somewhat as in *Tylonycteris* but with a longer, narrower rostrum; dentition with . . . three lower premolars on each side." Distinguishing features are the large adhesive disk on the foot and the three pairs of lower premolars.

The type specimen was collected at an elevation of 1,320 meters. Koopman (1972) noted that the flattened skull suggests that this bat must crawl through narrow crevices and that the foot pads indicate that it must cling to relatively smooth surfaces. The related genus *Tylonycteris* shares these two specializations and is known to roost inside hollow bamboo stems. *Eudiscopus,* evidently very rare and not reported for many years, is designated as near threatened by the IUCN.

CHIROPTERA; VESPERTILIONIDAE; **Genus**
PIPISTRELLUS
Kaup, 1829

Pipistrelles

There are 7 subgenera and 68 species (Aggundey and Schlitter 1984; Ansell and Dowsett 1988; Chasen 1940; Corbet 1978; Corbet and Hill 1992; De Vree 1972; Duckworth, Harrison, and Timmins 1993; Ellerman and Morrison-Scott 1966; Francis and Hill 1986; Gaisler 1983; Green and Rainbird 1984; E. R. Hall 1981; Flannery 1990a, 1995; Hanák and Elgadi 1984; D. L. Harrison 1979; Hayman and Hill *in* Meester and Setzer 1977; Hill 1972a, 1974b, 1976b, 1982a, 1983a; Hill and Francis 1984; Hill and Harrison 1987; Horacek and Hanák 1986; Ibáñez and Fernandez 1985a; Juste and Ibáñez 1994; Kitchener, Caputi, and Jones 1986; Kitchener, Jones, and Caputi 1987; Koch-Weser 1984; Kock 1981b; Koopman 1973, 1975, 1982b, 1984c, 1986, and *in* Wilson and Reeder 1993; Koopman, Mumford, and Heisterberg 1978; Laurie and Hill 1954; Lekagul and McNeely 1977; Mahoney and Walton *in* Bannister et al. 1988; Makin and Harrison 1988; McKean 1975; McKean, Richards, and Price 1978; Medway 1977, 1978; Qumsiyeh 1985; Qumsiyeh and Schlitter 1982; Ride 1970; Roberts 1977; Schlitter, Robbins, and Buchanan 1982; Soota and Chaturvedi 1980; Tate 1942; Taylor 1934; Thomas 1915; Thompson 1982; Thorn 1988; Varty and Hill 1988; Wang 1982; Wolton et al. 1982):

A. Pipistrelle *(Pipistrellus pipistrellus)* in flight, photo by Walter Wissenbach. B. *P. subflavus* resting on hand, photo by Ernest P. Walker. C. Pipistrelle *(P. pipistrellus)*, close-up of head.

subgenus *Pipistrellus* Kaup, 1829

P. pipistrellus, Europe, northern Africa, southwestern and Central Asia, possibly Korea and Japan;

P. nathusii, Europe, Caucasus, western Asia Minor;

P. ceylonicus, Pakistan to Indochina, Hainan, Sri Lanka, Borneo;

P. minahassae, northern Sulawesi;

P. permixtus, Tanzania;

P. abramus, eastern Siberia, Japan, China, Taiwan, Hainan, Viet Nam;

P. javanicus, eastern Afghanistan, northern Pakistan, northern and central India, southeastern Tibet, Burma, Thailand, Indochina, Malay Peninsula, Andaman and Nicobar islands, Sumatra, Bangka Island, Java, Madura Island, Borneo, Philippines, Talaud Islands, Sulawesi, Flores, possibly Aru Islands, two specimens from uncertain localities in Australia;

P. endoi, Honshu;

P. paterculus, northern India, Yunnan, northern Burma;

P. coromandra, Afghanistan, Pakistan, India, southeastern Tibet, Burma, northern Thailand, Sri Lanka, Nicobar Islands;

P. tenuis, Iran, Afghanistan, Pakistan, India, southern China, Burma, Thailand, Indochina, Malay Peninsula, Hainan, Sri Lanka, Christmas Island and possibly Cocos-Keeling Islands (Indian Ocean), Sumatra, Java, Bali, Borneo, Serasan Island, Negros, Mindanao, Sulawesi, South Molucca Islands, New Guinea,

Bismarck Archipelago, Solomon Islands, D'Entrecasteaux Islands, Louisiade Archipelago;

P. collinus, central highlands of Papua New Guinea;

P. angulatus, New Guinea and some nearby islands, Bismarck Archipelago, D'Entrecasteaux Islands, Louisiade Archipelago, Solomon Islands;

P. adamsi, northern Northern Territory, Cape York Peninsula of Queensland;

P. papuanus, South Molucca Islands, Aru Islands, New Guinea and several nearby islands, Bismarck Archipelago, D'Entrecasteaux Islands, Louisiade Archipelago;

P. wattsi, southeast Papua New Guinea and nearby Samari Island;

P. westralis, northern coastal Australia;

P. sturdeei, Bonin Islands (south of Japan);

P. ruepelli, Iraq, Algeria, Egypt, much of Africa south of the Sahara;

P. crassulus, Cameroon, southern Sudan, Zaire, northern Angola;

P. nanulus, Sierra Leone to Kenya, island of Bioko (Fernando Poo);

P. kuhli, southern Europe to Kazakhstan, southwestern Asia, most of Africa, Bioko; Canary Islands;

P. maderensis, Madeira and Canary islands;

P. aegyptius, Algeria, Libya, Burkina Faso, southern Egypt, Sudan;

P. rusticus, Liberia and Ethiopia south to Namibia and Transvaal;

P. aero, northwestern Kenya, possibly Ethiopia;

P. inexspectatus, Benin, Cameroon, Zaire, Uganda, Kenya, perhaps Sudan;

subgenus *Vespadelus* Troughton, 1943

P. pumilus, northern, central, and eastern Australia;

P. vulturnus, central and southeastern Australia, Tasmania;

P. baverstocki, central and southern Australia;

P. regulus, southwestern and southeastern Australia, Tasmania;

P. douglasorum, northern Western Australia;

P. sagittula, southeastern Australia, Tasmania, Lord Howe Island (southwestern Pacific);

subgenus *Perimyotis* Menu, 1984

P. subflavus, eastern North America from Minnesota and Nova Scotia to Honduras and Florida;

subgenus *Hypsugo* Kolenati, 1956

P. savii, southern Europe to Korea and Burma, Japan, northwestern Africa, Canary and Cape Verde islands;

P. pulveratus, southern China, Thailand;

P. bodenheimeri, Israel, Sinai, Yemen, Oman, possibly Socotra Island;

P. ariel, Israel, Egypt, Sudan;

P. anchietae, Zaire, Angola, Zambia;

P. arabicus, Oman;

P. nanus, Senegal to Ethiopia and south to South Africa, Madagascar;

P. musciculus, Cameroon, Gabon, Zaire;

P. hesperus, western North America from southern Washington to central Mexico;

P. eisentrauti, Liberia to Somalia and Kenya;

P. imbricatus, Java, Kangean Islands, Bali, Borneo, Luzon, Palawan, Negros, southern Sulawesi, Lombok;

P. macrotis, Malay Peninsula, Sumatra and nearby Enggano Island, Bali;

P. vordermanni, Borneo, Billiton Island between Sumatra and Borneo;

P. cadornae, northeastern India to northern Thailand;

P. lophurus, Tenasserim (southern Burma);

P. kitcheneri, Borneo;

P. joffrei, Burma;

P. stenopterus, Malay Peninsula, Riau Archipelago, Sumatra, Borneo, Mindanao;

P. anthonyi, northern Burma;

subgenus *Falsistrellus* Troughton, 1943

P. affinis, India, Nepal, southeastern Tibet, Yunnan-Burma border area, Sri Lanka;

P. petersi, northern Borneo, Luzon, Mindanao, northern Sulawesi, Buru and Amboina islands in the South Moluccas;

P. mordax, now thought to be restricted to Java;

P. tasmaniensis, southeastern Queensland, eastern New South Wales, southern Victoria, Tasmania, southern Western Australia;

P. mackenziei, southwestern Western Australia;

subgenus *Neoromicia* Roberts, 1926

P. brunneus, Liberia to Zaire;

P. capensis, throughout Africa south of the Sahara, Bioko; Madagascar;

P. somalicus, Guinea-Bissau to Somalia and south to South Africa, Madagascar;

P. guineensis, Senegal to Sudan and northeastern Zaire;

P. flavescens, Angola, Burundi;

P. rendalli, Gambia to northern Ethiopia and south to Botswana;

P. tenuipinnis, Senegal to Ethiopia and Angola, Bioko;

subgenus *Arielulus* Hill and Harrison, 1987

P. circumdatus, Yunnan (southern China), northern Burma, possibly Thailand, Malay Peninsula, Java;

P. societatis, Malay Peninsula;

P. cuprosus, Sabah (northern Borneo).

This genus is sometimes considered to be no more than subgenerically distinct from *Eptesicus, Vespertilio,* and *Nyctalus. Pipistrellus* usually differs from *Eptesicus* in having a pair of rudimentary upper premolars in addition to the normal pair, but this character is not constant. While acknowledging such difficulties, most recent authors have maintained *Pipistrellus* as a separate genus for purposes of convenience (see Corbet 1978; Ellerman and Morrison-Scott 1966; and Hayman and Hill *in* Meester and Setzer 1977). A new revision by Hill and Harrison (1987) based mainly on bacular morphology and cranial characters has clarified the situation to some extent and has resulted in the transfer of certain species placed above in the subgenera *Vespadelus* and *Neoromicia* from *Eptesicus* to *Pipistrellus.* Assessments of karyological data by McBee, Schlitter, and Robbins (1987) and Morales et al. (1991) tend to support this transfer, at least with respect to *Neoromicia.* Although *Perimyotis, Hypsugo,* and *Falsistrellus* were considered to be full genera by Menu (1984), Horacek and Hanák (1986), and Kitchener, Caputi, and Jones (1986, supported by Adams et al. 1987b), respectively, they were treated as subgenera by Hill and Harrison. Additional karyological analyses (Volleth and Heller 1994b; Volleth and Tidemann 1991) indicated that *Vespadelus* and *Falsistrellus* are full genera and are not closely related to *Pipistrellus.* An electrophoretic study by Ruedi and Arlettaz (1991) led to similar results with respect to *Hypsugo.* Koopman (*in* Wilson and Reeder 1993) retained *Falsistrellus, Hypsugo,* and *Perimyotis,* as well as *Scotozous,* as subgenera of *Pipistrellus;* he also did not accept the transfer of the species of *Vespadelus* and *Neoromicia* from *Eptesicus* to *Pipistrellus.*

Koopman's list (*in* Wilson and Reeder 1993) also differed from the above as follows: *P. abramus* was included in *P. javanicus; P. babu* of India and adjacent areas and *P. peguensis* of southern Burma were considered species distinct from *P. javanicus; P. mimus* of south-central and southeastern Asia was treated as a species distinct from *P. tenuis; P. collinus, P. angulatus, P. adamsi, P. papuanus, P. wattsi,* and *P. westralis* were included in *P. tenuis; P. inexspectatus* was put in the subgenus *Hypsugo; P. nanus* and *P. macrotis* were put in the subgenus *Pipistrellus; P. vordermanni* was included in *P. macrotis;* and *P. mackenziei* was included in *P. tasmaniensis.* Koopman also listed *P. melckorum* of southern Africa as a species but noted that it had not been clearly distinguished from *P. capensis* (it had been considered a synonym of *P. capensis* by Thorn 1988).

Still other subgenera have been used for *Pipistrellus,* but one, *Ia,* has been raised to generic level (Topal 1970a), and the single species of another, *P. (Megapipistrellus* Bianchi, 1916) *annectans,* is now considered to belong to *Myotis* (Topal 1970b). The following species were recently transferred from *Pipistrellus* to other genera: *P. rosseti* and *P.*

Pipistrelle *(Pipistrellus (Vespadelus) regulus)*, photo by Pavel German.

ridleyi to *Myotis* (Hill and Topal 1973) and *P. brachyoterus* to *Philetor* (Hill 1971c). Heller and Volleth (1984) considered *P. societas* a synonym of *P. circumdatus* and thought that the latter should be transferred to the genus *Eptesicus*, but Hill and Francis (1984) and Hill and Harrison (1987) disagreed. The name *P. deserti*, which was synonymized under *P. aegyptius* by Qumsiyeh (1985), was used by most authorities cited in this account, though Koopman (*in* Wilson and Reeder 1993) used *P. aegyptius*. Ansell and Dowsett (1988) recommended use of the name *P. africanus* in place of *P. nanus*. Das (1990) argued that *P. camortae* of the Andaman and Nicobar islands is a species distinct from *P. javanicus*, but neither Corbet and Hill (1991) nor Koopman (*in* Wilson and Reeder 1993) agreed. Yoshiyuki (1989) treated *P. coreensis* of Korea and Tsushima Island as a species distinct from *P. savii*. Studies of allozyme electrophoresis by Adams et al. (1987a) and of morphological characters by Kitchener, Jones, and Caputi (1987) would suggest substantial modification in the above content of the subgenus *Vespadelus*; *P. caurinus* of northwestern, *P. finlaysoni* of western and central, and *P. troughtoni* of eastern Australia would be species separate from *P. pumilus*, and *P. sagittula* would be replaced by the name *P. darlingtoni*. These modifications were accepted by Corbet and Hill (1991) but not by Koopman (*in* Wilson and Reeder 1993) or Flannery (1995). Based on karyological data, Rautenbach, Bronner, and Schlitter (1993) regarded *P. zuluensis* of southern Africa as a species distinct from *P. somalicus*.

Head and body length is 35–62 mm, tail length is 25–50 mm, forearm length is 27–50 mm, and adult weight is 3–20 grams. The ear is usually shorter and broader than in *Myotis*, and the tragus is not as sharply pointed. Pipistrelles are usually dark brown or blackish, but some are gray, chocolate brown, reddish brown, or pale brown.

There is considerable variation in habitat. In Japan, for example, *P. javanicus* forages in open areas and roosts both summer and winter in houses; *P. endoi*, in contrast, is found only in mountain deciduous forests and never roosts in houses (Wallin 1969). Several European species seem to be associated with woodlands, roosting either in hollow trees or in buildings (Kowalski 1955; Ognev 1962). In Botswana, Smithers (1971) reported *P. capensis* to have a wide habitat tolerance, occurring in open dry scrub, rich riverine wood-

land, and *Acacia* woodland and scrub. In the northeastern United States *P. subflavus* is often found during the summer in open woods near water; it roosts in rock crevices or, less frequently, in caves and buildings (Godin 1977). In the southern United States it commonly inhabits clusters of Spanish moss (Barbour and Davis 1969). *P. nanus* of Africa is commonly called the "banana bat" because of its habit of sheltering in the rolled leaves of bananas and plantains (Rosevear 1965).

Over most of their range pipistrelles are among the first bats to appear in the evening. Some members of this genus occasionally fly about in bright sunlight. Their early appearance and jerky, erratic flight are characteristic. According to Barbour and Davis (1969), the fluttery flight of *P. hesperus* is the slowest and weakest of any bat in North America. Some species make spring and autumn migrations between summer and winter ranges, and many apparently hibernate. In colder areas *P. subflavus* utilizes caves, mines, or deep crevices for hibernation. Around mid-October it enters a profound state of torpor from which it generally cannot be aroused until spring (Banfield 1974; Barbour and Davis 1969; Godin 1977; Schwartz and Schwartz 1959). In Louisiana, however, this species fre-

Pipistrelle *(Pipistrellus (Falsistrellus) tasmaniensis)*, photo by Klaus-Gerhard Heller.

quently moves about from one roost to another during the winter (Lowery 1974). In Britain *P. pipistrellus* goes into hibernation in late November or December, commonly spending the winter in crevices between the beams of country churches (Racey 1973). *P. mimus* of southeastern Asia also hibernates during the winter, even when temperatures are relatively warm (Lekagul and McNeely 1977). The diet of *Pipistrellus* consists of small insects caught in flight.

Pipistrellus is not as gregarious as some species of *Myotis*. *P. subflavus*, for example, roosts either alone or in small groups (Godin 1977). *P. hesperus* forms small maternity colonies of up to 12 individuals, including both females and young (Barbour and Davis 1969). In *P. pipistrellus*, however, the nursery colonies may contain hundreds of females (Kowalski 1955). These colonies are divided into groups of 1–10 bats associated with a single territorial male. This male defends a particular roosting site, allowing females to enter but driving away other males (Gerel and Lundberg 1985).

In India *P. ceylonicus* generally occurs in small colonies in old buildings, numbering from a couple dozen to a couple hundred bats (Madhavan 1971). There is a sharply defined breeding season in this species, with mating in the first week of June, ovulation and fertilization in the second week of July, and a gestation period of 50–55 days. There are usually 2 young per birth, lactation lasts 25–30 days, and sexual maturity is attained by both sexes before 1 year. *P. tenuis* breeds throughout the year in India, with the males and females remaining together at all times. Like most other species of *Pipistrellus* that have been studied, it normally produces 2 young per birth (Gopalakrishna, Thakur, and Madhavan 1975). In a study in Natal, South Africa, LaVal and LaVal (1977) found that *P. nanus* roosted within banana plants in groups usually comprising 2–3 individuals, sometimes of both sexes. Also observed, however, was one apparent maternity colony of 150 individuals roosting in the thatched roof of a hut. This species was found to be a monestrous, seasonal breeder producing 1–2 young in November or December. In the Northern Territory of Australia, Maddock and McLeod (1974) determined that *P. pumilus* was polyestrous, with periods of birth at least in August–October and March. Of 6 females with young that were examined, 5 had a single young and 1 had twins.

P. subflavus has two mating periods per year, one in the autumn, probably involving only individuals over 1 year old and followed by storage of the sperm in the uterus of the female during hibernation, and a second in the spring, in which the yearlings participate and which is followed by ovulation. Measured from implantation to parturition, gestation lasts at least 44 days. The young are born from late May to July; litter size is usually 2, rarely 1 or 3, but as many as 4 embryos have been reported; flying begins at about 3 weeks (Banfield 1974; Barbour and Davis 1969; Fujita and Kunz 1984; Schwartz and Schwartz 1959). Maximum known longevity in this species is 14.8 years (Walley and Jarvis 1971).

The IUCN classifies *P. joffrei* and *P. anthonyi*, of Burma, as critically endangered and *P. sturdeei*, of the Japanese Bonin Islands, as endangered; all occur in restricted habitat that is undergoing intensive environmental disturbance. Other IUCN classifications are: *P. maderensis*, *P. anchietai*, *P. arabicus*, *P. mackenziei*, and *P. cuprosus*, vulnerable; and *P. paterculus*, *P. papuanus*, *P. wattsi*, *P. caurinus* (found in Western Australia, recognized by the IUCN as a species distinct from *P. pumilus*, and also included by the IUCN in the genus *Eptesicus*; see above), *P. douglasorum* (included in *Eptesicus*), *P. pulveratus*, *P. bodenheimeri*, *P. ariel*, *P. musculus*, *P. macrotis*, *P. cadornae*, *P. kitcheneri*, *P. mordax*, *P. brunneus* (included in *Eptesicus*), and *P. guineensis* (included in *Eptesicus*), near threatened.

Stebbings and Griffith (1986) listed *P. pipistrellus*, *P. nathusii*, *P. kuhli*, and *P. savii* as vulnerable. These species are declining in Europe because of loss of remaining natural roosts in trees and the use of toxic chemicals to treat wood in the buildings on which the bats now have come to depend (see account of *Myotis*). Gerel and Lundberg (1993) attributed the decline of populations of *P. pipistrellus* in southern Sweden mainly to deteriorating feeding conditions because of drainage and water pollution.

CHIROPTERA; VESPERTILIONIDAE; **Genus SCOTOZOUS**
Dobson, 1875

Dormer's Bat

The single species, *S. dormeri*, is found in northeastern Pakistan and most of India (Corbet and Hill 1992). This species usually has been placed in the genus *Pipistrellus* (e.g., by Koopman *in* Wilson and Reeder 1993), but *Scotozous* was regarded as a distinct genus by Corbet and Hill (1991, 1992) and Hill and Harrison (1987).

According to Roberts (1977), head and body length in one specimen was 52 mm, tail length is about 35–38 mm, and forearm length is 34–36 mm. The dorsal fur is hoary with a scattering of silver-tipped hairs, and the ventral fur is very pale whitish gray. All species of *Pipistrellus* on the Indian subcontinent have brownish gray belly fur. *Scotozous* also differs from *Pipistrellus* in having only one pair of incisors in the upper jaw; when a second, outer pair is present the teeth are reduced in size and almost vestigial. Specimens have been collected near a grove of mango trees and roosting under roof tiles.

CHIROPTERA; VESPERTILIONIDAE; **Genus NYCTALUS**
Bowdich, 1825

Noctule Bats

There are apparently six species (Corbet 1978; Corbet and Hill 1992; Ellerman and Morrison-Scott 1966; Hanák and Elgadi 1984; Hanák and Gaisler 1983; Harrison and Jennings 1980; Hayman and Hill *in* Meester and Setzer 1977; Ibáñez 1988; Jones 1983; Maeda 1983; Medway 1978; Mitchell 1980; Neuhauser and DeBlase 1974; Palmeirim 1982, 1991; Qumsiyeh and Schlitter 1982; Speakman and Webb 1993; Trujillo, Barone, and González 1988):

N. noctula, Europe through most of temperate Asia to Japan and Burma, Oman, Viet Nam, Taiwan, Algeria, possibly Mozambique and Singapore;

N. lasiopterus, western Europe to the Urals and Iran, Morocco, Libya;

N. aviator, Japan, Korea, vicinity of mouth of Yangtze Kiang in China;

N. leisleri, Europe, Iran, eastern Afghanistan, northern India, western Himalayas, North Africa, Madeira and Canary islands;

N. azoreum, Azores (northeastern Atlantic);

N. montanus, eastern Afghanistan, northern India, Nepal.

The species *Pipistrellus joffrei* and *P. stenopterus* have sometimes been referred to *Nyctalus* and would extend the range of the latter genus considerably farther to the south-

Dormer's bat *(Scotozous dormeri)*, photo by Klaus-Gerhard Heller.

east. Corbet (1978) listed both *N. azoreum* of the Azores and *N. verrucosus* of Madeira as subspecies of *N. leisleri,* but *N. azoreum* was considered a separate species by Hanák and Gaisler (1983), Palmeirim (1991), and Speakman and Webb (1993). Yoshiyuki (1989) recognized *N. furvus* of Honshu as a species distinct from *N. noctula.*

Head and body length is 50–100 mm, tail length is 35–65 mm, and forearm length is 40–70 mm. *N. noctula,* the most widely distributed species, weighs 15–40 grams. The color is golden brown or yellowish brown to dark brown above and usually pale brown below.

These bats are generally associated with forests but may also forage in open areas and reside in or near human settlements. Roosting sites include hollow trees, buildings, and caves (Kowalski 1955; Wallin 1969). Noctule bats fly early, with quick turns. They resemble *Pipistrellus* in appearance

and in the habit of beginning to fly before or just after sunset. There may be two main feeding flights of one to two hours duration each, one in the early evening and the other ending shortly before sunrise. Moore (1975) reported that *N. azoreum* flies extensively during daylight in the spring and was often observed from 0900 to 1600 hours. The voice of *N. noctula* has been described as "bursts of piercing, staccato cries" and as "a very loud, shrill, prolonged trilling sound . . . similar to the prolonged song of a bird."

N. leisleri and *N. noctula* are migratory. Three individuals of the latter species banded during hibernation in a cave in Germany were recovered the following summer north and east of the banding site. The most distant recovery was about 750 km from the cave. Medway (1978) wrote that *N. noctula* has made journeys as long as 2,347 km.

These bats are fond of beetles; in fact, there seem to be

Noctule bat *(Nyctalus noctula)*, photo by Erwin Kulzer.

local movements of the males of *N. noctula* in the early summer in England in search of cockchafers. Noctule bats also eat winged ants, moths, and other insects. An unusual instance is the capture of house mice *(Mus musculus)* by an *N. noctula*, which then ate them.

Breeding usually takes place in September and again in the spring, but only a single litter is produced each year. The pregnant females generally assemble in groups of as many as 400 individuals in trees and buildings. Kozhurina (1993) noted that related females may roost together and sometimes nurse young other than their own; males roost separately. The 1 or 2, rarely 3, young are born in May and June. Kulzer and Schmidt (*in* Grzimek 1990) recorded that in *N. noctula* and *N. leisleri* the gestation period is about 70–75 days and the young weigh about 5 grams at birth. Adult dimensions are attained when the young are 6–7 weeks old and weaning occurs at about the same time. Kleiman and Racey (1969) found that some captive females mated during their first autumn, when only 3 months old, whereas males mated during their second autumn.

Stebbings and Griffith (1986) listed *N. leisleri, N. noctula,* and *N. lasiopterus* as rare or vulnerable. These species are declining in Europe as natural habitat, roosting trees, and large insect prey are eliminated. The IUCN designates *N. leisleri, N. lasiopterus, N. aviator,* and *N. montanus* as near threatened. Speakman and Webb (1993) found *N. azoreum* occurring almost throughout the Azores archipelago and to have a much higher population density than that of mainland *Nyctalus.* They estimated its numbers at between 1,750 (based on observations of daylight activity) and 23,650 (based on nocturnal surveys). The IUCN classifies *N. azoreum* as vulnerable because its range is relatively small and it is expected to decline in response to habitat loss.

CHIROPTERA; VESPERTILIONIDAE; **Genus**
GLISCHROPUS
Dobson, 1875

Thick-thumbed Bats

There are two species (Chasen 1940; Corbet and Hill 1992; Lekagul and McNeely 1977):

G. tylopus, Burma, Thailand, Malay Peninsula, Sumatra, Borneo, Palawan Island (Philippines), Bacan Island (North Moluccas);
G. javanus, Java.

This genus is sometimes considered a subgenus of *Pipistrellus.* The species *Pipistrellus tasmaniensis* and *Myotis rosseti* were formerly placed in *Glischropus.*

Head and body length is about 40 mm, tail length is 32–40 mm, and forearm length is 28–35 mm. Payne, Francis, and Phillipps (1985) reported weights of 3.5–4.5 grams. The coloration is dark reddish brown to blackish above and paler below. This genus resembles *Pipistrellus,* differing in that the pads on the thumb and foot are more developed, probably as a grasping modification. The longer and pointed tragus distinguishes *Glischropus* from *Tylonycteris,* another genus of bats with pads on the hand and foot.

G. tylopus is frequently associated with bamboo, roosting in dead or damaged stalks; it also roosts in rock crevices and in new banana leaves. It tends to fly rather high (Lekagul and McNeely 1977). The type specimen of *G. javanus* was taken from the hollow top of a broken-off and partially dead bamboo stem in cultivated country not far from the mountain forest. The diet consists of insects. The IUCN designates *G. javanus* as near threatened.

CHIROPTERA; VESPERTILIONIDAE; **Genus**
LAEPHOTIS
Thomas, 1901

African Long-eared Bats

Four species are currently recognized (Ansell and Dowsett 1988; Fenton 1975; Herselman and Norton 1985; Hill 1974c; Kock and Howell 1988; Peterson 1971b, 1973; Rautenbach, Fenton, and Braack 1985; Setzer 1971; Watson 1990a):

L. wintoni, Ethiopia, Kenya, Tanzania, Orange Free State and Cape Province of South Africa, Lesotho;
L. namibensis, Namibia;
L. angolensis, southern Zaire, northeastern Angola, Zimbabwe;
L. botswanae, southern Zaire, Zambia, Malawi, northwestern Zimbabwe, Botswana, eastern Transvaal.

The reported occurrence of *L. wintoni* in southern Africa long was based on a single specimen from southwestern Cape Province. Some authorities, including Corbet and Hill (1991) and Skinner and Smithers (1990), have regarded this specimen as actually representing *L. namibensis.* However, Herselman and Norton (1985) indicated that

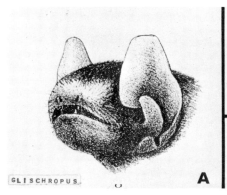
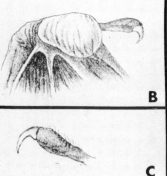

Thick-thumbed bat *(Glischropus tylopus):* A. Photo from *Catalogue of the Chiroptera in the Collection of the British Museum*, G. E. Dobson; B & C. Thumb; D. Foot, photos from *Proc. Zool. Soc. London.*

Thick-thumbed bat *(Glischropus tylopus)*, photos by Klaus-Gerhard Heller.

the measurements of the specimen are by far closest to *L. wintoni,* and Koopman (*in* Wilson and Reeder 1993) included South Africa in the range of that species. Watson (1990a) reported additional specimens of *L. wintoni* from Orange Free State and Lesotho. It may be that the genus *Laephotis* actually comprises only a single highly variable species.

Head and body length is 45–59 mm, tail length is 35–47 mm, and forearm length is about 32–38 mm. The weight of *L. botswanae* is 4.5–6.8 grams (Skinner and Smithers 1990). Two female *L. angolensis* averaged 7.7 grams (Fenton 1975). The upper parts are tawny olive or coppery brown and the underparts are pale brown, dark brown, or gray.

The bats of this genus are quite similar to those of *Histiotus* of South America, differing mainly in skull features. The teeth are similar to those of *Histiotus.* The ears of *Laephotis* are separate and 15–19 mm long, relatively shorter than in *Histiotus.*

A female has been found "under the bark of a dead tree

Laephotis wintoni, photo by John Visser.

Philetor brachypterus, photo by Klaus-Gerhard Heller.

limb with another (presumably a male) which flew away and was not collected." Thirteen specimens from Zaire, 12 of them females, were found under the bark of trees. The IUCN classifies *L. namibensis* as endangered, noting that it is expected to decline by at least 50 percent over the next decade in response to habitat destruction. *L. wintoni, L. angolensis,* and *L. botswanae* are designated as near threatened.

CHIROPTERA; VESPERTILIONIDAE; Genus PHILETOR
Thomas, 1902

The single species, *P. brachypterus,* now is known from Nepal, the Malay Peninsula, Sumatra, Borneo, the Philippines (Mindanao, Leyte, Negros), Sulawesi, New Guinea, and New Britain Island in the Bismarck Archipelago; records from Java and Bangka Island do not seem valid, though it is likely that the species occurs there (Corbet and Hill 1992; Hill 1974b; Hill and Francis 1984; Hill and Rozendaal 1989; Kock 1981a; Koopman 1983; Medway 1977; Rickart et al. 1993). This species formerly was known as *P. rohui,* but Hill (1971c) showed that the species then called *Pipistrellus brachypterus* actually is referable to *Philetor,* that it is conspecific with *P. rohui,* and that its name has priority. The species formerly called *Eptesicus verecundus* also now is considered part of *Philetor brachypterus* (Hill 1966, 1971c). Still another species that may be referable to *Philetor* is *Pipistrellus anthonyi* of northeastern Burma (Koopman 1983).

Head and body length is about 52–64 mm, tail length is 30–38 mm, forearm length is 30–38 mm, and weight is 8–13 grams. Coloration is reddish brown to dark brown. The wings are described as being relatively short. The muzzle is broad and the skull is short and rounded with a large, rounded braincase.

On Borneo, Lim, Shin, and Muul (1972) collected most specimens of *Philetor* from tree holes about 1.5–4.5 meters above the ground. The stomachs of three specimens contained insects such as Coleoptera and Hymenoptera. A pregnant female with a single embryo was in the period from late May to mid-June.

CHIROPTERA; VESPERTILIONIDAE; Genus HESPEROPTENUS
Peters, 1869

There are two subgenera and five species (Hill 1976a, 1983a; Hill and Francis 1984; Mitchell 1980):

subgenus *Hesperoptenus* Peters, 1869

H. doriae, known only by the holotype from Sarawak in Borneo and a single specimen from Selangor on the Malay Peninsula;

subgenus *Milithronycteris* Hill, 1976

H. tickelli, India, Nepal, Sri Lanka, Andaman Islands, southeastern Burma, Thailand, possibly China;
H. tomesi, Malay Peninsula, Borneo;
H. gaskelli, central Sulawesi;
H. blanfordi, Thailand, southeastern Burma, Malay Peninsula, Borneo.

Head and body length is 40–75 mm, tail length is 24–53 mm, and forearm length is 25 to about 60 mm. Payne, Francis, and Phillipps (1985) reported weights of 6.1–6.4 grams

Hesperoptenus blanfordi, photo by Klaus-Gerhard Heller.

Hesperoptenus doriae, photo by Klaus-Gerhard Heller.

for *H. blanfordi* and 30–32 grams for *H. tomesi*. *H. tickelli* is grayish yellow to golden brown, *H. blanfordi* is reddish brown, *H. tomesi* is dark brown, and *H. doriae* is light brown. The black wings of *H. tickelli* are usually marked with white.

Externally these bats are like *Eptesicus,* differing from that and similar genera in cranial and dental features. *H. tickelli* has friction pads on the thumbs.

Bats of the genus *Hesperoptenus* roost singly or in small groups of both sexes, often in the foliage of trees, and begin to fly early in the evening, soon after *Pipistrellus*. *H. tickelli* usually establishes a feeding territory, which it patrols night after night in search of insects. Except when food is abundant this species will chase away other individuals of the same species when they transgress the limits of the territory. Sometimes a male and a female share the circumscribed area. Most young of *H. tickelli* are born in May or June, but one pregnant female was noted in December. Litter size is one young (Hayssen, Van Tienhoven, and Van Tienhoven 1993). The IUCN classifies *H. doriae* as endangered and *H. gaskelli* as vulnerable; both are declining because of loss of their very limited habitat.

CHIROPTERA; VESPERTILIONIDAE; **Genus
CHALINOLOBUS**
Peters, 1866

Lobe-lipped Bats, Groove-lipped Bats, or Wattled Bats

There are six species (Koopman 1971, 1984c; Mahoney and Walton *in* Bannister et al. 1988; Ride 1970; Ryan 1966; Van Deusen and Koopman 1971):

C. gouldii, Australia, Tasmania, Norfolk Island, New Caledonia;
C. morio (chocolate bat), southern Australia, Tasmania;
C. picatus, southwestern Queensland, northwestern New South Wales, eastern South Australia, western Victoria;
C. nigrogriseus, Papua New Guinea, northeastern Western Australia to Queensland and northeastern New South Wales;

C. dwyeri, central New South Wales and adjacent part of Queensland;
C. tuberculatus, New Zealand.

The African genus *Glauconycteris* (see account thereof) sometimes is considered a subgenus of *Chalinolobus*. *C. neocaledonicus* of New Caledonia in the South Pacific was treated as a species distinct from *C. gouldii* by Flannery (1995), though Corbet and Hill (1991) and Koopman (*in* Wilson and Reeder 1993) followed Tidemann (1986) in regarding the two as conspecific. This population, as well as the one on Norfolk Island, may have become extinct in recent years through habitat loss.

Head and body length is 42–75 mm, tail length is 32–60 mm, and forearm length is 30–50 mm. Weight is 4–18 grams (Allison, Richards, and Dwyer, all *in* Strahan 1983). These bats are chocolate brown, dark brown, or black, sometimes with a reddish tinge. The hairs may be tipped with white. *C. picatus* is called the "pied wattled bat" because it is black above and has a whitish fringe of hairs below. This genus is characterized by wattlelike, fleshy, outwardly projecting lobes at the corners of the mouth. Glandular swellings are present on the short, broad muzzle.

These bats roost in the foliage of trees, hollow trees, caves, and mine tunnels. Like most other insectivorous bats, they become active at dusk and often visit water holes to drink and hunt insects. *C. gouldii* is active throughout the year in arid regions but becomes torpid during the winter in cooler climates (Dixon *in* Strahan 1983). Through radio tracking, Lunney, Barker, and Priddel (1985) determined that *C. morio* has a large home range, at least 5 km long.

Both *C. gouldii* and *C. morio* have been reported to roost in groups of 30 to several hundred individuals. Dixon and Huxley (1989) studied a relatively stable maternity colony of 20–30 *C. gouldii* in Melbourne for about seven years. The bats remained together through winter hybridization and at other times. All long-term residents were females up to five years of age; no adult males were recorded for longer than one year. In a study in Western Australia, Kitchener (1975) determined that female *C. gouldii* are monestrous. In the southern part of the state mating occurs at the beginning of winter (about July), sperm is stored by the females over the winter, ovulation and fertilization start at the end of winter (about September), gestation lasts about three months, and births occur from late November to mid-

tat ar
viron
Glau
now
and C
threa

CHIR
NYC1
Rafin

Eve

There

N. hu
   Sta
N. cub

Corbe
er 199
er aut
compr
accoun
genus,
liceius
(Sinha
   Hea
mm, a
humer
the un
more r
be disti
terden
   N. h
in hard
field 19
of tree
found
steady;

Wattled bat *(Chalinolobus nigrogriseus)*, photo by B. Thompson / National Photographic Index of Australian Wildlife.

January; in the central and northern parts of the state births begin around October. The normal litter size is 2 young. According to Kitchener and Coster (1981), *C. morio* also is monestrous, but there normally is a single young, and births take place from mid-September to mid-November. Dwyer (1966) reported that mating in *C. dwyeri* occurs in late summer or autumn, births take place in early December, and average litter size is 1.8. Menzies (1971) studied a maternity colony of *C. nigrogriseus* in the roof of a house in Port Moresby, Papua New Guinea. Several hundred bats were present in November 1969, roosting in tight clusters of 10–40 individuals. All 35 adults collected were lactating females, and 76 juveniles were also taken.

C. tuberculatus of New Zealand is very rare and may be in danger of extinction. For several decades only two small colonies were known, each containing 5–10 individuals, but Daniel and Williams (1983) reported discovery of a group of 200–300 on North Island. Observations indicate that these bats leave their cave at night, fly up to 2 km, and feed mainly on mosquitoes, moths, and midges caught in the air. They apparently hibernate during the winter. Their breeding season evidently extends from October to February (spring and summer), with most females giving birth to a single young in January or February.

The IUCN now designates *C. tuberculatus*, as well as *C. picatus*, as near threatened. *C. dwyeri* is classed as vulner-

Evening
Bat Cons

Chocolate bat *(Chalinolobus morio)*, photo by Pavel German.

Schlieffen's twilight bat *(Nycticeinops schlieffeni)*, photo by M. Brock Fenton.

erratic flight. Although usually solitary, numbers may congregate to forage.

Van der Merwe and Rautenbach (1986) reported *Nycticeinops* to be monestrous and to breed during the spring in the Transvaal. Pregnant females were taken there from September to November, and lactating females in November. Litter size evidently ranged from one to four young, with three being the most common number.

CHIROPTERA; VESPERTILIONIDAE; **Genus**
SCOTEANAX
Troughton, 1943

Greater Broad-nosed Bat

The single species, *S. rueppelli,* occurs in coastal eastern Australia from northeastern Queensland to southern New South Wales. *Scoteanax* long was considered by most authorities to be only a subgenus of *Nycticeius* but was given full generic rank by Kitchener and Caputi (1985). This position was supported by Baverstock et al. (1987) and Hill and Harrison (1987), but Koopman (*in* Wilson and Reeder 1993) continued to treat *Scoteanax* as a subgenus.

Head and body length is 63–73 mm, tail length is 44–58 mm, forearm length is 51–56 mm (Kitchener and Caputi 1985), and weight is 25–35 grams (Richards *in* Strahan 1983). The upper parts are hazel to cinnamon brown and the underparts are tawny olive. From *Nycticeius*, *Scoteanax* differs in its generally greater size, more robust skull with much more pronounced occipital helmet, and considerably reduced third upper molar tooth.

Scoteanax favors tree-lined creeks and the junction of woodland and cleared paddocks but also forages in rainforests. It usually roosts in tree hollows but also has been found in the roof spaces of old buildings. It emerges just after sundown and usually flies slowly and directly at a

Scoteanax rueppelli, photo by G. A. and M. M. Hoye / National Photographic Index of Australian Wildlife.

Lesser broad-nosed bat *(Scotorepens greyi)*, photo by Stanley Breeden.

height of 3–6 meters. The diet includes beetles and other large, slow-flying insects, as well as small vertebrates. A single young is produced in January (Richards *in* Strahan 1983). The IUCN (which treats *Scoteanax* as a subgenus of *Nycticeius*) designates *S. rueppelli* as near threatened.

CHIROPTERA; VESPERTILIONIDAE; **Genus SCOTOREPENS**
Troughton, 1943

Lesser Broad-nosed Bats

There are four species (Baverstock et al. 1987; Flannery 1990*b*; Kitchener, Adams, and Boeadi 1994; Kitchener and Caputi 1985):

S. orion, coastal southeastern Australia;
S. balstoni, Australia except extreme north and south;
S. sanborni, Timor, southeastern New Guinea, northern
    Western Australia, northern Northern Territory,
    northern Queensland;
S. greyi, throughout mainland Australia except
    southwestern Western Australia, Victoria, and northern
    Queensland.

Scotorepens long was considered by most authorities to be only a subgenus of *Nycticeius* but was given full generic rank by Kitchener and Caputi (1985). Hill and Harrison (1987) agreed and placed the two genera in separate tribes of the subfamily Vespertilioninae. Koopman (1978*a*, 1984*c*) considered *orion* and *sanborni* to be part of *S. bal-*

stoni and also recognized*S. influatus* of inland Queensland as a species separate from *S. balstoni.* Koopman (*in* Wilson and Reeder 1993) accepted the above arrangement of species but continued to treat *Scotorepens* as a subgenus of *Nycticeius.*

The information for the remainder of this account was taken from Hall (*in* Strahan 1983), Kitchener and Caputi (1985), and Richards (*in* Strahan 1983). Head and body length is 37–65 mm, tail length is 25–42 mm, forearm length is 27–40 mm, and weight is 6–18 grams. The upper parts are brown, gray brown, or tawny olive, and the underparts are paler brown or buff. *Scoteanax* resembles *Nycticeius* in having a broad, square muzzle and only a single upper incisor on each side but differs in having a considerably reduced third upper molar.

Lesser broad-nosed bats occur in a variety of habitats. They often are found in areas that are generally dry but have access to watercourses, lakes, and ponds. Small colonies roost by day in hollow trees and buildings and emerge at dusk to pursue small insects that congregate over water. Mating evidently occurs just before winter (about May in Australia), at least in temperate areas, and the young are born around November. In *S. balstoni* pregnancy lasts a total of 7 months and litters contain 1–2 young. They cling to their mother for 10 days and subsequently are left at the roost until they are ready to fly on their own, at about 5 weeks.

CHIROPTERA; VESPERTILIONIDAE; **Genus**
SCOTOECUS
Thomas, 1901

There are three species (Hayman and Hill *in* Meester and Setzer 1977; Hill 1974*d;* Kingdon 1974*a;* Koopman 1986; McLellan 1986; Robbins 1980):

S. pallidus, Pakistan, India;
S. albofuscus, Senegal to Malawi;
S. hirundo, Senegal to Ethiopia, and south to Angola and Zambia.

Scotoecus sometimes is treated as a subgenus of *Nycticeius.* The subspecies *S. hirundo hindei* sometimes is considered a separate species.

Head and body length is 46–68 mm, tail length is 28–40 mm, and forearm length is 28–38 mm. Weight is 4–5 grams (Hayssen, Van Tienhoven, and Van Tienhoven 1993). The pelage is fine and silky. The upper parts are brownish, reddish, or dark gray. In the African *S. hirundo* the belly is pale and the wings are dark, while in *S. albofuscus* the belly is dark and the wings are pale. *S. pallidus,* of Asia, has brownish white underparts. *Scotoecus* is distinguished from *Nycticeius* by its broader rostrum and the broad, flat anterior surface of its canines.

The African species occur mostly in open woodland, and *S. pallidus* occupies semidesert (Hill 1974*d*). According to Roberts (1977), the latter species has been collected in roof crevices, old tombs, and pump houses. It flies slowly and close to the ground. It is gregarious, with both sexes occurring together in mid-March but separately during the summer, when the young are born. Kingdon (1974*a*) wrote that in western Uganda *S. hirundo* appears to have a well-defined breeding season; pregnant females, each with two embryos, have been taken in March and lactating females have been taken in May. The IUCN designates *S. albofuscus* as near threatened.

CHIROPTERA; VESPERTILIONIDAE; **Genus**
EPTESICUS
Rafinesque, 1820

Big Brown Bats, House Bats, or Serotines

There are 2 subgenera and 19 species (Carter et al. 1981; Chasen 1940; Corbet 1978; Davis 1966*b;* Eisenberg 1989; Ellerman and Morrison-Scott 1966; E. R. Hall 1981; Hayman and Hill *in* Meester and Setzer 1977; Hill and Harrison 1987; Hollander and Jones 1988; Koopman 1975, 1978*b,* 1989*b,* and *in* Wilson and Reeder 1993; Lekagul and Mc-Neely 1977; Morgan 1989; Nader and Kock 1990*a;* Rautenbach and Espie 1982; Schlitter and Aggundey 1986; Strelkov 1986; Williams 1978*b*):

subgenus *Eptesicus* Rafinesque, 1820

E. bobrinskoi, Kazakhstan to northwestern Iran;
E. nilssonii, central Europe to Japan and Tibet;
E. gobiensis, Mongolia, Sinkiang, and adjacent parts of Central Asia and northern China;
E. nasutus, Arabian Peninsula to Pakistan;
E. tatei, reportedly from Darjeeling in northeastern India;
E. fuscus, southern Canada to Colombia and Venezuela, Bahamas, Cuba, Cayman Islands, Jamaica, Hispaniola, Puerto Rico, Dominica and Barbados (Lesser Antilles);
E. guadeloupensis, known only from Guadeloupe in the Lesser Antilles;
E. innoxius, Pacific coast of Ecuador and Peru;
E. brasiliensis, southern Mexico to northeastern Argentina and Uruguay, Trinidad;
E. furinalis, Tamaulipas (eastern Mexico) to northern Argentina;
E. diminutus, central Brazil to northern Argentina and Uruguay, also an isolated record in northern Venezuela;
E. serotinus, western Europe to Korea and Thailand, northern Africa, most islands of the Mediterranean, Taiwan;
E. platyops, Senegal, Nigeria, island of Bioko (Fernando Poo);
E. bottae, Central and southwestern Asia, northeastern Egypt;
E. kobayashii, reportedly from Korea;
E. hottentotus, Kenya, Angola, Zambia, Malawi, Zimbabwe, Mozambique, South Africa, Namibia;
E. demissus, peninsular Thailand;
E. pachyotis, Assam, northern Burma, Thailand;

subgenus *Rhinopterus* Miller, 1906

E. floweri, Mali, Sudan.

Rhinopterus sometimes has been treated as a distinct genus. A number of African and Australian species were transferred by Hill and Harrison (1987) from *Eptesicus* to the subgenera *Neoromicia* and *Vespadelus* of the genus *Pipistrellus* (see account thereof). Koopman (*in* Wilson and Reeder 1993), however, treated *Neoromicia* and *Vespadelus* as subgenera of *Eptesicus.* Koopman, as well as Corbet and Hill (1991), did not follow Ibáñez and Valverde (1985) in designating *E. platyops* a subspecies of *E. serotinus.* Koopman did not accept Strelkov's (1986) designation of *E. gobiensis* as a species distinct from *E. nilssonii,* though Corbet and Hill did. Yoshiyuki (1989) recognized *E. japonensis* of Honshu as a species distinct from *E. nilssonii.* E. R. Hall (1981) tentatively treated *E. lynni* of Jamaica as a species separate from *E. fuscus,* but Koopman (1989*b*) regarded the two as conspecific. Koopman (*in* Wilson and Reeder 1993) suggested that *E. fuscus* may be conspecific with *E. serotinus* and that *E. kobayashii* is probably conspecific with *E. bottae.* Schlitter and Aggundey (1986) showed *E. loveni* of Kenya to be a synonym of *Myotis tricolor.* Data compiled by Brosset and Charles-Dominique (1990) and Ochoa G. et al. (1993) suggest that *E. andinus,* found from the northern and central Andes to the Guianas, is a species distinct from *E. brasiliensis.*

Head and body length is 35–75 mm, tail length is about 34–60 mm, and forearm length is 28–55 mm. Weights in the species *E. fuscus* range from 14 to 30 grams and in *E. serotinus* and *E. nilssonii* from 8 to 18 grams. The usual color is dark brown to black above and paler below. A buffy wash is often present. Some African species have white or translucent membranes. In the subgenus *Rhinopterus* the upper parts are pale fawn, the underparts are buffy, and the arms, legs, and tail are covered with small, horny, raised areas, giving a scabby appearance.

In Venezuela, Handley (1976) collected several species of *Eptesicus* in a variety of habitats, but mainly in moist, wooded areas; most roosting bats were found in holes in trees or logs. For North America, Banfield (1974) wrote that *E. fuscus* was originally a forest dweller, using hollow trees for roosting during warmer months and hibernating in caves in the winter. Many individuals still follow this way

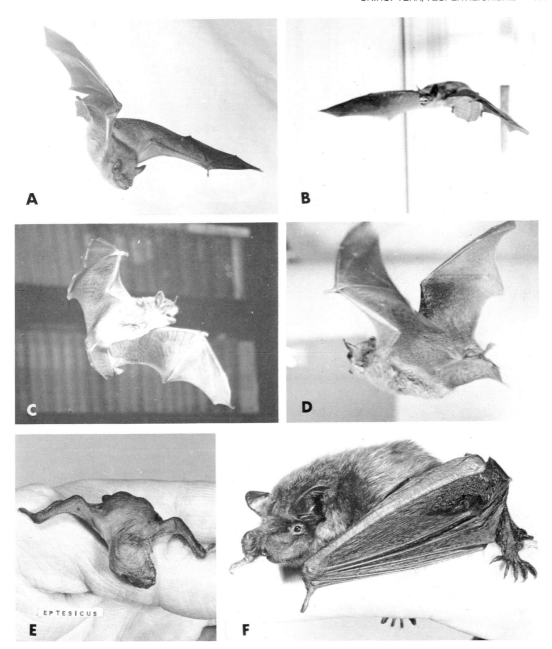

Big brown bat *(Eptesicus fuscus):* A. The upper arm and forearm are seen clearly from both above and below, and the legs are seen to be operating in unison with the wings; B. With tail and legs down to check momentum in coming in for a landing; C. Approaching for a landing using wings as well as tail membrane, but the mouth is still open, giving its ultrasonic sound, which is used in aerial navigation; D. Shows legs working in unison with wings in straightaway flight; E. Baby one day old; F. Adult in which the five equal-length toes with sharp, curved claws, the extremely long forearm, and the free thumb are clearly shown. Photos by Ernest P. Walker.

of life, but this species has become more closely associated with people than any other American bat (Barbour and Davis 1974). Colonies of *E. fuscus* are often found in attics and church belfries and behind shutters and loose boards of buildings. Hibernating sites include houses, tunnels, and storm sewers. Big brown bats usually emerge about sunset with slow, ponderous, fluttering flight and generally feed near the ground at lower levels than those of bats with a rapid, erratic flight. *E. fuscus* reportedly does not make substantial migrations or move very far from its place of birth.

Banfield (1974) reported the average distance traveled probably to be under 32 km. The record natural movement for the species is 288 km (Mills, Barrett, and Farrell 1975). Hibernation in *E. fuscus* is not very profound and may be relatively brief, lasting only from December to April in Canada (Banfield 1974). *E. nilssonii* also is nonmigratory and hibernates only from November or December until March or April. It may shift roosts several times during the summer and winter. It is usually nocturnal, but in the far north of Europe during the summer, when there is little or

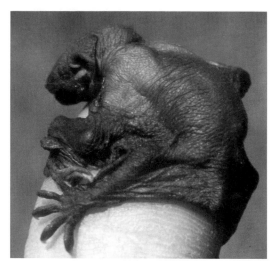

Big brown bat *(Eptesicus fuscus)*, newborn, photo by Mary E. Stewart and Allen Kurta.

no darkness, the bats fly in daylight (Rydell 1993). The diet of *Eptesicus* consists of insects.

Rydell (1986, 1993) reported that female *E. nilssonii* defended small feeding territories, even against members of their own nursery colony, by means of chases and shrill, clearly audible calls. The nursery colonies form during the summer and consist of 10–80 adult females. There are stable, age-related dominance relationships within these groups. Young females are recruited into their natal colonies and sometimes do not breed until they are 1–3 years old. Young males disperse in their first year of life and adult males roost alone in the summer. During winter hibernation individuals are found alone or in groups of 2–4.

The females of *E. fuscus* also form maternity colonies to rear young, during which time the males roost alone or in small groups. Later during the summer both sexes are found roosting together. In hibernating colonies apparently there are usually more males than females (Goehring 1972). According to Barbour and Davis (1974), summer colonies in Kentucky have ranged in size from 12 adults to about 300, the most common number being 50–100. In a study in Ohio, Mills, Barrett, and Farrell (1975) found nursery colonies to contain 8–700 bats, with an average of 154. In northern temperate regions the young of *Eptesicus* are usually born from April through July, following mating in the autumn and storage of sperm in the uterus over the winter. The number of young is usually 1 or 2. The species *E. fuscus* usually has a single young in the Rocky Mountains and westward and twins in that part of its range east of the Rockies. *E. nilssonii* usually has a single young in Scandinavia, but twins are more common further south in Europe (Rydell 1993). In Nicaragua, Jones, Smith, and Turner (1971) found a pregnant female *E. furinalis* with 2 embryos on 22 April and a lactating female on 5 July. Birth weight is about 3 grams in *E. fuscus* and 5 grams in *E. serotinus*, lactation lasts 1 or 2 months, and sexual maturity may be attained at about 1 year (Hayssen, Van Tienhoven, and Van Tienhoven 1993). *Eptesicus* has a potentially long life span, there being numerous records of wild *E. fuscus* recaptured after being banded more than 10 years earlier. Individuals that had survived at least 19 years were reported by Hitchcock (1965) and by Schowalter, Harder, and Treichel (1978).

According to Barbour and Davis (1974), *E. fuscus* is gen-

erally beneficial to people because of its insectivorous diet. Banfield (1974), writing on the mammals of Canada, stated that this aspect of the species is overbalanced by its nuisance value in occupying buildings and its menace to health as a carrier of rabies. This last point should be put in perspective, however, by noting that since 1925 there have been only 3 documented human deaths in Canada resulting from contact with a rabid bat (Rosatte 1987). For the United States, Tuttle (1995) stated that only 23 cases of bat-transmitted rabies had been recorded since 1946. An actual bite is not necessary to pass the disease from bats to humans, and in some cases respiratory transmission is thought to have occurred. According to Squires (1995), 10 persons are known to have contracted and died from rabies within the United States from 1993 to late 1995, and in 9 of these cases bats were involved (see also account of *Felis catus*). There also has been a recent increase of bat rabies in Europe, with *E. serotinus* being the most frequently involved species (Müller 1988).

Several species of *Eptesicus* have restricted ranges and are declining in response to environmental disruption. The IUCN classifies *E. guadaloupensis* as endangered, *E. nasutus, E. innoxius, E. platyops,* and *E. demissus* as vulnerable, and *E. floweri* as near threatened.

CHIROPTERA; VESPERTILIONIDAE; **Genus IA**
Thomas, 1902

Great Evening Bat

The single species, *Ia io,* is known from central and southern China, Assam, northern Thailand, Laos, and northern Viet Nam (Corbet and Hill 1992; Lekagul and McNeely 1977). Ellerman and Morrison-Scott (1966) and various other authors treated *Ia* as a subgenus of *Pipistrellus,* but Topal (1970*a*) showed that *Ia* should be considered a separate genus closely related to *Eptesicus.*

According to Lekagul and McNeely (1977), head and body length is 89–104 mm, tail length is 61–83 mm, and forearm length is 71.5–80 mm. The upper parts are uniform sooty brown, and the underparts are dark grayish brown. The face is relatively hairless, but the insides of the ears are densely haired near the tip. The tip of the tail extends slightly beyond the interfemoral membrane. According to Topal (1970*a*), *Ia* has the same dental formula as *Pipistrellus,* while *Eptesicus* differs from both in having only a single upper premolar. Nonetheless, in other critical characters, especially the structure of the baculum, *Ia* resembles *Eptesicus* and differs from *Pipistrellus.*

Data summarized by Topal (1970*a*) indicate that specimens of *Ia* have been taken in caves at altitudes of 400–1,700 meters. Some specimens were observed in the process of returning to caves at only about 1730 hours, thus indicating that their initial period of foraging is relatively early. Banding studies suggest the possibility that *Ia* undertakes lengthy migrations. It is designated as near threatened by the IUCN.

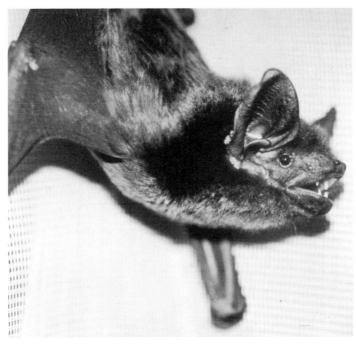

Great evening bat *(Ia io)*, photo by Gábor Csorba.

CHIROPTERA; VESPERTILIONIDAE; **Genus**
VESPERTILIO
Linnaeus, 1758

Frosted Bats, or Particolored Bats

There are three species (Corbet 1978):

V. murinus, Europe to southeastern Siberia and
 Afghanistan;
V. superans, northeastern China, Manchuria, Ussuri
 region of southeastern Siberia, Korea, Japan;
V. orientalis, eastern China, Japan, Taiwan.

Koopman (*in* Wilson and Reeder 1993) included *V. orientalis* within *V. superans,* but Corbet and Hill (1991, 1992) continued to treat the two as separate species.

Head and body length is 55–75 mm, tail length is 35–50 mm, and forearm length is 40–60 mm. *V. murinus* weighs about 14 grams. The coloration is reddish brown to blackish brown above and usually dark brown or gray below. White-tipped hairs may give these bats a "frosted" appearance. The ear in this genus is shorter and broader than in *Eptesicus,* and the facial part of the skull is flattened, with a deep pit on each side that is lacking in *Eptesicus.*

According to Wallin (1969), *V. orientalis* occurs mainly within broad-leaved deciduous forests in mountainous areas. This species has been found roosting in bushes during the day, in a more exposed manner than *V. murinus.* It apparently does not spend the winter in caves, but has been found in buildings at this time. *V. murinus* often inhabits large forest areas but sometimes occurs on steppes, and in some regions it is often found in towns and cities. When sleeping, *V. murinus* usually lies in the narrowest crevices.

This may be a species that prefers the vicinity of rocks for roosting sites; it has been suggested that the present extensive range of *V. murinus* is the result of adapting to the utilization of human structures as shelters. There is evidence that some populations of this species migrate for several hundred kilometers between summer and winter roosts. Other populations seem to move only a short distance, perhaps just from summer roosts in small suburban houses to autumn and winter retreats within large city buildings (Rydell and Baagoe 1994).

V. murinus appears late in the evening and flies at a considerable height, usually more than 20 meters above the ground. It feeds on small dipterans, beetles, moths, and other prey that are captured in the air. This species, in flight in the autumn and winter, "utters a strong, shrill, grinding or whistling . . . cry. This is sometimes combined with a continuous buzzing sound (which latter is possibly produced with the wings)" (Ryberg 1947). This is believed to be a calling and mating cry.

There have been reports that several thousand *V. murinus* may summer together in such localities as the protecting walls of old fortresses. However, according to Rydell and Baagoe (1994), summer maternity colonies usually are small, consisting of 10–100 females and their young. In central and northern Europe these groups form in May and disperse in August. Males roost alone or in small groups but occasionally form colonies of more than 200 individuals. Hibernating bats may be found singly or in groups. Mating presumably occurs in autumn or early winter and the sperm are stored until spring. Births take place in June or early July, with litter size 2 or occasionally 3 in central Europe and 1 or 2 in Scandinavia (Rydell and Baagoe 1994). The period of actual gestation has been reported to be 40–50 days. Kulzer and Schmidt (*in* Grzimek 1990) indicated that the young are weaned after 6–7 weeks and reach sexual maturity at 1 year.

The IUCN classifies *V. orientalis* as vulnerable. The

Frosted bat, or particolored bat *(Vespertilio murinus)*, photo by Liselotte Dorfmüller.

species is subject to severe environmental disruption and is expected to decline by at least 20 percent over the next decade.

CHIROPTERA; VESPERTILIONIDAE; **Genus**
HISTIOTUS
Gervais, 1855

Big-eared Brown Bats

Four species are recognized (Cabrera 1957; Koopman 1982*a*):

H. montanus, Colombia through western and southern
   South America to Tierra del Fuego;
H. macrotus, Peru, Bolivia, Argentina, Chile;
H. alienus, southeastern Brazil, Uruguay;
H. velatus, Brazil, Paraguay.

An unnamed species of this genus occurs in Venezuela (Handley 1976; Linares 1973).

   Head and body length is 48–70 mm, tail length is 39–55 mm, and forearm length is 42–52 mm. Weight is 11–14 grams (Redford and Eisenberg 1992). The upper parts are light brown, grayish brown, or dark brown and the underparts are grayish brown, whitish gray, or dark brown.

   This genus closely resembles *Laephotis* of Africa. *Histiotus* resembles *Eptesicus* in dental and skull characters but differs from that genus in its much larger ears, which are at least as long as the head. In *H. macrotus* the ears are connected by a low band of skin.

   These comparatively rare bats apparently occur over a wide area in a variety of habitats, sometimes being found in forests or at high elevations in the mountains. They may roost in buildings, and Baker (1974) discovered a specimen of *H. montanus* in Ecuador in a hole in a cliff at an altitude of 4,117 meters. In southeastern Brazil, Mumford and Knudson (1978) found *H. velatus* to roost in clusters of 6–12 individuals in the attics of buildings. In July the clusters comprised both males and nongravid females. On 31 October 6 females were taken, each with a suckling young. Peracchi (1968) also found a colony of this species in October; it comprised adults and young of various ages. In Colombia in July, Arata and Vaughan (1970) collected a pregnant female *H. montanus,* 7 lactating females, and 4 immatures. In Malleco Province, Chile, in December, Greer (1965) caught 6 immature *H. montanus* and 11 adult females, 4 of which were lactating. In Patagonia, pregnant *H. montanus* have been found from August to November, there is a single young, and reproductive maturity is attained at about 1 year (Redford and Eisenberg 1992). *H. alienus,* which is restricted to a region experiencing intensive human habitat destruction, is classified as vulnerable by the IUCN; *H. macrotus* and *H. velatus* are designated as near threatened.

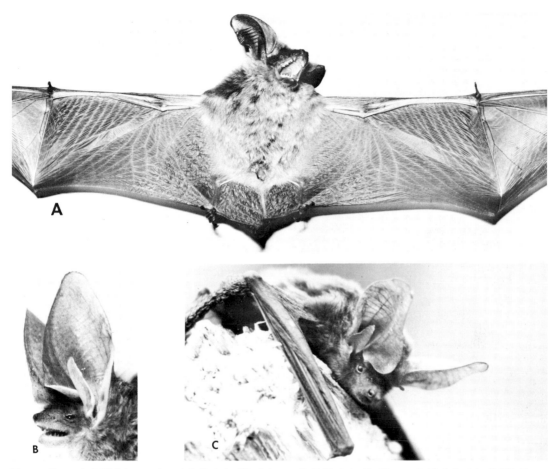

Big-eared brown bat *(Histiotus montanus):* A. Photo by Alfredo Langguth; B. Photo by Abel Fornes; C. Photo by John P. O'Neill.

CHIROPTERA; VESPERTILIONIDAE; **Genus**
TYLONYCTERIS
Peters, 1872

Club-footed Bats, or Bamboo Bats

There are two species (Corbet and Hill 1992; Heaney and Alcala 1986; Lekagul and McNeely 1977; Medway 1973):

T. pachypus, India to southern China and the Malay
  Peninsula, Sumatra, Java, Bali, Borneo, throughout the
  Philippines, Lombok;
T. robustula, Yunnan (southern China), Burma, Thailand,
  Indochina, Malay Peninsula, Sumatra, Java, Bali, Luzon
  and Calauit islands (Philippines), Sulawesi, Peleng
  Island, Lombok, Timor.

Head and body length is 35–50 mm, tail length is 24–33 mm, and forearm length is 22–33 mm. Medway (1972) listed weights of 3.5–5.8 grams for *T. pachypus* and 7.1–11.2 grams for *T. robustula*. However, Heaney and Alcala (1986) noted that *T. pachypus* weighs only about 2 grams and is a competitor with *Craseonycteris thonglongyai* for the title of world's smallest bat. Coloration is reddish brown or dark brown above and paler below.

The bats of this genus may be recognized by the greatly flattened skull and the presence of cushions or pads on the thumb and foot. The ears are about as long as the head, and the tragus is short and bluntly rounded at the tip. *Mimetillus,* from Africa, has been referred to as the African counterpart of *Tylonycteris,* as it also has a flattened and broadened skull. The special adaptations of *Tylonycteris* may be another case of similar or parallel development that fits the animal for a certain mode of life. These bats are remarkably adapted for gaining access to and roosting in the hollow joints of bamboo stems. The small size and flattened skull facilitate their entrance through cracks in the stem, and the suction pads enable them to hang up in the joint.

The following natural history information is taken from Medway (1972, 1978), Medway and Marshall (1972), and Lekagul and McNeely (1977). There appears to be considerable overlap in the ecological niches of the two species. Both characteristically roost in the internodal spaces of standing culms of the large bamboo, *Gigantochloa scortechinii,* access to which is provided by a narrow vertical slit originating as the pupation chamber and emergence hole of the chrysomelid beetle, *Lasiochila goryi*. The larger of the two bats, *T. robustula,* tends to use slightly larger roosting sites than those sometimes entered by *T. pachypus* and may be found in other kinds of bamboo and between rocks. The two species have been seen together feeding on swarms of termites. These bats are gregarious, sometimes roosting in groups of 40 or more individuals. Most groups apparently

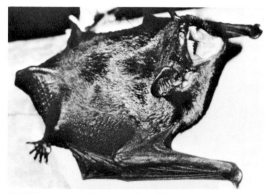

Club-footed bat: Top (*Tylonycteris* sp.), photo by H.-G. Heller; Middle and bottom *(T. robustula)*, photos by David Pye.

have a harem-type arrangement, though there are also lone individuals and groups composed of only a few males. One complete roosting group collected had a single adult male, 12 adult females, and 24 infants. In a study on the Malay Peninsula, however, average composition was 1.7 males and 3.2 females for *T. pachypus* and 1.4 males and 2.1 females for *T. robustula*. Marking investigations have shown that there are frequent changes in group composition and roosting sites. There seems to be a restricted annual breeding season, with births generally occurring over a 1-month period that may take place from February to May, but there is one record of a pregnant female *T. robustula* in August. The gestation period in both species is about 12–13 weeks, and there are usually twin births. The young are carried by the mother for the first few days and then are left at the roost until weaning and independence at about 6 weeks. Both sexes attain sexual maturity in their first year of life.

CHIROPTERA; VESPERTILIONIDAE; **Genus**
MIMETILLUS
Thomas, 1904

Moloney's Flat-headed Bat, or Narrow-winged Bat

The single species, *M. moloneyi*, is found from Sierra Leone to Ethiopia and south to Angola and Zambia (Hayman and Hill *in* Meester and Setzer 1977; Largen, Kock, and Yalden 1974).

Head and body length is 48–58 mm, tail length is 26–38 mm, and forearm length is 27–31 mm. The wingspan is only about 175 mm. Kingdon (1974*a*) gave the weight as 6.0–11.5 grams. The pelage is short and rather scanty. Coloration is dark brown to blackish, or somewhat brighter with a chestnut tinge.

This bat differs from all others in its greatly reduced wings relative to the size of its head and body. It has a remarkably flattened and broadened skull. The legs are short and stout. The short, triangular ears have rounded tips.

Kingdon (1974*a*) provided the following information. *Mimetillus* is found across tropical Africa; it is not restricted to forests and occurs in wooded country to an altitude of 2,300 meters. It roosts in cracks beneath the bark of dead trees. It forages at dusk, often while there is still light, with very rapid flight. This bat apparently needs to drop some distance before it can achieve flight, and it must return to its roost to rest every 10–15 minutes because it has to beat its wings so fast to fly. The diet consists of small flying insects. *Mimetillus* occurs in colonies of 9–12 individuals. Births take place biannually at the end of dry spells, in February–March and August. There is a single young (Hayssen, Van Tienhoven, and Van Tienhoven 1993).

CHIROPTERA; VESPERTILIONIDAE; **Genus**
RHOGEESSA
H. Allen, 1866

Rhogeëssa Bats, or Little Yellow Bats

There are two subgenera and eight species (Audet, Engstrom, and Fenton 1993; R. J. Baker 1984; Carter et al. 1981; Handley 1976; Jones, Arroyo-Cabrales, and Owen 1988; LaVal 1973*c*):

subgenus *Baeodon* Miller, 1906

R. alleni, mountains of western Mexico from Zacatecas and Jalisco to central Oaxaca;

subgenus *Rhogeessa* H. Allen, 1866

R. gracilis, mountains of western Mexico;
R. parvula, western Mexico from central Sonora to Isthmus of Tehuantepec;
R. mira, known only from the type locality in Michoacán (central Mexico);
R. tumida, Tamaulipas (northeastern Mexico) to Bolivia and southern Brazil, Trinidad;
R. genowaysi, Chiapas (southern Mexico);
R. aeneus, Yucatan Peninsula;
R. minutilla, northeastern Colombia, northern Venezuela.

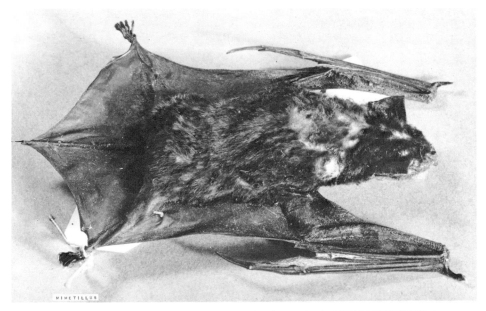

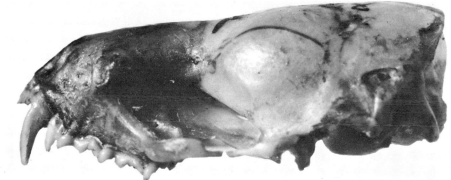

Moloney's flat-headed bat *(Mimetillus moloneyi)*, photos from British Museum (Natural History).

Baeodon was considered a distinct genus by Corbet and Hill (1991) but not by Jones, Arroyo-Cabrales, and Owen (1988) or Koopman (*in* Wilson and Reeder 1993). Smith and Genoways (1974) thought that *R. minutilla* eventually would prove to be a geographic race of what is now recognized as the species *R. tumida* (it was then usually considered part of *R. parvula*). However, morphological analyses by Ruedas and Bickham (1992) confirmed that *R. minutilla* is specifically distinct from *R. tumida*. Based on chromosomal data, there also have been suggestions that *R. tumida* actually comprises several biological species that are difficult, if not impossible, to distinguish by examination of skins and skulls (Baker, Bickham, and Arnold 1985; Honeycutt, Baker, and Genoways 1980). This view was confirmed with the description by R. J. Baker (1984) of *R. genowaysi*, which can be separated from *R. tumida* solely by chromosomal and genic means. This discovery serves as a warning that there may be numerous cryptic species of mammals that cannot be distinguished by classical systematic methods.

Head and body length is 37–50 mm, tail length is 28–48 mm, and forearm length is 25–35 mm. Adults usually weigh 3–10 grams. Coloration is yellowish brown or light brown above and slightly paler below. The genus is distinguished by features of the skull and dentition.

Little yellow bats probably occur in a variety of habitats. In Mexico *R. gracilis* has a known altitudinal range of 600–2,000 meters (J. K. Jones 1977). In Venezuela Handley (1976) found *R. tumida* mainly in humid lowlands in both open and forested sites and *R. minutilla* largely in arid thorn forest. These bats roost mostly in hollow trees but have also been found under palm fronds, in thatched roofs, and under boards. They have been observed in San Luis Potosi in slow, fluttery flight resembling that of *Pipistrellus*, usually from 1 to 4 meters above the ground. Definite hunting areas are generally established. When hunting over clearings they usually fly in circles from 15 to 30 meters in diameter, but when hunting over trails and roads they fly back and forth for distances of about 30 meters. A given area may be shared by 2–5 bats.

LaVal (1973c) listed the following reproductive data: *R. gracilis*, pregnant females taken on 15 May, two subadults taken on 27 July; *R. parvula*, pregnant females collected from late February to early June, lactating females taken from late April to early July, and flying young taken from June to September; and *R. tumida*, pregnancy and lactation

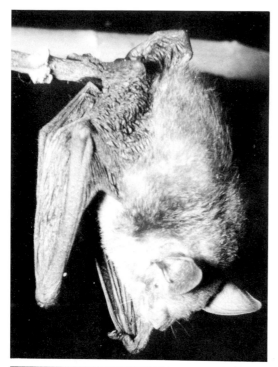

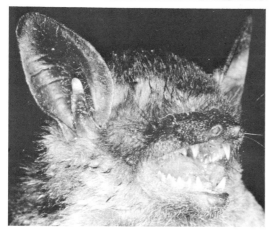

Rhogeëssa bat *(Rhogeessa parvula):* A. Photo by Bruce J. Hayward; skull photo by P. F. Wright of specimen in U.S. National Museum of Natural History; B. *R. tumida,* photo by David Pye.

extending from mid-February to mid-July in Mexico, Central America, and Venezuela. Apparently, the usual litter size for the genus is two.

Most species of *Rhogeessa* occur within a zone experiencing intensive human habitat destruction. The IUCN now classifies *R. alleni, R. gracilis, R. parvula, R. mira, R. genowaysi,* and *R. minutilla* as near threatened.

CHIROPTERA; VESPERTILIONIDAE; **Genus SCOTOMANES**
Dobson, 1875

Harlequin Bat

The single species, *S. ornatus,* occurs in northeastern India, southern China, northern Burma, northern Thailand, and northern Indochina (Corbet and Hill 1992; Lekagul and McNeely 1977; Sinha and Chakraborty 1971). A second named species, *S. emarginatus,* known only by the type specimen from an unknown locality in India, formerly was placed in the genus *Nycticeius* and the subgenus *Scoteinus* but was shown by Sinha and Chakraborty (1971) to belong to *Scotomanes.* Koopman (*in* Wilson and Reeder 1993) stated that *S. emarginatus* probably is a synonym of *S. ornatus,* and Corbet and Hill (1992) treated it as such. A specimen from Indochina given the subgeneric designation *Parascotomanes* Bourret, 1942, and the specific name *beaulieui,* which sometimes has been referred to *Scotomanes,* was shown by Topal (1970*a*) to belong to the species *Ia io.*

Head and body length is about 72–78 mm, tail length is 50–62 mm, and forearm length is 50–60 mm. *Scotomanes ornatus* can be identified by its unusual and attractive color pattern. In this species the upper parts are russet brown and the underparts are dark brown along the midline and whitish along the sides. There is a white stripe down the middle of the back, a patch of white hairs on the crown of the head, and two white spots just above the wing, behind each shoulder.

One specimen was taken in a cave, but the usual roosting site of *S. ornatus* is in the foliage and on the branches of trees, often from two to four meters above the ground. Because of their peculiar markings, a cluster of these bats hanging in a tree could be mistaken for fruit. One specimen was taken from a folded plantain leaf. Another individual flew into an inhabited cabin in the early evening, apparently in pursuit of insects. *S. ornatus* is designated as near threatened by the IUCN.

CHIROPTERA; VESPERTILIONIDAE; **Genus SCOTOPHILUS**
Leach, 1821

House Bats, or Yellow Bats

There are 10 species (Corbet 1978; Corbet and Hill 1992; De Vree 1973; Ellerman and Morrison-Scott 1966; Gaucher 1993; Harrison and Bates 1991; Hayman and Hill *in* Meester and Setzer 1977; Hill 1980*b;* Koopman 1986 and *in* Wilson and Reeder 1993;; Laurie and Hill 1954; Lekagul and McNeely 1977; McLellan 1986; Meester et al. 1986; Robbins 1978, 1980, 1984; Robbins, De Vree, and Van Cakenberghe 1985; Schlitter et al. 1986):

S. nigrita, Senegal, Ghana, Benin, Nigeria, Sudan, Zaire, Kenya, Zimbabwe, Malawi, Mozambique;
S. nux, Sierra Leone to Kenya;
S. nucella, high forest zone of Ghana and Uganda, and probably similar habitat in other parts of West, Central, and East Africa;
S. dinganii, Africa south of the Sahara;
S. robustus, Madagascar;
S. leucogaster, drier areas south of the Sahara from Mauritania and Senegal to Ethiopia and northern

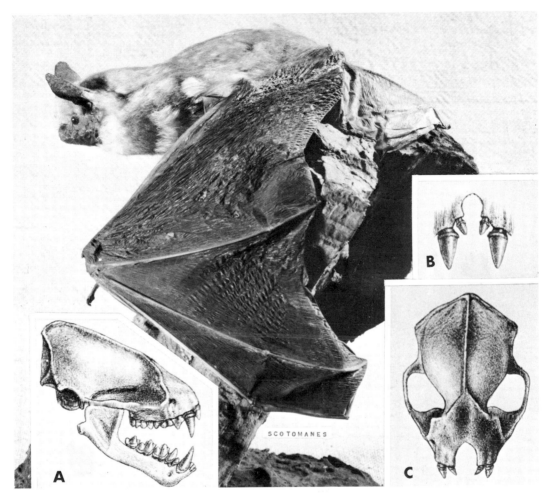

Harlequin bat *(Scotomanes ornatus)*, photo from Royal Scottish Museum through Tom Scott. A. Lateral view of skull; B. Upper incisors; C. Dorsal view of skull, photos from *Catalogue of the Chiroptera in the Collection of the British Museum*, G. E. Dobson.

Kenya, from southern Angola and Namibia to Zambia, southwestern Saudi Arabia, Yemen;

S. viridis, savannah zones from Senegal to Central African Republic and from southern Sudan to eastern South Africa;

S. borbonicus, Madagascar, Reunion in the Indian Ocean;

S. kuhlii, Pakistan to Indochina and Malay Peninsula, Riau Archipelago, Sri Lanka, Nicobar Islands, Taiwan, Hainan, Sumatra, Krakatau, Java, Bali, Borneo, Philippines, Sulawesi, Lombok, Flores, Sumba, Savu, Timor;

S. heathi, Afghanistan, Pakistan, India, Bangladesh, southern China, Burma, Thailand, Indochina, Sri Lanka, Hainan, possibly Sulawesi.

Koopman (*in* Wilson and Reeder 1993) included *S. nucella* in *S. leucogaster* and indicated that the species in Arabia is *S. dinganii.* Koopman also listed *S. celebensis,* reported to occur on Sulawesi, as a distinct species, though he stated that it probably is a subspecies of *S. heathi;* it was tentatively included in the latter species by Corbet and Hill (1992). *S. kuhlii* has been reported from the Aru and Banda islands southwest of New Guinea, but Flannery (1995) doubted the validity of the record. Kitchener and Caputi

(1985) indicated that *S. nigrita* does not belong in *Scotophilus.* Robbins, De Vree, and Van Cakenberghe (1985) did recognize *S. nigrita* as a very distinctive species of *Scotophilus,* though they noted that the name *S. gigas* often is applied to this species. There is much additional controversy regarding the systematics of this genus, especially as to whether various African mainland forms are conspecific with *S. borbonicus.* The originally described subspecies of the latter, *S. b. borbonicus* of Reunion, may have been extinct for over a century (Hill 1980*b*).

Head and body length is 60–117 mm, tail length is 40–65 mm, and forearm length is 42–89 mm. Lekagul and McNeely (1977) gave the weight of *S. kuhlii* as 15–22 grams. Fenton (1975) gave the average weight of *S. nigrita* as about 19 grams. The upper parts are often yellowish brown and the underparts buffy, yellow, or white, but the coloration is quite variable in a given species. In Java and perhaps throughout its range in Indonesia *S. kuhlii* has two color phases, one a brilliant rufous, the other a dull olive brown.

These are rather heavy-bodied, strongly built bats with powerful jaws and teeth. *Scotophilus* is distinguished by the structure of the molar teeth and the dental formula.

The members of this genus are common house-roosting bats over most of their range, usually sheltering in attics,

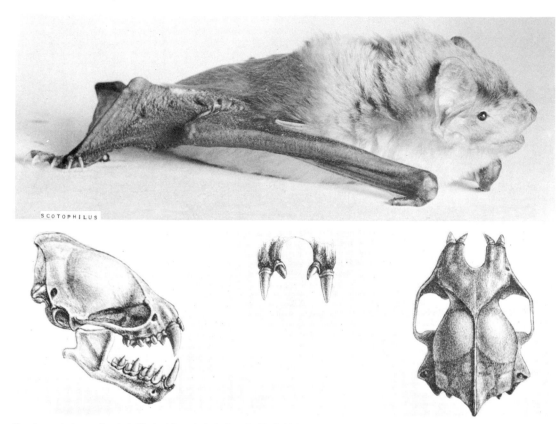

Top, house bat, or yellow bat *(Scotophilus nigrita)*, photo by Erwin Kulzer. Bottom, skull and teeth of house bat (*Scotophilus* sp.), from *Catalogue of the Chiroptera in the Collection of the British Museum*, G. E. Dobson.

often in those that are roofed with corrugated iron and those having extremely high temperatures. Hollow trees, often palms, are also used as roosting sites, and a colony was found among the leaf stalks of a fan palm. In the Philippines, Rickart, Heideman, and Utzurrum (1989) observed *S. kuhlii* roosting in palmate umbrella "tents" such as those described in the account of *Uroderma*, apparently constructed within fan palms. In southern Africa *Scotophilus* sometimes frequents abandoned woodpecker and barbet nests. These bats appear about dusk in fairly steady and strong flight and generally feed from 3 to 12 meters above the ground. The diet consists of beetles, termites, moths, and other insects.

In some areas the roosting colonies are fairly small, generally fewer than 20 individuals, while in other areas they may number in the hundreds. Female *S. kuhlii* in India are reported to assemble in maternity colonies comprising several hundred individuals. Breeding occurs there in March, the gestation period is about 105–15 days, and the young number 1–2. Pregnant female *S. kuhlii* have been captured in the Philippines in April and June, and lactating females from June to August (Ingle 1992).

Another monestrous Indian species, *S. heathi*, mates in January or early February; sperm is then stored 60–80 days, ovulation takes place in early March, gestation lasts about 116 days, parturition is in July, two young are usually produced, and lactation lasts 2–3 weeks (Krishna and Dominic 1981). In Zimbabwe pregnant or lactating female *S. nigrita* have been taken from October to December and the normal number of embryos is two (Smithers 1971). Births of this species, often twins, have been recorded in July in Su-

dan and in January–March in Tanzania and Uganda, there possibly being a second period of births in August–September. Pregnant female *S. leucogaster*, each with two embryos, have been taken in February–March (Kingdon 1974a). Pregnant female *S. dinganii*, also with two embryos each, were collected in Zambia in August and September, and a lactating female of that species was taken there in December (Ansell 1986).

The IUCN classifies *S. borbonicus* of Madagascar as critically endangered. It is thought to have declined by at least 80 percent over the past decade because of human habitat destruction. The IUCN also designates *S. robustus*, of Madagascar, and *S. nigrita*, of mainland Africa, as near threatened.

CHIROPTERA; VESPERTILIONIDAE; **Genus LASIURUS**
Gray, 1831

Hairy-tailed Bats

There are 15 species (Baker et al. 1988; Cabrera 1957; Carter et al. 1981; Dinerstein 1985; Fazzolari-Corrêa 1994; Genoways and Baker 1988; E. R. Hall 1981; Hill and Yalden 1990; Koopman *in* Wilson and Reeder 1993; Koopman and Gudmundsson 1966; Krzanowski 1977; Masson and Cosson 1992; Maunder 1988; Morales and Bickham 1995; Webster, Jones, and Baker 1980; Wilkins 1987; Wilson 1991):

Red bats *(Lasiurus borealis)*, mother with nursing baby, photo by Ernest P. Walker.

L. intermedius, Atlantic and Gulf coasts of United States, western and eastern coasts of Mexico, southern Mexico to Honduras;

L. insularis, Cuba and nearby Isle of Pines;

L. egregius, eastern Panama to southern Brazil;

L. ega (yellow bat), southern Texas and eastern Mexico to northeastern Argentina, Trinidad;

L. xanthinus, southwestern United States, central and western Mexico, Baja California;

L. cinereus (hoary bat), north-central and southern Canada and, rarely, east to Nova Scotia and Newfoundland, conterminous United States, Mexico, Guatemala, Colombia and Venezuela to Chile and Argentina, resident in Hawaii, regularly appears on Bermuda, occasionally on Iceland and Cuba, one record from the Orkney Islands north of Scotland;

L. brachyotis, Galapagos Islands;

L. castaneus, Costa Rica, eastern Panama, French Guiana;

L. seminolus, Pennsylvania, southeastern United States, possibly northeastern Mexico, occasionally Bermuda;

L. pfeifferi, Cuba;

L. degelidus, Jamaica;

L. minor, Bahamas, Hispaniola, Puerto Rico;

L. borealis (red bat), Alberta and Nova Scotia to Chihuahua (northern Mexico) and Florida;

L. blossevillii, southern British Columbia through Utah and Mexico to Chile and Argentina, Tres Marías Islands off western Mexico, Trinidad;

L. ebenus, known only by a single specimen from the vicinity of Sao Paulo in southeastern Brazil.

E. R. Hall (1981) used the name *Nycteris* Borkhausen, 1797, for this genus. Hall and most other American authors have not continued to recognize *Dasypterus* Peters, 1871, as a separate genus for the yellow bats, *L. intermedius, L. egregius,* and *L. ega,* though Corbet and Hill (1991) and Koopman (*in* Wilson and Reeder 1993) accepted *Dasypterus* as a subgenus. Some authorities, including Buden (1985), Corbet and Hill (1991), and Koopman (*in* Wilson and Reeder 1993), have included *L. pfeifferi, L.*

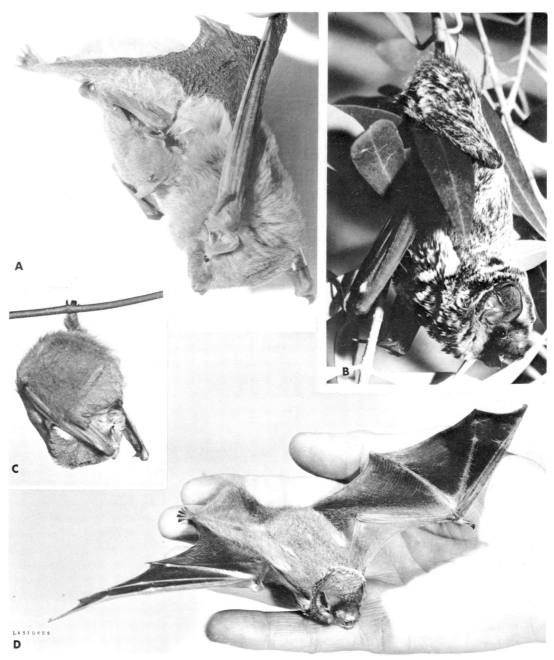

A. Hairy-tailed bat *(Lasiurus intermedius)*, with nursing young clinging to its mother, photo by David K. Caldwell. B. Hoary bat *(L. cinereus)*, photo by Bruce Hayward. Red bat *(L. borealis)*: C. In sleeping position; D. Resting on hand, photos by Ernest P. Walker.

degelidus, and *L. minor* within *L. borealis*, but Bickham (1987) disagreed on the basis of electrophoretic data. Chromosomal data led Baker et al. (1988) to recognize *L. degelidus* as a full species and to suggest that it, *pfeifferi*, and *minor* might actually be subspecies of *L. seminolus*. Koopman (*in* Wilson and Reeder 1993) also included *L. xanthinus* in *L. ega* and *L. blossevillii* in *L. borealis*, but Jones et al. (1992) listed both as full species. Koopman, as well as Corbet and Hill (1991), included *L. brachyotis* in *L. cinereus*. However, whereas Koopman included *L. insularis* within *L. intermedius*, Corbet and Hill listed each as a dis-

tinct species. Recent analyses of mitochondrial DNA (Morales and Bickham 1995) have supported the specific distinction of *L. insularis, L. xanthinus,* and *L. blossevillii* and the affinity of *L. pfeifferi* to *L. seminolus.*

Head and body length is 50–90 mm, tail length is 40–75 mm, and forearm length is 37–58 mm. Adults weigh 6–30 grams. The red bats are brick red to rusty red, usually washed with white; the hoary bats receive their common name from the silver frosting on the yellowish brown to mahogany brown hairs of the upper parts; and the yellow bats are whitish buff, yellowish, or orange, usually with a blackish wash. Male red bats tend to be more brightly colored than the females. The tail membrane is well furred above.

These bats generally occur in wooded areas and roost in foliage or occasionally in tree holes and buildings. The distribution of *L. seminolus* nearly coincides with that of Spanish moss, within clumps of which it roosts for most of the year. Hairy-tailed bats usually appear early in the evening. Feeding flights are usually 6–15 meters above the ground. Most insects are captured in flight, but *L. borealis* will alight on vegetation to pick off insects.

Populations of *L. seminolus* may shift southward in the autumn, and this species is known to become torpid in cold weather (Barbour and Davis 1969; Lowery 1974). Both *L. borealis* and *L. cinereus* are highly migratory, sometimes moving south many hundreds of kilometers during the autumn and occasionally landing on ships at sea or on oceanic islands, especially Bermuda. Migratory movements of these species are not well documented, but it appears likely that from September to November populations in Canada and the northern United States move south to the central and southern United States. Winter populations of *L. borealis* build up in Missouri and Kentucky but may not increase substantially farther to the south. During the winter limited hibernation may take place, but the bats awake and forage on warmer days. Northward migration begins as early as March or April and reaches Canada by late May (Banfield 1974; Barbour and Davis 1969, 1974; Lowery 1974; Schwartz and Schwartz 1959). The population density of *L. borealis* was estimated at 1 per 0.4 ha. in part of Iowa and 1 per ha. in Indiana (Mumford 1973).

Available evidence suggests that these bats are generally solitary but that the females of some species form small nursery colonies and that flocks of up to several hundred individuals form for migration. During the summer the sexes of *L. cinereus* apparently segregate, with males concentrated in the western United States and females and young in the east. Barbour and Davis (1969) gave the following information on *L. intermedius*. At least 50 females with their young were once found in less than 0.4 ha. of live oaks; during the summer female aggregations are formed, but males are rare therein; and during the winter males apparently congregate. In populations of *Lasiurus* in temperate regions mating takes place in the late summer or autumn, sperm is stored over the winter in the uterus, ovulation and fertilization occur in the spring, and births take place from late May to early July. Barbour and Davis (1969) gave the estimated gestation period of *L. borealis* as 80–90 days. Greer (1965) collected female *L. borealis,* each with two young, in Chile in December and January.

Lasiurus is the only genus of bats in which there are commonly more than two young per birth. *L. cinereus, L. intermedius,* and *L. seminolus* are known to produce litters of up to four young (Banfield 1974; Barbour and Davis 1969; Lowery 1974), and there are two records of female *L. borealis* found with five clinging young (Hamilton and Stalling 1972). The normal litter size in all species apparently is two or three, but single young have been reported for *L. ega, L. intermedius,* and *L. cinereus* (Birney et al. 1974; Bogan 1972). Old observations suggested that females initially carried their litters during foraging, but Davis (1970) found no evidence for the transportation of young by *Lasiurus* or any other North American bat except in cases of disturbance.

The newly described *L. ebenus* is known only from an area undergoing intensive human habitat destruction. The IUCN classifies it and *L. castaneus* as vulnerable and *L. egregius* as near threatened. The Hawaiian hoary bat *(L. cinereus semotus)* is listed as endangered by the USDI. Comparatively little is known about this subspecies, but it is thought to have declined because of loss of its forest habitat. A few thousand individuals may still survive. Recent observations by Fullard (1989), Jacobs (1994), and Kepler and Scott (1990) indicate a widespread but rare resident population on the large island of Hawaii, a small and rare population on Kauai, and occasional appearances on Maui and Oahu; these bats sometimes forage along the coast, possibly hibernate during winter, form aggregations of about 20 individuals, and apparently rear their young during the summer. The species *L. brachyotis,* one of the few mammals native to the Galapagos Islands, was observed regularly in the late nineteenth century (Allen 1942) but was listed as extinct by Goodwin and Goodwin (1973).

CHIROPTERA; VESPERTILIONIDAE; **Genus**
OTONYCTERIS
Peters, 1859

Desert Long-eared Bat

The single species, *O. hemprichi,* occurs in the desert zone from Morocco and northern Niger through Egypt and the Arabian Peninsula to Kazakhstan, Kyrgyzstan, and Pakistan (Aulagnier and Mein 1985; Corbet 1978; Fairon 1980;

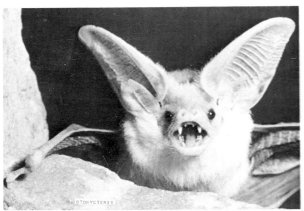

Hemprich's long-eared bat *(Otonycteris hemprichi)*, photos by Erwin Kulzer. The skin of the wings and feet of bats is so thin and tender that cages for newly captured bats should be lined with coarse mesh cloth, as in these pictures.

Horacek 1991; Roberts 1977; Shaimardanov 1982). The systematic position of *Otonycteris* is unclear, but chromosomal analysis by Qumsiyeh and Bickham (1993) suggests close affinity to *Barbastella* and *Plecotus*.

Head and body length is about 73–81 mm, tail length is about 47–70 mm, and forearm length is 57–67 mm. Gaisler, Madkour, and Pelikan (1972) gave the weight of two adult males as 18 and 20 grams. The coloration above is pale sandy to dark brown; the underparts are usually whitish.

The large ears, about 40 mm in length, are directed nearly horizontally and are connected across the forehead by a low band of skin. Five female specimens revealed two pairs of pectoral mammae, a unique condition in mammals. It is not known whether both pairs are functional. The skull and teeth resemble those of *Eptesicus*.

Harrison (1964) stated that this bat is capable of inhabiting extremely barren and arid regions. In the Negev Desert a pair was found roosting in a rocky crevice on a hill. This bat has also been found in buildings and reportedly has a slow, floppy flight. Based on an analysis of its body mass, low aspect ratio, and low relative wing loading, Norberg and Fenton (1988) predicted that *Otonycteris* will be found to be carnivorous in diet. Observations in Kyrgyzstan indicate that *Otonycteris* forages close to the ground, uses echolocation to detect large flying or surface-dwelling invertebrates, and feeds mostly on arachnids and orthopterans that are seized directly from the ground (Arletaz et al. 1995; Horacek 1991). Breeding colonies of 3–15 females have been found, and 7 pregnant females, most with two embryos, have been collected in Central Asia; one was taken on 12 June (Horacek 1991; Roberts 1977). Three pregnant females, each with two embryos, were found in a deserted hut in Jordan on 2 May (Atallah 1977).

CHIROPTERA; VESPERTILIONIDAE; **Genus BARBASTELLA** *Gray, 1821*

Barbastelles

There are two species (Corbet 1978; Corbet and Hill 1992; Ibáñez and Fernández 1985*a*):

B. barbastellus, Europe and the larger Mediterranean islands, Morocco, Canary Islands;
B. leucomelas, from the Caucasus, through Iran and Central Asia, to Japan, the Himalayas, south-central China, and possibly Indochina.

Corbet (1978) also mentioned old records from Sinai, Senegal, and Eritrea in Ethiopia, each of which is dubious or, if valid, might apply to either species. Qumsiyeh (1985), however, considered *leucomelas* to be only a subspecies of *B. barbastellus* and accepted the presence of this subspecies in Sinai and southern Israel. The occurrence of *leucomelas* in those two areas has since been confirmed (Harrison and Bates 1991).

Head and body length is 43–60 mm, tail length is 40–55 mm, and forearm length is 35–45 mm. Adults usually weigh 6–10 grams. The pelage is soft and long. In *B. barbastellus* the coloration is blackish, the tips of the hairs being whitish or yellowish to impart a "silvered" or "frosted" appearance, and the underparts are somewhat paler. *B. leucomelas* is uniformly dark brown above and below, the tips of the hairs being gray to buffy on the upper parts and whitish on the underparts.

The fur extends onto the tail membrane and the wings. The nostrils of *Barbastella* open upward and outward be-

Barbastelle *(Barbastella barbastellus)*, photo by Erwin Kulzer.

hind a median pad. The ears, conspicuously notched on their external margins, are short, broad, and united by a low band across the forehead. This genus resembles *Plecotus, Corynorhinus, Idionycteris,* and *Euderma* in most skull features and in dental characters but lacks the extreme auditory specializations of those genera. The auditory bullae of *Barbastella* are only slightly enlarged and are circular in outline, whereas in the other genera the bullae are substantially enlarged and elliptical or elongate. Like *Corynorhinus, Barbastella* has greatly enlarged muzzle glands (Frost and Timm 1992).

The flight is described as relatively slow and flapping. Barbastelles often fly near the ground but sometimes feed at fairly high altitudes, particularly during good weather. *B. barbastellus* is reported to appear early, not later than sunset, and *B. leucomelas* is said to emerge rather late in the evening. There is some indication that these bats wander considerably, and they sometimes emerge from hibernation and fly around during the winter. They hibernate in caves, usually in drier regions, from late September to early April. They have been found in caves with *Myotis, Pipistrellus,* and *Plecotus.* The common roosts in summer are trees, generally on and under bark, and buildings. One captive barbastelle picked houseflies off a ceiling and also fed on mealworms, and another individual readily accepted small pieces of meat. From one to about half a dozen individuals are usually found in a given roost, but during the breeding season barbastelles apparently congregate in fairly large numbers. According to Hayssen, Van Tienhoven, and Van Tienhoven (1993), births occur from May to early August in Germany, there is a single annual litter, and there are commonly two young.

Stebbings and Griffith (1986) designated *B. barbastellus* as vulnerable worldwide and as possibly endangered in western Europe. One of the rarest bats in Europe, it appears to be declining through habitat disturbance, pollution, and loss of hollow trees. The IUCN also now classifies *B. barbastellus* as vulnerable.

CHIROPTERA; VESPERTILIONIDAE; **Genus**
PLECOTUS
E. Geoffroy St.-Hilaire, 1818

Old World Long-eared Bats

There are four species (Corbet 1978; Corbet and Hill 1992; Ibáñez and Fernández 1985*b*; Koopman *in* Wilson and Reeder 1993; Nader and Kock 1990*b*; Palmeirim 1990; Yoshiyuki 1991*a*):

P. auritus, Europe to Sakhalin Island, Japan, and the Himalayas;
P. taivanus, Taiwan;
P. teneriffae, Canary Islands;
P. austriacus, southern England and Portugal to Mongolia and Sichuan (southwestern China), Arabian Peninsula, northern Africa, Ethiopia, Senegal, Cape Verde Islands.

The American genera *Corynorhinus* and *Idionycteris* (see accounts thereof) sometimes have been regarded as subgenera of *Plecotus.* Corbet (1978) listed *P. teneriffae* as part of *P. austriacus,* but Ibáñez and Fernández (1985*b*) regarded it as a distinct species with closer affinity to *P. auritus.*

Head and body length is 40–53 mm, tail length is 34–50 mm, forearm length is 34–43 mm, and adult weight is 5–14 grams (Kulzer and Schmidt *in* Grzimek 1990). The upper parts are brown or grayish brown, the underparts are paler,

and the ears and membranes are brownish. These bats may be distinguished by their large ears, up to 40 mm in length, which are joined at the base by a prominent interauricular septum. There are glandular masses on the muzzle, but these are not greatly enlarged as they are in *Corynorhinus* and *Barbastella* (Frost and Timm 1992).

Long-eared bats roost in caves, tunnels, buildings, and trees. In Europe *P. auritus* is usually found in buildings and trees in the summer and in caves in the winter, whereas *P. austriacus* usually occurs in buildings during the summer and in buildings or cellars in the winter (Horacek 1975). There also is some difference in general habitat, *P. auritus* seemingly preferring woodlands, though often in areas inhabited by people, and *P. austriacus* being found more in open, cultivated country (Van Den Brink 1968). The diet consists of small to medium-sized insects.

These bats are gregarious. Horacek (1975) found the usual number of individuals in summer colonies to be 5–10 in *P. auritus* and 10–20 in *P. austriacus. P. auritus* mates in the autumn and gives birth in the spring, about 60–70 days after arousing from hibernation (Hayssen, Van Tienhoven, and Van Tienhoven 1993). Kulzer and Schmidt (*in* Grzimek 1990) indicated that the usually single young is weaned after 6–7 weeks and attains sexual maturity at 1–3 years. Lehmann, Jenni, and Maumary (1992) recaptured a female *P. auritus* in the Swiss Alps 30 years after it had been banded, noted that it was in good condition, and suggested that longevity of the species may be considerably greater.

Because of its ability to utilize human structures for roosting, Horacek (1975) suggested that *P. austriacus* may have spread through Europe in historical time. Nonetheless, there is concern about the future of this and several other species of *Plecotus.* Piechocki (1966) stated that numbers of *P. austriacus* and *P. auritus* in central Germany had declined considerably in recent years because of environmental disruption. Stebbings and Griffith (1986) designated both species as vulnerable throughout their ranges. A new problem is the use of toxic chemicals to treat the wood in the buildings on which the bats have come to depend (see account of *Myotis*). The IUCN now classifies *P. taivanus* and *P. teneriffae* as vulnerable, noting that both are expected to decline by at least 20 percent over the next decade.

CHIROPTERA; VESPERTILIONIDAE; **Genus**
CORYNORHINUS
H. Allen, 1865

American Long-eared or Lump-nosed Bats

There are three species (E. R. Hall 1981; Tumlison 1991):

C. mexicanus, northern and central Mexico, northern Yucatan, Cozumel Island;
C. rafinesquii, Indiana, southeastern United States;
C. townsendii, southwestern Canada, western conterminous United States, Ozark and central Appalachian regions, Baja California, Mexico as far south as Oaxaca.

Corynorhinus long was considered a full genus, but most authorities, including Corbet and Hill (1991) and Koopman (*in* Wilson and Reeder 1993), eventually came to treat it as a subgenus of the Old World *Plecotus.* A recent series of morphological and karyological studies have demonstrated that *Corynorhinus* should be restored to generic rank

A & D. *Plecotus auritus,* photos by Walter Wissenbach. B & C. Eastern lump-nosed bat *(Corynorhinus rafinesquii),* photos by Ernest P. Walker.

Long-eared bat (*Corynorhinus townsendii*), photo by Scott D. Keefer.

(Frost and Timm 1992; Tumlison and Douglas 1992; Volleth and Heller 1994*a*).

Head and body length is 45–70 mm, tail length is 35–55 mm, forearm length is 35–52 mm, and adult weight is 5–20 grams. The upper parts are brown, the underparts are paler, and the ears and membranes are brownish. The large ears, up to 40 mm in length, are joined basally across the forehead. Two very large glandular masses on the dorsal surface of the rostrum rise above the muzzle as flaplike lumps. There also are large muzzle glands in *Barbastella*, but these glands are much smaller in the genera *Plecotus*, *Idionycteris*, and *Euderma*.

Frost and Timm (1992) also distinguished *Corynorhinus* as follows: rostrum arched without median concavity (the rostrum is flattened with varying degrees of median concavity in *Barbastella*, *Plecotus*, *Idionycteris*, and *Euderma*); braincase dorsally domed, as in *Barbastella* (the braincase is flattened in *Plecotus*, *Idionycteris*, and *Euderma*); fourth upper premolar much wider than long (subequal in *Plecotus*); metacone present on third upper molar (absent in *Plecotus*); basial pits present (absent in *Barbastella*, *Plecotus*, *Idionycteris*, and *Euderma*); tragus narrow, bladelike (tragus broader in *Plecotus*, *Idionycteris*, and *Euderma*).

Barbour and Davis (1969) listed the following habitat conditions: *C. rafinesquii*, forest; and *C. townsendii* desert scrub, pinon-juniper or pine forest. *C. townsendii* generally shelters in caves throughout the year except in parts of the west, where it inhabits buildings. *C. rafinesquii* most frequently roosts in partially lighted, mostly unoccupied buildings but is also found in caves, trees, and other natural shelters (C. Jones 1977). *C. mexicanus* typically is found in high, humid mountainous areas dominated by pine-oak forests and roosts in caves (Tumlison 1992). These bats do not begin flying until after dark. Their flight is slow, and they are able to hover like a butterfly at a point that interests them. The ears are thrown forward when they search for insects in foliage. These bats also hunt insects in flight. The prey is often picked off the foliage of trees and bushes and the walls of buildings while the bat is hovering, hummingbirdlike. All investigated species have been found to hibernate for at least part of the winter. While there may

be changes between summer and winter roosting sites, apparently there are not extensive migrations. Humphrey and Kunz (1976) found *C. townsendii* in western Oklahoma and Kansas to be relatively sedentary, with over 80 percent of 88 bats being recovered at the same site as banded. Overall population density in this region varied from 1 per 40 ha. to 1 per 53 ha.

C. Jones (1977) stated that *C. rafinesquii* roosts singly or in clusters of 2–100, with females greatly outnumbering males in most colonies. Barbour and Davis (1969) wrote that maternity colonies of *C. townsendii* contain up to 1,000 females; this species is known to roost in clusters. Based on a study in the karst region of western Oklahoma and Kansas, Humphrey and Kunz (1976) provided the following social information on *C. townsendii*: during the summer males roost alone or in groups of up to 6 bats, nursery colonies contain 17–40 adult females, and winter hibernacula have 1–62 individuals.

In *C. townsendii* most mating occurs in the winter roost, and the sperm then stored in the reproductive tract of the female remains motile for 76 days. The gestation period is 56–100 days, apparently being dependent upon body temperature, and the young are born from late May through July. Young *C. townsendii* weigh nearly 25 percent as much as their mothers, are capable of flight at 2.5–3.0 weeks of age, and are fully weaned by 6 weeks (Kunz and Martin 1982). In *C. rafinesquii* the females give birth to a single young in late May or early June, and the young is capable of nonagile flight at 15–18 days (C. Jones 1977). Pregnant female *C. mexicanus*, each with one embryo, have been collected in April, and lactating females in May and July (Tumlison 1992). The longevity record for the genus appears to be held by a female *C. rafinesquii* that had a life span of at least 10 years and 1 month (Paradiso and Greenhall 1967).

Humphrey and Kunz (1976) observed that the total population of *C. townsendii* on the southern plains of the United States is at most 14,000, that the species is barely persisting because of precise natural requirements, and that avoidance of human disturbance is essential for its continued existence. The subspecies *C. townsendii ingens* of the Ozark region and *C. t. virginianus* of the central Appalachians are listed as endangered by the USDI. The former subspecies may number only about 1,600 individuals (Harvey and Barkley 1990), the latter about 13,000 (*Endangered Species Tech. Bull.* 19, no. 5 [1994]: 14). These bats have declined because of direct killing by people and through abandonment of roosting caves when disturbed by explorers and vandals. Noting that declines are likely to continue, the IUCN now classifies the entire species *C. townsendii* and *C. rafinesquii* as vulnerable.

CHIROPTERA; VESPERTILIONIDAE; **Genus**
IDIONYCTERIS
Anthony, 1923

Allen's Big-eared or Lappet-eared Bat

The single species, *I. phyllotis*, is found in mountainous regions from the southwestern United States to central Mexico (Czaplewski 1983). Two subspecies now are recognized: *I. p. hualapaiensis* in southern Nevada, southern Utah, and northeastern Arizona; and *I. p. phyllotis* in Arizona south of the Grand Canyon, southwestern New Mexico, and Mexico from Chihuahua to northern Oaxaca (Bonilla, Cis-

Allen's big-eared bat *(Idionycteris phyllotis)*, photo by Merlin D. Tuttle, Bat Conservation International.

neros, and Sánchez-Cordero 1992; Tumlison 1993). Although E. R. Hall (1981) considered *Idionycteris* a subgenus of *Plecotus*, there now has been general acceptance of the suggestion by Williams, Druecker, and Black (1970), on the basis of karyotype, that the two are generically distinct (Corbet and Hill 1991; Koopman *in* Wilson and Reeder 1993; Jones et al. 1992). A recent analysis of both morphological and karyological characters indicated that *Idionycteris* is a synonym of *Euderma* (Frost and Timm 1992); this study was supported by a further evaluation of chromosomal data (Volleth and Heller 1994a). However, another study of the morphology of skulls and skins, while finding close affinity of *Idionycteris* and *Euderma*, maintained a generic distinction between the two (Tumlison and Douglas 1992). Except as noted, the information for the remainder of this account was taken from Czaplewski (1983).

Head and body length is about 50–60 mm, tail length is 44–55 mm, forearm length is 42–49 mm, and weight is 8–16 grams. The dorsal pelage is long, soft, and basally blackish with tips a contrasting yellowish gray. The ventral hairs are black basally with pale buffy tips. The ears, which are 34–43 mm long, are as large as those of *Plecotus* and *Corynorhinus*, but *Idionycteris* is distinguished by a pair of fleshy lappets projecting over the forehead from the anterior bases of the ears. *Plecotus* and *Corynorhinus* also differ in having keeled calcars and nostrils that are not elongated posteriorly.

Frost and Timm (1992) observed that both *Idionycteris* and *Euderma* are characterized by having the rostrum of the skull not flattened and with a median concavity, the braincase dorsally flattened, the auditory bullae very enlarged and elliptical, a paddlelike tragus, and muzzle glands that are not enlarged. *Idionycteris*, however, has a third lower premolar, while *Euderma* does not. Tumlison and Douglas (1992) pointed out the following additional characters, among others, that distinguish the two genera: the lateral borders of the pterygoids are angled medially in relation to the longitudinal axis of the skull in *Idionycteris* but are vertical in *Euderma*; *Idionycteris* has a spine at the anterior tip of the nasals, but *Euderma* does not; in lateral view the coronoid process of the dentary appears rounded in *Idionycteris* but has a hooklike process in *Euderma*; the medial aspect of the auditory bullae is smooth in *Idionycteris* but emarginated in *Euderma*; the infraorbital foramen is large in *Idionycteris* but small in *Euderma*; the upper ca-

nine tooth is longer than the fourth upper premolar in *Idionycteris* but shorter in *Euderma*; the posterior nares opens in the posterior third of the pterygoids in *Idionycteris* but in the middle third in *Euderma*; and the posterior basal lobe of the auricle is not attached to the base of the tragus in *Idionycteris* but is so attached in *Euderma*.

Allen's big-eared bat dwells primarily in forested mountainous areas, occasionally in desert scrub, and usually in the vicinity of rocks or cliffs. Maternity colonies have been found roosting in tunnels and piles of boulders. Open-air flight is fast, direct, and often accompanied by loud "peeps." Foraging flight is slow, highly maneuverable, and characterized by long-constant-frequency echolocation. The diet consists largely of moths.

Seasonal movements and winter location and activity are unknown. During the summer the sexes segregate, with females gathering into maternity colonies of about 25–100 individuals and males possibly remaining solitary. Pregnant females, each with a single embryo, have been collected in June in New Mexico, Arizona, and Durango. Lactating females have been reported from June until early August, and flying young as early as 31 July.

CHIROPTERA; VESPERTILIONIDAE; Genus EUDERMA
H. Allen, 1892

Spotted Bat, or Pinto Bat

The single species, *E. maculatum*, is found in southern British Columbia, the western conterminous United States, and northern Mexico (T. L. Best 1988; E. R. Hall 1981; Watkins 1977).

Head and body length is 60–77 mm, tail length is 47–51 mm, forearm length is 44–55 mm, and weight averages about 15 grams; the ears, larger than those of any other American bat, measure 34–50 mm from notch to tip (T. L. Best 1988; Watkins 1977). The color is dark reddish brown to black, with a characteristic white spot on each shoulder and a white spot at the base of the tail. The hairs of the underparts are tipped with white, and the ears and membranes are grayish. The color pattern of *Euderma* is unique. For additional characters that distinguish *Euderma* from related genera see the above accounts of *Corynorhinus* and *Idionycteris*.

The spotted bat is known from 57 meters below sea level to the high transition zone of Yosemite National Park in California. It has been reported from a wide variety of habitats but has been collected most often in dry, rough desert country (Watkins 1977). Several have been taken in and on houses; others have been obtained in caves or cavelike structures and around water. Easterla (1973) observed 13 banded *Euderma* after release to ignore trees and fly to cliffs, where many entered crevices or other retreats under loose rocks or boulders. He suggested that distribution is determined mainly by the availability of suitable cliff habitat. Poché and Bailie (1974) netted 4 male *Euderma* over scattered pools in Utah. When one was released, it was seen to drop to the ground, where it seized and ate a grasshopper. Other observations indicate that the spotted bat feeds mainly on moths and other insects caught in flight. Indeed, recent studies (Leonard and Fenton 1983; Woodsworth, Bell, and Fenton 1981) show that the spotted bat is not a ground feeder or gleaner but a fast-flying, high-level forager. One individual was seen to fly around a 50-ha. area for about an hour, always remaining at or above treetop height, 10–30 meters above the ground; it made six attempts to catch insects during a 44-minute period. In a ra-

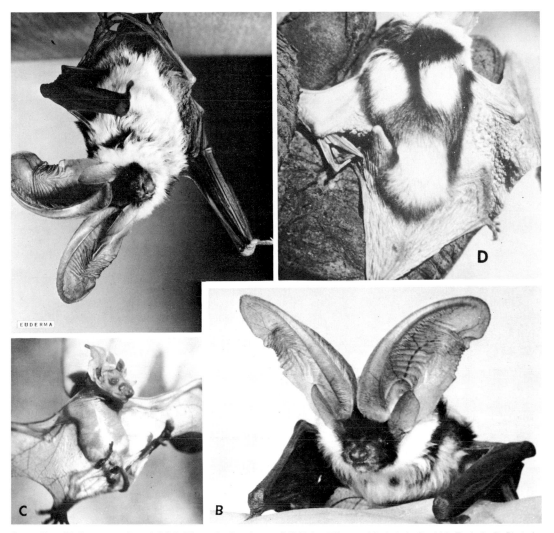

Spotted bat *(Euderma maculatum):* A & B. Photos by Sam Learned; C. Baby, 24 hours old, photo by David A. Easterla; D. Photo by David A. Easterla.

dio-tracking study, Wai-Ping and Fenton (1989) found that *Euderma* flies continuously when away from the day roost.

Although four individuals once were reported hibernating on the walls of a cave above a large pool of water in Utah, the recent studies confirm that *Euderma* is usually solitary and roosts alone in steep cliff faces. It also has been reported to defend an exclusive foraging territory, but radio-tracking data (Wai-Ping and Fenton 1989) showed extensive overlap of large, individual foraging areas. Apparently there is a single young. Births have been observed in western Texas in early June; lactating females have been taken in Texas and New Mexico in June and early July and in Utah in August (Barbour and Davis 1969; Easterla 1973, 1976).

The spotted bat was unknown until 1890; the second specimen was found 13 years later; and the first collections in Canada, Oregon, and Colorado have been reported only since 1980 (Finley and Creasey 1982; McMahon, Oakley, and Cross 1981; Woodsworth, Bell, and Fenton 1981). The species sometimes has been called America's rarest mammal. There have been an increasing number of observations

recently, and Easterla (1973) netted 54 individuals in Big Bend National Park, Texas. Nonetheless, a rangewide survey involving the monitoring of echolocation calls (Fenton, Tennant, and Wyszecki 1987) confirmed that the species is generally rare and perhaps numerous only in a few restricted localities.

CHIROPTERA; VESPERTILIONIDAE; **Genus**
ANTROZOUS
H. Allen, 1862

Pallid Bat, or Desert Bat

The single species, *A. pallidus,* is found from southern British Columbia and Montana to central Mexico and also is known from a few localities on Cuba (Hermanson and O'Shea 1983; Shryer and Flath 1980). *A. p. koopmani* of Cuba was listed as a full species by E. R. Hall (1981) but was shown to be a subspecies of *A. pallidus* by Martin and

Pallid bat, or desert bat *(Antrozous pallidus)*, photo by Robert T. Orr.

Schmidly (1982). *Bauerus* (see account thereof) sometimes is considered a subgenus or synonym of *Antrozous*.

Head and body length is 60–85 mm, tail length is 35–57 mm, forearm length is 45–60 mm, and adult weight is usually 17–28 grams. The woolly fur is creamy, yellowish, or light brown on the upper parts and paler, sometimes almost whitish, on the underparts. Distinctive features of this genus are the large ears and the presence of a small horseshoe-shaped ridge on the squarely truncate muzzle. The nostrils are located beneath the ridge on the front of the muzzle. The ears are large and separate, though not as large, proportionately, as in *Plecotus* and *Euderma*.

The pallid bat favors rocky outcrops with desert scrub but commonly ranges up to forested areas with oak and pine. It roosts in caves, rock crevices, mines, hollow trees, and buildings; emergence is fairly late in the evening (Barbour and Davis 1969). There is evidence for both migration and hibernation in various areas. In a study in central Arizona, Vaughan and O'Shea (1976) found *A. pallidus* to ar-

rive in March or April and to disappear in November. All daytime retreats were crevices or chambers in vertical or overhanging cliffs. During the hottest periods shelter was sought in deep crevices, where body temperatures of 30° C could be maintained passively. O'Shea and Vaughan (1977) found that *A. pallidus* had two nightly foraging periods with an intervening roosting period. This roosting interval was longer in the autumn than in warmer months. Foraging took place at a height of 0.5–2.5 meters above the ground, with a flight pattern consisting of dips, rises, and low gliding swoops. This flight style was well suited to the taking of relatively large, substrate-roving or slow-flying prey. Food items included Coleoptera, Orthoptera, Homoptera, Lepidoptera, arachnids, and a lizard. Barbour and Davis (1969) stated that *A. pallidus* takes food primarily from the ground but also forages in foliage and alights on flowers in search of insects.

These bats appear to be highly social. O'Shea and Vaughan (1977) stated that after the initial foraging period individual *A. pallidus* located one another through vocal communication and gathered in night roosting clusters, where they entered torpor. Brown (1976) listed four main kinds of adult vocalization: a directive call, used for orientating individuals to one another; squabble notes, used for spacing bats when roosting; an irritation buzz, used in agonistic intraspecific encounters; and ultrasonic orientation pulses, used to communicate exploratory activity to other individuals. According to Barbour and Davis (1969), maternity colonies of *A. pallidus* begin to form in early April and number about 12–100 bats. Males have been reported both from nursery colonies and in separate groups. Vaughan and O'Shea (1976) found that 95 percent of a studied population roosted in groups of 20 or more, the largest colony numbering 162 individuals. O'Shea and Vaughan (1977) observed the following conditions in central Arizona: during most of the summer adult males seemed to occur separately; in July and August the females and young appeared to forage together and maintain large colonies; and in mid-August a postbreeding dispersal occurred.

Barbour and Davis (1969) wrote that mating in *A. pallidus* begins in late October and probably takes place sporadically in the winter; sperm is retained in the uterus of the female through the winter; and gestation is 53–71 days.

Pallid bat, or desert bat *(Antrozous pallidus)*, photo by W. G. Winkler and D. B. Adams.

The young are born in May and June; usually there are twins, but 20 percent of births are single. Manning et al. (1987) reported that of 12 pregnant females collected in Texas on 11 May, 1 had a single fetus, 8 had two fetuses, 1 had three fetuses, and 2 had four fetuses. Grindal, Collard, and Brigham (1991) captured a lactating female, a reproductively active male, and a juvenile in southern British Columbia during August. Birth weight is around 3 grams (Hayssen, Van Tienhoven, and Van Tienhoven 1993). The young open their eyes by about 5 days, begin to fly at 4–5 weeks, are weaned after 6–8 weeks, and may breed in their first year (Hermanson and O'Shea 1983). A wild *A. pallidus* is known to have lived for at least 9 years and 1 month (Cockrum 1973). A captive lived for 11 years and 8 months (Marvin L. Jones, Zoological Society of San Diego, pers. comm., 1995).

CHIROPTERA; VESPERTILIONIDAE; Genus BAUERUS
Van Gelder, 1959

Van Gelder's Bat

The single species, *B. dubiaquercus*, is known from mainland Mexico in Jalisco, Chiapas, southern Veracruz, Guerrero, and Quintana Roo, from the Tres Marías Islands off the west coast of Nayarit, and from Belize, Honduras, Guatemala, and Costa Rica (Engstrom, Lee, and Wilson 1987; Engstrom, Reid, and Lim 1993; Juarez G., Jiménez A., and Navarro L. 1988). *Bauerus* originally was described as a subgenus of *Antrozous*, and such status was supported by Pine, Carter, and LaVal (1971). However, based on several morphological and karyological distinctions, Engstrom and Wilson (1981) and Martin and Schmidly (1982) considered the two taxa to be generically distinct. This view was accepted by Corbet and Hill (1991) and Jones, Arroyo-Cabrales, and Owen (1988), though not by Koopman (*in* Wilson and Reeder 1993).

Head and body length is about 57–75 mm, tail length is 46–57 mm, forearm length is 48–57 mm, and weight is 13–20 grams (Engstrom and Wilson 1981; Juarez G., Jiménez A., and Navarro L. 1988). The pelage is soft, lax, and dark brown. *Bauerus* is distinguished from most other North American vespertilionids by its darker coloration

and larger ears. It resembles *Antrozous* externally but is darker and has slightly smaller ears.

Cranially it is distinguished from *Antrozous* by its more pronounced sagittal crest, relatively small auditory bullae, and anteriorly inflected upper toothrow. Most specimens have a spiculelike third lower incisor, whereas *Antrozous* has only two lower incisors on each side (Engstrom, Lee, and Wilson 1987).

Bauerus has been found in a variety of tropical forest habitats at elevations from 370 to 1,450 meters. Most specimens have been taken on the Tres Marías Islands. Roosting sites are unknown. On the basis of morphology it has been speculated that this bat feeds on large insects and takes food exclusively in flight. A pregnant female with one fetus was collected in Honduras in April, lactating females were taken in Chiapas in April and in Quintana Roo in June, and two postlactating females were found in Costa Rica in July (Engstrom, Lee, and Wilson 1987; Juarez G., Jiménez A., and Navarro L. 1988). The IUCN includes *B. dubiaquercus* in the genus *Antrozous* (see above) and classifies it as vulnerable; a decline of at least 20 percent is anticipated over the next decade due to habitat loss.

CHIROPTERA; VESPERTILIONIDAE; Genus MINIOPTERUS
Bonaparte, 1837

Long-winged Bats, or Bent-winged Bats

There are 11 species (Aggundey and Schlitter 1984; Baeten, Van Cakenberghe, and De Vree 1984; Corbet 1978; Corbet and Hill 1992; Ellerman and Morrison-Scott 1966; Flannery 1995; Hayman and Hill *in* Meester and Setzer 1977; Hill 1971a, 1974b, 1982a, 1983a, 1993; Hill and Beckon 1978; Juste and Ibáñez 1992; Koopman 1982b, 1984c, and *in* Wilson and Reeder 1993; Laurie and Hill 1954; Lekagul and McNeely 1977; McKean 1972; Medway 1977; Qumsiyeh and Schlitter 1982; Ride 1970; Taylor 1934):

M. minor, coastal Kenya and Tanzania, lower Congo River, Sao Tomé Island, Madagascar, Comoro Islands;

Van Gelder's bat *(Bauerus dubiaquercus)*, photo by D. S. Rogers through Mark D. Engstrom and courtesy of Carnegie Museum.

Long-winged bat, or long-fingered bat (*Miniopterus* sp.), photo by Howard E. Uible. Inset: face *(M. minor)*, photo by David Pye.

M. *inflatus,* Liberia, Cameroon, Gabon, eastern and
  southern Zaire, western Kenya, Uganda, Rwanda,
  Burundi, Ethiopia, Somalia, Mozambique, Madagascar;
M. *fraterculus,* Angola, Zambia, Mozambique, Malawi,
  South Africa, possibly Kenya, Madagascar;
M. *schreibersi,* southern Europe to Japan and Indochina,
  Malay Peninsula, northern Africa, most of Africa south
  of the Sahara, Madagascar, Taiwan, Sri Lanka, Sumatra,
  Java, Borneo, throughout the Philippines, Sulawesi,
  Lombok, possibly Timor, South Molucca Islands, New
  Guinea, Bismarck Archipelago, Solomon Islands,
  Australia;
M. *magnater,* Yunnan (southern China), Hong Kong,
  Burma, Thailand, Malay Peninsula, Hainan, Sumatra,
  Java, Nusa Penida Island (near Bali), Borneo, possibly
  Timor, Molucca Islands, New Guinea, Bismarck
  Archipelago;
M. *fuscus,* Ryukyu Islands, Okinawa;
M. *medius,* Malay Peninsula, Java, Borneo, Anamba
  Island, New Guinea;
M. *australis,* Java, Borneo, Philippines, Sulawesi, Togian
  Islands, Timor, South Molucca Islands, New Guinea and
  some nearby islands, Bismarck Archipelago, Solomon
  Islands, Trobriand Islands, D'Entrecasteaux Islands,
  Louisiade Archipelago, Vanuatu (New Hebrides), New
  Caledonia, Loyalty Islands (South Pacific), eastern
  Australia;
M. *pusillus,* southern India, possibly Nepal, Thailand,
  Hong Kong, Nicobar Islands, Sumatra, Java, Sulawesi,
  Timor, Moluccas, New Guinea, Bismarck Archipelago,
  Woodlark Island, Solomon Islands, Vanuatu, New
  Caledonia, Loyalty Islands;
M. *tristis,* Philippines, Sulawesi, Togian Islands, New
  Guinea, Bismarck Archipelago, Woodlark Island,
  Solomon Islands, Vanuatu;
M. *robustior,* Loyalty Islands.

There is considerable disagreement regarding the classification of this genus. Maeda (1982) described and/or distinguished 11 species in addition to those listed above, but Hill
(1983a) accepted none of them. M. *blepotis* (Malay Peninsula, Borneo, Java, Amboina and Kei islands), M. *esch-
scholtzi* (Philippines, Australia), and M. *haradai* (Thailand)
were considered by Maeda to form a related group, but they
were regarded as components of M. *schreibersi* by Hill. M.
macrodens (Burma and southern China to Java and Borneo), M. *oceanensis* (eastern Australia), and M. *fuliginosus*
(Afghanistan and India to Japan and northern Viet Nam)
were considered by Maeda to form another group that also
included M. *magnater,* but while Hill did regard *macrodens*
as a subspecies of M. *magnater,* he considered *oceanensis*
and *fuliginosus* to be components of M. *schreibersi.* M.
solomonensis (Solomon Islands) and M. *paululus* (Philippines, Java, Rennell Island in the Solomons) were considered by Maeda to be species related to M. *australis,* but Hill
considered them to be only subspecies thereof. M. *macroc-
neme* (New Guinea to New Caledonia), considered by
Maeda to be a species related to M. *fuscus* and accepted as
such by Flannery (1995), was designated a subspecies of M.
pusillus by Hill. M. *melanesiensis* (Solomon Islands, New
Hebrides) and M. *bismarckensis* (Bismarck Archipelago)
were described by Maeda as species related to M. *tristis,* but
Hill synonymized *melanesiensis* under M. *tristis* and suggested that *bismarckensis* is part of M. *magnater.*

Koopman (*in* Wilson and Reeder 1993) generally accepted the species of *Miniopterus,* as listed above and by
Hill (1983a) and Corbet and Hill (1991), but included M.
medius in M. *fuscus.* Several reports of *Miniopterus* in
South Australia, cited by Maeda, were not accepted by Hill
or Koopman, but Mahoney and Walton (*in* Bannister et al.
1988) indicated that M. *schreibersi* occurs along the coast
of South Australia. Peterson (1981) described a new species,
M. *propitristis,* from Sulawesi, New Guinea, and other islands extending eastward to the New Hebrides; Flannery
(1995) accepted it, but Hill (1983a) considered it a subspecies of M. *tristis.* Peterson (cited by Wilson 1985) accepted M. *oceanensis* as a valid taxon.

Head and body length is 40–78 mm, tail length is 38–72

mm, forearm length is 34–59 mm, and adult weight is usually 6–24 grams. The coloration is reddish, reddish brown, dark brown, grayish brown, or grayish.

The second bone of the longest finger is about three times as long as the first bone. When the bat hangs by its hind feet this lengthened terminal part of the third finger folds back upon the wing. The tail is completely enclosed within the interfemoral membrane and is proportionately longer than in other bats of the same size.

The members of this genus usually roost in caves but have also been found in rock clefts, culverts, eaves and roofs of buildings, and crevices of trees. They are often associated with *Notopteris* and species of *Myotis* in their daytime retreats. They appear early in the evening, with a rapid, jerky flight. They feed on small beetles and other insects, usually at heights of 10–20 meters. *Miniopterus* hibernates in the cooler parts of its range.

In a study of *M. schreibersi* in South Africa, Van der Merwe (1975) found that seasonal migrations occurred. From late winter to late spring there was a movement of pregnant females from wintering caves in the southern Transvaal to maternity caves in the north. In late summer the females and weaned young moved back to the south. Studies of this species in India showed that the population of a given area tended to be centered in one large cave but that individuals spent part of their time in secondary roosts within a 70-km radius (Lekagul and McNeely 1977).

These bats may be highly gregarious. Van der Merwe (1978) reported that in one maternity cave of *M. schreibersi* in the Transvaal the juveniles alone numbered 110,000 (each female gives birth to a single young from early November to early December). His studies, and others in Asia (Lekagul and McNeely 1977), showed that young are not carried by the mother but are deposited in a large communal nursery, in clusters separate from those of the adults. Lekagul and McNeely stated that the typical roosting group of *M. australis* consists of only one male and six females and that such groups are transitory, breaking up when the bats go out to feed. Dwyer (1968), however, reported a nursery colony of 4,000 *M. australis*, including 1,800 young, associated with a much larger group of *M. schreibersi* in New South Wales. Males were found to disappear from the colony by December, when the young were born. In Europe mating occurs in August and September, fertilization occurs shortly thereafter, embryonic development is retarded through hibernation, and the young are not born until spring. Females are ready for breeding again after the young are weaned. In a study in eastern Australia Richardson (1977) found both *M. schreibersi* and *M. australis* to be monestrous. In *M. schreibersi* mating took place in the autumn (late May–early June), with fertilization and development to the blastocyst stage following immediately; implantation was then delayed until August and births occurred in December. In *M. australis* mating was in August, implantation occurred by mid-September with little or no delay, and births took place in December. Birth weight in *M. schreibersi* is about 2–3 grams, lactation lasts around two months, and sexual maturity evidently comes sometime between one and two years of age (Hayssen, Van Tienhoven, and Van Tienhoven 1993).

In a study on coastal Kenya McWilliam (1990) found evidence that *M. minor* utilizes a "lek" mating system such as described for *Hypsignathus*. Males seemed to compete for access to a particular site within a cave; in one area about 30 individuals, or 10 percent of the males in the area, were successful. Females then visited the site and mating occurred in June and July. Brosset and Saint Girons (1980)

found *M. inflatus* to be monestrous and strictly a seasonal breeder in Gabon; there was no delayed implantation, females gave birth to a single young in October, and pregnancy and lactation together lasted six months. The longevity record in the genus is nine years.

The IUCN classifies *M. robustior* as endangered, *M. fuscus* as vulnerable, and *M. minor*, *M. fraterculus*, and *M. schreibersi* as near threatened. Flannery (1995) indicated that *M. robustior* has not been recorded for many years. Stebbings and Griffith (1986) regarded *M. schreibersi* as endangered in western Europe and possibly throughout the world. Several colonies that formerly contained thousands of individuals have almost or entirely disappeared. This species is very sensitive and may be eliminated if its roosts in caves and mines are disturbed by human workers or tourists.

CHIROPTERA; VESPERTILIONIDAE; **Genus MURINA** *Gray, 1842*

Tube-nosed Insectivorous Bats

There are 2 subgenera and 14 species (Chasen 1940; Corbet 1978; Corbet and Hill 1992; Ellerman and Morrison-Scott 1966; Flannery 1995; Hill 1962, 1963a, 1972a; Hill and Francis 1984; Koopman and Danforth 1989; Laurie and Hill 1954; Maeda 1980; Medway 1977, 1978; Sly 1975; Taylor 1934; Yoshiyuki 1983):

subgenus *Murina* Gray, 1842

M. leucogaster, southern Siberia to central China and Japan;

M. tenebrosa, known only by a single specimen from Tsushima Island (Japan), possibly also on Yakushima Island;

M. aurata, mountains of Sichuan and Yunnan (south-central China), Nepal, Sikkim, northeastern India, Burma, northern Thailand;

M. ussuriensis, Ussuri region of southeastern Siberia, Manchuria, Korea, Sakhalin, Kuril Islands;

M. silvatica, Japan;

M. florium, Sulawesi, Sumbawa, Flores, South Molucca Islands, New Guinea and Umboi Island to northeast, northeastern Queensland;

M. suilla, Malay Peninsula, Sumatra and nearby Nias Island, Java, Borneo;

M. tubinaris, Kashmir, northern Pakistan, northeastern India, northern Thailand, Burma, Laos, northern Viet Nam;

M. rozendaali, Borneo;

M. huttoni, northern India to southeastern China, Malay Peninsula;

M. puta, Taiwan;

M. aenea, Malay Peninsula, Borneo;

M. cyclotis, southern and eastern India to Indochina and Malay Peninsula, Sri Lanka, Hainan, Borneo, Philippines, Lombok;

subgenus *Harpiola* Thomas, 1915

M. grisea, known only by the holotype from Kumaon in northern India.

Koopman (*in* Wilson and Reeder 1993) listed *M. fusca* of Manchuria as a distinct species, but it was included in *M.*

Tube-nosed insectivorus bat *(Murina leucogaster)*, photo by Modoki Masuda, Nature Productions, Japan.

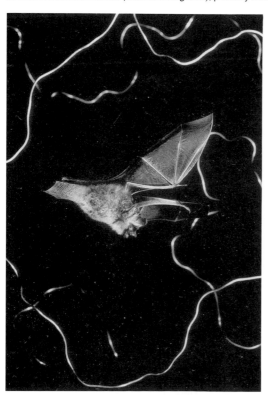

Tube-nosed insectivorus bat *(Murina leucogaster)*, photo by Modoki Masuda, Nature Productions, Japan.

leucogaster by Corbet (1978) and Corbet and Hill (1991). Yoshiyuki (1989) treated *M. hilgendorfi* of Japan as a species separate from *M. leucogaster.*

Head and body length is 33–60 mm, tail length is 30–42 mm, and forearm length is 28–45 mm. Medway (1978) listed the following weights: *M. suilla,* 3–4 grams; *M. cyclotis,* 9–10 grams; *M. huttoni,* 6.4 grams; and *M. aenea,* 7.5 grams. The fur is usually thick and woolly. The coloration is often brownish or grayish, but in some forms it is reddish and in *M. aurata* it is golden yellow above. The nostrils, which are located at the ends of tubes, are somewhat like those of *Harpiocephalus* and *Paranyctimene* of the family Pteropodidae.

Murina is often found in hilly areas. These bats seem to be low-flying, as they have been noted skimming over the surface of crops and grass in their feeding flights. Members of this genus have been found roosting in the dead dry leaves of cardamon plants and in caves. Several usually roost together. Medway (1978) wrote that pregnant *M. cyclotis* carrying two fetuses each were taken in February and May on the Malay Peninsula. *M. florium* is considered the rarest bat of Australia, but its continued presence recently was confirmed by Spencer, Schedvin, and Flick (1992). The IUCN classifies *M. ussuriensis* as endangered, forecasting a decline of at least 50 percent over the next decade because of habitat loss. *M. grisea* also is declining and classed as endangered, *M. puta* is designated as vulnerable, and *M. aurata, M. silvatica, M. rozendaali, M. huttoni,* and *M. aenea* are designated as near threatened.

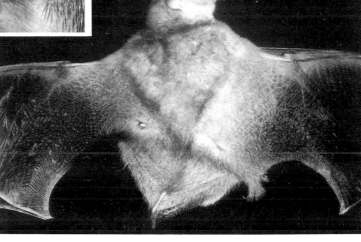

Hairy-winged bat *(Harpiocephalus harpia)*, photos by Paul D. Heideman.

CHIROPTERA; VESPERTILIONIDAE; **Genus**
HARPIOCEPHALUS
Gray, 1842

Hairy-winged Bat

There apparently are two species (Corbet and Hill 1992):

H. harpia, northeastern and southern India, southeastern
China, northern Indochina, Taiwan, Sumatra, Java,
Borneo, Lombok, Amboina Island in the Moluccas;
H. mordax, Burma, Thailand, Borneo.

Koopman (*in* Wilson and Reeder 1993) included *H. mordax*
in *H. harpia* but noted that it may be a separate species.
 Head and body length is 55–75 mm, tail length is 37–55
mm, and forearm length is 40–54 mm. One specimen of *H.
harpia* weighed 19 grams (Ingle and Heaney 1992). Coloration above is chestnut, orange-red, or rusty, usually
mixed with gray; the underparts are grayish buffy. As in
Murina, the fur is thick and woolly. The legs, tail membrane, and occasionally part of the wings are covered with
hair.
 Four genera of bats possess tubular nostrils—*Nyctimene* and *Paranyctimene* in the family Pteropodidae and
Murina and *Harpiocephalus* in the family Vespertilionidae. *Harpiocephalus* is distinguished from *Murina* mainly by dentition. The cheek teeth of *Harpiocephalus* are
massive, powerful, and supplied with blunt cusps.
 Little information has been recorded on the habits and
life history of these bats. However, they have been ob-

served to frequent hilly country, and they probably roost
in vegetation. Beetle remains were found in the stomach of
one individual, perhaps an indication that they feed on
chitinous insects, as their dentition suggests. The IUCN
designates *H. mordax* as near threatened.

CHIROPTERA; VESPERTILIONIDAE; **Genus**
KERIVOULA
Gray, 1842

Painted Bats, or Woolly Bats

There are 2 subgenera and 22 species (Bergmans and Van
Bree 1986; Chasen 1940; Corbet and Hill 1992; Ellerman
and Morrison-Scott 1966; Flannery 1995; Hayman and Hill
in Meester and Setzer 1977; Hill 1965, 1977*a;* Hill and
Francis 1984; Hill and Rozendaal 1989; Koopman *in* Wilson
and Reeder 1993; Laurie and Hill 1954; Lekagul and Mc-
Neely 1977; Lunney and Barker 1986; Medway 1977, 1978;
Ride 1970; Taylor 1934; Walton, Busby, and Woodside
1992):

subgenus *Kerivoula* Gray, 1842

K. argentata, southern Kenya to Namibia and eastern
South Africa;
K. lanosa, Liberia to Ethiopia, and south to South Africa;
K. eriophora, Ethiopia;
K. smithii, Liberia, Ivory Coast, Nigeria, Cameroon,
eastern Zaire, Kenya;

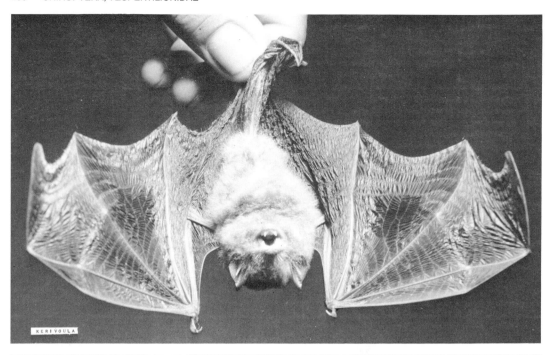

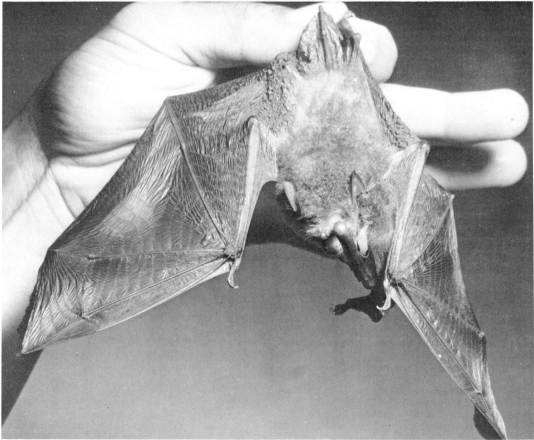

Woolly bats, or painted bats *(Kerivoula papillosa)*. A small baby is between the wing and head of the mother in the lower picture. This is not the normal position in which baby bats are carried; it probably resulted from the handling of the bats. Photos by Lim Boo Liat.

K. cuprosa, southern Cameroon, northern Zaire, Kenya;

K. phalaena, Liberia, Ghana, Cameroon, middle Congo River;

K. africana, Tanzania;

K. whiteheadi, Malay Peninsula, Borneo, Philippines;

K. picta, southern and eastern India to southern China and Indochina, Malay Peninsula, Sri Lanka, Hainan, Sumatra, Java, Bali, possibly Borneo, Molucca Islands;

K. muscina, Papua New Guinea;

K. agnella, Fergusson, Woodlark, Sudest (Tagula), and St. Aignan's (Misima) islands southeast of New Guinea;

K. minuta, Malay Peninsula, Borneo;

K. intermedia, Malay Peninsula, Borneo;

K. pellucida, Malay Peninsula, Sumatra, Java, Borneo, southern Philippines and nearby Sulu Islands;

K. papillosa, northeastern India, Thailand, Indochina, Malay Peninsula, Sumatra, Java, Borneo, Sulawesi;

K. flora, known from Borneo, Lombok, and Flores, with possible records from Bali, Sumbawa, and Sumba;

K. hardwickei, southern and eastern India to southern China and Indochina, Malay Peninsula, Sri Lanka, Sumatra, Mentawai Islands, Dua Island, Java, Kangean Islands, Bali and nearby Nusa Penida Island, Borneo, southern Philippines, Talaud Islands, Sulawesi, Peleng Island, Lombok, possibly Sumba;

K. myrella, Wetar Island north of Timor, Admiralty Islands and Bismarck Archipelago northeast of New Guinea;

subgenus *Phoniscus* Miller, 1905

K. atrox, Malay Peninsula, Sumatra, Borneo;

K. jagorii, Java, Bali, Borneo, Samar Island (Philippines), Sulawesi, Lombok;

K. papuensis, Biak and Supiroi islands northeast of New Guinea, Papua New Guinea, eastern Queensland, eastern New South Wales;

K. aerosa, reputedly from east coast of South Africa but possibly from Sulawesi.

Phoniscus was considered a distinct genus by Hill (1965) and Corbet and Hill (1991, 1992) but not by Koopman (1979, 1982*b*, and *in* Wilson and Reeder 1993) or Mahoney and Walton (*in* Bannister et al. 1988). The status of *K. eriophora* is doubtful since the type and only known specimen has been lost (Yalden and Largen 1992).

Head and body length is 31–57 mm, tail length is 32 to about 55 mm, and forearm length is 27–46 mm. Medway (1978) listed weights of 9–10 grams for *K. papillosa* and 4–6 grams for *K. hardwickei.* Flannery (1995) recorded weights of 8.1–9.5 grams for *K. papuensis.* Hill and Francis (1984) reported that the smallest examples of *K. minuta* are little larger than *Craseonycteris,* probably the smallest of bats. The painted bat, *K. picta,* is bright orange or scarlet with black wings and orange along the fingers. It thus resembles *Myotis formosus* in its color pattern, but the painted bat is a smaller species. *K. argentata* of Africa also has a striking appearance: this bat is bright orange rufous interspersed with black and frosted with whitish tips to the hairs. The other forms of *Kerivoula* are reddish brown, yellowish brown, olive brown, dark brown, or grayish; shining gray hairs simulating a tuft of moss are sometimes present in the pelage. This genus is characterized by the long, woolly, rather curly hair; the delicate form; the rather large and somewhat pointed, funnel-shaped ears; and the presence of 38 teeth. In distinguishing *Phoniscus* from *Kerivoula,* Corbet and Hill (1992) indicated that the former has the posterior margin of the tragus deeply notched near the base, the anterior palatal emargination usually wider than long, and

the upper canine tooth with a prominent lateral groove.

Painted bats are forest inhabitants. Some of the African species often roost and apparently raise their young in the abandoned hanging nests of weaverbirds and sunbirds. Tree hollows and trunks, foliage, huts, and buildings are also used as roosts by *Kerivoula. K. picta,* from Asia, shelters among the dry leaves of vines and other plants, in plantain fronds, and in flowers. One individual of this species was noted in the foliage of a longan tree, an evergreen that retains decaying orange and black leaves throughout the year; *K. picta* could readily be mistaken for just another russet leaf in such a roost. These bats emerge late in the evening, have a rather weak, fluttering flight, and usually fly in circles close to the ground.

Kerivoula roosts singly or in small groups. Four individuals of *K. argentata* were clinging together so tightly under the eaves of a hut in Africa that they resembled the mud nest of a wasp. A group of seven *Kerivoula* in the Philippines were noted flying together in the daytime and performing aerial gyrations between periodic returns to a roosting site in foliage; this behavior may possibly have been associated with breeding. Young *Kerivoula* have been found in October in Zaire and Sri Lanka. Medway (1978) stated that a lactating female *K. hardwickei* with a nearly full-grown young was taken in January on the Malay Peninsula. Lim, Shin, and Muul (1972) reported the collection of two female *K. papillosa,* each with one embryo, in mid-June on Borneo. On Biak Island in September 1992 Flannery (1995) discovered a group of *K. papuensis* consisting of one adult male, two lactating females, one pregnant female, three other females, and two juveniles. Kitchener et al. (1990) collected pregnant female *K. flora, K. hardwickei,* and *K. jagorii,* each with a single embryo, in September and October on Lombok Island.

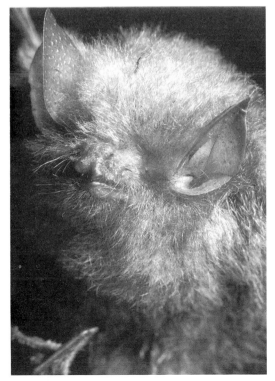

Woolly bat, or painted bat (*Kerivoula* sp.), photo by Lawrence R. Heaney.

The species *K. africana* was described in 1878 and is known only by a single specimen collected on the African coast near Zanzibar (Kingdon 1974*a*). It is probably extinct, and a number of other species are declining because of human environmental disturbance. The IUCN classifies *K. eriophora, K. muscina, K. agnella,* and *K. myrella* as vulnerable and *K. smithi, K. cuprosa, K. minuta,* and *K. intermedia* as near threatened.

CHIROPTERA; VESPERTILIONIDAE; **Genus**
NYCTOPHILUS
Leach, 1821

New Guinean Bats and Australian Big-eared Bats

There are nine living species (Churchill, Hall, and Helman 1984; Corbet and Hill 1992; Flannery 1990*b*; Hill and Pratt 1981; Kitchener, How, and Maharadatunkamsi 1991*a*; Koopman 1984*c*; Laurie and Hill 1954; Maddock *in* Strahan 1983; Mahoney and Walton *in* Bannister et al. 1988; Parnaby 1987; Richards *in* Strahan 1983; Ride 1970):

N. walkeri, northeastern Western Australia, northern Northern Territory;
N. arnhemensis, northeastern Western Australia, northern Northern Territory;
N. microtis, Salawati Island off western New Guinea, Papua New Guinea;
N. gouldi, southwestern Western Australia, southeastern Australia, Tasmania;
N. bifax, New Guinea, northern Western Australia to eastern Queensland;
N. timoriensis, New Guinea, Western Australia, northwestern Northern Territory, South Australia, eastern Queensland, eastern New South Wales, Victoria, Tasmania, possibly Timor;
N. heran, Lembata Island in the Lesser Sundas;
N. geoffroyi, throughout Australia except northern Queensland, Tasmania;
N. microdon, eastern New Guinea.

An additional living but unnamed species recently was discovered on New Caledonia (Flannery 1995). Still another species, *N. howensis* from Lord Howe Island, east of Australia, was described by McKean (1975). Although known only from fossil remains, it may have occurred in modern times, since there are reports of the former presence of a large bat on the island. *N. bifax* was included within *N. gouldi* by Koopman (1984*c* and *in* Wilson and Reeder 1993) but was considered a distinct species by Corbet and Hill (1991) and Parnaby (1987). The genus and species *Lamingtona lophorhina* were described by McKean and Calaby (1968) on the basis of six specimens from Mount Lamington in southeastern Papua New Guinea. *Lamingtona* was considered to be related to *Nyctophilus* but to be distinguished by several morphological characters, including the absence of a band of integument connecting the ears across the forehead and a basally more slender tragus. Hill and Koopman (1981) considered *Lamingtona lophorhina* to closely resemble *Nyctophilus microtis* in these and other characters, concluded that the two are conspecific, and provisionally recognized *lophorhina* as a subspecies of *N. microtis.*

Head and body length is 38–75 mm, tail length is 30–55 mm, forearm length is 31–47 mm, and weight is 4–20

Australian big-eared bat *(Nyctophilus timoriensis),* photo by Stanley Breeden.

grams (Allison *in* Strahan 1983; Churchill, Hall, and Helman 1984; Kitchener *in* Strahan 1983; Maddock *in* Strahan 1983; McKenzie *in* Strahan 1983; Richards *in* Strahan 1983). The coloration is light brown, cinnamon brown, orange brown, dark brown, or ashy gray. *Nyctophilus* and *Pharotis* are the genera in the family Vespertilionidae that combine long united ears with a small, horseshoe-shaped nose leaf. The ears, about 25 mm in length, usually are united by a low membrane. There is a fleshy disk behind the nose leaf, and the muzzle is abruptly truncate.

These bats have been observed in forested areas, scrub country, and arid regions. Daytime retreats are small caverns, crevices in rocks, tree hollows, and under the bark of trees. *N. geoffroyi* seems to have adapted well to human presence and may be found in cities roosting within the roofs of buildings (Maddock *in* Strahan 1983). Some populations are thought to be active throughout the year, but Phillips and Inwards (1985) concluded that in southeastern Australia *N. gouldi* undergoes lengthy bouts of hibernation during the colder months, from April to September. Most if not all of the species have a rather slow and fluttering flight. They are highly maneuverable and pursue a variety of insects in the open air but also glean from foliage and tree branches. Ride (1970) stated that *N. geoffroyi* will land on the ground to pick up beetles and other insects. He also said (pers. comm.) that this species will often come to an observer's hand to take a proffered insect, preferably one that is so held as to kick, flutter, or "buzz" and thus attract the bat's attention. The bat hovers in front of the hand and at times will even land gently on it to take the insect. Big-eared bats can also fly off a horizontal surface by leaping into the air with a sudden downward cupping of the outstretched wings, as does *Antrozous.*

Social structure seems to vary, with *N. timoriensis* reported to roost alone or in pairs and *N. geoffroyi* forming maternity colonies of 10 to more than 100 individuals (Maddock *in* Strahan 1983; Richards *in* Strahan 1983). On the mainland these colonies are made up almost entirely of pregnant females, but in Tasmania they contain adults of both sexes. Births, commonly of 2 young, take place in the colonies in late spring or early summer. In New South Wales, McKean and Hall (1964) found pregnant females or newborn young in October and November. Studies of both wild and captive *N. gouldi* in southeastern Australia (Phillips and Inwards 1985) showed that mating takes place within hibernating colonies during the winter. Females are monestrous and store the sperm until ovulation in September or early October. The 1 or 2 young are born in late October and November; they are never carried by the mother, begin practice flights at 4–5 weeks, and are weaned after 6 weeks. Females reach sexual maturity at 7–9 months, males at 12–15 months.

The IUCN classifies *N. howensis*, which may have existed in historical times (see above) as extinct. Another restricted island species, *N. heran*, is probably declining and is classed as endangered. The IUCN also designates *N. timoriensis* and *N. microdon* as vulnerable and *N. walkeri*, *N. sherrini* (found in Tasmania and recognized by the IUCN as a species distinct from *N. timoriensis*), and *N. daedalus* (found in Northern Territory and recognized as distinct from *N. hifax*) as near threatened.

CHIROPTERA; VESPERTILIONIDAE; **Genus**
PHAROTIS
Thomas, 1914

The single species, *P. imogene*, is known only by a series of 45 specimens taken in 1890 approximately at sea level in far southeastern New Guinea (Flannery 1990*b*; Koopman 1982*b*).

Measurements of two female specimens are as follows: head and body length, 47–50 mm; tail length, 42–43 mm; and forearm length, 37.5–38.6 mm (Flannery 1990*b*). The coloration is dark brown both above and below.

This bat is similar to the bats of the genus *Nyctophilus*, differing externally from that group in the shorter muzzle and the larger ears and nose leaf. *Pharotis* and *Nyctophilus* are the only genera of the family Vespertilionidae that combine long ears united at the base with a small, horseshoe-shaped nose leaf.

Nothing is known about *Pharotis* except that it must be near or at the point of extinction. The IUCN now formally classifies it as critically endangered, judging that any surviving population would contain fewer than 250 mature individuals and would probably continue to decline in response to habitat destruction.

CHIROPTERA; VESPERTILIONIDAE; **Genus**
TOMOPEAS
Miller, 1900

The single species, *T. ravus*, is restricted to the arid and semiarid coastal region of Peru (Davis 1970*b*; Koopman 1978*b*). The phylogenetic position of *Tomopeas* is questionable; recent protein electrophoretic and mitochondrial DNA analyses suggest that the genus is a primitive member of the family Molossidae (Sudman, Barkley, and Hafner 1994).

Davis (1970*b*) listed the following measurements: total length, 73–85 mm; tail length, 34–45 mm; forearm length, 31.2–34.5 mm; and weight, 2–3.5 grams. The coloration is pale brown above and dull buff to whitish cream buff below. The basal half of the fur is dull slaty gray. The face, ears,

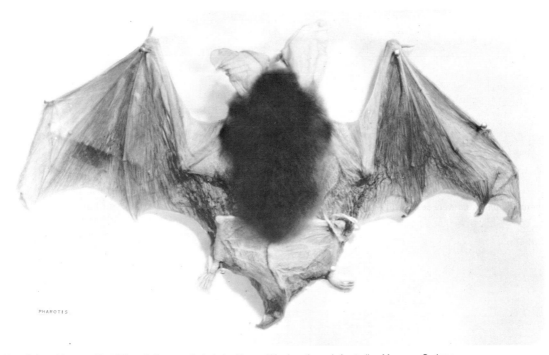

New Guinea big-eared bat *(Pharotis imogene)*, photo by Howard Hughes through Australian Museum, Sydney.

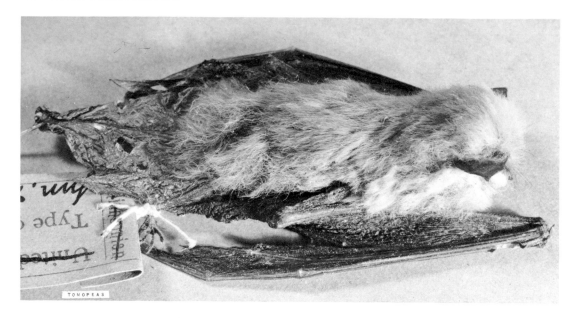

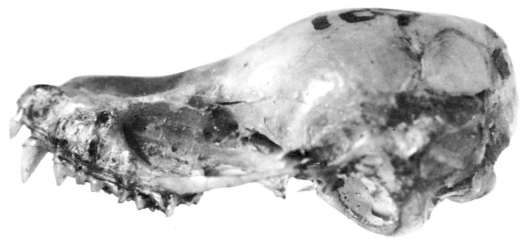

Tomopeas ravus, photos from British Museum (Natural History).

and membranes are black. The ears lack an anterior basal lobe but have a rudimentary keel, and the tragus is short and blunt.

Tomopeas is generally placed in a separate subfamily of the Vespertilionidae, but it has characters that are suggestive of both the families Vespertilionidae and Molossidae, and it may well belong in the latter. Allen (1939*a*:208) stated: "It has the delicate form and the long tail wholly included in the membrane as in the former, and the peculiar ear structure and the fusion of the last cervical with the first thoracic vertebra rather characteristic of the latter. It may thus be thought of as a very ancient and ancestral type, partly bridging the gap between the two families."

This bat has a known altitudinal range from sea level to 1,000 meters. Specimens have been taken in mist nets set among large mesquite trees and have been found roosting under the exfoliations of granite boulders and outcroppings. Juvenile females, as well as adults of both sexes, were collected in July and August (Davis 1970*b*). Based on the presence of juveniles in August but only nonpregnant females in August and November, Graham (1987) suggested that parturition was possible in the coastal dry season, from May to September. The IUCN classifies *Tomopeas* as vulnerable, noting that it could decline precipitously in response to human disturbance of its restricted habitat.

CHIROPTERA; **Family MOLOSSIDAE**

Free-tailed Bats and Mastiff Bats

This family of 16 genera and 86 species is found in the warmer parts of the world, from southern Europe and southern Asia south through Africa and Malaysia and east to the Fiji Islands, and from the central United States south through the West Indies, Mexico, and Central America to the southern half of South America. The sequence of gen-

Mastiff bat *(Eumops perotis)*, photo by Abel Fornes.

era presented here follows that suggested in Freeman's (1981) discussion of generic considerations for the Molossidae, which was adopted by Koopman (1984b), except that the highly distinctive *Cheiromeles* is placed last. A somewhat different sequence was proposed by Legendre (1984), who recognized three subfamilies: Molossinae, with the genera *Molossops* (including *Cynomops* and *Neoplatymops* as subgenera), *Myopterus, Molossus, Eumops,* and *Promops;* Tadaridinae, with *Mormopterus* (including *Micronomus, Platymops,* and *Sauromys* as subgenera), *Nyctinomops, Rhizomops* (including only the species here called *Tadarida brasiliensis*), *Otomops,* and *Tadarida* (including *Chaerephon* and *Mops* as subgenera); and Cheiromelinae, with the single genus *Cheiromeles.* Recent analyses of protein electrophoretic and mitochondrial DNA data (Sudman, Barkley, and Hafner 1994) tended to support the latter arrangement but also indicated that the genus *Tomopeas* (here included in the Vespertilionidae) is

a basal and phylogenetically distant member of the Molossidae.

Head and body length is 40 to approximately 130 mm, tail length is 14–80 mm, and forearm length is 27–85 mm. The tail projects far beyond the free edge of the narrow tail membrane, hence the common name "free-tailed bats" for this family. The short body hair has a velvetlike texture. In one genus *(Cheiromeles)* the hair is so short that the animal appears to be naked. Some species of *Tadarida* have an erectile crest of hairs on the top of the head. The coloration is usually brown, buff, gray, or black. Some species, particularly in the genus *Molossus,* have two color phases. The head is rather thick and the muzzle is broad and obliquely truncate, usually with a scattering of short hairs having spoon-shaped tips. The eyes are small. The ears are thick and rather leathery, variable in size and form, and often united across the forehead and directed forward. A tragus is present. The nostrils usually open on a pad the upper surface of which is often adorned with small hornlike projections. There is no nose leaf. The lips are large; in some genera the upper lip is furrowed by vertical wrinkles. Throat glands are often present. The wings are long, narrow, thick, and leathery. The legs are short and strong, and the foot is broad. Curved, spoon-shaped bristles on the outer toes of each foot are used by the bat in cleaning and grooming its fur.

The teeth are of the normal cuspidate insectivorous type. The dental formula varies from (i 1/1, c 1/1, pm 1/2, m 3/3) $\times$ 2 = 26 to (i 1/3, c 1/1, pm 2/2, m 3/3) $\times$ 2 = 32. The braincase is thick, flat, and broad.

Molossid bats roost in caves, tunnels, buildings, hollow trees, foliage (at least *Molossus*), the decayed wood of logs (reported for *Molossops*), the crevices of rock cliffs, and holes in the earth (reported for *Cheiromeles*). They also shelter under bark and rocks and often under the corrugated iron roofing of human-made structures; they prefer the high temperatures, up to 47° C and above, in such places. A

Mexican free-tailed bat *(Tadarida brasiliensis)* with right wing extended, showing that it is proportionately narrower than that of most bats except other members of the family Molossidae. The rate of wingbeat of these bats is faster than that of most other insectivorous bats. Photo by Ernest P. Walker.

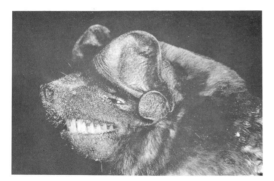

Molossus molossus, photo by David Pye.

characteristic strong musky odor generally permeates their retreats. Some species live in groups of hundreds of thousands or even millions, others associate in smaller groups, and some forms are solitary. The colonial species generally return to their favored haunts year after year. *Nyctinomops femorosaccus* and *Eumops perotis* have been found roosting together in a crack in a large granite boulder; the two genera were segregated, the *Nyctinomops* hanging in the higher, narrower part of the crevice and the *Eumops* sheltering in the lower, wider part.

The members of this family tend to be active throughout the year. The northernmost species may be inactive for short periods during the winter, but there is no definite evidence of true hibernation. Rather, these species usually make local movements or, as in *Tadarida brasiliensis* of the New World, a seasonal migration. Compared with the flight of many insectivorous bats, their flight is swift and relatively straight. They fly with their mouths open and send out ultrasonic sounds. The diet consists of insects, often hard-shelled forms.

These bats usually have one young per litter and one litter per year. Two young are born on rare occasions, and *Molossus ater*, in Trinidad, possibly has two litters per year. Two successive pregnancies have been reported in a female *Tadarida* from Africa. Breeding generally takes place just before ovulation in late winter and spring. There may be partial or complete segregation of the sexes in some species.

During the Civil War the Confederate Army used guano, probably produced by *Tadarida brasiliensis*, as a source of niter (sodium nitrate) for gunpowder. Guano for fertilizer was obtained in commercial quantities for 20 years from Carlsbad Caverns, New Mexico. Glover M. Allen stated that during the period of greatest activity, lasting about 15 years, this guano was obtained from September to March; one to three carloads were shipped daily, each weighing about 40 tons. Allen reported that the prices realized for it were said to be about $20–$80 per ton. The attempt of C. Campbell to increase the number of *Tadarida brasiliensis* by erecting wood "bat roosts" proved to be a failure. He had hoped to secure guano in commercial quantities and also thought that these bats would eradicate mosquitoes and malaria. Only small amounts of guano were obtained, and mosquitoes are seldom eaten by this species. The attempted utilization of this same species as a carrier of small incendiary bombs during World War II was also abandoned.

The geological range of this family is late Eocene to Recent in Europe and North America, late Oligocene to Recent in South America, middle Miocene to Recent in Africa and Australia, and Pleistocene to Recent in Asia and the East and West Indies (Hand 1990; Koopman 1984*a*).

CHIROPTERA; MOLOSSIDAE; **Genus MORMOPTERUS**
Peters, 1865

Little Goblin Bats

There are 2 subgenera and 10 species (Allison *in* Strahan 1983; Cabrera 1957; Flannery 1995; Freeman 1981; E. R. Hall 1981; Hayman and Hill *in* Meester and Setzer 1977; Hill 1983*b*; Koopman 1982*a*, 1982*b*, 1984*c*, and *in* Wilson and Reeder 1993; Legendre 1984; Peterson 1985; Richards *in* Strahan 1983):

subgenus *Mormopterus* Peters, 1865

M. acetabulosus, Ethiopia, Natal (South Africa), Madagascar, Mauritius and Reunion (Indian Ocean);
M. jugularis, Madagascar;
M. doriae, Sumatra;
M. kalinowskii, coastal Peru, northern Chile;
M. phrudus, Peru;
M. minutus, Cuba;

subgenus *Micronomus* Troughton, 1943

M. planiceps, central and southern Australia;
M. loriae, Papua New Guinea, northern and eastern Australia;
M. norfolkensis, southeastern Queensland, coastal New South Wales, possibly Norfolk Island (southwestern Pacific);
M. beccarii, Halmahera and Amboina islands in the Moluccas, New Guinea and Fergusson Island to southeast, northern Australia.

Mormopterus sometimes has been treated as a subgenus of *Tadarida*, but it is now almost universally accepted that the two are generically distinct. On the other hand, there has been disagreement concerning whether *Platymops* and *Sauromys* (see accounts thereof) are subgenera of *Mormopterus*. In addition, *Micronomus* was recognized as a subgenus by Legendre (1984) but was considered only a synonym of *Mormopterus* by Freeman (1981) and Koopman (*in* Wilson and Reeder 1993). *M. loriae* was treated as a subspecies of *M. planiceps* by Koopman (1982*b*, 1984*c*, and *in* Wilson and Reeder 1993) but was considered a distinct species by Corbet and Hill (1991) and Flannery (1990*b*). Peterson (1985) implied agreement that *M. loriae* and *M. planiceps* are conspecific and indicated further that the resulting species should have the name *M. petersi*. He also suggested that *M. astrolabiensis* of the Cape York Peninsula of northern Queensland is a species distinct from *M. beccarii* and that another species, with affinity to *M. jugularis*, is present at Hermannsburg in the south-central Northern Territory. Based on electrophoretic analyses, Adams et al. (1988) suggested that there are at least five species of *Mormopterus* in Australia, but they could not equate those species with the ones discussed above.

Based on the Australian and Cuban species, head and body length is 43–65 mm, tail length is 27–40 mm, forearm length is 29–41 mm, and weight is 6–19 grams; the upper parts are dark brown, grayish brown, or charcoal, and the underparts are usually paler (Allison *in* Strahan 1983; E. R. Hall 1981; Richards *in* Strahan 1983). *Mormopterus* is characterized by small overall size, separated and erect ears, unwrinkled lips, medium-thick jaws, a skull with a tall and posteriorly curving coronoid process, shallow basisphenoid pits, and a well-developed third upper molar tooth (Freeman 1981).

Little goblin bat *(Mormopterus planiceps)*, photo by G. B. Baker / National Photographic Index of Australian Wildlife.

The small amount of available natural history information is restricted mainly to the Australian species (Allison *in* Strahan 1983; Richards *in* Strahan 1983). These bats, found in tropical forests, woodlands, open areas, and cities, roost mainly in roofs and tree hollows. They usually forage for insects above the tree canopy, water holes, and creeks but sometimes scurry after prey on the ground. Flight is swift and direct. Colonies vary in number from fewer than 10 to several hundred individuals. A single young is produced in December or January (summer). Crichton and Krutzsch (1987) reported that female *M. planiceps* stored sperm for at least two months prior to ovulation, which took place in late July and August, and that gestation then lasted three to five months. Development of young was slow, but females reached sexual maturity before one year. Graham (1987) collected a lactating female *M. kalinowskii* in coastal Peru in May.

Little is known about the non-Australian species, except that they all appear to be in serious decline because of habitat loss. The IUCN classifies *M. phrudus* as endangered and *M. acetabulosus*, *M. jugularis*, *M. doriae*, *M. kalinowskii*, and *M. minutus* as vulnerable.

CHIROPTERA; MOLOSSIDAE; Genus SAUROMYS
Roberts, 1917

South African Flat-headed Bat

The single species, *S. petrophilus*, is found in Namibia, Zimbabwe, Botswana, South Africa, the Tete district of western Mozambique, possibly Ghana (Meester et al. 1986), and apparently the desert region of southwestern Angola (Crawford-Cabral 1989*a*). Both Freeman (1981) and Legendre (1984) concluded that *Sauromys* is a subgenus of *Mormopterus*. This position was followed by Koopman (1984*b* and *in* Wilson and Reeder 1993) but not by Corbet and Hill (1991), Meester et al. (1986), or Peterson (1985).

Head and body length is about 50–85 mm, tail length is 30–45 mm, forearm length is 38–50 mm, and weight is 9–22 grams (Skinner and Smithers 1990). The upper parts are brownish gray to tawny olive and the underparts are paler.

Sauromys differs from *Platymops* in the absence of wartlike granulations on the forearm, the absence of a well-developed throat sac, a well-developed anterior premolar (reduced or absent in *Platymops*), the strong development of a distinct hypocone on the first and second upper molars, and other cranial details. *Sauromys* differs from *Tadarida* in the flatness of the skull, the conformation of the anterior nasal aperture, the weakly bicuspidate upper incisors, the wide degree of separation of the ears, and other details. Freeman (1981) indicated that *Sauromys* shares the characters of *Mormopterus*, as given in the account thereof, and that the bats of both genera (or subgenera) are flatheaded to some degree. However, Meester et al. (1986) suggested that flattening of the skull is more conspicuous in *Sauromys*.

Specimens have been taken by overturning slabs of rocks in hilly regions. *Sauromys*, with its flattened skull, apparently is adapted to such situations. It rests during the day under slabs of exfoliated rock or in narrow rock crevices or fissures. It generally occurs in small numbers, most records being of up to four (Smithers 1983).

South African flat-headed bat *(Sauromys petrophilus)*, photo by M. Brock Fenton.

Flat-headed free-tailed bat *(Platymops setiger)*, photo by Bruce J. Hayward.

CHIROPTERA; MOLOSSIDAE; **Genus PLATYMOPS**
Thomas, 1906

Flat-headed Free-tailed Bat

The single species, *P. setiger,* is found in southeastern Sudan, southern Ethiopia, and Kenya (Hayman and Hill *in* Meester and Setzer 1977). Peterson (1985) recognized *P. macmillani* of Ethiopia as a separate species. Both Freeman (1981) and Legendre (1984) concluded that *Platymops* is a subgenus of *Mormopterus*. This position was followed by Koopman (1984b and *in* Wilson and Reeder 1993) but not by Corbet and Hill (1991), Meester et al. (1986), or Peterson (1985).

Head and body length is 50–60 mm, tail length is 22–36 mm, and forearm length is 29–36 mm. A small, tufted sac is present in the throat region of both sexes. This genus resembles some species of *Tadarida* in external appearance, but in *Platymops* the skull is flattened, so the depth of the braincase is equal to only about a third of its width. The ears are separate. In some individuals the forearm is roughened with warty protuberances. Freeman (1981) indicated that *Platymops* shares the general characters of *Mormopterus,* as given in the account thereof, but its lips appear to be wrinkled and are covered with unusually thick hairs; she considered it a highly derived subgenus.

Platymops does not appear to be gregarious. Usually one to five individuals shelter under rocks and slabs, the flattened skull being an adaption for hiding in crevices. Individuals have been seen over a small permanent water hole and marsh in rapid erratic flight at heights of nine meters and less. The bats appeared when it was almost dark and were feeding on small beetles. Data cited by Hayssen, Van Tienhoven, and Van Tienhoven (1993) indicate that a single young is produced in Kenya in January and that lactating females have been taken in February.

CHIROPTERA; MOLOSSIDAE; **Genus MOLOSSOPS**
Peters, 1865

Broad-faced Bats

There are two subgenera and five species (Alberico 1987; Anderson 1991; Ascorra, Wilson, and Handley 1991; Cabrera 1957; Eisenberg 1989; Goodwin and Greenhall 1961; E. R. Hall 1981; Handley 1976; Ibáñez and Ochoa G. 1985; Koopman 1978b; Myers and Wetzel 1983; Williams and Genoways 1980b, 1980c):

subgenus *Molossops* Peters, 1865

M. temminckii, Venezuela and Peru to central Brazil and northern Argentina;
M. neglectus, Surinam, northeastern Brazil, eastern Peru;

subgenus *Cynomops* Thomas, 1920

M. planirostris, Panama to Paraguay and northern Brazil;
M. greenhalli, southwestern Mexico to northeastern Brazil, Trinidad;
M. abrasus, Colombia, Venezuela, Guyana, Surinam, French Guiana, Peru, Bolivia, Brazil, Paraguay, northern Argentina.

Cynomops sometimes is considered a distinct genus (as by Emmons 1990), and *Neoplatymops* and *Cabreramops* (see accounts thereof) sometimes are designated subgenera of *Molossops.* Handley (1976) recognized an additional species in Venezuela, *M. paranus;* it was not considered distinct from *M. planirostris* by Koopman (1978b), though Ochoa G. and Ibáñez (1985) suggested that the name may be valid. Corbet and Hill (1991) did list *M. paranus* as a species and included central Mexico within its range, but Jones, Arroyo-Cabrales, and Owen (1988) did not mention the presence of such a species in Mexico or Central America. The name *M. brachymeles* sometimes has been used in place of *M. abrasus,* but Williams and Genoways (1980b) considered the latter to be correct.

Head and body length is 40 to about 95 mm, tail length is 14–37 mm, and forearm length is 28–51 mm. Eisenberg

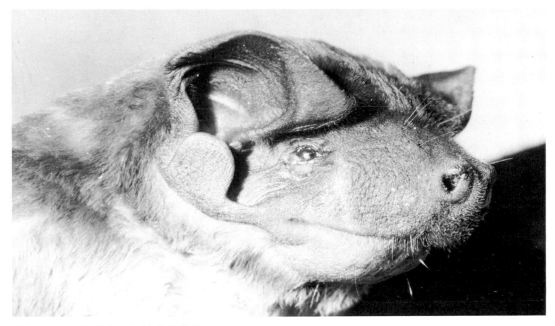

Molossops greenhalli, photo by Merlin D. Tuttle.

(1989) listed average weights of about 55 grams for *M. abrasus*, 19 grams for *M. greenhalli*, 14 grams for male *M. planirostris*, and 11 grams for female *M. planirostris*. The upper parts are yellowish brown, russet, chocolate brown, dark brown, or slate black and the underparts are usually yellowish white, gray, or slaty; some species have a white throat.

The bats of this genus resemble those of *Molossus* in external features, "but more conspicuous lines of fur diverging from the angle in the bend of the wing along the forearms and fourth finger are usually distinctive." According to Freeman (1981), both *Molossops* and *Myopterus* are characterized by a broad face, widely separated ears, no development of wrinkles on the lips, thickish dentaries, and no anterior palatal emargination. *Molossops* lacks basisphenoid pits, whereas *Myopterus* has very deep ones.

In Venezuela, Handley (1976) found most specimens of *M. planirostris* roosting in rotting snags in swamps. *M. temminckii* also has been found sheltering in decaying logs. *M. greenhalli* generally roosts in colonies of 50–75 individuals in the hollow branches of large trees. The males and females of this species appear to remain together throughout the year.

Vizotto and Taddei (1976) reported *M. temminckii* and *M. planirostris* in Brazil to roost mainly in hollow fence posts in groups of up to 8 individuals; pregnant females were found there from September to January and lactating females in February. Taddei, Vizotto, and Martins (1976) collected pregnant female *M. abrasus* from October to December in southeastern Brazil. Pregnant or lactating females of *M. greenhalli* have been taken in May, June, and July (Gardner, LaVal, and Wilson 1970; Valdez and LaVal 1971). There is a single young (Hayssen, Van Tienhoven, and Van Tienhoven 1993).

The IUCN classifies *M. neglectus* as endangered. It evidently is declining because of habitat loss. *M. abrasus* is designated as near threatened.

CHIROPTERA; MOLOSSIDAE; **Genus**
NEOPLATYMOPS
Peterson, 1965

South American Flat-headed Bat

The single species, *N. mattogrossensis*, is known from six localities in extreme eastern Colombia, east-central and southern Venezuela, southern Guyana, the Matto Grosso region of Brazil, and east-central Brazil, but it probably has a wide distribution in the Amazon Basin and the eastern Brazilian highlands (Linares and Escalante 1992; Ortiz-Von Halley and Alberico 1989; Willig 1985a). Although Peterson (1965b) originally described *Neoplatymops* as a full genus, it was regarded only as a subgenus of *Molossops* by both Freeman (1981) and Legendre (1984). Koopman (*in* Wilson and Reeder 1993) agreed, but not Corbet and Hill (1991). Except as noted, the information for the remainder of this account was taken from Willig (1985a) and Willig and Jones (1985), who also considered *Neoplatymops* generically distinct on the basis of morphological and karyological data.

Neoplatymops is one of the smallest molossids. Average measurements for males, followed by those for females, are: head and body length, 52.6 mm and 51.1 mm; tail length, 25.5 mm and 25.7 mm; forearm length, 30.1 mm and 30.2 mm; and weight, 6.1 grams and 5.4 grams. The pelage is short and sparse. The upper parts are generally pale brownish, the ears and membranes are dark brown, and the underparts are whitish to grayish. The ears are widely separated, there is a gular gland on the throat, the skull is conspicuously flattened, and there is no sagittal crest.

Neoplatymops has some resemblance to *Molossops* but has two (rather than one) well-developed upper premolar teeth, a flatter skull, and a series of small, wartlike granular structures on the dorsal surface of the forearm. These forearm granulations, which are not found in any other

South American flat-headed bat *(Neoplatymops mattogrossensis)*, photo by Merlin D. Tuttle.

New World molossid, may function to provide anchorage on the porous surfaces of the rocks on which *Neoplatymops* roosts, thereby reducing the likelihood of being pulled off by a predator.

The local distribution of *Neoplatymops* is restricted to areas containing rocky outcrops, where it roosts close to the ground in narrow horizontal crevices beneath granitic exfoliations. In this regard it is convergent with the Old World *Platymops* and *Sauromys*. Its morphology suggests that it is capable of maneuverable flight and the utilization of both soft and hard-bodied prey. Stomach contents of specimens show a wide variety of small insects, with beetles being most common.

Three or four individuals sometimes roost under the same granitic exfoliation. Linares and Escalante (1992) reported that such groups usually contain a single male and two to four females, thereby suggesting a harem–a polygynous social system. Females are monotocous and seasonally monestrous. In eastern Brazil pregnancies have been found initially in August, during the middle of the dry season. Synchronized births occur there in November and December, during the transition from the dry to the wet season, and lactation persists through the wet season, from December to April.

The IUCN designates *N. mattogrossensis* as near threatened but considers it part of the genus *Molossops*.

CHIROPTERA; MOLOSSIDAE; Genus CABRERAMOPS
Ibáñez, 1980

Ibáñez (1980) erected this genus for the species originally described as *Molossops aequatorianus,* known only by four specimens collected in 1864 at Babahoyo in west-central Ecuador. Koopman (*in* Wilson and Reeder 1993) treated *Cabreramops* as a subgenus of *Molossops,* though Corbet and Hill (1991) listed it as a full genus.

In two specimens head and body length was 50 and 52 mm, tail length was 29 and 31 mm, and forearm length was 37.6 and 35.9 mm. From *Molossops, Cabreramops* differs in having the anterior bases of the ears near together on the

forehead, a wrinkled upper lip (though more weakly than in *Tadarida*), and well-developed basisphenoid pits. There is no anterior palatal emargination in the skull, the sagittal crest is absent, and there are only two lower incisors and a single upper premolar on each side.

The IUCN classifies *C. aequatorianus* as vulnerable but considers it part of the genus *Molossops*. It is expected to decline by at least 20 percent over the next decade because of human destruction of its very restricted habitat.

CHIROPTERA; MOLOSSIDAE; Genus MYOPTERUS
E. Geoffroy St. Hilaire, 1818

There probably are only two species (Adam, Aellen, and Tranier 1993; Hayman and Hill *in* Meester and Setzer 1977):

M. whitleyi, Ghana, Nigeria, Cameroon, Gabon, Zaire, Uganda;
M. daubentonii, Senegal, Ivory Coast, Central African Republic, northeastern Zaire.

A third species, *M. albatus* of Ivory Coast and northeastern Zaire, was listed by Hayman and Hill (*in* Meester and Setzer 1977) as well as Corbet and Hill (1991), but it evidently is conspecific with *M. daubentonii* (Adam, Aellen, and Tranier 1993; Koopman 1989c). The name *Eomops* Thomas, 1905, has sometimes been used in place of *Myopterus.*

Head and body length is about 56–66 mm, tail length is about 25–33 mm, and forearm length is about 33–37 mm. The upper parts are dark brown and the underparts are light reddish yellow to white. This genus resembles some of the other molossid bats in external features, but it differs from them in cranial and dental characters. Some external features of *Myopterus* are as follows: the ears are shorter than the head; the ears and the short tragus are rounded above; the muzzle projects beyond the jaws; the end of the nose is separate from the upper lip; and the nostrils open almost laterally.

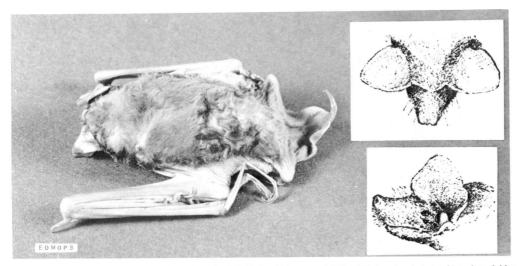

Myopterus whitleyi, photo by P. F. Wright of specimen in U.S. National Museum of Natural History. Inset photos from *Ann. & Mag. Nat. Hist.*

According to Happold (1987), *M. whitleyi* lives only in the rainforest zone. It seems to be solitary. Individuals have been found clinging to the bark of a tree, among the leaves of plantains, and in the roof of a shed. At night it flies in the forest, into gardens, and around houses. Freeman (1981) speculated that *Myopterus* is probably more maneuverable in flight than are most molossids, that *M. whitleyi* takes small, soft-bodied prey, and that *M. daubentonii* is able to consume large, hard-bodied prey.

CHIROPTERA; MOLOSSIDAE; Genus TADARIDA
Rafinesque, 1814

Free-tailed Bats

There are 2 subgenera and 8 species (Aggundey and Schlitter 1984; Ansell 1986; Ansell and Dowsett 1988; Cabrera 1957; Corbet 1978; Corbet and Hill 1992; Ellerman and Morrison-Scott 1966; Flannery 1990*b*; Freeman 1981; E. R. Hall 1981; Harrison 1975*b*; Harrison and Bates 1984; Hayman and Hill *in* Meester and Setzer 1977; Hill 1961*b*, 1982*a*, 1983*b*; Hill and Morris 1971; Kock 1975; Koopman 1975, 1982*b*, 1984*c*; Largen, Kock, and Yalden 1974; Laurie and Hill 1954; Legendre 1984; Lekagul and McNeely 1977; McKean and Calaby 1968; Medway 1977, 1978; Peterson 1971*c*, 1974*b*; Pienaar 1972; Rautenbach, Fenton, and Braack 1985; Ride 1970; Schlitter et al. 1986; Skinner and Smithers 1990; Wilkins 1989):

subgenus *Tadarida* Rafinesque, 1814

T. teniotis, Mediterranean zone of Europe and northern Africa to Japan and Taiwan, Madeira and Canary islands;
T. lobata, Kenya, Zimbabwe;
T. fulminans, eastern Zaire, Rwanda, Kenya, Tanzania, Zambia, Malawi, Zimbabwe, extreme northeastern Transvaal, Madagascar;

Free-tailed bat *(Tadarida aegyptiaca),* photo by Erwin Kulzer.

A. Free-tailed bat *(Tadarida aegyptiaca)* at rest on a horizontal surface with its two-day-old baby on its back. This position is normal for the young when these bats are on ledges in caves and tombs, but the babies of most bats cling beneath the mother so they do not interfere with her wings. B. Two-day-old young of an Egyptian tomb bat *(T. aegyptiaca)*. Photos by Erwin Kulzer.

T. ventralis, southern Sudan, Ethiopia, eastern Zaire, Kenya, Malawi, Zambia, Transvaal;

T. aegyptiaca, Algeria, Egypt, Nigeria and Sudan to South Africa, southern Arabian Peninsula, Iran, Pakistan, India, Sri Lanka;

T. australis, Western Australia to southern Queensland and Victoria;

T. kuboriensis, central (Kubor Range) and probably southeastern Papua New Guinea;

subgenus *Rhizomops* Legendre, 1984

T. brasiliensis, Oregon and North Carolina to Chile and Argentina, throughout the West Indies, occasionally as far north as South Dakota and Ohio.

Tadarida often is considered to include as subgenera what are here regarded as the separate genera *Mormopterus, Chaerephon, Mops,* and *Nyctinomops* (see accounts thereof). *Rhizomops,* described as a full genus by Legendre (1984), was not even accepted as a subgenus by Corbet and Hill (1991) or Koopman (*in* Wilson and Reeder 1993). Owen, Chesser, and Carter (1990) discussed the validity of *Rhizomops* and also provided evidence that *T. cynocephala* of the southeastern United States and possibly the West Indies may be specifically distinct from *T. brasiliensis.* Yoshiyuki (1989) recognized *T. insignis* of Japan as a species separate from *T. teniotis. T. kuboriensis* was considered a subspecies of *T. australis* by Koopman (1982b, 1984c, and

in Wilson and Reeder 1993) but a separate species by Corbet and Hill (1991), Flannery (1990b), Freeman (1981), and Legendre (1984). Koopman (*in* Wilson and Reeder 1993) listed an additional species, *T. espiritosantensis* of Brazil, but as both he and Corbet and Hill (1991) suggested, it eventually was shown to be a synonym of *Nyctinomops laticaudatus* (Zortéa and Taddei 1995). The generic name *Nyctinomus* E. Geoffroy St.-Hilaire, 1813, was used in place of *Tadarida* by Mahoney and Walton (*in* Bannister et al. 1988). The name *T. africana* often is used in place of *T. ventralis.*

In *T. australis,* one of the largest species, head and body length is 85–100 mm, tail length is 40–55 mm, forearm length is 57–63 mm, and weight is 25–40 grams (Richards *in* Strahan 1983). In African species head and body length is 65–100 mm, tail length is 30–59 mm, forearm length is 45–66 mm, and weight is 14–39 grams (Kingdon 1974a; Smithers 1983). In the New World *T. brasiliensis* head and body length is 46–66 mm, tail length is 29–44 mm, forearm length is 36–46 mm, and weight is 10–15 grams (E. R. Hall 1981; Hoffmeister 1986; Lowery 1974). Coloration in the genus ranges from reddish brown to almost black. *Tadarida* is characterized by wrinkled lips, deep anterior palatal emargination, relatively thin jaws, and a third upper molar with an N-shaped occlusal pattern. Most species have ears that are joined by a band of skin across the top of the head. In *T. brasiliensis* and *T. aegyptiaca* the ears are separated, but not as widely as in *Mormopterus,* and are large and forward-facing as in other *Tadarida* (Freeman 1981).

Habitat varies considerably. African species live in either forest or open country and generally are reported to roost in trees and buildings (Hayman and Hill *in* Meester and Setzer 1977; Rosevear 1965; Smithers 1971). In Venezuela Handley (1976) found most *T. brasiliensis* roosting in houses. Barbour and Davis (1969) stated that *T. brasiliensis* roosts in buildings on the west coast and in the southeastern United States and mainly in caves in the southwest. Prior to European settlement the populations in the southeastern United States probably roosted in the hollows of mangrove and cypress trees (Wilkins 1989). Colonies of *T. brasiliensis* may make spectacular mass exits from their caves after sunset and then fly up to 65 km to foraging areas. The diet consists mostly of small moths and beetles. Based on radar observations in the southwest, Williams, Ireland, and Williams (1973) determined the average speed of groups of *T. brasiliensis* to be 40 km/hr (7–105 km/hr) and the average maximum altitude to be 2,300 meters (600–3,100 meters). Some populations of this species also make lengthy migrations. In a banding study Cockrum (1969) found that maternity colonies began assembling in the southwestern United States in April and disappeared by mid-October. He determined that populations in California, western Arizona, and southern Arizona made only localized movements in the spring and autumn or relatively short migrations to the south or west. Populations from central Arizona to Kansas and Texas, however, migrated deep into Mexico, sometimes traveling more than 1,600 km. Less is known about southeastern populations. Lowery (1974) described a colony of 20,000 *T. brasiliensis* in New Orleans that existed for at least 35 years, moving to an unknown location in the autumn and then returning in the spring.

The large summer groups of *T. brasiliensis* are basically maternity colonies, consisting mainly of females. In the southeastern United States such colonies usually number more than 1,000 bats, but in the southwest there may be millions of individuals in a single cave. During the 1960s about 100 million *T. brasiliensis* occupied 13 caves in Texas, and an estimated 25 million to 50 million bats were in Eagle Creek Cave in Arizona (Barbour and Davis 1969; Cockrum 1969). In terms of sheer numbers, no larger concentration of mammalian life is known to have existed. Some males are always present in the large nursery colonies, but most tend to gather in relatively small groups nearby. Cockrum (1969) found summer male colonies usually to number only 10–300 bats in the southwestern United States. It is possible that some males do not even migrate northward for the summer, as a group of 40,000 was discovered in late June in Chiapas, Mexico. In late summer, after the young are full grown, the sexes begin to reassociate. Most other species are not known to be as gregarious as *T. brasiliensis*. For example, *T. australis* seldom is found in groups of more than 10; females of that species apparently mate during the austral spring or early summer and give birth to a single young toward the end of the year (Richards *in* Strahan 1983). In Zimbabwe females of *T. fulminans* are seasonally polyestrous, being potentially able to give birth during the wet season in November and the start of the dry season in May or June, following a gestation period of about 100 days (Cotterill and Fergusson 1993). In southern Africa female *T. aegyptiaca* apparently form maternity colonies of up to several hundred individuals and bear young during the warm, wet austral summer (Skinner and Smithers 1990).

In *T. brasiliensis* in North America mating takes place in February–March, ovulation occurs around late March, gestation lasts about 77–84 days, and the single young is usually born in June or July (Barbour and Davis 1969). Birth weight is about 3–4 grams (Hayssen, Van Tienhoven, and Van Tienhoven 1993). Since maternity colonies may consist of millions of tightly packed females and young, it once was believed that mothers nursed offspring indiscriminately during the 5-week lactation period. New investigations (McCracken 1984; McCracken and Gustin 1987, 1991), however, show that each female usually locates her own pup, perhaps through scent or vocalization, though occasionally another young will "steal" her milk as she searches through the colony. The mother finds her pup using vocal and olfactory cues, and apparently the young also has some ability to detect and move toward its mother. In a study in New Orleans, Pagels and Jones (1974) found that the gestation period was about 11 weeks, the young were capable of maneuverable flight at 38 days, and full size was attained by 60 days. Short (1961) reported that males reached sexual maturity at about 18–22 months. LaVal (1973a) reported that a banded *T. brasiliensis* lived at least 8 years.

In recent years there has been increasing concern about drastic declines in some populations of *T. brasiliensis* in the southwestern United States (Geluso, Altenbach, and Wilson 1976, 1981; Gosnell 1977; Humphrey 1975; Mohr 1972). The numbers in the most famous colony, that of Carlsbad Caverns National Park in New Mexico, fell from an estimated 8.7 million bats in 1936 to only 200,000 in 1973. The even larger group in Eagle Creek Cave, Arizona, was reduced to only about 600,000 individuals in 1970. The exact cause of such declines is unknown, but the most likely factor is poisoning through accumulation of organochlorine residues in the bats as a result of the spraying of their prey insects with DDT. There may have been a moderate recovery of the Carlsbad Caverns population since DDT was banned in the United States in 1972, but this insecticide is still being used in agricultural operations in Mexico, where the bats spend the winter. Fortunately, at least one of the giant summer colonies still exists: McCracken and Gustin (1987) reported the presence of about 20 million *T. brasiliensis* at Bracken Cave in central Texas. This site recently was purchased and placed under protection by Bat Conservation International.

The IUCN now designates *T. brasiliensis*, as well as *T. fulminans*, *T. ventralis*, and *T. australis*, as near threatened. *T. lobata*, restricted to a much smaller area and subject to severe human disturbance, is classified as vulnerable.

CHIROPTERA; MOLOSSIDAE; Genus CHAEREPHON
Dobson, 1874

Lesser Mastiff Bats

There are 14 species (Aggundey and Schlitter 1984; Ansell 1986; Chasen 1940; Corbet 1978; Corbet and Hill 1992; Eger and Peterson 1979; Fenton and Peterson 1972; Harrison 1975b; Hayman and Hill *in* Meester and Setzer 1977; Hill 1961b, 1982a, 1983b; Hill and Beckon 1978; Hill and Morris 1971; Hutterer, Dieterlen, and Nikolaus 1992; Juste and Ibáñez 1993c; Kock 1975; Koopman 1975, 1982b, and *in* Wilson and Reeder 1993; Koopman and Steadman 1995; Largen, Kock, and Yalden 1974; Lekagul and McNeely 1977; Medway 1978; Nader and Kock 1980; Peterson 1971a, 1972; Peterson and Harrison 1970; Ride 1970; Schlitter et al. 1986; Schlitter, Robbins, and Buchanan 1982):

C. bivittata, Ethiopia to Zambia and Mozambique;
C. ansorgei, eastern Nigeria and Ethiopia to Angola and Natal;
C. bemmelini, Liberia, Cameroon, eastern Zaire, southern Sudan, Kenya, Uganda, Tanzania;
C. nigeriae, southwestern Saudi Arabia, Ghana and Niger to Ethiopia and Botswana;
C. major, savannah zones from Mali and Liberia to southern Sudan and Tanzania;
C. pumila, southwestern Arabian Peninsula, most of Africa south of the Sahara, Sao Tomé and Bioko islands off West Africa, Madagascar, Comoro Islands, Aldabra and Amirante islands in the Seychelles;
C. chapini, Ethiopia, Zaire, Uganda, Zambia, Angola, Namibia, Botswana, Zimbabwe;
C. russata, Ghana, Cameroon, northeastern Zaire, Kenya;
C. aloysiisabaudiae, Ghana, Gabon, Zaire, Uganda, possibly Ethiopia;
C. gallagheri, eastern Zaire;
C. tomensis, known only from Sao Tomé Island off West Africa in the Gulf of Guinea;
C. johorensis, Malay Peninsula, Sumatra;
C. plicata, India to southern China and Indochina, Malay Peninsula, Sri Lanka, Hainan, Sumatra, Java, Bali, Borneo, Philippines, Lombok, Cocos Keeling Island in the Indian Ocean;
C. jobensis, Seram Island (Moluccas), New Guinea and nearby Japen (Jobi) Island, Solomon Islands, northern Australia, Vanuatu (New Hebrides), Fiji Islands, also skeletal remains from 'Eua in The Tonga Islands dating from less than 3,000 years ago.

Chaerephon was regarded as a full genus by Freeman (1981), Koopman (1984b and *in* Wilson and Reeder 1993), and Mahoney and Walton (*in* Bannister et al. 1988) but only as a subgenus of *Tadarida* by Corbet and Hill (1991), Legendre (1984), and Meester et al. (1986). The species *C. bivittata*, *C. ansorgei*, and *C. bemmelini* sometimes are placed in *Tadarida*, but Freeman (1981) assigned them to *Chaerephon*. *C. bregullae* of Vanuatu and Fiji and *C. solomonis* of the Solomon Islands were treated by Flannery (1995) as species distinct from *C. jobensis*.

In *C. plicata* of Southeast Asia head and body length is

Chaerephon pumila, photo by John Visser.

65–75 mm, tail length is 30–40 mm, forearm length is 40–50 mm, and weight is 17–31 grams. The pelage is dense and soft, the upper parts are dark brown, and the underparts are slightly paler. The face is covered with stiff, short, black bristles, and the ears are thick, round, broad, and joined on the front of the muzzle (Lekagul and McNeely 1977). In *C. jobensis* of Australia head and body length is 80–90 mm, tail length is 35–45 mm, forearm length is 46–52 mm, and weight is 20–30 grams. Coloration is chocolate to gray brown above and slightly grayer below (Richards *in* Strahan 1983). In various African species head and body length is about 50–80 mm, tail length is 25–45 mm, forearm length is 35–53 mm, and weight is 8–26 grams. The upper parts are dark brown or black, the underparts are slightly paler, and there are sometimes white markings on the sides or belly (Happold 1987; Kingdon 1974*a*; Smithers 1983). Several species have tufts of glandular hairs, arising from the crown, behind the ears, and several have a heavy crest of long straight hairs on the back of the membrane uniting the ears. These patches of specialized hairs are often restricted to the males, as is the saclike throat gland.

Chaerephon differs from *Tadarida* in having ears that are joined by a band of skin, usually a more elevated mandibular condyle, and broader wing tips. *Chaerephon* differs from *Mops* in having less robust jaws and more constricted anterior palatal emargination. There are usually five upper cheek teeth, and the last upper molar has an N-shaped occlusal surface (Freeman 1981; Kingdon 1974*a*).

These bats inhabit open forests, savannahs, and agricultural areas, sometimes in the mountains. They roost in hollow trees, crevices, and caves. Some species have adapted well to human presence and can be found in roofs, rafters, thatch, and other suitable parts of buildings (Richards *in* Strahan 1983; Smithers 1983). *C. pumila* is known to be a fast flier and to hunt above the forest canopy and buildings at heights of more than 70 meters; it takes a wide variety of small insects (Kingdon 1974*a*).

Social structure appears to vary. Rosevear (1965), for example, reported that *C. pumila* is usually found roosting singly. Smithers (1983), however, noted that whereas *C. bivittata* usually does not occur in groups of more than about 6 individuals, *C. pumila* may occur in colonies of hundreds in favorable areas. Kingdon (1974*a*) observed that the larger groups are found in buildings where there is much available roosting space. A colony of 350 *C. jobensis* was found in a building in a town (Richards *in* Strahan 1983). *C. plicata* of southeastern Asia generally is found in caves in groups of 200,000 or more individuals (Lekagul and McNeely 1977; Medway 1978).

In a detailed study of *C. pumila* in Ghana, McWilliam (1988*b*) found harems of up to 21 adult females and their young attended by a single adult male. Group composition was stable, with some individuals remaining together at the same site throughout the 16-month study period. Three cohorts of young were produced during the wet season, from April to October. A few young females stayed in their natal groups to take the place of older females, but most young of both sexes dispersed during the dry season, after they had attained reproductive maturity.

Young *C. jobensis* have been found in December and January (Richards *in* Strahan 1983), and a pregnant female *C. bivittata* was taken in March (Kingdon 1974*a*). In Uganda *C. pumila* is a continuous breeder, the peaks coinciding with maximum rainfall in October–November and April–May (Mutere 1973*b*). Studies of the same species in the Transvaal (Van der Merwe, Rautenbach, and Van der Colf 1986) have suggested less of a correlation with rainfall. In that area there is an extended breeding season of 8 months per year, with birth peaks in November, January, and April; this period evidently is associated with higher minimum temperatures and thus greater availability of in-

Chaerephon jobensis, photo by B. G. Thomson / National Photographic Index of Australian Wildlife.

sects. Females were found to be polyestrous and able to have three pregnancies per season. The gestation period is about 60 days and a single young is born. Weaning apparently occurs before 21 days, and females become sexually mature at 5–12 months.

Flannery (1995) reported that a colony of many thousands of *C. jobensis bregullae* occupies a cave roost on the island of Vanua Levu in Fiji. In December 1990 there were many pregnant females and young at various stages of growth in the cave; adult males also were present. In a cave colony on the island of Espiritu Santo in Vanuatu most females were lactating in April 1992. These observations suggest that there is synchronous seasonal breeding in the South Pacific, with young being born around December and weaned by April. Flannery noted that the bats were being killed for use as food, and he designated *bregullae* as vulnerable.

The IUCN recognizes *bregullae* and *solomonis* as species distinct from *C. jobensis* (see above) and designates them, as well as *C. chapini* and *C. johorensis*, as near threatened. *C. gallagheri*, restricted to one small area and declining in response to habitat destruction, is classified as critically endangered. *C. tomensis*, expected to decline by at least 20 percent over the next decade, is classed as vulnerable. *C. pusilla* of the Seychelles Islands, which already has been reduced to fewer than 1,000 individuals, is recognized by the IUCN as a species distinct from *C. pumila* and also is classified as vulnerable.

CHIROPTERA; MOLOSSIDAE; **Genus MOPS**
Lesson, 1842

Greater Mastiff Bats

There are 2 subgenera and 14 species (Aggundey and Schlitter 1984; Chasen 1940; Corbet 1978; Corbet and Hill 1992; El-Rayah 1981; Freeman 1981; Harrison 1975*b*; Hayman and Hill *in* Meester and Setzer 1977; Hill 1961*b*; Hill and Morris 1971; Jones 1971*a*; Kock 1975; Koopman 1975 and *in* Wilson and Reeder 1993; Koopman, Mumford, and Heisterberg 1978; Largen, Kock, and Yalden 1974; Laurie and Hill 1954; Medway 1977; Peterson 1972; Schlitter, Robbins, and Buchanan 1982):

subgenus *Xiphonycteris* Dollman, 1911

M. spurrelli, Liberia, Ivory Coast, Ghana, Togo, Benin, Equatorial Guinea (Rio Muni and island of Bioko), Zaire;
M. nanulus, forest zone from Sierra Leone to Ethiopia and Kenya;
M. petersoni, Ghana, Cameroon;
M. brachypterus, forest zone from Gambia to Kenya and Mozambique, island of Bioko (Fernando Poo);
M. thersites, forest zone from Sierra Leone to Rwanda, Bioko, possibly Mozambique and Zanzibar;

subgenus *Mops* Lesson, 1842

M. condylurus, most of Africa south of the Sahara, Madagascar;
M. niveiventer, Zaire, Rwanda, Burundi, Tanzania, Angola, Zambia, northern Botswana, Mozambique;
M. demonstrator, Burkina Faso, Sudan, northeastern Zaire, Uganda;
M. mops, Malay Peninsula, Sumatra, Borneo, possibly Java;
M. sarasinorum, Sulawesi and nearby Peleng Island, Mindanao;
M. trevori, northeastern Zaire, Uganda;
M. niangarae, northeastern Zaire;
M. congicus, Ghana, Nigeria, southern Cameroon, northeastern Zaire, western Uganda;
M. midas, southwestern Saudi Arabia, much of savannah zone of Africa south of the Sahara, Madagascar.

Mops was regarded as a full genus by Freeman (1981) and Koopman (1984*b* and *in* Wilson and Reeder 1993) but only as a subgenus of *Tadarida* by Corbet and Hill (1991), Legendre (1984), and Meester et al. (1986). *Xiphonycteris* was treated as a full genus, with only the one species, *M. spurrelli*, by Hayman and Hill (*in* Meester and Setzer 1977). The above use and composition of *Xiphonycteris* as a subgenus is based primarily on Koopman (1975); neither Freeman (1981) nor Legendre (1984) recognized this subgenus, but Corbet and Hill (1991) and El-Rayah (1981) did. Another name, *Philippinopterus* Taylor, 1934, is a synonym of *Mops*, and its only species, *M. lanei* of Mindanao, was considered conspecific with *M. sarasinorum* by Koopman (1975). *M. leonis*, found from Sierra Leone to eastern Zaire and Uganda, was listed as a separate species by Corbet and Hill (1991) but was included within *M. brachypterus* by Koopman (*in* Wilson and Reeder 1993).

The information for the remainder of this account was compiled from Freeman (1981), Happold (1987), Kingdon (1974*a*), Medway (1978), Rosevear (1965), and Smithers

Greater mastiff bats: Left, *Mops condylurus*, photo by Richard K. LaVal; Right, *Mops midas*, photo by M. Brock Fenton.

(1983). Head and body length is 52–121 mm, tail length is 34–56 mm, forearm length is 29–66 mm, and weight is 7–64 grams. The upper parts are often dark brown but vary in color both within and among species from reddish to almost black. The underparts are usually paler, and in some species there are white markings.

Mops is characterized by ears that are joined over the top of the head by a band of skin, very wrinkled lips, a robust skull, medium to thick jaws, a last upper molar that is reduced to a V pattern, and usually only two lower incisors on each side. From *Tadarida* and *Chaerephon* it is distinguished by having more developed sagittal and lambdoidal crests in the skull, a generally thicker dentary with a higher coronoid process, usually more anterior palatal emargination, and reduced dentition.

Habitat varies widely within species and includes forest, woodland, savannah, and dry brushland. These bats roost in caves, mines, culverts, hollow trees, crevices, attics, and thatched roofs. They emerge after sundown and fly high and fast in pursuit of insects. Their powerful jaws suggest a diet of hard-bodied prey such as beetles.

Colonies range in size from fewer than 10 to several hundred individuals. Studies in Uganda have indicated the following reproductive patterns: *M. nanulus*, two breeding seasons, with fertilization in January and July and synchronized births in March–April and September, when insect availability is greatest; *M. midas*, a well-defined breeding season, with pregnancy in January and lactation in March, and possibly a second season, as a lactating female was taken in October; *M. congica*, two breeding seasons, with pregnancies reported in March and September; and *M. condylurus*, two seasons, with most mating in April–May and November–December and birth peaks coinciding with maximum rainfall in July–August and February–March. Pregnant females of *M. midas* and *M. condylurus* have been taken in Botswana from December to February. The gestation period of the latter species is about two months, and a single young is born. Females are thought to be able to breed in the season following their birth.

M. niangarae, known only from a single specimen taken long ago in an area subject to environmental disruption, is classified by the IUCN as critically endangered. *M. petersoni*, *M. demonstrator*, *M. sarasinorum*, *M. trevori*, and *M. congicus* are designated as near threatened.

CHIROPTERA; MOLOSSIDAE; **Genus OTOMOPS**
Thomas, 1913

Big-eared Free-tailed Bats

There are six widely scattered species (Ansell 1974; Chasen 1940; Ellerman and Morrison-Scott 1966; Hayman and Hill *in* Meester and Setzer 1977; Hill 1983*b*; Hill and Morris 1971; Kitchener, How, and Maryanto 1992; Largen, Kock, and Yalden 1974; Laurie and Hill 1954; J. K. Long 1995):

O. martiensseni, Central African Republic, Congo, Djibouti, Ethiopia, Zaire, Uganda, Kenya, Tanzania, Angola, Zimbabwe, Malawi, Botswana, Mozambique, Natal (South Africa), Madagascar;
O. wroughtoni, southern India;
O. formosus, Java;
O. johnstonei, Alor Island in the Lesser Sundas;
O. papuensis, Papua New Guinea;
O. secundus, northeastern New Guinea.

Head and body length is 60–103 mm, tail length is 30–50 mm, and forearm length is 49 to about 70 mm. Ansell (1974) gave the weights of a male and a female *O. martiensseni* as 36 grams and 27 grams, respectively. The coloration is reddish brown, pale brown, or dark brown. Most species have a grayish or whitish area on the back of the neck and upper back.

There is a series of small spines along the anterior borders of the ears. The ears, 25–40 mm in length, are united by a low membrane. A glandular sac is sometimes located in the lower throat region.

These bats roost in caves, hollow trees, and human-made structures and reportedly are usually solitary or associate in small groups. According to Kingdon (1974*a*), however, the few known colonies of *O. martiensseni* contain many hundreds of bats packed close together. There are particularly large numbers in the lava tunnels on Mount Suswa in Kenya. Mutere (1973*a*) studied two populations of this species in Kenya. In the group at Ithunda, to the south, adult females showed evidence of pregnancy from October to January, and adult males had a peak in their sexual cycle in August. In the Suswa population, to the north, the same

Big-eared free-tailed bat *(Otomops martiensseni)*, photos by D. W. Yalden.

pattern was evident, but there were also a few isolated pregnancies in May and June. Brosset (1962) reported that female *O. wroughtoni* collected in December in India had newborn young, while others were on the verge of delivery. Specimens taken in May had no young, nor were any females pregnant. Brosset suggested that the breeding season was near the end of autumn. There is a single young (Hayssen, Van Tienhoven, and Van Tienhoven 1993).

O. wroughtoni, which has a range of less than 100 sq km and is declining because of habitat destruction, is classified as critically endangered by the IUCN. Most of the other species also have restricted distributions, and all are classified as vulnerable. Both *O. papuensis* and *O. secundus* have been recorded on only two occasions (Flannery 1990*b*).

CHIROPTERA; MOLOSSIDAE; **Genus NYCTINOMOPS**
Miller, 1902

New World Free-tailed Bats

There are four species (Cabrera 1957; Di Salvo, Neuhauser, and Mancke 1992; E. R. Hall 1981; Handley 1976; Husson 1978; Ibáñez and Ochoa G. 1989; Koopman 1978*b*, 1982*a*; Milner, Jones, and Jones 1990; Ochoa G. 1984; Taddei and Garutti 1981):

N. aurispinosus, east and west coasts of Mexico, southern Mexico, Colombia, Peru, Bolivia, Venezuela, eastern Brazil;

N. femorosaccus, southwestern United States, northern and western Mexico;

N. laticaudatus, eastern and southern Mexico to northern Argentina and southern Brazil, Cuba;

N. macrotis, southwestern United States, northern and central Mexico, most of South America, Cuba, Hispaniola, Jamaica, extralimital records as far north as southwestern British Columbia and Iowa and as far east as South Carolina.

Nyctinomops was regarded as a full genus by Freeman (1981), Koopman (1984*b* and *in* Wilson and Reeder 1993), and Legendre (1984) but only as a subgenus of *Tadarida* by Corbet and Hill (1991) and Jones, Arroyo-Cabrales, and Owen (1988). *N. europs* of Venezuela, Surinam, French Guiana, Brazil, and Trinidad and *N. gracilis* of Venezuela and Brazil were listed as synonyms of *N. laticaudatus* by Freeman (1981).

Head and body length is 54–84 mm, tail length is 34–57 mm, and forearm length is 41–64 mm (E. R. Hall 1981, Hoffmeister 1986). The average weight of nine specimens of *N. macrotis* from Cuba was 20.6 grams (Milner, Jones, and Jones 1990). Average weight of South American *N. laticaudatus* was 14.6 grams for males and 13.8 grams for females (Eisenberg 1989). The upper parts are brown or reddish and the underparts are paler. Although the species of *Nyctinomops* formerly were joined with *Tadarida brasiliensis* in the genus and subgenus *Tadarida*, Freeman (1981) indicated that *T. brasiliensis* actually has closer affinity to certain African *Tadarida*. From *T. brasiliensis*, *Nyctinomops* is distinguished by its short second phalanx of the fourth digit of the wing, narrower rostrum of the

A New World free-tailed bat *(Nyctinomops laticaudatus)* wearing an aluminum band on its right forearm. The band bears a number and the name of the organization to which it should be reported if captured. Note that the very delicate skin of the forearm is being injured by the band, which emphasizes the fact that bands must be very carefully placed; otherwise the bat may not survive and banding will be futile. Photo by Ernest P. Walker.

New World free-tailed bat *(Nyctinomops macrotis)*, photo by Merlin D. Tuttle, Bat Conservation International.

skull, loss of the third lower incisor tooth, well-joined (rather than separated) ears, and slightly narrower anterior palatal emargination.

In Venezuela, Handley (1976) found most *N. laticaudatus* roosting in rocks. In the Yucatan, Jones, Smith, and Genoways (1973) reported this species to frequent buildings. According to Hoffmeister (1986), *N. femorosaccus* inhabits rocky cliffs and slopes in the southern desert of Arizona but also makes use of buildings. *N. macrotis* has been taken in a variety of habitats in Arizona, including ponderosa pine, Douglas fir, and desert scrub, but apparently requires rocky cliffs with crevices and fissures for roosting. Both species apparently migrate from Mexico into the United States and form maternity colonies during the summer, but some groups overwinter. They emerge from their roosts late in the evening and fly fast and high in pursuit of insects, mainly large moths.

Nyctinomops is not as gregarious as *Tadarida brasiliensis*. According to Schmidly (1977), colonies of *N. femorosaccus* usually number fewer than 100 individuals and a nursery colony of *N. macrotis* contained 130. Such groups consist almost entirely of adult females and their young; the males roost separately. Barbour and Davis (1969) stated that both species give birth to one young during June or July. This information is corroborated by numerous records from the southwestern United States, northern Mexico, and Cuba cited by Kumirai and Jones (1990) and Milner, Jones, and Jones (1990). In the Yucatan, Jones, Smith, and Genoways (1973) found pregnant female *N. laticaudatus*, each with a single embryo, in April and lactating females in August. Based on collections from a colony of 500–1,000 *N. laticaudatus* roosting within a Mayan ruin, Bowles, Heideman, and Erickson (1990) suggested that parturition of *N. laticaudatus* in the Yucatan occurs synchronously in June. Pregnant female *N. aurispinosus* were collected in Bolivia in September (Jones and Arroyo-Cabrales 1990).

The IUCN continues to recognize the validity of the species *Tadarida espiritosantensis* of southeastern Brazil and classifies it as near threatened. However, as noted in the account of *Tadarida*, *T. espiritosantensis* now is considered a synonym of *Nyctinomops laticaudatus*. In any event, there would be concern for the survival of bat populations restricted to this rapidly developing region.

CHIROPTERA; MOLOSSIDAE; Genus EUMOPS
Miller, 1906

Mastiff Bats, or Bonneted Bats

There are eight species (Anderson 1991; Anderson, Koopman, and Creighton 1982; Brosset and Charles-Dominique 1990; Dolan and Carter 1979; Eisenberg 1989; Eger 1977; Graham and Barkley 1984; E. R. Hall 1981; Ibáñez and Ochoa G. 1989; Koopman 1978*b*, 1989*b*; McCarthy et al. 1993; Myers and Wetzel 1983; Polaco, Arroyo-Cabrales, and Jones 1992; Redford and Eisenberg 1992; Sanchez H., Ochoa G., and Ospino 1992):

E. auripendulus, southern Mexico to northern Argentina and southern Brazil, Jamaica, Trinidad;

E. underwoodi, southern Arizona to Nicaragua;

E. glaucinus, southern Florida, central Mexico to northern Argentina and southeastern Brazil, Cuba, Jamaica;

E. maurus, Venezuela, Guyana, Surinam;

E. dabbenei, Magdalena River Valley of Colombia, northern Venezuela, central Paraguay, Chaco Province of northern Argentina;

E. bonariensis, southern Mexico to central Argentina;

E. hansae, extreme southern Mexico, Honduras, Costa Rica, Panama, Venezuela, Guyana, French Guiana, north-central Peru, Bolivia, northern Brazil;

E. perotis, southwestern United States, northern and central Mexico, Colombia, Venezuela, Guianas, Ecuador, Peru, Bolivia, Paraguay, Brazil, northern Argentina, Cuba.

E. R. Hall (1981) considered *E. nanus,* found from southern Mexico to northern South America, to be specifically distinct from *E. bonariensis.*

Head and body length is 40–130 mm, tail length is 35–80 mm, and forearm length is 37–83 mm. *E. perotis* is the largest bat found in the United States (Barbour and Davis 1969). Adults of *E. underwoodi sonoriensis,* which have a head and body length of about 110 mm, weigh from 40 to about 65 grams. Eisenberg (1989) listed average weights of about 7 grams in *E. bonariensis,* 23 grams in *E. auripendulus,* and 34 grams in *E. glaucinus.* The upper parts are brownish, grayish, or black and the underparts are slightly paler.

Mastiff bat, or bonneted bat *(Eumops perotis)*, photo by Lloyd G. Ingles.

The large ears are rounded or angular in outline and usually connected across the head at the base. Throat sacs are present in some of these bats. When present, they are more developed in the males than in the females. Apparently, when the males are sexually active these sacs swell and secrete material with an odor that may attract females.

These bats usually roost in crevices in rocks, tunnels, trees, and buildings. Like most molossid bats, *Eumops* often shelters under the corrugated iron roofing of human-made structures. The roosts, at least in the larger species, are usually six meters or more above the ground. Because of their size and long narrow wings, these bats require considerable space to launch themselves into flight. Although *E. perotis* appears to be nonmigratory, it moves to different roosting sites with the changing seasons, at least in the northern parts of its range. Some individuals of this species become inactive for short periods in winter in the southwestern United States. *E. perotis* emits loud, high-pitched "peeps" while in flight. This species feeds on small insects, mainly members of the order Hymenoptera, probably catching them from near ground level to treetop height.

Groups of 10–20 bats are usual in this genus, though some individuals may roost alone and some colonies may consist of as many as 70 bats. The adult males and females do not segregate. In the United States births of *E. perotis* occur from June to August, and those of *E. underwoodi* and *E. glaucinus* take place in June and July. Usually a single young is produced by these species (Barbour and Davis 1969). In Costa Rica, Gardner, LaVal, and Wilson (1970) recorded pregnant female *E. glaucinus*, each with a single embryo, on 19 May and 30 December and lactating females in April, May, and August. In the Yucatan, Birney et al. (1974) collected 7 adult female *E. glaucinus* on 17 May; 1 was not pregnant, 5 had one embryo each, and 1 had two embryos. Based on further collections in the Yucatan, Bowles, Heideman, and Erickson (1990) concluded that there is synchronous parturition, with female *E. glaucinus* giving birth in the latter half of June and then lactating for at least 5–6 weeks and *E. bonariensis* giving birth in mid- to late June and then lactating for 6–8 weeks. Dolan and

Carter (1979) collected 3 lactating female *E. underwoodi* in Nicaragua on 27 July. In Peru, Graham (1987) collected juvenile *E. auripendulus* in September and *E. glaucinus* in August.

The isolated subspecies *E. glaucinus floridanus* was designated as threatened by Belwood (*in* Humphrey 1992). Once considered common in the vicinity of Miami and Coral Gables in southeastern Florida, it has declined drastically in recent years, perhaps because of destruction of its restricted natural habitat, loss of large roosting trees, and the adverse effects of pesticides. There had been fear that the subspecies was extinct, but on 30 August 1988 a pregnant female was found in an office building in downtown Coral Gables (Robson, Mazzotti, and Parrott 1989). The species *E. maurus* and *E. underwoodi* also may be declining through habitat loss and are classified by the IUCN as vulnerable and near threatened, respectively.

CHIROPTERA; MOLOSSIDAE; Genus **PROMOPS**
Gervais, 1855

Domed-palate Mastiff Bats

There are two species (Barquez and Lougheed 1990; Cabrera 1957; Eisenberg 1989; Freeman 1981; Genoways and Williams 1979*b*; E. R. Hall 1981; Handley 1976; Koopman 1978*b*, 1982*a*; Marinkelle and Cadena 1972; Redford and Eisenberg 1992):

P. centralis, southern Mexico to northern Argentina, Trinidad;
P. nasutus, Colombia and Surinam to northern Argentina and central Brazil, Trinidad.

P. pamana, known by a single incomplete specimen from central Brazil and sometimes designated a full species, was listed as a synonym of *P. nasutus* by Freeman (1981).

Head and body length is 60–90 mm, tail length is 45–75 mm, and forearm length is 43–63 mm. *P. centralis* is the largest species. A male and two females of this species from

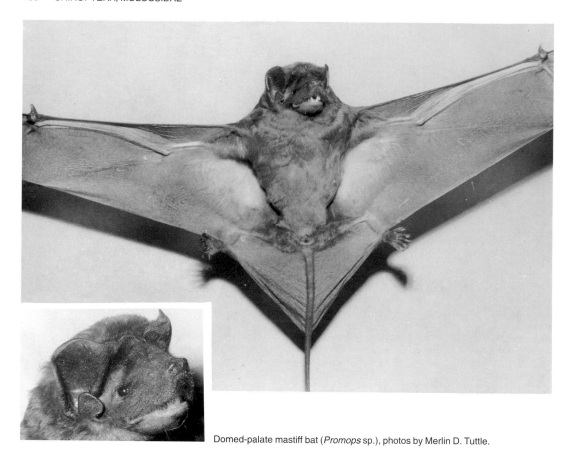

Domed-palate mastiff bat (*Promops* sp.), photos by Merlin D. Tuttle.

Trinidad weighed 14.4–17.2 grams. Coloration is drab brown to glossy black above and somewhat paler below. The short, broad skull, with its strongly domed palate, in combination with certain dental features, is characteristic. The short, rounded ears meet on the forehead, and a small muzzle pad (without processes) and throat sacs are present.

The members of this genus do not seem to be as gregarious as some of the other molossid bats. Colonies of approximately half a dozen individuals have been found roosting in hollow trees and on the underside of palm leaves. A colony of six *P. centralis,* all females, found on the underside of a palm leaf in Trinidad in April contained two lactating individuals. In a study of a small group of *P. nasutus* roosting in the roof of a house in southern Brazil, Sazima and Uieda (1977) found parturition and lactation periods to coincide with the rainy season (November–December). Based on the collection of pregnant and lactating female *P. centralis* in April and October, Graham (1987) considered parturition to be possible in both the wet and dry seasons of coastal Peru. As far as is known, the diet consists of insects. Collections in the Yucatan indicated that female *P. centralis* give birth in late June to a single young and that lactation lasts at least 6 weeks (Bowles, Heideman, and Erickson 1990).

CHIROPTERA; MOLOSSIDAE; **Genus MOLOSSUS**
E. Geoffroy St.-Hilaire, 1805

Velvety Free-tailed Bats

There are seven species (Anderson, Koopman, and Creighton 1982; Cabrera 1957; Carter et al. 1981; Dolan 1989; Dolan and Carter 1979; Freeman 1981; Goodwin and Greenhall 1961; E. R. Hall 1981; Handley 1976; Koopman 1978b, 1982a, 1989a; Marinkelle and Cadena 1972; McCarthy et al. 1993; Myers and Wetzel 1983):

M. rufus, northern Mexico to northern Argentina, Trinidad;

M. pretiosus, southern Mexico to Colombia and northern Guyana;

M. bondae, Honduras to southwestern Colombia and Venezuela;

M. sinaloae, western and southern Mexico to southwestern Colombia and Surinam, Trinidad;

M. aztecus, Guatemala, Honduras, Nicaragua, possibly southern Mexico;

M. coibensis, extreme southern Mexico to Peru and the Mato Grosso of central Brazil;

M. molossus, Mexico to northern Argentina and Uruguay, throughout the West Indies except Bahamas.

Velvety free-tailed bat *(Molossus molossus)*, photo by Bruce J. Hayward. Inset: *M. molossus*, photo by David Pye.

Most recent authorities, including Koopman (*in* Wilson and Reeder 1993), have used the name *M. ater* in place of *M. rufus*, but Dolan (1989) presented convincing evidence that the latter designation is appropriate. Koopman (*in* Wilson and Reeder 1993) did not accept Dolan's (1989) distinction of *M. aztecus* and *M. coibensis* from *M. molossus*. Dolan herself indicated some question about the ranges of *M. coibensis* and *M. molossus* in much of South America. Corbet and Hill (1991) followed Freeman (1981) in giving specific status to *M. trinitatus*, found from Costa Rica to Surinam and on Trinidad, but most other recent authorities (e.g., Dolan 1989; Jones, Arroyo-Cabrales, and Owen 1988; Koopman 1982*a* and *in* Wilson and Reeder 1993) have included it within *M. sinaloae.* Freeman (1981) also listed *M. macdougalli* of southern Mexico as a species apart from *M. rufus*, and Corbet and Hill (1991) listed *M. barnesi* of French Guiana and Brazil as a species distinct from *M. coibensis.*

Head and body length is 50–95 mm, tail length is 20–70 mm, forearm length is 33–60 mm, and adult weight is usually 10–30 grams. The general coloration is reddish brown, dark chestnut brown, dark brown, rusty blackish, or black. Many, perhaps all, of the species have two color phases. Some species have a long, bicolored pelage, whereas others have short, velvety, unicolored fur.

Externally these bats resemble *Tadarida* and *Molossops,* but they differ in cranial and dental characters. The bases of the ears meet on the forehead. Throat sacs may be present.

According to Dolan (1989), *M. bondae, M. aztecus,* and *M. sinaloae* are forest dwellers but the other species are found in nonforest habitats. In Venezuela, Handley (1976) collected several species of *Molossus,* mostly in moist areas but often at dry sites, within a variety of forested and open habitats; roosts included buildings, hollow trees and logs, holes in trees, and rocks. The genus is also known to roost in the fronds of palm trees and in caves. It is often found in attics with galvanized roofing, where the temperature may reach 55° C. In Trinidad *M. rufus* has been found resting in a horizontal rather than a vertical position, and individuals of other species probably do the same. Species of *Molossus* and *Noctilio albiventris* often are found roosting in the same trees and buildings. These bats begin to fly early in the evening, often before sunset. Observations by Chase et al. (1991) suggest that *M. molossus* has a bimodal pattern of foraging, being more strictly crepuscular than any other bat yet studied and entering torpor in the middle hours of the night. *M. rufus* is a fast and erratic flier, sometimes flying high in the air and sometimes near the ground, depending on where insects are to be found. Both sexes of *M. rufus* have large internal cheek pouches; when these are filled to capacity, the bat returns to the roost to chew and swallow its catch.

Molossus roosts in groups of up to hundreds of individuals. The adult males and females of *M. rufus* may segregate and live apart even when roosting at the same site. This species may produce two litters per year on Trinidad. Pregnant female *M. rufus* have been collected in Nayarit, Mexico, in July; in the Yucatan from April to August (Bowles 1972; Jones, Smith, and Genoways 1973); in Guatemala in March (Jones 1966); in El Salvador in November; in Nicaragua in March and July–August (Jones, Smith, and

Turner 1971); and in Costa Rica in February and March (LaVal and Fitch 1977). Of 7 female *M. rufus* taken in Hidalgo, Mexico, on 30 July, 4 were lactating (Watkins, Jones, and Genoways 1972). Of 10 female *M. rufus* taken in Coahuila on 14 May, 2 were pregnant and 5 were lactating (Ramirez-Pulido and Lopez-Forment 1979). In a study of a colony of about 500 *M. rufus* in a house in Manaus, Brazil, Marques (1986) found females to be polyestrous. Pregnancies occurred almost throughout the several years of study, but successful reproduction and lactation was limited to a period of favorable environmental conditions from July to November 1980. The gestation period was estimated to last 2–3 months.

Studies by Krutzsch and Crichton (1985) of a population of *M. molossus* on Puerto Rico (which they treated as a separate species, *M. fortis*) also indicate polyestry. Mating occurs in February or March, there is a birth season in June, and then a postpartum estrus, renewed mating in June or July, and another birth season in September. A single young is born, and lactation lasts about six weeks. Of 32 adult *M. molossus* caught in a church in Jalisco, Mexico, on 7 August, 31 were females, and 27 of these were pregnant (Watkins, Jones, and Genoways 1972). In Nicaragua *M. molossus* reportedly gives birth from early March to mid-July (Jones, Smith, and Turner 1971). In southwestern Colombia pregnant females and immatures of this species were taken in July and August (Arata and Vaughan 1970). Of 18 female *M. molossus* taken in late July on Montserrat, 7 were pregnant and 11 were lactating (Jones and Baker 1979). Of 42 female *M. molossus* taken in late July on Guadeloupe, 18 were pregnant (Baker, Genoways, and Patton 1978). Birth weight in *M. molossus*

is about 3 grams (Hayssen, Van Tienhoven, and Van Tienhoven 1993).

Pregnant or lactating female *M. pretiosus* have been taken in Nicaragua in March, April, June, July, and August (Jones, Smith, and Turner 1971) and in Costa Rica in May, July, August, and October (Dolan and Carter 1979; LaVal and Fitch 1977). Pregnant or lactating female *M. sinaloae* have been taken in the Yucatan from March to May (Birney et al. 1974; Bowles 1972), in Honduras in July (Dolan and Carter 1979), and in Nicaragua in February and July (Jones, Smith, and Turner 1971). LaVal and Fitch (1977) thought that *M. sinaloae* might breed throughout the year in Costa Rica. Pregnant female *M. bondae* have been collected in Costa Rica in January, March, and August (Gardner, LaVal, and Wilson 1970; LaVal and Fitch 1977) and in Nicaragua in late July (Dolan and Carter 1979). All reproductive records for the genus refer to a single young.

The IUCN classifies *M. aztecus* and *M. coibensis* as near threatened.

CHIROPTERA; MOLOSSIDAE; Genus CHEIROMELES
Horsfield, 1824

Naked Bats, or Hairless Bats

Two species were recognized by Corbet and Hill (1991, 1992), Flannery (1995), and Lekagul and McNeely (1977):

C. torquatus, Malay Peninsula and some nearby small
    islands, Sumatra and nearby Simeuleu and Bangka

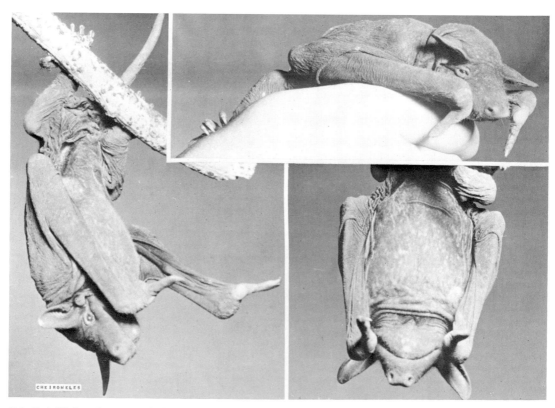

Naked bat *(Cheiromeles torquatus)*, photos by Lim Boo Liat.

islands, Java, Borneo, southern Philippines, doubtfully in Thailand and Indochina;

C. parvidens, Sulawesi and Sanana Island to the east, Mindoro Island in the Philippines.

There long have been suggestions that these two species are not more than subspecifically distinct, and Koopman (1989*b* and *in* Wilson and Reeder 1993) included *parvidens* within *C. torquatus*. Freeman (1981) regarded *Cheiromeles* as the most extreme member of the Molossidae but considered it phenetically closest to *Molossops*. Corbet and Hill (1991) listed *Cheiromeles* last among the Molossidae, and Legendre (1984) placed this genus in its own new subfamily.

Head and body length is 104–45 mm, tail length is 50–71 mm, and forearm length is 70–86 mm. Medway (1978) gave the weight of *C. torquatus* as 167–96 grams. Flannery (1995) listed weights of 75–100 grams for *C. parvidens*. These bats are nearly devoid of hair. Fine, short hairs are present on the head, tail membrane, and underparts, and a ruff of black bristles is present on the lower neck in the vicinity of a conspicuous glandular throat sac, where a strong-smelling secretion is produced. The skin is remarkably thickened and elastic. It is almost black in *C. torquatus* and dark brown in *C. parvidens*. The wings are attached to the back near the midline, and unlike in most Molossidae, the ears are separate and the lips are smooth.

This genus is unique among bats in several respects: the essentially naked appearance, the great development of the throat sacs, and the wing pouches. In both sexes a pouch is present along the sides of the body, formed by an extension of a fold of skin to the upper arm bone and to the upper leg bone. This pouch opens toward the rear and is 25–50 mm deep. The folded wings are pushed into these pouches by the hind feet; the first toe of each hind foot is opposable to the other toes and is supplied with a flattened nail instead of a claw. When the membrane portions of the wings are folded within the pouch the bat can move about relatively freely on all four limbs. According to Burton (1955), "The supposition is that, living in large, hollow trees, the bulldog bat must do a fair amount of climbing to find a suitable place to rest," and "perhaps the folding of the wings into pouches is to give elbow-room."

Naked bats roost in hollow trees, rock crevices, and holes in the earth. They are not rare; nearly a thousand have been noted in a hollow tree, and a colony of about 20,000 was observed in a cave in Borneo (Freeman 1981). *Cheiromeles* has been seen at dusk flying rapidly high in the air. Observations by both Flannery (1995) and Kirk-Spriggs (1989) suggest, however, that flocks of these bats sometimes fly close to the ground, where they can easily be caught. Medway (1978) reported that the diet consists of termites and other insects that are hunted in the open air, either above the forest canopy or over clearings or paddy fields. A captive individual was maintained for several weeks on a diet of grasshoppers and moths.

There are usually two offspring (Freeman 1981). The mammae are located near the opening of the pouch, and it was formerly thought that the young were carried and nursed in this pocket. The young are probably left in the roost by the parents when they leave on their evening flights.

The IUCN classifies both species of Cheiromeles as near threatened.

Order Primates

Primates

A single species of this order, *Homo sapiens*, is nearly worldwide in distribution. Otherwise the order is found in the Americas from eastern and southern Mexico to northern Argentina and southeastern Brazil, most of Africa, Madagascar, the southwestern part of the Arabian Peninsula, south-central and southeastern Asia, Japan, and the East Indies as far as Sulawesi and Timor. Included under the Primates here are 15 families (including 2 Recently extinct families in Madagascar), 77 genera, and 279 species. Two recent, exhaustive, authoritative surveys of the Primates (Groves 1989; Martin 1990) recognized somewhat different arrangements of families; further discussion is found below and in the appropriate familial accounts.

Since people themselves are primates, the order has attracted much interest and investigation, and there are numerous views on classification, especially between the order and family levels. A basic scheme that has been widely followed is Simpson's (1945) division of the Primates into two suborders: Prosimii, for the families Tupaiidae, Lemuridae, Indriidae, Daubentoniidae, Lorisidae, and Tarsiidae; and Anthropoidea, for the families Cebidae, Callitrichidae, Cercopithecidae, Pongidae, and Hominidae. Two major problems with this plan involve the status of the Tupaiidae and Tarsiidae. The tupaiids long were considered by some authorities to belong to the Insectivora but by others to be part of, or immediately related to, the Primates. Now, however, there has been general acceptance that the tupaiids represent an entirely separate order, the Scandentia, which may be part of a supraordinal grouping that includes the Primates but is not the closest living relative of the latter (Macphee 1993b). Additional discussion of this grouping (sometimes called the Archonta) and the affinities of its components is found in the accounts of the orders Chiroptera and Dermoptera.

Although the tarsiers have a number of primitive features linking them with the prosimians, several critical characters suggest a closer relationship to the monkeys and apes. Therefore, most authorities now appear to be following a systematic arrangement by which the prosimians, not including the Tarsiidae, are placed in one group, the Strepsirhini, and the anthropoids and tarsiers are joined in a second group, the Haplorhini. There is some question regarding the level of these two groups, though they seem usually to be called suborders (Goodman 1975; Groves 1989; Hershkovitz 1977; Martin 1975a; McKenna 1975; Petter and Petter-Rousseaux 1979; Thorington and Anderson 1984). In the Strepsirhini the nostrils are crescentic or comma-shaped, the rhinarium surrounding the nostrils is moist and glandular, and there is a median cleft down the middle of the rhinarium. In the Haplorhini the nostrils are ovate or elliptical, the rhinarium is not moist or glandular, and there is no median cleft (Hershkovitz 1977). Schwartz (1986) argued that *Tarsius* belongs in the Strepsirhini and thus that Simpson's classification of the Primates into the two suborders Prosimii and Anthropoidea is, after all, the most appropriate. Schwartz and Tattersall (1987) associated *Tarsius* with the prosimian infraorder Lemuriformes (see below). Martin (1990) provided a detailed analysis of the issue from several perspectives and concluded that there is overwhelming evidence of the validity of the Strepsirhini and Haplorhini but also used Simpson's basic classification. Groves (1989) and Szalay, Rosenberger, and Dagosto (1987), however, retained the tarsiers in the suborder Haplorhini.

The overall sequence of strepsirhine families presented here follows Petter and Petter-Rousseaux (1979), though there are various alternatives, as covered in detail by Groves (1989 and *in* Wilson and Reeder 1993) and Martin (1990). The living Strepsirhini sometimes (as by Martin 1990) are divided into two infraorders: the Lemuriformes, for what Simpson termed the Lemuridae, the Indriidae, and the Daubentoniidae; and the Lorisiformes, for the Lorisidae. Studies of the postcranial skeleton of *Daubentonia* (the sole genus of Daubentoniidae) indicated to Oxnard (1981) that this genus differs more from any primate than any primate does from any other and that its referral to the Lemuriformes is questionable. Based on karyotype, Rumpler et al. (1988) placed *Daubentonia* in a separate infraorder, the Daubentoniiformes. Groves (1989), considering a variety of evidence, also concluded that *Daubentonia* represented one main branch of the Strepsirhini, which he designated the infraorder Chiromyiformes. However, he thought that all other living strepsirhines, including the Lorisidae, belong in a single infraorder, the Lemuriformes. The genera *Microcebus, Mirza, Cheirogaleus, Allocebus,* and *Phaner,* all formerly assigned to the Lemuridae, are now placed in the family Cheirogaleidae and that family sometimes has been associated with the Lorisidae (Tattersall 1982; Tattersall and Schwartz 1975). However, Martin (1990), while accepting the plausibility of an origin of the Cheirogaleidae from the same ancestral stock that gave rise to the Lorisidae, still considered the two groups to belong in separate infraorders. A recent cladistic analysis by Yoder (1994) supports the position that all the primates of Madagascar, including the Daubentoniidae and the Cheirogaleidae, form a monophyletic group that excludes the Lorisidae.

Three other families distinguished here but not recognized by Simpson (1945) are the Megaladapidae, with the single living genus *Lepilemur,* which formerly was regard-

Humankind's closest living relative, the pygmy chimpanzee or bonobo *(Pan paniscus)*, photo by Russell A. Mittermeier.

ed as part of the Lemuridae, and the extinct Palaeopropithecidae and Archaeolemuridae of Madagascar, which formerly were included in the Indriidae. These three families, together with the Indriidae, were united in a superfamily, the Indrioidea, by Mittermeier et al. (1994). If this last taxonomic step is taken and if Martin (1990) is otherwise followed with respect to the content of the Lemuriformes, except for placing the Daubentoniidae in a separate infraorder, then it will be necessary to recognize another superfamily, the Lemuroidea, for the families Lemuridae and Cheirogaleidae.

An extensive review of dentition led Schwartz and Tattersall (1985) to suggest that the Megaladapidae, as well as the Indriidae and the Lemuridae, are associated phylogenetically with the fossil family Adapidae of the European Eocene and that the Cheirogaleidae and the Lorisidae are associated with certain Eocene non-adapid strepsirhine primates. Groves (1989) questioned the validity of trying to trace direct connections between Eocene fossils and modern groups. Subsequently, Martin (1993) reported that new fossil discoveries had provided suggestive indicators of a connection between the adapids and the strepsirhines. Both the Adapidae and the Omomyidae, a widespread tarsierlike group of the early Tertiary, have been proposed as the ancestral stock of the modern Haplorhini. Recent detailed analyses by Kay, Ross, and Williams (1997) support a divergence between the Strepsirhini and the Haplorhini by the earliest Eocene, with the former including the Adapidae and the latter including the Omomyidae as a basal stock.

The sequence of haplorhine families presented here generally follows that of Hershkovitz (1977), who divided the living members of this group into three infraorders: Tarsii,

for the tarsiers; Platyrrhini, for the New World monkeys; and Catarrhini, for the Old World monkeys, the apes, and humans. Martin (1990) recognized the same basic groupings but used the term Tarsiiformes for the tarsier infraorder and considered it prosimian. Groves (1989) accepted only two living haplorhine infraorders, the Tarsiiformes for the Tarsiidae and the Simiiformes for all other families. Hershkovitz recognized the New World family Callimiconidae, with the single genus *Callimico*, as being distinct from the Callitrichidae and Cebidae. In contrast, Groves included *Callimico* in the Callitrichidae and Martin included it in the Cebidae (though he noted that a shift to the Callitrichidae might eventually prove justifiable). Martin, after an exhaustive analysis of mainly morphological characters, concluded that the Callitrichidae are derived from cebidlike ancestors. Such a view has been supported by molecular analysis (Schneider et al. 1993). Those studies, rather than the arrangements of Groves and Hershkovitz, are reflected in the following sequence of families.

Hershkovitz divided the Catarrhini into two superfamilies: the Cercopithecoidea, with the single family Cercopithecidae; and the Hominoidea, with the families Hylobatidae (the gibbons being recognized as a distinct family), Pongidae, and Hominidae. Martin used the same arrangement, but Groves recognized two living catarrhine superfamilies: the Cercopithecoidea, with the Cercopithecidae and the separate family Colobidae (leaf monkeys); and the Hominoidea, with the Hylobatidae and the Hominidae (including the Pongidae).

There long has been speculation about the ancestral relationship of the Old and New World monkeys. For many years the most commonly held view was that the two groups arose independently, with the New World groups being descended from primitive North American stock. More recently, it has been widely accepted that the Catarrhini and the Platyrrhini had a common ancestor in Africa and that the latter or their precursors crossed the South Atlantic Ocean sometime between the late Eocene and the middle Oligocene (see Aiello 1993; and papers *in* Ciochon and Chiarelli 1980). At that time Africa and South America were much closer together than they are now, there were numerous islands between the two continents, and the relatively small oceanic gaps that remained could have been traversed when animals became isolated on large masses of drifting vegetation. Martin (1990) generally supported this view, though Groves (1989) found it unconvincing and encouraged investigation of alternative routes of origin.

The smallest living primate genera are *Microcebus* (mouse lemurs) and *Cebuella* (pygmy marmoset), which may weigh around 100 grams or less. The species *Microcebus myoxinus* weighs only about 30 grams. The largest living species is *Gorilla gorilla*, which sometimes weighs more than 200 kg, but the extinct *Gigantopithecus blacki* of the middle Pleistocene of Asia probably was even larger (Hershkovitz 1977). The pelage in primates varies in texture from velvety soft to harsh and coarse. The hair has various arrangements, sometimes forming crests on the head, fringing the face with whiskers, forming a mantle on the shoulders, or falling from the sides or over the rump. The hairy coats vary greatly in color; some primates are among the most brilliantly colored mammals.

In primates, except people, the first digits of the hind feet are opposable, and usually the thumbs are also opposable. This power of opposition makes the hands and feet good grasping organs and thus aids in climbing. Most primates have five digits on each limb. When they are walking or running, the sole of the foot and heel touch the ground, giving plantigrade locomotion, but there are exceptions to

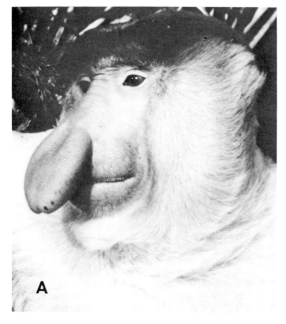

A. Proboscis monkey *(Nasalis larvatus)*. B. Woolly monkey *(Lagothrix lagotricha)*. C. Mouse lemur *(Microcebus murinus)*. D. Mandrill *(Mandrillus sphinx)*. Photos by Bernhard Grzimek.

this general rule. The fingers and toes of many forms have flat nails, but some species have modified clawlike nails on certain digits. A tail is absent in the Hominidae, Pongidae, and Hylobatidae and in a few species of other families. When a tail is present, it varies from a mere knob to an appendage considerably longer than the head and body. Some of the New World monkeys have prehensile tails.

An outstanding characteristic of the Primates is the trend toward great development of the cerebral hemispheres of the brain. This feature is most pronounced in *Homo*. The primate skull, with few exceptions, has a large braincase that approaches globular form. Usually the jaws

are relatively short and the face is flattened. Most primates have a short nose, with the sense of smell being secondary to the senses of vision, touch, and hearing, but some lemurs have an elongated snout. The orbit of the skull is encircled by bone and directed forward. The eyeballs generally face more forward than laterally and thereby facilitate true stereoscopic vision, which aids in seeing objects in three dimensions. The development of a special field of the retina of the eye allows for sharper definition of objects viewed in the direct line of the optic axis. The incisor teeth are usually 2/2. The third incisor is absent above, and the first premolar is absent below. The premolars vary from simple in form to somewhat molarlike. The molars are adapted for grinding, each usually having four main tubercles.

Daubentonia has one pair of abdominal mammae; females of the family Cheirogaleidae have one pectoral, one abdominal, and one inguinal pair; and the genus *Varecia* has one pectoral and two abdominal pairs. All other primates have a pair of mammary glands in the chest or axillary region. The penis of males is of the pendulous type (a rare pattern in mammals), and a baculum is present in most genera but absent in *Tarsius, Lagothrix, Homo,* and a few others. The testes are borne in a sac known as the "scrotum." The typical primate reproductive cycle differs in certain respects from the typical mammalian "heat" cycle and is known as the "menstrual cycle" in the female. The males of many primates, unlike many other mammals, do not exhibit a specific period of pronounced sexual activity, but are capable of breeding at any time. According to Austad and Fischer (1992), primates are among the longest-lived mammals, though as a group they are exceeded by several other orders. The exceptional longevity of primates may be associated less with brain size or evolutionary development than with a generalized low mortality risk due to arboreal habit and propensity for social grouping.

Many primates have been adversely affected by human activity. In addition to habitat loss and direct hunting, there has been an enormous commercial trade in live animals for use as pets and in entertainment and research. An estimated 200,000 primates were being imported to the United States annually during the 1950s. Subsequent national and international restrictions have greatly reduced this trade (Kavanagh, Eudey, and Mack 1987). Imports to the United States have fallen by about 90 percent. However, Dobson and Lyles (1989) pointed out that more than 50,000 primates still were being used in the nation for experiments each year, about half for the first time; of the other half, 8,000 are born annually in primate centers and around 17,000 per year are imported. There was concern that the wild populations supplying biomedical operations were not being properly monitored. It was predicted that any given population would tend to crash when the survival of adult females fell below around 70 percent per interbirth interval. The entire order Primates has been placed on appendix 2 of the CITES, except for those species on appendix 1.

The known geological range of this order is Paleocene to Recent (Petter and Petter-Rousseaux 1979). However, Martin (1990) pointed out that even though the first fossil currently recognized as a primate is just 67 million years old, the hypothetical progenitor of the order would have lived 90–100 million years ago in the Cretaceous. Martin (1993) calculated the time of origin of "primates of modern aspect" at about 80 million years ago.

PRIMATES; Family LORISIDAE

Lorises, Pottos, and Galagos

This family of 9 genera and 18 species is found in Africa south of the Sahara, southern India, Sri Lanka, southeastern Asia, and the East Indies. Petter and Petter-Rousseaux (1979) divided the family into two subfamilies: the Lorisinae for the lorises and pottos *(Arctocebus, Loris, Nycticebus, Perodicticus)* and the Galaginae for the galagos *(Euoticus, Galago, Otolemur, Galagoides).* This sequence of genera is followed here, with the newly described *Pseudopotto* (see account thereof) being added first to reflect its primitive nature. Some authorities (e.g., Goodman 1975; Schwartz and Tattersall 1985) consider both subfam-

ilies to be full families. Schwartz (1987) not only gave familial rank to the Lorisinae and the Galaginae but also placed *Nycticebus* and *Perodicticus* in a new subfamily, the Nycticebinae. Groves (1989) did not at first accept such a familial division, but subsequently (*in* Wilson and Reeder 1993) he did, and he also followed Jenkins (1987) in using the names Loridae (in place of Lorisidae) and Galagonidae (in place of the more commonly used Galagidae). Schwartz (1996) argued for maintenance of Lorisidae.

In the lorises and the pottos head and body length is 170–390 mm and the tail is short or absent. The head and eyes are round, and the small, rounded ears are nearly hidden by the surrounding fur. The forelimbs and hind limbs are nearly equal in length. The hands and feet and the digits, especially the first, are relatively stronger than in the galagos. The first digit of each limb is more widely opposable to the other digits than in the galagos, and the index finger is more reduced. All the digits of lorises and pottos have nails except the second digit of the foot, which has a claw. The wrist and ankle joints have more freedom of movement than in galagos. The grip by either hands or feet is powerful and can be maintained for long periods. Male lorises and pottos do not have a baculum, and females have two or three pairs of mammae.

The forms of galagos are very different from those of lorises and pottos. The bodies of galagos are slender; the tail is usually longer than the head and body and is almost bushy. The arms and legs are comparatively longer and more slender with rather long fingers and toes. Head and body length is 105–465 mm. The ears are large and mobile. The hind limbs are considerably longer than the forelimbs. Males have a baculum, and females usually have two pairs of mammae.

The dental formula for the family is: (i 2/2, c 1/1, pm 3/3, m 3/3) $\times$ 2 = 36. In lorises and galagos the lower incisors project forward and slightly upward, the canines are long and sharp, and the molars have three or four cusps. The orbits in lorises and pottos face more directly forward than in the galagos.

Lorises and pottos are essentially arboreal and nocturnal, sheltering by day in hollow trees, in crevices of trees, or on branches. They generally sleep in a flexed position, with the head tucked between the arms. They usually move with slow, deliberate, hand-over-hand movements. They move quickly at times but do not leap or jump. The forward progress of lorises and pottos is accompanied by a rhythmic sinuous movement of the spine, produced by placing a hand and then a foot in consecutive order on the surface they are traversing near the midline of the body. They travel along the underside of limbs or poles as readily as on the top. The powerful grasp is facilitated by the great development and wide angle of divergence of the first digit on the hand and foot, the mobility of the wrist and ankle joints, and the presence of storage channels *(reta mirabilia)* for blood in the vessels of the hands and feet. These channels enable the muscles to remain contracted over long periods without fatigue by aiding the exchange of oxygen and carbon dioxide in the muscle fibers.

The geological range of this family is early Miocene to Recent (Thorington and Anderson 1984).

PRIMATES; LORISIDAE; Genus PSEUDOPOTTO
Schwartz, 1996

The single species, *P. martini,* is known only by two specimens taken years ago and now in the Anthropological Institute and Museum, University of Zurich-Irchel. One, a

Slender loris *(Loris tardigradus)*, photo by Bernhard Grzimek.

nearly complete skeleton, represents an adult female that was captured in "the Cameroons" and subsequently lived at the Zurich Zoo. The other, consisting only of a skull and mandible, represents a subadult male collected somewhere in "Equatorial Africa." These geographic terms, in use until about 1960, indicate that the first specimen originated either in the current nation of Cameroon or in extreme eastern Nigeria and that the second came from either Cameroon or an immediately neighboring nation. The specimens had been cataloged as *Perodicticus potto* but recently were reexamined by Schwartz (1996) and described as a new genus and species.

The phylogenetic position of *Pseudopotto* is uncertain, and Schwartz did not even assign it definitely to any family. However, as noted above in the account of the Lorisidae, Schwartz (1987) treated the Galaginae as a full family (Galagidae). Since Schwartz (1996) indicated that *Pseudopotto* appears to be more closely related to what he had termed the Lorisidae than to what he had termed the Galagidae, and since those two groups are here considered only subfamilies of the family Lorisidae, it seems reasonable to assign *Pseudopotto* to that family here. Also, in keeping with Schwartz's indication that *Pseudopotto* is more primitive than any of the other living lorisids, it is here placed first among the group.

Pseudopotto bears a superficial resemblance to *Pero-*

dicticus, and its cranial and humeral measurements fall at the lower end of the size range of the latter genus. It is immediately distinguished from *Perodicticus* and all other living lorisids, however, by its much longer tail (obvious even though at least one caudal vertebra is missing from the specimen with postcranial elements) and in lacking a distinctly hooked ulnar styloid process.

Pseudopotto is dentally primitive relative to *Perodicticus, Arctocebus,* and *Nycticebus* in retaining a more bucally emplaced cristid obliqua and lacking deep hypoflexid notches on the lower molars, as well as in having relatively longer lower middle and last premolars. *Pseudopotto* is similar to *Perodicticus* and *Nycticebus* in having a bulky skull and mandible, expanding suprameatal shelves, and a broad snout, but differs in having a more hooked mandibular coronoid process and goneal region. It differs from *Perodicticus* and *Arctocebus* in having taller cheektooth cusps; more crisply defined cusp apices, crests, and margins; and an entepicondylar foramen. *Pseudopotto* is distinguished from most of the Lorisidae and Cheirogaleidae (but not *Phaner*) in having a very diminutive third upper premolar, from most extant prosimians (but not *Nycticebus*) in having the lacrimal fossa incorporated into the inferior orbital margin, and from all extant prosimians in having a diminutive third upper molar.

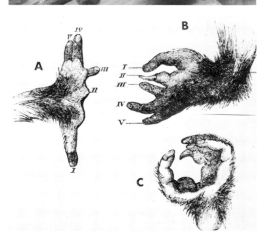

Golden potto *(Arctocebus calabarensis)*, photo from Zoological Society of London. A. Left hand spread; B. Left foot in natural position as seen from above; C. The same as seen from below, photos from *Proc. Zool. Soc. London.*

Golden Potto, or Angwantibo

Petter and Petter-Rousseaux (1979) recognized a single species, *A. calabarensis*, occurring in the tropical forest belt from southeastern Nigeria to southern Congo and western Zaire. Groves (1989 and *in* Wilson and Reeder 1993) accepted the distinction of a second, allopatric species, *A. aureus*, occupying that part of the range of the genus to the south of the Sanaga River, which runs through central Cameroon.

Head and body length is 229–305 mm and the tail is only a slight protuberance. Napier and Napier (1967) gave the weight as 266–465 grams. The fur has a golden sheen and is thick, long, and wool-like over the body, but the hands and feet are thinly haired. The species or subspecies *aureus* is generally bright golden red above and grayish beneath; the species or subspecies *calabarensis* is yellowish brown to fawn above and paler, almost whitish, on the underparts. The face is darker than the back, and the sides of the face are light. There is a white line from the brow to the nose. *Arctocebus* resembles *Perodicticus*, and the two are frequently confused. *Arctocebus* is slightly smaller and has a relatively longer face. The index finger is reduced to a mere stub. The feet of *Arctocebus* are well suited for grasping for long periods of time with little or no change of position.

Although found within the overall tropical forest region, the golden potto seems to prefer clearings where trees have been felled by storms or areas of young secondary growth with dense and intricate vegetation. In primary forest it has never been observed at a height above 15 meters, and it is generally found below 5 meters in the undergrowth. It is nocturnal, sleeping by day in thick foliage. It moves with a stealthy, slow climb in which three extremities are involved in gripping supports at any one time. If threatened, it will remain firmly attached to a branch with its body rolled into a ball, and it may reach out under an arm to bite. The diet consists largely of insects, mostly caterpillars in some areas, and also includes fruits. Insects may be located under leaves by smell, and this primate also may rear on its hind legs to catch moths in flight. Population density has been estimated at 2/sq km in primary forest and 7/sq km in clearings and secondary forest (Charles-Dominique 1977; Charles-Dominique and Bearder 1979; Hladik 1979; Walker 1979).

The golden potto appears to be a generally solitary species. Several vocalizations have been reported, including a deep, hoarse growl in agonistic encounters and a fairly powerful call for distant communication (Petter and Charles-Dominique 1979). Estrous cycles of 36–45 days have been recorded in a captive female (Van Horn and Eaton 1979). Females apparently can breed more than once a year, and in Gabon births have been observed throughout the year, but with a minimum number from June to August. The gestation period is 131–36 days, and weight at birth is 24–30 grams. There is one young per birth. It clings to the belly of the female for about 3–4 months, after which time it is weaned and begins to follow the mother or ride dorsally. Adult weight and sexual maturity are attained at 8–10 months (Charles-Dominique 1977). Record longevity is 13 years (Jones 1982).

Arctocebus is threatened by loss of forest habitat to logging and cultivation and by being hunted for its meat. Its ability to inhabit secondary forests, however, may give it a chance to survive (Lee, Thornback, and Bennett 1988). It is

Slender loris *(Loris tardigradus),* photo from New York Zoological Society.

on appendix 2 of the CITES. The IUCN distinguishes *A. aureus* and *A. calabarensis* as separate species and designates both as near threatened.

PRIMATES; LORISIDAE; Genus LORIS
E. Geoffroy St.-Hilaire, 1796

Slender Loris

Petter and Petter-Rousseaux (1979) and Corbet and Hill (1992) recognized a single species, *L. tardigradus,* occurring in southern India below about 15° N and in Sri Lanka. Groves (1989 and *in* Wilson and Reeder 1993) suggested that the three subspecies in Sri Lanka are sharply distinct and might represent full species. Goonan (1993) noted an apparent loss of reproductive viability when individuals from two of the Sri Lankan subspecies were joined in a captive group.

Head and body length is 175–264 mm, there is no external tail, and weight is 85–348 grams. The fur is soft, thick, and woolly and varies in coloration from yellowish gray to dark brown above, while the underparts are silvery gray to buff. Superficially this animal resembles the better-known slow loris *(Nycticebus),* but it is smaller and much more slenderly built. The great toes and the thumbs are opposable to the other toes and fingers, thereby producing highly efficient grasping organs, and the limbs are very slender. The eyes are round and large, and the prominent ears are thin, rounded, and naked at the edges.

In Sri Lanka this species occurs both in montane forest, where temperatures are relatively low, and in deciduous forest in thick vegetation (Hladik 1979). The slender loris is arboreal and nocturnal, spending daylight hours sleeping on branches. It moves with a slow, stealthy climb, at least three limbs always grasping a support. *Loris* is almost always found on the top side of branches, whereas *Arctocebus* sometimes rests by hanging below (Walker 1979). According to Roonwal and Mohnot (1977), *Loris* is capable of running at a fair pace on the ground but cannot swim. The diet is chiefly insectivorous but also includes shoots, young leaves, fruits with hard rinds, birds' eggs, and small vertebrates. When trying to catch prey, the method of attack is usually a cautious stalk followed by a quick grab with both hands. In Sri Lanka each individual utilizes approximately 1 ha. of forest (Hladik 1979).

According to Roonwal and Mohnot (1977), the slender loris usually lives alone or with a mate. Most reports indicate that it is aggressive, and when several individuals are kept together there initially is constant squealing and fighting, sometimes resulting in death. Eventually the animals become more tolerant of one another. Reported vocalizations are a "growl" indicating disturbance, a "screech" heard at night, a "whistle" probably expressing satisfaction upon meeting a mate, and loud distress and warning calls. Females apparently enter estrus twice a year and are receptive for about a week. In southern India mating reportedly takes place in April–May and October–November, and in Sri Lanka births have been recorded in May and December. There may be one or two young per birth. Studies of a captive colony (Izard and Rasmussen 1985) indicated no reproductive seasonality and the following parameters: estrous cycle, 29–40 days; gestation, 166–69 days; interbirth interval, 9.5 months; litter size, 1 young; lactation, 6–7 months; and sexual maturity, 10 months in females and about 18 months in males. According to Marvin L. Jones (Zoological Society of San Diego, pers. comm., 1995), one individual lived in captivity for 15 years and 5 months and another was still living after 14 years and 5 months.

Schulze and Meier (1994) indicated that *Loris* may be endangered in India because its forest habitat has been reduced to isolated fragments and it is hunted for body parts that have alleged medicinal use. Much of its habitat in Sri Lanka also has been destroyed and it is absent from most areas even where suitable habitat seems to remain. *Loris* currently is classified as vulnerable by the IUCN and is on appendix 2 of the CITES.

PRIMATES; LORISIDAE; **Genus NYCTICEBUS**
E. Geoffroy St.-Hilaire, 1812

Slow Lorises, or Cu Lan

There are two species (Corbet and Hill 1992; Eudey 1987; Groves 1971*b*; Petter and Petter-Rousseaux 1979; Wolfheim 1983; Zhang, Chen, and Shi 1993):

N. pygmaeus, east of the Mekong River in southern
    Yunnan, Laos, Viet Nam, and Cambodia;
N. coucang, eastern Bangladesh, Assam, Burma, southern

Yunnan and Guangxi, Thailand, Indochina, Malay Peninsula and the nearby islands of Penang and Tioman, Riau Archipelago, Sumatra, Bangka, Borneo, Natuna Islands, southwestern part of the Sulu Archipelago in the southern Philippines.

Groves (1971*b*) stated that *N. coucang* also occurred on Mindanao, but perhaps as a result of introduction. Fooden (1991*a*) concluded that records from Mindanao were completely erroneous; both he and Timm and Birney (1992) showed that the species is found no farther east than the small islands of Tawitawi, Sanga Sanga, Bongao, and Simunul, which politically are part of the Philippines though they are located less than 100 km off northeastern Borneo. Petter and Petter-Rousseaux (1979) considered *N. pygmaeus* a subspecies of *N. coucang,* but the two were regarded as specifically distinct by Corbet and Hill (1991), Eudey (1987), Groves (1971*b* and *in* Wilson and Reeder 1993), Napier and Napier (1985), Wolfheim (1983), and Zhang, Chen, and Shi (1993).

In *N. coucang* head and body length is 265–380 mm and weight is 375–2,000 grams (Roonwal and Mohnot 1977). In *N. pygmaeus* head and body length is about 180–210 mm. The tail is vestigial in this genus. The fur is short, thick, and woolly. The coloration above ranges from light brownish gray to deep reddish brown, sometimes with a hoary effect produced by the light tips of individual hairs. The coloration beneath is usually somewhat lighter, vary-

Slow loris *(Nycticebus coucang),* photo by Lim Boo Liat. Inset photo by Ernest P. Walker.

Slow loris *(Nycticebus coucang)*, two-day-old baby, photo by Ernest P. Walker.

ing from almost white to buffy or grayish. There is usually a dark midline along the neck and back and a light streak between the dark orbital rings.

The thumb is practically perpendicular to the other fingers, and the great toe is perpendicular or even points slightly backward. The muscular arrangement enables an effortless grasp, with the hands and feet rigidly clenched.

The remainder of this account, except as noted, is taken from Lekagul and McNeely (1977) and applies to *N. coucang*. The slow loris is found in primary or secondary forest or in groves of bamboo. It is nocturnal and arboreal, seldom descending to the ground. By day it sleeps curled up in the fork of a tree or in a clump of bamboo. It walks hand over hand along branches, alternately placing a hand, then bringing the foot of the same side up to the hand, then extending the other hand, then bringing the other foot up to the hand, and repeating the process. It generally climbs slowly and deliberately but climbs considerably faster when the day is windy. The diet consists of large mollusks and insects, lizards, birds, small mammals, and fruits. Although it appears to be a slow creature, *Nycticebus* can strike with amazing speed. It grips a branch with both hind feet, stands erect, throws its body forward, and seizes its prey with both hands.

Little is known of social structure, but pairs and family groups have been collected and successfully maintained in captivity. Births evidently sometimes occur in the open rather than in a nest. The infant is carried by the mother and perhaps also by the father but may be left clinging to a branch for brief periods. Adult males appear to be strongly territorial and will not tolerate the presence of another adult male. The territory is marked with urine. Reported vocalizations include a low buzzing hiss or growl when disturbed, a single high-pitched rising tone for making contact, and a high pure whistle produced by the female when in estrus. There also are whistles for making contact and emitted by either sex during courtship (Daschbach, Schein, and Haines 1981; Zimmermann 1985).

Examination of specimens indicates that breeding in the wild continues throughout the year. Such also occurred in one captive colony in North Carolina (Izard, Weisenseel, and Ange 1988), but there was a peak in estrous activity from August to December and a birth peak from March to May. In a captive group in Germany (Zimmermann 1989) mating occurred mainly from June to mid-September, and parturition from December to March. Observations of these captives indicated that the estrous cycle is 29–45 days, estrus lasts 1–5 days, and gestation is 185–97 days. Litter size was always one in these two colonies, but occasional twins have been reported elsewhere. The young weigh 30{endash}60 grams at birth and are weaned after 5–7 months. Females reach sexual maturity at 18–24 months. Males may be physiologically capable of reproduction at 17 months. Fathers become increasingly hostile when their male offspring are 12–14 months old and eventually seek to chase them out of the group. A number of individuals have lived in captivity for more than 20 years and one was still living in December 1994 at an estimated age of more than 26 years (Marvin L. Jones, Zoological Society of San Diego, pers. comm., 1995).

MacKinnon (1986) estimated that *N. coucang* once occupied 610,570 sq km of habitat in Indonesia, that this area had been reduced to 227,883 sq km through logging and clearing for agriculture, and that the number of animals remaining was 1,139,415. Hunting and deforestation have reduced the number in southeastern China to a few hundred (Tan 1985). The lesser slow loris, *N. pygmaeus*, is listed as vulnerable by the IUCN and as threatened by the USDI and was regarded as highly vulnerable by Eudey (1987). The restricted habitat of this species is said to have been subject to severe environmental disruption through military activity. However, Duckworth (1994) suggested that it still is common in Laos. Both species of *Nycticebus* are on appendix 2 of the CITES.

PRIMATES; LORISIDAE; Genus PERODICTICUS
Bennett, 1831

Potto

The single species, *P. potto*, is found in the tropical forest belt from Guinea to western Kenya and central Zaire (Petter and Petter-Rousseaux 1979).

Head and body length is 305–90 mm and tail length is 37–100 mm. Charles-Dominique (1977) gave the weight as 850–1,600 grams. The dense, woolly coat of adults varies from a brownish gray to a rich, dark reddish brown, slightly lighter on the underparts. The index finger is a mere rudiment, and the thumb is readily opposable to the three remaining fingers, thus producing an excellent grasping organ. The great toe is so placed that it readily opposes the other toes, making the foot equally efficient for grasping. The second toe is slightly shortened and has a long, sharp claw. The nails of all the other toes and fingers are flattened like those of humans. The modification of the hand is similar to, but not as extreme as, that of *Arctocebus*. The presence of a moderate tail distinguishes *Perodicticus* from *Arctocebus*.

The spinal processes of the neck vertebrae are very long and project above the general contour of the flesh of the neck. They are hidden by thick fur but can be felt on either living or dead specimens. Walker (1970) stated that it was not true, as sometimes claimed, that the spinal processes actually penetrate through the skin. Instead the spines are covered by sensitive hairless tubercles. This nuchal region was postulated to function in tactile stimulation rather

Potto *(Perodicticus potto)*, photo by Ernest P. Walker.

than for defense, aggression, abrasion, or muscular support. Charles-Dominique (1977), however, thought this region was used for defense. He stated that the potto lowers its head between its forelegs when threatened, thereby presenting an attacker with a scapular shield, and that it may also thrust forward with this shield to knock a predator off balance. Oates (1984) suggested that the scapular shield may provide defense while the potto forages along a branch with its head down to locate food by olfaction.

The remainder of this account, except as noted, is taken from Charles-Dominique (1977). The potto is an inhabitant of dense forest. It is nocturnal and arboreal, generally

Potto *(Perodicticus potto)*, hand and foot, photo by Bernhard Grzimek.

being found in trees at a height of 5–30 meters. It spends the day sleeping in foliage. It is generally slow, moving strictly by climbing rather than leaping, but it can make quick grabs with its hands and mouth. The diet consists largely of fruits but includes some gums and insects. The latter are located by smell, and ants are lapped up with the tongue. Small vertebrates, such as birds and bats, also may be killed and eaten.

In studies in Gabon population density was estimated to be 8–10/sq km in primary forest and 28/sq km in flooded forest. Home ranges were 6–9 ha. for females and 9–40 ha. for males. Both sexes are territorial. The females defend an area large enough to support themselves and their young. The older and larger males establish individual home ranges, from which other males are excluded but which overlap the home ranges of one or more females. A relationship between male and female is maintained throughout the year, but contact is made only at night, and the sexes do not sleep together. Vocalizations include a high-pitched "tsic" for communication between mother and young, a whistling call emitted by females in estrus, a groan used in threats, and a high-pitched distress call. The potto also leaves urine trails about 1 meter long on branches, apparently as a means of passing on information to conspecifics. Although studies in the wild indicate a solitary nature for this genus, long-term observations of captives (Cowgill, States, and States 1989) show that individuals, even adult males, form attachments and benefit from sociable activity.

The estrous cycle averages 39 days (Van Horn and Eaton 1979), estrus lasts less than 2 days (Hayssen, Van Tienhoven, and Van Tienhoven 1993), and the gestation period is 180–205 days (Cowgill, States, and States 1989; Müller *in* Grzimek 1990). In Gabon all births occurred from August to January, and weaning took place 4–5 months later, from January to March, when fruit was most abundant. Birth weight is 30–52 grams, and there is normally a single young, but twins were reported once in captivity. The young clings to the belly of the female for the first few days of life but subsequently is left suspended from a hidden branch during the night and is collected by the mother and returned to the nest in the morning. At 3–4 months the young begins to follow the mother or ride dorsally. Adult weight is attained at 8–14 months, and sexual maturity at 18 months. Young males leave the mother's home range when they are only 6 months old. Young females are independent after 8 months but still share the mother's range. Captive pottos have lived to be more than 26 years old (Cowgill, States, and States 1989). *Perodicticus* is on appendix 2 of the CITES.

PRIMATES; LORISIDAE; **Genus EUOTICUS**
Gray, 1863

Needle-clawed Bushbabies

There are three species (Groves 1989; Hill and Meester *in* Meester and Setzer 1977):

E. pallidus, north of the Sanaga River in Cameroon and southeastern Nigeria, island of Bioko (Fernando Poo) in Equatorial Guinea;
E. elegantulus, south of the Sanaga River in Cameroon, Rio Muni (mainland Equatorial Guinea), Gabon, Congo;
E. matschiei, northeastern Zaire, western Uganda.

Euoticus was regarded only as a subgenus of *Galago* by Meester et al. (1986), Olson (*in* Nash, Bearder, and Olson 1989), and Petter and Petter-Rousseaux (1979) but was considered a full genus by Corbet and Hill (1991), Groves (1974b, 1989, and *in* Wilson and Reeder 1993), and Napier and Napier (1985). The species *E. matschiei* (sometimes called *E. inustus*) was transferred to *Galago* by Groves but was retained in *Euoticus* by Corbet and Hill, Napier and Napier, Olson, and Petter and Petter-Rousseaux. The species *E. pallidus* was distinguished from *E. elegantulus* by Groves but not by any of the other authorities cited herein.

In *E. matschiei* head and body length is 147–95 mm, tail length is 195–279 mm, and weight is 170–250 grams (Kingdon 1974a). In *E. pallidus* and *E. elegantulus* head and body length is 182–210 mm, tail length is 280–310 mm, and weight is 270–360 grams (Happold 1987). The upper parts are orange-cinnamon to dull cinnamon-gray. The fur is thick, woolly, soft, and without luster. From *Galago, Euoticus* is distinguished by the presence of pointed and keeled nails on all digits except the first, which has flat nails.

Needle-clawed bushbabies are found in the closed canopy of tropical rainforest and are completely arboreal (Napier and Napier 1985). In studies in Gabon, Charles-Dominique (1977) found that *E. elegantulus* lived at heights of 5–50 meters above the ground, hardly ever descended, and slept by day rolled up in a ball in foliage. It was seen to make horizontal leaps of 2.5 meters and to cover up to 8 meters with some loss of height. This species spends the entire night searching for the gums of trees, its primary food. The nails on the feet of this species allow it to investigate large tree trunks and other areas inaccessible to most galagines. It locates gums by smell and visits

500–1,000 collection points per night. Insects and fruits also form part of its diet.

According to Charles-Dominique (1977), *E. elegantulus* has a population density of 15–20/sq km in favorable habitat. It sleeps in tight clusters of two to seven individuals, which break up at night. It does not urine wash, as some other bushbabies do, but deposits urine directly on branches. In Gabon births occur mostly from January to March, when fruits and insects are most abundant, and there are no births from June to August, when resources are minimal. Litter size is normally one, and it is cared for by the mother, as described below in the account of *Galagoides*. Kingdon (1974a) indicated that a breeding peak in *E. matschiei* may occur in November–December. Müller (*in* Grzimek 1990) listed a longevity of more than 15 years for captive *Euoticus*.

The IUCN classifies all three species as near threatened. They occupy relatively small ranges that are subject to severe disturbance by human activity. All species are on appendix 2 of the CITES.

PRIMATES; LORISIDAE; **Genus GALAGO**
E. Geoffroy St.-Hilaire, 1796

Galagos, or Bushbabies

There are three species (Hill and Meester *in* Meester and Setzer 1977; Meester et al. 1986; Nash, Bearder, and Olson 1989; Petter and Petter-Rousseaux 1979):

G. senegalensis, savannah and forest savannah zones from Senegal to Somalia and Tanzania;
G. gallarum, southern Ethiopia, southern Somalia, northern Kenya;
G. moholi, Zaire and Tanzania to northern Namibia and Transvaal.

Euoticus, Otolemur, and *Galagoides* (see accounts thereof) sometimes are considered subgenera or synonyms of *Galago.* Meester et al. (1986) summarized the different views relative to the status of these genera and the assignment of species thereto. Groves (*in* Wilson and Reeder 1993) included within *Galago* what are here treated as the species *Euoticus matschiei* and *Galagoides alleni.*

Head and body length is 88–210 mm, tail length is 180–303 mm, and weight is about 95–300 grams (Nash, Bearder, and Olson 1989). The fur is dense, woolly, rather long, slightly wavy, and without luster. Coloration is silvery gray to brown, being slightly lighter on the underparts. The large ears have four transverse ridges and can be independently or simultaneously bent back and wrinkled downward from the tips well toward the base. This furling and unfurling of the ears is frequent and produces a most quizzical expression. On the ends of all digits are flat disks of thickened skin that help in grasping limbs and slippery surfaces. The long front digits and flattened hind digits have flattened nails, except the second digit of the hind foot, which is armed with a curved grooming claw. Females have two pairs of mammae.

In contrast to most other galagines, which are restricted largely to dense forests, *Galago* occurs in open woodlands, scrub, wooded savannahs, and grasslands with thickets (Doyle and Bearder 1977; Happold 1987). Like the others, *Galago* is arboreal and nocturnal. In studies in South Africa Doyle and Bearder (1977) observed *G. moholi* to have a midnight rest period and to make use of two or three fa-

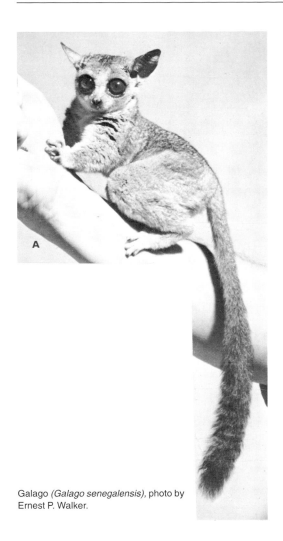

Galago *(Galago senegalensis)*, photo by Ernest P. Walker.

ful version of the same used by adults of some species to assemble at the sleeping site, a loud cry or bark used for distant communication and territorial encounters, and a high-pitched alarm call (Petter and Charles-Dominique 1979). Galagos also may communicate by olfactory methods, especially through urination. Individuals, especially dominant males, frequently "urine wash" their hands and feet, so that scent is spread over the entire three-dimensional space through which the animals move. This process apparently allows dissemination of information regarding the presence and condition of individuals. However, studies by C. Harcourt (1981) suggest that at least in some individuals urine washing functions primarily to facilitate grip.

Studies indicate that females experience at least two restricted periods of mating and two pregnancies per year, the resulting birth seasons being September–October and January–February in southern Africa, December–February and June in Uganda, and March and July in Sudan. The estrous cycle averages about 32 days, estrus lasts 2–7 days, and the gestation period is 120–42 days. Litter size is commonly one or two, occasionally three. Birth weight is about 9.5 grams in *G. moholi* and about 12 grams in *G. senegalensis*. The young are fully furred and have their eyes open at birth. They first leave the nest at 10–11 days, and they are

vorite sleeping sites during the day. The females made nests to give birth and shelter the family. According to Smithers (1983), this species may construct an open-topped, platformlike nest or shelter in an unused bird nest, a hollow tree, or a clump of dense foliage. Bushbabies are alert, sprightly, and agile, making long leaps from branch to branch. Doyle and Bearder (1977) found *G. moholi* to progress mainly by hops and leaps and only rarely to walk quadrupedally. On the average night it covered about 2.1 km. It fed exclusively on acacia gums and insects, catching its prey with rapid strikes of one hand.

According to Doyle and Bearder (1977), population density of *G. moholi* was 95–200/sq km in favorable habitat. This species often occurs in small family groups of about two to seven individuals. These groups may consist of an adult pair with or without young, two adult females plus infants, or an adult female with young. Such groups spend the day sleeping at the same site, but the adults split up at night to forage. Adult males are territorial and may fight viciously for control of a home range that overlaps the ranges of several females. Harcourt and Bearder (1989) reported home range sizes of 1.5–22.9 ha. for males and 4.4–11.7 ha. for females.

Vocalizations identified for *Galago* include a clicking sound by which the young calls its mother, a more power-

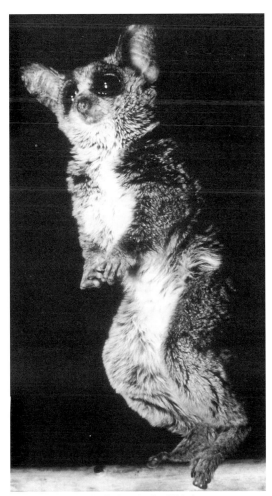

Galago *(Galago moholi)*, photo by David Haring.

able to catch insects at 4 weeks. Weaning was reported at 11 weeks of age in a laboratory study of *G. senegalensis*. Young males leave their mother when they reach physiological sexual maturity, at about 10 months, but young females may remain longer, even though they also mature at this time (Charles-Dominique 1977; Doyle and Bearder 1977; Hayssen, Van Tienhoven, and Van Tienhoven 1993; Kingdon 1974*a;* Nash, Bearder, and Olson 1989; Smithers 1983; Van Horn and Eaton 1979). Record longevity for galagines is held by a *G. senegalensis* that was still living after 18 years and 10 months in captivity (Marvin L. Jones, Zoological Society of San Diego, pers. comm., 1995).

Lee, Thornback, and Bennett (1988) did not regard *Galago* as threatened. Nonetheless, the newly distinguished species *G. gallarum* occupies a restricted area that may be undergoing severe habitat disruption because of human activity. It is designated as near threatened by the IUCN. All species of *Galago* are on appendix 2 of the CITES.

PRIMATES; LORISIDAE; **Genus OTOLEMUR**
Coquerel, 1859

Greater Bushbabies

There are two species (Hill and Meester *in* Meester and Setzer 1977; Masters 1986, 1988; Meester et al. 1986; Nash, Bearder, and Olson 1989):

O. garnettii, southern Somalia to southeastern Tanzania, and on the nearby islands of Pemba, Mafia, and Zanzibar;
O. crassicaudatus, eastern Zaire and southwestern Kenya to Angola and eastern South Africa.

Otolemur sometimes has been regarded as a synonym or subgenus of *Galago* but was recognized as a distinct genus by Corbet and Hill (1991), Groves (1989 and *in* Wilson and Reeder 1993), Meester et al. (1986), and Napier and Napier (1985).

Head and body length is 230–465 mm, tail length is 300–550 mm, and weight is about 600–2,000 grams (Charles-Dominique 1977; Kingdon 1974*a;* Nash, Bearder, and Olson 1989; Nash and Harcourt 1986). The upper parts are pale gray tinged with buff or brown and sometimes penciled with black; the underparts are paler. The tail is usually lighter and buffier than the upper parts, and the head is darker around the eyes (Smithers 1983). From *Galago, Otolemur* is distinguished by larger size, a pronounced postorbital constriction of the skull, a long and robust muzzle, and a foramen magnum that is directed backwards rather than downwards (Meester et al. 1986). Characteristics of the ears and digits are about the same as described above for *Galago.*

Greater bushbabies are found in forests, thickets, and well-developed woodlands. They may occupy urban areas if there are sufficient trees for shelter and food sources (Smithers 1983). In studies in South Africa Doyle and Bearder (1977) observed *O. crassicaudatus* to sleep in a nest 5–12 meters high during the day. Its nocturnal activity lasted from 9.5 hours in summer to about 12 hours in winter, with a rest period in the middle of the night. It generally walked quadrupedally when undisturbed and could make leaps of up to 2 meters without losing height. It moved over a distance of about 1 km per night. Although some fruits and insects were eaten, 62 percent of the diet throughout the year consisted of gums (tree exudates). Population density in favorable habitat was 72–125/sq km, and one ma-

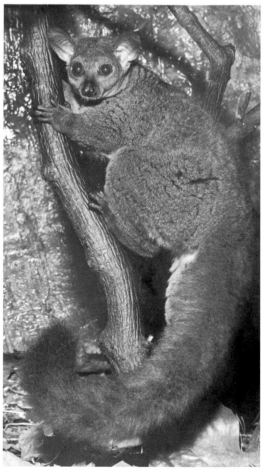

Greater bush baby (*Otolemur* sp.), photo from Zoological Society of London.

ternal group had a home range of 7 ha. In coastal Kenya, Nash and Harcourt (1986) found *O. garnettii* to move an average of 1.6 km per night. Population density in this species was 31–38/sq km, and home range averaged 17 ha. for males and 12 ha. for females.

O. crassicaudatus often occurs in small, stable family groups of about two to six individuals. These groups may comprise an adult pair with or without young, two adult females plus infants, or an adult female with young. Such groups spend the day sleeping at the same site, but the adults split up at night to forage. Adults may be territorial; the ranges of adjacent groups overlap only to a small extent, but the ranges of males overlap extensively with those of females. The males evidently fight one another to secure control of such areas (Doyle and Bearder 1977). *O. garnettii* has a similar social structure but more pronounced territorial behavior. Fully adult animals, even of opposite sexes, rarely share a nest. There is little overlap between the ranges of adults of the same age but extensive overlap between the ranges of individuals of different ages. Young females mature in their natal ranges, but males disperse to other areas (Nash and Harcourt 1986).

According to Kingdon (1974*a*), *Otolemur* announces its presence for several months of the year by a loud, croaking wail repeated at frequent intervals during the early part of the night. This vocalization appears to correspond to the

Greater bush baby *(Otolemur crassicaudatus)*, photo from San Diego Zoological Society.

breeding season and to have a territorial function. Its resemblance to the cry of a child is said to be the basis for the common name "bushbaby." Urine washing such as described above in the account of *Galago* also is characteristic of *Otolemur*.

Except as noted, the following information on natural history was taken from Doyle and Bearder (1977), Izard (1987), Kingdon (1974a), Nash and Harcourt (1986), Smithers (1983), and Van Horn and Eaton (1979). In the northern Transvaal, mating occurs mainly in June or July and births are restricted largely to November. Births have been reported in March in Somalia and between August and November in Tanzania and Zambia, at the end of the dry season. Pregnancies seem to peak in August on Zanzibar and Pemba. *O. garnettii* breeds once a year, from August to October, in coastal Kenya. The estrous cycle of *O. crassicaudatus* averages about 44 days, and gestation about 133 days. Litter size is commonly two young and occasionally three in *O. crassicaudatus* but usually only one in *O. garnettii*. The young weigh about 40 grams, have their eyes open at birth, and can crawl after about 30 minutes. They begin feeding on their own at around 1 month, though lactation may last for nearly 5 months. Both sexes of *O. garnettii* are able to breed by 20 months of age. Captives of both species have lived for about 18 years or more (Marvin L. Jones, Zoological Society of San Diego, pers. comm., 1995).

Lee, Thornback, and Bennett (1988) did not consider either species of *Otolemur* to be threatened. Nonetheless, *O. garnettii*, at least, could be of long-term concern, as it occupies a restricted area undergoing increasing human activity and habitat disruption. It is designated as near threatened by the IUCN. Both species are on appendix 2 of the CITES.

Dwarf Galagos

There are two subgenera and four species (Hill and Meester *in* Meester and Setzer 1977; Meester et al. 1986; Nash, Bearder, and Olson 1989):

subgenus *Sciurocheirus* Gray, 1873

G. alleni, southeastern Nigeria to Congo, Bioko (Fernando Poo);

subgenus *Galagoides* A. Smith, 1833

G. demidoff, forests from Senegal to Uganda and western Tanzania, Bioko (Fernando Poo);
G. thomasi, an extremely disjunctive range including the vicinity of Mount Cameroon in southwestern Cameroon, possibly Mount Marsabit in northern Kenya, the Kivu and Ituri regions of eastern Zaire and southwestern Uganda, the region between Dilolo and Kolwezi in southern Zaire, and the Loanda Highlands of central Angola;
G. zanzibaricus, southern Somalia, coastal Kenya and Tanzania, Zanzibar, Mozambique, Malawi, eastern Zimbabwe.

Galagoides sometimes is regarded as a subgenus or synonym of *Galago* but was given generic rank by Corbet and Hill (1991), Groves (1989 and *in* Wilson and Reeder 1993), Meester et al. (1986), Napier and Napier (1985), and Olson (*in* Nash, Bearder, and Olson 1989). However, Corbet and Hill, Groves, and Napier and Napier included *G. alleni* in

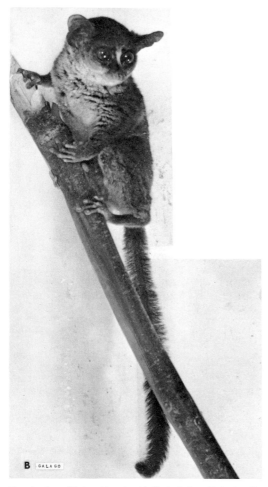

Dwarf galago *(Galagoides demidoff)*, photo from Paignton Zoo, Great Britain.

Galago rather than in *Galagoides*. *G. granti* of Mozambique was listed as a separate species by Corbet and Hill but was included within *G. zanzibaricus* by Groves, Meester et al., and Nash, Bearder, and Olson. *G. thomasi* was distinguished from *G. demidoff* by Nash, Bearder, and Olson but not by Groves or any of the other authorities cited above. Based on karyological analysis, Stanyon et al. (1992) suggested that *G. demidoff* may actually comprise several species; like some other authorities, they used the name *G. demidovii* for that taxon.

The following specific descriptions were taken largely from Nash, Bearder, and Olson (1989). In *G. alleni* head and body length is 155–240 mm, tail length is 205–300 mm, and weight is 200–445 grams. The upper parts are very dark brown to almost black, the underparts are paler, the flanks of the limbs are brightly colored, and the face has dark circumocular rings. The head is long and narrow, the muzzle is pointed, and the tail is bushy (Nash, Bearder, and Olson 1989). In *G. demidoff* head and body length is 73–155 mm, tail length is 110–215 mm, and weight is 44–97 grams. The upper parts vary in color from rufus to reddish brown, the underparts and flanks of the limbs are paler, and the circumocular markings are variable. The head is narrow, the muzzle is pointed and upturned, and the tail is not bushy. In *G. thomasi* head and body length is 123–66 mm, tail length is 150–233 mm, and weight is 55–149 grams. The upper parts are blackish brown, the underparts and flanks of the limbs are yellowish, and the circumocular markings are prominent. The head is narrow, the muzzle is pointed, and the tail is not bushy. In *G. zanzibaricus* head and body length is 120–90 mm, tail length is 170–265 mm, and weight is 104–203 grams. The upper parts are cinnamon reddish brown, the underparts are bright yellow, the flanks of the limbs are close in color to the upper parts, and the circumocular markings are prominent. The muzzle is long, pointed, and concave dorsally. The tail is moderate to bushy.

Galagoides resembles *Galago* but differs in that the upper parts are generally brownish or cinnamon rather than predominantly gray and in having a relatively longer muzzle. The nails of the feet are not pointed. Except in *G. alleni* there is a pronounced elongation of the premaxillae of the skull in front of the incisor teeth.

Dwarf galagos live in primary or secondary forests, usually where there is dense foliage, and are nocturnal and arboreal (Napier and Napier 1985). In studies in Gabon, Charles-Dominique (1977) found *G. alleni* only 1–2 meters above the ground; it spent each day sleeping in one of several favored tree hollows within its home range. *G. demidoff* occurred 5–40 meters above the ground and slept in leaf nests or thick vegetation. The latter species was seen to make horizontal leaps of 1.5–2 meters. *G. demidoff* fed largely on insects, mostly small beetles and nocturnal moths, and also ate some fruits and gums. Population densities in favorable habitat were 15–20/sq km for *G. alleni* and 50–117/sq km for *G. demidoff*. Harcourt and Nash (1986) determined that in the coastal forests of Kenya *G. zanzibaricus* moved about 1.5–2.0 km per night, had a population density of 170–80/sq km, and had a group home range of 1.6–2.8 ha. Charles-Dominique and Bearder (1979) listed the following home range sizes: *G. alleni*, females, 3.9–16.6 ha., and males, 17–50 ha.; *G. demidoff*, females, 0.6–1.4 ha., and males, 0.5–2.7 ha.

In *G. alleni* and *G. demidoff* usually a single animal occupies a sleeping site, but sometimes two or three adult females plus young are found together (Charles-Dominique 1977). Adult males are aggressive toward one another and apparently are territorial. Adults of opposite sexes, however, may share the same area, sometimes sleep together by day, and have some contact at night even though foraging separately. Each male seeks to control a home range that overlaps those of several females, and intense competition may result. Charles-Dominique (1977) distinguished four categories of male *G. demidoff*. The heaviest males (averaging 75 grams) have large home ranges that include at least one female's home range and are in a central position overlapping a number of female ranges. The home ranges of several such males converge at a common point of slight overlap, where interaction occurs. The lightest males (averaging 56 grams) are tolerated within the ranges of the heavy males and have small home ranges of their own. Medium-sized males (averaging 61 grams) occupy relatively large home ranges, but on the periphery of female ranges. These males may associate with other peripheral males and eventually may gain weight and shift into a central position. The last category, nomadic males, includes mostly young animals that do not remain long in any one area.

In *G. zanzibaricus* in coastal Kenya there is a closer regular association between the sexes, and both are probably territorial. Adult males usually have nonoverlapping ranges that are shared with one or two females and their offspring. The male usually sleeps with these other animals, but a mother will sleep alone for about three weeks after giving birth. Females can breed twice a year, the peak

A. Mother dwarf galago *(Galagoides demidoff)* carrying one of twins, aged 30 days; B. Baby, aged two days. Photos by A. S. Woodhall.

birth seasons in this area being February–March and August–October. Gestation is 120 days, and there usually is a single young, occasionally two. Young males disperse from their natal ranges, but females remain in the area and initially give birth there at about 12 months (C. Harcourt 1986; Harcourt and Nash 1986). One *G. zanzibaricus* lived at the San Antonio Zoo for 16 years and 6 months (Marvin L. Jones, Zoological Society of San Diego, pers. comm., 1995).

The following additional information on natural history was taken from Charles-Dominique (1977), Hayssen, Van Tienhoven, and Van Tienhoven (1993), and Van Horn and Eaton (1979). In *G. demidoff* there usually is one pregnancy per female per year; mating takes place in Congo in September–October and January–February; and births occur all year in Gabon, with a peak in January–April, when fruits and insects are most abundant. Gestation has been reported to be about 133 days in *G. alleni* and 111–14 days in *G. demidoff*. Birth weight is about 5–10 grams in *G. demidoff*. In *G. alleni* and *G. demidoff* the female takes the young out of the nest when it is a few days old, leaves it hidden in vegetation while she forages during the night, and carries it back to the nest in the morning. After about a month the young is able to follow the female, but it still is carried on occasion. It has been observed that if the young is unable to leap across a gap after its mother, it will emit a call, upon which the female will return, pick up the young, and make the jump. Weaning occurs after about 6 weeks and sexual maturity at 8–10 months. Wild *G. alleni* are known to have lived at least eight years.

The IUCN classifies *G. alleni* and *G. zanzibaricus* as near threatened (placing both in the genus *Galago*). The indigenous forests on which the species depend are being cleared for development purposes or cut and replaced with exotic conifers (Lee, Thornback, and Bennett 1988). All species are on appendix 2 of the CITES.

PRIMATES; Family CHEIROGALEIDAE

Dwarf Lemurs and Mouse Lemurs

This family of five genera and eight species is found only in Madagascar. Petter and Petter-Rousseaux (1979) divided the family into two subfamilies: the Cheirogaleinae, for *Microcebus, Mirza, Cheirogaleus*, and *Allocebus*; and the Phanerinae, for *Phaner*. Tattersall (1982, 1986), however,

did not consider *Phaner* to warrant subfamilial separation. In any case, all of these genera once were placed within the family Lemuridae but now are thought to represent a distinct family on the basis of new anatomical, behavioral, ecological, and cytogenetic information. Moreover, Tattersall (1982) united the Cheirogaleidae with the Lorisidae in the superfamily Lorisoidea of the strepsirhine infraorder Lorisiformes, while he assigned all other Malagasy primates to the superfamily Lemuroidea of the infraorder Lemuriformes. Dene, Goodman, and Prychodko (1980) considered the Cheirogaleidae to represent a distinct superfamily, Cheirogaleoidea, within the Lemuriformes. Groves (1989 and *in* Wilson and Reeder 1993) did not accept a close affinity between the Cheirogaleidae and the Lorisidae.

The members of the Cheirogaleidae are smaller than those of the Lemuridae. Head and body length is about 125–275 mm and tail length is about 125–350 mm. The hind legs are substantially longer than the forelegs. The fur is soft, thick, and woolly. In this family the eyes are large and set closely together, whereas in the Lemuridae the eyes are smaller and placed more laterally. Female cheirogaleids have three pairs of mammae located pectorally, abdominally, and inguinally, while female lemurids, except in *Varecia*, have only a single pair located pectorally. The dental formula of the Cheirogaleidae is: (i 2/2, c 1/1, pm 3/3, m 3/3) × 2 = 36. The incisors are well developed, not small and peglike as in the Lemuridae or absent as in the Megaladapidae.

The Cheirogaleidae are arboreal and nocturnal. *Cheirogaleus* and *Microcebus* are known to accumulate fat reserves on the hind legs and base of the tail and then to become torpid during the dry season.

PRIMATES; CHEIROGALEIDAE; Genus MICROCEBUS
E. Geoffroy St.-Hilaire, 1828

Mouse Lemurs

There are three species (Mittermeier et al. 1994; Petter and Petter-Rousseaux 1979; Schmid and Kappeler 1994):

M. murinus, western and southern Madagascar;
M. myoxinus, west-central Madagascar;
M. rufus, northern and eastern Madagascar.

The distinction of *M. myoxinus* from *M. murinus* was recognized only recently. Still another species, *M. coquereli*, is

Mouse lemur *(Microcebus murinus)*, photo by Howard E. Uible.

here assigned to the separate genus *Mirza* (see account thereof) but has been retained in *Microcebus* by Groves (*in* Wilson and Reeder 1993) and some other authorities.

In *M. murinus* and *M. rufus* head and body length is usually 125–50 mm, tail length is about the same, and weight is 39–98 grams. *M. myoxinus*, with a total length of 178–219 mm and a weight of 24–38 grams, is evidently the smallest of primates. All species usually have soft fur, a short snout, a rounded skull, prominent eyes and ears, long hind limbs, and a long tail. The upper parts are generally grayish in *M. murinus*, rufous brown with an orange tinge in *M. myoxinus*, and brownish in *M. rufus*. There is a dorsal stripe along the middle of the back, which is not always distinct, and a distinct white median nasal stripe. The underparts are white.

Mouse lemurs live in forests and are associated with trees. They have been found to be most common in secondary forest and to construct spherical leaf nests in foliage or nest in hollow trees (Martin 1973). Mouse lemurs are agile and active at night, usually traveling along branches on all four legs, leaping at times, and using the tail as a balancing organ. On the ground *M. murinus* moves with froglike hops (Martin 1973). *Microcebus* does not undergo true hibernation, but there is a decline in activity during the winter (June–September). From July to December fat accumulates in the tail, and then in the dry season the animals experience short periods of torpor with a decline in body temperature. The diet consists of insects, spiders, occasionally small frogs and lizards, a large amount of fruit, flowers, nectar, some gums and insect secretions, and leaves (Hladik 1979).

Petter (1978) reported population densities of about 300–400/sq km for *M. murinus*. Pollock (1979) listed densities of up to 262/sq km for *M. rufus* and 1,300–2,600/sq

km for *M. murinus*. Martin (1973) stated that the individual home range of both species is not more than 50 meters in diameter. Hladik, Charles-Dominique, and Petter (1980) reported home range not to exceed 2 ha. in dry forest and to be smaller in humid areas. Females form relatively stable groups, usually comprising two to nine individuals, that share a sleeping nest but forage alone by night. A male occasionally may be found with a group of females even outside of the breeding season. Males usually nest alone or in pairs (Martin 1973; Pollock 1979). A male's home range may overlap those of several females; ranges often are scent-marked with urine and feces (Mittermeier et al. 1994). The vocalizations of *M. murinus* are very high-pitched but are known to include a variety of calls for seeking contact, mating, distant communication, alarm, and distress (Petter and Charles-Dominique 1979).

Reproduction seems to be well known only for *M. murinus* (Doyle 1979; Hayssen, Van Tienhoven, and Van Tienhoven 1993; Van Horn and Eaton 1979). There is a restricted breeding season from August to March in Madagascar, but it shifts to the other half of the year among captive animals in the Northern Hemisphere. The estrous cycle usually lasts about 45–55 days, estrus 1–5 days. Gestation averages 60 (54–68) days. Litter size is normally two or three, the newborn weigh 5 grams each, and weaning occurs after 25 days. Sexual maturity is attained at 10–29 months by females and at 7–19 months by males. A captive *M. murinus* reportedly lived to an age of 15 years and 5 months (Marvin L. Jones, Zoological Society of San Diego, pers. comm., 1995).

Because of the general concern about the pace of habitat destruction in Madagascar, all species of *Microcebus* are listed as endangered by the USDI and are on appendix 1 of the CITES. The IUCN classifies *M. myoxinus* as vulnerable. Although *M. murinus* and *M. rufus* are still relatively common and may be the only Malagasy lemurs not undergoing a general decline, they are apparently losing habitat in some areas through logging and excessive grazing by cattle and goats (Richard and Sussman 1975). Information compiled by Mittermeier et al. (1992, 1994) suggests that *M. murinus* and *M. rufus* may still number in the hundreds of thousands, if not millions, of individuals but that *M. myoxinus*, which occupies a much smaller known range, may be vulnerable to extinction.

PRIMATES; CHEIROGALEIDAE; **Genus MIRZA**
Gray, 1870

Coquerel's Dwarf Lemur

The single species, *M. coquereli*, occurs on the northwest and west-central coasts of Madagascar. Until recently this species generally was assigned to *Microcebus*, and Groves (*in* Wilson and Reeder 1993) continued to do so, but *Mirza* was recognized as a distinct genus by Corbet and Hill (1991), Mittermeier et al. (1994), and Tattersall (1982).

Head and body length is about 250 mm, tail length is about 280 mm, and weight is 280–335 grams. The upper parts are dark gray washed with rufous, being darkest along the midline; many of the hairs have golden tips, producing an olive brown effect. The underparts are yellowish gray. Facial markings are indistinct or absent. The ears are long and hairless. Females have two pairs of mammae.

Mirza is much larger than *Microcebus*. The two genera are both characterized by a relatively small second upper premolar, a small pericone on the first and second upper molars, the reduction or loss of the paraconid on the posterior lower premolar, and a complete buccal cingulum on all lower molars (Tattersall 1982). *Mirza*, however, lacks the greater bullar development, obliteration of the interincisal gap, and toothrow curvature of *Microcebus* (Groves 1989).

Mirza generally lives near rivers or ponds in thick forest, where it is usually observed 1–6 meters above the ground (Hladik 1979; Pages 1978). A spherical nest 50 cm in diameter is constructed of interlaced vines, branches, and leaves, usually in a tree fork (Pages 1980). *Mirza* is nocturnal, is active all year except on very cold nights, and is not known to accumulate fat reserves (Petter 1978). During the wet season it has a varied diet consisting of fruits, flowers, gums, insects, spiders, frogs, lizards, small birds, and eggs. In the dry season it feeds largely on a sweet liquid secreted onto branches by the larvae of the homopteran insect family Flatidae and also eats the insects themselves (Pages 1980; Petter 1978).

Reported population densities have varied from about 30/sq km to as high as 385/sq km (Mittermeier et al. 1994). Pages (1980) reported that each individual has a home range with up to 12 nests in the central portion. Male ranges consist of a core area of about 1.5 ha. and a peripheral, less used area of up to 4.0 ha. Females have a core area of 2.5–3.0 ha. and a peripheral zone of up to 4.5 ha. Males are solitary and seem to defend their core areas from other males. However, a male's core area may overlap that of a female. Marking is accomplished through urination, salivation, and anogenital dragging. Vocalizations include a "hum" that accompanies all movements and meetings between individuals.

Observations of a captive colony at the Duke University Primate Center in North Carolina (Stanger, Coffman, and Izard 1995) indicate that reproduction occurs throughout the year. Females there usually had one litter per year

Greater mouse lemur *(Mirza coquereli)*, photo by David Haring.

but sometimes had a second if the first did not survive or was removed. The estrous cycle averaged 22.1 days, estrus lasted no longer than a day, and the gestation period averaged 89.2 days. Of 51 litters, 21 contained a single infant and 60 had twins. Mean birth weight was 17.5 grams. Females first gave birth at an average age of 22.5 months. According to Hayssen, Van Tienhoven, and Van Tienhoven (1993), lactation lasts 134–38 days. Jones (1982) listed a record longevity of 15 years and 3 months.

Mirza is designated as vulnerable by the IUCN and as endangered by the USDI and is on appendix 1 of the CITES. It occupies a restricted area of humid forest and is threatened both by a long-term drying trend in the climate and by human destruction of the forests. Mittermeier et al. (1992, 1994) suggested that a reasonable estimate of numbers would be 10,000–100,000 individuals.

PRIMATES; CHEIROGALEIDAE; **Genus**
CHEIROGALEUS
E. Geoffroy St.-Hilaire, 1812

Dwarf Lemurs

There are two species (Petter and Petter *in* Meester and Setzer 1977; Petter and Petter-Rousseaux 1979):

Dwarf lemur *(Cheirogaleus medius)*, photo by David Haring.

C. major, northern and eastern Madagascar, possibly one restricted area in west-central Madagascar;
C. medius, western and extreme southern Madagascar.

Groves (1989) also recognized *C. crossleyi* of northeastern Madagascar as a separate species, but subsequently (*in* Wilson and Reeder 1993) he included it within *C. major.*

Head and body length is 167–264 mm, tail length is 195–310 mm, and weight is 177–600 grams (Tattersall 1982). The fur is soft, woolly, and sometimes silky. The up-

Dwarf lemur *(Cheirogaleus major)*, photo by Howard E. Uible. Insets: hand and foot of dwarf lemur, photos from *Proc. Zool. Soc. London.*

per parts vary from buffy or gray to reddish brown; the underparts are whitish, often tinged with yellow. The ears are thin, membranous, and naked. The eyes are large, lustrous, and surrounded by dark rings. Females have two pairs of mammae.

These dwarf lemurs live in forests. They are nocturnal, spending the daylight hours in globular nests that they construct of twigs and leaves in hollow tree trunks or in the tops of trees. They move about their arboreal heights on all fours, but they sit up to eat, holding the food in their hands. The diet consists mainly of fruits, flowers, nectar, and insects (Hladik 1979). Those living in areas subject to periodic dry seasons store up quantities of fat in the basal portion of the tail during the wet season and estivate when food is scarce. In a study in western Madagascar Petter (1978) found that *C. medius* gained weight from November to March, went into torpor within hollow tree trunks in April, and emerged in October or November, at the start of the rainy season. Individuals could lose 100 grams of body weight during estivation.

Pollock (1979) reported a population density of 75–110/sq km for *C. major.* Hladik, Charles-Dominique, and Petter (1980) found densities of 200/sq km and 350/sq km for *C. medius* and determined that most individuals did not have ranges of more than 200 meters in diameter. There was much overlap of these areas and no evidence of territoriality; groups of 3–5 estivated together. Nonetheless, these animals are usually solitary, and captives are generally intolerant of others of the same sex. They are relatively quiet but have a number of weak calls for contact and a louder cry in agonistic situations (Petter and Charles-Dominique 1979). Captive individuals reportedly are easy to tame, and at least one became quite affectionate, enjoyed being handled, and would come when called by name. Limited data indicate that in the wild mating occurs in October or November, births take place in January or February, and litter size is 2–3 (Klopfer and Boskoff 1979; Mittermeier et al. 1994; Van Horn and Eaton 1979). Estrus lasts 2–3 days in *C. major,* gestation 70 days, and lactation about 45 days (Hayssen, Van Tienhoven, and Van Tienhoven 1993). In a captive colony of *C. medius* in North Carolina mating takes place from April to early June, births occur from June to August, the estrous cycle averages 19.7 days, and males compete fiercely for the estrous females. Gestation lasts 61 days, and litter size is one to four. The newborn are fully furred and have their eyes open. Females reach sexual maturity at 10–14 months (Foerg 1982). One captive specimen was still living at 19 years and 2 months (Marvin L. Jones, Zoological Society of San Diego, pers. comm., 1995).

According to Richard and Sussman (1975), both species of *Cheirogaleus* are declining because of loss of forest habitat. Both are listed as endangered by the USDI and are on appendix 1 of the CITES. Mittermeier et al. (1992, 1994) suggested that each species probably numbers more than 100,000 individuals.

PRIMATES; CHEIROGALEIDAE; Genus ALLOCEBUS
Petter-Rousseaux and Petter, 1967

Hairy-eared Dwarf Lemur

The single species, *A. trichotis*, is known with certainty only from a few specimens collected in lowland rainforest in the vicinity of Mananara in northeastern Madagascar,

though there have also been recent reports from the Masoala Peninsula farther north (Mittermeier et al. 1994).

Petter-Rousseaux and Petter (1967) described a specimen with a head and body length of 133 mm and a tail length of 170 mm. Gunther, who originally described *trichotis* as a species of *Cheirogaleus*, gave the measurements of the type specimen as: head and body length, 152 mm, and tail length, 149 mm. He emphasized that the tail was shorter than the head and body, but Petter-Rousseaux and Petter (1967) thought that Gunther's specimen may have been damaged. In four recently collected living adults head and body length was 125–45 mm, tail length was 150–95 mm, and weight was 75–98 grams (Meier and Albignac 1991). The upper parts are brownish gray, the underparts are whitish gray, and the tail is reddish brown. The ears are short with tufts of long hair in front and on the internal side of the lobe.

This genus resembles *Microcebus* in size and color but has a noticeably longer tail and better-developed hind limbs. Its muzzle is elongated and rounded off at the end, whereas that of *Microcebus* is short and pointed. The hands and feet are larger than in *Microcebus*. According to Groves (1989), *Allocebus* differs from *Microcebus*, *Mirza*, and *Cheirogaleus* but resembles *Phaner* in lacking enlarged molars, convergent toothrows, and a molariform fourth upper premolar. *Allocebus* is unique in having caniniform

Hairy-eared dwarf lemur *(Allocebus trichotis)*, photos by Bernhard Meier.

second and third upper premolars, a greatly enlarged first upper incisor, and a deflated tympanic bullar region, hardly defined from the mastoid inflation.

This genus is extremely rare, and little is known of its habits in the wild. Meier and Albignac (1991) found *Allocebus* only in primary lowland rainforest. The specimen described by Petter-Rousseaux and Petter (1967) was taken from a hole in a tree. Walker (1979) stated that *Allocebus* apparently progresses in a quadrupedal fashion, like a large *Microcebus murinus*. Coimbra-Filho and Mittermeier (1977*b*) observed that *Allocebus* has large upper incisors like those of *Phaner* and may also use these teeth to scrape tree bark to obtain exudates.

Allocebus had not been seen in more than 20 years when, in March 1989, it was rediscovered in the vicinity of the Mananara River in northeastern Madagascar. In a survey for the World Wildlife Fund, Yoder (1989) obtained considerable new information through interviews with native people in the Mananara area. *Allocebus* is found only in primary forest, probably at low densities. It is nocturnal and arboreal. It makes nests of fresh leaves in small holes in either living or dead trees, usually 3–5 meters above the ground. The people encounter it only from October to March, when they cut down nest trees in the course of slash-and-burn agriculture. During the entire cold season, from early May to mid-October, *Allocebus* evidently hibernates deep within tree holes. The diet includes new leaves and small fruits. *Allocebus* is found only in pairs consisting of one male and one female or in pairs with a single infant.

Observations of four captive individuals have confirmed part of Yoder's report and provided some new information (Meier and Albignac 1991). The animals take a wide variety of foods but seem to prefer large insects, which they capture with both hands. They are less active during the dry season, June–September, but fat reserves that would facilitate dormancy are less obvious than in *Cheirogaleus* and are distributed across the body, not accumulated in the tail. The captive *Allocebus* use straw to construct nests in holes, where they sleep together as a group during the day. Various evidence suggests that estrus occurs at the beginning of the wet season, in November–December, with births in January–February.

Yoder (1989) indicated that *Allocebus* is being killed and eaten regularly by people, and this factor, along with rapid deforestation, places its survival in doubt. Meier and Albignac (1991) stated that it is certainly declining rapidly. Mittermeier et al. (1992) gave the main threat as slash-and-burn agriculture, estimated the total population at only 100–1,000 individuals, and considered the genus of the highest conservation priority. *Allocebus* is classified as critically endangered by the IUCN and the USDI and is on appendix 1 of the CITES.

PRIMATES; CHEIROGALEIDAE; **Genus PHANER**
Gray, 1870

Fork-marked Dwarf Lemur

The single species, *P. furcifer*, occupies scattered areas across northern Madagascar, a section of the west-central part of the island, and a small area in the far southeast; four distinctive subspecies now are recognized (Groves and Tattersall 1991; Mittermeier et al. 1994).

Head and body length is about 227–85 mm, tail length is 285–370 mm, and weight is 300–500 grams. The thickly furred body is reddish gray or brownish gray; the color is

brightest on the head and neck. Black streaks extend from the eyes to the top of the head, where they converge in a middorsal stripe to about the hips. This particular color pattern on the top of the head results in the vernacular name. The throat and underparts are pale rufous or yellowish. The hands and feet are dark brown, and the bushy tail is dark reddish brown with a black or white tip. *Phaner* differs from other members of the Cheirogaleidae in chromosomal features, cranial and dental characters, the presence of distinctive dermatoglyphs and a specific marking gland on the surface of the neck, and its unique behavior (Petter and Petter-Rousseaux 1979). The second upper premolar tooth is uniquely long, resembling a second canine, and the first upper incisor is long, procumbent, and separated from the second by a gap (Groves 1989).

Phaner is primarily a forest inhabitant but also frequents narrow lines of trees that project into or through savannahs. During the day some individuals use nests made by *Mirza coquereli*, but most rest in holes in tree trunks or branches. *Phaner* is active from evening to early morning. It runs quadrupedally along branches and jumps from one to another. It is capable of horizontal leaps of 4–5 meters but can cover 10 meters with some loss of height (Petter, Schilling, and Pariente 1975). In a study in an area of dry forest it was found to forage at heights of 8–10 meters and often in the treetops (Hladik, Charles-Dominique, and Petter 1980). Apparently, it neither accumulates fat reserves nor estivates (Petter 1978). It is a specialized gum eater. In October–November it sometimes feeds on nectar, but its diet consists primarily of vegetable resins and insect secretions. Its keeled nails allow it to descend along smooth tree trunks to favored feeding places; there it scrapes off the bark with its incisor teeth and laps up sap (Hladik 1979; Petter, Schilling, and Pariente 1975).

Population densities of 40–870/sq km have been reported (Hladik, Charles-Dominique, and Petter 1980; Pollock 1979). According to Petter and Charles-Dominique (1979), females occupy well-defined territories that may overlap to some extent. The males have individual territories, separate from those of other males but including the territories of one or more females. An associated male and female often move around together by night, with the female taking the lead. Territorial encounters are purely vocal and do not involve urine marking. Small groups of animals regularly assemble in encounter zones, all calling at once, and then return to normal foraging activity. Identified vocalizations include a mild "hong" sound emitted every few seconds by an adult in motion; a series of "tia" calls exchanged by an associated pair, which may stimulate a territorial encounter with other individuals; a powerful call emitted by males for territorial purposes; and a series of staccato calls and grunts expressing agitation.

Charles-Dominique and Petter (1980) stated that territory size averages 4 ha. and that breeding occurs during the austral spring. Petter, Schilling, and Pariente (1975) wrote that according to local informants, *Phaner* gives birth about 15 November, around the time of the first rains. The single young is initially sheltered in a tree hole, then clings to the belly of the mother, and finally rides on her back. Several individuals reportedly have lived about 12 years in captivity (Marvin L. Jones, Zoological Society of San Diego, pers. comm., 1995).

The fork-marked dwarf lemur is declining because of loss of its forest habitat to human logging and agricultural activity. It is listed as endangered by the USDI and is on appendix 1 of the CITES. Each of the four recognized subspecies of *P. furcifer* was given a high-priority rating for conservation action by Mittermeier et al. (1992), and each was estimated to number 1,000–10,000 individuals. Of the

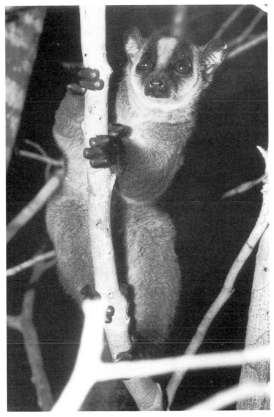

Fork-marked mouse lemur *(Phaner furcifer)*, photos by Russell A. Mittermeier.

four, the IUCN classifies *P. f. furcifer*, of northeastern Madagascar, as near threatened and *P. f. electromontis*, of the extreme north, *P. f. parienti*, of the northwest (south of Ambanja), and *P. f. pallescens*, of the west and possibly the southeast, as vulnerable.

PRIMATES; Family LEMURIDAE

Lemurs

This family of 4 living genera, one Recently extinct genus *(Pachylemur)*, and 11 species is confined to Madagascar and (probably through introduction) the Comoro Islands. Several genera that once were usually placed in this family are now considered to belong to the families Cheirogaleidae and Megaladapidae (see accounts thereof). With regard to the remaining genera there is disagreement about relationships. Some authorities consider *Varecia, Pachylemur, Lemur,* and *Eulemur* to be closely related or even congeneric but place *Hapalemur* in a separate subfamily (Schwartz and Tattersall 1985; Tattersall and Schwartz 1991). Others see close affinity of *Lemur* and *Hapalemur,* with *Varecia, Pachylemur,* and *Lemur* being more distantly related (Groves 1989; Groves and Eaglen 1988; Simons and Rumpler 1988).

In the four living genera head and body length is 280–458 mm and tail length is 280–600 mm. Lemurs have long, heavily furred tails and slender bodies and limbs. The hind limbs are longer than the forelimbs. The pelage is usu-

ally soft, thick, woolly, and solid-colored, but sometimes there are color patterns. In *Lemur catta,* the ring-tailed lemur, the tail is ringed with black and white; in all other species it is unicolored. The face is elongated and foxlike in *Lemur, Eulemur,* and *Varecia* but shortened in *Hapalemur.* The ears, which are at least partially haired, are short or of only moderate length. A clawlike nail is present on the second toe of the hind foot of some forms, and the naked areas of the palms and soles are generally marked by large, ridged pads. In *Lemur, Eulemur,* and *Hapalemur* there is only one pair of mammae, located pectorally, but *Varecia* has one pectoral pair and two abdominal pairs.

The dental formula is: (i 2/2, c 1/1, pm 3/3, m 3/3) × 2 = 36. The upper incisors are small and peglike. The first upper incisors are separated from each other by a wide space. The canines are somewhat elongate, sharp, and separated from the premolars by a space. The lower incisors and canines project forward and somewhat upward. The first lower premolar resembles a canine tooth. The molars have three tubercles.

Although the geological range of the lemurs often is said to extend back into the Pleistocene, the oldest known remains are actually subfossils dating from about 2,850 B.P. The subfossil material evidently represents several of the species that are still extant, as well as the extinct genus *Pachylemur.*

Bamboo lemur *(Hapalemur griseus)*, photo by David Haring.

Gentle lemur *(Hapalemur griseus)*, photo by David Haring.

PRIMATES; LEMURIDAE; **Genus HAPALEMUR**
I. Geoffroy St.-Hilaire, 1851

Bamboo Lemurs, or Gentle Lemurs

There are three species (Meier et al. 1987; Petter and Petter-Rousseaux 1979; Rumpler and Dutrillaux 1978; Vuillaume-Randriamanantena, Godfrey, and Sutherland 1985; Wilson et al. 1989):

H. griseus (bamboo lemur), eastern, northwestern, and west-central Madagascar;
H. simus (broad-nosed gentle lemur), found in much of eastern Madagascar until at least 100 years ago, also known by subfossil material from central and far northern Madagascar;
H. aureus (golden bamboo lemur), southeastern Madagascar.

Head and body length is 260–458 mm, tail length is 240–560 mm, and weight is about 1.0–2.5 kg. The fur is soft and of moderate length. Coloration is brownish gray, reddish gray, orangish brown, grayish green, or reddish green above; it is darkest on top of the head. The underparts are whitish, buffy, gray, or yellowish. The head is globose, the muzzle is short, and the ears are short and hairy. The hands and feet are short and broad with large pads under the tips of the toes and fingers. All the teeth except the molars have a serrated cutting edge adapted for tough foods such as bamboo and other coarse vegetation.

On the inner side of the wrist of *H. griseus* is a rough tract of skin over a gland. In the male this area is covered with spinelike processes, but in the female the processes are hairlike. *H. simus* does not possess such glandular areas.

Bamboo lemurs seem to be restricted to humid forests and marshes where bamboos and reeds are abundant. They have been observed at all hours of the day but are most active in the evening and early morning. They can run quickly on the ground and jump considerable distances from branch to branch in the trees. *H. simus* is more terrestrial than *H. griseus* and often runs across the ground in the forest. The subspecies *H. griseus alaotrensis*, found at Lake Alaotra, apparently is semiaquatic and can swim well. The natural diet reportedly consists almost entirely of bamboo shoots and leaves (J.-J. Petter 1975; Petter and Peyrieras 1975).

Population densities of 47–120/sq km and defended group territories of 6–15 ha. have been reported for *H. griseus*, while a group of *H. aureus* maintained an exclusive territory of about 80 ha. (Harcourt and Thornback 1990). These animals are seen most frequently in groups of 3–5 individuals thought to be family units, comprising an adult male, 1 or 2 adult females, and 1 or 2 juveniles (Petter and Peyrieras 1975). *H. griseus alaotrensis* has been reported to form aggregations of up to 40 individuals when Lake Alaotra is at high water, though permanent group size may be no greater than that of other populations (Feistner and Rakotoarinosy 1993). Group sizes of 2–6 individuals and a mean group home range of 2 ha. for that subspecies were observed by Mutschler and Feistner (1995). There are numerous vocalizations, including a weak grunt to maintain group cohesion, a strong call for distant communication, and a very powerful "creee" in threat situations (Petter and Charles-Dominique 1979).

Gentle lemur *(Hapalemur griseus)*, photo by David Haring.

Births of *H. griseus* in Madagascar have been reported to occur from late October to February and to always be of a single young (Harcourt and Thornback 1990; Petter and Peyrieras 1975). Records from Duke University, however, show that of four births in captivity one was of twins; gestation lasted 135–50 days (Klopfer and Boskoff 1979). One newborn weighed 32 grams (Hayssen, Van Tienhoven, and Van Tienhoven 1993). The young initially may be carried ventrally, but by 3 weeks of age they are riding on the mother's back. One captive infant was carried until 11 weeks and was weaned at 20 weeks (Steyn and Feistner 1994). A captive specimen of *H. griseus* was still living at 17 years (Marvin L. Jones, Zoological Society of San Diego, pers. comm., 1995).

All species of *Hapalemur* are listed as endangered by the USDI and are on appendix 1 of the CITES. The IUCN classifies *H. simus, H. aureus,* and *H. griseus alaotrensis* as critically endangered and *H. griseus occidentalis,* found in a small part of west-central Madagascar, as vulnerable. Mittermeier et al. (1994) estimated a population of more than 10,000 individuals for *H. g. occidentalis,* which is jeopardized by human destruction of forest habitat. The population of *H. g. alaotrensis,* consisting of about 7,500 animals, is threatened by burning of the reed beds in which it lives, draining of parts of Lake Alaotra for agricultural purposes, and hunting (Harcourt and Thornback 1990; Mutschler and Feistner 1995). *H. g. alaotrensis* also seems to be a popular pet (Feistner and Rakotoarinosy 1993). *H. g. griseus,* which occurs throughout the eastern forests of Madagascar, numbers more than 100,000; it is generally declining through loss of forest habitat and excessive hunting by people but may now occur at higher than original densities in certain areas where bamboo has replaced the original forest. A fourth subspecies of *H. griseus (H. g.*

meridionalis) was recently reported to occur at one locality in the extreme southeast but was not accepted as a distinct form by Mittermeier et al. (1992, 1994) or Harcourt and Thornback (1990).

Skeletal remains indicate that *H. simus* may still have been present in the Ankarana region at the northern tip of Madagascar until less than 50 years ago, and there are some indications that it still exists there (Wilson et al. 1989). Otherwise *H. simus* and *H. aureus* are known to survive only in a few small areas east of Fianarantsoa in southeastern Madagascar, where they are critically endangered by logging, clearing of forest habitat for agriculture, and direct hunting by people (Meier and Rumpler 1987). Both species are estimated to number only about 1,000 individuals and are considered to be of the highest priority for conservation attention (Mittermeier et al. 1992).

PRIMATES; LEMURIDAE; **Genus LEMUR**
Linnaeus, 1758

Ring-tailed Lemur

The single species, *L. catta,* is found in southern Madagascar south and west of a line running approximately from the vicinity of Morondava on the west coast, east to Fianarantsoa, and then south to Taolanaro (Harcourt and Thornback 1990; Tattersall 1982). Several other species usually placed in *Lemur* are here assigned to the separate genera *Eulemur* and *Varecia* (see accounts thereof). Simons and Rumpler (1988) considered *L. catta* to be more closely related to *Hapalemur* than to *Eulemur.*

Head and body length is 385–455 mm, tail length is 560–624 mm, and weight is about 2.3–3.5 kg (Tattersall 1982). The upper parts are brownish gray, the underparts are whitish, and the tail is ringed with black and white. The palms and soles are long, smooth, and leatherlike, affording a firm footing on slippery rocks, and the great toe is considerably smaller than in the various species formerly assigned to *Lemur,* which are more arboreal.

According to Simons and Rumpler (1988), *Lemur* resembles *Hapalemur* but differs from *Eulemur* in having an antebrachial (carpal) gland on the forearm as well as brachial glands. Males rub fatty secretions from these various glands onto the tail in order to disperse a scent during agonistic interaction. The karyotype of *Lemur* also resembles that of *Hapalemur* and differs from that of *Eulemur.* Tattersall and Schwartz (1991) suggested that the resemblance of *Lemur* and *Hapalemur* was the result of parallel evolution and that numerous cranial and dental characters indicate that *Lemur* is more closely related to *Eulemur.* Among the characters shared by *Lemur, Eulemur,* and *Varecia* are a laterally swollen, prenasopalatine portion of the palatine bone and at least some posterior expansion of this element; a paranasal sinus; at least some obscuring of the maxilla in the lateral part of the orbital floor; an infraorbital foramen lying anterior to the lacrimal foramen; the maxilla unexposed in the medial orbital wall; and some cingular development on the first and second upper molars.

The ring-tailed lemur is capable of arboreal activity but is partly terrestrial and sometimes is found in thinly wooded country. It ranges farther into the interior highlands of Madagascar than any other lemur. In a study in southwestern Madagascar, Sussman (1975, 1977) observed *L. catta* to spend more time on the ground than within any one level of the trees. Groups foraged intensively, covering 900–1,000 meters a day, and took only a short rest at midday. Jolly (1966) reported that *L. catta* may be moderately

Ring-tailed lemur *(Lemur catta)*, photo by Bernhard Grzimek.

active at any time of the night, though like most species of *Lemur,* it is primarily diurnal. The diet includes mostly fruits, some leaves and other plant parts, and only rarely insects (Hladik 1979).

Most reported population densities are from about 150/sq km to 350/sq km, though there is one estimate of only 17.4/sq km in an area of disturbed forest (Harcourt and Thornback 1990; Pollock 1979; Sussman 1975, 1977; Sussman and Richard 1974). Jolly (1966) found a group of 20 to have a home range of 5.7 ha. In southwestern Madagascar groups used home ranges of 6.0–8.8 ha. (Sussman 1977; Sussman and Richard 1974). Budnitz and Dainis (1975) reported that troops inhabiting closed canopy and open forest had home ranges of 6 ha. and 8.1 ha., whereas a troop occupying an area of brush, scrub, and open forest utilized 23.1 ha. Sauther and Sussman (1993) stated that known home range size is 6–35 ha., with those in wet habitats averaging about twice the size of those in drier areas. Group home ranges overlap, with few or no areas of exclusive use.

In her study of *L. catta* Jolly (1966) found group sizes of 12–24, no consistent group leadership, and considerable agonistic activity, including some fighting. The sexes had separate dominance hierarchies, and the females dominated the males. Budnitz and Dainis (1975) found group size to range from 5 to 22, averaging 12.8 before the birth season and 17.3 after. They reported the basic troop to be organized around a core group of adult females and their infants, young juveniles, and sometimes 1 or more dominant males. The average troop comprised 6 adult females, 4 adult males, and 4 young. There was no constant leader of the troop, but adult females dominated others and seemed responsible for territorial defense, while males were general-

ly peripheral to group activity. Females remained in the troop of their birth, but males moved among troops. Sauther and Sussman (1993) reported that males emigrate from their natal group upon reaching adulthood, that adult males usually change groups every 3–5 years, and that each group appears to have a single "central male" that interacts with the females at a greater rate than do other males and that is the first to mate.

Jolly (1966) reported that troops of *L. catta* have well-defined, nonoverlapping territories within their overall home ranges. However, Sauther and Sussman (1993) questioned whether the species is truly territorial; there is almost total overlap of the ranges used by different groups, though certainly there is vigorous defense of areas being used at a given time. Resulting disputes generally involve two opposing groups of females running at each other and vocalizing, but direct physical contact is rare (Budnitz and Dainis 1975). Jolly (1966) identified 15 different vocalizations, including a howl audible to humans at a distance of 1,000 meters.

Like other lemurs, *L. catta* mates from about April to June and gives birth from August to October, just before or at the beginning of the rainy season (Petter 1965; Pollock 1979). In any one area, however, mating and births may be synchronized to occur within a period of a few days (Jolly 1966). Most females give birth annually in the wild (Sauther and Sussman 1993). The estrous cycle averages 40 days, estrus lasts 1 day, and weight at birth is 50–80 grams (Hayssen, Van Tienhoven, and Van Tienhoven 1993). Doyle (1979) listed a precise gestation period of 136 days. According to Van Horn and Eaton (1979), females usually produce a single young, though twins are not rare. For the first two weeks of life the young clings to the underside of the mother and rides longitudinally; subsequently it rides on her back (Tattersall 1982). Infants may suckle until they are 5 months old, but they begin to take some solid food during the second month of life. Female *L. catta* first conceive at an average age of 19.56 months (Van Horn and Eaton 1979). Males are sexually mature at 2.5 years but may not be allowed to mate then by the older males (Budnitz and Dainis 1975). A captive specimen was still living at about 33 years (Marvin L. Jones, Zoological Society of San Diego, pers. comm., 1995).

According to Mittermeier et al. (1994), *L. catta* generally has been considered common, but its preferred habitats—gallery forests along rivers and *Euphorbia* bush—are disappearing rapidly because of fires, overgrazing by livestock, and cutting of trees for charcoal production. The species can only survive in primary vegetation. Recent surveys indicate that suitable habitat has been diminished alarmingly through human activity, that hunting pressure is severe, and that the species may be more threatened than was thought previously (Harcourt and Thornback 1990). Mittermeier et al. (1992) considered its conservation of high priority and estimated its total numbers at only 10,000–100,000 individuals. *L. catta* now is classified as vulnerable by the IUCN and as endangered by the USDI and is on appendix 1 of the CITES.

Ruffed lemur *(Varecia variegata)*, photo from San Diego Zoological Society.

PRIMATES; LEMURIDAE; **Genus VARECIA**
Gray, 1863

Variegated Lemur, or Ruffed Lemur

The single species, *V. variegata,* is found in the forests of eastern Madagascar. There are two distinctive subspecies, *V. v. ruber* (red ruffed lemur) on the Masoala Peninsula east of the Antainambalana River and *V. v. variegata* (black and white ruffed lemur) from the Antainambalana River in the north to the Mananara River in the south (Harcourt and Thornback 1990; Mittermeier et al. 1994; Tattersall 1982). Until recently this species usually was placed in the genus *Lemur,* but because of various distinctive anatomical and behavioral characteristics it now is considered to represent a separate genus (Groves and Eaglen 1988; Petter and Petter-Rousseaux 1979; Rumpler 1974). The extinct genus *Pachylemur* (see account thereof) often has been included in *Varecia.*

V. variegata is the largest living member of the family Lemuridae. Head and body length is 510–60 mm, tail length is 560–650 mm, and weight is about 3.2–4.5 kg (Tattersall 1982). The fur is long and soft, and the ears are hid-den by a ruff of hair. The color pattern varies and may be different on the right and left sides of a specimen. In the subspecies *V. v. variegata* much of the pelage is black, but there are large white areas on the limbs, back, and head. In the subspecies *V. v. ruber* the pelage is mostly red, there may be white markings on the limbs, and the tail and bel-ly are black.

Varecia differs from *Lemur* and *Eulemur* in dermato-glyphic pattern and the presence of a marking gland on the neck (Rumpler 1974). Whereas the other genera have only a single pair of mammae, *Varecia* has three pairs. Accord-ing to Tattersall and Schwartz (1991), *Varecia* shares nu-merous cranial and dental characters with *Lemur* (see ac-count thereof) and *Eulemur.* From the others, however, *Varecia* can be distinguished by its elongate talonid basins in the lower molars, the absence of entoconids on the low-er molars, the protocone fold of the first upper molar, the posterolingual opening of the talonid basin of the second lower molar, and the anterior expansion of the lingual cin-gulum of the first and second upper molars.

The ruffed lemur is an arboreal forest dweller. It nor-mally progresses by walking or running on larger branch-es and makes leaps from tree to tree, but its locomotion is more labored and cautious than that of *Lemur* (Walker 1979). It is crepuscular, being most active from 1700 to 1900 hours, when its peculiar calls are commonly heard (Petter

Black and white ruffed lemur *(Varecia variegata variegata)*, photo by David Haring.

of rubbing the chest, chin, and neck onto the substrate (Pereira, Seeligson, and Macedonia 1988).

The reproductive pattern of this genus differs from those of *Lemur* and *Eulemur* in a number of ways (Bogart, Cooper, and Benirschke 1977; Boskoff 1977; Harcourt and Thornback 1990; Hayssen, Van Tienhoven, and Van Tienhoven 1993; Klopfer and Boskoff 1979; Mittermeier et al. 1994; Pollock 1979). The reported gestation period of *Varecia* is only 90–102 days, considerably shorter than that of any species of the other genera. The estrous cycle lasts about 30 days, with estrus averaging 6.25 days. On Nosy Mangabe Island most mating occurs in June and July, births in September and October. More than half of the births are of twins, and the remainder are mostly single or of triplets, but litters with as many as six young have been reported in captivity. Birth weight is 80–100 grams. The young are initially carried in the mouth of the female and are often deposited in a convenient place while she forages. By 5 weeks the young can climb to the tops of trees. Weaning occurs at around 135 days. Females may become pregnant at 20 months, but average age at first reproduction in captivity is 3.4 years. According to Marvin L. Jones (Zoological Society of San Diego, pers. comm., 1995), a number of specimens have lived in captivity for more than 25 years and one was still living at about 33 years.

The ruffed lemur is listed as endangered by the USDI and is on appendix 1 of the CITES. It reportedly is declining be-

and Charles-Dominique 1979; Pollock 1979). The diet consists largely of fruits. On the Masoala Peninsula, animals moved an average of 436 meters per day but generally remained in one area with large fruit trees for several weeks before shifting to another area (Rigamonti 1993). Population densities of 20–30/sq km have been reported on Nosy Mangabe Island, a protected reserve, but apparently are much lower elsewhere (Harcourt and Thornback 1990).

There apparently is some variation in social organization. In southeastern Madagascar an adult male and female formed a cohesive pair and foraged through a home range of 197 ha. (Harcourt and Thornback 1990). Two groups on the Masoala Peninsula consisted of 5 and 6 members, used home ranges of about 25 ha., and during the cool wet season (May–August) fragmented into subgroups that used different core areas (Rigamonti 1993). A study on Nosy Mangabe Island indicated that groups consist of 8–16 individuals, all members use a common home range, groups are aggressive toward one another, and females form the core of the group and defend its home range; subgroups of 2–5 individuals formed during the cool season, possibly representing mated pairs and offspring (Morland 1991). There are a variety of vocalizations, the most characteristic being an intense roar of alarm and a powerful plaintive call for territorial expression (Petter and Charles-Dominique 1979). Both sexes scent-mark, females only with the anal-genital region but males mainly with a unique process

Red ruffed lemur *(Varecia variegata ruber)*, photo by David Haring.

cause of human destruction of its forest habitat, hunting for use as food, and commercial exportation (Richard and Sussman 1975; Wolfheim 1983). Although the species occupies a relatively large range, there are few well-protected areas therein (Mittermeier et al. 1992). Each of the two subspecies is thought to number only 1,000–10,000 individuals in the wild (Mittermeier et al. 1994), and there are another 859 in captivity (Olney, Ellis, and Fisken 1994). The IUCN now classifies *V. v. ruber* as critically endangered, forecasting a decline of at least 80 percent over the next decade, and *V. v. variegata* as endangered.

PRIMATES; LEMURIDAE; Genus PACHYLEMUR
Lamberton, 1948

The single species, *P. insignis,* is known only by subfossil material from various sites in northern, central, and southern Madagascar (Jenkins 1987; Simons et al. 1990; Tattersall 1982). *P. jullyi,* named on the basis of subfossil material from central Madagascar, sometimes is treated as a separate species but is here considered at most subspecifically distinct from *P. insignis.* These species commonly have been referred to the genus *Varecia,* together with the living species *V. variegata. Pachylemur,* originally described as a subgenus for *Varecia insignis* and *V. jullyi,* was accepted as a full genus by Mittermeier et al. (1994), Ravosa (1992), and Simons et al. (1990).

Based on skeletal remains, *Pachylemur* resembles *Varecia* but is somewhat larger, having a cranial length of 114.5–126.0 mm, compared with 97.2–110.7 mm in the latter genus. According to Mittermeier et al. (1994), the skull structure of *Pachylemur* is very close to that of *Varecia,* but its postcranial bones have a heavier build. This condition may suggest a more terrestrial way of life than that of the highly arboreal *Varecia.* Tattersall (1982) noted that the skull of *Pachylemur* also can be distinguished by the presence of sagittal and nuchal cresting and by the more forward orientation of its orbits. In a statistical analysis of cranial and dental measurements, Ravosa (1992) found substantial differences between the two genera. The skull of *Pachylemur* is relatively broader, the jaws more massive, and the molar teeth larger. These adaptations may indicate that the diet of *Pachylemur* was more obdurate or fibrous than that of *Varecia.*

Pachylemur has been found at sites dated at about 2,000–1,000 years ago, but there is little information about when or why it finally disappeared. Except for *Mesopropithecus* (see account thereof), it was the smallest of the subfossil Malagasy prosimians and it may have survived until about the same time. If it was terrestrial, it probably was more susceptible than *Varecia* to hunting by people.

PRIMATES; LEMURIDAE; Genus EULEMUR
Simons and Rumpler, 1988

Lemurs

There are five species (Harcourt and Thornback 1990; Mittermeier et al. 1994; Petter and Petter-Rousseaux 1979; Tattersall 1982, 1993):

E. coronatus (crowned lemur), extreme northern
   Madagascar;
E. rubriventer (red-bellied lemur), eastern rainforest zone
   of Madagascar;

Black lemurs *(Eulemur macaco),* photo by Bernhard Grzimek.

E. macaco (black lemur), northwestern Madagascar north
   of Narinda Bay;
E. mongoz (mongoose lemur), northwestern Madagascar
   south of Narinda Bay, Moheli and Anjouan islands in
   the Comoros (probably introduced);
E. fulvus (brown lemur), Madagascar, Mayotte Island in
   the Comoros (probably introduced).

These species usually have been placed in the genus *Lemur,* but Simons and Rumpler (1988) concluded that they should be put in a genus different from the type species of *Lemur, L. catta,* and proposed the name *Eulemur* for such a genus. Almost simultaneously, Groves and Eaglen (1988) independently came to a similar conclusion, proposing the name *Petterus* for the same group of species. Most subsequent authorities, including Groves (*in* Wilson and Reeder 1993), have accepted *Eulemur* as the appropriate name, though Corbet and Hill (1991) used *Petterus.* Based on detailed analyses of craniodental characters, Tattersall (1993)

Crowned lemur *(Eulemur coronatus)*, photo by David Haring.

and Tattersall and Schwartz (1991) recommended that the species of *Eulemur* be restored to the genus *Lemur*. However, *Eulemur* was maintained as a separate genus by Crovella, Montagnon, and Rumpler (1993), Shedd and Macedonia (1991), and Mittermeier et al. (1992, 1994). Matings between *E. fulvus* and some other species of *Eulemur* have resulted in fertile offspring (Ratomponirina, Andrianivo, and Rumpler 1982; Tattersall 1993).

The species are allopatric, except for *E. fulvus,* which overlaps all the others and is well differentiated from them. According to Tattersall (1982), its head and body length is 380–500 mm, tail length is 465–600 mm, and weight is 2.1–4.2 kg. There is much variation in color, and in some subspecies there is sexual dichromatism. The upper parts are usually gray or brown, the underparts are paler, and the tail often darkens distally. The face is usually dark, though there may be light patches above the eyes. In some subspecies the head is mostly white or pale gray, and in some the ears are tufted. All subspecies have mystacial, submental, superciliary, and carpal vibrissae, and in all the face is clothed in short hair except at the tip of the muzzle. The circumanal region is distinguished by an area of naked, wrinkled, glandular skin. Females have one or two pairs of mammae, though only the anterior pair are functional.

In the other four species head and body length is 300–450 mm, tail length is 400–640 mm, and weight is about 2–3 kg. The tail is slightly shorter than the head and body in some forms but longer in others. The fur is soft and relatively long, and there is a pronounced ruff about the neck and ears. The coloration varies considerably. Some forms are speckled reddish brown or gray; others are more or less reddish, brownish, or blackish throughout. In *E. macaco* males are usually black, and females, brown.

According to Simons and Rumpler (1988), *Eulemur* differs from *Lemur* and *Varecia* in having a hairy, not naked, scrotum. *Eulemur* lacks the antebrachial and brachial glands of *Lemur* and *Hapalemur* but possesses perianal glands not seen in *Varecia, Lemur,* or *Hapalemur.* It is fur-

ther distinguished from *Lemur* by a suite of dental characters that include smaller third upper and lower molars, distinct protostyle development on the first and second molars, a more developed anterior basin in the posterior lower premolars, and more continuous crests in the trigonids and talonids of the molars. Groves and Eaglen (1988) added that *Eulemur* is unique among the Lemuridae in having the paranasal air sinuses much dilated and the interorbital region projecting above the plane of the remainder of the skull roof, creating a bubblelike effect in its cranial profile. *Eulemur* resembles *Varecia* in the reduction of the anterior or upper premolar and in lacking any lingual outlet for the talonid basin. Tattersall and Schwartz (1991) pointed out that *Eulemur* shares numerous other cranial and dental characters with *Varecia* and *Lemur* (see account thereof).

All species are arboreal forest dwellers. *L. coronatus* occurs mainly in dry forest and *L. rubriventer* in rainforest, but all species show some adaptability to various habitats, including moderately disturbed areas (Harcourt and Thornback 1990). Lemurs are active, quadrupedal animals that run and walk on horizontal and diagonal branches and are capable of leaping to and from vertical and horizontal supports. Resting postures range from sitting upright to lying sprawled on a horizontal branch. Movement on the ground is normally by quadrupedal walking, running, or galloping, but short bouts of bipedal running have been observed (Walker 1979). Most activity is diurnal, but all species are sometimes active by night. *E. mongoz* has been found to be exclusively nocturnal in some places in Madagascar and the Comoros but largely diurnal in others; its shift from day to night activity may coincide with the transition from the rainy to the dry season (Harcourt and Thornback 1990; Harrington 1978; Tattersall 1978a). The diet of *Eulemur* consists largely of flowers, fruits, and leaves (Tattersall 1982). *E. macaco* also is known to eat bark, and *E. mongoz* feeds extensively on nectar (Harcourt and Thornback 1990). Captive *E. macaco* will accept meat (Kolar *in* Grzimek 1990).

Red-bellied lemur *(Eulemur rubriventer)*, photo by Russell A. Mittermeier.

In a study in southwestern Madagascar, Sussman (1975, 1977) found *E. fulvus* to contrast sharply with *Lemur catta* in that it spent 95 percent of its time in the tops of trees, rarely descending to the ground. It was relatively sedentary, staying within only a few trees and covering about 125–50 meters in its daily foraging. It rested most of the afternoon and had peaks of feeding activity from 0600 to 0930 and 1700 to 1825 hours. In a study on Mayotte Island in the Comoros, Tattersall (1977a) also found the brown lemur to prefer the upper levels of the forest but found it to be active both day and night and to range 450–1,150 meters per day. The diet consisted predominantly of kily leaves in southwestern Madagascar and fruits on Mayotte Island.

Population densities as great as 900–1,000/sq km have been reported for *E. fulvus* (Sussman 1975, 1977), though such figures would apply only in small remaining areas of quality habitat, and densities as low as 40–60/sq km have been reported for the species in some areas (Harcourt and Thornback 1990). Other recorded densities are 58–200/sq km for *E. macaco* (Colquhoun 1993; Tattersall 1982), 350/sq km for *E. mongoz* (Wolfheim 1983), and 50–500/sq km for *E. coronatus* (Harcourt and Thornback 1990; Mittermeier et al. 1994). Lemurs appear to be territorial. A group of 12 *E. fulvus* utilized an area of 7 ha. in northwestern Madagascar (Harrington 1975). Other studies of that species have indicated home ranges as large as 14 ha. (Mittermeier et al. 1994). Groups of 5–14 *E. macaco* occupied ranges of 3.5–7.0 ha. (Colquhoun 1993). In *E. fulvus* and *E. mongoz* home ranges overlap extensively, but troops avoid one another or engage in disputes over feeding areas through vocalization and gestures (Harrington 1975; Sussman 1975; Tattersall 1978a). There is a rich repertoire of

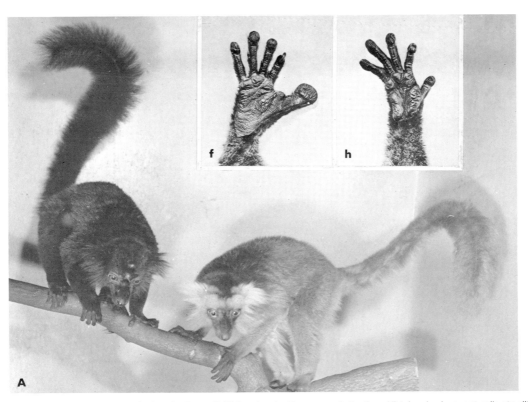

Black lemurs *(Eulemur macaco)*, photo by Ernest P. Walker. Insets: *E. mongoz*, photos from *Histoire physique, naturelle et politique de Madagascar*, Alfred Grandidier.

Mongoose lemur *(Eulemur mongoz)*, photo by David Haring.

calls for contact, greeting, territorial expression, threats, and alarm (Petter and Charles-Dominique 1979).

Social organization in *Eulemur* is variable. In southwestern Madagascar 18 groups of *E. fulvus* contained an average of 9.4 (4–17) individuals. These groups appeared

Brown lemur *(Eulemur fulvus)*, photo by David Haring.

cohesive and peaceful, agonistic activity was rare, and no dominance hierarchy was observed (Harrington 1975; Sussman 1975, 1977). On Mayotte Island, Tattersall (1977*a*) found group size to average 9.1 (2–29) individuals, but the groups changed in composition from day to day. Investigation of *E. macaco* indicates a group size of 4–15 individuals, leadership by a female, and some exchange of adults between troops (Pollock 1979). At least in some areas, *E. mongoz* has been found to live in small family groups comprising a permanently bonded adult pair and not more than 2 young (Pollock 1979; Tattersall 1978*a*). *E. rubriventer* has been observed in groups of 5 or fewer individuals (Pollock 1979). *L. coronatus* and *L. rubriventer* generally occur in groups of 2–10 animals, including several adults of both sexes (Harcourt and Thornback 1990).

It seems likely that all species of *Eulemur* throughout Madagascar mate from about April to June, have a gestation period of about 4.5 months, and give birth from August to October, just before or at the beginning of the rainy season (Petter 1965; Pollock 1979). In any one area, however, mating and births may be synchronized to occur within a period of a few days (Jolly 1966). Doyle (1979) listed the following precise gestation periods: *E. fulvus,* 117 days; *E. macaco,* 127 days; *E. mongoz,* 128 days; and *E. rubriventer,* 127 days. Hayssen, Van Tienhoven, and Van Tienhoven (1993) listed birth weights of 60–70 grams, estrous cycles of about 1 month, and estrus lengths of 1–2 days. There usually is a single young, though twins are not rare. For the first four weeks of life the young grips the mother's fur and rides ventrally in a transverse position; subsequently it rides on her back (Tattersall 1982). Young *E. rubriventer* frequently are carried by the father (Mittermeier et al. 1994). Doyle (1979) reported weaning to occur at 135 days and sexual maturity to be attained at 548 days in *E. fulvus* and *E. macaco.* In *L. coronatus,* which seems to have a few unusual characteristics, gestation is 125 days, some births evidently occur in January and May, twins appear to be as common as singletons, and both males and females reach sexual maturity at 20 months (Harcourt and Thornback 1990). Lemurs thrive in captivity and are often exhibited in zoos. A specimen of *E. fulvus* lived to more than 36 years and a captive hybrid *E. macaco* × *E. fulvus* survived for 39 years (Marvin L. Jones, Zoological Society of San Diego, pers. comm., 1992).

Red-bellied lemur (Eulemur rubriventer), photo by David Haring.

area is rapidly being reduced and fragmented by logging, burning, and grazing, and poaching is rampant; *L. macaco flavifrons* has an even smaller area of natural habitat, most of which has been converted to agriculture; the mainland population of *L. mongoz* is declining as forests are cleared for pastures and charcoal production. The populations of *E. mongoz* and *E. fulvus* on the Comoro Islands, which probably were introduced through human agency long ago, face the same threats as those faced by the lemurs in Madagascar (Tattersall 1977b). In addition, there appears to have been a recent increase in the frequency of cyclones striking the islands, which have a devastating effect on the remaining natural vegetation (Harcourt and Thornback 1990). The estimated number of *E. fulvus* on Mayotte fell from about 50,000 in 1975 to less than half that in 1991 (Mittermeier et al. 1994).

PRIMATES; **Family MEGALADAPIDAE**

Sportive Lemurs, or Weasel Lemurs, and Koala Lemurs

This family of one living and one Recently extinct genus and 10 species is found only in Madagascar. The living genus, *Lepilemur*, long was placed in the family Lemuridae. Recent systematic investigations involving especially cytogenetics indicated that it represents a separate family that took the name Lepilemuridae (Petter and Petter Rousseaux 1979; Rumpler 1975; Rumpler and Albignac 1975); however, it now seems that *Lepilemur* belongs in the same family as the subfossil *Megaladapis* and that the correct name for the resulting group is Megaladapidae (Buettner Janusch and Tattersall 1985; Tattersall 1982, 1986). Some authorities, including Corbet and Hill (1991), have continued to place *Lepilemur* in the Lemuridae, but new studies by Groves and Eaglen (1988) indicate that *Lepilemur* is more closely related to the Indriidae than to the Lemuridae. Other authorities, including Jenkins (1987), while acknowledging affinity between *Lepilemur* and *Megaladapis*, consider the latter to warrant placement in a separate family. Tattersall (1982) recognized three subfamilies: the Lepilemurinae, with *Lepilemur*; the Hapalemurinae, with *Hapalemur*; and the Megaladapinae, with *Megaladapis*. However, Schwartz and Tattersall (1985) and Groves and Eaglen (1988) regarded *Hapalemur* to be a member of the Lemuridae, to which it also has been assigned by most other authorities.

Lepilemur and *Megaladapis* differ markedly in size and certain other morphological characters, and the latter genus probably lived in a manner unlike that of modern lemurs. In the skull of both genera there is a median vertical articular area in addition to the transverse articular surface of the mandible (Jenkins 1987). The close relationship between the two genera also is indicated by the dentition (Tattersall 1982). Each has the same dental formula as that of the Lemuridae, but in both *Lepilemur* and *Megaladapis* the adults lose the upper incisors, so that there are only 32 teeth. The cheek teeth of both genera show much resemblance. The premolars broaden posteriorly, and the upper molars display buccal cingula and parastyles, lingual buttressing of the paracones and metacones, and distal displacement of their lingual moieties. Differences between *Lepilemur* and the Lemuridae have been found in studies of dermatoglyphics, hematology, parasitology, behavior, and cytogenetics (Petter and Petter-Rousseaux 1979).

Geologically the Megaladapidae are known only from

All species of *Eulemur* are listed as endangered by the USDI and are on appendix 1 of the CITES. The IUCN classifies most species and subspecies of the genus. Those classifications, together with population estimates by Mittermeier et al. (1992, 1994), are: *E. fulvus albifrons* (found in northeast Madagascar), more than 100,000; *E. fulvus albocollaris* (southeast), endangered, 1,000–10,000; *E. fulvus collaris* (southeast), vulnerable, 10,000–100,000; *E. fulvus fulvus* (northwest and east-central), more than 100,000; *E. fulvus mayottensis* (Mayotte Island, not recognized as a valid taxon by Mittermeier et al.), fewer than 25,000; *E. fulvus rufus* (west and southeast), more than 100,000; *E. fulvus sanfordi* (extreme north), vulnerable, 10,000–100,000; *E. macaco flavifrons*, critically endangered, 100–1,000; *E. macaco macaco*, vulnerable, 10,000–100,000; *E. mongoz*, vulnerable, 1,000–10,000; *E. coronatus*, vulnerable, 10,000–100,000; *E. rubriventer*, vulnerable, 10,000–100,000.

All of these populations are considered to be declining through loss of forest habitat to industrial activity, logging for local use and export, plantations, and slash-and-burn agriculture (Mittermeier et al. 1992, 1994; Richard and Sussman 1975; Wolfheim 1983). The relatively high population densities reported in field studies may be misleading in that they generally apply only to small, protected reserves. In addition to habitat destruction, certain forms reportedly are threatened by intensive hunting and trapping for use as food and because of their alleged raids on crops. Harcourt and Thornback (1990) reported the following for some of the most critically endangered forms: *L. coronatus* has less than 1,300 sq km of suitable habitat remaining, the

the Recent of Madagascar, subfossil remains having been found at sites dated nearly as far back as 3,000 years ago. However, Tattersall (1982) indicated close affinity between the Megaladapidae and the prosimian family Adapidae of the Eocene of Europe.

PRIMATES; MEGALADAPIDAE; Genus LEPILEMUR
I. Geoffroy St.-Hilaire, 1851

Sportive Lemurs, or Weasel Lemurs, and Lemurs

There are seven species (Petter and Petter *in* Meester and Setzer 1977; Petter and Petter-Rousseaux 1979; Rumpler 1975; Rumpler and Albignac 1975, 1977):

L. dorsalis, extreme northwestern Madagascar and Nosy-Be Island;
L. ruficaudatus, southwestern Madagascar;
L. edwardsi, west-central Madagascar;
L. leucopus, extreme southern Madagascar;
L. mustelinus, northern part of eastern forests of Madagascar;
L. microdon, southern part of eastern forests of Madagascar;
L. septentrionalis, extreme northern tip of Madagascar.

Compiled from the sources cited above, this list of seven species is based to a large extent on studies involving cytogenetics. Tattersall (1982) questioned the validity of such work and tentatively followed the more traditional view that there is only a single species, *L. mustelinus.* That position also was taken by Corbet and Hill (1991), but the above seven species of *Lepilemur* were accepted by Groves (1989 and *in* Wilson and Reeder 1993), Harcourt and Thornback (1990), and Mittermeier et al. (1992, 1994).

Head and body length is 300–350 mm, tail length is 255–305 mm, and weight is about 500–900 grams. The upper parts are rufous, brown, or gray, and the underparts are white or yellowish. The head is conical and short; the ears are large, round, and membranous; the feet are only slightly elongated, and the fourth and fifth toes are the longest. Except for the nail of the great toe, which is large and flat, the nails are keeled. Females have a single pair of mammae located pectorally.

These lemurs are arboreal forest dwellers and strictly nocturnal (Pollock 1979). They normally move by rapid leaps from one vertical support to another using a powerful extension of the hind limbs. On the ground they usually progress with bipedal hops, as do kangaroos (Walker 1979). During the day they sleep rolled up in a ball in a hollow tree or thick foliage. Their diet is folivorous, consisting mostly of leaves and flowers (Hladik 1979; Klopfer and Boskoff 1979). The species *L. leucopus* feeds mainly on thick, juicy leaves and reingests part of its fecal material, as do some lagomorphs (Hladik 1978).

Most species are not well known, but Harcourt and Thornback (1990) compiled population density estimates of 57/sq km for *L. edwardsi,* 13–100/sq km for *L. microdon,* 60–564/sq km for *L. septentrionalis,* and 180–350/sq km for *L. ruficaudatus.* Several species have been found to occupy defended home ranges about 1 ha. in size. Two or three individual *L. edwardsi* may sleep together in a tree hole by day, but they move about separately by night.

Field studies in southern Madagascar have provided substantial information on *L. leucopus* (Charles-Do-

Lemur (*Lepilemur* sp.), photo by Jean-Jacques Petter.

minique 1974; Hladik 1978; Hladik and Charles-Dominique 1974; Petter and Charles-Dominique 1979; Pollock 1979). Population densities vary from about 200/sq km to 810/sq km. Home ranges are small and apparently coincide with stable, well-defined territories; those of adult females average 0.18 (0.15–0.32) ha. and those of males average 0.30 (0.20–0.46) ha. The territory of a large male may overlap those of up to five females, but that of a small male overlaps those of only one or two females. The animals are basically solitary, except that there may be some association between related females. Territorial defense against members of the same sex is very pronounced and occupies the greater part of an individual's time during the night. Since a territory is relatively small, the entire area can be surveyed from a high branch by the resident, and animals may spend hours observing one another. Defense involves visual displays, vocalization, chases, and sometimes severe fighting. A variety of calls ranging from weak squeals to powerful, high-pitched sounds function in distant communication and territoriality.

The mating season is May–August. A single young is born in the period from mid-September to December and is raised in a nest within a hollow tree. The gestation period has been calculated as 120–50 days (Klopfer and Boskoff 1979). Birth weight is about 50 grams. The female often leaves the young clinging to a branch when she forages. Weaning occurs at about 4 months, but the young may follow its mother until it is more than 1 year old. Sexual maturity is attained at around 1.5 years. Captives have lived for 12 years (Doyle 1979; Kolar *in* Grzimek 1990).

Like the other Malagasy prosimians, all species of *Lepilemur* are listed as endangered by the USDI and are on appendix 1 of the CITES. The IUCN classifies *L. dorsalis* and *L. septentrionalis* as vulnerable. These animals are threatened by loss of forest habitat to slash-and-burn agriculture, clearing for settlement, burning to encourage growth of pastures, and overgrazing by cattle and goats (Harcourt and Thornback 1990). Most species also are subject to hunting for use as food (Mittermeier et al. 1994). Total populations have been estimated to number more than 100,000 individuals for *L. edwardsi* and *L. leucopus* and 10,000–100,000 for each of the other five species (Mittermeier et al. 1992).

PRIMATES; MEGALADAPIDAE; Genus **MEGALADAPIS**
Forsyth Major, 1894

Koala Lemurs

Vuillaume-Randriamanantena et al. (1992) recognized two subgenera and three species, all extinct:

subgenus *Peloriadapis* Grandidier, 1899

M. edwardsi, known by subfossil material from
    southwestern and extreme southeastern Madagascar;

subgenus *Megaladapis* Forsyth Major, 1894

Koala lemur *(Megaladapis edwardsi)*, reconstruction by Stephen D. Nash.

M. madagascariensis, known by subfossil material from
    southwestern and perhaps northern Madagascar;
M. grandidieri, known by subfossil material from central
    and perhaps northern Madagascar.

Remains referable to the subgenus *Megaladapis* have been found at four sites in northern Madagascar but have not yet been distinguished by species.

The three species are estimated to have ranged in weight from about 40 kg to 80 kg (Mittermeier et al. 1994). *M. edwardsi,* with a cranial length of 277–317 mm, is among the largest of the known prosimians. It also is distinguished by its huge molar teeth, far larger than would be expected in an animal of its overall body size. Cranial length is 235–44 mm in *M. madagascariensis* and 273–300 mm in *M. grandidieri.* The cranium is relatively narrow and greatly elongated. The facial region is long, the orbits divergent, the braincase remarkably small, and the auditory bullae flat. The nasal bones are long and project well beyond the anterior end of the palate, possibly indicating the presence of a mobile snout in life. The zygomatic arches are massive, and there are strong nuchal and sagittal crests. The foramen magnum is backward-facing, and the occipital condyles are oriented perpendicularly to the cranial base. Adults have no upper incisor teeth; in their place are bony ridges suggesting the presence in life of a horny pad such as is found in some ruminants. The molars have complex cusps and increase strikingly in size from front to rear. The skull is disproportionately large in relation to the short, stocky postcranial skeleton. The forelimbs are longer than the hind limbs, and all four extremities are long and somewhat curved, clearly having been powerful grasping organs (Jenkins 1987; Tattersall 1978*b*, 1982; Tat-

tersall and Schwartz 1975; Vuillaume-Randriamanantena et al. 1992).

Both the cranial and the postcranial morphology of *Megaladapis* indicate that its locomotion and lifestyle paralleled that of *Phascolarctos,* the living koala of Australia. It evidently clung to tree trunks and branches with all four limbs, moved upward by a series of short hops, and crossed to nearby trees by short leaps. It presumably fed by cropping leaves pulled by the forelimbs within reach of the mouth (Jenkins 1987; Preuschoft 1971; Tattersall 1982). The species *M. grandidieri* and *M. madagascariensis* had postcranial specializations suggesting greater flexibility of limbs and perhaps more pronounced arboreal adaptation than in *M. edwardsi.* Newly discovered pedal remains suggest also that hind limb suspension may have been an important behavior of the former two species (Vuillaume-Randriamanantena et al. 1992).

Megaladapis is known from sites with radiocarbon dates of about 2,850 to 600 years ago (Culotta 1995). Humans invaded Madagascar around the middle of this period, probably then moved across the island with flocks of domestic livestock, and also introduced various suids that became feral (Dewar 1984). The resulting hunting pressure and environmental disruption, especially the elimination of natural forest habitat, would have been disastrous for a huge, arboreal, and slow lemur that had evolved without predators. There have been recent suggestions that severe drought may already have reduced the lemur population, which thus would have been especially vulnerable to human disturbances, and that people also may have brought a lethal disease to Madagascar (Culotta 1995). *Megaladapis* and the other Malagasy primates known only as subfossils probably became extinct prior to the arrival of Europeans about 500 years ago. There is, however, a seventeenth-century description of a living animal that could fit any of several of the larger extinct genera (Tattersall 1982).

PRIMATES; Family INDRIIDAE

Avahi, Sifakas, and Indri

This family of three genera and five species is confined to forests and scrublands of Madagascar. Groves (*in* Wilson and Reeder 1993) followed Jenkins (1987) in spelling the name of this family Indridae. The sequence of genera presented here follows that suggested by Tattersall (1982), who, however, considered the extinct families Palaeopropithecidae and Archaeolemuridae (see accounts thereof) to be subfamilies of the Indriidae.

The three genera may be distinguished as follows: *Avahi*, size small, tail long, pelage uniformly brown and woolly; *Propithecus*, size moderate, tail long, pelage largely white and more silky than woolly; and *Indri*, size large, tail short, and pelage silky. The muzzle is shortened and bare, so that these primates have a somewhat monkeylike appearance rather than the foxlike appearance of certain members of the family Lemuridae. The eyes are large; the orbits of the skull are large and well separated, being directed more forward than laterally in *Avahi* but about midway between forward and laterally in the other genera. As in some genera of Lemuridae, the external ears are largely concealed by the pelage. The hand is more elongated and narrowed than in the Lemuridae, and all fingers have pointed nails. The thumb is short and only slightly opposable. A fold of skin along the arm that extends to the side of the chest represents a vestigial parachutelike membrane. The legs are about one-third longer than the arms, and the foot is larger than the hand but also elongated and narrow, differing in the relatively greater development of the first toe. The other four toes are united at the base by a web of skin, and they function as a single unit in opposition to the first toe, providing good grasping ability for the feet. The palms and soles are padded. Females have a single pair of mammae in the chest region, and the males have a penis bone.

The salivary glands are greatly enlarged, as in the langurs and colobus monkeys. The stomach is large, and the intestine is relatively long, due partly to the elongated caecum.

The members of this family have fewer teeth than other primates, the dental formula being: (i 2/1, c 1/1, pm 2/2, m 3/3) × 2 = 30. Some workers interpret the arrangement of the teeth as: (i 2/2, c 1/0, pm 2/2, m 3/3) × 2 = 30. The upper incisors are often large, and the upper canines are elongate and sharp. The lower front teeth project forward and slightly upward as in the Lemuridae. The cusps and ridges of the molars are arranged in alternating V's or crescents.

Avahis are nocturnal, whereas the other members of this family are diurnal. These animals are sometimes solitary, but they usually associate in pairs or in groups of 10 or more individuals. They are arboreal, though they also often descend to the ground. In both arboreal and terrestrial locomotion they differ from the Lemuridae. When climbing they utilize a slow, deliberate hand-over-hand movement. Generally they leap from one upright to another, gaining footholds by means of the powerful grasping action of the hind limbs. In trees they usually cling in an erect position to vertical branches. They descend awkwardly, tail first. On the ground they stand upright with their arms held out in front of their body and usually progress by a series of short leaps or hops. Common resting positions are clinging to an upright branch, sprawling horizontally on a lateral branch with limbs dangling over the sides, and sitting upright in a crotch, sometimes with their arms held outward so as to expose their underparts to the sun. This latter resting position has given rise to a Malagasy superstition that these animals worship the sun.

These primates are strictly vegetarian in the wild, feeding on leaves, buds, fruits, nuts, bark, and flowers. They thus occupy a dietary position in the tropical forests of Madagascar similar to that of the leaf-eating monkeys of Africa and Asia and the howler monkeys *(Alouatta)* of tropical America. When feeding on the ground they often pick up food directly with the mouth; however, they often use their hands to convey food to the mouth and then lick their palms.

Geologically the Indriidae are known only from the Recent of Madagascar. Subfossil remains of some living species, together with those of extinct members of related families, have been found at various sites dated between about 2,000 and 1,000 years ago.

PRIMATES; INDRIIDAE; Genus INDRI
E. Geoffroy St.-Hilaire and G. Cuvier, 1796

Indri

The single species, *I. indri*, now occurs in northeastern Madagascar from the vicinity of Sambava to the Mangoro River. Subfossil remains indicate that the range of this species once extended as far west as the Itasy Massif in central Madagascar and north to the Ankarana Range (Mittermeier et al. 1994).

This is the largest living prosimian. Head and body length is 610–900 mm, tail length is 50–64 mm, and weight is about 6–10 kg. The body is thickly covered with long, silky fur above, which becomes shorter beneath. Coloration is usually patterns of grays, browns, and blacks, but there is variation. Some individuals are black, while others are almost or entirely white. The head, ears, shoulders, back, and arms are usually black. On the rump there is a large triangular patch of white, usually surrounded by black. The great toe is long and opposable, and the remaining toes are joined by a web as far as the ends of the first joint. The muzzle is small and nearly naked, the eyes are large, and the skin is black. Females have two mammae.

The indri inhabits coastal and montane rainforest from sea level to about 1,800 meters. It is diurnal and arboreal. It moves mainly by powerful leaps between large vertical stems and trunks (Walker 1979). On the ground it stands erect and usually progresses by jumps, holding its relatively short arms above its body. It is generally found in trees at heights of 2–40 meters, and 30–60 percent of its activity involves feeding. Activity begins two to three hours after dawn and continues until one to three hours before dusk, the period of daily activity being shorter in winter than in summer. Groups move about 300–700 meters in a day. The diet consists of leaves, flowers, and fruits (Pollock 1977, 1979).

Population densities have been estimated at 9–16/sq km but are thought to be lower in some areas. The ranging areas of two groups were 17.7 ha. and 18.0 ha. A large central part of each ranging area constitutes a defended territory from which other groups are excluded. Defense is by adult males, which mark territories with urine and secretions from glands on the muzzle. The most characteristic vocalization of the indri is a melodious song that can be heard

Indris *(Indri indri)*, photos by Russell A. Mittermeier.

by humans up to 2 km away. Often there are loud singing sessions by several members of a group, with each song lasting 40–240 seconds and consisting of a series of cries or howls. These calls probably function to unite groups, express territoriality, and convey information relative to age, sex, and reproductive availability of individuals (Pollock 1975, 1977, 1979, 1986).

According to Pollock (1977), groups comprise two to five individuals, apparently small families with a mated adult pair and their offspring. There is frequent intragroup agitation and competition for food, and it has been observed that the female and young are dominant in such situations, easily displacing the adult male. Nonetheless, it is the male that is responsible for territorial defense, and during encounters with intruders the female moves to a safe location. A single young is born to the female at two- to three-year intervals. It is carried ventrally by the mother for a month or more and then rides on her back until about eight months. Adult size, and presumably full reproductive maturity, is attained at seven to nine years. Thalmann et al. (1993) indicated that births occur in May in the southern part of the range but in December and onward at one site near the northern edge of the range. Doyle (1979) listed the gestation period as 137 days and the time of weaning as 180 days after the birth.

The indri is classified as endangered by the IUCN and the USDI and is on appendix 1 of the CITES. Its range has been reduced and fragmented through the spread of slash-and-burn agriculture and the commercial logging of the forests on which it depends. Like other prosimians, it is protected by law in the Malagasy Republic, but unlike most others, it is also still protected through local custom in some areas (Harcourt and Thornback 1990). With a total wild population estimated at only 1,000–10,000 individuals and no captives being maintained, the indri was considered to be of the highest priority for conservation attention by Mittermeier et al. (1992).

PRIMATES; INDRIIDAE; **Genus AVAHI**
Jourdan, 1834

Avahi, or Woolly Lemur

Groves (1989 and *in* Wilson and Reeder 1993), Harcourt and Thornback (1990), Mittermeier et al. (1992), Petter and Petter-Rousseaux (1979), and Tattersall (1982) recognized a single species, *A. laniger,* with two subspecies: *A. l. laniger,* in the eastern forests of Madagascar from near the northern tip of the island to the extreme southeast and, based on subfossil remains, formerly in central Madagascar at least as far west as Analavory; and *A. l. occidentalis,* now in northwestern Madagascar to the north and east of the Betsiboka River but possibly once found as far south as Morondava. Based on karyological analysis by Rumpler et al. (1990), Mittermeier et al. (1994) recognized these two taxa as separate species. Rakotoarison, Mutschler, and Thalmann (1993) reported the discovery of an isolated population of *Avahi,* probably referable to *occidentalis,* still living in west-central Madagascar, not far north of Morondava.

Head and body length is 300–450 mm, tail length is

Avahi (*Avahi laniger*); Top, photo by R. D Martin; Bottom, photo by Chris Raxworthy.

about 325–400 mm, and weight is about 600–1,200 grams. The fur is thick and woolly, not silky as in *Indri* and *Propithecus*. The face is covered with short hairs, and the small ears are concealed by the fur on the head. The most common coloration is brown gray with a lighter rump. The legs, forearms, hands, and feet are white and the tail is reddish orange. There is considerable color variation in this genus, however, some individuals being almost white and others reddish. The head is almost spherical in shape, the snout is short, and the eyes are large.

The avahi is found in rainforests. It spends most of its time in trees but sometimes descends to the ground. In trees it maintains a nearly vertical position and clings to upright limbs and trunks. It is nocturnal, spending the daylight hours sleeping in hollow trees or thick vegetation. On the ground it stands upright and leaps like an indri or sifaka, in a vertical position with arms held upward and legs close together. The diet in the wild seems to be exclusively vegetarian, consisting chiefly of leaves and also including, buds, bark, and fruit.

Harcourt and Thornback (1990) cited estimated population densities of 72–100/sq km and home range sizes of up to 4 ha. They noted also that the ranges of the subspecies *A. l. occidentalis* are larger than those of *A. l. laniger* and overlap more and that *occidentalis* shows less territorial aggression. Ganzhorn, Abraham, and Razanahoera-Rakotomalala (1985) reported that 10 groups of *A. l. laniger* occupied nonoverlapping home ranges of 1–2 ha. each and that a baby was born to four of these groups in August or September. There is some question regarding social structure. Groups consist of up to five individuals, and it has been suggested that these represent either a mated pair and young or an adult female with several generations of offspring (Klopfer and Boskoff 1979; Pollock 1979). Based on a radio-tracking study and other observations, Harcourt (1991) concluded that *Avahi* is monogamous, being usually found in male-female pairs that sleep and forage together. The various vocalizations include a high-pitched whistle for distant communication and perhaps territorial expression (Petter and Charles-Dominique 1979). The gestation period is four to five months, and there is normally a single young per birth. The young are born in the dry season, August–November (Hayssen, Van Tienhoven, and Van Tienhoven 1993), and are transported on the back of the mother for several months.

The avahi is classified as endangered by the USDI and is on appendix 1 of the CITES. It is declining through loss of forest habitat, especially because of clearing and burning to create pastures and cropland. Unlike many primates, the avahi is extremely difficult to keep in captivity, none currently known to be held. The number of individuals in the wild may be more than 100,000 for *A. l. laniger* and 10,000–100,000 for *A. l. occidentalis* (Mittermeier et al. 1994). The IUCN recognizes the latter as a separate species and classifies it as vulnerable.

PRIMATES; INDRIIDAE; Genus PROPITHECUS
Bennett, 1832

Sifakas

There are three species (Petter and Petter-Rousseaux 1979; Simons 1988):

P. verreauxi, western and southern Madagascar;
P. diadema, eastern Madagascar;
P. tattersalli, extreme northeastern Madagascar.

Head and body length is 450–550 mm, tail length is 432–560 mm, and weight is 3–7 kg. The fur is rather long and soft with woolly hair above but is sparse below. Coloration varies greatly within each species, from white, often tinged with yellowish, to black, gray, or reddish brown, arranged in various patterns. The face is hairless and black. The short arms are limited in their movement by small gliding membranes; the thumbs are scarcely opposable to the fingers; and the hind limbs are large and strong. *Propithecus* is distinguished from *Indri* by its long tail and smaller ears, which are largely concealed by hair.

Sifakas inhabit deciduous and evergreen forests and are

Sifaka *(Propithecus verreauxi)*, photo from San Diego Zoological Society.

diurnal. They stay mainly in large trees and sleep at a height of about 13 meters. They can make leaps of about 10 meters from tree to tree using the strength of the hind legs. On the ground they stand on their hind feet and progress by bipedal leaps, throwing their arms above their heads for balance. When picking up food from the ground they usually stoop and seize the food in the mouth, seldom using their hands. The diet consists mainly of leaves, flowers, bark, and fruits. In a study of *P. verreauxi*, Richard (1977) found that in the wet season there was an increase in the take of flowers and fruits, whereas in the dry season there was an increase in the amount of leaves eaten. Reported estimates of population density are up to 20/sq km for *P. diadema*, 50–500/sq km for *P. verreauxi*, and 60–70/sq km for *P. tattersalli* (Harcourt and Thornback 1990; Mittermeier et al. 1994).

Comparatively little is known of *P. diadema* and *P. tattersalli*, but *P. verreauxi* is one of the most intensively investigated of the Malagasy prosimians, and the remainder of this account, except as noted, deals with the latter species. Jolly (1966) reported that *P. verreauxi* woke up at about dawn, fed until 0800–0900 hours, moved to a sunning site for a while, and then fed again. In warm weather the sifakas slept at noon, had a second period of activity, and then settled for sleep at dusk. Richard (1977, 1978) found peak feeding activity during the wet season at 0700–0900 and 1300–1400 hours. She studied sifakas in both the northern and the southern parts of their range. The mean distance moved in a day in the north was 1,100 meters in the wet season and 750 meters in the dry season. In the south the respective distances were 1,000 and 550 meters. Group home range in these studies was 6.75–8.50 ha., and the entire range was visited by the troop about every 10–20 days. Jol-

ly (1966) reported home ranges of only 2.2–2.6 ha., through which the troop moved in a 7- to 10-day period. Estimated home ranges of 20 ha. to more than 250 ha. have been reported for *P. diadema* (Harcourt and Thornback 1990) and of 9–12 ha. for *P. tattersalli* (Mittermeier et al. 1994).

Harcourt and Thornback (1990) compiled additional available information on *P. diadema* and *P. tattersalli* but suggested that their social behavior may be much like that of *P. verreauxi*. Groups of the latter species are territorial in at least some areas. Jolly (1966) reported that there was an area of 0.6–1.8 ha. within the home range of each troop that other troops did not enter. Richard (1977) stated that each group had exclusive use of 24–51 percent of its home range. Adult males mark territories with urine and with a scent gland located on the front of the throat. Territorial confrontations between two troops may involve growling, scent marking, and ritualistic leaping toward the enemy territory but not direct physical contact. In addition to growls at other troops, vocalizations include barks at aerial predators, a resonant bark for distant communication and spacing, and the sound "sifaka," which is made when there are intruders on the ground (Jolly 1966; Petter and Charles-Dominique 1979).

Groups include 2–13 individuals, there usually being 2–3 adult males, 2–3 adult females, and several young (Pollock 1979). Groups are compact, with the various individuals generally keeping in sight of one another (Jolly 1972). According to Jolly (1966), intragroup social life is peaceful, with the young solicitously cared for and the adults playing together and grooming one another. During the breeding season, however, there are fights and sometimes serious injuries. Richard (1974) observed that at this time there is a breakdown in group structure, with adult males under-

Sifaka *(Propithecus tattersalli)*, photo by David Haring.

Sifaka *(Propithecus verreauxi)*, photo by Martin E. Nicoll.

All three species are listed as endangered by the USDI and are on appendix 1 of the CITES. The IUCN also classifies all species and subspecies. Those classifications, together with population estimates by Mittermeier et al. (1992), are: *P. diadema diadema* (found in eastern Madagascar from the Antanambalana to the Mangoro River), endangered, 1,000–10,000; *P. d. edwardsi* (in the east, south of the Mangoro River), endangered, 1,000–10,000; *P. d. candidus* (to the north and east of the Antanambalana River), critically endangered, 100–1,000; *P. d. perrieri* (extreme north), critically endangered, 2,000; *P. verreauxi verreauxi* (in the south and west, as far north as the Tsiribihina River), vulnerable, more than 100,000; *P. v. coquereli* (in the northwest, to the east and north of the Betsiboka River), endangered, 1,000–10,000; *P. v. coronatus* (in the northwest, between the Betsiboka and the Mahavavy du Sud River), critically endangered, 100–1,000; *P. v. deckeni* (in the west, from the Mahavavy du Sud to the Tsiribihina River), vulnerable, 1,000–10,000; and *P. tattersalli*, critically endangered, 8,000. The primary threat to all of these taxa is destruction and fragmentation of natural forest habitat because of slash-and-burn agriculture, commercial logging, charcoal production, fires to stimulate growth of pasture, and overgrazing by domestic livestock. Enforcement of protective laws is often ineffective, even in parks and reserves, and hunting for use as food also jeopardizes certain populations (Harcourt and Thornback 1990). All of these problems are especially severe for the very narrowly restricted *P. diadema perrieri, P. d. candidus, P. verreauxi,* and *P. tattersalli,* which are considered to be among the most critically endangered primates. Although *P. verreauxi* has been successfully maintained and bred in captivity, most attempts to keep *P. diadema* have failed (J.-J. Petter 1975);

Sifaka *(Propithecus diadema)*, photo by Gustav Peters.

going "roaming" behavior and there being competition for females, food, and resting places. The dominance hierarchy that existed in the nonmating season may change. A formerly subordinate male may take over his own group or leave his group and attain dominant status in another group. Females permit mating only by males that retain or achieve dominance during the mating season. Based on such behavior, Richard (1985) suggested that sifaka groups are primarily foraging units of closely related females and a varying number of males.

The mating season extends from January to March. Females in a given group enter estrus only once a year and for a relatively short time (Van Horn and Eaton 1979). Gestation periods of 130 and 141 days have been reported (Doyle 1979; Jolly 1966). The single young usually is born during June–September and weighs about 40 grams (Kolar *in* Grzimek 1990; Mittermeier et al. 1994). Jolly (1966) wrote that young born in July were carried on the belly of the mother until October, rode on her back until December–January, and reached full size at 21 months. Doyle (1979) reported that weaning occurred at 180 days, and sexual maturity at 913 days. A captive *P. verreauxi* was still living at 23 years and 4 months (Marvin L. Jones, Zoological Society of San Diego, pers. comm., 1995).

a few individuals of *P. d. diadema* and *P. tattersalli* are currently maintained at the Duke University Primate Center (Mittermeier et al. 1994).

PRIMATES; Family PALAEOPROPITHECIDAE

Sloth Lemurs

This family of four Recently extinct genera and five species is known only by subfossil remains found in Madagascar. The Palaeopropithecidae were treated as a subfamily of the Indriidae by Groves (1989) and Tattersall (1982) but as a full family by Jenkins (1987), Jungers et al. (1991), Mittermeier et al. (1994), and Simons et al. (1992).

This family usually had been thought to comprise only the genera *Palaeopropithecus* and *Archaeoindris*. However, substantial postcranial remains of *Mesopropithecus*, only recently discovered, suggest that the genus is allied with the Palaeopropithecidae rather than the Indriidae, to which it was assigned previously. Moreover, the newly described genus *Babakotia* is a somewhat intermediate form with affinity to both *Mesopropithecus* and *Palaeopropithecus*. In all genera of the family the skull is heavily built, the dental formula is the same as in the Indriidae, and the forelimbs are substantially longer than the hind limbs. According to Tattersall (1982), *Palaeopropithecus* and *Archaeoindris* share unusual conditions of the nasal and auditory regions. The premaxillae are enlarged bilaterally in their superior portions to produce paired protuberances projecting from the nasal aperture. Auditory bullae, which are prominent in the Indriidae and Archaeolemuridae, are entirely lacking. *Mesopropithecus* and *Babakotia* do not have the nasal projections and do retain inflated bullae. However, Simons et al. (1992) noted that all four genera of palaeopropithecids are characterized by relatively small orbits, robust zygomatic arches with convex superior margins, and a bony palate more or less rectangular in outline, its posterior rim bearing prominent tubercles for the attachment of the muscles of the soft palate.

Although most of the extinct Malagasy lemurs were described about a century ago, much information on their likely overall appearance, natural history, and systematic affinity has become available only in the last decade or is still unpublished. Simons et al. (1992) used the term "sloth lemurs" for the Palaeopropithecidae because of their many extraordinary convergences with the South American sloths. Mittermeier et al. (1994) suggested that the Palaeopropithecidae evolved from and are closely related to the Indriidae. They were medium-sized to large primates, apparently slow-moving, and at least to some extent slothlike in suspensory habit. *Mesopropithecus*, the smallest and most primitive genus, probably weighed about 10 kg. It and *Babakotia* seem to represent stages, in size and adaptation for arboreal suspension, that culminate in *Palaeopropithecus*. *Archaeoindris*, which weighed up to 200 kg, probably was largely terrestrial and somewhat analogous to a giant ground sloth. The prominent shearing crests on the molar and premolar teeth of the Palaeopropithecidae suggest that all members were highly folivorous (Simons et al. 1992).

Geologically the family is known only from Recent sites dated as about 8,000–1,000 years old, though *Mesopropithecus*, at least, could have survived to within 500 years ago. All genera probably disappeared as a direct result of the human invasion of Madagascar (Tattersall 1971, 1978b; Tattersall and Schwartz 1975). Simons et al. (1992) stated that they were probably diurnal, noisy, slow to reproduce, and slow-moving and would have been easy to hunt.

PRIMATES; PALAEOPROPITHECIDAE; Genus MESOPROPITHECUS
Standing, 1905

There are two species, both extinct (Jenkins 1987; Mittermeier et al. 1994; Simons et al. 1990, 1992; Tattersall 1982):

M. globiceps, known only by subfossil remains from the type locality in central Madagascar;

M. pithecoides, known by subfossil remains from northern, central, western, and southwestern Madagascar.

M. globiceps sometimes has been placed in a separate genus, *Neopropithecus* Lamberton, 1936.

Until recently *Mesopropithecus* was known mainly from cranial material, on the basis of which it usually was assigned to the family Indriidae. Cranial length is 83.3–94.4 mm in *M. globiceps* and 94.0–103.5 mm in *M. pithecoides.* The skull is similar in size to that of *Indri* (cranial length 97.1–117.7 mm) but closely resembles that of *Propithecus* in proportions and dentition. It differs from that of *Propithecus* in having a more robust build, slightly smaller and more convergent orbits, more massive zygomatic arches, a broader snout, a more pronounced postorbital constriction, a steeper facial angle, a more rounded braincase, a sagittal crest or temporal lines that are anteriorly confluent, a distinct nuchal ridge confluent with the posterior root of the zygoma, and larger upper incisors and canines. *M. pithecoides* is the more divergent of the two species, being characterized by larger size, sagittal and nuchal crests, massive zygomatic arches, and a broad muzzle. *M. globiceps,* with its more gracile build and narrower snout, is more primitive and more like *Propithecus.*

Mesopropithecus is the smallest of the subfossil prosimians of Madagascar and may have been the last to become extinct. A recently discovered postcranial skeleton apparently referable to *M. pithecoides* shows that the species probably weighed about 10 kg and had elongated forelimbs. These and other characters indicate adaptation for arboreal suspension and affinity to the family Palaeopropithecidae rather than to the Indriidae. The dentition of *Mesopropithecus* suggests that the genus was folivorous in diet. It may have survived until about 500 years ago and then disappeared through direct hunting and/or environmental disruption by people.

PRIMATES; PALAEOPROPITHECIDAE; Genus BABAKOTIA
Godfrey, Simons, Chatrath, and Rakotosamimanana, 1990

The single species, *B. radofilai,* is known only by subfossil remains from Antsiroandoha and one other cave in the Ankarana Range in extreme northern Madagascar (Godfrey, Simons, et al. 1990; Jungers et al. 1991; Mittermeier et al. 1994; Simons et al. 1992).

Cranial length in one specimen is 114.5 mm. The skull has a superficial resemblance to that of *Indri* because of the elongation of the face, buccolingual compression and mesiodistal expansion of the maxillary and mandibular premolars, and the limited depth of the corpus of the mandible. *Babakotia* also resembles the indriids, as well as *Mesopropithecus,* in possessing inflated bullae, an unfused mandibular symphysis, and a relatively long and procumbent tooth comb. Details of the dental morphology, however, do not suggest a close relationship with any other genus. *Babakotia* has especially elongated upper premolars with shallow paracristas and metacristas; a mesial projection of the crown of the anterior upper premolar well beyond the mesial root of the tooth; fine enamel crenulations on the occlusal surfaces of the molars; incipient bilophodonty of the second maxillary molar; and a relatively large upper third molar with four distinct cusps.

A nearly complete postcranial skeleton indicates that *Babakotia* was a mid-sized primate weighing about 15–20 kg. It was adapted for climbing and hanging, as were the other palaeopropithecids, rather than leaping, a characteristic of the indriids. Its forelimbs are elongated, being about 20 percent longer than the hind limbs. The forefeet and hind feet are also long and adapted for strong grasping. The hind feet are reduced, like those of animals adapted for suspension and in contrast to those of leapers. These and other structural characters indicate specialized suspensory behavior involving frequent use of all four limbs and simultaneous feeding. The living arboreal sloths, *Bradypus* and *Choloepus*, possess similar features. *Babakotia* itself appears to have become extinct within the last 1,000 years, probably as a result of human hunting.

PRIMATES; PALAEOPROPITHECIDAE; Genus **PALAEOPROPITHECUS**
G. Grandidier, 1899

Jenkins (1987) and Tattersall (1982) recognized a single species, *P. ingens*, known only by subfossil remains from central, southwestern, and southern Madagascar. Mittermeier et al. (1994) reported that there is at least one other

extinct species and that the range of the genus extends into northern Madagascar.

Weight probably was about 40–60 kg. Cranial length is 181–211 mm. The skull is long, relatively low, and robustly built. It resembles that of *Indri* but is much larger, and its braincase and orbits are relatively smaller. The orbits are forwardly and dorsally directed and are heavily ringed by bone. There is a pronounced postorbital constriction and a small degree of nuchal cresting. The structure of the face and the occipital condyles indicate that the head was habitually held at a considerable angle to the neck. In contrast to *Mesopropithecus* and *Babakotia*, as well as to the members of the Archaeolemuridae and the Indriidae, *Palaeopropithecus* has no auditory bullae. Much of the space they would have occupied is taken up by a massive structure consisting of the partially coalesced and highly developed styloid, mastoid/postglenoid, and paroccipital processes. The cheek teeth are similar to those of the Indriidae, and the dental formula is the same. The medial upper incisor is large and subcylindrical, while the lateral is greatly reduced.

The forelimbs are long relative to the hind limbs, and the feet of each are long and hooklike. These and other postcranial specializations indicate that *Palaeopropithecus* was exclusively arboreal and practiced a form of suspensory locomotion similar to that of *Pongo*. Recently collected postcranial remains suggest that of the extinct Malagasy lemurs, *Palaeopropithecus* was the most highly specialized for arboreal suspension and was slothlike in habit. The diet probably was like that of living indriids. *Palaeopropithecus* is known from sites dated at about 2,300 to 1,000 years ago, and it evidently became extinct not long after the end of this period. Humans arrived in Madagascar around the middle of this same period. They and their domestic stock

Extinct Madagascar prosimians (with living *Indri* for comparison), reconstructions by Stephen D. Nash.

probably eliminated much of the forest habitat on which *Palaeopropithecus* depended, and they also would have hunted this vulnerable giant lemur whenever possible.

PRIMATES; PALAEOPROPITHECIDAE; **Genus ARCHAEOINDRIS**
Standing, 1908

The single species, *A. fontoynonti*, is known only by sub-fossil remains found at Ampasambazimba in central Madagascar (Jenkins 1987; Mittermeier et al. 1994; Tattersall 1982; Vuillaume-Randriamanantena 1988).

Archaeoindris apparently was the largest of the Malagasy lemurs and, with an estimated weight of 160–200 kg, about as large as an adult male gorilla. The length of the single known skull is 269 mm. It is similar to that of *Palaeopropithecus* but larger and more robust, the facial region is shorter, the orbits are less dorsally oriented, and there is a broad sagittal crest. There are no auditory bullae. The few known postcranial bones have some resemblance to those of the genus *Megaladapis* and had suggested a similar form of locomotion. However, recent analysis indicates that in its postcranial morphology *Archaeoindris* actually shows affinity to *Palaeopropithecus*. As in the latter genus, the forelimbs of *Archaeoindris* were longer than the hind limbs, but there was far less disparity between the humeral and femoral lengths.

Archaeoindris probably was largely terrestrial, having evolved from a more arboreal ancestor, and may have been ecologically comparable to the extinct ground sloths of the New World. The diet is thought to have been folivorous. There is some question about the period during which *Archaeoindris* lived, but certain specimens were collected at a stratigraphic level recently radiocarbon dated at about 8,000 years B.P.. Considering also the apparent rarity of the genus, it is possible that it disappeared even before the human invasion of Madagascar. In any event, it would certainly have been extremely vulnerable to environmental disruption and hunting by people.

PRIMATES; **Family ARCHAEOLEMURIDAE**

Baboon Lemurs

This family of two Recently extinct genera and three species is known only by subfossil remains found in Madagascar. The Archaeolemuridae were treated as a subfamily of the Indriidae by Groves (1989) and Tattersall (1982), but as a full family by Jenkins (1987), Mittermeier et al. (1994), and Simons et al. (1992).

The members of this family were medium-sized primates weighing about 15–25 kg. Their skulls resemble those of the Indriidae but are much more massive with especially robust mandibles, fused and less oblique mandibular symphyses, and sagittal and nuchal cresting. The dental formula is the same as that of the Indriidae and Palaeopropithecidae except that there are three, rather than two, upper and lower premolars. The teeth are highly specialized, as described below in the generic accounts. Examination of the postcranial skeleton indicates that the archaeolemurids were short-limbed and powerfully built and that all four feet were strikingly short.

The Archaeolemuridae seem to have evolved from an ancestral indriid that still had three premolar teeth. The Palaeopropithecidae represent an evolutionary branch that

has closer affinity to the living indriids. Although the skulls of the indriids resemble those of the archaeolemurids, the latter show considerable divergence in their dentition and in a highly developed suite of locomotor adaptations for terrestrial living. The two genera have been called the ecological equivalents of the African baboons. Geologically, the Archaeolemuridae are known only from Recent sites dated at about 3,000–1,000 years ago. Like the other extinct Malagasy prosimians, they probably disappeared as a direct result of the human invasion of Madagascar.

PRIMATES; ARCHAEOLEMURIDAE; **Genus ARCHAEOLEMUR**
Filhol, 1895

Jenkins (1987), Simons et al. (1990), and Tattersall (1982) recognized two species, both extinct:

A. majori, known by subfossil remains from southern Madagascar;
A. edwardsi, known by subfossil remains from northern and central Madagascar.

Godfrey and Petto (1981) suggested that there is only a single, geographically variable species. In contrast, Godfrey, Sutherland, et al (1990), while finding extensive variation in *Archaeolemur,* thought that the genus probably comprises more than two species. Mittermeier et al. (1994) noted that *Archaeolemur* is perhaps the most widely distributed of all the Malagasy subfossil lemurs, but they did not comment on specific differentiation. All of the sources cited above have been used in compiling the remainder of this account.

Archaeolemur is much larger than any living prosimian. Maximum weight probably was more than 22 kg. Cranial length is 122.3–135.0 mm in *A. majori* and 139.6–153.5 mm in *A. edwardsi.* Nonetheless, the skull is similar to that of living indriids, especially *Propithecus,* except for the presence of a sagittal crest. This crest is particularly pronounced in *A. edwardsi.* The skull also is similar to that of the extinct *Palaeopropithecus* and *Archaeoindris* except that the bullae are not reduced. It differs from the skull of these other genera in having more forwardly directed orbits, as in monkeys. There are three upper and three lower premolar teeth, compared with only two of each in living indriids. These unique premolars are buccolingually compressed and together form a single longitudinal shearing blade. The molars are subsquare and bilophodont, the paracone-protocone and metacone-hypocone pairs being united by transverse crests. The upper canines are reduced, the upper incisors are greatly expanded, and the lower canines and incisors are angled forward at only about 45°, though they apparently were derived from the procumbent condition found in the Indriidae.

The postcranial skeleton is well known and bears some resemblance to that of living terrestrial cercopithecid monkeys. *Archaeolemur* appears to have been a powerfully built and rather short-legged quadruped with reduced leaping power compared with that of some other lemurs. The foot seems to have been modified for grasping, and thus it is likely that *Archaeolemur* retained the ability to exploit arboreal food resources. Its general locomotor and postural characters evidently closely parallel those of *Papio.* Its cheek teeth also are similar to those of cercopithecids and may have been used for cropping, husking, and pulping a frugivorous or mixed diet.

Contrary to an earlier view that Madagascar was covered largely by dense forest prior to human modification, there now is evidence that the original vegetation over much of the island was woodland, savannah, bushland, or grassland. Such areas would have been suitable for baboonlike primates. *Archaeolemur* evidently occurred in a number of these habitats, there being a general pattern of larger individuals in wetter (mainly northern and central) zones and smaller animals in more arid (mainly southern and western) areas. Altogether the genus is known from more than 20 sites with radiocarbon dates ranging from about 2,850 to 1,035 years ago, though it may have survived several hundred years longer. People first arrived in Madagascar about 1,500–2,000 years ago. Although the disappearance of *Archaeolemur* and the other subfossil lemurs sometimes is attributed to a variety of factors (Dewar 1984), there is little doubt that human hunting and environmental disruption were directly responsible. A large and primarily terrestrial herbivore such as *Archaeolemur* would have been especially vulnerable to an expanding pastoral culture.

PRIMATES; ARCHAEOLEMURIDAE; Genus HADROPITHECUS
Lorenz Von Liburnau, 1899

The single species, *H. stenognathus,* is known only by subfossil remains from central, southwestern, and southern Madagascar (Jenkins 1987; Mittermeier et al. 1994; Tattersall 1982).

Hadropithecus is about the same size as *Archaeolemur.* Cranial length in two specimens is 128.2 and 141.8 mm. The skull differs from that of *Archaeolemur* in having a much shorter snout, a deeper facial region, more forwardly directed orbits, a more marked interorbital constriction, very robust and widely projecting zygomatic arches, larger auditory bullae, and even more specialized dentition. The incisor and canine teeth are greatly reduced. The posterior lower premolar is molariform, with a cruciform arrangement of rounded ridges, and the upper premolars increase in size and complexity posteriorly. The anterior molars, upper and lower, are large and subsquare, with high, rounded enamel folds replacing the transverse crests of *Archaeolemur.* This entire crowded battery of cheek teeth rapidly is worn flat, with thick enamel ridges enclosing shallow basins in the softer dentine, as occurs in various ungulates. The postcranial skeleton is not well known; identified bones are similar to those of *Archaeolemur* but more gracile.

Hadropithecus is thought to have been quadrupedal and terrestrial and more specialized in this regard than *Archaeolemur.* Its skull and dentition show many characters that are shared by the living gelada baboon *(Theropithecus)* of East Africa. Like the gelada, *Hadropithecus* apparently had an abrasive diet consisting of grass and other small, tough, low-growing items that require heavy grinding by the cheek teeth. This genus is known from sites dated at about 2,000–1,000 years ago and disappeared along with *Archaeolemur* in the face of the human invasion of Madagascar. As a large, terrestrial grazer, it would have been highly susceptible to the spread of people, their domestic flocks, and introduced suids (Dewar 1984).

PRIMATES; Family DAUBENTONIIDAE; Genus DAUBENTONIA
E. Geoffroy St.-Hilaire, 1795

Aye-ayes

The single genus, *Daubentonia*, contains two species (Harcourt and Thornback 1990; MacPhee and Raholimavo 1988; Petter and Petter *in* Meester and Setzer 1977; Tattersall 1982):

D. madagascariensis, formerly found in much of eastern, northern, and west-central Madagascar;
D. robusta, known only by subfossil remains found in southwestern Madagascar.

Groves (*in* Wilson and Reeder 1993) listed *robusta* as a synonym of *D. madagascariensis* but noted that it may be a distinct species.

In *D. madagascariensis* head and body length is 360–440 mm, tail length is 500–600 mm, and weight is about 2–3 kg. *D. robusta,* which is known only from postcranial material and a few teeth, is thought to have been about 30 percent larger. The pelage is coarse and straight, and the bushy tail has hairs up to 100 mm long. The coloration is dark brown to black, the pale bases of the individual hairs showing through to some extent. The nose, cheeks, chin, throat, and spots over the eyes are yellowish white. The hands and feet are black.

Aye-aye *(Daubentonia madagascariensis)*, photo by David Haring.

Aye-aye *(Daubentonia madagascariensis)*, from *Bull. Acad. Malgache*. Insets: right hand and left foot photos from *Zoologie de Madagascar*, G. Grandidier and G. Petit, photo (right) by Jean-Jacques Petter.

D. madagascariensis has a rounded head with a short face, large eyes, and large, naked, membranous ears. The body and limbs are slender. The fingers are long and narrow, the third finger extremely so. The thumb is flexible but not truly opposable; however, the first toe is opposable. All the digits have pointed, clawlike nails, except the first toe, which has a flat nail. Females have two mammae located abdominally, and males have a penis bone.

The incisors are large, curved, and similar to those in rodents, that is, chisel-like with enamel on only the front surface and ever growing. Canines are absent in the permanent dentition but present in the deciduous dentition. In the adult aye-aye there is a large space between the incisors and the premolars. The modified cheek teeth have flattened crowns with indistinct cusps. The formula for the permanent dentition is: (i 1/1, c 0/0, pm 1/0, m 3/3) $\times$ 2 = 18. The skull resembles that of a squirrel, but primate features are prevalent (see also above account of the order Primates).

The modern aye-aye lives mostly in forests but appears to be adaptable, having been found in secondary growth, mangroves, bamboo thickets, and cultivated areas, particularly coconut groves; it may, however, depend on large trees for nesting (Harcourt and Thornback 1990). It is arboreal and nocturnal. During the day it sleeps in a nest constructed amidst dense foliage in a strong fork of a large tree, about 10–15 meters above the ground. The nest, about 50 cm in diameter, probably requires 24 hours to build and is very complex. Made of twigs or interlaced leaves, it is closed at the top and has a lateral opening; the bottom is covered by a layer of shredded leaves. The 5-ha. home range of two an-

imals contained 20 such nests. Each nest was used for several days in succession. The aye-aye leaves its nest at nightfall to forage and returns at first light. Its locomotion through the branches is much like that of a lemur, but it is less adept in horizontal movements. Vertical climbing is by rapid successive leaps. It frequently descends to the ground and can make long trips there (J.-J. Petter 1977; Petter and Petter 1967). When moving about, it carries its tail in a curve. Occasionally the aye-aye, like a loris, will suspend itself by its hind feet, using its hands for feeding or cleaning. In the latter operation it uses the long third finger in combing, scratching, and cleansing; the other fingers are flexed during this performance.

On Nosy Mangabe Island, Sterling (1993) found individuals to move about 800–4,400 meters per night. The main foods there were seeds from the fruit of *Canarium*, cankerous growths on *Intsia*, nectar from *Ravenala*, and larvae from several families of insects. The aye-aye apparently listens carefully for the sound of larvae in decaying wood and often taps the surface of the wood with its long third finger. Smelling may also be involved in finding the larvae. Erickson (1991, 1994) found that the tapping serves to locate the galleries made by larvae, as well as adult ants and termites, within the wood. The aye-aye also usually can determine which cavities are actually occupied by prey. This capability could result from a combination of perceptual factors, including a cutaneous sense in the third finger and an echolocation system involving triangulation of the large ears of the aye-aye and the tones emitted in response to the tapping. When the prey is located, the aye-aye bites through the wood with its powerful incisors and inserts its

Aye-aye *(Daubentonia madagascariensis)*, photo by David Haring.

third finger to crush and extract the larvae. The aye-aye also eats coconuts by gnawing a hole with its incisors and extracting the juice and pulp with the third finger (Hladik 1979; J.-J. Petter 1977; Petter and Petter 1967). Before biting into the coconut the aye-aye may tap the surface with its third finger, perhaps to evaluate the milk content (Winn 1989). Other studies on Nosy Mangabe Island, where the aye-aye was introduced in 1967, indicate that the diet also includes bamboo shoots, tree exudates, large insects, and possibly small vertebrates (Pollock et al. 1985).

The aye-aye appears to be basically solitary, but six captive adults kept in the same cage did not show signs of aggression (J.-J. Petter 1977). The animals in a captive colony at the Jersey Wildlife Preservation Trust are kept in separate cages except when mixed for breeding purposes (Carroll and Beattie 1993). Some observations of captive animals suggest that females may be dominant to males (Rendall 1993). An adult female and a male not fully grown shared a home range of about 5 ha. (Petter and Petter 1967). A male, a female, and a juvenile ranged over an area about 5 km long (Hladik 1979). In a radio-tracking study on Nosy Mangabe Island, Sterling (1993) found home range size for two males to be 126 ha. and 215 ha. and that for two females to be 32 ha. and 40 ha. Male ranges overlapped greatly with one another and with those of several females, but female ranges were well separated from one another. Social interaction varied, with some animals avoiding one another but others forming foraging units consisting of two or three males or of a male and a female. Adults slept separately, though observations elsewhere indicate that two males sometimes share a nest. Reproductive activity was asynchronous and evidently occurred throughout the year. During a three-day period of estrus a female called repetitively and was subsequently surrounded by up to three males.

According to J.-J. Petter (1977), several vocalizations have been distinguished, including the sounds "rontsit" for

alarm or danger and "creee" possibly as a contact signal. Both sexes frequently mark with urine. Reproduction in any one female may occur only once every two or three years, and apparently there is a single young, which is raised in a nest. A captured 1-year-old male and 2-year-old female were still with the mother and did not appear sexually mature. Limited observations in the wild suggest that breeding may occur through much of the year (Sterling 1993). Beattie et al. (1992) reported that a female, apparently already pregnant when captured, gave birth at the Duke University Primate Center in April. The first fully captive-bred aye-aye was born at the Jersey Wildlife Preservation Trust in August after a gestation period of 158 days; it weighed about 140 grams. Another infant, subsequently born at Duke after a gestation period of 172 days, weighed only 103 grams. According to Winn (1989), a young male at the Paris Zoo was weaned at 7 months of age but still had not reached adult size by about 3 years. Although captive specimens are rare, one lived for 23 years and 3 months (Jones 1982).

D. robusta apparently disappeared less than 1,000 years ago (J.-J. Petter 1977). As with other large Malagasy prosimians, its extinction was probably brought on by human agency. Specimens of its teeth, evidently modified by people for ornamental purposes, have been recovered (MacPhee and Raholimavo 1988). *D. madagascariensis* still survives but is designated as endangered by the IUCN and the USDI and is on appendix 1 of the CITES. It has declined mainly through cutting of the large forest trees upon which it depends. It also is killed on sight by villagers, who believe it to be a harbinger of misfortune. Based on its rarity and systematic uniqueness, Mittermeier et al. (1986) considered its survival to be one of the highest primate conservation priorities, if not the highest, in the world. At that time only a few scattered individuals were thought to remain on the northeastern and possibly northwestern coasts of Madagascar. A number of animals also had been captured alive in 1967 and released on Nosy Mangabe Island, off northeastern Madagascar, which is a protected reserve, and evidently a population became established there (Constable et al. 1985).

Subsequently, considerable attention was devoted to locating and investigating the aye-aye. In 1985 a small population was discovered in a rainforest about 900–1,000 meters above sea level in east-central Madagascar (Ganzhorn and Rabesoa 1986). Several were observed and captured in the northeast in 1986 (Albignac 1987). In 1991 another population was discovered in the northwest to the east of Narinda Bay (Simons 1993). Mittermeier et al. (1994) thought the many new sightings over a large region to be an incredible turn of events since the aye-aye had been feared to be on the verge of extinction; nonetheless, they cautioned that the species remained rare and of much conservation concern. Only 1,000–10,000 individuals are estimated to survive in the wild (Mittermeier et al. 1992). There also are 17 in captivity, 3 of which were born in that state (Olney, Ellis, and Fisken 1994).

PRIMATES; Family TARSIIDAE; Genus TARSIUS
Storr, 1780

Tarsiers

The single Recent genus, *Tarsius*, contains five species (Groves 1976; Musser and Dagosto 1987; Niemitz 1984*b*; Niemitz et al. 1991; Petter and Petter-Rousseaux 1979):

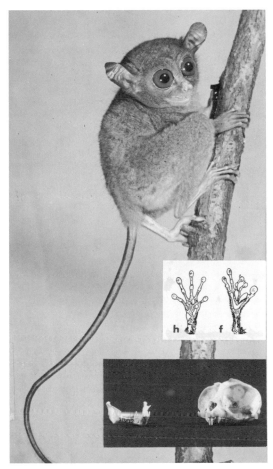

Mindanao tarsier *(Tarsius syrichta)*, photo by Ernest P. Walker. Insets: hand and foot (*Tarsius* sp.), photo from *Proc. Zool. Soc. London;* skull (*Tarsius* sp.), photo by P. F. Wright of specimen in U.S. National Museum of Natural History.

T. spectrum, Sulawesi and the nearby islands of Sangihe, Peleng, and Salayar;

T. pumilus, central Sulawesi;

T. dianae, known only from the type locality in central Sulawesi;

T. bancanus, southern Sumatra and the nearby islands of Bangka and Billiton, Borneo and the nearby islands of Karimata and Serasan, possibly Java;

T. syrichta, the islands of Samar, Leyte, Dinagat, Siargao, Bohol, Mindanao, and Basilan in the Philippines.

The forms *T. pelengensis,* of Peleng Island, and *T. sangirensis* of Sangihe Island are sometimes regarded as species distinct from *T. spectrum,* but they were not regarded as such by Corbet and Hill (1992), Flannery (1995), or Groves (*in* Wilson and Reeder 1993).

Head and body length is 85–160 mm, tail length is 135–275 mm, and adult weight is 80–165 grams. The individual hairs have a wavy, silky texture, and certain parts of the body are sparsely haired. The coloration above ranges from buff or grayish brown to dark brown, and the underparts are buff, grayish, or slate. The tail is naked except for a few short hairs on the tip. The very large eyes are the most outstanding structural feature, the diameter of the eyeball being approximately 16 mm. The head is round

with a reduced muzzle, and the neck is short. The ears are thin, membranous, and nearly naked. Members of this genus differ from all lemurs in that the nasal region is clothed with short hairs to the margins of the nostrils and there is a narrow strip of naked skin around the nostrils. Lemurs have a moist muzzle with a central prolongation dividing the upper lip. The forelimbs of *Tarsius* are short, but the hind limbs are greatly elongated—the name "tarsier" refers to the elongated tarsal, or ankle, region. The digits are long and tipped with rounded pads that enable the animal to grip almost any surface. The thumb is not truly opposable, but the first toe is well developed and widely opposable. Except for the second and third toes, which have clawlike nails used in grooming, the digits have flattened nails.

The dental formula is: (i 2/1, c 1/1, pm 3/3, m 3/3) × 2 = 34. The upper incisors are large and pointed and the lower incisors point upward, not forward. The upper canines are relatively small. The cheek teeth are adapted for a diet of insects. The orbits of the skull are directed forward.

Tarsiers seem to prefer secondary forest, scrub, and clearings with thick vegetation, but they have also been found in primary forest and mangroves (IUCN 1978). They are nocturnal or crepuscular and mainly arboreal. They spend the day sleeping in dense vegetation on a vertical branch or, rarely, in a hollow tree. They are not believed to build nests. When clinging to an upright branch they rigidly apply their tail as a support, and when they are asleep the head may drop downward between the shoulders. If disturbed while resting, a tarsier moves up or down its support and faces the suspected enemy. Its mouth may be opened and the teeth bared at the same time.

A tarsier can rotate its head nearly 360°, giving the animal an extremely wide field of vision. Tarsiers are notable acrobats in trees and shrubs, making quick leaps of several meters with no apparent effort. MacKinnon and MacKinnon (1980) reported an average leap of 1.4 meters and a maximum leap of 5–6 meters for *T. spectrum.* The tail trails behind as they leap from one support to another. Their movements are much like those of a tree frog, except that in *Tarsius* grasping is also involved. On any flat surface they also leap froglike, with their tail arched over their back, but they can walk on all fours with the tail hanging down. Their leaps on the ground are 1,200–1,700 mm long and up to 600 mm high. The ears of tarsiers are in almost constant motion during the waking period, being furled or crinkled frequently.

Tarsiers prey mainly on insects and readily accept small lizards and crustaceans, such as shrimps, in captivity. A tarsier watches its moving prey, adjusts its position and focus, then suddenly leaps forward and seizes the prey with both hands. The prey is chewed with side-to-side movements of the jaw while the tarsier sits upright on its hindquarters. These primates drink water by lapping.

MacKinnon (1986) estimated that 9,912,500 individual *T. bancanus* occupied 198,250 sq km of suitable remaining habitat in Indonesia, at an average density of 50/sq km, and that 14,146,000 *T. spectrum* occupied 70,730 sq km at a density of 200/sq km. However, such figures may not account for widely varying habitat conditions and human disturbances and thus could be highly excessive.

Niemitz (1984b) reported that an adult pair of *T. bancanus* inhabits a home range of 1–2 ha. This area apparently is a territory, which is marked by urine and the scent of various glands. Fogden (1974), also working in Borneo, found home range to be 2–3 ha., larger for males than for females. Ranges of individuals of the same sex seemed mostly exclusive, but there was extensive overlap between the ranges of males and females. Despite reports that tar-

Tarsier *(Tarsius syrichta)*, photo by Bernhard Grzimek.

siers are usually found in pairs, Fogden saw two together only eight times during his study. He recorded immature males mainly in primary forest, which is not considered good habitat, and suggested that these young animals were forced to disperse into marginal areas on attaining independence. Females, however, established ranges near those of their parents. *T. syrichta* is usually seen in male-female pairs (IUCN 1978). Wright et al. (1987) reported maintaining captive *T. bancanus* in male-female pairs and *T. syrichta* in pairs or in groups of one male with two or three females.

In studies on northern Sulawesi, MacKinnon and MacKinnon (1980) found *T. spectrum* to occupy a wider range of habitats than that reported for *T. bancanus*. Family groups occupied an average home range of approximately 1 ha. and regularly slept together at the same sites each day. They were territorial, actively chasing others out of their range and marking the area by rubbing branches with urine and epigastric glands. Vocalizations included loud shrieks during territorial battles and loud duets, with the male and female of a pair having separate parts. Pairs formed a close and stable bond and were seen to remain together for more than 15 months. Young females remained with their parents until adulthood, whereas young males departed as juveniles. One group had 3 males and 3 females but appeared to be splitting up. There were two breeding seasons 6 months apart, at the beginning and the end of the rainy season. Births occurred in May and November–December. The precocious young could travel with the group at 23 days and hunt alone at 26 days.

Fogden (1974) reported that *T. bancanus* has a sharply defined breeding season, with mating in October–December and births in January–March. Van Horn and Eaton (1979), however, questioned Fogden's data and pointed out that earlier investigation had found pregnant females in every month of the year on Bangka Island. These authorities also cited studies indicating that litter size in the genus is one and that the estrous cycle is about 25–28 days in *T. syrichta*, but they questioned a report that the gestation period in a specimen of *T. syrichta* lasted 6 months. Subsequent studies, however, have shown that gestation in *T. bancanus* is 178 days, an unusually long period for such a small mammal, and that the estrous cycle in this species is 18–27 days, with a 1- to 3-day estrus (Izard, Wright, and Simons 1985; Wright, Izard, and Simons 1986). The young is born in a fairly well-developed state—well furred, its eyes open, and capable of climbing and making short hops on a level surface. It is unable to leap until 1 month old. The weight at birth is approximately 20–31 grams. The young usually clings to the mother's abdomen but sometimes is carried in her mouth. It begins to capture prey at 42 days and is weaned shortly thereafter (Wright et al. 1987). Fogden (1974) stated that adult weight was attained at 15–18 months. A captive specimen of *T. syrichta* lived for 13 years and 5 months (Jones 1982).

T. syrichta is classified as threatened by the USDI and as conservation dependent by the IUCN. All species are on appendix 2 of the CITES. Tarsiers are jeopardized by destruction of forest habitat and capture by people (Wolfheim 1983).

Simons and Bown (1985) tentatively assigned fossils from the Oligocene of Egypt to the Tarsiidae, and Ginsburg and Mein (1987) described the species *Tarsius thailandica* from the lower Miocene of northwestern Thailand. Otherwise, fossils referable to the family have not been found, but the Omomyidae, a related family of the infraorder Tarsii (or Tarsiiformes), has a geological range of early Eocene to late Oligocene in North America and early Eocene to early Oligocene in Eurasia. This diverse group contains about 30 known genera, and there is increasing evidence that *Tarsius* is a living descendant (Beard and Wang 1991; Gingerich 1984).

PRIMATES; Family CEBIDAE

New World Monkeys

This family of 11 living genera, 3 Recently extinct genera (in the West Indies), and 65 species inhabits forests from northeastern Mexico to northern Argentina. There is almost universal acceptance of these genera, but the number of ways in which they have been assigned to subfamilies or otherwise arranged to show phylogenetic affinity is nearly equal to the number of authorities who have attempted such a task. Groves (1989) divided the Cebidae into five full families: the Cebidae, for *Cebus* and *Saimiri*; the Aotidae, for *Aotus*; the Callicebidae, for *Callicebus* and the extinct *Xenothrix*; the Pitheciidae, for *Pithecia*, *Chiropotes*, and *Cacajao*; and the Atelidae, with two subfamilies, the Atelinae for *Lagothrix*, *Ateles*, and *Brachyteles* and the Alouattinae for *Alouatta* (the extinct *Paralouatta* and *Antillothrix* had not yet been described). Subsequently Groves (*in* Wilson and Reeder 1993) reduced these families to subfamilial level. Hershkovitz (1977) recognized a seventh subfamily for *Saimiri*, the Saimirinae, whereas Martin (1990) accepted only five subfamilies, placing *Aotus* (see account thereof) in the Callicebinae. There has been a recent trend toward dividing the traditional cebids into two full families, the Atelidae and the Cebidae, and including the Callitrichidae in the latter, but there is further disagreement about the details of the division. Ford (1986*b*)

considered the Atelidae to comprise the Alouattinae, the Atelinae, and the Pithecinae; Schneider et al. (1993) also included the Callicebinae; and Rosenberger, Setoguchi, and Shigehara (1990) added the Aotinae (thereby restricting the Cebidae to *Cebus*, *Saimiri*, and also the callitrichids).

Most of the authorities cited above, along with Kay (1990) and Thorington and Anderson (1984), agree that there is affinity between the genera *Lagothrix*, *Ateles*, *Brachyteles*, and *Alouatta* and between the genera *Pithecia*, *Chiropotes*, and *Cacajao*; a looser connection between both groups also seems to be recognized. There is a further partial consensus on some phylogenetic association of *Callicebus* and *Aotus*, and of those two genera and *Cebus*, *Saimiri*, and the callitrichids. These factors are taken into account in the following sequence of genera. The extinct *Xenothrix* is something of an enigma, having been assigned to various subfamilies and families over the years. MacPhee and Fleagle (1991) supported an early view that *Xenothrix* belongs in its own family, the Xenotrichidae, though they also recognized the Atelidae and the Cebidae as distinct families. Since such a division is not made here, *Xenothrix* is retained in the Cebidae and placed near *Callicebus* in keeping with the relationships suggested by Groves (1989) and Rosenberger, Setoguchi, and Shigehara (1990). The also extinct *Paralouatta* and *Antillothrix* are here placed adjacent to *Callicebus* in accordance with the finding by MacPhee et al. (1995) that they form a sister group to the latter (and also may have some phylogenetic association with *Xenothrix*). However, Rivero and Arredondo (1991) had concluded that *Paralouatta* is closely related to *Alouatta*.

Most cebids are much larger than any callitrichid. They range in size from *Aotus*, with a head and body length of 240–370 mm, to *Alouatta*, with a head and body length of up to 915 mm (*Brachyteles* may sometimes exceed *Alouatta* in weight). The tail is well haired in all members and long in all genera except *Cacajao*. The tail is fully prehensile in four genera: *Ateles*, *Brachyteles*, *Lagothrix*, and *Alouatta*. In these genera the underside of the tail is naked near the tip and soft and sensitive. The prehensile tail is used as though it were an additional hand—for grasping objects beyond the reach of the arms, in swinging and checking falls, and in steadying the animal by retaining a grasp with the tail when the hands and feet are being used in progression. The naked underside of the tip of the tail also serves as an organ of touch. In the squirrel monkeys *(Saimiri)* the tail is exceedingly mobile, but it is not prehensile. Monkeys of the genus *Cebus* carry the tail coiled and use it to brace or steady themselves, but it is not regularly used to grasp objects or to hang by.

Some zoologists use the general term Platyrrhine for the American monkeys, whose nostrils are widely separated and open to the sides, and the term Catarrhine for the Old World monkeys, in which the nostrils are set close together and open to the front. However, a number of species exhibit intermediate appearances.

In cebid monkeys the orbits are large and forwardly directed with large eyes and well-developed eyelids. The eyes of the nocturnal *Aotus* are very large. Except for this genus, the eye of cebid monkeys has the typical diurnal type of retina; that is, rods and cones are present and are arranged as in the human eye. The retina in *Aotus* contains rods only. None of the American monkeys possesses cheek pouches or rump callosities such as are present in many of the Old World monkeys. The digits, which are usually long and thin, bear flattened or curved nails. The thumb in cebid monkeys is not opposable, but the great toe is well developed and widely opposable to the other toes. A baculum is

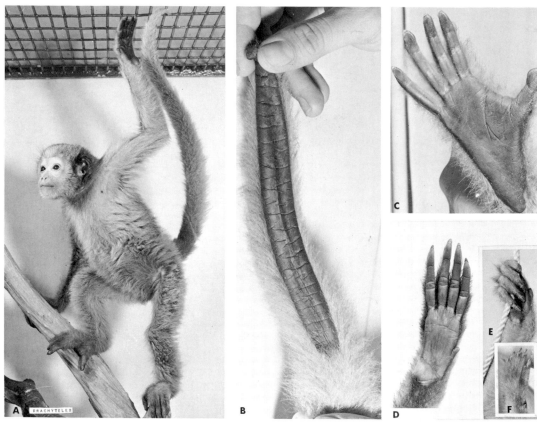

A. Woolly spider monkey *(Brachyteles arachnoides);* B. Underside of tail; C. Plantar surface of right foot; photos from New York Zoological Society. D. Palmar surface of right hand, photo by G. E. Erikson, Harvard Medical School. E. Grasping hand; F. Dorsal view of hand; photos from New York Zoological Society.

usually present but is lacking in *Aotus, Ateles,* and *Lagothrix.*

The dental formula is: (i 2/2, c 1/1, pm 3/3, m 3/3) × 2 = 36. The upper molars have four cusps, and the lower molars have four or sometimes five.

Most cebids are agile jumpers and runners, swinging and leaping through trees, but some forms are rather deliberate in their movements. Their movements tend to be smoother and less jerky than those of marmosets and tamarins. The claim of early writers that monkeys made living chains and bridges to travel between trees too far apart for leaping has generally been discredited, but Dr. Carl Lovelace, of Waco, Texas, stated in a letter to William Mann that he had seen such a performance in 1905 in South America: "These monkeys made a living bridge . . . formed by one monkey swinging tail down from a limb and wrapping his tail around the head of another monkey, and so on until five successive monkeys made a pendant chain which then swung back and forth until the ultimate monkey attached himself to the tree across the interval. The remaining monkeys, including two females with little baby monkeys around their necks, then crossed this bridge. The last monkey to cross was . . . evidently the leader of the band. . . . The initial end of the bridge then turned loose and the whole band proceeded in the original direction in less time than it has taken me to describe it."

The known geological range of the Cebidae is upper Oligocene to Recent. Rosenberger, Setoguchi, and Shigehara (1990) listed 10 extinct genera but assigned all to living subfamilies. MacPhee and Iturralde-Vinent (1995a) reported the first specimen of a New World monkey of Tertiary age to be discovered outside of South America. It is an astragalus found at an early Miocene site in central Cuba.

PRIMATES; CEBIDAE; **Genus LAGOTHRIX**
E. Geoffroy St.-Hilaire, 1812

Woolly Monkeys

There are two species (Fooden 1963):

L. lagotricha, eastern slope of the Andes in Colombia to the Rio Tapajós and the Mato Grosso in central Brazil, eastern Ecuador and Peru, possibly Bolivia;
L. flavicauda, eastern slope of Cordillera Central in northern Peru.

The name of the first species sometimes is spelled *L. lagothricha.*

Among American monkeys this genus is exceeded in weight only by some *Alouatta* and *Brachyteles* and in length by some individuals of *Ateles,* with their very long tails. Head and body length is about 508–686 mm, tail length is about 600–720 mm, and weight is 5.5–10.8 kg. Captives may be much heavier. The hair is short, thick, and woolly, with a good growth of underfur. In some individu-

Woolly monkey *(Lagothrix lagotricha)*. This photo, by Ernest P. Walker, illustrates the use of the prehensile tail.

als the color of the head is distinctly darker than that of the back; in others it is almost the same color. The upper parts are hoary gray, blue gray, tawny, dark brown, or blackish brown, and the underparts are only slightly lighter. Some individuals are yellowish buff. The naked face is almost black. *L. flavicauda* differs from *L. lagotricha* in having mahogany-colored pelage, a white or buffy circumbuccal patch on the muzzle, a long tuft of yellow scrotal hair, and a yellow band on the posteroventral surface of the tail (Mittermeier et al. 1977).

Woolly monkeys have a round and massive head, a relatively heavy body for tree monkeys, and a tail that is bare on the undersurface for several millimeters near the tip and prehensile. The thumbs and toes are well developed, and the fingers are short and thick with long pointed nails.

In Colombia *Lagothrix* lives in gallery forests, palm forests, flooded and nonflooded primary forest, and cloud forest as high as 3,000 meters but not in secondary forest. It is found in the crowns of the tallest trees as well as in the shrub layer (IUCN 1978). It seems to be less active in the wild than most American tree monkeys, and it definitely is less active in captivity. It is diurnal and arboreal but often comes to the ground, where it walks upright on its hind legs, using its arms to help maintain balance. When standing on its hind legs it sometimes uses the tail as a brace. Mittermeier and Coimbra-Filho (1977) reported the diet to consist mainly of fruits, supplemented by leaves, seeds, and some insects. Hernandez-Camacho and Cooper (1976), however, stated that leaves were not eaten.

In Peru, Neville et al. (1976) estimated a population density of 12–46/sq km. Studies in the upper Amazon region of Colombia showed that home range was at least 4 sq km

for four groups of *L. lagotricha* and that one group of 42–43 individuals used a nomadic range of 11 sq km, moving about 1 km per day (Izawa 1976; Nishimura and Izawa 1975). Overall group size range in this region was 20–70 individuals, there usually being 30–40. Groups sometimes included as many or more adult males than adult females. Nishimura (1990) added that group ranges overlapped extensively; two and occasionally three groups were observed to move in proximity for hours, usually without agonistic interaction. Individuals, especially subadult females, sometimes temporarily left their own group and spent a few hours or days with another. Males evidently maintained an intragroup dominance hierarchy through aggressive behavior. Hernandez-Camacho and Cooper (1976) observed groups of 4–6 *L. lagotricha* in Colombia, always including a rather old adult male. Available information on *L. flavicauda* indicates a group size of 4–35, with an average of 13 individuals (Mittermeier et al. 1977).

Female *L. lagotricha* carrying newborn young on their backs and bellies have been observed in the Amazon region of Brazil in November. In this species the estrous cycle is 12–49 days, estrus lasts 3–4 days, the gestation period is about 225 days, normal litter size is 1, weight at birth is 140 grams, lactation is estimated to continue for 9–12 months, and full sexual maturity is attained at 6–8 years by females and more than 5 years by males (Eisenberg 1977; Hayssen, Van Tienhoven, and Van Tienhoven 1993). A specimen of *L. lagotricha* lived for 24 years and 9 months in captivity (Jones 1982).

L. flavicauda, the yellow-tailed woolly monkey, is designated as critically endangered by the IUCN and as endangered by the USDI and is on appendix 1 of the CITES.

This species, restricted to a small area of montane rainforest on the eastern slope of the Andes in northern Peru, where fewer than 250 individuals now survive, was discovered by Alexander von Humboldt in 1802. There was no new knowledge of this monkey until two specimens were collected in 1925 and three in 1926, and then nothing was heard again until the species was rediscovered by an expedition in 1974 (Mittermeier et al. 1977). Despite legal protection, this monkey is subject to intensive hunting by people for food. The area where it lives is being opened through construction of new roads, and this increased accessibility will lead to more hunting pressure, habitat destruction, and probably extinction (Butchart et al. 1995; Thornback and Jenkins 1982). The subspecies *L. lagotricha lugens* of Colombia also numbers fewer than 250 animals and is classified as critically endangered by the IUCN. Two other subspecies of *L. lagotricha*—*cana* and *poeppigii* of Ecuador, Peru, and Brazil—are designated as vulnerable by the IUCN. *L. lagotricha* also is on appendix 2 of the CITES, and Peres (1991*a*) argued that the entire species should be reclassified as endangered as it is being rapidly reduced in range and numbers by human encroachment. It is perhaps more susceptible to hunting than any other primate in the New World tropics. Its meat is esteemed as food, and its large size makes it a choice target. It is also avidly sought for local use as a pet, fetching up to U.S. $80 each. Generally the mother must be killed in order to capture her infant, and it is estimated that at least 10 females are sacrificed for every live young that reaches the market. In addition, this species seems unable to adapt to secondary forest and so is especially vulnerable to human habitat modification. It and *Ateles* tend to be the first primates to disappear following human encroachment (Thornback and Jenkins 1982).

PRIMATES; CEBIDAE; Genus ATELES
E. Geoffroy St.-Hilaire, 1806

Spider Monkeys

Based on the provisional conclusions of Groves (1989 and *in* Wilson and Reeder 1993) and on information from Cabrera (1957), Eisenberg (1989), and E. R. Hall (1981), the following six allopatric species are recognized to exist at present on the mainland of Middle and South America:

A. geoffroyi, Tamaulipas (northeastern Mexico) and Jalisco (west coast of Mexico) to Panama;

A. fusciceps, eastern Panama, Colombia and Ecuador west of the Andes;

A. belzebuth, eastern Colombia and Ecuador, Venezuela, northeastern Peru, northwestern Brazil;

A. paniscus, north of the Amazon River in the Guianas and northeastern and central Brazil;

A. marginatus, Brazil south of the lower Amazon River;

A. chamek, south of the Amazon River in eastern Peru, northern and central Bolivia, and western Brazil as far south as the Mato Grosso.

Several authors (e.g., Hernandez-Camacho and Cooper 1976 and Hershkovitz 1972) consider all of the above species to be intergrading subspecies of the single species *A. paniscus.* An additional named species, *A. anthropomorphus,* is represented by a series of teeth found in a cave in central Cuba, supposedly in direct association with pre-Columbian Amerindian remains. Although there is a possibility that the specimens are from an endemic Cuban

Spider monkey *(Ateles geoffroyi)*, photo by Ernest P. Walker.

species that became extinct before or shortly after the arrival of European explorers (Ford 1990), recent morphological analysis and radiometric dating indicate that they most likely represent a captive *A. fusciceps* that was brought to Cuba by people within the last 300 years (MacPhee and Rivero 1996).

Head and body length is about 382–635 mm and tail length is about 508–890 mm. The weight is as much as 8 kg for captives, but it is probably nearer 6 kg for animals in the wild. Most forms have coarse, stringy hair, but some have a soft and fine pelage. Underfur is lacking. The coloration above is yellowish gray, darker grays, reddish brown, darker browns, or almost black; the sides may be golden yellow to rufous; and the underparts are lighter, usually whitish or yellowish. Most forms have a black face with white eye rings, but some forms have a flesh-colored face.

Spider monkeys have an exceptionally long tail and legs in relation to body length, a prehensile and extremely flexible tail, and thumbs that are poorly developed or lacking. The head is small, and the muzzle prominent. The clitoris of the female is greatly elongated, so that it is often mistaken for the penis of the male.

Spider monkeys are found in rain and montane forests and tend to occupy small branches of the high strata of the canopy. They are entirely diurnal, may feed intensively early in the morning, and rest for most of the remainder of the day (Hall and Dalquest 1963; Klein and Klein 1977).

The diet consists largely of fruits, supplemented by nuts, seeds, buds, flowers, leaves, insects, arachnids, and eggs (Mittermeier and Coimbra-Filho 1977). In Surinam, Van Roosmaien (1985) found groups of *A. paniscus* to move as much as 5,000 meters per day during the wet season (March–June), when fruit was abundant, and to have prolonged feeding periods at many feeding sites, apparently thereby building up energy reserves. During the subsequent dry season (July–September), when fruit was scarce, foraging covered as little as 500 meters per day, feeding times were short, and resting periods were long.

Probably only the gibbons exceed *Ateles* in agility in trees. Spider monkeys move swiftly through trees and use their tails as a fifth arm or leg. They pick up objects with the tail and can hang by a single hand or foot or by the tail alone. The normal movement through trees is along the upper surfaces of limbs, with the tail arched over the back. Only rarely do they descend to the ground. When approached in the wild, spider monkeys sometimes break off dead branches weighing nearly 5 kg and drop them, attempting to hit the observer. They also emit terrierlike barks when approached. The most frequently heard call resembles the whinnying of a horse and is made when the monkeys are separated.

Population densities of 1–35/sq km and home ranges of 1–4 sq km have been recorded (Eisenberg 1989; Wolfheim 1983). In studies of *A. paniscus* in Peru, Symington (1988) found group home ranges of 150–250 ha. Overlap with neighboring groups was only 10–15 percent. Females, particularly those with infants, tended to restrict their movements to a core area, 20–33 percent of the total group range, while males used the entire area. Apparently, the males were cooperatively defending this area as a joint territory. Although there generally was little contact between groups, at least some of the young females evidently emigrated to other groups. Males tended to stay with their natal unit. Van Roosmaien (1985) reported a similar pattern in Surinam and noted that older females apparently lead the foraging groups.

It is difficult to describe the social structure of these monkeys. Group size seems to vary considerably, partly with respect to habitat conditions, and there is a tendency for large bands regularly to divide into subgroups. Symington (1988) suggested that *Ateles* has a fission-fusion society such as has been described for the chimpanzee. Aggregations of up to 100 individuals have been reported, but the usual size range is 2–30 (Freese 1976; Hall and Dalquest 1963; Hernandez-Camacho and Cooper 1976; Izawa 1976; Leopold 1959; Mittermeier and Coimbra-Filho 1977; Napier and Napier 1967; Vessey, Mortenson, and Muckenhirn 1978). Probably the higher figures represent sightings when groups are united and the lower figures represent observations of subgroups. In a study in Colombia, Klein and Klein (1977) determined that in an 8-sq-km area there were three distinct and mutually exclusive social networks of 20 adults, almost always dispersed into subgroups averaging about 3.5 individuals each. In southeastern Peru, Durham (1971) found that average group size decreased sharply from 18.5 individuals at an elevation of 275 meters to 11 individuals at 576 meters and then less sharply to 7 at 889 meters and 4.5 at 1,424 meters. At higher altitudes there was only a single adult male in each group, but at 275 meters the number of adult males averaged 3.75.

There appears to be no regular breeding season. The estrous cycle is 24–27 days, estrus is 2 days, gestation is 200–232 days, the normal litter size is 1, birth weight is 300–500 grams, the interbirth interval averages about 34.5 months, lactation has been reported to last from 10 weeks to more than 2 years, and the age of sexual maturity is

about 5 years in males and 4 years in females (Eisenberg 1977; Hayssen, Van Tienhoven, and Van Tienhoven 1993; Symington 1988). A captive *A. geoffroyi* lived for 48 years and a specimen of *A. chamek* was still living at about 40 years (Marvin L. Jones, Zoological Society of San Diego, pers. comm., 1995).

In many parts of their range spider monkeys are avidly hunted for use as food. They are easily located because of their large size and noisy habits, and populations have been eliminated in some accessible areas (Hernandez-Camacho and Cooper 1976; Mittermeier and Coimbra-Filho 1977). Although they are relatively tolerant of a limited amount of logging and land clearing, they still require large tracts of tall forest and thus are vulnerable to human colonization (Leopold 1959). They formerly occurred throughout the tropical parts of Veracruz, Mexico, but now are found only in the south of that state (Hall and Dalquest 1963). Even in that area their range has become fragmented, and many populations have been exterminated (Estrada and Coates-Estrada 1984). The habitat of the two subspecies of *A. geoffroyi* in southern Mexico, *vellerosus* and *yucatanensis*, has now been reduced by at least 90 percent (Estrada and Coates-Estrada 1988). The subspecies *A. geoffroyi frontatus* and *A. geoffroyi panamensis*, of Central America, are listed as endangered by the USDI and are on appendix 1 of the CITES; all other species and subspecies are on appendix 2.

The IUCN classifies the species *A. marginatus* as endangered and the species *A. fuscipes* and *A. belzebuth* generally as vulnerable. However, the IUCN also classifies the following subspecies individually: *A. geoffroyi yucatanensis* (Mexico, Belize, Guatemala), vulnerable; *A. g. frontatus* (Nicaragua, Costa Rica), vulnerable; *A. g. ornatus* (Costa Rica), vulnerable; *A. g. panamensis* (Costa Rica, Panama), endangered, fewer than 2,500 individuals surviving; *A. g. azurensis* (Azuero Peninsula and adjacent parts of Panama), critically endangered, fewer than 250 individuals surviving; *A. g. grisescens* (eastern Panama and adjacent parts of Colombia), endangered, fewer than 2,500 individuals surviving; *A. fuscipes fuscipes* (Pacific slope forests of northern Ecuador), critically endangered, fewer than 250 individuals surviving, continuing to decline precipitously; *A. belzebuth brunneus* (Colombia), endangered; *A. b. hybridus* (northern Colombia and Venezuela), endangered. All of these species and subspecies are threatened by habitat destruction and hunting by people (Mittermeier et al. 1986; Thornback and Jenkins 1982). The same problems are leading to the rapid disappearance of *A. paniscus* in Surinam (Van Roosmaien 1985).

PRIMATES; CEBIDAE; **Genus BRACHYTELES**
Spix, 1823

Muriqui, or Woolly Spider Monkey

The single species, *B. arachnoides*, occurs in southeastern Brazil, formerly being found from the state of Bahia to the northern part of the state of Paraná (Cabrera 1957; Martuscelli, Petroni, and Olmos 1994). There are two distinct subspecies, *B. a. arachnoides* to the north and *B. b. hypoxanthus* to the south in the states of Sao Paulo and Paraná. Recent morphological assessments led Rylands, Mittermeier, and Luna (1995) to recognize *hypoxanthus* as a full species, though Strier (1995) continued to treat it as a subspecies.

Head and body length is 460–630 mm and tail length is 650–800 mm. Milton (1984) indicated that weight is 12–15

Woolly spider monkey *(Brachyteles arachnoides)*, photo by Andrew Young.

kg, which would make *Brachyteles* the largest New World monkey. Peres (1994) questioned this size range, noting that recent weights of live-captured animals have been about 7–10 kg, but added that individuals of a Pleistocene population may have weighed up to 20 kg. Males and females of the living population are about equal in size (Strier 1992). The pelage is woolly, and the hairs of the head are short and directed backward. There may be a sexual difference in coloration, but this may be only individual and geographic variation. Most muriquis have a prevailing color of yellowish gray or ashy brown, but some are reddish. The head is blackish brown tinged with yellow, or dark gray washed with brown, or the forehead and back of the neck may be orange rufous and the top of the head chestnut. The flat, naked face is often brilliant red in color, particularly when the animal is excited.

The head is round, the body is heavy, the limbs are long and slender; the tail is longer than the body, naked beneath near the tip, and prehensile. *Brachyteles* resembles *Ateles* in such features as the long limbs. Its general appearance is even more like that of *Lagothrix*. The thumb in *Brachyteles* is vestigial or lacking, and the nails of the fingers and toes are sharp and laterally compressed. Lemos de Sá et al. (1990) found that *B. a. arachnoides* has a small but still obvious thumb, whereas *B. a. hypoxanthus* has no thumb at all.

The muriqui is found only in undisturbed high forest, including both lowland tropical and montane rainforest, and is predominantly arboreal (IUCN 1978). It long had been among the least known primates, but a series of recent studies have provided considerable new information. Strier (1992) found that during the hot summer groups begin to forage just after sunrise, rest in the shade during the middle of the day, and then move again in the late afternoon until sunset. During the winter they sun themselves

for hours before beginning to forage and then are active until about an hour before sunset. Daily movements cover about 1,000 meters in the winter and 1,400 meters in the summer. Most locomotion is through the trees, using the limbs and tail, but individuals occasionally come to the ground to drink or eat certain items. *Brachyteles* is strongly folivorous, spending about 51 percent of its annual foraging time feeding on leaves, 32 percent on fruits and seeds, and 11 percent on pollen and nectar. Fruit increases in importance at times, and individuals may remain for several days in areas with abundant fruit.

Strier et al. (1993) reported the results of a nine-year study of two groups of muriquis inhabiting an 800-ha. private reserve. At the start of the study the main study group had about 24 members, including 6 adult males and 8 adult females, and used a home range of 168 ha. The other group contained 18 individuals. The study group eventually grew in size to 42 members, mainly through the birth of 23 young, and its range nearly doubled in size. This group evidently formed an intact social unit for six years, but subsequently there was an increasing subdivision into two groups, which stayed apart for days at a time and increasingly interacted with the other original group on the reserve. The following additional observations of group and range sizes were listed by Torres de Assumpcao (1983): 7 or 8 individuals on 170 ha., 13 on 105 ha., 12 on 217 ha., and 30 on 580 ha. The group with 13 monkeys had 4 adult males, 4 adult females, and 5 young. Milton (1984, 1985) indicated that home ranges of females cover an estimated 70 ha., including a core area of 4 ha. where the animal often rests and sleeps, and that there is no territorial defense, though there is a dominance hierarchy among females. Milton's observations indicated somewhat smaller group size and looser social organization than found in Strier's study, but both agreed on a remarkable lack of aggression.

Communication may be carried out through the depositing of urine on the hands and tail, and there are also eight known vocalizations, including a loud, piercing call for ascertaining the location of other individuals.

Strier (1990, 1992) reported that males and females are equal in rank and that there is little agonistic interaction between males. Receptive females attract numerous males and mate with several of them. In contrast to the behavior of most social primates, young male *Brachyteles* tend to remain in their natal group, but young females emigrate at about 5–6 years and join other groups. Females apparently are polyestrous and have an estrous cycle of about 4 weeks, an estrus of 2–3 days, and a gestation period of 7–8.5 months. Births are highly seasonal, being concentrated in the dry winter months (May–September). Females seem to give birth about once every 3 years. There usually is a single young, but twins have been observed on rare occasions. The infant initially is carried on its mother's belly or flanks but begins to ride on her back by 6 months. It is weaned at 18–30 months. The earliest known age of reproduction in a female is 7.5 years.

Although *Brachyteles* once presumably occupied all Atlantic coastal forests of eastern and southeastern Brazil, it is now only sparsely distributed in the states of Sao Paulo, Minas Gerais, Rio de Janeiro, and Paraná. Mittermeier (1987) estimated its total numbers at not over 400, compared with perhaps 400,000 in 1500. Martuscelli, Petroni, and Olmos (1994) reported recent surveys showing the current figure to be nearly twice as great and also cited still higher estimates but indicated that the species is continuing to decline because of hunting for food and clearing of forests for agriculture and settlement even within supposedly protected areas. More than 95 percent of the kind of forest on which it depends has been eliminated (Milton 1986). It is classified as endangered by the USDI and is on appendix 1 of the CITES. The IUCN recognizes *arachnoides* and *hypoxanthus* as separate species and designates both as endangered.

PRIMATES; CEBIDAE; Genus ALOUATTA
Lacépède, 1799

Howler Monkeys

There are nine species (Cabrera 1957; Froehlich and Froehlich 1987; E. R. Hall 1981; Hershkovitz 1972; Horwich 1983; Rylands, Mittermeier, and Luna 1995; Wolfheim 1983):

A. palliata, southern Mexico to northwest coast of South America, possibly as far south as Peru;
A. coibensis, Azuero Peninsula and Coiba Island of Panama;
A. pigra, Yucatan, Guatemala, Belize;
A. seniculus, Colombia, Venezuela, Guianas, eastern Ecuador, northeastern Peru, Amazonian Brazil, Bolivia;
A. arctoidea, northern Colombia and Venezuela;
A. sara, southeastern Bolivia;
A. belzebul, Amazonian Brazil and adjacent regions;
A. fusca, Bolivia, eastern Brazil, extreme northeastern Argentina;
A. caraya, eastern Bolivia, southern Brazil, Paraguay, northern Argentina.

Groves (*in* Wilson and Reeder 1993) included *A. arctoidea* in *A. seniculus*.

Howlers are among the largest of New World monkeys: head and body length is 559–915 mm, tail length is 585–915 mm, and adult weight generally ranges from 4 to 10 kg. Data compiled by Peres (1994) indicate that males are substantially larger than females and that those of *A. pigra* weigh up to 11.6 kg. The hair is coarse, and in some species it is long on the head and shoulders; the face is naked. Coloration is of three types—yellowish brown, deep reddish brown, and black—but there is considerable age and individual variation. Males are generally larger than females, and sometimes the sexes are dichromatic (Crockett and Eisenberg 1987). In *A. caraya* males average 6.7 kg and are all black, and females average 4.4 kg and are yellow-brown (Thorington, Ruiz, and Eisenberg 1984).

The angle of the lower jaw and the hyoid bone are both greatly enlarged, making it possible for the animal to produce such a remarkable voice. The lower jaw and neck are quite large, which, with other characteristics, such as the low facial angle, gives the animals a forbidding expression. The legs are shorter and stouter than those of the spider monkeys (*Ateles*). The strongly prehensile tail is naked on the underside for the terminal third. The tail is so powerful that the animal may jump from a limb but check its flight by not releasing its tail hold. Also, if a monkey falls, it can keep from hitting the ground by grabbing onto a limb with its tail.

Howlers are arboreal, mainly diurnal forest dwellers. Mittermeier and Coimbra-Filho (1977) reported that *A. seniculus* inhabits cloud forest, rainforest, secondary forest, and mangroves. Hernandez-Camacho and Cooper (1976) stated that *A. seniculus* is an able swimmer. Freese (1976) observed that *A. palliata* prefers primary forest and usually avoids scrub forest. According to Carpenter (1965), the daily pattern of *A. palliata* on Barro Colorado Island in the Panama Canal Zone is as follows: a bout of roaring at dawn, rest, feeding in the immediate vicinity of the sleeping area, moving out to food trees in the midmorning, active feeding, rest till midafternoon, and then movement, with simultaneous vocalization, to the night lodging trees. C. C. Smith (1977) found that adults on Barro Colorado Island rested for 74 percent of the day and all of the night, fed for most of the remaining time, and used only 4 percent of their time for social activity. Glander (1978) reported a daily movement of 207–1,261 meters for *A. palliata*. Other reported day ranges for *Alouatta* average 123–706 meters (Crockett and Eisenberg 1987). The diet of *Alouatta* is varied but consists mainly of leaves, fruits, and other vegetable matter (Hernandez-Camacho and Cooper 1976; C. C. Smith 1977). According to Mittermeier and Coimbra-Filho (1977), this genus eats more leaves than does any other New World monkey. Crockett and Eisenberg (1987), however, suggested that *Alouatta* should be labeled a folivore-frugivore rather than simply a folivore.

In Venezuela, Sekulic (1982a) found that the abundance of fig trees was a principal determinant of group size for *A. seniculus*, with smaller groups being maintained when fewer figs were available. Four groups moved an average of 340–445 meters per day and slept in the tallest available trees at night. Population density in the area was more than 115/sq km, and yearly home range was 3.88–7.44 ha. Territoriality was not evident, with ranges overlapping by 32–63 percent. Groups contained an average of 8.9 (4–17) individuals, including 1–3 adult males and 2–4 adult females. Young of both sexes dispersed from their natal groups.

Klein and Klein (1976) reported population densities of about 12–30/sq km for *A. seniculus* and 12–15/sq km for *A. belzebul* in La Macarena National Park in Colombia. Thorington, Ruiz, and Eisenberg (1984) determined that a population of *A. caraya* in northern Argentina had a den-

Howler monkey *(Alouatta palliata),* photo from New York Zoological Society.

sity of 130/sq km and troop sizes of 3–19; groups included 1–3 adult males and 2–7 adult females. Baldwin and Baldwin (1976) reported a density of 1,040/sq km for *A. palliata* in southwestern Panama and noted that this figure was 12–21 times greater than densities found earlier on Barro Colorado Island. Crockett and Eisenberg (1987), however, pointed out that this figure was based on a relatively brief and unnatural situation and that other reported densities include 16–90/sq km for *A. palliata,* 8–22/sq km for *A. pigra,* and 15–118/sq km for *A. seniculus.* Group home ranges determined in Baldwin and Baldwin's study were 3.2–6.9 ha., and Crockett and Eisenberg cited other reported ranges of 10–60 ha. for *A. palliata,* 11–125 ha. for *A. pigra,* and 4–25 ha. for *A. seniculus.* In another investigation of *A. palliata,* Neville et al. (1976) found home ranges of 4.89–76.00 ha. Izawa (1976) reported that one group of *A. seniculus* used a nomadic range of 3–4 sq km. He also observed that group size averaged 5.4 and that there appeared to be 1 dominant male in the group. Various other reports indicate an overall range of 2–15 individuals in groups of *A. seniculus,* with 4–7 being most common (Hernandez-Camacho and Cooper 1976; Izawa 1976; Klein and Klein 1976; Mittermeier and Coimbra-Filho 1977; Neville et al. 1976; Vessey, Mortenson, and Muckenhirn 1978).

There have been a number of studies of social structure of *A. palliata* (Baldwin and Baldwin 1972, 1976; Carpenter 1965; Freese 1976; Heltne, Turner, and Scott 1976; Mittermeier 1973; Mittermeier and Coimbra-Filho 1977; C. C. Smith 1977). Group size in this species varies from 2 to 45, there generally being about 10–20 individuals. Each group contains about 2–4 adult males and 5–10 adult females. The males have a dominance hierarchy and appear to act in concert during travel and vocal battles with other groups. Females seem to maintain a peaceful relationship among themselves. A few males live alone but not in complete isolation. Groups generally are closed units, but sometimes they split up and new groups are formed. Occasionally several groups join in an aggregation of 38–65 animals. Most troops have little or no area of exclusive use, but they do defend the place where they happen to be at a given time. The dawn chorus of loud calls seems to function as a means of preventing direct contact between groups.

New studies (Crockett and Eisenberg 1987) indicate that altough troops of *A. palliata* have several adult males, other species of *Alouatta* usually have only 1–2 adult males and 2–3 adult females. Competition between males is severe, with invasions and infanticide by adult males reported for *A. palliata* and *A. seniculus.* Maturing individuals of both sexes usually emigrate from troops of these species, perhaps because of the limited opportunities for reproduction in an established hierarchy. As a result, the adults of both old and new groups have few or no close relatives therein.

All the species of this genus are particularly noted for their remarkably loud and persistent calls, described by some observers as deep howls and by others as deep, carrying growls comparable to the roars of lions. These calls have been heard by people 3 km away through the jungle and 5 km away across lakes. The cries are made by the males and may be emitted at any time the animals are active, though Horwich and Gebhard (1983) reported *A. pigra* to have intense periods of roaring at dawn and dusk during the dry season. Sekulic (1982*b*) suggested that the roars of *A. seniculus* elicit responses that allow it to assess neighboring groups, as an alternative to energetically expensive chases and fights. In addition to these calls, Carpenter (1965) identified a "deep cluck" used by a leader to coordinate group movement, "grunts" of adults when disturbed, and "wailing" by females to solicit help in retrieving separated young.

Alouatta appears to breed throughout the year, though Crockett and Rudran (1987*a*, 1987*b*) found *A. seniculus* in two habitats of Venezuela to have a consistently reduced birth frequency during the early wet season, in May–July; females there had an average interbirth interval of 17 months. Shoemaker (1982*a*) reported the interbirth interval in captive *A. caraya* to be 12–15 months for newly matured females and 7–10 months for older mothers. The estrous cycle in *A. palliata* is 13–24 days and gestation is 180–94 days (Eisenberg 1977). Figures for other species fall within the same range (Crockett and Eisenberg 1987). There normally is one young. That of *A. palliata* weighs 275–400 grams at birth (Hayssen, Van Tienhoven, and Van Tienhoven 1993). It clings to its mother's fur, and as it gets older it makes its permanent riding position on her back. This continues for about 1 year. Crockett and Rudran (1987*b*) found that 80 percent of the young of *A. seniculus* survived at least 1 year but that 44 percent of infant mortality was the result of infanticide by older males. Studies on Barro Colorado Island (Carpenter 1965; Froehlich, Thorington, and Otis 1981) indicate that the young of *A. palliata* are weaned at around 10 months and that sexual maturity is achieved at 3–4 years by females and 5 years by males but that several more years may pass before males attain social maturity and are allowed to mate. The average adult life span in this population is about 16 years, and maximum known longevity there is over 20 years. A captive *A. caraya* was still living at about 23 years (Marvin L. Jones, Zoological Society of San Diego, pers. comm., 1995).

The USDI lists *A. palliata* as endangered and *A. pigra* as threatened. Those two species also are on appendix 1 of the CITES, and all other species are on appendix 2. The IUCN classifies the species *A. fusca* and the subspecies *A. palliata mexicana* (southern Mexico, Guatemala) and *A. seniculus insulans* (Trinidad) as vulnerable, the subspecies *A. coibensis coibensis* (Coiba Island off Panama) as endangered, and the subspecies *A. coibensis trabeata* (Azuero Peninsula of Panama), *A. fusca fusca* (southeastern Brazil), and *A. belzebul ululata* (southeastern Brazil) as critically endangered. The last three of these subspecies have each been reduced to fewer than 250 individuals and are continuing to decline because of rapid human destruction of their limited habitat. Mittermeier et al. (1986) also considered the subspecies *A. fusca clamitans* of southeastern Brazil and adjacent Argentina to be endangered. Oliver and Santos (1991) reported that it still occupied an extensive range, but Di Bitetti et al. (1994) indicated that its range had been drastically reduced and fragmented. In addition, all species of *Alouatta* are hunted for food and are subject to commercial export. Mittermeier and Coimbra-Filho (1977) observed that *A. seniculus* was easy to locate because of its loud voice and had become rare in some areas but was still abundant in Amazonian Brazil.

Investigations by Estrada and Coates-Estrada (1984, 1988) and Horwich and Johnson (1986) indicate that habitable areas for *A. palliata* and *A. pigra* are rapidly being lost in Mexico and Central America and that troop sizes are falling as forests become fragmented. There was relatively little destruction of such habitat in southern Mexico until the 1940s, but in the last 40 years at least 90 percent of the original rainforests in the region have been converted to agriculture and pasture. The basic cause is rapid growth of the human population and a consequent demand for land to raise crops and graze livestock. Remaining monkey populations are small and scattered and could disappear entirely by the end of the century.

PRIMATES; CEBIDAE; **Genus PITHECIA**
Desmarest, 1804

Sakis

There are five species (Bodini and Pérez-Hernández 1987; Hershkovitz 1979, 1987*b*):

P. pithecia, southern and eastern Venezuela, Guianas, northeastern Brazil;

P. monachus, Amazon Basin of southern Colombia, Ecuador, Peru, and northwestern Brazil;

P. irrorata, south of the Amazon River in western Brazil, southeastern Peru, and northern Bolivia;

P. aequatorialis, known only from three areas in the upper Amazon Basin of eastern Ecuador and northeastern Peru;

P. albicans, south bank of the Amazon between the Jurua and Purús rivers in western Brazil.

According to Hershkovitz (1987*b*), the species now known as *P. monachus* was long called *P. hirsuta*, and the species now known as *P. aequatorialis* was previously identified as *P. monachus*.

Head and body length is 300–705 mm, tail length is 255–545 mm, and adult weight is generally 700–1,700 grams. *P. monachus*, *P. irrorata*, and *P. aequatorialis* are generally dark agouti in color with pale hands and feet. The underparts of *P. aequatorialis* vary from pale yellowish brown to rufescent; those of *P. monachus* and *P. irrorata* are predominantly blackish. The crown pelage of *P. aequatorialis* is short and stiff in males and long and lax in females, and the hood partly conceals the ears and face. *P. albicans* has a predominantly black back and tail, and the rest of the body is buffy to reddish. Male *P. pithecia* are uniformly blackish except that the crown, face, and throat are whitish to reddish; the hood does not cover the face. Female *P. pithecia* are predominantly blackish or brownish agouti and have a bright stripe extending from beneath each eye to the corner of the mouth or chin.

Sakis inhabit forests, usually at elevations of 210–750 meters. They are essentially diurnal and wholly arboreal but sometimes descend to the lower limbs of trees or even to bushes in search of food. Happel (1982) reported *P. monachus* to rest and feed mainly in the upper, emergent layers of the forest canopy, at heights of 15–24 meters, but to travel mostly at 10–19 meters. Sakis descend vertical supports tail first. They usually move about on all fours, but they are capable of making long leaps in trees and also have been observed running on horizontal branches in an erect posture with their arms held upward and their fingers extended. When alarmed, they can travel through trees with considerable speed. The pale-headed saki in its sleeping position coils up on a branch like a cat. Sakis, like many other monkeys, sometimes hang by their hind limbs when feeding. The diet consists of berries and other fruits, honey, leaves, flowers, small mammals such as mice and bats, and small birds. Sakis tear mammals and birds apart with their hands before eating them. In the Guianas, sakis have been observed going into tree hollows to collect bats, which they tore apart and skinned before eating.

Neville et al. (1976) reported a population density of 7.5–30.0/sq km in Peru. Oliveira et al. (1985) reported that *P. pithecia* occurred at a density of 40/sq km in Brazil and that groups used home ranges of 8–13.5 ha. These authorities and others (Hernandez-Camacho and Cooper 1976; Izawa 1976; Mittermeier and Coimbra-Filho 1977) indicate that sakis may occur alone or in small family groups com-

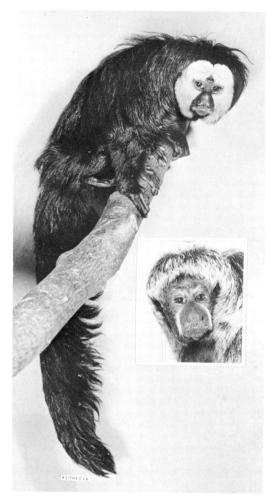

Saki *(Pithecia pithecia)*, male, photo by Ernest P. Walker.
Inset: saki *(P. monachus)*, female, photo from Zoological
Society of Philadelphia.

prising an adult pair and several young. In Guyana, Vessey, Mortenson, and Muckenhirn (1978) found group size of *P. pithecia* to average 3.3 and range from 1 to 5 individuals. In Peru, Happel (1982) observed four groups of *P. monachus* with a mean size of 4.5. Sakis are usually silent in captivity but emit a loud and penetrating call in the wild. Nervous and melancholy in appearance, they are essentially gentle in disposition and have made affectionate pets.

Well-developed young have been observed in Surinam at the beginning of the rainy season. Oliveira et al. (1985) indicated that a birth of *P. pithecia* evidently took place in late November or December. Johns (1986) recorded births of *P. albicans* in March, June, November, and December and noted that the young became totally independent after 6 months. Shoemaker (1982*b*) reported births of *P. pithecia* to occur throughout the year in captivity. He estimated the estrous cycle to last 18 days. A single young is produced. One infant was born to a male about 35 months old and a female about 26 months old. The gestation period of *P. pithecia* is 163–76 days (Eisenberg 1989). A specimen of *P. pithecia* living in the Pittsburgh Zoo at the end of 1994 was approximately 35 years old (Marvin L. Jones, Zoological Society of San Diego, pers. comm., 1995).

These monkeys are hunted for food and captured for use as pets but do not appear to be endangered in Brazil (Mittermeier and Coimbra-Filho 1977). Even *P. albicans*, with its very limited range, is reported to be common and not threatened (Johns 1986), though Mittermeier et al. (1986) warned that it should be carefully watched. All species are on appendix 2 of the CITES, and the IUCN classifies the subspecies *P. monachus milleri*, of Colombia, as vulnerable.

PRIMATES; CEBIDAE; **Genus CHIROPOTES**
Lesson, 1840

Bearded Sakis

There are two species (Hershkovitz 1985):

C. albinasus, Brazil from the Amazon and Xingu rivers to the northern Mato Grosso;
C. satanas, southern Venezuela, Guianas, Brazil north of the Amazon, Brazil south of the Amazon and east of the Xingu.

Head and body length is 327–511 mm, tail length is 300–507 mm, and weight is 2,000–4,000 kg (Hershkovitz 1985). The species *C. albinasus* is shiny black in color. The nose and upper lip are covered with downward-growing, stiff, yellowish white hairs; the red skin in this area contrasts sharply with the blackness of the remainder of the body. The species *C. satanas* is colored as follows: the head and beard are black; the shoulders and back, including the flanks, vary from dark reddish chestnut to blackish brown; the limbs are black or chestnut; the hands and feet are

Red-backed saki *(Chiropotes satanas)*, photo from New York Zoological Society.

blackish or reddish; and the tail is black. The beard of *Chiropotes* is well developed, and the tail is so heavily furred that it is full and thick.

In Venezuela, Handley (1976) collected all 64 specimens of *C. satanas* in trees near streams within evergreen or deciduous forest. *C. albinasus* is reported to inhabit tropical rainforest in unflooded areas and sometimes flooded forest along riverbanks (IUCN 1978). The diet of the genus consists of fruits, nuts, flowers, leaves, apparently insects, and possibly small vertebrates (Ayres and Nessimian 1982; Mittermeier and Coimbra-Filho 1977; Mittermeier et al. 1983). Population densities of about 4–82/sq km have been recorded for *C. satanas* (Wolfheim 1983). In Guyana, Vessey, Mortenson, and Muckenhirn (1978) found group size in *C. satanas* to average 13.1 and range from 4 to 20 individuals. Eisenberg (1989) referred to troops of up to 25 and to one group with a home range of nearly 100 ha. Welker and Schäfer-Witt (*in* Grzimek 1990) wrote that the gestation period is about 5 months, there is a single young per birth, sexual maturity is attained at 4 years, and longevity is more than 18 years.

C. albinasus is classified as endangered by the USDI and is on appendix 1 of the CITES. It is declining because of the destruction of its forest habitat and excessive hunting by people (Thornback and Jenkins 1982). The subspecies *C. sa-tanas satanas,* found in Brazil to the south of the Amazon and east of the Tocantins, is classified as endangered by the IUCN. Although reportedly intolerant of human disturbance, its range now is densely populated by people and subject to logging and other environmental disruption (Johns 1985). Mittermeier et al. (1986) regarded it as perhaps the most endangered Amazonian primate. The subspecies *C. satanas utahicki,* found from the Tocantins west to the Xingu, is classified as vulnerable by the IUCN.

PRIMATES; CEBIDAE; Genus CACAJAO
Lesson, 1840

Uakaris, or Uacaris

There are two species (Barnett and Da Cunha 1991; Boubli 1993; Cabrera 1957; Hernandez-Camacho and Cooper 1976; Hershkovitz 1972, 1987*c*):

C. melanocephalus, southeastern Colombia, southern Venezuela, northwestern Brazil;

C. calvus, upper Amazonian region of western Brazil, eastern Peru, and southern Colombia.

Red uakaris *(Cacajao calvus rubicundus)*, photo from New York Zoological Society.

Head and body length is 300–570 mm, tail length is 125–210 mm, and weight is about 2.7–3.5 kg (Hershkovitz 1987c). Uakaris are the only short-tailed American monkeys, the tail being less than half the length of the head and body, often much less. The face and cheeks are almost naked, but the jaws and throat are moderately haired, sometimes with a well-developed beard. In this genus the ears are large and may resemble those of *Homo*. In *C. calvus* the top of the head may appear nearly bald, having only a few scattered silky hairs, and the ears are exposed. In *C. melanocephalus* the top of the head is fully haired and the ears are concealed by fur.

There are several rather distinctive patterns of coloration (Hershkovitz 1987c). The subspecies *C. calvus calvus* of northwestern Brazil, known as the "white uakari," is whitish, yellowish, or buffy in color, except for the face, which is scarlet. The subspecies *C. calvus ucayali* of eastern Peru and immediately adjacent Brazil is entirely reddish orange or golden. *C. calvus rubicundus* of the Brazil-Colombia border region, sometimes called the "red uakari," is bright reddish brown to chestnut brown throughout, except for the strikingly paler nape of the neck, the hands, feet, and ears, which are brown, and the face and forehead, which are almost vermillion red. The subspecies *C. calvus novaesi*, found farther to the east, has the same general appearance, but the pale color of the nape extends down the middle of the back. *C. melanocephalus melanocephalus* of southern Venezuela and adjacent Brazil, the "black-headed uakari," is chestnut brown, except for the face, shoulders, arms, hands, feet, and lower surface of the tail, which are black. *C. melanocephalus ouakary* of northwestern Brazil and adjacent Colombia looks about the same, but the middle of the back is a contrasting pale orange, golden, or buffy.

Uakaris are more restricted in habitat than most South American primates; they seem to be found mostly along small rivers and lakes within forests and to avoid the margins of large rivers (Mittermeier and Coimbra-Filho 1977). They frequent the tops of large trees and rarely descend to the ground, at least during the part of the year when the forest floor is flooded. They are most active during the daytime and progress quite rapidly and nimbly on all fours but do not leap. Uakaris subsist chiefly on fruits but probably also eat leaves, insects, and other small animals.

In Brazil, Mittermeier and Coimbra-Filho (1977) observed groups of 10 to more than 30 individuals and heard reports that groups of 30–50 were not rare and that some troops contained as many as 100 animals. Fontaine and Du Mond (1977) reported that the members of a captive colony of *C. calvus rubicundus* in Florida established a linear dominance hierarchy through noisy fighting but without apparent injury. The breeding season in this and other captive groups in the Northern Hemisphere lasts from May to October. Females in the Florida colony were able to breed at 3 years, and one produced viable young until 11 years. Males became proven breeders at 6 years. Females gave birth to one young at a time, at intervals of about 2 years. One female appeared to be about 12 years old when brought into the colony and then lived for 11 more years. Hayssen, Van Tienhoven, and Van Tienhoven (1993) listed an estrous cycle of 14–48 days and a lactation period of about 21 months. According to Marvin L. Jones (Zoological Society of San Diego, pers. comm., 1995), a captive specimen of *C. calvus rubicundus* lived to at least 31 years.

Both species of *Cacajao* are listed as endangered by the USDI and are on appendix 1 of the CITES. The IUCN classifies the subspecies *C. calvus ucayali* as vulnerable and the subspecies *C. c. calvus*, *C. c. rubicundus*, and *C. c. novaesi* as endangered. All are declining because of human disruption of their limited habitat. *C. melanocephalus* is not now classified by the IUCN. *C. melanocephalus ouakary* of northwestern Brazil and adjacent Colombia is the most widely distributed subspecies. Barnett and Da Cunha (1991) reported that it still is common, even in the vicinity of long-established villages, but that it is subject to heavy hunting pressure and is declining in those areas now being settled and disrupted. *C. calvus* has become rare in Peru because of excessive hunting and collection for use as a pet. The subspecies *C. calvus calvus* is naturally restricted to a very small area surrounded by rivers in northwestern Brazil. Although it is not often hunted at present, increased access to its range could lead to greater pressure by people and rapid extermination. Ayres and Johns (1987) reported the principal threat to *C. c. calvus* to be the growth of the timber industry.

PRIMATES; CEBIDAE; Genus XENOTHRIX
Williams and Koopman, 1952

Jamaican Monkey

The single species, *X. mcgregori*, is definitely known only by a subfossil partial mandible found in 1920 in Long Mile Cave in northwestern Jamaica. This specimen has been subject to much speculation; it sometimes was thought to be nothing more than the remains of a modern captive monkey, but Hershkovitz (1977) considered it to represent a distinctive New World family, the Xenotrichidae. The original describers (Williams and Koopman 1952) assigned it to the family Cebidae, and most subsequent authorities have agreed. Rosenberger (1977) and Rosenberger, Setoguchi, and Shigehara (1990) indicated affinity with *Aotus* and *Callicebus*. Ford (1986a, 1990), however, suggested that *Xenothrix* might be a callitrichid. Recently, MacPhee and Fleagle (1991) reported the existence of several postcranial bones apparently assignable to *Xenothrix*, found these specimens to show no evidence of affinity between *Xenothrix* and any other primate, and revived the position that the genus does indeed represent a separate family.

As in the living Callitrichidae (except *Callimico*), *Xenothrix* lacks a third lower molar, but the first and second molars are unlike those of marmosets, being bunodont, with the individual cusps greatly enlarged and the trigonid and talonid basins correspondingly reduced. The mandible is large and stout relative to that of the Callitrichidae, and the angle is expanded as in *Callicebus*. The mandible deepens posteriorly, also as in *Callicebus*, and the canine apparently was reduced, as in *Callicebus* and *Aotus*. The newly described postcranial elements indicate that the hind limbs were relatively short and robust.

The type specimen was discovered in detritus beneath an aboriginal kitchen midden in a rock shelter. The layer in which it was found has been dated by radiometric means at 2,145 years before the present (MacPhee 1984). The time of extinction and the natural history of the represented population are unknown. However, considering the morphology of the associated postcranial remains, MacPhee and Fleagle (1991) stated that *Xenothrix* was probably a heavy, slow-moving quadruped or climber. Those authorities also suggested that certain descriptions of living Jamaican animals by early European naturalists offer the possibility that *Xenothrix* externally resembled *Eulemur* or *Perodicticus* and that it may have survived at least until the eighteenth century.

PRIMATES; CEBIDAE; **Genus PARALOUATTA**
Rivero and Arredondo, 1991

Cuban Monkey

The single species, *P. varonai,* is known definitely only by a well-preserved and nearly complete skull found in 1987 and by a portion of a mandible of a different animal found in 1992 in a cave deposit in Cueva del Mono Fósil, Pinar del Río Province, western Cuba (Ford 1990; MacPhee 1993a; MacPhee et al. 1995; Rivero and Arredondo 1991; Salgado et al. 1992). Several additional teeth and postcranial elements found in the same vicinity are probably also referable to this species. Inclusion of *Paralouatta* here is provisional as the specimens have not yet been radiometrically dated, but associated fauna and conditions of discovery indicate a late Quaternary age.

The skull of *Paralouatta* is within the upper size range of those of living *Alouatta* and also superficially resembles those in the form of hafting of the neurocranium and face, the deep malar body, the marked lateral flaring of the maxillary root of the zygomatic process, and the low, posteriorly tapering cranial vault. The upper teeth of *Paralouatta* bear some resemblance to those of *Alouatta* and differ greatly from those of the possibly contemporary Cuban spider monkey, *Ateles anthropomorphus.* However, details of cusp morphology of the lower teeth appear to show less than expected correspondence with *Alouatta. Paralouatta* also differs from living *Alouatta* in having the foramen magnum more downwardly directed, the nuchal plane less vertically oriented, inconspicuous stylar shelves on the molars and premolars, three-rooted third and fourth upper premolars, and a completely incisiform lower canine. The cranial height is relatively great and the orbits relatively wide, being almost as large, relatively, as those of *Aotus.* Recent studies show that *Paralouatta* most closely resembles *Antillothrix* (see account thereof) in critical characters.

Mammalian remains found together with the new genus include almost exclusively those of living genera, such as *Solenodon* and *Capromys,* and very Recently extinct forms, such as *Nesophontes* and *Boromys.* It is therefore reasonable to think that, like those other mammals, *Paralouatta* survived until less than 5,000 years ago and became extinct in association with the human invasion and environmental disruption of the West Indies.

PRIMATES; CEBIDAE; **Genus ANTILLOTHRIX**
MacPhee, Horovitz, Arredondo, and Jiménez Vasquez, 1995

Hispaniolan Monkey

The single species, *A. bernensis,* is known only by skeletal remains found at Cueva de Berne and another site in eastern Dominican Republic and at and around Trou Woch Sa Wo in southwestern Haiti. This species, described in 1977, was then thought to represent the genus *Saimiri* and subsequently was associated with *Cebus* (see accounts of those two genera and of the family Cebidae), but detailed new morphological analyses by MacPhee et al. (1995) show that it represents a separate genus most closely related to the

also extinct *Paralouatta* and the living *Callicebus.* These new studies also indicate that the living *Alouatta* is not a close relative of any of the other genera mentioned here.

The outstanding diagnostic feature of *Antillothrix* is a crest on the upper first and second molar teeth, not consistently present in any other anthropoid. This "distal crest" runs directly distally from the protocone but is not distinct from it. *Antillothrix* also differs (1) from *Saimiri* and *Paralouatta* in lacking a continuous distal wall of the trigon running from protocone to metacone on the second upper molar; (2) from *Saimiri, Cebus,* and *Callicebus* in displaying a relatively lingual intersection of the protolophid and oblique cristid on the first lower molar; (3) from all of the other genera mentioned except *Paralouatta* in possessing a mesially projecting lobe of the lingual cingulum on the fourth upper premolar and an oblique alignment (relative to the midsagittal plane) of the protocone and hypocone of the first upper molar; and (4) from all of the other genera mentioned in the positioning of the infraorbital foramen above the interval between the third and fourth upper premolars.

The exclusive resemblances between *Antillothrix* and *Paralouatta* in the above critical dental characters are indicative of close phylogenetic relationship. In contrast, the resemblance between *Paralouatta* and *Alouatta* in cranial features is probably the result of convergence. The original association of the latter two genera was based largely on a skull with such worn teeth that no details of cusp morphology could be discerned.

Radiocarbon dates associated with the remains of *Antillothrix* range from about 3,850 to 9,850 years before the present. The genus still would have been present at least 2,000 years into the period of human occupation of Hispaniola. Details of its natural history and the exact time and cause of its extinction are not known.

PRIMATES; CEBIDAE; **Genus CALLICEBUS**
Thomas, 1903

Titi Monkeys

Hershkovitz (1988a, 1990c) recognized 13 species:

C. modestus, upper Beni River Basin in western Bolivia;

C. donacophilus, central and southern Bolivia, northern and western Paraguay;

C. olallae, upper Beni River Basin in western Bolivia;

C. oenanthe, upper Mayo River Valley in north-central Peru;

C. cinerascens, Madeira River Basin in central Brazil;

C. hoffmannsi, south of the Amazon and between the Canuma and Tapajós rivers in central Brazil;

C. moloch, south of the Amazon and between the Tapajós and Tocantins-Araguaia rivers in central Brazil;

C. brunneus, upper Madeira and upper Purús river basins in western Brazil and eastern Peru;

C. cupreus, upper Amazon and Orinoco basins in central and southern Colombia, eastern Ecuador, eastern Peru, and western Brazil;

C. caligatus, south of the Solimoes and between the Ucayali and Madeira rivers in western Brazil, eastern Peru, and extreme northern Bolivia;

C. dubius, south of the Amazon and between the lower Purús and Madeira rivers in central Brazil;

C. personatus, states of Bahia, Espirito Santo, Minas Gerais, Rio de Janeiro, and Sao Paulo in southeastern Brazil;

CALLICEBUS

C. torquatus, upper Amazon and southern Orinoco basins in Venezuela, southern Colombia, eastern Ecuador, northeastern Peru, and northwestern Brazil.

In an earlier assessment of the genus, Hershkovitz (1963*c*) recognized only three species, *C. personatus* and *C. torquatus,* distributed largely as given above, and *C. moloch,* which essentially comprised the ranges of all the other species listed above.

Head and body length is 230–460 mm and tail length is 260–560 mm. Jones and Anderson (1978) stated that the weight range of *C. moloch* is 510–730 grams, and Hershkovitz (1990*c*) indicated that the genus can weigh up to 2,000 grams. The nonprehensile tail is well haired and sometimes tufted at the tip. The pelage consists of long, soft hairs. Coloration is variable; it ranges from reddish gray or yellowish through reddish brown to black on the upper parts and is usually paler on the underparts. A light-colored collar band is present on some species. Titi monkeys may have a black or white band on the forehead. These monkeys, with their harmonious colors and thick, soft fur, are attractive animals. They have a small, rounded head and a somewhat flattened, high face. Among the diagnostic characters pointed out by Hershkovitz (1990*c*) are: supraorbital ridges well defined, frontal bone above and between ridges depressed or flattened; interorbital septum broad, pneumatized, not perforated; dental (palatal) arcade more nearly V-shaped than U-shaped; mandibular angle greatly expanded; first upper incisor tooth with cutting edge bluntly pointed; second upper incisor staggered behind first; and molars heavy and brachyodont, their cusps high and sharp, buccal and secondary lingual cingula usually well developed. In general form and behavior titi monkeys bear some resemblance to *Aotus.*

Titis are arboreal and diurnal forest dwellers. In Venezuela, Handley (1976) found all specimens of *C. torquatus* in trees near streams or in other moist parts of evergreen forest. Jones and Anderson (1978) stated that *C. moloch* inhabits low and dense tropical forests, especially near riverbanks, and frequently occurs in understory vegetation. Hernandez-Camacho and Cooper (1976) reported *C. moloch* to occur mostly at heights of 3–8 meters, but it

Dusky titi monkey *(Callicebus moloch),* photo from New York Zoological Society.

is said to descend to the ground occasionally. When progressing on all four limbs *Callicebus* hunches its body, and the hind limbs are rotated outward. In the characteristic resting position the body is hunched and all four limbs are brought together on the branch, with the tail hanging vertically down. From this position a titi monkey can jump by rapid extension of the hind limbs while it stretches its arms outward to enable the hands to grasp another branch. In the usual sleeping position the body is extended, the side of the head resting on the folded hands. The hands are used to convey food to the mouth.

There is an intensive feeding period in the early morning, a major resting period at midday, considerable additional feeding after 1500 hours, and selection of a sleeping tree before dark (Jones and Anderson 1978). The diet consists mostly of fruits but also includes leaves, other vegetation, insects, other small invertebrates, birds' eggs, and small vertebrates (Kinzey 1977; Mittermeier and Coimbra-Filho 1977). A captive specimen of *C. personatus* exhibited considerable ingenuity; it was fond of cockroaches, and if these hid in a crevice, it used pieces of straw to pry them out. This individual also deliberately moved his sleeping box, presumably to obtain cockroaches that accumulated beneath.

Population densities of 2–31/sq km have been recorded (Wolfheim 1983). Mason (1968) reported a daily movement of 315–570 meters for *C. moloch*, and home ranges of three groups were 3,201, 5,093, and 4,762 sq meters. There was seldom more than 20 percent overlap between the ranges of different groups. Mason (1971) noted that each group of *C. moloch* occupied a small, well-defined, stable home range. There were regular territorial confrontations between groups where their ranges overlapped. These engagements involved displays, vocalizations, and vigorous chasing, but physical fighting was rare and never severe. Kinzey et al. (1977) reported that *C. torquatus*, like *C. moloch*, was territorial but that *C. torquatus* had a much larger group home range, about 20 ha.

Titi groups consist of two to seven individuals, including a strongly bonded pair of adults and their offspring (Jones and Anderson 1978; Kinzey 1977; Mittermeier and Coimbra-Filho 1977). The adult male searches for food, leads group movements, and also usually carries the infant when it is not being nursed by the female. There is a wide range of visual signals and vocalizations for communication. The intertwining of tails occurs frequently when two or more animals of the same group sit side by side. Litter size in this genus is one. Births in wild *C. moloch* occur from December to April. Birth weight is about 70 grams, and adult weight is attained at around 10 months. Hayssen, Van Tienhoven, and Van Tienhoven (1993) listed a gestation period of 5–6 months, a lactation period of 12–16 weeks, and an interbirth interval of 1 year. According to Marvin L. Jones (Zoological Society of San Diego, pers. comm., 1995), a specimen of *C. moloch* was still living in captivity at an age of at least 25 years.

All species of *Callicebus* are probably declining (Wolfheim 1983) and are on appendix 2 of the CITES. The IUCN gives an overall classification of vulnerable to the species *C. personatus* and singles out the subspecies *C. p. barbarabrownae* as critically endangered. *C. personatus* is restricted to the Atlantic coastal forests of eastern Brazil, at least 95 percent of which have been cut down for lumber and charcoal production and to make way for agriculture and pasture. The monkeys occur at low densities in widely scattered forest fragments that continue to decline as human settlement and activity proceeds (Mittermeier et al. 1986; Oliver and Santos 1991). The IUCN also now classifies *C. dubius*, *C. oenanthe*, *C. cupreus ornatus* of central Colombia, and *C. torquatus medemi* of southern Colombia as vulnerable. Each of these taxa occupies a relatively small area and is immediately jeopardized by human activity.

PRIMATES; CEBIDAE; Genus AOTUS
Humboldt, 1811

Douroucoulis, or Night Monkeys

Rylands, Mittermeier, and Luna (1995), based largely on the work of Hershkovitz (1983), recognized two species groups and ten species:

primitive gray-neck group

A. lemurinus, Panama, northern and western Colombia, western Venezuela;
A. brumbacki, eastern Colombia;
A. hershkovitzi, known only from the type locality on the east side of the Andes in central Colombia;
A. trivirgatus, southern Venezuela, north-central Brazil;
A. vociferans, southern Colombia, eastern Ecuador, northeastern Peru, northwestern Brazil;

derived red-neck group

A. miconax, north-central Peru;
A. nigriceps, eastern Peru, western Brazil;
A. nancymaae, northeastern Peru, northwestern Brazil;
A. infulatus, east-central Brazil;
A. azarai, Bolivia, Paraguay, northeastern Argentina.

Groves (*in* Wilson and Reeder 1993) accepted the above and used the spellings *A. nancymaae* and *A. azarai*, respectively, in place of the more commonly used *A. nancymae* and *A. azarae*. Based on analysis of cranial morphology, pelage characters, karyology, and blood protein variation, Ford (1994) generally accepted the above arrangement but concluded that *A. lemurinus* and *A. brumbacki* are part of *A. vociferans* and that the population of *A. azarai* in Bolivia actually is referable to *A. infulatus*; she also questioned whether *A. infulatus* is even specifically distinct from *A. azarai* and whether *A. nancymaae* is distinct from *A. miconax*. Pieczarka et al. (1993) reported chromosomal evidence that *A. azarai* and *A. infulatus* form a single intergrading species. Corbet and Hill (1991) considered the entire gray-neck group to compose only one species, *A. trivirgatus*, and the entire red-neck group to compose only one species, *A. azarai*. Other authorities, including Rathbun and Gache (1980) and Thorington and Vorek (1976), have recognized only the single species *A. trivirgatus*. E. R. Hall (1981) listed an additional species, *A. bipunctatus*, of the Azuero Peninsula of Panama, but stated that it was almost certainly no more than a subspecies of *A. trivirgatus*. Hall also noted that an old record of *Aotus* from Nicaragua probably involved a pet monkey that had been brought from elsewhere. On the basis of studies of chromosomes, Brumback (1974) suggested that there were three isolated species: *A. griseimembra* in Central America and northern South America, *A. trivirgatus* in the Amazon Valley, and *A. azarai* in the Parana-Paraguay River Valley in south-central South America.

In addition to this diversity of opinion regarding the intrageneric composition of *Aotus*, there is profound disagreement with respect to the proper phylogenetic position of the genus itself. Most authorities place it in the Cebidae, or sometimes in the Atelidae if the latter is recognized as a separate family, and affinity with *Callicebus* is often suggested. Rosenberger, Setoguchi, and Shigehara (1990) regarded *Aotus* as part of a tribe, together with *Callicebus* and *Xenothrix*, that in turn is part of the subfamily Pitheciinae. In contrast, Groves (*in* Wilson and Reeder 1993),

Douroucoulis, or night monkeys (*Aotus* sp.), inset showing baby riding on mother's back, photos by Ernest P. Walker.

Hershkovitz (1977), and Schneider et al. (1993) considered *Aotus* to represent a monotypic subfamily, the Aotinae. Earlier, Groves (1989) had treated this group as a full family, the Aotidae, and suggested that it might be the most primitive component of all the Platyrrhini (New World primates). Utilizing various morphological, immunological, and behavioral evidence, Tyler (1991*a*) suggested that *Aotus* actually diverged from both the Platyrrhini and Catarrhini (Old World monkeys, apes, and humans) before the latter two groups separated from one another and that it is thus the most primitive member of the entire Anthropoidea.

Head and body length is 240–370 mm and tail length is 316–400 mm. Weight is about 0.6–1.0 kg. The pelage is short, dense, semiwoolly, and soft. Coloration varies but is usually silvery gray to dark gray above and gray, buff, or brownish beneath. The markings on the face are usually three dark brown or black lines separated and bordered by grayish areas, but in some this pattern is only faintly indicated or is lacking. The head is relatively round and the eyes are very large. The body is so heavily covered with fur that it appears to be short and thick. The densely furred, slightly club-shaped tail is not prehensile. In most forms the ears are small and almost completely concealed in the fur. A sac under the chin can be inflated at will and gives resonance to the voice.

Douroucoulis live in primary and secondary forests from sea level to about 3,200 meters (Aquino and Encarnación 1994). In Colombia, Hernández-Camacho and Cooper (1976) found them to be present at all levels in every major forest type except mangrove. Their day nests were located in tree hollows and/or woody climbing vines in accumulations of dry leaves and twigs. Aquino and Encarnación (1994) found *A. vociferans* only in tree holes but found *A. nancymaae* in a variety of sleeping sites, including tree hollows, thick vegetation on branches, and termite nests. *Aotus* is nocturnal; it has good vision at night and is adept at running on tree limbs, leaping, and performing remarkable acrobatics. Wright (1994) reported that the average nightly foraging path of a group was 708 meters, being about twice as long on moonlit nights as on nights with no moon. Mittermeier and Coimbra-Filho (1977) stated that the diet consists mainly of fruits, nuts, leaves, bark, flowers, gums, insects, and small vertebrates.

In Peru, Aquino and Encarnación (1986) reported an average population density of 8.75 groups or 25 individuals per sq km. Aquino and Encarnación (1994) listed other estimates of density, ranging from about 8 to 242 individuals and 2 to 70 groups per sq km, and indicated that the lower figures resulted from hunting pressure, deforestation and decreasing food resources, and cold temperatures at higher elevations. Thorington, Muckenhirn, and Montgomery (1976) determined that a radiotracked individual on Barro Colorado Island spent 72 percent of its time in an area of 800 sq meters. Wright (1994) indicated that groups maintain territories averaging 9.2 ha. in rainforests and 5 ha. in dry forest and that aggressive interaction sometimes is seen along boundaries.

Douroucoulis are usually seen in family groups of two to five individuals, including an adult pair and their young of up to three reproductive seasons (Aquino and Encarnación 1986, 1994; Izawa 1976). Subadults disperse from the natal group at about three years (Wright 1994). Any attempt to manage captives other than mated pairs invariably results in stress, conflict, and injury (Meritt 1980). The alarm or danger call is "wook," and individuals give a deep resonant call of several seconds' duration by inflating the throat sac. The night calls are referred to variously as squeaks, hisses, and barks and are sometimes likened to the calls of cats *(Felis)*. Ernest P. Walker learned from a pet that *Aotus* utters perhaps as many as 50 vocal sounds.

In northern Argentina, Rathbun and Gache (1980) found no evidence of a marked birth season. Most births in Peru occur from November to January (Aquino and Encarnación 1994). Observations of individuals in captivity (Dixson 1994; Gozalo and Montoya 1990; Meritt 1977) indicate that the estrous cycle lasts about 16 days, the gestation period is 133–41 days, females are capable of giving birth every eight months but average interbirth interval is about 12.7 months, there is almost always a single young, birth weight is about 96 grams, and sexual maturity is usually attained at about 3 years in both sexes. A female more than 13 years old and a male more than 11 years old were still reproductively active. Fathers play a major role in carrying and otherwise providing for the infant (Wright 1994). According to Marvin L. Jones (Zoological Society of San Diego, pers. comm., 1995), a captive *A. trivirgatus* was still living at about 27 years.

Although *Aotus* seems more tolerant of human activity than many other New World primates, its habitat is being rapidly reduced in some areas through the total elimination of forests. It also is killed for its meat and fur and has been collected extensively for the pet trade and biomedical research (Wolfheim 1983). *Aotus* has proved to be of particular value in studies of malaria and diseases of the eyes (Baer 1994). Importation to the United States declined substantially after federal restrictions were implemented in the 1970s, but it continues and most animals that are captured actually die before leaving the country of origin (Aquino and Encarnación 1994). All species are on appendix 2 of the CITES, but none are listed as endangered or threatened by the USDI. The IUCN now classifies the species *A. lemurinus, A. brumbacki,* and *A. miconax* as vulnerable and the subspecies *A. lemurinus griseimembra,* of Colombia and western Venezuela, as endangered. Aquino and Encarnación (1994) expressed concern for the restricted habitat of those taxa, as well as that of *A. nigriceps* and *A. trivirgatus.*

PRIMATES; CEBIDAE; Genus CEBUS
Erxleben, 1777

Capuchins, or Ring-tail Monkeys

There are six currently recognized species (Cabrera 1957; Cameron et al. 1989; Goodwin and Greenhall 1961; E. R. Hall 1981; Hershkovitz 1972; McCarthy 1982*b*; Ottocento et al. 1989; Queiroz 1992; Redford and Eisenberg 1992; Rylands and Luna 1993; Rylands, Mittermeier, and Luna 1995; Wolfheim 1983):

C. olivaceus, northern Colombia, Venezuela, Guianas, northern Brazil to north of the Amazon;
C. kaapori, small area of northeastern Brazil south of the lower Amazon along the Gurupi River;

C. capucinus, Belize and Honduras to western Colombia and Ecuador;
C. albifrons, the west coast of Colombia and Ecuador, parts of eastern Colombia, western and southern Venezuela, Trinidad, eastern Peru, northern Bolivia, much of Amazonian Brazil;
C. apella, eastern and southern Colombia, southern Venezuela, Margarita Island, the Guianas, eastern Ecuador and Peru, most of Brazil, Bolivia, Paraguay, northern Argentina;
C. xanthosternos, state of Bahia in southeastern Brazil.

Groves (*in* Wilson and Reeder 1993) included *C. xanthosternos* in *C. apella* and did not distinguish *C. kaapori* from *C. olivaceus.* Subfossil fragments, some of which have radiocarbon dates as recent as 3,860 years ago, suggested the former presence of still another species of *Cebus* on the island of Hispaniola (Ford 1990; Hershkovitz 1988*b;* MacPhee and Woods 1982) but recently have been assigned to the new genus *Antillothrix* (see account thereof). Use of the name *C. olivaceus* in place of *C. nigrivittatus* is in accordance with Corbet and Hill (1991).

Head and body length is 305–565 mm, tail length is 300–560 mm, and weight is about 1,100–4,300 grams. Capuchins can be divided into two groups with respect to pelage. Those in the first group *(C. apella* and *C. xanthosternos)* have tufts, that is, horns of hair over the eyes or ridges of hair along the sides of the top of the head. These monkeys are usually without pronounced contrasting color or pattern, the color normally ranging through grayish browns. The monkeys of the other group *(C. capucinus, C. olivaceus, C. kaapori,* and *C. albifrons)* do not have tufts on the head. The more characteristically marked are white on the face, throat, and chest and black elsewhere. From this pattern they vary toward pale yellowish grays and grayish browns, some clearly showing the basic pattern and others having but little evidence of the color pattern. In all the body and tail are well haired, some having silky or shiny hair. A well-defined thumb is present. The tail is slightly prehensile and often carried coiled at the tip, hence the name "ring-tail." The tail is strong enough to support the weight of the animal.

Like most other New World primates, capuchins are arboreal and diurnal. According to information summarized by Hernandez-Camacho and Cooper (1976) and Mittermeier and Coimbra-Filho (1977), all *Cebus* are flexible in choice of habitat. *C. apella,* which has the largest range of any New World monkey, is especially adaptable and occurs in nearly all kinds of humid forest, both seasonally flooded and nonflooded, up to an elevation of 2,700 meters. It is found in broken forest and secondary growth and crosses open ground when traveling from one forest segment to another. *C. albifrons* also can live in secondary forest and even has been collected in mangroves. *C. olivaceus* is found in dry forest on the llanos of Venezuela as well as in mature tropical forest in the Guianas. *C. capucinus* lives in a variety of forest types and in the western Andes occurs to an elevation of 2,100 meters. In Costa Rica it has been seen in every kind of forest and even in mangroves and sparsely forested areas (Freese 1976). The diet of *Cebus* is more varied than that of any other New World monkey. Foods include fruits, nuts, berries, seeds, flowers, buds, shoots, bark, gums, insects, arachnids, eggs, small vertebrates, and even certain kinds of marine life, such as oysters and crabs, found in coastal areas. Antinucci and Visalberghi (1986) reported that a captive *C. apella* used various objects as tools to crack nuts open. Visalberghi (1990) also cited cases of *Cebus* using sticks and other tools for various purposes but concluded that this genus does not actually have the abili-

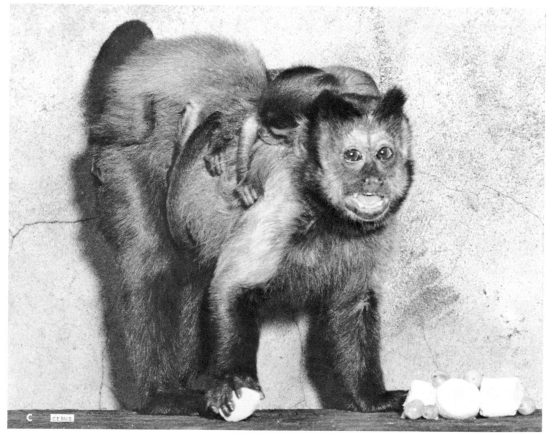

A. White-fronted capuchin *(Cebus albifrons)*, photo from New York Zoological Society. B. White-faced sapajou *(C. capucinus)*, photo from New York Zoological Society. C. Weeping capuchins *(C. apella)*, photo from San Diego Zoological Garden.

ty, as found in *Pan*, to understand a problem and mentally devise a tool that will provide a solution.

Population densities reported vary from about 1/sq km to 111/sq km (Klein and Klein 1976; Neville et al. 1976; Wolfheim 1983). Izawa (1976) estimated that a nomadic group of *C. albifrons* in the upper Amazon used a range of 3 sq km. A group of 35 *C. albifrons* occupied a home range of 110–20 ha. in eastern Colombia and moved up to 5 km per day (Defler 1979a). Groups of this species are strongly intolerant of one another and defend largely exclusive territories. In contrast, home ranges of *C. apella* have been found to overlap by at least 40 percent, and groups may feed side by side with no apparent antagonism (Defler 1982). In Panama, Baldwin and Baldwin (1976) determined that a group of 27–30 *C. capucinus* had a home range of 24–40 ha. They also observed that in coastal areas where *C. capucinus* was not hunted or harassed it traveled in groups of 20–30 individuals that would make bold displays at humans but that in disturbed inland forests the monkeys lived in small bands that fled silently from people. Overall group size ranges reported for the four species of *Cebus* are: *C. capucinus*, 6–50; *C. apella*, 1–20; *C. olivaceus*, 6–15; and *C. albifrons*, 10–35 (Defler 1979a, 1979b, 1982; Freese 1976; Hernandez-Camacho and Cooper 1976; Klein and Klein 1976; Neville et al. 1976; Oppenheimer 1969; Vessey, Mortenson, and Muckenhirn 1978). Several hundred individuals of *C. albifrons* were once observed to congregate to feed on fruits.

In general, groups appear to contain more adult females than males but to be dominated by a large older male. Defler (1982) reported that groups of *C. albifrons* were multimale and that one group had 7 adult and subadult males, but two groups of *C. apella* had only 1 adult male. Izawa (1980) found groups of the latter species to have a strong dominance hierarchy for both sexes and indicated that rank order was less rigid in other species. Capuchins have a variety of chatters, squeaks, shrieks, and other sounds for communication. Oppenheimer and Oppenheimer (1973) noted 11 different vocalizations for *C. olivaceus* in Venezuela. These calls were divided into four fundamental categories according to their purpose: (1) gaining or maintaining contact with the troop, (2) gaining or maintaining contact with an individual, (3) terminating a potential or actual attack, and (4) alerting the troop to potential danger from without.

Hernandez-Camacho and Cooper (1976) stated that *Cebus* evidently is not a seasonal breeder in Colombia. Oppenheimer (1969), however, reported most births of *C. capucinus* on Barro Colorado Island, Panama Canal Zone, to occur in the dry season or the early rainy season. In *C. apella* the estrous cycle is 16–20 days, estrus lasts 2–4 days, gestation is 180 days, weight at birth is about 200–250 grams, and lactation lasts 9 months (Eisenberg 1977; Hayssen, Van Tienhoven, and Van Tienhoven 1993). Female *Cebus* normally give birth to a single young; immediately thereafter the infant clings tightly to its mother's hair with both hands and feet. As it gains strength it moves about on the mother or leaves her for short periods to explore nearby, but it nurses for at least several months. When a helpless young loses its mother or becomes separated from her, other members of the troop often come in response to its distress cries and assist it. Oppenheimer (1969) observed that full size and sexual maturity were reached by female *C. capucinus* at 4 years and by males at 8 years. According to Marvin L. Jones (Zoological Society of San Diego, pers. comm., 1995), a captive *C. capucinus* lived about 55 years and a specimen of *C. apella* lived in captivity for more than 45 years.

Capuchins are such vivacious, intelligent little creatures

that they have become the most numerous monkeys in captivity in both the United States and Europe. They were the monkeys most often used by itinerant organ grinders. More recently they have been trained to assist human quadripeligics. They must be kept fairly warm and free from drafts, receive a wide variety of food, and have an opportunity for plenty of exercise and sunlight. They are so active and mischievous that they not only are first-class entertainers but usually become serious nuisances if allowed freedom in a home. Importation for such purposes now is illegal.

Four of the species of *Cebus* are generally common, but all are on appendix 2 of the CITES and the survival of certain subspecies is in question. The IUCN classifies *C. kaapori* as vulnerable and *C. xanthosternos* as critically endangered; fewer than 250 individuals of the latter are thought to survive. *C. alibifrons trinitatis*, of Trinidad, and *C. apella margaritae*, of Margarita Island, also have populations of fewer than 250 and also are classed as critically endangered. *C. capucinus curtus*, of Colombia, and *C. apella robustus*, of southeastern Brazil, are designated as vulnerable. The forms *robustus* and *xanthosternos* are restricted to remnants of the Atlantic forest region, are jeopardized by widespread habitat destruction and hunting, and may soon be entirely eliminated outside of protected reserves (Mittermeier et al. 1986; Oliver and Santos 1991). The other forms classified by the IUCN all have restricted ranges in areas that are undergoing severe habitat disruption. *C. o. kaapori* may be in particular jeopardy as its range is small, isolated, and likely to be affected by a massive mining project that will be fueled through extensive local charcoal production (Ferrari and Queiroz 1994).

PRIMATES; CEBIDAE; **Genus SAIMIRI**
Voigt, 1831

Squirrel Monkeys

Based on the studies of Ayres (1985), Hershkovitz (1984b), and Thorington (1985), there are at most five living species:

S. boliviensis, upper Amazon Basin in western Brazil, eastern Peru, and Bolivia;

S. vanzolinii, known only from the vicinity of the confluence of the Amazon and Japura rivers in west-central Brazil;

S. sciureus, east of the Andes from Colombia and northern Peru to northeastern Brazil;

S. ustus, south of the Amazon in central Brazil;

S. oerstedii, a small area on the Pacific coast of Costa Rica and Panama.

Groves (*in* Wilson and Reeder 1993) accepted the content of *Sciureus* as listed above. However, based on analyses of pelage characters, dental morphology, biochemical data, and behavior, Costello et al. (1993) concluded that *S. oerstedii*, with a range as given above, is the only species distinct from *S. sciureus*. Thorington (1985) considered *S. boliviensis* and *S. oerstedii* to be subspecies of *S. sciureus* and used the name *S. madeirae* in place of *S. ustus*. Silva et al. (1992) reported natural hybridization between *S. sciureus* and *S. boliviensis* in the Amazon Basin of Peru. Ayres's (1985) description of *S. vanzolinii* was published subsequent to the work of Hershkovitz and Thorington. Another species, *S. bernensis*, originally was based on a few subfossil fragments with an estimated age of 3,860 years B.P. found on the island of Hispaniola. Subsequent studies sug-

Squirrel monkey *(Saimiri sciureus)*, photo by Ernest P. Walker. Inset photo by John G. Vandenburgh.

gested that this species may actually belong to the genus *Cebus* (Ford 1986a, 1986b; MacPhee and Woods 1982), and it recently was assigned to a new genus, *Antillothrix* (see account thereof).

Head and body length is usually 260–360 mm, tail length is 350–425 mm, and weight is 750–1,100 grams. The pelage is short, thick, soft, and brightly colored. The skin of the lips, including the area around the nostrils, is black and

nearly devoid of hair. The most common coloration is: white around the eyes and ears and on the throat and sides of the neck; the top of the head black to grayish; the back, forearms, hands, and feet reddish or yellow; the shoulders and hind feet suffused with gray; the underparts whitish or light ochraceous; and the tail bicolored like the body except that the terminal part is black. The markings and coloration of the sexes and of animals of different ages are similar.

The eyes are large and set close together, and the large ears are shaped about as in people. The thumb is short but well developed. The tail is covered with short hair, slightly tufted at the tip, and not prehensile. There is no laryngeal pouch. The facial part of the skull is small in proportion to the cranial part.

These monkeys are found in primary and secondary forests and in cultivated areas, usually along streams. They are diurnal and arboreal but occasionally descend to the ground. Thorington (1968) reported that a wild group was most active from early morning to midmorning and from midafternoon to late afternoon, resting for one or two hours at midday. The diet consists of fruits, berries, nuts, flowers, buds, seeds, leaves, gums, insects, arachnids, and small vertebrates (Mittermeier and Coimbra-Filho 1977).

Reported population densities are 130/sq km in Panama (Baldwin and Baldwin 1976), 20–30/sq km in Colombia (Klein and Klein 1976), and 151–528/sq km in Peru (Neville et al. 1976). In the upper Amazon, Izawa (1976) found that one group of 42 or 43 individuals had a nomadic range of about 3 sq km and made daily movements of 0.6–1.1 km. In Colombia, Mason (1971) observed that groups occupied an ill-defined home range and showed no evidence of territorial encounters with adjacent groups. In Panama, Baldwin and Baldwin (1976) determined that a group of 23 monkeys used a home range of 17.5 ha., of which 1.8 ha. was exclusive to that group. A group of 27 monkeys in a coastal marsh had a home range of 24–40 ha., but troops of 10–20 animals lived in isolated patches of forest as small as 0.8–2.0 ha.

Squirrel monkeys form larger groups than do any other New World monkeys. Based on a survey of areas of natural habitat, Baldwin and Baldwin (1971) reported troop size to range from 10 to 35 individuals in Panama and Colombia and from 120 to 300 or more in unaltered Amazonian rainforests. Mittermeier and Coimbra-Filho (1977) found group size to be 20–70 in the Amazon and mentioned a report of 550 individuals in one troop. Hernandez-Camacho and Cooper (1976), however, were unaware of any reliable accounts of more than 40–50 monkeys in a group.

A wild group observed on the llanos of Colombia was seen to split daily into subgroups composed of pregnant females, females with young, and adult males (Thorington 1968). A group in a 6-ha. enclosure in Florida appeared to be divided into socially independent subgroups (Du Mond 1968). The adult males generally had little to do with the other animals and had no relationship with the young. During the mating season they established a dominance hierarchy through fierce fighting, and the higher-ranking individuals then interacted with the females. The female subgroup began to form a separate identity, probably with its own dominance hierarchy, when pregnancy was evident. Following birth of the young, the adult males were excluded by the females. A third subgroup comprised young adult or preadult males. As these males matured they could move to the adult male subgroup, but they sometimes had to fight to win a place. In the course of this study squirrel monkeys were found to be among the most vocal of primates. The multiplicity of sounds included chirps and peeps for contact or alarm, squawks and purrs during the mating and birth seasons, barks of aggression, and screams of pain.

Winter (1968) identified 26 different calls in these monkeys.

Observations of enclosed animals in Florida showed that there was a discrete, 2- to 4-month mating period during the dry season and a period of births from late June to early August (Baldwin 1970; Du Mond 1968). The estrous cycle in *Saimiri* has been reported to last 6–25 days, with estrus itself being 12–36 hours (Hayssen, Van Tienhoven, and Van Tienhoven 1993; Rosenblum 1968). Reported gestation periods have ranged from 152 to 172 days (Goss et al. 1968; Lorenz, Anderson, and Mason 1973). The newborn weigh about 100 grams, normally cling to the back of the mother for the first few weeks of life, and are independent at 1 year. Reproductive status is attained at about 3 years in females and about 5 years in males (Du Mond 1968). Several individuals have lived in captivity for about 30 years (Marvin L. Jones, Zoological Society of San Diego, pers. comm., 1995).

The United States Department of the Interior (USDI 1976) reported that squirrel monkeys were being captured extensively for export. From 1968 to 1972 more than 173,000 were brought into the United States alone. In 1968 about half of those imported were for use in biomedical research, but in 1969 fewer than 20 percent were so used. The remainder went into the pet market or to zoos. Ultimately, the United States Department of Health, Education and Welfare (1975) issued regulations essentially ending the importation of all primates into the United States for use as pets and considerably restricting importation for all purposes. Despite the heavy commercial take and local use for food, bait, and pets, Mittermeier and Coimbra-Filho (1977) considered *S. sciureus* to be perhaps the most abundant and least threatened primate in Brazilian Amazonia. Long-term habitat alteration, however, could jeopardize this and other species in the region. Myers (1987b) suggested that expansion of colonization and beef production could deforest much of the Amazon Valley and Central America. The species *S. oerstedii* of Panama and Costa Rica has already declined drastically through clearing of the forests. It is listed as endangered by the IUCN and the USDI and is on appendix 1 of the CITES; all other species are on appendix 2. Boinski (1985) estimated the population in Costa Rica to number 3,000.

However, the IUCN now singles out the subspecies *S. o. citrinellus*, known only from the vicinity of the Pirris River in Costa Rica, as critically endangered and notes that both that subspecies and *S. o. oerstedii*, which occupies the remainder of the range of the species, have populations of fewer than 250 mature individuals. The IUCN also classifies *S. vanzolinii* as vulnerable, noting that it numbers fewer than 10,000 individuals and is continuing to decline because of habitat disturbance through logging. Ayres (1985) had suggested that it probably has the smallest geographic range of any South American primate and estimated the population at 50,000.

PRIMATES; **Family CALLITRICHIDAE**

Marmosets, Tamarins, and Goeldi's Monkey

This family of 5 Recent genera and 35 species is found in the tropical forests of Central and South America, mainly in the Amazon region. Although the family name sometimes is spelled Callithricidae, most authorities now are following Hershkovitz (1977, 1984a) in accepting Callitrichi-

dae as the correct version. Much disagreement remains, however, relative to the phylogenetic position of this group. Traditionally the Callitrichidae have been considered a distinct family and the more primitive of the New World monkeys. That position basically was taken by Hershkovitz (1977), Groves (1989 and *in* Wilson and Reeder 1993), and Corbet and Hill (1991). However, there has been an increasing trend toward treating the Callitrichidae only as a subfamily of the family Cebidae (Rosenberger and Coimbra-Filho 1984; Rosenberger, Setoguchi, and Shigehara 1990; Schneider et al. 1993; Thorington and Anderson 1984). Martin (1990) continued to accept the Callitrichidae as a full family but presented substantial evidence that the family is a highly specialized group derived from cebidlike ancestral stock. Other recent authorities giving familial status to the Callitrichidae include Ford (1986b), Rylands, Coimbra-Filho, and Mittermeier (1993), and Sussman and Kinzey (1984). An additional question involving the group is the affinities of the genus *Callimico* (Goeldi's monkey); it was placed in a separate family by Hershkovitz (1977), tentatively included in the Cebidae by Martin (1990), allocated to a subtribe of what was regarded as the subfamily Callitrichinae by Schneider et al. (1993), and considered to represent a separate subfamily of the family Callitrichidae by Groves (1989), Napier and Napier (1985) and Sussman and Kinzey (1984). Recently, Martin (1992) summarized extensive morphological, chromosomal, and biochemical evidence and concluded that *Callimico* and the four callitrichid genera form a monophyletic group but that *Callimico* branched away from the group prior to the divergence of the other genera. Based on that conclusion and a review of the various other authorities cited above, the *Callitrichidae* are treated here as a family, the components being arranged phylogenetically as follows: the subfamily Callimiconinae, with the genus *Callimico*; and the subfamily Callitrichinae, with *Saguinus, Callithrix, Cebuella* (included in *Callithrix* by some of the authorities), and *Leontopithecus*.

These are among the smallest of primates. Head and body length is 130–370 mm and tail length is 150–420 mm. Adults weigh from approximately 100 to 900 grams. The pelage is soft, dense, and in some forms silky. Often there are hair tufts and other adornments on the head. The face is naked or only sparsely haired. The coloration is quite variable in the family. In some species the tail is marked with alternate dark- and light-colored bands. The tail is not prehensile in any of the species.

The forelimbs are shorter than the hind limbs. The thumb is not opposable. The hand and the foot are elongate, and all the digits bear pointed, sickle-shaped nails, except the great toe, which has a flat nail. A baculum is present in the males, and the females have two mammae in the chest region.

The dental formula of the Callitrichinae is: (i 2/2, c 1/1, pm 3/3, m 2/2) × 2 = 32; the Callimiconinae have m 3/3. In *Callithrix* and *Cebuella* the lower canines barely extend beyond the adjacent incisors, the "short-tusked" condition, but in *Saguinus* and *Leontopithecus* the lower canines are longer than the incisors, the "long-tusked" condition. This shared characteristic is not necessarily indicative of phylogenetic affinity; *Callithrix* and *Cebuella* actually are considered more closely related to *Leontopithecus* than any of those genera is to *Saguinus* (Groves 1989; Rosenberger and Coimbra-Filho 1984).

Callitrichids sometimes are considered primitive, squirrel-like primates, but new evidence indicates that they are both morphologically and ecologically advanced (Sussman and Kinzey 1984; Thorington and Anderson 1984). Unlike squirrels, which range through the forest canopy mainly

A mother pygmy marmoset *(Cebuella pygmaea)* with two babies clinging to her. The young of all the primates but *Homo* ride on either the mother or the father during their long period of dependence. Photo from San Diego Zoological Garden.

by ascending and descending large vertical trunks, callitrichids move by running quadrupedally along horizontal branches and leaping between thin terminal supports. Squirrels are adapted to feed on hard nuts, while callitrichids eat mainly insects, soft fruits, nectar, and plant exudates.

Callitrichids have acute sight, good hearing, and appar-

ently a good sense of smell. Facial expression is indicated mainly by lip movements. Emotions are expressed by movements of the eyelids, ears, and the hairy adornments on the head if present. The voice has a variety of high-pitched trilling and staccato calls. These animals are diurnal, sheltering at night in tree holes and cavities. Most of their time is spent in trees or shrubs, though *Callithrix ar-*

gentata in the Matto Grosso sometimes goes into tall grass. They are active and agile, running, jumping, and occasionally leaping among trees and shrubs. *Leontopithecus* is said to bound from tree to tree with remarkable rapidity. Movements are usually quick and jerky. At rest, these animals sometimes draw the fingers inward, so that the clawlike nails pierce the bark of the support, and rest with the belly in contact with the branch, all four limbs hanging on either side. They are fastidious in the care of their pelage and engage in both individual and mutual grooming, the former being more common. The hands are used in securing food but not invariably for conveying it to the mouth.

Callitrichids live in small groups. Contrary to what is sometimes reported, they do not have a monogamous mating system. A female apparently mates with more than one male during the breeding season, and all raise her young cooperatively (Sussman and Kinzey 1984). Females give birth to one to three young after a gestation period of approximately 130–50 days. It has been reported that a female giving birth in captivity sometimes kills and eats its first litter or the third member of a set of triplets. Whether this occurs in the wild is not known. It is probable that this behavior is similar to that of many captive animals and is caused by anxiety. The males assist in birth and carry the offspring on their backs. The males transfer the young to the mother at feeding time and then accept them from the mother again after feeding. This takes place every two to three hours. Sexual maturity is attained at 12–18 months, and longevity in captivity has been as high as 28 years.

Callitrichids are usually docile and gentle in captivity but bite if handled against their will, especially in the presence of strangers. Sudden noises or movements sometimes cause them to panic. They do quite well in captivity if fed, housed properly, and exposed to ultraviolet radiation to replace the lack of sunshine. It has been reported that the members of this family do not seem to be as intelligent as monkeys of the family Cebidae; however, it is very difficult to determine relative intelligences of animals of different groups.

The name "marmoset" dates back to Middle English and is said to be adapted from the Old French word *marmouset*, meaning "a grotesque image" or "manikin." Members of the genera *Saguinus* and *Leontopithecus* are usually called "tamarins," a word of obscure origin.

There was no fossil record of this family until Setoguchi and Rosenberger (1985) reported specimens of a new genus, *Micodon*, from the middle Miocene of Colombia. Rosenberger, Setoguchi, and Shigehara (1990) indicated that a subsequently described genus from the same site, *Mohanamico*, may have affinity to *Callimico*. However, Kay (1990) suggested that *Mohanamico* is probably a primitive member of the subfamily Pitheciinae of the family Cebidae.

PRIMATES; CALLITRICHIDAE; Genus CALLIMICO
Miranda Ribeiro, 1911

Goeldi's Monkey

The single species, *C. goeldii*, occurs in the upper Amazonian rainforests of southern Colombia, eastern Ecuador, eastern Peru, western Brazil, and northern Bolivia. Although here referred to the Callitrichidae, Hershkovitz (1977) observed that it is not a callitrichid or a cebid, nor a link between the two, and thus should be placed in a separate family. A recent analysis of DNA (Montagnon, Crovella, and Rumpler 1993), however, indicates that *Callimico*

is part of the group here designated the family Callitrichidae; most other authorities now seem to agree (see the family account, above).

Head and body length is 210–34 mm, tail length is 255–324 mm, and adult weight is 393–860 grams (Hershkovitz 1977). Adults are brownish black with buffy markings on the back of the neck and two or three light buff-colored rings on the basal part of the tail. Young animals lack the buffy tail rings and sometimes the buff on the back. The head or even the entire dorsal surface may be spotted with white. The hair is thick and soft. There are no ear tufts. A mane drapes from the neck and shoulders, and the elongated hairs of the rump extend skirtlike over the base of the tail.

Although the facial appearance, foot structure, and clawlike nails resemble those of other callitrichids, the dental formula and cranial configuration are like those of the Cebidae. *Callimico* differs from the Cebidae in the structure of the molar teeth, for example, in the absence of external cingula on the lower molars (Hershkovitz 1977).

Information available to Hershkovitz (1977) indicated that *Callimico* prefers deep, mature forests and is relatively rare in areas generally accessible to people. Mittermeier and Coimbra-Filho (1977) observed that *Callimico* occurs in bamboo forest, secondary forest, and primary forest but is naturally rare and very sparsely distributed. Hernandez-Camacho and Cooper (1976) stated that this monkey is found most frequently in the understory and on the ground. The diet includes fruits, insects, and some vertebrates.

In a study in a seasonally dry rainforest in northwestern Bolivia, Pook and Pook (1981) located only a single group within an area of 4 sq km. Groups were largely isolated from one another, and no interaction was seen. The main study group had a home range of 30–60 ha. It traveled about 2 km per day, usually at a height of a few meters, but occasionally came to the ground or climbed to the tops of trees to feed. Leaps of up to 4 meters were made without loss of height. Average group size was 6 individuals, but the main study group had 8, including an adult male and more than 1 breeding female. Members usually remained well within 15 meters of one another and maintained contact by a shrill call. Studies by Masataka (1981a, 1981b) in the same area suggest a similar social structure and support other new evidence that callitrichids are not monogamous. The main study group contained about 6 individuals, including 1 adult male and 2 breeding females, and membership was not exclusive. Also in northern Bolivia, Christen and Geissmann (1994) observed groups of 2–5 individuals occupying adjacent home ranges, one of which was 80 ha. Old reports of troops of up to 40 individuals may have been exaggerated or mistaken. Masataka (1982) reported 40 different vocalizations in his wild study group. These sounds included a whistle for long-distance contact, trills for alarm and warning, and a "truuu" during agonistic behavior.

Births have taken place throughout the year in captivity (Beck et al. 1982), but in northwestern Bolivia they occur mainly in the early wet season, September–November (Masataka 1981a; Pook and Pook 1981). Females are polyestrous, the estrous cycle is about 22–24 days, estrus lasts about 7 days, and gestation averages about 155 days, but the reported range is 139–80 days. Twins occur rarely in captivity, but normally there is a single young weighing 30–60 grams at birth. It is handled solely by its mother for about the first 10–20 days of life, but subsequently there is increasing participation in its care by other members of the group. At 7 weeks the young is able to move about and forage on its own. Weaning is completed by 12 weeks and sex-

CALLIMICO

Goeldi's marmoset *(Callimico goeldii)*, photo by Ernest P. Walker.

ual maturity is attained at 18–24 months (Altmann, Warneke, and Ramer 1988; Hayssen, Van Tienhoven, and Van Tienhoven 1993; Heltne, Turner, and Scott 1976; Hershkovitz 1977; Mallinson 1977). Maximum recorded longevity is 17.9 years (Ross 1991).

Goeldi's monkey is classified as endangered by the USDI and as vulnerable by the IUCN and is on appendix 1 of the CITES. Its natural rarity and apparently specialized habitat make it vulnerable to such adverse factors as habitat destruction and hunting, and parts of its range may soon come under development (Thornback and Jenkins 1982).

PRIMATES; CALLITRICHIDAE; **Genus SAGUINUS**
Hoffmannsegg, 1807

Long-tusked Tamarins

There are 12 species (Hershkovitz 1977; Moore and Cheverud 1992; Natori and Hanihara 1988; Rylands, Coimbra-Filho, and Mittermeier 1993; Skinner 1991; Thorington 1988; Wolfheim 1983):

S. nigricollis, eastern Ecuador, southern Colombia, northeastern Peru, a small part of northwestern Brazil;
S. fuscicollis, upper Amazonian region of eastern Ecuador, southern Colombia, eastern Peru, western Brazil, and northern Bolivia;

S. tripartitus, eastern Ecuador, northeastern Peru, possibly Colombia;

S. labiatus, middle Amazonian region of Brazil and adjacent fringes of southeastern Peru and northwestern Bolivia;

S. mystax, eastern Peru, western Brazil, northwestern Bolivia, possibly southern Colombia;

S. imperator, southeastern Peru, northwestern Bolivia, and adjacent parts of western Brazil;

S. midas, Guianas, northern Brazil;

S. inustus, southeastern Colombia and adjacent part of Brazil;

S. bicolor, an area north of the Amazon in north-central Brazil;

S. leucopus, north-central Colombia;

S. oedipus, northwestern Colombia;

S. geoffroyi, southeastern Costa Rica to extreme northwestern Colombia.

Until recently this genus sometimes was known as *Tamarin* Gray, 1870, or *Leontocebus* Wagner, 1840. Rylands, Coimbra-Filho, and Mittermeier (1993) suggested that *S. melanoleucus* of the upper Juruá River is specifically distinct from *S. fuscicollis* and that *S. niger* to the south of the Amazon River is specifically distinct from *S. midas* to the north, but they did not actually proceed with such designations. Morphological and genetic differentiation between the various subspecies of *S. fuscicollis* was discussed by Cheverud, Jacobs, and Moore (1993), Cheverud and Moore (1990), and Moore and Cheverud (1992) and was found to be pronounced in some cases, but no divisions at the specific level were indicated. Additional modifications of the above list as suggested by the work of Natori (1988) and Natori and Hanihara (1988, 1992) are discussed below.

Head and body length is 175–310 mm, tail length is 250–440 mm, and weight is usually 225–900 grams. Hershkovitz (1977) divided the species into a number of sections and groups depending largely on pelage. The following brief descriptions are based mainly on his detailed analysis.

In the hairy-faced tamarin section the forehead, crown, cheeks, and temples are covered with long hairs. In the *S. nigricollis* group of this section there is no development of a mustache, but an area around the mouth is covered with short white hairs. The species of this group include *S. nigricollis,* which has uniformly dark-colored upper parts, and *S. fuscicollis,* which, except for extreme albinistic races, has upper parts with a trizonal pattern, there being a sharp division between the colorations of the mantle, saddle, and rump. According to Thorington (1988), *S. tripartitus,* which Hershkovitz considered to be a subspecies of *S. fuscicollis,* is distinguished from the latter by a golden, rather than dark red, mantle and a prominent chevron between and above the eyes. The *S. mystax* group of the hairy-faced tamarin section has a mustache as well as a white area around the mouth. The three species of this group are *S. labiatus,* with a poorly developed mustache, a black back marbled with silver, a golden or reddish crown, and mostly reddish underparts; *S. mystax,* with a well-developed but not particularly elongate mustache and a black crown and underparts; and *S. imperator,* the emperor tamarin, with a long white mustache that extends to the shoulders when laid back. The *S. midas* group of this section, with the one species *S. midas,* lacks the white area around the mouth and has a blackish face and orange or yellowish hands and feet.

The mottled-face tamarin section includes only the species *S. inustus.* Its crown is densely haired, but the cheeks and sides of the face are naked, or nearly so. The face is mostly unpigmented and the pelage is melanistic.

The bare-faced tamarin section also has a densely haired crown. The mostly naked face is black. The *S. bicolor* group of this section contains two species: *S. bicolor,* with yellowish or white forequarters sharply defined from the grayish brown hindquarters; and *S. leucopus,* with silvery hairs on the head, buffy brown upper parts, and white limbs. The *S. oedipus* group of this section, with the cotton-top marmosets *S. oedipus* and *S. geoffroyi,* has a general agouti pelage but a prominent crest of white hairs extending back over the head.

Somewhat different groupings than those set forth by Hershkovitz are suggested by analyses of cranial and dental characters (Natori 1988; Natori and Hanihara 1988, 1992). These analyses place *S. oedipus, S. geoffroyi, S. bicolor,* and *S. midas* in one group of related species. Within that group, *S. oedipus* and *S. geoffroyi* have affinity with one another and *S. bicolor* and *S. midas* seem immediately related. The other eight species are placed in another group, with *S. leucopus, S. imperator, S. nigricollis, S. fuscicollis,* and *S. tripartitus* forming one monophyletic subgroup and *S. inustus, S. labiatus,* and *S. mystax* forming a second such subgroup of related taxa.

Saguinus inhabits tropical forests, open woodlands, and secondary growth. It is diurnal and arboreal, moving through trees with rather quick and jerky movements. An individual *S. midas* jumped 20 meters from the top of a tree to the ground without injury. Neyman (1977) reported that *S. oedipus* slept in broad tree forks, began moving about 80 minutes after dawn, and entered the sleeping tree at 1630 to 1830 hours. Data from studies of different species summarized by Buchanan-Smith (1991) and Garber (1993) indicate that groups move about 1,000–2,000 meters per day. Garber also reported that the primary components of the diet of *Saguinus* now are known to be insects (especially large orthopterans), ripe fruits, plant exudates (sap, gums, resin), and nectar. Other foods include some tender vegetation, spiders, small vertebrates, and probably birds' eggs. With the exception of some of the smaller prey, animals are killed by a bite on the head.

Freese, Freese, and Castro (1977) reported a population density of 30/sq km for *S. fuscicollis* in eastern Peru. Density of *S. nigricollis* in the upper Amazon of Colombia was 10–13/sq km, and groups had extensively overlapping home ranges of about 0.3–0.5 sq km (Izawa 1978). In a study of *S. geoffroyi* in the Panama Canal Zone, Dawson (1977) found a density of 20–30/sq km and home ranges of 26 ha. and 32 ha. In Colombia, Neyman (1977) estimated densities of 30–180/sq km for *S. oedipus* and found well-defined home ranges of 7.8, 7.8, and 10 ha. Neighboring home ranges overlapped substantially, but contact between groups was agonistic. Hershkovitz (1977) observed that each troop of *S. oedipus* has a defended territory. Defense of an area is associated with its importance as a feeding site; overall reported group home range size for the various species is 8–120 ha., with about 10–83 percent overlap between such ranges (Garber 1993).

The investigations of Neyman (1977) indicate that groups of *S. oedipus* do not represent extended families. Such groups may consist of a dominant mated pair, their young of the year, and a transient complement of subordinate, probably young animals of both sexes. These subordinates sometimes form small groups of their own within the home range of the main groups. They enter and leave the main groups and possibly sometimes remain and rise to breeding position. The overall size range reported for groups of *S. oedipus* is 1–19, with about 3–9 being most

Long-tusked marmosets, or tamarins *(Saguinus):* A. Pied marmoset *(S. bicolor),* photo by Harald Schultz; B. White-lipped marmoset *(S. nigricollis),* photo by Ernest P. Walker; C. Geoffroy's marmoset *(S. geoffroyi),* photo from New York Zoological Society; D. Emperor marmoset *(S. imperator),* photo from New York Zoological Society; E. Yellow-handed marmoset *(S. midas),* photo by Ernest P. Walker; F. Cotton-top marmoset *(S. oedipus),* photo by Ernest P. Walker.

Long-tusked tamarin *(Saguinus nigricollis)*, photo by Russell A. Mittermeier.

common. Izawa (1978) reported much the same social arrangement for *S. nigricollis* and observed that 2–3 family groups often joined to form assemblies of 10–20 individuals that usually remained together for less than a day.

Reported group sizes in some other species of *Saguinus* are: *S. nigricollis*, 5–10; *S. fuscicollis*, 2–40; *S. labiatus*, 5–10; *S. imperator*, 1–5; *S. midas*, 1–20, and *S. leucopus*, 3–12 (Buchanan-Smith 1991; Freese, Freese, and Castro 1977; Hernandez-Camacho and Cooper 1976; Neyman 1977). Castro and Soini (1977) stated that *S. mystax* lives in parental groups of 2–6 individuals, the largest comprising an adult pair, 2 subadults, and 2 dorsally carried young. Garber, Moya, and Malaga (1984) and Garber (1993) reported a somewhat different social structure for this species, with groups of 3–11 individuals, including a single breeding female and 1 or 2 other adult females but up to 3 reproductively active males. The female may mate with each of the males, and all of the latter assist in caring for the young. There is frequent migration of adults and subadults between groups. Most troops of *S. mystax* have been observed to travel together with bands of *S. fuscicollis* in apparently stable associations. Garber and Teaford (1986) suggested that such mixed troops were able to utilize information obtained by both species relative to location and productivity of food sources and thus to increase foraging efficiency. Epple et al. (1993) reported that *Saguinus* possesses specialized scent glands in the gular-sternal region of the mid-chest and the area surrounding the genitalia. Secretions from these glands, together with urine and other discharges, are rubbed or deposited in various places to mark territories and convey information about identity, social status, sexual receptivity, and other factors.

A common characteristic of tamarin groups is the suppression of reproductive activity in all but the dominant female member. This effect results from a combination of inhibitory behavior by the dominant and loss of ovulatory capacity in the subordinate (Abbott, Barrett, and George 1993). Nonetheless, tamarins generally display minimal intragroup aggression, with a marked degree of cooperation and tolerance, even by sexually active males towards one another (Caine 1993).

Hernandez-Camacho and Cooper (1976) stated that

there is a tendency to seasonal reproduction in the species *S. leucopus* and *S. oedipus*. Hershkovitz (1977) noted that in many years of observation of *S. geoffroyi* in Panama and Colombia pregnant females and suckling young were found only from January to June. Eisenberg (1977) cited studies showing that the estrous cycle in this species is about 15 days, gestation lasts about 140 days, litter size is usually 2, and sexual maturity is attained at about 18 months in females and 24 months in males. According to Gengozian, Batson, and Smith (1977), gestation in *S. fuscicollis* is estimated at 140–50 days; births in captivity have occurred in every month, with a peak from March to May; and 75 percent of births have been of twins, 20 percent single, and 5 percent triplets. Birth weight in *Saguinus* is 25–55 grams (Hayssen, Van Tienhoven, and Van Tienhoven 1993).

In *Saguinus* the father and sometimes other adult members of a group assist at birth, receiving and washing the young. The newborn have a coat of short hair and are helpless. They cling tightly with their hands and feet to the body of the mother or father. The father transfers the young to the mother at feeding time and then accepts them from the mother again after feeding. This behavior, which is common among callitrichids, takes place every two to three hours and lasts about half an hour. At about 21 days the young begin to explore nearby surroundings, but they continue to ride on the backs of the parents until they are about 6–7 weeks of age. At 4 weeks they begin to accept soft food in addition to their mother's milk. Several members of a group besides the mother and father may help carry and provision the young, such communal roles apparently being more important in *Saguinus* than in *Callithrix* (Tardif, Harrison, and Simek 1993). According to Marvin L. Jones (Zoological Society of San Diego, pers. comm., 1995), a captive specimen of *S. imperator* lived for more than 20 years, one of *S. fuscicollis* lived for about 24 years, and one of *S. oedipus* lived for nearly 25 years.

The USDI lists the species *S. bicolor*, *S. oedipus*, and *S. geoffroyi* as endangered and *S. leucopus* as threatened. Those four species also are on appendix 1 of the CITES, and all other species are on appendix 2. The IUCN classifies the species *S. oedipus* and the subspecies *S. bicolor bicolor*, of central Brazil, as endangered, noting that each has a population of fewer than 2,500 individuals and is continuing to decline. The IUCN also designates the species *S. leucopus* and the subspecies *S. nigricollis hernandezi*, of southern Colombia, and *S. imperator imperator*, found in a small area where Bolivia, Brazil, and Peru come together, as vulnerable; each has an estimated surviving population of fewer than 10,000. Clearing of forest habitat by people is the main problem for all these forms, and populations of *S. oedipus* also were depleted through taking for the animal trade. *S. oedipus* already has lost most of its habitat, and it is unlikely that many remaining tracts of forest are large enough to maintain populations (Thornback and Jenkins 1982). Rylands, Coimbra-Filho, and Mittermeier (1993) expressed particular concern for *S. bicolor bicolor*, which occupies a very small range near the large city of Manaus. They also considered the subspecies *S. fuscicollis melanoleucus* and *S. f. acrensis*, confronted with rapid development in the Brazilian state of Acre, as vulnerable.

PRIMATES; CALLITRICHIDAE; **Genus CALLITHRIX**
Erxleben, 1777

Marmosets

There are 17 named species (Coimbra-Filho and Mittermeier 1973*b*; Ferrari and Lopes 1992; Mittermeier, Schwarz, and Ayres 1992; Muskin 1984; Natori 1990, 1994; Rylands, Coimbra-Filho, and Mittermeier 1993; Rylands, Mittermeier, and Luna 1995):

C. jacchus, eastern Brazil;

C. penicillata, inland east-central Brazil;

C. kuhli, southern Bahia and northeastern Minas Gerais in coastal eastern Brazil;

C. geoffroyi, states of Espirito Santo and Minas Gerais in coastal eastern Brazil;

C. flaviceps, southern Espirito Santo and Minas Gerais in coastal southeastern Brazil;

C. aurita, states of Minas Gerais, Rio de Janeiro, and Sao Paulo in coastal southeastern Brazil;

C. argentata, south of the western part of the Amazon River in eastern Brazil;

C. leucippe, a small area between the Cuparí and Tapajós rivers in central Brazil;

C. melanura, eastern Bolivia, northeastern Paraguay, and adjacent parts of southwestern Brazil;

C. emiliae, states of Pará and Matto Grosso in central Brazil;

C. nigriceps, along the Madeira River in the states of Amazonas and Rondônia in west-central Brazil;

C. marcai, known only from near the confluence of the Aripuana and Roosevelt rivers in the state of Amazonas in central Brazil;

C. intermedia, Guariba River Basin in the state of Amazonas in west-central Brazil;

C. humeralifer, south of the Amazon and between the Tapajós and Canuma rivers in central Brazil;

C. chrysoleuca, between the Aripuana and Canuma rivers in the state of Amazonas in central Brazil;

C. mauesi, between the Urariá and Maués rivers in the state of Amazonas in central Brazil;

C. saterei, central Amazon region of Brazil.

Rylands, Coimbra-Filho, and Mittermeier (1993) pointed out that an additional species, unnamed but related to *C. melanura* and *C. emiliae,* occurs in much of the state of Rondônia in west-central Brazil. They noted also that a large part of the state of Bahia on the coast of east-central Brazil is apparently occupied by a hybrid zone involving *C. jacchus, C. penicillata,* and *C. kuhli.* Coimbra-Filho, Pissinatti, and Rylands (1993) reported that there also has been natural hybridization involving *C. geoffroyi, C. flaviceps,* and *C. aurita* and that experiments in captivity have produced fertile hybrids between a number of the eastern Brazilian species. Partly on the basis of such hybridization, Hershkovitz (1975*a*, 1977) recognized only three valid species: *C. jacchus,* which would include *C. penicillata, C. geoffroyi, C. flaviceps, C. aurita,* and *C. kuhli; C. argentata,* which would include *C. leucippe, C. melanura, C. emiliae,* and probably *C. nigriceps;* and *C. humeralifer,* which would include *C. intermedia, C. chrysoleuca,* and probably *C. mauesi.* Groves (*in* Wilson and Reeder 1993) included *emiliae, intermedia, leucippe,* and *melanura* in *C. argentata,* and *chrysoleuca* in *C. humeralifer.* Groves, like some other authorities, also included *Cebuella* (see account thereof) in *Callithrix.*

Head and body length is 180–300 mm, tail length is 172–405 mm, and weight is usually 230–453 grams. The general coloration of *C. jacchus* is agouti gray; its tail has alternating broad blackish and narrow pale bands. There are long tufts in front of, and often above and behind, the base of the ears, and these tufts, along with other areas of the head, vary among species as follows: *C. penicillata,* blackish circumauricular tufts, crown, and temples; *C. geoffroyi,* blackish circumauricular tufts, white or creamy crown and sides of face; *C. jacchus,* mostly white or grayish circumauricular tufts, blackish or brown forehead and temples; *C. aurita,* small white ear tufts, black throat, head, and cheeks; and *C. flaviceps,* small yellow to ochraceous ear tufts, other areas of head the same color. In *C. argentata,* the black-tailed or silvery marmoset, the pelage is fine and silky and silvery white, sometimes washed with gray or yellowish gray, especially on the back. The tail is black, contrasting sharply with the color of the body. The face and ears are devoid of hair and reddish in color. There is a complete absence of tufts, plumes, or other adornments. In *C. humeralifer* there are pronounced tassels on the ears. The entire pelage is predominantly whitish in the western part of the range of the species. In the east the tassels are silvery or buffy, the back is black flecked with white, the tail is black banded with silver, and the underparts are orange.

In this genus and in *Cebuella* the lower canines are incisiform, barely extending beyond the adjacent incisors. *Callithrix* differs from *Cebuella* in its larger size and in cranial and dental features.

These marmosets generally live in tropical or subtropical forests. They run and hop in trees and bushes and are capable of leaping. Their movements are usually quick and jerky. They are diurnal, sleeping at night in tree holes or other shelters. The diet includes insects, spiders, small vertebrates, birds' eggs, fruits, and tree exudates. Like *Cebuella,* they have specialized short lower canines for perforating tree bark and inducing the flow of gums and sap (Coimbra-Filho and Mittermeier 1977*b*; Mittermeier and Coimbra-Filho 1977). Reportedly more than 70 percent of the foraging time of *C. jacchus* involves the collection of such exudates (Bouchardet da Fonseca and Lacher 1984). *C. penicillata* also depends heavily on exudates; however, other species are less adapted for exploitation of such resources, and *C. humeralifer* and *C. argentata* are highly frugivorous (Rylands and de Faria 1993).

Information summarized by Sussman and Kinzey (1984) indicates that these marmosets live in groups of about 4–15 individuals that have overlapping home ranges of 0.5–28.0 ha. and do not defend territories. Studies in captivity suggest that groups are dominated by a monogamous pair, but the mating system in the wild may be comparable to that described above in the account of *Saguinus.* Groups may contain several adults of each sex; generally only the dominant female breeds, but groups of *C. jacchus* with two reproductively active females have been observed (Digby and Ferrari 1993). Observations of *C. jacchus* show that ovulation in subordinate females is inhibited through suppression of necessary hormone secretions, perhaps in response to pheromones from the dominant female (Abbott, Barrett, and George 1993). Groups of *C. jacchus* appear to be extended families, and emigration seems less significant than it is in *Saguinus* (Digby and Barreto 1993). Dominance is maintained through aggressive interaction. The vocal repertoire includes a soft "phee" for contact, a louder "phee" for territorial expression, a "tsee-tsee-tsee" as an aggressive threat, and a high-pitched whistle as a warning signal. *Callithrix* has scent glands and utilizes them in the same manner described above in the account of *Saguinus* (Epple et al. 1993).

Data cited by Hayssen, Van Tienhoven, and Van Tien-

Short-tusked marmosets *(Callithrix):* A. Black ear-tufted marmoset *(C. penicillata),* photo by Ernest P. Walker; B. Silky marmoset *(C. chrysoleuca),* photo by Harald Schultz; C. Geoffroy's marmoset *(C. geoffroyi),* photo by Bernhard Grzimek; D. Black-tailed marmoset *(C. argentata);* E & F. White ear-tufted marmosets *(C. jacchus),* photos by Ernest P. Walker.

hoven (1993) indicate that breeding is year-round in captivity but that births of *C. jacchus* in Brazil occur in late October–November and late March–April. Interbirth interval in that species is about 6 months, the estrous cycle is about 30 days, estrus lasts 2–3 days, the gestation period is around 130–50 days, birth weight is 20–35 grams, lactation lasts up to 100 days, and sexual maturity is attained at 11–15 months by males and 14–24 months by females. Litter size in *C. jacchus* is 1–4, with more than half of recorded births resulting in twins (Hershkovitz 1977). *Callithrix* carries and provisions its young for a shorter period than does *Saguinus*, and there is less involvement in this function by adults other than the mother (Tardif, Harrison, and Simek 1993). Life span in the genus seems to be about 10 years in the wild and up to 16 years in captivity.

The species *C. aurita* and *C. flaviceps*, found only along the rapidly developing southeastern coast of Brazil and each numbering fewer than 2,500 individuals, are classified as endangered by the IUCN and the USDI and are on appendix 1 of the CITES; all other species are on appendix 2. The IUCN also classifies *C. geoffroyi*, *C. nigriceps*, *C. leucippe*, and *C. chrysoleuca* as vulnerable. All of these species are jeopardized through destruction of their habitat by people. Oliver and Santos (1991) reported that *C. flaviceps* has the most limited distribution of any marmoset but that *C. kuhli* and *C. geoffroyi* remain relatively widespread. The small ranges of *C. leucippe*, *C. nigriceps*, and *C. humeralifer* are bisected by the new Trans-Amazonian Highway in Brazil and are being subjected to much clear-cutting to make way for cattle ranches (Ferrari and Queiroz 1994; Rylands, Coimbra-Filho, and Mittermeier 1993; Thornback and Jenkins 1982). Recent studies of *C. flaviceps* indicate that although it has lost most of its original habitat, it is adaptable to secondary forest and can be expected to survive with proper protection (Ferrari and Mendes 1991).

PRIMATES; CALLITRICHIDAE; **Genus CEBUELLA**
Gray, 1865

Pygmy Marmoset

The single species, *C. pygmaea*, inhabits tropical forests of the upper Amazon region in western Brazil, southeastern Colombia, eastern Ecuador, eastern Peru, and northwestern Bolivia (Hershkovitz 1977; Rylands, Coimbra-Filho, and Mittermeier 1993). Groves (1989 and *in* Wilson and Reeder 1993) and Rosenberger and Coimbra-Filho (1984) included this species in the genus *Callithrix*, but *Cebuella* was maintained as a full genus by Schneider et al. (1993).

This is the smallest New World primate. Head and body length is 117–52 mm, tail length is 172–229 mm, and adult weight is 85–140 grams (Hershkovitz 1977; Soini 1993). The head and neck are dark brown and gray or dark brown and buff; the back is grayish, black mixed with buff, or brownish tawny, sometimes with a greenish cast; the hands and feet are yellowish or orangish; the tail is indistinctly ringed with black and tawny; and the underparts are often orangish but vary from white to tawny. From *Callithrix*, *Cebuella* differs in its smaller size, multibanded agouti pelage, penis shaft lacking spines, sessile scrotum, second upper incisor tooth and both lower incisors having a mesial as well as a distal capsule, and lower premolars being at least as long as they are wide (Groves 1989).

The pygmy marmoset prefers low second growth along streams where cover is thick, visibility is good, and insects are most abundant (Hershkovitz 1977). It is arboreal and completely diurnal, being most active on cool mornings and in late afternoons. Movement usually is by running along horizontal or diagonal branches with a galloping gait. It can make long, horizontal leaps of a meter or more and also clings and leaps vertically. Movements are sometimes exceedingly slow but more often occur in spurts; an alarmed individual continually turns its head in all directions (Moynihan 1976). The diet consists of fruits, buds, insects, and sap or exudates from trees, the latter item being an especially important food source. The short lower canine teeth are specialized for gouging holes in trees and inducing exudate flow (Mittermeier and Coimbra-Filho 1977). Each pygmy marmoset group has one or more trees in its range, which are riddled with small holes for feeding on the sap (Moynihan 1976). There may be hundreds of such holes per square meter of tree surface (Ramirez, Freese, and Revilla C. 1977).

In a floodplain forest of Amazonian Peru, Soini (1982) found a density of 51.5 independently locomoting individuals per sq km, but along river edges density increased to 274/sq km. Of these animals, 50 percent were adults; 83 percent were in stable troops, and the rest were incipient pairs and lone individuals. Forest troops lived in home ranges of 0.2–0.4 ha., while the ranges of riverside troops covered 70–90 meters of shoreline and extended 20–60 meters inland. Home ranges were contiguous but nonoverlapping. Stable troops contained a mated adult pair and their mature offspring of up to four generations. All members of a troop roosted together, usually huddled in a clump on a leafy branch 7–10 meters high.

In another Peruvian study, Castro and Soini (1977) calculated a population density of 5.6/ha. Of 4 troops of 5–10 animals each, 3 occupied contiguous, nonoverlapping home ranges along a river and 1 inhabited a single large "home tree" in which the marmosets foraged and slept. Also in Peru, Ramirez, Freese, and Revilla C. (1977) found that troops numbered 7–9 animals and utilized home ranges measuring about 75 × 40 meters. About a third of this area formed a core zone containing all of the troop's exudate trees, in which the group spent 80 percent of its time. The overall reported size range for pygmy marmoset groups is 2–15 (Mittermeier and Coimbra-Filho 1977). Although sometimes considered a relatively quiet primate, *Cebuella* has a variety of vocalizations, including a trill for communication over distance, a high, sharp warning whistle, and a clicking sound for threats (Pola and Snowdon 1975).

According to Soini (1993), some troops contain more than one adult of each sex, but in such cases only one female appears to be reproductively active and one male maintains exclusive mating access to her. Soini (1982, 1993) reported that births in the wild occur throughout the year but that in Amazonian Peru there are two peaks, in October–January and May–June. Hershkovitz (1977) wrote that no reproductive seasonality was evident in captivity, that gestation lasted about 20 weeks, and that sexual maturity was attained at 18–24 months. Of 21 recorded births, 14 were of twins, 6 were single, and 1 was of triplets. Hayssen, Van Tienhoven, and Van Tienhoven (1993) listed birth weights of 14–27 grams, a minimum interbirth interval of about 150 days, and a lactation period of 3 months. One captive specimen in Japan lived for 18 years (Marvin L. Jones, Zoological Society of San Diego, pers. comm., 1995).

Moynihan (1976) found that in the Putumayo region of Colombia the pygmy marmoset was subject to intensive collection for use as a pet. Mittermeier and Coimbra-Filho (1977), however, stated that the species was in no danger in Brazil as it was highly adaptable to change and could inhabit degraded as well as virgin forest. It is on appendix 2 of the CITES.

Pygmy marmosets *(Cebuella pygmaea)*, photo by Howard E. Uible. Inset photo by Ernest P. Walker.

PRIMATES; CALLITRICHIDAE; **Genus**
LEONTOPITHECUS
Lesson, 1840

Lion Tamarins

There are four species, all in the lowland rainforests of southeastern Brazil (Coimbra-Filho and Mittermeier 1977a; Lorini and Persson 1990; Pinto and Tavares 1994; Rosenberger and Coimbra-Filho 1984; Rylands, Coimbra-Filho, and Mittermeier 1993):

L. chrysopygus (black lion tamarin), state of Sao Paulo;
L. caissara (black-headed lion tamarin), Superagui Island
    and nearby mainland of states of Paraná and Sao Paulo;
L. chrysomelas (golden-headed lion tamarin), state of
    Bahia and extreme northeastern Minas Gerais;
L. rosalia (golden lion tamarin), state of Rio de Janeiro.

The name *Leontideus* Cabrera, 1956, often has been used for this genus. Hershkovitz (1977) considered the species to be color grades of an otherwise morphologically uniform species, *L. rosalia*, but Rosenberger and Coimbra-Filho (1984) showed that they are distinct in cranial and dental characters as well as in color.

Head and body length is about 200–336 mm, tail length is 315–400 mm, and adult weight is 600–800 grams (Kleiman 1981). The pelage is long and silky; the common name "lion" refers to the mane on the shoulders. *L. chrysopygus* is mostly black with a gold rump and thighs; *L. caissara* is mostly gold with black head, hands, feet, and tail; *L. chrysomelas* also is mostly black but has a golden mane; and *L. rosalia* is entirely gold, reddish, or buffy to white (Hershkovitz 1977; Lorini and Persson 1990).

Lion tamarins prefer primary tropical forest but have been reported in secondary forest and areas under partial cultivation. They are usually found at a height of 3–10 meters in the trees, where interlacing branches, vines, and epiphytes provide optimum shelter and an abundance of insect and small vertebrate prey. They are diurnal, sleeping at night in tree holes or occasionally in vines or epiphytes (Coimbra-Filho and Mittermeier 1977a; Hershkovitz 1977). Groups forage along a path of about 1,300–2,600 meters each day (Rylands 1993a). The animals leap from branch to branch with great agility. They are primarily in-

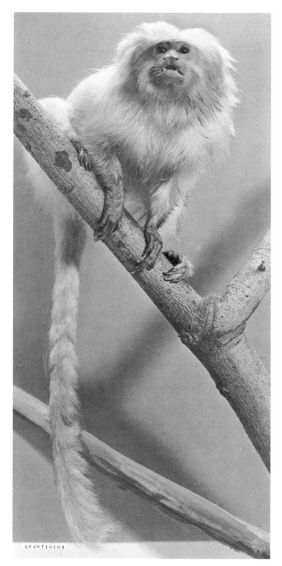

Golden lion tamarin (Leontopithecus rosalia), photo from New York Zoological Society.

sectivorous and frugivorous but also eat spiders, snails, small lizards, birds' eggs, small birds, and plant exudates.

Leontopithecus occurs at population densities of about 1–17/sq km and groups occupy home ranges of up to 200 ha. (Rylands 1993a). A portion of the home range appears to be a territory that is defended from other groups (Peres 1989). Territory size has been found to average 75 ha. for *L. chrysomelas* and 42 ha. for *L. rosalia* (Dietz, De Sousa, and Da Silva 1994). Groups consist of 2–11, usually 3–7, related individuals, but temporary associations of 15–16 have been observed. The basic group is thought to consist of a mated pair plus their young of one or more years. Although adults of the same sex have been reported to be extremely aggressive toward one another, even fighting to the death if kept together, numerous observations now indicate that natural groups frequently contain more than one adult of each sex. The dominant adult male and female of a group form a permanent pair bond, are equal in rank, and share

responsibility for raising the young. In the rate of dispersal of subordinate adults, groups of *Leontopithecus* probably are comparable to those of *Saguinus* rather than to the more stable groups of *Callithrix* (Coimbra-Filho and Mittermeier 1973a, 1977a; Rylands 1993a; Snyder 1974). McLanahan and Green (1977) identified 17 types of vocalization, grouped into several classes with such associations as trills (for solo activity), clucks (for foraging), long calls (for vigilance), and whines (for contact).

The dominant female of a group may inhibit reproductive activity in other females, but unlike the process discussed in the account of *Callithrix*, this one seems to be strictly behavioral and does not involve physiological mechanisms. This results in a more flexible mating system, perhaps associated with resource availability, and in some groups there is mating by more than one female (Dietz and Baker 1993; Rylands 1993a). *Leontopithecus* apparently is a seasonal breeder. In Brazil births occur from September to March, the warmest and wettest period of the year (Coimbra-Filho and Mittermeier 1973a). In captivity in the Northern Hemisphere most litters are born from January to June. The estrous cycle averages about 2–3 weeks and gestation averages 128 days, ranging from 125 to 132 days. Of the recorded births in the National Zoo in Washington, D.C., 116 were of twins, 41 were single, and 8 were of triplets (Kleiman 1977b). The young are born fully furred with their eyes open and weigh an average of 60.6 grams (Kleiman 1981). Within a few days after a birth the father begins to carry the infant at times, and he is the principal carrier after the third week. Juvenile members of the family also carry and care for the newborn (Kleiman 1981). Weaning is complete at 12 weeks (Hayssen, Van Tienhoven, and Van Tienhoven 1993). Sexual maturity is reached by males at about 24 months and by females at 18 months (Eisenberg 1977). Life span in captivity, when the animals are well cared for, can be 15 years or longer. One specimen of *L. rosalia* was still living in December 1995 at an age of 28 years and 2 months (Marvin L. Jones, Zoological Society of San Diego, pers. comm., 1995).

Lion tamarins are among the world's most critically endangered mammals. They are classified as endangered by the USDI and are on appendix 1 of the CITES. The IUCN now designates *L. chrysomelas* as endangered and the other three species as critically endangered. They are found in the part of Brazil that has the densest human population. They have declined largely because of destruction of their forest habitat for lumber, agriculture, pasture, and housing. Until the 1960s they also were subject to considerable exportation for use in zoos, laboratories, and the pet trade.

In the nineteenth century the species *L. rosalia* occurred across most of the present state of Rio de Janeiro. Attempts to establish reserves for *L. rosalia* failed in the 1960s, and two prime sites were deforested in 1971 when money could not be found for purchase (Coimbra-Filho and Mittermeier 1977a). *L. rosalia* now is restricted to the Sao Joao Basin in Rio de Janeiro, where the habitat it currently occupies totals about 105 sq km and is continuing to decline. A 5,500-ha. reserve for *L. rosalia* was established in the mid-1970s and is now the site of a reintroduction program. Intensive efforts to establish a captive breeding pool for *L. rosalia* were begun in the 1960s. By 1983 the captive population of that species contained about 370 animals and was increasing at a rate of 20–25 percent annually. Efforts to reintroduce some of these captives were initiated in Brazil in 1984 and eventually achieved a high degree of success, including substantial reproduction in the wild (Beck et al. 1991; Kleiman et al. 1986). The captive population has now leveled off at just over 500 and is being intensively managed for genetic viability (Rylands 1994). The total wild population of *L. rosalia* is now estimated at 560 individuals, including about 39 surviving reintroduced animals and 53 of their offspring (Mallinson 1994).

L. chrysomelas has always been restricted to a small coastal area of the state of Bahia and extreme northeastern Minas Gerais, but its range is now fragmented. A reserve of about 11,400 ha. was authorized for its protection, but only about 7,000 ha. has been purchased and much of that has been taken over by squatters (Mittermeier et al. 1986; Oliver and Santos 1991; Rylands 1994). The species also is able to use some of the cocoa plantations within its range since cocoa is grown in the shade and many large forest trees are left standing for this purpose. The great majority of wild individuals thus do not live in officially protected areas. Although the total remaining range is about 37,500 sq km, suitable habitat is fragmented and declining precipitously. During the 1980s a substantial number of *L. chrysomelas* were exported illegally from Brazil to various other countries. An international project to recover these animals was largely successful and contributed to the start of a captive breeding pool of *L. chrysomelas* (Mallinson 1987). Although once thought to number only about 200 individuals, intensive conservation efforts have resulted in a wild population of the species estimated at 6,000–14,000 and to at least 575 more in captivity (De Bois 1994; Mallinson 1994; Pinto and Tavares 1994; Rylands 1994).

L. chrysopygus, which once occupied a large forested part of the state of Sao Paulo, now has been reduced in range to a 37,157-ha. state park at the western edge of the state, two nearby private ranches, and two small areas in central Sao Paulo, one private and one state-owned. There were once estimated to be only 50–100 individuals left in the wild (Mittermeier et al. 1985), but conservation efforts seem to have led to a moderate improvement in status. There now are about 1,000 in the wild and another 80 in captivity (Mallinson 1994; Valladares-Padua, Padua, and Cullen 1994).

The newly discovered *L. caissara* probably is the rarest and most endangered member of the Callitrichidae. Its entire range may be less than 30,000 ha., though about a third of this area is within protected parks. There are only about 260 individuals in the wild, divided into three isolated groups, and none in captivity (Lorini and Persson 1994; Mallinson 1994; Rylands, Coimbra-Filho, and Mittermeier 1993).

PRIMATES; Family CERCOPITHECIDAE

Old World Monkeys

This family of 21 living genera and 96 species is found in Africa (and, possibly through introduction, in Gibraltar in extreme southern Spain), the southwestern part of the Arabian Peninsula, south-central and southeastern Asia, Japan, and the East Indies as far as Sulawesi and Timor. The sequence of genera presented here is an attempt to express phylogenetic relationships with consideration of the arrangements of both Hershkovitz (1977) and Groves (1989). Two subfamilies are recognized: the Cercopithecinae, with the genera *Erythrocebus, Chlorocebus, Cercopithecus, Miopithecus, Allenopithecus, Cercocebus, Lophocebus, Macaca, Papio, Mandrillus,* and *Theropithecus;* and the Colobinae, with *Nasalis, Simias, Pygathrix, Rhinopithecus, Presbytis, Semnopithecus, Trachypithecus, Colobus, Piliocolobus,* and *Procolobus.* Groves (1989) actually had divided the Cercopithecidae into two full families: the Cercopithecidae, with the subfamily Cercopithecinae

for the genera *Erythrocebus, Chlorocebus,* and *Cercopithecus* and Papioninae for all other genera; and Colobidae, with the subfamily Nasalinae for *Nasalis* and *Simias* and Colobinae for all other genera. Subsequently, Groves (*in* Wilson and Reeder 1993) accepted only the single family Cercopithecidae with the two subfamilies Cercopithecinae and Colobinae. The same basic subfamilial divisions were given by Dandelot (*in* Meester and Setzer 1977), Delson (1994), and Thorington and Groves (1970). Hill (1970), however, placed the genera *Cercocebus, Lophocebus, Macaca, Papio, Mandrillus,* and *Theropithecus* in a separate subfamily, the Cynopithecinae.

Head and body length is 325–1,100 mm. The tail, when present, is 20–1,030 mm long; it is absent in the Barbary ape *(Macaca sylvanus).* The heaviest members of this family weigh up to 54 kg *(Mandrillus).* The tail is slightly prehensile in young guenons *(Cercopithecus).* Adult mangabeys *(Cercocebus)* sometimes coil their tails around a branch for support, but the tail is not prehensile. With these exceptions, the tail, when well developed, serves as a balancing organ. The pelage varies from short to long, and a cap or crest on the head, a mane, or a beard is present on some. The face is bare or nearly so and is often almost black, but adult male *Mandrillus sphinx* have brilliant red, blue, or purple streaks on the face. The palms and soles are naked, and the underparts are usually sparsely haired. Naked buttock pads are often brightly colored. The muzzle is elongate to rounded, usually longer in male baboons than in females. The ears are rounded. The space between the nostrils is narrow, the nostrils being higher than wide and directed forward or downward. The members of the genera *Rhinopithecus* and *Simias* have a peculiar upturned nose, and in old males of *Nasalis* the nose is long and pendulous. Cheek pouches inside the lips are present in the macaques, mangabeys, guenons, and baboons but absent in the langurs, proboscis monkey, and colobus monkeys. The stomach in the latter group is large and has folds or pockets in the walls that are adaptations for handling bulky leafy food. The females of this family have one pair of mammae in the chest region.

The forelimbs are generally shorter than the hind limbs. The hands and feet are adapted for grasping and have five digits, but the thumb is absent, or nearly so, in colobus monkeys. The thumb, if present, is opposable, as is the large toe. The nails are flattened.

The dental formula is: (i 2/2, c 1/1, pm 2/2, m 3/3) $\times$ 2 = 32. The lower incisors are nearly upright and in contact with the canines, whereas the upper incisors are pressed close together and separated from the canines by a small space. The upper canines are elongate and tusklike, and the lower canines are curved inward and backward slightly. The premolars are smaller than the molars, and the molars are supplied with tubercles. The highly concave palate extends beyond the last molar, and the skull is rounded or flattened, with a large braincase.

The members of this family have good sight, hearing, and sense of smell, a moderate amount of facial expression, and a variety of calls. Nearly all Old World monkeys and baboons are diurnal. Baboons are mainly terrestrial; macaques are at home either in trees or on the ground; and the other members of the family are mainly arboreal. Cercopithecids can stand upright, but they generally progress on all four limbs. Langurs and colobus monkeys, perhaps the most arboreal members of the family, jump and leap through trees with amazing agility and speed. Sentinels may be posted to detect danger; the usual reaction when they are threatened is to flee, but male baboons often stay at the rear of a retreating group to fight an enemy that comes too close. Most forms are good swimmers.

Some members of this family raid crops or forage near villages. Some are hunted, usually illegally, for their meat and fur. The rhesus monkey *(Macaca mulatta)* and the crab-eating macaque *(M. fascicularis)* are used extensively in laboratory research. The constitution of most cerco-pithecids makes them better suited for a life in captivity than the New World monkeys. However, their manners are less gentle, and when they grow old they usually are not satisfactory pets. The capture of primates for use as pets is usually illegal in the countries where the animals occur, and importation into the United States for such purposes is now prohibited by federal law.

The geological range of this family is early Miocene to Recent in Africa and late Miocene to Recent in Europe and Asia (Thorington and Anderson 1984). There was an an-cestral subfamily, the Victoriapithecinae, and the diver-gence of the modern subfamilies Cercopithecinae and Colobinae apparently occurred in the middle to late Miocene (Delson 1994).

PRIMATES; CERCOPITHECIDAE; Genus ERYTHROCEBUS
Trouessart, 1897

Patas Monkey, or Red Guenon

The single species, *E. patas,* inhabits open country from Senegal to Ethiopia and south to Tanzania. Dandelot *(in* Meester and Setzer 1977) and Lernould (1988), among oth-ers, treated *Erythrocebus* as a subgenus of *Cercopithecus,* but Groves (1989), Hershkovitz (1977), Napier and Napier (1985), and Thorington and Groves (1970) regarded it as a separate genus.

Head and body length is 600–875 mm and tail length is 500–750 mm. The adult males, weighing about 7–13 kg, are usually much larger than the females, which weigh about 4–7 kg. On the adult males the hairs of the nape and shoul-ders may be long and manelike, and there is a mustache of white hairs, along with whiskers that are usually white. The upper parts are usually reddish mixed with gray and the underparts are whitish, grayish, buff pink, or orange ru-fous. The thighs are often whitish, and the tail is dark above and light below. There may be a light forehead band and a dark dorsal stripe. The young are almost entirely red. The long limbs of patas monkeys are of about equal length. The digits are considerably reduced, but the more proximal bones of the feet are elongated (Kingdon 1988). According to Groves (1989), the limb specializations and unusual red coloration of *Erythrocebus* set it apart from *Cercopithecus.* The latter also is distinguished from *Erythrocebus,* as well as *Chlorocebus* (see account thereof), by various cranial characters.

Much of what is known about the natural history of this genus has resulted from Hall's studies (1967, 1968) in Uganda. The patas monkey occupies grass and woodland savannahs and avoids the more densely treed areas within its range. It is diurnal and largely terrestrial. It can climb trees when alarmed but usually relies on its speed on the ground to escape from danger. Perhaps the fastest of all pri-mates, it has been timed running at about 55 km/hr. Dur-ing a 10- to 12-hour day a group travels from as little as 500 meters, when food is abundant, to as much as 12,000 meters, when resources are scarce. During the 2–3 hottest hours of the day the animals rest in a large shade tree. At night each individual goes up a separate tree to sleep, ex-cept that females retain their infants at this time. A group may be spread over an area of 250,000 sq meters at night

and thus be protected from severe loss to predators. The bulk of the diet consists of grasses, berries, fruits, beans, and seeds, occasionally supplemented by mushrooms, ants, grasshoppers, lizards, and birds' eggs. In the 311,200-ha. area studied by Hall there were 110 patas. Each group was found to occupy an extensive home range, that of one troop of 31 animals being 51.8 sq km. There is no core area, and almost any part of the range may be used at night for sleep-ing. Chism and Rowell (1988) reported that two groups used home ranges of 23.4 and 32.0 sq km.

Hall observed group size to average 15 and to range from 9 to 31 individuals. Each heterosexual group included only 1 adult male and 2–8 adult females. Isolated adult males and one group of 4 males were also seen. Presumably, the domi-nant male of a heterosexual group drives the other males out. In a study in Cameroon that in most respects con-firmed Hall's findings, Struhsaker and Gartlan (1970) found mean group size to be 21 (7–34) patas. Although typically there was a single adult male per heterosexual group, one troop apparently had 2 adult males, and several all-male groups were also seen. In Kenya, Harding and Ol-son (1986) found as many as 10 males to enter a group of 74 individuals during the mating period and to mate with the females, thus suggesting that the patas monkey's social structure is more variable than previously thought. Per-haps the situation can be explained by the work of Chism and Rowell (1988), also in Kenya, who found the patas population to consist both of heterosexual groups, usually with 1 or a few adult males and numerous females, and of other males that lived alone or in all-male associations. These other males appeared to have large home ranges that overlapped those of several heterosexual groups. Males fre-quently moved between their own associations and the het-erosexual groups, their presence in the latter coinciding with the mating season.

Whereas Hall found spacing mechanisms to be such that intergroup encounters were rare, Struhsaker and Gartlan reported much contact and agonistic behavior between groups concentrated around water holes at the peak of the dry season. Perhaps such concentrations are responsible for earlier reports of groups comprising as many as 200 indi-viduals. Chism and Rowell also found aggressive interac-tion between heterosexual groups and observed females and juveniles to be the chief antagonists. According to Hall, however, the adult male is the guardian of the group. He is constantly watchful and often moves several hundred me-ters away from the group to survey a new feeding area or check for danger. Since he is often away, the adult females establish their own hierarchy to assure group order, and one or more of the females may lead routine movement. Chism and Rowell found that adult females usually initi-ate group movement, determine the daily route, and select sleeping sites; they maintain contact with a "moo" vocal-ization. Though usually silent, the male may bark upon en-countering another patas group or create a noise to distract a predator. Laboratory investigation has shown that *Ery-throcebus* has many vocalizations but that most are muted and can be heard by humans only within 100 meters.

The birth season apparently is restricted to Novem-ber–January in Cameroon and to around February in Uganda. In Kenya the reproductive period also is limited, with mating occurring from June to September (Harding and Olson 1986) and most young being born in December and January (Chism and Rowell 1988). Studies of captives indicate that the estrous cycle averages about 33 days, es-trus 13.5 days, gestation 167 days, and interbirth interval 384 days (Loy 1981). There is normally a single young. In a study of captives, Rowell (1977*b*) determined that fe-males usually can first conceive at 2.5 years and that males

Patas monkeys *(Erythrocebus patas):* A. Photo by Ernest P. Walker. B. Photo from U.S. National Zoological Park.

reach sexual maturity 1–2 years later. Record longevity is 23 years and 11 months (Marvin L. Jones, Zoological Society of San Diego, pers. comm., 1995).

The subspecies *pyrrhonotus* of Nubia and Somaliland, known as the Nisnas, was probably the Cebus of the ancients, which is pictured so frequently on their monuments and which was described by Aelian as being of a bright flame color with white whiskers and underparts. The Teso natives formerly thought that the flesh of the patas monkey would cure leprosy.

Erythrocebus sometimes is destructive to crops. For that reason and for its meat, it often is hunted by people. Its habitat is being reduced in some areas by heavy cattle grazing and conversion of the savannahs to farmland (Wolfheim 1983). Although it is not dependent on large trees, it does require some woodland and will disappear from an area if such habitat is lost (Chism and Rowell 1988).

PRIMATES; CERCOPITHECIDAE; Genus CHLOROCEBUS
Gray, 1870

Savannah Guenons

Dandelot (*in* Meester and Setzer 1977) and Lernould (1988) recognized four allopatric species:

C. sabaeus (green monkey), Senegal to the Volta River in Ghana;
C. tantalus, savannah zone from the Volta River in Ghana to Uganda;
C. aethiops (grivet), Sudan, Ethiopia;
C. pygerythrus (vervet), southern Ethiopia to Angola and South Africa.

Sineo, Stanyon, and Chiarelli (1986) treated *C. cynosurus* of Angola and some adjacent areas as a species distinct from

C. pygerythrus. All of these species were considered only subspecies or synonyms of *C. aethiops* by Corbet and Hill (1991), Groves (1989 and *in* Wilson and Reeder 1993), and Thorington and Groves (1970). Of these various authorities only Groves actually recognized *Chlorocebus* as a genus separate from *Cercopithecus.*

Head and body length is 350–660 mm, tail length is 420–720 mm, and weight is 2.5–9.0 kg (Kingdon 1971; Van Hoof *in* Grzimek 1990). Average adult weight is 7.0 kg in males and 5.6 kg in females (Fedigan and Fedigan 1988). The fur is predominantly gray- to yellow-green and the face is black with a white fringe. Males have a light red penis and a blue scrotum. Groves (1989) pointed out that both *Chlorocebus* and *Erythrocebus* lack certain cranial specializations of *Cercopithecus.* In the latter genus the inferior suborbital region is regularly curved toward the dental arcade, the pyriform aperture is round to oval (not angled in the middle), the temporal lines anteriorly follow the posterior borders of the orbits, the nasal bones usually run straight across inferiorly instead of being pointed at the midline, and the second upper incisor tooth is small and pointed. *Chlorocebus* and *Erythrocebus* also share the following specializations: in side view the orbits do not slope forward inferiorly, but their lower borders are situated behind the upper margins; the auditory tube has a V-shaped lower margin; and the orbits themselves are angular instead of oval. The hand and foot of *Chlorocebus* exhibit some of the terrestrial adaptations of *Erythrocebus* but in less extreme form (Kingdon 1988).

According to Fedigan and Fedigan (1988), these monkeys are found in a variety of habitats, including savannah with minimal tree cover, open woodland, and gallery and rainforest edge. They prefer riverine woodland but seem limited only by the need for water and sleeping trees. They seem more adaptable than *Cercopithecus,* being able to utilize such marginal habitats as secondary growth, agricultural areas, and mangrove swamps. They are partly terrestrial, readily forage and run about on the ground, and are good swimmers. Like other guenons, however, they depend on trees as sleeping sites at night, and they take shelter in

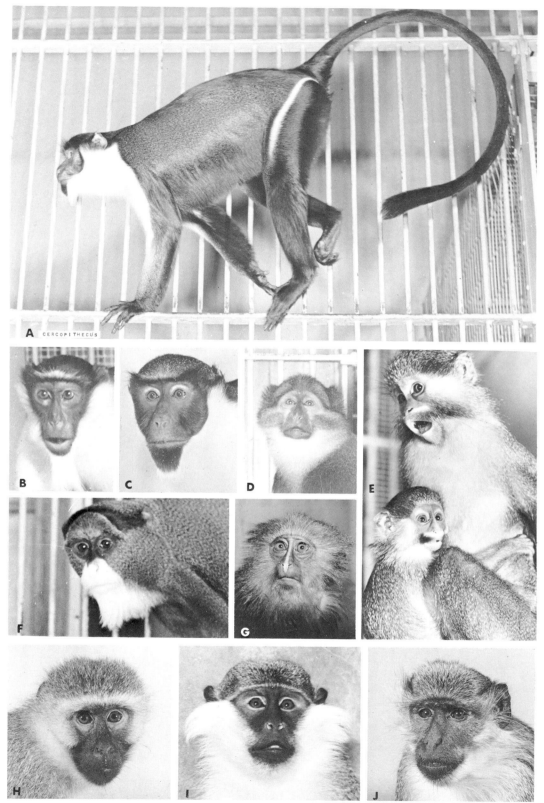

African guenons: A & C. Diana monkey *(Cercopithecus diana)*; B. Roloway monkey *(Cercopithecus diana roloway)*; D. Preuss's monkey *(Cercopithecus preussi)*; E. Moustached monkey and baby *(Cercopithecus cephus)*; F. De Brazza's monkey *(Cercopithecus neglectus)*, photos by Ernest P. Walker. G. Owl-faced monkey *(Cercopithecus hamlyni)*; H. Vervet monkey *(Chlorocebus pygerythrus)*, photos from San Diego Zoological Garden. I. Grivet monkey *(Chlorocebus aethiops)*, photo by Bob McIntyre through Cheyenne Mountain Zoo. J. Green monkey *(Chlorocebus sabaeus)*, photo from San Diego Zoological Garden.

trees when alarmed. *C. aethiops* may even gather and hold well-leaved branches around itself. Savannah guenons eat a great variety of natural and cultivated fruits, other plant parts, insects, crustaceans, birds' eggs, and small vertebrates. In the mangroves of Senegal more than half of the feeding time of *C. sabaeus* involved hunting for fiddler crabs (Galat and Galat-Luong 1976).

According to Fedigan and Fedigan (1988), group home ranges have been calculated at 13–178 ha. and group size at 5–76, averaging 25, individuals; there usually are several adults of each sex in a group. Young males generally disperse from their natal groups, but young females remain and become part of an established social hierarchy. In some populations a number of groups seem to form a larger community, in which there is an interchange of males and other friendly interaction. The degree of territoriality displayed by groups seems to vary widely, depending on such factors as history of association with neighboring groups and availability and concentration of resources. In Kenya, Struhsaker (1967*b*) found groups of *C. pygerythrus* occupying home ranges of 18–96 ha., most of which was a defended territory. Galat and Galat-Luong (1976) found groups of *C. sabaeus* to defend their territories through loud barking and displays. *C. aethiops* has a creaking cry and a staccato bark that enable the members of a troop to keep in touch with one another. Other calls of this species express alarm, excitement, pain, rage, and melancholy. Struhsaker (1967*a*) found *C. pygerythrus* to make 36 sounds that conveyed at least 21–23 different messages.

In the Sahel region of Senegal, Galat and Galat-Luong (1977) found that over a three-year period a group of *C. sabaeus* numbered 33–41 animals, including 2–5 adult males and 10–15 adult females. Another troop in this region had 140–74 members (Galat and Galat-Luong 1978), but in a separate study of *C. sabaeus* in Senegal, Dunbar (1974) found group size to be only 4–16. In Uganda, Gartlan and Brain (1968) found group size in *C. pygerythrus* to range from 4 to 51 and to average 9–11. In contrast with what was observed in some other investigations, the adult males seemed peripheral to the group and may even have moved between groups. The nucleus of a group was formed by the adult females and their young. Rowell (1971), studying captive *C. aethiops*, observed that adult males were only equal to or lower than adult females in dominance. In an investigation of an introduced population of *C. sabaeus* on St. Kitts Island in the West Indies, however, McGuire (1974) found that all groups, each numbering 4–65 animals, had a single dominant adult male.

Dunbar (1974) observed that in Senegal there was an apparent birth peak in *C. sabaeus* from January to March. Fedigan and Fedigan (1988) reported that in any given population of *Chlorocebus* there is usually a synchronized peak of parturition, with approximately 80 percent of births occurring in a period of 2–3 months. This period seems usually to coincide with the beginning of the season during which resources are most plentiful. Females commonly produce young at intervals of 1 year. Weaning is usually completed when the young is 8.5 months old. Females are capable of giving birth at 2.5 months. Males are physiologically capable of reproduction at about 3 years but are not behaviorally mature until after 5 years. The estrous cycle lasts about 30 days (Hayssen, Van Tienhoven, and Van Tienhoven 1993). Van Hoof (*in* Grzimek 1990) noted that there normally is a single young, birth weight is 300–400 grams, gestation is 5.5 months, and the life span is more than 30 years.

All species of *Chlorocebus* are on appendix 2 of the CITES but are thought to be common and to be relatively adaptable to the presence of human activity (Fedigan and Fedigan 1988). Populations of *C. sabaeus* are established on St. Kitts, Nevis, and Barbados islands in the West Indies. They are probably descended from pets brought by African slave traders in the seventeenth century and have been the subject of a number of field studies (Horrocks 1986; McGuire 1974; Poirier 1972). Denham (1982) suggested that numbers initially increased rapidly on Barbados, then crashed in the eighteenth century because of loss of forest habitat and bounty hunting, and finally increased again in the 1950s after some areas had become reforested.

PRIMATES; CERCOPITHECIDAE; **Genus CERCOPITHECUS**
Linnaeus, 1758

Guenons

There are 20 species (Colyn, Gautier-Hion, and Thys van den Audenaerde 1991; Dandelot *in* Meester and Setzer 1977; Dutrillaux et al. 1988; Harrison 1988; Kuroda, Kano, and Muhindo 1985; Lernould 1988; Oates 1988; Sineo, Stanyon, and Chiarelli 1986; Thorington and Groves 1970):

C. nictitans (white-nosed guenon), Liberia, Ivory Coast, Nigeria to Congo and Zaire north of Congo River, island of Bioko (Fernado Poo);

C. mitis (blue monkey), forested areas west of Rift Valley in southern Sudan, Ethiopia, Uganda, Kenya, Zaire, Angola, and Zambia;

C. albogularis (Sykes monkey), east of Rift Valley from Somalia south to eastern Cape Province in South Africa, islands of Phylax, Zanzibar, and Mafia;

C. ascanius (black-cheeked white-nosed monkey), Central African Republic to western Kenya and Angola;

C. petaurista (lesser white-nosed guenon), Guinea to Benin;

C. erythrogaster (red-bellied guenon), southern Nigeria to west of Niger River;

C. sclateri, southeastern Nigeria between Niger and Cross rivers;

C. erythrotis (russet-eared guenon), southeastern Nigeria east of Cross River, Cameroon, island of Bioko;

C. cephus (mustached monkey), southern Cameroon, Equatorial Guinea, Gabon, Congo, western Central African Republic;

C. campbelli, Gambia to Ghana;

C. mona (mona monkey), Ghana to southwestern Cameroon;

C. pogonias, southeastern Nigeria, Cameroon, Equatorial Guinea, Gabon, Congo, Central African Republic, northwestern Zaire;

C. wolfi, Zaire, western Uganda;

C. diana (Diana monkey), Sierra Leone to Ghana;

C. dryas, central Zaire;

C. neglectus (De Brazza's monkey), forested zones from southeastern Cameroon to southern Ethiopia and southern Zaire;

C. hamlyni (owl-faced monkey), eastern Zaire, Rwanda;

C. lhoesti (L'hoest's monkey), northeastern Zaire, Rwanda, Burundi, Uganda;

C. solatus, central Gabon;

C. preussi, southeastern Nigeria, Cameroon, island of Bioko.

Chlorocebus (see account thereof) often is included within *Cercopithecus*. Various authorities also have included *Al-*

lenopithecus, Cercocebus, Erythrocebus, Miopithecus, Papio, and *Theropithecus.* Thorington and Groves (1970) considered *C. erythrotis* part of *C. cephus* and thought *C. albogularis* might be a subspecies of *C. mitis.* Corbet and Hill (1991), Groves (*in* Wilson and Reeder 1993), and Lawes (1990) agreed with such allocation for *C. albogularis,* but Lernould (1988) retained it as a separate species. Groves, as well as Lernould, listed *C. sclateri* as a separate species, though Lernould indicated that it might be a hybrid between *C. erythrotis* and *C. erythrogaster.* Lernould also showed the presence of a hybrid zone between *C. erythrotis* and *C. cephus* along the Sanaga River in southwestern Cameroon, and he listed numerous other records of wild and captive hybrids between various species and subspecies of *Cercopithecus.*

Head and body length is about 325–700 mm and tail length is about 500–1,000 mm. Chiarelli (1972) listed the following weights: *C. ascanius,* 1.8–6.4 kg; *C. mona,* 3–6 kg; *C. neglectus,* 4.5–7.8 kg; and *C. mitis,* 6–12 kg. Some forms have long white whiskers associated with a blackish or dark red belly. There may be a white brow band and a tuft of hair on the chin; others have variously colored whiskers with or without a nasal spot and a patch of yellow or grayish yellow hair on the cheek defined by a black stripe above and below. The nasal spot may be white, red, or blue. Some forms have white or buff stripes on the thighs. Some guenons are known for the beauty of their pelage—the individual hairs ringed with different colors and combined with speckled black and gray. The upper parts are often greenish gray, but browns, grays, and intermediate shades are the common colors. The color of underparts also varies considerably. The skin of part of the face in some forms is blue or violet.

Cercopithecus is quite variable cranially but generally can be distinguished from *Chlorocebus* by the characters discussed in the account of the latter. Cheek pouches are present. Guenons may be characterized by their roundish head, slender body, long hind limbs, and long tail. The nostrils are close together, a beard is often present, and the side whiskers are well developed. The common name comes from the French word *guenon,* meaning "fright," and refers to the fact that these animals grimace and expose teeth when they are excited or angry.

Guenons inhabit forests, woodlands, and savannahs, usually near rivers and streams. *C. mona* has been found to live successfully in mangroves (Gartlan and Struhsaker 1972). Guenons are diurnal; they are most active in the early morning and again in the late afternoon or evening. Some forms spend nearly all of their time in trees, jumping from one tree to another. The tail in young guenons is prehensile, but adults normally use the tail as a balancing organ. Some guenons often run about on the ground, but all take shelter in trees when alarmed. They sleep in trees, probably in a sitting position holding onto branches or to each other. The diet is varied but consists mainly of fruits and the seeds thereof (Gautier-Hion 1988). A substantial amount of leaves also is eaten, some grain and roots, and, at least on occasion, young birds, birds' eggs, small reptiles, and insects.

Home range size varies greatly. In the coastal forests of the Ivory Coast a troop of *C. campbelli* stayed within 3 ha. from 1967 to 1970 and spent most of its time in an even smaller core area (Bourliere, Hunkeler, and Bertrand 1970). In a degraded forest of the Central African Republic a group of 17–23 *C. ascanius* used a 15-ha. home range with a 5-ha. core area (Galat-Luong 1975). The same species, however, has been reported to have a home range of up to 130 ha. (Chalmers 1973). In Uganda a group of *C. mitis* used a 72.5-ha. home range but moved through only a small part of it

De Brazza's monkeys *(Cercopithecus neglectus),* photo from San Diego Zoological Society.

on any given day (Rudran 1978). In Kenya troops of *C. neglectus* were found to use home ranges of 4.1–6.0 ha. and to move 330–1,001 meters per day (Wahome, Rowell, and Tsingalia 1993).

There appears to be considerable variation in the social structure of *Cercopithecus,* but data compiled by Cords (1988) suggest that permanent groups generally comprise a single dominant male, who may maintain his position for some years, and a number of adult females. Other males in the vicinity may sometimes move into the group and even mate with the females. Although one troop of about 200 *C. ascanius* has been reported, Struhsaker and Leland (1988) stated that this species typically is found in social groups of 30–35 individuals. There is a single adult male, who usually has a rather short tenure, while the females and their young form the core of the group and are the most active defenders of the territory; one large group was observed to gradually divide into two new permanent groups. Galat-Luong (1975) determined that a group of *C. ascanius* in the Central African Republic numbered only 17–23 individuals; 4 were adult males, 1 of whom was much larger than the others. The group often split into subgroups composed of either adult pairs and their young, females, or juveniles. Rudran (1978) reported groups of *C. mitis* in Uganda to average 20.8 (13–27) individuals and to include only 1 adult male but 4–12 adult females and 7–16 immatures. Apparently, most males dispersed from the groups upon reaching maturity and then lived alone. Tsingalia and Rowell (1984), however, observed 16 adult males in association with a troop of *C. mitis* during a six-month period. Although there was a dominant male, most of the others participated in mating with the group's females. Bourliere, Hunkeler, and Bertrand (1970) observed that a group of *C. campbelli* in the Ivory Coast had 9 members, including a single adult male. This male had a central role in the group and was the only member to emit loud calls, but it was not strongly dominant over the other animals. Younger males seemed to leave the group willingly at about 3–4 years and to take females of the same age with them. Wahome, Rowell, and Tsingalia (1993) found three troops of *C. neglectus* in Kenya to number 11, 13, and 16 individuals; each had a single adult male and at least 3 adult females.

Guenons apparently are territorial but generally avoid serious conflicts. DeVos and Omar (1971) reported *C. mitis* to have territories of 13.2–16.0 ha. Rudran (1978) stated that aggressive interaction in this species was rare as home ranges did not greatly overlap. Different species of guenons may travel together. In fact, guenons and monkeys of the genera *Cercocebus* and *Colobus* may associate. Baboons and chimpanzees, however, do not seem to mix with these smaller monkeys, though they may feed in the same area. Gautier (1988) identified 22 different vocalizations in *Cercopithecus*, including a variety of low- and high-pitched sounds for maintaining group cohesion, whistles and chirps for warning, and loud whoops and booms by adult males.

Data compiled by Butynski (1988) show that all guenon populations have peak mating and birth seasons but that breeding may occur year-round in areas of high rainfall. Most populations have synchronized mating seasons centered on July, August, or September and birth seasons centered on December, January, or February. Interbirth interval has been about a year in some of the populations studied but up to five years in others. A study of *C. mitis* in Kenya (Omar and DeVos 1971) indicated that conception occurred from July to November and that birth were confined to a dry period from November to March. Most births took place at the end of this dry season, thereby allowing lactation to proceed when rainfall was high. Rudran (1978) also found seasonality for this species in Uganda, with births occurring from December to May. In the Ivory Coast *C. campbelli* was observed to give birth only from November to January, at the end of the rains or the beginning of the dry season, following a gestation period of about 6 months (Bourliere, Hunkeler, and Bertrand 1970). Gestation lasted approximately 5.6 months in most of the other species investigated but was only 4.7 months in *C. mitis* and up to 7.0 months in *C. badius* (Butynski 1988). Twins are born occasionally, but a single young is usual. Birth weight is around 400 grams (Hayssen, Van Tienhoven, and Van Tienhoven 1993). A young guenon clings to the fur of the underparts of the mother and entwines its tail with its mother's as they travel. Females generally produce their first young at about 4–5 years (Cords 1988). Several specimens of *Cercopithecus* are thought to have lived in captivity for more than 30 years, including one *C. campbelli* that may have reached an age of 38 years (Marvin L. Jones, Zoological Society of San Diego, pers. comm., 1995).

These guenons generally are less adaptable than *Chlorocebus* and have proved more vulnerable to human disruption of forests. The IUCN assigns the following specific and subspecific classifications: *C. mitis kandti* (Uganda, Rwanda, eastern Zaire), endangered; *C. erythrogaster,* vulnerable; *C. sclateri*, endangered; *C. erythrotis* (in general), vulnerable; *C. erythrotis erythrotis* (island of Bioko), endangered; *C. pogonias pogonias* (Cameroon, Equatorial Guinea), vulnerable; *C. diana diana* (Sierra Leone, eastern Guinea, Liberia), vulnerable; *C. diana roloway* (Ivory Coast, Ghana), endangered; *C. hamlyni*, near threatened; *C. lhoesti*, near threatened; *C. solatus*, vulnerable; and *C. preussi*, endangered. The species *C. diana, C. erythrotis, C. erythrogaster,* and *C. lhoesti* (including *C. preussi*) are listed as endangered by the USDI. *C. diana* is on appendix 1 of the CITES and all other species are on appendix 2. All of the species classified by the IUCN are thought to be declining because of destruction of forest habitat through agricultural expansion and logging and/or excessive hunting for meat by people (Lee, Thornback, and Bennett 1988). For the same reasons, Brennan (1985) regarded *C. neglectus* to be seriously endangered in Kenya, though Muriuki and Tsingalia (1990) reported that the species was somewhat more widely distributed in that country than had been previously thought. *C. solatus*, which was not described until 1988, and *C. sclateri*, a wild population of which was not found

until that same year, both occupy highly restricted ranges that are undergoing severe disturbance (Gautier et al. 1992; Oates and Anadu 1989; Oates et al. 1992). *C. mona* was introduced on the island of Grenada in the West Indies, but Fedigan and Fedigan (1988) indicated that it did not adapt as well to the region as *Chlorocebus sabaeus* and may now have disappeared.

PRIMATES; CERCOPITHECIDAE; **Genus**
MIOPITHECUS
I. Geoffroy St.-Hilaire, 1842

Talapoin

The single named species, *M. talapoin,* is found in southwestern Zaire, northwestern Angola, and possibly the Ruwenzori region farther east. *Miopithecus* also occurs in southern Cameroon, Equatorial Guinea, Gabon, Congo, and possibly the Central African Republic, but the populations in those countries are thought to represent a second, unnamed species (Groves 1989; Lernould 1988; Wolfheim 1983). *Miopithecus* was listed as a subgenus of *Cercopithecus* by Dandelot (*in* Meester and Setzer 1977), Lernould (1988), and a number of other recent authorities, but Groves (1989) tentatively placed it in a subfamily entirely separate from that containing *Cercopithecus*.

This is the smallest Old World monkey. Head and body length is about 320–450 mm and tail length is 360–525 mm. Napier and Napier (1967) listed weights of 1,230 and 1,280 grams for two males and 745 and 820 grams for two females. The coloration above ranges from greenish gray or greenish black to greenish yellow or greenish buff. The underparts are white or grayish white. The face is naked except for a few blackish hairs on the lips (upper lip sometimes yellow) and nose. The hairs on the cheeks are golden yellow with black tips; a black streak extends from the corner of the eye halfway to the ear, which is also black. The skin around the eyes is sometimes yellowish or orangish. White hairs with black tips in front of the ear radiate from a point like a fan. The top of the head is ochraceous black. The color of the tail varies from grayish black to brownish black above; beneath, it is yellowish or yellowish gray at the base, blending into buff, yellowish black, or black at the tip. The outer side of the limbs is pale or chrome yellow sometimes tinged with red, and the hands and feet are chrome yellow with a buffy or reddish yellow tint. There is considerable color variation among individuals; some are nearly olive, while others have more of a true green tint.

The limbs and tail of the talapoin monkey are relatively long, but the hands and feet are short. There is some webbing between the fingers and toes. The facial part of the skull is small, and the cranium is relatively large.

Studies by Gautier-Hion (1971, 1973) in Gabon have added considerably to our knowledge of this monkey. All 25 troops observed were centered in inundated rainforest along rivers, and 17 of these troops were located near human settlements. At night the members of each group slept at a communal site in trees along the edge of the water. By day the animals made foraging trips of 1,500–2,950 meters, which sometimes took them into secondary growth or even plantations but never more than 450 meters from a river. There were two daily peaks of activity: from early morning to 1030 hours and from 1330 to 1830 hours. The diet was insectivorous and frugivorous.

On the average there was one troop along every 4,835 meters of river, and mean group home range was 1,216 sq meters. Gautier-Hion distinguished between "parasitic"

Talapoin monkey *(Miopithecus talapoin)*, photo from San Diego Zoological Garden.

mean estrous cycle to be 35 days and gestation to be 158–66 days. The newborn were relatively enormous; one female weighing 1,100 grams gave birth to a 230-gram infant. Rowell (1977*b*) found that females usually were able to conceive at 4.5 years and that males reached sexual maturity 1–2 years after the females. One captive talapoin lived for 27 years and 8 months (Jones 1982).

Miopithecus is on appendix 2 of the CITES, but Lee, Thornback, and Bennett (1988) did not consider it to be threatened. Being so small, it is hardly worth hunting, and its northern population is centered in a region where human habitat disruption has been somewhat less intense than in most of Africa. However, the southern population is more restricted and susceptible to forest disturbance, and recognition of it as a separate species could cause increased conservation concern.

PRIMATES; CERCOPITHECIDAE; **Genus**
ALLENOPITHECUS
Lang, 1923

Allen's Monkey

The single species, *A. nigroviridis,* inhabits northeastern Congo and western Zaire. Dandelot (*in* Meester and Setzer 1977), Lernould (1988), and some other authorities regarded *Allenopithecus* as a subgenus of *Cercopithecus,* but Gautier (1985), Hershkovitz (1977), and Thorington and Groves (1970) treated it as distinct, and Groves (1989) placed it in a subfamily entirely separate from that containing *Cercopithecus.*

Head and body length is 400–510 mm and tail length is 350–520 mm (Van Hoof *in* Grzimek 1990). Gautier (1985) reported strong sexual dimorphism in size, with an average weight of about 6 kg for males and 3.5 kg for females. The hair is slightly longer on the nape and shoulders than on the back and scant on the underparts, hands, and feet. The skin of the face is dark grayish brown to black but lighter around the eyes. The hairs on the sides of the face extend outward to form a ruff from the ears to the mouth. The general coloration above is grayish to almost black, sometimes with a yellowish tinge, and the underparts are light gray or whitish, sometimes speckled with black and yellow. The tail is dark above and lighter below.

Groves (1989) observed that *Allenopithecus* differs from *Cercopithecus,* and shows some resemblance to *Papio,* as follows: the sexual area of the female's skin—vagina, perineum, anus—undergoes cyclic enlargement, being maximally swollen around the time of ovulation (there is no sexual swelling in *Cercopithecus*); the ischial callosities of adult males are fused across the midline (they are smaller and discrete in male *Cercopithecus*); there is some facial elongation; the molar teeth are large and have well-developed flare, a convexity of the buccal and lingual surfaces; and there is some facial elongation. Also unlike in *Cercopithecus,* the first and third lower molars of *Allenopithecus* have a small buccal cusplet between the anterior and posterior cusps. The body of *Allenopithecus* is rather heavy; in appearance it is similar to *Cercopithecus,* but anatomically it bears some resemblance to the baboons.

Allenopithecus apparently forages on the ground as the baboons do. Its diet is primarily frugivorous and insectivorous (Lee, Thornback, and Bennett 1988). It frequents swampy areas and probably goes into water freely. In Zaire, Gautier (1985) observed troops in swamp forests, with regular sleeping sites on riverbanks. These groups comprised more than 40 individuals, including several adult males.

troops, which lived near human settlements and depended in part on cultivated crops, and nonparasitic troops, which were found at a considerable distance from villages. Nonparasitic troops had on average about 66 members and lived at a population density of 30/sq km. Parasitic troops averaged 112 individuals and had a density of 92/sq km. Each group included a number of adult males that acted as leaders during daily movements and as sentinels at night. The adult males of a group were outnumbered by adult females, and the latter, along with their infants and year-old juveniles, formed a definite subgroup, at least during the night. No displays or interaction suggesting group territoriality were observed. Gautier-Hion noted that *Miopithecus* differed from *Cercopithecus* in its lower population density, larger troops, and lack of territoriality.

Based on a study of captives, Dixson, Scruton, and Herbert (1975) favored retaining the talapoin in a genus separate from *Cercopithecus.* Unlike *Cercopithecus,* the talapoin was found to be able to raise its eyebrows and move its ears during facial expression. Also, talapoins that groomed each other frequently sat or huddled with their tails entwined. Animals adjacent in rank groomed one another far more than was the case in *Cercopithecus.* Earlier observations showed that the call of the talapoin is a short, explosive "k-sss!" like the splash of a rock in water, and is different from the calls of most guenons.

In Gabon, Gautier-Hion (1971, 1973) found that the birth season of any given troop lasted for about two months during the period November–April and that most adult females gave birth to a single young every year. In a study of captive talapoins, Rowell (1977*a*) determined the

Allen's monkey *(Allenopithecus nigroviridis)*, photo by Ernest P. Walker.

Captives appeared to enjoy wading in shallow water at the National Zoo in Washington, D.C. The animals there were friendly and comparable to *Cercopithecus* in behavior. The single young weighs about 200 grams at birth (Van Hoof *in* Grzimek 1990). Weaning occurs at 2.5 months (Hayssen, Van Tienhoven, and Van Tienhoven 1993). The record known longevity for *Allenopithecus* is 23 years (Jones 1982). The genus is classified as near threatened by the IUCN and is on appendix 2 of the CITES. Its numbers are thought to have declined in recent years, and remaining populations are threatened by hunting and habitat loss (Lee, Thornback, and Bennett 1988).

PRIMATES; CERCOPITHECIDAE; **Genus CERCOCEBUS**
E. Geoffroy St.-Hilaire, 1812

Mangabeys

There are three species (Dandelot *in* Meester and Setzer 1977; Groves 1978a, 1989, and *in* Wilson and Reeder 1993; Homewood and Rodgers 1981; Struhsaker 1971):

C. agilis, Cameroon, Equatorial Guinea, Gabon, Congo, Central African Republic, Zaire, southern Tanzania;
C. galeritus, lower Tana River in eastern Kenya;
C. torquatus, Senegal to Congo.

Dandelot (*in* Meester and Setzer 1977) considered these species to form the superspecies *torquatus* and considered the two species here placed in *Lophocebus* (see account thereof) to form another superspecies of *Cercocebus.* Groves (1989) is followed here in recognizing *C. agilis* as a species separate from *C. galeritus,* though the two frequently are treated as conspecific. *C. atys* (sooty mangabey), found from Senegal to Benin, sometimes is regarded as a species distinct from *C. torquatus* (Lee, Thornback, and Bennett 1988).

Head and body length is about 382–888 mm, tail length is about 434–764 mm, and weight is about 3–20 kg. The height when seated is 35–40 cm. In *C. galeritus* the upper parts are golden brown speckled with black or dark grayish brown speckled with gold yellow; the underparts are ochraceous orange or whitish. There is neither a crown patch nor a dorsal line. This species has a whorl, or parting of the hair, on top of the head. In *C. agilis* this whorl is less developed and the pelage is strongly agouti-banded (Groves 1989). In *C. torquatus* the upper parts are slaty gray or grayish brown to dark brown and the underparts are generally grayish. There usually is no speckling in the pelage. The top of the head usually has a conspicuous patch of mixed olive and black, blackish brown, or chestnut red, and the individual hairs are tipped with black. There is no whorl on top of the head. The crown patch is surrounded by a black or white band that is continued behind into a neck patch. This gradually leads into a black line along the middle of the back that extends to the tail. The tip of the tail of *C. torquatus* is paler than the rest of the tail. A conspicuous cheek beard is present.

Mangabeys, though similar in most respects, are longer, taller, and of more slender proportions than are guenons *(Cercopithecus).* They also differ in that they have pale upper eyelids, the buttock pads are fused in the males, and the third lower molar has five tubercles (as in *Macaca*) rather than four (as in *Cercopithecus*). Mangabeys have an oval-shaped head with a fairly long muzzle. Large cheek pouches are present and constantly used. There are webs between the fingers; the second and third toes are united by a web for most of their length, and the fourth is united to the third and fifth as far as the middle joints.

According to Lee, Thornback, and Bennett (1988), *C. torquatus* is found mostly in high-canopy primary forest but also in mangrove, gallery, and inland forests, as well as in young secondary forests and around cultivated areas. In a study in Equatorial Guinea, Jones and Sabater Pi (1968) found *C. torquatus* occupying a broad range of habitats, including swamp and agricultural areas, and to be partly terrestrial. Groups had 14–23 individuals, of which several were large males. This species is mainly frugivorous but may eat significant amounts of animal matter.

In Gabon, Quris (1975) found *C. agilis* apparently to be restricted to periodically flooded areas and to prefer the lower forest strata. Fruit was the favorite food, and leaves, grass, and mushrooms were also eaten. Population density in the study area was 6.7–12.5/sq km, and one troop had a home range of 1.98 sq km. The number of individuals in a troop ranged from 8–9 to 17–18, but there was frequent division into subgroups. Each of three observed groups contained only a single adult male, who functioned as leader and defender. The loud calls of the adult males could be heard by people up to 1,000 meters away and perhaps played a role in regulation of group spacing. However, groups sometimes met one another with no apparent antagonism and temporarily united. Exchange of members between troops sometimes occurred. The presumed season of birth in the area was December–February.

In a study of the Tana River mangabey, *C. galeritus*, in eastern Kenya, Homewood (1975) found population density to vary from about 3/ha., with a home range of as little as 15 ha., in optimal diverse evergreen forest to fewer than 1/2 ha., with a home range of 50–100 ha., on poor, low-diversity deciduous woodland. These mangabeys seemed equally at home 30 meters up in trees and down on the ground, where they traveled between adjacent forest patches and spent much of the day foraging in leaf litter. The diet consisted mainly of fruits, seeds, and insects. Group size was 13–36 individuals, usually including at least 2 adult males and about twice as many adult females. Some males lived alone. During the dry season, when food was limited, the groups maintained discrete territories, with spacing maintained by the long-range vocalizations, displays, and occasional active combat between the dominant males. During the rains, however, when food was abundant, territories seemed to break down and groups became more tolerant of one another, sometimes joining in temporary associations of 40–50 mangabeys.

Gray-cheeked mangabey and baby *(Lophocebus albigena)*, photo from New York Zoological Society. A. Red-crowned mangabey *(Cercocebus torquatus)*; B. Sooty mangabey *(Cercocebus torquatus atys)*; C. Black-crested mangabey *(Lophocebus aterrimus)*, photos by Ernest P. Walker.

Apparently, a single young is born at a time. The gestation periods in three births of *C. torquatus* were 164, 173, and 175 days (Stevenson 1973). At the Yerkes Regional Primate Center, in Atlanta, Georgia, births of captive *C. torquatus atys* occurred in all months of the year but predominantly from March to August. Females first gave birth at an average age of 56.5 months and then had an average interbirth interval of 16.6 months. Females more than 22 years old were still reproductively active (Gust, Busse, and Gordon 1990).

All species of mangabeys are losing forest habitat to logging and other human activity (Wolfheim 1983). All are on appendix 2 of the CITES, except for the isolated Tana River mangabey *(C. galeritus)*, which is on appendix 1. *C. galeritus* (not including *C. agilis*) is designated as endangered by the IUCN and the USDI. It is restricted to a forested area of about 730 ha. along the Tana River in eastern Kenya. Suitable habitat has been reduced and severely modified through agriculture, logging, and construction of hydroelectric dams upstream. Recent population estimates of this species have ranged from about 600 to 1,200 individuals (Butynski and Mwangi 1995; Medley 1993). A subspecies of *C. agilis* discovered in the Uzungwa Mountains of Tanzania and not yet formally described also is classified as endangered by the IUCN (under the name *C. galeritus*) because of hunting and habitat loss (Lee, Thornback, and Bennett 1988). The IUCN designates the entire species *C. agilis* (under the name *C. galeritus*) as near threatened.

The species *C. torquatus* is listed as endangered by the USDI because large-scale logging apparently has greatly reduced its range in West Africa. Schlitter, Phillips, and Kemp (1973), however, stated that this species is adaptable and could survive in areas of cutover forest and agriculture. The IUCN recognizes *C. torquatus* and *C. atys* as separate species (see above) but designates both generally as near threatened. It singles out as vulnerable the subspecies *C. atys lunulatus*, of Ivory Coast and Ghana, which is threatened by habitat loss, hunting, and conflicts resulting from raids on fruit plantations (Lee, Thornback, and Bennett 1988).

**PRIMATES; CERCOPITHECIDAE; Genus
LOPHOCEBUS**
Palmer, 1908

Black Mangabeys

In designating *Lophocebus* as a genus distinct from *Cercocebus*, with consideration of morphological and immunological evidence, Groves (1978*a* and *in* Wilson and Reeder 1993) indicated that it comprised only the species *L. albigena* of southeastern Nigeria, Cameroon, Congo, Gabon, Equatorial Guinea, northeastern Angola, Central African Republic, Zaire, Burundi, Uganda, western Kenya, and western Tanzania. The populations to the south of the Congo River, in Zaire and Angola, were considered to represent a subspecies, *L. albigena aterrimus*. On the basis of sociological data, Horn (1987*b*) supported the subspecific designation of *aterrimus* (he did not comment on the generic status of *Lophocebus*). However, Groves (1989) indicated that despite evidence of some very slight gene flow across the Congo River, it would be appropriate to recognize *aterrimus* as a separate species. Dandelot (*in* Meester and Setzer 1977) treated *L. albigena* and *L. aterrimus* as full species that together formed the superspecies *albigena* of the genus *Cercocebus* (see account thereof). *Lophocebus* also was not accepted as a distinct genus by Corbet and Hill

(1991) or Napier and Napier (1985). Such distinction, however, was supported by recent molecular and chromosomal analyses (Disotell 1994).

Head and body length is 400–720 mm, tail length is 550–1,000 mm, and weight is about 4–11 kg (Van Hoof *in* Grzimek 1990). In *L. albigena* the hair is soft and dull black. The cheeks have short hairs, a long brow fringe and crest are present, and there is a mantle on the shoulders. *L. aterrimus* has coarse and glossy black hairs, long grayish brown whiskers partly concealing the ears, and a high conical crest, but the long brow fringe and mantle are not present. Groves (1989) wrote that *Lophocebus* differs from *Cercocebus* in the form of the nasal bones and interorbital region, cranial flexion, the form of the suborbital fossa, smaller cheek teeth and reduced molar flare, orbit shape, malar foramen size, and characters of the mandible. *Lophocebus* also differs in not having pale eyelids that contrast with an otherwise dark face.

Dandelot (*in* Meester and Setzer 1977) stated that *L. albigena* and *L. aterrimus* (which he considered to compose the superspecies *albigena* of the genus *Cercocebus*) are arboreal and have "supple" movements, very dark pelage, and a long, ruffled tail. In contrast, the species *Cercocebus torquatus* and *C. galeritus* (which he considered to compose the superspecies *torquatus*) are semiterrestrial and have a "stiff" gait, rather light pelage, and a medium-long, not very supple tail. *Lophocebus* has a throat sac that gives resonance to its calls. This sac is small in the female but large in the male. The call is a shrieking, howling sound resembling "hu, hu." According to Kingdon (1971), there are a variety of other vocalizations, including barks, chuckles, twitters, and grunts.

Kingdon noted also that black mangabeys are never found far from water or at elevations above 1,700 meters. In a study in Equatorial Guinea, Jones and Sabater Pi (1968) found *L. albigena* to inhabit mainly primary and secondary forest, whereas *Cercocebus torquatus* occupied a broad range of habitats. *L. albigena* was entirely arboreal and *C. torquatus* was partly terrestrial. Groups of *L. albigena* contained 9–11 animals, including only a single adult male, whereas groups of *C. torquatus* had 14–23 individuals, of which several were large males.

In Uganda, Chalmers (1968*a*, 1968*b*) observed *L. albigena* to spend most of its time in the high canopy of the forest but occasionally to come down to raid cultivated fields. Unlike most monkeys, these mangabeys had three, rather than two, daily feeding periods: 0800–0900, 1200–1300, and 1600–1700 hours. Food consisted mainly of fruits, buds, and shoots and was eaten mainly high in the trees. Population density in the study area was about 78/sq km and group home range was about 0.13–0.26 sq km. Groups often moved through their entire home range in a single day. They generally avoided one another, perhaps being spaced by the loud vocalizations of the males. Groups had 7–25 members, with an equal number of adult males and females. Males were aggressive but were solicitous of the young. Breeding seemed to continue throughout the year. Age of sexual maturity in males was about 5–7 years.

Additional studies of *L. albigena*, in the Kibale Forest Reserve of Uganda, showed population density to increase from 6–7/sq km in 1971 to 13–14/sq km in 1991, apparently in response to regeneration of forest cover. Groups moved about 1,000 meters per day and utilized home ranges of approximately 1 sq km (Olupot et al. 1994). In another investigation of *L. albigena* in Uganda, Waser (1977) observed group movement pattern apparently to center on large fruiting trees. A group would repeatedly visit a single such tree for days or weeks but also would spread out over an enormous area to search for other resources. One group had a home range of 4.1 sq km. Group size ranged from 6 to 28 animals. A particular group in this area studied by Waser and Floody (1974) was made up of 15–16 individuals, of which 6–7 were adult or subadult males.

In Zaire, Horn (1987a) found L. aterrimus to be completely arboreal and to use all levels of the forest canopy, but especially the middle layers. Foraging activity was concentrated in the early morning, sometimes even before daybreak. The diet consisted mainly of fruits and nuts. Two groups had home ranges of 48 ha. and 70 ha., which mostly overlapped. The groups contained 14–19 individuals, including several adults of both sexes. There was often division into subgroups, each with at least 1 adult male. Vocalizations were frequent and loud. Among them was the "whoopgobble," audible to humans more than 500 meters away and apparently given only by adult males for purposes of intergroup location and spacing.

Gevaerts (1992) found births in northern Zaire to occur mainly in July and August, at the start of the wet season. Kingdon (1971), however, indicated that in East Africa black mangabeys do not show any marked seasonal breeding and that in most troops there are usually babies and young of all sizes; the gestation period is about 174 days. A captive specimen of L. albigena lived for 32 years and 8 months (Jones 1982).

According to Lee, Thornback, and Bennett (1988), L. albigena is not threatened, but L. aterrimus is subject to intensive, uncontrolled hunting for use as food and also is vulnerable to loss of its restricted forest habitat. Both species are on appendix 2 of the CITES, and L. aterrimus is designated as near threatened by the IUCN.

PRIMATES; CERCOPITHECIDAE; Genus MACACA
Lacépède, 1799

Macaques

There are 20 species in 5 species groups (Corbet 1978; Corbet and Hill 1992; Cronin, Cann, and Sarich 1980; Delson 1980; Ellerman and Morrison-Scott 1966; Eudey 1987; Fa 1989; Fooden 1969, 1971, 1975, 1976b, 1988, 1989, 1990, 1991b; Fooden, Mahabal, and Saha 1981; Fooden et al. 1985, 1994; Groves 1980c; Lekagul and McNeely 1977; MacKinnon 1986; Tan 1985):

sylvanus group

M. *sylvanus* (Barbary ape), northern Africa, a possibly introduced population at Gibraltar in extreme southern Spain;

silenus group

M. *silenus* (liontail macaque), southwestern peninsular India;
M. *nemestrina* (pigtail macaque), Assam, Burma, southern Yunnan, Thailand, Indochina, Malay Peninsula, Sumatra, Bangka, Borneo and some small nearby islands;
M. *pagensis*, Mentawai Islands off western Sumatra;
M. *tonkeana*, central Sulawesi and nearby Togian Islands;
M. *hecki*, western part of northern peninsula of Sulawesi;
M. *maura*, southwestern peninsula of Sulawesi;
M. *ochreata*, southeastern peninsula of Sulawesi;
M. *brunnescens*, Muna and Butung islands off southeastern peninsula of Sulawesi;

M. *nigra* (Celebes ape), eastern part of northeastern peninsula of Sulawesi;
M. *nigrescens*, central part of northeastern peninsula of Sulawesi;

fascicularis group

M. *fascicularis* (crab-eating macaque), southeastern Burma, southern Thailand and Indochina, Malay Peninsula, Nicobar Islands, Sumatra, Java, Borneo, throughout the Philippines, Lesser Sunda Islands as far east as Timor, but not Sulawesi;
M. *mulatta* (rhesus monkey), eastern Afghanistan and possibly formerly southeastern Pakistan, through much of India and Nepal, to northeastern China, Indochina, and Hainan;
M. *cyclopis*, Taiwan;
M. *fuscata* (Japanese macaque), Honshu, Shikoku, Kyushu, Yakushima;

sinica group

M. *sinica* (toque macaque), Sri Lanka;
M. *radiata* (bonnet macaque), peninsular India;
M. *assamensis*, mountainous areas from Nepal and Bangladesh to northern Indochina and western Thailand;
M. *thibetana* (Père David's stump-tailed macaque), Sichuan to Fujian and Guangxi provinces in southeastern China;

arctoides group

M. *arctoides* (stump-tailed macaque), Brahmaputra River of northeastern India to Guangxi (southeastern China), Viet Nam, and northern Malay Peninsula.

Fooden (1976b) combined the *sylvanus* and *silenus* groups, and Delson (1980) thought that M. *arctoides* belonged in the *sinica* group. Bernstein and Gordon (1980) and Fa (1989) reported that many species of *Macaca* are capable of interbreeding and producing offspring no less fertile than the parents. Groves (*in* Wilson and Reeder 1993) and Corbet and Hill (1992) included M. *pagensis* in M. *nemestrina*, though the latter authorities noted its distinctiveness and thought that it might warrant specific status. There is substantial disagreement regarding the systematic situation on Sulawesi, with some authorities recognizing a single intergrading species and others dividing various species into groups or subgenera (Fa 1989). Most recent workers have followed either Fooden (1976b), who recognized the seven Sulawesian species listed above, or Groves (1980c), who on the basis of evident interbreeding between certain populations considered M. *nigrescens* a subspecies of M. *nigra*, M. *hecki* a subspecies of M. *tonkeana*, and M. *brunnescens* a subspecies of M. *ochreata*. A recent series of more detailed investigations have suggested that although there is gene flow between the species of Sulawesi along most lines where their ranges meet, such interaction is more indicative of sporadic hybridization between normally separate species than of total intergradation between subspecies (Camperio Ciani et al. 1989; Watanabe, Lapasere, and Tantu 1991; Watanabe and Matsumura 1991; Watanabe et al. 1991).

The smallest species, M. *sinica*, has a head and body length of 367–528 mm and weighs about 2.5–6.1 kg (Fooden 1979). Other species range up to 764 mm in head and body length and 18 kg in weight. Males generally are about

A. Rhesus macaque *(Macaca mulatta)*; B. Pigtail macaque *(M. nemestrina)*; C. Barbary ape *(M. sylvanus)*; D. Crab-eating macaque *(M. fascicularis)* with left cheek pouch filled; E. Liontail macaque *(M. silenus)*, photos by Ernest P. Walker. F. Japanese macaques *(M. fuscata)*, photo from U.S. National Zoological Park.

Barbary ape *(Macaca sylvanus)*, photo by Russell A. Mittermeier.

50 percent heavier than females. In *M. sylvanus* there is no tail, and in the following species the tail length ranges from only about 10 mm to approximately 20 percent of the head and body length: *M. nigra, M. nigrescens, M. ochreata, M. brunnescens, M. maura, M. tonkeana, M. hecki, M. arctoides, M. thibetana,* and *M. fuscata.* In the remaining species the tail length is equal to at least 25 percent of the head and body length. In *M. sinica* the tail generally is longer than the head and body (Fooden 1976*b*, 1979). The coloration is usually yellowish brown above and lighter below, but some forms are olive and others almost black. Some species have a "cap" of hairs on the head. *M. silenus* has a ruff of long grayish hair on each side of the face, while *M. fuscata* has long whiskers, a beard, and long dense fur. *M. nigra,* of Sulawesi, has a conspicuous cone-shaped mass of long hairs on the crown of the head. Macaques are medium-sized monkeys with stout bodies and strong limbs. The snout is somewhat elongated.

M. mulatta has adapted to a wide range of habitats, from sea level to elevations of 2,500 meters, from snow to intense heat, and from near-desert situations to dense forests (Lindburg 1971). More than half of the rhesus monkeys in the northern Indian state of Uttar Pradesh actually live in cities and towns, where there is ideal habitat—shelter, large trees, abundant food, and water from wells (Southwick, Beg, and Siddiqi 1965). A group of 62 of these monkeys lived successfully in an area of less than 4 ha. within urban Calcutta, among large buildings, motor vehicles, crowds of people, and domestic animals (Southwick et al. 1974). In contrast, *M. silenus* is an obligate rainforest dweller, and *M. nemestrina* is restricted to dense, inland evergreen and deciduous forests. Both of these species seem unable to adapt to human encroachment and are declining in numbers

(Green and Minkowski 1977; Roonwal and Mohnot 1977). Most species of macaques usually inhabit forests, sometimes at relatively high altitudes. *M. assamensis, M. arctoides,* and *M. sylvanus* are often present in upland areas or mountains at elevations of 2,000 meters or more (Lekagul and McNeely 1977; Taub 1977). *M. fascicularis* may also be found in such habitat and even in urban situations (Medway 1978), but it seems to prefer coastal mangroves; like *M. mulatta,* it is an expert swimmer (Lekagul and McNeely 1977; Roonwal and Mohnot 1977). *M. cyclopis,* of Taiwan, may once have been associated with the seacoast but now has been restricted largely to inland hills because of human activity (Kuntz and Myers 1969).

All macaques are primarily diurnal, and all have an arboreal capability, but most species come down from the trees at least on occasion to forage or move over long distances. *M. radiata* has been observed to spend about one-third of its time on the ground (Simonds 1965), and *M. arctoides* is primarily terrestrial, sometimes making seasonal migrations from one mountain range to another (Lekagul and McNeely 1977). Groups of this species generally leave their sleeping trees at dawn, feed until about 1000 hours, rest till 1400 hours, and then feed again until 1800 hours (Roonwal and Mohnot 1977). *M. sinica,* of the tropical forests of Sri Lanka, spends about 75 percent of its time in the trees, always sleeping high in the canopy at night, and seems cautious and fearful when it does descend to the ground (Dittus 1977; Fooden 1979). *M. silenus* spends less than 1 percent of its time on the ground (Green and Minkowski 1977). Studies in Borneo indicated that one group of *M. fascicularis* slept in trees by night and came down to forage by day (Fittinghoff and Lindburg 1980) but that another group spent more than 97 percent of its time in the trees (Wheatley 1980). *M. mulatta* readily leaves the trees if attracted by food supplies and moves freely among and through human habitations if tolerated. The time and area of its activity may shift according to season. Lindburg (1971) found that in warmer periods it began feeding before dawn but took rest sessions during the day; in the cool season it began to feed later and lacked well-defined intervals of resting. Daily distance averaged 1,428 (350–2,820) meters. Like most other species of macaques, *M. mulatta* is largely vegetarian, feeding on wild and cultivated fruits, berries, grains, leaves, buds, seeds, flowers, and bark. All species, however, eat insects and other small invertebrates when they are available and occasionally take eggs and small vertebrates (Fooden 1979; Lekagul and McNeely 1977; Lindburg 1971; Roonwal and Mohnot 1977). *M. fascicularis* is known to feed on crabs, other crustaceans, shellfish, and other littoral animals exposed by the tide; and some of the macaques of Sulawesi may have a comparable diet (Fooden 1969; Lekagul and McNeely 1977). Captive *M. silenus* were found to manufacture and use tools for the extraction of syrup from a container experimentally provided to the animals (Westergaard 1988).

Fooden (1982) provided a summary of the distribution of the eight species of macaques found on the mainland of Asia, explaining that all are segregated either ecologically or geographically. The species *M. assamensis* and *M. thibetana* are found in subtropical broad-leaved evergreen forest, while *M. silenus* and *M. nemestrina* occur in tropical broad-leaved evergreen forest; none of these four species is sympatric. *M. arctoides* is present both in tropical and subtropical broad-leaved evergreen forest and is sympatric with *M. assamensis* and *M. nemestrina.* However, *M. arctoides* is largely terrestrial, *M. assamensis* is arboreal, and *M. nemestrina* is arboreal in those areas where its range overlaps that of *M. arctoides.* The other three mainland species—*M. mulatta, M. radiata,* and *M. fascic-*

Macaque *(Macaca tonkeana)*, photo by Russell A. Mittermeier.

ularis—are rare or absent in broad-leaved evergreen forest but widely distributed in secondary, deciduous, coniferous, riverine, and mangrove forests and in disturbed habitats. They are not sympatric, M. mulatta being subtropical and the other two tropical.

MacKinnon (1986) listed the following population estimates and densities in Indonesia: *M. fascicularis*, 2,176,860 individuals occupying 73,371 sq km of primary habitat, at a density of 30/sq km, and 1,550,000 occupying 38,750 sq km of secondary habitat, at a density of 40/sq km; *M. nemestrina*, 895,700 individuals in 179,140 sq km, density 5/sq km; *M. pagensis*, 9,000 individuals in 4,500 sq km, density 2/sq km; *M. nigra* (including *M. nigrescens*), 144,000 individuals in 4,800 sq km, density 30/sq km; *M. tonkeana* (including *M. hecki*), 385,000 individuals in 38,500 sq km, density 10/sq km; *M. maura*, 56,000 individuals in 2,800 sq km, density 20/sq km; and *M. ochreata* (including *M. brunnescens*), 277,500 individuals in 18,500 sq km, density 15/sq km. Some other reported population densities of macaques are: *M. fascicularis*, 36–90/sq km (Kurland 1973); *M. mulatta*, 5–15/sq km in high forest, 57/sq km in low forest, and 753/sq km in towns (Roonwal and Mohnot 1977); *M. silenus*, 0.25/sq km (Green and Minkowski 1977); *M. nemestrina*, 5.5/sq km (Rodman 1978); *M. sylvanus*, 43–70/sq km (Deag 1977); and *M. sinica*, from 0.3/sq km in marginal habitats to 100/sq km in optimal habitats (Dittus 1977; Fooden 1979). It should be understood that these figures are speculative and based largely on assumptions regarding the overall extent and condition of the remaining habitat.

Estimated group home ranges are: *M. fascicularis*, generally 0.5–1.0 sq km but as small as 0.25 sq km in mangroves (Crockett and Wilson 1980); *M. mulatta*, up to 16 sq km in sub-Himalayan forests, 1–3 sq km in other forests, and 0.05 sq km in towns (Lindburg 1971; Roonwal and Mohnot 1977); *M. silenus*, 1–2 sq km during a period of 1–2 months and 5 sq km over a year (Green and Minkowski 1977); *M. nemestrina*, 1–3 sq km (Crockett and Wilson 1980); *M. sinica*, from 0.17 sq km for a group of 8 to 1.15 sq km for a group of 37 (Dittus 1977); and *M. radiata*, 0.4–5.2 sq km (Roonwal and Mohnot 1977). There is generally extensive and sometimes nearly complete overlap between the home ranges of groups, and there is little evidence of territoriality. However, there may be defense of the area occupied at a given time, mutual avoidance by groups, and dominance of large groups over smaller ones. Lindburg (1971) found that most intergroup encounters of *M. mulatta* resulted in withdrawal of one group, but occasionally there was peaceful proximal feeding or agonistic displays and stalemate. Green and Minkowski (1977) indicated that a study group of *M. silenus* had a 300-ha. core area that was rarely entered by other groups. Roonwal and Mohnot (1977) stated that groups of *M. sinica* spend considerable time in the company of *Semnopithecus entellus* and that the two often play, feed, rest, and even groom together.

Overall heterosexual group size in *M. mulatta* is 8–180. Generally there are from two to four times as many adult females as adult males since most of the males leave to live alone or in small groups of their own (Roonwal and Mohnot 1977). The average composition of 399 groups in Uttar Pradesh was 17.6 animals, including 3.7 adult males, 7.7 adult females, 4.5 infants, and 1.7 juveniles (Southwick, Beg, and Siddiqi 1965). Another study in this state found 32 groups to number 8–98 individuals and to include 3 adult females for every adult male (Lindburg 1971). In *M. mulatta* there are dominance hierarchies in both sexes, but that of the females is less evident. Relationships among adult males range from peaceful to hostile, but the females generally live together in harmony. There is a tendency for mating between high-ranking adults. Although males appear to lead and defend the group, the females and infants form a central subgroup within which the young are raised. According to Sade (1967), the social status of the young is dependent on the rank of their mother. In *M. fuscata* of Japan, groups are known to be matrilineal. Young males emigrate, but females remain within the group of their birth and form close relationships among themselves (Kurland 1977). Groups of this species become very large; in one case a troop divided three times, in each instance a group of 70–100 animals separating after the mother troop had reached a population of 600–700 (Masui et al. 1975). Like most other macaques, *M. mulatta* has a wide variety of vocalizations. Lindburg (1971) identified a shrill bark for alarm, barking or screeching as a response to aggression, a scream when under attack, a growl of aggression, and a squawk for surprise.

The social structure of *M. fascicularis* appears close to that of *M. mulatta*. Reported group size is 6–100, and there are about 2.5 adult females for every adult male. Roonwal and Mohnot (1977) stated that in this species groups are divided into a central component, comprising a male leader and most of the females and young, and a peripheral component, made up of the younger males and perhaps a few females. Angst (1975) found dominance hierarchies based on age, kinship, and coalitions in each sex, with senile individuals dropping in rank. Dittus (1977) reported an average group size of 24.7 (8–43) in *M. sinica*. He observed that natural mortality was largely the result of lower-ranking animals' being prevented from foraging in areas of high

Moor macaque *(Macaca maura)*, photo by Ernest P. Walker.

quality. In *M. radiata,* the common performing monkey of southern India, group size is about 7–76, but usually there are more adult males than females (Roonwal and Mohnot 1977). Although there is a dominance hierarchy among the males, they seem more tolerant of one another and much less aggressive than males of *M. mulatta* (Simonds 1965). This situation may be attributable to the rich tropical habitat of *M. radiata,* which allows more large individuals (males) to live at close quarters without intense competition for resources (Caldecott 1986).

Some other reported information on group structure in macaques is: *M. nemestrina,* group size 3–15 in disturbed forest and 30–47 in undisturbed areas; *M. silenus,* group size 4–34, usually about 10–20, including 1–3 adult males; *M. assamensis,* group size 10–100; and *M. arctoides,* group size 20 to more than 100, usually, but not always, led by an adult male (Green and Minkowski 1977; Lekagul and Mc-Neely 1977; Roonwal and Mohnot 1977). According to Fooden (1969), the Sulawesi species form troops of 5–25 individuals, usually with an old male as leader; several groups may occasionally merge into a temporary aggregation of more than 100 animals. Ménard and Vallet (1993) observed a group of *M. sylvanus* to more than double in size over a four-year period, to 76 individuals, and to then undergo fission into three independent groups largely at the instiga-

tion of adult females. Maternally related individuals stayed together in the new groups. Most of the original males stayed in one group, immigrant males moved into another group, and the third new group had only one adult male. Wu and Lin (1992) also observed group fission in *M. cyclopis,* of Taiwan. Groups of that species are small, usually containing 10–20 individuals, and only one male, though solitary males sometimes move into the permanent units and mate with the females.

According to Roonwal and Mohnot (1977), there is mounting evidence that reproductive seasonality is the rule in all wild populations of *M. mulatta.* In south-central Asia births occur mainly in February–May, with some also in September–October. The season is about the same in *M. radiata,* of southern India, but seems more variable in *M. sinica,* of Sri Lanka, perhaps being synchronized with the rainfall pattern (Fooden 1979). There apparently is a spring birth peak in *M. fascicularis* and *M. assamensis* in Thailand (Fooden 1971). In peninsular Malaysia *M. fascicularis* may give birth at any time of the year but is most likely to do so from May to July (Kavanagh and Laursen 1984). *M. cyclopis,* of Taiwan, mates mostly from November to January and gives birth mostly from April to June (Wu and Lin 1992). Female *Macaca* may enter estrus several times during the mating season, with receptivity being signaled by

Celebes ape *(Macaca nigra)*, photo by Osman Hill.

swelling and reddening of the genital region and vaginal discharge. Some reported estrous cycle lengths (in days) are: *M. nigra,* mean of 33.5 (Dixson 1977); *M. silenus,* mean of 31 days (Lindburg, Lyles, and Czekala 1989); *M. arctoides,* mean of 30.7; *M. fascicularis,* 24–52; *M. mulatta,* mean of 28; and *M. nemestrina,* mean of 42 (Roonwal and Mohnot 1977). Estrus commonly lasts about 9 days (Hayssen, Van Tienhoven, and Van Tienhoven 1993). In the laboratory a female *M. mulatta* has conceived as early as 45 days after giving birth, but in the wild usually only one birth occurs every two years. Interbirth interval in a captive colony of *M. arctoides* averaged about 619 days when weaning occurred naturally and outdoors but only 292 days following an abortion or stillbirth (Nieuwenhuijsen et al. 1985).

Roonwal and Mohnot (1977) listed the following gestation periods (in days): *M. nemestrina,* 162–86; *M. arc-* *toides,* mean of 177.5; *M. fascicularis,* 160–70; *M. mulatta,* 135–94 (mean of 166); and *M. radiata,* 150–70. Ross (1992) added these means: *M. fuscata,* 173; *M. silenus,* 180; and *M. sylvanus,* 165. Based on 700 pregnancies in captivity, Silk et al. (1993) reported a range in gestation of 133–200 days (mean 166.5) for *M. mulatta,* with older females having significantly longer periods and heavier infants than younger mothers. Normally there is a single young, but twins have been reported on rare occasion for several species. A newborn weighs about 400–500 grams (Ross 1992) and nurses for about 1 year. It initially clings to the belly of the female and later may ride ventrally or dorsally. Sexual maturity is usually attained at around 2.5–4 years in females and 2–3 years later in males (Roonwal and Mohnot 1977; Simonds 1965). In *M. silenus,* however, females do not have their first offspring until they are 5 years old and males do not reach sexual maturity until 8 years

Japanese macaques *(Macaca fuscata)*, photo by Carl B. Koford.

(Green and Minkowski 1977). Most reproductive activity of female *M. fuscata* occurs between 6 and 18 years (Koyama, Norikoshi, and Mano 1975). Perhaps females of that species must attain maximum size before they can successfully bear young in the harsh climate of Japan (Ross 1992). Female *M. mulatta* reach menopause at about 25 years (Roonwal and Mohnot 1977). Although the average life span for wild *M. sinica* is less than 5 years, a captive individual lived for 33–35 years (Fooden 1979). Several other species have survived for more than 25 years in captivity, and record longevity for the genus appears to be held by a specimen of *M. fascicularis* that died at 37 years and 1 month (Jones 1982).

Several species of macaques are closely associated with people. *M. mulatta,* the rhesus monkey, is sacred to the Hindu religion and is commonly found in the vicinity of temples and urban areas in northern India and Nepal. This species has been used extensively in the West for experiments involving biology, behavior, medicine, and even space flight. It served in the studies that first demonstrated the Rh factor in blood. It also does well and is widely exhibited in zoos and circuses. The demand for *M. mulatta* was such that nearly 200,000 were exported each year to the United States alone in the late 1950s (Mohnot 1978). Subsequent restrictions by the governments of both India and the United States have greatly reduced this commercial traffic, but major problems continue. *M. mulatta* seems to have been adversely affected both by the loss of juveniles for export and by changing environmental and cultural conditions. After a lengthy study of population trends, Southwick and Siddiqi (1977) cautioned that food shortages in India were leading to a shift in villagers' attitude from traditional reverence and protection of the rhesus to the view that the species is a threat to crops and a nuisance to be eliminated. Southwick and Siddiqi stated that unless the rhesus received protection from local people, it would be extirpated from most agricultural areas of India within 25 years. Further investigations confirmed that habitat changes, intensification of agriculture, and loss of traditional protection were resulting in drastic overall declines

and the disappearance of most populations along roads and canals, in villages, and even around temples (Southwick et al. 1982; Southwick, Siddiqi, and Oppenheimer 1983). The number of *M. mulatta* in India was estimated to be as high as 20 million in the 1940s, less than 2 million in 1960, and 180,000 by 1980 (Southwick and Lindburg 1986). Remarkably, a 1978 ban on exportation, a renewed interest in protection for both religious and bioconservation reasons, and a striking increase in agricultural production, apparently in association with greater tolerance of the monkeys, evidently led to recovery in some areas. In 1985 the total rhesus population of the country was estimated at 410,000–460,000, and more recent surveys indicate that the number continues to grow (Southwick and Siddiqi 1988, 1994a). This information is tempered, however, by the finding that more than 85 percent of the rhesus in India are now living in a commensal or semicommensal status with people (Southwick and Siddiqi 1994b).

The population of *M. mulatta* in China, estimated at 150,000, also is declining through habitat loss and commercial trade (Wang and Quan 1986). A colony of about 50 individuals northeast of Beijing, once thought to be the result of an introduction, now is known to have been the last relict of a population that occupied much of northeastern China during the historical period (Tan 1985; Zhang et al. 1991). Unfortunately, this group, the northernmost nonhuman primates on the mainland of Eurasia, evidently was extirpated in the late 1980s through hunting and habitat destruction by people (Zhang et al. 1989). Another population, with more than 2,000 monkeys, survives about 500 km south of Beijing, in the Taihang Mountains on the Shanxi-Henan border (Qu et al. 1993). That population is nonetheless considered seriously endangered and has been found to represent a genetically distinctive subspecies, *M. m. tcheliensis* (Zhang and Shi 1993). *M. fascicularis,* which replaces *M. mulatta* to the southeast, has the reputation of being a pest in some areas because of its raids on fields and gardens (Medway 1978). This species, which is frequently referred to in medical terminology as the cynomolgus monkey, was used extensively in studies that led to the development of vaccine for the control of polio. The IUCN now designates both *M. fascicularis* and *M. mulatta* as near threatened.

A number of other species of macaques are of concern to the conservation community. Zhang et al. (1991) considered all six species that occur in China to be endangered or vulnerable. Eudey (1987) listed all the species and subspecies of Sulawesi as either highly vulnerable or endangered through human usurpation of habitat; the IUCN now classifies *M. maura* and *M. nigra* as endangered, *M. brunnescens* as vulnerable, *M. nigrescens* as conservation dependent, *M. tonkeana* and *M. hecki* as near threatened, and *M. thibetana* as conservation dependent. Supriatna et al. (1992) found the habitat of *M. maura* to be restricted to small and fragmented nature reserves, though *M. tonkeana* still occupied large and protected areas of forest. Hamada, Oi, and Watanabe (1994) indicated that *M. nigra* is critically endangered on Sulawesi, though a population apparently introduced long ago on Bacan Island, in the Northern Moluccas, seems to be in good condition. *M. nemestrina*, a species often domesticated and trained to harvest coconuts, is reportedly declining rapidly throughout its range and is classified as vulnerable by the IUCN. Its flesh is sought by people for food and medicinal purposes. Because of intensive shooting, it has become rare in parts of Borneo (Medway 1970, 1977; Roonwal and Mohnot 1977). Khan (1978) estimated that the population in peninsular Malaysia had declined from 80,000 in 1958 to 45,000 in 1975. *M. sinica*, of Sri Lanka, is also declining because of the logging of its forest habitat and is listed as threatened by the USDI and as near threatened by the IUCN. Dittus (1977) estimated that the most threatened subspecies, *M. sinica opisthomelas*, numbered only 1,469 animals.

Also listed as threatened by the USDI are *M. arctoides*, *M. cyclopis*, and *M. fuscata*. The IUCN classifies *M. cyclopis*, *M. assamensis*, and *M. arctoides* as vulnerable, *M. fuscata* as endangered, and *M. pagensis* of the Mentawai Islands as critically endangered. Eudey (1987) already had regarded this last species, which sometimes is treated only as a subspecies of *M. nemestrina* (see above), as highly endangered. *M. pagensis* is estimated to have declined by at least 80 percent over about the past two decades and is forecast to decline by at least another 80 percent in the next two decades. The habitat of the species, which is restricted to a few small islands off southwestern Sumatra, is rapidly being destroyed by logging. In addition, the monkeys are being hunted by the native people for food and use as pets. *M. arctoides* is thought to have declined because of deforestation and environmental disruption resulting from military activity. Roonwal and Mohnot (1977) referred to *M. arctoides* as rare throughout its range. *M. cyclopis*, now restricted to the remote highlands of Taiwan by an expanding human population, is jeopardized by taking for use as food, pets, medicinal preparations, and research. Poirier (1982) estimated that at least 1,000–2,000 *M. cyclopis* are lost annually for such purposes but stated that the main threat is loss of forest habitat; every locality that was found to support the species in a 1940 survey is now occupied by a village or city. More recent surveys have found that these problems are continuing but that *M. cyclopis* is still widespread on Taiwan (Lee and Lin 1990). *M. fuscata*, while traditionally protected in Japan, is declining through cutting of the forests on which it depends and through persecution by people who consider it a threat to crops (Eudey 1987). Iwano (1975) warned that there had been a rapid population decrease throughout Japan since 1923 and that free-living monkeys could disappear within the next few decades. Maruhashi, Sprague, and Matsubayashi (1992) reported that about 5,000 *M. fuscata* were being removed annually

in response to claims of agricultural damage. Hill and Sprague (1991) indicated that more than 500 of those taken each year represented the subspecies *M. f. yakui*, of Yakushima Island, which had begun to raid orange groves after their natural habitat had been reduced.

One of the most seriously jeopardized macaques is *M. silenus* of the Western Ghats Mountains in southern India. This species seems to have a low reproductive rate and to be unable to adapt to human encroachment. Suitable habitat has greatly diminished through clearing of the forests for agriculture. The largest remaining population contains about 300 animals and occupies about 160 sq km of forest, just about the minimum area needed to maintain viability. The total estimated number in the wild in 1975 was 405, and there were another 275–300 in captivity (Green and Minkowski 1977). A subsequent investigation resulted in an overall estimate of about 915 individuals, but it also revealed numerous new threats, including the construction of dams and roads in the habitat of the largest populations (Ali 1985). More recent reports indicate that the area occupied by the largest population has been incorporated into a biosphere reserve, that the total number in the wild has increased to more than 1,700, and that there is a rapidly growing captive population (Kurup 1988; Lindburg and Gledhill 1992). *M. silenus* is classified as endangered by the IUCN and the USDI and is on appendix 1 of the CITES; all other species of *Macaca* are on appendix 2.

The only macaque to occur naturally outside of Asia is *M. sylvanus* of northern Africa. It may have survived in Libya and Egypt in the early nineteenth century and as far east as Tunisia around 1900, but currently it is restricted to high cedar and oak forests in the Atlas and nearby ranges of Morocco and northern Algeria. It is continuing to decline because of the commercial logging of the forests. In 1975 the total population was estimated to number about 14,000–23,000 individuals, 77 percent of which were in Morocco (Deag 1977; Fa et al. 1984; Taub 1977, 1984). In 1992 the total estimate was 10,000–16,000 (Lilly and Mehlman 1993). The IUCN classifies *M. sylvanus* as vulnerable.

A small number of *M. sylvanus* also occupy the Rock of Gibraltar, a British possession at the tip of southern Spain. This population is fed and protected by the British army and long was artificially maintained, through capture and export to zoos, at about 33 individuals. Recently numbers have increased to approximately 105 and there are concerns that the health of the population is being adversely affected through excessive feeding by people and exposure to human diseases. The monkeys are a popular tourist attraction, and there is a superstition that if they should disappear from the Rock, British rule would end. Aside from that, the most interesting question about these animals is whether they are descended from stock originally introduced from Africa through human agency or are the natural remnant of a formerly widespread European monkey population. It is known that macaques were present at Gibraltar when the British arrived in 1704, but numerous introductions have occurred since then. Fossil remains indicate that *M. sylvanus* occurred in much of Europe during the late Pleistocene, and some animals may possibly have survived in southern Spain, outside of Gibraltar, as late as the 1890s. More recently, some *M. sylvanus* escaped from captivity in Spain and began to live and reproduce in the wild (Deag 1977; O'Leary 1993; Taub 1977, 1984).

Successful introductions of macaques are not unusual. In 1763 a troop of *M. sylvanus* was set loose in Germany, where it thrived until deliberately exterminated in 1784 (Grzimek 1975). There is currently a large, free-ranging colony of *M. sylvanus* in France (R. W. Thorington, Jr., U.S. National Museum of Natural History, pers. comm., 1981).

M. fascicularis apparently was brought to Mauritius in the Indian Ocean by Dutch or Portuguese sailors in the early sixteenth century, and the monkey population on the 1,865-sq-km island now numbers 25,000–35,000 (Sussman and Tattersall 1986). *M. mulatta* has been established in Puerto Rico and on an island near Rio de Janeiro, Brazil (Hausfater 1974; Roonwal and Mohnot 1977). A feral population of *M. mulatta* in central Florida, probably established originally as a tourist attraction, has existed since the 1930s (Wolfe and Peters 1987). M. cyclopis has been introduced on the Izu Islands, south of Tokyo (Corbet 1978), *M. nemestrina* on Singapore and the Natuna Islands (Corbet and Hill 1992), and *M. nigra* on Bacan Island in the Moluccas (Flannery 1995).

PRIMATES; CERCOPITHECIDAE; Genus PAPIO
Erxleben, 1777

Baboons

There are five species (Corbet 1978; Dandelot *in* Meester and Setzer 1977; Harrison and Bates 1991; Yalden, Largen, and Kock 1977):

P. hamadryas (hamadryas baboon), upper Egypt, northeastern Sudan, eastern Ethiopia, northern Somalia, western Arabian Peninsula north to about 24° N;

P. anubis (olive baboon), savannah zone from Mali to Ethiopia and northern Tanzania, several mountainous areas in the Sahara Desert;

P. papio (western baboon), Senegal, Gambia, Guinea;

P. cynocephalus (yellow baboon), Angola, southern Zaire, Tanzania, Zambia, Malawi, northern Mozambique, Kenya, Somalia;

P. ursinus (chacma baboon), southwestern Angola, southern Zambia, southern Mozambique, Namibia, Botswana, Zimbabwe, South Africa, Swaziland, Lesotho.

The authority for the name *Papio* sometimes is given as Müller, 1773. *Mandrillus* (see account thereof) sometimes is regarded as a subgenus of *Papio*. The ranges of the above five species are largely, if not entirely, allopatric, and it may be that all intergrade and are no more than subspecies of a single widespread species (De Vore and Hall 1965; Jolly 1993; Maples 1972; Thorington and Groves 1970). Groves (1989) indicated that the five taxa are strongly differentiated and that at least some might be specifically distinct, but subsequently (*in* Wilson and Reeder 1993) he recognized only the species *P. hamadryas*, with the other four included therein. Corbet and Hill (1991), Meester et al. (1986), and Napier and Napier (1985) treated all five as separate species. *P. anubis* and *P. hamadryas* have formed an apparently stable hybrid zone where their ranges meet in Ethiopia (Gabow 1975; Kummer, Goetz, and Angst 1970; Nagel 1971). Limited hybridization also evidently occurs between *P. anubis* and *P. cynocephalus* and between *P. cynocephalus* and *P. ursinus*, but a recent odontomorphometric analysis by Hayes, Freedman, and Oxnard (1990) indicates that all are valid species. In contrast, Jolly (1993) concluded that the degree of gene flow between the various populations implies that they should be called a single species. Jolly also questioned whether *P. hamadryas* ever occurred naturally in Egypt and suggested that the population in Arabia is the result of introduction by human agency within the last 4,000 years.

In *P. hamadryas* head and body length is 610–762 mm and tail length is 382–610 mm. One individual weighed 18 kg. Young animals are brown, becoming ashy gray with age; females retain the brown coat longer than the males. The heavy mane around the neck and shoulders of the older males is lacking in the younger males and in the females. In the other four species head and body length is 508–1,143 mm, tail length is 456–711 mm, and weight is 14–41 kg. As with other baboons, the males are usually considerably larger than the females. The hair is rather coarse. The face and neck are thinly haired, but long hairs develop around the neck and shoulders in males. General coloration is olive brown in *P. anubis*, olive rufous in *P. papio*, yellowish in *P. cynocephalus*, and dark olive brown in *P. ursinus*. These four species lack the heavy mane of the hamadryas baboon and are not as brilliantly colored as *Mandrillus*.

Baboons inhabit open woodland, savannahs, grassland, and rocky hill country. All species are primarily diurnal and terrestrial but capable of climbing. If available, trees may be used for sleeping, but populations in Ethiopia and South Africa generally spend the night on the faces of cliffs (Davidge 1978; De Vore and Hall 1965; Kummer 1968). The mean daily distance traveled by *P. cynocephalus, P. ursinus,* and *P. anubis* is about 3–4 km (Altmann and Altmann 1970). They walk with an awkward swaggering gait and run in a "rocking horse" gallop. They are said to be good swimmers and to have excellent vision. All five species are omnivorous, taking a variety of vegetable matter but seeming to concentrate on whatever is easily available at a given time (De Vore and Hall 1965). During the dry season in Kenya, grass may make up 90 percent of the diet. On the coast of South Africa, mollusks, crabs, and other marine creatures are regularly eaten. Altmann and Altmann (1970) observed *P. cynocephalus* chasing and capturing hares *(Lepus capensis)* and digging out small fossorial mammals. Hamilton, Buskirk, and Buskirk (1978) found that when there were massive outbreaks of grasshoppers or scale insects *P. ursinus* ignored other food sources.

Wolfheim (1983) listed the following recorded population densities: *P. anubis,* 1.12–63.00/sq km; *P. papio,* 2–15/sq km; *P. cynocephalus,* 9.65–60.00/sq km; and *P. ursinus,* 2.31–43.2/sq km. These species reportedly have definite home ranges from which it is difficult to drive them. Group home range size is 2–40 sq km, though there may be smaller core areas where most time is spent. There is extensive or total overlap between the ranges of different groups and little evidence of territorial defense (Altmann and Altmann 1970; Chalmers 1973; Davidge 1978; De Vore and Hall 1965).

P. hamadryas generally occurs at low densities, recorded at 1.8–3.4/sq km, and groups move across a considerable distance, 6.5–19.6 km, each day (Wolfheim 1983). The detailed studies of Kummer (1968) in Ethiopia demonstrate that *P. hamadryas* has developed a trilevel social organization enabling the species to adapt to an environment with (1) sparse food resources, thereby favoring dispersed, small groups; and (2) scarce sleeping sites, thereby favoring group concentration. The largest social unit is the "troop," which comprises all the animals sleeping at a particular cliff. Troop size generally is 100 or more individuals, and one sleeping party of 750 animals was observed. In the morning the troops assemble above the cliffs and then move out in separate foraging "bands" of about 30–90 animals. The largest moving group seen had 494 baboons. Each band is composed of several single-male units, containing an adult male and 1–9 (usually 2–5) adult females with their young. Often the male leaders of two units will cooperate as a team, the younger individual initiating movement for both units and the older male bringing up

Chacma baboon *(Papio ursinus)*, photo by Bernhard Grzimek.

the rear and accepting or rejecting the indicated direction. All females live in single-male units, and young males or adult males with no unit to lead may either maintain a loose association with a unit or live alone. Of the various social groupings, the single-male unit has the most stable membership. In six such units studied from one to six months there was no change in the male leader. He consistently defends the same set of females from the advances of other males. If a female strays from a unit, the adult male will drive her back, though if she is in estrus, the male may force the entire unit to follow her.

Of all the intragroup social activities, grooming takes up most of the adults' time. If two females simultaneously try to groom the adult male, they will soon begin to scream and fight until one withdraws. Aggressive displays by females seem always to be associated with winning the attention of the unit leader and are never directed against females of other units. Adult males sometimes fight one another for possession of females. Such fighting involves rapid fencing with open jaws and hitting out with the hands, but there is almost never actual physical contact. Although the adult males of a given band seem to tolerate one another, there may be general conflicts between the males of different bands. When danger threatens, female *P. hamadryas* gather up the young and move along at a fast pace; the males usually lag behind, often turning around in threatening poses and uttering grunts and deep-throated sounds.

Sigg et al. (1982) reported a significant fourth level of organization, the "clan," between the band and the single-male unit. A clan is an association of interacting single-male units. Although young males may leave their natal unit, they always remain within their original clan. Females also usually transfer to a unit within the clan, though sometimes they switch clans and even bands. Age of departure from the natal unit is about 2 years in males and 3.5 years in females.

Studies of *P. cynocephalus, P. anubis,* and *P. ursinus* in eastern and southern Africa indicate a somewhat different social structure (Altmann and Altmann 1970; De Vore and Hall 1965; Hall and De Vore 1965). Group size is about 8–198, usually 30–60. These groups often contain several adults of each sex, though usually more females than males. There is no division into single-male units; rather, all the males compete for estrous females. The males form a dominance hierarchy, with the position of each animal depending on how well he fights and on his ability to form alliances with other males. Females have a less evident hierarchy, but several may sometimes join in aggression against others. Group movement is directed by the adult males, though females may walk in front. A young or

Hamadryas baboons *(Papio hamadryas)*, photo by Erik Parbst through Zoologisk Have, Copenhagen.

subadult male moves well ahead of the column and sounds an alarm bark in case of danger, whereupon an adult male will move forward to investigate. Numerous other vocalizations have been identified in these baboons.

The differences in the social life of *P. anubis* and *P. hamadryas* have stimulated special interest in a zone of hybrids formed by these two species where their ranges meet along the Awash River in Ethiopia (Gabow 1975; Kummer, Goetz, and Angst 1970; Nagel 1971, 1973; Phillips-Conroy, Jolly, and Brett 1991). Since male *P. anubis* lack the ability to herd females, they are unable to control and mate with female *P. hamadryas*. Male *P. hamadryas*, however, can bring female *P. anubis* under their domination, and this process apparently was responsible for establishment of the hybrid zone. Male *P. hamadryas* have been observed to enter groups of *P. anubis*, establish "harems" therein, and sire young, though there is no evidence that they actually abduct females. The hybrid young are viable, but the zone has not expanded beyond 20 km, evidently because male offspring do not inherit the herding ability of their fathers. Mori and Belay (1990) reported another zone of extensive hybridization between *P. anubis* and *P. hamadryas* along the Wabi-Shebeli River, about 100 km south of the Awash.

The following reproductive data on *P. hamadryas* were summarized by Kummer (1968): average estrous cycle, 30 days; gestation, 170–73 days; average lactation period, 239 days; and normally a single young. Breeding occurs throughout the year, but there are birth peaks in Ethiopia in May–June and November–December. The age of sexual maturity is 5 years in females and 7 years in males. Hayssen, Van Tienhoven, and Van Tienhoven (1993) listed birth weights of 600–900 grams and lactation periods of 6–15 months. According to Marvin L. Jones (Zoological Society of San Diego, pers. comm., 1995), a captive specimen of *P. hamadryas* lived 37 years and 6 months.

In the other species of *Papio* there seems to be seasonal reproduction in some areas but not in others. In Kenya, births may occur at any time, but most take place from Oc-

tober to December. The interbirth interval is at least 15 months, the gestation period is 6 months, and lactation lasts 6–8 months. There is usually a single young, rarely two. At first the young is carried clinging to the mother's breast, but it soon begins to ride on her back. Females have their first estrus at 3.5–4.0 years and reach full size and bear their first offspring at 5 years. Males are sexually mature at 5 years but are not fully developed and able to mate successfully until 7–10 years (De Vore and Hall 1965; Hall and De Vore 1965). A specimen of *P. ursinus* lived 45 years in captivity.

The hamadryas was the sacred baboon of the ancient Egyptians and was often pictured on temples and monoliths as the attendant or representative of Thoth, the god of letters and scribe of the gods. It was also mummified, entombed, and associated with sun worship. Today *P. hamadryas* is generally tolerated and even actively fed by people in Saudi Arabia, and a number of populations have become commensal in the vicinity of cities and garbage dumps (Biquand et al. 1992*a*, 1992*b*). The species reportedly has recently extended its range northward on the Arabian Peninsula and increased in numbers, sometimes reaching pest proportions, with troops of up to 1,000 baboons moving freely through settled areas and showing little fear of people (Harrison and Bates 1991). *P. hamadryas*, however, has been exterminated in Egypt and reduced in numbers in some other areas. Irrigation projects in Ethiopia, home to 90 percent of the hamadryas population, have brought parts of its range under cultivation, led to conflicts with people, and often resulted in the baboons' being killed or harassed (Lee, Thornback, and Bennett 1988). The other four species reportedly are more adaptable to human environmental disturbance; nonetheless, they are losing habitat to agricultural expansion in many areas. They become subject to organized extermination efforts when they turn to raiding crops, and they also are hunted for food and sport (Wolfheim 1983). All species are on appendix 2 of the CITES. *P. hamadryas* and *P. papio* are designated as near threatened by the IUCN.

Chacma baboons *(Papio ursinus)*, photos by Ernest P. Walker.

PRIMATES; CERCOPITHECIDAE; **Genus**
MANDRILLUS
Ritgen, 1824

Drill and Mandrill

There are two species (Dandelot *in* Meester and Setzer 1977; Grubb 1973; Wolfheim 1983):

M. sphinx (mandrill), south of Sanaga River in Cameroon, Equatorial Guinea, Gabon, Congo;
M. leucophaeus (drill), from Cross River in southeastern Nigeria to Sanaga River in Cameroon, island of Bioko (Fernando Poo).

Mandrillus was regarded as a subgenus of *Papio* by Dandelot (*in* Meester and Setzer 1977) and Wolfheim (1983) but as a distinct genus by Corbet and Hill (1991), Groves (1989), Grubb (1973), Hill (1970), and Napier and Napier (1985). Recent molecular and chromosomal analyses suggest that *Mandrillus* is less closely related to *Papio* than is *Theropithecus* (Disotell 1994).

Head and body length is about 610–764 mm and tail length is about 52–76 mm. Weight is usually about 25 kg in males and 11.5 kg in females (Napier and Napier 1985). *M. sphinx*, the heavier of the two species, is the largest of all monkeys, having a shoulder height of about 508 mm and a weight of up to 54 kg. Both species have prominent ridges on each side of the nasal bones; these are outgrowths of ridged and grooved bone. There are usually six grooves in each main ridge on the mandrill's face. In the adult male mandrill the skin is purple in the grooves and blue on the small ridges. The median tract between the outgrowths is brilliant scarlet. The scarlet extends forward to the muzzle and on the area around the nostrils. Female and young male

mandrills have less pronounced ridges and lack the purple color in the grooves, and their noses are black, not scarlet. In the drill the face is black, lacking bright coloration, and each outgrowth has only two grooves. In both species the buttock pads and the skin around them have a lilac tinge, which becomes reddish purple at the edges. The red color of the skin is the result of the distribution and richness of blood vessels. The bright color of the skin of these animals becomes more pronounced when they are excited. Jouventin (1975*b*) suggested that the rear coloration serves as a signal facilitating group movement through thick vegetation.

Both species have a beard, a crest, and a mane. The hair on the upper parts of the mandrill is tawny greenish, and that on the underparts is yellowish. The pelage of the drill is brownish.

The drill and mandrill usually are found in thick rainforest. In a study of *M. sphinx* in Gabon, Jouventin (1975*a*) generally encountered the large adults on the ground and smaller females and juveniles at midlevel in the trees. These animals were observed to begin activity at dawn, rest at midday, and enter sleeping trees at 1700–1800 hours. On the average they foraged over about 8 km per day. In Cameroon, Hoshino (1985) determined that *M. sphinx* moved 2.5–4.5 km per day. Both *M. sphinx* and *M. leucophaeus* feed mainly on the ground, searching for fruits, nuts, other plant material, mushrooms, invertebrates, and occasionally small vertebrates. They turn over stones and debris when seeking food. Kudo and Mitani (1985) observed a pair of mandrills attacking and killing a duiker *(Cephalophus)*.

In a study in Gabon, Jouventin (1975*a*) found *M. sphinx* to use very large home ranges, two such areas being estimated at 40 and 50 sq km. The basic social unit comprises 1 large adult male, 5–10 adult females with or without infants, and approximately 10 juveniles. During the dry season, however, 6–7 such harems occasionally associate to form troops of 200 or more individuals. Excess males live alone. When a group is quietly foraging, the adult male keeps to the rear, but in the event of danger he moves out to the front. Three vocalizations were identified: a contact call by the females and young, an alarm call by females and subadults, and the rallying call of the male leader. Studies of the same species in Cameroon (Hoshino et al. 1984) indicated a generally similar social structure, with an average of 13.9 individuals per adult male, sometimes larger aggregations with several adult males and some solitary males. Observations of a group of *M. sphinx* in a large enclosure in Gabon (Dixson, Bossi, and Wickings 1993) revealed a multimale organization with a strict dominance hierarchy. The highest-ranking males had the most prominent facial coloration and the heaviest rumps, and they associated almost exclusively with the females and sired young. Other males tended to be peripheral and solitary. Gartlan (1970) observed troops of 14–170 *M. leucophaeus* in western Cameroon but suggested that the larger groups were temporary aggregations and that social structure in this species is much like that of *M. sphinx*.

Jouventin (1975a) reported apparent seasonal reproduction for *M. sphinx* in Gabon, with most births occurring from January to April. Sabater Pi (1972) stated that young less than 3 months old had been taken in Equatorial Guinea in December, February, March, and June. Carman (1979) reported gestation periods of 168–76 days for captive *M. sphinx*. Based on observations of a captive group of *M. sphinx* in an enclosure of 5.3 ha., Feistner (1992) calculated an average gestation period of 176 days, an interbirth interval of 13–14 months for multiparous females, and age of sexual maturity in females as about 3.5 years. Hadidian and Bernstein (1979) listed the median estrous cycle as 33.5

days in *M. sphinx* and 35 days in *M. leucophaeus.* According to Hill (1970), mating in *M. leucophaeus* takes place throughout the year and the gestation period is 7.5 months. He also reported that record longevity for *Mandrillus* is 46 years.

Both Hill (1970) and Sabater Pi (1972) indicated that despite their formidable appearance and reputation for ferocity, specimens of *Mandrillus* are gentle and adapt readily to captivity. In the wild both *M. sphinx* and *M. leucophaeus* occupy relatively small ranges and are declining because of habitat loss and excessive hunting by people for food. Recent surveys indicate that hunting is mainly commercial, the meat being resold in urban centers (Blom et al. 1992). Numbers of *M. leucophaeus* are estimated to be fewer than 10,000 (Cox and Boer *in* Bowdoin et al. 1994). Considering both degree of threat and taxonomic distinctiveness, Oates (1996) ranked *M. leucophaeus* the African primate species most in need of conservation action. Both species are listed as endangered by the USDI and are on appendix 1 of the CITES. The IUCN classifies *M. leucophaeus,* the more narrowly distributed and seriously jeopardized of the two, as endangered and *M. sphinx* as near threatened. Blom et al. (1992) indicated that the range of *M. sphinx* is now more restricted in Gabon than previously reported but that there had been recent unconfirmed reports of *M. leucophaeus* in the country.

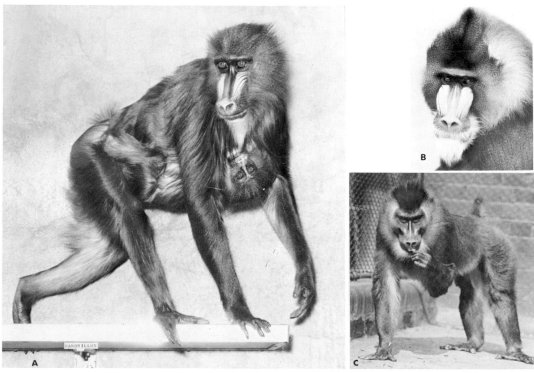

A. Mandrills *(Mandrillus sphinx)*, female with baby riding in usual position, photo from Cheyenne Mountain Zoo. B. Mandrill *(M. sphinx)*, male, photo by Bernhard Grzimek. C. Drill *(M. leucophaeus)*, photo from Zoological Society of London.

PRIMATES; CERCOPITHECIDAE; **Genus**
THEROPITHECUS
I. Geoffroy St.-Hilaire, 1843

Gelada

The single living species, *T. gelada,* is restricted to the mountains of central Ethiopia (Yalden, Largen, and Kock 1977). A previously unknown population about 300 km south of the others was discovered in 1989 (Mori and Belay 1990). During the Pliocene and Pleistocene various extinct species occurred across much of Africa and southwestern Asia (Pickford 1993). *Theropithecus* is sometimes considered a subgenus of *Papio.*

Head and body length is about 500–725 mm and tail length, including the tuft, is about 700–800 mm. Napier and Napier (1967) recorded weights of 20.5 kg for a male and 13.6 kg for a female. The whiskers are long, the head is crested, a mane is present on the shoulders, and the tail is long and tufted. Coloration is brown and black, but there is a bare red area on the chest that is conspicuous when movements of the animal bring it into full view. The underparts are lighter than the upper parts, usually being grayish or grayish white. The legs are grayish, or pale yellowish from the knees to the ankles. The buttock pads are black, not red as in *Papio hamadryas.* The gelada also may be distinguished from all other baboons by its nostrils, which open on the side of the nose, not terminally as in the other species. The body form is massive.

Jablonski (1993*a*) pointed out that *Theropithecus* is distinguished by a large number of skeletal characters associated mostly with adaptation to a terrestrial, grazing lifestyle and the stresses of mastication. These features include a skull with large infratemporal fossa, marked postorbital constriction, deep posterior maxilla, and deep mandibular body and symphysis; cheek teeth high-crowned with steep sides and increased relief and showing a distinctive pattern of wear with the exposure of completely curved infoldings of enamel and dentinal lakes as cusps become worn; and in the postcranial skeleton relative elongation of the forearm, greater flexibility of the elbow and wrist joints, relative elongation of the thumb, and relative shortening of the index finger.

The gelada is strictly associated with rocky gorges and precipices at altitudes between about 1,700 meters and the montane moorland zone at 4,400 meters (Iwamoto 1993; Yalden, Largen, and Kock 1977). It is dependent on the gorge systems for sleeping sites and refuge from predators and is never found more than 2 km from the rim of a gorge. About an hour after dawn groups climb out of the gorges to the plateaus above. They spend about two hours in social activity, and then around 0900–1000 hours they move out to feed. By 1600 hours they are back at the sleeping cliffs. *Theropithecus* is the only true grazing primate. One study showed that 90–95 percent of its diet consists of the leaves, roots, and seeds of grasses; the remainder includes other vegetable matter and occasionally insects. Specialization in eating grass means that the gelada does not have to disperse very far to find food. Small groups may average as

Gelada baboons *(Theropithecus gelada),* females, photo from Los Angeles Zoo.

Gelada baboon *(Theropithecus gelada)*, male, photo by P. R. McCann.

little as 600 meters a day, and even large herds of 300–400 animals rarely travel more than 2 km per day. This situation has resulted in population densities as great as 75–80/sq km in undisturbed areas (Dunbar 1977a, 1977b).

The complex social structure of *Theropithecus* has been the subject of several recent major studies, the results of which do not entirely coincide (Dunbar 1983a, 1983b, 1993b; Dunbar and Dunbar 1975; Kawai 1979; Ohsawa and Kawai 1975). Some of the differences were resolved by Kawai et al. (1983). The overall gelada population seems to be divided into "bands" of about 30–350 animals. Each band utilizes a home range of about 1.5–2.0 sq km, which overlaps to some extent with the ranges of neighboring bands. When grazing conditions are good, several bands may join in temporary aggregations of as many as 670 individuals. Bands are relatively stable, socially cohesive entities, but they are divided into several "reproductive units" and "all-male units" that may act independently.

All-male units, averaging about eight individuals each, contain mainly young animals, though an older male may also be present. Most reproductive units have only a single adult male, but about 25 percent have two or three. These units usually have about three to six adult females and their young. There seems to be general agreement that the main factor in keeping a reproductive unit together is the strength of bonds between the females rather than male dominance and herding, the case in *Papio hamadryas*. Through this female relationship and its descent from mother to daughter the reproductive unit persists generation after generation even if males are not present at times. Since there also is movement of females between groups, some or most of these animals may not be related. There may be little or no social interaction, particularly groom-

ing, between unrelated individuals. The females take at least partial responsibility for group defense, and they may unite in aggression against the male. If the unit has more than one adult male, the females sometimes fight amongst themselves regarding which of the males is to be allowed to mate.

There is some disagreement regarding the role of the males, but it seems likely that at least certain males are capable of controlling group movements. This may be especially true if the females are young and have not yet developed strong bonds. In established groups, however, the male may have only a peripheral role. He does not necessarily interact with all of the females but has a single predominant social partner who, in turn, is likely to be an individual without female relatives in the unit. Males of reproductive units fight younger males attempting to move in from all-male units. Sometimes, however, a second or third male attaches itself to a reproductive unit. Such "follower" males may be tolerated and may even cooperate with the first male, but they appear to be merely waiting for the opportunity to obtain the females for themselves. A follower male may eventually try to take over the entire reproductive unit, which usually involves fierce combat with the first male, or he may gradually develop a relationship with one or two of the younger females, ultimately departing with them to form a new unit. If combat occurs, the actual issue to be decided is which male will have interaction with the females, and it is the latter who may choose the winner. If the first male is defeated, he sometimes remains in the unit as a follower. *Theropithecus* has a large number of visual signals and calls for communication; about 25 different vocalizations have been recognized.

Dunbar and Dunbar (1975) found birth peaks in January and June in their study area but noted that such peaks differed in other areas. They also stated that there was an evident interbirth interval of 24 months. Hayssen, Van Tienhoven, and Van Tienhoven (1993) listed estrous cycle lengths of 30–39 days, estrus lengths of 7–10 days, and lactation periods of 12–18 months. Mallinson (1974) reported gestation periods of 147, 171, and 192 days. Males are considered to be adults at 8 years and females, at 3.5 years. An individual in the San Antonio Zoo in December 1994 was probably well over 30 years old (Marvin L. Jones, Zoological Society of San Diego, pers. comm., 1995).

Dunbar (1977b) estimated the total gelada population at about 600,000 animals but warned that suitable natural habitat was being lost to agriculture. Population densities had been reduced to only about 15–20/sq km in areas disturbed by cultivation and topsoil erosion. In addition, in the southern part of the range of the species natives intensively hunt adult male geladas to obtain their manes for use in headdresses and capes. Such hunting was thought to reduce average life expectancy of males and to upset the social structure of reproductive units, though Dunbar (1993a) suggested that it has little or no impact on overall numbers. Dunbar and Dunbar (1974a) stated that human activity had resulted in a breakdown of the natural ecological separation of *Theropithecus* and *Papio anubis* and that hybridization between the two species might have occurred. Lee, Thornback, and Bennett (1988) stated that numbers may have fallen during the droughts of the 1980s, though Dunbar (1993a) considered any marked reduction to be unlikely. The gelada is listed as near threatened by the IUCN and as threatened by the USDI and is on appendix 2 of the CITES.

PRIMATES; CERCOPITHECIDAE; **Genus NASALIS**
E. Geoffroy St.-Hilaire, 1812

Proboscis Monkey

The single species, *N. larvatus*, occurs on Borneo (Groves 1970*a*). *Simias* (see account thereof) sometimes is considered part of *Nasalis*.

Males have a head and body length of 660–762 mm and weigh 16.0–22.5 kg; females have a head and body length of 533–609 mm and weigh 7–11 kg. Tail length is 559–762 mm. The upper parts vary from chestnut brown to cinnamon rufous, and the lighter underparts are usually creamy, creamy buff, or grayish. The skin of the face is entirely rufous.

The outstanding feature of the genus is the protruding nose, which becomes long and pendulous in old males; it is less developed in females. In young animals the nose is turned upward as in *Simias*. The large nostrils open downward. Bennett (1987) suggested that the nose may be involved with attracting females or radiating excess body heat. She stated also that the feet are partially webbed, thereby facilitating swimming and perhaps walking on mangrove mud.

The proboscis monkey is usually found near fresh water in lowland rainforest or mangrove swamp. Kawabe and Mano (1972) noted that it seemed to depend on mangrove trees for resting and sleeping. Salter et al. (1985) found it to sleep primarily along the edge of a river by night and to move an average of 1,312 meters during the day. Kern (1964) reported it to be active throughout the day, with a peak extending from late afternoon to dark. He observed it swimming both on the surface and underwater and considered it the best swimmer among the primates. A dive of an entire group from a height of about 16 meters was witnessed. Kern found that 95 percent of the diet consisted of leaves, pedada leaves being favored. Fruits and flowers were also eaten. In contrast, Yeager (1989) recorded leaves being eaten in 52 percent of observations and fruits in 40 percent, and seed-eating was a specialization.

Kern (1964) reported a population density of about 9/sq km. Three troops each had a home range of about 2 sq km, and they moved through these areas randomly, with the animals spread out over about 0.5 km. At night each individual slept alone in a tree. Group size averaged 20 (12–27) animals. Kawabe and Mano (1972) reported troops to contain 11–32 monkeys, of which several were adult males and considerably more were adult females and young. The adult males would confront intruders while the others departed. Macdonald (1982) observed group sizes of 2–63 animals and suggested that the larger parties were temporary foraging aggregations. The call of the male is a drawn-out, resonant "honk" or "kee-honk"; that of the female "has a milder, petulant, rather resounding cry faintly suggestive of a goose."

In forests along the Samunsam River in Sarawak, Bennett and Sebastian (1988) found a population density of 5.93/sq km. Most of the animals were organized into groups that generally stayed within 600 meters of the river and had a daily average movement of 483 meters. The basic social unit was a relatively stable harem that used an average length of 6.4 km of river and had an average size of 9 (6–16) individuals. Each harem had a single adult male

Proboscis monkeys *(Nasalis larvatus)*, photo by Hilmi Oesman.

Proboscis monkey *(Nasalis larvatus)*, photo by Bernhard Grzimek.

and 2–7 adult females. Different groups frequently joined together by day to forage and often spent the night in close proximity. The harems were never observed to split into subgroups. Young females commonly transferred from one group to another, either before or after reaching reproductive age. Juvenile males regularly left their natal groups and moved about alone for a while or joined to form bachelor bands.

Based on a year of observations in the Indonesian portion of Borneo, Yeager (1989, 1990, 1991, 1992) reported an average population density of 63/sq km, an average group home range of 130.3 ha., and an average group size of 12 (3–23) animals. Social organization was found to closely resemble that of *Theropithecus*, there being "bands" comprising a number of associated groups that moved and rested in proximity and both "all-male" and "one-male" groups. The latter units, aside from the single male, contained 1–9 adult females and their young. Female bonding was the most important factor in holding the groups together, with young females remaining in their mother's unit and young males leaving to join an all-male group. Although agonistic displays were common, serious fighting was rare and bands tended to avoid one another. The females of a one-male unit seemed to initiate movement, and the male sometimes acted to stop agonistic behavior between other group members.

There appears to be no restricted breeding season; a single young is the rule, after a gestation period of about 166 days (Schultz 1942). Weaning is complete by 7 months (Hayssen, Van Tienhoven, and Van Tienhoven 1993). For many years the proboscis monkey was considered difficult or impossible to maintain in captivity, but according to Olney, Ellis, and Fisken (1994), zoos held 16 specimens, of which 14 had been born in captivity. One captive individual lived for about 23 years (Marvin L. Jones, Zoological Society of San Diego, pers. comm., 1995).

N. larvatus is classified as vulnerable by the IUCN and as endangered by the USDI and is on appendix 1 of the CITES. This species was long considered safe from severe human disturbance because of its prime habitat in inaccessible mangrove swamps. Now, however, improved technology is allowing the economical clear-cutting of mangroves, and the proboscis monkey is declining. MacKinnon (1986)

estimated that suitable habitat has fallen from 19,622 to 10,438 sq km and that the latter area is occupied by 260,950 individuals.

PRIMATES; CERCOPITHECIDAE; **Genus SIMIAS**
Miller, 1903

Pig-tailed Langur

The single species, *S. concolor*, occurs on Siberut, Sipura, North Pagai, and South Pagai in the Mentawai Islands off western Sumatra. Groves (1970*a*) included *Simias* within *Nasalis* based on the close resemblance of the two in cranial features, limb proportions, and hair pattern. The differences in the nose and tail were regarded as superficial. Groves (*in* Wilson and Reeder 1993), as well as Brandon-Jones (1984) and Corbet and Hill (1992), continued to follow that arrangement, but Corbet and Hill (1991), Eudey (1987), Napier and Napier (1985), and Oates, Davies, and Delson (1994) all regarded *Simias* as a distinct genus.

Head and body length is usually 450–525 mm and tail length is only about 130–80 mm. A female specimen weighed 6.9 kg (Groves 1970*a*). The general coloration throughout is dusky brown; this may be lightened slightly on the nape, shoulders, back, and upper arms by buffy rings

Pig-tailed langur *(Simias concolor)*, mounted specimen, photo by H. Unte.

on the hairs. The hands and feet, as well as the naked palms, soles, and large buttock pads, are black. Creamy buff individuals have been noted. The short tail is bare except for an inconspicuous tuft of brownish hair at the tip. Unlike in *Nasalis*, the nose shows no trace of tubular elongation. The openings of the nostrils are on the surface of the wide upper lip, and only the upper nasal margin is lengthened to give a snub-nosed appearance.

Like *Nasalis*, *Simias* is a large, stoutly built monkey. Both genera have arms nearly as long as the legs. The hair of the cheeks is brushed back toward the midline of the neck and does not have the "untidy" appearance found in *Pygathrix*. In males of both *Simias* and *Nasalis* the ischial callosities are united across the midline (Groves 1970a).

An investigation by Tilson (1977) on Siberut Island added considerably to our knowledge of this rare primate. *Simias* is found mainly in hilly primary forests of the interior, not in swamp or mangrove. It is totally arboreal, descending only when disturbed. The diet consists predominantly of leaves, but fruits and berries are also taken. The group foraging area is about 25–30 ha. The social unit is an adult pair with one to three immature offspring. Noticeable vocalizations are rare. Newborn young were observed only in June and July.

S. concolor is listed as endangered by the USDI and the IUCN and is on appendix 1 of the CITES. It is thought to have declined by at least 50 percent over the last decade and is expected to decline by at least another 50 percent during the next decade. It is intensively hunted for food, and its conspicuous size makes it especially vulnerable in this regard. More significantly, it is restricted to a few islands and is dependent on primary forests, which are now being logged. No specimens are known to be in captivity. MacKinnon (1986) estimated that the remaining 4,500 sq km of suitable habitat is occupied by 31,500 individuals, though Eudey (1987) indicated that the total population numbered fewer than 10,000.

PRIMATES; CERCOPITHECIDAE; **Genus PYGATHRIX**
E. Geoffroy St.-Hilaire, 1812

Douc Langur

The single species, *P. nemaeus*, is found to the east of the Mekong River in central and southern Viet Nam, central Laos, and eastern Cambodia (Groves 1989). *P. nigripes* of southern Viet Nam and adjacent parts of Cambodia was treated as a separate species by Brandon-Jones (1984), but was considered a subspecies of *P. nemaeus* by Corbet and Hill (1991, 1992), Groves (*in* Wilson and Reeder 1993), Jablonski (1995), Jablonski and Peng (1993), and Oates, Davies, and Delson (1994). *P. nemaeus* also has been reported to occur on Hainan, but Groves (1970a) stated that there is no evidence for its existence there. *Rhinopithecus* (see account thereof) sometimes is treated as a subgenus of *Pygathrix*.

Head and body length is usually 610–762 mm and tail length is 558–762 mm. *Pygathrix* is among the most strikingly colored of all mammals. It is marked with sharply contrasting patches and areas of color. The head is brown with a bright chestnut band below the ears. The body is mottled grayish, darker above than below, and the rump, tail, and forearms are white. The upper parts of the arms and legs, and the hands and feet, are black. In the subspecies *P. n. nemaeus* the face is bright yellow, the whiskers are white, and the lower legs are reddish chestnut. In *P. n. nigripes* the face is dark and the lower legs are black. One or

two other distinctively colored subspecies may exist (Wirth, Adler, and Thang 1991), though Jablonski (1995) recognized only *nemaeus* and *nigripes*.

The douc langur inhabits tropical rainforests from sea level to 2,000 meters (IUCN 1972). Most of the available information about this monkey was obtained by Lippold (1977), who studied *P. nemaeus* both in captivity in the San Diego Zoo and in the wild on Mount Son Tra, near Da Lat, Viet Nam, from June to August 1974. The species is diurnal and arboreal. It eats a wide variety of leaves and also certain abundant fruits. Early reports suggested a group size of 30–50 individuals, but recent observations indicate a usual range of 4–15. Groups contain 1 or more adult males and generally about twice as many adult females. Each sex apparently has its own dominance hierarchy, and males are dominant over females. In the wild a peak in births probably occurs between February and June, being correlated with maximum seasonal availability of fruit. The estrous cycle is 28–30 days long, and estimates of gestation have ranged from 165 to 190 days. An individual born at the San Diego Zoo was still living in December 1995 at an age of 25 years and 9 months (Marvin L. Jones, Zoological Society of San Diego, pers. comm., 1995).

The douc langur is classified as endangered by the IUCN and the USDI and is on appendix 1 of the CITES. There has been recent confirmation that both the subspecies *nemaeus* and *nigripes* survive in small numbers, but they still are being actively collected for the pet market and hunted for use as food (Eames and Robson 1993; Wirth, Adler, and Thang 1991). They also are thought to have declined because of environmental disruption brought on by military activity during the Viet Nam War. The population studied by Lippold has now disappeared (Eudey 1987).

PRIMATES; CERCOPITHECIDAE; **Genus RHINOPITHECUS**
Milne-Edwards, 1872

Snub-nosed Langurs

There are two subgenera and four species (Brandon-Jones 1984; Eudey 1987; Groves 1970a; Jablonski and Peng 1993):

subgenus *Presbytiscus* Pocock, 1924

R. avunculus, mountains of northern Viet Nam;

subgenus *Rhinopithecus* Milne-Edwards, 1872

R. roxellana, mountainous region on southeastern slopes of Tibetan Plateau in Chinese provinces of Hubei, Shaanxi, Gansu, Sichuan, and Yunnan;
R. bieti, Yun-ling Range of eastern Tibet and Yunnan;
R. brelichi, Fan-jin Range south of Middle Yangtze in Guizhou Province of China.

Groves (1970a, 1989, and *in* Wilson and Reeder 1993) considered *Rhinopithecus* not more than subgenerically distinct from *Pygathrix* on the basis of similarities in the skull and in limb proportions and in this regard was followed by Corbet and Hill (1991, 1992), Oates, Davies, and Delson (1994), and various other authorities. However, exhaustive analyses of characters of the skeleton, internal anatomy, and pelage of many specimens by Jablonski (1995) and Jablonski and Peng (1993) supported the validity of the above arrangement. A study of cranial morphology by Peng, Pan, and Jablonski (1993) even suggested that *Pres-*

Douc langur *(Pygathrix nemaeus)*, photo by Bernhard Grzimek.

bytiscus might best be treated as a full genus with closest affinity to *Pygathrix*.

Known measurements for head and body length and tail length, respectively, are as follow: *R. roxellana*, 570–760 mm and 510–720 mm; *R. bieti*, 740–830 mm and 510–720 mm; *R. brelichi*, 730 mm and 970 mm; and *R. avunculus*, 510–620 mm and 660–920 mm. The males of *R. roxellana* are grayish black on the top of the head, nape, shoulders, upper parts of the arms, back, and tail (the back being more or less overlaid with long silvery hairs). The forehead, sides of the head, sides of the neck, and underparts are golden. The coloration of the females is much the same except that the head and upper parts are brownish black. *R. bieti* has mostly blackish gray upper parts, long yellow-gray hairs scattered on the shoulders, and white underparts. *R. brelichi* has a slaty gray back with a patch of white in the midline between the shoulders. The crown is yellowish, the forelimbs are yellowish and whitish, the hind limbs are grayish, and the belly is gray. In *R. avunculus* the forehead

and cheeks are creamy, the sides of the neck are orange buff, and the back and limbs are black. There is a buffy white patch on the rump on either side of the tail. The hairs are long, measuring 150–80 mm in some forms. A crest of hair may be present on the crown.

The nose is flat in *Rhinopithecus*, whereas it is turned upward in *Pygathrix*. According to Napier and Napier (1985), who also regarded the two as generically distinct, the nose of *Rhinopithecus* is strongly retroussé and set well back from the large rounded muzzle. The nostrils are widely open and face forward, and two flaps of skin on the upper rim of the nostrils form two little peaks that almost touch the forehead. In *Pygathrix* these flaps lie flat.

Very little is known about snub-nosed langurs. *R. avunculus* lives at relatively low elevations in tropical monsoon forests. The other three species inhabit high mountain forests up to about 4,500 meters but may descend to lower elevations in winter. Part of their range is covered by snow for more than half the year. Recent studies in China (Chen

Snub-nosed langurs *(Rhinopithecus roxellana)*, photos by Wang Sung.

et al. 1983; Happel and Cheek 1986; Kirkpatrick 1995; Li et al. 1982; Long et al. 1994; Mu and Yang 1982; Tan 1985) have added to our knowledge. Most activity takes place in trees, but some feeding and social interaction occurs on the ground. If frightened, the animals flee through the upper levels of the canopy at great speed. Normal daily foraging movement is about 1,000–1,500 meters. The diet seems to consist largely of leaves, grass, lichens, bamboo shoots, buds, and fruits. Troops of more than 100 and even up to 600 individuals have been reported, though such groups

may be divided into many units, each containing one dominant male, 3–5 females, and young. The male secures his position by fighting other males. The social pattern is thought to be comparable to the fission and fusion system of *Pan troglodytes*. Groups have nonoverlapping home ranges of about 4–50 sq km, through which they move about once every month. If two groups meet, they quickly go off in different directions. The most common vocalization is "ga-ga," uttered loudly upon the discovery of a rich food source. Births take place during the spring and summer, following a reported gestation period of about 200 days. The number of young is usually 1, occasionally 2. Males reach sexual maturity at 7 years, females at 4–5 years.

Formerly the pelage of *Rhinopithecus* was believed to ward off rheumatism and could be worn only by Manchu officials. There still is intensive hunting to obtain pelts and other parts for supposed medicinal purposes. In some cases entire communes have joined in mass roundups of these monkeys in China (Tan 1985). Environmental disruption also is a problem. The IUCN classifies *R. roxellana* as vulnerable, *R. bieti* and *R. brelichi* as endangered, and *R. avunculus* as critically endangered. All species now are listed as endangered by the USDI and are on appendix 1 of the CITES. *R. avunculus* evidently has declined because of military activity and hunting. It occupies a restricted habitat, and fewer than 300 individuals are thought to survive (Nisbett and Ciochon 1993). Population sizes of the other species are estimated at 10,000–15,000 individuals for *R. roxellana*, 2,000 for *R. bieti*, and 800–1,200 for *R. brelichi* (Bleisch 1991; Eudey 1987; Long et al. 1994; Wang and Quan 1986). Eudey (1987) rated *R. avunculus*, *R. bieti*, and *R. brelichi* among the four primate species in Asia with the highest priority for conservation action (the fourth is *Simias concolor*).

PRIMATES; CERCOPITHECIDAE; **Genus PRESBYTIS**
Eschscholtz, 1821

Langurs, or Leaf Monkeys

There are eight species (Aimi and Bakar 1992; Chasen 1940; Corbet and Hill 1992; Eudey 1987; Fooden 1976a; Lekagul and McNeely 1977; Medway 1977; Weitzel and Groves 1985):

P. femoralis, Malay Peninsula (including southern
    Thailand), Riau Archipelago, east-central Sumatra,
    western Borneo, North Natuna Islands;
P. melalophos, southern and west-central Sumatra;
P. thomasi, northern Sumatra;
P. potenziani, Siberut, Sipura, North Pagai, and South
    Pagai in the Mentawai Islands off western Sumatra;
P. comata, western Java;
P. hosei, northern and eastern Borneo;
P. frontata, Borneo;
P. rubicunda, Borneo and nearby Karimata Island.

Several species formerly assigned to *Presbytis* now are placed in the genera *Semnopithecus* and *Trachypithecus* (see accounts thereof). The species listed above are the same as those accepted by Corbet and Hill (1992) and Groves (*in* Wilson and Reeder 1993). Oates, Davies, and Delson (1994) included *P. femoralis* in *P. melalophos*. Brandon-Jones (1993) considered *P. thomasi* and *P. hosei* to be subspecies of *P. comata*.

Head and body length is 420–610 mm, tail length is

Stripe-crested langur *(Presbytis comata)*, photo by Russell A. Mittermeier.

500–850 mm, and weight is 5.0–8.1 kg (Brandon-Jones 1984). The upper parts generally are brownish, grayish, or blackish, and the underparts are paler. Some forms have light-colored markings on the head or stripes on the thighs. Reddish and whitish mutants also occur. The newborn are generally whitish. This genus is distinguished by poorly developed or absent brow ridges, prominent nasal bones, and a crest on the crown, which takes a variety of shapes in the different species (Brandon-Jones 1984; Napier and Napier 1967). The body is slender, the tail is long, the hands are long and slender, and the thumb is small but the other fingers are well developed and strong.

Langurs are diurnal, arboreal forest dwellers (Medway 1970). In Borneo *P. hosei* and *P. rubicunda* are found in tall and, less abundantly, secondary forests (Payne, Francis, and Phillipps 1985). They are active throughout the tree canopy, occasionally descending to the ground to visit natural mineral sources. Daily foraging range is about 500–800 meters; although the diet includes a substantial amount of foliage, the greater part consists of fruits and seeds (Bennett and Davies 1994).

MacKinnon (1986) listed population densities of 2–30/sq km for the various species in Indonesia. Ruhiyat (1983) found densities of *P. comata* to be 11–12/sq km in one part of Java and 35/sq km in another. Respective home range sizes were 35–40 ha. with little overlap and 14 ha. with much overlap. On the Malay Peninsula, Chivers (1973) found group territories of about 13 ha. for *P. femoralis*. According to Bennett and Davies (1994), a group of *P. rubicunda* occupied about 84 ha. of forest in Borneo, but ranges usually are smaller and mostly constitute actively defended territories. Groups usually include a single adult male and two or more adult females, but *P. potenziani* sometimes is found in monogamous pairs. Heterosexual group size averaged about 15 individuals for *P. melalophos* on the Malay Peninsula, 7 for *P. rubicunda* on Borneo, and 6 for *P. thomasi* on Sumatra; there also are lone males and small all-male units. Outside males may attempt to take over a heterosexual group or split off some of its females. Juvenile males disperse from their natal groups. According

to Medway (1978), the alarm call of *P. femoralis* is a harsh rattle followed by a loud "chak-chak-chak-chak." There normally is a single young. A specimen of *P. melalophos* lived in captivity for 18 years and 4 months (Marvin L. Jones, Zoological Society of San Diego, pers. comm., 1995).

The IUCN classifies *P. comata* as endangered, noting that it has declined by at least 50 percent in the last decade because of habitat destruction and that fewer than 2,500 mature individuals survive. The IUCN also classifies *P. potenziani* as vulnerable and *P. femoralis* and *P. thomasi* as near threatened. *P. potenziani* is listed as threatened by the USDI and is on appendix 1 of the CITES. It is endemic to the Mentawai Islands, which are now being logged. It also is hunted extensively for food by the increasing human population. An estimated 45,000 individuals occupy the remaining 4,500 sq km of suitable habitat (MacKinnon 1986). All other species of *Presbytis* are on appendix 2 of the CITES. Most are suffering to some extent because of habitat loss. Khan (1978) estimated that populations of *P. femoralis* in peninsular Malaysia declined from 962,000 in 1958 to 554,000 in 1975. Eudey (1987) designated *P. comata* as highly endangered because of deforestation and fragmentation of remnant populations. It also now is listed as endangered by the IUCN.

PRIMATES; CERCOPITHECIDAE; Genus SEMNOPITHECUS
Desmarest, 1822

Hanuman Langur

The single species, *S. entellus,* is found in extreme southern Tibet, Nepal, Sikkim, northern Pakistan, Kashmir, India, Bangladesh, and Sri Lanka (Corbet and Hill 1992; Ellerman and Morrison-Scott 1966). Although this species often is considered to be simply a distinctive member of the genus *Presbytis,* Eudey (1987), based on studies by V. Weitzel and C. P. Groves, regarded it as the representative of a separate genus. Eudey did not accept Brandon-Jones's (1984) recognition of *S. hypoleucos* of southwestern India as a species distinct from *S. entellus. Semnopithecus* was treated as a monotypic genus by Groves (1989 and *in* Wilson and Reeder 1993) and Oates, Davies, and Delson (1994) but was considered by Corbet and Hill (1992) to include *Trachypithecus* (see account thereof).

According to Brandon-Jones (1984), head and body length is 410–780 mm, tail length is 690–1,080 mm, and weight is 5.4–23.6 kg. The upper parts are various shades of gray, brown, or buff, and the crown and underparts are white, orange-white, or yellowish. The coat of the newborn is blackish brown. From *Presbytis, Semnopithecus* is distinguished by its prominent, shelflike brow ridges. Napier and Napier (1967) stated also that the brow hairs are directed forward, and the crown hairs backward, by a whorl on the forepart of the crown. Postcranial morphology is like that of *Presbytis.*

The Hanuman langur is found in many environments, from desert edge to wet tropical forest and alpine scrub, and from sea level to more than 4,000 meters. In some areas where trees are scarce it has adapted well and spends most of its time on the ground (Oppenheimer 1977). On the ground it walks or runs on all four feet. In the trees it is remarkably agile; it can make horizontal leaps of 3–5 meters or cover up to 13 meters with some loss of height (Roonwal and Mohnot 1977). It is generally most active in the early morning and late afternoon, and it sleeps at midday. Like all langurs, it is almost entirely a vegetarian; leaves make up most of its diet, but fruits, flowers, and cultivated crops are also eaten. Groups may forage over several kilometers in the course of a day.

Population densities of up to 904/sq km have been reported (Roonwal and Mohnot 1977), but more usual density seems to range from about 3/sq km in grass and cropland to 130/sq km in forest (Oppenheimer 1977). Home range is about 0.05–13.00 sq km for groups made up of both males and females and 7–22 sq km for all-male groups. Bennett and Davies (1994) listed home range sizes of 2–12 sq km in the Himalayas but only 17–18 ha. in Sri Lanka. A portion of the home range is a core area where most time is spent, but there may be much overlap of entire ranges and there appears to be no strictly territorial defense. The most outstanding vocalization is a booming whoop, frequently given in a morning chorus and apparently functioning to space groups.

Social structure is variable (Bennett and Davies 1994; Oppenheimer 1977; Roonwal and Mohnot 1977). In areas where populations are below carrying capacity of the habitat, groups containing both sexes may include more than a single adult male. In areas where the environment is subject to at least seasonal stress heterosexual troops may have only a single adult male, and there may also be separate all-male groups. Reported size for heterosexual groups is 5–125 individuals, but perhaps the larger figures involve temporary aggregations of several social units. The more normal group size range is 13–37, and there are usually about 2 adult females for each adult male. Additional males may live alone or in groups of 2–32 individuals, including adults, subadults, and a few juveniles. Intragroup structure is generally stable once the males establish a dominance hierarchy through fighting. Both males and females some-

times migrate between heterosexual groups, but males do so more often. The leading male may eventually be ousted from his position by a younger male, or the younger individual may succeed in splitting off some of the females to form a new group of his own. Females also have a rank order, but it is not pronounced, and a dominant male is the leader and defender of the group. One male is known to have remained in control of a group for 10 years. Relations between heterosexual troops are generally peaceful, but in some areas these troops are subject to attack from all-male groups. If an outside male succeeds in defeating the dominant male of a heterosexual group, his next action may be to kill the infants of the group. This measure will bring the females into estrus within two weeks and allow the new group leader to mate and father progeny of his own.

Births may occur at any time of the year but are concentrated in the dry season during April–May in northern India and in December–March in southern India. There is much intergroup movement of males during the breeding season, and birth periods in groups have been reported to shift or to cease altogether in the face of persistent attacks by outside males (Bennett and Davies 1994; Oppenheimer 1977). In western India there is a definite season of births, November–March, and in Sri Lanka the young are born from March to May and in September (Roonwal and Mohnot 1977). The estrous cycle is about 24 days long (Winkler, Loch, and Vogel 1984). Estrus lasts 5–7 days (Hayssen, Van Tienhoven, and Van Tienhoven 1993). If an infant is lost, cycles can resume within 8 days (Oppenheimer 1977). The normal interbirth interval is about 15–24 months, and the gestation period is about 190–210 days (Roonwal and Mohnot 1977). There usually is a single young, but twins are said to occur frequently in the Hi-

Hanuman langur *(Semnopithecus entellus)*, photo from Zoological Society of San Diego.

malayas (Roonwal and Mohnot 1977). A birth weight of 500 grams was listed by Hayssen, Van Tienhoven, and Van Tienhoven (1993). Weaning is completed at about 10–12 months. Adulthood is reached by females at 3–4 years and by males at 6–7 years. Subadult males may be driven from a group by the dominant male. A captive specimen was still alive after 25 years (Marvin L. Jones, Zoological Society of San Diego, pers. comm., 1995).

The Hanuman langur is classified as endangered by the USDI and as near threatened by the IUCN and is on appendix 1 of the CITES. Although considered the most widely distributed nonhuman primate in India, it is declining through loss of habitat to agriculture and forest clearance. It is regarded as sacred to the Hindu religion, but its raids on crops may no longer be tolerated in the face of human population increases and food shortages. The total population in India is estimated at only 233,800 individuals (Southwick and Lindburg 1986).

PRIMATES; CERCOPITHECIDAE; **Genus**
TRACHYPITHECUS
Reichenbach, 1862

Brow-ridged Langurs, or Leaf Monkeys

There are two subgenera and nine species (Chasen 1940; Corbet and Hill 1992; Ellerman and Morrison-Scott 1966; Eudey 1987; Fooden 1976*a;* Lekagul and McNeely 1977; Medway 1977; Morris 1991; Weitzel and Groves 1985):

subgenus *Kasi* Reichenbach, 1862

T. vetulus, Sri Lanka;
T. johnii, southwestern India;

subgenus *Trachypithecus* Reichenbach, 1862

T. geei, Nepal, Bhutan, northeastern India;
T. pileatus, Assam, Bangladesh, western and northern Burma;
T. phayrei, eastern Assam and Bangladesh, Burma, Thailand, Laos, northern Viet Nam, extreme western Yunnan (southern China);
T. francoisi, southern Guangxi (southeastern China), northern Viet Nam, west-central Laos;
T. cristatus, eastern Assam, parts of southeastern Burma and southern Thailand north of Malay Peninsula, southern Indochina, western mainland Malaysia, Riau and Lingga archipelagoes, Sumatra, Bangka, Java, Bali, Borneo, Belitung Island, South Natuna Islands;
T. auratus, Java, Bali, Lombok;
T. obscurus, Malay Peninsula (including extreme southern Burma and Thailand) and nearby islands.

Trachypithecus often is considered a subgenus or synonym of *Presbytis* (or of *Semnopithecus*) but was recognized as a separate genus by Eudey (1987), Groves (1989 and *in* Wilson and Reeder 1993), Oates, Davies, and Delson (1994), and Weitzel (1992). Corbet and Hill (1992) included *Trachypithecus* in *Semnopithecus.* All of those authorities accepted the species listed above and did not accept the following views set forth by Brandon-Jones (1984): that *T. leucocephalus* of southwestern Guangxi and *T. delacouri* of northern Viet Nam are species distinct from *T. francoisi;* that populations of *T. cristatus* and *T. phayrei* in southern

China and northern Burma and Indochina actually represent another species, *T. barbei;* and that the remaining populations of *T. phayrei* are referable to *T. cristatus.* The species *T. vetulus* sometimes has been called *T. senex.*

Head and body length is 400–760 mm, tail length is 570–1,100 mm, and weight is 4.2–14.0 kg (Brandon-Jones 1984). The overall coloration is usually brown, dark gray, or black. Most species have various white, yellowish, or black markings on the head, shoulders, rump, limbs, or tail.

T. obscurus and *T. phayrei* have conspicuous white circles around the eyes. Some species have a pointed crest on top of the head. The coat of the newborn is bright yellow or orange-red. From *Presbytis, Trachypithecus* is distinguished by its prominent brow ridges, which resemble raised eyebrows (Brandon-Jones 1984). The postcranial morphology is like that of *Presbytis,* but the thumb is particularly short (Napier and Napier 1985) and the hind limbs are relatively shorter (Oates, Davies, and Delson 1994).

The well-studied species *T. johnii* inhabits upland forests and is more arboreal than *Semnopithecus entellus.* It may shift its range in the course of a year in response to the fruiting and flowering seasons of various plants. It is also known to raid potato and cauliflower crops and occasionally to feed on insects (Poirier 1970; Roonwal and Mohnot 1977). The species *T. pileatus* is considered entirely arboreal and apparently never leaves the trees except when forced to find water during the dry season (Roonwal and Mohnot 1977). In a study of that species in Bangladesh, Stanford (1992) found groups to move an average of 325 meters per day in moist deciduous forest habitat and 485 meters in wet semievergreen forest; the respective population densities were 53/sq km and 13/sq km, and the diet consisted mostly of leaves and fruit. *T. phayrei* feeds almost entirely on leaves and also eats some flowers and fruits but never was observed to invade cultivated fields (Mukherjee 1982). Bennett and Davies (1994) reported that foliage amounts to about 60 percent of the diet for *Trachypithecus.*

Population densities of 116–327/sq km have been reported for *T. vetulus,* of Sri Lanka. According to MacKinnon (1986), about 2 million *T. cristatus* occur in 133,167 sq km of suitable habitat in Indonesia, at a density of 15/sq km. Group home range averages smaller in *T. johnii* than in *Semnopithecus,* about 5–50 ha., but territories are maintained through the vigilance, loud vocalizations, and visual displays of the adult males (Bennett and Davies 1994; Poirier 1970). No intergroup aggression has been noted in *T. geei,* but *T. vetulus* is highly territorial and adult males defend the entire home range of just 1–10 ha. through powerful calls and spectacular displays of running and leaping (Roonwal and Mohnot 1977). *T. vetulus* usually lives in cohesive groups of about 8 individuals, including a single adult male, but on average an outside male takes over the group every three years; at such times there is much violent behavior, with the new male driving out or killing all the young in the group (Bennett and Davies 1994). On the Malay Peninsula, Chivers (1973) found group territories of 5–12 ha. for *T. obscurus.*

In *T. johnii* reported heterosexual group size is 6–30 and most units contain only a single adult male. Additional males live alone or in small groups. There is a strong, stable dominance hierarchy among the females, who appear to be responsible for most group socialization. Intragroup relations are generally peaceful, but the leading male has frequent territorial encounters with the males of other groups and may eventually be ousted by an outside male (Bennett and Davies 1994; Poirier 1970; Roonwal and Mohnot 1977). Observations of most other species of *Trachypithecus* suggest a comparable group size and social structure (Bennett

Brow-ridged langurs *(Trachypithecus obscurus)*, photo from Amsterdam Zoo.

and Davies 1994; Chivers 1973; Medway 1970; Mukherjee 1982; Roonwal and Mohnot 1977; Stanford 1992).

No seasonality has been reported for *T. johnii* or *T. geei*. For *T. vetulus* Rudran (1973) reported that in one area of Sri Lanka high rainfall and abundant food stimulated a peak mating period in October–January that resulted in a peak birth season in May–August. In another area, where rainfall and food supplies were high all year, no reproductive peaks were observed. Lekagul and McNeely (1977) stated that births in wild *T. obscurus* usually occur from January to March and that the estrous cycle lasts about 3 weeks. The normal interbirth interval is 2 years. Gestation is 200–220 days in *T. vetulus* (Rudran 1973) and 140–50 days in *T. obscurus* (Lekagul and McNeely 1977). Litter size is usually 1 in the genus, but twins are born occasionally. Hayssen, Van Tienhoven, and Van Tienhoven (1993) listed a birth weight of 113 grams for *T. obscurus* and an age at sexual maturity of around 3–4 years for *T. vetulus*. A captive *T. cristatus* lived to 31 years and 1 month (Marvin L. Jones, Zoological Society of San Diego, pers. comm., 1995).

Some species of *Trachypithecus* are of serious concern from a conservation standpoint. The USDI lists *T. geei, T. pileatus,* and *T. francoisi* as endangered and *T. vetulus* as threatened. The IUCN classifies *T. delacouri* (recognized as a species distinct from *T. francoisi;* see above) as critically endangered, *T. poliocephalus* (recognized as a species distinct from *T. francoisi;* see below) as endangered, *T. vetulus, T. johnii, T. pileatus, T. francoisi,* and *T. auratus* as vulnerable, and *T. cristatus* as near threatened. In addition, *T. geei* and *T. pileatus* are on appendix 1 of the CITES and all other species are on appendix 2. *T. johnii* is restricted to a small part of southern India, where its range continues to shrink through deforestation and other environmental disturbances. Most other species of *Trachypithecus* also are suffering to some extent because of habitat loss (Wolfheim 1983). Khan (1978) estimated that populations in peninsular Malaysia declined from 1958 to 1975 as follows: *T. cristatus,* from 6,000 to 4,000 individuals, and *T. obscurus,* from 305,000 to 155,000. Eudey (1987) designated *T. francoisi* as endangered and noted that the subspecies *T. f. delacouri* of central Viet Nam may be the most endangered monkey in Asia. Fewer than 250 mature individuals are thought to survive. Weitzel (1992) noted that *T. f. poliocephalus* is now restricted to Cát Bà Island, just off Haiphong, and Canh and Campbell (1994) estimated that fewer than 200 individuals survived in a protected national park on the island. Tan (1985) reported that the total population of *T. francoisi leucocephalus* of Guangxi

Brow-ridged langur *(Trachypithecus cristatus)*, photo from New York Zoological Society.

Province, southeastern China, had fallen to only about 400, at least in part because of hunting by people who believe its parts have medicinal value.

PRIMATES; CERCOPITHECIDAE; **Genus COLOBUS**
Illiger, 1811

Black-and-white Colobus Monkeys

There are five species (Dandelot *in* Meester and Setzer 1977; Oates and Trocco 1983; Rahm 1970a):

C. satanas (black colobus), southern Cameroon, Equatorial Guinea, Gabon, island of Bioko (Fernando Poo);
C. angolensis, northern Angola, Zaire, Uganda, Rwanda, Burundi, southern Kenya, Tanzania, northern Zambia;
C. polykomos, Gambia to Ivory Coast;
C. vellerosus, Ivory Coast to southwestern Nigeria;
C. guereza (guereza), eastern Nigeria to Ethiopia and Tanzania.

Some authorities regard *Piliocolobus* and/or *Procolobus* (see accounts thereof) as subgenera of *Colobus.* Most recent authorities, including Groves (*in* Wilson and Reeder 1993), have regarded *C. vellerosus* as a subspecies of *C. polykomos,* but Oates and Trocco (1983), on the basis of vocalizations and morphology, considered *C. vellerosus* to be a full species most closely related to *C. guereza.* Groves, Angst, and Westwood (1993) explained further that the supposed subspecies *dollmani* between the Sassandra and Bandama rivers in Ivory Coast, which has been reported to be morphologically intermediate to *C. polykomos* and *C. vellerosus,* actually represents a hybrid swarm formed when the latter species extended its range westward, genetically swamping the original population of *C. polykomos.*

Head and body length is 450–720 mm, tail length is 520–1,000 mm, and weight is 5.4–14.5 kg (Brandon-Jones 1984). *C. satanas* is entirely black; the other four species have some white markings. In *C. polykomos* the whiskers and chest are white, the rest of the body is black, and the tail is entirely white and not tufted. In *C. vellerosus* most of the body is black, but there are white thigh patches, the face is framed in a white mane, and the tail is entirely white and not tufted. In *C. guereza* there is a white beard, a con-

Colobus monkeys *(Colobus guereza)*, photo from San Diego Zoological Garden. Inset: *C. guereza*, photo from *Bull. Amer. Mus. Nat. Hist.*

Guereza *(Colobus guereza)*, photo by Bernhard Grzimek.

spicuous white mantle extending from the shoulders to the lower back, and a large white tuft on the end of the tail. And in *C. angolensis* there are long white hairs around the face and on the shoulders, and the tail ends in a small white tuft. Infants are pure white at birth but gain the full adult markings by about 3.5 months of age (Napier and Napier 1985).

Colobus has a slender body, a long tail, and prominent rump callosities. As in the langurs of Asia, a complex stomach is present and cheek pouches are absent. The thumb, which is small but present in the langurs, is suppressed and represented only by a tubercle in *Colobus*. The skull is somewhat prognathous, and the orbits are oval with narrow superciliary ridges (Napier and Napier 1967). The nostrils are more or less lengthened by an extension of the nasal skin and may nearly reach the mouth (Dandelot *in* Meester and Setzer 1977)

Colobus monkeys are generally diurnal, highly arboreal residents of deep forests, but there is some variation. *C. guereza* may be found in dry, moist, or riparian forest either in lowlands or up to 3,300 meters; it is most abundant in secondary forest or along rivers. Where trees are not densely packed it frequently feeds and travels on the ground (Oates 1977*b*, 1977*c*). *Colobus* spends much of the day resting, and groups generally have a daily foraging path of only about 500 meters (Oates 1994). The diet consists mostly of leaves, but fruits and flowers are at least seasonally important. Population densities of 100–500/sq km have been reported for *C. guereza* (Oates 1977*a*).

According to Oates (1994), most populations of *Colobus* live in relatively small social groups, normally with a single fully adult male and 2–6 adult females. Groups with more than one male tend to be larger. Temporary aggregations of more than 300 *C. angolensis* have been observed. Permanent groups use a home range of up to 84 ha., at least part of which is a defended territory.

Oates (1977*b*, 1977*c*) found *C. guereza* to live in small, highly cohesive social groups. There were usually 3–15 individuals per group, averaging about 9 in large forest blocks and 7 in riparian forest and small patches of forest. Most groups had a single fully adult male, but sometimes more than 1 male was present. There generally were also 3–4 adult females, plus young. The female membership seemed stable, but adult males sometimes were ousted by younger males that had either grown up within the group or moved in from the outside. Intragroup relations normally were friendly and were reinforced by considerable mutual grooming. Infants often were handled by individuals other than the mother. Relatively small home ranges were utilized, that of one group being about 15 ha. Either the entire home range or a core area therein was vigorously defended as a territory. Intergroup relations were usually agonistic, with most hostility being expressed by the adult males through gestures, vocalizations, displays of leaping, and occasional chasing and fighting. Some groups, however, did share water holes and other resources. There were loud nocturnal and dawn roaring choruses by adult males, probably as a means of spacing groups. McKey and Waterman (1982) reported an absence of early morning roaring in *C. satanas*. The annual home range in that species was found to be about 60 ha., and groups averaged about 15 individuals.

There seems to be little or no reproductive seasonality in most populations of *Colobus* that have been studied. Sabater Pi (1973), however, reported that in Equatorial Guinea there apparently was a birth season in *C. satanas* extending from December to early April, coincident with the period of greatest fruit production. Oates (1977*b*) stated that each adult female *C. guereza* produced one young every 20 months. Hayssen, Van Tienhoven, and Van Tienhoven (1993) listed an estrus of 1–3 days, a gestation period of about 175 days, a birth weight of 820 grams, a lactation period of 6 months, and an age at sexual maturity of about 2 years for *C. polykomos*. Oates (1977*c*) determined that the age of full sexual maturity in *C. guereza* was at least 6 years for males and 4 years for females. C. Hill (1975) reported that a specimen of *C. polykomos* died after spending 23 years and 6 months in the San Diego Zoo and that the animal probably had been brought into captivity when it was 4–7 years old.

Oates (1977*b*) reported that all species of *Colobus* had declined drastically over the last 100 years, though *C. guereza* was not as seriously threatened as *C. polykomos* and *C. satanas*. The decline was caused initially by the international fur trade in the nineteenth century and subsequently by rapid human population growth and habitat destruction in Africa. All species are on appendix 2 of the CITES; the IUCN classifies *C. satanas, C. vellerosus,* and *C. angolensis ruwenzorii* (Rwanda, Uganda, eastern Zaire) as vulnerable and *C. polykomos* as near threatened; and the USDI lists *C. satanas* as endangered. This last species, according to Lee, Thornback, and Bennett (1988), has the most limited distribution of any species in the genus and is declining as a result of hunting for its meat and skin and loss of habitat to logging and agriculture. It seems unable to survive in secondary forest. According to Oates (1996), both *C. satanas* and *C. vellerosus* are among the most threatened primate species in Africa.

PRIMATES; CERCOPITHECIDAE; **Genus PILIOCOLOBUS**
Rochebrune, 1887

Red Colobus Monkeys

Following Lee, Thornback, and Bennett (1988) and Oates (1985), five species are tentatively recognized here:

P. badius, Senegal to Ghana;
P. pennantii, southeastern Nigeria, Cameroon, Congo, Bioko (Fernando Poo);
P. rufomitratus, northern and eastern Zaire, Uganda, mouth of Tana River in eastern Kenya, western Tanzania;
P. gordonorum, Uzungwa Mountains in south-central Tanzania;
P. kirkii, Zanzibar.

There is considerable disagreement regarding the taxonomy of red colobus monkeys at both the generic and the specific level. Some authorities, including Dandelot (*in* Meester and Setzer 1977), consider *Piliocolobus* to be a subgenus of *Colobus*. Others, including Brandon-Jones (1984), Lee, Thornback, and Bennett (1988), and Oates (1985), place *Piliocolobus* in the genus *Procolobus* (see account thereof). Groves (1989 and *in* Wilson and Reeder 1993) treated *Piliocolobus* as a subgenus of *Procolobus*.

Struhsaker (1981) noted general agreement that there are 14 named kinds of the genus *Piliocolobus,* with different authorities accepting from 1 to 6 of those kinds as full species. Oates, Davies, and Delson (1994), Rahm (1970*a*), Wolfheim (1983), and various other recent authors have recognized only the single widely distributed species *P. badius*. Oates (1985), followed by Lee, Thornback, and Bennett (1988), thought the best arrangement to be ranking of *P. badius* as a "superspecies" that includes the 5 allopatric species listed above. Dandelot (*in* Meester and Setzer 1977) listed *P. badius, P. pennantii, P. kirkii, P. rufomitratus* (including *P. gordonorum*), and also *P. tholloni* of east-central Zaire, which is here considered a subspecies of *P. rufomitratus*. Dandelot also listed three other "potential species" in *Piliocolobus: P. ellioti* of the Ituri Forest in Zaire, *P. preussi* of the Nigeria-Cameroon border area, and *P. waldroni* of the Ivory Coast–Ghana region. Groves (*in* Wilson and Reeder 1993) recognized *P. badius, P. pennantii* (including *kirkii, gordonorum,* and *tholloni*), *P. preussi* (here considered a subspecies of *P. pennantii*), and *P. rufomitratus*. Corbet and Hill (1991) and Napier and Napier (1985) regarded only *P. kirkii* as a species distinct from *P. badius.*

Red colobus *(Piliocolobus badius)*, photo by Dawn Starin.

Head and body length is 450–670 mm, tail length is 520–800 mm, and weight is 5.1–11.3 kg (Brandon-Jones 1984). These monkeys have black, slaty, or brownish upper parts and red or chestnut brown arms, legs, and head; tail tufts and long fringes of hair are lacking. According to Napier and Napier (1967), the body form of *Piliocolobus* is similar to that of *Colobus*, but the fingers are relatively longer and the hallux shorter. The thumb tubercle, apparent in *Colobus* and *Procolobus*, is totally absent. The orbits of the skull are angular and have thick superciliary ridges with a distinct supraorbital groove or foramen, which is lacking in *Colobus* and *Procolobus*. Dandelot (*in* Meester and Setzer 1977) noted also that in contrast to *Colobus*, *Piliocolobus* has nostrils that are V-shaped and not lengthened by an extension of the nasal skin.

Most populations of red colobus are found in rainforests, but in Senegal and Gambia they often inhabit savannah woodland (Struhsaker 1975). Some inhabit mangroves and floodplains. They are diurnal and arboreal. Red colobus are generally considered to be clumsy climbers, but they only infrequently descend to the ground and seem to favor heights of 16.5–27.0 meters. They appear to lack a precise pattern of activity, and their day alternates between periods of feeding and resting. During the course of 54 days one group of *P. badius* had an average daily movement of 648.9 (222.5–1,185.0) meters (Struhsaker 1975). The diet consists principally of leaves and also includes some fruits and seeds. A population density of 80–100/sq km has been reported for *P. badius* (Nishida 1972a).

Apparently there is a pronounced difference in social structure between the genera *Colobus* and *Piliocolobus*. According to information collected or cited by Struhsaker (1975), the red colobus, like the black-and-white, form stable groups, but these range in size from 12 to 82 members, averaging about 50. There are usually several adult males and about 1.5–3 times as many adult females. One group of 75–80 individuals included 10 adult males and 21 adult females. There were also solitary animals, mostly immature males driven out of groups by older males. There is a pronounced dominance hierarchy within a group, expressed by priority access to food, space, and grooming. Most mating is done by the highest-ranking male. Rank order is maintained by aggressive interaction, but actual physical fighting is rare. Infants are handled by the mother alone until they are 1–3.5 months old. Red colobus have a great variety of vocalizations, perhaps the most outstanding of which is the alarm bark of the adult males. Reported group home range size has varied from about 8.5 to 132.1 ha., there evidently usually being about 1–2 ha. for each individual in the group. There is no territoriality, but the area being used at a given time may be defended through aggressive displays and vocalizations.

Reproductive seasonality is not generally evident but has been reported in Sierra Leone, where most births occur in the early dry season, from October to December (Oates 1994). Struhsaker (1975) gave the estimated gestation period of red colobus as 4.5–5.5 months, and Oates (1994) cited interbirth intervals of around 26 months. A single offspring is born. Studies of *P. badius* in Gambia (Starin 1981) indicate that the young of both sexes leave their natal group prior to maturity. Females depart voluntarily and are readily accepted into other troops, but males are forcibly expelled and then may again encounter hostility, and even be killed, when they try to join another group.

Struhsaker (1975) warned that red colobus monkeys were being adversely affected by timber exploitation, not only through loss of required forest habitat but also because logging made the animals more accessible to hunters. He considered red colobus to be the easiest monkeys in Africa

to hunt. The IUCN includes *Piliocolobus* in the genus *Procolobus* and recognizes only the single species *P. badius*, which it designates generally as near threatened. It also assigns the following additional designations (technical names adjusted in accordance with the taxonomic classification used above): *P. badius waldroni*, Ivory Coast and Ghana, critically endangered; *P. b. temmincki*, Senegal to northwestern Guinea, endangered; *P. pennantii pennantii*, island of Bioko (Fernando Poo), endangered; *P. p. epieni*, Nigeria, endangered; *P. p. preussi*, Cameroon, endangered; *P. p. bouvieri*, Congo, endangered; *P. rufomitratus rufomitratus*, floodplains of Tana River in eastern Kenya, endangered; *P. gordonorum*, endangered; and *P. kirkii*, endangered. *P. r. rufomitratus* and *P. kirkii* also are listed as endangered by the USDI and are on appendix 1 of the CITES; all other species and subspecies are on appendix 2. Oates (1996) reported that *P. badius waldroni* appeared to be on the verge of extinction because of intensive hunting and habitat destruction. Lee, Thornback, and Bennett (1988) expressed particular concern for the following: *P. r. rufomitratus*, numbers declined from about 1,860 in 1972 to 200–300 and range greatly reduced; *P. gordonorum*, extremely rare with a patchy distribution in a restricted range; *P. pennantii bouvieri*, probably found only along the Lefini River in southern Congo, numbers perilously low; *P. pennantii preussi*, eliminated from southeastern Nigeria and now confined to a small part of adjacent Cameroon, fewer than 8,000 survive; and *P. kirkii*, about 1,500 individuals remaining in isolated patches of forest, all of which are subject to cutting. More recently, Butynski and Mwangi (1995) reported that numbers of *P. r. rufomitratus* were somewhat higher, about 1,100–1,300, but that the total occupied habitat was considerably less than 13 sq km and was immediately jeopardized by planned construction of hydroelectric dams. The one population of *P. gordonorum* with a reasonable chance for long-term survival, comprising about 544 individuals, inhabits a 6-sq-km forested area, legal ownership of which is currently being disputed between commercial interests and the Selous Game Reserve (Decker 1994).

PRIMATES; CERCOPITHECIDAE; **Genus
PROCOLOBUS**
Rochebrune, 1887

Olive Colobus Monkey

The single species, *P. verus*, is found from Sierra Leone and southeastern Guinea to Benin; there also is an isolated population to the east of the Niger River in southeastern Nigeria, which suggests that the overall range once was more widespread (Napier and Napier 1985). *Procolobus* sometimes has been treated as a subgenus of *Colobus* but was regarded as a separate monotypic genus by Corbet and Hill (1991) and Napier and Napier (1985). In contrast, Brandon-Jones (1984), Oates (1985), and Lee, Thornback, and Bennett (1988) also placed the red colobus monkeys (here considered to belong to the genus *Piliocolobus*) in *Procolobus*. Oates, Davies, and Delson (1994), while recognizing "that the olive colobus is a very distinctive animal," treated *Piliocolobus* as a subgenus of *Procolobus*. Groves (1989 and *in* Wilson and Reeder 1993) also accepted the genus *Procolobus* with the two subgenera *Procolobus* and *Piliocolobus*.

Except as noted, the information for the remainder of this account was taken from Napier and Napier (1967, 1985). Head and body length is 430–90 mm, tail length is 570–640 mm, and weight is 2.9–4.4 kg. The upper parts are olive gray or olive brown and the underparts and limbs are grayish. There is a small longitudinal crest on the crown. The morphological characters of *Procolobus* are similar to those of *Colobus*, but in the former there are five cusps on the lower third molar tooth, while in the latter there are six. Unique among primates, the glans penis of *Procolobus* bears minute horny papillae (Oates, Davies, and Delson 1994). Dandelot (*in* Meester and Setzer 1977) noted also that like *Piliocolobus*, but in contrast to *Colobus*, *Procolobus* has V-shaped nostrils that are not lengthened by an extension of the nasal skin. Unlike either of the other genera, *Procolobus* has two partings of the frontal hairs separated by a median crest.

The olive colobus is restricted to rainforest and is diurnal and arboreal. It inhabits thickets and the low stratum of the canopy, often near riverbanks and swamps. It moves into the middle stratum to sleep but never ascends to the upper stratum. The diet is strictly vegetarian and probably consists mainly of leaves. A population density of 21/sq km has been reported (Davies 1994). One series of observations indicates that young foliage is eaten most often and that fruit and seeds are also important (Oates 1994). Groups contain 5–20 individuals, commonly 10–15, and there may be more than a single adult male. One group used a home range of about 28 ha. but did not appear to engage in active territorial defense (Oates 1994). The voice is little used, but there is a characteristic complex call, alto in pitch and ending in a scream. In Sierra Leone most matings have been seen to occur from March to August and very small infants were seen only from November to April, suggesting a distinct breeding season and a gestation period of about 6 months (Oates 1994). A newborn infant is carried in its mother's mouth for several months, a practice unique among monkeys and apes.

The olive colobus is classified as near threatened by the IUCN and is on appendix 2 of the CITES. Its presence in Nigeria was unknown until a single specimen was taken in 1967. Oates (1982) observed a few living individuals there but reported the population to be threatened by intensive hunting and habitat destruction. Anadu and Oates (1988) received reports of the possible presence of *Procolobus* in the Niger River Delta.

PRIMATES; **Family HYLOBATIDAE; Genus HYLOBATES**
Illiger, 1811

Gibbons, or Lesser Apes

The single living genus, *Hylobates*, contains 4 subgenera and 11 species (Chivers 1977; Chivers and Gittins 1978; Corbet and Hill 1992; Geissmann 1994; Groves 1972*b*, 1984, 1993; Groves and Wang 1990; Haimoff et al. 1982; Ma and Wang 1986; Marshall 1981*a*; Marshall and Sugardjito 1986; Prouty et al. 1983):

subgenus *Nomascus* Miller, 1933

H. concolor (crested gibbon), southeastern China, Hainan, extreme northwestern corner of Laos east of Mekong River, northern Viet Nam northeast of Song Bo and Song Ma rivers;
H. leucogenys, extreme southern Yunnan (southern China), northern Laos, northern Viet Nam southwest of the Song Bo and Song Ma rivers;
H. gabriellae, southern Laos, eastern Cambodia, central and southern Viet Nam;

subgenus *Symphalangus* Gloger, 1841

H. syndactylus (siamang), mainland Malaysia, Sumatra;

subgenus *Bunopithecus* Matthew and Granger, 1923

H. hoolock (white-browed gibbon), east of the Brahmaputra River in Assam and Bangladesh, Burma and adjacent border area of Yunnan (China);

subgenus *Hylobates* Illiger, 1811

H. lar (white-handed gibbon), extreme southern Yunnan (southern China), eastern and southern Burma, Thailand south to below the Isthmus of Kra, eastern and southern mainland Malaysia, northern Sumatra;
H. pileatus (capped gibbon), southeastern Thailand, extreme southwestern Laos, Cambodia;
H. agilis (dark-handed gibbon), extreme southern peninsular Thailand, northwestern mainland Malaysia, Sumatra except north, southwestern Borneo;
H. moloch (silvery gibbon), Java;
H. muelleri (gray gibbon), Borneo except southwest;
H. klossii (Kloss's gibbon), Siberut, Sipura, North Pagai, and South Pagai in the Mentawai Islands off western Sumatra.

This family often has been considered to be only a subfamily of the Pongidae, but anatomical, immunological, karyological, paleontological, and behavioral evidence now support its recognition as a separate though closely related family (Andrews and Groves 1976; Chivers and Gittins 1978). *Symphalangus* often is treated as a distinct genus, and Lekagul and McNeely (1977) treated both *Symphalangus* and *Nomascus* as full genera. All of the above species of *Hylobates* are allopatric, except that the range of *H. syndactylus* overlaps parts of those of *H. lar* and *H. agilis* (Corbet and Hill 1992), there evidently was an area of sympatry between *H. lar* and *H. pileatus* to the east of Bangkok (Geissmann 1991*b*), and *H. concolor* and *H. leucogenys* seem to occur at the same localities in parts of extreme southern Yunnan, northern Viet Nam, and northwestern Laos (Groves 1993; Groves and Wang 1990). Fooden, Quan, and Luo (1987) considered *H. leucogenys* to be only a subspecies of *H. concolor*, but both it and *H. gabriellae* were recognized as distinct species by Corbet and Hill (1992) and Groves and Wang (1990). The species of the subgenus *Hylobates* readily interbreed in captivity, and small hybrid populations now are known at points of contact between some of these species in the wild, though there is no evidence of massive introgression (Brockelman and Gittins 1984; Marshall and Sugardjito 1986).

H. syndactylus is by far the largest species, with an armspread of as much as 1.5 meters. Its head and body length is 750–900 mm, there is no tail, and weight is 8–13 kg. The coat is longer and less dense than in the other species. The body and limbs are black, and there is a large gray or pink throat sac that is inflated during calls. This species also is distinguished from most other gibbons by a webbing uniting the second and third toes.

In the other species head and body length is 440–635 mm, there is no tail, and weight is usually 4–8 kg. Chivers and Gittins (1978) listed the following diagnostic features: *H. concolor*, spidery form, conical shape to crown, males black, females golden or gray brown; *H. hoolock*, longer hair than in species of subgenus *Hylobates*, males black with white eyebrows; *H. klossii*, both sexes completely black, like *H. syndactylus*, and sometimes with interdigital webbing, but much smaller and with shorter and denser

Gibbon (*Hylobates* sp.), photo by Bernhard Grzimek. Inset: *Hylobates agilis*, photo from West Berlin Zoo (Gerhard Budich).

hair; *H. pileatus,* males black with white hands and feet, females ash blond with black cap and chest; *H. muelleri,* coloration varied, from mouse gray to brown, cap and chest darker; *H. moloch,* both sexes blue gray, cap and chest darker; *H. agilis,* coloration varied, from very dark brown to light buff, often with reddish tinge, males with white brows and cheeks, females with white eyebrows; and *H. lar,* coloration varied, from black to light buff, hands and feet white. Groves and Wang (1990) added that in *H. leucogenys* the pelage is longer and coarser than in *H. concolor,* the males being black with some silvery hairs intermixed and with white patches on the cheeks and the females tending to be more richly colored than those of *H. concolor* and having no conical tuft on the crown. Corbet and Hill (1992) indicated that the pelage of *H. gabriellae* is fine, like that of *H. concolor,* but the males have pinkish cheeks and the females have only a short crown patch. Many gibbons are almost white at birth and do not attain their final color until two to four years old.

Prouty et al. (1983) characterized the four subgenera as follows: *Nomascus,* diploid chromosome number 52, late maturation color change in female, nasal bones flat, prominent laryngeal sac not in both sexes, male component of the "great call" vocalization concentrated at end; *Symphalangus,* diploid number 50, no late color change in female, nasal bones flat, prominent laryngeal sac in both sexes, male and female parts of great call interspersed; *Bunopithecus,* diploid number 38, late color change in female, nasal bones convex, no prominent laryngeal sac, male and female parts of great call interspersed; and *Hylobates,* diploid number 44, no late color change in female, nasal

bones flat, no prominent laryngeal sac, male component of great call concentrated at end (unless call is solo).

Gibbons resemble the Pongidae in lacking a tail and in having the same dental formula but differ in their much smaller size, more slender form, relatively longer arms, longer canine teeth, and the presence of buttock pads. Gibbons do not have the pronounced sexual dimorphism in size of body, skull, and teeth found in the great apes, but as noted above, several species of gibbons are sexually dichromatic. The thumb of *Hylobates* is unique among higher primates in having the basal part freed from the palm and extending from near the wrist, allowing a wide range of movement (Lekagul and McNeely 1977).

Hylobates means "dweller in the trees," and gibbons fully justify the name. They exceed all other animals in agility. They move primarily by brachiation, in which the long arms are extended above the head to suspend and propel the body. Gibbons do most of their traveling through the trees by swinging alternately from branch to branch, using the hands like hooks, not grasping the limb, and often making long swings, like leaps, in which neither arm supports them. By this method of rapid locomotion gibbons may cover 3 meters in a single swing. When traveling on a large limb or on the ground they usually walk upright with their arms held high for balance. Gibbons can leap from branch to branch for a distance of 9 meters or more. Roonwal and Mohnot (1977) reported that *H. hoolock* can swim well, but Marshall and Sugardjito (1986) stated that gibbons do not swim, avoid open water, and do not walk long distances on the ground. These characteristics were factors in the speciation and largely allopatric distribution of gibbons.

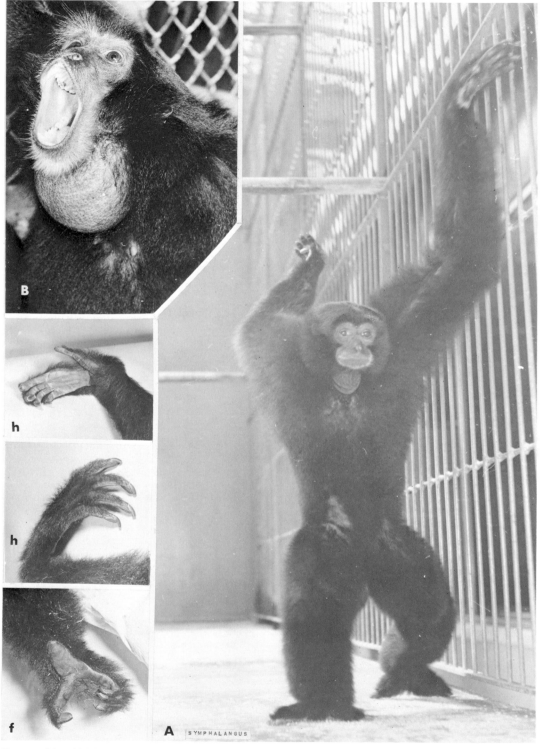

Siamang gibbon *(Hylobates syndactylus):* A. Photo by Ernest P. Walker; B. Head showing throat sac inflated while calling, photo by Marcel Langer. Hands and foot photos by Ernest P. Walker.

Crested gibbon *(Hylobates concolor)*, photo by B. E. Joseph.

Gibbons occur throughout the deciduous monsoon and evergreen rainforests of the islands and mainland of Southeast Asia (Chivers 1977, 1984). Most inhabit lowlands and hills, though *H. syndactylus* is sometimes found in submontane forests at altitudes of up to 1,800 meters. All species are primarily diurnal and almost exclusively arboreal. *H. syndactylus* is usually found at a height of 25–30 meters in the trees. It is active for about 10.5 hours a day, during which it alternates periods of feeding and rest. Its average daily movement is nearly 1 km, being more during the dry season and less in the wet season. At around 1600–1800 hours groups settle in a regularly used sleeping tree for the night (Chivers 1972, 1974, 1977). *H. lar* generally is seen higher in the trees than the siamang and moves faster and farther during the day. Srikosamatara (1984) listed the following mean distances (in meters) traveled per day: *H. lar,* 1,490; *H. agilis,* 1,150; *H. muelleri,* 890; *H. klossii,* 1,514; and *H. pileatus,* 833. About 50 percent of the diet of *H. syndactylus* consists of leaves, 40 percent is fruits, and the remainder consists of flowers, buds, and insects (Chivers 1977). The other species are mostly frugivorous but also eat leaves, other vegetable matter, a variety of insects, and perhaps eggs and small vertebrates. Raemaekers (1984) suggested that the difference in dietary requirements between *H. syndactylus* and the smaller gibbons has allowed sympatry but that the members of the latter group exclude one another geographically through resource competition.

The following account of social structure is based on the studies of Chivers (1972, 1974, 1977, 1984), Ellefson (1974), Tilson (1981), and Whitten (1982). All species of gibbons are monogamous. A mated adult pair and their offspring occupy a small, stable home range, most or all of which is defended as a territory. The usual group size observed is three or four individuals, but solitary animals are often seen. Apparently, the lone individuals are subadults that have been forced out of the family group and have not yet established a territory of their own.

There is a density of 0.8 groups of *H. syndactylus* per sq km, and reported densities of the smaller species range from about 0.7 to 6.5 groups per sq km. Surprisingly, densities are highest in the drier and more seasonal forests of the northern part of the range of *Hylobates,* where there are usually at least 4 groups per sq km; on Sumatra, Java, and Borneo there tend to be only 3 per sq km. Where two species are sympatric, as on the Malay Peninsula, there may be only 1 group of each species per sq km.

Groups of *H. syndactylus* include up to three immature animals, separated in age by two or three years. These groups are very cohesive, with the animals being an average of only eight meters apart during the day. Social bonds are reinforced by considerable mutual grooming. Departure of young siamangs from the group after they reach adolescence at around five or six years is brought about both by their motivation to mate and by the increasing intolerance of the adult pair. Young of both sexes begin to call and move separately from their parents and may spend several years searching for a mate. Social behavior in *H. lar* is much the same, except that groups seem somewhat less cohesive and may include up to four offspring. Adults of *H. klossii* may assist their offspring to obtain territories by accompanying them into unoccupied areas and intimidating potential rivals or by forcefully usurping land belonging to other families.

All species of *Hylobates* are strongly territorial. In most species about 75 percent of the group home range is defended. The overall average home range size for the genus is 34.2 ha., and modal territory size is about 25 ha. Home ranges are largest on the Malay Peninsula, being about 50 ha. there for *H. lar.* The smallest reported ranges are 22 ha. for *H. hoolock* in Assam and Bangladesh; 17 ha. for *H. moloch* in Java; and about 10–20 ha. for *H. klossii.* Average size for *H. syndactylus* is about 30 ha. Defense rarely involves physical contact, but groups range near the borders of their territories and call at, display to, and chase intruders. Displays include acrobatics and breaking of branches in the trees. Spacing of groups also apparently is accomplished through loud vocalizations given even when groups are not in contact. Such calls, which may carry for several kilometers to the human ear, are emitted on about 30 percent of days by *H. syndactylus* and on 80–90 percent of days by *H. lar.*

The calls or songs of gibbons have been the subject of considerable study, and pertinent information was summarized by Chivers and Gittins (1978). The vocalizations vary from one species to another, being valuable in identification, and also usually differ markedly between sexes. In some species males alone give a morning chorus, but group calls generally are composed of a duet between the male and female, with some accompaniment from their offspring. The female emits the longest and most distinctive sounds. The call of *H. klossii* is the most spectacular, with a long series of slowly ascending pure notes leading into a bubbling trill. In *H. syndactylus* there are alternate booms and barks, made louder by resonance across the greatly inflated throat sac. Cowlishaw (1992) hypothesized that the male-female duets are unrelated to pair bonding but are mainly for intimidation of neighbors; the female's song is associated strongly with territorial defense, the male's with keeping potential rivals away.

Siamang gibbon *(Hylobates syndactylus)*, photo by Bernhard Grzimek.

Roonwal and Mohnot (1977) stated that the young of *H. hoolock* are probably born from November to March, but otherwise there seems to be little information on reproductive seasonality in *Hylobates*. The interbirth interval is 2–3 years in *H. syndactylus* (Chivers 1977) and 2–2.5 years in *H. lar* (Ellefson 1974). The estrous cycle averages 28 days in *H. hoolock* and 30 days in *H. lar* (Roonwal and Mohnot 1977). Reported gestation periods are 230–35 days in *H. syndactylus* (Hill 1967), 7–7.5 months in *H. lar* (Roonwal and Mohnot 1977), and 7.5 months in *H. pileatus* (Lekagul and McNeely 1977). Normally a single young is born, and it initially clings around the mother's body like a belt. Weaning is gradual, not being completed in *H. lar* until the young is about 1 year and 8 months old (Ellefson 1974). In *H. syndactylus* early in the young animal's second year the father takes over most of its care. The age of full sexual maturity is about 8–9 years, though Geissmann (1991*a*) reported cases in which both sexes bred successfully in captivity at around 4–6 years. The animals in one pair of wild *H. syndactylus* were known to be nearly 25 years old (Chivers 1977). A captive specimen of *H. syndactylus* lived to about 40 years, and individuals of both *H. agilis* and *H. leucogenys* are reported to have lived in zoos for 44 years (Marvin L. Jones, Zoological Society of San Diego, pers. comm., 1995).

The following quotation from Rumbaugh (1972) on maintaining *Hylobates* in captivity may apply to a number of other primates as well: "Infant gibbons are a delight and can be tractable, loving and friendly. But not always! In-termixed with these treasured attributes are episodes of tantrums and blind fury that render them quite unpredictable. This unpredictability almost invariably increases with maturity and the day comes when as adults, equipped with formidable canines, they attack without either warning or obvious cause even their most trusted human affiliates."

Wild gibbons have suffered severe losses through the increase of people and consequent hunting and destruction of forest habitat. About 1,000 years ago gibbons, probably of the species *H. concolor*, could be found as far north as the Yellow River in eastern China. *Hylobates* now occurs in China only in the extreme southern border region and on the island of Hainan. The subspecies in the latter area, *H. concolor hainanus*, is restricted to a reserve of less than 2,000 ha. and may be on the verge of extinction (Haimoff 1984). Because of poaching and deforestation within the reserve the population declined from perhaps 30–40 individuals in the 1980s to only 15 in 1993 (Fooden, Quan, and Luo 1987; Zhang and Sheeran 1994). Bleisch and Nan (1990) estimated that 1,750–5,350 *H. concolor* survived in reserves on the mainland of China and noted that the species was otherwise extremely rare. Even the so-called reserves were subject to deforestation, mining activity, and hunting.

Populations in several other countries have been greatly reduced and fragmented. In his excellent summary of data on the natural history and conservation of gibbons Chivers (1977) listed the following estimated numbers: *H. concolor* (including *H. leucogenys* and *H. gabriellae*),

228,000; *H. syndactylus,* 167,000; *H. hoolock,* 532,000; *H. klossii,* 84,000; *H. pileatus,* 100,000; *H. muelleri,* 1,825,000; *H. moloch,* 20,000; *H. agilis,* 744,000; *H. lar,* 250,000; total, 3,950,000. He observed that while some species might remain numerous, present levels of logging and agriculture and of taking for food and commerce would lead within 15–20 years to the overall destruction of 80–90 percent of then current gibbon populations, with the extinction of *H. moloch* and possibly of *H. klossii, H. pileatus,* and *H. concolor.*

Brockelman and Chivers (1984) reported that the anticipated declines were being realized as forests were destroyed and opened to human access and that probably all species were headed eventually for relict status. Indeed, Kool (1992) thought that the habitat of *H. moloch,* of Java, had been reduced by 96 percent and that only 4,800 individuals survived, 1,800 of which were in protected areas. In peninsular Malaysia alone deforestation is resulting in an annual loss of 31,000 *H. agilis* and *H. lar* and 3,400 *H. syndactylus* (Chivers 1986). MacKinnon and MacKinnon (1987) provided the following revised estimates for some other species: *H. concolor* (including *H. leucogenys* and *H. gabriellae*), 131,000; *H. hoolock,* 169,134; and *H. pileatus,* 33,600. More recently, however, Eames and Robson (1993) calculated that the total world population of *H. concolor* (including *H. leucogenys* and *H. gabriellae*) could be as low as 10,500–14,000. All species of *Hylobates* are listed as endangered by the USDI and are on appendix 1 of the CITES. The IUCN now classifies *H. moloch* as critically endangered and *H. concolor* (including *H. leucogenys* and *H. gabriellae*) as endangered and estimates the surviving number of mature individuals at fewer than 250 for the former and 2,500 for the latter. The IUCN also designates *H. pileatus* and *H. klossii* as vulnerable and *H. syndactylus, H. lar, H. agilis,* and *H. muelleri* as near threatened.

Fossil evidence suggests that the Hylobatidae diverged from the Pongidae prior to the separation of the Pongidae and the Hominidae. Gibbonlike remains are known from the Oligocene and Miocene of Africa, the Miocene of Europe, and the Miocene, upper Pliocene, and Pleistocene of Asia. Possibly, however, the similarity of the earlier material represents parallel development rather than direct relationship (Chivers 1977; Fleagle 1984; Lekagul and McNeely 1977). The family Pliopithecidae of the Oligocene and Miocene sometimes is considered to have given rise to the Hylobatidae. Both Groves (1989) and Tyler (1991*b*) rejected that view but indicated that there are fossil genera from the same period in India and China that appear to be ancestral hylobatids.

PRIMATES; Family PONGIDAE

Great Apes

This family of three living genera and four species is found in equatorial Africa and on Sumatra and Borneo. Some authorities, such as E. R. Hall (1981), place the great apes in the family Hominidae, along with humans. Andrews and Cronin (1982) suggested an alternative classification by which the living Pongidae would be restricted to *Pongo* and the Hominidae would comprise two subfamilies: the Gorillinae, with *Pan* and *Gorilla;* and the Homininae, with *Homo* (and the fossil *Australopithecus*). In the arrangement of Groves (1986*a,* 1989, and *in* Wilson and Reeder 1993) the family Hominidae comprises the subfamily Ponginae, with *Pongo,* and the subfamily Homininae, with *Gorilla, Pan,* and *Homo.* The phylogeny thus indicated is

supported by a variety of genetic data showing that *Gorilla, Pan,* and *Homo* are more similar to one another than any is to *Pongo* (Marks 1988). Schwartz (1986) divided the involved taxa into three full families: the Panidae, with *Pan* and *Gorilla;* the Pongidae, with *Pongo* (and the fossil *Gigantopithecus* and *Sivapithecus*); and the Hominidae, with *Homo* (and *Australopithecus*). Martin (1990) argued for continuing to combine all three great apes in the Pongidae and to place humans in the distinctive family Hominidae (see account thereof).

All three genera of great apes contain many individuals that exceed most human beings in size. The gorilla may be 1.75 meters tall when standing on two feet and may weigh well over 200 kg. Like humans, the Pongidae lack a tail. The pelage is short, shaggy, and coarse. The face is nearly naked, and the ears are round and naked. Cheek pouches are not present. Unlike in humans, the arms are longer than the legs. The second to fifth fingers are long and the thumb is short. The big toe is opposable, allowing at least some individuals to carry objects with their feet.

The dental formula is: (i 2/2, c 1/1, pm 2/2, m 3/3) × 2 = 32. The incisors are nearly equal in size, the canines are large, the premolars resemble the molars, and the molars are wide-crowned and supplied with tubercles. The form of the skull is variable, but it is always longer than broad. In general form and structure the brain closely resembles that of *Homo.* Average brain capacities in adult Pongidae are: *Pongo,* 411 cu cm; *Pan,* 394 cu cm; and *Gorilla,* 506 cu cm. In *Hylobates* this capacity is 95 cu cm; in an 11-month-old *Homo* it is 850 cu cm; and in an adult human it is 1,350 cu cm (Rumbaugh 1970).

The great apes generally support themselves on all four limbs, but they have a limited bipedal capability. They do not take naturally to water and are not known to be able to swim. They can distinguish colors, and their best-developed senses are vision and hearing. They have a wide range of vocalizations, and some have laryngeal sacs that can be inflated for resonance. A variety of facial expressions are evidenced, especially in the chimpanzee.

The great apes are intelligent, at least from the human viewpoint. Chimpanzees are used extensively in laboratory research (often with little consideration of the needs of the individual or the conservation of the species) and are capable of learning many complicated tricks. Orangutans may be as intelligent as chimpanzees, but since they tend to be introverts by nature, testing their intelligence is more difficult than testing that of the extrovert chimpanzees, who love to show off. D. P. Erdbrink (pers. comm.) reported that a newly captured young male orangutan almost succeeded in unscrewing the bars of his travel cage; having observed how workers used a shifting spanner, he grasped the spanner through the bars when the workers left. In the zoo at Bandung, Java, before World War II an orangutan, seeing tame chickens wandering near his cage, captured one by throwing a path of cornseeds (part of his food) toward the cage. The size, strength, and relative rarity of gorillas has precluded intensive training and mental analysis, though some recent work suggests a remarkable ability to learn from and communicate with people (Patterson 1986).

According to Andrews (1978), the family Pongidae originated in the Oligocene, at which time the genera *Propliopithecus* and *Oligopithecus* were present in Africa. By the early Miocene, about 22 million years ago, the genus *Proconsul* had developed in Africa, and this genus is thought to represent the ancestral line of both the modern great apes and humans. The last common ancestor of all living Pongidae and Hominidae is probably the genus *Kenyapithecus,* which lived about 15 million years ago in the middle Miocene of eastern Africa (Andrews and Martin 1987; Mc-

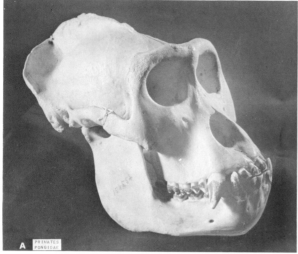
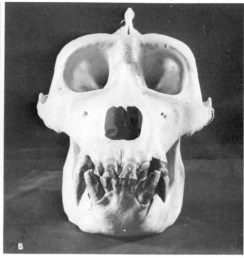

Skull of gorilla *(Gorilla gorilla)*, photos by P. F. Wright of specimen in U.S. National Museum of Natural History.

Crossin and Benefit 1993). There once was a general belief that a somewhat later Miocene descendant, *Sivapithecus*, represented the start of the evolutionary line leading to *Homo*. More recent studies, however, have indicated that *Sivapithecus* is not a direct ancestor of *Homo* and that the divergence between the lines leading to modern humans and apes did not occur until the early Pliocene (Kelley and Pilbeam 1986; Zihlman and Lowenstein 1979). Andrews and Cronin (1982) suggested that *Sivapithecus* and *Pongo* are closely related and in an evolutionary line completely separate from that leading to *Pan, Gorilla,* and *Homo.* The latter view has now been widely accepted, but Schwartz

Chimpanzee *(Pan troglodytes)*, photo by Bernhard Grzimek.

Nest of a Bornean orangutan *(Pongo pygmaeus)*
accommodating two babies, photo by Barbara Harrison.

(1986, 1987, 1988a) argued that the closest living relative of *Homo* is *Pongo.*

Andrews (1978) also considered that in the late Miocene *Sivapithecus* gave rise to *Gigantopithecus,* a huge ape that survived at least until the middle Pleistocene in Asia. This ape may have been 2.5–3.0 meters tall and probably was the largest primate ever to exist. It was first discovered by von Koenigswald, the Dutch geologist, who located its teeth during his travels in China and Java from 1934 to 1939 while wandering through Chinese pharmacies to study fossil bones used as medicines (see also account of the family Hominidae).

PRIMATES; PONGIDAE; Genus PONGO
Lacépède, 1799

Orangutan

The single species, *P. pygmaeus,* is now found only on the islands of Sumatra and Borneo. The two subspecies, *P. p. abelii,* of Sumatra, and *P. p. pygmaeus,* of Borneo, have many morphological and behavioral differences (Courtenay, Groves, and Andrews 1988; Groves 1986a; Groves and Holthius 1985). The population in southwestern Borneo may also warrant subspecific designation (Groves, Westwood, and Shea 1992). Pleistocene fossils from southern China, northern Viet Nam, Laos, and Java indicate that the genus once had a much greater distribution. Additional remains discovered in northern Viet Nam, as well as traditional beliefs in peninsular Malaysia, suggest that the orangutan still occurred on the mainland of Southeast Asia in early Recent times (Groves 1986a; Groves and Holthius 1985; Kahlke 1973).

Head and body length is about 1.25–1.50 meters (according to D. P. Erdbrink, an unusually large Bornean male

measured 1.80 meters) and the sitting height is about 0.70–0.90 meters. The arms, which reach to the ankles when the animal is erect, have a spread of about 2.25 meters. Rijksen (1978) estimated adult weight at 30–50 kg for females and 50–90 kg for males. Most of those that have survived in captivity have become fatter and heavier than those in the wild. Fooden and Izor (1983) reported full adult weight in captivity to average 65 kg for females and 144 kg for males and indicated that several of the latter had reached about 200 kg. Coloration is dark rufous or reddish brown. The hairy coat is rather thin and shaggy.

The forehead is high, and the snout bulging. The skull profile is more sloping than that of *Pan* and *Gorilla,* and the skull exhibits little of the brow ridging so prominent in those genera. Cheek pads are present in adults, especially in old males. These represent localized deposits of subcutaneous fat bound by connective tissue and covered by unmodified skin with scant and irregularly distributed hairs. The mouth projects and the lips are thin. The small ears are devoid of hair. The legs are short and relatively weak, but the hands and arms are powerful. According to Courtenay, Groves, and Andrews (1988), adult males from Sumatra are characterized by cheek pads that lie flat against the face, giving it a very wide appearance; the pads are thickly covered with hair, and there also is a pronounced beard. In males from Borneo the pads bulge outward and have little or no hair and the beard is short and scruffy.

A number of recent field studies have helped to clarify our understanding of *Pongo* (Horr 1975; MacKinnon 1971, 1973, 1974; Rijksen 1978; Rodman 1973, 1977, 1988). The orangutan appears to be well adapted to several different types of primary forest, ranging from swamps and other areas near sea level to mountainous forests at around 1,500 meters. The species is primarily arboreal and diurnal. It uses vegetation to construct a large platform nest in a tree, in which it sleeps at night. Usually a new nest is made for each night, but occasionally the same one is reused. Smaller nests may be made for rests during the day. The orangutan also shelters itself from rain and sun by holding leafy branches over its head or draping large leaves around its head and neck.

Tool use by fully wild orangutans is much more limited than has been reported for the chimpanzee, but released captive individuals use sticks for digging, fighting, prying, eating, scratching, and many other purposes. Some of these ex-captives learned (without human assistance) to untie complex knots securing boats and rafts and then to shove off, board, and ride the vessels across rivers (Galdikas 1982).

Normal movements are mainly by climbing and walking through large trees and swinging from branch to branch. Terrestrial movement is rare, cautious, and usually only to get from one tree to another. Surface locomotion is mainly quadrupedal, with the clenched fist, but not the knuckles, being placed on the ground. Movements generally are much less hurried than those of gibbons. In a lowland area of Borneo, Galdikas (1988) found daily movement usually to be 200–1,000 meters. There are morning and late afternoon peaks of activity and a rest period around the middle of the day. The diet is largely frugivorous and seems to include a high proportion of wild figs. Various species of these figs ripen at different times of the year, and this sequence is responsible for considerable local movement by the orangutan. Many other kinds of vegetation, as well as mineral-rich soil, insects, and perhaps small vertebrates and birds' eggs, are also eaten. Captive orangutans have accepted meat readily, and Sugardjito and Nurhuda (1981) observed a wild individual to consume a gibbon carcass.

Reported population densities have ranged from 0.2 to

Orangutan *(Pongo pygmaeus)*, photo from New York Zoological Society. A. Adult male showing great width of face produced by deposits of tissue beneath the skin and heavy beard and mustache, photo by Bernhard Grzimek. B. Photo from San Diego Zoological Garden. C. Photo by Bernhard Grzimek.

5.0 individuals per sq km. Local variation has been reported in the social structure of the orangutan. Observations generally indicate, however, that individuals usually occur in very small groups or alone, that most animals in a given area maintain a loose relationship, that temporary aggregations are sometimes formed, and that adult males oc-cupy larger home ranges than adult females and are hostile to one another. Perhaps some of the reported variation results from inconsistent interpretation by different investigators.

Studies in Borneo (Galdikas 1981, 1985*a*, 1985*b*; Horr 1975; MacKinnon 1971, 1973; Rodman 1973, 1977) showed that the number of orangutans seen together averaged about 1.8. Adult females, usually with dependent young, occupied overlapping home ranges of about 0.65 sq km or less. Adult males were generally alone and used home ranges of about 2–6 sq km that overlapped the ranges of several adult females. In Sabah (Malaysian Borneo) male home ranges also were reported to overlap one another, but in Kalimantan (Indonesian Borneo) these ranges were considered discrete. Although infrequent, contact between mature males invariably involved aggression or marked avoidance. Adult females, however, frequently paired and traveled and fed together for up to three days. Several immature animals sometimes associated with one another or with an adult of either sex. Adults of opposite sex came together only for a brief period of courtship. For purposes of mating, males preferred fully adult females and females preferred dominant adult males. Choice of sexual partners was very much a prerogative of females, who could not be monopolized by a particular male, though "rapes" did occur.

For Sumatra, MacKinnon (1973, 1974) reported a closer relationship than in Borneo among the various orangutans in a particular area. Although he thus recognized dispersed groups of up to 17 individuals apparently centered on the leadership of an adult male, he again found the number of animals seen together to average fewer than 2; adult males, however, were sometimes observed in the presence of females with young. Rijksen (1978), also working in Sumatra, determined group size to average only 1.47; adult females were usually accompanied by offspring, and adult males were usually solitary. Both sexes lived in home ranges of 2–10 sq km that overlapped considerably. Upon reaching adolescence, at about four or five years of age, young animals became increasingly independent of their mothers and formed small groups of their own. There also was temporary association of adults of either sex with immature individuals.

Rijksen (1978) described 13 vocalizations, and MacKinnon (1971) listed 15. The orangutan is generally quieter than other higher primates. The most notable sound is the long call of the adult male, a series of groans that may be heard up to 1 km away by a human, which probably serves as a spacing mechanism.

The following reproductive information was taken largely from Graham (1988), MacKinnon (1971, 1974), Nadler (1988), Rijksen (1978), and Rodman (1988). The estrous cycle of *Pongo* averages about 28 days, receptivity usually lasts 5–6 days, gestation averages 245 (227–75) days, and a single young is normal, but twins occur rarely. Mean birth weight is 1.56 kg for females and 1.93 kg for male when born singly (Fooden and Izor 1983). The young clings to the ventral surface of the female until it is nearly 1 year old and may still ride on the mother at 2.5 years. Weaning is usually completed at about 3.5 years. The interbirth interval is generally about 3–4 years but has been reported to be 8 years in an area subject to considerable human disturbance. Young individuals become increasingly independent of the mother after a second young is born but may still seek protection from the female until about 7–8 years. Physiological sexual maturity occurs at around 7 years in females and shortly thereafter in males; the females, however, usually do not give birth in the wild until at least 12 years, and the males do not attain full physical and social maturity, and hence reproductive capability, until 13–15 years or more. Young females generally remain in the vicinity of their birth, but males emigrate to other areas. According to Marvin L. Jones (Zoological Society of San Diego, pers. comm., 1995), a pair of orangutans that died in 1976 and 1977, after 48 and 49 years, respectively, in captivity, were estimated to have been 58 and 59 years old. These two animals hold the known record longevity for nonhuman primates.

The range of the orangutan has been shrinking since the Pleistocene, largely because of excessive hunting and envi-

Orangutan *(Pongo pygmaeus)*, photo by Polly McCann.

Orangutan *(Pongo pygmaeus)*, photo by Bernhard Grzimek.

ronmental disruption by people. Distribution in Sumatra has declined by 20–30 percent since 1935–38, and the species is now found only in the northern part of that island. Both MacKinnon (1971) and Rijksen (1978) suggested that the orangutan's arboreal mode of life and dispersed social structure may have resulted in part from many thousands of years of human hunting pressure. The other large primates *(Papio, Mandrillus, Gorilla, Pan, Homo)* are mainly terrestrial and highly gregarious. Pleistocene remains from China and Sumatra suggest that some orangutans were considerably larger than those of today, possibly even exceeding the gorilla in size. Galdikas (1988), however, considered the solitary nature of *Pongo* to result mainly from the association of its large size and frugivorous diet and the consequent competition at opportunistic food sources; she doubted that Pleistocene orangutans had been terrestrial.

The killing of the orangutan may have been stimulated in part by its resemblance to humans and may have served as an alternative to headhunting. Through much of the twentieth century a major threat to surviving populations was considered to be the collection of young animals, inevitably requiring the killing of the mother, for exhibition in zoos and circuses. The species has long been protected by the laws of the countries where it lives, but it is still sometimes taken and exported illegally. Indeed, in the late 1980s as many as 1,000 young orangutans were being smuggled from Indonesia to Taiwan (Eudey 1991).

By far the greatest current problem is the logging of the forests upon which the orangutan depends. Reports had circulated that fewer than 5,000 individuals survived, but apparently there are still more than that. MacKinnon

(1986) calculated that there could be as many as 179,000 in the Indonesian portion of the range. Rijksen (1986), after considering the large amounts of unsuitable and destroyed habitat, estimated populations to number about 6,000 on Sumatra and 37,000 on Borneo. More recent and precise estimates are that Sumatra has about 11,700 sq km of remaining habitat with 9,200 orangutans and that Borneo has about 22,400 sq km of habitat with 10,282–15,546 animals; there are another 901 in captivity (Perkins 1993; Tilson and Eudey 1993). The orangutan is classified as vulnerable by the IUCN and as endangered by the USDI and is on appendix 1 of the CITES. Recent conservation efforts have included the live capture of individuals isolated by logging and their relocation to more suitable areas of forest (Andau, Hiong, and Sale 1994).

PRIMATES; PONGIDAE; Genus GORILLA
I. Geoffroy St.-Hilaire, 1852

Gorilla

The single species, *G. gorilla,* is found mainly in two widely separated regions of equatorial Africa (Dandelot *in* Meester and Setzer 1977; Goodall and Groves 1977; Groves 1970*b*, 1971*b*; Schaller 1963; Vedder 1987). The subspecies *G. g. gorilla* (western lowland gorilla) occurs in southeastern Nigeria, west-central and southern Cameroon, extreme southwestern Central African Republic, mainland Equatorial Guinea, Gabon, Congo, and the extreme western tip of Zaire, as well as in the Cabinda enclave of Angola at the

Gorillas *(Gorilla gorilla):* A & C. Photos from San Diego Zoological Society; B. Baby, not more than 36 hours old, photo from Zoologischer Garten, Basel, through E. M. Lang.

Gorilla *(Gorilla gorilla)*, photo by Bernhard Grzimek.

mouth of the Congo River. An isolated population of this subspecies may also have existed in the Bondo district of north-central Zaire. The subspecies *G. g. graueri* (eastern lowland gorilla) is separated by about 1,000 km from the western lowland gorilla. *G. g. graueri* occurs in east-central Zaire to the southeast of Kisangani (Stanleyville), between the Lualaba (Congo) River, Lake Edward, and the northern part of Lake Tanganyika. A third subspecies, *G. g. beringei* (mountain gorilla), is restricted to the six extinct volcanoes of the Virunga Range straddling the Zaire-Rwanda-Uganda border and may also be represented by a small population in the Bwindi (Impenetrable) Forest of southwestern Uganda. However, a recent morphological analysis indicates that this population is subspecifically distinct from that of the Virungas and may be referable either to *G. g. graueri* or to an undescribed taxon (Sarmiento, Butynski, and Kalina 1995). A population in the Mount Kahuzi vicinity, just west of Lake Kivu, in Zaire, also sometimes has been assigned to *G. g. beringei*, but Casimir (1975) suggested that this population actually represents *G. g. graueri*, and most subsequent authorities have agreed. A recent study of mitochondrial DNA indicates a striking divergence between *G. g. gorilla*, on the one hand, and *G. g. graueri* and *G. g. beringei*, on the other, and has raised the possibility that the western and eastern populations may constitute separate species (Morell 1994*b*). There also have been reports of a pygmy species of *Gorilla*, but Groves (1985*a*) showed that such reports were based on misunderstanding. Goodall and Groves (1977), Tuttle (1986), and some other authors have taken the position that *Gorilla* is only a subgenus of *Pan*, but Groves (1986*a*) diagnosed the two as being generically distinct.

When standing on two feet the gorilla has a height of 1.25–1.75 meters; the total length is somewhat greater since the animal normally stands with its knees slightly bent. There is no tail. The span of the outstretched arms, about 2.00–2.75 meters, is far greater than the standing height. The shoulder breadth of this powerful primate is nearly twice that of the chimpanzee, and the chest is as much as 508 mm across. The circumference of the chest of male *Gorilla* is 1.25–1.75 meters. According to Grzimek (1975), weight is 70–140 kg in females and 135–275 kg in males, but occasionally it reaches 350 kg in individuals that become fat in captivity. The face, ears, hands, and feet are bare, and the chest in old males lacks hair. There is no beard. A pad of skin and connective tissue relatively dense and fibrous in nature is present on the crown. According to Groves (1970*b*), the coloration of both the skin and hair of *G. g. beringei* and *G. g. graueri* is jet black, and the adult male develops a silvery white saddle on the back between the shoulders and rump; *G. g. beringei* can easily be distinguished, however, by its much longer, silkier fur, especially on the arms. In *G. g. gorilla* the hair is short as in *graueri* but more grayish or brownish, and the male's saddle of whitish hair extends onto the thighs and grades more into the body color.

The gorilla has a short muzzle and an extremely stocky body. The nostrils are large, the eyes are small, and the small ears lie close to the head. The forearm is much shorter than the upper arm, and the hand is very large. The thumbs are larger than the fingers. The only animal that resembles the gorilla at all is the chimpanzee, which is smaller, usually has large, conspicuous ears, and is quicker in its movements. According to Napier and Napier (1967), the skull of *Gorilla* is characterized by a markedly prognathic face, rectangular and widely separated orbits, sagittal and nuchal crests in males and sometimes in females, and lower molar teeth, which increase in size toward the rear.

Most gorillas inhabit lowland tropical rainforests, but in the Cross River region of southeastern Nigeria and southern Cameroon and in extreme eastern Zaire and adjacent parts of Uganda and Rwanda some populations are found in montane rainforest between 1,500 and 3,500 meters and in bamboo forest from about 2,500 to 3,000 meters. The gorilla is primarily terrestrial but is fully capable of climbing. Goodall and Groves (1977) reported that some animals in the Mount Kahuzi area of Zaire fed in trees at heights of 40 meters and that even the largest male, weighing at least 200 kg, frequently climbed to 20 meters. Surface locomotion is by a quadrupedal walk, with the soles of the feet and the middle phalanges of the fingers placed on the ground. A few bipedal steps are often taken, but Schaller (1963) never observed an animal to progress in this manner for more than about 18 meters. The collection and preparation of food is done almost entirely with the hands. The gorilla does not seem to be able to swim and, indeed, is sometimes reluctant to wade across shallow streams; this factor may partially explain the unusual distribution of the species. Every adult and juvenile builds its own crude platform nest in which to sleep at night and occasionally for rest during daylight. Construction time is less than five minutes. In some areas more than 90 percent of nests are made on the ground, but in other areas, apparently because of availability of materials, most nests are built in trees. A nest is not used for more than a single night (Goodall and Groves 1977; Schaller 1963, 1965).

According to Schaller (1963), the gorilla is diurnal, with nearly all activity occurring between 0600 and 1800 hours. After awakening, the animals feed intensively for a while, then rest from about 1000 to 1400 hours, and then travel and feed until bedding down for the night. Groups observed by Schaller moved 90–1,800 meters per day. Other observations (Casimir and Butenandt 1973; Fossey and Harcourt 1977; Jones and Sabater Pi 1971) suggest that daily movement is usually about 400–1,000 meters. Goodall and

Groves (1977) stated that during the wet season (April–June) in eastern Zaire groups moved an average of 596 meters per day but that in the dry months (July–August) they covered 1,240 meters per day. A. G. Goodall (1977) determined that movements in the course of a year were not random but followed an established migratory pattern with respect to availability of food. Casimir and Butenandt (1973) found that one gorilla group in the Mount Kahuzi area spent October and November in bamboo forest, because of the abundance then of young bamboo shoots, and the rest of the year in other kinds of forest. Harcourt and Stewart (1984) determined that the mountain gorilla spends about 45 percent of the day feeding. The gorilla is almost entirely a vegetarian in the wild, though captives have been known to eat meat. Fossey and Harcourt (1977) reported that 85.8 percent of the diet of *G. g. beringei* consists of leaves, shoots, and stems. Small amounts of wood, roots, flowers, fruits, and grubs also are eaten. Harcourt and Harcourt (1984) noted that many small insects are eaten inadvertently while feeding on plants. Tutin and Fernandez (1983) found that in Gabon the gorilla frequently breaks into termite nests to feed on the insects therein. A. G. Goodall (1977) observed that gorillas had ample opportunity to eat eggs, helpless young birds, and the honey of stingless bees but never did.

Population densities of 0.35–0.75/sq km have been recorded in various parts of Africa where gorillas occupy areas of at least 100 sq km; if unfavorable habitat within such areas is not considered, maximum density would be nearly 1/sq km (Harcourt, Fossey, and Sabater Pi 1981). Annual group home range probably varies from 4 sq km to 25 sq km and averages about 8 sq km (Fossey 1974; Fossey and Harcourt 1977). A. G. Goodall (1977), however, reported the home range of a group to be 34 sq km, and Casimir and Butenandt (1973) stated that over a period of years 40–50 sq km probably would be used by one group. Schaller (1963) found much overlap of home range in the Virunga Volcanoes region, with several groups frequenting the same parts of the forest. Fossey and Harcourt (1977) reported that groups in the same region had core areas that were not entered by outsiders but that there was no overt territorial defense. A. G. Goodall (1977) stated that some territoriality apparently did exist in the Mount Kahuzi area.

When two groups approach, they may ignore each other, temporarily associate, or express hostility through vocalizations and displays (Fossey 1974; Schaller 1963). It is possible that there is a larger community structure beyond the group level and that certain groups interact peacefully among themselves but not with other groups (Goodall and Groves 1977). A group may seemingly go far out of its way to avoid contact with another group. Fossey (1974) thought that the main determinant in overall group movement was the location of other groups. A. G. Goodall (1977), however, considered food availability to be the chief determinant.

Intragroup relations seem to be much more cohesive in the gorilla than in the chimpanzees. Schaller (1963) found the splitting of groups into smaller units to be infrequent and temporary. In his study area in the Virunga Volcanoes region group size ranged from 2 to 30 individuals. On the average there were 16.9 animals, including 1.7 fully adult males (silverbacks), 1.5 subadult males (blackbacks), 6.2 adult or subadult females, 2.9 juveniles (3–6 years old), and 4.6 infants (under 3 years). In other regions average group size has varied from 6 to 13 (Goodall and Groves 1977). Median group composition throughout the range of *Gorilla* is 1 silverback, 1 blackback, 3 adult females, and 2 or 3 immatures (Harcourt, Fossey, and Sabater Pi 1981). The smallest groups, averaging 4–8 members, are found in the

western lowland subspecies (Dixson 1981). If there is more than a single adult male, one is dominant and is the leader of the group, and only that one normally breeds. There is a general rank order for the group based mainly on size, with silverback males being dominant over all other animals.

Groups are highly stable, and the dominant male retains leadership for years. Certain of the other males are only temporarily associated with the group and eventually leave to live alone or to join other groups (Goodall and Groves 1977). The adult males that do remain are probably the sons of the dominant male and eventually will take over leadership (Harcourt 1978; Harcourt, Stewart, and Fossey 1976). Males that leave a group wander alone for a number of years, then sometimes establish a range adjacent to or overlapping that of their former group, and may eventually take one or more females to begin a new group (Caro 1976; Fossey 1974). Unlike the situation in most primates, nearly all female gorillas emigrate from their parental group at maturity. Departure usually seems to be of the female's own volition, and she almost immediately joins with another group or a lone young silverback male (Harcourt 1978, 1979; Harcourt, Fossey, and Sabater Pi 1981; Harcourt, Stewart, and Fossey 1976; Watts 1992). A female may join, and reproduce within, several groups in succession (Watts 1991). However, the long-term female residents of a large group sometimes form cliques that may try to prevent the arrival of, or remain hostile to, immigrant females (Watts 1994).

Fossey (1972) identified 16 vocalizations emitted by *Gorilla*, and Schaller (1963) described 22. These include roars and growls of aggression, grunts and barks for group coordination, and loud hoots that can be heard 1 km away by people and may serve to space different groups. Schaller also described the elaborate chest-beating display that may be elicited by the presence of outsiders, either other gorillas or humans. It begins with a series of hoots, after which the animal stands on two legs and slaps its chest with slightly cupped hands, first one, then the other, 2–20 times. The gorilla then runs about and breaks vegetation.

According to Schaller (1963), there is no evidence of a particular breeding season in the wild. Females give birth every 3.5–4.5 years unless the infant dies. The estrous cycle lasts about 27–28 days, and estrus 1–3 days (Harcourt, Stewart, and Fossey 1981; Watts 1991). Estimates of gestation range from 251 to 295 days, though Dixson (1981) gave the average as precisely 257.6 days. There is normally a single young, but twins occur rarely. The weight at birth is about 2 kg. Young are weaned at 3–4 years (Watts 1991). The age of physiological sexual maturity is about 6–8 years in females and 10 years in males, but breeding in the wild normally does not begin until females are 9–10 and males 15 years old, about the time when emigration occurs. Mortality in stable populations is 42 percent for immatures, mostly in the first year of life, and 5 percent for adults. Females generally give birth to only 2–3 surviving young during their reproductive life (Harcourt, Fossey, and Sabater Pi 1981; Watts 1991). Life span in the wild may be up to 50 years. A captive gorilla lived for exactly 54 years (*Washington Post*, 1 January 1985). Although once considered impossible, breeding of *Gorilla* in captivity now is accomplished regularly; Olney, Ellis, and Fisken (1994) indicated that 362 of the 633 individuals held by the world's zoos had been bred in captivity.

The gorilla is classified as endangered by the USDI and is on appendix 1 of the CITES. The IUCN designates the species in general as endangered but designates the subspecies *G. g. beringei* as critically endangered. That subspecies is found in the Virunga Volcanoes region at the point where Zaire, Uganda, and Rwanda come together.

Most of this region is included within national parks or reserves, but there have been encroachments for agriculture, grazing, and wood gathering. During the 1970s poaching intensified in the Virunga region, especially to obtain gorilla heads for commercial sale (Harcourt and Curry-Lindahl 1978). The number of gorillas in the region fell from about 450 in 1960 to about 225 two decades later (Harcourt and Fossey 1981). A conservation program implemented in Rwanda in 1978 and funded in part by tourism may have helped to stabilize the population and reduce threats to the habitat (A. H. Harcourt 1986; Weber and Vedder 1983).

G. g. beringei also may be represented by the gorilla population of the Bwindi (Impenetrable) Forest in southwestern Uganda. As noted above, however, the most recent assessment (Sarmiento, Butynski, and Kalina 1995) indicates that the Bwindi population is referable to *G. g. graueri* or to an undescribed subspecies. Although this area is a reserve, the gorilla population was thought to have been reduced to only about 100 animals because of intense human activity (A. H. Harcourt 1981). Recent surveys indicate, however, that there are approximately 320 gorillas in the Bwindi Forest (Butynski, Werikhe, and Kalina 1990). Apparently based on the earlier figure, Vedder (1987) estimated total numbers of *G. g. beringei* as 370–440. By 1995 there was recognition that the Virunga and Bwindi gorilla populations each had grown to about 320 animals, but protection had become more difficult and poaching had increased in association with the terrible social unrest sweeping Rwanda and adjoining areas (Morris 1995).

Comparatively little is known about the overall status of *G. g. graueri* and *G. g. gorilla*. Schaller (1963) estimated the entire population of eastern gorillas as 5,000–15,000 individuals. More recent estimates were only 2,500–4,500 (Vedder 1987) and 3,000–5,000 (Lee, Thornback, and Bennett 1988). A population just west of Lake Kivu in the vicinity of Mount Kahuzi, Zaire, has been estimated at about 275 animals (Yamagiwa et al. 1993). The western subspecies has been estimated to number 40,000 animals (A. H. Harcourt 1986), of which about 35,000 are located in the nation of Gabon (Tutin and Fernandez 1984). There also may be a large number in Congo (Fay and Agnagna 1992). In contrast, the species had not been recorded at all in Nigeria for 30 years, but a few isolated groups totaling perhaps 150 individuals were located there recently by Harcourt, Stewart, and Inahoro (1989). This Nigerian population is classified as critically endangered by the IUCN. Jones and Sabater Pi (1971) considered the western lowland gorilla to be declining because of habitat destruction and killing by people for food and allegedly to protect crops. A small population of *G. g. gorilla* that once may have existed in the Bondo district of north-central Zaire has not been reported since 1908 (Goodall and Groves 1977).

PRIMATES; PONGIDAE; **Genus PAN**
Oken, 1816

Chimpanzees

There are two species (Dandelot *in* Meester and Setzer 1977; Kano 1992; Lee, Thornback, and Bennett 1988):

P. troglodytes (chimpanzee), originally found in the forest zone from Gambia to Uganda, and south to Lake Tanganyika, but not including the central forests of Zaire south and west of the Congo–Lualaba River;
P. paniscus (pygmy chimpanzee, bonobo), central Zaire to

the south and west of the Congo–Lualaba River and north of the Kasai–Sankuru River.

There has been considerable controversy regarding the technically proper generic name for the chimpanzees. *Pan* is used here because of its familiarity and in the hope of promoting nomenclatural stability. Van Gelder (*Bull. Zool. Nomencl.* 32 [1975]: 69–73) presented a detailed argument in favor of using the name *Chimpansee* Voigt, 1831. *P. paniscus* sometimes is considered a subspecies of *P. troglodytes* (see, e.g., Horn 1979), but recent evidence suggests not only that the two are distinct but that *P. paniscus* bears some resemblance to the ancestral stock that gave rise to the Pongidae and Hominidae (see, e.g., papers *in* Susman 1984*b*). Three other subspecies of *P. troglodytes* usually are recognized: *P. t. verus,* found from Gambia to the Niger River in central Nigeria; *P. t. troglodytes,* from the Niger River to the Congo–Ubanghi River along the eastern border of Congo; and *P. t. schweinfurthi,* from the northwestern corner of Zaire to western Uganda and Tanzania (Lee, Thornback, and Bennett 1988; Shea, Leigh, and Groves 1993). Tuttle (1986) followed some older authorities in listing a fourth, *P. t. koolokamba,* a large, "gorilla-like" animal. Shea (1984), however, explained that *koolokamba* is based in part on local folklore and that purported specimens thereof have generally turned out to be either large male chimpanzees or small female gorillas. Recent analyses of DNA (Morell 1994*a*; Morin et al. 1994) have suggested such a great genetic distance between *P. t. verus,* on the one hand, and *P. t. troglodytes* and *P. t. schweinfurthi,* on the other, that *verus* may warrant elevation to full species rank. This work also indicates the following times of divergence: between *P. t. troglodytes* and *P. t. schweinfurthi,* 440,000 years ago; between *P. t. verus* and the other two subspecies, 1,580,000 years ago; between the species *P. troglodytes* and *P. paniscus,* 2,500,000 years ago; and between the genera *Pan* and *Homo,* 4,700,000 years ago.

According to Napier and Napier (1967), *Pan* is characterized by prominent ears, protrusive lips, arms that are longer than the legs, a long hand but short thumb, and no tail. The facial skeleton is moderately prognathic. The orbits are frontally directed and surmounted by prominent supraorbital crests. A small sagittal crest is seen only occasionally in large males and females, and there is no nuchal crest. The molar teeth decrease in size toward the rear. Additional information is provided separately for each species, and the account of *P. troglodytes* was prepared with the assistance of Geza Teleki, of the Committee for the Conservation and Care of Chimpanzees.

Pan troglodytes (chimpanzee)

Head and body length is about 635–940 mm, and the height when erect is about 1.0–1.7 meters. The arms, which extend below the knees when the animal is standing, have a spread that is about 50 percent greater than the animal's height. Weight in the wild is 34–70 kg for males and 26–50 kg for females (Jungers and Susman 1984), but respective figures for captives are as much as 80 kg and 68 kg. The face is usually bare and generally black in color. Younger animals have flesh-colored ears, nose, hands, and feet and a white patch near the rump. At maturity the overall skin color is dark and the pelage varies from deep black to light brown. The hair on the head may grow in any direction, and both sexes are prone to partial baldness early in maturity.

The distribution of *Pan troglodytes* is centered in tropical rainforest but also extends into forest-savannah mosaic and montane forest up to 3,000 meters (Dandelot *in*

Chimpanzees *(Pan troglodytes):* A. Mother and baby, photo from San Diego Zoological Garden; B. Young chimpanzee, photo by Ernest P. Walker. Inset: photo from *Bull. Amer. Mus. Nat. Hist.*

Meester and Setzer 1977). During daylight hours most time is spent in trees, and chimpanzees, especially young individuals, sometimes brachiate from branch to branch. Movement through the trees, however, seldom proceeds very far, and most travel takes place on the ground (Reynolds and Reynolds 1965; Van Lawick–Goodall 1968). The most common means of locomotion is a quadrupedal walk with the hind legs slightly flexed, the body inclined forward, and the arms straight. The soles of the feet and the backs of the middle phalanges of the fingers are placed on the ground. In this position chimpanzees may proceed at a rapid gallop. They also are capable of a bipedal walk, but with the toes turned inward, and may proceed for more than 1 km in this manner. The hands are used to eat and may also be used to throw objects at enemies. For sleeping at night, each individual, except infants, constructs a nest of vegetation in a tree, usually at a height of 9–12 meters. Nests normally are occupied for only one night but occasionally are reused. Nests also may be constructed for rest during daytime in the rainy season (Van Lawick–Goodall 1968).

The chimpanzee has achieved some renown for its manufacture and utilization of tools. It frequently feeds by carefully poking a stick or vine, which it may have modified for the purpose, into an entrance of a termite nest and then withdrawing the tool, which has become covered with the insects (McGrew and Collins 1985). Sticks also have been used as hooks to pull down fruit-laden branches, as weapons in inter- and intraspecific fighting, and for various other purposes (Goodall 1986). Stones are used as hammers and anvils to crack nuts (Kortlandt and Holzhaus 1987). The spread of this nut-cracking ability through a group,

with one animal after another learning, was documented by Hannah and McGrew (1987). Chimpanzees also have been observed to use leaves as sponges to soak up drinking water and as tissues for cleaning the body. Apparently, tool use and nest construction are not instinctive skills but are culturally acquired by younger animals observing more experienced ones (Van Lawick–Goodall 1968). Tool use has been found in all populations of *P. troglodytes,* but there are regional differences; it is especially extensive in far West Africa and nut-cracking seems to be confined mainly to that region (Boesch and Boesch 1990; McGrew 1992).

Chimpanzees are largely diurnal but sometimes move about by night. During the day they feed for 6–8 hours and forage over a distance of 1.5–15.0 km. Peaks of activity occur in the early morning and at around 1530–1830 hours (Van Lawick–Goodall 1968). They may shift their activity seasonally depending on the fruiting times of various kinds of plants (Izawa 1970). The diet includes fruits, leaves, blossoms, seeds, stems, bark, resin, honey, insects, eggs, and meat. Food intake varies by season, consisting on an annual basis of about 60 percent fruits, 30 percent other vegetation, and 10 percent animal matter. Termites are the most important insect. Chimpanzees occasionally stalk, kill, and eat young artiodactyls, baboons, and other monkeys. Individuals can hunt some prey successfully, but cooperative hunting seems essential in taking baboons and the large young of bush pigs. Cannibalism on infant chimpanzees has been observed, and there have been a few reports of the carrying off of young human children (J. Goodall 1977, 1986; Nishida, Uehara, and Nyundo 1979; Sugiyama 1973; Suzuki 1971; Van Lawick–Goodall 1968).

Reported population densities for *P. troglodytes* are

0.05–26.00/sq km, usually about 1–10/sq km (IUCN 1972; Jones and Sabater Pi 1971). Populations are divided into "communities," loose and flexible associations of males and females with a shared home range (Tutin, McGrew, and Baldwin 1983). Subgroups of a community vary from solitary individuals to large mixed parties with all ages of both sexes. Only rarely, if ever, do all members of a community congregate in one place. Subgroups are generally unstable, except for females with immature offspring. There is little difference in community range size between forest and woodland; eight studies in such habitats have produced estimates of 5–40 sq km, with an average of 12 sq km. Savannah ranges are much larger, size estimates being 120–560 sq km. Communities of about 50 individuals each have been reported in all three kinds of habitat, but overall size range is around 15–80.

A community of 28 animals investigated by Tutin, McGrew, and Baldwin (1983) occurred in dry savannah, and this marginal habitat apparently resulted in a large home range of 228 sq km and the formation of large, relatively long-lasting traveling parties. Izawa (1970) investigated two communities of 40 animals each. Both groups used a nomadic home range of about 120 sq km, but there was only a 20 percent overlap in these ranges and the groups avoided one another. Sugiyama (1973) reported that a group of 70–80 animals maintained itself in a home range of 7–8 sq km for several years. He added that another group of 50–60 had a total range of 7–8 sq km and that a group of 60–80 used 17 sq km.

The chimpanzee is thought to be territorial, and intercommunity relationships sometimes appear to resemble those between enemy human tribes (Goodall 1986). Borders are patrolled, and outsiders may be attacked and killed.

Chimpanzee *(Pan troglodytes)*, photo from West Berlin Zoo.

Males even have been seen to attack nonestrous females of another community. Long-term studies have revealed that an entire community may be destroyed as its range is gradually occupied by a stronger neighboring group and its members are systematically killed or driven away. Manson and Wrangham (1991) noted that "lethal raiding," in which several individuals make unprovoked aggressive invasions of the range of another group, where they deliberately seem to hunt and kill members of that group, is unknown in primates except for *Pan* and *Homo*. According to Goodall (1986:534), "The chimpanzee, as a result of a unique combination of strong affiliative bonds between adult males on the one hand and an unusually hostile and violently aggressive attitude toward nongroup individuals on the other, has clearly reached a stage where he stands at the very threshold of human achievement in destruction, cruelty, and planned intergroup conflict."

Goodall (1986) described the chimpanzee community as a "fusion-fission" society. Individuals of either sex have almost complete freedom to come and go as they wish. They associate with one another for various lengths of time, depending on the intensity of the relationship, reproductive status, and resource distribution. Each animal has its own network of social contacts. A community has a dominant male leader, and there also are power-wielding coalitions among both males and females. The ability of a male to enlist support during conflict is perhaps the most crucial factor in attaining and maintaining his rank. Subgroups of a community are temporary and may change in composition within a matter of hours or days. They can include any category of animals—all adult males, all adult females, both sexes, adults and young, or all young. Friendly relationships, even between adult males, appear to predominate within these groupings, though animals can suddenly turn violent. When two parties meet at a fruiting tree there is considerable greeting behavior, and large adult males sometimes embrace and groom one another (Sugiyama 1973). Wrangham (1977) observed that when an adult male arrived at a major food source, it would give a "food call" that tended to attract other individuals of both sexes.

Of 498 parties seen by Van Lawick–Goodall (1968), 44 percent had 2–4 animals each, and only 1 percent had more than 20. However, there was a recognized dominance hierarchy, and the highest-ranking individual present would lead the party. Because associations were continuously changing, most mature and adolescent animals had an opportunity to be dominant. Rank depended partly on size and age but also on relationship. A young animal might show dominance over an older one if its mother was high-ranking and present. Aggressive behavior within communities was frequent but seldom violent, and even serious fights resulted in no evident injury. Communication was complex and involved constantly varying facial expression, numerous gestures, and 24 identified vocalizations. The most common adult call is the "pant-hoot," which is given in many different situations and often seems to identify the caller and solicit information from other animals. Mitani and Nishida (1993) suggested that pant-hooting serves to space potential rivals and bring together allies. Grooming, particularly the removal of external parasites from one individual by another, is an important factor in maintaining friendly relationships but also can be used as a means of exchange to gain favors from other individuals (Goodall 1986).

The chimpanzee once was thought to be largely promiscuous, with several males, showing little aggression toward one another, following and mating with a female in estrus. Such activity does occur, but a male may enter a short-term relationship with a receptive female and prevent lower-ranking males from mating with her. An adult pair may also establish a temporary consortship, leaving the vicinity of other animals or even moving beyond the community range and mating exclusively with one another (Tutin and McGinnis 1981). Consortships may last up to three months. Older females are more popular with the males of their community than are younger females. Females, especially those that have not reached full adulthood, may leave their natal community either temporarily or permanently and mate with the males of neighboring communities. According to Goodall (1986), the presence or absence of an estrous female is the single most significant factor in the overall patterning of a chimpanzee community. Such an individual, unless involved in a consortship, may be surrounded by most or all of the community males and many of the females.

The following information on reproduction and life history was taken in part from Goodall (1986) and Van Lawick–Goodall (1968, 1973). Breeding occurs throughout the year. The estrous cycle lasts about 36 days, and females are receptive for about 6.5 days during a period of maximum genital swelling. There is a minimum interbirth interval of 3 years in the wild, but the period usually is 5–6 years if the first young survives. A female can give birth every year if the young is removed when only a few months old. The gestation period is 202–261 days, averaging about 230 days. The normal number of young is one; twins are rare, though perhaps more common than in *Homo sapiens*. Average birth weight is about 1.9 kg. For the first 3 months of life the infant is cradled by the mother when she is sitting. Until around 6 months the young clings ventrally when the female moves. For the next several years the young rides on the mother's back. Weaning takes place at 3.5–4.5 years, but the young remains dependent on the mother for a longer period and may still travel with her at 10 years. Bonds between young, especially females, and the mother sometimes persist throughout life. Puberty in both sexes occurs at about 7 years, but females do not usually give birth until 13–14 years and males are not fully integrated into the social hierarchy until 15–16 years. Reproductive capability in the female apparently lasts until at least age 40, and maximum life span in the wild may be 60 years. According to Marvin L. Jones (Zoological Society of San Diego, pers. comm., 1995), a female died in 1992 at the Yerkes Primate Center at an age of about 59 years, another female was still living in 1976 at the Mysore Zoo in India at an age of 53 years, and a male ("Cheeta" of the *Tarzan* movies) was still living in 1990 at 56 years.

There may have been several million *P. troglodytes* in the early twentieth century, but current estimates generally range from 150,000 to 235,000, and the species has disappeared entirely from large areas (Lee, Thornback, and Bennett 1988; Luoma 1989). The largest numbers remaining are found in the Central African forests of Cameroon, Gabon, Congo, and Zaire. A survey during the early 1980s suggested the presence of about 64,000 chimpanzees in Gabon alone (Tutin and Fernandez 1984). More recently, however, Blom et al. (1992) indicated that numbers may be declining in Gabon, and J. Goodall (1994) warned that alarming declines were occurring even in Central Africa. Populations have been reduced and fragmented through loss of forest habitat to agriculture and logging, hunting by people for food and to protect crops, and commercial exportation for the animal trade. In 1976 the USDI classified *P. troglodytes* as threatened, and consequent regulations reduced legal importation to the United States, though illicit trade continued. Recently there has been widespread concern that a demand for animals for biomedical research, especially involving work related to Acquired Immune De-

ficiency Syndrome, or AIDS, would lead to large-scale collection of young animals in the wild and consequent devastation of remaining populations. There are several thousand animals in captivity, many of them in colonies that are being managed with the objective of their becoming self-sustaining. In 1990 the USDI reclassified *P. troglodytes* as endangered in the wild and also strengthened regulations covering animals in captivity. *P. troglodytes* is on appendix 1 of the CITES. The IUCN also now classifies the species as endangered.

Pan paniscus (pygmy chimpanzee, bonobo)

Despite the common name, *P. paniscus* is often equal in size to *P. troglodytes*. Head and body length is 700–828 mm and weight is 37–61 kg for males and 27–38 kg for females (Jungers and Susman 1984). Coloration of *P. paniscus* is about the same as in *P. troglodytes*. *P. paniscus* has longer limbs than *P. troglodytes*, and the chest girth of the former is smaller relative to its height (Zihlman 1984). *P. paniscus* has long hair on the sides of the head, relatively small ears, and a relatively high forehead (Kano 1992).

According to Kano and Mulavwa (1984), *P. paniscus* has a narrower ecological range than *P. troglodytes*, being found exclusively in lowland, especially primary, forest. Activity is largely diurnal; the animals feed in trees shortly after waking, rest, move leisurely on the ground to the next food trees, become less active toward the middle of the day, and then repeat the pattern in the afternoon. Average daily movement is about 2.4 km. Nishida (1972b) reported *P. paniscus* to build sleeping nests in trees in the same manner as *P. troglodytes*. Tool use in the wild has not been observed, but Jordan (1982) reported captives to construct ropes to swing from, to wipe themselves with leaves, and to use sticks to probe, rake, and pole-vault over water. Movements in the wild include quadrupedal knuckle walking on the ground, arm swinging through the trees, and leaping from branches (Susman 1984a). The natural diet consists mostly of fruits and also includes leaves, seeds, and, rarely, invertebrates and small vertebrates. Badrian and Malenky (1984) reported predation on young duikers *(Cephalophus)*.

The following information on population structure and social life was taken from Badrian and Badrian (1984), Kano (1992), and Kano and Mulavwa (1984). *P. paniscus* has a "fusion-fission" community organization comparable to

Pygmy chimpanzee *(Pan paniscus)*, photo by Bernhard Grzimek.

that of *P. troglodytes* but seems to be more gregarious, more mutually tolerant, and less agonistic. Each known community or "unit group" contains about 40–120 individuals occupying a home range of about 22–68 sq km, at a density of approximately 2/sq km. Membership is fluid, mainly because young, nulliparous females transfer freely from group to group, but is more stable than in groups of *P. troglodytes*. Subgroups, or "parties," usually contain 2–15 individuals but may contain as many as 40 and are commonly based on a female and her immature male offspring, as well as on adult female associations, and also usually include adult males. Larger subgroups sometimes are found in the vicinity of rich food sources. Upon discovery of such a site, individuals may emit a loud call, perhaps both to alert members of their community and to warn other groups. There is extensive overlap between community ranges, but parties of one community avoid those of the other. When there is an encounter, there may be loud vocalizing, agonistic displays, and occasionally serious fighting, but the deliberate "lethal raids" of *P. troglodytes* have not been observed.

Adolescent female *P. paniscus* emigrate from their natal groups and have no further ties with their mothers. Adult females, however, appear to have a greater tendency to come together in strongly bonded units than do female *P. troglodytes*. Such female association seems to form the primary basis of the community, and it is the females that generally have the leadership role. In contrast to young adult male *P. troglodytes*, which are incorporated into a male hierarchy and then have little to do with their mothers, male *P. paniscus* stay in their mothers' units and maintain a permanent association with them. Adult male *P. paniscus* seem to lack the dominant role of male *P. troglodytes* and are about equal in rank to the females. The males are sometimes aggressive toward one another but are only rarely aggressive toward the females.

Reproductive seasonality is not known. The estrous cycle lasts about 46 days (Thompson-Handler, Malenky, and Badrian 1984). The period of receptivity is about twice as long as that of *P. troglodytes*, but interbirth interval usually is about the same. There normally is a single young and interaction with the mother is similar to that seen in *P. troglodytes* (Kano 1992). Based on observations in captivity, the gestation period is 220–30 days, birth weight is 1–2 kg, and females reach sexual maturity at about 9 years (Neugebauer 1980). There are far fewer captive *P. paniscus* than captive *P. troglodytes*, the number of the former being about 100 (Reinartz and Van Puijenbroeck *in* Bowdoin et al. 1994).

The pygmy chimpanzee is classified as endangered by the IUCN and the USDI and is on appendix 1 of the CITES. Although it occurs over a large region of dense jungle, its distribution is discontinuous because of both natural ecological requirements and fragmentation of habitat through increasing human populations, agricultural expansion, and logging. It also is hunted extensively by people for meat and for parts used in religious rituals. Its numbers are unknown, but extrapolation of estimates of known populations to potential remaining habitat suggests an overall figure of about 15,000 (Lee, Thornback, and Bennett 1988). Assessments of continued habitat deterioration and expansion of human activity in Zaire suggest that numbers in the wild may be fewer than 5,000 (Thompson-Handler, Malenky, and Reinartz 1995). Recently, Kortlandt (1995) suggested that the pygmy chimpanzee may still inhabit a vast area of relatively undisturbed habitat and could number more than 100,000. Susman (1995) agreed that remaining distribution and population size may be large but cautioned that human activity could quickly jeopardize

survival of the species. Both authorities emphasized the need for more intensive and careful field surveys.

PRIMATES; Family HOMINIDAE; Genus HOMO
Linnaeus, 1758

People, or Human Beings

The single living genus and species, *Homo sapiens,* maintains permanent residence in nearly all terrestrial parts of the earth's surface. Exceptions include extremely arid regions, such as certain sections of Australia and the central Sahara Desert, and extensive ice-covered regions, especially most of Greenland and Antarctica. Their knowledge and technical skills allow humans to temporarily occupy even those inhospitable areas, move freely on and under most water-covered portions of the world, and fly through and orbit above the atmosphere. A Russian space station has been maintained about 350 km above the earth since February 1986, usually occupied by several persons, and in 1995–98 was visited by American crews. From December 1968 to December 1972 a series of rocket-powered spacecraft carried people from the United States to the vicinity of the moon and back to earth; three of these expeditions merely circled the moon, but six landed teams of two men each on the lunar surface.

There is controversy regarding the content of the family Hominidae. Although it long was common practice to treat *Homo sapiens* as the only living species, some authorities have suggested that the great apes (here considered to compose the Pongidae) should be placed within the same family as humans. Groves (1986*a*, 1989, and *in* Wilson and Reeder 1993), for example, divided the Hominidae into two living subfamilies: the Homininae, with the living genera *Pan, Gorilla,* and *Homo*; and the Ponginae, with *Pongo*. Based largely on DNA evidence, Goodman (1992) went even further, including the gibbons in the Hominidae as the subfamily Hylobatinae and placing the great apes and humans in a second subfamily, the Homininae, with *Pan* and *Homo* united at the subtribal level. The relegation of humans to a common family with the apes is partly a consequence of a principle of the taxonomic school known as cladistics. This view holds (1) that all components of one systematic grouping must be considered to have descended from a common ancestor and (2) that the grouping must include all the descendants of that ancestor. With regard to the first point, there is general agreement that apes and humans did have a common ancestor. Nonetheless, a problem develops relative to the currently favored version of subsequent phylogenetic history, as discussed below. If *Pongo* diverged from the ancestral stock prior to the split between the two African apes and *Homo,* and if *Pongo* is placed in the same family as the African apes, then in accordance with the second point of cladistics *Homo* must also be included in that family. Not all taxonomists accept such an arrangement even if they agree with its evolutionary basis. Martin (1990) discussed this matter in some detail and concluded that the great morphological distinction between the three living great apes, on the one hand, and *Homo,* on the other, warrants placing the latter in a separate family. That position has been followed here.

Until about 1970 there was a general view, based primarily on fossil evidence, that the evolutionary line leading to *Homo* diverged from that leading to the great apes in the middle of the Miocene epoch, about 15 million years ago. It was thought that the first clearly hominid genera were *Sivapithecus* and *Ramapithecus,* of southern Asia and

Homo sapiens Astronaut Edward H. White II makes America's first "space walk" during the *Gemini IV* mission, 3 June 1965. He was accompanied by James A. McDivitt. On 27 January 1967, White, along with astronauts Virgil I. Grissom and Roger B. Chaffee, died in a fire during a test of an Apollo spacecraft. Photo courtesy of National Aeronautics and Space Administration.

Africa. Subsequently, based in large part on biochemical analyses of humans and apes, there has been a growing consensus that *Sivapithecus* (which may include *Ramapithecus*) is close to the ancestral line of only *Pongo*. That line is thought to have split off in the middle Miocene from the line leading to humans and the two African apes. The latter line, however, did not begin to branch until the late Miocene, around 8 million years ago, with the split between the ancestors of *Pan* and *Homo* taking place in the Pliocene, perhaps only 4–5 million years ago (Andrews and Cronin 1982; Hasegawa 1992; Johanson and White 1979; Pilbeam 1984; Simons 1989; Zihlman and Lowenstein 1979). It is commonly thought that *Pan* is the closest living relative of *Homo* and that the ancestral stock of *Gorilla* split off considerably earlier, but Rogers (1993) suggested that the lines leading to all three genera diverged over a relatively brief period. There is a minority view, supported by certain morphological evidence, that *Pongo* is the closest living relative of *Homo* (Schwartz 1987, 1988a). There also have been some suggestions that *Gigantopithecus*, the huge ape of the Pliocene and Pleistocene of southeastern Asia, which usually is thought to have been a relative of *Pongo*, may actually represent an aberrant branch of the line leading to *Homo* (Ciochon 1988; Gelvin 1980; Oxnard 1985).

If the possible affinities of *Pongo* and *Gigantopithecus* are disregarded, and if the Hominidae are restricted to creatures that developed subsequent to their divergence from the ancestors of the modern African apes, then the current consensus is that the family comprises only *Homo*, the extinct *Australopithecus*, and perhaps one or two other extinct genera closely related to the latter; it also is thought

that all major developments in the evolution of the family took place in Africa (Lewin 1993; Martin 1990; Pilbeam 1984; Simons 1989; Wood 1992). Some fragmentary remains, possibly referable to *Australopithecus*, date from around 5 million years ago, but the earliest described hominid species is based mostly on dentition from sites about 4.4 million years old in Ethiopia. This species, *A. ramidus*, originally was assigned to *Australopithecus* (White, Suwa, and Asfaw 1994; Wood 1994) but subsequently was placed in an entirely new genus, *Ardipithecus* (White, Suwa, and Asfaw 1995). In several critical morphological characters it is closer to *Pan* than to *Australopithecus*, but it evidently represents a basal line leading to the latter.

Fossilized bones and bipedal footprints suggest that *Australopithecus* itself arose about 4 million years ago. The oldest relatively complete specimen, popularly known as "Lucy" and representing a species designated *A. afarensis*, was found in Ethiopia. There is some controversy about the content of this taxon, with Ferguson (1992) arguing that it is a composite species. In any case, Lucy is thought to have been about 1 meter in height and 30 kg in weight and to have had a brain capacity of 400 cu cm, about equal to that of a modern chimpanzee. The males of her species may have been up to 1.5 meters tall and averaged 45 kg in weight. Later fossil material from eastern and southern Africa suggests that *A. afarensis*, either directly or after having evolved into another species, *A. africanus* (Skelton and McHenry 1992), gave rise to two evolutionary lines. One seems to have consisted of relatively lightly built creatures that probably were ancestral to *Homo*.

The other line is represented by a specimen known as the "black skull," which was found in Kenya in deposits about 2.5 million years old. This line apparently gave rise to two heavily built species, *A. robustus* and *A. boisei*, which some authorities, including Groves (1989) and Wood (1992), place in the separate genus *Paranthropus*. These species survived for a substantial period, were contemporary in Africa with the earliest known populations of *Homo*, and did not become extinct until about 1.2 million years ago (Ward 1991). The remains of both *Australopithecus* and *Homo* have been found in the lowest strata of Olduvai Gorge in Tanzania, so these two genera of hominids may have occurred together in this region. Both may also have used stone implements and tools, since some of these have been found in association with the skeletal remains.

The earliest definitely known specimens of *Homo* itself are from Kenya and Malawi and have been dated to about 2.4 million years ago (Hill et al. 1992; Schrenk et al. 1993). These occurrences suggest that the genus arose in tropical Africa during a period of climatic cooling and subsequently spread southward. There is some question about the number of early species, with certain authorities recognizing only one, *H. habilis*, and others accepting at least one contemporary, *H. rudolfensis*. The brain capacity of *H. habilis* was about 700 cu cm, considerably greater than that of any modern ape. These early forms of *Homo* disappeared, but from among them evolved *H. erectus*. The latter is first represented by specimens about 1.6 million years old from eastern Africa, which had a brain capacity of around 900 cu cm. At least 1 million years ago *H. erectus* expanded its range from Africa to Eurasia. From among the resulting widespread populations there seems to have gradually evolved an archaic form of *Homo sapiens*. In Europe and southwestern Asia this form became established about 150,000 years ago as the subspecies *H. sapiens neanderthalensis*. Some genetic studies suggest that modern *H. sapiens* meanwhile originated about 100,000–200,000 years ago in Africa and eventually spread out to replace other human populations. There also is substantial evidence that the replacement of *H. s. neanderthalensis* by modern human beings occurred 30,000–40,000 years ago. Strahan (1983) noted that people first entered Australia at least 40,000 years ago. According to Owen (1984), there is no substantial evidence that *Homo* occupied the Americas, except possibly the area immediately east of the Bering Strait, until about 12,000 years ago.

There is substantial disagreement about the phylogeny, biogeography, and nomenclature of the populations involved in the development of *Homo* (Sussman 1993). Some authorities, for example, have argued that *H. erectus* represents an early offshoot that developed in Asia and was not directly involved in the evolution of *H. sapiens*. Instead, they say, another species, *H. ergaster*, would have arisen in Africa subsequent to the emigration of the forebears of *H. erectus* and represented the ancestral stock of modern humans (Groves 1989; Harrison 1993; Turner and Chamberlain 1989; Wood 1992). In contrast, Kramer (1993) concluded that all the variation demonstrated by *Homo* in the period from 1.7 million to 0.5 million years ago can be accommodated by the single species *H. erectus*. There are questions concerning how distinct archaic *H. sapiens* was from the modern populations and whether the disappearance of the former involved interbreeding with, or total replacement by, the latter. Tattersall (1992) argued that *H. s. neanderthalensis* should be regarded as a distinctive species. A highly publicized analysis of mitochondrial DNA suggesting that all modern humans can trace their ancestry to a single female that lived in Africa about 200,000 years ago was refuted by Pickford (1991).

The family Hominidae, as accepted here, is distinguished from the Pongidae by skeletal modifications that adapt it to erect bipedalism; such characters in the Pongidae are directed primarily to arboreal brachiation. Thus, in the Hominidae there is a proportional lengthening of the lower extremities, as well as modifications of the pelvis, femur, and musculature to reflect erect posture, whereas in the Pongidae there is lengthening of the upper extremities and modification of morphological details to reflect brachiation. The two families also differ in a number of other minor skeletal and dental details (Buettner-Janusch 1966).

The genus *Australopithecus* has been described as having, among others, the following characters: (1) a large masticatory apparatus reflecting an adaptation to a diet that required heavy chewing; (2) a small cranial capacity (350–530 cu cm); (3) strong supraorbital ridges; (4) a low sagittal crest; (5) a skeleton resembling that in *Homo* but differing in detail (a forward prolongation of the region of the anterior superior spine of the ilium and a relatively small sacroiliac surface); (6) large, long-fingered hands; and (7) arms shorter in relation to torso length than in apes of comparable size (Johanson and White 1979; Le Gros Clark 1955; Simons 1989). The primary characters of the genus *Homo* have been described as (1) no specialization relating to a heavily masticated diet; (2) a skeleton modified for habitual erect posture; and (3) a large cranial capacity (700 to almost 2,000 cu cm) (Johanson and White 1979; Leakey, Tobias, and Napier 1964; Le Gros Clark 1955). Mayr (1951) perceived the genus *Homo* as being characterized by progressive brain enlargement associated with increasing cultural elaboration. The species *H. sapiens* presents the ultimate in the tendencies toward brain expansion and reduction of the masticatory apparatus characteristic of earlier forms of hominids.

There is considerable variation in size of *Homo*. Recently there were announcements of the death of both the world's smallest man, who was about 74 cm tall, and the world's tallest man, 243 cm tall (*Washington Post*, 13 February 1988, C-4, and 24 January 1990, B-7, respectively). According to Thorington and Anderson (1984), the standing height of adult males averages 163–72 cm; weight averages about 75 kg in males and 52 kg in females; pelage is sparse, usually of one color, black to blond or reddish; and dental formula is: (i 2/2, c 1/1, pm 2/2, m 3/3) × 2 = 32.

Living populations of *H. sapiens* differ from one another only in minor details of skin coloration, hair texture and color, facial features, and so on. Using standard criteria as applied to other mammals, four or five possible geographic races of people can be recognized, with a number of minor variants.

1. Negroid—originally distributed in tropical parts of Africa and Southeast Asia; now also widely found in North, Central, and South America and in the Caribbean area. This race contains some of the shortest human beings (Ituri Forest pygmies in Zaire, whose height is usually under 1.4 meters) as well as some of the tallest (the Tutsi of Burundi and Rwanda, whose height is as great as 2 meters and over). Negroids are characterized by dark skin coloration and curly, stiff hair that is only sparsely distributed on the body.
2. Caucasoid—originally distributed in Europe, North Africa, and extreme western Asia; now widely scattered throughout much of the world. The subspecies is characterized by light or dark skin coloration and wavy, soft hair that forms a heavy beard in the males and is usually densely distributed over the body.
3. Mongoloid—originally distributed in eastern Asia; Mongoloid stock also migrated in late Pleistocene and

The transcendency and diversity of *Homo sapiens*, as represented by the crew of the space shuttle *Challenger*. Front row, left to right: Michael J. Smith, Francis R. Scobee, Ronald E. McNair; Back row, left to right: Ellison S. Onizuka, Sharon Christa McAuliffe, Gregory Jarvis, Judith A. Resnik. All died when their spacecraft exploded during launch on 28 January 1986. Photo courtesy of National Aeronautics and Space Administration.

early Recent times to the Americas by way of the Bering Land Bridge or some other route and gave rise to the American Indians. Mongoloids tend to be intermediate in skin color to Negroids and Caucasoids, though many are very pale. They are characterized by straight, very dark hair sparsely distributed over the body, a flat face with protruding cheekbones, and, in some varieties at least, the Mongolian eye folds.

4. Australoid and Khoisanic—possibly distinct from the three major races listed above. The Australoids are represented by the Australian aborigines (and perhaps by the Ainus of Japan, as well as others) and are characterized in very general terms as having dark skin coloration, wavy hair, which may be thick over the body and face, and heavy brow ridges. The true Khoisanics are represented today by only a few Bushman tribes living in the deserts of southern Africa but may at one time have been very widely distributed across the African continent. In general, they can be described as having yellowish skin coloration; sparse, wiry hair that grows in clumps over the head ("peppercorn hair"); steatopygia, especially marked in females; short stature; and delicate facial features with little development of a brow ridge. Both Australoids and Khoisanics have suffered greatly from persecution by other peoples and today are reduced in numbers and distribution.

It must be emphasized that at least in recent years *Homo sapiens* has become a case apart from all other mammals.

The extreme mobility of the species has thoroughly mixed the races enumerated above in many areas and has distributed at least the three major races widely throughout the world. Even before modern means of rapid transportation, humans were wide-ranging animals and there was extensive interbreeding between the races. Because of the movements of human populations, zones of intergradation tended to cover vast areas, thus further confusing the picture.

Three calculations of average length of the menstrual cycle in *Homo sapiens* are 28.7, 30.4, and 33.9 days (Hayssen, Van Tienhoven, and Van Tienhoven 1993). The gestation period of the species varies from 243 to 298 days, usually being about 280 days. Generally a single baby is born, with twins occurring once in 88 births, triplets once in 7,600 births, and quadruplets once in 670,000 births. Very rarely there are five, six, or more offspring. In the United States, apparently in association with a societal trend toward delaying childbearing, the rate of twins recently has increased to once in 43 births, triplets to once in 1,341 (*Washington Post Health,* 26 July 1994, 5). The weight of single babies averages 3.25 kg and ranges from 2.5 to 6 kg. Humans have one of the longest periods of development following birth of any living creature. Human females are fertile from about 13 to 49 years of age. Maximum longevity may be as great as 100 years or more, but average life expectancy is much lower and differs widely from population to population, depending mainly on such factors as the availability of modern medical expertise, equipment, and drugs and also, apparently, on the lifestyle of the people. The average life expectancy at birth of peo-

The first record of *Homo sapiens*, or of any mammal, on the surface of a world other than the Earth. Astronaut Edwin E. Aldrin stands on the moon during the *Apollo XI* mission, 20 July 1969. He was photographed by the mission commander, Neil A. Armstrong, while pilot Michael Collins remained in lunar orbit. Photo courtesy of National Aeronautics and Space Administration.

ple in some of the developing countries may be as low as 30 years, mainly because of high infant mortality, but it is well above 70 years in some of the more developed countries. In the United States, life expectancy at birth is 78.9 years for females and 72.0 years for males (*Washington Post*, 1 September 1993, A-3).

All three of the principal racial groups of people (Negroid, Caucasoid, and Mongoloid) have populations that are simple hunters/gatherers, farmers, and herders, and all three have produced populations that have advanced civilizations, that is, having written languages, highly developed arts and sciences, and complex social and cultural developments. Hunters and gatherers perhaps represent the basic form of human society, and the one that comes closest to representing the type of life lived by early human beings. According to Hayden (1981), existing hunter/gatherer societies still occupy many habitats, including tropical forests, semideserts, temperate woodlands, and arctic tundra. The diet is omnivorous but may include more meat than is generally thought. Reported population densities range from about 0.4/sq km to 75.0/sq km. Group size

varies from 9 to 1,500 but usually is about 25–50. For food gathering, members of such groups forage up to 8 km per day during a period of 2–5 hours, and hunting for meat involves movements as great as 16–24 km. The maximum yearly home range for groups that lack transportation aids is about 2,500 sq km.

The Negroid Mbuti pygmies of the Ituri Forest in Zaire are an example of a hunting and gathering people and have been well studied. According to Turnbull (1976), their economy requires a minimal technology and is still at the Stone Age level, though Harako (1981) explained that they have been substantially influenced by contact with village peoples. They domesticate neither plants nor animals. They gather mushrooms, roots, fruits, berries, and nuts, but hunting is the primary influence on their society. They live in small bands of 3–14 family units, and each band is territorial, with a large no man's land in the center of the forest. Harako (1981) indicated that originally each band had an exclusive hunting area of 100–200 sq km. When a band grows to a size of more than 30 individuals, it subdivides, since no band can grow in size beyond what local game and vegetable supplies can feed. To prevent excessive consumption in any one area, the bands move from camp to camp, never staying more than a month in one place.

The Mbuti family is the basic unit, but in a sense the entire band considers itself a single family. A band does not necessarily consist of families related to each other in one line or another, and the composition may change with every monthly move. The families within a band are extremely cooperative with one another, and great affection and intimacy are usually evident between all members. The entire band will participate in a hunt, with men setting up nets and guarding with spears, youths standing farther back to shoot with bow and arrow any game that escapes, and women and children forming an opposing semicircle at some distance to drive the game into the net. Usually the band can obtain more than enough game with several casts of the nets, whereupon everyone returns to camp, and by noon each family is cooking its meal over an open fire in front of the small leaf hut in which it lives. Everyone, even the children, has a role to play in this sort of society.

The pygmies apparently have a belief in God as a universal creator; although they enjoy meat as food, they still somehow regard it as wrong to take life. At an early age, children are introduced to a concept of dependence on and trust in their forest world and made to feel a part of it. They personalize it and refer to it as father and mother and say that it provides them with all that they need, with life itself. Thus, they adapt to their environment instead of trying to control it.

Since pygmies are highly mobile, their social organization must be fluid. Because bands are constantly changing in size and composition, there can be no chiefs or individual leaders, for they would be as likely to move somewhere else as anyone, leaving the band without a leader. Elders, however, act as arbitrators and make decisions on major issues facing the band, and they are highly respected by all; there is no lineage system among the Mbuti.

The pygmies' technology is simple but more than adequate for their needs. Their clothing consists only of a barkcloth pubic covering, and their houses and furniture can be made within a matter of minutes from saplings and leaves (they obtain metal machetes and knives from neighboring village peoples). They place themselves completely under the forest's protection, and in times of trouble they sing songs to "awaken the forest" and draw its attention to the plight of its children; then all will be well.

The Mbuti are a society of human beings who live in harmony with their environment. The lifestyle of these people may be as close to that of early *H. sapiens* as can be found in the world today. They and other hunting and gathering societies have been under great pressure in recent years as vast areas of forest and woodland have been cut over for agricultural and pastoral purposes; nevertheless, such cultures still exist widely in all subspecies of humans.

Agricultural and herding peoples represent the next step in the ladder leading to the "technologically advanced" societies that control much of the world today. It was a revolutionary step when people first learned that they could plant crops and reap a steady harvest year after year and not have to depend on the availability and abundance of wild nuts, fruits, and tubers. It was also a major achievement when people learned to raise their own meat in the form of cattle, pigs, and sheep, so that they no longer had to be constantly on the move searching for game. Then, according to Turnbull (1976), there was the dramatic development of true civilization. Apart from the trend toward industrialization, however, herding and cultivating cultures still characterize most human societies today.

With crops and herds providing a ready and dependable source of food, people were able to devote more of their time and energies to other affairs. Humans thus began to control nature rather than simply live in harmony with it. As better methods of herding and farming were developed in certain areas, fewer and fewer people could supply food for more and more people, so that the mass of humanity no longer needed to be tied closely to the soil. Cities arose and people devoted their minds to such matters as art, literature, medicine, and science. Great technologies developed that enabled humans to change and control more and more of their natural environment. Rapid means of transportation were developed, roads built, and diseases conquered, and populations grew and expanded. Except for some of the social insects, no other animal has a society as complex and varied as humans', and none has been able to exert such control over the environment.

Along with these advances have come some serious problems, not only for humans themselves but for the earth as a whole. Until recently, human population grew slowly. It is thought that there were only 250 million people on earth in the first year A.D. but 1 billion by 1830. It took a century to add the second billion, 30 years to add the third, and only 15 years to add the fourth. The rate of increase recently has declined dramatically in some areas, particularly because women in developing countries have become better educated and have been able to reduce their number of births. Nonetheless, the world population now is 5.6 billion people, and it is expected to reach at least 10 billion by the year 2050 (*Washington Post*, 4 September 1994, A-1). To provide for the needs of these people, forests are being rapidly cleared, coal and other minerals are being torn from the earth, oil resources are being exploited and probably depleted, industrialization and development are spreading into even remote regions, fisheries are being overharvested, and other animals and plants are being exterminated at an unprecedented rate. The waste and incidental products of human activity are poisoning freshwater and even the oceans, devastating the natural and cultural environment through pollution and acid rain, and threatening the integrity of the atmosphere and climate.

In a recent overview, Vitousek et al. (1997) pointed out that between one-third and one-half of the earth's land surface has been transformed by human action, the carbon dioxide concentration in the atmosphere has increased by nearly 30 percent since the beginning of the Industrial Revolution, more atmospheric nitrogen is fixed by humanity than by all natural terrestrial sources combined, more than half of all accessible surface fresh water is put to human use, two-thirds of major marine fisheries are fully or overexploited or depleted, and the rate of species extinction is now 100 to 1,000 times that before human dominance of the earth. Never has a mammal been so destructive to the habitat upon which it depends for survival. That there is still hope, however, may be seen in the accompanying photographs illustrating the remarkable capabilities and courage of the human species.

Order Carnivora

Dogs, Bears, Raccoons, Weasels, Civets, Mongooses, Hyenas, and Cats

This order of 8 Recent families, 97 genera, and 246 species occurs naturally throughout the world except in Australia, New Guinea, New Zealand, Antarctica, and many oceanic islands. One species, *Canis familiaris*, apparently was introduced into Australia by human agency in prehistoric time and subsequently established wild populations on that continent. Although recently there has been much reassessment of carnivore systematics, most authorities (e.g., E. R. Hall 1981; Hunt and Tedford 1993; Stains 1984; Wayne et al. 1989; Werdelin and Solounias 1991; Wozencraft 1989*b*; Wyss and Flynn 1993) continue to recognize two basic phylogenetic divisions: (1) the suborder Caniformia or superfamily Arctoidea (or Canoidea), with the families Canidae, Ursidae, Procyonidae, and Mustelidae; and (2) the suborder Feliformia or superfamily Aeluroidea (or Feloidea), with the families Viverridae, Herpestidae, Hyaenidae, and Felidae. A major point of contention involves the interrelationships of the Aeluroidea; some evidence indicates that the Hyaenidae and Felidae are most closely related and that the other two families represent earlier evolutionary divergences; other data suggest a dichotomy, with the Herpestidae and Hyaenidae forming one related group and the Viverridae and Felidae forming another.

Many, perhaps most, authorities also now include the Pinnipedia within the Carnivora. Simpson (1945) treated the Pinnipedia as a suborder of the Carnivora and united all the terrestrial families in another suborder, the Fissipedia. Other authorities (e.g., Rice 1977; Tedford 1976; Wozencraft 1989*b* and *in* Wilson and Reeder 1993; Wyss and Flynn 1993) place the pinnipeds in the suborder Caniformia and/or the superfamily Arctoidea. The Pinnipedia are here treated as a full order, the account of which should be seen for further information on questions regarding its classification.

The smallest living carnivore is the least weasel *(Mustela nivalis)*, which has a head and body length of 135–85 mm, a tail length of 30–40 mm, and a weight of 35–70 grams. The largest is the grizzly or brown bear, some individuals of which, particularly along the coast of southern Alaska, attain a head and body length of 2,800 mm and a weight of 780 kg.

Carnivores have four or five clawed digits on each limb. The first digit (pollex and hallux) is not opposable and sometimes is reduced or absent. Some carnivores, including canids and felids, are digitigrade, walking only on their toes. Others, such as ursids, are plantigrade, walking on their soles with the heels touching the ground. The brain has well-developed cerebral hemispheres, and the skull is heavy, with strong facial musculature. The articulation of the lower jaw permits only open-and-shut (not side-to-side) movements. The stomach is simple. Males have a baculum. The number of mammae in females is variable; they are located on the abdomen, except that in the Ursidae some are pectoral.

The teeth are rooted. The small, pointed incisors number 3/3 in all species except *Ursus ursinus*, which has 2/3, and *Enhydra lutris*, which has 3/2. The first incisor is the smallest, and the third is the largest, the difference in size being most marked in the upper jaw. The canine teeth are strong, recurved, pointed, elongate, and round to oval in section. The premolars are usually adapted for cutting, and the molars usually have four or more sharp, pointed cusps. The last upper premolar and the first lower molar, the carnassials, often work together as a specialized shearing mechanism. The carnassials are most highly developed in the Felidae, which have a diet consisting almost entirely of meat, and are least developed in the omnivorous Ursidae and Procyonidae.

Most carnivores are terrestrial or climbing animals. Two genera, *Potos* and *Arctictis,* have prehensile tails. Apparently, all carnivores can swim if necessary, but the polar bear *(Ursus maritimus)* and the river otters *(Lutra, Lontra, Lutrogale, Aonyx, Pteronura)* are semiaquatic, and the sea otter *(Enhydra)* spends practically its entire life in the water. Land-dwelling carnivores shelter in caves, crevices, burrows, and trees. They may be either diurnal or nocturnal.

Most species of the Canidae, Mustelidae, Viverridae, Herpestidae, and Felidae live solely or mainly on freshly killed prey. Their whole body organization and manner of living are adapted for predation. The diet may vary by season and locality. Hunting is done by scent and sight, and the prey is captured by a surprise pounce from concealment *(Panthera pardus)*, a stalk followed by a swift rush *(Mustela frenata)*, or a lengthy chase *(Canis lupus)*. Some species regularly eat carrion. *Arctictis* is largely frugivorous, and the diet of certain other viverrids consists partly of fruit. The Hyaenidae include one genus *(Crocuta)* that is primarily a hunter of large animals, two *(Hyaena, Parahyaena)* that feed mainly on carrion, and one *(Proteles)* that is largely insectivorous. The Procyonidae and Ursidae, except for the carnivorous polar bear, are omnivorous, eating a wide variety of plant and animal life.

Carnivores are solitary or associate in pairs or small

A. Maned wolf *(Chrysocyon brachyurus)*. B. Linsang *(Prionodon linsang)*. C. Leopard *(Panthera pardus)*. D. Giant panda *(Ailuropoda melanoleuca)*. Photos by Bernhard Grzimek.

groups. Females commonly produce a single litter each year, but those of a few species may give birth two or three times annually, and those of some large species usually mate at intervals of several years. Most species have gestation periods of about 49–113 days. Delayed implantation of the fertilized egg occurs in ursids and some mustelids, so the period from mating to birth is considerably longer than average. Litter size commonly ranges from 1 to 13. The offspring usually are born blind and helpless but with a covering of hair. They are cared for solicitously by the mother and, in some species, by the father. There is often a lengthy period of parental care and instruction.

The Carnivora were once considered to include the Creodonta, an extinct group dating back to the late Cretaceous, as a suborder. It now appears, however, that the Carnivora evolved independently, perhaps from ancestral insectivores or from the same basal stock that gave rise to the Primates and Chiroptera (Novacek 1992; Wozencraft 1989a; Wyss and Flynn 1993). The oldest known groups usually referred to the Carnivora are the Viverravidae, an arboreal family that lived from the early Paleocene to the middle Eocene, and the Miacidae, another small, viverridlike family of the Eocene (L. D. Martin 1989). Heinrich and Rose (1995) reported the oldest known miacid skeleton, from the early Eocene of Wyoming, and indicated that it represented a highly arboreal animal that weighed about 1.3 kg. Wyss and Flynn (1993) suggested that the Viverravidae and Miacidae be included not within the Carnivora but in a larger, supraordinal grouping called the Carnivoramorpha.

CARNIVORA; Family CANIDAE

Dogs, Wolves, Coyotes, Jackals, and Foxes

This family of 16 Recent genera and 36 species has a natural distribution that includes all land areas of the world except the West Indies, Madagascar, Taiwan, the Philippines, Borneo and islands to the east, New Guinea, Australia, New Zealand, Antarctica, and most oceanic islands. There are wild populations of the species *Canis familiaris* in Australia and New Guinea, but these apparently origi-

nated through introduction by human agency. The living Canidae traditionally have been divided, mainly on the basis of dentition, into three subfamilies: the Caninae, with the genera *Canis, Alopex, Vulpes, Fennecus, Urocyon, Nyctereutes, Dusicyon, Cerdocyon, Atelocynus,* and *Chrysocyon*; the Simocyoninae, with *Speothos, Cuon,* and *Lycaon*; and the Otocyoninae, with *Otocyon*.

Recent studies have indicated that subfamilial distinction for the Simocyoninae and the Otocyoninae is not warranted and have revealed considerable controversy regarding the systematics of the Caninae. Langguth (1975a) referred most species of the South American *Dusicyon* to a subgenus *(Pseudalopex)* of *Canis*, referred one species *(D. vetulus)* to the genus *Lycalopex*, and retained only one species *(D. australis)* in *Dusicyon*. Clutton-Brock, Corbet, and Hills (1976) considered the genus *Vulpes* to include *Fennecus* and *Urocyon* and the genus *Dusicyon* to include *D. australis*, the species referred by Langguth to *Pseudalopex* and *Lycalopex*, and also *Cerdocyon* and *Atelocynus*. Van Gelder (1978) expanded the genus *Canis* to contain the following as subgenera: *Dusicyon*, with the single species *D. australis*; *Pseudalopex*, with most species traditionally assigned to *Dusicyon*; *Lycalopex*, with the species formerly called *Dusicyon vetulus*; *Cerdocyon*; *Atelocynus*; *Vulpes*, including *Fennecus* and *Urocyon*; and *Alopex*. Berta (1987) considered *Otocyon* and *Urocyon* to be closely related to *Vulpes*, *Lycalopex* to be part of *Pseudalopex*, *D. australis* to be the only modern species of *Dusicyon*, *Nyctereutes* to be the closest living relative of *Cerdocyon*, and *Speothos* to be the closest living relative of *Atelocynus*. Based on analyses of mitochondrial DNA, Geffen, Mercure, et al. (1992) considered *Alopex* to be very closely related to *Vulpes velox* and *Fennecus* to form a mono-

A. Maned wolf pup *(Chrysocyon brachyurus)*, photo from Los Angeles Zoological Society. B. Red fox pup *(Vulpes vulpes)*, photo by Leonard Lee Rue III.

Bush dogs *(Speothos venaticus)*, photo by Bernhard Grzimek.

phyletic unit with *V. cana*. They recommended that all be included in the single genus *Vulpes* but also concluded that *Urocyon* is not closely related to any of the other genera.

Because of the controversy, a conservative position has been taken here, and all traditionally recognized genera have been maintained. In addition, *Pseudalopex* and *Lycalopex* have been given generic rank. Wozencraft (*in* Wilson and Reeder 1993) took much the same position but included *Lycalopex* within *Pseudalopex*. The sequence of genera presented here is based partly on information given by Langguth (1975a) and Nowak (1978, 1979) indicating that *Vulpes* and the other foxes are more primitive than *Canis*. The sequence also gives some consideration to the molecular analyses of Wayne (1993) and Wayne and O'Brien (1987), which suggest the existence of a wolflike (including all *Canis*), a foxlike, and a South American group of genera. That scheme, however, indicates that *Lycaon* and possibly *Speothos* are associated with the wolflike group and that the genera *Urocyon, Nyctereutes,* and *Otocyon* show no close affinity with any of the groups. Also by that approach *Chrysocyon* is associated distantly with the South American group, but a chromosomal analysis (Vitullo and Zuleta 1992) suggested northern affinity for *Chrysocyon*. Further systematic comments are given in the generic accounts. Two additional genera, *Cubacyon* and *Paracyon*, described from subfossil material found on Cuba, almost certainly represent domestic *Canis familiaris* introduced by Amerindian peoples (Morgan and Woods 1986) and are not discussed further here.

In wild species head and body length is 357–1,600 mm, tail length is 125–560 mm, and weight is 1–80 kg. *Fennecus zerda* is the smallest species and *Canis lupus* is the largest.

In a given population males generally are larger than females. Most species are uniformly colored or speckled, but one species of jackals *(Canis adustus)* has stripes on the sides of its body, and *Lycaon* is covered with blotches.

Canids have a lithe, muscular, deep-chested body; usually long, slender limbs; a bushy tail; a long, slender muzzle; and large, erect ears. There are four digits on the hind foot and five on the forefoot, except in *Lycaon,* which has four on both the front and the back foot. The claws are blunt. Males have a well-developed baculum, and females generally have three to seven pairs of mammae.

The skull is elongate. The bullae are prominent but usually are not highly inflated. The dental formula in all but three species is: (i 3/3, c 1/1, pm 4/4, m 2/3) × 2 = 42. The molars are 1/2 in *Speothos,* 2/2 in *Cuon,* and 3/4 or 4/4 in *Otocyon.*

Canids occur from hot deserts *(Fennecus)* to arctic ice fields *(Alopex).* For dens they may use burrows, caves, crevices, or hollow trees. These alert, cunning animals may be diurnal, nocturnal, or crepuscular. They are generally active throughout the year. They walk, trot tirelessly, amble, or canter, either entirely on their digits or partly on more of the foot. At full speed they gallop. The gray foxes *(Urocyon)* often climb trees, an unusual habit for canids. The senses of smell, hearing, and sight are acute. Prey is captured by an open chase or by stalking and pouncing. The diet may vary by season, and vegetable matter is important to some species at certain times.

Some canids, especially the larger species, occur in packs of up to 30 members and seek prey animals that are larger than themselves. Most smaller canids hunt alone or in pairs, preying on rodents and birds. There is usually a regular home range, part or all of which may be an exclusive territory. Females generally give birth once a year. Litters usually contain 2–13 young. Gestation averages around 63 days. The offspring are blind and helpless at birth but are covered with hair. They are cared for solicitously by the mother and often by the father and other group members as well. Sexual maturity comes after 1 or 2 years. Potential longevity is probably at least 10 years in all species.

The geological range of this family is late Eocene to Recent in North America and Europe, early Oligocene to Recent in Asia, late Pliocene to Recent in South America, Pliocene to Recent in Africa, and late Pleistocene to Recent in Australia (Berta 1987; Langguth 1975a; Macintosh 1975).

CARNIVORA; CANIDAE; Genus VULPES
Frisch, 1775

Foxes

There are 10 species (Clutton-Brock, Corbet, and Hills 1976; Coetzee *in* Meester and Setzer 1977; Corbet 1978; Corbet and Hill 1992; Ellerman and Morrison-Scott 1966; Geffen et al. 1993; E. R. Hall 1981; Mendelssohn et al. 1987; Roberts 1977):

V. vulpes (red fox), Eurasia except the southeastern tropical zone, northern Africa, most of Canada and the United States;

V. corsac (corsac fox), dry steppe and subdesert zone from the lower Volga River to Manchuria and Tibet;

V. ferrilata (Tibetan sand fox), high plateau country of Tibet, Nepal, and north-central China;

V. cana (Blanford's fox), dry mountainous regions of southern Turkmenistan, Iran, Pakistan, Afghanistan, southern and western Arabian Peninsula, Israel, and Sinai;

V. velox (swift fox), southern Alberta and North Dakota to northwestern Texas;

V. macrotis (kit fox), southern Oregon to Baja California and north-central Mexico;

V. bengalensis (Bengal fox), Pakistan, India, Nepal, Bangladesh;

V. rueppellii (sand fox), desert zone from Morocco and Niger to Afghanistan and Somalia;

V. pallida (pale fox), savannah zone from Senegal to northern Sudan and Somalia;

V. chama (Cape fox), dry areas of southern Angola, Namibia, Botswana, western Zimbabwe, and South Africa.

The reason for using Frisch, 1775, rather than Bowdich, 1821, as the authority for this generic name was explained by Corbet and Hill (1992). Treatment of *Vulpes* as a distinct genus that does not include *Fennecus, Urocyon,* or *Alopex* is in keeping with the arrangements of such authorities as Coetzee (*in* Meester and Setzer 1977), E. R. Hall (1981), Jones et al. (1992), and Rosevear (1974). Other views have been to consider *Vulpes* a full genus that includes *Alopex* (Youngman 1975), a full genus that includes *Fennecus* and *Urocyon* but not *Alopex* (Clutton-Brock, Corbet, and Hills 1976), a full genus that includes *Fennecus* but not *Urocyon* and *Alopex* (Wozencraft *in* Wilson and Reeder 1993), a full genus that includes *Fennecus* and *Alopex* but not *Urocyon* (Geffen, Mercure, et al. 1992), and a subgenus of *Canis* that includes *Fennecus* and *Urocyon* but not *Alopex* (Van Gelder 1978). The North American red fox has sometimes been designated a separate species, *V. fulva,* but most authorities now consider it to be conspecific with the Palaearctic *V. vulpes.*

E. R. Hall (1981) treated *V. macrotis* as being conspecific with *V. velox* because Rohwer and Kilgore (1973) had reported interbreeding between these two kinds of fox where their ranges meet in eastern New Mexico and western Texas. Combining the two species was supported by Dragoo et al. (1990) based on genic data, though morphometric data were inconclusive, and Wozencraft (*in* Wilson and Reeder 1993) took the same position. Additional studies using morphological characters (Stromberg and Boyce 1986; Thornton and Creel 1975) and analyses of mitochondrial DNA (Mercure et al. 1993) have argued for continued recognition of the two as separate species.

Vulpes is characterized by a rather long, low body; relatively short legs; a long, narrow muzzle; large, pointed ears; and a bushy, rounded tail that is at least half as long, and often fully as long, as the head and body. The pupils of the eyes generally appear elliptical in strong light. Some species have a pungent "foxy" odor, arising mainly from a gland located on the dorsal surface of the tail, not far from its base. Females usually have six or eight mammae. Additional information is provided separately for each species.

Vulpes vulpes (red fox)

Head and body length is 455–900 mm, tail length is 300–555 mm, and weight is 3–14 kg. Average weights in North America are 4.1–4.5 kg for females and 4.5–5.4 kg for males (Ables 1975). The usual weight in central Europe is 8–10 kg (Haltenorth and Roth 1968). The typical coloration ranges from pale yellowish red to deep reddish brown on the upper parts and is white, ashy, or slaty on the underparts. The lower part of the leg is usually black, and the tail is generally tipped with white or black. Color variants, known as the "cross fox" and the "silver fox," represent, respectively, about 25 percent and 10 percent of the species. The cross fox is reddish brown in color and gets its name from the cross formed by one black line down the middle of the back and another across the shoulders. The color of the silver fox, whose fur is the most prized among foxes, ranges from strong silver to nearly black. The general color effect depends on the proportion of white or white-tipped hairs to black hairs. An individual with only a few white hairs is sometimes called a "black fox." In Europe such black individuals occur only in the north and represent at most 1 percent of the population; in North America they are more common (Krott 1992). Aberrant individuals known as "samson foxes" sometimes appear in a population and may occur in substantial numbers for various periods (Voipio 1990). They lack the guard hairs of the normal pelage and have other unusual morphological and behavioral characters.

The red fox rivals the gray wolf *(Canis lupus)* for having the greatest natural distribution of any living terrestrial mammal besides *Homo sapiens.* Habitats range from deep forest to arctic tundra, open prairie, and farmland, but the red fox prefers areas of highly diverse vegetation and avoids large homogeneous tracts (Ables 1975). Elevational range is sea level to 4,500 meters (Haltenorth and Roth 1968). Daily rest may be taken in a thicket or any other protected spot, but each individual or family group usually has a main earthen den and one or more emergency burrows within the home range. An especially large den may be constructed during the late winter and subsequently used to give birth and rear the young. Some dens are used for many years by one generation of foxes after another. The preferred site is a sheltered, well-drained slope with loose soil. Often a marmot burrow is taken over and modified. Tunnels are up to 10 meters long and lead to a chamber 1–3 me-

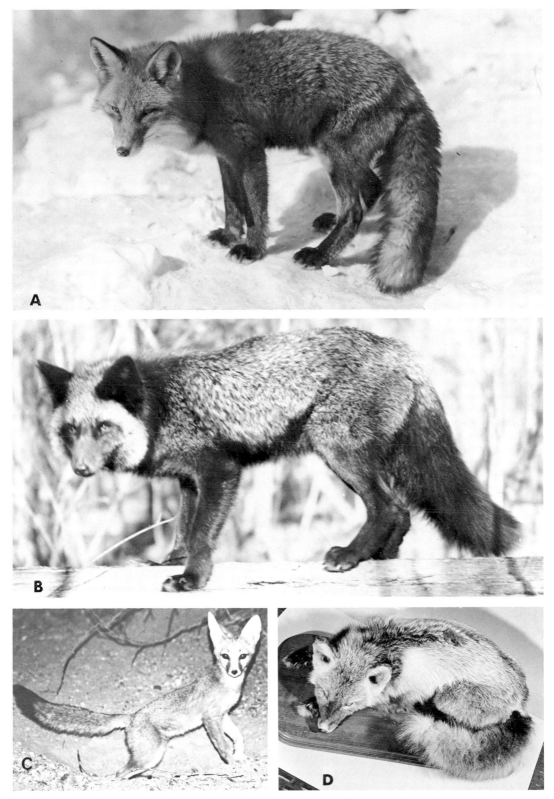

A. Red fox *(Vulpes vulpes)*, photo from New York Zoological Society. B. Silver fox *(V. vulpes)*, photo from Fromm Brothers, Inc. C. Kit fox *(V. macrotis)*, photo by O. J. Reichman. D. Cross fox *(V. vulpes)*, photo by Howard E. Uible of mounted specimen in U.S. National Museum of Natural History.

ters below the surface. There is sometimes only a single entrance, but there may be as many as 19. A system of pathways connects the dens, other resting sites, favored hunting areas, and food storage holes (Ables 1975; Banfield 1974; Haltenorth and Roth 1968; Stanley 1963).

The red fox is terrestrial, normally moving by a walk or trot. It has great endurance and can gallop for many kilometers if pursued. It can run at speeds of up to 48 km/hr, can leap fences 2 meters high, and can swim well (Haltenorth and Roth 1968). It has keen senses of sight, smell, and hearing. Its ability to survive in the close proximity of people and often to elude human hunters and their dogs has given it a reputation for cunning and intelligence. Most activity is nocturnal and crepuscular. Individuals cover up to 8 km per night as they move on circuitous routes through the home range (Banfield 1974). During the autumn the young born the previous spring disperse from the parental home range. The usual distance traveled at this time is about 40 km for males and 10 km for females; the maximum known distance traveled is 394 km (Ables 1975; Storm et al. 1976). Once the young animals establish themselves in a new area, they generally remain there for life.

The diet is omnivorous, consisting mostly of rodents, lagomorphs, insects, and fruit. To hunt mice, the red fox stands motionless, listens and watches intently, and then leaps suddenly, bringing its forelegs straight down to pin the prey. Rabbits are stalked and then captured with a rapid dash (Ables 1975). Daily consumption is around 0.5–1.0 kg. Sometimes a hole is dug and excess prey placed therein and covered over, to be eaten at a later time (Haltenorth and Roth 1968).

The most favorable areas usually support an average of one or two adults per sq km (Ables 1975; Haltenorth and Roth 1968; Insley 1977). Home range size varies with habitat conditions and food availability; it becomes larger in winter and smallest around the time of the arrival of newborn (Ables 1975). According to Zimen (in Grzimek 1990), the home range under natural conditions is usually 1–10 sq km but may be as small as 10 ha. in suburban areas. Jones and Theberge (1982) reported home range to average about 16 sq km in tundra habitat of northwestern British Columbia, larger than in temperate environments. V. vulpes is apparently territorial. There is little overlap of home ranges, and individuals on different ranges avoid one another (Storm and Montgomery 1975). Captive males were found continually to harass and chase foxes newly introduced to the enclosure, but females seldom became involved in such interaction (Preston 1975). In the breeding season, however, females do exhibit territorial behavior (Haltenorth and Roth 1968).

A home range is typically occupied by an adult male, one or two adult females, and their young (Storm and Montgomery 1975). Occasionally two females have litters in the same den (Pils and Martin 1978). Males may fight one another during the breeding season. A vixen sometimes mates with several males, but she later establishes a partnership with just one of them (Haltenorth and Roth 1968). For a period extending from shortly before birth to several weeks thereafter the female remains in or very near the den. The male then brings her food but does not actually enter the maternal den.

The mating season varies with latitude. In Europe it is December–January in the south, January–February in central regions, and February–April in the north (Haltenorth and Roth 1968). In North America mating occurs over about the same period (Ables 1975; Storm et al. 1976). Females are monestrous; estrus is 1–6 days and gestation is 49–56, usually 51–53, days. Litter size is 1–13 young but averages about 5 throughout the fox's range (Ables 1975;

H. G. Lloyd 1975). The young weigh 50–150 grams at birth, open their eyes after 9–14 days, emerge from the den at 4–5 weeks, and are weaned at 8–10 weeks. They may be moved to a new den at least once. The family remains together until the autumn. Sexual maturity is reached at about 10 months. Potential natural longevity is around 12 years, though few individuals live more than 3–4 years, at least where the species is heavily hunted and trapped (Ables 1975). The oldest individual reliably recorded was a male aged 10 years and 8 months taken in Labrador (Chubbs and Phillips 1996).

The red fox is killed by people for sport, to protect domestic animals and game, to prevent the spread of rabies, and to obtain the valuable pelt. Sport hunting may involve an elaborate daytime chase by large numbers of riders and dogs or a nocturnal effort by one person to lure the fox with a call imitating that of a wounded rabbit. In Great Britain V. vulpes is traditionally valued as a game animal, but H. G. Lloyd (1975) noted that it was also the only mammal in the country subject to a government-approved bounty. It has become common in parts of London and other cities, and control efforts there have not substantially reduced its numbers (Harris and Smith 1987). The red fox is often considered to be a threat to poultry, but depredations are generally localized and many of the birds eaten are taken in the form of carrion. Studies have indicated that the red fox has little effect on wild pheasant populations (Ables 1975). Rabid foxes are said to be a serious menace in some areas, especially Europe, and intensive persecution there may be threatening the species in certain parts of its range; 200,000 individuals are taken every year in Germany alone. Such direct killing has had little effect in preventing the disease, but Switzerland, which has an extensive program of spreading oral vaccine baits, has become virtually free of rabies (Zimen in Grzimek 1990). A rabies epizootic, the main vector of which is the fox, spread from Poland across much of Europe from the 1940s to 1970s; however, only 5–10 percent of reported cases of rabies in domestic animals in the involved region have resulted from this epizootic (Steck and Wandeler 1980). The epizootic reached a peak in 1989 but subsequently declined in association with greater use of oral vaccine baits (Barrat and Aubert 1993). Most cases of rabies in Canada from 1958 to 1986 were reported from Ontario, and most of those occurrences (17,982) were in the red fox (Rosatte 1988b). Problems caused by the fox are perhaps more than balanced by its control of rodent populations, which might otherwise multiply and damage human interests.

From 1900 to 1920 in North America, and to some extent in other parts of the world, catching wild foxes and raising them in captivity developed into an important industry. In the early stages of the breeding effort choice animals often sold for more than $1,000 each. Through selective breeding, strains were developed that nearly always produce silver-colored offspring. The number of foxes being raised for their fur now exceeds that of all other normally wild animals except possibly the mink (Mustela vison). One fox farm permanently employed about 400 people and sold pelts worth more than $18 million annually. The fur is used in coats, stoles, scarves, and trimming. The value of fox pelts has varied widely depending on fashions, availability, and economic conditions. According to Banfield (1974), the average price of a silver fox skin was $246.46 in 1919–20 but only $17.94 in 1971–72. The average price of a wild-caught U.S. red fox skin rose from $12.00 in the 1970–71 season to about $48.00 in 1976–77. The reported number of red foxes trapped for their fur during the latter season in the United States and Canada was 421,705 (Deems and Pursley 1978). The number of pelts taken annually rose to more than 500,000 during the early 1980s, with average prices peaking at more than $60.00, but had declined to less than $20.00 by 1984 (Voigt 1987). In the 1991–92 season 217,257 skins were taken in the United States and sold for an average price of only $10.75 (Linscombe 1994). Considering inflation, this price was far less than what it had been one to two decades earlier. The drop may have been associated with growing social disdain for the use of wild-caught furs in fashion and with a pending ban by the European Community on the importation of pelts derived from the use of leghold traps.

Despite human persecution, *V. vulpes* has maintained or even increased its numbers in many parts of its range. There are now probably more in Great Britain than there were in medieval times because of improved habitat conditions resulting from the establishment of hedgerows and crop rotation (H. G. Lloyd 1975). This species is able to carry on its mode of life in intensively farmed areas and sometimes even in large cities (Ables 1975; Grzimek 1975). It also has been successfully introduced in some areas, especially by persons of English background who desired to continue traditional fox hunting. The species was brought to Australia in 1868 and subsequently spread over much of that continent, to the lasting detriment of the native fauna (Clutton-Brock, Corbet, and Hills 1976; Ride 1970).

Introductions from England also were made in eastern North America in colonial times. The species was naturally present in this region but apparently was not abundant. It subsequently increased in numbers and became established in areas not previously occupied, mainly because of the breaking up of the homogeneous forests by people and continuous introduction by hunting clubs. In the twentieth century the red fox has greatly extended its range in the southeastern United States and has occupied Baffin Island and moved as far north as the southern coast of Ellesmere Island. It has spread westward across the Great Plains, possibly in response to a human-caused reduction of coyote *(Canis latrans)* numbers (Banfield 1974; E. R. Hall 1981; Hatcher 1982; Lowery 1974). The only major North American population that may be in trouble from a conservation viewpoint is that of the Sierra Nevada of California, where surveys indicate that the native subspecies *(V. v. necator)* is very rare and evidently declining (Schempf and White 1977).

Vulpes corsac (corsac fox)

Head and body length is 500–600 mm and tail length is 250–350 mm. The fur is thick and soft. The general coloration of the upper parts is pale reddish gray, or reddish brown with silvery overtones. The underparts are white or yellow. *V. corsac* is externally similar to *V. vulpes* but has relatively longer legs. Its ears are large, pointed, and very broad at the base (Novikov 1962).

The corsac fox is a typical inhabitant of steppes and semidesert. It avoids forests, thickets, plowed fields, and settled areas. It lives in a burrow, often taken over from another mammal, such as a marmot or badger. Self-excavated burrows, sometimes found in groups, are simple and usually very shallow (Novikov 1962). Although usually reported to be nocturnal in the wild, *V. corsac* is active by day in captivity; it is said to be an excellent climber (Grzimek 1975). It runs with only moderate speed and can be caught by a slow dog, but it has excellent senses of vision, hearing, and smell (Stroganov 1969). Most reports indicate that it is nomadic and does not keep to a fixed home range (Ognev 1962). It may migrate southward when deep snow and ice make hunting difficult (Stroganov 1969). The diet consists mostly of small rodents but also includes pikas, birds, insects, and plant material.

This species is more social than other foxes, with sever

Corsac foxes *(Vulpes corsac)*, photo from Amsterdam Zoo.

al individuals sometimes living together in the same burrow (Ognev 1962). Small hunting packs are said to form in the winter (Stroganov 1969), though perhaps these represent mated pairs and their grown young of the previous spring. Males fight one another during the breeding season but then remain with the family (Grzimek 1975; Novikov 1962). Mating occurs from January to March, gestation lasts 50–60 days, and litters usually contain 2–11 young (Stroganov 1969). Females in the Berlin Zoo did not reach sexual maturity until their third year of life (Grzimek 1975).

The corsac fox lacks the penetrating odor of most *Vulpes* and was frequently kept as a pet in eighteenth-century Russia (Grzimek 1975). Its warm and beautiful fur led to large-scale commercial trapping; up to 10,000 pelts were sold annually in the western Siberian city of Irbit in the late nineteenth century. For this reason, and also because of the settlement and plowing of the steppes, the corsac fox has disappeared in much of its range (Ognev 1962; Stroganov 1969).

Vulpes ferrilata (Tibetan sand fox)

Head and body length is 575–700 mm, tail length is 400–475 mm, and males weigh up to 7 kg (Mitchell 1977). The fur is soft and thick and the tail is bushy. The upper parts are pale gray agouti or sandy, with a tawny band along the dorsal region. The underparts are pale, the front of the leg is tawny, and the tip of the tail is white. The skull is peculiarly elongated and has a very narrow maxillary region (Clutton-Brock, Corbet, and Hills 1976).

Mitchell (1977) found this fox on barren slopes and in streambeds at 3,000–4,000 meters in the Mustang district of Nepal. In this area dens are made in boulder piles or in burrows under large rocks. The diet consists of rodents, lagomorphs, and ground birds. Mitchell observed pairs hunting along streambeds, on boulder heaps, and in wheat fields. Mating occurs in late February, and two to five young are born in April or May.

Vulpes cana (Blanford's fox)

Specimens from Afghanistan and Iran had a head and body length of 400–500 mm and a tail length of 330–410 mm (Geffen 1994). A series from Israel averaged about 420 mm in head and body length, 320 mm in tail length, and 1 kg in weight (Geffen, Hefner, et al. 1992*b*). Clutton-Brock, Corbet, and Hills (1976:155) described *V. cana* as "a small fox with extremely soft fur and a long very bushy tail. The colouring is blotchy black, grey and white with a dark tip to the tail and a dark patch over the tail gland. There is an almost black mid-dorsal line and the hind legs may be dark. . . . The underparts are white, the ears are grey, and there is a small dark patch between the eyes and nose."

According to Roberts (1977), the habitat of *V. cana* is mountain steppe. It is reportedly more frugivorous than the other foxes of Pakistan, being fond of ripe melons and seedless grapes, and sometimes damages crops. Geffen et al. (1993) found it relatively common in the hot, rocky habitats of the Negev and Judean deserts. Mendelssohn et al. (1987) added that it has an astonishing jumping ability and can move upward among cliffs by pushing itself from one vertical wall to another. It is strictly nocturnal and shows little change in temporal activity throughout the year (Geffen and Macdonald 1993). Its diet in Israel consists mainly of arthropods, fruit, and other plant material (Geffen, Hefner, et al. 1992*a*), and it apparently has no need to drink free water (Geffen, Degen, et al. 1992).

Reported population densities in Israel are 0.5/sq km and 2.0/sq km (Geffen 1994). A radio-tracking study there showed that individuals moved 7–11 km per night and uti-

lized home ranges of about 1.6 sq km. These ranges were occupied by strictly monogamous pairs and overlapped only minimally with the ranges of other individuals (Geffen and Macdonald 1992). Females captured in that area gave birth to litters of 1 and 3 young in February and April (Mendelssohn et al. 1987). The gestation period is 50–60 days, pups weigh about 29 grams at birth, lactation lasts 30–45 days, and age at sexual maturity is 8–12 months (Geffen 1994).

The skin of *V. cana* is valued in commerce and is heavily hunted. Novikov (1962) called this species one of the rarest predators of the old Soviet Union. It is on appendix 2 of the CITES.

Vulpes velox (swift fox)

Head and body length is 375–525 mm, tail length is 225–350 mm, and weight is 1.8–3.0 kg. Males average larger than females. The winter coat is long and dense; the upper parts are dark buffy gray; the sides, legs, and lower surface of the tail are orange tan; and the underparts are buff to pure white. In summer the coat is shorter, harsher, and more reddish. *V. velox* differs from the closely related *V. macrotis* in having smaller ears, a broader snout, and a shorter tail (Egoscue 1979).

The swift fox inhabits prairies, especially those with grasses of short and medium height. For shelter it depends on burrows, which are either self-excavated or taken over from another mammal. The burrows are usually simple and located on high, well-drained ground. The tunnels may be 350 cm long and lead to a chamber as much as 150 cm below the surface. There are one to seven or more entrances. The swift fox is primarily nocturnal but sometimes suns itself near the den. Its diet consists mostly of lagomorphs and also includes rodents, birds, lizards, and insects (Egoscue 1979; Kilgore 1969).

A radio-tracking study in Nebraska indicated an average nightly movement of about 13 km and found the average home range of 11 individuals to be about 32 sq km, larger than that of any other species of *Vulpes* (Hines and Case 1991). The usual social unit is a mated pair and their young, but occasionally a male will live with two adult females. The mating season in Oklahoma is late December to early January, and most young are born in March or early April. Females are monestrous. Litters consist of three to six young. Their eyes open after 10–15 days, weaning occurs after 6–7 weeks, and they probably remain with the parents until August or early September. A captive lived for 12 years and 9 months (Egoscue 1979; Kilgore 1969).

The swift fox is not as cautious as *V. vulpes* and seems to take poison baits readily. In the mid- and late nineteenth century intensive poisoning was carried out on the Great Plains, mainly to eliminate wolves, coyotes, and other predators, and many swift foxes were accidentally killed. Subsequently much habitat was lost as the prairies were converted to agriculture. The swift fox was also taken for its fur. From 1853 to 1877 in Canada the Hudson's Bay Company reportedly sold more than 100,000 pelts. By the 1920s the northern subspecies, *V. velox hebes*, apparently had disappeared, though occasional reports continued in Canada, and the southern subspecies, *V. velox velox*, survived only in Colorado, New Mexico, western Texas, and possibly western Kansas. For reasons not fully understood the species reappeared in Oklahoma, much of Kansas, Nebraska, and Wyoming in the 1950s and in South Dakota, North Dakota, and Montana in the 1960s and 1970s (Carbyn, Armbruster, and Mamo 1994; Egoscue 1979; Floyd and Stromberg 1981; Kilgore 1969; Moore and Martin 1980; Zumbaugh and Choate 1985). In the last few years it has been recognized that *V. velox* is again declining, mainly because of destruction and fragmentation of native habitat by agricultural activity and perhaps excessive fur trapping; the entire species is estimated to occupy only 10 percent of its original range, and numbers are very low in the northern part of the range (Allardyce 1995; Smeeton 1993). *V. velox* is classified as conservation dependent by the IUCN.

V. velox hebes is on appendix 1 of the CITES. The USDI lists the subspecies as endangered, but this designation officially applies only in Canada. The swift fox populations now on the northern plains of the United States may be descended from animals that moved north from the range of *V. velox velox*. There has been considerable controversy during the last decade regarding both the systematic status of the swift fox subspecies and a reintroduction program in Canada. The latter project, which began in 1983, involves the capture of foxes in Colorado, Wyoming, and South Dakota and the release of them or their offspring in Alberta and Saskatchewan. Stromberg and Boyce (1986) argued that *hebes* probably is not a valid subspecies but that there is significant geographic variation in *V. velox* and that gene flow from the transplanted animals might adversely affect the viability of natural populations on the U.S. side of the border. Herrero, Schroeder, and Scott-Brown (1986) replied that the transplanted stock was taken from the northernmost readily available populations and that more northerly animals in the United States may be recently descended from the same stock. Despite the USDI listing, the original swift fox population of Canada evidently had disappeared by the 1930s. Carbyn, Armbruster, and Mamo (1994) reported that the program had released 569 animals by the end of 1992, that perhaps 13 percent were known to have survived at least 1–2 years, and that reproduction was occurring. In late 1991 an estimated breeding population of 150–225 foxes occupied about 1,200 sq km of southern Alberta and Saskatchewan.

Vulpes macrotis (kit fox)

The account of this species is based in large part on McGrew (1979). Head and body length is 375–500 mm and tail length is 225–323 mm. Average weight is 2.2 kg for males and 1.9 kg for females. The back is generally light grizzled or yellowish gray, the shoulders and sides are buffy to orange, and the underparts are white. The ears, which are proportionally the largest among the North American canids, are set close together.

The kit fox is closely associated with steppe and desert habitat, generally with a covering of shrubs or grasses. Dens usually have multiple entrances, the number varying from 2 to 24. There are groups of dens in favorable areas, and a fox family may move from one to another during the year, leaving most vacant at any given time. *V. macrotis* is nocturnal and may travel several kilometers per night during hunts. Its diet consists largely of rodents, such as kangaroo rats, and lagomorphs.

Optimal habitat in Utah was found to support two adults per 259 ha. Other areas had densities of one fox per 471–1,036 ha. In the San Joaquin Valley, Morrell (1972) found that each fox apparently spent its entire life in an area of 260–520 ha. Home ranges overlap extensively, and there apparently is no definite territory. Usually an adult male and female live together, though not necessarily permanently, and a second female is sometimes present. When a female is nursing young it rarely leaves the den, and the male supplies it with food. Several vocalizations are known, including a bark by mothers to recall the young. Females are monestrous. Mating occurs from December to February, and the young are born in February and March. There are usually four or five offspring weighing about 40 grams each. The young emerge from the den at 1 month and begin to accompany their parents at 3–4 months. The family splits up in the autumn, with the young dispersing beyond the parental home range. According to Jones (1982), one specimen was still living after 20 years in captivity.

The kit fox is not a particularly cautious animal, and its numbers have been greatly reduced in some areas by poisoning, trapping, and shooting. Habitat disruption also has led to declines, especially in California. The subspecies *V. m. macrotis*, of the southwestern corner of the state, had disappeared by 1910. The San Joaquin Valley subspecies, *V. m. mutica*, is classified as endangered by the USDI and as rare by the California Department of Fish and Game (1978). It has declined because of conversion of areas of natural veg-

Kit fox *(Vulpes macrotis)*, photo from San Diego Zoological Society.

etation to irrigated agriculture. Most states classify *V. macrotis* as a fur bearer and allow trapping. O'Farrell (1987) reported that the annual legal kill was 5,400–7,800 and that pelts sold for $12 to $23. In the 1991–92 season 1,519 skins were taken in the United States and sold for an average price of only $4.24 (Linscombe 1994).

Vulpes bengalensis (Bengal fox)

The account of this species is based largely on Roberts (1977). Head and body length is 450–600 mm and tail length is 250–350 mm. Males weigh 2.7–3.2 kg and females weigh less than 1.8 kg. The upper parts are yellowish gray or silvery gray, the underparts are paler, and the backs of the ears and the tip of the tail are dark.

The Bengal fox is generally found in open country with a scattering of trees. It avoids deserts and mountains. It digs its own burrow and hunts mainly at night. A captive could climb low trees and a vertical wire net. The diet is omnivorous and includes small vertebrates, insects, and fruit. *V. bengalensis* is adept at catching frogs and digging lizards out of their burrows. The gestation period is 50–51 days (Hayssen, Van Tienhoven, and Van Tienhoven 1993). The young, usually four to a litter, are born from February to April.

Although once extremely common in India, the Bengal fox is rapidly declining and has disappeared from much of its range. It is not an agricultural pest and its pelt has no commercial value; it is being killed simply for sport (Ginsberg and Macdonald 1990).

Vulpes rueppellii (sand fox)

Head and body length is 400–520 mm and tail length is 250–350 mm (Roberts 1977). Weight is about 1.5–3.0 kg. The coat is very soft and dense. The upper parts are silvery gray, the sides are grayish buff, and the underparts are whitish. *V. rueppellii* is much more lightly built than *V. vulpes*; it has rather short legs and broad ears (Dorst and Dandelot 1969).

The usual habitat is stony or sandy desert. Activity is mainly nocturnal. In a study in Oman, Lindsay and Macdonald (1986) found individuals to spend the day in underground dens and to change their den site on an average of once every 4.7 days. By night these animals hunted over a large home range calculated to average 30.4 sq km by one method and 69.1 sq km by another. The diet in that area consisted mostly of small mammals and also included lizards, insects, and grass. There were three monogamous pairs; each pair shared a den site and a portion of the hunting range, but there also was some overlap in the ranges of different pairs. Two of the pairs had cubs during January–April. Other reported information (Dorst and Dandelot 1969; Roberts 1977) is that a captive female had a litter of three young, and a litter of two young evidently was born in March.

Vulpes pallida (pale fox)

Head and body length is 406–55 mm, tail length is 270–86 mm, and weight is 1.5–3.6 kg. The upper parts are pale sandy fawn variably suffused with blackish, the flanks are paler, and the underparts are buffy white. The tail is long and bushy (Dorst and Dandelot 1969; Rosevear 1974).

The habitat is savannah. Burrows are large, with tunnels extending 10–15 meters and opening into chambers lined with dry vegetation. Activity is mainly nocturnal. The diet includes rodents, small reptiles, birds, eggs, and vegetable matter. The pale fox is gregarious (Dorst and Dandelot 1969; Coetzee *in* Meester and Setzer 1977). Three captive adults, a female and two males, seemed to get along amicably. The female gave birth to a litter of four young in June 1965 (Bueler 1973).

Vulpes chama (Cape fox)

Head and body length is about 560 mm, tail length is about 330 mm, and weight is about 4 kg. The upper parts are silvery gray and the underparts are pale buff. The very bushy tail has a black tip. The ears are pointed. The muzzle is short but pointed (Dorst and Dandelot 1969).

The Cape fox inhabits dry country, mainly open plains

Sand fox *(Vulpes rueppelli)*, photo from Antwerp Zoo.

Cape fox *(Vulpes chama)*, photo from San Diego Zoological Society.

and karroo. It is nocturnal, hiding by day under rocks or in burrows in sandy soil. The diet consists mainly of small vertebrates and insects. This species lives alone or in pairs. Average population density is 0.3/sq km in the Orange Free State. Home range is 1.0–4.6 sq km and there may be overlap where prey density is high. The call is a yell followed by several yaps. The breeding season is September–October, gestation lasts 51–52 days, and litters contain three to five young. Human persecution has been responsible for numerical and distributional declines, though studies have shown that *V. chama* is not a harmful predator (Bekoff 1975; Bothma 1966; Dorst and Dandelot 1969; Ginsberg and Macdonald 1990).

CARNIVORA; CANIDAE; Genus FENNECUS
Desmarest, 1804

Fennec Fox

The single species, *F. zerda*, occurs in the desert zone from southern Morocco and Niger to Egypt and Sudan (Coetzee *in* Meester and Setzer 1977). There also are records from Sinai, southern Iraq, Kuwait, and the southeastern Arabian Peninsula (Gasperetti, Harrison, and Büttiker 1985). *Fennecus* was included in the genus *Vulpes* by Clutton-Brock, Corbet, and Hills (1976), Corbet (1978), Geffen, Mercure, et al. (1992), and Wozencraft (*in* Wilson and Reeder 1993) and in the subgenus *Vulpes* of the genus *Canis* by Van Gelder (1978).

With the possible exception of *Vulpes cana* this is the smallest canid, but it has the proportionally largest ears in the family. Head and body length is 357–407 mm, tail length is 178–305 mm, and weight is about 1.0–1.5 kg. The ears are 100–150 mm long. The coloration of the upper parts, the palest of any fox's, is reddish cream, light fawn, or almost white. The underparts are white, and the tip of the tail is black. The coat is thick, soft, and long. The tail is heavily furred. The bullae are exceedingly large, and the dentition is weak (Clutton-Brock, Corbet, and Hills 1976). The feet have hairy soles, enabling the animal to run in loose sand (Bekoff 1975).

Fennecus occurs in arid regions and usually lives in burrows several meters long in the sand. It digs so rapidly that it has gained the reputation of being able to sink into the ground. It is nocturnal and quite agile; a captive could spring 60–70 cm upward from a standing position and could jump about 120 cm horizontally. Some food apparently is obtained by digging, as evidenced by the pronounced scratching or raking habit of captives. The diet includes plant material, small rodents, birds and their eggs, lizards, and insects, such as the noxious migratory locusts. Although an abundance of fennec tracks around some water holes indicates that this fox drinks freely when the opportunity arises, travelers also report tracks in the desert far from oases. Laboratory studies suggest that *Fennecus* can survive without free water for an indefinite period (Banholzer 1976).

The fennec lives in groups of up to 10 individuals. Males mark their territory with urine and become aggressive during the breeding season. One captive male became dominant over a group at the age of 4 years and then killed its 8-year-old father. Females are aggressive and defend the nest site when they have newborn offspring. Males remain with their mate after the young are born and defend them but do not enter the maternal den. Mating occurs in January and February in captivity, and the young are born in late winter and early spring. Females normally give birth once a year, but if the first litter is lost, another may be produced 2.5–3 months later. The gestation period is about 50–52 days, and there are 2–5 young per litter. The young are weaned at 61–70 days and become sexually mature at 11 months (Bekoff 1975; Koenig 1970). One captive lived 14 years and 7 months (Jones 1982).

Although it does no harm to human interests, the fennec is intensively hunted by the native people of the Sahara. It has disappeared or become rare in many parts of northern Africa (Zimen *in* Grzimek 1990). It is on appendix 2 of the CITES.

Fennec foxes *(Fennecus zerda)*, photo from New York Zoological Society.

CARNIVORA; CANIDAE; **Genus ALOPEX**
Kaup, 1829

Arctic Fox

The single species, *A. lagopus,* occurs on the tundra and adjacent lands and ice-covered waters of northern Eurasia, North America, Greenland, and Iceland (Banfield 1974; Chesemore 1975; Corbet 1978; E. R. Hall 1981). *Alopex* was included within the genus *Vulpes* by Geffen, Mercure, et al. (1992) and Youngman (1975) and was considered a subgenus of *Canis* by Van Gelder (1978) but was treated as a full genus by Clutton-Brock, Corbet, and Hills (1976) and all other authorities cited in this account. Although *Alopex* once was thought to have some distant affinity to *Lycalopex,* the work of Geffen, Mercure, et al. (1992), Wayne (1993), Wayne and O'Brien (1987), and other authorities suggests that *Alopex* is closely related to *Vulpes,* far more so than is *Urocyon.* In any event, *Lycalopex* is really not like *Alopex.*

Head and body length is 458–675 mm, tail length is 255–425 mm, shoulder height is about 280 mm, and weight is 1.4–9.0 kg. The dense, woolly coat gives *Alopex* a heavy appearance. There are two color phases. Individuals of the "white" phase are generally white in winter and brown in summer but may remain fairly dark throughout the year in areas of less severe climate. Individuals of the "blue" phase are pale bluish gray in winter and dark bluish gray in summer. The blue phase constitutes less than 1 percent of the arctic fox population of most of mainland Canada and less than 5 percent of that of Baffin Island but makes up 50 percent of the population of Greenland. In Iceland most foxes are blue, perhaps because of increased camouflage value in coastal areas (Hersteinsson 1989). The winter pelage develops in October and is shed in April. In addition to its coloration, *Alopex* differs from *Vulpes* in having short, rounded ears and long hairs on the soles of its feet.

Alopex is found primarily in arctic and alpine tundra, usually in coastal areas. It generally makes its den in a low mound, 1–4 meters high, on the open tundra. Dens usual-ly have 4–12 entrances and a network of tunnels covering about 30 sq meters. Some dens may be used for centuries, by many generations of foxes, and eventually become very large, with up to 100 entrances. The organic matter that accumulates in and around a den stimulates a much more extensive growth of vegetation than is found on most of the tundra. *Alopex* sometimes dens in a pile of rocks at the base of a cliff. It may use its den throughout the year but often seems not to have a fixed home site except when rearing young. It is active at any time of the day and throughout the year. It moves easily over snow and ice and swims readily (Banfield 1974; Chesemore 1975; Macpherson 1969; Stroganov 1969). During blizzards it may shelter in burrows dug in the snow. Several individuals were noted in winter on the Greenland icecap, more than 450 km from the nearest ice-free land and with the temperature below −50° C. Captives have survived experimental temperatures as low as −80° C.

According to Prestrud (1991), the adaptations of *Alopex* to low temperatures and winter food scarcity include: the best insulative fur of any mammal, with a seasonal increase in fur depth of nearly 200 percent; short muzzle, ears, and legs and a short, rounded body; increased blood flow to a capillary rete in the skin of the pads, which keeps the feet from freezing, and probably a countercurrent vascular exchange mechanism to avoid heat loss when warm blood flows into the legs; storage of substantial body fat reserves, with individuals commonly having a fat content of 20–40 percent during winter; and capability to reduce the basal metabolic rate by a remarkable 40–50 percent during periods of limited food availability. At such times the fox possibly seeks shelter for an extended period but remains relatively alert and able to immediately resume foraging when conditions improve. Hersteinsson and Macdonald (1992) observed that the fur of the arctic fox provides 50 percent better insulation than that of the red fox, but they did not think the differing distributions of the two species resulted directly from relative adaptation to extreme cold. Rather, the northern limit of the red fox seems to be determined by resource availability, and the southern limit of the arctic fox is a factor of interspecific competition with the red fox.

Arctic foxes *(Alopex lagopus):* A. White and blue phases; B. Blue phase; C. Summer coat; photos by A. Pedersen. D. Winter coat, photo by Ernest P. Walker.

The arctic fox makes the most extensive movements of any terrestrial mammal other than *Homo sapiens*. Over much of its range, including Alaska, there is a basic seasonal pattern, probably associated with food availability (Chesemore 1975). In the autumn and early winter the animals shift toward the shore and out onto the pack ice. They are capable of remarkably long travels across the sea ice, having been sighted 640 km north of the coast of Alaska. One individual reportedly reached a latitude of 88° N, at a point 800 km from the nearest land in Russia. A fox tagged on 8 August 1974 on Banks Island was trapped on 15 April 1975 on the northeastern mainland of the Northwest Territories, having covered a straight-line distance of 1,530 km (Wrigley and Hatch 1976). And a fox tagged in northeast-

ern Alaska reached a point on Banks Island, 945 km away (Eberhardt and Hanson 1978). Some foxes have been carried by ice floes as far as Cape Breton and Anticosti islands in the Gulf of St. Lawrence (Banfield 1974).

In some areas, though apparently not in Alaska, there are inland migrations to the forest zone during the winter, in addition to or instead of a shift onto the sea ice. These movements sometimes involve large numbers of foxes and seem especially extensive following a crash in populations of lemmings, a major food source (Banfield 1974; Chesemore 1975; Pulliainen 1965). *Alopex* has occasionally traveled overland to south-central Ontario and the St. Lawrence River. The deepest known penetration in North America was made by an individual taken in December

1974 in southern Manitoba, nearly 1,000 km south of the tundra. In Russia, however, inland movements have extended as far as 2,000 km (Wrigley and Hatch 1976).

Alopex takes any available animal food, alive or dead, and frequently stores it for later use. It may follow polar bears on the pack ice and wolves on land in order to obtain carrion (Banfield 1974). Its winter diet includes the remains of marine mammals, invertebrates, sea birds, and fish (Chesemore 1975). It also is known to be an important predator of the ringed seal *(Phoca hispida)*, digging through the snow to reach the pups in their subnivean lairs (Smith 1976).

In winter for those populations that move inland and in summer throughout most of the range of *Alopex* the diet consists mainly of lemmings. Other small mammals, ground-nesting birds, and stranded marine mammals are also utilized, but arctic fox populations seem to be associated with those of lemmings in a three- to five-year cycle (Chesemore 1975; Wrigley and Hatch 1976). Following a lemming crash, population density has been estimated at only 0.086 foxes per sq km (Banfield 1974).

Although a number of dens may be found in a favorable area, at any one time most are not utilized. Occupied dens are at least 1.6 km apart and are usually distributed such that there is only one family per every 32–70 sq km (Macpherson 1969). In a nonmigratory population on the coast of Iceland home ranges varied from 8.6 to 18.5 sq km and evidently represented defended territories. The foxes there lived in social groups comprising one adult male, two vixens, and young of the year. In such situations one of the vixens apparently is a nonbreeding animal born the previous year that stays on to help care for the next litter before dispersing (Hersteinsson and Macdonald 1982). *Alopex* is monogamous and may mate for life (Chesemore 1975). A number of adults sometimes gather temporarily around a food source, such as a stranded whale, but they then may fight one another. Vocalizations include barks, screams, and hisses (Banfield 1974).

Mating occurs from February to May, and births take place from April to July. Females are monestrous and have an estrus of 12–14 days and an average gestation period of 52 (49–57) days. The number of young per litter varies depending on environmental conditions but ranges from 2 to as many as 25. The usual number seems to be about 6–12. The young weigh an average of 57 grams at birth. They emerge from the den and are weaned at around 2–4 weeks. Subsequently they are brought food by both parents for a brief period, but by autumn they have dispersed. They are capable of breeding at 10 months. Most young do not survive their first 6 months of life, and few animals live more than several years in the wild (Banfield 1974; Chesemore 1975; Macpherson 1969; Stroganov 1969). One individual, however, lived in captivity for 16 years and 2 months (Marvin L. Jones, Zoological Society of San Diego, pers. comm., 1995).

Raising blue-phase foxes has been an important industry; the undressed skins have sometimes sold for up to $300 each. Some operations have pens, where selective breeding is practiced, but most were on small islands, where the animals ran at liberty. The history of the industry in Alaska has been covered in detail by Bailey (1993). Foxes originally did not occur on most islands of the state, and despite some assertions to the contrary, *Alopex* evidently is not indigenous to any part of the Aleutian chain. The Russians began releasing arctic foxes in the Aleutians in 1750, and the process continued after the American purchase in 1867. Introductions also were made on islands off the southern coast of Alaska and in the Alexander Archipelago, but they were generally unsuccessful in the latter region. There are known records of introduction on 455 Alaskan islands, with formal leasing for the purpose of fox farming starting in 1882. The industry grew rapidly as fur prices rose in the early twentieth century, and by 1925 there were 391 farms with over 36,000 blue foxes valued at about $6 million. The Great Depression effectively killed the industry, with average prices falling from $185 in 1919 to $108 in 1929 and only $26 in 1945. Although a few people still raise foxes in pens, no farms remain on the islands. Feral foxes persist, however, on 46 islands, mostly in the Aleutians and off the Alaska Peninsula. Even in the early nineteenth century these animals were known to be adversely affecting the native avifauna of the islands, and as the foxes increased in numbers many populations of waterfowl and seabirds were nearly or entirely exterminated. Some of these populations recovered after the foxes died out or were eliminated by trappers.

White-phase foxes also have desirable furs and have been subject to intensive trapping. The skins may be left the natural white or dyed one of many colors, especially "platinum" or "blue" imitation. The arctic fox has long been important to the economy of the native people living within its range. Fluctuations in fox numbers and fur prices have frequently combined to cause hardship for native trappers. Trade in white fox skins developed in northern Alaska during the nineteenth century, in conjunction with the whaling industry. In the 1920s fox trapping was the most important source of income in the area, the price per pelt averaging $50. The Depression destroyed the market, and by 1931 skins sold for $5 or less (Chesemore 1972). Recently prices began to rise again. During the 1976–77 season 4,261 skins were marketed in Alaska at an average price per pelt of $36; the figures for Canada were 36,482 and $54.20 (Deems and Pursley 1978). The kill fell to 720 in Alaska and 16,405 in Canada during the 1983–84 season, when the average price in Canada was only $6 (Novak, Obbard, et al. 1987).

The arctic fox still is generally common in the wild, there being a total annual harvest of 100,000–150,000 pelts throughout the North American and Eurasian parts of its range (Prestrud 1991). However, there has been a general northward contraction in distribution during the twentieth century, perhaps in association with long-term climatic fluctuation (Ginsberg and Macdonald 1990; Hersteinsson and Macdonald 1992). In northern Alaska there also is concern about the effects of petroleum exploitation and other human development on the limited number of long-term favorable den sites (Garrott, Eberhardt, and Hanson 1983). In Iceland the arctic fox is persecuted because of its reputed depredations on sheep and lambs. Legislation there has promoted its destruction since at least as early as A.D. 1295. State-subsidized hunting has been so efficient that large parts of the island are now devoid of the species (Hersteinsson and Macdonald 1982). Overall numbers in Iceland fell to about 1,000 in the early 1970s but subsequently rebounded. A drastic decline occurred during the early twentieth century in Scandinavia, probably because of overhunting, and only a few hundred individuals now are present throughout Norway, Sweden, and Finland. Although the arctic fox has been legally protected in those three countries for decades, it has failed to recover, perhaps because of competition and predation from red fox populations that were able to occupy its range (Hersteinsson and Macdonald 1992; Hersteinsson et al. 1989; Frafjord, Becker, and Angerbjörn 1989).

Gray fox *(Urocyon cinereoargenteus)*, photo from San Diego Zoological Society.

CARNIVORA; CANIDAE; **Genus UROCYON**
Baird, 1858

Gray Foxes

There are two species (Cabrera 1957; E. R. Hall 1981):

U. cinereoargenteus, Oregon and southeastern Canada to
western Venezuela;
U. littoralis, San Miguel, Santa Rosa, Santa Cruz, Santa
Catalina, San Nicolas, and San Clemente islands off
southwestern California.

Urocyon was included in the genus *Vulpes* by Clutton-
Brock, Corbet, and Hills (1976) and in the subgenus *Vulpes*
of the genus *Canis* by Van Gelder (1978). All other authorities cited in this account treated *Urocyon* as a full
genus. *U. littoralis* often is considered to be conspecific with
U. cinereoargenteus, but a series of recent investigations
involving morphology, genetics, biochemical analysis, paleontology, and biogeography indicate that it is a distinct
species that originated in the later Pleistocene (Collins
1991*a,* 1991*b,* 1992; Moore and Collins 1995; Wayne,
George, et al. 1991). It also is possible that the population
now assigned to *U. cinereoargenteus* on Tiburon Island in
the Gulf of California is a separate species (Collins 1992).

In *U. cinereoargenteus* head and body length is 483–685
mm and tail length is 275–445 mm. In *U. littoralis* head and
body length is 480–500 mm and tail length is 110–290 mm.
Usual weight in the genus is 1.8–7.0 kg. The face, upper part
of the head, back, sides, and most of the tail are gray. The
throat, insides of the legs, and underparts are white. The
sides of the neck, lower flanks, and ventral part of the tail
are rusty. The hairs along the middle of the back and top of
the tail are heavily tipped with black, which gives the effect
of a black mane. Black lines also occur on the legs and face

of most individuals. A concealed mane of stiff hairs occurs
on top of the tail. The pelage is coarse. The skull of *Urocyon* is distinguished from that of *Vulpes* in having a deeper
depression above the postorbital process, much more pronounced and more widely separated ridges extending from
the postorbital processes to the posterior edges of the parietals, and a conspicuous notch toward the rear of the lower edge of the mandible (Lowery 1974).

Gray foxes frequent wooded and brushy country, often
in rocky or broken terrain, and are possibly most common
in the arid regions of the southwestern United States and
Mexico. The preferred habitat in Louisiana is mixed pine-
oak woodland bordering pastures and fields with patches of
weeds (Lowery 1974). Several sheltered resting sites may
be used on different days. The main den is in a pile of brush
or rocks, a crevice, a hollow tree, or a burrow that is either
self-excavated or taken over from another animal. Dens in
hollow trees have been found up to 9.1 meters above the
ground. Dens used for giving birth may be lined with vegetation (Trapp and Hallberg 1975).

Urocyon is sometimes called a "tree fox" because it frequently climbs trees, a rather unusual habit for a canid. It
lacks the endurance of the red fox, and when pursued it often will seek refuge in a tree. It also climbs without provocation, shinnying up the trunk and then leaping from
branch to branch. Most activity is nocturnal and crepuscular. Nightly movements cover about 200–700 meters
(Trapp and Hallberg 1975). *U. littoralis* tends to be diurnal
(Moore and Collins 1995). Juvenile gray foxes have dispersed as far as 84 km (Fritzell and Haroldson 1982). The
diet includes many kinds of small vertebrates as well as insects and vegetable matter. *Urocyon* seems to take plant
food more than do other foxes, and its diet may consist
mostly of fruits and grains at certain seasons and places.

Reported population densities are about 0.4–10.0/sq km
(Trapp and Hallberg 1975). Reported home range size
varies from 0.13 to 7.7 sq km. Four females in central Cal-

ifornia followed by radio tracking for various periods from January to July used ranges of 0.3–1.85 sq km (Fuller 1978). Range size increases in autumn and winter. Apparently, each family group uses a separate area and the normal social unit is an adult pair and their young, though there is some conflicting evidence on these points (Fritzell and Haroldson 1982). There is one litter per year. Mating occurs from late December to March in the southeastern United States and from mid-January to late May in New York. The gestation period, never precisely determined, has been variously estimated at 50–63 days. Litters usually contain about 4 young but may have from 1 to 10. The young weigh around 100 grams at birth, are blackish in color, and open their eyes after 9–12 days. They can climb vertical tree trunks after 1 month and begin to take solid food in 6 weeks. They forage independently by late summer or early autumn but apparently remain in the parental home range until January or February. Most females breed in their first year of life (Trapp and Hallberg 1975). A captive lived for 13 years and 8 months (Jones 1982).

If captured when small, gray foxes tame readily, are as affectionate and playful as domestic dogs, and make more satisfactory pets than do red foxes. *U. littoralis*, in particular, is said to be docile and to readily approach humans (Moore and Collins 1995). However, deliberate removal from the wild is often illegal and may be dangerous to both the animals and the persons involved. The attractive skins of *Urocyon* are used commercially but are not classed as fine furs. For much of the twentieth century the price per pelt averaged around $0.50, but it began to rise in the 1960s (Lowery 1974). For the 1970–71 season the reported harvest in the United States was 26,109 skins with an average value of $3.50. By the 1976–77 season the reported take had grown to 225,277 pelts, and the average price to $34 (Deems and Pursley 1978). In the 1991–92 season 90,604 skins were taken in the United States and sold at an average price of $9.09 (Linscombe 1994).

The island fox, *U. littoralis*, is classified as conservation dependent by the IUCN and as threatened by the California Department of Fish and Game (Crooks 1994) and is fully protected by California state law. Each of the six Channel Islands has its own recognized subspecies, whose validity recent studies have upheld. Effective population size ranges from 150 individuals on the smallest island, San Miguel, to approximately 1,000 on the largest island, Santa Cruz. It is thought that *Urocyon* initially reached one of the three northern islands (Santa Cruz, Santa Rosa, and San Miguel) prior to 24,000 years ago, perhaps by accidental rafting, and subsequently spread to all three islands when they were joined at certain times during the late Pleistocene. A fossil fox collected on Santa Rosa shows that *U. littoralis* had evolved its diminutive size by at least 16,000 years ago. Native Americans arrived on the Channel Islands about 10,000 years ago, and apparently it was they who transported foxes from the northern to the southern islands (Santa Catalina, San Clemente, and San Nicolas). The first occurrence in the south dates from about 3,400–4,300 years ago on San Clemente. Numerous remains of foxes have been found at archaeological sites and indicate that the animals were kept as pets and had an important role in religious and ceremonial practices. They probably were brought to the various islands in the course of trade and subsequently became feral. Their very small population sizes and lack of genetic variability now make them especially vulnerable to environmental disruption and introduction of disease (Collins 1991*a*, 1991*b*, 1992; Garcelon, Wayne, and Gonzales 1992; George and Wayne 1991; Wayne, George, et al. 1991).

Hoary fox *(Lycalopex vetulus)*, photo by Luiz Claudio Marigo.

Hoary Fox

Langguth (1975*a*) considered *Lycalopex* to be a full genus with a single species, *L. vetulus*, occurring in the states of Mato Grosso, Goias, Minas Gerais, and Sao Paulo in south-central Brazil. *Lycalopex* was treated as a subgenus of *Dusicyon* by Cabrera (1957) and as a subgenus of *Canis* by Van Gelder (1978). *L. vetulus* was included within *Dusicyon* by Clutton-Brock, Corbet, and Hills (1976), who noted, however, that it was the most foxlike member of that genus. Berta (1987) included *Lycalopex* within *Pseudalopex* and suggested a relationship to *P. sechurae*. Wozencraft (*in* Wilson and Reeder 1993) also included *Lycalopex* in *Pseudalopex*.

Head and body length is about 585–640 mm, tail length is 280–320 mm, and weight is about 2.7–4.0 kg. The coat is short. The upper parts are a mixture of yellow and black, giving an overall gray tone. The ears and the outsides of the legs are reddish or tawny. The tail has a black tip and a marked dark stripe along the dorsal line. The underparts are cream to fawn. Compared with *Dusicyon*, *Lycalopex* has a short muzzle, a small skull and teeth, reduced carnassials, and broad molars (Bueler 1973; Clutton-Brock, Corbet, and Hills 1976; Röhrs *in* Grzimek 1990).

The habitat is grassy savanna on smooth uplands, or savannahs with scattered trees. A deserted armadillo burrow may be used for shelter and for the natal nest. The diet consists of small rodents, birds, and insects, especially grasshoppers. The dentition suggests substantial dependence on insects, but *Lycalopex* is persecuted by people because of presumed predation on domestic fowl (Langguth 1975*a*). It is usually timid but courageously defends itself and its young. Births occur in the austral spring (September), and there are usually two to four young per litter (Bueler 1973; Röhrs *in* Grzimek 1990).

South American Foxes

There are four species (Berta 1987; Clutton-Brock, Corbet, and Hills 1976; Duckworth 1992; Ginsberg and Macdonald

South American fox *(Pseudalopex culpaeus)*, photo by Ernest P. Walker.

1990; Langguth 1975a; Pacheco et al. 1995; Redford and Eisenberg 1992; Van Gelder 1978):

P. gymnocercus, humid grasslands of southern Brazil, Paraguay, northern Argentina, and Uruguay;

P. culpaeus, Andes and adjacent highlands from Ecuador and possibly Colombia to Patagonia and Tierra del Fuego;

P. griseus, plains and low mountains of Chile and Argentina, including Patagonia, and possibly up the Pacific coast to Peru and southwestern Ecuador;

P. sechurae, arid coastal zone of southwestern Ecuador and northwestern Peru.

Pseudalopex was considered a subgenus of *Canis* by Langguth (1975a) and Van Gelder (1978). The four species listed above were included in the genus *Dusicyon* along with the type species of that genus, *D. australis,* by Clutton-Brock, Corbet, and Hills (1976). Berta (1987) recognized *Pseudalopex* as a full genus comprising the above four species and considered *D. australis* to be the only living species of *Dusicyon.* Information provided by these various authorities suggests that *D. australis* is at least as different from the species of *Pseudalopex* as it is from *Canis* and that *Pseudalopex* is at least as different from *Canis* as it is from *Vulpes.* Therefore, since *Dusicyon* (for the species *D. australis*) and *Vulpes* are here being maintained as separate genera, it is also advisable to give generic rank to *Pseudalopex.* Medel et al. (1990) and Miller et al. (1983) treated *P. fulvipes,* known from Chiloe Island and a mainland area of Chile about 600 km to the north, as a species distinct from *P. griseus.*

Head and body length is 530–1,200 mm, tail length is 250–500 mm, and weight is 4–13 kg. *P. culpaeus* is the largest species and *P. sechurae* is the smallest. The coat is usually heavy, with a dense underfur and long guard hairs. The upper parts are generally gray agouti with some ochraceous or tawny coloring (Clutton-Brock, Corbet, and Hills 1976). The head, ears, and neck are often reddish. The underparts are usually pale. The tail is long, bushy, and black-tipped. There is some resemblance to a small coyote *(Canis latrans).* The dentition, however, is more foxlike than doglike; the molars are well developed, and the carnassials are relatively short (Clutton-Brock, Corbet, and Hills 1976).

Habitats include sandy deserts for *P. sechurae;* low, open grasslands and forest edge for *P. griseus;* pampas, hills, deserts, and open forests for *P. gymnocercus;* and dry rough country and mountainous areas up to 4,500 meters in elevation for *P. culpaeus* (Crespo 1975; Langguth 1975a; Röhrs in Grzimek 1990). Dens are usually among rocks, under bases of trees and low shrubs, or in burrows made by other animals, such as viscachas and armadillos. Most activity is nocturnal, but some individuals are occasionally active during the day. *P. gymnocercus* sometimes collects and stores objects, such as strips of leather and cloth. This species may freeze and remain motionless upon the appearance of a human being; in one case it reportedly did not move even when approached and struck with a whip handle. The voice of *Pseudalopex* has been described as a howl or a series of barks and yaps. It is heard mainly at night, especially during the breeding season.

The omnivorous diet of *Pseudalopex* includes rodents, lagomorphs, birds, lizards, frogs, insects, fruit, and sugar cane. Studies of stomach contents indicate that *P. gymnocercus* takes an equal amount of plant and animal food. *P. culpaeus* seems to be more carnivorous than other species and reportedly sometimes preys heavily on introduced sheep and European hares. In western Argentina during the spring, part of the population of *P. culpaeus* shifts 15–20 km into the higher mountains in response to the seasonal movements of the sheep and hares. Its normal home range in that area is 4 km in diameter (Crespo 1975). In areas where *P. culpaeus* is sympatric with the smaller *P. griseus* there is no overlap of local territories, and the latter species is able to survive in poorer habitat by supplementing its diet with beetles and plant material, especially from spring to autumn (Johnson 1992; Johnson and Franklin 1994). *P. griseus* has been found to occur at an average density of 1/43 ha. in southern Chile (Durran, Cattan, and Yáñez 1985). Observations of *P. gymnocercus* in the Paraguayan Chaco usually are of a single individual and indicate a density of about 1/100 ha. (Brooks 1992). In a study of *P. sechurae* in northwestern Peru, Asa and Wallace (1990) found mostly nocturnal activity, a diet consisting largely of plant matter and invertebrates, and two home ranges, one occupied by a single male, the other apparently by an adult female and two juveniles.

Observations of *P. griseus* in Patagonia (Johnson 1992) indicate that a monogamous pair maintains an exclusive year-round territory, though occasionally a second female is present in the area and assists in rearing the young; births, usually of four to five pups, occur by mid-October. According to Crespo (1975), *P. culpaeus* and *P. gymnocercus* mate from August to October and give birth from October to December (the austral spring). Females are monestrous. The gestation period is 55–60 days. Embryo counts range from one to eight, with averages of about four in *P. gymnocercus* and five in *P. culpaeus*. The male helps to provide food to the family. At 2–3 months the young begin to hunt with the parents. Sexual maturity apparently is attained by 1 year in *P. culpaeus* and *P. griseus* (Ginsberg and Macdonald 1990). Few individuals live more than several years in the wild, but a captive *P. gymnocercus* lived 13 years and 8 months (Jones 1982).

These canids are killed by people because they are alleged to prey on domestic fowl and sheep and because their fur is desirable. Populations of *Pseudalopex* have thus declined in some areas, such as in Buenos Aires Province, Argentina. *P. griseus* was introduced to Tierra del Fuego in 1951 to control the previously introduced European rabbit *(Oryctolagus)* and also has been released on several small islands in the Falklands (Ginsberg and Macdonald 1990). In southern Chile *P. griseus* is legally protected but is threatened by persecution for alleged depredations and environmental disturbance (Durran, Cattan, and Yáñez 1985; Miller et al. 1983). The subspecies (or possibly species) *P. griseus fulvipes* of Chiloe Island is probably the rarest canid in South America (Medel et al. 1990). Crespo (1975) noted, however, that *P. culpaeus* occurred in relatively low numbers in Neuquen Province, western Argentina, until the early twentieth century and then greatly increased in response to the introduction of sheep and European hares *(Lepus europaeus)*. *P. griseus*, *P. gymnocercus*, and *P. culpaeus* are on appendix 2 of the CITES.

Records of the CITES indicate an annual export during the 1980s of about 100,000 skins of *P. griseus* and several thousand of *P. culpaeus*, virtually all from Argentina (Broad, Luxmoore, and Jenkins 1988). Such uncontrolled hunting is thought to have resulted in a decline of about 80 percent in populations of *P. griseus* since 1970 (Roig 1991). Numbers of *P. culpaeus* seem to have been less seriously affected overall but are declining on Tierra del Fuego (Novaro 1993). *P. gymnocercus* also is heavily hunted and trapped for its fur; up to 30,000 skins may be exported annually from Paraguay even though the species is officially protected there (Ginsberg and Macdonald 1990).

CARNIVORA; CANIDAE; Genus DUSICYON
Hamilton-Smith, 1839

Falkland Island Wolf

Berta (1987) considered this genus to include a single modern species, *D. australis*, which formerly occurred on West and East Falkland Islands, off the southeastern coast of Argentina. She added that another species, *D. avus*, occurred in the late Pleistocene of southern Chile and survived into the Recent of southern Argentina. Additional species usually have been assigned to *Dusicyon*. Cabrera (1957) recognized two subgenera: *Dusicyon*, with the species here placed in the genus *Pseudalopex*; and *Lycalopex*, which is here treated as a full genus. Clutton-Brock, Corbet, and Hills (1976) included within *Dusicyon* all of those species that are here placed in the genera *Lycalopex, Pseudalopex, Dusicyon, Cerdocyon*, and *Atelocynus*. Those authorities indicated, however, that *D. australis* is at least as close, systematically, to some species of *Canis* as it is to the species here assigned to *Pseudalopex*. They even discussed the possibility that *D. australis* is a form of *Canis familiaris*. Both Langguth (1975a) and Van Gelder (1978) considered *Dusicyon* to be a subgenus of *Canis* and to include only *D. australis*.

Only 11 specimens of *D. australis* are known, and not all include skins. In one specimen head and body length was 970 mm and tail length was 285 mm. The upper parts are brown with some rufous and a speckling of white, and the underparts are pale brown. The coat is soft and thick. The tail is short, bushy, and tipped with white. The face and ears are short and the muzzle is broad. The skull is large and has inflated frontal sinuses, more like the situation of *Canis* than that of *Pseudalopex* (Allen 1942; Clutton-Brock, Corbet, and Hills 1976).

Dusicyon was the only terrestrial mammal found on the Falkland Islands by the early explorers. Its natural diet consisted mainly of birds, especially geese and penguins, and also included pinnipeds. Its presence on the islands, about 400 km from the mainland, is something of a mystery.

Clutton-Brock, Corbet, and Hills (1976) thought it most likely that *Dusicyon* had been taken to the Falklands as a domestic animal by prehistoric Indians. Possible descent from either *Pseudalopex* or *Canis* was suggested. Berta (1987), however, pointed out that a lowered sea level during the Pleistocene would have facilitated natural movement and that the distinguishing characters of *Dusicyon* probably result from subsequent isolation rather than from domestication.

Dusicyon demonstrated remarkable tameness toward people. Individuals waded out to meet landing parties. Later they came into the camps in groups, carried away articles, pulled meat from under the heads of sleeping men, and stood about while their fellow animals were being killed. The dogs were often killed by a man holding a piece of meat as bait in one hand and stabbing the dog with a knife in the other hand.

Although *Dusicyon* was discovered in 1690, it was still common, and still behaving in a very tame manner, when Darwin visited the Falklands in 1833. In 1839, however, large numbers were killed by fur traders from the United States. In the 1860s Scottish settlers began raising sheep on the islands. *Dusicyon* preyed on the sheep and was therefore intensively poisoned. The genus was very rare by 1870, and the last individual is said to have been killed in 1876 (Allen 1942). The IUCN classifies *D. australis* as extinct.

Crab-eating fox *(Cerdocyon thous)*, photo from San Diego Zoological Garden.

CARNIVORA; CANIDAE; Genus CERDOCYON
Hamilton-Smith, 1839

Crab-eating Fox

The single species, *C. thous*, has been recorded from Colombia, Venezuela, Guyana, Surinam, eastern Peru, Bolivia, Paraguay, Uruguay, northern Argentina, and most of Brazil outside of the lowlands of the Amazon Basin (Grimwood 1969; Husson 1978; Langguth 1975a). *Cerdocyon* was recognized as a distinct genus by Berta (1982, 1987), Cabrera (1957), and Langguth (1975a), as a part of *Dusicyon* by Clutton-Brock, Corbet, and Hills (1976), and as a subgenus of *Canis* by Van Gelder (1978).

Head and body length is 600–700 mm, tail length is about 300 mm, and weight is 5–8 kg. The coloration is variable, but the upper parts are usually grizzled brown to gray, often with a yellowish tint, and the underparts are brownish white. The short ears are ochraceous or rufous. The tail is fairly long, bushy, and either totally dark or black-tipped (Clutton-Brock, Corbet, and Hills 1976). The relatively short and robust legs may be tawny (Brady 1979).

Except as noted, the information for the remainder of this account was taken from Brady (1978, 1979). *Cerdocyon* inhabits woodlands and savannahs and is mainly nocturnal. On the llanos of Venezuela during the wet season it uses high ground and shelters by day under brush. During the dry season it occupies lowlands and spends the day in clumps of matted grass. These grass shelters have several entrances, are used repeatedly, and may serve as natal dens.

Cerdocyon forages from about 1800 hours to 2400 hours. It stalks and pounces on small vertebrates and apparently listens for crabs in tussocks of grass. In the dry season the percentage composition of its diet is: vertebrates, 48; crabs, 31; insects, 16; carrion, 3; and fruit, 2. In the wet season the percentage breakdown is: insects, 54; vertebrates, 20; fruit, 18; and carrion, 7.

Population densities of 4/sq km and group territory sizes of 5–10 sq km have been reported (Ginsberg and Macdonald 1990). However, three pairs studied by Brady occupied home ranges of approximately 54, 60, and 96 ha. The ranges overlapped to some extent but were regularly marked with urine. Tolerance of neighbors was shown in the wet season, but aggression increased in the dry season. Mated pairs form a lasting bond and commonly travel together but do not usually hunt cooperatively. There is a variety of vocalizations, including a siren howl for long-distance communication between separated family members (Brady 1981). Breeding may take place throughout the year, but births peak in January and February on the llanos of Venezuela. Captive females produced two litters annually at intervals of about 8 months. The gestation period is 52–59 days and litter size is three to six young. The young weigh 120–60 grams at birth, open their eyes at 14 days, begin to take some solid food at 30 days, and are completely weaned by about 90 days. Both parents guard and bring food to the young. Independence comes at 5–6 months, and sexual maturity at about 9 months. According to Smielowski (1985), one specimen lived in captivity for 11 years and 6 months.

Cerdocyon is now on appendix 2 of the CITES. Ginsberg

and Macdonald (1990), however, noted that it is widespread and common. It is hunted, but its pelt has little value.

CARNIVORA; CANIDAE; **Genus NYCTEREUTES**
Temminck, 1839

Raccoon Dog

The single species, *N. procyonoides*, originally occurred in the woodland zone from southeastern Siberia to northern Viet Nam as well as on all the main islands of Japan (Corbet 1978). Although their present distributions may suggest otherwise, *Nyctereutes* and *Cerdocyon* evidently are closely related. This view is supported by the fossil record, which suggests the presence of a common ancestor that ranged over Eurasia and North America 4–10 million years ago (Berta 1987). Molecular analyses show no affinity of *Nyctereutes* with any other genus (Wayne 1993; Wayne and O'Brien 1987).

Head and body length is 500–680 mm and tail length is 130–250 mm. Weight is 4–6 kg in the summer but 6–10 kg prior to winter hibernation (Novikov 1962). The pelage is long, especially in winter. The general color is yellowish brown. The hairs of the shoulders, back, and tail are tipped with black. The limbs are blackish brown. The facial markings resemble those of raccoons *(Procyon)*. There is a large dark spot on each side of the face, beneath and behind the eye. *Nyctereutes* is somewhat like a fox in external appearance but has proportionately shorter legs and tail.

According to Novikov (1962), the raccoon dog occurs mainly in forests and in thick vegetation bordering lakes and streams. It usually dens in a hole initially made by a fox or badger or in a rocky crevice but sometimes digs its own burrow. *Nyctereutes* is the only canid that hibernates, though the process is neither profound for individuals nor

universal for the species. In northern parts of the range well-nourished animals hibernate from as early as November to as late as March but may awaken occasionally to forage on warm days. In the southern parts of the range there is no winter sleep. Poorly nourished individuals do not hibernate, even in the north. Successful hibernation may be preceded by a period of intensive eating that increases weight by nearly 50 percent. The summer diet may consist largely of frogs and also includes rodents, reptiles, fish, insects, mollusks, and fruit. In the autumn, vegetable matter such as berries, seeds, and rhizomes becomes important. Northern individuals that do not hibernate have a difficult time in the winter but may subsist on small mammals, carrion, and human refuse. *Nyctereutes* usually is reported to be primarily nocturnal, but Ward and Wurster-Hill (1989) found it active both by day and by night at two study sites in Japan. The greatest distance traveled in one night was 8 km.

In the Ussuri region of southeastern Siberia, Kucherenko and Yudin (1973) found population density to average 1–3/1,000 ha. and to reach 20/1,000 ha. in the best habitat. According to Kauhala, Helle, and Taskinen (1993), reported home range size of the introduced European populations has varied from 0.4 sq km to 20.0 sq km. Studies in Finland found an average range size of 9.5 sq km, including a core area of 3.4 sq km. Such ranges were generally shared by an evidently monogamous adult male-female pair. The ranges, especially those of the males, were larger in autumn than in summer. The core areas of most adjacent pairs did not overlap in the pup-rearing season but did overlap to some extent in the autumn. Juveniles of both sexes left their natal ranges in their first autumn, many traveling more than 10 km.

Much smaller ranges have been recorded in Japan. Ward and Wurster-Hill (1989) found averages of 49 ha. on Kyushu and 59 ha. on Honshu during the autumn. The ranges were not exclusive, though there was no evidence of

Raccoon dog *(Nyctereutes procyonoides)*, photo by Ernest P. Walker.

social grouping. During the spring on Kyushu, Ikeda (1986) found home ranges of only 8–48 ha., which overlapped extensively, and observed *Nyctereutes* to undergo much amicable social activity. At major food sources, such as fruiting trees and garbage dumps, up to 20 individuals were seen to intermingle without belligerence. There also was communal use of latrines, which may serve as sites to pass on information. Antagonism increased during the breeding season. Male-female pairs formed in late winter, and both parents cared for the young, though each sex foraged separately. The family dissolved at the end of summer, but pair bonds may have persisted from year to year. Vocalizations include growls and whines but not barks.

Ward and Wurster-Hill (1990) indicated some question concerning whether *Nyctereutes* is monogamous or polygamous, noting that an estrous female may be courted by several males. Mating occurs from January to April, estrus lasts about 4 days, and the gestation period is usually 59–64 days. There are commonly 4–10 young, but as many as 19 have been reported. They weigh 60–90 grams at birth, open their eyes after 9–10 days, and nurse for up to 2 months. Both parents, however, begin to bring them solid food after only 25–30 days. They are capable of an independent existence by 4–5 months and attain sexual maturity at 9–11 months (Helle and Kauhala 1993; Novikov 1962; Valtonen, Rajakoski, and Mäkelä 1977). Maximum longevity reported in a study of wild individuals in Finland was 7–8 years (Helle and Kauhala 1993). A captive specimen lived 10 years and 8 months (Jones 1982).

In Japan the flesh of the raccoon dog has been used for human consumption, and the bones for medicinal preparations. The skin, known commercially as "Ussuri raccoon," is used widely in the manufacture of such items as parkas, bellows, and decorations on drums. Although about 70,000 are taken every year in Japan for the fur trade, *Nyctereutes* is still common there and sometimes is found within large cities (Ikeda 1986). Populations have declined in southeastern Siberia through overhunting and habitat disturbances (Kucherenko and Yudin 1973).

From 1927 to 1957 more than 9,000 raccoon dogs were released by people in regions to the west of the natural range, especially in European Russia. The hope was to create new and valuable fur-producing populations. *Nyctereutes* did become established, eventually spreading almost throughout European Russia, and its importance in the fur trade increased. It also, however, extended its range westward, reaching Finland in 1935, Sweden in 1945, Romania in 1951, Poland in 1955, Slovakia in 1959, Germany and Hungary in 1962, and France in 1979 (Artois and Duchêne 1982; Kubiak 1965; Mikkola 1974; Novikov 1962). It now is very common in the Baltic states and many parts of eastern Europe (Kauhala 1994) and occurs throughout Germany (Roben 1975). The rate of expansion has slowed in recent years, though the prognosis is that the Balkans, Scandinavia, and France will be colonized (Nowak 1984). An individual was recently caught in England (*Oryx* 13 [1977]: 434), and the USDI designates the genus as "injurious," thereby prohibiting importation of live specimens. *Nyctereutes* is generally considered a nuisance to the west of Russia. It destroys small game animals and fish, and its fur, which does not become as long as in the native habitat, is nearly worthless in most areas. However, in Finland about 80,000 pelts are derived annually from farmed animals and another 40,000–60,000 from wild-trapped individuals (Ginsberg and Macdonald 1990; Helle and Kauhala 1990).

CARNIVORA; CANIDAE; Genus ATELOCYNUS
Cabrera, 1940

Small-eared Dog

The single species, *A. microtis*, inhabits the Amazon, upper Orinoco, and upper Parana basins in Brazil, Peru, Ecuador, Colombia, and probably Venezuela. *Atelocynus* was recognized as a distinct genus by Berta (1986, 1987), Cabrera

Small-eared dog *(Atelocynus microtis)*, photo from Field Museum of Natural History.

(1957), and Langguth (1975a), as a part of *Dusicyon* by Clutton-Brock, Corbet, and Hills (1976), and as a subgenus of *Canis* by Van Gelder (1978).

Head and body length is 720–1,000 mm, tail length is 250–350 mm, shoulder height is about 356 mm, and weight is around 9–10 kg. The ears, rounded and relatively shorter than those of any other canid, are only 34–52 mm in length. The upper parts are dark gray to black, and the underparts are rufous mixed with gray and black. The thickly haired tail is black except for the paler basal part on the underside. It has been reported to sweep the ground when hanging perpendicularly; however, when captives at the Brookfield Zoo, in Chicago, were standing, they curved the tail forward and upward against the outer side of a hind leg, so that the terminal hairs did not drag on the ground. The temporal ridges of the skull are strongly developed, and the frontal sinuses and cheek teeth are relatively large (Clutton-Brock, Corbet, and Hills 1976).

Atelocynus inhabits tropical forests from sea level to about 1,000 meters. It moves with a catlike grace and lightness not observed in any other canid. Observations in the wild indicate that *Atelocynus* hunts individually and captures primarily small and sometimes medium-sized rodents, whereas the sympatric *Speothos* hunts in packs and commonly kills larger animals (Peres 1991b). A male in captivity in Bogota ate raw meat, shoots of grass, and foods commonly eaten by people. This male and a female were brought to the Brookfield Zoo. They proved to be completely different in temperament: the male was exceedingly friendly and docile, but the female exhibited constant hostility. Although the male was shy in captivity before being sent to Brookfield, and growled and snarled when angry or frightened, it did not show any unfriendly actions at the Chicago zoo and, in fact, became very tame. It permitted itself to be hand fed and petted by persons it recognized,

responding to petting by rolling over on its back and squealing. This male came to react to attention from familiar people by a weak but noticeable wagging of the back part of its tail. The female, on the other hand, emitted a continuous growling sound, without opening its mouth or baring its teeth, when under direct observation.

The odor from the anal glands was strong and musky in the male but scarcely detectable in the female. In both sexes the eyes glowed remarkably in dim light. The male, though smaller, was dominant in most activities. Although some snapping was observed between the two animals, no biting or fighting was noted, and they occupied a common sleeping box when not active. According to Jones (1982), one captive lived 11 years.

According to Thornback and Jenkins (1982), *Atelocynus* seems to be rare in some countries, though its status is difficult to assess because of its nocturnal, solitary habits. It may be at risk from indirect influence of human intrusion throughout its range. It is protected by law in Brazil and Peru.

CARNIVORA; CANIDAE; **Genus SPEOTHOS**
Lund, 1839

Bush Dog

The single living species, *S. venaticus,* occurs in Panama, Colombia, Ecuador, Venezuela, the Guianas, eastern Peru, Brazil, eastern Bolivia, Paraguay, and extreme northeastern Argentina (Cabrera 1957; Ginsberg and Macdonald 1990; E. R. Hall 1981; Thornback and Jenkins 1982). *Speothos* was first described from fossils collected in caves in Brazil. Although *Speothos* sometimes has been united with *Cuon*

Bush dog *(Speothos venaticus)*, photo from San Diego Zoological Garden.

and *Lycaon* in the subfamily Simocyoninae, Berta (1984) showed that its true affinities lie with other South American canids, especially *Atelocynus*. She reported also that a second species, *S. pacivorus*, is known from late Pleistocene and early Recent deposits in Minas Gerais, Brazil.

Head and body length is 575–750 mm, tail length is 125–50 mm, and shoulder height is about 300 mm. Two males weighed 5 and 7 kg. The head and neck are ochraceous fawn or tawny, and this color merges into dark brown or black along the back and tail. The underparts are as dark as the back, though there may be a light patch behind the chin on the throat (Clutton-Brock, Corbet, and Hills 1976). The body is stocky, the muzzle is short and broad, and the legs are short. The tail is short and well haired but not bushy.

Speothos inhabits forests and wet savannahs, often near water. It seems to be mainly diurnal and to retire to a den at night, in either a burrow or a hollow tree trunk. It is reportedly semiaquatic: one captive could dive and swim underwater with great facility. Wild individuals have been observed to swim across large rivers and to chase prey into water. The bush dog captures mainly relatively large rodents, such as *Agouti* and *Dasyprocta* (Husson 1978; Kleiman 1972; Langguth 1975a; Strahl, Silva, and Goldstein 1992).

Speothos is a highly social canid, living and hunting cooperatively in packs of up to 10 individuals. Several captives of the same or opposite sex can stay confined together without fighting, though a dominance hierarchy may be established. There are a number of vocalizations, the most common of which is a high-pitched squeak that appears to help maintain contact as the group moves about in dense forest (Clutton-Brock, Corbet, and Hills 1976; Kleiman 1972).

Husson (1978) wrote that litters of 2–3 young are produced during the rainy season in the wild. Extended observations in captivity (Porton, Kleiman, and Rodden 1987), however, indicate that the reproductive pattern of *Speothos* is aseasonal and uninterrupted and is partly influenced by social factors. Young females typically do not experience estrus when living with their mother or sisters but quickly do so when paired with males. Estrus and births have been recorded throughout the year. Estrus periods average 4.1 days, and some females show polyestrous activity, which has not been recorded in any other canid. The interbirth interval averages 238 days, and gestation, 67 days. Litter size averages 3.8 young and ranges from 1 to 6. The earliest age of conception in a female was 10 months. Hayssen, Van Tienhoven, and Van Tienhoven (1993) listed newborn weights of 130–90 grams and a nursing period of 4–5 months. According to Marvin L. Jones (Zoological Society of San Diego, pers. comm., 1995), a captive lived 13 years and 4 months.

Speothos is classified as vulnerable by the IUCN and is on appendix 1 of the CITES. It is still widespread but is scarce throughout its range, and it seems to disappear as settlement progresses and forests are cleared (Thornback and Jenkins 1982).

CARNIVORA; CANIDAE; **Genus CANIS**
Linnaeus, 1758

Dogs, Wolves, Coyotes, and Jackals

There are eight species (Clutton-Brock, Corbet, and Hills 1976; Coetzee *in* Meester and Setzer 1977; Corbet 1978; Ferguson 1981; Ginsberg and Macdonald 1990; E. R. Hall 1981; Kingdon 1977; Nowak 1979; Sillero-Zubiri and Gottelli 1994; Thomas and Dibblee 1986; Vaughan 1983):

C. adustus (side-striped jackal), open country from Senegal to Somalia, and south to northern Namibia and eastern South Africa;

Simien jackal or Ethiopian wolf *(Canis simensis)*, photo by Claudio Sillero-Zubiri.

Dingos *(Canis familiaris)*, photos by Lothar Schlawe.

C. mesomelas (black-backed jackal), open country from Sudan and Ethiopia to Tanzania and from southwestern Angola to Mozambique and South Africa;

C. aureus (golden jackal), Balkan Peninsula to Thailand and Sri Lanka, Morocco to Egypt and northern Tanzania;

C. simensis (Simien jackal or Ethiopian wolf), mountains of central Ethiopia;

C. latrans (coyote), Alaska to Nova Scotia and Panama;

C. rufus (red wolf), central Texas to southern Pennsylvania and Florida;

C. lupus (gray wolf), Eurasia except tropical forests of southeastern corner, Egypt and Libya, Alaska, Canada, Greenland, conterminous United States except southeastern quarter and most of California, highlands of Mexico;

C. familiaris (domestic dog), worldwide in association with people, extensive feral populations in Australia and New Guinea.

Van Gelder (1977*b*, 1978) considered *Canis* to comprise *Vulpes* (including *Fennecus* and *Urocyon*), *Alopex*, *Lycalopex*, *Pseudalopex*, *Dusicyon*, *Cerdocyon*, and *Atelocynus* as subgenera. *C. simensis* sometimes has been placed in a separate genus or subgenus, *Simenia* Gray, 1868, but recent mitochondrial DNA analysis indicates that it is phylogenetically closer to the coyote and wolves than to the other jackals (Sillero-Zubiri and Gottelli 1994). There also is molecular evidence that *C. aureus* has closer affinity to the coyote-wolf group than to the jackal group even though it has demonstrated some morphological convergence with *C. adustus* and *C. mesomelas* in East Africa (Van Valkenburgh and Wayne 1994). Lawrence and Bossert (1967, 1975) suggested that *C. rufus* is not more than subspecifically distinct from *C. lupus*. In contrast, recent analyses of mitochondrial and nuclear DNA have led to arguments that *C. rufus* is a hybrid population that originated subsequent to European settlement through interbreeding between *C. latrans* and *C. lupus* (Jenks and Wayne 1992; Roy et al. 1994; Roy, Girman, and Wayne 1994; Wayne 1992; Wayne and Gittleman 1995; Wayne and Jenks 1991).

That view has been questioned based on other interpretations of genetic data (Cronin 1993; Dowling, DeMarais, et al. 1992; Dowling, Minckley, et al. 1992), behavioral and ecological data (Phillips and Henry 1992), and morphological and paleontological evidence indicating that *C. rufus* is a primitive kind of wolf that has been present in the Southeast since the mid-Pleistocene (Nowak 1992). The cited presence of *C. lupus* in Egypt and Libya is taken from Ferguson's (1981) remarkable determination that the population formerly known as *C. aureus lupaster* actually consists of small wolves rather than large jackals. However, Spassov (1989) argued that *lupaster* shows closer affinity to *C. aureus* than to *C. lupus* and may even be a separate species. *C. familiaris* was included in *C. lupus* by Wozencraft (*in* Wilson and Reeder 1993), and that view has been widely accepted. The feral populations of *C. familiaris* in Australia and apparently related populations in parts of southern Asia and the East Indies sometimes are considered to represent a distinct species, *C. dingo*.

Canis is characterized by a relatively high body, long legs, and a bushy, cylindrical tail. The pupils of the eyes generally appear round in strong light. Although most species have a scent gland near the base of the tail, it does not produce as strong an odor as that of *Vulpes*. The skull has large frontal sinuses and temporal ridges that are close together, often uniting to form a sagittal crest. The facial region of the skull, except in *C. simensis*, is relatively shorter than in *Vulpes* and *Pseudalopex* (Clutton-Brock, Corbet, and Hills 1976). Females have 8 or 10 mammae. Additional information is provided separately for each species.

Canis adustus (side-striped jackal)

Head and body length is 650–810 mm, tail length is 300–410 mm, shoulder height is 410–500 mm, and weight is 6.5–14.0 kg (Kingdon 1977). Males are larger than females. The coat is long and soft. The upper parts are generally mottled gray; on each side of the body is a line of white hairs below which is a line of dark hairs; and the underparts and tip of the tail are white (Clutton-Brock, Corbet, and Hills 1976).

According to Kingdon (1977), the side-striped jackal is

Side-striped jackal *(Canis adustus)*, photo by Cyrille Barrette.

widespread in the moister parts of savannahs, thickets, forest edge, cultivated areas, and rough country up to 2,700 meters in elevation. Favored denning sites include old termite mounds, abandoned aardvark holes, and burrows dug into hillsides. The animal is strictly nocturnal in areas well settled by people. It is a more omnivorous scavenger than other kinds of jackals, taking a variety of invertebrates, small vertebrates, carrion, and plant material. Social groups are well spaced and usually consist of a mated pair and their young. In East Africa births occur mainly in June–July and September–October, gestation lasts 57–70 days, and litters consist of three to six young. Ginsberg and Macdonald (1990) stated that lactation lasts 8–10 weeks, sexual maturity is attained at 6–8 months, and longevity is 10–12 years. They noted that the species is rare throughout its range. In a recent radio-tracking study of the three species of jackals in East Africa, Fuller et al. (1989) were able to obtain data on only a single *C. adustus;* in 24 days of monitoring it was found to use a home range of 1.1 sq km.

Canis mesomelas (black-backed jackal)

Head and body length is 450–900 mm, tail length is 260–400 mm, shoulder height is 300–480 mm, and weight is 6.0–13.5 kg. Males average about 1 kg heavier than females. A dark saddle extends the length of the back to the black tip of the tail. The sides, head, limbs, and ears are rufous, and the underparts are pale ginger. The build is slender, and the ears are very large (Bekoff 1975; Clutton-Brock, Corbet, and Hills 1976; Kingdon 1977).

The black-backed jackal is found mainly in dry grassland, brushland, and open woodland. It generally dens in an old termite mound or aardvark hole. It is partly diurnal and crepuscular in undisturbed areas but becomes nocturnal in places intensively settled by people. The diet includes a substantial proportion of plant material, insects, and carrion, but *C. mesomelas* frequently hunts rodents and is capable of killing antelopes as large as *Gazella thomsoni.* Some food may be cached (Bothma 1971*b;* Kingdon 1977; Smithers 1971).

In a study in southern Africa average daily movement was found to be about 12 km, but one radio-collared indi-

vidual covered 87 km in four nights. Average home range size in this area was about 11 sq km for adults, but some subadults wandered over much greater areas (Ferguson, Nel, and de Wet 1983). In the same region, Rowe-Rowe (1982) estimated a population density of 1/2.5–2.9 sq km and an average home range size of 18.2 sq km. According to these and other investigations (Bekoff 1975; Kingdon 1977; Moehlman 1978, 1983, 1987), all or a major part of the adult home range is a territory that is marked with urine and defended by both sexes. Each sex seems to repel mainly members of the same sex. Although 20–30 individuals may gather around a lion kill, the basic social unit is a mated pair and their young. Some young remain with the parents, do not breed, and assist in feeding, grooming, and guarding the next litter. This tendency seems to be stronger in *C. mesomelas* than in *C. aureus,* perhaps because the former maintain larger territories and, because of a dependency on smaller game, must make a greater effort to provide for the newborn. Final dispersal of the young may not occur until they are about 2 years old. Mated pairs are known to have maintained a monogamous relationship for at least 6 years.

Births occur mainly from July to November. The gestation period is about 60 days. Average litter size is four young, ranging from one to eight. Weight at birth is about 159 grams (Hayssen, Van Tienhoven, and Van Tienhoven 1993). The young emerge from the den after 3 weeks and may then be moved several times to new sites. They are weaned at 8–9 weeks, and the adults then regurgitate food for them. After about 3 months they no longer use the den, and by the age of 6 months they are hunting on their own. Sexual maturity may come at 10–11 months, though young that remain with the parents do not breed. Captives have lived up to 14 years (Bekoff 1975; Bothma 1971*a;* Kingdon 1977; Moehlman 1978, 1987).

Unlike *C. adustus,* the black-backed jackal is considered a serious predator of sheep and is therefore intensively hunted and poisoned (Clutton-Brock, Corbet, and Hills 1976). According to Bothma (1971*b*), *C. mesomelas* causes more problems than any other animal in the sheep-farming areas of the Transvaal.

Black-backed jackal *(Canis mesomelas)*, photo by Ernest P. Walker.

Canis aureus (golden jackal)

Head and body length is 600–1,060 mm, tail length is 200–300 mm, shoulder height is 380–500 mm, and weight is 7–15 kg (Kingdon 1977; Lekagul and McNeely 1977). The fur is generally rather coarse and not very long. The dorsal area is mottled black and gray; the head, ears, sides, and limbs are tawny or rufous; the underparts are pale ginger or nearly white; and the tip of the tail is black (Clutton-Brock, Corbet, and Hills 1976).

The golden jackal is found mainly in dry, open country. It is strictly nocturnal in areas inhabited by people but may be partly diurnal elsewhere. Its opportunistic diet includes young gazelles, rodents, hares, ground birds and their eggs, reptiles, frogs, fish, insects, and fruit. It takes carrion on occasion but is a capable hunter (Kingdon 1977; Moehlman 1987; Van Lawick and Van Lawick–Goodall 1971).

The basic social unit is a mated pair and their young. Monogamy is the rule, though pair bonds do not seem to

Golden jackal *(Canis aureus)*, photo by Lothar Schlawe.

be as strong as in *C. mesomelas*. Sometimes the young of the previous year remain in the vicinity of the parents and even mate and bear their own litters. However, a more usual procedure is for one or two of the young to stay with the parents, refrain from breeding, and help care for the next litter. The resulting associations are probably responsible for the reports of large packs hunting together. The usual hunting range of a family is about 2–3 sq km. A portion of this area is a territory marked with urine and defended against intruders. The young are born in a den within the territory. Estrus lasts 3–4 days and newborn weight is about 201–14 grams (Hayssen, Van Tienhoven, and Van Tienhoven 1993). Births occur mainly in January–February in East Africa and in April–May in Central Asia but take place throughout the year in tropical Asia. The gestation period is 63 days. Litters contain one to nine, usually two to four, young. Their eyes open after about 10 days, and they begin to take some solid food at about 3 months. Both parents provide food and protection. Sexual maturity comes at 11 months. Captives have lived up to 16 years (Kingdon 1977; Lekagul and McNeely 1977; Moehlman 1983, 1987; Novikov 1962; Rosevear 1974; Van Lawick and Van Lawick–Goodall 1971).

As with other wild canids, the relationship of the golden jackal with people is controversial. In Bangladesh, for example, *C. aureus* plays an important scavenging role by eating garbage and animal carrion around towns and villages and may benefit agriculture by preventing increases in the numbers of rodents and lagomorphs. However, this jackal also raids such crops as corn, sugar cane, and watermelons. It may be involved in the spread of rabies, and in 1979 two young children were attacked and killed by jackals (Poché et al. 1987). Since about 1980 *C. aureus* has extended its European range into eastern Austria and northeastern Italy; it also has occupied parts of the Balkan Peninsula where it had not previously been seen, perhaps in association with a decline of the native wolf population

(Ginsberg and Macdonald 1990; Hoi-Leitner and Kraus 1989; Krystufek and Tvrtkovic 1990). On the other hand, the intensive spreading of poison baits for wolves after World War II evidently also eliminated the jackal in some areas, such as Macedonia, where no more were collected until 1989 (Krystufek and Petkovski 1990).

Canis simensis (Simien jackal or Ethiopian wolf)

Head and body length is 841–1,012 mm, tail length is 270–396 mm, shoulder height is 530–620 mm, and weight is 11–19 kg; males average about 20 percent larger than females (Sillero-Zubiri and Gottelli 1994). The general coloration of the upper parts is tawny rufous with pale ginger underfur. The chin, insides of the ears, chest, and underparts are white. There is a distinctive white band around the ventral part of the neck. The tail is rather short, the facial region of the skull is elongated, and the teeth, especially the upper carnassials, are relatively small (Clutton-Brock, Corbet, and Hills 1976).

The present habitat is montane grassland and moorland at elevations of approximately 3,000–4,000 meters. *C. simensis* may have been found at lower elevations before becoming subject to severe human persecution (Yalden, Largen, and Kock 1980). In the Bale Mountains of south-central Ethiopia, Morris and Malcolm (1977) found the Simien jackal to be relatively numerous in grasslands supporting large rodent populations. It was uncommon in scrub and was not seen in forests at low altitudes. It was primarily diurnal, but there was some evidence of activity at night, especially in moonlight. Farther north, in the Simien Mountains, the few remaining jackals were reported to be almost entirely nocturnal, probably because of human disturbance. The animals made little effort to find shelter, and individuals slept in the open or in places with slightly longer grass than usual even when temperatures were as low as −7° C. Sillero-Zubiri and Gottelli (1994) noted that dens are dug in open ground or located in a rocky crevice

Simien jackal, or Ethiopian wolf *(Canis simensis)*, photo by Claudio Sillero-Zubiri.

and are used only by females to give birth. Rodent prey also commonly is dug out of the ground. and kills often are cached for future use. Morris and Malcolm (1977) found the diet of *C. simensis* in the Bale Mountains to consist mainly of diurnal rodents, especially *Tachyoryctes* and *Otomys*. Hunting is done mainly by walking slowly through areas of high rodent density, investigating holes and listening carefully, and then stealthily creeping toward the prey and capturing it with a final dash. Individuals generally were seen to hunt alone and to not come together during the day, though there was apparently considerable overlap in hunting ranges.

According to Gottelli and Sillero-Zubiri (1994a) and Sillero-Zubiri and Gottelli (1994), *C. simensis* also is a facultative, cooperative hunter. Small packs occasionally have been seen chasing and killing young antelopes, lambs, and hares. The species lives in cohesive packs of 3–13 adults that share and defend an exclusive territory. These animals include all males born into the pack during consecutive years and 1 or 2 females. Annual home ranges of eight packs monitored for four years had an average size of 6 sq km, and ranges in an area of less prey averaged 13.4 sq km. Population density has been found to vary from about 0.1 to 1.0 adults per sq km, also depending on prey availability. Pack members congregate for friendly social interaction and territorial patrols at dawn and noon and in the evening and rest together at night but commonly forage individually in the morning and afternoon. Territories are marked with urine and feces and sometimes are contested through aggressive behavior and chases. There are various vocalizations, including yelps and barks during interaction and howls that can be heard 5 km away. The males and females of a pack form separate dominance hierarchies, and only the highest-ranking pair mates. Births in the Bale Mountains occur from October to January. The gestation period evi-

dently is 60–62 days and litters contain 2–6 young. They are regularly shifted between dens up to 1,300 meters apart and finally emerge on their own after 3 weeks. They begin to take some solid food at about 5 weeks, are fully weaned at 10 weeks, begin to hunt with the pack at 6 months, and attain sexual maturity in their second year.

If *C. simensis* actually is a wolf, as suggested by genetic analysis, its distribution may not seem so incongruous if consideration is given to the presence of another small wolf, *Canis lupus arabs*, just across the Red Sea to the east. *C. simensis* was reported from most Ethiopian provinces in the nineteenth century. It subsequently declined because of agricultural development in its range and environmental disruption that reduced prey populations. Also, it was incorrectly believed to be a predator of sheep and so was frequently shot. A newly recognized and potentially devastating threat is hybridization with domestic dogs *(C. familiaris)*. The latter species also is a source of harassment and diseases, such as rabies. At present only about 70–150 individuals of the subspecies *C. s. simensis* remain in the Simien Mountains and nearby parts of north-central Ethiopia. It is estimated that 270–370 individuals of the subspecies *C. s. citerni* remain in the Bale Mountains and that there are a few others in neighboring highlands. There are national parks in the Simien and Bale mountains, but these areas are still being used for grazing (Ginsberg and Macdonald 1990; Gottelli and Sillero-Zubiri 1992, 1994a, 1994b; Gottelli et al. 1994; Sillero-Zubiri and Gottelli 1994). *C. simensis* is protected by law in Ethiopia and is classified as critically endangered by the IUCN and as endangered by the USDI.

Canis latrans (coyote)

Head and body length is 750–1,000 mm and tail length is 300–400 mm. In a given population males are generally

Coyote *(Canis latrans)*, photo from U.S. Fish and Wildlife Service.

larger than females. Bekoff (1977) listed weights of 8–20 kg for males and 7–18 kg for females. Northern animals are usually larger than those to the south. Gier (1975) gave average weight as 11.5 kg in the deserts of Mexico and 18 kg in Alaska. The largest coyotes of all are found in the northeastern United States; there is considerable debate whether this development has resulted from hybridization with *C. lupus* or other genetic factors or is simply a phenotypic response to enhanced nutrition (Larivière and Crête 1993; Nowak 1979; Peterson and Thurber 1993). Pelage characters are variable, but usually the coat is long, the upper parts are buffy gray, and the underparts are paler. The legs and sides may be fulvous, and the tip of the tail is usually black (Clutton-Brock, Corbet, and Hills 1976). From *C. lupus* the coyote is distinguished by its smaller size, narrower build, proportionally longer ears, and much narrower snout.

The coyote is found in a variety of habitats, mainly open grasslands, brush country, and broken forests. The natal den is located in such places as brush-covered slopes, thickets, hollow logs, rocky ledges, and burrows made by either the parents themselves or other animals. Tunnels are about 1.5–7.5 meters long and lead to a chamber about 0.3 meter wide and 1 meter below the surface. Activity may take place at any time of day but is mainly nocturnal and crepuscular. The average distance covered in a night's hunting is 4 km. The coyote is one of the fastest terrestrial mammals in North America, sometimes running at speeds of up to 64 km/hr. There may be a migration to the high country in the summer and a return to the valleys in the autumn. Some of the young disperse from the parental range in the autumn and winter, generally covering a distance of 80–160 km (Banfield 1974; Bekoff 1977; Gier 1975). In a tagging study in Iowa, Andrews and Boggess (1978) found individuals to travel an average of about 31 km, but up to 323 km, from the point of capture. Carbyn and Paquet (1986) reported that a tagged individual in south-central Canada moved a record 544 km.

About 90 percent of the diet is mammalian flesh, mostly jack rabbits, other lagomorphs, and rodents. Such small animals are usually taken by stalking and pouncing (Bekoff 1977; Gier 1975). Coyotes have been seen to enter shallow streams and snatch fish from the water (Springer 1980). Larger mammals, especially deer, are also eaten, often in the form of carrion but sometimes after a chase in which several coyotes work together. In northwestern Wisconsin deer represent 21.3 percent of the diet (Niebauer and Rongstad 1977), and in northern Minnesota deer is the coyote's main food, mostly as carrion (Berg and Chesness 1978). In Alberta half of the food is carrion (Nellis and Keith 1976). Analyses of coyote predation on large mammals generally indicate that most of the individuals taken are immature, aged, or sick (Bekoff 1977). Livestock is killed by some coyotes, but the diet of the species seems to be largely neutral or beneficial with respect to human interests. *C. latrans* sometimes forms a "hunting partnership" with the badger (*Taxidea*). The two move together, the coyote apparently using its keen sense of smell to locate burrowing rodents and the badger digging them up with its powerful claws. Both predators then share in the proceeds. Recent observations by Minta, Minta, and Lott (1992) have confirmed that such associations do commonly exist, that they are mutually beneficial, and that the two species interact in a friendly manner.

Population density is generally 0.2–0.4/sq km but may be as high as 2.0/sq km under extremely favorable conditions (Bekoff 1977; Knowlton 1972). Reported home range size is 8–80 sq km. The ranges of males tend to be large and to overlap one another considerably; the ranges of females

are smaller and do not overlap (Bekoff 1977). In northern Minnesota Berg and Chesness (1978) found home range size to average 68 sq km for males and 16 sq km for females. In Alberta, however, Bowen (1982) found home ranges to be shared by an adult male and female and to average about 14 sq km for both sexes. Laundré and Keller (1984) reviewed various reports of home range size of about 10–100 sq km and concluded that most were based on inadequate data.

Social structure and territoriality vary depending on habitat conditions and food supplies. *C. latrans* is found alone, in pairs, or in larger groups but generally is less social than *C. lupus* (Bekoff 1977). Several males may court a female during the mating season, but apparently she eventually selects one for a lasting relationship. The pair then live and hunt together, sometimes for years. Gier (1975) suggested that pair territories tend to be bordered by natural features, cover about 1 sq km or less, and are defended only during the denning season. More recent work suggests that territories are larger and are exclusive throughout the year, though the ranges of young or transient animals may be superimposed (Andelt 1985).

Some of the offspring of a pair disperse, but others may remain with the parents and thus form the basis of a pack. These helpers, whose main function apparently is to provide increased protection to the next litters, may not leave until they are 2–3 years old. Studies in northwestern Wyoming indicate that packs are larger and more stable if a major, clumped food resource, such as ungulate carrion, is readily available (Bekoff and Wells 1980, 1982). In this area, on the National Elk Refuge, Camenzind (1978) found that 61 percent of the coyotes composed resident packs of 3–7 members, 24 percent were resident mated pairs, and 15 percent were nomadic individuals. The latter ranged through the territories of the resident animals and were continually chased away. The packs had well-defined social hierarchies and marked and defended territories. The sizes of two pack territories were 5 sq km and 7.2 sq km. Sometimes more than one pair of a pack mated and bore young. Occasionally up to 22 individuals would come together around carrion, but such aggregations usually lasted less than an hour.

The coyote has a great variety of visual, auditory, olfactory, and tactile forms of communication (Lehner 1978). At least 11 different vocalizations have been identified; these are associated with alarm, threats, submission, greeting, and contact maintenance. The best-known sounds are the howls given by one or more individuals apparently to announce location. Group howls may have a territorial and spacing function.

Mating usually occurs from January to March, and births take place in the spring. Females are monestrous. Estrus averages 10 (4–15) days, gestation averages 63 (58–65) days, and litter size averages about 6 (2–12) young. Larger numbers of offspring are sometimes found in one den, but these apparently represent the litters of more than one female. The young weigh about 250 grams at birth, open their eyes after 14 days, emerge from the den at 2–3 weeks, and are fully weaned at 5–6 weeks. The mother, and perhaps the father and older siblings, begins to regurgitate some solid food for the young when they are about 3 weeks old. Adult weight is attained at about 9 months. Some individuals mate in the breeding season following their birth, but others wait another year. Maximum known longevity in the wild is 14.5 years, but few animals survive that long. In one unexploited population 70 percent of the animals were less than 3 years old (Bekoff 1977; Gier 1975; Kennelly 1978). A captive lived 21 years and 10 months (Jones 1982).

Coyote *(Canis latrans)*, photo by David M. Shackleton.

The coyote long has been considered a serious predator of domestic animals, especially sheep. It is also sometimes alleged to spread rabies and to damage populations of game birds and mammals. It generally is considered harmless to people, though during the 1980s several children reportedly were seriously injured, and one killed, in attacks in California and western Canada (Carbyn 1989). Bounties against the coyote were first offered in the United States in 1825; they are still paid in some areas but have never achieved lasting prevention of depredations. In 1915 the U.S. government initiated a large-scale program of predator control, mainly in the West. The resulting known kill of *C. latrans* often exceeded 100,000 individuals per year. From 1971 to 1976 the annual take in 13 western states averaged 71,574 (Evans and Pearson 1980). Some state governments and private organizations also have control programs. Killing methods include trapping, poisoning, shooting from aircraft, and destroying the young in dens. In 1972 the federal government banned the general use of poison on its lands and in its programs.

Few wildlife issues have been as controversial as coyote control (Bekoff 1977; Bekoff and Wells 1980; Hall 1946; Sterner and Shumake 1978; Wade 1978). No authorities deny that predation is a problem to some farmers and ranchers, but almost since the beginning of the federal program there have been serious doubts concerning the extent of claimed livestock losses and the appropriate response. There was particular resentment against the widespread application of poison on western ranges, which was thought to kill not only far more coyotes than necessary but also many other carnivorous mammals. It has been argued that sheep raising is declining in the United States regardless of the coyote and its control and that damage by predators represents a relatively small part of overall losses to the industry. Surveys by trained biologists generally have indicated that losses are substantial in some areas but lower than often claimed. The stated intention of most current control operations is elimination of the specific individual coyotes that may

be causing problems, not the destruction of entire populations.

Information compiled by the U.S. Department of Agriculture (Gee et al. 1977) indicates that in 1974 losses attributed to the coyote in the western states amounted to 728,000 lambs (more than 8 percent of those born and one-third of the total lamb deaths) and 229,000 adult sheep (more than 2 percent of inventory and one-fourth of all adult deaths), valued at $27 million. In that same year the amount expended for coyote control in the West was $7 million (Gum, Arthur, and Magleby 1978). In 1974, 71,522 coyotes were killed by the federal control program. Data compiled by Pearson (1978) suggest that the total kill in the 17 western states in 1978 (including the take by fur trappers and sport hunters) was around 300,000 individuals, approximately the same as in 1946. Data compiled by Terrill (1986) indicate that the rate of loss of sheep and lambs to coyote predation through 1985 generally remained about the same as given above, though there was a temporary decline in 1980, apparently caused by the spread of canine parovirus in the coyote population. Subsequent losses have been subject to conflicting reports, though one estimate sets the value of sheep and lambs killed by coyotes in the 17 western states during 1989 at $18 million (Connolly 1992). About 89,000 coyotes were killed during fiscal year 1995 by the federal control program (unpublished data from U.S. Department of Agriculture, 1996).

Coyote fur has varied in price. In 1974 skins taken in the United States sold for an average of $17.00 each. For the 1976–77 season 30 states reported a total harvest of 320,323 pelts with an average value of $45.00; in 6 Canadian provinces the take was 65,819 pelts with an average value of $59.76 (Deems and Pursley 1978). Coyote numbers evidently rebounded in the 1980s, but fur prices dropped. The total known kill in the United States during the 1983–84 season was 439,196. In Canada 68,975 pelts with an average value of $15.90 were taken (Novak, Obbard, et al. 1987). In the 1991–92 season 158,001 skins were taken in the United States and sold at an average price of $13.53 (Linscombe 1994).

It is sometimes said that *C. latrans,* as a species, can easily sustain human-inflicted losses because of its adaptability, cunning, and high reproductive potential. That may not necessarily be the case. In the past, intensive control programs eliminated the species from central Texas, much of North Dakota, and certain large sheep-raising sections of Colorado, Wyoming, Montana, Utah, and Nevada (Gier 1975; Nowak 1979).

It does appear that *C. latrans* has substantially extended its range since the arrival of European colonists in North America (Brady and Campbell 1983; Gier 1975; Hill, Sumner, and Wooding 1987; Hilton 1978; Holzman, Conroy, and Pickering 1992; Monge-Nájera and Brenes 1987; Moore and Millar 1984; Nowak 1978, 1979; Sabean 1989; Thomas and Dibblee 1986; Vaughan 1983; Weeks, Tori, and Shieldcastle 1990; Wooding and Hardisky 1990). The newly occupied regions may include Alaska, the Yukon, and the southern part of Central America, though some evidence suggests that the species was already there, at least intermittently, in prehistoric times. It is known to have moved through Costa Rica and into western Panama since 1960. The main expansion was eastward, the two main factors apparently being (1) the creation of favorable habitat through the breaking up of the forests by settlement and (2) human elimination of the wolves *(C. rufus* and *C. lupus)* that might have competed with the coyote. In the late nineteenth and early twentieth centuries the coyote moved from its prairie homeland, through the Great Lakes region, and into southeastern Canada. From the 1930s to the 1960s the species established itself in New England and New York. It now has occupied Ohio and Kentucky, moved as far east as Nova

Scotia and Prince Edward Island, and spread down the Appalachians to southern Virginia. Other coyote populations pushed into southern Missouri and Arkansas in the 1920s, overran Louisiana in the 1950s, crossed into Mississippi by the 1960s, and subsequently became established in Tennessee and as far as Florida and the Carolinas. The expanding coyote populations were modified by hybridization with the remnant pockets of wolves that they encountered, *C. lupus* in southeastern Canada and *C. rufus* in the south-central United States, with the result that the coyotes of eastern North America are larger and generally better adapted for forest life than their western relatives.

Canis rufus (red wolf)

Head and body length is 1,000–1,300 mm, tail length is 300–420 mm, shoulder height is 660–790 mm, and weight is 20–40 kg. While the red element of the fur is sometimes pronounced, the upper parts are usually a mixture of cinnamon buff, cinnamon, or tawny with gray or black; the dorsal area is generally heavily overlaid with black. The muzzle, ears, and outer surfaces of the limbs are usually tawny. The underparts are whitish to pinkish buff, and the tip of the tail is black. There was reported to be a locally common dark or fully black color variant in the forests of the Southeast. From *C. lupus* the red wolf is distinguished by its narrower proportions of body and skull, shorter fur, and relatively longer legs and ears.

Habitats include upland and bottomland forests, swamps, and coastal prairies. Natal dens are located in the trunks of hollow trees, stream banks, and sand knolls. Dens, which the animal either excavates itself or takes over from

Red wolf *(Canis rufus),* photo from the *Washington Post.*

some other animal, average about 2.4 meters in length and usually extend no further than 1 meter below the surface. The red wolf is primarily nocturnal but may increase its daytime activity during the winter. It hunts over a relatively small part of its home range for about 7–10 days and then shifts to another area. Reported foods include nutria, muskrats, other rodents, rabbits, deer, hogs, and carrion (Carley 1979; Nowak 1972; Riley and McBride 1975).

Home range in southeastern Texas has been reported: (1) to average 44 sq km for 7 individuals (Shaw and Jordan 1977); (2) to cover 65–130 sq km over 1–2 years (Riley and McBride 1975); and (3) to average 116.5 sq km for males and 77.7 sq km for females (Carley 1979). Extensive observations of the reintroduced population in North Carolina indicate that pack home range is 50–100 sq km (Phillips 1994). The basic social unit is apparently a mated, territorial pair. Groups of 2–3 individuals are most common, though larger packs have frequently been reported. The normal age of dispersal appears to be 16–22 months (Phillips 1995). Vocalizations are intermediate to those of *C. latrans* and *C. lupus* (Paradiso and Nowak 1972). Mating occurs from January to March and offspring are produced in the spring. The gestation period is 60–63 days, and litters contain up to 12 young, usually about 4–7. Several of the individuals reintroduced in North Carolina are known to have lived more than 4 years in the wild (Phillips 1995), and potential longevity in captivity is at least 14 years (Carley 1979).

Like most large carnivores, the red wolf was considered to be a threat to domestic livestock, if not to people themselves. It was therefore intensively hunted, trapped, and poisoned after the arrival of European settlers in North America. In addition, disruption of its habitat and reduction of its numbers allowed *C. latrans* to invade its range from the west and north and evidently stimulated interbreeding between the two species. This process led to the genetic swamping of the small pockets of red wolves that survived human persecution. By the 1960s the only pure populations of *C. rufus* were in the coastal prairies and swamps of southeastern Texas and southern Louisiana. Conservation efforts, especially by the U.S. Fish and Wildlife Service, were not successful. The hybridization process continued to spread, and by 1975 the prevailing view was that the species could be saved only by securing and breeding some of the remaining animals that appeared to represent unmodified *C. rufus* (Carley 1979; McCarley and Carley 1979; Nowak 1972, 1974, 1979).

Many animals were captured and examined, and eventually 14 were selected for the breeding program. Since 1977 offspring have been produced from this stock, and by 1995 there were 289 living descendants (the last wild-caught wolf died in 1989). Most of these animals are still in captivity, the majority at the Point Defiance Zoo in Tacoma, Washington, but experimental reintroductions have been made on Bulls Island off South Carolina, St. Vincent Island off Florida, and Horn Island off Mississippi. Starting in 1987, pairs have been released as part of a large-scale effort to reestablish a permanent red wolf population at the Alligator River National Wildlife Refuge and adjacent lands in eastern North Carolina. By 1995 this population contained about 50 animals, which were behaving normally, reproducing freely, and readily taking deer and other natural prey. Another small group had been released to the west in Great Smoky Mountains National Park (Henry 1993, 1995; Phillips 1995; Rees 1989; Waddell 1995a, 1995b).

From a management standpoint the reintroductions have been highly successful; indeed, they serve as a model for future efforts involving other large predatory mammals. However, the projects are jeopardized by intense political controversy evidently generated by the concerns of a small minority of landowners in the affected areas (Bourne 1995a). Widely publicized views that the red wolf may have originated as a hybrid (see above), though not generally accepted by the scientific community, have provided further grounds for challenging the reintroductions and related conservation work. In August 1995 a measure that would have cut off funding for this work was brought before the U.S. Senate; it was narrowly defeated, largely through the efforts of one environmentally oriented Republican senator (Chafee 1995). Notwithstanding the controversy, the red wolf continues to be classified as endangered by the USDI and now is classified as critically endangered by the IUCN.

Canis lupus (gray wolf)

This species includes the largest wild individuals in the family Canidae. Head and body length is generally about 1,000–1,600 mm and tail length is 350–560 mm. However, there are several small subspecies along the southern edge of the range of the gray wolf, especially *C. lupus arabs* of the Arabian Peninsula; an adult male of that subspecies had a head and body length of only 820 mm and a tail length of 320 mm (Harrison 1968). Ferguson (1981) listed the following data: *C. lupus arabs* of Israel and Arabia, head and body length 1,140–1,440 mm, tail length 230–360 mm, and weight 14–19 kg; *C. lupus lupaster* of Egypt and Libya, head and body length 1,058–1,300 mm, tail length 283–355 mm, and weight 10–16 kg. In North America the largest animals are found in Alaska and western Canada, the smallest in Mexico. In a given population males are larger on average than females. Overall mean (and extreme) weights are about 40 (20–80) kg for males and 37 (18–55) kg for females (Mech 1970, 1974a). The pelage is long; the upper parts are usually light brown or gray, sprinkled with black, and the underparts and legs are yellow white. Entirely white individuals occur frequently in tundra regions and occasionally elsewhere. Black individuals are also common in some populations.

The gray wolf has the greatest natural range of any living terrestrial mammal other than *Homo sapiens*. It is found in all habitats of the Northern Hemisphere except tropical forests and arid deserts. Dens, used only for the rearing of young, may be located in rock crevices, hollow logs, or overturned stumps but are usually in a burrow, either dug by the parents themselves or initially made by another animal and enlarged by the wolves. Sometimes several such burrows, perhaps as far apart as 16 km, are excavated in the same season. A den may be used year after year. The animals prefer an elevated site near water. Tunnels are about 2–4 meters long and lead to an enlarged underground chamber with no bedding material. There may be several entrances, each marked by a large mound of excavated soil. When the young are about 8 weeks old they are moved to a rendezvous site, an area of about 0.5 ha., usually near water and marked by trails, holes, and matted vegetation. Here the young romp and play, and the other members of the pack gather for daily rest during the summer. Such sites are frequently changed, but some may be used for one or two months (Mech 1970, 1974a; Peterson 1977).

Movements are extensive and usually take place at night, but diurnal activity may increase in cold weather. During the summer the pack usually sets out in the early evening and returns to the den or rendezvous site by morning. In winter the animals wander farther and do not necessarily return to a particular location. They tend to move in single file along regular pathways, roads, streams, and

Gray wolf (*Canis lupus*), photo from Zoological Society of London

ice-covered lakes. The daily distance covered ranges from a few to 200 km (Mech 1970). In Finland the mean daily movement was determined to be 23 km (Pulliainen 1975). On Isle Royale, in Lake Superior, Peterson (1977) found that packs averaged 11 km per day or 33 km per kill. When individuals permanently disperse from a pack, they move much farther than normal; one traveled 206 km in 2 months (Mech 1974a), and another covered a straight-line distance of 670 km in 81 days (Van Camp and Gluckie 1979). Recent record movements are of a single animal that dispersed 886 km, from Minnesota to Saskatchewan (Fritts 1983), and a group of at least 2 and probably 4 wolves that moved at least 732 km across Alaska (Ballard, Farnell, and Stephenson 1983). Five dispersing wolves tracked by Fritts and Mech (1981) emigrated about 20–390 km, though others remained in the vicinity of the parental groups. Packs that depend on barren ground caribou make seasonal migrations with their prey and move as far as 360 km (Kuyt 1972; Mech 1970, 1974b).

The gray wolf usually moves at about 8 km/hr but has a running gait of 55–70 km/hr. It can cover up to 5 meters in a single bound and can maintain a rapid pursuit for at least 20 minutes. Prey is located by chance encounter, direct scenting, or following a fresh scent trail. Odors up to 2.4 km away can be detected. A careful stalk may be used to get as close as possible to the prey. If the objective is large and healthy and stands its ground, the pack usually does not risk an attack. Otherwise a chase begins, usually covering 100–5,000 meters. If the wolves cannot quickly close with the intended victim, they generally give up. There seems to be a continuous process of testing the individual members of the prey population to find those that are eas-

ily captured. Most hunts are unsuccessful. On Isle Royale, for example, only about 8 percent of the moose tested are actually killed (Mech 1970, 1974b). In one exceptional case the chase of a deer went for 20.8 km (Mech and Korb 1978). Once the wolves overtake an ungulate, they strike mainly at its rump, flanks, and shoulders.

The gray wolf is primarily a predator of mammals larger than itself, such as deer, wapiti, moose, caribou, bison, muskox, and mountain sheep. The smallest consistent prey is beaver. Following a drastic decline of deer in central Ontario, beaver remains were found in most wolf droppings (Voight, Kolenosky, and Pimlott 1976). Kill rates vary from about one deer per individual wolf every 18 days to one moose per wolf every 45 days (Mech 1974a). Usually a pack of several wolves will make a kill, consume a large amount of food, and then make another kill some days later. An adult can eat about 9 kg of meat in one feeding. Studies in Minnesota indicate that an average daily consumption is about 2.5 kg of deer per wolf and that pack members generally remain in the vicinity of a kill for several days, eventually utilizing nearly the entire carcass, including much of the hair and bones (Mech et al. 1971). Most analyses of predation on wild species show that immature, aged, and otherwise inferior individuals constitute most of the prey taken by wolves (Mech 1974a).

There long has been controversy regarding the wolf's effect on overall prey populations. It was once generally thought, even by some experienced zoologists, that the wolf could and did eliminate its prey. For example, Bailey (1930) wrote that "wolves and game animals can not be successfully maintained on the same range." Subsequently a popular view developed that the wolf and the animals

B

Gray wolf *(Canis lupus)*, photo from New York Zoological Society.

on which it depends are in precise balance, with the wolf taking just enough to prevent substantial population increases. Mech (1970) tentatively concluded that wolf predation is the major controlling factor where prey-predator ratios are about 11,000 kg of prey per wolf or less, as on Isle Royale. More recently it became apparent that the wolf is not regulating the moose herd on Isle Royale but merely cropping part of the annual surplus production (Mech 1974b; Peterson 1977). On the other hand, studies by Mech and Karns (1977) indicate that wolf predation was a major contributing factor in a serious decline of deer in the Superior National Forest of Minnesota from 1968 to 1974. Primary causes in the decline there, as well as in other parts of the Great Lakes region, were maturation of the forests that had been cut around the turn of the century and a series of severe winters.

Wolf population density varies considerably, being as low as 1/520 sq km in parts of Canada. In Alberta, Fuller and Keith (1980) found densities of 1/73 sq km to 1/273 sq km. Several studies in the Great Lakes region found apparently stable densities of around 1/26 sq km and suggested that this level was the maximum allowed by social tolerance (Mech 1970; Pimlott, Shannon, and Kolenosky 1969). Subsequent investigations determined that the wolf could attain densities nearly twice as great in areas where prey concentrate for the winter (Kuyt 1972; Parker 1973; Van Ballenberghe, Erickson, and Byman 1975). On Isle Royale, an area of 544 sq km, the number of wolves remained relatively stable at a mid-winter average of 22 from 1959 to 1973. There was an increase to 44 in 1976 evidently in response to rising moose numbers, thus indicating that food supply is the main determinant of density (Peterson 1977). The Isle Royale wolf population peaked at 50 in 1980 and then underwent a crash in association with falling moose numbers (Peterson and Page 1988). By 1990 only 14 wolves

survived there even though the moose population again was on the rise, which led to concern about genetic viability (Wayne, Gilbert, et al. 1991).

Home range size depends on food availability, season, and number of wolves. The largest pack range, found in Alaska during winter, was 13,000 sq km, and the smallest, in southeastern Ontario during summer, was 18 sq km (Mech 1970; Pimlott, Shannon, and Kolenosky 1969). In Minnesota, ranges of 52–555 sq km have been reported (Fritts and Mech 1981; Harrington and Mech 1979; Van Ballenberghe, Erickson, and Byman 1975). One radio-collared pack in Minnesota used summer ranges of 117–32 sq km and winter ranges of 123–83 sq km (Mech 1977a). Alberta packs used ranges of 195–629 sq km in summer and 357–1,779 sq km in winter (Fuller and Keith 1980). Pack range in Russia and Kazakhstan varies from 30 sq km in areas of abundant food and good cover to more than 1,000 sq km in deserts and tundra (Bibikov, Filimonov, and Kudaktin 1983).

The home range of a wolf pack usually corresponds to a defended territory. There is generally little or no overlap between the ranges of neighboring packs. Lone wolves that split off from a pack may disperse a considerable distance or they will be continuously pursued by the resident packs and forced to shift about (Mech 1974a; Peterson 1977; Rothman and Mech 1979). Territories are relatively stable; some are known to have been used for at least 10 years (Mech 1979). Buffer zones, areas of little use, tend to develop between territories; lone wolves may sometimes center their activities there. In Minnesota deer have been found to have higher densities in such border areas and to survive longer there than elsewhere during times of general population declines (Mech 1977c, 1979; Rogers et al. 1980). Under normal conditions packs avoid areas where they might encounter other packs. When food shortages induce stress, however, wolves move into the buffer zones to hunt and eventually trespass into the territories of other packs. Meetings between packs are agonistic and may result in chases, savage fighting, and mortality (Harrington and Mech 1979; Mech 1970, 1977b; Van Ballenberghe and Erickson 1973).

Although packs are hostile to one another, the gray wolf is among the most social of carnivores. Groups usually contain 5–8 individuals but have been reported to have as many as 36 (Mech 1974a). Such associations are probably essential for consistent success in the pursuit and overpowering of large prey. The number of wolves in a pack tends to increase with the size of the usual prey, being 7 or less where deer is the only important food, 6–14 where both deer and wapiti are eaten, and 15–20 on Isle Royale, where moose is the primary prey (Peterson 1977). In certain southern parts of its range, such as Mexico, Italy, and Arabia, C. lupus seems to be less gregarious than in the north (Harrison 1968; McBride 1980; Zimen 1981). This situation may be partly unnatural, resulting from intensive human persecution and forced dependence on easily captured domestic animals and garbage. However, it has been demonstrated that a single female, and sometimes even an adult male, can successfully rear a litter of young (Boyd and Jimenez 1994).

A pack is essentially a family group, comprising an adult pair, which may mate for life, and their offspring of one or more years (Mech 1970). A few exceptions have been revealed by both field and genetic study (Lehman et al. 1992; Mech and Nelson 1990). The leader of a normal pack is usually a male, often referred to as the alpha male. He initiates activity, guides movements, and takes control at critical times, such as during a hunt. The males and females of a pack may have separate dominance hierarchies reinforced by aggressive behavior and elaborate displays of greeting and submission by subordinate members. Generally only the most dominant pair mate, and they inhibit sexual activity in the others. Social status is rather consistent, and a leader may retain its position for years, but roles can be reversed. Intragroup strife, perhaps resulting from increasing membership or declining food supplies, can result in a division into two packs or the splitting off of individuals. The latter may maintain a loose association with the parental group, sometimes following at a distance and feeding on scraps left behind, or disperse to a new area to seek a mate and begin a new pack (Mech 1970, 1974a; Peterson 1977; Wolfe and Allen 1973; Zimen 1975).

Studies by Fritts and Mech (1981) have contributed to our understanding of new pack formation in an area of an expanding wolf population. A young individual evidently does not remain with its parental pack past breeding age (about 22 months) unless it becomes a breeder itself upon the death of an alpha animal of the same sex. As it approaches maturity it may actively explore the fringes of the parental territory. After dispersal it either joins another lone wolf to search for a new territory or establishes itself in an area and awaits the arrival of an animal of the opposite sex. A new pack territory is relatively small but may incorporate a portion of the parental territory. Therefore, as population size increases, average territory size decreases. Data summarized by Gese and Mech (1991) indicate that most wolves disperse from their natal pack during February–April, when they are 11–12 months old, or during October–November, when they are 17–19 months old.

The gray wolf has a variety of visual, olfactory, and auditory means of communication. Vocalizations include growls, barks, and howls—continuous sounds usually lasting 3–11 seconds (Mech 1974a). Individuals have distinctive howls (Peterson 1977). Humans can hear howls 16 km away on the open tundra, and wolves probably respond to howls at distances of 9.6–11.2 km. Howling functions to bring packs together and as an immediate, long-distance form of territorial expression (Harrington and Mech 1979).

Territories are also maintained by scent marking via scratching, defecation, and especially urination. Scent marking differs from ordinary elimination in that there is a regular pattern of deposition at certain repeatedly used points. Peters and Mech (1975) determined that as a pack moves through its territory—visiting most parts at least every three weeks—signs are left at average intervals of 240 meters. Rothman and Mech (1979) found that scent marking also is important in bringing new pairs together for breeding and in helping established pairs to achieve reproductive synchrony.

Threats and attacks by the dominant members of a pack probably prevent sexual synchronization in subordinates; thus only the highest-ranking female normally bears a litter during the reproductive season (Zimen 1975). Some exceptions were reported by Van Ballenberghe (1983), and Mech and Nelson (1989) documented that during a single breeding season a male mated with two females, both of which raised litters. Mating may occur anytime from January, in low latitudes, to April, in high latitudes. Births take place in the spring. Courtship may extend for days or months, estrus lasts 5–15 days, and the gestation period is usually 62–63 days. Mean litter size is 6 young, and the range is 1–11. The young weigh about 450 grams at birth and are blind and deaf. Their eyes open after 11–15 days, they emerge from the den at 3 weeks, and they are weaned at around 5 weeks. The mother usually stays near the den for a period, during which time the father and other pack members hunt and bring food for both her and the pups. The young are commonly fed by regurgitation. At 8–10 weeks the young are shifted to the first in a series of rendezvous sites, each up to 8 km from the other. If they are in good condition, the young begin to travel with the pack in early autumn. Sexual maturity generally comes at around 22 months, but social restrictions often prevent mating at that time. A captive pair successfully bred at only 10 months. Mortality is highest among the young. In times of sharply declining food supplies all pups may be severely underweight or die from malnutrition, and reproduction may even cease. For adults in such a situation the primary mortality factor has been found to be intraspecific strife. Annual survival of adults in a population not under nutritional stress or human exploitation has been calculated at 80 percent. Wild females are known to have given birth when 10 years old and to have lived to an age of 13 years and 8 months. Potential longevity is at least 16 years (Mech 1970, 1974a, 1977a, 1977b, 1989; Medjo and Mech 1976; Van Ballenberghe, Erickson, and Byman 1975; Van Ballenberghe and Mech 1975).

The wolf is often believed to be a direct threat to people. In Eurasia attacks are unusual but evidently have occurred, sometimes resulting in death (Pulliainen 1980; Ricciuti 1978). In North America there appear to have been only four well-documented attacks by wild wolves (none resulting in death): (1) an injury inflicted by a female that may have been attracted to a camp by male dogs (Jenness 1985); (2) a prolonged assault involving a probably rabid individual (Mech 1970); (3) simply an aggressive leap that made contact with the person but caused no injury (Munthe and Hutchison 1978); and (4) several lunges by members of a pack at a group of three people (Scott, Bentley, and Warren 1985). Mech (1990) discussed a number of less detailed reports of attacks as well as several incidents involving apparently accidental wolf-human contact and rabid wolves (again without death or serious injury to people).

A far more substantive basis for the age-old warfare between people and the gray wolf is depredation by the latter on domestic animals, notably cattle, sheep, and reindeer. The wolf also has been persecuted, especially in the twentieth century, because of its alleged threat to populations of the wild ungulates that are desired by some persons for sport and subsistence hunting. The wolf was long taken by various kinds of traps and snares as well as by pursuit with packs of specially trained dogs. In the nineteenth century poison came into widespread use, and in the mid–twentieth century hunting from aircraft became popular, especially on the open tundra.

The last wolves in the British Isles were exterminated in the eighteenth century. By the early twentieth century the species, except for occasional wandering individuals, had disappeared in most of western Europe and in Japan. Modest comebacks occurred in Europe during World Wars I and II, but currently the only substantial populations on the Continent west of Russia are in the Balkans. Recent conservation efforts have prevented extirpation in Poland, Spain, and Italy, even allowing increases in distribution and numbers (Blanco, Reig, and de la Cuesta 1992; Boitani 1992; Okarna 1993). There are also remnant groups in Portugal, Slovakia, Hungary, and Scandinavia. A few individuals recently have entered Germany and France (*Oryx* 28 [1994]: 6). *C. lupus* survives over much of its former range in southwestern and south-central Asia but is generally rare. Only 500–800 survive in India.

There were estimated to be 150,000–200,000 wolves in the old Soviet Union after World War II, but these became subject to an intensive government control program. The annual kill was 40,000–50,000 individuals from 1947 to 1962 and subsequently dropped to about 15,000. In the mid-1970s the estimated number of wolves in the old Soviet Union was 50,000, about two-thirds of them in the Central Asian republics (Grzimek 1975; Pimlott 1975; Pulliainen 1980; Roberts 1977; Shahi 1982; Smit and Van Wijngaarden 1981; Zimen 1981). Wolf numbers subsequently again increased in the Soviet Union, leading to implementation of a bounty of up to 100 rubles (Bibikov 1980). There also is aerial hunting and poisoning by the government. In 1980 the number of wolf pelts taken there was 35,573 (Bibikov, Ovsyannikov, and Filimonov 1983).

The decline of the gray wolf was even more sweeping in the New World than in the Old. The species was largely eliminated along the east coast and in the Ohio Valley by the mid–nineteenth century. By 1914 the last gray wolves had been killed in Canada south of the St. Lawrence River and in the eastern half of the United States, except for northern parts of Minnesota and Wisconsin and the upper peninsula of Michigan. In the following year the federal government began a large-scale program to destroy predators. Partly as a result of this campaign resident populations of the gray wolf apparently had disappeared from the western half of the conterminous United States by the 1940s. The rate of decline subsequently slowed, mainly because the wolf had been eliminated from nearly all parts of the continent that were conducive to the raising of livestock. Conflict with agricultural interests does continue all along the lower edge of the current major range of the gray wolf, which extends across southern Canada, from coast to coast, and dips down into northern Minnesota (Nowak 1975a). Regular government control operations are still carried out in parts of Canada and even have intensified recently in some areas in response to reported predation on livestock and big game (Gunson 1983; Stardom 1983; Tompa 1983).

The wolf still occurs over much of the northern part of Minnesota, primarily in boreal forest and bog country that is not suitable for agriculture but also in adjoining lands that are well settled. The species is protected by federal law, except that individuals preying on domestic animals may be taken by government agents. Many persons in Minnesota have argued that current protective regulations are overly restrictive and do not allow adequate control of depredations and that the wolf should be subject to sport hunting and commercial trapping. Recent studies in northwestern Minnesota, however, indicate that wolves infrequently attack domestic livestock even when living nearby (Fritts 1982; Fritts and Mech 1981; Fritts et al. 1992; Mech 1977d). In 1983 the U.S. Fish and Wildlife Service actually attempted to open the Minnesota wolf to sport hunting, but this effort was defeated through litigation. The species thus remained under protection, and perhaps as a result, it continued to occupy more of Minnesota and even to extend its range into northern Wisconsin, upper Michigan, and the Dakotas (see below).

In Alaska the wolf is legally open to sport hunting and fur trapping, but there is also controversy (Harbo and Dean 1983). Shooting from aircraft was a major hunting method for many years. In 1972 the practice was banned by federal law, except for government predator-control programs. Starting in the mid-1970s the Alaska Department of Fish and Game authorized aerial hunting of wolves in an effort to increase the numbers of moose and caribou in certain parts of the state. There followed a series of legal battles involving opposing conservationist organizations, the state government, and the federal agencies that owned much of the land where the activity was to take place. Some of the hunts were halted, but others were allowed to proceed. The taking of the wolf in Alaska has been stimulated in part by rising fur prices. For the 1976–77 season the reported state harvest was 1,076 pelts with an average value of $200 (Deems and Pursley 1978). In the 1991–92 season 1,162 pelts were taken in Alaska and sold for an average price of $275 (Linscombe 1994).

The wolf population of the Alexander Archipelago and adjacent mainland of southeastern Alaska represents a subspecies distinct from that of the rest of Alaska and western Canada. This population preys largely on the black-tailed deer herds of the area, which in turn are dependent on the understory of old-growth forests (Ingle 1994). Planned logging of these forests could result in a drastic decline of the deer and ultimately extinction of the wolf. Nonetheless, the U.S. Fish and Wildlife Service recently rejected a petition to classify the wolf population as threatened (Beattie 1995), a measure that would have allowed protection of the forests, nearly all of which are on federal lands.

The total number of gray wolves in North America can only be guessed at. Overall populations do not appear to have declined since the 1950s, but there is concern that excessive hunting, as well as oil and mineral exploitation, could adversely affect prey species, especially the caribou of tundra regions. The species was exterminated in Greenland by the 1930s, but individuals subsequently wandered back in from arctic Canada. There now is a resident population of perhaps 50 animals, and breeding has occurred well down the east coast of the island (Dawes, Elander, and Ericson 1986; Higgins 1990). However, the wolves of the arctic islands of Canada may themselves be in jeopardy because of a crash in native caribou populations and increasing human harassment. There also is some suggestion that the wolves of that region have been affected by hybridization with domestic dogs (Clutton-Brock, Kitchener, and Lynch 1994). There are about 4,000–7,000 wolves in Alaska, and there are estimates of 30,000–60,000 in Canada (Carbyn 1983, 1987; Theberge 1991). The fur take in Canada was 5,000–7,000 annually in the late 1970s but fell to about 4,000 in 1984 (Carbyn 1987). Wolf numbers have continued to fall in Mexico, mainly because of persecution by cattle ranchers. There may still have been hundreds in the 1950s, but there were no more than 50 in the 1970s and only 10 by the 1990s (Ginsberg and Macdonald 1990; McBride 1980). Occasional individuals from Mexico crossed into the United States until the 1970s, but they usually were quickly killed. There is a captive breeding pool of about 140 Mexican wolves, and consideration is being given to reintroduction projects in Arizona and New Mexico (Parsons 1996; Siminski *in* Bowdoin et al. 1994).

In Minnesota the number of wolves has grown steadily under protection, from about 600 in the 1950s, to 1,200 in the 1970s, and to more than 1,800 in the early 1990s (Fuller, Berg, et al. 1992; Mech 1977d; Mech, Pletscher, and Martinka 1995). Numerous individuals from this expanding population evidently have dispersed into the Dakotas (Licht and Fritts 1994). Resident populations may have been eliminated in Wisconsin and Michigan in the 1950s, and an attempted reintroduction in Michigan failed in 1974 when all four of the involved wolves were quickly killed by human agency (Weise et al. 1975). Nonetheless, legal protection subsequently allowed dispersing wolves from Minnesota to move back into both northern Wisconsin and the adjacent upper peninsula of Michigan, and by the early 1990s a number of packs were established, with at least 50 animals in each state (Jensen, Fuller, and Robinson 1986; Mech 1995; Mech and Nowak 1981; Mech, Pletscher, and Martinka 1995; Robinson and Smith 1977; Thiel 1985).

There is increasing concern about the effects of hybridization on wolves in the Great Lakes region and possibly some other areas. There long has been morphological evidence indicating that the expanding population of *C. latrans* hybridized with the small wolf subspecies *C. lupus lycaon* in southeastern Ontario and southern Quebec (Nowak 1979). Recent studies have shown that all wolves in those areas and on Isle Royale have mitochondrial DNA derived from coyotes and that most wolves in Minnesota and southwestern Ontario have been similarly affected (Lehman et al. 1991). Additional investigation indicates a general decline in the genetic viability of the small, isolated population of wolves on Isle Royale (Wayne, Gilbert, et al. 1991). Fortunately, there is as yet no evidence suggesting any morphological, behavioral, or ecological modification of *C. lupus* in Minnesota, western Ontario, Isle Royale, or Algonquin National Park (Nowak 1992). Hybridization between *C. lupus* and *C. familiaris* has not yet been shown to be detrimental to wild populations of the former, despite the far greater abundance of the latter. Of possible concern is the growing popularity of captive wolf-dog hybrids, now estimated to number about 300,000 in the United States. If such animals become feral, they could facilitate a breakdown of genetic integrity, especially in sparse or recolonizing wolf populations. A few of these hybrids have escaped and attacked children, thereby contributing to a negative image of *C. lupus* in general (Willems 1995).

Since the 1940s there have been regular reports of wolves from the northwestern conterminous United States, especially along the Rocky Mountains between Glacier and Yellowstone national parks (Ream 1980; Weaver 1978). Such records probably represented individuals that wandered from Canada. Legal protection, research, and conservation measures provided pursuant to the U.S. Endangered Species Act of 1973 allowed some of these wolves to establish resident packs. A breeding population of about 100 animals now utilizes land on both sides of the border in and around Glacier National Park (Chadwick 1995a). Some individuals from this population may have spread as far as Wyoming and Idaho. However, development of a federal program to reintroduce wolves directly into those states was initiated in 1988. In early 1995, following an expensive and controversial process, several packs of wolves that had been captured in western Canada were released in Yellowstone National Park, Wyoming, and wilderness areas of central Idaho. As of mid-1996 the reintroduced populations contained more than 100 wolves, breeding was occurring regularly, and predation on domestic livestock had not been a significant problem (Bangs and Fritts 1996; Cook 1993; Gerhardt 1995). And recently there has been a natural movement of wolves into the state of Washington (Mech 1995).

The IUCN classifies the population of *C. lupus* in Mexico as extinct in the wild, the population in Italy as vulnerable, and the population in Spain and Portugal as conservation dependent. The USDI lists all populations of *C. lupus* in the conterminous United States and Mexico as endangered, except for that in Minnesota, which is designated as threatened. There have been repeated claims that the Minnesota population should be completely removed from the U.S. List of Endangered and Threatened Wildlife. However, it now is known that in addition to direct human persecution and long-term habitat disturbance, this population is jeopardized by parasitic heartworms and the deadly disease canine parovirus, both evidently spread by domestic dogs (Mech and Fritts 1987). *C. lupus* is on appendix 2 of the CITES, except for the populations of Pakistan, India, Nepal, and Bhutan, which are on appendix 1.

Canis familiaris (domestic dog)

There are approximately 400 breeds of domestic dog, the chihuahua being the smallest and the Irish wolfhound the largest (Grzimek 1975). According to the National Geographic Society (1981), head and body length is 360–1,450 mm, tail length is 130–510 mm, shoulder height is 150–840 mm, and the normal weight range is 1–79 kg. The heaviest individual on record, a St. Bernard, weighs about 150 kg (*Washington Post*, 29 November 1987). In the wild subspecies of Australia, *C. f. dingo*, head and body length is 1,170–1,240 mm, tail length is 300–330 mm, shoulder height is about 500 mm, and weight is 10–20 kg. The dingo is usually tawny yellow in color, but some individuals are white, black, brown, rust, or other shades. The feet and tail tip are often white (Clutton-Brock, Corbet, and Hills 1976). From other forms of *C. familiaris* of comparable size and shape the dingo can be distinguished by its longer muzzle, larger bullae, more massive molariform teeth, and longer, more slender canine teeth (Newsome, Corbett, and Carpenter 1980).

There have been various ideas regarding the origin of the domestic dog. The current consensus is that it was derived from one of the small south Eurasian subspecies of *C. lupus* and subsequently spread throughout the world in association with people (Nowak 1979). Some authorities, including Clutton-Brock, Corbet, and Hills (1976), think that the direct ancestor is probably the Indian wolf, *C. lupus pallipes*, but Olsen and Olsen (1977) argued in favor of the Chinese wolf, *C. l. chanco*. The wolf and dog still hybridize readily, but only when brought together in captivity or under very unusual natural conditions. The oldest well-documented remains of *C. familiaris*, dating from about 11,000 and 12,000 years ago, were found in Idaho and Iraq, respectively. Beebe (1978) reported a specimen from the northern Yukon with a minimum age of 20,000 years, but Olsen (1985) considered that record doubtful.

At present, from the Balkans and North Africa to Japan and the East Indies dogs known as pariahs lead a semidomestic or even feral existence around villages (Bueler 1973; Fox 1978; Trumler *in* Grzimek 1990). They may take food from people, scavenge, or actively hunt deer and other an-

Dingo *(Canis familiaris)*, photo from New York Zoological Society.

imals. They are socially flexible, being found either alone or in packs, but there seems to be a dominance hierarchy in any given area. These dogs, generally primitive in physical appearance, are probably closely related to the earliest dogs as well as to the Australian dingo. In New Guinea and Timor also there are wild dog populations, which are related to the primitive pariah-dingo group (Troughton 1971). The oldest definitely known fossils of the dingo in Australia date from about 3,500 years ago. People arrived in Australia at least 30,000 years ago. The dingo evidently was brought in long afterward but before true domestication had been achieved, and it was able to establish wild populations (Macintosh 1975). It is possible that these dogs were spread by maritime peoples of south-central Asia rather than by migrating aboriginal peoples (Gollan 1984). There are many other populations of feral dogs, notably on islands and in Italy, but these animals are descendants of fully domesticated individuals (Lever 1985).

In a recent reassessment of the early history of dogs Corbett (1995) identified the entire complex of primitive, dingolike animals across southern Asia, the East Indies, New Guinea, and Australia as the single taxon *dingo.* This revisionist view, which was followed by Ginsberg and Macdonald (1990), holds that *dingo* evolved on mainland southern Asia in association with people and then about 3,500–4,000 years ago was spread by seafarers to other regions. Some of these animals may even have been brought across the Bering Strait to North America and persist today in the form of the Carolina hunting dog. Others are thought to have reached some of the Pacific islands, Madagascar, and southern Africa. It is difficult to determine how to approach these populations from the standpoint of research and conservation as they are not fully natural entities but may have been established components of their ecosystems for thousands of years. Outside of Australia little is known about *dingo;* it probably is declining in association with environmental disruption and is threatened by hybridization with true domestic dogs, but there apparently still are some pure, viable, and wild-living populations in Southeast Asia, especially Thailand.

The dingo population of New Guinea long has been recognized as a distinctive wild population and even has been given the specific name *Canis hallstromi.* It once was thought to be ancestral to the Australian dingo, but Flannery (1990*b*) noted that remains from New Guinea are not as old as those found in Australia and that introduction to the latter region thus probably occurred first. The New Guinea dingo sometimes is called the "singing dog" because of its unique vocalizations, including a form of howling marked by an extraordinary degree of frequency modulation and a number of signals, such as a high-pitched rapid trill, not reported for other canids. Social behavior seems to be limited, and females show a single annual estrus. The wild population is now restricted to a few isolated sites and is jeopardized by hybridization with true domestic dogs (Brisbin et al. 1994).

The dingo was once found throughout Australia, in forests, mountainous areas, and plains (Bueler 1973). Studies by Corbett and Newsome (1975) were carried out in an arid region of deserts and grasslands in central Australia, where the dingo depends in part on water holes made for cattle. Natal dens were found in caves, hollow logs, and modified rabbit warrens, usually within 2–3 km of water. Most activity in this region is nocturnal. In good seasons the diet consists mainly of small mammals, especially the introduced European rabbit *(Oryctolagus).* In times of drought the dingo takes kangaroos and cattle, mostly calves. Studies by Robertshaw and Harden (1985) in a forested area of northeastern New South Wales found the diet to consist mostly of larger native mammals, particularly wallabies. In an open area of Western Australia, Thomson (1992*b*) found the major prey to be large kangaroos and wallaroos *(Macropus rufus* and *Macropus robustus).*

Harden (1985) found daily movements to be about 10–20 km. Periods of movement usually were short and separated by even shorter periods of rest. On the average, individuals were active for 15.25 hours and at rest for 8.75 hours each day. Average home range for adults was 27 sq km. Thomson (1992*c*) reported that in Western Australia packs occupy permanent territories of 45–113 sq km. These areas have very little overlap and there is minimal interaction between neighboring packs, the animals evidently being kept apart by howling and scent marking. According to

Domestic dogs *(Canis familiaris)*, photo from J. Nowak and E. Nowak.

Corbett (1995), territory size averages 39 sq km in the tropical north and is 10–21 sq km in the forested east. Most individuals remain in their natal area, but some disperse, especially young males. The longest recorded movement for a tagged individual is 250 km over a 10-month period.

The dingo may be basically solitary in some areas (Corbett and Newsome 1975), but where it is not subject to human persecution there are discrete and stable packs of 3–12 individuals (Corbett 1995). Rank order is determined and maintained largely by aggressive interaction, especially among males. The dingo is not a particularly vocal canid but has a variety of sounds. Howls, which probably function to locate friends and repel strangers, are heard frequently in the single annual breeding season. Corbett (1988) reported that a captive group had separate male and female dominance hierarchies; all the females became pregnant, but the alpha female killed the newborn of the others. Wild females have a single annual estrus and bear their pups mostly during the winter, following a gestation period of 63 days. Litter size is usually 4–5 young but ranges from 1 to 10 (Bueler 1973; Corbett 1995; Thomson 1992a). Yearlings may assist an older pair to raise their pups. Independence is generally achieved by 3–4 months, but the young animals often then associate with a mature male (Corbett and Newsome 1975). Maximum known longevity is 14 years and 9 months (Grzimek 1975).

Although the dingo is said to be regularly captured and tamed by the natives and other people of Australia, Macintosh (1975) argued that it has never been successfully domesticated. He noted that it does have a close relationship with some groups of aborigines but evidently is not used in hunting nor intentionally fed and that its main function may be to sleep in a huddle with persons and thereby provide protection from the cold. Corbett (1995), however, cited several records of the dingo's being used for the pur-

suit of game and otherwise living in close association with people.

Agriculturalists in Australia generally have a low opinion of the dingo mainly because of its predation on sheep. Intensive persecution began in the nineteenth century and has continued to the present. During the 1960s about 30,000 dingos were killed for bounty annually in Queensland alone. Nearly 10,000 km of fencing has been constructed in eastern Australia in an effort to keep the wild dogs off the sheep ranges. Nonetheless, Macintosh (1975) suggested that depredations have been greatly exaggerated, and Corbett (1995) pointed out that the dingo has actually aided agriculture by destroying introduced rabbits. Examination of stomach contents in Western Australia indicated that domestic stock was not a significant part of the dingo's diet even though sheep and cattle were common in the study area (Whitehouse 1977). Human persecution has caused a decline in dingo populations, but another serious problem is hybridization with the domestic dog, a phenomenon that evidently is spreading with human settlement (Newsome and Corbett 1982, 1985). Dingo populations in the tropical north and in the northwest are still pure, but those of southeastern Australia now seem to have hybridized to some extent (Corbett 1995; Jones 1990; Thomson 1992a).

Aside from the dingo, *C. familiaris* long was one of the least-known canids with respect to its behavior and ecology under noncaptive conditions, but there has been increasing study in recent years. Beck (1973, 1975) estimated that up to half of the 80,000–100,000 dogs in Baltimore, Maryland, are free-ranging, at least at times, with an average density of about 230/sq km. They shelter in vacant buildings and garages and under parked cars and stairways. They are active mainly from 0500 to 0800 and from 1900 to 2200 hours in the summer, remaining out of sight during the midday heat. Their diet, consisting mostly of garbage but also including rats and ground-nesting birds, seems adequate to maintain weight and good health. Studies of unrestrained pets (Berman and Dunbar 1983; Rubin and Beck 1982) indicate that movement is concentrated in the early morning and late evening but usually covers a very small area, probably because of the lack of a need to secure food. The distance a dog tends to move away from its home is positively correlated with the amount of time that it is allowed freedom.

Fully feral dogs sometimes occur in the countryside. Scott and Causey (1973) found Alabama packs to use moist floodplains in warm weather and dry uplands in cool weather and to cover distances of 0.5–8.2 km per day. Nesbitt (1975) determined that the females on a wildlife refuge in Illinois did not dig a den but gave birth in heavy cover. Activity in that area occurred anytime but was mainly nocturnal and crepuscular. The dogs traveled single file along roads, trails, and crop rows; if frightened, they took cover among trees and bushes. They fed on crippled waterfowl and deer, road kills and other carrion, small animals, some vegetation, and garbage.

In a study of free-ranging urban dogs in Newark, New Jersey, Daniels (1983a, 1983b) calculated a population density of about 150/sq km. Home range varied from 0.2 to 11.1 ha. in summer and from 0.1 to 5.7 ha. in winter. The relatively few groups that formed rarely contained more than two animals and often lasted only a few minutes; aggression was unusual, and there was no evidence of territoriality. However, presence of an estrous female in an area led to a congregation of males, usually about five or six, and to an increase in aggression and formation of a dominance hierarchy. Familiarity was observed to be an important basis for all social activity and for acceptance of a male by a female. At several sites in Mexico and the southwestern United States, Daniels and Bekoff (1989) found feral dogs to be more social than those in urban and rural areas. It was suggested that the formation of permanent groups offers little advantage to dogs dependent on people but that pack living facilitates feeding and protection for feral animals.

Home range was found to be 1.74 ha. for temporarily unrestrained suburban pets in Berkeley, California (Berman and Dunbar 1983); about 2.6 ha. each for 2 full-time free-ranging individuals in Baltimore (Beck 1975); 61 ha. for a group of 3 unowned dogs that lived in an abandoned building and used the streets and parks of St. Louis (Fox, Beck, and Blackman 1975); 444–1,050 ha. each for three feral packs of 2–5 dogs (Scott and Causey 1973); and 28,500 ha. for a feral pack of 5–6 (Nesbitt 1975). The last group was observed to have a dominance hierarchy and to be led by a female. Pets tend to be solitary and form social groups only randomly, whereas the three unowned urban animals maintained a long-term relationship, the single female member initiating movements. In the Baltimore study, half of the animals seen were solitary, 26 percent were in pairs, and the rest were in groups of up to 17 members. Spring and autumn breeding peaks there were suggested by fluctuations in reports of unwanted dogs.

Female dogs enter estrus twice a year, usually in late winter or early spring and in the autumn. Heat lasts about 12 days. At such times males tend to leave their owners' homes, mark territories, and fight rivals. The gestation period averages 63 days, litter size is usually 3–10 young, and nursing lasts about 6 weeks. Males often remain with the females and young. Sexual maturity comes after 10–24 months, and old age generally after 12 years, but a few individuals live for 20 years (Asdell 1964; Bueler 1973; Grzimek 1975).

There are an estimated 50 million owned dogs in the United States, and many more lack owners. Although these animals have abundant uses and values, they may cause problems for some people. In Baltimore, for example, dogs have been implicated in the spread of several diseases (in addition to rabies), they may benefit rats by overturning garbage cans, and there are about 7,000 reported attacks on people each year (Beck 1973, 1975). There are 1–3 million reported attacks annually for the whole United States (Rovner 1992). In 1986, 13 people, most of them children, were killed in the United States; in 7 of these cases the breed known as the pit bull was responsible (*Washington Post*, 1 September 1987, B-1, B-5). In 1990, 18 people were killed, 8 by pit bulls (ibid., 22 May 1991, A-29). With respect to rabies, dogs are traditionally viewed as the primary threat to people, but in recent decades other animals have been of far greater significance, at least in the United States and Europe (see accounts of *Lasionycteris*, *Eptesicus*, *Vulpes vulpes*, *Procyon*, and *Felis catus*). *C. familiaris* also often is considered to be a serious predator of livestock and game animals, especially deer; however, field studies have indicated that feral dogs do not significantly affect deer populations and may even have a sanitary function in eliminating carrion and crippled animals (Gipson and Sealander 1977; Nesbitt 1975; Scott and Causey 1973).

CARNIVORA; CANIDAE; Genus **CHRYSOCYON**
Hamilton-Smith, 1839

Maned Wolf

The single species, *C. brachyurus*, originally occurred in open country of central and eastern Brazil, eastern Bolivia, Paraguay, Argentina, and Uruguay (Cabrera 1957; Langguth 1975a; Mones and Olazarri 1990; Roig 1991). There may be disjunct populations on the llanos of Colombia (Dietz 1985) and in southeastern Peru (Ginsberg and Macdonald 1990).

Head and body length is 950–1,320 mm, tail length is 280–490 mm, shoulder height is 740–900 mm, and weight is 20–26 kg. There is a general impression of a red fox *(Vulpes vulpes)* on stilts (Clutton-Brock, Corbet, and Hills 1976). The general coloration is yellow red. The hair along the nape of the neck and middle of the back is especially long and may be dark in color. The muzzle and lower parts of the legs are also dark, almost black. The throat and tail tuft may be white. The coat is fairly long, somewhat softer than that of *Canis*, and has an erectile mane on the back of the neck and top of the shoulders. The ears are large and erect, the skull is elongate, and the pupils of the eyes are round.

The maned wolf inhabits grasslands, savannahs, and swampy areas (Langguth 1975a). The natal nest is located in thick, secluded vegetation (Bueler 1973). Activity is mainly nocturnal and crepuscular (Dietz 1984). It has been suggested that the remarkably long legs are an adaptation for fast running or for movement through swamps, but the actual function is probably to allow seeing above tall grass. *Chrysocyon* is not an especially swift canid, does not pursue prey for long distances, and generally stalks and pounces like a fox (Kleiman 1972). Its omnivorous diet includes rodents, other small mammals, birds, reptiles, insects, fruit, and other vegetable matter.

Chrysocyon is monogamous, with mated pairs sharing defended territories averaging 27 sq km. However, most activity is solitary; the male and female associate closely only during the breeding season (Dietz 1984). Captives some-

Maned wolf *(Chrysocyon brachyurus)*, photo by Bernhard Grzimek.

times can be kept together without apparent strife, though there is usually an initial period of fighting and then establishment of a dominance hierarchy (Brady and Ditton 1979). The father is known to regurgitate food for the young in captivity (Rasmussen and Tilson 1984) and may also have a significant parental role in the wild (Dietz 1984). The three main vocalizations are: a deep-throated single bark, heard mainly after dusk; a high-pitched whine; and a growl during agonistic behavior (Kleiman 1972).

Births in captivity have occurred in July and August in South America and in January and February in the Northern Hemisphere. Females are monestrous, heat lasts about 5 days, the gestation period is 62–66 days, and litter size is two to four young. The young weigh about 350 grams at birth, open their eyes after 8 or 9 days, begin to take some regurgitated food at 4 weeks, and are weaned by 15 weeks (Brady and Ditton 1979; Da Silveira 1968; Faust and Scherpner 1967). One captive lived 15 years and 8 months (Marvin L. Jones, Zoological Society of San Diego, pers. comm., 1995).

The maned wolf is not extensively hunted for its fur, but it is sometimes persecuted because of an unjustified belief that it kills domestic livestock. It often is killed because of alleged depredations on chickens. Resident populations disappeared from most of Argentina and from Uruguay in the nineteenth century, though one individual recently was taken in the latter country. It now is threatened in other regions by the annual burning of its grassland habitat, hunting, live capture, and disease (Ginsberg and Macdonald 1990; Roig 1991; Thornback and Jenkins 1982). It also, however, recently extended its range into a deforested zone of central Brazil (Dietz 1985). It is classified as near threatened by the IUCN and as endangered by the USDI and is on appendix 2 of the CITES.

CARNIVORA; CANIDAE; **Genus OTOCYON**
Müller, 1836

Bat-eared Fox

The single species, *O. megalotis,* is found from Ethiopia and southern Sudan to Tanzania and from southern Angola and Zimbabwe to South Africa (Coetzee *in* Meester and Setzer 1977). According to Ansell (1978), it does not occur in Zambia. *Otocyon* sometimes has been placed in a separate subfamily, the Otocyoninae, on the basis of its unusual dentition. Clutton-Brock, Corbet, and Hills (1976), however, considered *Otocyon* to be simply an aberrant fox with systematic affinities to *Urocyon* (which they included in *Vulpes*) and some behavioral similarities to *Nyctereutes*.

Head and body length is 460–660 mm, tail length is 230–340 mm, shoulder height is 300–400 mm, and weight is 3.0–5.3 kg. The upper parts are generally yellow brown with gray agouti guard hairs. The throat, underparts, and insides of the ears are pale. The outsides of the ears, mask, lower legs, feet, and tail tip are black. In addition to coloration, distinguishing characters include the relatively short legs and the enormous ears (114–35 mm long).

Otocyon has more teeth than any other placental mammal that has a heterodont condition (the teeth being differentiated into several kinds). Whereas in all other canids there are no more than two upper and three lower molars, *Otocyon* has at least three upper and four lower molars. This condition is sometimes held to be primitive, but it more likely represents the results of a mutation that caused the appearance of the extra molar teeth in what had been a fox population with normal canid dentition (Clutton-Brock, Corbet, and Hills 1976).

Bat-eared fox *(Otocyon megalotis)*, photo by Bernhard Grzimek.

The bat-eared fox is found in arid grasslands, savannahs, and brush country. It seems to prefer places with much bare ground or where the grass has been kept short by burning or grazing. When the grass again grows high, *Otocyon* may depart and wander about in search of a new place of residence. A capable digger, it either excavates its own den or enlarges the burrow of another animal. A family may have more than one den in its home range, each with multiple entrances and chambers and several meters of tunnels. In the Serengeti 85 percent of activity occurs at night, but in South Africa *Otocyon* is mainly diurnal in winter and nocturnal in summer. One female was observed to forage over about 12 km per night. The diet consists predominantly of insects, most notably termites, and also includes other arthropods, small rodents, the eggs and young of ground-nesting birds, and vegetable matter (Kingdon 1977; Lamprecht 1979; Nel 1978; Smithers 1971). In one part of the central Karoo about half of the food was plant material, especially wild fruits (Kuntzsch and Nel 1992).

In the Masai Mara Reserve of Kenya, Malcolm (1986) found an overall population density of 0.8–0.9/sq km. Group home ranges there averaged 3.53 sq km and overlapped; in one case there was no area of exclusive use. Groups consisted of 2–5 members and generally had amicable relations with other groups. In South Africa, Nel (1978) also found home ranges to overlap extensively and observed no territorial defense or marking. Up to 15 individuals, representing four groups, were seen foraging within less than 0.5 sq km. In the Serengeti, however, Lamprecht (1979) found resident families to occupy largely exclusive home ranges of 0.25–1.5 sq km and to mark them with urine. Groups usually consisted of a mated adult pair accompanied by their young of the year for a lengthy period. A few observations suggested that there may sometimes be 2 adult females with a male. Strangers of the same sex generally were hostile. Contact between members of a group was maintained by soft whistles. Nel and Bester

(1983) reported that most communication involves visual signaling, mainly with the large ears and tail.

In both the Serengeti and Botswana, births occur mainly from September to November (Lamprecht 1979; Smithers 1971), but pups have been recorded in Uganda in March (Kingdon 1977), and reproduction may be year-round in some parts of East Africa (Malcolm 1986). Gestation is usually reported as 60–70 days, but Rosenberg (1971) calculated the period at 75 days for a birth in captivity. Litters contain 2–6 young. According to Skinner and Smithers (1990), the average litter size is 5, there sometimes are two litters per year, newborn weight is about 100–140 grams, and weaning occurs at about 10 weeks. The young then begin to forage with the parents; regurgitation is evidently rare. The young are probably full grown by 5 or 6 months and separate from the parents prior to the breeding season. According to Jones (1982), a captive lived 13 years and 9 months.

Otocyon has declined in settled parts of South Africa (Coetzee *in* Meester and Setzer 1977). Nonetheless, it is apparently extending its range eastward into Mozambique and into previously unoccupied parts of Zimbabwe and Botswana (Pienaar 1970). Rabies has recently been identified as a major problem to the genus in East Africa and was the main known cause of mortality in Serengeti National Park in the late 1980s (Maas 1994).

CARNIVORA; CANIDAE; **Genus CUON**
Hodgson, 1838

Dhole

The single species, *C. alpinus,* is found from southern Siberia and eastern Kazakhstan to India and the Malay Peninsula and on the islands of Sumatra and Java but not

Dhole *(Cuon alpinus)*, photo from New York Zoological Society.

Sri Lanka. In the Pleistocene, *Cuon* did occur in Sri Lanka and probably Borneo and also had a vast range extending from western Europe to Mexico (Corbet and Hill 1992; Kurten 1968; Kurten and Anderson 1980). Recently, Serez and Eroglu (1994) reported the presence of a living population of *C. alpinus* in northeastern Turkey, far from the remainder of the known modern range of the species. Except as otherwise noted, the information for this account was taken from the review papers by Cohen (1977, 1978) and Davidar (1975).

Head and body length is 880–1,130 mm, tail length is 400–500 mm, and shoulder height is 420–550 mm. Males weigh 15–21 kg and females, 10–17 kg. The coloration is variable, but generally the upper parts are rusty red, the underparts are pale, and the tail is tipped with black. In the northern parts of the range the winter pelage is long, soft, dense, and bright red; the summer coat is shorter, coarser, sparser, and less vivid in color. *Cuon* resembles *Canis* externally, but the skull has a relatively shorter and broader rostrum. Females have 12–16 mammae.

The dhole occupies many types of habitat but avoids deserts. In the Soviet Union it occurs mainly in alpine areas, and in India it is found in dense forest and thick scrub jungle. The preferred habitat in Thailand is dense montane forest at elevations of up to 3,000 meters (Lekagul and McNeely 1977). *Cuon* may excavate its own den, enlarge a burrow made by another animal, or use a rocky crevice. M. W. Fox (1984) found one earth den with six entrances leading to a labyrinth of at least 30 meters of interconnected tunnels and four large chambers; many generations of dholes probably had developed this complex. *Cuon* may be active at any time but mainly in early morning and early evening. Cohen et al. (1978) reported a major peak of activity at 0700–0800 hours and a lesser peak at 1700–1800 hours.

The dhole hunts in packs and is primarily a predator of mammals larger than itself. Prey is tracked by scent and then pursued, sometimes for a considerable distance. When the objective is overtaken, it is surrounded and attacked from different sides. Prey animals include deer, wild pigs, mountain sheep, gaur, and antelopes. The chital *(Axis axis)* is probably the major prey in India, though Cohen et al. (1978) found remains of this deer to occur less frequently than those of *Lepus* in the droppings of *Cuon*. The diet also includes rodents, insects, and carrion. Reports of predation on tigers, leopards, and bears generally are not well documented, but those carnivores are sometimes driven from their kills by packs of dholes. There are numerous records of leopards being treed.

In a study in southern India, Johnsingh (1982) found population densities of 0.35–0.90/sq km. A pack in this area contained an average of 8.3 adults and used a home range of 40 sq km. Other work suggests that there are usually 5–12 dholes in a pack, but up to 40 have been reported. This discrepancy was perhaps explained by M. W. Fox (1984), who pointed out that the larger groups are actually clans comprising several related packs. In parts of India the clans keep together during that part of the year when only juvenile and adult chital are available as prey but divide into smaller hunting packs when the chital have fawns. A pack apparently consists of a mated pair and their offspring. Although the social structure has not been closely studied, there seems to be a leader, a dominance hierarchy, and submissive behavior by lower-ranking animals. Intragroup fighting is rarely observed. More than one female sometimes den and rear litters together. Vocalizations include nearly all of those made by the domestic dog except loud and repeated barking; the most distinctive sound is a peculiar whistle that probably serves to keep the pack together during pursuit of prey.

In India mating occurs from September to November, and births from November to March. At the Arignar Anna Zoo in India captives first showed signs of sexual maturity at 11 months of age, females had estrus periods of 14–39 days, and normal gestation lasted 60–67 days (Paulraj et al. 1992). In the Moscow Zoo mating occurred in February and gestation lasted 60–62 days (Sosnovskii 1967). Litters usually contain four to six young, but up to nine embryos have been recorded. The young weigh 200–350 grams at birth and are weaned after about 2 months (Hayssen, Van Tienhoven, and Van Tienhoven 1993). In the wild both mother and young are provided with regurgitated food by other pack members. The pups leave the den at 70–80 days and participate in kills at 7 months (Johnsingh 1982). Longevity in the Moscow Zoo is 15–16 years.

Although the dhole only rarely takes domestic livestock and was only once reported to attack a person, it has been intensively poisoned and hunted throughout its range. This situation seems based mainly on dislike by hunters, who see *Cuon* as a competitor for game and who are repulsed by its method of predation. Some of the village people of India actually welcome the dhole, following it in order to expropriate its kills. Because of direct persecution, elimination of natural prey, and destruction of its forest habitat, the dhole has declined seriously in range and numbers. Recent surveys indicate that remnant populations in southern Asia are small, widely scattered, and difficult to locate even in protected parks (Stewart 1993, 1994). *Cuon* has nearly disappeared in Siberia and Kazakhstan, possibly because it is inadvertently killed in poisoning campaigns against *Canis lupus* (Ginsberg and Macdonald 1990). It is classified as vulnerable by the IUCN, which estimates total numbers at fewer than 10,000, and as endangered by the USDI and Russia and is on appendix 2 of the CITES.

CARNIVORA; CANIDAE; Genus LYCAON
Brookes, 1827

African Hunting Dog

The single species, *L. pictus,* originally occurred in most of Africa south of the Sahara Desert as well as in suitable parts of the Sahara in Egypt (Coetzee *in* Meester and Setzer 1977; Kingdon 1977).

Head and body length is 760–1,120 mm, tail length is 300–410 mm, shoulder height is 610–780 mm, and weight is 17–36 kg (Kingdon 1977). Males and females are about the same size (Frame et al. 1979). There is great variation in pelage, the mottled black, yellow, and white occurring in almost every conceivable arrangement and proportion. In most individuals, however, the head is dark and the tail has a white tip or brush. The fur is short and scant, sometimes so sparse that the blackish skin is plainly visible. The ears are long, rounded, and covered with short hairs. The legs are long and slender, and there are only four toes on each foot. The jaws are broad and powerful. *Lycaon* has a strong, musky odor. Females have 12–14 mammae (Van Lawick and Van Lawick–Goodall 1971).

Lycaon inhabits grassland, savannah, and open woodland (Coetzee *in* Meester and Setzer 1977). Its den, usually an abandoned aardvark hole, is occupied only to bear young (Kingdon 1977). The pack does not wander very far from the den when pups are present (Frame et al. 1979). Otherwise movements are generally correlated with hunting success: if prey is scarce, the entire home range may be traversed in two or three days. Hunts take place in the morning and early evening. Schaller (1972) recorded peaks of activity from 0700 to 0800 and from 1800 to 1900 hours. Prey is apparently located by sight, approached silently, and then pursued at speeds of up to 66 km/hr for 10–60 minutes (Kingdon 1977). Van Lawick and Van Lawick–Goodall (1971) observed a pack to maintain a speed of about 50 km/hr for 5.6 km. Their investigation indicated this distance to be about the maximum that *Lycaon* would usually follow before giving up. They observed 91 chases, 39 of which were successful. In all but 1 of the latter cases the quarry was killed within five minutes of being caught. In his study, Schaller (1972) observed 70 percent of chases to be successful. Groups of *Lycaon* generally cooperate in hunting large mammals, but individuals sometimes pursue hares, rodents, or other small animals. The main prey seems to vary by area, being bushduiker and reedbuck in the Kafue Valley of Zambia, impala in Kruger National Park of South Africa, and Thomson's gazelle and wildebeest in the Serengeti of Tanzania (Kingdon 1977). Certain packs in the Serengeti, however, specialize in the capture of zebra (Malcolm and Van Lawick 1975). Some food may be cached in holes (Malcolm 1980*b*).

Reported population densities vary from 2 to 35 per 1,000 sq km (Fuller, Kat, et al. 1992). From 1970 to 1977 density in the Serengeti declined from 1 adult *Lycaon* per 35 sq km to 1/200 sq km. Pack home range in this area is generally 1,500–2,000 sq km (Frame et al. 1979). A pack in South Africa reportedly used a home range of about 3,900 sq km (Van Lawick and Van Lawick–Goodall 1971). Range contracts to only about 50–200 sq km when there are small pups at a den (Fuller, Kat, et al. 1992); at such time one Serengeti pack used an area of 160 sq km for 2.5 months (Schaller 1972). The home range of a pack overlaps by about 10–50 percent with those of several neighboring packs (Frame and Frame 1976). Territoriality does not seem to be well developed, and hundreds of individuals may once have gathered temporarily in response to migrations of the formerly vast herds of springbok in southern Africa (Kingdon 1977).

Studies on the Serengeti Plains of Tanzania have revealed that *Lycaon* has an intricate and unusual social structure (Frame and Frame 1976; Frame et al. 1979; Malcolm 1980*a*; Malcolm and Marten 1982; Schaller 1972; Van Lawick and Van Lawick–Goodall 1971). Groups were found to contain averages of 9.8 (1–26) individuals, 4.1 (0–10) adult males, and 2.1 (0–7) adult females. This sexual proportion is unlike the usual condition in social mammals. Some packs have as many as 8 adult males with only a single adult female. Moreover, in a reversal of the usual mammalian process, females emigrate from their natal group far more than do males. Commonly, several sibling females 18–24 months old leave their pack and join another that

African hunting dog *(Lycaon pictus)*, photo by Bernhard Grzimek.

lacks sexually mature females. Following the transfer, one of the females achieves dominance, whereupon her sisters may depart. Whereas no female seems to stay in its natal pack past the age of 2.5 years, about half of the young males do remain; the other males emigrate, usually in sibling groups. The typical pack thus consists of several related males, often representing more than one generation, and 1 or more females that are genetically related to each other but not to the males. Some male lineages within a pack are known to have lasted at least 10 years.

There are separate dominance hierarchies for each sex. Normally, only the highest-ranking male and female breed, and they inhibit reproduction by subordinates. There is intense rivalry among the females for the breeding position. If a subordinate female does bear pups, the dominant one may steal them. Females sometime fight savagely, and the loser may leave the group and perish. Aside from this aspect of the social life of *Lycaon*, packs are remarkably amicable, with little overt strife. Food is shared, even by individuals that do not participate in the kill. An animal with a broken leg was allowed to feed, when it hobbled up after the others, throughout the time required for its leg to mend. Pups old enough to take solid food are given first priority at kills, eating even before the dominant pair. Subordinate animals, especially males, help feed and protect the pups. There are several vocalizations, the most striking of which is a series of wailing hoots that probably serves to keep the pack together during pursuit of prey.

Studies at Masai Mara Reserve in Kenya and Kruger National Park in South Africa suggest some differences in behavior (Fuller, Kat, et al. 1992). Both males and females regularly disperse from their natal packs at 1–2 years, usually in groups composed only of one sex. They move up to 250 km, eventually joining with groups of the opposite sex that have dispersed from elsewhere to form a new breeding pack. Sometimes they join an established pack. Occasionally a female does not leave her natal pack and may

even replace her mother. It is not unusual for two females of a pack to mate and successfully rear young.

Births may occur at any time of year but peak from March to June, during the second half of the rainy season. The interval between births is normally 11–14 months but may be as short as 6 months if all the young perish. The gestation period is 79–80 days. There is some confusion about litter size. Although up to 23 young have been seen at den sites, and although single females have given birth to as many as 21 pups, litter sizes at various captive facilities have averaged only 4–8. It is likely that the very large litters observed in the wild represent the synchronous offspring of two or even three females in the pack. The newborn weigh about 300 grams and open their eyes after 13 days. At about 3 weeks the young emerge from the den and begin to take some solid food. Weaning is normally completed by 11 weeks. All adult pack members regurgitate food to the young. Once, when a mother died, the males of the pack were able successfully to raise her 5-week-old pups. When the pack is hunting, 1 or 2 adults remain at the den to guard the pups. After 3 months the young begin to follow the pack, and at 9–11 months they can kill easy prey, but they are not proficient until about 12–14 months. Social restrictions blur the actual time of sexual maturity. Five males were observed to first mate at 21, 33, 36, 36, and 60 months. The youngest female to give birth was 22 months old at the time. Maximum observed longevity in the wild is 11 years (Frame et al. 1979; Fuller, Kat, et al. 1992; Kingdon 1977; Malcolm 1980a; Schaller 1972; Van Heerden and Kuhn 1985; Van Lawick and Van Lawick–Goodall 1971). A captive lived for about 17 years (Marvin L. Jones, Zoological Society of San Diego, pers. comm., 1995).

Lycaon has been reported only rarely to attack people but is intensively hunted and poisoned because of a largely undeserved reputation as an indiscriminate killer of livestock and valued game animals. Remnant and fragmented populations also are jeopardized by habitat loss, diseases spread by domestic animals, loss of genetic viability, and possibly even the stress caused by well-meaning research workers. The genus has been almost entirely extirpated in the Saharan region, only small and probably inviable populations survive in West and Central Africa, and it has been wiped out in South Africa except in the vicinity of Kruger National Park. It is thought to have vanished entirely even in Zaire and Congo. There still may be viable populations present from southern Ethiopia, through Kenya, Tanzania, Zambia, Zimbabwe, and Botswana, to Namibia. Even in those countries, however, there has been a great decline in distribution. The largest remaining population, with perhaps more than 1,000 animals, is located in the Selous Game Reserve and adjacent lands of southern Tanzania. Numbers in Zimbabwe have been estimated at 300–600 individuals, and persecution continues there except in national parks. Numbers throughout all of Africa were thought to be fewer than 7,000 in 1980. Estimates issued in the early 1990s varied from only 2,000 to about 5,000 (Buk 1994; Burrows, Hofer, and East 1994; Childes 1988; Fanshawe, Frame, and Ginsberg 1991; Fuller, Kat, et al. 1992; Ginsberg 1993; Ginsberg and Macdonald 1990; Kingdon 1977; Lensing and Joubert 1977; Malcolm 1980a; Skinner, Fairall, and Bothma 1977). There are another 300 in captivity (Brewer *in* Bowdoin et al. 1994).

Lycaon now is classified as endangered by the IUCN and USDI. With the establishment in 1992 of the Licaone Fund by concerned Italian biologists, as well as other international efforts, its plight is finally receiving substantive attention. Considering its immense former distribution and its scientific, cultural, and behavioral interest, the prospective disappearance of this genus from the wild at a time of supposed increasing emphasis on conservation values must rank as one of the great wildlife tragedies of the late twentieth century.

CARNIVORA; Family URSIDAE

Bears

This family of three Recent genera and eight species occurred historically almost throughout Eurasia and North America, in the Atlas Mountains of North Africa, and in the Andes of South America. E. R. Hall (1981) recognized three living subfamilies: Tremarctinae, with the genus *Tremarctos*; Ursinae, with *Ursus*; and Ailuropodinae, with *Ailuropoda*. O'Brien (1993) and Wayne et al. (1989), using molecular and karyological data, supported the same arrangement. Wozencraft (*in* Wilson and Reeder 1993) included *Tremarctos* in the Ursinae and placed *Ailuropoda*, together with *Ailurus*, in another ursid subfamily, the Ailurinae. The sequence of genera presented here basically follows that of Simpson (1945), though he, like many other authorities, recognized additional genera.

Head and body length is 1,000 to approximately 2,800 mm, tail length is 65–210 mm, and weight is 27–780 kg. Males average about 20 percent larger than females. The coat is long and shaggy, and the fur is generally unicolored, usually brown, black, or white. Some species have white or buffy crescents or semicircles on the chest. *Tremarctos*, the spectacled bear of South America, typically has a patch of white hairs encircling each eye. *Ailuropoda*, the giant panda, has a striking black and white color pattern.

Bears have a big head; a large, heavily built body; short, powerful limbs; a short tail; and small eyes. The ears are small, rounded, and erect. The soles are hairy in species that are mainly terrestrial but naked in species that climb con-

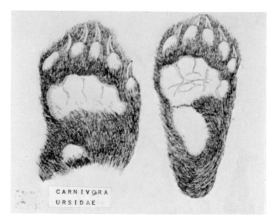

American black bear *(Ursus americanus)*, right forepaw and right hind foot, photo from *Proc. Zool. Soc. London.*

siderably, such as *Ursus malayanus*. All limbs have five digits. The strong, recurved claws are used for tearing and digging. The lips are free from the gums.

The skull is massive and the tympanic bullae are not inflated. In most genera the dental formula is: (i 3/3, c 1/1, pm 4/4, m 2/3) $\times$ 2 = 42. The species *Ursus ursinus*, however, has only 2 upper incisors and a total of 40 teeth. Ursid incisors are not specialized, the canines are elongate, the first 3 premolars are reduced or lost, and the molars have broad, flat, and tubercular crowns. The carnassials are not developed as such.

Habitats range from arctic ice floes to tropical forests. Those populations that occur in open areas often dig dens in hillsides. Others shelter in caves, hollow logs, or dense vegetation. Bears have a characteristic shuffling gait. They walk plantigrade, with the heel of the foot touching the ground. They are capable of walking on their hind legs for short distances. When need be, they are surprisingly agile and careful in their movements. Their eyesight and hearing are not particularly good, but their sense of smell is excellent. Bears are omnivorous, except that the polar bear *(Ursus maritimus)* feeds mainly on fish and seals.

During the autumn bears in most parts of the range of the family become fat. With the approach of cold weather they cease eating and go into a den that they have prepared in a protected location. Here they sleep through the winter, living mainly off stored fat reserves. With certain exceptions, especially pregnant females, the polar bear does not undergo winter sleep. Some authorities prefer not to call this process hibernation since body temperature is not substantially reduced, body functions continue, and the animals can usually be easily aroused. Sometimes they awaken on their own during periods of mild weather. Folk, Larson, and Folk (1976) found, however, that the heart rate of a hibernating bear drops to less than half of normal and that other physiological changes occur. They concluded that bears do experience true mammalian hibernation.

Except for courting pairs and females with young, bears live alone. Litters are produced at intervals of 1–4 years. In most regions births occur from November to February, while the mother is hibernating. The period of pregnancy is commonly extended 6–9 months by delayed implantation of the fertilized egg. Litter size is one to four young. The young are relatively tiny at birth, ranging from 225–680 grams each. They remain with the mother at least through their first autumn. They become sexually mature at 2.5–6 years and normally live 15–30 years in the wild.

Bears are usually peaceful animals that try to avoid con-

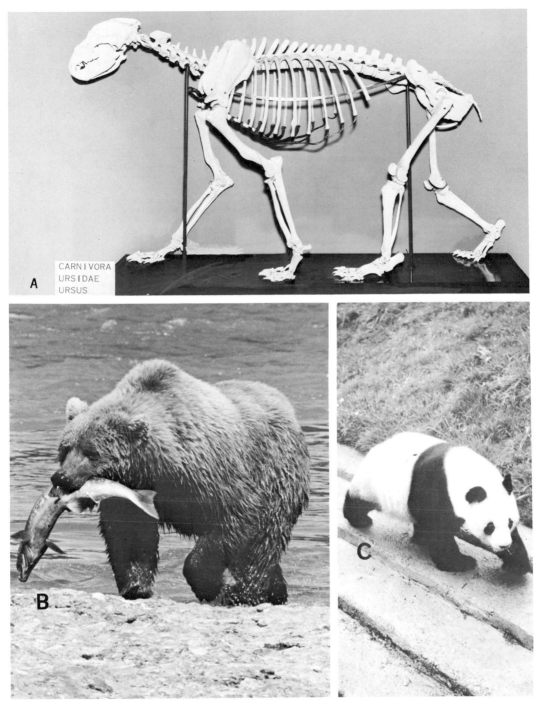

A. Brown bear *(Ursus arctos)*, skeleton from U.S. National Museum of Natural History. B. Photo of *U. arctos* by Leonard Lee Rue III. C. Giant panda *(Ailuropoda melanoleuca)*, photo by Elaine Anderson.

flict. However, if they consider themselves, their young, or their food supply threatened they can become formidable adversaries. Only a small proportion of the stories of unprovoked attacks by bears on people are true. When such cases are carefully investigated, it is usually found that there was provocation. Nonetheless, bears have been persecuted almost throughout their range because of alleged

danger to humans and because they are sometimes considered to be serious predators of domestic livestock. These problems tend to increase dramatically when the natural habitat of bears is invaded and fragmented by people for purposes of agriculture, recreation, settlement, and logging, which inevitably leads to conflicts and also facilitates access by human hunters. Another longstanding problem that only recently has achieved widespread recognition is the killing of bears for their body parts, especially the gallbladder, for use as medicines and food. Most species of bears in Asia are now directly threatened by such exploitation, and bear populations around the world have been increasingly subject to consequent illegal hunting and commercial trade (Rose and Gaski 1995).

The geological range of the Ursidae is late Miocene to Recent in North America, late Pliocene to Recent in South America, late Eocene to Recent in Europe, early Miocene to Recent in Asia, Pleistocene to Recent in North Africa, and Pliocene to Recent in South Africa (Hendey 1977; Stains 1984). The cave bear *(Ursus spelaeus)*, which was common in the late Pleistocene of Europe, was equal in size to the modern brown bear of Alaska. Although generally thought to have vanished about 10,000 years ago, certain unfossilized remains suggest that the species persisted until more recent times (Geist 1993).

CARNIVORA; URSIDAE; Genus TREMARCTOS
Gervais, 1855

Spectacled Bear

The single species, *T. ornatus,* is known to inhabit the mountainous regions of western Venezuela, Colombia, Ecuador, Peru, and western Bolivia (Cabrera 1957). It also has been reported from eastern Panama and northern Argentina (Peyton 1986).

Head and body length is usually 1,200–1,800 mm, tail length is about 70 mm, and shoulder height is 700–800 mm. One male 1,740 mm in length weighed 140 kg. Peyton (1980) reported that a male 2,060 mm in total length weighed 175 kg. Grzimek (1975) gave the weight of females as 60–62 kg. The entire body is uniformly black or dark brown except for large circles or semicircles of white around the eyes and a white semicircle on the lower side of the neck, from which lines of white extend onto the chest. The common name is derived from the white around the eyes. The head and chest markings are variable, however, and may be completely lacking in some individuals.

In the Andes of Peru, Peyton (1980) found *Tremarctos* to occupy a wide variety of habitats from 457 to 3,658 meters in elevation. The preferred habitats are humid forests between 1,900 and 2,350 meters and coastal thorn forests when water is available. High-altitude grasslands are also utilized. *Tremarctos* apparently is mainly nocturnal and crepuscular. During the day it beds down between or under large tree roots, on a tree trunk, or in a cave. It frequently climbs large trees to obtain fruit. In a tree it may assemble a large platform of broken branches, on which it positions itself to eat and to reach additional fruit. Peyton found one such platform at a height of 15 meters. Goldstein (1991) observed similar platforms apparently used as resting places.

Tremarctos feeds extensively on fruit, moving about in response to seasonal ripening. It also depends on plants of the family Bromeliaceae, especially when ripe fruit is not available. It tears off the leaves of large bromeliads to feed on the white bases and obtains the edible hearts of small bromeliads by ripping the entire plant off the substrate. In addition, this bear climbs large cacti to get the fruits at the top, tears into the green stalks of young palms to eat the unopened inner leaves, and strips bark off trees to feed on the cortex. The diet also includes bamboo hearts, corn, rodents, and insects. Only about 4 percent of the food was found to be animal matter (Peyton 1980).

In Peru, Peyton (1980) received reports that a male of-

Spectacled bear *(Tremarctos ornatus)*, photo from San Diego Zoological Garden.

ten enters a cornfield with one or more females, and sometimes with yearling animals, during the months of March–July. According to Grzimek (1975), *Tremarctos* has a "striking shrill voice." In the Buenos Aires Zoo young were produced in July, while in European zoos births have occurred from late December to March. Pregnancy lasts 6.5–8.5 months and apparently involves delayed implantation. Litters contain one to three young weighing about 320 grams each (Bloxam 1977; Gensch 1965; Grzimek 1975). One captive lived 38 years and 8 months (Marvin L. Jones, Zoological Society of San Diego, pers. comm., 1995).

Mittermeier et al. (1977) reported that the meat of *Tremarctos* is highly esteemed in northern Peru and that this bear also is killed by people for its skin and fat. Servheen (1989) wrote that the body parts of bears are commonly used for alleged medicinal purposes in South America and that legal protection is ineffective. Grimwood (1969) warned that *Tremarctos* had become rare and endangered in Peru through intensive hunting by sportsmen and landowners, who consider it to be a predator of domestic livestock. Mondolfi (1989) regarded it as endangered and still declining in Venezuela, with remnant populations subject to hunting and habitat disruption even in national parks. Peyton (1980) did not believe *Tremarctos* to be in immediate danger of extinction because it is adapted to a diversity of habitats, some of which are largely inaccessible to people. He did note, however, that some bears become habituated to raiding cornfields and that these animals are frequently shot. Later, Peyton (1986) noted that vast parts of the original range have been replaced by agriculture and that surviving bear populations are fragmentary. Thornback and Jenkins (1982) stated that the spectacled bear is declining through much of its range because of habitat loss due to settlement, to human persecution that subsequently results from raids on crops and livestock, and to hunting for its meat and skin. *Tremarctos* is classified as vulnerable by the IUCN and is on appendix 1 of the CITES.

CARNIVORA; URSIDAE; Genus URSUS
Linnaeus, 1758

Black, Brown, Polar, Sun, and Sloth Bears

There are six species (Corbet 1978; Corbet and Hill 1992; Ellerman and Morrison-Scott 1966; E. R. Hall 1981; Kurten 1973; Laurie and Seidensticker 1977; Lay 1967; Lekagul and McNeely 1977; Ma 1983a; Simpson 1945):

U. thibetanus (Asiatic black bear), Afghanistan, southeastern Iran, Pakistan, Himalayan region, Burma, Thailand, Indochina, China, Manchuria, Korea, extreme southeastern Siberia, Japan (except Hokkaido), Taiwan, Hainan;

U. americanus (American black bear), Alaska, Canada, conterminous United States, northern Mexico;

U. arctos (brown or grizzly bear), western Europe and Palestine to eastern Siberia and Himalayan region, Atlas Mountains of northwestern Africa, Hokkaido, Alaska to Hudson Bay and northern Mexico;

U. maritimus (polar bear), primarily on arctic coasts, islands, and adjacent sea ice of Eurasia and North America;

U. malayanus (Malayan sun bear), Assam southeast of Brahmaputra River, Sichuan and Yunnan in south-central China, Burma, Thailand, Indochina, Malay Peninsula, Sumatra, Borneo;

U. ursinus (sloth bear), India, Nepal, Bangladesh, Sri Lanka.

Each of these species has often been placed in its own genus or subgenus: *Selenarctos* Heude, 1901, for *U. thibetanus*; *Euarctos* Gray, 1864, for *U. americanus*; *Ursus* Linnaeus, 1758, for *U. arctos*; *Thalarctos* Gray, 1825, for *U. maritimus*; *Helarctos* Horsfield, 1825, for *U. malayanus*; and *Melursus* Meyer, 1793, for *U. ursinus*. E. R. Hall (1981), however, placed all of these names in the synonymy of *Ursus*. The latter arrangement is supported by molecular and karyological data (O'Brien 1993; Wayne et al. 1989) and in part by the captive production of viable offspring through hybridization between several of the above species (Van Gelder 1977b). Wozencraft (in Wilson and Reeder 1993) treated *Helarctos* and *Melursus* as valid genera.

The systematics of the brown or grizzly bear have caused considerable confusion. Old World populations long have been recognized as composing a single species, with the scientific name *U. arctos* and the general common name "brown bear." In North America the name "grizzly" is applied over most of the range, while the term "big brown bear" is often used on the coast of southern Alaska and nearby islands, where the animals average much larger than those inland. E. R. Hall (1981) listed 77 Latin names that have been used in the specific sense for different popu-

lations of the brown or grizzly bear in North America. No one now thinks that there actually are so many species, but some authorities, such as Burt and Grossenheider (1976), have recognized the North American grizzly (*U. horribilis*) and the Alaskan big brown bear (*U. middendorffi*) as species distinct from *U. arctos* of the Old World. Other authorities (Erdbrink 1953; Kurten 1973; Rausch 1953, 1963), based on limited systematic work, have referred the North American brown and grizzly to *U. arctos*. This procedure is being used by most persons now studying or writing about bears and is followed here. Kurten (1973) distinguished three North American subspecies: *U. a. middendorffi*, on Kodiak and Afognak islands; *U. a. dalli*, on the south coast of Alaska and the west coast of British Columbia; and *U. a. horribilis*, in all other parts of the range of the species. Hall (1984), however, recognized nine North American subspecies of *U. arctos*.

From *Tremarctos*, *Ursus* is distinguished by its masseteric fossa on the lower jaw not being divided by a bony septum into two fossae. From *Ailuropoda* it is distinguished in having an alisphenoid canal (E. R. Hall 1981). Additional information is provided separately for each species.

Ursus thibetanus (Asiatic black bear)

Head and body length is 1,200–1,800 mm and tail length is 65–106 mm. Stroganov (1969) listed weight as 110–50 kg for males and 65–90 kg for females. Roberts (1977) stated that an exceptionally large male weighed 173 kg but that an adult female weighed only 47 kg. The coloration is usually black but is sometimes reddish brown or rich brown. There is some white on the chin and a white crescent or V on the chest.

The Asiatic black bear frequents moist deciduous forests and brushy areas, especially in the hills and mountains. It ascends to elevations as high as 3,600 meters in the summer and descends in the winter. It swims well. According to Lekagul and McNeely (1977), this bear is generally nocturnal, sleeping during the day in hollow trees, caves, or rock crevices. It is also seen abroad by day when favored fruits are ripening. It climbs expertly to reach fruit and beehives. It usually walks on all fours but often stands on its hind legs so that its forepaws can be used in fighting. The diet includes fruit, buds, invertebrates, small vertebrates, and carrion. Domestic livestock is sometimes taken, and animals as large as adult buffalo are killed by breaking their necks. Individuals become fat in late summer and early autumn before hibernation, but some populations either do not undergo winter sleep or do so only during brief periods of severe weather. Roberts (1977) stated that in the Himalayas *U. thibetanus* hibernates, sometimes in a burrow of its own making, but that in southern Pakistan there is no evidence of hibernation. According to Stroganov (1969), hibernation in Siberia begins in November and lasts four or five months. Dens in that area are usually in tree holes. The bears are easily aroused during the first month but sleep more deeply from December to February.

A radio-tracking study in Sichuan, central China (Reid et al. 1991), found the ecology of *U. thibetanus* to compare closely to that of *U. americanus* in Tennessee and to differ in a few ways from the pattern indicated above. Bears were generally active in daylight and far less active at night. Daily movements decreased from about 650 meters per day during the summer to 300 meters per day in the autumn, though there also was an increase in nocturnal activity in the autumn. The diet was almost exclusively vegetarian. In the autumn there was a shift to lower elevations to obtain acorns and other mast foods. Winter denning took place from late November to early April. Population density in

Asiatic black bear *(Ursus thibetanus)*, photo from New York Zoological Society.

the study area was 0.1–1.3/sq km, and minimum home range estimates for two individuals during a 30-month period were 16.4 and 36.5 sq km. Seasonal ranges were smaller, about 5–6 sq km.

In Siberia individual home range is 500–600 ha., only one-third or one-fourth the size of that of *U. arctos*. Mating in Siberia occurs in June or July, and births take place from late December to late March, mostly in February (Stroganov 1969). In Pakistan mating is thought to occur in October, and the young are born in February (Roberts 1977). Newborn weight is 300–450 grams (Hayssen, Van Tienhoven, and Van Tienhoven 1993). According to Lekagul and McNeely (1977), pregnancy lasts 7–8 months, and usually two cubs are born in a cave or hollow tree in early winter. The young open their eyes after about 1 week and shortly thereafter begin to follow the female as she forages. They are weaned at about 3.5 months but remain with the mother until they are 2–3 years old. Females have been seen with two sets of cubs. Sexual maturity comes at about 3 years. An individual living at the zoo in Portland, Oregon, in December 1994 was approximately 36 years old (Marvin L. Jones, Zoological Society of San Diego, pers. comm., 1995).

The Asiatic black bear sometimes raids cornfields and attacks domestic livestock. Occasionally it has been reported to kill humans. Hayashi (1984), for example, reported that from 1970 to 1983 in Fukui Prefecture of Japan there were attacks on 16 people, one of whom was killed. For these reasons *U. thibetanus* is hunted by people, and it also has declined because of the destruction of its forest habitat (Cowan 1972). The subspecies *U. t. gedrosianus,* of southern Pakistan and possibly adjacent parts of Iran, is classified as critically endangered by the IUCN and as endangered by the USDI. There is no recent information on its status, but any remaining population would be very small and isolated. In neighboring Afghanistan *U. thibetanus* evidently has been completely extirpated, and populations in

Bangladesh and Korea may be on the verge of disappearing. The subspecies *U. t. japonicus* is threatened by deforestation and subsistence and commercial hunting. It was wiped out on Kyushu in the 1950s, has been reduced to fewer than 100 individuals on Shikoku, and survives in viable numbers only in western Honshu (Servheen 1989). The same problems jeopardize *U. t. formosanus,* which is endemic to Taiwan, where probably fewer than 200 bears remain, and Hainan, where there are fewer than 50 (Garshelis 1995). Wild populations of *U. thibetanus* in mainland China are estimated to number about 12,000–18,000 and are generally declining (Guo 1995).

The species *U. thibetanus* has been given an overall classification of vulnerable by the IUCN and also is now on appendix 1 of the CITES. This species is probably the one most severely affected by the commercial trade in bear parts. Although bears long have been used in traditional Oriental foods and medicines, only recently has the extent of the resulting international trade and the devastating impact on bear populations been generally recogized (Rose and Gaski 1995). Nearly all parts of a bear's body are utilized (Highley and Highley 1995; Mills 1993, 1994, 1995). The meat, fat, and paws are popular foods, and their consumption is believed to give strength and ward off colds and other illnesses. With rising affluence in such countries as Singapore, Hong Kong, and South Korea, bear-paw dishes have become a status symbol and can cost hundreds of U.S. dollars per serving at some restaurants. The bones, brain, blood, and spinal cord also are reputed to have medicinal uses, and the whole head and skin are prized decorative items.

By far the most valued part, and the one that decidedly has made bears an international commodity, is the gallbladder. The bile salts found within that organ are used as a medicine to treat diseases of the liver, heart, and digestive system and for many other purposes, such as relieving pain, improving vision, and cleaning toxins from the blood.

There apparently is some scientific basis for these uses, as bears are the only mammals that manufacture the bile salt ursodeoxycholic acid. This substance has been shown in Western laboratory tests to be effective in treating some liver diseases, and a synthesized form is widely used to dissolve gallstones without surgery. A recent survey of traditional doctors in South Korea indicated that most sold bear bile for at least U.S. $37.50 per gram and would pay more than $1,000 for a gallbladder from a wild bear. Customs records show that the amount of bear bile imported by South Korea alone from 1970 to 1993 represented 2,867 bears annually; by far the greatest supplier was Japan. Bear populations in the Russian Far East reportedly also are being decimated by this trade. In China more than 10,000 bears are maintained in captivity for purposes of bile production. The animals are kept in small cages and "milked" of their bile by a tube surgically implanted in the gallbladder. Although this activity has been extolled as a conservation mechanism, wild bears still are being killed and sold regularly in China and their gallbladders considered superior to those of captives for medicinal purposes.

Ursus americanus (American black bear)

Head and body length is 1,500–1,800 mm, tail length is about 120 mm, and shoulder height is up to 910 mm. Banfield (1974) listed weights of 92–140 kg for females and 115–270 kg for males. The most common color phases are black, chocolate brown, and cinnamon brown. Different colors may occur in the same litter. A white phase is generally rare and never in the majority but seems to be most common on the Pacific coast of central British Columbia. A blue black phase is also generally rare but occurs frequently in the St. Elias Range of southeastern Alaska. Rounds (1987) reported that coloration varies geographically, with nearly all bears in eastern North America being black but most in some southwestern populations being non-black. Compared with U. arctos, U. americanus has a shorter and more uniform pelage, shorter claws, and shorter hind feet. Females have three pairs of mammae (Banfield 1974).

The American black bear occurs mainly in forested areas. It may originally have avoided open country because of the lack of trees in which to escape U. arctos. The latter species is known to be a competitor with, and sometimes a predator upon, U. americanus (Jonkel 1978). The black bear now appears to have extended its range northward onto the tundra, possibly in response to the decline of the barren ground grizzly (Jonkel and Miller 1970; Veitch 1993). Following the extermination of the grizzly in the mountains of southern California, U. americanus moved into the area (E. R. Hall 1981).

The usual locomotion is a lumbering walk, but U. americanus can be quick when the need arises. It swims and climbs well. It may move about at any hour but is most active at night (Banfield 1974). Like other bears that sleep through the winter, it becomes fat with the approach of cold weather, finally ceases eating, and goes into a den in a protected location. The shelter may be under a fallen tree, in a hollow tree or log, or in a burrow. In the Hudson Bay area individuals may burrow into the snow. During hibernation body temperature drops from 38° C to 31°–34° C, the respiration slows, and the metabolic rate is depressed (Banfield 1974). The winter sleep is interrupted by excursions outside during periods of relatively warm weather. Such emergences are more numerous at southern latitudes. Hibernation begins as early as late September and may last until May. When dens are entered may depend on how much fat has been accumulated through feeding, and even after emergence a bear may remain lethargic and in the vicinity of its den until food again is plentiful. In Washington the average period of hibernation is 126 days, but three

A

American black bear (Ursus americanus), photo from San Diego Zoological Garden.

American black bear *(Ursus americanus)*, photo by J. Perley Fitzgerald.

Louisiana bears each slept for 74–124 days (Lindzey and Meslow 1976; Lowery 1974; Rogers 1987). At least 75 percent of the diet consists of vegetable matter, especially fruits, berries, nuts, acorns, grass, and roots. In some areas sapwood is important; to reach it the bear peels bark from trees, thereby causing forest damage (Poelker and Hartwell 1973). The diet also includes insects, fish, rodents, carrion, and occasionally large mammals.

Banfield (1974) suggested an overall population density of about 1/14.5 sq km. Field studies in Alberta, Washington, and Montana, however, indicate a usual density of 1/2.6 sq km (Jonkel and Cowan 1971; Kemp 1976; Poelker and Hartwell 1973). Still higher densities have been reported: 1/1.3 sq km in southern California (Piekielek and Burton 1975) and 1/0.67 sq km on Long Island off southwestern Washington (Lindzey and Meslow 1977b). In the latter area home range was found to average 505 ha. for adult males and 235 ha. for adult females (Lindzey and Meslow 1977a). Farther north in Washington, however, Poelker and Hartwell (1973) determined home range to average about 5,200 ha. for males and 520 ha. for females. The ranges of males did not overlap one another, but the ranges of females overlapped with those of males and occasionally with those of other females. On the tundra of the Ungava Peninsula adult males commonly range over 50,000–100,000 ha., females over 5,000–20,000 ha. (Veitch 1993). In Idaho, Amstrup and Beecham (1976) found home ranges to vary from 1,660 to 13,030 ha., to remain stable from year to year, and to overlap extensively. Despite such overlap, in-

dividuals tend to avoid one another and to defend the space being used at a given time. A number of bears sometimes congregate at a large food source, such as a garbage dump, but they try to keep out of one another's way. Dominance hierarchies may be formed in such situations, but more tolerance is shown to familiar individuals than to strangers (Banfield 1974; Jonkel 1978; Jonkel and Cowan 1971; Rogers 1987). There are a variety of vocalizations. When startled, the ordinary sound is a "woof." When cubs are lonely or frightened they utter shrill howls.

Perhaps the most intensive study of *U. americanus* was carried out from 1969 to 1985 in northeastern Minnesota by Rogers (1987), who made use of live capture and radio tracking. The annual cycle of social behavior was found to be closely tied to plant growth, fruiting, and availability of food. The population density of bears, including cubs, was calculated at 1/4.1–6.3 sq km. Mature males used overlapping home ranges that averaged 75 sq km and appeared to be arranged so as to allow access to the maximum number of potentially estrous females. Females occupied territories averaging 9.6 sq km and vigorously chased intruders. However, after fruit and nuts disappeared in late summer, 67 percent of the males and 40 percent of the females foraged beyond their ranges or territories. Mothers tolerated their independent offspring within their territories and eventually allowed their daughters to take over a portion. Young males dispersed an average distance of 61 km from their natal areas.

The sexes come together briefly during the mating season, which generally peaks from June to mid-July. During this period individual females apparently are in estrus only 1–3 days. They usually give birth every other year but sometimes wait 3–4 years. Pregnancy generally lasts about 220 days, but there is delayed implantation. The fertilized eggs are not implanted in the uterus until the autumn, and embryonic development occurs only in the last 10 weeks of pregnancy. Births occur mainly in January and February, commonly while the female is hibernating. The number of young per litter ranges from one to five and is usually two or three. At birth the young weigh 225–330 grams and their eyes are tightly closed. They may appear naked but have a coat of short grayish hair (Frederick A. Ulmer, Jr., Academy of Natural Sciences, Philadelphia, pers. comm., 1994). They usually are weaned at around 6–8 months but remain with the mother and den with her during their second winter. Upon emergence in the spring they usually depart in order to avoid the aggression of the adult males in the breeding season. Females reach sexual maturity at 4–5 years, and males about a year later. One female is known to have lived 26 years and to have been in estrus at that age (Banfield 1974; Jonkel 1978; Poelker and Hartwell 1973; Rogers 1987). A captive reportedly lived for 31 years (Marvin L. Jones, Zoological Society of San Diego, pers. comm., 1995).

Except when wounded or attempting to protect its young, the black bear is generally harmless to people. In areas of total protection, such as national parks, the species has become accustomed to humans. It thus can be easily seen and is a popular attraction but is sometimes a nuisance, raiding campsites or begging for food along roads. Physical attacks are rare but occur with some regularity, often because the persons involved disregard safety regulations (Cole 1976; Jonkel 1978; Pelton, Scott, and Burghardt 1976). Black bears have killed people on occasion, most recently an adult woman in Alaska in July 1992 (*Washington Post*, 8 September 1992) and a four-year-old boy in British Columbia in September 1994 (*International Bear News* 41[1] [1995]: 19).

People have intensively killed *U. americanus* because of fear, to prevent depredations on domestic animals and crops, for sport, and to obtain fur and meat. According to Lowery (1974), attacks on livestock are negligible, but the bear does serious damage to cornfields and honey production. The economic loss caused to beekeepers in the Peace River Valley of Alberta was estimated at $200,000 in 1973, and a government control program is directed against the bear in that area (Gilbert and Roy 1977). In most of the states and provinces occupied by the black bear it is treated as a game animal, subject to regulated hunting. Approximately 40,000 individuals are killed legally each year in North America (McCracken, Rose, and Johnson 1995). Relatively few skins go to market now, as regulations sometimes forbid commerce and there is no great demand. The average price per pelt in the 1976–77 season was about $44 (Deems and Pursley 1978), but it had fallen to less than $20 by 1983–84 (Novak, Obbard, et al. 1987).

A more significant factor in recent years has been the killing of *U. americanus* and the exportation of its body parts for medicinal use or food in the Orient, as described above in the account of *U. thibetanus*. This problem may have intensified in association with a decline in the native bears of Asia. Indeed, the United States now is the second largest supplier of bear gallbladders to South Korea (Mills 1995). Other markets include China, Japan, Hong Kong, Taiwan, and Asian communities in the United States and Canada. In 1992 *U. americanus* was placed on appendix 2 of the CITES. According to McCracken, Rose, and Johnson (1995), this measure was taken primarily to prevent illegal

commerce in the parts of Asian bears, which were being falsely labeled as those of American black bears. However, investigations have revealed an extensive trade in gallbladders and other parts of *U. americanus* itself, together with poaching operations in the United States and Canada. This activity appears to be intensifying, and CITES controls have thus far had only a very limited effect.

The distribution of the black bear has declined substantially, but the species is still common in Alaska, Canada, the western conterminous United States, the upper Great Lakes region, northern New England and New York, and parts of the Appalachians. Small native populations also survive in coastal lowlands from the Dismal Swamp of Virginia to the Okefenokee of Georgia and in the bottom land forests of southern Alabama and southeastern Arkansas (Wooding, Cox, and Pelton 1994). *U. americanus* evidently had been totally extirpated in Texas by the 1940s but recently has been reestablished through the apparent movement of animals from the Sierra del Carmen of northern Mexico into Big Bend National Park (Hellgren 1993). Data compiled by Cowan (1972) indicated the presence of about 170,000 black bears in the conterminous United States, and Raybourne (1987) gave an estimate of 400,000–500,000 for all of North America. A new survey (McCracken, Rose, and Johnson 1995) developed estimates of approximately 200,000 for the contiguous United States, 150,000 for Alaska, and 330,000 for Canada. Leopold (1959) thought the species was still widespread in Mexico, but more recent information indicates that it is critically endangered there (Ceballos and Navarro L. 1991).

The subspecies *U. a. floridanus*, of Florida and adjacent areas, is considered to be threatened through habitat loss, fragmentation of remnant populations, and persecution by beekeepers (Brady and Maehr 1985; Layne 1978). About 500–1,000 individuals are thought to survive (Maehr in Humphrey 1992). The subspecies *U. a. luteolus*, formerly found from eastern Texas to Mississippi, was by the mid-twentieth century reduced to a few individuals along the Mississippi and lower Atchafalaya rivers in eastern Louisiana and possibly neighboring Mississippi. In 1992 it was classified as threatened by the USDI, a measure that provided a basis for protection of substantial portions of its habitat (Neal 1993). During the 1960s the wildlife agencies of both Louisiana and Arkansas imported a number of bears from Minnesota (within the range of the subspecies *U. a. americanus*) to their respective states (Lowery 1974; Sealander 1979), thus further jeopardizing the genetic viability of the native populations. The Arkansas introduction has been spectacularly successful from a management standpoint, there now being about 2,100 bears in the Ozark and Ouachita highlands of the state; this population has expanded into adjacent parts of Missouri and Oklahoma, where there now may be an additional 300 bears (Smith and Clark 1994; Smith, Clark, and Shull 1993).

Ursus arctos (brown or grizzly bear)

Head and body length is 1,700–2,800 mm, tail length is 60–210 mm, and shoulder height is 900–1,500 mm. In any given population adult males are larger on average than adult females. The largest individuals—indeed, the largest of living carnivores—are found along the coast of southern Alaska and on nearby islands, such as Kodiak and Admiralty. In this area weight is as great as 780 kg. Size rapidly declines to the north and east. In southwestern Yukon, for example, Pearson (1975) found average weights of 139 kg for males and 95 kg for females. In the Yellowstone region, Knight, Blanchard, and Kendall (1981) found weights of full-grown animals to average 181 kg and to range from 102 to 324 kg. In Siberia and northern Europe weight is

Alaskan brown bear *(Ursus arctos)*, photo by Ernest P. Walker.

usually 150–250 kg. In parts of southern Europe average weight is only 70 kg (Grzimek 1975). Coloration is usually dark brown but varies from cream to almost black. In the Rocky Mountains the long hairs of the shoulders and back are often frosted with white, thus giving a grizzled appearance and the common name "grizzly" or "silvertip." From *U. americanus, U. arctos* is distinguished in having a prominent hump on the shoulders, a snout that rises more abruptly into the forehead, longer pelage, and longer claws.

The brown bear has one of the greatest natural distributions of any mammal. It occupies a variety of habitats but in the New World seems to prefer open areas such as tundra, alpine meadows, and coastlines. It was apparently common on the Great Plains prior to the arrival of European settlers. In Siberia the brown bear occurs primarily in forests (Stroganov 1969). Surviving European populations are restricted mainly to mountain woodlands (Van Den Brink 1968). Even when living in generally open regions *U. arctos* needs some areas with dense cover (Jonkel 1978). It shelters in such places by day, sometimes in a shallow excavation, and moves and feeds mainly during the cool of the evening and early morning. Egbert and Stokes (1976) noted that activity in coastal Alaska occurs throughout the day but peaks from 1800 to 1900 hours. Seasonal movements are primarily toward major food sources, such as salmon streams and areas of high berry production (Jonkel 1978). In Siberia individuals may travel hundreds of kilometers during the autumn to reach areas of favorable food supplies (Stroganov 1969).

According to Banfield (1974), the usual gait is a slow walk. *U. arctos* is capable of moving very quickly, however, and can easily catch a black bear. Its long foreclaws are not adapted for climbing trees. It has excellent senses of hearing and smell but relatively poor eyesight. The brown bear has great strength. Banfield saw one drag a carcass of a horse about 90 meters. In another case, a 360-kg grizzly killed and dragged a 450-kg bison.

Hibernation begins in October–December and ends in March–May. The exact period depends on the location, weather, and condition of the animal. In certain southerly areas hibernation is very brief or does not take place at all. In most cases the brown bear digs its own den and makes a bed of dry vegetation. The burrow is often located on a sheltered slope, either under a large stone or among the roots of a mature tree. The bed chamber has an average volume of around two cubic meters. A den is sometimes used year after year. During winter sleep there is a marked depression in heart rate and respiration but only a slight drop in body temperature. The animal can be aroused rather easily and can make a quick escape, if necessary (Craighead and Craighead 1972; Grzimek 1975; Stroganov·1969; Slobodyan 1976; Ustinov 1976).

The diet consists mainly of vegetation (Jonkel 1978). Early spring foods include grasses, sedges, roots, moss, and bulbs. In late spring succulent, perennial forbs become important. During the summer and early autumn berries are essential and bulbs and tubers are also taken. Banfield (1974) wrote that *U. arctos* consumes insects, fungi, and roots at all times of the year and also digs mice, ground squirrels, and marmots out of their burrows. In the Canadian Rockies the grizzly is quite carnivorous, hunting moose, elk, mountain sheep and goats, and even black bears. In Mount McKinley National Park, Alaska, Murie (1981) found *U. arctos* to feed mostly on vegetation but also to eat carrion whenever available and occasionally to capture young calves of caribou and moose. During the summer, when salmon are moving upstream along the Pacific coasts of Canada, southern Alaska, and northeastern Siberia, brown bears gather to feed on the vulnerable fish (Banfield 1974; Egbert and Stokes 1976; Kistchinski 1972). Perhaps because of this abundant food supply the bears of these areas are larger and are found at greater densities than anywhere else.

Some approximate reported population densities are: Carpathian Mountains, one bear per 20 sq km; Lake Baikal area, one per 60 sq km; coast of Sea of Okhotsk, one per 10 sq km; Kodiak Island, one per 1.5 sq km; Mount McKinley National Park, one per 30 sq km; northern parts of Alaska and Northwest Territories, one per 150 sq km; and Glacier National Park, Montana, one per 21 sq km (Dean 1976; Harding 1976; Kistchinski 1972; Martinka 1974, 1976; Slobodyan 1976; Ustinov 1976). In the Yellowstone region of the western United States overall average density is about one bear per 88 sq km. In summer, however, individuals have concentrated by night at feeding sites, so densities have reached about one per 0.05 sq km. Daytime dispersal has reduced density to about one per 0.36 sq km. In the Yellowstone ecosystem individual home range averages about 80 sq km and varies from about 20 to 600 sq km, with respect to the area used in the course of a year (Craighead 1976; Knight, Blanchard, and Kendall 1981). The home ranges of males are generally substantially larger than those of females. Lifetime individual ranges in Yellowstone have averaged 3,757 sq km for males and 884 sq km for females (Blanchard and Knight 1991). In the northern Yukon, Pearson (1976) found averages of 414 sq km for males and 73 sq km for females.

Home ranges overlap extensively and there is no evidence of territorial defense (Craighead 1976; Murie 1981). Although generally solitary, the grizzly is the most social of North American bears, occasionally gathering in large numbers at major food sources and often forming family foraging groups with more than one age class of young (Jonkel 1978). At a salmon stream in southern Alaska, Egbert and Stokes (1976) sometimes observed more than 30 bears at one time. Considerable intraspecific tolerance was demonstrated in such aggregations, but dominance hierarchies were enforced by aggression. The highest-ranking animals were the large adult males, which most other bears attempted to avoid. The most aggressive animals were females with young, and the least aggressive were adolescents. Overt fighting was usually brief, and no infliction of serious wounds was observed, but the researchers suspected that killing of young individuals by adult males was a factor in population regulation.

Except as noted, the following life history information was taken from Craighead, Craighead, and Sumner (1976), Egbert and Stokes (1976), Glenn et al. (1976), Jonkel (1978), Murie (1981), Pearson (1976), and Slobodyan (1976). The only lasting social bonds are those between females and young. During the breeding season males may fight over females. Successful males attend one or two females for 1–3 weeks. Mating takes place from May to July, implantation of the fertilized eggs in the uterus is usually delayed until October or November, and births generally occur from January to March, while the mother is in hibernation. The total period of pregnancy may last 180–266 days. Females have been reported to remain in estrus throughout the breeding season until mating, but data listed by Hayssen, Van Tienhoven, and Van Tienhoven (1993) suggest an estrus period of less than a month. After mating, females do not again enter estrus for at least 2, usually 3 or 4, years. The number of young in a litter averages about two and ranges from one to four. They weigh 340–680 grams at birth and their eyes are closed. They may appear nearly naked but are actually covered with short grayish brown hair (Frederick A. Ulmer, Jr., Academy of Natural Sciences, Philadelphia, pers. comm., 1994). They are weaned at about 5 months. They remain with the mother at least until their

second spring of life and usually until their third or fourth. Litter mates sometimes maintain an association for 2–3 years after leaving the mother. They reach puberty at around 4–6 years but continue to grow. Males in southern Alaska may not reach full size until they are 10–11 years old. Females in the Yellowstone region are known to have lived 25 years and still be capable of reproduction. A female at the Leipzig Zoo bore 54 young in 19 litters and lived to the age of 39 years and 4 months, the known record life span for the species (Marvin L. Jones, Zoological Society of San Diego, pers. comm., 1995). Potential longevity in captivity may be as great as 50 years (Stroganov 1969).

The brown bear is reputedly the most dangerous animal in North America. If we disregard venomous insects, disease-spreading rodents, domestic animals, and people themselves, this may be true. Three persons were killed by grizzlies in Glacier National Park, Montana, in 1980 (*Washington Post*, 25 July 1980, A-14; 9 October 1980, A-54; 20 October 1980, A-5). Another was killed in Canada (J. R. Gunson, Alberta Fish and Wildlife Division, pers. comm., 1981). That was an unusually tragic year, but injuries and an occasional death have been reported from the western national parks since around 1900 (Cole 1976; Herrero 1970, 1976). During the nineteenth century, attacks on people apparently occurred with some regularity in California (Storer and Tevis 1955). Two people were killed in a single attack in Alaska in July 1995 (*Washington Post*, 3 July 1995), and another died in August 1996 when he inadvertently disturbed a mother bear with a young cub (ibid., 27 August 1996). Three men recently were killed by the bears they were photographing, one in Yellowstone in October 1986, one in Glacier Park in April 1987, and one on the Kamchatka Peninsula in August 1996 (*Audubon* 89[4]: 16–17; Mitchell 1987; *New York Times*, 22 September 1996). According to Ustinov (1976), more than 70 attacks and 17 deaths have been attributed to *U. arctos* in the Lake Baikal area of Siberia. Cicnjak and Ruff (1990) reported 4 deaths in Bosnia, Croatia, and Slovenia from 1986 to 1988. Many such incidents have probably been provoked by an effort to shoot or harass the animal, as the brown bear normally tries to avoid humans. It is unpredictable, however, if startled at close quarters, especially when accompanied by young or engrossed in a search for food. Jonkel (1978) cautioned that there may be more difficulties as recreational and commercial activity increases in areas occupied by the grizzly. He suggested that problems could be reduced by improved education and planning and by not locating campsites, trails, and residential facilities in places regularly used by bears.

The brown bear long has been persecuted as a predator of domestic livestock, especially cattle and sheep. Those parts of North America from which it has been eliminated correspond closely to areas of intensive ranching and grazing. In the nineteenth and early twentieth centuries there apparently were some remarkably destructive bears (Hubbard and Harris 1960; Storer and Tevis 1955), and their activities earned the entire species the lasting enmity of cattle ranchers and sheepherders. The brown bear also has been widely sought as a big-game trophy, and it is currently subject to regulated sport hunting in most of its range. During the 1983–84 season 1,441 brown bears were killed legally in the United States and Canada, and pelts sold in Canada for an average price of $162.05 (Novak, Obbard, et al. 1987).

The original eastern limits of *U. arctos* in North America are not certain. A skull found in a Labrador Eskimo midden dating from the late eighteenth century supports earlier stories that the grizzly used to occur to the east of Hudson Bay (Spiess 1976). Pelts, reportedly those of *U. arc-*

tos, were taken on the Ungava Peninsula as late as 1927 (Veitch 1993). The decline of the species on the Great Plains may have begun when the Indians of that region obtained the horse and hence an improved hunting capability. A precipitous drop in grizzly numbers came in the nineteenth century as settlers and livestock filled the West, setting up confrontations that usually ended to the detriment of the bear. This process was intensified by logging, mining, and road construction, which increased human presence in remote areas. The distinctive subspecies of California and northern Baja California, *U. a. californicus* (Hall 1984), evidently had disappeared by the 1920s.

In the early nineteenth century there may have been 100,000 grizzlies in the western conterminous United States. Now there are probably fewer than 1,000. Of these, about 200 are in Glacier National Park (Martinka 1974), 300 more are in nearby parts of northwestern Montana, and perhaps another 100 occupy isolated mountain ecosystems westward into northern Idaho and Washington. The population in Yellowstone National Park and vicinity has been estimated at 136 (Craighead, Varney, and Craighead 1974), 247 (Knight, Blanchard, and Kendall 1981), and 329 (Mattson et al. 1995). Although numbers have appeared to be stable or increasing for the last two decades and research and conservation efforts have intensified, this isolated population may not be able to survive anticipated long-term environmental perturbations and levels of mortality resulting from increased human habituation (Mattson, Blanchard, and Knight 1992; Mattson and Reid 1991). Some authorities have warned that the Yellowstone population may not be reproductively viable and that current management practices in and around the park are leading to conflicts between human interests and bears, often resulting in the death of the latter (Chase 1986). Such problems probably caused the recent disappearance of very small remnant groups in south-central Colorado and northwestern Mexico. The grizzly has been extirpated from the Great Plains of Canada, except for an isolated group in west-central Alberta (Banfield 1974). The species also has declined on the barrens of the Northwest Territories (Macpherson 1965). In the mountainous regions of western Canada and in Alaska *U. arctos* is still relatively common, perhaps numbering about 55,000 individuals (Cowan 1972; Jonkel 1987; Schoen, Miller, and Reynolds 1987; Servheen 1989).

Chestin et al. (1992) estimated that the number of brown bears in the countries of the old Soviet Union had increased from about 105,000 in the early 1960s to 131,000 in 1989. The annual legal kill was approximately 3,500 and poaching was thought to account for several thousand more. It must be noted that these figures were derived prior to a reported relaxation of controls and intensification of illegal killing that followed the breakup of the Soviet Union and is associated in part with the Far Eastern commerce in bear parts (see account of *Ursus thibetanus*). The bears of the Kamchatka Peninsula, comparable in size to those of southern Alaska, are said to have been especially hard hit (*Oryx* 28 [1994]: 224). The great majority of the bears estimated to be present in 1989 were in European Russia and Siberia, 3,500 were in the Caucasus, and 3,500 were in the Central Asian republics.

Vereschagin (1976) estimated that there were about 30,000 brown bears in Eurasia outside of the Soviet Union, though there may have been a reduction since then. There now are 4,500–7,600 in China (Guo 1995) and possibly some viable populations in Mongolia, Hokkaido, the Himalayas, and Turkey, but all appear to be declining in the face of uncontrolled hunting and habitat disruption (Servheen 1989). To the west of Russia there are about 700 in northern Scandinavia, 700 in southern Poland, 700 in

Slovakia, 6,000 in Romania, 700 in Bulgaria, and as many as 2,000 in the countries originally comprised by Yugoslavia. There are small, isolated groups in Albania, Greece, Austria, Italy, southern France, northern Spain, and southern Norway (Elgmork 1978; Pasitschniak-Arts 1993; Servheen 1989; Smit and Van Wijngaarden 1981). A recent analysis of mitochondrial DNA, suggesting a phylogenetic divergence between the bears of Russia and northern Scandinavia, on the one hand, and those surviving in southern Scandinavia and southern Europe, on the other, may have effects on reintroduction planning (Taberlet and Bouvet 1994). *U. arctos* apparently disappeared from Great Britain in the tenth century (Pasitschniak-Arts 1993) and from northwestern Africa around the mid–nineteenth century (Harper 1945).

Appendix 2 of the CITES includes all North American populations of *U. arctos* except *U. a. nelsoni*, the Mexican grizzly bear (treated as a synonym of *U. a. arctos* by Hall 1984), and now also includes all Eurasian populations except those on appendix 1. On appendix 1 are *U. a. nelsoni*, *U. a. pruinosus* of Tibet and Mongolia, and *U. a. isabellinus* of the mountains of Central Asia. Schaller, Tulgat, and Navantsatsvalt (1993) considered the range of *isabellinus* to extend as far as Great Gobi National Park in western Mongolia, where there is a unique, isolated, and arid-adapted population of only about 30 bears. The IUCN classifies *U. a. nelsoni* as extinct. The USDI lists *U. a. nelsoni*, *U. a. pruinosus*, and the Italian populations of *U. arctos* as endangered. Those populations of *U. arctos* in the conterminous United States are listed as threatened by the USDI.

Ursus maritimus (polar bear)

Head and body length is 2,000–2,500 mm, tail length is 76–127 mm, and shoulder height is up to 1,600 mm. DeMaster and Stirling (1981) gave the weight as 150–300 kg for females and 300–800 kg for males. Banfield (1974), however, wrote that males usually weigh 420–500 kg. The color is often pure white following the molt but may become yellowish in the summer, probably because of oxidation by the sun. The pelage also sometimes appears gray or almost brown depending on the season and light conditions. The neck is longer than that of other bears, and the head is relatively small and flat. The forefeet are well adapted for swimming, being large and oarlike. The soles are haired, probably for insulation from the cold and traction on the ice. Females have four functional mammae (DeMaster and Stirling 1981).

The polar bear is often considered to be a marine mammal. It is distributed mainly in arctic regions around the North Pole. The southern limits of its range are determined by distribution of the pack ice. It has been recorded as far north as 88° N and as far south as the Pribilof Islands in the Bering Sea, the island of Newfoundland, the southern tip of Greenland, and Iceland. There also are permanent populations in James Bay and the southern part of Hudson Bay. Although found generally in coastal areas or on ice hundreds of kilometers from shore, individuals have wandered up to 200 km inland (Stroganov 1969).

According to DeMaster and Stirling (1981), the preferred habitat is pack ice that is subject to periodic fracturing by wind and sea currents. The refreezing of such fractures provides places where hunting by the bear is most successful. Some animals spend both winter and summer along the lower edge of the pack ice, perhaps undergoing extensive north-south migrations as this edge shifts. Others move onto land for the summer and disperse across the ice as it forms along the coast and between islands during winter. The bears of the Labrador coast sometimes move north to Baffin Island, and some individuals have traveled

Polar bears *(Ursus maritimus):* Top, photo from New York Zoological Society. Insets: A. Forefoot; B. Hind foot; photos from *Proc. Zool. Soc. London;* C. Young, 24 hours old, photo by Ernest P. Walker. Bottom, photo by Sue Ford, Washington Park Zoo, Portland, Oregon.

as far as 1,050 km, to the islands of northern Hudson Bay (Stirling and Kiliaan 1980). Individuals in the population of the Beaufort Sea off northern Alaska move several thousand kilometers annually and may use an area exceeding 500,000 sq km over a period of years (Amstrup, Garner, and Durner 1995). The population that summers along the southern shore of Hudson Bay spreads all across the partly ice-covered bay in November and returns to shore in July or August (Stirling et al. 1977). During the latter, ice-free period adult males occupy coastal areas and family groups and pregnant females move farther inland (Derocher and Stirling 1990a). Despite such movements, the polar bear is not a true nomad. There are a number of discrete populations, each with its own consistently used areas for feeding and breeding (Stirling, Calvert, and Andriashek 1980).

The polar bear can outrun a reindeer for short distances on land and can attain a swimming speed of about 6.5 km/hr. It swims rather high, with head and shoulders above the water. If killed in the water, it will not immediately sink. According to DeMaster and Stirling (1981), it has been reported to swim for at least 65 km across open water. It is capable of diving under the ice and surfacing in holes utilized by seals. It seems to be most active during the first third of the day and least active in the final third. From July to December in the James Bay region, when a lack of ice prevents seal hunting, *U. maritimus* spends about 87 percent of its time resting, apparently living off of stored fat (Knudsen 1978). The bears sometimes excavate depressions or complete earthen burrows on land during the summer in order to avoid the sun and keep cool (Jonkel et al. 1976).

Any individual bear may make a winter den for temporary shelter during severe weather, but only females, especially those that are pregnant, generally hibernate for lengthy periods. As with other bears, winter sleep involves a depressed respiratory rate and a slightly lowered body temperature but not deep torpor. Most pregnant females evidently do not spend the winter along the pack ice but hibernate on land from October or November to March or April. Maternal dens are usually found within 8 km of the coast, but in the southern Hudson Bay region they are concentrated 30–60 km inland. They are excavated in the snow to depths of 1–3 meters, often on a steep slope. They usually consist of a tunnel several meters long that leads to an oval chamber of about 3 cubic meters. Some dens have several rooms and corridors (DeMaster and Stirling 1981; Harington 1968; Larsen 1975; Stirling, Calvert, and Andriashek 1980; Stirling et al. 1977; Uspenski and Belikov 1976).

The polar bear feeds primarily on the ringed seal *(Phoca hispida)* (DeMaster and Stirling 1981). The bear either remains still until a seal emerges from the water or stealthily stalks its prey on the ice (Stirling 1974). It may also dig out the subnivean dens of seals to obtain the young (Stirling, Calvert, and Andriashek 1980). During summer and autumn in the southern Hudson Bay region *U. maritimus* often swims among sea birds and catches them as they sit on the water (Russell 1975). The diet also includes the carcasses of stranded marine mammals, small land mammals, reindeer, fish, and vegetation. Berries become important for some individuals during summer and autumn (Jonkel 1978).

Reported population densities range from 1/37 sq km to 1/139 sq km (DeMaster and Stirling 1981). Home ranges are not well defined but are thought to vary from 150 to 300 km in diameter and to overlap extensively (Kolenosky 1987). Although *U. maritimus* is generally solitary, large aggregations may form around a major source of food (Jonkel 1978). As many as 40 individuals have been seen at one time in the vicinity of the Churchill garbage dump, on the southern shore of Hudson Bay (Stirling et al. 1977). Of the adult males in the population that inhabits that region, more than half occur in aggregations during the ice-free season. These groups, with an average size of 4 bears, tend to occur at environmentally favorable sites and may help develop familiarity that will avoid future conflicts for resources (Derocher and Stirling 1990b). Wintering females evidently tolerate one another well, as dens on Wrangel Island are sometimes found at densities of one per 50 sq meters (Uspenski and Belikov 1976). High concentrations of summer dens also have been reported (Jonkel et al. 1976). Adult females with young are not subordinate to any other age or sex class but tend to avoid interaction with adult males, presumably because the latter are potential predators of the cubs (DeMaster and Stirling 1981).

Estrus lasts 3 days (Hayssen, Van Tienhoven, and Van Tienhoven 1993). The sexes usually come together only briefly during the mating season, March–June. Delayed implantation apparently extends the period of pregnancy to 195–265 days. The young are born from November to January, while the mother is in her winter den. Females give birth every 2–4 years. The number of young per litter averages about two and ranges from one to four. They weigh about 600 grams at birth and are well covered with short white fur, but their eyes are tightly closed. Upon emergence from the den in March or April the cubs weigh 10–15 kg. They usually leave the mother at 24–28 months. The age of sexual maturity averages about 5–6 years. Adult weight is attained at about 5 years by females but not until 10–11 years by males. Wild females apparently have a reduced natality rate after the age of 20. Annual adult mortality in a population is about 8–16 percent. Potential longevity in the wild is estimated at 25–30 years (DeMaster and Stirling 1981; Ramsay and Stirling 1988; Stirling, Calvert, and Andriashek 1980; Ulmer 1966; Uspenski and Belikov 1976). A female at the Detroit Zoo gave birth at 36 years and 11 months (Latinen 1987) and was still living at 45 years (Marvin L. Jones, Zoological Society of San Diego, pers. comm., 1995).

The polar bear often is considered to be dangerous to people, though usually the two species are not found in close proximity. An exception developed during the 1960s in the vicinity of the town of Churchill, on the southern shore of Hudson Bay (Stirling et al. 1977). Bears apparently increased in this area because of a decline in hunting. At the same time, more people moved in and several large garbage dumps were established. A number of persons were attacked and one was killed. Many bears were shot or translocated by government personnel. There also have been a number of confrontations in Svalbard, and one person was killed by a bear there in 1977 (Gjertz and Persen 1987).

The native peoples of the Arctic have long hunted the polar bear for its fat and fur. Sport and commercial hunting increased in the twentieth century. Of the regularly marketed North American mammal pelts, that of *U. maritimus* is the most valuable. During the 1976–77 season 530 skins from Canada were sold at an average price of $585.22 (Deems and Pursley 1978). Some individual prime pelts have brought more than $3,000 each (Smith and Jonkel 1975). There subsequently appears to have been a decline in demand, and other parts of the polar bear, particularly the gallbladder, seem to lack the market value of those of other bears (Frampton 1995a).

The use of aircraft to locate polar bears and to land trophy hunters in their vicinity developed in Alaska in the late 1940s. The annual kill by such means reached about 260 bears by 1972. In that year, however, the killing of *U. mar-*

itimus, except for native subsistence, was prohibited by the United States Marine Mammal Protection Act. Canada and Denmark (for Greenland) also limit hunting to resident natives, and Russia and Norway (for Svalbard) provide complete protection. In 1973 these five nations drafted an agreement calling for the restriction of hunting, the protection of habitat, and cooperative research on polar bears. The agreement was ratified by the United States in 1976. The yearly worldwide kill is now estimated at around 1,000 animals. The total number of polar bears in the wild in 1993 was estimated at 21,470–28,370, and populations are generally thought to be stable or increasing. *U. maritimus*, however, may be threatened by the exploitation of oil and gas reserves in the Arctic, especially by development in the limited areas suitable for denning by pregnant females (DeMaster and Stirling 1981; Stirling and Kiliaan 1980; U.S. Fish and Wildlife Service 1980; Wiig, Born, and Garner 1995).

There now also is concern that the influx of cash from oil and gas development will stimulate increased hunting by native peoples and that such hunting, not being restricted to adult males, could damage the relatively small and vulnerable polar bear populations off northern Alaska and northwestern Canada (Amstrup, Stirling, and Lentfer 1986). Another long-term concern is that global warming, resulting from the greenhouse effect of atmospheric polluting gases, will reduce the southerly extent of sea ice and thereby deny accessibility to seals. Even now the polar bear population of southwestern Hudson Bay is showing signs of nutritional stress (Stirling and Derocher 1993). The species is classified as conservation dependent by the IUCN and is on appendix 2 of the CITES. The provisions of both CITES and the United States Marine Mammal Protection Act do allow for limited importation under certain circumstances. The U.S. Fish and Wildlife Service recently proposed allowing importation of polar bear trophies taken in accordance with carefully regulated sport-hunting programs in the Northwest Territories of Canada (Frampton 1995*a*).

Ursus malayanus (Malayan sun bear)

This is the smallest bear. Head and body length is 1,000–1,400 mm, tail length is 30–70 mm, shoulder height is about 700 mm, and weight is 27–65 kg. The general coloration is black. There is a whitish or orange breast mark and a grayish or orange muzzle, and occasionally the feet are pale in color. The breast mark is often U-shaped but is variable and sometimes wholly lacking. The body is stocky, the muzzle is short, the paws are large, and the claws are strongly curved and pointed. The soles are naked.

The sun bear inhabits dense forests at all elevations (Lekagul and McNeely 1977). It is active at night, usually sleeping and sunbathing by day in a tree, two to seven meters above the ground. Tree branches are broken or bent to form a nest and lookout post. *U. malayanus* has a curious gait in that all the legs are turned inward while walking. The species is usually shy and retiring and does not hibernate. An expert tree climber, it is cautious, wary, and intelligent. A young captive observed the way in which a cupboard containing a sugar pot was locked with a key. It then later opened the cupboard by inserting a claw into the eye of the key and turning it. Another captive scattered rice from its feeding bowl in the vicinity of its cage, thus attracting chickens, which it then captured and ate.

The diet is omnivorous, and the front paws are used for most of the feeding activity. Trees are torn open in search of nests of wild bees and for insects and their larvae. The soft growing point of the coconut palm, known as "palmite," is ripped apart and consumed. After digging up termite colonies, the animal places its forepaws alternately in the nest and licks the termites off. Jungle fowl, small rodents, and fruit juices also are included in the diet.

Births may occur at any time of the year. In the East Berlin Zoo a female produced one litter on 4 April 1961 and another on 30 August 1961. The gestation period for six births at that zoo was 95–96 days (Dathe 1970). At the Fort Worth Zoo, however, three pregnancies lasted 174, 228, and 240 days, evidently because of delayed fertilization or im-

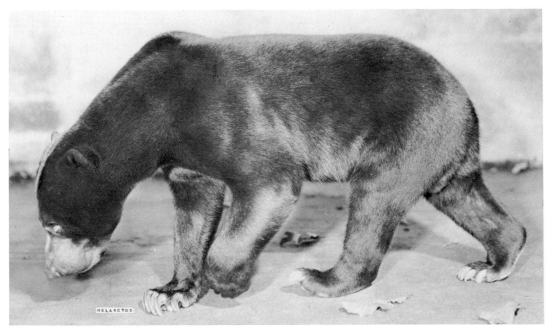

Malayan sun bear *(Ursus malayanus)*, photo by Ernest P. Walker.

plantation (McCusker 1974). Litters usually contain one or two young weighing about 325 grams each. They remain with the mother until nearly full grown (Lekagul and Mc-Neely 1977). Two captives lived to an age of approximately 31 years (Marvin L. Jones, Zoological Society of San Diego, pers. comm., 1995).

In the wild, *U. malayanus* sometimes does considerable damage to coconut plantations and is said to be one of the most dangerous animals within its range (Lekagul and Mc-Neely 1977). Young individuals make interesting pets but become unruly within a few years. According to Mills (1993, 1994, 1995), sun bear cubs are highly popular as pets in the Orient, but when they reach adult size they commonly are sold for their parts, which are used in the medicinal trade. International commerce involving such utilization also is adversely affecting wild populations (see account of *Ursus thibetanus*). Servheen (1989) reported a general decline because of hunting for gallbladders and other body parts, lack of effective regulation, and logging and conversion of vast areas of forest to cropland and rubber plantations. *U. malayanus* may now have been extirpated in India and is very rare in Bangladesh and China. It is on appendix 1 of the CITES.

Ursus ursinus (sloth bear)

Head and body length is 1,400–1,800 mm, tail length is 100–125 mm, shoulder height is 610–915 mm, and weight is 55–145 kg. The shaggy black hairs are longest between the shoulders. The overall black coloration is often mixed with brown and gray, but cinnamon and red individuals also have been noted. The chest mark, typically shaped like a V or Y, varies from white or yellow to chestnut brown.

The sloth bear has a number of structural modifications associated with an unusual method of feeding. The lips are protrusible, mobile, and naked; the snout is mobile; the nostrils can be closed at will; the inner pair of upper incisors is absent, thus forming a gap in the front teeth; and the palate is hollowed. These features enable the bear to feed on termites (white ants) in the following manner: the nest is dug

up, the dust and dirt blown off, and the occupants sucked up in a "vacuum cleaner" action. The resulting noises can be heard more than 185 meters away and often lead to the bear's detection by hunters.

The sloth bear inhabits moist and dry forests, especially in areas of rocky outcrops. It may be active at any hour but is mainly nocturnal. During cool weather it spends the day in dense vegetation or shallow caves. The sense of smell is well developed, but sight and hearing are relatively poor. Hibernation is not known to occur. Termites are the most important food for most of the year, but the diet also includes other insects, grubs, honey, eggs, carrion, grass, flowers, and fruit.

In the Royal Chitawan National Park of Nepal, Sunquist (1982) found an adult male to move over an area of at least 10 sq km during a period of about 2 years. In the same area, Laurie and Seidensticker (1977) found a minimum density of about 0.1/sq km. Most observations were of lone bears or of females with cubs. Vocalizations, heard mainly in association with intraspecific agonistic encounters, included roars, howls, screams, and squeals. Births apparently occurred mostly from September to January.

Previous observations indicate that breeding takes place mainly in June in India and during most of the year in Sri Lanka. Pregnancy lasts about 6–7 months. The young, usually one or two and rarely three, are born in a ground shelter. They leave the den at 2–3 months and often ride on the mother's back. They remain with the mother until they are almost full grown, possibly 2 or 3 years. Captives have lived for 40 years.

The sloth bear normally is not aggressive but is held in great respect by some of the people that inhabit its range. Apparently because of its poor eyesight and hearing, it is sometimes closely approached by humans. It may then attack in what it considers to be self-defense and inflict severe wounds. Since it is thought to be dangerous, and since it sometimes damages crops, it has been extensively hunted. It also seems not to tolerate regular human disturbance and is losing habitat to agriculture, logging, settlement, and

Sloth bear *(Ursus ursinus):* A. Photo by Hans-Jürg Kuhn. B. Photo from New York Zoological Society.

hydroelectric projects (IUCN 1978). Total numbers surviving throughout the range of the species are estimated at 7,500–10,000 individuals (Nobbe and Garshelis 1994). A major problem that only recently has received general recognition is the killing of *U. ursinus* for its body parts, which then are sold into international commerce and eventually used for food, alleged medicines, and decorative purposes, much the same as described above in the account of *Ursus thibetanus*. Bile from the gallbladder is widely used in the Orient to treat diseases of the liver, heart, and stomach, and bear paws are prized for consumption (Rose and Gaski 1995). Based on data showing the number of bear gallbladders exported from India to Japan from 1978 to 1988, Servheen (1989) calculated an annual kill of 728–1,548 sloth bears, probably a substantial part of the remaining populations. Such information contributed to the placing of *U. ursinus* on appendix 1 of the CITES in 1988, and the species now is also classified as vulnerable by the IUCN.

CARNIVORA; URSIDAE; **Genus AILUROPODA**
Milne-Edwards, 1870

Giant Panda

The single species, *A. melanoleuca*, now is known from the central Chinese provinces of Gansu, Shaanxi, and Sichuan; there also are historical records from much of eastern China to the south of the Huang River (Schaller et al. 1985).

There is much controversy regarding the systematic position of this genus. It once was considered to be a close relative of *Ailurus* and was placed with that genus in the family Procyonidae, but most authorities eventually came to treat *Ailuropoda* as a bear. E. R. Hall (1981) put it in the ursid subfamily Ailuropodinae, but Chorn and Hoffmann (1978) referred it to the otherwise extinct ursid subfamily Agriotheriinae. Thenius (1979) recognized the giant panda as an offshoot of the Ursidae but suggested that it represents a distinct family, the Ailuropodidae. A recent trend has been to again treat *Ailuropoda* and *Ailurus* as close relatives. Schaller et al. (1985) stated that on the basis of anatomical, biochemical, and paleontological evidence the giant panda's position remains equivocal but that certain characters of morphology, reproduction, and behavior indicate that *Ailuropoda* and *Ailurus* are related to both bears and raccoons, but more closely to the former, and that the two genera belong either in separate but closely related families of their own or together in their own family, the Ailuridae. Wozencraft (*in* Wilson and Reeder 1993) combined *Ailuropoda* and *Ailurus* in the subfamily Ailurinae of the family Ursidae.

Head and body length is 1,200–1,500 mm, tail length is about 127 mm, and weight is 75–160 kg. The coat is thick and woolly. The eye patches, ears, legs, and band across the shoulders are black, sometimes with a brownish tinge. The remainder of the body is white but may become soiled with age. There are scent glands under the tail.

Ailuropoda resembles other bears in general appearance but is distinguished by its striking coloration and certain characters associated with its diet. The head is relatively

Giant panda *(Ailuropoda melanoleuca)*, photo from New York Zoological Society. Insets: A. Right hind foot; B. Right forefoot; photos from *Proc. Zool. Soc. London.* C. Skull showing dentition, photo by P. F. Wright of specimen in U.S. National Museum of Natural History.

massive because of the expanded zygomatic arches of the skull and the well-developed muscles of mastication. The second and third premolar teeth and the molars are relatively larger and broader than those of other bears. The forefoot has an unusual modification that is thought to aid in the grasping of bamboo stems. The pad on the sole of each forepaw has an accessory lobe, and the pad of the first digit—and to a lesser extent the pad of the second digit—can be flexed onto the summit of this accessory lobe and its supporting bone.

The giant panda is found in montane forests with dense stands of bamboo. Its usual elevational range is 2,700–3,900 meters, but it may descend to as low as 800 meters in the winter. It does not make a permanent den but takes shelter in hollow trees, rock crevices, and caves. It lives mainly on the ground but evidently can climb trees well. Activity is largely crepuscular and nocturnal. Schaller et al. (1985) found animals to be active about 14.2 hours per day. *Ailuropoda* does not hibernate but descends to lower elevations in the winter and spring (Chorn and Hoffmann 1978). Average daily movement has been reported to cover 596 meters (Johnson, Schaller, and Hu 1988).

The diet consists mainly of bamboo shoots, up to 13 mm in diameter, and bamboo roots. *Ailuropoda* spends 10–12 hours a day feeding, usually in a sitting position with the forepaws free to manipulate the bamboo. Average daily consumption is about 12.5 kg of bamboo (Johnson, Schaller, and Hu 1988). One radio-tracked individual used an area of 3.8 sq km during a 9-month period, except for a 15-day foray during the spring bamboo shooting season, which increased the overall area to 6.8 sq km (Johnson, Schaller, and Hu 1988). Other plants, such as gentians, irises, crocuses, and tufted grasses, are also taken. *Ailuropoda* occasionally hunts for fish, pikas, and small rodents.

Schaller et al. (1985) studied wild individuals by radio tracking in the Wolong Natural Reserve of China. About 145 pandas occurred in this 2,000 sq km area. Within the study area, which measured 35 sq km, there were 7 adult males, 5 or 6 adult females, 4 independent subadults, and 2 infants. Home range size varied from 3.9 to 6.4 sq km, with male ranges being only as large as or slightly larger than those of females. However, males were found to occupy greatly overlapping ranges lacking well-defined core areas. They shift frequently within their ranges, show no evidence of territorial behavior, and spend considerable time within the core areas of females and subadults. Females also may have overlapping ranges, but they spend most of their time within a discrete core area of only 30–40 ha. They evidently do not tolerate other females and subadults within their core areas. As the animals move about, they mark their routes by spraying urine, clawing tree trunks, and rubbing against objects. Captives are known to scent-mark with secretions from glands in the genital region (Kleiman 1983). The vocal repertoire consists of bleats, honks, squeals, growls, moans, barks, and chirps (Peters 1982).

Except as noted, the following information on reproduction and life history was taken from Schaller et al. (1985). *Ailuropoda* is generally solitary, but during the breeding season several males may compete for access to a female. Mating generally occurs from March to May. Females have a single estrous period of 12–25 days, but peak receptivity lasts only 1–5 days. Births usually take place during August or September in a cave or hollow tree. The overall period of pregnancy is 97–163 days; the variation evidently results from a delay in implantation of 45–120 days. The number of young per litter is one or two, and occasionally three, but normally only a single cub is raised. The neonatal/maternal weight ratio may be the smallest among the eutherian mammals (Kleiman 1983). At birth the off-spring weighs 90–130 grams, is covered with sparse white fur, and has a tail that is about one-third as long as the head and body; adult coloration is attained by the end of the first month, and the eyes open after 40–60 days (Chorn and Hoffmann 1978). The young begin to walk at 3–4 months and to eat bamboo at 5–6 months. They may be fully weaned at 8–9 months, usually leave their mothers at about 18 months, and attain sexual maturity after 5.5–6.5 years. A captive specimen lived to an age of approximately 34 years (Marvin L. Jones, Zoological Society of San Diego, pers. comm., 1995).

The range of the giant panda began to decline in the late Pleistocene because of both climatic changes and the spread of people (Wang 1974). In the last 2,000 years the species has disappeared from Henan, Hubei, Hunan, Guizhou, and Yunnan provinces (Schaller et al. 1985). At present about 1,000 individuals are thought to survive, apparently divided into three isolated groups. In the mid-1970s about 100 pandas starved when an important food plant died over a large area. The species receives complete legal protection, and cooperative field investigations have been carried out by the Chinese government and the World Wildlife Fund (Schaller 1981, 1993). From 1985 to 1991, Chinese courts convicted 278 persons for panda poaching or pelt smuggling, with 16 sentenced to life imprisonment and 3 to death! Nonetheless, poaching to obtain the valuable skin continues to be a major threat (Schaller 1993). The giant panda is classified as endangered by the IUCN and the USDI and is on appendix 1 of the CITES. It is among the most popular of zoo animals but has been extremely difficult to breed. As of June 1993 there were 98 giant pandas in captivity in China and 15 in other countries (Frampton 1995*b*). A Chinese program of loaning pandas for exhibit in Western zoos became subject to intense controversy in the late 1980s, with supporting parties arguing that the fees charged would be used for panda conservation and opponents claiming that the animals should be kept together in China for breeding purposes (Drew 1989).

The West's foremost authority on the giant panda, George B. Schaller (1993), wrote that he initially had favored strictly regulated loans to foster cooperative research and conservation but had changed his mind after observing the resulting "undisciplined scramble for pandas." In response to political and financial interests, the involved countries ignored guidelines and CITES provisions designed to ensure that exportation would not be detrimental to maintenance of viable wild and captive breeding populations in China. The U.S. Fish and Wildlife Service was said to have buckled under political pressure, though more recently that agency proposed a new policy by which all imports would have to be part of an international effort to benefit panda conservation and not interfere with China's breeding and research program (Frampton 1995*b*). However, Schaller, like some other zoologists whose research has helped to publicize and stimulate the extensive capturing of an endangered species, wondered whether the giant panda would have been best left in "obscurity."

CARNIVORA; Family PROCYONIDAE

Raccoons and Relatives

This family contains 7 Recent genera and 19 species. Two subfamilies are traditionally recognized: Ailurinae, with the single genus *Ailurus* (lesser panda), found in the Himalayas and adjacent parts of eastern Asia; and Procyoninae, with the other 6 genera, which occur in temperate and tropical areas of the Western Hemisphere. Such an arrangement is supported by recent molecular and biochemical analyses (Wayne et al. 1989). However, Corbet and Hill (1991), Glatston (1994), and Roberts and Gittleman (1984) elevated the Ailurinae to full familial rank. Wozencraft (*in* Wilson and Reeder 1993) considered the Ailurinae to also include *Ailuropoda* (giant panda) and placed the group in the family Ursidae. On the basis of morphology, the remaining procyonids were divided into two subfamilies: the Procyoninae, with *Procyon, Bassariscus, Nasua,* and *Nasuella;* and the Potosinae, with *Potos* and *Bassaricyon* (Decker and Wozencraft 1991; Wozencraft *in* Wilson and Reeder 1993).

Head and body length is 305–670 mm, tail length is 200–690 mm, and weight is about 0.8–12.0 kg. Males are about one-fifth larger and heavier than females. The pelage varies from gray to rich reddish brown. Facial markings are often present, and the tail is usually ringed with light and dark bands. The face is short and broad. The ears are short, furred, erect, and rounded or pointed. The tail is prehensile in the arboreal kinkajou *(Potos)* and is used as a balancing and semiprehensile organ in the coatis *(Nasua).* Each limb bears five digits, the third being the longest. The claws are short, compressed, recurved, and in some genera semiretractile. The soles are haired in several genera. Males have a baculum.

The dental formula is usually: (i 3/3, c 1/1, pm 4/4, m 2/2) × 2 = 40. The premolars, however, number 3/3 in *Potos* and 3/4 in *Ailurus* (Stains 1984). The incisors are not specialized, the canines are elongate, the premolars are small and sharp, and the molars are broad and low-crowned. The carnassials are developed only in *Bassariscus.*

Procyonids walk on the sole of the foot with the heel touching the ground or partly on the sole and partly on the digits. The gait is usually bearlike. They are good climbers, and one genus *(Potos)* spends nearly its entire life in trees. Most procyonids shelter in hollow trees, on large branches, or in rock crevices. Most become active in the evening, but *Nasua* may be primarily diurnal. The diet is omnivorous, though *Potos* and *Bassaricyon* seem to depend largely on fruit and *Ailurus* feeds mainly on bamboo. Most genera travel in pairs or family groups and give birth in the spring.

The geological range of the Procyonidae is early Oligocene to Recent in North America, late Miocene to Recent in South America, late Eocene to early Pleistocene in Europe, and early Miocene to Recent in Asia (Stains 1984). Although *Ailurus* is now geographically far removed from the other procyonids, a related Pliocene genus, *Parailurus,* occurred in both Europe and North America (L. D. Martin 1989).

CARNIVORA; PROCYONIDAE; Genus AILURUS
F. Cuvier, 1825

Lesser Panda

The single species, *A. fulgens,* is known to occur in Nepal, Sikkim, Bhutan, northern Burma, and the provinces of Yunnan and Sichuan in south-central China and probably also exists along the border of Tibet and Assam (Ellerman and Morrison-Scott 1966; Roberts and Gittleman 1984).

Head and body length is 510–635 mm, tail length is 280–485 mm, and weight is usually 3–6 kg. The coat is long and soft and the tail is bushy. The upper parts are rusty to deep chestnut, being darkest along the middle of the back. The tail is inconspicuously ringed. Small, dark-colored eye patches are present, and the muzzle, lips, cheeks, and edges of the ears are white. The back of the ears, the limbs, and the underparts are dark reddish brown to black. The head is rather round, the ears are large and pointed, the feet have hairy soles, and the claws are semiretractile. The nonprehensile tail is about two-thirds as long as the head and body. There are glandular sacs in the anal region. Females have four mammae.

The lesser panda inhabits mountain forests and bamboo thickets at elevations of 1,800–4,000 meters. It seems to prefer colder temperatures than does the giant panda *(Ailuropoda).* It is usually said to be nocturnal and crepuscular (Roberts and Gittleman 1984) and to sleep by day in a tree.

Lesser panda *(Ailurus fulgens),* photo by Arthur Ellis, *Washington Post.*

However, Reid, Hu, and Huang (1991) found it to be much more active in daylight, especially during summer coincident with arboreal foraging; it apparently rested in direct sunlight during winter to minimize heat loss. When sleeping, it generally curls up like a cat or dog, with its tail over its head, but it may also sleep while sitting on top of a limb, its head tucked under the chest and between the forelegs, as the American raccoon *(Procyon)* does at times. Although *Ailurus* is a capable climber, it seems to do most of its feeding on the ground. The diet consists mostly of bamboo sprouts, grasses, roots, fruits, and acorns. It also occasionally takes insects, eggs, young birds, and small rodents. A female radio tracked by Johnson, Schaller, and Hu (1988) for 9 months had an average daily movement of 481 meters and a home range of 3.4 sq km. Reid, Hu, and Huang (1991) estimated two home ranges to be 0.9 and 1.1 sq km and population density to be 1/2–3 sq km.

In the wild the lesser panda sometimes travels in pairs or small family groups. Such groups probably represent a consorting male and female or a mother with cubs. The young seem to stay with the mother for about a year, or until the next litter is about to be born. The disposition of *Ailurus* is mild; when captured, it does not fight, tames readily, and is gentle, curious, and generally quiet. The usual call is a series of short whistles or squeaking notes; when provoked, it utters a sharp, spitting hiss or a series of snorts while standing on its hind legs. A musky odor is emitted from the anus when the animal is excited. According to Roberts and Gittleman (1984), *Ailurus* scent-marks its territory by urine, feces, and secretions from anal and circumanal glands. In the wild, births take place in the spring and summer, mainly in June, in a hollow tree or rock crevice.

Studies at the U.S. National Zoo in Washington, D.C. (Roberts 1975, 1980), indicate that adult males can be kept with females and young but that adult females are not tolerant of one another. Reproduction is most successful when a single adult male and female are placed together, though they will sleep and rest apart except during the breeding season. Females apparently have only one estrus annually and are then receptive for just 18–24 hours. They may begin to build a nest of sticks and leaves several weeks before giving birth. Mating occurs from mid-January to early March, and births from mid-June to late July. Recorded

gestation periods at the National Zoo are 114–45 days; however, there also is a record of only 90 days at the San Diego Zoo. It thus may be that there is delayed implantation in temperate zones but not in subtropical zones. Litters contain one to four, usually two, young. The cubs weigh about 200 grams at 1 week, open their eyes after 17–18 days, attain full adult coloration by 90 days, and take their first solid food at 125–35 days. The young are removed from the parents at 6–7 months and placed in small peer groups. Both sexes attain sexual maturity at about 18 months. According to Marvin L. Jones (Zoological Society of San Diego, pers. comm., 1995), a captive in China was still living at 17 years and 6 months.

The lesser panda is a very popular zoo animal and is frequently involved in the animal trade. It now is on appendix 1 of the CITES and is classified as endangered by the IUCN, which estimates that fewer than 2,500 mature individuals survive. There recently has been increasing evidence that the species is rare and continuing to decline because of the destruction of its forest habitat, killing for its pelt, and illegal trade in live animals (Glatston 1994). The species was found to be scarce even in a national park in Nepal; fecundity there was low, the habitat was deteriorating through intensive cattle grazing, and most young were killed as a result of human disturbance or attack by domestic dogs (Yonzon and Hunter 1991*a*, 1991*b*).

CARNIVORA; PROCYONIDAE; **Genus BASSARISCUS** *Coues, 1887*

Ringtails, or Cacomistles

There are two species (E. R. Hall 1981):

B. astutus, southwestern Oregon and eastern Kansas to Baja California andsouthern Mexico;
B. sumichrasti, southern Mexico to western Panama.

The latter species formerly was often placed in a separate genus, *Jentinkia* Trouessart, 1904.

In *B. astutus* head and body length is 305–420 mm, tail length is 310–441 mm, and shoulder height is about 160

Ringtail *(Bassariscus astutus)*, photo by Woodrow Goodpaster.

Central American cacomistle *(Bassariscus sumichrasti)*, photo by I. Poglayen-Neuwall.

mm. Armstrong, Jones, and Birney (1972) listed weights of 824–1,338 grams. The upper parts are buffy with a black or dark brown wash, and the underparts are white or white washed with buff. The eye is ringed by black or dark brown, and the head has white to pinkish buff patches. The tail is bushy, longer than the head and body, and banded with black and white for its entire length. Females have four mammae.

In *B. sumichrasti* head and body length is 380–470 mm and tail length is 390–530 mm. One individual weighed 900 grams. The color is usually buffy gray to brownish, and the tail is ringed with buff and black. From *B. astutus, B. sumichrasti* is distinguished in having pointed (rather than rounded) ears, a longer tail, naked (rather than hairy) soles, nonretractile (rather than semiretractile) claws, and low (rather than high) ridges connecting the cusps of the molariform teeth.

B. astutus utilizes a varied habitat but seems to prefer rocky, broken areas, often near water. It dens in rock crevices, hollow trees, the ruins of old Indian dwellings, and the upper parts of cabins. Activity occurs mainly at night.

B. sumichrasti is found in tropical forests and appears to be more arboreal than *B. astutus,* denning exclusively in trees and almost never coming to the ground (Poglayen-Neuwall and Poglayen-Neuwall 1995). The latter, however, is a good climber and travels quickly and agilely among cliffs and along ledges. In a study of captive *B. astutus,* Trapp (1972) determined that the hind foot can rotate at least 180°, permitting a rapid, headfirst descent and great dexterity. One individual traveled upside down along a cord 5 mm in diameter. Ringtails sometimes climb in a crevice by pressing all four feet on one wall and the back against the other. They also maneuver by ricocheting off of smooth surfaces to gain momentum to continue to an objective. The diet includes insects, rodents, birds, fruit, and other vegetable matter.

According to Poglayen-Neuwall (*in* Grzimek 1990), population density of *B. astutus* reaches 20/sq km in central California. Territories cover up to 136 ha., and those of animals of the same sex do not overlap. *Bassariscus* is generally solitary except during the mating season, and individuals may be aggressive toward one another. Nonetheless, aggregations of 5–9 *B. sumichrasti* have been observed in favored fruit trees (Poglayen-Neuwall and Poglayen-Neuwall 1995). *B. astutus* scent-marks its territory by regularly urinating at certain sites. *B. sumichrasti,* however, eliminates at random (Poglayen-Neuwall 1973). Both species have a variety of vocalizations. Adult *B. astutus* may emit an explosive bark, a piercing scream, and a long, plaintive, high-pitched call.

Female *B. sumichrasti* may enter estrus at any time but usually do so from February to June. Average length of the estrous cycle is 44 days, receptivity lasts 1 day, the gestation period is 63–66 days, and there commonly is a single young (Poglayen-Neuwall 1991, 1993; Poglayen-Neuwall and Poglayen-Neuwall 1995). *B. astutus* is seasonally monestrous, usually mating from February to May and giving birth from April to July. Heat usually does not last more than 24 hours, and the gestation period is about 51–54 days, the shortest among procyonids. The young number one to four, usually two or three, and are born in a nest or den. The female is mainly responsible for care, though the father may be tolerated in the vicinity and may play with the young as they grow older. They weigh about 25 grams at birth, open their eyes fully by 34 days, begin taking some solid food after 6–7 weeks, begin to forage with the adults at 2 months, and usually are completely weaned at 3 months. Both sexes attain sexual maturity and disperse at approximately 10 months (Poglayen-Neuwall and

Central American cacomistle *(Bassariscus sumichrasti)*, immature, photo from Jorge A. Ibarra through Museo Nacional de Historia Natural, Guatemala City.

Poglayen-Neuwall 1980, 1993). A captive *B. astutus* lived for more than 16 years (Poglayen-Neuwall *in* Grzimek 1990). A captive *B. sumichrasti* lived for 23 years and 5 months (Marvin L. Jones, Zoological Society of San Diego, pers. comm., 1995).

Ringtails, especially females obtained when young, make charming pets. They were sometimes kept about the homes of early settlers as companions and to catch mice. The coat is not of particularly high quality. It is known commercially as "California mink" or "civet cat," but there is no scientific basis for either name. In the 1976–77 trapping season 88,329 pelts of *B. astutus* were reported taken in the United States and sold for an average price of $5.50 (Deems and Pursley 1978). The harvest peaked at about 135,000 in 1978–79 but subsequently declined (Kaufmann 1987). In the 1991–92 season only 5,638 skins were taken in the United States and the average price was $3.62 (Linscombe 1994). *B. astutus* apparently extended its range into Kansas, Arkansas, and Louisiana in the twentieth century and has even been reported from Alabama and Ohio (E. R. Hall 1981). These occurrences might result partly from the habit of boarding railroad cars (Sealander 1979). *B. sumichrasti* now is classified as near threatened by the IUCN. It is completely dependent on forests, yet its entire range is in a region subject to intensive deforestation. It also is hunted for its fur and meat. The species has become rare or has disappeared over large areas (Glatston 1994).

Raccoons

Two subgenera and seven species are currently recognized (Cabrera 1957; Gardner 1976; E. R. Hall 1981; Redford and Eisenberg 1992):

subgenus *Procyon* Storr, 1780

P. lotor, southern Canada to Panama;
P. insularis, Tres Marías Islands off western Mexico;
P. maynardi, New Providence Island (Bahamas);
P. pygmaeus, Cozumel Island off northeastern Yucatan;
P. minor, Guadeloupe Island (Lesser Antilles);
P. gloveralleni, Barbados (Lesser Antilles);

subgenus *Euprocyon* Gray, 1865

P. cancrivorus (crab-eating raccoon), eastern Costa Rica to eastern Peru, northern Argentina, andUruguay.

Lotze and Anderson (1979) suggested that several of the designated species of the subgenus *Procyon* might be conspecific with *P. lotor.* It was Olson and Pregill's (1982) view that *P. maynardi* probably is not a valid species and represents an introduced population of *P. lotor.* The latter species also now is established on Grand Bahama Island (Buden 1986). Based on lack of fossil evidence, Morgan and Woods (1986) suggested that all *Procyon* in the West Indies represent human introductions.

Head and body length is 415–600 mm, tail length is 200–405 mm, shoulder height is 228–304 mm, and weight

Raccoon *(Procyon lotor)*, photo from Zoological Society of Philadelphia.

is usually 2–12 kg. Generally, males are larger than females and northern animals are larger than southern ones. Five adult males in the Florida Keys averaged 2.4 kg. Mean weights in Alabama were 4.31 kg for males and 3.67 kg for females. Means in Missouri were 6.76 kg for males and 5.94 kg for females (Johnson 1970; Lotze and Anderson 1979). In Wisconsin the normal weight range is about 6–11 kg, but there is one record of a male weighing 28.3 kg (Jackson 1961).

The general coloration is gray to almost black, sometimes with a brown or red tinge. There are 5–10 black rings on the rather well-furred tail and a black "bandit" mask across the face. The head is broad posteriorly and has a pointed muzzle. The toes are not webbed, and the claws are not retractile. The front toes are rather long and can be widely spread. The footprints resemble those of people. Females have four pairs of mammae (Banfield 1974).

Raccoons frequent timbered and brushy areas, usually near water. They are more nocturnal than diurnal and are good climbers and swimmers. The den is usually in a hollow tree, with an entrance more than 3 meters above the ground (Banfield 1974). The den may also be in a rock crevice, an overturned stump, a burrow made by another animal, or a human building. Urban (1970) found most raccoons in a marsh to den in muskrat houses. Except when sequestered during severe winter weather or in cases of females with newborn, each den is usually occupied for only one or two days. The average distance between dens has been reported as 436 meters, and the general movements of Procyon are not extensive, but one individual was found to have traveled 266 km (Lotze and Anderson 1979).

Raccoons do not hibernate. In the southern parts of their range they are active throughout the year. In northern areas they may remain in a den for much of the winter but will emerge during intervals of relatively warm weather. While they are in winter sleep their heartbeat does not decline, their body temperature stays above 35° C, and their metabolic rate remains high. They do, however, live mostly off fat reserves accumulated the previous summer and autumn and may lose up to 50 percent of their weight (Lotze and Anderson 1979). The remarkable success of P. lotor, compared with other procyonids, in occupying a broad range of habitats and climates has been attributed to its well-defined cyclic changes in fat content and thermal conductance, high capacity for evaporative cooling, high level of heat tolerance, high basal metabolic rate, extraordinarily diverse diet, and high reproductive potential (Mugaas, Seidensticker, and Mahlke-Johnson 1993).

Procyon has a well-developed sense of touch, especially in the nose and forepaws. The hands are regularly used almost as skillfully as monkeys use theirs. Food is generally picked up with the hands and then placed in the mouth. Although raccoons have sometimes been observed to dip food in water, especially under captive conditions, the legend that they actually wash their food is without foundation (Lowery 1974). The omnivorous diet consists mainly of crayfish, crabs, other arthropods, frogs, fish, nuts, seeds, acorns, and berries.

As many as 167 raccoons have been found in an area of 41 ha., but more typical population densities are 1/5–43 ha. (Lotze and Anderson 1979). Reported home range size varies from 0.2 to 4,946 ha. but seems to be typified by the situation found by Lotze (1979) on St. Catherine's Island, Georgia: he reported an annual average of 65 ha. for males and 39 ha. for females but indicated that there was much variation and that different study methods might give different results. About the smallest population density (0.5–1.0/100 ha.) and the largest home range (means of 1,139 ha. for males and 806 ha. for females) were reported

by Fritzell (1978) for the prairies of North Dakota. His study indicated that the ranges of adult males are largely exclusive of one another but do commonly overlap the ranges of 1–3 adult females and up to 4 yearlings. This and other studies (Lotze and Anderson 1979; Schneider, Mech, and Tester 1971) suggest that female ranges often are not exclusive and that territorial defense is not well developed in Procyon but that unrelated animals tend to avoid one another. Nonetheless, as many as 23 individuals have been found in the same winter den, and about the same number have congregated around artificial feeding sites (Lotze and Anderson 1979; Lowery 1974). Raccoons have a variety of vocalizations, most with little carrying power.

In the United States the reproductive season extends from December to August. Mating peaks in February and March, and births from April to June (Johnson 1970; Lotze and Anderson 1979). The breeding season of P. cancrivorus is July–September (Grzimek 1975). If a female P. lotor loses a newborn litter, she may ovulate a second time during the season (Sanderson and Nalbandov 1973). The estrous cycle has been reported to last 80–140 days (Hayssen, Van Tienhoven, and Van Tienhoven 1993). The gestation period averages 63 days and ranges from 60 to 73. The number of young per litter is one to seven, usually three or four. Captives in New York weighed 71 grams at birth. The eyes open after about 3 weeks (Banfield 1974), and weaning takes place at from 7 weeks to 4 months (Lotze and Anderson 1979). In Minnesota, Schneider, Mech, and Tester (1971) found that the young were kept in a den in a hollow tree until they were 7–9 weeks old and then were moved to one or a series of ground beds. At 10–11 weeks they were taken on short trips by the mother, and after another week the family began to move together. In November the members denned either together in one hollow tree or individually in nearby trees. The young usually separate from the mother at the end of winter. They may attain sexual maturity at about 1 year, but most do not mate until the following year. Few wild raccoons live more than 5 years, but some are estimated to have survived for 13–16 years (Lotze and Anderson 1979). One captive was still living after 20 years and 7 months (Jones 1982).

Raccoons sometimes damage corn and other crops but usually not to a serious extent (Jackson 1961). They make good pets and are interesting to observe in the wild; however, they carry pathogens known to cause such human diseases as leptospirosis, tularemia, and rabies. In 1992, 4,311 rabid P. lotor were reported in the United States, considerably more than any other animal, and there were expanding epizootics in the northeast and southeast (Krebs, Strine, and Childs 1993. This same species is currently the most valuable wild fur bearer in the United States, though prices have fallen since the peak in the late 1970s. The harvest in 44 states during the 1976–77 season was 3,832,802 skins, which sold at an average price of $26.00 (Deems and Pursley 1978). In the 1983–84 season the take in the United States and Canada was 3,410,548 pelts with an average value of $5.54 (Novak, Obbard, et al. 1987). During the 1991–92 season, 1,417,198 skins were taken in the United States and sold for an average price of $5.82 (Linscombe 1994). Because of its commercial value, P. lotor was introduced in France, the Netherlands, Germany, and various parts of the old Soviet Union, but now it is sometimes considered a nuisance in those areas (Corbet 1978; Glatston 1994). P. lotor seems to have extended its range and increased in numbers in certain parts of North America since the nineteenth century (Lotze and Anderson 1979).

In contrast, the IUCN now classifies the insular species P. insularis, P. maynardi, and P. minor as endangered. According to Glatston (1994), little is known about those three, but all are apparently rare and subject to hunting and habitat disruption. *P. gloveralleni*, the last living specimen of which was seen in 1964, is classed as extinct by the IUCN. Wilson (1991) indicated that the subspecies *P. insularis vicinus* of María Magdalena Island in the Tres Marías also may be extinct. Navarro L. and Suarez (1989) observed that *P. pygmaeus* is being killed for alleged depredations on fruit and also is jeopardized by construction of tourist facilities; it now is designated as endangered by the IUCN.

CARNIVORA; PROCYONIDAE; Genus NASUA
Storr, 1780

Coatis, or Coatimundis

There are two species (Decker 1991):

N. narica, Arizona to Gulf of Uraba in northwestern Colombia;

N. nasua, Colombia to northern Argentina and Uruguay.

E. R. Hall (1981) and many other authorities have recognized another species, *N. nelsoni*, on Cozumel Island off northeastern Yucatan, but Decker (1991) considered it a subspecies of *N. narica*. Glatston (1994) accepted *N. nelsoni* as a species but noted suggestions that it had been introduced to Cozumel by the Mayans.

Head and body length is 410–670 mm, tail length is 320–690 mm, and shoulder height is up to 305 mm. Poglayen-Neuwall (*in* Grzimek 1990) gave the weight as 3.5–6.0 kg. Males are usually larger than females. *N. narica nelsoni* has short, fairly soft, silky hair, but in other populations of both species the fur is longer and somewhat harsh. *Nelsoni* also is relatively small, but its size has been found to overlap that of mainland populations. The two recognized species can be distinguished consistently by a suite of cranial characters, the most easily discernible of which is that the palate is flat in *N. nasua* and depressed along the midline in *N. narica*. In both species the general color is reddish brown to black above and yellowish to dark brown below. Usually the muzzle, chin, and throat are whitish and the feet are blackish. Black and gray markings are present on the face, and the tail is banded. The muzzle is long and pointed, and the tip is very mobile. The forelegs are short, the hind legs are long, and the tapering tail is longer than the head and body.

Coatis are found mainly in wooded areas. They forage in trees as well as on the ground using the tail as a balancing and semiprehensile organ. While moving along the ground the animals usually carry the tail erect, except for the curled tip. The long, highly mobile snout is well adapted for investigating crevices and holes. Adult males are often active at night, but coatis are primarily diurnal. They move about 1,500–2,000 meters a day in their search for food and usually retire to a roost tree at night (Kaufmann 1962). The diet includes both plant and animal matter. When fruit is abundant coatis are almost exclusively frugivorous. At other times females and young forage for invertebrates on the

Coatimundi *(Nasua nasua)*, photo from New York Zoological Society.

forest floor and adult males tend to prey on large rodents (Smythe 1970).

Reported population densities are 26–42/100 ha. on Barro Colorado Island in the Panama Canal Zone, 15–20/ha. in Guatemala, and 1.2–2.0/100 ha. in Arizona (Gompper 1995; Lanning 1976). Home ranges of 4 solitary males in Arizona were 70–270 ha. each (Kaufmann, Lanning, and Poole 1976). On Barro Colorado Island groups had overlapping, undefended home ranges of 35–45 ha. Each group, however, spent about 80 percent of its time in an exclusive core area within its home range (Kaufmann 1962).

The social system of *N. narica*, as studied on Barro Colorado Island, is an interesting example of the interrelationship of ecology and behavior (Kaufmann 1962; Russell 1981; Smythe 1970). All females, as well as males up to two years old, are found in loosely organized bands usually comprising 4–20 individuals. Males more than two years old commonly become solitary except during the breeding season. They are usually excluded from membership in the bands by the collective aggression of the adult females, sometimes supported by the juveniles. Several cases of adult males associating with female bands outside of the breeding season and also of two males in nonagonistic association were reported by Gompper and Krinsley (1992). In the breeding season an adult male is accepted into each of the female groups, but he is completely subordinate to the females. The breeding season corresponds with the period of maximum abundance of fruit. At this time there is thus a minimum of competition for food between the large males and the other animals. Moreover, at other times of the year, when the males become carnivorous, they may attempt to prey on young coatis and thus threaten the survival of the group. According to Gompper (1995), *N. narica* is highly vocal, with a repertoire of sounds associated with aggression, appeasement, alarm, and contact maintenance.

In the populations studied mating took place synchronously once a year during a period of 2–4 weeks. Births occurred in Arizona in June and in Panama in April or early May, at the beginning of the rains. The gestation period lasts 10–11 weeks. Pregnant females separate from the group and construct a tree nest, where they give birth to a litter of two to seven young. When the young are 5 weeks old, they leave the nest, and they and the mother join the group. The young weigh 100–180 grams at birth, open their eyes after 11 days, are weaned at 4 months, reach adult size at 15 months, and attain sexual maturity at 2 years (Gompper 1995; Kaufmann 1962; Poglayen-Neuwall *in* Grzimek 1990). A captive coati was still living after 17 years and 8 months (Jones 1982).

Coatis seem to have extended their range northward in the twentieth century. Numbers reached a peak in Arizona in the late 1950s, crashed in the early 1960s, and then began a slow recovery (Kaufmann, Lanning, and Poole 1976). According to Glatston (1994), however, *N. narica* now is uncommon in the United States and is subject to year-round hunting in Arizona, where the only substantial populations remain. Moreover, suitable habitat is being disrupted throughout its range, especially in northern Mexico, and there is a possibility that the U.S. populations will become isolated from those farther south. *N. narica nelsoni*, which is restricted to rapidly developing Cozumel Island, is classified as endangered by the IUCN. Coatis rarely damage crops and only infrequently take chickens. They are hunted for their meat by natives, who sometimes have dogs trained for this purpose. If cornered by a dog on the ground, a coati can inflict serious wounds with its large, sharp canine teeth. Coatis can be tamed and make interesting and inquisitive pets, but collection for such purposes can be damaging to wild populations.

CARNIVORA; PROCYONIDAE; Genus NASUELLA
Hollister, 1915

Mountain Coati

The single species, *N. olivacea*, is known from the Andes of western Venezuela, Colombia, and Ecuador (Cabrera 1957). Glatston (1994) indicated that the species also may occur in extreme northern Peru and noted that some authorities include it in the genus *Nasua*, to which it originally was assigned.

Nasuella resembles *Nasua* but is usually smaller and has a shorter tail. A specimen of a male from Colombia in the U.S. National Museum of Natural History has a head and body length of 383 mm and a tail length of 242 mm. Respective measurements for a female are 394 mm and 201 mm (John Miles, U.S. National Museum of Natural History, pers. comm., 1980). The general color is grayish sooty brown. The tail is ringed with alternating yellowish gray and dark brown bands. The skull of *Nasuella* is smaller and more slender than that of *Nasua*, the middle part of the facial portion is greatly constricted laterally, and the palate extends farther posteriorly.

Handley (1976) collected seven specimens in Venezuela at elevations of 2,000–3,020 meters. All were taken on the ground, four at dry and three at moist sites, and four were taken in cloud forest and three in paramo. Like *Nasua*, *Nasuella* probably feeds on insects, small vertebrates, and fruit. Glatston (1994) observed that the genus appears to be uncommon and that its upland forest habitat is being converted to agriculture and pine plantations.

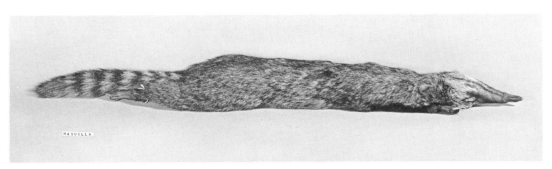

Mountain coati *(Nasuella olivacea)*, photo by Howard E. Uible of skin in U.S. National Museum of Natural History.

Kinkajou *(Potos flavus)*, photo from New York Zoological Society.

CARNIVORA; PROCYONIDAE; **Genus POTOS**
E. Geoffroy St.-Hilaire and G. Cuvier, 1795

Kinkajou

The single species, *P. flavus,* is found from southern Tamaulipas in eastern Mexico to the Mato Grosso of central Brazil (Cabrera 1957; E. R. Hall 1981). Confusion sometimes results from the application of the vernacular name "potto" to this species because the same term is used for the African primate *Perodicticus.*

Head and body length is 405 to about 760 mm, tail length is 392–570 mm, shoulder height is as much as 254 mm, and weight is 1.4–4.6 kg. Males are generally larger than females (Kortlucke 1973). The upper parts and upper surface of the tail are tawny olive, yellow tawny, or brownish; some individuals have a black middorsal line. The underparts and undersurface of the tail are tawny yellow, buff, or brownish yellow, and the muzzle is dark brown to blackish. The hair is soft and woolly.

The kinkajou has a rounded head, a short face, a long, prehensile tail, and short, sharp claws. The hind feet are longer than the forefeet. The tongue is narrow and greatly extensible. *Potos* is similar to *Bassaricyon* but differs in its round, tapering, short-haired, prehensile tail; its stockier body form; and its face, which is not grayish. Females have two mammae (Grzimek 1975).

The kinkajou inhabits forests and is almost entirely arboreal. It spends the day in a hollow tree, sometimes emerging on hot, humid days to lie out on a limb or in a tangle of vines. By night it forages among the branches. Although it moves rapidly through a single tree, progress from one tree to another is made cautiously and relatively slowly. An individual probably returns to the same trees night after night. The long tongue of *Potos* is an adaptation for a frugivorous diet. It eats mainly fruit but also takes honey, flowers, insects, and small vertebrates (Husson 1978).

Population densities of 12.5–74.0/sq km have been reported in various parts of the range of *Potos,* and an individual home range of 8 ha. was estimated in Veracruz (Ford and Hoffmann 1988). Studies in French Guiana (Julien-Laferriere 1993) indicated that a male moved an average of 2,540 meters per night and a female averaged 1,495 meters; respective estimated home range sizes were 26.6–39.5 ha. and 15.7–17.6 ha. Individuals were considered to be territorial and solitary except during the mating season. Male territories evidently overlapped those of several females, and both sexes defended their areas against other animals of the same sex. Scent marking probably is used in territorial behavior or for intersexual communication. Small groups may form, but usually only temporarily in a fruit-bearing tree. *Potos* barks when disturbed and emits a variety of other vocalizations, but the usual call, given while feeding during the night, seems to be "a rather shrill, quavering scream that may be heard for nearly a mile" (Dalquest 1953:182).

According to Poglayen-Neuwall (*in* Grzimek 1990), females enter estrus every three months, but Husson (1978) wrote that in Surinam births reportedly take place in April and May. Hayssen, Van Tienhoven, and Van Tienhoven (1993) listed an estrous cycle range of 46–92 days and an estrus range of 12–24 days. The gestation period is 112–20 days. The number of young is usually one, rarely two. Born in a hollow tree, they weigh 150–200 grams at birth, open their eyes at 7–19 days, and begin to take solid food, and can hang by the tail, at 7 weeks. Sexual maturity comes at 1.5 years for males and 2.25 years for females. One pair lived together for 9 years and had their first litter at 12.5 years of age. One individual lived to an age of 32 years at the Bronx Zoo (Marvin L. Jones, Zoological Society of San Diego, pers. comm., 1995).

If captured when young and treated kindly, the kinkajou becomes a good pet. It is often sold under the name "honey bear." Its meat is said to be excellent (Husson 1978), and its pelt is used in making wallets and belts. It does little damage to cultivated fruit.

CARNIVORA; PROCYONIDAE; Genus BASSARICYON
J. A. Allen, 1876

Olingos

Five species are currently recognized (Cabrera 1957; Glatston 1994; E. R. Hall 1981; Handley 1976):

B. gabbii, Nicaragua to Ecuador and Venezuela;
B. pauli, known only from the type locality in western Panama;

B. lasius, known only from the type locality in central Costa Rica;
B. beddardi, Guyana and possibly adjacent parts of Venezuela and Brazil;
B. alleni, Ecuador, Peru, western Bolivia, possibly Venezuela.

All of the above species may be no more than subspecies of *B. gabbii* (Cabrera 1957; Grimwood 1969; E. R. Hall 1981).

Head and body length is 350–475 mm and tail length is 400–480 mm. Grzimek (1975) gave the weight as 970–1,500 grams. The fur is thick and soft. The upper parts are pinkish buff to golden mixed with black or grayish above, and the underparts are pale yellowish. The tail is somewhat flattened and more or less distinctly annulated along the median portion. The general body form is elongate, the head is flattened, the snout is pointed, and the ears are small and rounded. The limbs are short, the soles are partly furred, and the claws are sharply curved. The tail of *Bassaricyon,* unlike that of *Potos,* is long-haired and not prehensile. Females have a single pair of inguinal mammae.

Bassaricyon is found in tropical forests from sea level to 2,000 meters. It is primarily arboreal and nocturnal. It is thought to spend the day in a nest of dry leaves in a hollow tree. According to Grzimek (1975), it is an excellent climber and jumper and can leap 3 meters, from limb to limb, with-

Olingo *(Bassaricyon gabbii),* photo from New York Zoological Society. Inset: olingo *(B. pauli),* photo by Hans-Jürg Kuhn.

out difficulty. It feeds mainly on fruit but also hunts for insects and warm-blooded animals. Poglayen-Neuwall (1966) found that in captivity *Bassaricyon* required considerably more meat than *Potos*.

In the wild, *Bassaricyon* lives alone or in pairs. It is sometimes found in association with *Potos* as well as with opossums and douroucoulis *(Aotus)*. It scent-marks with urine, perhaps for its own orientation or to attract the opposite sex (Poglayen-Neuwall *in* Grzimek 1990). Studies of captives by Poglayen-Neuwall (1966, 1989) found that *Bassaricyon* is less social than *Potos* and that two males could not be kept together. There was no definite breeding season, gestation lasted 73–74 days, and each of 16 births yielded a single offspring. The young weighed about 55 grams at birth, opened their eyes after 27 days, began taking solid food after 2 months, and attained sexual maturity at 21–24 months. One female remained reproductively active until 12 years and ultimately lived to be more than 25 years old.

The IUCN now classifies *B. pauli* and *B. lasius* as endangered, estimating that each numbers fewer than 250 mature individuals, and *B. gabbii* and *B. beddardi* as near threatened. Glatston (1994) stated that those species, as well as *B. alleni*, are highly dependent on intact tropical forest and do not adapt readily to disturbed areas or secondary forest. They generally are found in a region where deforestation is rampant and seem to have become rare over much of their range.

Weasels, Badgers, Skunks, and Otters

This family of 25 Recent genera and 67 species occurs in all land areas of the world except the West Indies, Madagascar, Sulawesi and islands to the east, most of the Philippines, New Guinea, Australia, New Zealand, Antarctica, and most oceanic islands. One genus, *Enhydra,* inhabits coastal waters of the North Pacific. The sequence of genera presented here follows basically that of Simpson (1945), who recognized five subfamilies: the Mustelinae (weasels), with *Mustela, Vormela, Martes, Eira, Galictis, Lyncodon, Ictonyx, Poecilictis, Poecilogale,* and *Gulo;* the Mellivorinae (honey badger), with *Mellivora;* the Melinae (badgers), with *Meles, Arctonyx, Mydaus, Taxidea,* and *Melogale;* the Mephitinae (skunks), with *Mephitis, Spilogale,* and *Conepatus;* and the Lutrinae (otters), with *Lutra, Lutrogale, Lontra, Pteronura, Aonyx,* and *Enhydra.* A cladistic analysis of 46 morphological characters (Bryant, Russell, and Fitch 1993) generally supported these basic groupings but found the Melinae to be polyphyletic, with *Melogale* and perhaps *Taxidea* showing affinity with the Mustelinae and *Mydaus* showing some approach to the Mephitinae. Wozencraft (1989*a*) pointed out that *Taxidea* differs from

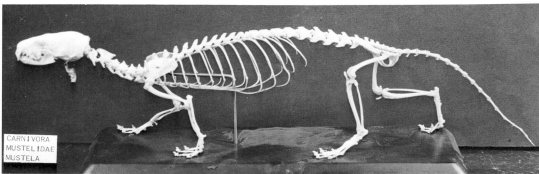

Giant otter *(Pteronura brasiliensis)*, photo by Bernhard Grzimek. Skeleton of mink *(Mustela vison)*, photo by P. F. Wright of specimen in U.S. National Museum of Natural History.

the other badgers in critical basicranial characters, and subsequently (*in* Wilson and Reeder 1993) he placed that genus in a separate subfamily, the Taxidiinae.

The smallest member of this family is the least weasel *(Mustela nivalis)*, which has a head and body length of 114–260 mm and a weight as low as 25 grams. The largest members are the otters *Pteronura* and *Enhydra,* which have a head and body length of around 1 meter or more and a weight of 22–45 kg. Male mustelids are about one-fourth larger than females. The pelage is uniformly colored, spotted, or striped. Some species of *Mustela* turn white in winter in the northern parts of their range. The body is long and slender in most genera but stocky in the wolverine *(Gulo)* and in badgers. The short ears are either rounded or pointed. The limbs are short and bear five digits each. The claws are compressed, curved, and nonretractile. The claws of badgers are large and heavy for burrowing. The digits of otters are usually webbed for swimming. Well-developed anal scent glands are present in most genera. Males have a baculum.

The skull is usually sturdy and has a short facial region. The dental formula is: (i 3/2–3, c 1/1, pm 2–4/2–4, m 1/1–2) × 2 = 28–38. *Enhydra* is the only genus with 2 lower incisors. The incisors of mustelids are not specialized, the canines are elongate, the premolars are small and sometimes reduced in number, and a constriction is usually present between the lateral and medial halves of the upper molar. The carnassials are developed. The second lower molar, if present, is reduced to a simple peg.

Mustelids are either nocturnal or diurnal and often shelter in crevices, burrows, and trees. Badgers usually dig elaborate burrows. Mustelids move about on their digits or partly on their digits and partly on their soles. The smaller slender forms usually travel by means of a scampering gait interspersed with a series of bounds. The larger, stocky forms proceed in a slow, rolling, bearlike shuffle. Mustelids often sit on their haunches to look around. Many genera are agile climbers, and otters and minks are skillful swimmers. Mustelids are mainly flesh eaters. They hunt by scent, though the senses of hearing and sight are also well developed. Some species occasionally feed on plant material, a few are omnivorous, and otters subsist mainly on aquatic animals. *Mustela, Poecilogale, Gulo,* and certain badgers reportedly store food.

Many mustelids use the secretions of their anal glands as a defensive measure. Some genera have a contrasting pattern of body colors; for example, skunks have black and white stripes. This pattern is thought to be a form of warning coloration associated with the fetid anal gland secretion and a reminder that the animal is better left alone. Some of the forms with contrasting body colors, such as the marbled polecat *(Vormela)* and certain skunks, expose and emphasize this contrasting pattern by means of bodily movements when they are alarmed.

Delayed implantation of the fertilized eggs in the uterus occurs in many genera. The actual period of developmental gestation is about 30–65 days, but with delayed implantation the total period of pregnancy is as long as 12.5 months in *Lutra canadensis.* There is usually a single litter per year. The offspring are usually tiny and blind at birth. The young of *Enhydra,* however, are born with their eyes open and in a more advanced stage than the young of other mustelids. The young of most genera can care for themselves at about 2 months and are sexually mature after a year or 2. The potential longevity in the wild is generally 5–20 years.

Mustelids sometimes kill poultry, but they also help to keep rodents in check. Many species are widely hunted for their fur.

The geological range of this family is early Oligocene to Recent in North America, Eocene to Recent in Europe, middle Oligocene to Recent in Asia; early Pliocene to Recent in Africa; and late Pliocene to Recent in South America (Stains 1984).

CARNIVORA; MUSTELIDAE; Genus **MUSTELA** *Linnaeus, 1758*

Weasels, Ermines, Stoats, Minks, Ferrets, and Polecats

There are 5 subgenera and 17 species (Alberico 1994; Coetzee *in* Meester and Setzer 1977; Corbet 1978; Corbet and Hill 1992; Ellerman and Morrison-Scott 1966; E. R. Hall 1951, 1981; Izor and de la Torre 1978; Izor and Peterson 1985; Lekagul and McNeely 1977; Schreiber et al. 1989; Van Bree and Boeadi 1978; Youngman 1982):

subgenus *Grammogale* Cabrera, 1940

M. *felipei* (Colombian weasel), known by five specimens from western Colombia and northern Ecudaor;
M. *africana* (tropical weasel), Amazon Basin of Brazil, eastern Ecuador, and northeastern Peru;

subgenus *Mustela* Linnaeus, 1758

M. *erminea* (ermine, or stoat), Scandinavia and Ireland to northeastern Siberia and the western Himalayan region, Japan, Alaska and northern Greenland to northern New Mexico and Maryland;
M. *nivalis* (least weasel), most of the Palaearctic region from western Europe and Asia Minor to northeastern Siberia and Korea, parts of China and northern Viet Nam, Great Britain, several Mediterranean islands, Japan, northwestern Africa, Egypt, Alaska, Canada, north-central conterminous United States, Appalachian region;
M. *frenata* (long-tailed weasel), southern Canada to Guyana and Bolivia;
M. *altaica* (mountain weasel), southern Siberia to the Himalayan region and Korea;
M. *kathiah* (yellow-bellied weasel), Himalayan region to southern China and northern Viet Nam, Hainan;

subgenus *Lutreola* Wagner, 1841

M. *lutreola* (European mink), France to western Siberia and the Caucasus;
M. *sibirica* (Siberian weasel), eastern European Russia to eastern Siberia and Thailand, Japan, Taiwan;
M. *lutreolina* (Indonesian mountain weasel), southern Sumatra, Java;
M. *nudipes* (Malaysian weasel), Malay Peninsula, Sumatra, Borneo;
M. *strigidorsa* (back-striped weasel), Nepal to northern Viet Nam;

subgenus *Vison* Gray, 1843

M. *vison* (American mink), Alaska, Canada, conterminous United States except parts of Southwest;
M. *macrodon* (sea mink), formerly on Atlantic coast from New Brunswick to Massachusetts;

Weasel *(Mustela* sp.), photo by Ernest P. Walker.

subgenus *Putorius* Cuvier, 1817

M. putorius (European polecat), western Europe to Ural Mountains;

M. eversmanni (steppe polecat), steppe zone from Austria to Manchuria andTibet;

M. nigripes (black-footed ferret), plains region from Alberta and Saskatchewan to northeastern Arizona and Texas.

Grammogale was considered to be a full genus by Cabrera (1957) and Stains (1967) but was given only subgeneric rank by Hall (1951) and Izor and de la Torre (1978). Although E. R. Hall (1981) also recognized *Lutreola* and *Putorius* as subgenera, he noted that some members of these two taxa closely resemble some members of the subgenus *Mustela* and that distinction from the latter may not be warranted. Corbet (1978) did not use any subgeneric designations. Anderson (1989) and Youngman (1982) recognized the five subgenera with content as given above. A morphometric analysis by Van Zyll de Jong (1992) suggested that *M. subpalmata* of Egypt is a species distinct from *M. nivalis. M. macrodon* was considered a subspecies of *M. vison* by Manville (1966) but was given specific rank by E. R. Hall (1981). Several authorities have suggested that *M. nigripes* of North America is conspecific with *M. eversmanni* of the Old World (Anderson 1973; Corbet 1978), but recent electrophoretic evidence indicates that the two are distinct species (Miller et al. 1988); late Pleistocene remains indicate that the range of each extended into Alaska and Yukon about 35,000 years ago (Youngman 1994). Both *M. erminea* and *M. putorius* have been reported, with much doubt, from northwestern Africa (Coetzee *in* Meester and Setzer 1977).

There is considerable variation in size, but *Mustela* includes the smallest species in the Mustelidae. The body is usually long, lithe, and slender. The tail is shorter than the head and body, often less than half as long. The legs are short, and the ears are small and rounded. The skull has a relatively shorter facial region than that of *Martes*. The bullae are long and inflated. Additional information is given separately for each species.

Mustela felipei (Colombian weasel)

In the type specimen, an adult male, head and body length is 217 mm and tail length is 111 mm. In a female head and body length is 225 mm, tail length is 122 mm, and weight is 138 grams (Alberico 1994). The fur is relatively long, soft, and dense. The upper parts and the entire tail are uniformly blackish brown, and the lower parts are light orange buff. All four plantar surfaces are naked, and there is extensive interdigital webbing. These features are thought to be adaptations for semiaquatic life. Most of the known specimens were collected in riparian areas at elevations of 1,750 and 2,700 meters (Izor and de la Torre 1978). However, the most recent collection was at a rugged site not near riparian habitat (Alberico 1994). The species occurs in a limited area where deforestation is rampant. Schreiber et al. (1989) noted that it probably is the rarest carnivore in South America. It now is classified as endangered by the IUCN.

Mustela africana (tropical weasel)

Head and body length is 240–380 mm and tail length is about 160–210 mm. The upper parts are reddish to chocolate; the underparts are pale and have a longitudinal median stripe of the same color as the upper parts. All four plantar surfaces are nearly naked, and there is extensive interdigital webbing. The tropical weasel has been reported mainly from humid riparian forests (Izor and de la Torre 1978) and is reported to be a good swimmer and climber. However, an observation of a group of four by Ferrari and Lopes (1992) indicates that the species may also occur on high ground and be primarily terrestrial. Schreiber et al. (1989) were aware of only 30 specimens collected in the 170 years since the discovery of the species and noted that it may be of the highest priority for conservation-related research on carnivores in South America.

Mustela erminea (ermine, or stoat)

Except as noted, the information for the account of this species was taken from Banfield (1974), Jackson (1961), C. M. King (1983), Novikov (1962), and Stroganov (1969). Head and body length is 170–325 mm, tail length is 42–120 mm, and weight is 42–365 grams. Old World animals average larger than those of North America, and males average larger than females. Except in certain southern parts of its range the ermine changes color during three- to five-week molts in April–May and October–November. In summer the back, flanks, and outer sides of the limbs are rich chocolate brown, the tip of the tail is black, and the underparts are white. In winter the coat is white except for the tip of

Ermine, or stoat *(Mustela erminea)* in winter coat, photo by Bernhard Grzimek.

the tail, which remains black. The winter pelage is longer and denser than the summer pelage. Females have eight mammae.

The ermine is found in many habitats, from open tundra to deep forest, but seems to prefer areas with vegetative or rocky cover. It makes its den in a crevice, among tree roots, in a hollow log, or in a burrow taken over from a rodent. It maintains several nests within its range, which are lined with dry vegetation or the fur and feathers of its prey. It is primarily terrestrial but climbs and swims well. It generally hunts in a zigzag pattern, progressing by a series of leaps of up to 50 cm each. It can easily run over the snow, and if pursued, it may move under the snow. It may travel 10–15 km in a night, though the average hunt covers 1.3 km. Activity takes place at any hour but is primarily nocturnal. The ermine is swift, agile, and strong and has keen senses of smell and hearing. Its slender body allows it to enter and move quickly through the burrows of its prey. It generally kills by biting at the base of the skull. It sometimes attacks animals considerably larger than itself, such as adult hares. The diet consists mainly of small rodents and also includes birds, eggs, frogs, and insects. Food may be stored underground for the winter.

Population density fluctuates with prey abundance. Under good conditions there may be an ermine for every 10 ha. Individual home range is up to about 200 ha., usually 10–40 ha., and is generally larger for males than for females. There are several vocalizations, including a loud and shrill squeaking. In a radio-tracking study in southern Sweden, Erlinge (1977) found home range sizes of 2–3 ha. for females and 8–13 ha. for males. The ranges of the males included portions of those of the females. Resident animals of both sexes maintained exclusive territories. Boundaries were regularly patrolled and scent-marked, and neighbors usually avoided one another. Adult males were dominant over females and young. Females usually spent their lives in the vicinity of their birthplace, but juvenile males wandered extensively in the spring to find a territory.

Females are polyestrous but produce only one litter per year. The estrous cycle is 4 weeks (Hayssen, Van Tienhoven, and Van Tienhoven 1993). Mating occurs in late spring or early summer, but implantation of the fertilized eggs in the uterus is delayed until around the following March and birth takes place in April or May. Pregnancy thus lasts about 10 months, but embryonic development only a little more than 1 month. Litter size is 3–18 young, averaging about 6 in the New World and 8–9 in the Old. The young weigh about 1.5–3.0 grams at birth, are blind and helpless, and are covered with fine white hair. They grow rapidly and by 8 weeks are able to hunt with the mother. Females attain sexual maturity at 2–3 months and sometimes can mate in their first summer of life. Males do not reach full size and sexual maturity until 1 year.

The ermine rarely molests poultry and is valuable to human interests because it destroys mice and rats. Its white winter fur has long been used in trimming coats and making stoles. The reported number of ermine pelts taken in eight Canadian provinces during the 1976–77 trapping season was 55,216, with an average price of $1.03 (Deems and Pursley 1978).

Mustela nivalis (least weasel)

Except as noted, the information for the account of this species was taken from Banfield (1974), Jackson (1961), and Stroganov (1969). The least weasel is the smallest carnivore, though there is considerable variation in size across its vast range. Head and body length is 114–260 mm, tail length is 17–78 mm, and weight is 25–250 grams. Old World animals average larger than those of North America, and males average larger than females. Except in certain southern parts of its range the least weasel changes color during the spring and autumn. In summer the upper parts are brown and the underparts are white. In winter the entire coat is white, though there may be a few black hairs at the tip of the tail.

The least weasel is much like *M. erminea* in details of habitat, nest construction, and movements. It is, however, even more agile and does not travel over such large areas.

A. Least weasel *(Mustela nivalis)*. B. Least weasel changing into winter coat. Photos by Ernest P. Walker.

It feeds almost entirely on small rodents and may store food for the winter. In an area of 27 ha. in England, King (1975) determined that not more than 4 adult males were resident at one time. Home range size was 7–15 ha. for males and 1–4 ha. for females. For the species as a whole Sheffield and King (1994) reported population densities of 1/1–100ha., male home ranges of 0.6–26.2 ha., and female ranges of 0.2–7.0 ha. Male ranges commonly include those of 1 or more females but do not overlap with those of other males. Both sexes scent-mark their ranges and defend them vigorously against individuals of the same sex.

According to Sheffield and King (1994), breeding may continue throughout the year but is concentrated in spring and late summer. Even in the Arctic, when small rodents are abundant *M. nivalis* breeds during winter under the snow. Females experience a postpartum estrus and can bear more than one litter annually. Delayed implantation does not occur, and the gestation period is 34–37 days. The number of young per litter averages about 5 and ranges from 3 to 10. The offspring weigh about 1.5 grams at birth, open their eyes at 26–30 days, are weaned at 42–56 days, leave their mother at 9–12 weeks, and attain adult size at 12–15 weeks. Females born in the spring are sexually mature at 3 months and may produce a litter in their first summer. Most wild individuals do not survive even 1 year, but captives have lived up to 10 years.

The least weasel is rare and of little or no commercial value. It is not known to prey on domestic animals and is beneficial to people through its destruction of mice and rats. It evidently has been introduced by human agency on Malta and Crete in the Mediterranean, as well as in the Azores, New Zealand, and Sao Tomé off west-central Africa (Corbet 1978; Sheffield and King 1994).

Mustela frenata (long-tailed weasel)

In Canada and the United States females have a head and body length of 203–28 mm, a tail length of 76–127 mm, and a weight of 85–198 grams; males have a head and body length of 228–60 mm, a tail length of 102–52 mm, and a weight of 198–340 grams (Burt and Grossenheider 1976). In Mexico the head and body length is 250–300 mm, tail length is 140–205 mm, and one male weighed 365 grams (Leopold 1959). Except as noted, the information for the remainder of this account was taken from Banfield (1974), Jackson (1961), and Lowery (1974). In Canada and the northern United States a color change occurs in the course of 25- to 30-day molts from early October to early December and from late February to late April. During the summer in these regions, and throughout the year farther south, the upper parts are brown, the underparts are ochraceous or buff, and the tip of the tail is black. During the winter in these regions the entire coat is pure white, except for

Immature long-tailed weasel *(Mustela frenata)*, photo by Ernest P. Walker.

the terminal quarter of the tail, which is black. Females have eight mammae.

The long-tailed weasel occurs in a variety of habitats but shows preference for open, brushy or grassy areas near water. It dens in a hollow log or stump, among rocks, or in a burrow taken over from a rodent. Its den is lined with the fur of its victims. It is primarily nocturnal but is frequently active by day. It can climb and swim, but apparently not as well as *M. erminea*. Its long and slender shape allows it to follow a mouse to the end of a burrow or to enter a chicken coop through a knothole. This shape also prevents curling into a spherical resting posture to conserve heat, and thus the metabolic rate of *M. frenata* (and presumably of other weasels) is 50–100 percent higher than that of "normally" shaped mammals of the same weight (Brown and Lasiewski 1972). It thus has a voracious appetite but also speed, agility, and determination. *M. frenata* seizes its prey with its claws and teeth and usually kills by a bite to the back of the neck. It may kill animals larger than itself and has even been known to attack humans who get between it and its prey. The diet, however, consists mainly of rodents and other small mammals. Although weasels are sometimes said to suck blood, this behavior has not been scientifically documented.

Population density varies from about 1/2.6 ha. to 1/260 ha., and home range from 4 to 120 ha. Home ranges may overlap, but individuals seldom meet except during the reproductive season. The voice of *M. frenata* has been reported to consist of a "trill, screech, and squeal" (Svendsen 1976).

Females are monestrous. Mating occurs in July and August, but implantation of the fertilized eggs in the uterus is delayed until around the following March. Embryonic development then proceeds for approximately 27 days until birth in April or May. The total period of pregnancy is 205–337 days. The mean number of young per litter is six, and the range is three to nine. The young weigh about 3.1 grams at birth, open their eyes after 35–37 days, and are weaned at 3.5 weeks. The father has been reported to assist in the care of the offspring. Females attain sexual maturity at 3–4 months, but males do not mate until the year following their birth.

The long-tailed weasel is more prone to raid henhouses than are other species of *Mustela* but is generally beneficial in the vicinity of poultry farms because it destroys the rats that prey on young chickens. The white winter pelage, known collectively with that of *M. erminea* as "ermine," has varied in value, with prime pelts sometimes bringing as much as $3.50 each. The reported number of pelts taken in the United States and Canada during the 1976–77 trapping season was 61,175, with an average price of about $1.00 (Deems and Pursley 1978). The reported take in the United States declined to 18,218 pelts in 1987–88 and to only 3,957 in 1991–92, but prices have held steady at around $1.00 (Linscombe 1994).

Mustela altaica (mountain weasel)

The information for the account of this species was taken from Stroganov (1969). In males head and body length is 224–87 mm, tail length is 108–45 mm, and weight is 217–350 grams. In females head and body length is 217–49 mm, tail length is 90–117 mm, and weight is 122–220 grams. *M. altaica* resembles *M. sibirica* but is smaller and has shorter fur and a less luxuriant tail. There are spring and autumn molts. The winter pelage is yellowish brown above and pale yellow below. In summer the coat is gray to grayish brown.

This weasel occurs in highland steppes and forests at elevations up to 3,500 meters. It nests in rock crevices, among tree roots, or in expropriated rodent burrows. It is quick, agile, and chiefly nocturnal or crepuscular. It feeds mainly on rodents, pikas, and small birds. In Kazakh mating occurs in February and March. The gestation period is 40 days, litters contain two to eight young, and lactation lasts 2 months. Following independence, litter mates remain together until autumn. Of little importance in the fur trade, *M. altaica* is considered to be beneficial to agricultural interests.

Mustela kathiah (yellow-bellied weasel)

The information for the account of this species was taken from Mitchell (1977). Head and body length is 250–70 mm and tail length is 125–50 mm. The tail is more than half and sometimes nearly two-thirds as long as the head and body. The upper parts are dark brownish and the underparts are deep yellow. This weasel inhabits pine forests and also occurs above the timber line. The elevational range is 1,800–4,000 meters. The diet includes birds, rodents, and other small mammals.

Mustela lutreola (European mink)

Except as noted, the information for the account of this species was taken from Novikov (1962) and Stroganov (1969). In males head and body length is 280–430 mm, tail length is 124–90 mm, and weight is up to 739 grams. In females head and body length is 320–400 mm, tail length is 120–80 mm, and weight is up to 440 grams. The general coloration is reddish brown to dark cinnamon. The underparts are somewhat paler than the back, and there may be some white on the chin, chest, and throat. The dense pelage is short even in winter.

The European mink inhabits the densely vegetated banks of creeks, rivers, and lakes. It is rarely found more than 100 meters from fresh water. It may excavate its own burrow, take one from a water vole (Arvicola), or den in a crevice, among tree roots, or in some other sheltered spot. It swims and dives well. Activity is mainly nocturnal and crepuscular. The summer is generally spent in an area of 15–20 ha., but there may be extensive autumn and winter movements to locate swift, nonfrozen streams. The chief prey is the water vole. The diet also includes other small rodents, amphibians, mollusks, crabs, and insects. Food is often stored.

In Russia reported population densities are 2–12 individuals per 10 km of shoreline and average territory size is 32 ha. for males and 26 ha. for females (Youngman 1990). Mating occurs from February to March, with births in April and May. Pregnancy lasts 35–72 days, the variation probably resulting from delayed implantation in some females. The number of young per litter is two to seven, usually four or five. The young open their eyes after 4 weeks, are weaned at 10 weeks, disperse in the autumn, and attain sexual maturity the following year. Longevity is 7–10 years.

The European mink appears to be in a precipitous decline, and it now is classified as endangered by the IUCN. Although its fur is not as valuable as that of M. vison, it has been widely trapped for commercial purposes. An annual average of 49,850 pelts was taken from 1922 to 1924 in the old Soviet Union (Youngman 1990). It also has been killed as a predator, has lost much habitat through hydroelectric developments and water pollution, and has suffered badly from competition with the introduced M. vison. Tumanov and Zverev (1986) reported that in the old Soviet Union its numbers had declined to 40,000–45,000 individuals. However, according to Maran (1992, 1994), recent surveys show that there are only about 25,000 in that region, the great majority in Russia, and that the estimated total for the world is less than 30,000. Introductions have been carried out in the Kuril Islands and Tajikistan, far from the natural range of the species. Outside of Russia, there are very small and declining populations in Georgia, Ukraine, Estonia, Belarus, Moldava, and possibly Finland and the Danube Delta of Romania. There also are about 2,000 in western France and 1,000 in northern Spain. Interestingly, the species may have spread through France as recently as the eighteenth and nineteenth centuries during a period of climatic amelioration at the same time that it was vanishing from central Europe, and it appears to have entered Spain only about 1950 (Youngman 1982).

Mustela sibirica (Siberian weasel)

Except as noted, the information for the account of this species was taken from Stroganov (1969). In males head and body length is 280–390 mm, tail length is 155–210 mm, and weight is 650–820 grams. In females head and body length is 250–305 mm, tail length is 133–64 mm, and weight is 360–430 grams. In winter the upper parts are bright ochre to straw yellow and the flanks and underparts are somewhat paler. The summer pelage is darker, shorter, coarser, and sparser. Females have four pairs of mammae.

The Siberian weasel dwells mainly in forests, especially along streams, but sometimes enters towns and cities. It dens in tree hollows, under roots or logs, between stones, in modified rodent burrows, or in buildings. It lines its nest with fur, feathers, and dried vegetation. It is swift and agile, has good senses of smell and hearing, and can climb and swim well. It is mainly nocturnal and crepuscular and has been observed to cover a distance of 8 km in one night. In the autumn it may move from upland areas to valleys. There are reports of mass migrations associated with food shortages. The diet consists mainly of small rodents but also includes pikas, birds, eggs, frogs, and fish. Food may be stored for winter use.

Several males may pursue and fight over a single female. Mating occurs in late winter and early spring, and births take place from April to June. The gestation period is 28–30 days, litter size is 2–12 young, the offspring open their eyes after 1 month, and lactation lasts 2 months. The young leave their mother by the end of August, but litter mates may travel together through the autumn. A captive lived 8 years and 10 months (Jones 1982).

The Siberian weasel is important in the fur trade. It occasionally attacks domestic fowl but is generally considered beneficial because of its destruction of noxious rodents. It has been introduced on Sakhalin and Iriomote islands (Corbet 1978).

Mustela lutreolina (Indonesian mountain weasel)

The information for the account of this species was taken from Van Bree and Boeadi (1978). Based on 6 male specimens, head and body length is 297–321 mm, tail length is 136–70 mm, and weight is 295–340 grams. The overall color is glossy dark russet, and there is no mask or other facial markings. M. lutreolina bears a striking resemblance to M. lutreola in size and color but resembles M. sibirica in characters of the skull. The 11 known specimens have been collected at elevations of 1,000–2,200 meters in the mountains of Java and southern Sumatra. M. lutreolina is a close relative of M. sibirica and sometimes is considered conspecific with the latter. Its lifestyle probably is much the same. Apparently, during a glacial phase of the Pleistocene a cooler climate and lower sea level allowed the ancestral stock of M. lutreolina to spread southward. Changing conditions subsequently isolated it in its present mountain habitat. This area now is again declining because of human disruption, and M. lutreolina is classified as endangered by the IUCN.

Mustela nudipes (Malaysian weasel)

The information for the account of this species was taken from Lekagul and McNeely (1977). Head and body length is 300–360 mm and tail length is 240–60 mm. The body coloration varies from pale grayish white to reddish brown, and the head is much paler than the body. The soles of the feet are naked around the pads. Females have two pairs of mammae. This species is not well known, but its habits are thought to be like those of other weasels. A litter of four young has been recorded.

Mustela strigidorsa (back-striped weasel)

The information for the account of this species was taken from Lekagul and McNeely (1977) and Mitchell (1977). Head and body length is 250–325 mm and tail length is 130–205 mm. The general coloration is dark brown, but the upper lips, cheeks, chin, and throat are pale yellow. There is a narrow whitish stripe down the middle of the back and another along the venter. As in *M. nudipes,* the area around the foot pads is entirely naked. Females have two pairs of mammae.

The back-striped weasel inhabits evergreen forest at elevations of 1,200–2,200 meters. One was taken in a tree hole 3–4 meters above the ground. An individual was observed to attack a bandicoot rat three times its own size. According to Schreiber et al. (1989), this species is probably rare, though it now is known from at least 31 specimens. The IUCN classifies it as vulnerable, noting that fewer than 10,000 mature individuals survive and a decline is continuing.

Mustela vison (American mink)

Except as noted, the information for the account of this species was taken from Banfield (1974), Burt and Grossenheider (1976), Jackson (1961), and Lowery (1974). In males head and body length is 330–430 mm, tail length is 158–230 mm, and weight is 681–2,310 grams. In females head and body length is 300–400 mm, tail length is 128–200 mm, and weight is 790–1,089 grams. The pelage is soft and luxurious. Its general color varies from rich brown to almost black, but the ventral surface is paler and may have some white spotting. Captive breeding has produced a number of color variants. Anal scent glands emit a strong, musky odor that some persons consider to be more obnoxious than that of skunks. Females have three pairs of mammae.

The mink is found along streams and lakes as well as in swamps and marshes. It prefers densely vegetated areas. It dens under stones or the roots of trees, in expropriated beaver or muskrat houses, or in self-excavated burrows. Such burrows may be about 3 meters long and 1 meter beneath the surface and have one or more entrances just above water level. The mink is an excellent swimmer, can dive to depths of 5–6 meters, and can swim underwater for about 30 meters. It is primarily nocturnal and crepuscular but is sometimes active by day. It normally is not wide-ranging but may travel up to about 25 km in a night during times of food shortage. The most important dietary components are small mammals, fish, frogs, and crayfish; other foods include insects, worms, and birds.

Population densities of about 1–8/sq km have been recorded. Females have home ranges of about 8–20 ha. The ranges of males are larger, sometimes up to 800 ha. Individuals are generally solitary and hostile to one another except when opposite sexes come together for breeding. Females are polyestrous but have only one litter per year. Mating occurs from February to April, with births in late April and early May. Because of a varying period of delay in implantation of the fertilized eggs, pregnancy may last from 39 to 78 days. Actual embryonic development takes 30–32 days. The number of young per litter averages 5 and ranges from 2 to 10. The young are born in a nest lined with fur, feathers, and dry vegetation. They are blind and naked at birth, open their eyes after 5 weeks, are weaned at 5–6 weeks, leave the nest and begin to hunt at 7–8 weeks, and separate from the mother in the autumn. Females reach adult weight at 4 months and sexual maturity at 12 months; males reach adult weight at 9–11 months and sexual maturity at 18 months. Potential longevity is 10 years.

Most of the mink fur used in commerce is produced on farms. The preferred breeding stock results from crossing the large Alaskan and dark Labrador forms. Selective breeding has led to development of strains that regularly yield such colors as black, white, platinum, and blue (Grzimek 1975). In Canada during the 1971–72 season 72,674 wild and 1,155,020 farm-raised mink pelts were sold (Banfield 1974). During the 1976–77 season the reported number of wild mink taken and the average price per pelt were 116,537 and $19.67 in Canada and 320,823 and $14.00 in the United States (Deems and Pursley 1978). In 1983–84 the total harvest of wild mink was 392,122, with an average value of $9.71 (Novak, Obbard, et al. 1987). In the 1991–92 season 129,106 skins were taken in the United States and sold for an average price of $18.47 (Linscombe 1994). *M. vison* was introduced deliberately in many parts of the old Soviet Union, and escaped animals have established populations in Iceland, Ireland, Great Britain, France, Spain, Portugal, Scandinavia, Germany, and Poland (Braun 1990; Corbet 1978; Romanowski 1990; Ruiz-Olmo and Palazon 1991; Smit and Van Wijngaarden 1981). The subspecies *M. v. evergladensis,* of southern Florida, appears to be rare and may be jeopardized by human water diversion projects (Smith and Cary 1982). Humphrey and Setzer (1989) have suggested that *evergladensis* actually is a disjunct population of the subspecies *M. vison mink,* which otherwise occurs no farther south than central Georgia. Populations of other subspecies are found in coastal areas of northwestern and extreme northeastern Florida.

Mustela macrodon (sea mink)

The information for the account of this extinct species was taken from Allen (1942), Campbell (1988), and Manville (1966). The sea mink was said to resemble *M. vison* but to

American mink *(Mustela vison)*, photo by Ernest P. Walker.

European polecat *(Mustela putorius)*, photo from Zoological Society of London.

be much larger, to have a coarser and more reddish fur, and to have an entirely different odor. Head and body length has been estimated at 660 mm and tail length at 254 mm. No complete specimen is known to exist, and descriptions are based only on recorded observations and numerous bone fragments and teeth found at Indian middens along the New England coast.

The sea mink reportedly made its home among the rocks along the ocean. Its den had two entrances. It is thought to have been nocturnal and solitary. The diet consisted mainly of fish and probably also included mollusks. An adult and four young estimated to be three or four weeks old were seen along a beach in August. Because of the large size of *M. macrodon*, its pelt brought a higher price than that of *M. vison* and was persistently sought. Some persons pursued the species from island to island, using dogs to locate individuals on ledges and in rock crevices and then digging or smoking them out. The sea mink apparently had been exterminated by about 1880, and its range seems subsequently to have been occupied by *M. vison*.

Mustela putorius (European polecat)

Except as noted, the information for the account of this species was taken from Blandford (1987), Grzimek (1975), Novikov (1962), and Stroganov (1969). Males have a head and body length of 295–460 mm, a tail length of 105–90 mm, and a weight of 405–1,710 grams. Females have a head and body length of 205–385 mm, a tail length of 70–140 mm, and a weight of 205–915 grams. The general coloration is dark brown to black; the underfur is pale yellow and is clearly seen through the guard hairs; and the area between the eye and the ear is silvery white.

The European polecat is most common in open forests and meadows. It dens in such places as crevices, hollow logs, and burrows made by other animals. It sometimes enters settled areas and buildings occupied by people. It is nocturnal and terrestrial but is capable of climbing. A radio-tracking study by Brzezinski, Jedrzejewski, and Jedrzejewska (1992) showed individuals to occupy stretches of about 0.65–3.05 km along a stream, to have an average daily movement of 1.1 km, and to use several dens each. The diet consists of small mammals, birds, frogs, fish, and invertebrates.

Population density has been calculated to be 1/1,000 ha., and home range about 100–150 ha. *M. putorius* is usually solitary and silent but has a variety of squeals, screams, and other sounds. Mating occurs from March to June, heat lasts 3–5 days, and the gestation period is about 42 days. The number of young per litter is 2–12, usually 3–7. The young

weigh 9–10 grams at birth, open their eyes and are weaned after about 1 month, and become independent at around 3 months. Sexual maturity may come in the first or second year of life. Longevity in the wild is usually as high as 5–6 years, but captives have lived as long as 14 years.

The domestic ferret sometimes is given the subspecific name *M. putorius furo*. It generally is thought to be a descendant of the European polecat, but Lynch (1995) indicated that *M. eversmanni* may also be involved in its ancestry. It was bred in captivity as early as the fourth century B.C. It is usually tame and playful and is used to control rodents and to drive rabbits from their burrows. It is now found in captivity in much of the world. An estimated 1 million are kept as pets in the United States, though there have been a few reported cases of rabid individuals and savage attacks on children (*Washington Post*, 5 April 1988). Unlike the wild polecat, it is generally white or pale yellow in color. According to Asdell (1964), females may have two or three litters annually. Both the domestic ferret and the polecat apparently were introduced in New Zealand, and large feral populations are now established there. These animals are trapped for their fur in New Zealand as well as in their original range. The pelt is sometimes called "fitch." The wild polecat generally has been persecuted as a predator of small domestic mammals and birds. By the early twentieth century it had been eliminated throughout Great Britain, except in Wales, but there has been some recovery in recent years. Recently, Lynch (1995) expressed concern about possible hybridization between wild and domestic polecats in Britain.

Mustela eversmanni (steppe polecat)

The information for the account of this species was taken from Stroganov (1969). Males have a head and body length of 370–562 mm, a tail length of 80–183 mm, and a weight of up to 2,050 grams. Females have a head and body length of 290–520 mm, a tail length of 70–180 mm, and a weight of up to 1,350 grams. There is much variation in color pattern, but generally the body is straw yellow or pale brown, somewhat darker above than below. There is a dark mask across the face. The chest, limbs, groin area, and terminal third of the tail are dark brown to black. Some individuals bear a striking resemblance to the North American black-footed ferret *(M. nigripes)*.

The steppe polecat is found in open grassland and semi-desert. It usually expropriates the burrow of a ground squirrel or some other animal and modifies the home for its own use. Some burrow systems, especially those of females, may be occupied for several years and become rather

Steppe polecat *(Mustela eversmanni)*, photo by Ernest P. Walker.

complex. *M. eversmanni* is quick and agile and has keen senses, especially of smell and hearing. It moves by leaps of up to 1 meter and constantly changes direction during hunts. It is nocturnal and has been known to cover up to 18 km in the course of a winter night. Local migrations may occur in response to extreme snow depth or food shortage. The diet consists mainly of pikas, voles, marmots, hamsters, and other rodents. Food is sometimes stored for later use.

Mating usually occurs from February to March, with births from April to May. If a litter is lost, however, the female may produce a second later in the year. The gestation period is 38–41 days. The average number of young per litter is about 8–10 and the range is 4–18. The young weigh 4–6 grams at birth, open their eyes after 1 month, are weaned and start hunting with the mother at 1.5 months, disperse at 3 months, and attain sexual maturity at 9 months.

The steppe polecat is considered to be beneficial to agriculture because of its destruction of rodents. Its fur has commercial importance but is not as valuable as that of *M. putorius*. The subspecies *M. e. amurensis,* of southeastern Siberia and Manchuria, is classified as vulnerable by the IUCN because of excessive human exploitation and habitat disruption.

Mustela nigripes (black-footed ferret)

Head and body length is 380–500 mm and tail length is 114–50 mm. The linear measurements of males are about 10 percent greater than those of females. Two males weighed 964 and 1,078 grams, and two females weighed 764 and 854 grams. The body color is generally yellow buff, being palest on the underparts. The forehead, muzzle, and throat are nearly white. The top of the head and middle of the back are brown. The face mask, feet, and terminal fourth of the tail are black. Females have three pairs of mammae (Burt and Grossenheider 1976; Henderson, Springer, and Adrian 1969; Hillman and Clark 1980).

The black-footed ferret is found mainly on short and midgrass prairies. It is closely associated with prairie dogs *(Cynomys)* and utilizes their burrows for shelter and travel; it may modify such burrows for its own use. It is primarily nocturnal and is thought to have keen senses of hearing, smell, and sight. It depends largely on prairie dogs for food, but captives have readily accepted other small

Black-footed ferret *(Mustela nigripes)*, photo by Luther C. Goldman.

mammals. Studies of a recently discovered population near Meeteetse, Wyoming (Richardson et al. 1987), indicate that the ferret confines its activity primarily to prairie dog colonies and seldom moves from one colony to another. Average nightly movement during winter was 1,406 meters, and areas of activity for periods of 3–8 nights were 0.4–98.1 ha. Some food was stored in many small caches.

Studies in South Dakota (Hillman, Linder, and Dahlgren 1979) indicate that *M. nigripes* kills only enough to eat. A prairie dog town of 14 ha. was occupied by a single ferret for six months, but the prey population was not severely reduced. Females raising litters require relatively large prairie dog towns. Whereas overall average town size was found to be 8 ha., the average town size occupied by females with young was 36 (10–120) ha. The mean distance between a town and the nearest neighboring town was 2.4 km. The mean distance between two towns occupied by ferrets was 5.4 km. Investigation of the wild population near Meeteetse, Wyoming (Clark 1987*b*), found a density of about 1 ferret per 50 ha. of prairie dog colonies. This population consisted of 67 percent juveniles and 33 percent adults each August.

M. nigripes is solitary except during the breeding season, and males apparently do not assist in the rearing of young (Henderson, Springer, and Adrian 1969; Hillman and Clark 1980). Individuals kept in a captive colony in Wyoming from 1986 to 1990 mated mostly in March and April and gave birth in May and June. Estrus lasted 32–42 days, the gestation period was 42–45 days, and litter size averaged 3.0 kits and ranged from 1 to 6 (Williams et al. 1991). In the wild the number of young per litter has been found to average 3.3. The young emerge from the burrow in early July and separate from the mother in September or early October. Young males then disperse for a considerable distance, but young females often remain in the vicinity of their mother's territory. Sexual maturity is attained by 1 year (Forrest et al. 1988; Hillman and Clark 1980; Miller et al. 1988). Captives are estimated to have lived up to about 12 years (Hillman and Carpenter 1980).

The black-footed ferret may once have been common on the Great Plains, but its subterranean and nocturnal habits made it difficult to locate and observe. According to Clark (1976), there had been about 1,000 reports of the species since 1851, and there are about 100 specimens in museums. In 1920 ferret numbers were estimated at more than 500,000 (Clark 1987*a*). During the twentieth century *M. nigripes* apparently declined in association with the extermination of prairie dogs by human agency. In Kansas, for example, the area occupied by prairie dog towns has been reduced by 98.6 percent since 1905. The ferret became so rare that some persons considered it to be extinct. Reports continued in several states, however, and in 1964 a population was discovered in South Dakota. This group was regularly observed and studied until 1974, when confirmed records ceased. Several captives were taken, but eventually all died without leaving surviving offspring. There again was fear that the species was extinct, though Clark (1978) gathered a number of reliable reports in Wyoming, and Boggess, Henderson, and Choate (1980) found the skull of a ferret that may have died as recently as 1977 in Kansas.

Then on 26 September 1981 a ferret was found that had been killed by dogs on a ranch near Meeteetse, Wyoming. Agents of the U.S. Fish and Wildlife Service quickly live-captured, radio-collared, and released another individual in the same area and found evidence of several additional animals. It soon was realized that a substantial ferret population, indeed the largest ever scientifically observed, was present. Intensive field studies were initiated, and by July 1984 the population was estimated to contain 129 individ-

uals. The following year, however, the number of ferrets began to fall, apparently as a result of a decline in their major prey, white-tailed prairie dogs, caused by sylvatic plague. Amidst efforts to control the plague and live-capture a portion of the remaining ferrets, canine distemper somehow was introduced into the wild ferret population, which promptly collapsed (the first 6 ferrets to be caught also died of the deadly disease). From late 1985 to early 1987 all of the ferrets known to survive, a total of 18 individuals, were brought into captivity and used as the basis of a breeding program. By 1989 there were 58 ferrets in a captive facility operated by the Wyoming Game and Fish Department and another 12 in the National Zoological Park at Front Royal, Virginia (Bender 1988; Clark 1987*b*; Collins 1989; Forrest et al. 1988; Miller et al. 1988; Solt 1981). There has been considerable controversy regarding the manner in which the ferret study and conservation program was handled, especially with respect to the length of time taken to establish a proper captive breeding program (Carr 1986; Clark 1987*a*, 1987*b*; May 1986; T. Williams 1986).

By the start of the 1991 breeding season there were 180 individuals in captivity, and during the season about 150 surviving kits were produced. It was then decided that the population was large enough to sustain removal of some animals for reintroduction. In the autumn of 1991, 49 ferrets were released in the Shirley Basin of southeastern Wyoming; 91 more were released there in 1992, and another 48 in 1993. Most of these animals are thought to have died, but some are known to have survived and reproduced; at least six litters were born in the wild in 1993 (Russell et al. 1994; Thorne and Russell *in* Bowdoin et al. 1994). By 1996 there were about 400 individuals in captivity at seven separate facilities, and the U.S. Fish and Wildlife Service had carried out additional reintroductions in north-central Montana, southwestern South Dakota, and northwestern Arizona (Christopherson and Torbit 1993; Reading et al. 1996; Searls and Torbit 1993; Wada 1995). At present there are no known nonintroduced wild populations, though reports continue to come from various areas, and two skulls were found in Montana in 1984 that evidently were from animals that had died not more than 10 years previously (Clark, Anderson, et al. 1987). The species is thought to be present in Canada, beyond the range of prairie dogs but in areas of native grasslands with high densities of ground squirrels (Laing and Holroyd 1989). *M. nigripes* is classified as extinct in the wild by the IUCN (because there has been no confirmation of a surviving nonintroduced population) and as endangered by the USDI and is on appendix 1 of the CITES. Prospects for the long-term survival of the species in the context of human manipulation and restricted size and genetic viability is covered in detail in Seal et al. (1989).

CARNIVORA; MUSTELIDAE; **Genus VORMELA**
Blasius, 1884

Marbled Polecat

The single species, *V. peregusna,* occurs in the steppe and subdesert zones from the Balkans and Palestine to Inner Mongolia and Pakistan (Corbet 1978).

Head and body length is 290–380 mm, tail length is 150–218 mm, and weight is 370–715 grams (Grzimek 1975; Stroganov 1969). *Vormela* resembles *Mustela putorius* but differs in its broken and mottled color pattern on the upper parts and in its long claws. The mottling on the back is red-

Marbled polecat *(Vormela peregusna)*, photo by Bernhard Grzimek.

dish brown and white or yellowish, and the tail is usually whitish with a dark tip. The underparts are dark brown or blackish, and the facial mask is dark brown. Females have five pairs of mammae (Roberts 1977).

Like most mustelids, *Vormela* possesses anal scent glands, from which a noxious-smelling substance is emitted. When this animal is threatened, it throws its head back, bares its teeth, erects its body hairs, and bristles and curls its tail over its back. This behavior results in the fullest display of the contrasting body colors, and the pattern thus exposed is thought to be a warning associated with the fetid anal gland secretion.

The marbled polecat seems to prefer steppes and foothills. With its strong paws and long claws it excavates deep, roomy burrows. It may also shelter in the burrows of other animals. It is chiefly nocturnal and crepuscular but is sometimes active by day. A radio-tracking study (Ben-David, Hellwing, and Mendelssohn 1988*a*) determined that *Vormela* moved up to 1 km per night, seldom used the same path repeatedly, and changed its den and activity area every 2–3 days. It is a good climber but feeds mainly on the ground. It preys on rodents, birds, reptiles, and other animals.

Vormela is solitary except during the breeding season. Investigations in Israel found each animal to occupy a home range of 0.5–0.6 sq km; there was some overlap of ranges and there were some encounters between animals, but each foraged and rested alone (Ben-David, Hellwing, and Mendelssohn 1988*a*). Observations in captivity suggest that males form a dominance hierarchy. In March, just prior to the mating season, all males that were housed alone underwent a conspicuous color change, the yellow spots of the fur changing to bright orange patches. However, of the males that were housed in groups only dominant individuals changed color (Ben-David, Mendelssohn, and Hellwing 1989).

The mating season in Israel extends from mid-April to early June, and some wild males undergo the color change described above at that time. Females kept in captivity gave birth 8–11 months after mating, meaning that delayed implantation is involved. Litter size was 1–8 young. They did not open their eyes until they were 40 days old, but they began eating solid food at 30 days. Females attained adult size and sexual maturity at only 3 months of age, but males continued to grow for another 2 months and reached sexual maturity at 1 year (Ben-David, Hellwing, and Mendelssohn 1988*b*). Only the mother cares for the young, which are reared in a nest of grass and leaves within a burrow. A captive specimen was still living after 8 years and 11 months (Jones 1982).

The fur of the marbled polecat has been sought at certain times but is not of major commercial importance (Grzimek 1975; Stroganov 1969). *Vormela* sometimes preys on poultry and has been eliminated in parts of its range. The subspecies *V. p. peregusna*, of Europe and Asia Minor, is classified as vulnerable by the IUCN and Russia. Schreiber et al. (1989) suggested that the major decline of this subspecies on the steppes of the Balkans and Ukraine, like the disappearance of the North American *Mustela nigripes*, has resulted from usurpation of grassland habitat by agriculture and the elimination of rodent prey.

CARNIVORA; MUSTELIDAE; **Genus MARTES**
Pinel, 1792

Martens, Fisher, and Sable

There are three subgenera and eight living species (Anderson 1970; Chasen 1940; Cholley 1982; Corbet 1978; Ellerman and Morrison-Scott 1966; Graham and Graham 1994; E. R. Hall 1981; Hsu and Wu 1981; Lekagul and McNeely 1977; Pilgrim 1980; Schreiber et al. 1989):

subgenus *Martes* Pinel, 1792

M. foina (beech marten, or stone marten), Denmark and
   Spain to Mongolia and the Himalayas, Crete, Rhodes,
   Corfu;
M. martes (European pine marten), western Europe to
   western Siberia and the Caucasus, Ireland, Great
   Britain, Balearic Islands, Corsica, Sardinia, Elba, Sicily;
M. zibellina (sable), originally the entire taiga zone from
   Scandinavia to eastern Siberia and North Korea,
   Sakhalin, Hokkaido;
M. melampus (Japanese marten), South Korea, Honshu,
   Kyushu, Shikoku, Tsushima;
M. americana (American pine marten), Alaska to
   Newfoundland, south in mountainous areas to central
   California and northern New Mexico, Great Lakes
   region, New England, historically as far south as Iowa
   and southern Ohio;

subgenus *Pekania* Gray, 1865

M. pennanti (fisher, or pekan), southern Yukon to
   Labrador, south in mountainous areas to central
   California and Utah, Great Lakes region, New England,
   Appalachian region, historically as far south as
   northeastern Alabama;

subgenus *Charronia* Gray, 1865

M. flavigula (yellow-throated marten), southeastern
   Siberia to Malay Peninsula, Himalayan region, Hainan,
   Taiwan, Sumatra, Bangka Island, Java, Borneo;
M. gwatkinsi (Nilgiri marten), Nilgiri Hills of extreme
   southern India.

An additional species, *M. nobilis* (noble marten), has been described from subfossil and fossil specimens from the Yukon, Idaho, Wyoming, Colorado, Nevada, and California and may have survived until about 1,200 years ago (Graham and Graham 1994). Youngman and Schueler (1991) concluded that *M. nobilis* is a synonym of *M. americana*, but Anderson (1994) questioned that view and treated *nobilis* as a subspecies of *M. americana*. In any event, *nobilis* was a large form, related to modern *M. americana* and evidently adapted to a wider variety of environments; the reason for its extinction is unknown but could have involved human impacts (Graham and Graham 1994; Grayson 1984). Anderson (1970, 1994) stated that additional study might show that *M. martes*, *M. zibellina*, *M. melampus*, and *M. americana* are conspecific. Corbet (1978) suggested that *M. gwatkinsi* is conspecific with *M. flavigula*, and Corbet (1984) cited information suggesting that *M. latinorum* of Sardinia is a species distinct from *M. martes*.

From *Mustela*, *Martes* is generally distinguished by a larger and heavier body, a longer and more pointed nose, larger ears, longer limbs, and a bushier tail. *Martes* has typical mustelid anal sacs, but they are poorly developed compared with those of other genera (Buskirk 1994). Additional information is provided separately for each species.

Martes foina (beech marten, or stone marten)

Except as noted, the information for the account of this species was taken from Anderson (1970), Grzimek (1975), Stroganov (1969), and Waechter (1975). Head and body length is about 400–540 mm, tail length is 220–300 mm, and weight is 1.1–2.3 kg. The general coloration is pale grayish brown to dark brown, and most specimens have a prominent white or pale yellow neck patch. The fur is coarser than that of *M. martes*, and the tail is relatively longer. The soles are covered with sparse hairs, through which the pads stand out markedly.

The stone marten is less dependent on forests than *M. martes* and prefers rocky and open areas. It is found in mountains at elevations of up to 4,000 meters. It often enters towns and may occupy buildings. Natural nest sites include rocky crevices, stone heaps, abandoned burrows of other animals, and hollow trees. *M. foina* is a good climber but rarely goes high in trees. It is nocturnal and crepuscular. The diet consists of rodents, birds, eggs, and berries. Vegetable matter forms a major part of the summer food in some areas.

American pine marten *(Martes americana)*, photo by Howard E. Uible.

A radio-tracking study in western Germany determined seasonal home range size to vary widely, from 12 to 211 ha. In general, ranges were largest during summer and smallest in winter, the smallest ranges were in areas of highest-quality habitat, male ranges were significantly larger than those of females, and adult ranges were larger than those of immature animals, especially during the mating season (Herrmann 1994). Mating occurs in midsummer, but because of delayed implantation of the fertilized eggs in the uterus, births do not occur until the following spring. The total period of pregnancy is 230–75 days, though only about a month is true gestation. Litters usually contain three to four young but may have as many as eight. Canivenc et al. (1981) reported that in southwestern France, where births take place in late March and early April, lactation lasts until mid-May. Sexual maturity has been reported to come at 15–27 months for both sexes (Mead 1994). A captive was still living after 18 years and 1 month (Jones 1982).

The stone marten is considered common in most of Eurasia. Its pelt is hunted but never reaches the quality of that of *M. martes* (Kruska *in* Grzimek 1990). A distinctive but undescribed form of *M. foina* found on Ibiza in the Balearic Islands apparently had been hunted to extinction by about 1960 (Clevenger 1993a, 1993b; Schreiber et al. 1989). An apparently breeding population of *M. foina* has been established in southeastern Wisconsin for at least 25 years (C. A. Long 1995).

Martes martes (European pine marten)

Except as noted, the information for the account of this species was taken from Grzimek (1975) and Stroganov (1969). Head and body length is 450–580 mm, tail length is 160–280 mm, and weight is 800–1,800 grams. The general coloration is chestnut to gray brown. There is a light yellow patch on the chest and lower neck. The winter pelage is luxuriant and silky, the summer pelage shorter and coarser. The paws are covered by dense hair. Females have four mammae.

The pine marten dwells in forests, both coniferous and deciduous. It is better adapted than *M. zibellina* for an arboreal life. An individual has several nests, located preferably in hollow trees. Activity is mainly nocturnal, and 20–30 km may be covered in a night's hunting. The diet consists of murids, sciurids, other small mammals, honey, fruit, and berries. There are several food storage sites within the home range, which may be 5 km in diameter.

Average home range size is 23 sq km for males and 6.5 sq km for females; there is little or no overlap between the ranges of individuals of the same sex, but the ranges of males greatly overlap those of one or more females (Powell 1994). Independent subadults are tolerated within the otherwise exclusive ranges of adult animals of the same sex, suggesting that the social system of this species (as well as that of many others in the Mustelinae) is determined not strictly by availability of resources but also by an inherent intolerance (Balharry 1993). Mating occurs in midsummer, but because of delayed implantation births do not take place until March or April. Captive females reportedly exhibit one to four periods of sexual receptivity, which usually last 1–4 days and recur at intervals of 6–17 days during the breeding season (Mead 1994). Pregnancy lasts 230–75 days. The number of young per litter is two to eight, usually three to five. The young weigh about 30 grams at birth, open their eyes after 32–38 days, are weaned at 6–7 weeks, separate from the mother in the autumn, and usually attain sexual maturity at 2 years. Maximum known longevity is 17 years.

The fur of the pine marten is more valuable than that of *M. foina*. Wild populations have been excessively trapped and greatly reduced during the twentieth century, but the species is not in immediate danger of extinction. Efforts at captive breeding have had only limited success. *M. martes* occurs in the Balearic Islands, perhaps through human introduction, but a large subspecies, *M. m. minoricensis*, has been described from Menorca. Trappers had nearly exterminated it by 1970, but subsequent legal protection has allowed recovery (Clevenger 1993a, 1993b). Populations on Mallorca, Elba, Sardinia, and Corsica also are of conservation concern (Clevenger 1993a, 1993b; Schreiber et al. 1989).

Martes zibellina (sable)

Except as noted, the information for the account of this species was taken from Grzimek (1975), Novikov (1962), and Stroganov (1969). Males have a head and body length of 380–560 mm, a tail length of 120–90 mm, and a weight of 880–1,800 grams. Females have a head and body length of 350–510 mm, a tail length of 115–72 mm, and a weight of 700–1,560 grams. The winter pelage is long, silky, and luxurious. Coloration varies but is generally pale gray brown to dark black brown. The summer pelage is shorter, coarser, duller, and darker. The soles are covered with extremely dense, stiff hairs. The body is very slender, long, and supple.

The sable dwells in both coniferous and deciduous forests, sometimes high in the mountains and preferably near streams. It is mainly terrestrial but can climb. An individual may have several permanent and temporary dens located in holes among or under rocks, logs, or roots. A burrow several meters long may lead to the enlarged nest chamber, which is lined with dry vegetation and fur. The sable hunts either by day or by night. It tends to remain in one part of its home range for several days and then move on. It sometimes stays in its nest for several days during severe winter weather. There may be migrations to higher country in summer and also large-scale movements associated with food shortage. The diet consists mostly of rodents but also includes pikas, birds, fish, honey, nuts, and berries.

Reported population densities vary from 1/1.5 sq km in some pine forests to 1/25 sq km in larch forests. Individual home range is usually several hundred hectares but may be as great as 3,000 ha. in more desolate parts of Siberia. At least part of the home range is defended against intruders, but a male may sometimes share its territory with a female. Mating occurs from June to August, with births usually in April or May. In contrast to *M. martes, M. zibellina* usually has a single estrus period of 2–8 days (Mead 1994). Actual embryonic development takes perhaps 25–40 days, but because of delayed implantation the total period of pregnancy is 250–300 days. The number of young per litter ranges from one to five and is usually three or four. The young weigh 30–35 grams at birth, open their eyes after 30–36 days, emerge from the den at 38 days, are weaned at about 7 weeks, and attain sexual maturity at 15–16 months. Maximum known longevity is 15 years.

The sable is one of the most valuable of fur bearers. Its pelt has been avidly sought since ancient times. Several hundred thousand skins were traded annually during the late eighteenth century in the western Siberian city of Irbit. Because of excessive trapping, the take dropped to 20,000–25,000/year by 1910–13, and the sable then had disappeared from much of its range. Subsequently, through programs of protection and reintroduction, the species increased in numbers and distribution. Total numbers in the wild in Russia are now estimated at 1,000,000–1,300,000, and 300,000–350,000 are taken each year (Bakeyev and

Sinitsyn 1994); approximately 27,000 captive-bred sables also are harvested annually. About half of the pelts obtained are released to the world market, and these may sell for more than $500 each (Daniloff 1986). China also was once a major source of sable fur, but the species is now rare in that country (Ma and Li 1994).

Martes melampus (Japanese marten)

Head and body length is 470–545 mm, tail length is 170–223 mm, and weight averaged 1,563 grams for nine males and 1,011 grams for four females (Anderson 1970; Tatara 1994). The general coloration is yellowish brown to dark brown, and there is a whitish neck patch. Little is known about the habits of the Japanese marten. It reportedly is decreasing in numbers through excessive hunting for its fur and because of the harmful effects of agricultural insecticides. Tatara (1994) noted that the hunting season is 1 December–31 January but that a possibly introduced population in southern Hokkaido is fully protected.

The subspecies M. m. tsuensis, of Tsushima Island, is classified as vulnerable by the IUCN. Schreiber et al. (1989) noted that this subspecies now is legally protected. According to Tatara (1994), it is found mainly in broad-leaved forests, dens in trees or ground burrows, and is nocturnal. The diet consists largely of fruit and insects but also includes frogs, birds, and small mammals. Both sexes occupy home ranges of about 0.5–1.0 sq km and are territorial. Predation by feral dogs and highway mortality appear to be major threats.

Martes americana (American pine marten)

Males have a head and body length of 360–450 mm, a tail length of 150–230 mm, and a weight of 470–1,300 grams. Females have a head and body length of 320–400 mm, a tail length of 135–200 mm, and a weight of 280–850 grams. The pelage is long and lustrous. The upper parts vary in color from dark brown to pale buff, the legs and tail are almost black, the head is pale gray, and the underparts are pale brown with irregular cream or orange spots. The body is slender, the ears and eyes are relatively large, and the claws are sharp and recurved. Females have eight mammae (Banfield 1974; Burt and Grossenheider 1976; Clark, Grensten, et al. 1987).

The American pine marten is found mainly in coniferous forest. It dens in hollow trees or logs, in rock crevices, or in burrows. The natal nest is lined with dry vegetation. The marten is primarily nocturnal and partly arboreal but spends considerable time on the ground. It can swim and dive well. It is active all winter but may descend to lower elevations if living in a mountainous area. The diet consists mostly of rodents and other small mammals and also includes birds, insects, fruit, and carrion.

Population density varies from about 0.5/sq km to 1.7/sq km of good habitat (Banfield 1974). In Maine, Soutiere (1979) found densities of 1.2 resident martens per sq km in undisturbed and partly harvested forests but only 0.4/sq km in commercially clear-cut forests. Average home range size throughout North America is 8.1 sq km for males and 2.3 sq km for females, and degree of overlap varies (Powell 1994). In a radio-tracking study in northeastern Minnesota, Mech and Rogers (1977) determined home ranges to be 10.5, 16.6, and 19.9 sq km for 3 males and 4.3 sq km for 1 female. There was considerable overlap between the ranges of 2 of the males. The marten is primarily a solitary species, but Herman and Fuller (1974) regularly observed what seemed to be an adult male and female together, sometimes apparently with two of their offspring, even though it was not the breeding season.

Mating occurs from June to August. Captive females reportedly exhibit one to four periods of sexual receptivity, which last 1–4 days and recur at intervals of 6–17 days during the breeding season, though possibly wild individuals have a longer estrus (Mead 1994). Implantation of the fertilized eggs in the uterus is delayed until February, and embryonic development then proceeds for about 28 days. The total period of pregnancy is 220–75 days. The average number of young per litter is 2.6 (1–5). The young weigh 28 grams at birth, open their eyes after 39 days, are weaned at 6 weeks, reach adult size at 3.5 months, and attain sexual maturity at 15–24 months. Wild females have still been capable of reproduction at 12 years (Banfield 1974; Clark, Grensten, et al. 1987). Captives have lived up to 17 years.

The fur of the marten is valuable and is sometimes referred to as "American sable." In the 1940s marten pelts were worth as much as $100 each. In the mid–nineteenth century the Hudson's Bay Company traded as many as 180,000 Canadian skins each year (Banfield 1974). By the early twentieth century excessive trapping had severely depleted M. americana in Alaska, Canada, and the western conterminous United States. Protective regulations subsequently allowed the species to make a comeback in some areas, but in the eastern United States the marten survives only in small parts of Minnesota, New York, and Maine (Blanchard 1974; Mech 1961; Mech and Rogers 1977; Yocom 1974). Reintroduction projects have been carried out in northern Michigan and Wisconsin (Knap 1975), and it appears that a self-sustaining population has been restored in that region (Slough 1994). Reintroduction also has been attempted in New Hampshire (Strickland and Douglas 1987) and in various other parts of the northeastern United States and southeastern Canada. However, continued absence or only very low populations in most of this region seem attributable to forestry practices that leave relatively poor habitat for the species (Thompson 1991). During the 1976–77 trapping season 27,898 marten skins were taken in the United States, mostly in Alaska, and sold for an average price of $14.00. The respective figures for Canada were 102,632 skins and $19.92 (Deems and Pursley 1978). In 1983–84 the total kill was 188,647 and the average price was $18.61 (Novak, Obbard, et al. 1987). In 1991–92 the kill in the United States was 22,827 and the average price was $44.06 (Linscombe 1994).

Martes pennanti (fisher, or pekan)

Head and body length is 490–630 mm and tail length is 253–425 mm. Males weigh 2.6–5.5 kg and females weigh 1.3–3.2 kg. The head, neck, shoulders, and upper back are dark brown to black. The underparts are brown, sometimes with small white spots. The thick pelage is coarser than that of M. americana. The body is slender but rather stocky for a weasel. Females have eight mammae.

Except as noted, the information for the remainder of this account was taken from Powell (1981, 1982). The fisher inhabits dense forests with an extensive overhead canopy and avoids open areas. It generally lacks a permanent den but seeks temporary shelter in hollow trees and logs, brush piles, abandoned beaver lodges, holes in the ground, and snow dens. All dens known to have been used for raising young were located high up in hollow trees. Activity may take place at any hour. The fisher is adapted for climbing but is primarily terrestrial. It is capable of traveling long distances; one individual covered 90 km in three days. Usual daily movement in New Hampshire, however, was found to be 1.5–3.0 km. The fisher generally forages in a zigzag pattern, constantly investigating places where prey might be concealed.

The diet consists mainly of small to medium-sized birds and mammals, and carrion. It has been calculated that a

Fisher *(Martes pennanti)*, photo from San Diego Zoological Garden.

fisher requires either 1 porcupine every 10–35 days, 1 snowshoe hare every 2.5–8.0 days, 1 kg of deer carrion every 2.5–8.0 days, 1–2 squirrels per day, or 7–22 mice per day. Hares are killed with a quick rush and a bite to the back of the neck. Porcupines are taken only on the ground and are killed by repeatedly circling and biting at the face.

Population density in preferred habitat is 1/2.6–7.5 sq km, but in other areas it may be as low as 1/200 sq km. Individual home range size averages 38 sq km for males and 15 sq km for females (Powell 1994). There is little overlap between the ranges of animals of the same sex but extensive overlap between the ranges of opposite sexes. *M. pennanti* is solitary except during the breeding season, at which time, according to Leonard (1986), normal spacing mechanisms seem to break down, with males leaving their territories to seek as many females as possible and perhaps coming into physical conflict with one another.

Mating occurs from March to May, but implantation of the fertilized eggs in the uterus is delayed until the following January to early April, and births occur from February to May. Postimplantation embryonic development lasts about 30 days, but the total period of pregnancy is nearly 1 year. Females probably mate within 10 days of giving birth and thus are pregnant almost continually (estrus is the same as described for *M. americana*). Litters contain an average of about three young, but there may be as many as six. The newborn weigh less than 40 grams (LaBarge, Baker, and Moore 1990). They are blind and only partly covered with fine hair. They do not open their eyes for about 7 weeks, and they do not walk until about 8–9 weeks. Weaning begins at 8–10 weeks, and separation from the mother occurs in the fifth month. Females reach adult weight after 6 months, and males after 1 year. Sexual maturity comes at 1–2 years. Longevities of about 10 years have been recorded for both wild and captive animals.

In the nineteenth and early twentieth centuries the fisher declined over most of its range because of excessive fur trapping and habitat destruction through logging. The species was almost totally eliminated in the United States and greatly reduced in eastern Canada. As a result, porcupines increased in numbers and began to do considerable forest damage. Widespread closed seasons and other protective regulations were initiated in the 1930s, and reintroductions were made in various areas in the 1950s and 1960s. To the south of Canada, populations of *M. pennanti* are now present in Washington, Oregon, northern California, Idaho, Montana, Minnesota, Wisconsin, Michigan, New York, Vermont, New Hampshire, Maine, Massachusetts, and West Virginia (Cottrell 1978; Coulter 1974; Handley 1980*b*; Mohler 1974; Mussehl and Howell 1971; Olterman and Verts 1972; Penrod 1976; Petersen, Martin, and Pils 1977; Schempf and White 1977; Yocom and McCollum 1973). There also have been several recent records in North and South Dakota (Gibilisco 1994). In some of these states during the past two decades the fisher has been taken legally for the fur market. The average price of a fisher pelt was $100 in the 1920s. The price subsequently declined but then again increased. During the 1976–77 trapping season 12,557 skins were taken in the United States and Canada, selling for an average of about $95 each (Deems and Pursley 1978). In 1983–84 the take was 20,248 and the average price was $49.45 (Novak, Obbard, et al. 1987), but in 1986 in Ontario the highest-quality pelts were selling for $450 (Douglas and Strickland 1987). According to Strickland (1994), there was a reduction in the popularity of furs in the 1990s, and trapping pressure on the fisher eased. During the 1991–92 season 3,336 skins were taken in the United States and sold for an average price of $34.43 (Linscombe 1994).

Martes flavigula (yellow-throated marten)

The information for the account of this species was taken from Lekagul and McNeely (1977) and Stroganov (1969). Head and body length is 450–650 mm, tail length is

Yellow-throated marten *(Martes flavigula)*, photo from Zoological Society of London.

370–450 mm, and weight is 2–3 kg. The fur is short, sparse, and coarse. There is much variation in color, but generally the top of the head and neck, the tail, the lower limbs, and parts of the back are dark brown to black. The rest of the body is pale brown, except for a bright yellow patch from the chin to the chest. Females have four mammae.

The yellow-throated marten is generally found in forests. It climbs and maneuvers in trees with great agility but often comes to the ground to hunt. Activity is primarily diurnal. The diet includes rodents, pikas, eggs, frogs, insects, honey, and fruit. In the northern parts of its range *M. flavigula* evidently preys heavily on the musk deer *(Moschus)* and the young of other ungulates. Individuals often hunt in pairs or family groups, and lifelong pair bonding has been suggested. Births occur in April (Mead 1995). The number of young per litter is usually two or three and may be as many as five. Maximum known longevity is about 14 years. The pelt of this marten has little commercial value. However, the subspecies on Taiwan, *M. f. chrysospila*, has become very rare because of hunting for use of its inner organs for food and through deforestation (Schreiber et al. 1989); it is classified as endangered by the USDI. The subspecies *M. f. robinsoni* of Java has not been collected since 1959, though it was sighted as late as 1979 (Schreiber et al. 1989); it is classified as endangered by the IUCN.

Martes gwatkinsi (Nilgiri marten)

According to Wirth and Van Rompaey (1991), a male had a head and body length of 515 mm, a tail length of 419 mm, and a weight of about 2 kg. As in *M. flavigula*, the general coloration is dark brown and there is a yellowish patch from the chin to the chest. The dorsal profile of the skull of *M. gwatkinsi* is flat, not convex as in *M. flavigula*. The Nilgiri marten is found only in a small isolated tract of hill forest, seldom occurring below 900 meters. It long has been considered rare, only 5–10 specimens are known, and it sometimes is persecuted by beekeepers (Schreiber et al. 1989). It is classified as vulnerable by the IUCN. Madhusudan (1995) reported a recent observation of an individual with an estimated total length of about 1,200 mm sleeping in and moving through the tree canopy in a small patch of forest.

CARNIVORA; MUSTELIDAE; **Genus EIRA**
Hamilton-Smith, 1842

Tayra

The single species, *E. barbara*, is found from southern Sinaloa (west-central Mexico) and southern Tamaulipas (east-central Mexico) to northern Argentina and on the island of Trinidad (Cabrera 1957; Goodwin and Greenhall 1961; E. R. Hall 1981).

Head and body length is 560–680 mm, tail length is 375–470 mm, and weight is 4–5 kg (Grzimek 1975; Leopold 1959). The short, coarse pelage is gray, brown, or black on the head and neck, has a yellow or white spot on the chest, and is black or dark brown on the body. There also is a rare, light-colored form, which is pale buffy and has a darker head. The tayra has a long and slender body, short limbs, and a long tail. The head is broad, the ears are short and rounded, and the neck is long. The soles are naked, and the strong claws are nonretractile.

The tayra is a forest dweller. It nests in a hollow tree or log, a burrow made by another animal, or tall grass. It can climb, run, and swim well. When pursued by dogs it may run on the ground for some distance, then climb a tree and leap through the trees for about 100 meters before descending to the ground again. It thus gains time while the dogs are trying to pick up the trail. Leopold (1959) observed a group of 4 animals springing swiftly through the trees with incredible agility. *Eira* is active both at night and, especially when there is cloud cover, in the morning. The diet seems to consist mostly of rodents but also includes rabbits, birds, small deer *(Mazama)*, honey, and fruit.

A radio-collared female on the llanos of Venezuela was found to use a home range of 9 sq km (Eisenberg 1989). *Eira* is often seen alone, in pairs, or in small family groups. Hall and Dalquest (1963) wrote that the genus is not social to any extent, but Leopold (1959) referred to an old record of hunting troops made up of 15–20 individuals. He also cited a report that 2 young are born in February, that they open their eyes after 2 weeks, and that they forage with their mother by the time they are 2 months old. Other reports suggest that the young may be born in any season.

Tayra *(Eira barbara)*, photo by Ernest P. Walker.

Poglayen-Neuwall (1992) and Poglayen-Neuwall et al. (1989) reported that *Eira* is a nonseasonal, polyestrous species, with an estrous cycle averaging 17 days and receptivity lasting only 2–3 days. Poglayen-Neuwall (1975) and Poglayen-Neuwall and Poglayen-Neuwall (1976) determined that six gestation periods lasted 63–65 days, that newborn weigh about 74–92 grams, that they do not open their eyes until they are 35–58 days old, and that they suckle for 2–3 months. According to Vaughn (1974), litters of 2 young each were produced in captivity on 4 March and 22 July, following gestation periods of about 67–70 days.

The tayra can live 18 years in captivity, loves to play, and can be tamed. It reportedly was used long ago by the Indians to control rodents (Grzimek 1975). It is of no particular importance as a fur bearer or predator of game. It may occasionally eat poultry but does no substantial damage (Leopold 1959). It also has been accused of raiding corn and sugar cane fields (Hall and Dalquest 1963). Schreiber et al. (1989) reported that the range of the tayra has been greatly reduced in Mexico because of the destruction of tropical forests and spread of agriculture. Remaining populations are small and threatened by habitat loss and hunting. The subspecies *E. b. senex*, of Mexico, Guatemala, Belize, and northern Honduras, is now classified as vulnerable by the IUCN.

CARNIVORA; MUSTELIDAE; Genus GALICTIS
Bell, 1826

Grisóns

There are two species (Cabrera 1957; E. R. Hall 1981):

G. vittata (greater grisón), southern Mexico to central Peru and southeastern Brazil;
G. cuja (little grisón), central and southern South America.

Galictis has often been referred to as *Grison* Oken, 1816. Cabrera placed *G. cuja* in the subgenus *Grisonella* Thomas, 1912.

In *G. vittata* the head and body length is 475–550 mm, tail length is about 160 mm, and weight is 1.4–3.3 kg. In *G. cuja* the head and body length is 280–508 mm, tail length is 120–93 mm, and weight is 1.0–2.5 kg (Redford and Eisenberg 1992). The color pattern is striking. In both species the black face, sides, and underparts, including the legs and feet, are sharply set off from the back. The back is smoky gray in *G. vittata* and yellow gray or brownish in *G. cuja*. A white stripe extends across the forehead and down the sides of the neck, separating the black of the face from the gray

Greater grisón *(Galictis vittata)*, photo by Ernest P. Walker.

or brown of the back. Short legs and slender bodies give the animals of this genus an appearance somewhat like that of *Mustela,* but the color pattern immediately distinguishes them.

Grisóns are found in forests and open country from sea level to 1,200 meters. They live under tree roots or rocks, in hollow logs, or in burrows made by other animals, such as the viscachas of South America. They are probably capable burrowers themselves. Quick and agile, they are are good climbers and swimmers and are active both by day and by night. The diet includes small mammals, birds and their eggs, cold-blooded vertebrates, invertebrates, and fruit.

A radio-collared female on the llanos of Venezuela was found to use a home range of 4.2 sq km (Eisenberg 1989). Grisóns are sometimes seen in pairs or groups and playing together. They have a number of vocalizations, including sharp, growling barks when threatened. Various reports indicate that offspring have been produced in March, August, September, and October (Grzimek 1975; Leopold 1959). The gestation period is 39 days (Eisenberg 1989). Litter size is two to four young. A captive *G. vittata* was still living after 10 years and 6 months (Jones 1982).

Young grisóns tame readily and make affectionate pets. In early-nineteenth-century Chile grisóns reportedly were domesticated by natives and used in the same manner as ferrets to enter the crevices and holes of chinchillas to drive the latter out (Osgood 1943).

CARNIVORA; MUSTELIDAE; **Genus LYNCODON**
Gervais, 1844

Patagonian Weasel

The single species, *L. patagonicus,* is found in Argentina and southern Chile (Cabrera 1957).

Head and body length is usually 300–350 mm and tail length is 60–90 mm. The coloration on the back is grayish brown with a whitish tinge. The top of the head is creamy or white, and this color extends as a broad stripe on either side to each shoulder. The nape, throat, chest, and limbs are dark brown, and the rest of the lower surface is lighter brown varied with gray. The color pattern is characteristic and quite attractive. *Lyncodon* is somewhat similar externally to *Galictis* but differs in such features as coloration and the shorter tail. Internally *Lyncodon* has fewer teeth than *Galictis,* there being only two upper and two lower premolars, and one upper and one lower molar, on each side. Like most mustelids, *Lyncodon* has a slender body and short legs.

The Patagonian weasel inhabits the pampas. Its habits are little known, but Ewer (1973:177) wrote: "The reduced molars and cutting carnassials of *Lyncodon* strongly suggest that it is a highly carnivorous species." Redford and Eisenberg (1992:163)) added: "It is reported to be nocturnal or crepuscular and to enter the burrows of *Ctenomys* and *Microcavia* in pursuit of prey." This weasel reportedly was sometimes kept in the houses of ranchers for the purpose of destroying rats. Miller et al. (1983) considered it to be rare in Chile.

Patagonian weasel *(Lyncodon patagonicus)*, photo by Tom Scott of mounted specimen in Royal Scottish Museum, Edinburgh.

Zorilla *(Ictonyx striatus)*, photo from Zoological Society of London.

CARNIVORA; MUSTELIDAE; **Genus ICTONYX**
Kaup, 1835

Zorilla, or Striped Polecat

The single species, *I. striatus*, occurs from Mauritania to Sudan and south to South Africa (Coetzee *in* Meester and Setzer 1977).

Head and body length is 280–385 mm, tail length is 200–305 mm, and weight is 420–1,400 grams. Males are generally larger than females. The body is black with white dorsal stripes, the tail is more or less white, and the face has white markings. The appearance is somewhat like that of the spotted skunks *(Spilogale)* of North America, and early writers sometimes confused the two genera. *Ictonyx* also bears some resemblance to the African genera *Poecilictis* and *Poecilogale* but may be recognized by its color pattern, long hair, and bushy tail. Females have two pairs of mammae (Rowe-Rowe 1978a).

The zorilla is found in a variety of habitats but avoids dense forest. It is mainly nocturnal, resting during the day in rock crevices or in burrows excavated by itself or some other animal. It occasionally shelters under buildings and in outhouses in farming areas. It is terrestrial but can climb and swim well. The usual pace is an easy trot, slower than that of a mongoose, with the back slightly hunched. The diet consists mainly of small rodents and large insects but also includes eggs, snakes, and other kinds of animals. The zorilla may take a chicken on occasion but is often useful in eliminating rodents from houses and stables.

At the sight of an enemy, such as a dog, a zorilla may erect its hair and tail and perhaps emit its anal gland secretion. Such behavior would seem to make the zorilla more formidable than it really is. When actually attacked, it usually emits fluid into the face of the enemy and then feigns death. The ejected fluid may vary in potency with the individual animal and perhaps with age and the time of year. Some writers have remarked that the fluid is much less pungent than that of the American skunks, but others have stated that it is most repulsive, acrid, and persistent.

Ictonyx is solitary. Captive males are totally intolerant of one another, and even adults of opposite sexes are amicable only at the time of mating (Rowe-Rowe 1978b). There are several adult vocalizations associated with threat, defense, and greeting (Channing and Rowe-Rowe 1977). In a study of captives, Rowe-Rowe (1978a) determined the reproductive season to extend from early spring to late summer. All births occurred from September to December. There was usually a single annual litter, but if all young died at an early stage, the female sometimes mated and gave birth a second time. Gestation periods of 36 days were recorded. Litter size was one to three young. They weighed about 15 grams at birth, began to take solid food after about

32 days, opened their eyes at 40 days, were completely weaned at 18 weeks, and were almost full grown at 20 weeks. A male first mated at 22 months, and a female bore its first litter at 10 months. According to Jones (1982), a captive zorilla lived 13 years and 4 months.

North African Striped Weasel

The single species, *P. libyca,* occurs from Morocco and Senegal to the Red Sea (Rosevear 1974). There is some question whether *Poecilictis* warrants generic distinction from *Ictonyx.* The two taxa are sometimes confused in northern Nigeria and Sudan, where their ranges overlap. Wozencraft (*in* Wilson and Reeder 1993) chose to place *Poecilictis* in the synonymy of *Ictonyx,* but the two were treated as distinct genera by Corbet and Hill (1991).

Head and body length is 200–285 mm and tail length is 100–180 mm. Three males weighed 200–250 grams each, though Sitek (1995) reported weights of 600 grams for males and 500 grams for females. The snout is black, the forehead is white, and the top of the head is black. The back is white with a variable pattern of black bands. The tail is white but becomes darker toward the tip. The underparts and limbs are black. From *Ictonyx, Poecilictis* differs in color or pattern, in smaller size, in larger bullae, and in having hairy soles except for the pads. The well-developed anal glands are capable of ejecting a malodorous fluid (Rosevear 1974).

According to Rosevear (1974), this weasel is restricted to the edges of the Sahara and the contiguous arid zones. It is nocturnal, sheltering throughout the daylight hours in single subterranean burrows. These it digs for itself, either down into level surfaces or in the sides of dunes. Maternal dens consist of a single gallery ending in an unlined chamber. The diet apparently consists of rodents, young ground birds, eggs, lizards, and insects.

Births reportedly occur in the wild from January to March. Rosevear (1974) wrote that gestation may be as short as 37 days or as long as 77. Litters usually contain two or three young. At birth they are blind and covered with short hair. Sitek (1995) reported that cubs born in captivity weighed 5 grams at birth, took some solid food after 5 weeks, weighed 250 grams at 2 months, and were separated from their mother at 3 months. Some captives have lived for 5 years or more. *Poecilictis* probably does not make a good pet, as it constantly has a disagreeable smell and is aggressive toward people.

African Striped Weasel

The single species, *P. albinucha,* is found from Zaire and Uganda to South Africa (Coetzee *in* Meester and Setzer 1977).

Head and body length is 250–360 mm and tail length is 130–230 mm. Rowe-Rowe (1978*b*) listed weights of 283–380 grams for males and 230–90 grams for females. From the white on the head and nape four whitish to orange yellow stripes and three black stripes extend on the black back toward the tail, which is white. The legs and underparts are black. There is little variation in pattern.

Poecilogale is smaller and more slender than *Ictonyx* and has narrower back stripes. It resembles *Mustela* in its remarkably slender and elongate body and short legs. Like other weasels, it is able to enter any burrow that it can get its head into. Females have two pairs of mammae (Rowe-Rowe 1978*a*).

This weasel is found in a variety of habitats, including forest edge, grassland, and marsh. It is almost entirely nocturnal, usually spending the day in a burrow that is either self-excavated or taken over from another animal (Kingdon 1977). It can climb well but spends most of its time on the ground. It cannot run rapidly, and patiently trails prey by scent. The diet consists mainly of small mammals and birds but also includes snakes and insects.

Like most weasels, *Poecilogale* attacks by grabbing the throat or neck and hanging on and chewing until the victim is dead. Fair-sized prey, such as springhare, may run some distance before dropping from exhaustion, with the weasel retaining its hold. *Poecilogale* kills venomous snakes in much the same manner as the mongoose *Her-*

North African striped weasel *(Poecilictis libyca),* photo by Robert E. Kuntz.

African striped weasel *(Poecilogale albinucha):* Top, photo of mounted specimen from Field Museum of Natural History; Bottom, photo by D. T. Rowe-Rowe.

pestes. It repeatedly provokes the snake to strike until the reptile is tired and slower in recovery and then seizes the snake by the back of the head.

Poecilogale is generally solitary, but groups of two to four individuals—apparently family parties—have been observed on occasion. Adults of opposite sexes lived together quite amicably in captivity, but adult males fought each other at every encounter (Rowe-Rowe 1978*b*). The area around the den is marked by defecation (Kingdon 1977).

When attacked or under stress *Poecilogale* emits a noxious odor from its anal glands. Although nauseating, this odor is not as strong or persistent as that of the American skunks or the African *Ictonyx*. Normally *Poecilogale* is silent, but when alarmed it utters a loud sound described as being between a growl and a shriek. Several adult vocalizations, associated with threat, defense, and greeting, were recorded by Channing and Rowe-Rowe (1977).

In studies of captives and wild-caught animals in South Africa, Rowe-Rowe (1978*a*) determined that births oc-

curred from September to April. Females were polyestrous and would mate a second time in the season if their first litter was lost. Gestation periods of 31–33 days were recorded. Litters contained one to three young. They weighed four grams at birth, started taking solid food after 35 days, opened their eyes at 51–54 days, were completely weaned at about 11 weeks, and were nearly full grown at 20 weeks. A male first mated at 33 months, and a female had her first litter at 19 months. According to Smithers (1971), one individual lived for 5 years and 2 months after capture.

Meester (1976) considered this weasel to be rare but not endangered in South Africa. However, Rowe-Rowe (1990) indicated that it is at risk because of loss of habitat to agriculture and overgrazing and through predation by and competition from increasing numbers of domestic dogs. Although accused of killing poultry on occasion, *Poecilogale* is generally beneficial to human interests because of its destruction of rats, mice, and springhares. It also eats quantities of locusts when available and digs their larvae out of the ground. Some African tribes use the skins of *Poecilo-*

gale in ceremonial costumes or as ornaments, and the Zulu people reportedly use its skin and other parts for medicinal purposes.

CARNIVORA; MUSTELIDAE; **Genus GULO**
Pallas, 1780

Wolverine

The single species, *G. gulo,* originally occurred from Scandinavia and Germany to northeastern Siberia, throughout Alaska and Canada, and as far south as central California, southern Colorado, Indiana, and Pennsylvania (Corbet 1978; E. R. Hall 1981). Although Hall treated the North American populations as a separate species, *G. luscus,* he indicated that there was evidence supporting their recognition as a subspecies of *G. gulo.*

Head and body length is 650–1,050 mm, tail length is 170–260 mm, shoulder height is 355–432 mm, and weight is 7–32 kg. Females average 10 percent less than males in linear measurements and 30 percent less in weight (E. R. Hall 1981). The fur is long and dense. The general coloration is blackish brown. A light brown band extends along each side of the body from shoulder to rump and joins its opposite over and across the base of the tail. *Gulo* resembles a giant marten, with a heavy build, a large head, relatively small and rounded ears, a short tail, and massive limbs (Stroganov 1969). Females have four abdominal and four inguinal mammae (Pasitschniak-Arts and Larivière 1995).

The wolverine occurs in the tundra and taiga zones. It may be found in forests, mountains, or open plains. For shelter it may construct a rough bed of grass or leaves in a cave or rock crevice, in a burrow made by another animal, or under a fallen tree (Banfield 1974; Stroganov 1969). Most maternal dens in Finland were found in holes under the snow (Pulliainen 1968). *Gulo* is mainly terrestrial, the usual gait being a sort of loping gallop, but it can climb trees with considerable speed and is an excellent swimmer. It has a keen sense of smell but apparently poor eyesight and indifferent hearing. It seems to be unexcelled in strength among mammals of its size and has been reported to drive bears and cougars from their kills. It is largely nocturnal but is occasionally active in daylight. In the far north, where there are periods of extended light or darkness, the wolverine reportedly alternates 3- to 4-hour periods of activity and sleep. In winter it has been known to cover up to 45 km in a day. It can maintain a gallop for a lengthy period, sometimes moving 10–15 km without rest (Stroganov 1969). In mountainous areas it moves to lower elevations during winter (Pasitschniak-Arts and Larivière 1995). Juveniles normally disperse 30–100 km from their natal range, and an individual marked on 6 March 1981 was found 378 km away on 29 November 1982 (Gardner, Ballard, and Jessup 1986).

The diet includes carrion, the eggs of ground-nesting birds, lemmings, and berries. Large mammals such as reindeer, roe deer, and wild sheep are taken mainly in winter, when the snow cover allows *Gulo* to travel faster than its prey. Most large mammals, however, are obtained in the form of carrion. Small rodents may be chased, pounced upon, or dug out of the ground (Pasitschniak-Arts and Larivière 1995). Caches of prey or carrion are covered with earth or snow or, sometimes, wedged in the forks of trees.

Gulo occurs at relatively low population densities (Van Zyll de Jong 1975). In Scandinavia the estimates vary from 1/200 sq km to 1/500 sq km. An adult male there has a territory that may be as large as 2,000 sq km in winter, and individuals of the same sex are not tolerated within it. Territories are regularly marked, mainly by secretions from anal

Wolverine *(Gulo gulo)*, photo by James K. Drake.

scent glands but also with urine. Within the territory of each male are the territories of three or four adult females and their young (Kruska *in* Grzimek 1990; Pulliainen and Ovaskainen 1975).

Somewhat different information was derived from studies in Montana (Hornocker 1983; Hornocker and Hash 1981; Koehler, Hornocker, and Hash 1980). A study area of 1,300 sq km there contained a minimum population of 20 wolverines. Average yearly home range was 422 sq km for males and 388 sq km for females; however, individuals sometimes left their ranges, and females occupied much smaller areas while they were nursing young. There was extensive overlap between the ranges of both the same and the opposite sexes, and no territorial defense was observed. The extensive marking was considered not to define a territory but rather to identify the area being utilized at a given time.

Investigations in arctic Alaska indicated that females generally exclude one another from their home ranges but males tolerate the presence of females (Gipson 1985). In south-central Alaska, Whitman, Ballard, and Gardner (1986) determined annual home range to average 535 sq km for males and 105 sq km for females with young. Annual ranges in Yukon were found to be 209–69 sq km for males and 76–269 sq km for females (Banci and Harestad 1990).

The wolverine is solitary except during the breeding season. Females are monestrous and apparently give birth about every two years. Mating usually occurs from late April to July, but implantation of the fertilized eggs in the uterus is delayed until the following November to March, with births from January to April. The usual number of young per litter is two to four, and the range is one to five. Active gestation lasts 30–40 days. The young weigh 90–100 grams at birth, nurse for 8–10 weeks, separate from the mother in the autumn, and attain adult size after 1 year. Sexual maturity comes in the second or third year of life (Banci and Harestad 1988; Banfield 1974; Grzimek 1975; Pulliainen 1968; Rausch and Pearson 1972; Stroganov 1969). A captive female approximately 10 years of age gave birth after a pregnancy of 272 days (Mehrer 1976). Another captive wolverine lived 17 years and 4 months (Jones 1982).

The pelt of the wolverine is not used widely in commerce but is valued for parkas by persons living in the Arctic because it accumulates less frost than other kinds of fur. During the 1976–77 trapping season 1,922 skins were reported taken in Canada, Alaska, and Montana. The average selling price was about $182 (Deems and Pursley 1978). In 1983–84 the take was 1,377, with an average value of $77.16 (Novak, Obbard, et al. 1987). In 1991–92, 591 skins were taken in Alaska and sold at an average price of $235 (Linscombe 1994). Fur trapping has contributed to a decline in the numbers and distribution of the wolverine, but a more important factor may be human consideration of the genus as a nuisance. In Scandinavia it was intensively hunted, often for bounty, because of alleged predation on domestic reindeer. Throughout its range the wolverine came into conflict with people by following traplines and devouring the fur bearers it found and by breaking into cabins and food caches and spraying the contents with its strong scent.

In Europe *Gulo* is now found only in parts of Scandinavia, where it is very rare, and northern Russia (Pasitschniak-Arts and Larivière 1995; Smit and Van Wijngaarden 1981). Only 120–50 individuals remain in Norway; although legally protected, they are regularly killed in response to claims of attacks on the increasing numbers of sheep being allowed to graze in mountainous areas (Be-

vanger 1992). The genus has disappeared over most of eastern and south-central Canada (Van Zyll de Jong 1975). It is classified as vulnerable by the IUCN and has been designated as endangered in eastern Canada. By the early twentieth century the wolverine also had been nearly eliminated in the conterminous United States. One population, however, held out and subsequently increased in the mountains of northern and eastern California, and another population reestablished itself in western Montana. Since 1960 there have been numerous reliable reports from Washington, Oregon, Idaho, Wyoming, and Colorado and a few questionable records in Minnesota, Iowa, and South Dakota (Birney 1974; Blus, Fitzner, and Fitzner 1993; Field and Feltner 1974; Johnson 1977; Nead, Halfpenny, and Bissell 1985; Nowak 1973; Schempf and White 1977). The U.S. Fish and Wildlife Service recently was petitioned to add the wolverine in the conterminous United States to the List of Endangered and Threatened Wildlife but decided against such a measure (Nordstrom 1995), seemingly in contradiction of the IUCN classification and the whole history of the species. The subspecies *G. g. katschemakensis* of Alaska's Kenai Peninsula now numbers only about 50 individuals and reportedly continues to decline because of an excessively long hunting season (Schreiber et al. 1989).

CARNIVORA; MUSTELIDAE; Genus **MELLIVORA**
Storr, 1780

Honey Badger, or Ratel

The single species, *M. capensis,* originally occurred from Palestine and the Arabian Peninsula to Turkmenistan and eastern India and from Morocco and lower Egypt to South Africa (Coetzee *in* Meester and Setzer 1977; Corbet 1978).

Head and body length is 600–770 mm, tail length is usually 200–300 mm, and shoulder height is usually 250–300 mm. Kingdon (1977) listed weight as 7–13 kg. The upper parts, from the top of the head to the base of the tail, vary from gray to pale yellow or whitish and contrast sharply with the dark brown or black of the underparts. Completely black individuals, however, have been found in Africa, particularly in the Ituri Forest of northern Zaire. The color pattern of the honey badger has been interpreted as being a warning coloration because it makes the animal easily recognizable. Females have two pairs of mammae.

The body is heavily built, the legs and tail are relatively short, the ears are small, and the muzzle is blunt. The large forefeet are armed with very large, strong claws. The hair is coarse and quite scant on the underparts. The skin is exceedingly loose on the body and very tough. The skull is massive and the teeth are robust. Anal glands secrete a vile-smelling liquid. This combination of characters provides an effective system of deterrence and defense.

Mellivora is very difficult to kill. Except on the belly the skin is so tough that a dog can make little impression. The ratel can twist about in its skin, so it can even bite an adversary that has seized it by the back of the neck. Porcupine quills and bee stings have little effect, and snake fangs are rarely able to penetrate. *Mellivora* seems to be devoid of fear, and it is doubtful that any animal of equivalent size can regularly kill it. It may rush out from its burrow and charge an intruder, especially in the breeding season. Horses, antelopes, cattle, and even buffalo have been attacked and severely wounded in this manner.

The honey badger occupies a variety of habitats, mainly in dry areas but also in forests and wet grasslands. It lives among rocks, in hollow logs or trees, and in burrows. Its

Honey badger, or ratel *(Mellivora capensis)*, photo by Bernhard Grzimek.

powerful limbs and large claws make it a capable and rapid digger. It is primarily terrestrial but can climb, especially when attracted by honey. It travels by a jog trot but is tireless and trails its prey until the prey is run to the ground. The diet includes small mammals, the young of large mammals, birds, reptiles, arthropods, carrion, and vegetation. Honey and bees are important foods at certain times of the year.

A remarkable association has developed, at least in tropical Africa, between a bird—the honey guide *(Indicator indicator)*—and *Mellivora*. The association is mutually beneficial in the common exploitation of the nests of wild bees. In the presence of any mammal, even people, the honey guide has the unusual habit of uttering a series of characteristic calls. If a honey badger hears these calls, it follows the bird, which invariably leads it to the vicinity of a beehive. The badger breaks open the nest, and the bird obtains enough of the honey and insects to pay for its work.

Mellivora may travel alone or in pairs. It is generally silent but may utter a very harsh, grating growl when annoyed. The sparse data on reproduction were summarized by Hancox (1992*b*), Kingdon (1977), Rosevear (1974), and Smithers (1971). Mating has been noted in South Africa in February, June, and December. A lactating female was found in Botswana in November, and newborn were recorded in Zambia in December. Seasonal breeding has been reported in Turkmenistan, with mating occurring in autumn and births in the spring. The gestation period is thought to be about 5–6 months. The number of young in a litter is commonly two, and the range is one to four. The young are born in a grass-lined chamber and evidently remain close to the burrow for a long time. *Mellivora* appears to thrive in captivity: one specimen lived 26 years and 5 months (Jones 1982).

If captured before it is half grown, the honey badger can become a satisfactory pet, as it is docile, affectionate, and active. It is, however, incredibly strong and energetic and can wreck cages and damage property in its explorations. Wild individuals sometimes prey on poultry, tearing through wire netting to effect entry (Smithers 1971). Destruction of commercial beehives in East Africa seems to be a significant problem, and *Mellivora* has thus been intensively poisoned, trapped, and hunted (Kingdon 1977). The overall range of the genus has declined, probably through human persecution. The ratel is classified as vulnerable in South Africa (Smithers 1986) and as rare by Russia.

CARNIVORA; MUSTELIDAE; **Genus MELES**
Storr, 1780

Old World Badger

The single species, *Meles meles*, originally occurred throughout Europe, including the British Isles and several Mediterranean islands (Sicily, Crete, Rhodes), and in Asia as far east as Japan and as far south as Palestine, Iran, Tibet, and southern China (Corbet 1978; Long and Killingley 1983). The authority for the generic name *Meles* sometimes is given as Brisson, 1762.

Head and body length is 560–900 mm, tail length is 115–202 mm, and weight is usually 10–16 kg; Novikov (1962), however, wrote that old males attain weights of 30–34 kg in the late autumn. The upper parts are grayish, and the underparts and limbs are black. On each side of the face is a dark stripe that extends from the tip of the snout to the ear and encloses the eye; white stripes border the dark stripe. Like other badgers, *Meles* has a stocky body, short limbs, and a short tail. It is distinguished from *Arctonyx* by its black throat and shorter tail, which is the same color as the back. Females have three pairs of mammae (Grzimek 1975).

The Old World badger is found mainly in forests and densely vegetated areas. It usually lives in a large communal burrow system that covers about 0.25 ha. There are numerous entrances, passages, and chambers. Nests may be located 10 meters from an entrance and 2–3 meters below the surface of the ground and have a diameter of 1.5 meters. A burrow system may be used for decades or centuries, by one generation of badgers after another; it continually increases in complexity and may eventually cover several hectares (Grzimek 1975; Novikov 1962). The animals occupying a system may utilize one nest for several months and then suddenly move to another part of the burrow. The living quarters are kept quite clean. Bedding material, in the form of dry grass, brackens, moss, or leaves, is dragged backward into the den. Occasionally this bedding is brought up and strewn around the entrance to air for an hour or so in the early morning. Around the burrows are dung pits, sunning grounds, and areas for play (badgers play all sorts of games, including leapfrog). Well-defined foraging paths may extend outward for 2–3 km (Novikov 1962). Kruuk (1978*b*) distinguished two kinds of burrows:

Old World badger *(Meles meles)*, photo from Paignton Zoo, Great Britain.

"main setts," with an average of 10.5 entrances; and small "outliers," usually having only a single entrance.

Meles usually does not emerge from its burrow until after sundown. During periods of very cold weather and high snow it may spend days or weeks in the den. Such intervals of winter sleep extend to several months in northern Europe and up to seven months in Siberia. There is no substantial drop in body temperature, and the badger can be aroused easily, but the animal lives off of fat reserves accumulated in the summer and autumn. The omnivorous diet includes almost any available food—small mammals, birds, reptiles, frogs, mollusks, insects, larvae of bees and wasps, carrion, nuts, acorns, berries, fruits, seeds, tubers, rhizomes, and mushrooms. Earthworms have been found to be of major dietary importance in some areas (Kruuk 1978*a;* Skoog 1970).

In a study in England, Kruuk (1978*b*) found badgers to be organized into clans of as many as 12 individuals. The minimum distance between the main burrows of clans was 300 meters. Most clans used ranges of 50–150 ha., and there was little overlap. The ranges were marked by defecation and secretions from subcaudal glands. Several fights were observed along territorial boundaries. Most clans had more females than males, but one, which used a range of only 21 ha., consisted solely of males (it was suggested that in other parts of the range of *Meles* the bachelors may be nomadic). Individuals moved around alone within the clan range. Adult males always slept in the main setts, but females sometimes slept in the outliers, especially in the summer.

Mating is possible throughout the year but occurs mostly from late winter to midsummer. Pregnancy sometimes proceeds without interruption, with birth 2 months after mating. However, development of the fertilized eggs commonly stops at the blastocyst stage, and implantation in the uterus is delayed for about 10 months. The time of implantation seems to be controlled by conditions of light and temperature. Following implantation, embryonic development proceeds for 6–8 weeks, and births occur mainly from February to March. The total period of pregnancy may thus be about 9–12 months. Females may experience a postpartum estrus. The number of young in a litter is two to six, usually three or four. The young weigh 75 grams at birth, open their eyes after about 1 month, nurse for 2.5 months, and usually separate from the female in the autumn. Both males and females apparently attain sexual maturity at the age of 1 year (Ahnlund 1980; Canivenc and Bonnin 1979; Hancox 1992*b;* Kruska *in* Grzimek 1990; Novikov 1962). One specimen lived in captivity for 16 years and 2 months (Jones 1982).

Meles sometimes damages ripening grapes, corn, and oats. It also has been widely accused of killing small game. Its hair is used to make various kinds of brushes, and its skin has been used in northern China to make rugs. For these reasons and because of habitat disruption *Meles* had declined in much of its European range by the late nineteenth century, and the decline continues in some areas. However, a recent survey of the continent west of Russia, Belarus, and Ukraine (Griffiths and Thomas 1993) indicates that populations are stable or increasing in most countries. About half of the countries provide year-round legal protection, though many animals are killed illegally and by vehicles. The minimum number of badgers in the region surveyed was estimated at 1,220,000 and there was an annual legal kill of 118,000. Sweden, with 350,000–400,000, had the largest number of any country. Previous surveys in Great Britain (Clements, Neal, and Yalden 1988; Cresswell et al. 1989) had shown a general recovery from a low point before World War I, when the species was intensively persecuted by gamekeepers. The species now occupies most of the island and has been estimated to number about 250,000 in 43,000 social groups. About 50,000 are killed by vehicles annually, however, and there are concerns about the effects of future urbanization. Evidence of transmission of bovine tuberculosis to cattle from badgers led to efforts to reduce numbers of the latter in parts of Britain and Ireland during recent decades, but such programs are now thought to be scientifically and economically unjustified (Hancox 1992*a*).

Hog badger *(Arctonyx collaris)*, photo by Ernest P. Walker.

CARNIVORA; MUSTELIDAE; **Genus ARCTONYX**
F. Cuvier, 1825

Hog Badger

The single species, *A. collaris*, occurs from Sikkim and northeastern China to peninsular Thailand and on the island of Sumatra (Corbet 1978; Lekagul and McNeely 1977).

Head and body length is 550–700 mm, tail length is 120–70 mm, and weight is usually 7–14 kg. The back is yellowish, grayish, or blackish, and there is a pattern of white and black stripes on the head. The dark stripes run through the eyes and are bordered by white stripes that merge with the nape and with the white of the throat. The ears and tail also are white, and the feet and belly are black. The body form is stocky. This badger is distinguished from *Meles* by its white, rather than black, throat and by its long and mostly white tail, as distinct from the short tail, colored the same as the back, in *Meles*. Another external difference is that in *Arctonyx* the claws are pale in color, whereas in *Meles* they are dark. The common name "hog badger" refers to the long, truncate, mobile, and naked snout, which is often compared to that of a pig *(Sus)*.

The general habits probably resemble those of *Meles*. As with *Meles* and *Mellivora*, the color pattern has been interpreted as a means of warning potential enemies that the animal so marked is best left alone. Like these other genera, *Arctonyx* is a savage and formidable antagonist. It has thick and loose skin, powerful jaws, fairly strong teeth, well-developed claws, and a potent anal gland secretion. The snout is thought to be used in rooting for the various plants and small animals that compose the diet.

According to Lekagul and McNeely (1977), the hog badger is usually found in forested areas at elevations of up to 3,500 meters. It is nocturnal, spending the day in natural shelters, such as rock crevices, or in self-excavated burrows. A study in Shaanxi Province, central China (Zheng et al. 1988), found *Arctonyx* to inhabit agricultural areas and mountain grassland as well as forests. There were two peaks in activity, 0300–0500 and 1900–2100, and earthworms were the favorite food. The genus hibernated from November to February or March and was normally solitary. Mating occurred in May and parturition in the following February or March, suggesting a period of delayed implantation as in *Meles*. Females with litters of three to five young were captured from March to July. The young were weaned at about 4 months and then became independent.

Parker (1979) reported that a captive pair from China was received at the Toronto Zoo in July 1976. The female gave birth to two cubs in February 1977; one cub survived and reached approximate adult size at 7.5 months. In February 1978 the same female gave birth to four young. Matings had been observed from April to September 1977, but delayed implantation was suspected, and true gestation was postulated at less than 6 weeks. According to Jones (1982), a captive lived 13 years and 11 months.

CARNIVORA; MUSTELIDAE; **Genus MYDAUS**
F. Cuvier, 1821

Stink Badgers

Long (1978) recognized two subgenera and two species:

subgenus *Mydaus* F. Cuvier, 1821

M. javanensis, Sumatra, Java, Borneo, Natuna Islands;

subgenus *Suillotaxus* Lawrence, 1939

M. marchei, Palawan and Calamian Islands (Philippines).

Suillotaxus often has been given generic rank.

In *M. javanensis* the head and body length is 375–510 mm, tail length is usually 50–75 mm, and weight is usually 1.4–3.6 kg. The coloration is blackish except for a white crown and a complete or partial narrow white stripe down

Stink badger *(Mydaus javanensis)*, photo from Michael Riffel.

the back onto the tail. In *M. marchei* the head and body length is 320–460 mm, tail length is 15–45 mm, and weight is about 2.5 kg. The upper parts are brown to black, with a scattering of white or silvery hairs on the back and sometimes on the head, and the underparts are brown.

Both species have a pointed face, a somewhat elongate and mobile snout, short and stout limbs, and well-developed anal scent glands. Compared with *M. javanensis, M. marchei* has smaller ears, a shorter tail, and larger teeth.

M. javanensis reportedly is a montane species, often being found at elevations over 2,100 meters (Long and Killingley 1983). It is nocturnal, residing by day in holes in the ground dug either by itself or by the porcupines with which it sometimes lives. The burrows usually are not more than 60 cm deep. On Borneo this species reportedly inhabits caves. Captives have consumed worms, insects, and the entrails of chickens.

M. marchei inhabits grassland-thicket and cultivated areas (Long and Killingley 1983). Grimwood (1976) wrote that it is active both by day and by night. It is common, leaving its tracks and scent along roads and paths. It moves with a rather ponderous, fussy walk. One individual shammed death when first touched and allowed itself to be carried but finally squirted a jet of yellowish fluid from its anal glands into the lens of a camera about 1 meter away.

M. javanensis may growl and attempt to bite when handled. If molested or threatened, it raises its tail and ejects a pale greenish fluid. This vile-smelling secretion is reported by natives sometimes to asphyxiate dogs or even to blind them if they are struck in the eye. The old Javanese sultans used this fluid, in suitable dilution, in the manufacture of perfumes.

Some natives eat the flesh of *Mydaus*, removing the scent glands immediately after the animals are killed. Others mix shavings of the skin with water and drink the mixture as a cure for fever or rheumatism. The IUCN classifies *M. marchei* as vulnerable because its restricted habitat is being lost to human encroachment.

CARNIVORA; MUSTELIDAE; **Genus TAXIDEA**
Waterhouse, 1839

American Badger

The single species, *T. taxus,* is found from northern Alberta and southern British Columbia to Ohio, central Mexico, and Baja California (E. R. Hall 1981).

Head and body length is 420–720 mm, tail length is 100–155 mm, and weight is 4–12 kg. The upper parts are grayish to reddish, and a dorsal white stripe extends rearward from the nose. In the north this stripe usually reaches only to the neck or shoulders, but in the south it usually extends to the rump (Long 1972c). Black patches are present on the face and cheeks; the chin, throat, and midventral region are whitish; the underparts are buffy; and the feet are dark brown to black. The hairs are longest on the sides. *Taxidea* can be recognized by its flattened and stocky form, large foreclaws, distinctive black and white head pattern, long fur, and short, bushy tail. There are anal scent glands. Females have eight mammae (Jackson 1961).

The American badger is usually found in relatively dry, open country. It is a remarkable burrower and can quickly dig itself out of sight. The usual signs of its presence are the large holes that it digs in pursuit of rodents. For shelter it either excavates a burrow or modifies one initially made by another animal. The burrow can be as long as 10 meters and can extend as far as 3 meters below the surface. A bulky nest of grass is located in an enlarged chamber, and the entrances are marked by mounds of earth (Banfield 1974).

Taxidea may be active at any hour but is mainly nocturnal. Its movements are restricted, especially in winter, and it shows a strong attachment to a home area. In the summer, however, the young animals disperse over a considerable distance; one traveled 110 km. The badger is active all year, but it may sleep in its den for several days or weeks during severe winter weather. One female was found to have emerged only once during a 72-day period (Messick and Hornocker 1981).

American badger *(Taxidea taxus)*, photo by E. P. Haddon from U.S. Fish and Wildlife Service.

Most food is obtained by excavating the burrows of fossorial rodents, such as ground squirrels. Also eaten are other small mammals, birds, reptiles, and arthropods (Banfield 1974; Messick and Hornocker 1981). Food is sometimes buried and used later. If a sizable meal, such as a rabbit, is obtained, the badger may dig a hole, carry in the prey, and remain below ground with it for several days. There are reports that a badger sometimes forms a "hunting partnership" with a coyote (see account of *Canis latrans*).

On the basis of a radio-tracking study in southwestern Idaho, Messick and Hornocker (1981) estimated a population density of up to 5/sq km and found average home ranges of 2.4 sq km for males and 1.6 sq km for females. The ranges overlapped, but individuals were solitary except during the reproductive season. A female radio-tracked in Minnesota used an area of 752 ha. during the summer. She had 50 dens within this area and was never found in the same den on two consecutive days. In the autumn she shifted to an adjacent area of 52 ha. and often reused dens; in the winter she used a single den and traveled only infrequently within an area of 2 ha. (Sargeant and Warner 1972).

Mating occurs in summer and early autumn, but implantation of the fertilized eggs in the uterus is delayed until December–February and births take place in March and early April (Long 1973b). The total period of pregnancy is thus about 7 months, but actual embryonic development lasts only about 6 weeks (Ewer 1973). The litter of one to five young, usually two, is born and raised in a nest of dry grass within a burrow. Newborn weight is about 94 grams (Hayssen, Van Tienhoven, and Van Tienhoven 1993). The young are weaned at about 6 weeks and disperse shortly thereafter. The study by Messick and Hornocker (1981) indicated that 30 percent of the young females mate in the first breeding season following birth, when they are about 4 months old, but that males wait until the following year. The oldest wild badger caught in this study had attained an age of 14 years. A captive badger lived 26 years (Jones 1982).

Taxidea is generally beneficial to human interests as it destroys many rodents and its burrows provide shelter for many kinds of wildlife, including cottontail rabbits. Because the badger's burrows and holes may constitute a hazard to cattle, horses, and riders, ranchers have often killed badgers. Many badgers also have been killed by poison put out for coyotes (Hall 1955). Although *Taxidea* has declined in numbers in some areas, it has extended its range eastward in the twentieth century. It now occupies most of Ohio and is invading southeastern Ontario (E. R. Hall 1981; Long 1978). There also are recent records from New York, New England, and southern Yukon, but it is not known whether these represent natural occurrences (Messick 1987).

For many years badger fur was used to make shaving brushes, but it has now been largely replaced by synthetic materials. The fur is still used for trimming garments. In the 1976–77 trapping season 49,807 pelts were reported taken in the United States and Canada, selling at an average price of about $38 (Deems and Pursley 1978). By 1983–84 the take had fallen to only about 20,000 and the average price to $10.00 (Messick 1987). In the 1991–92 season 10,825 skins were taken in the United States and sold for an average price of $6.45 (Linscombe 1994).

CARNIVORA; MUSTELIDAE; **Genus MELOGALE**
I. Geoffroy St.-Hilaire, 1831

Ferret Badgers

There are four species (Corbet and Hill 1992; Long 1978; Riffel 1991; Zheng and Xu 1983):

M. moschata, Assam to central China and northern Indochina, Taiwan, Hainan;
M. personata, Nepal to Indochina;
M. orientalis, Java, Bali;
M. everetti, Borneo.

Long (1978) noted that *M. moschata sorella*, of southeastern China, may be a full species.

Ferret badger *(Melogale moschata)*, photo by Gwilym S. Jones.

Head and body length is 330–430 mm, tail length is 145–230 mm, and weight is 1–3 kg. The general coloration of the upper parts is gray brown to brown black; the underparts are somewhat paler. A white or reddish dorsal stripe is usually present. *Melogale* is distinguished by the striking coloration of the head, which combines black with patches of white or yellow. The tail is bushy, the limbs are short, and the feet are broad and have long, strong claws for digging (Lekagul and McNeely 1977).

Ferret badgers are found in wooded country and grassland. They reside in burrows and natural shelters during the day and are active at dusk and during the night. They climb on occasion. *M. moschata* on Taiwan is reported to be a good climber and often to sleep on the branches of trees. Ferret badgers are savage and fearless when provoked or pressed and have an offensive odor. The conspicuous markings on the head have been interpreted as a warning signal. The omnivorous diet is known to include small vertebrates, insects, earthworms, and fruit. A ferret badger is sometimes welcome to enter a native hut because of its destruction of insect pests.

The young, usually one to three per litter, are born in a burrow in May and June. They apparently are dependent on the milk of the mother for some time, as two nearly full-grown suckling animals and their mother were once found in a burrow. According to Marvin L. Jones (Zoological Society of San Diego, pers. comm., 1995), a specimen of *M. moschata* lived for more than 17 years in captivity.

The species *M. everetti* is classified as vulnerable by the IUCN because its restricted habitat is being lost to human encroachment. *M. orientalis* (treated by the IUCN as a subspecies of *M. personata*) is designated as near threatened.

CARNIVORA; MUSTELIDAE; **Genus SPILOGALE**
Gray, 1865

Spotted Skunks

There are three species (Dragoo et al. 1993; E. R. Hall 1981):

S. pygmaea, Pacific coast of Mexico from southern Sinaloa to Oaxaca;
S. gracilis, southern British Columbia and central Wyoming to Costa Rica and central Texas, Baja California, Channel Islands;
S. putorius, eastern Wyoming, Minnesota, and southern Pennsylvania to northeastern Mexico and Florida.

E. R. Hall (1981) and Wozencraft (*in* Wilson and Reeder 1993) included *S. gracilis* in *S. putorius,* but the two were regarded as distinct species by Dragoo et al. (1993) and Jones et al. (1992). The association of these two taxa is unusual. They are entirely allopatric, their ranges meeting all along a line through Wyoming, Colorado, Texas, and northeastern Mexico. On a morphological basis they do not appear to be more than subspecifically separable, but molecular analysis indicates a greater distinction. These data support earlier investigations showing a profound difference between the breeding patterns and seasons of the two species, effectively making them reproductively isolated from one another (see below).

Head and body length is 115–345 mm, tail length is 70–220 mm, and weight is usually 200–1,000 grams. Of the three genera of skunks *Spilogale* has the finest fur. The hairs are longest on the tail and shortest on the face. The basic color pattern consists of six white stripes extending along the back and sides; these are broken into smaller stripes and spots on the rump. There is a triangular patch in the middle of the forehead, and the tail is usually tipped with white. The variations are infinite as no two individuals have been found to have exactly the same pattern. This genus may be distinguished from the other two genera by its small size, forehead patch, and pattern of stripes and spots, the white never being massed. There is a pair of scent glands under the base of the tail, from which a jet of strong-

Spotted skunk *(Spilogale putorius)*, photo by Ernest P. Walker.

smelling fluid can be emitted through the anus. Females have 10 mammae (Lowery 1974).

Spotted skunks occur in a variety of brushy, rocky, and wooded habitats but avoid dense forests and wetlands. They generally remain under cover more than striped skunks *(Mephitis)*. They usually den underground but can climb well and sometimes shelter in trees. The dens are lined with dry vegetation. Spotted skunks are largely nocturnal and are active all year. The omnivorous diet consists mainly of vegetation and insects in the summer and of rodents and other small animals in the winter.

The white-plumed tail is used to warn other animals that spotted skunks should not be molested. If, however, the sudden erection of the tail is not sufficient deterrence, *Spilogale* may stand on its forefeet and sometimes even advance toward its adversary. Finally, the fluid from the anal glands is discharged at the enemy, usually after the skunk has returned its forefeet to the ground and assumed a horseshoe position (Lowery 1974).

According to Banfield (1974), population densities reach 5/sq km in good agricultural land and winter home range is approximately 64 ha. In spring the males wander over an area of about 5–10 sq km, but females have smaller ranges. Spotted skunks are very playful with one another. As many as eight individuals sometimes share a den.

The reproductive pattern is not the same in all parts of the range (Foresman and Mead 1973; Mead 1968a, 1968b). Populations in South Dakota and Florida *(S. putorius)* apparently mate mainly in March and April. Implantation of the fertilized eggs in the uterus occurs only 14–16 days lat-

er, and births take place in late May and June. Pregnancy is estimated to last 50–65 days. Populations farther to the west *(S. gracilis)* mate in September and October, but implantation is delayed until the following March or April, with births from April to June. The total period of pregnancy thus lasts 230–50 days, but actual embryonic development takes only 28–31 days. The number of young per litter is two to nine, usually three to six. The young weigh about 22.5 grams at birth, have adult coloration after 21 days, open their eyes at 32 days, can spray musk at 46 days, are weaned at about 54 days, and attain adult size at about 15 weeks. A captive specimen lived 9 years and 10 months (Egoscue, Bittmenn, and Petrovich 1970).

Spotted skunks have been reported to carry rabies and occasionally to take poultry and eggs but generally benefit people through their destruction of rodents and insects. The pelts are very attractive and durable, but they generally sold for well under $1.00 each until about 1970 (Jackson 1961; Lowery 1974). In the 1976–77 trapping season the reported harvest in the United States was 41,952 skins, with an average selling price of $4.00 (Deems and Pursley 1978). The subspecies *S. putorius interrupta*, which apparently expanded into Minnesota and Wisconsin only in the early twentieth century in response to favorable agricultural development, now is in decline throughout the northern part of its range as its habitat undergoes further human modification (Boppel and Long 1994). The species *S. pygmaea* occupies a restricted area of generally deteriorating habitat and apparently has become rare (Schreiber et al. 1989). The subspecies *S. gracilis amphiala*, found only on

Santa Cruz and Santa Rosa in the Channel Islands off southern California, also appears to have become very rare (Crooks 1994).

CARNIVORA; MUSTELIDAE; **Genus MEPHITIS**
E. Geoffroy St.-Hilaire and G. Cuvier, 1795

Striped Skunk and Hooded Skunk

There are two species (E. R. Hall 1981; Janzen and Hallwachs 1982):

M. mephitis (striped skunk), southern Canada to northern Mexico;
M. macroura (hooded skunk), Arizona and southwestern Texas to Costa Rica.

Head and body length is 280–380 mm, tail length is 185–435 mm, and weight is 700–2,500 grams. Both species have black and white color patterns, but with considerable variation. *M. mephitis* usually has white on top of the head and on the nape extended posteriorly and separated into stripes. In some individuals of this species the top and sides

of the tail are white, whereas in others the white is limited to a small spot on the forehead. The white areas are composed entirely of white hairs, with no black hairs intermixed. *M. macroura* has a white-backed color phase and a black-headed color phase. In the former there are some black hairs mixed with the white hairs of the back; in the latter the two white stripes are widely separated and are situated on the sides of the animal instead of being narrowly separated and situated on the back as in *M. mephitis*. Female *Mephitis* have 10–14 mammae (Jackson 1961; Leopold 1959).

According to Lowery (1974), the well-known scent of *Mephitis* is expelled from two tiny nipples located just inside the anus, which mark the outlets of the two ducts leading from glands lying adjacent to the anus. This musk is discharged either as an atomized spray or as a short stream of rain-sized drops. The skunk usually employs this weapon only after much provocation. When confronted by an antagonist, it arches its back, elevates its tail, erects the hairs thereon, and sometimes stamps its feet on the ground. Finally, it makes its body into a U, with the head and tail facing the intruder. The musk usually travels 2–3 meters, but the smell can be detected up to 2.5 km downwind. Lowery observed that one squirt is sufficient to send the most ferocious dog yelping in agony from burning eyes and nostrils and retching with nausea.

These skunks are found in a variety of habitats, includ-

Striped skunk *(Mephitis mephitis):* A. Facing danger that does not appear to be imminent. B. Aimed toward the enemy ready to spray its scent. Photos by Ernest P. Walker.

ing woods, grasslands, and deserts. They generally are active at dusk and through the night and spend the day in a burrow, under a building, or in any dry, sheltered spot. In Minnesota, Houseknecht and Tester (1978) found a general shift from underground, upland dens in winter to aboveground, lowland dens in summer. Although skunks tended to remain at a single den for a long time in winter, females with young able to travel changed dens every one or two days. Bjorge, Gunson, and Samuel (1981) reported that females moved a minimum daily distance of 1.5 km between dens. Juveniles were found to disperse up to 22 km. in summer.

In northern parts of its range *M. mephitis* stays in one den and sleeps through much of the winter. Males tend to sleep for shorter periods than females and to become active more readily during intervals of mild weather (Banfield 1974). The degree of lethargy achieved during the winter is not well understood. It does not appear to be deep torpor, but some skunks are known to have remained underground for more than 100 consecutive days (Sunquist 1974). In Alberta the overall period of female hibernation is 120–50 days, whereas in Illinois it is 62–87 days (Gunson and Bjorge 1979). A striped skunk may become very fat in the autumn before hibernation. The diet is omnivorous and includes rodents, other small vertebrates, insects, fruit, grains, and green vegetation.

Density estimates for striped skunk populations have ranged from 0.7/sq km to 18.5/sq km, but most are 1.8–4.8/sq km (Wade-Smith and Verts 1982). The home ranges of 6 females radio-tracked for 45–105 days each in Alberta (Bjorge, Gunson, and Samuel 1981) averaged 208 ha. and varied from 110 to 370 ha. Males wandered over larger areas, most notably in the autumn. *Mephitis* is generally solitary, but there is a tendency for individuals to den together, especially in the north, as a means of optimizing winter survival and reproductive success. In Alberta, Gunson and Bjorge (1979) found that only males, both adults and juveniles, denned alone during the winter. Communal winter dens contained an average of 6.7 (2–19) individuals. Usually there was only a single adult male per den and an average of 5.8 females. A male apparently wanders in search of a group of females during the autumn and then keeps other males away. *Mephitis* is usually silent but makes several sounds, such as low churrings, shrill screeches, and birdlike twitters (Lowery 1974).

Mating takes place from mid-February to mid-April, with births in May and early June. The period of pregnancy is 59–77 days, and delayed implantation may be involved. Females are usually monestrous but sometimes have a second estrus and parturition subsequent to the normal period if their first pregnancy is not successful. Litters contain 1–10 young, usually about 4–5. The young weigh about 30 grams at birth, open their eyes after 3 weeks, are weaned at 8–10 weeks, and separate from the mother by the autumn. Females may bear their first litter at 1 year (Lowery 1974; Wade-Smith and Richmond 1975, 1978). The average longevity in captivity is about 6 years, but one individual was still living after 12 years and 11 months (Jones 1982).

Striped skunks are generally beneficial to human interests because of their destruction of rodents and insects. *Mephitis*, however, sometimes attacks poultry and is reportedly the principal carrier of rabies among North American wildlife (Wade-Smith and Richmond 1975). The fur is durable and of good texture, but demand and value have varied widely (Lowery 1974; Jackson 1961). During the 1976–77 season 175,884 skins were reported taken in the United States and Canada, selling at an average price of about $2.25 (Deems and Pursley 1978). In 1983–84 the take

was about the same, but the price fell to less than $1.00 (Rosatte 1987). In the 1991–92 season 45,148 skins were taken in the United States and sold at an average price of about $2.00 (Linscombe 1994).

CARNIVORA; MUSTELIDAE; Genus CONEPATUS
Gray, 1837

Hog-nosed Skunks

Five species are now recognized (Cabrera 1957; Ewer 1973; E. R. Hall 1981; Kipp 1965; Manning, Jones, and Hollander 1986; Pine, Miller, and Schamberger 1979; Redford and Eisenberg 1992):

C. mesoleucus, southern Colorado and eastern Texas to Nicaragua;
C. leuconotus, southern Texas, eastern Mexico;
C. semistriatus, southern Mexico to northern Peru and eastern Brazil;
C. chinga, central and southern Peru, Bolivia, southern Brazil, Chile, northern and western Argentina, Uruguay;
C. humboldti, southern Chile and Argentina.

Statements by the various authorities cited above suggest that some, perhaps all, of the listed species are conspecific.

Head and body length is 300–490 mm, tail length is 160–410 mm, and weight is usually 2.3–4.5 kg. *Conepatus* has the coarsest fur of all skunks. There are two main color patterns, with variations. In one the top of the head, the back, and the tail are white, and the remainder of the animal is black. This coloration occurs most commonly in areas where the ranges of *Conepatus* and *Mephitis* overlap. In the other pattern the pelage is black except for two white stripes, beginning at the nape and extending on the hips, and a mostly white tail. This coloration resembles that of *Mephitis mephitis* and seems to be most common in areas where *Conepatus* is the only kind of skunk present. In all cases hog-nosed skunks lack the thin white stripe down the center of the face that is present in *Mephitis*. *Conepatus* may be distinguished from the other two genera of skunks by its nose, which is bare, broad, and projecting. Females have three pairs of mammae (Leopold 1959).

Hog-nosed skunks are found in both open and wooded areas but avoid dense forests. They occur at all elevations up to at least 4,100 meters (Grimwood 1969). Dens are located in rocky places, hollow logs, or burrows made by other animals. Like other skunks, *Conepatus* is mainly nocturnal, is generally slow-moving, does not ordinarily climb, and defends itself by expelling musk from anal scent glands.

The diet may consist principally of insects and other invertebrates, though fruit and small vertebrates, including snakes, probably are also eaten. Hog-nosed skunks may turn over the soil in a considerable area with their bare snout and their claws when in search of food. Like *Mephitis*, they also pounce on insects. At least in the Andes hognosed skunks are resistant to the venom of pit vipers. There is some evidence that the spotted skunks *(Spilogale)* also are resistant to rattlesnake venom. Since the musk of skunks produces an alarm reaction in rattlers (the same reaction that they exhibit in the presence of king snakes, which prey on them), it may be that skunks feed on rattlesnakes quite extensively.

In southern Chile Fuller et al. (1987) found *C. humboldti* to be solitary but to occupy overlapping home ranges of

Hog-nosed skunk (*Conepatus mesoleucus*), photo by Lloyd G. Ingles.

7–16 ha. According to Davis (1966a), *C. mesoleucus* is not as social as *Mephitis*, and usually only one individual lives in a den. The breeding season in Texas begins in February, most mature females are pregnant by March, and births occur in late April or early May. Gestation lasts approximately 2 months. Of six pregnant females on record three contained three embryos each and three contained two each. By August most of the young are weaned and foraging for themselves. Available evidence indicates that in Mexico the young are also born in the spring (Hall and Dalquest 1963; Leopold 1959). The gestation period of one South American species is 42 days, and litter size is usually two to five young. Sexual maturity in *C. mesoleucus* has been reported to come at 10–11 months (Hayssen, Van Tienhoven, and Van Tienhoven 1993). A captive of that species lived 8 years and 8 months (Marvin L. Jones, Zoological Society of San Diego, pers. comm., 1995).

The pelt of *Conepatus* is inferior in quality to that of *Mephitis*, but large numbers have been marketed from Texas (Davis 1966a). Perhaps because of this commerce, the subspecies *C. mesoleucus telmalestes*, isolated in the Big Thicket area of southeastern Texas, has declined to near the point of extinction (Schmidly 1983). It now is classified as extinct by the IUCN. *C. chinga rex* of northern Chile also is hunted for its pelt and now seems to be rare (Miller et al. 1983). Some natives use the skins for capes or blankets, and others consider the meat to have curative properties. *C. humboldti* is on appendix 2 of the CITES. Records of the CITES indicate that during the 1970s about 155,000 skins of *Conepatus* were exported annually, each with a value of about U.S. $8.00; the trade may subsequently have declined (Broad, Luxmoore, and Jenkins 1988).

CARNIVORA; MUSTELIDAE; Genus LUTRA
Brünnich, 1771

Old World River Otters

There are two subgenera and three species (Chasen 1940; Coetzee *in* Meester and Setzer 1977; Corbet 1978; Corbet and Hill 1992; Ellerman and Morrison-Scott 1966; Lekagul and McNeely 1977; Rosevear 1974; Van Zyll de Jong 1972, 1987):

subgenus *Lutra* Brünnich, 1771

L. lutra, western Europe to northeastern Siberia and Korea, Asia Minor and certain other parts of southwestern Asia, Himalayan region, southern India, China, Burma, Thailand, Indochina, northwestern Africa, British Isles, Sri Lanka, Sakhalin, Japan, Taiwan, Hainan,Sumatra, doubtfully Java;

L. sumatrana (hairy-nosed otter), Indochina, Thailand, Malay Peninsula, Sumatra, Bangka, Java,Borneo;

subgenus *Hydrictis* Pocock, 1921

L. maculicollis (spotted-necked otter), Sierra Leone to Ethiopia, and south to South Africa.

The New World river otters (genus *Lontra*) and the smooth-coated otter (genus *Lutrogale*) of southern Asia (see accounts thereof) often are considered only subgenerically distinct from *Lutra*, but Van Zyll de Jong (1972, 1987) considered all three to be full genera. However, he did not agree with Rosevear (1974), who stated that there is a strong case for regarding *Hydrictis* as a full genus. With respect to the possibly extinct otter populations of Japan, Imaizumi and Yoshiyuki (1989) restricted *L. lutra* to

European river otter *(Lutra lutra)*, photo by Annelise Jensen.

Hokkaido and described a separate species, *L. nippon,* from Shikoku, Honshu, and Kyushu.

Head and body length is 500–820 mm, tail length is 330–500 mm, and weight is 5–14 kg. Males average larger than females (Van Zyll de Jong 1972). The upper parts are brownish and the underparts are paler; the lower jaw and throat may be whitish. The fur is short and dense. The head is flattened and rounded; the neck is short and about as wide as the skull; the trunk is cylindrical; the tail is thick at the base, muscular, flexible, and tapering; the legs are short; and the digits are webbed. The small ears and the nostrils can be closed when the animal is in the water.

These aquatic mammals inhabit all types of inland waterways as well as estuaries and marine coves. Otters are excellent swimmers and divers and usually are found no more than a few hundred meters from water. They may shelter temporarily in shallow burrows or in piles of rocks or driftwood, but they also have at least one permanent burrow beside the water (Stroganov 1969). The main entrance may open underwater and then slope upward into the bank to a nest chamber that is above the high-water level. Erlinge (1967) found *L. lutra* to utilize the following types of facilities in southern Sweden: "dens," generally with several passages and a chamber lined with dry leaves and grass; "rolling places," bare spots near water where the otters roll and groom themselves; "slides," either on slopes or in level places but most common on winter snow; "feeding places," including holes kept open through winter ice; "runways," well-defined paths on land that connect waterways and other facilities; and "sprainting spots and sign heaps," prominent points of land where the animals mark by scratching and elimination.

Otters swim by movements of the hind legs and tail and usually dive for one or two minutes, five at the most (Kruska *in* Grzimek 1990). When traveling on ground, snow, or ice they may use a combination of running and sliding. Although normally closely associated with water, river otters sometimes move many kilometers overland to reach different river basins and to find ice-free water in winter (Stroganov 1969). They may be either diurnal or nocturnal but are generally more active at night. With the possible exception of the Old World badger *(Meles),* river otters are the most playful of the Mustelidae. Some species engage in the year-round activity of sliding down mud and snow banks, and individuals of all ages participate. Sometimes they tunnel under snow to emerge some distance beyond. The diet consists largely of fish, frogs, crayfish, crabs, and other aquatic invertebrates. Birds and land mammals, such as rodents and rabbits, are also taken. Studies indicate that the fish consumed are mainly nongame species. River otters capture their prey with the mouth, not the hands (Rowe-Rowe 1977).

Investigations by Erlinge (1967, 1968) in southern Sweden indicate that *L. lutra* occurs at population densities of 0.7–1.0/sq km of water area, or 1 for every 2–3 km of lakeshore or 5 km of stream. The straight-line length of a home range, including land area, was found to average about 15 km for adult males and 7 km for females with young. The ranges of males constitute territories, which may overlap the ranges of 1 or more adult females and from which other males are excluded. Females also defend their ranges against individuals of the same sex. Territories are marked with scent, and fights occasionally take place. Apparently, males form a dominance hierarchy, with the highest-ranking animals occupying the most favorable ranges. Erlinge noted that the males generally are solitary and ignore the females and young. Studies in Scotland (Mason and Macdonald 1986) suggest a somewhat less rigid social system, with females occupying overlapping home ranges, though again the dominant males have relatively exclusive territories. Still another variation evidently exists on Shetland Island, where up to five reproductively active females were found to share a group range of about 5–14 km of coastline; male ranges were larger and overlapped those of two or more female groups (Kruuk and Moorhouse 1991). On Lake Victoria *L. maculicollis* may undergo a regular cycle of aggregation and dispersal, with males and females each forming their own groups. The males' groups grow larger after the mating season, when males are not tolerated by the females with young. These groups may contain 8–20 individuals from January to May but then become smaller from June to August, when the older males leave to pair with the females (Kingdon 1977).

Within the wide geographic range of *Lutra* there is considerable variation in reproductive pattern (Duplaix-Hall 1975; Ewer 1973; Kingdon 1977; Liers 1966; Mason and Macdonald 1986; Stroganov 1969). Female *L. lutra* are polyestrous, with the cycle lasting about 4–6 weeks and estrus about 2 weeks. In some areas (e.g., most of England) mating and birth may occur at any time of the year. In areas with a more severe climate (e.g., Sweden and Siberia) mating takes place in late winter or early spring, with births in April or May. The gestation period of this species is 60–63 days. The length of pregnancy is about the same in *L. maculicollis*. Litter size in the genus is one to five young, usually two or three. The young weigh about 130 grams at birth, open their eyes after 1 month, emerge from the den and begin to swim at 2 months, nurse for 3–4 months, separate from the mother at about 1 year, and attain sexual maturity in the second or third year of life. Captives have lived up to 22 years (Kruska *in* Grzimek 1990).

River otters have been intensively hunted for their excellent fur and also have proved highly vulnerable to human environmental disruption and pollution of their aquatic habitat. The major problem in Europe is thought to be contamination of fish prey by bioaccumulating organochlorine pollutants, both pesticides and polychlorinated biphenyls (Macdonald and Mason 1994). The Eurasian *L. lutra* has declined drastically in such diverse places as Great Britain (Chanin and Jefferies 1978), Germany (Roben 1974), southeastern Siberia (Kucherenko 1976), and Japan (Mikuriya 1976). Since the 1980s there have been signs of recovery in England following a crash caused by hunting and insecticide pollution, and populations in Scotland and Ireland are now widespread (Foster-Turley, Macdonald, and Mason 1990; Jefferies 1989). There also is a thriving population in Latvia, Lithuania, and Belarus (Baranauskas et al. 1994). Extensive recent surveys of *L. lutra* in the Mediterranean region and southern Europe have found the species to be still relatively common in parts of Spain, Portugal, Greece, various other parts of the Balkans, the Jordan River Valley, and North Africa but rare or absent in Italy, Austria, Switzerland, and most of France (Macdonald and Mason 1994; Mason and Macdonald 1986). It has nearly or completely disappeared in Belgium, the Netherlands, Switzerland, and most of Germany and Poland. It is still common in Finland and northern Norway but rare in the rest of Scandinavia. It is rare in all parts of Asia where its status is known, has not been sighted in Japan since 1986, and may also have been extirpated in Thailand (Foster-Turley, Macdonald, and Mason 1990; Macdonald and Mason 1994).

L. lutra is on appendix 1 of the CITES, and the other species of *Lutra* are on appendix 2. *L. maculicollis* has become very rare in most of West Africa because of destruction of riparian forests and contamination of waterways (Foster-Turley, Macdonald, and Mason 1990) and is considered endangered in South Africa (Stuart 1985). The IUCN classifies *L. sumatrana* as vulnerable; it also is subject to habitat disruption and is still regularly hunted for its pelt. It appears to be rare throughout its range and may have disappeared completely from Thailand and mainland Ma-laysia (Foster-Turley, Macdonald, and Mason 1990). Neither it nor *L. lutra* was found during recent surveys on Java (Melisch, Asmoro, and Kusumawardhani 1994).

CARNIVORA; MUSTELIDAE; Genus LUTROGALE
Gray, 1865

Smooth-coated Otter

The single species, *L. perspicillata*, is found in southern Iraq, from Pakistan to Indochina and the Malay Peninsula, and on the islands of Sumatra, Java, and Borneo. *Lutrogale* often has been considered a subgenus or synonym of *Lutra* but was treated as a distinct genus by Corbet and Hill (1992), Van Zyll de Jong (1972, 1987), and Wozencraft (*in* Wilson and Reeder 1993).

Head and body length is 650–790 mm, tail length is 400–505 mm, and weight is 7–11 kg. The upper parts are raw umber to smoky gray-brown; the underparts are a lighter drab color. The cheeks, upper lip, throat, neck, and upper chest are whitish. The coat is short and very smooth and sleek rather than coarse, and the large footpads are smooth rather than granular. The general external form is similar to that of *Lutra* except that the dorsoventral flattening of the tail is much more pronounced and there are distinct lateral keels distally. Compared with that of *Lutra*, the skull of *Lutrogale* is less depressed, its muzzle shorter with larger orbits set lower and farther forward, its braincase deeper and more inflated, and the teeth larger and flatter (Foster-Turley, Macdonald, and Mason 1990; Harrison and Bates 1991; Lekagul and McNeely 1977).

According to Lekagul and McNeely (1977), *Lutrogale* is a plains otter, inhabiting mostly areas of low elevation. It is found in lakes, streams, reservoirs, canals, and flooded fields and will enter the open sea. It is quite active on land, often traveling long distances in search of suitable streams, and during the dry season may become a jungle hunter. It is a capable burrower and may dig its own breeding den. Family groups may combine when fishing, swimming in a semicircle and driving fish before them. Villagers in India have taken advantage of this habit by using *Lutrogale* to drive fish into nets. On the Malay Peninsula this genus typically occurs in groups consisting of a mated adult pair and as many as four young, which have a territory of 7–12 km of river. Breeding may occur in the early part of the year, and delayed implantation may be a factor. Both sexes bring nesting material into the den and carry food to the cubs. The true gestation period is 63 days, and a captive female first mated at 3 years of age. Chakrabarti (1993) reported that a captive male lived to an age of about 20 years.

The smooth-coated otter is classified as vulnerable by the IUCN (under the genus *Lutra*) and is on appendix 2 of the CITES. It still is generally common but is declining because of destruction of riparian habitat, deforestation, water pollution by industrial wastes and agricultural pesticides, damming and impoundment of streams, and killing by people who seek its pelt or consider it a threat to fisheries. The population on Java may already be extinct (Foster-Turley, Macdonald, and Mason 1990). At the opposite end of the range is the isolated subspecies *L. p. maxwelli*, known only from a few sites in the southern marshes of Iraq (Harrison and Bates 1991). A captive individual of this subspecies was featured in Gavin Maxwell's book *Ring of Bright Water*, published in 1960. Unfortunately, no information on its status has since become available. Its restricted habitat has been the scene of much recent military activity and also probably is being severely affected by current projects to drain the marshes (Oryx 28 [1944]:8), reportedly as a means of extending government control over the indigenous human population.

New World River Otters

There are four species (E. R. Hall 1981; Redford and Eisenberg 1992; Van Zyll de Jong 1972, 1987):

L. canadensis, Alaska, Canada, conterminous United States;
L. longicaudis (neotropical river otter), northwestern Mexico to Uruguay and Buenos Aires Province of Argentina;
L. provocax, central and southern Chile, southern Argentina, Tierra del Fuego;
L. felina (marine otter), Pacific coast from northern Peru to Tierra del Fuego.

Lontra was considered a subgenus or synonym of *Lutra* by E. R. Hall (1981) and Jones et al. (1992) but was treated as a distinct genus by Van Zyll de Jong (1972, 1987) and Wozencraft (*in* Wilson and Reeder 1993). The basis on which various authors rejected generic status for *Lontra* was questioned by Kellnhauser (1983).

Head and body length is 460–820 mm, tail length is 300–570 mm, and weight is 3–15 kg. Males average larger than females (Van Zyll de Jong 1972). The upper parts are various shades of brown, the underparts are light brown or grayish, and the muzzle and throat may be whitish or silvery gray. The fur is short and sleek, with dense underfur overlaid by glossy guard hairs. The general morphology is much like that of *Lutra*, but in the skull the posterior palatine foramina are located more posteriorly, the vomer-ethmoid partition of the nasal cavity extends posteriorly to or behind the first upper molar tooth (in *Lutra* it extends only to between the third and fourth premolars), the first upper molar has a prominent cingulum and expanded talon (in *Lutra* the cingulum is little developed and the talon is small), and the sectorial fourth upper premolar has a talon extending more than two-thirds the length of the tooth (in *Lutra* the talon extends less than two-thirds the length). In both genera the toes are webbed to the terminal digit pads or beyond, the proximal part of the tail is broad and moderately dorsoventrally flattened, and females have four mammae, except in *Lontra provocax*, which has more than four (Van Zyll de Jong 1987).

The natural history of New World river otters, includ-

Canadian river otters *(Lontra canadensis)*, photos by Ernest P. Walker.

ing their aquatic habits and playful behavior, is much like that of *Lutra* (see account thereof). However, there are some particularly different aspects of the ecology of *L. felina* (Estes 1986; Redford and Eisenberg 1992). That species is found largely or exclusively along the exposed seashore, though it may enter freshwater estuaries and large rivers. It stays within about 500 meters of the coast, mainly in areas characterized by rocky outcroppings, heavy seas, and strong winds. It shelters in caves that open at water level, is active mostly in the afternoon, and makes food dives of 15–45 seconds. It feeds mostly on crustaceans, mollusks, and fish. *L. provocax* also sometimes is found along rocky coastlines, whereas *L. longicaudis* depends on permanent streams or lakes with ample riparian vegetation, and shelters in a self-excavated burrow.

According to Melquist and Dronkert (1987), *L. canadensis* is found in both marine and freshwater environments and from coastal areas to high mountains. Densities appear highest in food-rich habitats such as estuaries, lower stream drainages, coastal marshes, and interconnected small lakes and swamps. Favored locations include those with riparian vegetation or rocks that can be used for dens. An individual may use numerous dens and temporary shelters in the course of a year. Beaver lodges are frequently occupied, sometimes simultaneously with the builder. Otters are mainly nocturnal and crepuscular, but daytime activity is not unusual. The diet consists primarily of fish and also includes crustaceans, reptiles, amphibians, and occasionally birds and mammals. Most reported population densities have been about 1/1–10 km of shoreline or waterway length, and home range lengths have been 4–78 km.

In Idaho, Melquist and Hornocker (1983) found *L. canadensis* to have an overall population density of 1/3.9 km of waterway. Seasonal home range length was 8–78 km, with males generally having larger ranges than females; however, there was extensive overlap between the ranges of both the same and opposite sexes. There was some mutual avoidance and defense of personal space but no strong territorial behavior. The basic social group consisted of an adult female and her juvenile offspring. Such families broke up before the female again gave birth, though yearlings occasionally associated with the group. Fully adult males were not observed to accompany family groups. Observations in other areas, however, suggest that although the male is excluded from the vicinity of the female when the latter has small young, he joins the family when the cubs are about 6 months old (Banfield 1974; Jackson 1961). *Lontra* has a variety of vocalizations, and like *Lutra*, it communicates through scent marking with urine, feces, and anal gland secretions (Melquist and Dronkert 1987).

In the southern part of the range of *L. felina* there seems to be a birth peak in September and October, the gestation period is 60–65 days, and litter size is two young; they are born in an earthen den or rocky crevice and stay with the female for approximately 10 months (Redford and Eisenberg 1992). Little is known about reproduction in *L. longicaudis*, but recently a captive pair in southern Brazil was observed to produce litters on 1 April 1992, 21 July 1992, and 14 February 1993; the young of the first two litters died shortly after parturition (Blacher 1994). In most populations of *L. canadensis*, of North America, there is delayed implantation of the fertilized eggs in the uterus. Mating occurs in the winter or spring, and births take place the following year, usually from January to May. The total period of pregnancy has been reported to vary from 290 to 380 days, though actual embryonic development is about 60–63 days, the same as in other kinds of river otters. Populations in southern Florida may not experience delayed implanta-

tion. The female does not excavate her own den but uses that of another animal or some natural shelter. The male does not assist in rearing the young. Litter size is one to six cubs, usually two or three. They weigh about 120–60 grams at birth, emerge from the den at about 2 months, are weaned at 5–6 months, and usually leave their mother just before she again gives birth. Both females and males attain sexual maturity at about 2 years, but males generally cannot successfully breed until 5–7 years of age. Wild individuals up to 14 years old have been taken, and captives have lived about 25 years.

As a group, river otters have suffered severely through habitat destruction, water pollution, misuse of pesticides, excessive fur trapping, and persecution as supposed predators of game and commercial fish. *L. canadensis* has disappeared or become rare throughout the conterminous United States except in the Northwest, the upper Great Lakes region, New York, New England, and the states along the Atlantic and Gulf coasts. The southwestern subspecies *L. c. sonora* has nearly disappeared, though there have been several recent reports in Arizona (Polechla *in* Foster-Turley, Macdonald, and Mason 1990). Otters of another subspecies were released in Arizona in 1981, perhaps inadvisably considering the possibility of genetic modification of the native population. Since 1976 there also have been efforts to reintroduce otters in Colorado, Iowa, Kansas, Kentucky, Missouri, Nebraska, Oklahoma, Pennsylvania, and West Virginia (Melquist and Dronkert 1987).

The IUCN classifies *L. felina* as endangered, noting that it has declined by at least 50 percent over the past decade, and *L. provocax* as vulnerable (both are included in the genus *Lutra*). *L. longicaudis*, *L. provocax*, and *L. felina* are listed as endangered by the USDI and are on appendix 1 of the CITES, and *L. canadensis* is on appendix 2. There may now be fewer than 1,000 individuals of *L. felina* (Estes 1986); it has declined in Chile because of excessive hunting for its fur and in Peru because of persecution for alleged damage to prawn fisheries (Thornback and Jenkins 1982). *L. provocax* remains common at a few isolated sites in extreme southern Chile and Argentina but has disappeared from most of its range because of overhunting and habitat alteration. *L. longicaudis* is still widespread but has disappeared from the highlands of Mexico and is threatened in the rest of that country by habitat destruction and fragmentation (Foster-Turley, Macdonald, and Mason 1990).

The beautiful and durable fur of river otters is used for coat collars and trimming. During the 1976–77 trapping season 32,846 pelts of *L. canadensis* were reported taken in the United States, and the average selling price was $53.00. Respective figures for Canada that season were 19,932 pelts and $69.04 (Deems and Pursley 1978). In 1983–84 the total take was 33,135, and the average selling price was $18.71 (Novak, Obbard, et al. 1987). In the 1991–92 season 10,916 pelts were taken in the United States and sold at an average price of $22.34 (Linscombe 1994). *L. longicaudis* also was taken in large numbers for its valuable skin, with probably about 30,000 killed annually during the early 1970s in Colombia and Peru alone. Continued illegal hunting, along with habitat loss and water pollution, jeopardizes the survival of this species and *L. provocax* (Mason and Macdonald 1986; Thornback and Jenkins 1982).

Giant otter *(Pteronura brasiliensis brasiliensis)*, photo from New York Zoological Society.

CARNIVORA; MUSTELIDAE; **Genus PTERONURA**
Gray, 1867

Giant Otter

The single species, *P. brasiliensis*, originally was found in Colombia, Venezuela, the Guianas, eastern Ecuador and Peru, Brazil, Bolivia, Paraguay, Uruguay, and northeastern Argentina (Cabrera 1957; Thornback and Jenkins 1982).

The remainder of this account is based largely on Duplaix's (1980) study of *Pteronura* in Surinam. Head and body length is 864–1,400 mm and tail length is 330–1,000 mm. Males weigh 26–34 kg and females, 22–26 kg. The short fur generally appears brown and velvetlike when dry and shiny black chocolate when wet. On the lips, chin, throat, and chest there are often creamy white to buff splotches, which may unite to form a large white "bib." The feet are large, and thick webbing extends to the ends of the five clawed digits. The tail is thick and muscular at the base but becomes dorsoventrally flattened with a noticeable bilateral flange. There are subcaudal anal glands for secretion of musk.

The giant otter is found mainly in slow-moving rivers and creeks within forests, swamps, and marshes. It prefers waterways with gently sloping banks that have good cover. At certain points along a stream, areas of about 50 sq meters are cleared and used for rest and grooming. Some of these sites have dens, which consist of one or more short tunnels leading to a chamber about 1.2–1.8 meters wide. *Pteronura* seems clumsy on land but may move a considerable distance between waterways. When swimming slowly or remaining stationary in the water it paddles with all four feet. When swimming at top speed it depends largely on undulations of the tail and uses the feet for steering. It is entirely diurnal. During the dry season, when cubs are being reared, activity is generally restricted to one portion of a waterway. In the wet season movements are far more extensive, and spawning fish are followed into the flooded forest. Prey is caught with the mouth and then may be held in the forepaws while being consumed. Small fish may be eaten in the water, but larger prey is taken to shore. The diet consists mainly of fish and crabs.

Group home range, including land area, measures about 12 km in both length and width. During the dry season, at least, several kilometers of stream form a defended territory. Both sexes regularly patrol and mark the area, but groups tend to avoid one another, and fighting is evidently rare. *Pteronura* is more social than *Lutra*. A population includes both resident groups and solitary transients. As many as 20 individuals have reportedly been seen together, but groups of 4–8 are usually observed. A group consists of a mated adult pair, 1 or more subadults, and 1 or more young of the year. There is a high degree of pair bonding and group cohesiveness. A male and female stay together and share the same den even when cubs are present. *Pteronura* is much noisier than *Lutra*. Nine vocalizations have been distinguished, including screams of excitement, often given while swimming with the forepart of the body steeply out of the water, and coos upon close intraspecific contact.

Although data are scanty, in the wild the young apparently are born from late August to early October, at the start of the dry season. If the first litter is lost, a second is sometimes produced from December to April. The gestation period is 65–70 days. The number of young per litter is one to five, usually one to three. The cubs weigh about 200 grams at birth and are able to eat solid food by 3–4 months. They remain with the parents at least until the birth of the next litter and probably for some time afterward. In a captive colony in Germany births occurred throughout the year, sexual maturity was attained at about 2 years, and one individual lived at least 14 years and 6 months (Hagenbeck and Wunnemann 1992).

The giant otter is classified as vulnerable by the IUCN and as endangered by the USDI and is on appendix 1 of the CITES. It has become very rare or has entirely disappeared over vast parts of its range. The main factor in its decline is excessive hunting by people for its large and valuable pelt. Because of its noise, diurnal habits, and tendency to approach intruders, it is relatively easy to locate and kill. In the early 1980s a skin could be sold by a hunter for the equivalent of U.S. $50 and on the market in Europe for $250. Such trade is now prohibited, but illegal hunting continues, facilitated by the opening of wilderness habitat (Thornback and Jenkins 1982). Recent surveys indicate that *Pteronura* survives in viable numbers in several parts of South America but has nearly or completely disappeared from Argentina, Uruguay, and southeastern Brazil (Foster-Turley, Macdonald, and Mason 1990).

CARNIVORA; MUSTELIDAE; Genus AONYX
Lesson, 1827

Clawless Otters

There are three subgenera and three species (Chasen 1940; Coetzee *in* Meester and Setzer 1977; Corbet and Hill 1992; Ellerman and Morrison-Scott 1966; Lekagul and McNeely 1977; Rosevear 1974):

subgenus *Aonyx* Lesson, 1827

A. capensis, Senegal to Ethiopia, and south to South Africa;

subgenus *Paraonyx* Hinton, 1921

A. congica, southeastern Nigeria and Gabon to Uganda and Burundi;

subgenus *Amblonyx* Rafinesque, 1832

A. cinerea, northwestern India to southeastern China and Malay Peninsula, southern India, Hainan, Sumatra, Java, Borneo, Riau Archipelago, Palawan.

Amblonyx was treated as a separate genus by Wozencraft (*in* Wilson and Reeder 1993) but not by Corbet and Hill (1991, 1992).

In the African species, *A. capensis* and *A. congica,* head and body length is 600–1,000 mm, tail length is 400–710 mm, and weight is 13–34 kg. In the smaller *A. cinerea,* of Asia, head and body length is 450–610 mm, tail length is 250–350 mm, and weight is 1–5 kg. The general coloration is brown, with paler underparts and sometimes white markings on the face, throat, and chest.

Aonyx differs from *Lutra* and *Pteronura* in having webbing that either does not extend to the ends of the digits or is entirely lacking and in having much smaller claws. In *A. congica* all the toes bear small, blunt claws; in *A. cinerea* the claws of adults are only minute spikes that do not project beyond the ends of the digital pads; and in *A. capensis* the only claws are tiny ones on the third and fourth toes of the hind feet. In association with these adaptations, *Aonyx* has developed very sensitive forepaws and considerable digital movement. *A. capensis* and *A. cinerea* have relatively large, broad cheek teeth, apparently for purposes of crushing the shells of crabs and mollusks. *A. congica* has lighter and sharper dentition, more adapted to cutting flesh.

The general habitat of *A. capensis* varies from dense rainforest to open coastal plain and semiarid country. The species is usually found near water, preferring quiet ponds and sluggish streams, but may sometimes wander a considerable distance overland. In coastal areas it has been seen to forage both in the sea and in adjoining freshwater streams and marshes (Verwoerd 1987). It is mainly nocturnal but may be active by day in areas remote from human disturbance. It dens under boulders or driftwood, in crannies under ledges, or in tangles of vegetation. It apparently does not dig its own burrow. *A. cinerea* occurs in rivers, creeks, estuaries, and coastal waters (Lekagul and McNeely 1977). *A. congica* is seemingly found only in small, torrential mountain streams within heavy rainforest.

These otters use their sensitive and dexterous forepaws to locate prey in mud or under stones. Captive *A. capensis* usually take food with the forepaws and do not eat directly off the ground. In the wild this species also catches most of its food with the forefeet, not with the mouth as do *Lutra* and *Pteronura* (Rowe-Rowe 1978b). The diet of both *A. capensis* and *A. cinerea* seems to consist mainly of crabs, other crustaceans, mollusks, and frogs; fish are relatively unimportant. There is thus apparently little competition for food between *Aonyx* and the fish-eating *Lutra* where both genera occur together. Piles of cracked crab and mollusk shells are signs of the presence of *A. capensis*. Donnelly and Grobler (1976) observed this species to use hard objects as anvils on which to break open mussel shells.

Little has been recorded about the habits of *A. congica*. Because of its scanty hair, weakly developed facial vibrissae, digital structure, and dental features, there has been speculation that it is more terrestrial than other otters. It is thought to feed mainly on relatively soft matter, such as small land vertebrates, eggs, and frogs. If this supposition is correct, there would be little competition for food between *A. congica* and the other kinds of otters that may occur in the same region.

Arden-Clarke (1986) studied a population of *A. capensis* inhabiting a rugged, densely vegetated environment along the coast of South Africa. The mean population density there was 1/1.9 km of coast, and dens were spaced at intervals of 470 meters. A radio-tracked adult male had a minimum home range of 19.5 km of coast, with a core area of 12.0 km, where it spent most of its time. An adult female had a 14.3-km home range with a 7.5-km core area. There apparently was a clan-type social organization, with groups of related animals defending joint territories. The home ranges of 4 adult males overlapped completely, and some of these animals were seen foraging together. In another

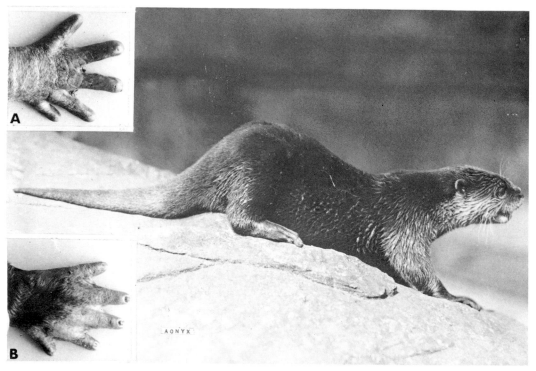

African clawless otter *(Aonyx capensis)*, photo from Zoological Society of London. Insets: A. Forefoot; B. Hind foot; photos by U. Rahm.

coastal area, Verwoerd (1987) found that an adult male might maintain a loose association with a female and cubs. *A. capensis* emits powerful, high-pitched shrieks when disturbed or trying to attract attention. According to Timmis (1971), *A. cinerea* lives in loose family groups of about 12 individuals and has a vocabulary of 12 or more calls, not including basic instinctive cries.

Births of *A. capensis* have been recorded in July and August in Zambia, and young have been found in March and April in Uganda (Kingdon 1977). There is probably no set breeding season in West Africa (Rosevear 1974). Most births in a coastal area of South Africa occurred in December and January (Verwoerd 1987). This species has a gestation period of 63 days and a litter size of two young. The

Oriental small-clawed otter *(Aonyx cinerea)*, photo by Lim Boo Liat.

young remain with the parents for at least 1 year. Female *A. cinerea* have an estrous cycle of 24–30 days, with an estrus of 3 days. They may produce two litters annually. The gestation period is 60–64 days, and litters contain one to six young, usually one or two. They open their eyes after 40 days, first swim at 9 weeks, and take solid food after 80 days (Duplaix-Hall 1975; Lekagul and McNeely 1977; Leslie 1971; Timmis 1971). Little is known of the reproduction of *A. congica*, but it probably has a gestation period of about 2 months, gives birth to two or three young, and attains sexual maturity at about 1 year. A captive specimen of *A. capensis* lived 14 years and a captive *A. cinerea* lived about 16 years (Marvin L. Jones, Zoological Society of San Diego, pers. comm., 1995).

If captured when young, these otters make intelligent and charming pets, though such activity may be illegal and detrimental to wild populations. *A. cinerea* has been trained to catch fish by Malay fishermen. The fur of *Aonyx* is not as good as that of *Lutra*; nonetheless, the subspecies *A. congica microdon* of Nigeria and Cameroon has declined seriously through uncontrolled commercial hunting. It is on appendix 1 of the CITES (other subspecies are on appendix 2) and is listed as endangered by the USDI. Data compiled by Foster-Turley, Macdonald, and Mason (1990) indicate that it is extremely rare. These data also show that the species *A. capensis* has become rare throughout West Africa, mainly in association with deforestation. Both *A. congica* and *A. cinerea* are designated as near threatened by the IUCN, and both *A. capensis* and *A. cinerea* are on appendix 2 of the CITES. *A. cinerea* apparently remains more common in southeastern Asia than the species of *Lutra* that occur there, but it is declining because of habitat loss and pollution and seems rare in most mainland parts of its range (Foster-Turley, Macdonald, and Mason 1990).

CARNIVORA; MUSTELIDAE; Genus ENHYDRA
Fleming, 1822

Sea Otter

The single species, *E. lutris*, was originally found in coastal waters off Hokkaido, Sakhalin, Kamchatka, the Commander Islands, the Pribilof Islands, the Aleutians, southern Alaska, British Columbia, Washington, Oregon, California, and western Baja California (Estes 1980). There are three subspecies: *E. l. lutris*, Hokkaido to the Commander Islands; *E. l. kenyoni*, Aleutians to Oregon; and *E. l. nereis*, California and Baja California (Wilson et al. 1991). The recent confirmation of the validity of these subspecies is important, both for consideration in reintroduction planning and to dispel claims that certain populations of bioconservation concern are of no systematic significance.

Head and body length is usually 1,000–1,200 mm and tail length is 250–370 mm. Males weigh 22–45 kg and females, 15–32 kg (Estes 1980). The color varies from reddish brown to dark brown, almost black, except for the gray or creamy head, throat, and chest. Albinistic individuals are rare. The head is large and blunt, the neck is short and thick, and the legs and tail are short. The ears are short, thickened, pointed, and valvelike. The hind feet are webbed and flattened into broad flippers; the forefeet are small and have retractile claws. *Enhydra* is the only carnivore with only four incisor teeth in the lower jaw. The molars are broad, flat, and well adapted to crushing the shells of such prey as crustaceans, snails, mussels, and sea urchins. Unlike most mustelids, the sea otter lacks anal scent glands. Females have two abdominal mammae (Estes 1980).

The sea otter differs from most marine mammals in that it lacks an insulating subcutaneous layer of fat. For protection against the cold water it depends entirely on a layer of air trapped among its long, soft fibers of hair. If the hair becomes soiled, as if by oil, the insulating qualities are lost and the otter may perish. The underfur, about 25 mm long, is the densest mammalian fur, averaging about 100,000 hairs per sq cm (Rotterman and Simon-Jackson 1988). It is protected by a scant coat of guard hairs.

Although the sea otter is a marine mammal, it rarely ventures more than 1 km from shore. According to Estes (1980), it forages in both rocky and soft-sediment communities on or near the ocean floor. Off California *Enhydra* seldom enters water of greater depth than 20 meters, but in the Aleutians it commonly forages at depths of 40 meters or more; the maximum confirmed depth of a dive was 97 meters. The usual period of submergence is 52–90 seconds, and the longest on record is 4 minutes and 25 seconds. The sea otter is capable of spending its entire life at sea but sometimes rests on rocks near the water. Such hauling-out behavior is more common in the Alaskan population than in that of California. *Enhydra* walks awkwardly on land. When supine on the surface of the water, it moves by paddling with the hind limbs and sculling with the tail. For rapid swimming and diving it uses dorsoventral undulations of the body. It can attain velocities of up to 1.5 km/hr on the surface and 9 km/hr for short distances underwater. The sea otter is generally diurnal, with crepuscular peaks and a midday period of rest (Riedman and Estes 1990). It often spends the night in a kelp bed, lying under strands of kelp to avoid drifting while sleeping. It sometimes sleeps with a forepaw over the eyes. Daily movements usually extend over a few kilometers, and there may be local seasonal movements but no extensive migrations (Riedman and Estes 1990).

The diet consists mainly of slow-moving fish and marine invertebrates, such as sea urchins, abalone, crabs, and mollusks (Estes 1980). Prey is usually captured with the forepaws, not the jaws. *Enhydra* floats on its back while eating and uses its chest as a "lunch counter." It is one of the few mammals known to use a tool. While floating on its back, it places a rock on its chest and then employs the rock as an anvil for breaking the shells of mussels, clams, and large sea snails in order to obtain the soft internal parts. This activity is most frequent in the population off California, and recent research there has shown considerable variation, such as using the rock as a hammer or using one rock as a hammer and another as an anvil (Riedman and Estes 1990). The sea otter requires a great deal of food: it must eat 20–25 percent of its body weight every day. It obtains about 23 percent of its water needs from drinking sea water and most of the rest from its food.

According to Estes (1980), *Enhydra* is basically solitary but sometimes rests in concentrations of as many as 2,000 individuals. Males and females usually come together only briefly for courtship and mating. At most times there is sexual segregation, with males and females occupying separate sections of coastline. Males usually occur at higher densities. Recent studies (Garshelis and Garshelis 1984; Garshelis, Johnson, and Garshelis 1984; Jameson 1989; Loughlin 1980; Rotterman and Simon-Jackson 1988) indicate that during the breeding season (which may be for most of the year) some males move into the areas occupied by females and establish territories. Such behavior has been documented in both Alaska and California, and males have been observed to return to the same place for up to seven years. The most favorable territories—those that seem to attract the most females—are characterized by availability of food; density of canopy-forming kelp, to which the ani-

Sea otters *(Enhydra lutris):* A. Juvenile walking. B. *E. lutris* floating on its back eating the head of a large codfish. C. Mother floating on her back with newborn pup. Photos by Karl W. Kenyon.

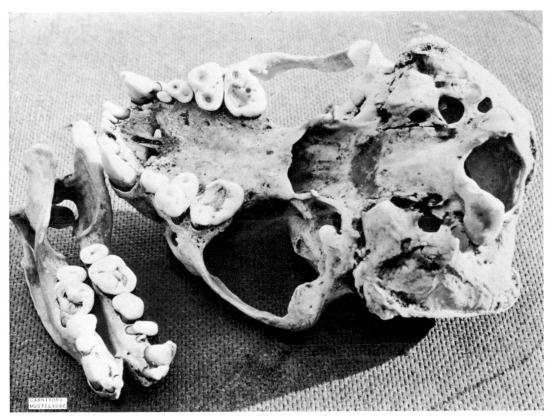

Skull and lower jaw of a sea otter *(Enhydra lutris)*, showing the cavities that develop in the teeth of old animals because of the hard, rough materials that they eat, photo by H. Robert Krear.

mals can attach themselves for secure resting; and associated shoreline features that provide a degree of shelter from the open sea (Riedman and Estes 1990). The boundaries of these territories are vigorously patrolled and intruding males are repulsed, but serious fighting is rare. The owner seeks to mate with any female that enters, though sometimes a pair bond is formed for a few days or weeks. Male territories are usually about 20–50 ha. and are smaller than female home ranges in the same area. Annual movements of both sexes frequently cover 50–100 km, considering foraging, breeding, and the passage of males between their territories and the all-male areas. McShane et al. (1995) described 10 vocalizations of *Enhydra*, including screams of distress, heard especially when mothers and young were separated, and coos, heard mostly when individuals were content or in familiar company.

Reproductive data have been summarized by Estes (1980) and Riedman and Estes (1990). Breeding occurs throughout the year, but births peak in late May and June in the Aleutians and from January to March off California; in the latter area there apparently is a secondary peak in late summer and early autumn. Males may mate with more than one female during the season. Pair bonding, and presumably estrus, lasts about 3 days. Females are capable of giving birth every year but usually do so at greater intervals. If a litter does not survive, the female may experience a postpartum estrus. Females are known to adopt and nurse orphaned pups. Reports of the period of pregnancy range from 4 to 12 months, and delayed implantation is probably involved. Estimates of the period of actual implanted ges-

tation vary from about 4.0 to 5.5 months. Births probably occur most often in the water. There is normally a single offspring. About 2 percent of births are multiple, but only one young can be successfully reared. The pup weighs 1.4–2.3 kg at birth. While still small it is carried, nursed, and groomed on the mother's chest as the mother swims on her back. The pup begins to dive in the second month of life. It may take some solid food shortly after birth but may nurse almost until it attains adult size. The period of dependency on the mother is thought to be about 5–8 months. Females become sexually mature at about 4 years. Males are capable of mating at 5–6 years but usually do not become active breeders until several more years have passed. Wendell, Ames, and Hardy (1984) concluded that the reproductive cycle of the California population is shorter than elsewhere, with some females giving birth each year. It now is known that Alaskan females also are capable of annual reproduction (Garshelis, Johnson, and Garshelis 1984). According to Rotterman and Simon-Jackson (1988), a captive male fathered young when at least 19 years old, and maximum estimated longevity for wild females is 23 years. Data from Alaska indicate that sea otter populations have the potential to increase by about 20 percent annually, but the population of California has tended to increase by only 5 percent a year, probably because of high preweaning mortality of young that may be associated with pollutants imparted through lactation (Estes 1990; Riedman et al. 1994).

The fur of the sea otter may be the most valuable of any mammal's. During the 1880s prices on the London market ranged from $105 to $165 per skin. By 1903, when the species had become scarce, large, high-quality skins sold for up to $1,125. Pelts taken in Alaska in the late 1960s, during a brief reopening of commercial activity, sold for an average of $280 each (Kenyon 1969).

Estimates of the original numbers of *Enhydra* are 150,000–300,000 (Riedman and Estes 1990). Intensive exploitation of the genus was begun by the Russians in 1741. Hunting was uncontrolled until 1799, when some conservation measures were established. Unregulated killing resumed in 1867, when Alaska was purchased by the United States. By 1911, when the sea otter was protected by a treaty among the United States, Russia, Japan, and Great Britain, probably only 1,000–2,000 of the animals survived worldwide (Kenyon 1969). Under protection of the treaty, state and national laws, and finally the United States Marine Mammal Protection Act of 1972, the sea otter has steadily increased in numbers and distribution. There are now probably 100,000–150,000 individuals in the major populations off southwestern and south-central Alaska and another 17,000 off Kamchatka and the Kuril and Commander islands in Russia. Alaskan populations are subject to limited killing for native subsistence purposes, may come into conflict with shellfisheries, and are potentially jeopardized by oil spills. Reintroduced populations (from Alaskan stock) apparently have been established off southeastern Alaska (now numbering 4,500 animals), Vancouver Island (about 350), and Washington (280). Reintroduced groups off Oregon and in the Pribilof Islands do not seem to have done well and have all but disappeared (Estes 1980; Jameson 1993; Jameson et al. 1982; Riedman and Estes 1990; Rotterman and Simon-Jackson 1988).

The magnitude of the threat posed by oil spills from damaged tanker ships was tragically demonstrated by the *Exxon Valdez* disaster in 1989. As many as 5,000 sea otters are estimated to have been killed directly (*Oryx* 26 [1992]: 195). Hence, a significant part of the entire world's population of *Enhydra* was destroyed in this single incident. More than 1,000 dead or dying sea otters were actually recovered. However, an intensive effort was made by the U.S. Fish and Wildlife Service and cooperating organizations to save as many of the affected animals as possible. Several hundred were rescued and treated, and 197 were released back into the wild (Bayha and Kormendy 1990).

The southern sea otter (subspecies *E. l. nereis*), which originally ranged from Baja California to at least Oregon was generally considered extinct by 1920. Apparently, however, a group of 50–100 individuals survived off central California in the vicinity of Monterey. In 1938 the presence of this population became generally known. By the 1970s it had grown to include about 1,800 animals, but subsequently numbers stabilized or even declined. The sea otter now regularly occurs along about 350 km of the central California coast, and there have been scattered reports of individuals from southern California and northern Baja California. As this population increased, there was concern that stocks of abalone and other shellfish were being depleted. Some parties with a commercial or recreational interest in these stocks have advocated control of the sea otter population, and there have been cases of illegal killing. Some otters also are being drowned accidentally in fishing nets. Another fear is that an oil spill, associated with either the extensive tanker traffic in the area or offshore drilling, could devastate the population (Armstrong 1979; Carey 1987; Estes and VanBlaricom 1985; Leatherwood, Harrington-Coulombe, and Hubbs 1978; U.S. Fish and Wildlife Service 1980).

There has been concern that the genetic viability of *E. l. nereis* was severely reduced when it approached extinc-

tion earlier in the century, but Ralls, Ballou, and Brownell (1983) calculated that the existing population should theoretically retain about 77 percent of the original diversity and that transplanted colonies should also be viable. An effort to establish such a colony was started in 1987 when 63 otters taken from the main California population were released around San Nicolas Island (Brownell and Rathbun 1988). By June of 1990, 137 animals had been brought there, but only 15 were known to have remained in the area (Riedman and Estes 1990) and the experiment was called unsuccessful (*Oryx* 28 [1994]: 95). The southern sea otter is listed as threatened by the USDI and is on appendix 1 of the CITES (other subspecies of *E. lutris* are on appendix 2).

CARNIVORA; Family VIVERRIDAE

Civets, Genets, and Linsangs

This family of 19 Recent genera and 35 species is found in southwestern Europe, southern Asia, the East Indies, Africa, and Madagascar. Certain genera have been introduced to areas in which the family does not naturally occur. The following sequence of genera, grouped in 5 subfamilies, is based on the classifications of Coetzee (*in* Meester and Setzer 1977), Corbet and Hill (1992), Ellerman and Morrison-Scott (1966), and Ewer (1973):

Subfamily Viverrinae (civets, genets, linsangs), genera *Viverra, Civettictis, Viverricula, Genetta, Osbornictis, Poiana, Prionodon;*

Subfamily Nandiniinae (African palm civet), genus *Nandinia;*

Subfamily Paradoxurinae (Asian palm civets), genera *Arctogalidia, Paradoxurus, Paguma, Macrogalidia, Arctictis;*

Subfamily Hemigalinae (banded palm and otter civets), genera *Hemigalus, Diplogale, Chrotogale, Cynogale;*

Subfamily Euplerinae (Malagasy civets), genera *Fossa, Eupleres.*

The Viverridae often have been considered to include the family Herpestidae (see account thereof). Wozencraft (*in* Wilson and Reeder 1993) placed the subfamily Cryptoproctinae in the Viverridae, but it is here included in the Herpestidae. Wozencraft is followed here in the use of the name Euplerinae rather than Fossinae. Wozencraft also is followed in placing *Nandinia* in its own subfamily. This is essentially a compromise position between that of Meester et al. (1986) and numerous other authorities who treated *Nandinia* as a member of the Paradoxurinae and that of Hunt and Tedford (1993), who suggested that this genus is so strikingly different from other viverrids that it actually represents a separate family. The latter view is based largely on the structure of the basicranium, a highly complex region of the skull thought to express phylogenetic relationships. In particular, *Nandinia* has a primitive auditory bulla, differing from that of other viverrids and herpestids in that it does not inflate during ontogeny and has a single (rather than double) chamber, no septum bullae, and a cartilaginous (rather than ossified) caudal entotympanic that uniquely intervenes between the ectotympanic and rostral entotympanic.

Viverrids are characteristic small and medium-sized carnivores of the Old World. Head and body length is 350–950 mm, tail length is 130–900 mm, and adult weight is 0.6–20.0 kg. There are various striped, spotted, and uniform color patterns. In some genera the tail is banded or ringed. The body is long and sinewy with short legs and generally a long, bushy tail. One genus *(Arctictis)* has a truly prehensile tail. The head is elongate and the muzzle is pointed. Most genera have five toes on each foot. The claws are retractile or semiretractile in some genera. Female viverrids usually have two or three pairs of abdominal mammae. Males have a baculum.

Most viverrids have scent glands in the anal region that secrete a nauseous-smelling fluid as a defensive measure. The conspicuous pattern of pelage in some genera has been interpreted as being a warning that the fetid secretion is present. Such a color pattern is also found in skunks and certain other members of the family Mustelidae. Rubbed on various objects, the secretion of the scent glands is recognized by other individuals of the same species and is probably used to communicate various information.

The skull is usually long and flattened. The dental formula is: (i 3/3, c 1/1, pm 3–4/3–4, m 1–2/1–2) × 2 = 32–40. The second lower incisor is raised above the level of the first and third, the canines are elongate, and the carnassials are developed.

Viverrids are essentially forest inhabitants, but they also live in dense brush and thick grass. They are either diurnal or nocturnal and shelter in any convenient retreat, usually a hole in a tree, a tangle of vines, ground cover, a cave, a crevice, or a burrow. A few species dig their own burrows. Those species living near people sometimes seek refuge under rafters or in the drains of houses. Those viverrids that walk on their digits (such as *Genetta*) have a gait described as "a waltzing trot," whereas the members of the family that walk on their soles, with their heels touching the ground (such as *Arctictis*), have a bearlike shuffle. Many genera are agile and extremely graceful in their movements. A number of species are skillful climbers; some apparently spend most of their lives in trees. Some genera take to water readily and swim well; two, *Osbornictis* and *Cynogale*, are semiaquatic. Sight, hearing, and smell are acute.

Viverrids may fight when cornered. They seek their prey in trees and on the ground either by stalking or by pouncing from a hiding place. They eat small vertebrates and various invertebrates and occasionally consume vegetable matter such as fruit, bulbs, and nuts. Carrion is taken by some species. Viverrids are solitary or live in pairs or groups. Breeding may occur seasonally or throughout the year. A number of genera have two litters annually. The one to six offspring are born blind but haired. Most species probably have a potential longevity of 5–15 years.

The secretion of the scent glands, civet, is obtained from several genera *(Civettictis, Viverra,* and *Viverricula)* for both perfumery and medicinal purposes. Some viverrids are tamed and kept to extract the musk. They may also be kept as pets. Viverrids occasionally kill poultry but also prey on rodents.

The geological range of this family is late Eocene to Recent in Europe, early Oligocene to Recent in Asia, early Miocene to Recent in Africa, and Pleistocene to Recent in Madagascar (Stains 1984).

CARNIVORA; VIVERRIDAE; Genus VIVERRA
Linnaeus, 1758

Oriental Civets

There are two subgenera and three species (Chasen 1940; Corbet and Hill 1992; Ellerman and Morrison-Scott 1966; Medway 1978; Taylor 1934):

subgenus *Viverra* Linnaeus, 1758

V. zibetha, Nepal and Bangladesh to southeastern China and Malay Peninsula, Hainan;
V. tangalunga, Malay Peninsula, Riau Archipelago, Sumatra, Bangka, Borneo, Belitung and Karimata islands, Palawan and Calamian Islands (Philippines);

Oriental civet *(Viverra tangalunga)*, photo by Ernest P. Walker.

subgenus *Moschothera* Pocock, 1933

V. megaspila, extreme southwestern peninsular India, extreme southeastern China, Burma, Thailand, Indochina, Malay Peninsula and some nearby small islands, possibly Sumatra.

Civettictis sometimes is regarded as a third subgenus of *Viverra* (see Coetzee *in* Meester and Setzer 1977). Corbet and Hill (1992) followed Wozencraft (1989*a* and *in* Wilson and Reeder 1993) in listing *V. civettina* of southwestern India as distinct from *V. megaspila* but noted that specific separation is very doubtful based on examination of the characters of the few available specimens.

Head and body length is 585–950 mm, tail length is 300–482 mm, and weight is 5–11 kg. The fur, especially in winter, is long and loose. It is usually elongated in the median line of the body, forming a low crest or mane. The color pattern of the body is composed of black spots on a grayish or tawny ground color. The sides of the neck and throat are marked with black and white stripes—usually three black and two white collars. The crest is marked by a black spinal stripe that runs from the shoulders to the tail, and the tail is banded or ringed with black and white. The feet are black. In *V. zibetha* and *V. tangalunga* the third and fourth digits of the forefeet are provided with lobes of skin that act as protective sheaths for the retractile claws. *Viverra* is distinguished from *Viverricula* by its larger size, by the presence of a dorsal crest of erectile hairs, and by the insertion of the ears, the inner edges of which are set farther apart on the forehead.

Oriental civets occur in a wide variety of habitats in forest, brush, and grassland. They stay in dense cover by day and come out into the open at night. They are mainly terrestrial and often live in holes in the ground dug by other animals. They apparently can climb readily but seldom do so. Like *Viverricula*, they are often found near villages and are common over most of their range. Like most civets, they are easily trapped. They are vigorous hunters, killing small mammals, birds, snakes, frogs, and insects and taking eggs, fruit, and some roots. The species *V. zibetha* has been observed fishing in India, and the remains of crabs have been found in the stomachs of two individuals from China.

Viverra is generally solitary. An adult male *V. zibetha*, radio-tracked by Rabinowitz (1991) in Thailand, moved within an area of 12 sq km in a period of 7 months. Its average monthly range was 5.4 sq km and its average daily movement was 1.7 km. *V. zibetha* is said to breed all year and to bear two litters annually (Lekagul and McNeely 1977). The number of young per litter is one to four, usually two or three. The young are born in a hole in the ground or in dense vegetation. The young of *V. zibetha* open their eyes after 10 days, and weaning begins at 1 month (Medway 1978). There are captive specimens of *V. zibetha* in the Ahmedabad Zoo in India that are more than 20 years old (Smielowski 1986).

Viverra is one of the sources of civet, a substance used commercially in producing perfume. Because of this function, *V. tangalunga* has been introduced through much of the East Indies, including most of the Philippines, Sulawesi, the Sangihe Islands, and the Moluccan islands of Buru, Seram, Amboina, and Halmahera (Corbet and Hill 1992; Groves 1976; Laurie and Hill 1954). The Malabar civet, *V. civettina*, known only from an isolated belt of rainforests in the Western Ghats of southwestern India, is recognized by the IUCN as a species distinct from *V. megaspila* and classified as critically endangered; fewer than 250 mature individuals are thought to survive. This form also is listed as endangered by the USDI. It evidently has become very rare through hunting by people and loss of habitat to agriculture. No specimens had been known to science since the nineteenth century, but in 1987 three were captured at Elayur (Schreiber et al. 1989). A survey in 1990 determined that a number of other individuals had been killed or captured in the last few decades and that a small population survives but is immediately jeopardized by land clearing for rubber plantations (Ashraf, Kumar, and Johnsingh 1993).

CARNIVORA; VIVERRIDAE; **Genus CIVETTICTIS** *Pocock, 1915*

African Civet

The single species, *C. civetta*, is found from Senegal to Somalia and south to northern Namibia and northeastern South Africa. Recognition of *Civettictis* as a distinct genus is in keeping with Ewer (1973), Kingdon (1977), Ray (1995), and Rosevear (1974). Some other authorities, such as Coetzee (*in* Meester and Setzer 1977) and Rowe-Rowe (1978*b*), have included *C. civetta* in the genus *Viverra*.

Head and body length is 670–890 mm, tail length is 340–470 mm, and weight is 7–20 kg (Kingdon 1977; Ray 1995). The color is black with white or yellowish spots, stripes, and bands. There is much variation in the pattern of markings, and some individuals are melanistic. The long and coarse hair is thick on the tail. The perineal glands under the tail contain the oily scented matter used commercially in making perfume. All the feet have five claws and the soles are hairy. From *Viverra*, *Civettictis* is distinguished by much larger molar teeth and a far broader lower carnassial (Rosevear 1974).

The African civet is widely distributed in both forests and savannahs, wherever long grass or thickets are sufficient to provide daytime cover (Ewer and Wemmer 1974). It seems to use a permanent burrow or nest only to bear young. It is nocturnal and almost completely terrestrial but takes to water readily and swims well. The omnivorous diet includes carrion, rodents, birds, eggs, reptiles, frogs, crabs, insects, fruits, and other vegetation. Poultry and young lambs are sometimes taken (Rosevear 1974).

Civettictis is generally solitary but has a variety of visual, olfactory, and auditory means of communication. Individuals may have defined and well-marked territories. The scent glands have a major social role, leaving scent along a path to convey information, such as whether a female is in estrus (Kingdon 1977). There are three agonistic vocalizations—the growl, cough-spit, and scream—but the most commonly heard sound is the "ha-ha-ha" used in making contact (Ewer and Wemmer 1974).

Available data on reproduction (Ewer and Wemmer 1974; Hayssen, Van Tienhoven, and Van Tienhoven 1993; Kingdon 1977; Mallinson 1973, 1974; Ray 1995; Rosevear 1974) suggest that breeding occurs throughout the year in West Africa, from March to October in East Africa, and in the warm, wet summer from August to January in South Africa. Females are polyestrous and there may be two or even three litters annually. The gestation period is usually 60–72 days but is occasionally extended to as many as 81 days, perhaps because of delayed implantation. The number of young per litter is one to four, usually two or three. The young are born fully furred, weigh about 300 grams at birth, open their eyes within a few days, cease suckling at 14–20 weeks, and attain sexual maturity at about 1 year. According to Jones (1982), a captive lived for 28 years.

In Ethiopia, and to a lesser extent in other parts of Africa,

African civet *(Civettictis civetta)*, photo from New York Zoological Society.

the natives keep civets in captivity and remove the musk from them several times a week. An average animal yields 3–4 grams weekly. The natives do not raise the civets, however, but merely capture wild ones. In 1934 Africa produced about 2,475 kg of musk with a value of U.S. $200,000. In that same year the United States imported 200 kg of musk. The production of civet musk is an old industry; King Solomon's supply came from East Africa. Rosevear (1974) reported that the trade in civet musk now has diminished considerably. However, Schreiber et al. (1989) indicated that in 1988 there still were more than 2,700 captive civets in Ethiopia and that their musk, exported mainly to France, was selling for about $438 per kg.

CARNIVORA; VIVERRIDAE; **Genus VIVERRICULA**
Hodgson, 1838

Lesser Oriental Civet, or Rasse

The single species, *V. indica,* occurs naturally from Pakistan, through most of India, to southeastern China and the Malay Peninsula as well as in Sri Lanka, Taiwan, Hainan, Sumatra, Java, the Kangean Islands, and Bali (Corbet and Hill 1992; Ellerman and Morrison-Scott 1966; Roberts 1977). The name *V. malaccensis* was used for this species by Medway (1978) and Lekagul and McNeely (1977).

Head and body length is 450–630 mm, tail length is 300–430 mm, and weight is usually 2–4 kg. The fur is harsh, rather coarse, and loose. The body color is buffy, brownish, or grayish and the feet are black. Small spots are present on the forequarters, and larger spots, tending to run into longitudinal lines, are present on the flanks. There are six to eight dark stripes on the back, and the tail is ringed black and white by six to nine rings of each color.

Viverricula is distinguished from *Viverra* by its smaller size, the absence of a dorsal crest of erectile hairs, and the insertion of the ears, the inner edges of which are set closer together on the forehead than those of *Viverra*. The muzzle is also shorter and more pointed. Internally the two genera differ in a number of cranial and dental features.

The rasse inhabits grasslands or forests. It probably ex-

cavates its own burrow (Roberts 1977) but may also shelter in thick clumps of vegetation, buildings, or drains (Lekagul and McNeely 1977). It is generally nocturnal but may be seen hunting by day in areas not populated by humans. It is mainly terrestrial but is said to climb well. It usually tries to escape from dogs by dodging and twisting through the underbrush. The diet consists of small vertebrates, carrion, insects and their grubs, fruits, and roots.

An adult male radio-tracked for six months in Thailand by Rabinowitz (1991) moved within a total area of 3.1 sq km, had an average monthly range of 0.83 sq km, and had an average daily movement of 500 meters. *Viverricula* is usually solitary but occasionally associates in pairs. It breeds throughout the year in Sri Lanka. Captives in Shanghai mate mostly from February to April and to some extent in August–September (Xu and Sheng 1994). The two to five young are born in a shelter on the ground. They are weaned after 4–4.5 months (Hayssen, Van Tienhoven, and Van Tienhoven 1993). A captive lived 10 years and 6 months (Jones 1982).

The rasse is kept in captivity by natives for the purpose of extracting the civet that is secreted and retained in sacs close to the genitals in both sexes. The removal of this secretion is accomplished by scraping the inside of the sac with a spoonlike implement. In India this secretion is used as a perfume to flavor the tobacco that is smoked by some natives. *Viverricula* was introduced by people to Socotra, the Comoro Islands, Madagascar, and perhaps the Philippines, probably for the production of civet. Its presence on the islands of Lombok and Sumbawa, to the east of Bali, also is thought to have resulted from introduction (Corbet and Hill 1992; Laurie and Hill 1954).

CARNIVORA; VIVERRIDAE; **Genus GENETTA**
Oken, 1816

Genets

There are 3 subgenera and 10 species (Ansell 1978; Coetzee *in* Meester and Setzer 1977; Crawford-Cabral 1981; Harrison and Bates 1991; Lamotte and Tranier 1983;

Lesser oriental civet *(Viverricula indica)*, photo by Ernest P. Walker.

Meester et al. 1986; Schlawe 1980; Schreiber et al. 1989; Skinner and Smithers 1990):

subgenus *Pseudogenetta* Dekeyser, 1949

G. thierryi, savannah zone from Senegal to the area south of Lake Chad;
G. abyssinica, Ethiopia, possibly Djibouti and northern Somalia;

subgenus *Paragenetta* Kuhn, 1960

G. johnstoni, southern Guinea, Liberia, Ivory Coast, possibly Ghana;

subgenus *Genetta* Oken, 1816

G. servalina, southern Nigeria to western Kenya;
G. victoriae, northern and eastern Zaire, Uganda;
G. genetta, France, Spain, Portugal, Balearic Islands, southwestern Saudi Arabia, Yemen, southern Oman, northwestern Africa, savannah zone of Africa from Senegal to northeastern Sudan and south to Tanzania, savannah and desert zone from southwestern Angola to southern Mozambique and South Africa;
G. angolensis, southern Zaire, central and northeastern

Angola, western Zambia, northern Mozambique, probably southern Tanzania, possibly northern Zimbabwe;
G. pardina, Gambia to Cameroon;
G. maculata, Senegal to Somalia, and south to Namibia and Natal (South Africa);
G. tigrina, Cape Province and Natal (South Africa), Lesotho.

Corbet (1978) suggested that the European populations of *G. genetta* are the result of introduction by human agency. Schlawe (1980) restricted the name *G. genetta* to Europe and northwestern Africa. He referred the populations to the south of the Sahara and on the southwestern Arabian Peninsula to a separate species, *G. felina,* and he showed that the reported presence of *Genetta* in Palestine is not correct. Later, Schlawe (1981) indicated that much work remains to be done before the systematics of this genus can be reasonably well understood. Corbet (1984) recognized the specific distinction of *G. felina,* but Corbet and Hill (1991), Meester et al. (1986), and Wozencraft (*in* Wilson and Reeder 1993) did not. Wozencraft included *G. pardina* within *G. maculata,* but the two were listed as separate species by Corbet and Hill (1991). Meester et al. (1986) considered *G. maculata* and *G. tigrina* to be conspecific, noting that the two intergraded over much of Natal. That same po-

Genets *(Genetta tigrina):* Top, photo by John Markham;
Bottom, photo by John Visser.

sition was taken by Coetzee (*in* Meester and Setzer 1977),
but the two taxa were treated as separate species by Corbet
and Hill (1991), Crawford-Cabral (1981), Crawford-Cabral
and Pacheco (1992), Schlawe (1981), and Wozencraft
(1989*a* and *in* Wilson and Reeder 1993). Some of those authorities used the name *G. rubiginosa* in place of *G. maculata.* Heard and Van Rompaey (1990) suggested that *G.
cristata* of southeastern Nigeria and southwestern
Cameroon may be a species distinct from *G. servalina.*

Head and body length is usually 420–580 mm, tail
length is 390–530 mm, and weight is 1–3 kg. Coloration is
variable, but the body is generally grayish or yellowish
with brown or black spots and blotches on the sides that
tend to be arranged in rows. A row of black erectile hairs is
usually present along the middle of the back. The tail has

black and white rings. Melanistic individuals seem to be
fairly common. Genets have a long body, short legs, a pointed snout, prominent and rounded ears, short and curved retractile claws, and soft and dense hair. They have the ability to emit a musky-smelling fluid from their anal glands.
Females have two pairs of abdominal mammae.

Genets inhabit forests, savannahs, and grasslands. They
are active at night, usually spending the day in rock
crevices, in burrows excavated by other animals, in a hollow
trees, or on large branches. They seem to return daily to the
same shelter. They climb trees to prey on nesting and roosting birds, but much of their food is taken on the ground.
They are silent and stealthy hunters; when stalking prey
they crouch until their body and tail seem to glide along
the ground. At the same time, the body seems to lengthen.
The genet's slender and loosely jointed body allows it to go
through any opening its head can enter. Nightly movements in Spain were found to average 2.78 km (Palomares
and Delibes 1994). The diet consists of any small animals
that can be captured, including rodents, birds, reptiles, and
insects. Genets sometimes take game birds and poultry.

Radio-tracking studies of *G. genetta* in Spain over a period of more than two years indicated an average maximum
home range size of 7.8 sq km (Palomares and Delibes 1994),
though one adult male wandered across 50 sq km during
about five months (Palomares and Delibes 1988). Home
ranges of adult males and females overlapped greatly, but
those of animals of the same sex were exclusive. Radio-
tracking studies of *G. maculata* in Kenya during June–August found average home ranges of 5.9 sq km for three
males and 2.8 sq km for two females (Fuller, Biknevicius,
and Kat 1990). Genets travel alone or in pairs. They communicate with one another by a variety of vocal, olfactory,
and visual signals. Breeding seems to correspond with the
wet seasons in both West and East Africa. In Kenya, for example, pregnant and lactating females have been taken in

May and from September to December. A pair of *G. genetta* in the National Zoo in Washington, D.C., regularly produced two litters per year, one in April–May and another in July–August. Gestation periods ranging from 56 to 77 days have been reported. The number of young per litter is one to four, usually two or three. The young weigh 61–82 grams at birth, begin to take solid food at 2 months, and attain adult weight at 2 years. One female *G. genetta* became sexually mature at about 4 years and produced young regularly until she died at the age of 13 years (Kingdon 1977; Rosevear 1974; Wemmer 1977). Another captive *G. genetta* lived 21 years and 6 months (Jones 1982).

According to Smit and Van Wijngaarden (1981), the genet has declined in Europe because of persecution for alleged depredations on game birds and poultry. In addition, its winter pelt is highly esteemed. A recently described subspecies on Ibiza Island in the Balearics, *G. genetta isabelae*, is classified as vulnerable by the IUCN. Although sometimes considered to represent an introduction by people long ago, this subspecies is clearly distinguishable from the nearby mainland populations of Europe and Africa (Schreiber et al. 1989). *G. cristata*, restricted to Nigeria and Cameroon, is recognized by the IUCN as a species distinct from *G. servalina* and classified as endangered. Although its survival has recently been confirmed, its only known habitat is rapidly being degraded (Heard and Van Rompaey 1990). There also is concern for *G. abyssinica* and *G. johnstoni*, both of which are apparently very rare, have not been definitely recorded from the field for many years, and oc-

cur in a limited area of habitat that is increasingly subject to human disturbance (Schreiber et al. 1989).

CARNIVORA; VIVERRIDAE; Genus OSBORNICTIS
J. A. Allen, 1919

Aquatic Genet

The single species, *O. piscivora*, is known only by about 30 specimens taken in northeastern Zaire (Coetzee *in* Meester and Setzer 1977; Hart and Timm 1978; Van Rompaey 1988).

An adult male had a head and body length of 445 mm, a tail length of 340 mm, and a weight of 1,430 grams; an adult female weighed 1,500 grams (Hart and Timm 1978). The body is chestnut red to dull red and the tail is black. There is a pair of elongated white spots between the eyes. The front and sides of the muzzle and the sides of the head below the eyes are whitish. Black spots and bands are absent, and the tail is not ringed. The pelage is long and dense, especially on the tail. The palms and soles are bare, not furred as in *Genetta* and other related genera. The skull is long and lightly built, and the teeth are relatively small and weak.

Osbornictis is among the rarest genera of carnivores. All specimens originated in areas of dense forest at elevations of 500–1,500 meters (Hart and Timm 1978). The genus is generally thought to be semiaquatic, as several

Aquatic genet *(Osbornictis piscivora)*, photo of mounted specimen by M. Colyn.

specimens have been taken in or near streams, and available evidence suggests that fish constitute a major part of the diet. Hart and Timm, for example, noted the following: the stomach of one specimen contained the remains of fish; natives of the area indicated that fish is the favored prey; the dentition of *Osbornictis* seems to be adapted to deal with slippery vertebrate prey, such as fish and frogs; and the bare palms may be an adaptation allowing the genet to feel for fish in muddy holes and then handle the prey. *Osbornictis* is apparently solitary. A pregnant female with a single embryo 15 mm long was taken on 31 December.

CARNIVORA; VIVERRIDAE; **Genus POIANA**
Gray, 1864

African Linsang, or Oyan

The single species, *P. richardsoni*, occurs from Sierra Leone to northern Zaire and on the island of Bioko (Fernando Poo) (Coetzee *in* Meester and Setzer 1977); the generic name reflects the occurrence on this island. Rosevear (1974) recognized the populations in the western part of the range of *Poiana* as a distinct species, *P. leightoni*.

The average head and body length is 384 mm and the average tail length is 365 mm (Rosevear 1974). The general color effect is light brownish gray to rusty yellow; dark brown to black spots and rings are present. Some individuals have alternating broad and narrow black bands on the tail, whereas others have only the broad bands. This genus differs from the Asiatic linsangs *(Prionodon)* in that the spots are smaller and show no tendency to run into bands or stripes except in the region of the head and shoulder. It also differs from them, and resembles *Genetta*, in having a narrow bare line on the sole of each hind foot.

The oyan is a forest animal and is nocturnal. According to Hans-Jürg Kuhn (Anatomisches Institut der Universität Frankfurt am Mein, pers. comm.), *Poiana* builds a round nest of green material, in which several individuals sleep for a few days, and then moves on and builds a new nest. The nests are at least two meters above the ground, usually higher. Although the oyan has been reported to sleep in the abandoned nests of squirrels, reliable hunters say that the reverse is true: the squirrels sleep in abandoned nests of *Poiana*. The diet includes cola nuts, other plant material, insects, and young birds. In the Liberian hinterland natives make medicine bags from the skins of *Poiana*.

Observations by Charles-Dominique (1978) in northeastern Gabon indicate a population density of 1/sq km. A lactating female has been noted in October. As in some other genera of viverrids, there may be two litters per year. The number of young per birth is 2 or 3. A captive oyan lived five years and four months (Jones 1982). The subspecies *P. r. liberiensis*, of Liberia, Ivory Coast, and Sierra Leone, is rare, isolated from the more easterly populations, and known only from about 14 museum specimens, including 2 killed in 1987 (Schreiber et al. 1989; Taylor 1989).

POIANA

African linsang *(Poiana richardsoni)*, photo from *Proc. Zool. Soc. London.*

CARNIVORA; VIVERRIDAE; **Genus PRIONODON**
Horsfield, 1822

Oriental Linsangs

There are two subgenera and two species (Chasen 1940;
Corbet and Hill 1992; Ellerman and Morrison-Scott 1966;
Lekagul and McNeely 1977):

subgenus *Prionodon* Horsfield, 1822

P. linsang (banded linsang), western and southern
Thailand, Tenasserim, Malay Peninsula, Sumatra,
Bangka, Java, Borneo, Belitung Island;

subgenus *Pardictis* Thomas, 1925

P. pardicolor (spotted linsang), eastern Nepal to northern
Indochina and nearby parts of southern China.

Head and body length is 310–450 mm and tail length is
304–420 mm. Medway (1978) listed the weight of *P. linsang* as 598–798 grams; on the average, *P. pardicolor* is
slightly smaller. In *P. linsang* the ground color varies from
whitish gray to brownish gray and becomes creamy on the
underparts. The dark pattern consists of four or five broad,
transverse black or dark brown bands across the back; there
is one large stripe on each side of the neck. The sides of the
body and legs are marked with dark spots, and the tail is
banded. Some individuals of *P. pardicolor* have a ground
color of orange buff, whereas others are pale brown. Black
spots on the upper parts are arranged more or less in longitudinal rows, and the tail has 8–10 dark rings.
    These animals are extremely slender, graceful, and beau-
tiful. The fur is short, dense, and soft; it has the appearance
and feel of velvet. The claws are retractile; claw sheaths are
present on the forepaws, and protective lobes of skin are
present on the hind paws. The skull is long, low, and narrow, and the muzzle is narrow and elongate. Unlike many
viverrids, *Prionodon* seems to be free from odor.
    Oriental linsangs dwell mainly in forests. They are noc-
turnal and generally arboreal but frequently move to the
ground in search of food (Lekagul and McNeely 1977). *P.
linsang* constructs a nest of sticks and leaves; in one case a
nest was located in a burrow at the base of a palm. This
species is also said to live in tree hollows. The diet includes
small mammals, birds, eggs, and insects.
    The limited data on reproduction suggest that *P. linsang*
has no clear breeding season (Lekagul and McNeely 1977).

Banded linsang *(Prionodon linsang)*, photo by Ernest P.
Walker.

Two pregnant females, one with two embryos and the other with three, were collected in May, and two lactating females were found in April and October. *P. pardicolor* is said
to breed in February and August and to have litters of two
young. Hayssen, Van Tienhoven, and Van Tienhoven
(1993) listed a newborn weight of 40 grams and an estrus
length of 11 days for *P. linsang*. A captive of that species
lived 10 years and 8 months (Jones 1982).
    The species *P. pardicolor* is listed as endangered by the
USDI and is on appendix 1 of the CITES. *P. linsang* is on
appendix 2 of the CITES. Schreiber et al. (1989) noted that
the former seems to be very rare but the latter is still relatively numerous in certain areas.

CARNIVORA; VIVERRIDAE; **Genus NANDINIA**
Gray, 1843

African Palm Civet

The single species, *N. binotata*, occurs from Guinea-Bissau
to southern Sudan and south to northern Angola and eastern Zimbabwe (Coetzee *in* Meester and Setzer 1977).
    Head and body length is 440–580 mm and tail length is
460–620 mm. Kingdon (1977) listed weight as 1.7–2.1 kg,
but Charles-Dominique (1978) reported that males
weighed as much as 5 kg. Coloration is quite variable but is
usually grayish or brownish tinged with buffy or chestnut.
Often two creamy spots are present between the shoulders,
and obscure dark brown spots are present on the lower back
and top of the tail. The tail, which is somewhat darker than
the body, is the same color above and below and has a variable pattern of black rings. The throat tends to be grayish,

African palm civet *(Nandinia binotata)*, photo by Ernest P. Walker.

and the underparts are grayish tinged with yellow. The pelage is short and woolly but coarse-tipped. The ears are short and rounded, the tail is fairly thick, the legs are short, and the claws are sharp and curved. There are scent glands on the palms, between the toes, on the lower abdomen, and possibly on the chin (Kingdon 1977). *Nandinia* has a number of unique cranial characters (see account of family Viverridae).

In a radio-tracking study in Gabon, Charles-Dominique (1978) found *Nandinia* to be largely arboreal and to occur mainly 10–30 meters above the ground in various types of forest. It was nocturnal, sleeping by day in a fork, on a large branch, or in a bundle of lianas. Stomach contents consisted of 80 percent fruit, on the average, but also included remains of rodents, bird eggs, large beetles, and caterpillars.

Charles-Dominique found a population density of 5/sq km in his study area. Adult females established territories averaging 45 ha. They allowed immature females on these areas but did not tolerate trespassing by other adult females. Large, dominant adult males had territories averaging about 100 ha., which overlapped a number of female territories. The large males drove away other animals of the same size and sex but allowed smaller adult males to remain; however, the small adult males were not permitted access to the females. Territories were marked with scent. Fighting was severe, sometimes resulting in death. Loud calls were exchanged during courtship.

In West Africa breeding apparently can occur during the wet or dry season (Rosevear 1974). Records from East Africa suggest that there are two birth peaks or seasons, May and October. The gestation period is 64 days. The number of young is usually two but up to four (Kingdon 1977). As soon as they are weaned, young males leave the territory of their mother. Sexual maturity is attained in the third year of life (Charles-Dominique 1978). One individual was still alive after 15 years and 10 months in captivity (Jones 1982).

The African palm civet is easily tamed and will drink milk in captivity. It is said to be quite clean and to keep houses free of rats, mice, and cockroaches.

CARNIVORA; VIVERRIDAE; **Genus ARCTOGALIDIA**
Merriam, 1897

Small-toothed or Three-striped Palm Civet

The single species, *A. trivirgata,* is found from Assam to Indochina and the Malay Peninsula and on Sumatra, Bangka, Java, Borneo, and numerous small nearby islands of the East Indies (Chasen 1940; Ellerman and Morrison-Scott 1966).

Head and body length is 432–532 mm, tail length is 510–660 mm, and weight is usually 2.0–2.5 kg. The color of the upper parts, proximal part of the tail, and outside of the limbs varies from dusky grayish tawny to bright orangish tawny. The head is usually darker and grayer, and the paws and distal part of the tail are brownish. There is a median white stripe on the muzzle, and there are three brown or black longitudinal stripes on the back. The median stripe is usually complete and distinct, whereas the laterals may be broken up into spots or almost absent. The undersides are grayish white or creamy buff with a whitish patch on the chest.

Only the females of this genus possess the civet gland, which is located near the opening of the urinogenital tract. *Arctogalidia* closely resembles *Paradoxurus* in external form as well as in the length of the legs and tail but differs externally in characters of the feet. Internally, the skull differs from that of *Paradoxurus,* and the back teeth are smaller.

Arctogalidia inhabits dense forests. In some areas it fre-

Small-toothed palm civet (*Arctogalidia trivirgata*), photo by Lim Boo Liat.

quents coconut plantations, though Lekagul and McNeely (1977) reported that it avoids human settlements. It is nocturnal, resting by day in the upper branches of tall trees (Medway 1978). It is arboreal, climbing actively and leaping from branch to branch with considerable agility. The omnivorous diet includes squirrels, birds, frogs, insects, and fruit.

Three animals, representing both sexes, occupied an empty nest of *Ratufa bicolor* about 20 meters above the ground in a tree. Mewing calls and light snarls, accompanied by playful leaps and chases, have been noted for a male and female at night in the wild. The young are reared in hollow trees. According to Lekagul and McNeely (1977), breeding probably continues throughout the year, there may be two litters annually, and litter size is two or three young. Batten and Batten (1966) reported that a female about 2 weeks old was captured in Borneo in August 1961. It entered estrus for the first time in December 1962 and then again at intervals of 6 months. In August 1964 it gave birth to its first litter (three young) after a gestation period of approximately 45 days. The young opened their eyes at 11 days and were suckled for more than 2 months. The father was reintroduced to the family when the young were 2.5 months old and was soon accepted by the others. According to Marvin L. Jones (Zoological Society of San Diego, pers. comm., 1995), one individual lived in captivity for 16 years and 1 month.

The subspecies *A. t. trilineata*, of Java, is classified as endangered by the IUCN. Schreiber et al. (1989) indicated that it was already rare 50 years ago; the last confirmed record was in 1978.

CARNIVORA; VIVERRIDAE; Genus PARADOXURUS
F. Cuvier, 1821

Palm Civets, Musangs, or Toddy Cats

There are four species (Chasen 1940; Corbet and Hill 1992; Ellerman and Morrison-Scott 1966; Laurie and Hill 1954; Schreiber et al. 1989):

P. hermaphroditus, occurs naturally from Kashmir and peninsular India to southeastern China and the Malay Peninsula and in Sri Lanka, Hainan, Sumatra and the nearby Simeulue and Enggano islands (but not the intervening Mentawai Islands), Java, Kangean Islands, Borneo, Palawan, and many small nearby islands of the East Indies;

P. lignicolor, Siberut, Sipura, North Pagai, and South Pagai in the Mentawai Islands off western Sumatra;

P. zeylonensis (golden palm civet), Sri Lanka;

P. jerdoni, extreme southwestern India.

Taylor (1934) referred populations of *Paradoxurus* in the Philippines to three species—*P. philippinensis, P. torvus,* and *P. minax*—but Corbet and Hill (1992) indicated that these populations, other than that on Palawan, probably were introduced and that the only species in the Philippines is *P. hermaphroditus.* Wozencraft (*in* Wilson and Reeder 1993) included *P. lignicolor* in *P. hermaphroditus,* but Corbet and Hill (1992) found them to be distinct species.

Head and body length is 432–710 mm, tail length is 406–660 mm, and weight is 1.5–4.5 kg. The ground color is grayish to brownish but is often almost entirely masked by the black tips of the guard hairs. There is a definite pattern of dorsal stripes and lateral spots, at least in the new coat, but this is sometimes concealed by the long black hairs. The pattern is most plainly shown in the species *P. hermaphroditus,* where it consists of longitudinal stripes on the back and spots on the shoulders, sides, and thighs and sometimes on the base of the tail. A pattern of white patches and a white band across the forehead may be present on the head of this species.

The species *P. hermaphroditus* can always be distinguished from *P. zeylonensis* and *P. jerdoni* by the backward direction of the hairs on the neck. In the other species the hairs on the neck grow forward from the shoulders to the head.

Paradoxurus differs from *Arctogalidia* and *Paguma* in color pattern and in characters of the skull and teeth. According to Lekagul and McNeely (1977), the teeth of *Paradoxurus* are less specialized for eating meat than those of most viverrids, having low, rounded cusps on rather square molars. Both sexes have well-developed anal scent glands. Females have three pairs of mammae.

Palm civet *(Paradoxurus hermaphroditus)*, photo by Ernest P. Walker.

Musangs are nocturnal forest dwellers. They are expert climbers and spend most of their time in trees, where they utilize cavities or secluded nooks. They are often found about human habitations, probably because of the presence of rats and mice. Under such conditions they shelter in thatched roofs and in dry drain tiles and pipes. They eat small vertebrates, insects, fruits, and seeds. They are fond of the palm juice, or "toddy," collected by the natives—thus one of the vernacular names.

An older male radio-tracked for 12 months in Thailand by Rabinowitz (1991) moved within a total area of 17 sq km and had an average monthly range of 3.2 sq km and an average daily movement of 1 km; respective figures for a younger male tracked for 7 months were 4.25 sq km, 0.74 sq km, and 660 meters. A female with five kittens radio-tracked for about a month in Nepal used a home range of 1.2 ha. (Dhungel and Edge 1985). Hayssen, Van Tienhoven,

and Van Tienhoven (1993) listed a gestation period of 60 days and newborn weights of 69–102 grams for *P. hermaphroditus*. Reproduction occurs throughout the year, though Lekagul and McNeely (1977) stated that the young of *P. hermaphroditus* seem to be seen most often from October to December. Litter size is two to five young. Sexual maturity is attained at 11–12 months in *P. hermaphroditus*. One individual was still living after 25 years in captivity (Marvin L. Jones, Zoological Society of San Diego, pers. comm., 1995).

Groves (1976) noted that *P. hermaphroditus* has been carried about from island to island by people for use as a rat catcher. Such activity is probably responsible for the presence of this species on most islands of the Philippines, Sulawesi, the Lesser Sundas as far east as Timor, Seram, the Sula and Kei islands, and Aru (Corbet and Hill 1992). *P. hermaphroditus* is generally widespread and often lives in close association with people; however, there is concern for the other three species, which have restricted ranges and depend on natural forest habitat. The IUCN classifies *P. lignicolor* (treated as a subspecies of *P. hermaphroditus*) and *P. jerdoni* as vulnerable. Schreiber et al. (1989) pointed out that the former is known only by four specimens and that its island habitat is being destroyed by commercial logging; the latter also occurs in a limited area and reportedly has been sighted only twice in the last 20 years. *P. zeylonensis* may still be common in some parts of Sri Lanka and is respected by older villagers, who realize that it spreads seeds of the valuable kitul palm, but it has lost much of its habitat and is trapped and eaten by younger people. The subspecies *P. hermaphroditus kangeanus* is restricted to the Kangean Islands off northeastern Java and may be jeopardized through interbreeding with introduced subspecies.

CARNIVORA; VIVERRIDAE; **Genus PAGUMA**
Gray, 1831

Masked Palm Civet

The single species, *P. larvata*, occurs from northern Pakistan and Kashmir to Indochina and the Malay Peninsula,

Masked palm civet *(Paguma larvata)*, photo from New York Zoological Society.

in much of eastern and southern China, and on the Andaman Islands, Taiwan, Hainan, Sumatra, and Borneo (Chasen 1940; Ellerman and Morrison-Scott 1966). In 1993 an individual was sighted in western Java, but this may have resulted from human introduction (Brooks and Dutson 1994).

Head and body length is 508–762 mm, tail length is usually 508–636 mm, and weight is usually 3.6–5.0 kg. In the facial region there is generally a mask, which consists of a median white stripe from the top of the head to the nose, a white mark below each eye, and a white mark above each eye extending to the base of the ear and below. The general color is gray, gray tinged with buff, orange, or yellowish red. There are no stripes or spots on the body and no spots or bands on the tail. The distal part of the tail may be darker than the basal part, and the feet are blackish. This genus differs externally from *Paradoxurus* and *Arctogalidia* in the absence of the striping and spotting. Like *Paradoxurus,* *Paguma* has a potent anal gland secretion, which it uses to ward off predators. The conspicuously marked head has been interpreted as a warning signal of the presence of the secretion. Females have two pairs of mammae (Lekagul and McNeely 1977).

The masked palm civet frequents forests and brush country. It reportedly raises its young in tree holes, at least in Nepal. It is arboreal and nocturnal (Roberts 1977). The omnivorous diet includes small vertebrates, insects, and fruits. An adult female radio-tracked for 12 months in Thailand by Rabinowitz (1991) moved within a total area of 3.7 sq km, had an average monthly range of 0.93 sq km, and moved an average of 620 meters per day.

Paguma is solitary, and apparently most young in the western parts of its range are born in spring and early summer (Roberts 1977). Births in Borneo have taken place in October (Banks 1978). Data on captives indicate that there may be two breeding seasons, in early spring and late autumn. Litters contain one to four young. They open their eyes at 9 days and are almost the size of adults by 3 months (Lekagul and McNeely 1977; Medway 1978). An individual lived at the London Zoo for more than 20 years (Marvin L. Jones, Zoological Society of San Diego, pers. comm., 1995).

In Tenasserim *Paguma* is reported to be a great ratter and not to destroy poultry. Medway (1978), however, wrote that it has been known to raid hen runs. The genus has been introduced on Honshu and Shikoku, Japan (Corbet and Hill 1992).

CARNIVORA; VIVERRIDAE; **Genus MACROGALIDIA**
Schwarz, 1910

Sulawesian Palm Civet

The single species, *M. musschenbroeki,* occurs only on Sulawesi (Laurie and Hill 1954).

Wemmer et al. (1983) reported that an adult male had a head and body length of 715 mm, a tail length of 540 mm, and a weight of 6.1 kg and that respective measurements for two adult females were 650 and 680 mm, 480 and 445 mm, and 3.8 and 4.5 kg. The upper parts are light brownish chestnut to dark brown. The underparts range from fulvous to whitish, with a reddish breast. The cheeks and a patch above the eye are usually buffy or grayish. Faint brown spots and bands are usually present on the sides and lower back, and the tail is ringed with dark and pale brown. The tail has more bands than that of *Arctogalidia* or *Paradoxurus.* Other distinguishing characters are the short, close fur and a whorl in the neck with the hairs directed forward. Both of the females noted above had two pairs of inguinal mammae.

According to Wemmer and Watling (1986), *Macrogalidia* occurs in both lowland and montane forests up to

Sulawesian palm civet *(Macrogalidia musschenbroeki)*, photo by Christen Wemmer.

about 2,600 meters. It seems to be dependent on primary forests but also moves onto adjacent grasslands and farms. It is a skillful climber and sometimes moves through trees but probably forages mainly on the ground. It feeds mainly on small mammals and fruits, especially palm fruit, and also raids domestic chickens and pigs. It is probably solitary, though a female and young may share a home range for a period after weaning. Although once thought to be extinct or restricted to the northern peninsula of Sulawesi and currently classified as vulnerable by the IUCN, recent observations indicate that *Macrogalidia* occurs in most parts of the island. It now seems to be neither abundant nor scarce but could be potentially jeopardized in some areas by logging and agricultural activity.

CARNIVORA; VIVERRIDAE; **Genus ARCTICTIS**
Temminck, 1824

Binturong

The single species, *A. binturong*, occurs from Sikkim, and probably originally from Nepal, to Indochina and the Malay Peninsula and on the Riau Archipelago, Sumatra, Bangka, Java, Borneo, and Palawan (Corbet and Hill 1992).

Head and body length is 610–965 mm and tail length is 560–890 mm. Weight is usually 9–14 kg, but a 19-kg individual reportedly was taken in Viet Nam (Rozhnov 1994). The fur is long and coarse, that on the tail being longer than that on the body. The lustrous black hairs often have a gray, fulvous, or buff tip. The head is finely speckled with gray

and buff, and the edges of the ears and the whiskers are white. The ears have long hairs on the back that project beyond the tips and produce a fringed or tufted effect. The tail is particularly muscular at the base and prehensile at the tip. The only other carnivore with a truly prehensile tail is the kinkajou *(Potos)*, which the binturong resembles in habits to some extent. Females have two pairs of mammae.

The binturong lives in dense forests and is nowhere abundant. It is mainly arboreal and nocturnal. It usually lies curled up, with the head tucked under the tail, when resting. It has never been observed to leap; rather, it progresses slowly but skillfully, using the tail as an extra hand. Its movements, at least during daylight hours, are rather slow and cautious, the tail slowly uncoiling from the last support as the animal moves carefully forward. According to Medway (1978), the binturong is reported to dive, swim, and catch fish. The diet also includes birds, carrion, fruit, leaves, and shoots.

Medway (1978) stated that *Arctictis* occurs either alone or in small groups of adults with immature offspring. Captives are very vocal, uttering high-pitched whines and howls, rasping growls, and, when excited, a variety of grunts and hisses. Lekagul and McNeely (1977) wrote that breeding seems to occur throughout the year. Numerous observations in captivity (Bulir 1972; Gensch 1963b; Grzimek 1975; Kuschinski 1974; Xanten, Kafka, and Olds 1976) indicate that females are nonseasonally polyestrous and may give birth to two litters annually. Wemmer and Murtaugh (1981) reported the following data for reproduction in captivity: breeding occurs year-round, but there is a pronounced birth peak from January to March; the estrous cycle averages 81.8 days; the mean gestation period is 91.1

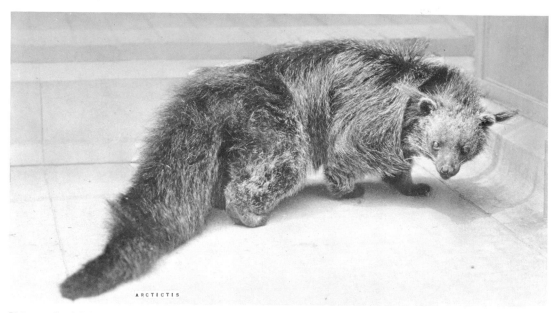

Binturong *(Arctictis binturong)*, photo from New York Zoological Society.

days (84–99); the number of young per litter averages 1.98 (1–6); the young weigh an average of 319 grams at birth and begin to take solid food at 6–8 weeks; the mean age of first mating is 30.4 months in females and 27.7 months in males; and both sexes can remain fertile until at least 15 years. According to Marvin L. Jones (Zoological Society of San Diego, pers. comm., 1995), a binturong lived at the Stuttgart Zoo for 25 years and 11 months.

Arctictis is sometimes kept as a pet. It is said to be easily domesticated, to become quite affectionate, and to follow its master like a dog. The subspecies *A. b. whitei* of Palawan is classified as vulnerable by the IUCN; it is declining because of habitat destruction by people.

CARNIVORA; VIVERRIDAE; Genus HEMIGALUS
Jourdan, 1837

Banded Palm Civet

The single species, *H. derbyanus*, occurs on the Malay Peninsula, including Tenasserim and southern Thailand, and on Sumatra, the Mentawai Islands (Siberut, Sipura, South Pagai), and Borneo (Chasen 1940; Corbet and Hill 1992; Ellerman and Morrison-Scott 1966; Lekagul and Mc-Neely 1977; Medway 1977). The genera *Diplogale* and *Chrotogale* (see accounts thereof) sometimes are included in *Hemigalus*.

Head and body length is 410–620 mm, tail length is 255–383 mm, and weight is usually 1.75–3.0 kg. On the head a narrow, median dark streak extends from the nose to the nape; on either side of this streak a broader dark stripe encircles the eye and passes backward over the base of the ear. Two broad stripes, sometimes more or less broken into shorter stripes or spots, run backward from the neck and curve downward to the elbow. Behind these are two shorter stripes. The back behind the shoulders is marked with four or five broad transverse stripes separated by pale, usually narrower spaces, and there are two imperfect stripes at the base of the tail. The ground color is whitish to orange buff, usually lighter and more buffy below, and the tail is usually black. The tips of the hairs on the back of the neck point forward. The five-toed feet have strongly curved claws that are retractile like those of cats. Small scent glands are present.

On the Malay Peninsula *H. derbyanus* is restricted to tall forest and apparently is largely terrestrial (Medway 1978); however, it is at least partly arboreal and climbs well (Lekagul and McNeely 1977). All specimens of *Hemigalus* taken by Davis (1962) in Borneo were collected on the ground. Davis observed that the animals were exclusively nocturnal and apparently foraged on the forest floor, picking food up from the surface. Orthopterans and worms made up 80 percent of the contents of 12 stomachs, the remaining 20 percent consisted mostly of other invertebrates.

Gangloff (1975) reported that captive *H. derbyanus* were fond of fruit, did not construct nests, and marked with scent. A pregnant female *H. derbyanus* with one embryo was taken in Borneo in February. A captive female in the Wassenaar Zoo, Netherlands, had two young. They weighed 125 grams at birth, opened their eyes after 8–12 days, and first took solid food at about 70 days (Ewer 1973). One specimen lived in captivity for about 18 years (Marvin L. Jones, Zoological Society of San Diego, pers. comm., 1995).

Schreiber et al. (1989) expressed concern for the subspecies *H. derbyanus minor* and *H. d. sipora* of the Mentawai Islands. These are the only small islands on which *Hemigalus* is found, which suggests that its occurrence there is natural and that the genus was not widely carried about by people as some other viverrids of the East Indies were. The status of these subspecies is not well known, but they probably are being adversely affected by commercial logging and direct killing by people who consider them predators on domestic chickens. The species *H. derbyanus* is on appendix 2 of the CITES.

Banded palm civet *(Hemigalus derbyanus)*, photo from Duisburg Zoo.

CARNIVORA; VIVERRIDAE; **Genus DIPLOGALE**
Thomas, 1912

Hose's Palm Civet

The single species, *D. hosei*, is known only from the mountains of northern Borneo. *Diplogale* sometimes has been included in *Hemigalus* but was considered a separate genus by Corbet and Hill (1992), Schreiber et al. (1989), and Wozencraft (*in* Wilson and Reeder 1993).

Head and body length is about 470–540 mm and tail length is about 280–335 mm. Coloration is dark brown or black above and grayish, yellowish white, or slightly rufescent below. The ears are thinly haired and white inside. A buffy gray patch extends from above the eye to the cheek and terminates where it meets the white of the lips and throat. The inner side of the limb near the body is grayish, while the remainder of each limb is black. The tail is not banded but is dark throughout. From *Hemigalus*, *Diplogale* differs in more uniform coloration, lack of a median foramen between the paired incisive foramina of the skull, lack of strongly curved incisor tooth rows, and more prominent accessory cusps on the anterior premolars (Corbet and Hill 1992).

This genus is found mainly in montane forest and is largely terrestrial (Medway 1977). It probably is nocturnal, and its long facial whiskers and hairs between the footpads suggest that it may specialize in foraging for small animals among mossy boulders and in streams; one stomach contained various small insects (Payne, Francis, and Phillips 1985). It is known only from 15 museum specimens, the last of which was collected in 1955 (Schreiber et al. 1989). It is classified as vulnerable by the IUCN.

CARNIVORA; VIVERRIDAE; **Genus CHROTOGALE**
Thomas, 1912

Owston's Palm Civet

The single species, *C. owstoni*, now is known from about 40 specimens taken in Yunnan and Guangxi provinces in extreme southern China, Laos, and northern and central Viet Nam (Rozhnov, Kuznetzov, and Anh 1992; Schreiber et al. 1989). *Chrotogale* was included within *Hemigalus* by Corbet and Hill (1992) but not by Wozencraft (*in* Wilson and Reeder 1993).

Head and body length is 560–720 mm, tail length is 350–470 mm, and weight is 2.5–4.0 kg (Dang, Anh, and Huynh 1992). The body and base of the tail have alternating and sharply contrasting dark and light transverse bands, and longitudinal stripes are present on the neck. The pattern of stripes and bands resembles that of *Hemigalus derbyanus* but it is supplemented by black spots on the

Owston's palm civet *(Chrotogale owstoni)*, photo from Hanoi Zoo through Jorg Adler.

sides of the neck, the forelimbs and thighs, and the flanks. Four seems to be the maximum number of dorsal bands. It has been suggested that this striking pattern serves as a warning signal as *Chrotogale* is thought to possess a particularly foul-smelling anal gland secretion. The underparts are pale buffy, and a narrow orange midventral line runs from the chest to the inguinal region. The terminal two-thirds of the tail is completely black.

Form and markings are strikingly similar to those of *Hemigalus derbyanus*, but the hairs on the back of the neck of *Chrotogale* are not reversed in direction. *Chrotogale* is also distinguished by cranial and dental characters. The incisor teeth are remarkable in that they are broad, close-set, and arranged in practically a semicircle, a type unique among carnivores and only approached in certain of the marsupials. The other teeth and the skull are also peculiar, indicating habits and a mode of life different from that of other genera in the family Viverridae.

The natural history of this genus was almost unknown until considerable new information was provided by Dang, Anh, and Huynh (1992). *Chrotogale* prefers forests and other areas of dense vegetation in the vicinity of streams and lakes. It makes simple dens under large tree trunks or in thick brush and also frequently uses natural holes in trees, rocks, or the ground. It is terrestrial but can climb well and often enters trees in search of food. Activity is mainly nocturnal, usually commencing at dusk and ending early in the morning. The natural diet consists mostly of earthworms but also includes small vertebrates, insects, and fruit. Captives readily accept beef, chicken, and bananas.

Although the genus is reported to be solitary in the wild, several individuals of both sexes have lived together peacefully in captivity and have accepted new members to the group without any show of aggression. They are generally silent and mark their home area with secretions from the anal-genital region. Mating seems to occur mainly from January to March but may continue until November, and there are one or two litters annually. The gestation period

is about 60 days, there are one to three young, and newborn weight is 75–88 grams.

Habitat loss and intensive hunting during the last few decades have greatly reduced the range and numbers of *Chrotogale*, and it now is legally protected as an endangered species in Viet Nam. Schreiber et al. (1989), however, indicated that it can survive close to villages and even approaches houses in search of kitchen wastes. It is classified as vulnerable by the IUCN.

CARNIVORA; VIVERRIDAE; **Genus CYNOGALE**
Gray, 1837

Otter Civets

Corbet and Hill (1992) and Schreiber et al. (1989) recognized two species:

C. bennettii, peninsular Malaysia and possibly adjacent southern Thailand, Sumatra, Borneo;
C. lowei, known with certainty only by a single specimen from Bac Can in northern Viet Nam, but likely records from southern Yunnan and northern Thailand.

Wozencraft (*in* Wilson and Reeder 1993) included *C. lowei* in *C. bennettii.*

Head and body length is 575–675 mm, tail length is 130–205 mm, and weight is usually 3–5 kg. The form is somewhat like that of the otters *(Lutra).* The underfur is close, soft, and short. It is pale buff near the skin and shades to dark brown or almost black at the tip. The longer, coarser guard hairs are usually partially gray, which gives a frosted or speckled effect on the head and body. The lower side of the body is lighter brown and not speckled with gray. The whiskers are remarkably long and plentiful; those on the snout are fairly long, but those on a patch under the ear are the longest. The newly born young lack dorsal

Otter civet *(Cynogale bennettii)*, photo from San Diego Zoological Society.

speckling; they have some gray on the forehead and ears and two longitudinal stripes down the sides of the neck extending under the throat.

Because of the deepening and expansion of the upper lip, the rhinarium occupies a horizontal position, with the nostrils opening upward on top of the muzzle. The nostrils can be closed by flaps, an adaptation for aquatic life. The ears also can be closed. Although the webbing on the feet does not extend farther toward the tips of the digits than in such genera as *Paradoxurus*, it is quite broad, and the fingers are capable of considerable flexion. A glandular area, merely three pores in the skin, is located near the genitals and secretes a mild scent material. The premolar teeth are elongate and sharp, adapted for capturing and holding prey, and the molars are broad and flat, for crushing. Females have four mammae.

Cynogale is usually found near streams and swampy areas. It can climb well and often takes refuge in a tree when chased by dogs. While walking it usually carries its head and tail low and arches its back. Although it is partly adapted for an aquatic life, its tail is short and lacks special muscular power, and the webbing between the digits is only slightly developed. *Cynogale* thus is probably a slow swimmer and cannot turn quickly in the water. It probably captures aquatic animals only after they have taken shelter from the chase and catches some birds and mammals as they come to drink. It cannot be seen by its prey because it is submerged with only the tip of the nose exposed above the surface of the water. The diet includes crustaceans, perhaps mollusks, fish, birds, small mammals, and fruits. There are records of pregnant females with two and three embryos. Young still with the mother have been noted in May in Borneo. A captive specimen lived five years (Jones 1982).

Cynogale bennettii (including *C. lowei*) now is classified as endangered by the IUCN and is on appendix 2 of the CITES. It is thought to have declined by at least 50 percent in the past decade and to now number fewer than 2,500 mature individuals. Schreiber et al. (1989) noted that *C. bennettii* is probably jeopardized by expanding human settlement and agriculture along rivers. The single known specimen of *C. lowei* was collected in 1926; however, there have been a few recent reports.

CARNIVORA; VIVERRIDAE; **Genus FOSSA**
Gray, 1864

Malagasy Civet

The single species, *F. fossana*, originally was found throughout the rainforests of eastern and southern Madagascar (Schreiber et al. 1989). Although the generic name is the same as the vernacular term for another Malagasy carnivore *(Cryptoprocta)*, the two animals are not closely related. The specific name sometimes is given as *fossa*.

Head and body length is 400–450 mm and tail length is 210–30 mm (Albignac 1972). Males weigh up to 2 kg and females, up to 1.5 kg (Coetzee *in* Meester and Setzer 1977). The ground color is grayish, washed with reddish. There are four rows of black spots on each side of the back and a few black spots on the backs of the thighs. These spots may merge to form stripes, and the gray tail is banded with brown. The underparts are grayish or whitish and more or less obscurely spotted. The limbs are slender, perhaps being adapted for running. There are no anal scent glands, but there probably are marking glands on the cheeks and neck (Albignac 1972).

The Malagasy civet inhabits evergreen forests and shelters in hollow trees or crevices. It is nocturnal and may occur in trees or on the ground. The preferred foods are crustaceans, worms, small eels, and frogs. Other kinds of animal matter and fruit are also taken (Coetzee *in* Meester and Setzer 1977).

Fossa lives in pairs, which share a territory. Vocalizations include cries, groans, and a characteristic "coq-coq," heard only in the presence of more than one individual. Births have been recorded from October to January. The gestation period is 3 months and a single offspring is born. It weighs 65–70 grams at birth, is weaned after 2 months, and prob-

Malagasy civets *(Fossa fossa)*, photo from U.S. National Zoological Park.

ably attains adult weight at 1 year (Albignac 1970*b*, 1972; Coetzee *in* Meester and Setzer 1977). A captive specimen lived 11 years (Jones 1982).

The Malagasy civet is classified as vulnerable by the IUCN and is on appendix 2 of the CITES. It still is not uncommon, but its habitat is rapidly declining because of human encroachment and now has been reduced to isolated patches; it also is subject to hunting and trapping (Schreiber et al. 1989).

CARNIVORA; VIVERRIDAE; Genus EUPLERES
Doyère, 1835

Falanouc

The single species, *E. goudotii,* is found in the coastal forests of Madagascar (Coetzee *in* Meester and Setzer 1977).

Head and body length is 450–650 mm, tail length is 220–50 mm, and weight is 2–4 kg (Albignac 1974). The subspecies *E. g. goudotii* is fawn-colored above and lighter below. In the subspecies *E. g. major* the males are brownish and the females are grayish. The pelage is woolly and soft, being made up of a dense underfur and longer guard hairs. The tail is covered by rather long hairs that give a bushy appearance.

Eupleres has a pointed muzzle, a narrow and elongate head, and short, conical teeth. It resembles the civets in some structural features and the mongooses in others. The small teeth are similar to each other and resemble those of insectivores rather than carnivores. Indeed, *Eupleres* was classified as an insectivore before its somewhat obscure relationship with the mongooses was detected. The claws are relatively long and not retractile, or are only imperfectly

so. The feet are peculiar in the comparatively large size and low position of the great toe and thumb.

The falanouc inhabits humid, lowland forests. It is crepuscular and nocturnal, resting by day in crevices and burrows. It is terrestrial, and if threatened, it may either run or remain motionless. In autumn up to 800 grams of fat can be accumulated in the tail, and it has been suggested that *Eupleres* hibernates during winter; however, active individuals have been observed in winter. The diet consists mainly of earthworms and also includes other invertebrates and frogs but apparently not reptiles, birds, rodents, or fruit (Albignac 1974; Coetzee *in* Meester and Setzer 1977).

Eupleres may live alone or in small family groups. It has several vocalizations and other means of communication. Mating probably takes place in July or August (winter), and a birth was observed in November. Litters contain one or two young. Weight of the newborn is 150 grams, and weaning occurs at nine weeks (Albignac 1974; Coetzee *in* Meester and Setzer 1977).

The falanouc is classified as endangered by the IUCN and is on appendix 2 of the CITES. It has declined in numbers and distribution because of deforestation, drainage of marshes, excessive hunting for use as food, predation by domestic dogs, and possibly competition from the introduced *Viverricula indica*. It still occurs over a large area but is nowhere common (Schreiber et al. 1989).

CARNIVORA; Family HERPESTIDAE

Mongooses and Fossa

This family of 19 Recent genera and 39 species is found in southern Asia, the East Indies, Africa, and Madagascar. One

Falanoucs *(Eupleres goudotii)*, photos by R. Albignac.

genus *(Herpestes)* occurs in Spain and Portugal, probably brought there by people in ancient times, and also has been introduced in other parts of Europe and on many islands around the world. The following sequence of genera, grouped in 3 subfamilies, is based on the classifications of Coetzee (*in* Meester and Setzer 1977), Ellerman and Morrison-Scott (1966), and Ewer (1973):

Subfamily Galidiinae (Malagasy mongooses), genera *Galidia, Galidictis, Mungotictis, Salanoia;*

Subfamily Herpestinae (mongooses), genera *Herpestes, Galerella, Mungos, Crossarchus, Liberiictis, Helogale, Dologale, Bdeogale, Rhynchogale, Ichneumia, Atilax, Cynictis, Paracynictis, Suricata;*

Subfamily Cryptoproctinae (fossa), genus *Cryptoprocta.*

The above three subfamilies often have been considered components of the family Viverridae, but there seems to be a growing consensus that the Herpestidae warrant full familial distinction (Corbet and Hill 1991, 1992; Hunt and Tedford 1993; Wozencraft (1989*a*, 1989*b*, and *in* Wilson and Reeder 1993). Some evidence even suggests that the Herpestidae are more closely related to the Hyaenidae than to the Viverridae, with the latter family being closest to the Felidae. There still is disagreement about the affinity of the Cryptoproctinae. Wozencraft (1989*a* and *in* Wilson and Reeder 1993) placed that subfamily in the Viverridae, but recent DNA analyses by Véron and Catzeflis (1993) indicate that it is more closely related to the Herpestidae.

The Herpestidae resemble the Viverridae in general appearance, size, distribution, and many aspects of natural history. They are on average somewhat smaller; excluding *Cryptoprocta*, head and body length is 180–710 mm, tail length is 150–530 mm, and weight is 230–5,200 grams. In *Cryptoprocta* head and body length is as much as 800 mm and weight is as much as 12 kg. The pelage of herpestids tends to be more uniform in color than that of viverrids, though some genera have stripes, bands, or contrasting patterns. The body is long, the limbs are rather short, and the tail usually is relatively shorter than in the Viverridae. Most genera have five toes on each foot; *Cynictis* has only four digits on the hind foot, and *Bdeogale, Paracynictis*, and *Suricata* have only four digits on all feet (Meester et al. 1986). The claws usually are not retractile. Female herpestids have two or three pairs of abdominal mammae. Males have a baculum. Like viverrids, herpestids have scent glands in the anal region. In herpestids, however, these glands open into a pouch or saclike depression outside the anus proper in which the secretion is stored. The skull is commonly shorter and broader than that of the Viverridae (Schliemann *in* Grzimek 1990). The dental formula is: (i 3/3, c 1/1, pm 3–4/3–4, m 2/2) × 2 = 36–40.

The presence of prominent anal scent glands and a large associated sac is one of the critical characters distinguishing the Herpestidae as a family from the Viverridae, in which the scent glands are simple. Other diagnostic characters are found in the skull, especially in the basicranial region (Hunt and Tedford 1993; Wozencraft 1989*b*; Wyss and Flynn 1993). In the Herpestidae the caudal entotympanic does not penetrate into the anterior chamber of the auditory bulla (as it does in the Viverridae), the ectotympanic contributes to an external auditory meatal tube (it does not in the Viverridae), and the internal carotid artery is enclosed in an osseous tube within the bony wall of the middle ear prior to entering the cranial cavity (in the Viverri-

dae this artery is not enclosed in a bony tube and actually enters the internal space of the bulla). In these characters the Viverridae resemble the Felidae more than they do the Herpestidae.

Mongooses tend to live in more open country than do civets, to be more diurnal, and to form larger social groups. Therefore, some genera have been studied extensively. Several genera, including *Cynictis* and *Suricata*, live in colonies in ground burrows. Some genera of mongooses associate in bands and take refuge as a group in any convenient shelter. Mongooses, particularly of the genus *Herpestes*, have been introduced into several areas to check the numbers of rodents and venomous snakes. Such introductions, however, generally have not proven beneficial because the mongooses quickly multiply and destroy many desirable forms of mammals and birds.

The geological range of this family is Oligocene to Recent in Europe, Miocene to Recent in Asia and Africa, and Pleistocene to Recent in Madagascar.

CARNIVORA; HERPESTIDAE; Genus GALIDIA
I. Geoffroy St.-Hilaire, 1837

Malagasy Ring-tailed Mongoose

The single species, *G. elegans*, is found in eastern, west-central, and extreme northern Madagascar (Albignac 1969).

Head and body length is about 380 mm and tail length is about 305 mm. Weight is 700–900 grams (Coetzee *in* Meester and Setzer 1977). The general coloration is dark chestnut brown, and the tail is ringed with dark brown and black. *Galidia* has some of the structural features of civets and some of mongooses. The feet differ from those of *Galidictis* in having shorter digits, fuller webbing, shorter claws, and more hairy soles. The lower canine teeth are smaller. From *Salanoia, Galidia* is distinguished by its ringed tail and very small second upper premolar. Dissection of two individuals indicates the presence of a scent gland, closely associated with the external genitalia, in males but not in females.

Galidia occurs in humid forests. It shelters in burrows, which it digs very rapidly, and probably also in hollow trees. It is mainly diurnal but may also be active at night. More arboreal than most mongooses, it is able to climb and descend on vertical trunks only 4 cm in diameter. It can also swim. The diet consists mainly of small mammals, birds and their eggs, and frogs and also includes fruits, fish, reptiles, and invertebrates (Albignac 1972; Coetzee *in* Meester and Setzer 1977).

Galidia is less social than most mongooses, being found alone or in pairs. Mating in Madagascar occurs from April to November, with births from July to February. The gestation period is 79–92 days, and a single young weighing 50 grams is produced. Physical maturity is reportedly attained at 1 year, and sexual maturity at 2 years (Coetzee *in* Meester and Setzer 1977; Larkin and Roberts 1979). One animal lived in captivity for 24 years and 5 months (Marvin L. Jones, Zoological Society of San Diego, pers. comm., 1995).

Malagasy ring-tailed mongoose *(Galidia elegans)*, photo from U.S. National Zoological Park.

CARNIVORA; HERPESTIDAE; **Genus GALIDICTIS**
I. Geoffroy St.-Hilaire, 1839

Malagasy Broad-striped Mongooses

There are two species (Coetzee *in* Meester and Setzer 1977; Wozencraft 1986, 1987):

G. fasciata, eastern Madagascar;
G. grandidieri, southwestern Madagascar.

Head and body length is 320–40 mm and tail length is 280–300 mm (Albignac 1972). The general body color is pale brown or grayish. In the subspecies *G. fasciata striata* there are usually 5 longitudinal black bands or stripes on the back and sides, and the tail is whitish. In *G. fasciata fasciata* there are usually 8–10 stripes, and the tail is bay-colored and somewhat bushy. *G. grandidieri* is larger than *G. fasciata* and has 8 dark brown longitudinal stripes on the back and sides that are narrower than the intervening spaces.

Galidictis differs from other viverrids in color pattern and in cranial and dental characters. The feet differ from those of *Galidia* in having longer digits, less extensive webbing, and longer claws. A scent pouch is present in females.

The species *G. fasciata* occurs in forests and is nocturnal and crepuscular. The diet consists mainly of small vertebrates, especially rodents, and also includes invertebrates. The three known specimens of *G. grandidieri* were taken in an area of spiny desert vegetation. *Galidictis* is found in

pairs or small social groups. It apparently produces one young per year, during the summer (Albignac 1972; Coetzee *in* Meester and Setzer 1977). Young individuals are said to tame readily, to follow their masters, and even to sleep in their masters' laps. *G. fasciata* is classified as vulnerable by the IUCN, *G. grandidieri* as endangered. Schreiber et al. (1989) reported the former to be locally common but threatened by habitat destruction. *G. grandidieri* had not been recorded since 1929, when a report of its possible survival in southwestern Madagascar was received from villagers in 1987.

CARNIVORA; HERPESTIDAE; **Genus MUNGOTICTIS**
Pocock, 1915

Malagasy Narrow-striped Mongoose

The single species, *M. decemlineata*, is found in western and southwestern Madagascar (Coetzee *in* Meester and Setzer 1977).

Head and body length is 250–350 mm, tail length is 230–70 mm, and weight is 600–700 grams. The fur is rather dense and generally gray beige in color. There are usually 8–10 dark stripes on the back and flanks. The underparts are pale beige. The soles of the feet are naked and the digits are partly webbed. Glands on the side of the head and on the neck appear to be for marking with scent. Females have a

Malagasy broad-striped mongoose *(Galidictis fasciata)*, photo by Peter Schachenmann through Steven M. Goodman.

single pair of inguinal mammae (Albignac 1971, 1972, 1976).

Mungotictis is found on sandy, open savannahs. It is diurnal and both arboreal and terrestrial. It spends the night in tree holes during the wet summer and in ground burrows during the dry winter. It can swim well. The diet consists mostly of insects but also includes small vertebrates, birds' eggs, and invertebrates. To break an egg or the shell of a snail, *Mungotictis* lies on one side, grasps the object with all four feet, and throws it abruptly until it breaks and the contents can be lapped up (Albignac 1971, 1972, 1976).

In a radio-tracking study Albignac (1976) determined that 22 individuals inhabited 300 ha. The animals were divided into two stable social units that sometimes engaged in agonistic encounters where their ranges met. Each group contained several adults of both sexes, as well as juveniles

Malagasy narrow-striped mongoose *(Mungotictis decemlineata)*. Inset: head *(M. decemlineata)*. Photos by Don Davis.

and young of the year. Group cohesiveness and intrarelationships varied. Generally, adult males and females came together in the summer. In the winter there was division into small units, such as temporary pairs, maternal family parties, all-male groups, and solitary males.

The breeding season extends from December to April, peaking in February and March (summer). The gestation period is 90–105 days. There is usually a single offspring weighing 50 grams at birth. Weaning occurs at 2 months, but the young remain with their mothers until they are 2 years old (Albignac 1972, 1976).

This mongoose is now classified as vulnerable by the IUCN. Schreiber et al. (1989) reported that it appears to be subject to very little direct human persecution but that its habitat is being burned and cleared at an alarming rate.

CARNIVORA; HERPESTIDAE; **Genus SALANOIA**
Gray, 1864

Malagasy Brown-tailed Mongoose, or Salano

The single species, *S. concolor,* is found in northeastern Madagascar (Coetzee *in* Meester and Setzer 1977).

Head and body length is 250–300 mm and tail length is 200–250 mm (Albignac 1972). The general coloration is brown, and there are either dark or pale spots. The tail is the same color as the body and is not ringed. The claws are not strongly curved, the ears are broad and short, and the muzzle is pointed.

The salano occurs only in the evergreen forests on the northeastern part of the central plateau of Madagascar. It is typically diurnal, sheltering at night in tree trunks or burrows. It feeds mainly on insects and fruits but also on frogs, small reptiles, and rodents. It occurs individually or in pairs

depending on the season. The young are born mainly during the summer (Albignac 1972; Coetzee *in* Meester and Setzer 1977). A captive lived 4 years and 9 months (Jones 1982). *Salanoia* is declining because of habitat loss and now is classified as vulnerable by the IUCN.

CARNIVORA; HERPESTIDAE; **Genus HERPESTES**
Illiger, 1811

Mongooses

There are 2 subgenera and 10 species (Chasen 1940; Coetzee *in* Meester and Setzer 1977; Corbet 1978; Corbet and Hill 1992; Ellerman and Morrison-Scott 1966; Medway 1977; Sanborn 1952; Wells 1989; Wells and Francis 1988):

subgenus *Herpestes* Illiger, 1811

H. ichneumon, southern Spain and Portugal (probably through human introduction), Asia Minor to Palestine, Morocco to Tunisia, Egypt, possibly eastern Libya, most of Africa south of the Sahara;

H. javanicus, Iraq through northern India to extreme southern China and Indochina, Malay Peninsula, Hainan, Java;

H. edwardsii, eastern Arabian Peninsula through Afghanistan and India to Assam, Sri Lanka;

H. smithii, peninsular India, Sri Lanka;

H. fuscus, southwestern India, Sri Lanka;

H. vitticollis, southwestern India, Sri Lanka;

H. urva, Nepal to southeastern China and peninsular Malaysia, Taiwan, Hainan;

H. semitorquatus, Sumatra, Borneo;

H. brachyurus, Malay Peninsula, Sumatra, Borneo, Palawan;

Malagasy brown-tailed mongoose *(Salanoia concolor),* photo of mounted specimen in Field Museum of Natural History.

A. African mongoose *(Herpestes ichneumon)*, photo by Ernest P. Walker. B. Crab-eating mongoose *(H. urva)*, photo by Robert E. Kuntz.

subgenus *Xenogale* J. A. Allen, 1919

H. naso, southeastern Nigeria to eastern Zaire.

H. auropunctatus, found from Iraq to Southeast Asia, sometimes has been treated as a species distinct from *H. javanicus,* but the two were considered conspecific by Corbet and Hill (1992) and Lekagul and McNeely (1977). *H. palustris,* described from the vicinity of Calcutta in eastern India (Ghose 1965; Ghose and Chaturvedi 1972), was listed as a separate species by Wozencraft (*in* Wilson and Reeder

1993) but placed within *H. javanicus* by Corbet and Hill (1992). Wozencraft also included *H. fuscus* in *H. brachyurus,* but Corbet and Hill argued that such a procedure is unjustified. Rosevear (1974) treated *Xenogale* as a distinct genus, but it was regarded as a subgenus or synonym of *Herpestes* by Corbet and Hill (1991), Meester et al. (1986), and Wozencraft (*in* Wilson and Reeder 1993). *Galerella* (see account thereof) also sometimes is included in *Herpestes.*

Head and body length is 250–650 mm, tail length is 200–510 mm, and weight is 0.5–4.0 kg. Coloration varies considerably. Some forms are greenish gray, yellowish brown, or grayish brown. Others are finely speckled with white or buff. The underparts are generally lighter than the back and sides and are white in some species. The fur is short and soft in some species and rather long and coarse in others. The body is slender, the tail is long, there are five digits on each limb, the hind foot is naked to the heel, and the foreclaws are sharp and curved. Small scent glands are situated near the anus, and some species can eject a vile-smelling secretion. Females have four or six mammae.

Mongooses occupy a wide variety of habitats ranging from densely forested hills to open, arid plains. They shelter in hollow logs or trees, holes in the ground, or rock crevices. They may be either diurnal or nocturnal. They are basically terrestrial but are very agile, and some species can climb skillfully (Grzimek 1975). During the morning they frequently stretch out in an exposed area to sun themselves. The diet includes insects, crabs, fish, frogs, snakes, birds, small mammals, fruits, and other vegetable matter. Some species kill cobras and other venomous snakes. Contrary to popular belief, mongooses are not immune to the bites of these reptiles. Rather, they are so skillful and quick in their movements that they avoid being struck by the snake and almost invariably succeed in seizing it behind the head. The battle usually ends with the mammal's eating the snake.

Population densities of 1–14/sq km have been recorded for *H. javanicus* in the West Indies (Hoagland, Horst, and Kilpatrick 1989). Data cited by Ewer (1973) indicate that in Hawaii the individual home range diameter of *H. javanicus* is about 1.6 km for males and 0.8 km for females. Mongooses are found alone, in pairs, or in groups of up to 14 individuals. In Spain, Palomares and Delibes (1993) observed most *H. ichneumon* alone, but pairs consisting of an adult male and female also were sighted year-round. Males were territorial, and the core areas of their ranges overlapped with those of one or more females. Ben-Yaacov and Yom-Tov (1983) found that in Israel family groups of *H. ichneumon* occupied permanent home ranges and consisted of 1 adult male, 2–3 females, and young. The group produced a single annual litter in the spring, and the young remained with the family for a year or more.

Some species breed throughout the year, with females giving birth two or three times annually. One female *H. edwardsii* produced five litters in 18 months. In *H. javanicus* the estrous cycle lasts about 3 weeks, estrus 3–4 days, and gestation 42–49 days. In *H. ichneumon* gestation has been variously reported to last 60–84 days. Litters contain one to four young. The young of *H. javanicus* are weaned after 4–5 weeks and attain sexual maturity at 1 year. Captive *H. ichneumon* have lived more than 20 years (Ewer 1973; Grzimek 1975; Kingdon 1977; Lekagul and McNeely 1977; Nellis 1989; Roberts 1977).

Mongooses have been widely introduced by people to kill rats and snakes. The presence of *H. ichneumon* in Spain and Portugal probably results from introduction in ancient times. That species also has been introduced to Madagascar. *H. javanicus* was introduced in the West Indies during the early 1870s, and populations now are present on 29 islands there, including all of the Greater Antilles. *H. javanicus* also has been established on the northeastern coast of South America, Mafia Island off East Africa, Mauritius, islands in the Adriatic off Croatia, and the Hawaiian and Fiji islands. Several individuals of this species have been taken on the mainland of North America. *H. edwardsii* has been introduced in central Italy and on the Malay Peninsula (though the population there probably is now extinct), Mauritius, and the Ryukyu Islands (Carpaneto 1990; Corbet 1978, 1984; Corbet and Hill 1992; Gorman 1976;

Krystufek and Tvrtkovic 1992; Nellis et al. 1978; Van Gelder 1979; Wells 1989). In addition to killing rats and snakes, mongooses have destroyed harmless birds and mammals and have contributed to the extinction or endangerment of many desirable species of wildlife. They also have become pests by preying on poultry. The importation or possession of mongooses is therefore now forbidden by law in some countries.

CARNIVORA; HERPESTIDAE; Genus GALERELLA
Gray, 1865

Gray and Slender Mongooses

There are four species (Skinner and Smithers 1990; Taylor and Goldman 1993; Watson 1990; Watson and Dippenaar 1987):

G. pulverulenta (Cape gray mongoose), South Africa to south of 28° S, Lesotho;

G. nigrita, southern Angola, northern and central Namibia;

G. sanguinea (slender mongoose), throughout Africa from Senegal and possibly Western Sahara to Sudan, and south to northern and eastern South Africa, but not the central forest zone of Gabon, Congo, and some adjacent areas.

G. ochracea, Somalia.

Galerella was treated as a subgenus of *Herpestes* by Taylor and Goldman (1993) but was considered a full genus by the other authorities cited above, as well as Corbet and Hill (1991), Meester et al. (1986), and Wozencraft (in Wilson and Reeder 1993). Watson and Dippenaar (1987) indicated that another named species, *G. swinnyi*, may have occurred in southeastern South Africa but that it could not be properly evaluated based on the limited available material. Wozencraft included *swinnyi* in *G. sanguinea* and noted that it is believed to have been extirpated from the type locality. Wozencraft followed Crawford-Cabral (1989b) in using the name *G. flavescens* in place of *G. nigrita* and followed Watson (1990b) in listing *G. swalius* of central and southern Namibia as a species distinct from *G. sanguinea*. Taylor and Goldman (1993) are followed here with regard to both those issues and also in accepting *G. ochracea* as a species separate from *G. sanguinea*.

Galerella generally is smaller than *Herpestes*. Head and body length is 268–425 mm, tail length is 205–340 mm, and weight is 373–1,250 grams. Males are about 9 percent larger than females. The pelage may be short or shaggy, and the color varies extensively. The upper parts are dark brown or even black, drab or pale brown, grayish, or grizzled reddish or yellowish. The hairs of the underparts lack annulations, being grayer in gray specimens, redder or yellow in the more brightly colored. The tip of the tail is dark or completely black. Females have been reported to have two or three pairs of mammae. From that of *Herpestes*, the skull of *Galerella* differs in being smaller, usually lacking the lower first premolar in adults, and in having the anterior chamber of the auditory bulla inflated and comparable in size to the posterior portion, rather than flattish and much smaller than the inflated posterior portion (Kingdon 1977; Meester et al. 1986; Skinner and Smithers 1990; Watson 1990b; Wozencraft 1989a).

Skinner and Smithers (1990) wrote that these mongooses are found in savannah, woodland, arid country, and a variety of other wet and dry habitats but not in dense forests or deserts. The elevational range is sea level to 3,600 meters. They are diurnal and terrestrial but good climbers and will readily take to trees. The diet consists mostly of insects and also includes a substantial amount of small vertebrates and some plant material.

In a radio-tracking study of *G. pulverulenta* in Cape Province, Cavallini and Nel (1990) found home ranges of 21–63 ha., with much overlap both between and within sexes. Ewer (1973) noted that family parties of *G. pulverulenta* den together but forage individually. Taylor (1975) wrote that *G. sanguinea* travels alone or in pairs, that its home range may be as small as 1 sq km but is much larger in desert areas, that it may be territorial, and that it is generally silent. Skinner and Smithers (1990), however, reported a variety of vocalizations and indicated that animals usually are solitary. In southern Africa *G. sanguinea* apparently gives birth in the wet summer months, from about October to March. *G. pulverulenta* has been reported to give birth from about August through December in Cape Province. The young are born in holes in the ground, crevices in rocks, hollow logs, or other suitable shelters. Litter sizes of one to three young have been recorded (Cavallini 1992). The gestation period has been reported as 58–62 days, the lactation period as 50–65 days (Hayssen, Van Tienhoven, and Van Tienhoven 1993).

CARNIVORA; HERPESTIDAE; Genus MUNGOS
E. Geoffroy St.-Hilaire and G. Cuvier, 1795

Banded and Gambian Mongooses

There are two species (Coetzee *in* Meester and Setzer 1977):

M. gambianus (Gambian mongoose), savannah zone from Gambia to Nigeria;
M. mungo (banded mongoose), Gambia to northeastern Ethiopia and south to South Africa.

The name *Mungos* formerly was sometimes used for many other species of mongooses, including those now assigned to *Herpestes*.

Head and body length is 300–450 mm, tail length is 230–90 mm, and weight is 1.0–2.2 kg. *M. mungo* is brownish gray with dark brown and well-defined yellowish or whitish bands across the back. The banded pattern is produced by hair markings of the same type that produce the ground color of the remainder of the body. The hairs are alternately ringed with dark and light bands; the color rings of the individual hairs coincide with like colors on adjacent hairs. *M. gambianus* lacks the transverse bands but has a dark streak on the side of the neck (Coetzee *in* Meester and Setzer 1977).

The pelage is coarse; compared with other mongooses, *Mungos* has little underfur. Although the tail is not bushy, it is covered with coarse hair and is tapered toward the tip. The foreclaws are elongate, the soles are naked to the wrist and heel, and there are five digits on each limb. There is no naked grooved line from the tip of the nose to the upper lip. Females have six mammae.

The information in the remainder of this account applies to *M. mungo* and was taken from Kingdon (1977), Neal (1970), Rood (1974, 1975), Rosevear (1974), and Simpson (1964). The banded mongoose is found in grassland, brushland, woodland, and rocky, broken country. It dens mainly in old termite mounds but also in such places as erosion gullies, abandoned aardvark holes, and hollow logs. Dens are communal and consist of one to nine entrance holes, a central sleeping chamber of about 1–2 cubic meters, and perhaps several smaller chambers. Most dens are used only for a few days, but some favorite sites may be occupied for as long as 2 months. *M. mungo* is terrestrial and diurnal and has excellent senses of vision, hearing, and smell. A group generally emerges from the den around 0700–0800 hours, forages for several hours, rests in a shady spot during the hottest part of the day, forages again, and returns to a den before sunset. More time is spent in the vicinity of the den if young are present therein. A group generally covers 2–3 km per day, moving in a zigzag pattern and searching among rocks and vegetation for food. The diet consists largely of invertebrates, especially beetles and millipedes, and also includes small vertebrates. To break an egg and obtain the contents, *M. mungo* grasps the item with its forefeet and propels it backward between its hind feet and against a hard object.

In the Ruwenzori National Park of Uganda population density was found to be about 18/sq km, and group home range varied from about 38 to 130 ha. In the Serengeti

Banded mongoose *(Mungos mungo)*, photo from Zoological Garden Berlin-West through Ernst von Roy.

Banded mongooses *(Mungos mungo)*, photo by J. P. Rood.

home range may be more than 400 ha. Ranges overlap, but intergroup encounters are generally noisy and hostile and sometimes involve chasing and fighting. Groups contain as many as 40 individuals, usually about 10–20, including several adults of both sexes. Captive females have been seen to dominate males. The animals sleep in contact and forage in a fairly close, but not bunched, formation. Groups are cohesive, but some splitting off has been observed, and mating between members of different groups sometimes occurs. Individuals may mark one another with scent from anal glands. The most common vocalization is a continuous birdlike twittering that probably serves to keep the group together during foraging. There are also various agonistic growls and screams and an alarm chitter.

In East Africa, at least, reproduction continues throughout the year. Breeding is synchronized within a given group, with several females bearing litters at approximately the same time. Groups breed as many as four times per year, though it cannot be said that any one female has that many litters annually. A period of mating often begins within 1–2 weeks after the birth of young. The gestation period is about 2 months. The number of young per litter seems usually to be about two or three but may be as high as six. The young weigh about 20 grams at birth but grow rapidly. They apparently are kept together and raised commonly by the group. They suckle indiscriminately from any lactating female and are usually guarded by one or two adult males while the rest of the group forages. They begin to travel with the others at about 1 month. Females attain sexual maturity at 9–10 months. According to Van Rompaey (1978), a captive *M. mungo* lived to an age of approximately 12 years.

CARNIVORA; HERPESTIDAE; Genus **CROSSARCHUS**
F. Cuvier, 1825

Cusimanses

There are four species (Colyn and Van Rompaey 1990; Colyn et al. 1995; Goldman 1984; Schreiber et al. 1989):

C. obscurus, Sierra Leone to Ghana;
C. platycephalus, southern Benin to southeastern
    Cameroon and possibly Congo;
C. alexandri, Zaire, Uganda, possibly Central African
    Republic and Zambia;
C. ansorgei, central Zaire, one specimen from
    northwestern Angola.

C. platycephalus was considered conspecific with *C. obscurus* by Wozencraft (1989a and *in* Wilson and Reeder 1993) but not by Corbet and Hill (1991), Goldman (1984), or Van Rompaey and Colyn (1992).

Head and body length is 305–450 mm, tail length is 150–255 mm, and weight is 450–1,450 grams. The body is covered by relatively long, coarse hair that is a mixture of browns, grays, and yellows. The head is usually lighter-colored than the remainder of the body, while the feet and legs are usually the darkest. The legs are short, the tail is tapering, the ears are small, and the face is sharp.

Cusimanses live in forests and swampy areas. Various reports suggest that they may be active either by day or by night (Kingdon 1977). They travel about in groups, seldom remaining in any one locality longer than two days and taking temporary shelter in any convenient place. While seeking food they scratch and dig in dead vegetation and in the soil. The diet consists principally of insects, larvae, small reptiles, crabs, tender fruits, and berries. It is said that cusimanses crack the shells of snails and eggs by hurling them

Cusimanse *(Crossarchus obscurus)*, photo by Ernest P. Walker.

with the forepaws back between the hind feet and against some hard object.

Groups contain 10–24 individuals. These probably represent 1–3 family units, each with a mated pair and the surviving members of 2–3 litters. There is no evidence of seasonal breeding in either East or West Africa. Observations in captivity indicate that there may be several litters annually (Kingdon 1977; Rosevear 1974). According to Goldman (1987), the gestation period averages 58 days and litters contain two to four, usually four, young; the young open their eyes after 12 days, take solid food at 3 weeks, and attain sexual maturity at about 9 months. Captives have been estimated to live for 9 years.

Cusimanses tame easily and make good pets. They are affectionate, playful, clean, and readily housebroken, but they sometimes mark objects with their anal scent glands (Rosevear 1974). The subspecies *C. ansorgei ansorgei* is known only by a single specimen collected in 1908 in a relict forest in northern Angola (Schreiber et al. 1989).

CARNIVORA; HERPESTIDAE; **Genus LIBERIICTIS**
Hayman, 1958

Liberian Mongoose

The single species, *L. kuhni*, now is known by 27 specimens from northeastern Liberia (Carnio 1989; Carnio and Taylor 1988) and has been reported from southwestern Ivory Coast (Hoppe-Dominik 1990; M. E. Taylor 1992).

Head and body length of an adult male was 423 mm, tail length was 197 mm, and weight was 2.3 kg. In an adult female head and body length was 478 mm and tail length was

Liberian mongoose *(Liberiictis kuhni)*, photo by F. Faigal, Metropolitan Toronto Zoo.

Liberian mongoose *(Liberiictis kuhni)*, photo of dead specimen by Lynn Robbins.

205 mm (Goldman and Taylor 1990). The predominant color of the pelage is dark brown. A dark stripe bordered above and below by a pale stripe is present on the neck. The throat is pale, the tail is slightly bicolored, and the legs are dark. From *Crossarchus, Liberiictis* is distinguished externally by the presence of neck stripes, a more robust body, and apparently longer ears (Schlitter 1974). The skull of *Liberiictis* is larger than that of *Crossarchus,* the rostrum and nasals are more elongate, the teeth are proportionally smaller and weaker, and there is an additional premolar in both the upper and lower jaws.

Schlitter (1974) noted that the long claws of the front feet, the long mobile snout, and the weak dentition of *Liberiictis* indicate that it is a terrestrial animal with a primarily insectivorous diet. Carnio and Taylor (1988) stated that the diet also includes worms, eggs, and small vertebrates. Goldman and Taylor (1990) reported that a captive was fed ground meat, dog food, young chickens, and fish. Schlitter's two specimens were taken in a densely forested area traversed by numerous streams. People native to the area said that *Liberiictis* is diurnal and found only on the ground. An adult male was taken in a snare on the ground. A juvenile female was excavated from a burrow associated with a termite mound on 29 July. An adult of unknown sex was also in the burrow but was not preserved. Goldman and Taylor (1990) indicated that breeding probably coincides with the rainy season, May–September. Schlitter stated that *Liberiictis* is eaten by human hunters. In a letter of 26 June 1963 to Ernest P. Walker, Dr. Hans-Jürg Kuhn, who had traveled in Liberia, wrote that native people say that *Liberiictis* is usually found in tree holes and lives in groups of 3–5 individuals. Carnio and Taylor (1988) reported that a group of 15 had been seen foraging. They added, however, that the genus seems to be very rare and may be jeopardized by human hunting and habitat destruction. In January 1989 a live specimen was obtained and placed in the Metro Toronto Zoo (Carnio 1989). The IUCN now classifies *Liberiictis* as endangered.

CARNIVORA; HERPESTIDAE; **Genus HELOGALE**
Gray, 1861

Dwarf Mongooses

There are two species (Coetzee *in* Meester and Setzer 1977; Yalden, Largen, and Kock 1980):

H. parvula, Ethiopia to Angola and eastern South Africa;
H. hirtula, southern Ethiopia, southern Somalia, northern Kenya.

Head and body length is 180–260 mm, tail length is 120–200 mm, and weight is 230–680 grams (Kingdon 1977). Coloration is variable, but generally the upper parts are speckled brown to grayish. The lower parts are only slightly paler, and the tail and lower parts of the legs are dark. In some individuals there is a rufous patch on the throat and breast and the basal portion of the lower side of the tail is reddish brown. Other individuals are entirely black.

Dwarf mongooses are found in savannahs, woodlands, brush country, and mountain scrub, from sea level to elevations of about 1,800 meters. They are mainly terrestrial and diurnal. They seek shelter at dusk in deserted or active termite mounds, among gnarled roots of trees, and in crevices. Their slender bodies enable them to squeeze into small openings. On occasion they may excavate their own burrows. Dens are changed frequently, sometimes daily (Rood 1978). Most of the day is spent in an active and noisy search for food among brush, leaves, and rocks. The diet consists mainly of insects (Rasa 1977; Rood 1980) and also includes small vertebrates, eggs, and fruit.

Helogale is found in organized groups that are thought to use a definite home range, though there is conflicting information on group movements and relationships. According to Rasa (1977), a portion of the range is occupied for two or three months and then there is a shift to another part, presumably because of depleted food supplies. Kingdon (1977) wrote that a group observed for seven years stayed mainly within an area of 2 ha.; another group used 2 ha. during the dry season but moved away with the coming of the rains. In the Serengeti, Rood (1978) found group home ranges to average 30 ha., to overlap by 5–40 percent with the ranges of one to four neighboring packs, and to contain 10–20 dens each. Rasa (1973, 1977) reported that

Dwarf mongoose *(Helogale parvula)*, photo by Bernhard Grzimek.

captives have been observed to mark the vicinity of dens with secretions from cheek and anal glands and to show considerable intergroup aggression. Rasa (1986) stated that groups inhabit home ranges of 0.65–0.96 sq km that show little overlap with one another and are traversed every 20–26 days, the length of time it takes for the marking secretions to decay.

The social organization of *Helogale* is unique among mammals (Kingdon 1977; Rasa 1972, 1973, 1975, 1976, 1977, 1983; Rood 1978, 1980). There are as many as 40 individuals in a group but usually about 10–12. The groups are matriarchal families founded and led by an old female. She initiates movements and has priority regarding food. The second highest ranking member of the group is her mate, an old male. These two dominant animals are monogamous and usually the only members of the group to produce offspring; they suppress sexual activity in other group members. The latter form a hierarchy, with the *youngest* individuals ranking highest. This arrangement probably serves to allow the young animals to obtain sufficient food, without competition from older and stronger mongooses, during the unusually long period of growth in this genus. Within any age class in the group females are dominant over males.

Despite the rigid class structure, or perhaps because of it, intragroup relations are generally harmonious, and severe fights are rare. Subordinate adults clean, carry, warm, and bring food to helpless young and take turns "baby-sitting" while the rest of the group forages. Females in addition to the mother sometimes nurse the young. The youngest mobile animals seem to have the role of watching for danger and alerting the others by means of visual signals or a shrill alarm call. Often a single animal occupies an exposed position, where it serves as a group guard. One series of observations showed that when a low-ranking male became sick, it was allowed a higher than normal feeding priority and was also warmed by other group members. In another case, a group restricted its normal movements in order to provide care and food for an injured member. Individuals communicate by depositing scent from the cheek and anal glands, and they also sometimes mark one another as an apparent sign of acceptance. As surviving subordinate animals grow older, they seem not to leave the group, even though they are not allowed to mate. If the dominant female dies, however, the group may split up.

In the Serengeti National Park of Tanzania, births occur mainly in the rainy season, from November to May, and the alpha female usually has three litters per year (Rood

1978, 1980). In a captive colony in Europe the young are born regularly in spring and autumn (Rasa 1972). According to Rasa (1977), females there normally give birth twice a year, entering estrus 4–7 days after lactation ceases. If the newborn young die, however, females may quickly remate, and they thus have the potential of producing five litters annually. The gestation period is 49–56 days. The number of young per litter averages about four and ranges from one to seven. Nursing lasts for at least 45 days, but group members begin to bring solid food to the young before weaning is complete. The young start to forage with the group by the time they are 6 months old. Females may reach physiological sexual maturity at as early as 107 days (Zannier 1965), but social restrictions normally delay breeding for several years. Apparently, full physical maturity is not attained until 3 years (Rasa 1972). One dominant pair was observed to mate first when about 3 years old and then to continue to breed for 7 years (Kingdon 1977). An individual lived at the Basel Zoo for 12 years and 3 months (Marvin L. Jones, Zoological Society of San Diego, pers. comm., 1995).

CARNIVORA; HERPESTIDAE; **Genus DOLOGALE**
Thomas, 1926

African Tropical Savannah Mongoose

The single species, *D. dybowskii*, is found in the Central African Republic, northeastern Zaire, southern Sudan, and western Uganda (Coetzee *in* Meester and Setzer 1977). This species was originally assigned to *Crossarchus* and has sometimes been referred to *Helogale*.

Head and body length is about 250–330 mm and tail length is 160–230 mm. Kingdon (1977) listed the weight as approximately 300–400 grams. Stripes are lacking. The head and neck are black, grizzled with grayish white. The back, tail, and limbs are lighter in color and have brownish spots. The underparts are reddish gray. The fur is short, even, and fine, in contrast to the loose, coarse pelage of *Crossarchus*. The snout is not lengthened like that of *Crossarchus*. *Dologale* closely resembles *Helogale* but does not have a groove on the upper lip and has weaker teeth (Coetzee *in* Meester and Setzer 1977).

Kingdon (1977) stated that the few known specimens of

African tropical savannah mongoose *(Dologale dybowskii)*, photo by F. Petter of mounted specimen in Museum National d'Histoire Naturelle.

Dologale suggest adaptation to a variety of habitats—thick forest, savannah-forest, and montane forest grassland. Some evidence indicates that the genus is at least partly diurnal. It has robust claws, suggesting digging habits as in *Mungos*. Asdell (1964) wrote that a litter of four young had been noted in Zaire. Schreiber et al. (1989) noted that there had been no records or sightings of *Dologale* for at least 10 years.

CARNIVORA; HERPESTIDAE; Genus BDEOGALE
Peters, 1850

Black-legged Mongooses

There are two subgenera and three species (Coetzee *in* Meester and Setzer 1977; Nader and Al-Safadi 1991):

subgenus *Bdeogale* Peters, 1850

B. crassicauda, Yemen, southern Kenya to central Mozambique;

subgenus *Galeriscus* Thomas, 1894

B. nigripes, southeastern Nigeria to northern Zaire and northern Angola;
B. jacksoni, southeastern Uganda, central Kenya.

Rosevear (1974) considered *Galeriscus* to be a separate genus. Kingdon (1977) suggested that *B. jacksoni* is only a subspecies of *B. nigripes*.

Head and body length is 375 to at least 600 mm and tail length is 175–375 mm. Kingdon (1977) listed weight as 0.9–3.0 kg. There is considerable variation in color both within and between species. The predominant general coloration is some shade of gray or brown, and the legs are usually black. The fur of adults is rather close, dense, and short; that of the young is nearly twice as long and lighter in color.

Bdeogale resembles *Ichneumia* in having black feet, soft underfur, and long, coarse hair over the upper parts of the body. It differs in lacking the first or inner toe on each foot and in having larger premolar teeth. *Bdeogale* differs from *Rhynchogale* in having a naked groove from the nose to the upper lip. The foreparts of the feet of *Bdeogale* are naked, but the hind parts are well haired.

Taylor (1986, 1987) described *B. crassicauda* as rare, unspecialized, nocturnal, insectivorous, and solitary. According to Kingdon (1977), *B. crassicauda* inhabits woodland and moist savannah, and the other species live in tropical forest. *B. crassicauda* feeds almost entirely on insects, especially ants and termites, but may also take crabs and rodents. *B. nigripes* seems to prefer ants but also eats small vertebrates and carrion. These mongooses are frequently seen in pairs. A female and a quarter-grown young were taken in December on the coast of Kenya, a pregnant female with a large fetus was also taken in December, and a female with a newborn infant was found in southeastern Tanzania in late November. Information compiled by Rosevear (1974) suggests that adults are basically solitary in the wild but are not quarrelsome when kept together in captivity; that births in West Africa occur from November to January; and that litters normally contain a single young.

Black-legged mongoose *(Bdeogale* sp.), photo by Don Davis.

According to Jones (1982), a captive *B. nigripes* lived 15 years and 10 months.

The IUCN classifies *B. jacksoni* as vulnerable and the subspecies *B. crassicauda omnivora,* of coastal Kenya and Tanzania, as endangered, and *B. c. tenuis,* of Zanzibar, as indeterminate. Schreiber et al. (1989) noted that *B. c. omnivora* is very rare and that its restricted forest habitat is rapidly being cut over. In 1988 a specimen of *B. crassicauda* was collected near Sanaa in central Yemen, the first record of *Bdeogale* outside of Africa. If human introduction is not involved, this record would be one in a series of recent discoveries demonstrating the faunal affinity of East Africa and the Arabian Peninsula and suggesting that the southern part of the peninsula holds one of the world's largest reservoir of mammalian surprises.

CARNIVORA; HERPESTIDAE; **Genus RHYNCHOGALE**
Thomas, 1894

Meller's Mongoose

The single species, *R. melleri,* occurs from southern Zaire and Tanzania to eastern South Africa and possibly northeastern Angola (Coetzee *in* Meester and Setzer 1977).

Head and body length is 440–85 mm and tail length is usually 300–400 mm. Kingdon (1977) listed weight as 1.7–3.0 kg. The general coloration is grayish or pale brown, the head and undersides are paler, and the feet are usually darker. *Rhynchogale* resembles *Ichneumia* in having coarse guard hairs protruding from the close underfur and the same dental formula. *Rhynchogale* differs from *Ichneumia* in the frequent reduction of the hallux and the lack of a naked crease from the nose to the upper lip. *Rhynchogale* has hind soles that are hairy to the roots of the toes. Females have two abdominal pairs of mammae.

According to Kingdon (1977), the habitat appears to be restricted to the woodland belt and possibly to moister and more heavily grassed or wooded areas, such as drainage lines and rock outcrops. Available information suggests that *Rhynchogale* is terrestrial, nocturnal, and solitary. The diet includes wild fruit, termites, and probably small vertebrates. A litter of two newborn young with eyes still unopened was found in a small cave on a rocky hill in Zambia in December. A pregnant female containing two embryos was found in the same area, also in December. In Zimbab-

we births occur around November and litters contain as many as three young.

CARNIVORA; HERPESTIDAE; **Genus ICHNEUMIA**
I. Geoffroy St.-Hilaire, 1837

White-tailed Mongoose

The single species, *I. albicauda,* occurs all across the southern part of the Arabian Peninsula and in Africa from Senegal to southeastern Egypt and south to northeastern Namibia and eastern South Africa, but not in the dense forest zone of West and Central Africa or in the deserts of the north or southwest (Coetzee *in* Meester and Setzer 1977; Corbet 1978; Gallagher 1992; Harrison and Bates 1991; Nader 1979; Skinner and Smithers 1990).

Head and body length is 470–710 mm, tail length is 355–470 mm, and weight is 1.8–5.2 kg (Kingdon 1977; Taylor 1972). Long, coarse, black guard hairs protrude from a yellowish or whitish close, woolly underfur, producing a grayish general body color. The four extremities, from the elbows and knees, are black. The basal half of the tail is of the general body color. The terminal portion is usually white but occasionally black. *Ichneumia* is characterized by large size; its bushy, tapering tail; having the soles of the forelimbs naked to the wrist; and the division of the upper lip by a naked slit from the nose to the mouth. Females have four mammae.

The white-tailed mongoose is found mainly in savannahs and grassland. It prefers areas of thick cover, such as forest edge and bush-fringed streams. It is basically terrestrial and nocturnal, sheltering by day in porcupine or aardvark burrows, termite mounds, or cavities under roots or rocks. The diet consists mainly of insects and also includes snakes, other small vertebrates, and fruit. *Ichneumia* breaks eggs by hurling them back between its hind legs and against some hard object. It may defend itself by ejecting a particularly noxious secretion from its anal scent glands (Kingdon 1977).

In Kenya, Taylor (1972) found home range of one individual to be about 8 sq km. Detailed studies on the Serengeti of Tanzania (Waser and Waser 1985) revealed average home range to be 0.97 sq km for adult males and 0.64 sq km for females. The male ranges did not overlap, but there was complete overlap between the ranges of opposite

Meller's mongoose *(Rhynchogale melleri)*, photo from *Proc. Zool. Soc. London.*

White-tailed mongoose *(Ichneumia albicauda)*, photo by Ernest P. Walker.

sexes. Some female ranges were exclusive, but in other cases several females and their offspring used a common range, though they foraged separately. Such an arrangement was thought to represent a matrilineal clan of related individuals. Reported pairs and family groups probably reflect observations of consorting individuals and mothers with young, respectively. *Ichneumia* is highly vocal, the most unusual sound being a doglike yap that may be associated with sexual behavior. Kingdon (1977) wrote that the two to four young are born in a burrow. All litters observed by Waser and Waser (1985) on the Serengeti contained one or two young, seen most frequently from February to May. No young were seen during the August–November dry season. The young usually were independent by 9 months. A captive was still living after 10 years (Jones 1982).

When the white-tailed mongoose dwells near a poultry raiser it may prove to be a pest. In captivity it is the shyest of mongooses, but it is said to become a pleasing pet if captured young.

CARNIVORA; HERPESTIDAE; Genus ATILAX
F. Cuvier, 1826

Marsh Mongoose, or Water Mongoose

The single species, *A. paludinosus,* is found from Senegal to Ethiopia, and south to South Africa (Baker 1992).

Head and body length is 440–620 mm, tail length is 250–430 mm, and weight is 2.5–4.1 kg (Baker 1992; Kingdon 1977). The pelage is long, coarse, and generally brown in color. A sprinkling of black guard hairs often gives a dark effect. In some individuals light rings on the hairs impart a grayish tinge. The head is usually lighter than the back, and the underparts are still paler.

Atilax is fairly heavily built. Although it is the most aquatic mongoose, it is the only one with toes that completely lack webbing. This feature may be associated with the habit of feeling for aquatic prey in mud or under stones. There are five digits on each limb, the soles are naked, and

the claws are short and blunt. The anal area is large and naked, and there is a narrow, naked slit between the nose and the upper lip. Females usually have two pairs of mammae.

Atilax is found in a variety of general habitat types, but its basic requirement seems to be permanent water bordered by dense vegetation (Rosevear 1974). Favored haunts are marshes, reed-grown streambeds, and tidal estuaries. Grassy patches and floating masses of vegetation often serve as feeding places and as dry resting spots. Like other mongooses, *Atilax* does little or no climbing but does run up leaning tree trunks or other inclines easy of access. It is an excellent swimmer and diver. When hard pressed, it submerges, leaving only the tip of the nose exposed for breathing. Normally the head and part of the back are exposed when it is swimming. Food is sought in the water and in travels on regular pathways along the borders of streams and marshes. Activity is usually said to be nocturnal and crepuscular, but Rowe-Rowe (1978b) referred to *Atilax* as diurnal and stated that it does much of its hunting while walking in shallow water.

The diet consists of almost any form of animal life that can be caught and killed. Regular foods include insects, mussels, crabs, fish, frogs, snakes, eggs, small rodents, and fruit (Kingdon 1977; Rosevear 1974). *Atilax* may throw such creatures as snails and crabs against hard surfaces to break their shells. A captive was seen to take a piece of beef rib between its forefeet, rear onto its hind feet with its forefeet held high, and then forcefully throw the bone to the floor of the cage in an effort to break it.

Kingdon (1977) wrote that *Atilax* is usually seen alone and that individuals are widely spaced and undoubtedly highly territorial. The young are said to be born in burrows in stream banks or on masses of vegetation gathered into heaps among reed beds. There is no evidence of a particular breeding season in West Africa (Rosevear 1974). Data obtained in southern Africa indicate that young are born during the warm, wet months from about October to February (Skinner and Smithers 1990). The gestation period is 69–80 days (Baker 1992). The number of young per litter has been reported as one to three and usually seems to be two or three. The young weigh about 100 grams at birth, open their eyes after 9–14 days, are weaned at 30–46 days,

Marsh mongoose, or water mongoose *(Atilax paludinosus)*, photo by Ernest P. Walker.

and reach adult size at approximately 27 weeks (Baker 1992; Baker and Meester 1986). One water mongoose lived in captivity for just over 19 years (Marvin L. Jones, Zoological Society of San Diego, pers. comm., 1995).

Rosevear (1974) observed that within the last 50 years *Atilax* has probably declined substantially in the drier parts of its range because of human destruction of the available riverine habitat. *Atilax* is also widely hunted by people because it is reputed to be a poultry thief.

CARNIVORA; HERPESTIDAE; **Genus CYNICTIS**
Ogilby, 1833

Yellow Mongoose

The single species, *C. penicillata*, is found in extreme southern Angola, Namibia, Botswana, extreme western Zimbabwe, and South Africa (Taylor and Meester 1993).

Head and body length is 270–380 mm and tail length is 180–280 mm. Smithers (1971) listed weights of 440–797 grams, though Cavallini (1993) found some males weighing up to 1,000 grams. The hair is fairly long, especially on the tail, which is somewhat bushy. The general color is dark orange yellow to light yellow gray. The underfur is rich yellow, the chin is white, the underparts and limbs are lighter than the back, and the tail is tipped with white. The individual guard hairs are usually yellowish in the basal half, followed by a black band and a white tip. There is a seasonal color change in the coat: the summer pelage, typical in January and February, is reddish, short, and thin; the winter coat, typical from June to August, is yellowish, long, and thick; and the transitional coat, typical in November and December, is pale yellow. The forefeet have five digits and the hind feet have four. The first, or inner, digit on the forefoot is small and above the level of the other four, so it does not touch the ground. Females have three pairs of mammae (Ewer 1973).

The yellow mongoose frequents open country, prefer-ably with loose soil, but may take refuge among rocks or in brush along the banks of streams when disturbed. It sometimes uses holes made by other animals, such as *Pedetes*, but usually excavates its own burrows. It is an energetic digger, constructing extensive underground systems that may cover 50 sq meters or more. These burrows may have 40 or more entrance holes, interconnecting tunnels on two or three levels to a depth of 1.5 meters, and enlarged nest chambers at intervals along the tunnels (Taylor and Meester 1993). Certain places within the burrow system are used for deposit of body wastes.

Cynictis is mainly diurnal but may be active at night when living near people. It basks in the sun and sits up on its haunches to obtain a better view of the surroundings. It is agile and capable of traveling at considerable speed. It seldom wanders more than 1 km from the burrow. Apparently, pairs and perhaps entire colonies seek new homes when food becomes scarce in the vicinity of a burrow. The diet consists mainly of insects and other invertebrates (Herzig-Straschil 1977; Smithers 1971). Other reported foods of *Cynictis* include lizards, snakes, birds, the eggs of birds and turtles, small rodents, and even mammals as large as itself.

Groups have been reported to include as many as 50 or more individuals, but Taylor and Meester (1993) indicated that such records result from confusion of *Cynictis* with *Suricata* and that colonies of the former usually have only 4–8 animals. There apparently is considerable variation in social structure. In a coastal area of limited resources, Cavallini (1993) found population density to be only about 1.2/100 ha. and observed little group organization. Adult males maintained separate home ranges of about 100 ha., about four times as large as the female ranges and encompassing several of the latter. Females in different dens maintained nonoverlapping ranges.

In South Africa, Earlé (1981) found the mean size of 5 colonies to be 8 individuals. The nucleus of each group consisted of an adult pair, their newest offspring, and 1 or 2 young or very old individuals. The other members had a loose, unclear association. The group hunting range of about 5–6 ha. corresponded to a territory, which was pa-

Yellow mongoose *(Cynictis penicillata)*, photo by Hans-Jürg Kuhn.

trolled each day by the adult male and marked with urine and secretion from the anal and cheek glands. Young animals from other groups were allowed to cross territorial boundaries if they showed submission by laying on their side and uttering high-pitched screams. Within the group, order of rank was: the adult male, the adult female, the youngest offspring, and then the other animals. The young showed reduced dominance by the age of 10 months. The mating season lasted from July to late September, and births occurred in October and November. Wenhold and Rasa (1994) found that most territorial defense and marking actually was carried out by younger, subordinate individuals. Subordinate males eventually dispersed to other colonies, and subordinate females temporarily left their natal territories to mate with males from other colonies. It was suggested that marking by these animals was partly a means of advertising for mates.

Long-term studies by Rasa et al. (1992) in South Africa and Namibia demonstrate that females commonly produce two litters annually within a period of 2–4 months. They are polyestrous and initiate a new cycle while still lactating. The first litter is born in October, the second from February to December. The gestation period is 60–62 days, mean litter size is 1.9 young (range 1–3), and lactation lasts 6–8 weeks. Smithers (1971) stated that there may be sporadic breeding throughout the year in Botswana, with a peak at some interval from October to April, and that the number of embryos per pregnant female averages 3.2 and ranges from 2 to 5. Taylor and Meester (1993) indicated that sexual maturity is attained after 1 year. One yellow mongoose lived in captivity for 15 years and 2 months (Jones 1982).

CARNIVORA; HERPESTIDAE; **Genus PARACYNICTIS**
Pocock, 1916

Gray Meerkat, or Selous's Mongoose

The single species, *P. selousi*, occurs in Angola, Zambia, Malawi, northern Namibia, Botswana, Zimbabwe, Mozambique, and eastern South Africa (Coetzee *in* Meester and Setzer 1977). This species originally was referred to *Cynictis*.

Head and body length is 390–470 mm and tail length is 280–400 mm. Smithers (1971) listed weights of 1.4–2.2 kg. The upper parts are dull buff gray, the belly is buffy, the feet are black, and the tail is white-tipped. There is no rufous in the coloring, nor are there any spots or stripes. Unlike *Cynictis*, *Paracynictis* has only four digits on each limb. The claws are long and slightly curved. These features are associated with a strong digging ability. *Paracynictis* can defend itself by expelling a strong-smelling secretion from its anal glands; its white-tipped tail, which makes the animal visible at night, may serve as a warning of this capability.

Paracynictis seems to prefer open scrub and woodland (Smithers 1971). It resides in labyrinthine burrows of its own construction. It is terrestrial and nocturnal but has been seen above ground by day. It has been described as shy and retiring. The diet consists of insects, other arthropods, frogs, lizards, and small rodents (Smithers 1971).

Apparently, each individual constructs its own burrow system, and there is less social activity than in *Cynictis*. On the basis of limited data, Smithers (1983) suggested that the births occur in the warm, wet months, probably from August to March, and that litter size is two to four young.

CARNIVORA; HERPESTIDAE; **Genus SURICATA**
Desmarest, 1804

Suricate, or Slender-tailed Meerkat

The single species, *S. suricatta*, occurs in southwestern Angola, Namibia, Botswana, and South Africa (Coetzee *in* Meester and Setzer 1977).

Head and body length is 250–350 mm and tail length is 175–250 mm. Smithers (1971) listed weights of 626–797 grams for males and 620–969 grams for females. The coloration is a light grizzled gray. The rear portion of the back is marked with black transverse bars, which result from the alternate light and black bands of individual hairs coinciding with similar markings of adjacent hairs. The head is almost white, the ears are black, and the tail is yellowish with a black tip. The coat is long and soft, and the underfur is dark rufous in color. The body is quite slender, though this feature is difficult to see because of the long fur. Scent glands peripheral to the anus open into a pouch that presumably stores the secretion (Ewer 1973). The forefeet have very long and powerful claws. Females have six mammae (Grzimek 1975).

The suricate inhabits dry, open country, commonly with hard or stony ground (Smithers 1971). It is an efficient digger. Colonies on the plains may excavate their own burrows or share the holes of African ground squirrels *(Xerus)*. Self-excavated burrow systems average about 5 meters in di-

Selous's mongoose *(Paracynictis selousi)*, photo by Don Carter.

ameter, have approximately 15 entrance holes, and consist of two or three levels of tunnels extending to a depth of 1.5 meters and interconnected with chambers about 30 cm across. The home range of a colony may contain up to five such burrows (Van Staaden 1994). Colonies in stony areas live in crevices among the rocks. Outside activity is almost entirely diurnal. The suricate seems to enjoy basking in the sun, lying in various positions or sitting up on its haunches like a prairie dog *(Cynomys)*. Individuals generally forage near the burrow, turning over stones and rooting in crevices. However, a colony may travel up to 6 km during

a day and use several burrows for sleeping (Skinner and Smithers 1990). The diet is primarily insectivorous (Ewer 1973) but also includes small vertebrates, eggs, and vegetable matter.

Home range of a colony is as large as 15 sq km (Van Staaden 1994). *Suricata* is highly social and occurs in troops of as many as 30 individuals. Usually, however, groups contain 2–3 family units and a total of 10–15 individuals. Each family contains a pair of adults and their young. The female may be larger than the male and may dominate him. Groups mark the vicinity of their burrows

Suricates *(Suricata suricatta)*, photo by Bernhard Grzimek.

with feces and secretions from the anal glands. Outside individuals are vigorously repulsed, and encounters between two groups are highly agonistic. Nonetheless, there is recruitment of new members, especially in the early summer, and both sexes occasionally change groups. At least 10 vocalizations have been identified, including a threatening growl and a shrill alarm bark. This call may be given by an individual that acts as a sentry and will cause the other members of the group to dive for their burrows or other cover (Ewer 1973; Skinner and Smithers 1990).

Breeding in captivity has been recorded throughout the year, with an interbirth interval of only about four months. Births in the wild occur during the warm, wet seasons, August–November and January–March, and there is no evidence of more than a single litter per year. The gestation period is 77 days, possibly less. In captivity litters number as many as seven young, but in the wild they number two to five, usually three or four (Van Staaden 1994). The young weigh 25–36 grams at birth, open their eyes after 10–14 days, and are weaned at 7–9 weeks (Ewer 1973). They attain sexual maturity by 1 year (Schliemann *in* Grzimek 1990). Two suricates living at the San Diego Zoo in December 1995 were 12 years and 8 months old (Marvin L. Jones, Zoological Society of San Diego, pers. comm., 1995).

Suricata tames readily, is affectionate, and enjoys the warmth of snuggling close to its master. It is sensitive to cold. It is often kept about homes in South Africa to kill mice and rats. It should not be confused with the yellow or thick-tailed meerkat *(Cynictis)*, with which it often associates. *Cynictis* is not as winning in its ways nor as pleasing as a pet.

Fossa

The single species, *C. ferox*, is found in Madagascar (Coetzee *in* Meester and Setzer 1977). *Cryptoprocta* sometimes has been placed in the cat family (Felidae), but recent debate has mostly been concerned with whether the genus should be assigned to the family Viverridae or to Herpestidae (see account thereof).

This is the largest carnivore of Madagascar. Head and body length is 610–800 mm, tail length is approximately the same, and shoulder height is about 370 mm. Weight is 7–12 kg (Albignac 1975). The fur is short, smooth, thick, soft, and usually reddish brown in color. Some black individuals have been captured. The mustache hairs are as long as the head. Like some other herpestids, *Cryptoprocta* has scent glands in the anal region that discharge a strong, disagreeable odor when the animal is irritated.

The general appearance is much like that of a large jaguarundi *(Felis yagouaroundi)* or small cougar *(Felis concolor)*. The curved claws are short, sharp, and retractile like those of a cat, but the head is relatively longer than in the Felidae. *Cryptoprocta* walks in a flat-footed manner on its soles, like bears, rather than on its toes, like cats.

The fossa dwells in forests and woodland savannahs, from coastal lowlands to mountainous areas at elevations of 2,000 meters (Coetzee *in* Meester and Setzer 1977). It is mainly nocturnal and crepuscular, is occasionally active by day, and often shelters in caves (Albignac 1972). A maternal den was located in an old termite mound and contained an unlined chamber about 70 cm deep, 100 cm wide, and 30 cm high (Albignac 1970a). The fossa is an excellent climber and pursues lemurs through the trees. Its diet consists mainly of small mammals and birds but also includes reptiles, frogs, and insects (Albignac 1972; Coetzee *in* Meester and Setzer 1977).

Cryptoprocta is solitary except during the reproductive

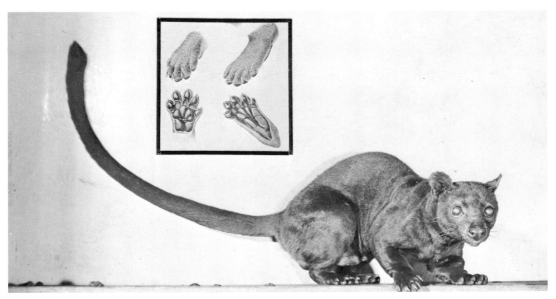

Fossa *(Cryptoprocta ferox)*, photo by Ernest P. Walker. Inset: forefeet and hind feet, top and bottom, photos from *Zoologie de Madagascar*, G. Grandidier and G. Petit.

season. Each individual has been estimated to require about 1 sq km. Mating occurs in September and October, and births take place in the austral summer. The gestation period lasts about 3 months. There are usually two young but sometimes three or four, each weighing about 100 grams at birth. The young leave the den after 4.5 months and are then weaned. Observations in captivity indicate that sexual maturity is not attained until 4 years. One specimen was still living after 20 years in captivity (Albignac 1972, 1975; Coetzee *in* Meester and Setzer 1977; Köhncke and Leonhardt 1986).

The fossa is a powerful predator but has an exaggerated reputation for savagery and destructiveness. It normally flees at the sight of a human, though it may be dangerous if wounded. It sometimes preys on poultry, and some accounts claim that it attacks wild hogs and even oxen. It is widely hunted by people and is now depleted in numbers. The IUCN now classifies it as vulnerable, noting that its habitat is declining because of human encroachment. It also is on appendix 2 of the CITES.

CARNIVORA; Family HYAENIDAE

Aardwolf and Hyenas

This family of four Recent genera and four species is found in Africa and southwestern and south-central Asia. The sequence of genera presented here follows that of Werdelin and Solounias (1991). Simpson (1945) and Wozencraft (*in* Wilson and Reeder 1993) recognized two subfamilies: the Protelinae, with the single genus *Proteles* (aardwolf); and the Hyaeninae, with *Parahyaena* (brown hyena), *Hyaena* (striped hyena), and *Crocuta* (spotted hyena). Some authorities, such as Coetzee (*in* Meester and Setzer 1977), Köhler and Richardson (1990), and Yalden, Largen, and Kock (1980), treat the Protelinae as a separate family.

Head and body length is 550–1,658 mm, tail length is 187–470 mm, and weight is 9–86 kg. The color pattern is striped in *Proteles* and *Hyaena*, spotted in *Crocuta*, and unmarked in *Parahyaena*, except for its barred feet and pale head. A well-developed mane is present in *Proteles, Parahyaena,* and *Hyaena.* In all genera the guard hairs are coarse and the tail is bushy.

The head and forequarters are large, but the hindquarters are rather weak. The forelimbs, which are slender in *Proteles* but powerfully built in the other genera, are longer than the hind limbs. *Proteles* has five digits on the forefoot and four on the hind foot; the other genera have four digits on each foot. The blunt claws are not retractile. Scent glands are present in the anal region. Males do not have a baculum, and females have one to three pairs of mammae.

All genera have strongly built jaws. In *Proteles* the cheek teeth are reduced and widely spaced and the carnassials are not developed, but the canines are sharp and fairly powerful. This genus may have as few as 24 teeth (Ewer 1973). In *Parahyaena, Hyaena,* and *Crocuta* the skull and jaws are massive and the teeth, including the carnassials, are powerfully developed for crushing bones. The dental formula of those three genera is: (i 3/3, c 1/1, pm 4/3, m 1/1) × 2 = 34. In all genera of the family the incisors are not specialized and the canines are elongate.

Hyenas and the aardwolf generally inhabit grassland or bush country, but they may also occur in open forest. They live in caves, dense vegetation, or the abandoned burrows of other animals. They are mainly nocturnal. They move about on their digits and seem to trot tirelessly. The aard-

wolf feeds primarily on insects, whereas hyenas are efficient scavengers and predators of large mammals. Individuals may occur alone, in pairs, or in groups. Females normally bear a single litter per year.

The phylogenetic origins of the family are uncertain; it may have evolved from a branch of the Viverridae or Herpestidae, or it could represent an entirely separate line of the aeluroid carnivores. The known geological range of the Hyaenidae is Miocene to Pleistocene in Europe, Miocene to Recent in Asia and Africa, Miocene to Pleistocene in Europe, and early Pleistocene in North America (Werdelin and Solounias 1991).

CARNIVORA; HYAENIDAE; Genus PROTELES
I. Geoffroy St.-Hilaire, 1824

Aardwolf

The single species, *P. cristatus,* is found from the southern border of Egypt to central Tanzania and from southern Angola and southern Zambia to the Cape of Good Hope (Coetzee *in* Meester and Setzer 1977; Corbet 1978; Kingdon 1977).

Head and body length is 550–800 mm, tail length is usually 200–300 mm, and shoulder height is usually 450–500 mm. Kingdon (1977) listed weight as 9–14 kg. The underfur is long, loose, soft, and wavy and is interspersed with larger, coarser guard hairs. The body is yellow gray with black stripes. The legs are banded with black, and the part below the knee and hock is entirely black. The tail is bushy and tipped with black, and the hair along the back is long and crestlike, as evidenced in the Boer vernacular name *manhaarjakkal,* or "maned jackal." The external ears and the auditory bullae are very large. The cheek teeth are vestigial and widely spaced, but the canine teeth are sharp and reasonably powerful. Köhler and Richardson (1990) noted that the large canines and relatively powerful jaws are associated more with intraspecific fighting and self-defense than with diet.

The aardwolf is most commonly found on open, sandy plains or in bush country. It dens in holes in the ground, usually abandoned burrows of the aardvark *(Orycteropus).* It is primarily nocturnal and has been observed to cover just over 1 km/hr while foraging (Bothma and Nel 1980). Nightly movement is 8–12 km in summer and 3–8 km in winter (Köhler and Richardson 1990). The diet consists almost entirely of termites and insect larvae. The aardwolf lacks the powerful digging claws of other myrmecophagous mammals and seems to be specialized for feeding on nasute harvester termites *(Trinervitermes),* which forage in dense concentrations on the surface of the soil and secrete a chemical that repels most other predators (Richardson 1987; Richardson and Levitan 1994). Although a captive juvenile killed a number of birds (Kingdon 1977), there are no substantive data on such behavior in the wild. *Proteles* has been accused of preying on chickens and lambs, but evidence to the contrary is overwhelming and it is known that the aardwolf can scarcely be induced to eat meat unless it is finely ground or cooked. When seen near carrion, *Proteles* is usually there to pick up carrion beetles, maggots, and other insects.

Skinner and Van Aarde (1988) reported a population density of about one adult per sq km and also that an adult female used a home range of 3.8 sq km over a period of 16 months. Köhler and Richardson (1990) stated that home range size is 1–4 sq km, varying inversely with termite density. The ranges are equivalent to defended territories

Aardwolf *(Proteles cristatus)*, photo by R. Pucholt.

and are commonly occupied year-round by a mated pair and their young, though there seems to be some variation in social structure. Kingdon (1977) noted that several females, together with their young, have been found in a single den. Bothma and Nel (1980) usually found a single animal per den but noted that dens could be less than 500 meters apart. Individuals from different territories sometimes form temporary aggregations, and there may even be mating between a female from one territory and an outside male, though generally adult males are hostile to one another and may fight viciously (Köhler and Richardson 1990). Both sexes mark their territories with secretions from anal scent glands. The hairs of the mane can be erected to make the aardwolf look twice its normal size for purposes of intraspecific display or defense against predators. Surprisingly loud growls and roars can be produced under stress (Smithers 1971). If attacked by dogs, the aardwolf ejects musky fluid from its anal glands and may fight effectively with its formidable canine teeth.

The following data on reproduction are available: estrus lasts 1–3 days and lactation up to 145 days (Hayssen, Van Tienhoven, and Van Tienhoven 1993); gestation is thought to last 90–110 days, an animal several months old was taken in Uganda in July, a birth in Kenya occurred in late May, the number of young per litter is usually two or three and ranges from one to five, and reports of larger litters may represent the utilization of one den by two mothers (Kingdon 1977); litters of two to four young are born in November and December in southern Africa (Ewer 1973); in Botswana lactating females have been taken in January and April, and pregnant females in July and October (Smithers 1971); and the young first emerge from the den at 1 month, begin to accompany the adults at about 4 months, and leave the natal territory at about 13 months (Köhler and Richardson 1990). One aardwolf lived at the Frankfurt Zoo

for 18 years and 11 months (Marvin L. Jones, Zoological Society of San Diego, pers. comm., 1995).

Proteles is threatened in at least some areas by human hunting and habitat destruction. It was classified as rare in South Africa by Smithers (1986).

CARNIVORA; HYAENIDAE; **Genus PARAHYAENA**
Hendey, 1974

Brown Hyena

The single species, *P. brunnea,* occurs in Namibia, Botswana, western and southern Zimbabwe, southern Mozambique, and South Africa (Coetzee *in* Meester and Setzer 1977). *Parahyaena* often has been included in *Hyaena,* the name originally having been proposed as a subgenus, but the two were recognized as separate genera by Werdelin and Solounias (1990, 1991) and Wozencraft (*in* Wilson and Reeder 1993).

Head and body length is 1,100–1,356 mm, tail length is 187–265 mm, shoulder height is 640–880 mm, and weight is 37.0–47.5 kg (Skinner 1976; Smithers 1971). Males average slightly larger than females. The coat is long, coarse, and shaggy. The general coloration is dark brown. The head is gray, the neck and shoulders are tawny, and the lower legs and feet are gray with dark brown bars. There are erectile manes of long hairs up to 305 mm in length on the neck and back. Like *Hyaena, Parahyaena* has scent glands leading to an anal pouch that can be extruded for depositing secretions. The powerful jaws and teeth can crush the largest bones of cattle, and the digestive system can effectively utilize all parts of a carcass. Each front and hind foot has four digits.

Brown hyenas *(Hyaena brunnea)*, photo by Bernhard Grzimek.

According to Werdelin and Solounias (1991), the following skeletal characters serve to distinguish *Parahyaena:* first upper molar tooth smaller, relative to fourth upper premolar, than in *Hyaena;* metastyle of fourth upper premolar longer than paracone (shorter than or equal to paracone in *Hyaena*); palate continues beyond last upper molar (ends at level of last upper molar in *Hyaena*); suture present between premaxillary and frontal on snout (absent in *Hyaena*); nasal wings of premaxilla diverge dorsad (vertically placed and parallel in *Hyaena*); sphenoid foramen and postpalatine located together in single depression (well separated in *Hyaena,* each in its own depression); processes for attachment of nuchal ligaments beneath terminus of sagittal crest much larger in *Parahyaena* than in *Hyaena;* supramastoid crest strongly developed (weak in *Hyaena*); axis and atlas have relatively long overlap (short overlap in *Hyaena*).

The brown hyena is found largely within arid habitat—open scrub, woodland savannah, grassland, and semidesert. It is mainly nocturnal and crepuscular, sheltering by day in a rocky lair, dense vegetation, or an underground burrow (Eaton 1976). It commonly uses an aardvark hole as a den but is capable of digging its own burrow. A favored maternal den site may be utilized for years (Owens and Owens 1979a). When foraging, *Parahyaena* travels at a walk of approximately 4 km/hr, but it is capable of running at 40–50 km/hr (M. G. L. Mills 1982). Foraging is done in a zigzag pattern along regular pathways. Nightly movements in the Kalahari average 31 km, range up to 54 km, and are usually longer in the dry season. Food is located mainly by scent, but hearing and night vision are also keen (Mills 1978, 1990; Owens and Owens 1978).

The brown hyena is primarily a scavenger of the remains of large mammals killed by other predators (Mills 1990; Mills and Mills 1978; Owens and Owens 1978; Skinner 1976). Other important foods include rodents, insects, eggs, and fruit. At certain times of the year vegetation may constitute up to half of the diet of some individuals. *Parahyaena* also is known to frequent shorelines and to feed on dead crabs, fish, and seals. It often stores excess food in shrubs or holes and usually recovers it within 24 hours. Individuals have been seen to raid ostrich nests and to cache the eggs at scattered locations.

Eaton (1976) wrote that maximum population density is about 1/130 sq km, but Skinner (1976) found at least six adults and two cubs within 20 sq km. Mills (1990) calculated a density of 1.8/100 sq km in the southern Kalahari. Social behavior evidently varies to some extent according to season (Mills 1978, 1982, 1990; Owens and Owens 1978, 1979a, 1979b). The animals in a given area are organized into a "clan" and are recognized by one another. They usually forage alone. In the dry season individual home range averages about 40 sq km. Individuals do not maintain territories but use common hunting paths and frequently meet and exchange greetings. At times, especially during the wet season, up to six clan members join to exploit carrion. Each clan has a central breeding den and defends a surrounding territory of approximately 300 sq km. Hyenas from neighboring clans behave aggressively toward one another.

A clan may include only one adult of each sex and associated young, but in some areas as many as four males and six females have been reported. In the latter situation, one of the males is thought to be dominant and the others subordinate. A clan is not a strictly closed system, and emigration is common, especially among the young. All males seem eventually to leave their natal clans, though sometimes not until after reaching adulthood, and they then may join another group. About a third of all adult males become nomadic, and these individuals frequently move into clans and mate with the females. Some females remain in their natal clans for life. There is a stable rank order maintained by ritualized displays of aggression that commonly involve biting at the back of the neck, where the skin is tough. Individuals regularly deposit scent from their anal glands as they move about. This activity seems to be mainly for communication of information to other clan members rather than demarcation of territory. Identified vocalizations include squeals, growls, yells, screams, and squeaks, all of which seem to be associated with conflict or

Top, brown hyena *(Parahyaena brunnea)* with a tumor near the left eye, photo from U.S. National Zoological Park. Bottom, striped hyena *(Hyaena hyaena),* photo by Klaus Kussmann.

submissive behavior. Sounds are infrequent and mostly soft, there being no loud "whoop" as in *Crocuta*.

Females are apparently seasonally polyestrous. Mating is thought to occur mainly from May to August, with births from August to November. The gestation period is approximately 97 days. There are one to five, usually two or three, young per litter. They weigh about 1 kg at birth, open their eyes after 8 days, and emerge from the den after 3 months (Eaton 1976; M. G. L. Mills 1982, 1990; Skinner 1976). There evidently is some variation in rearing cubs. According to Owens and Owens (1979*a*, 1979*b*), the young of several litters are raised together in a communal den, each suckles from any lactating female, and all clan females participate in bringing food to the den. Mills (1990) reported that there usually is but one family per den, though the females of a clan tend to den within a few square kilometers of one another, and after about four months both the mother and other clan members begin to bring food to the cubs. There is no regurgitation. The cubs are not fully weaned and do not leave the vicinity of the den until they are about 14 months old. Females usually produce their first litter late in their second year, the interbirth interval is 12–41 months, natural mortality rates are relatively low until old age, and wild individuals are known to have lived at least 12 years (M. G. L. Mills 1982). A captive specimen was still living at about 29 years (Marvin L. Jones, Zoological Society of San Diego, pers. comm., 1995).

The brown hyena is classified as endangered by the USDI and is on appendix 1 of the CITES. Although protected and still widely distributed in Botswana, it has declined drastically in range and numbers in Namibia and South Africa (Eaton 1976). The main problem is killing by people, who consider it to be a predator of domestic animals. The damage that it does seems to have been greatly exaggerated; for example, on one large cattle ranch in the Transvaal where *Parahyaena* was studied no depredations were known to have occurred in 15 years (Skinner 1976).

CARNIVORA; HYAENIDAE; **Genus HYAENA**
Brünnich, 1771

Striped Hyena

The single species, *H. hyaena,* is found in open country from Morocco and Senegal to Egypt and Tanzania and from Asia Minor and the Arabian Peninsula to the Caucasus, southern Turkmenistan and Tajikistan, and eastern India (Coetzee *in* Meester and Setzer 1977; Corbet 1978; Heptner and Sludskii 1992). *Parahyaena* (see account thereof) often is included in *Hyaena*. Werdelin and Solounias (1991) considered the proper authority for the name *Hyaena* to be Zimmermann, 1777.

Except as noted, the information for the remainder of the account of this species was taken from the review papers by Rieger (1979, 1981). Head and body length is 1,036–1,190 mm, tail length is 265–470 mm, shoulder height is 600–942 mm, and weight is 25–55 kg. Males and females are about the same size. The general coloration is gray to pale brown, with dark brown to black stripes on the body and legs. The mane along the back is more distinct than in *Parahyaena*. The hairs of the mane are up to 200 mm long, while those of the rest of the body are about 70 mm long. Females have two or three pairs of mammae.

The striped hyena prefers open or rocky country and has an elevational range of up to 3,300 meters. It avoids true deserts and requires the presence of fresh water within 10 km. It is mainly crepuscular or nocturnal, resting by day in

Young striped hyena *(Hyaena hyaena)*, photo by Reginald Bloom.

a temporary lair, usually under overhanging rocks (Kruuk 1976). Cubs are reared in natural caves, rocky crevices, or holes dug or enlarged by the parents. When searching for food the striped hyena moves in a zigzag pattern at 2–4 km/hr (Kruuk 1976). The diet varies by season and area but seems to consist mainly of mammalian carrion. The larger, more northerly subspecies prey on sheep, goats, donkeys, and horses. Other important foods are small vertebrates, insects, and fruit. Some food is stored in dense vegetation for later use. Food is often brought to the den, and eventually a large number of bones may accumulate (Skinner, Davis, and Ilani 1980).

The striped hyena may have a small defended territory around its breeding den, surrounded by a larger home range. In a radio-tracking study in the Serengeti, Kruuk (1976) determined home range to be about 44 sq km for a female and 72 sq km for a male. This species seems to be mainly solitary in East Africa, but there are indications of greater social activity farther north, where the animals are more likely to have a predatory life. In more than half of his sightings of *H. hyaena* in Israel, Macdonald (1978) saw more than one animal. An adult pair and their offspring may sometimes forage together, and a family unit may persist for several years. Adult females evidently are intolerant of one another and are dominant over males. Scent marking with anal glands is an important means of communication. Aggressive displays involve erection of the long hairs of the mane and tail. *H. hyaena* is much less vocal than *Crocuta* but growls, whines, and makes several other sounds.

Females are polyestrous and breed throughout the year. The estrous cycle may be 40–50 days long. Estrus lasts 1 day and may follow birth by 20–21 days. The gestation period is 88–92 days. The number of young per litter is 1–5, averaging 2.4. The young weigh about 700 grams at birth, open their eyes after 5–9 days, first take solid food at 30 days, and nurse for at least 4–5 months. Food is brought to the den for the young (Skinner and Ilani 1979). Mendelssohn (1992) found two wild females pregnant at an age of only about 15 months. However, sexual maturity usually has been reported at 2–3 years, and maximum longevity in captivity is 23–24 years.

The striped hyena has been known to attack and kill people, especially children. It can be tamed, however, and is said to become loyal and affectionate. Many of its parts are believed by some persons to have medicinal value. It is notorious in Israel for its destruction of melons, dates, grapes, apricots, peaches, and cucumbers (Kruuk 1976). It has been nearly or entirely eliminated in the Caucasus (Heptner and Sludskii 1992). The North African subspecies, *H. h. barbara,* is classified as endangered by the USDI. It has declined through habitat loss and persecution as a predator. Its range is now restricted to the highlands of Morocco, Algeria, and Tunisia.

CARNIVORA; HYAENIDAE; **Genus CROCUTA**
Kaup, 1828

Spotted Hyena

In historical time the single species, *C. crocuta,* was found throughout Africa south of the Sahara except in equatorial rainforests (Coetzee *in* Meester and Setzer 1977). Through the end of the Pleistocene the same species occurred in much of Europe and Asia (Kurten 1968).

Head and body length is 950–1,658 mm, tail length is 255–360 mm, shoulder height is 700–915 mm, and weight is 40–86 kg. On the average, females are 120 mm longer and 6.6 kg heavier than males (Kingdon 1977). The hair is coarse and woolly. The ground color is yellowish gray, and the round markings on the body are dark brown to black.

There is no mane, or only a slight one. The jaws are probably the most powerful, in proportion to size, of any living mammal's. Each front and hind foot has four digits. From *Parahyaena* and *Hyaena, Crocuta* differs in its larger size, shorter and more rounded ears, paler and spotted coat, lack of a mane, larger and more swollen braincase, and greatly reduced upper molar tooth. In *Parahyaena* and *Hyaena* the diameter of the upper molar tooth is at least twice that of the first upper premolar, and there is a small metaconid on the first lower molar, whereas in *Crocuta* the upper molar is much smaller than the first premolar and there is no metaconid on the first lower molar.

The external genitalia of the female *Crocuta* so closely resemble those of the male that the two sexes are practically impossible to distinguish in the field (Kingdon 1977; Kruuk 1972): the clitoris looks like the penis, occupies the same position, and is capable of elongation and erection; in addition, the female has a pair of sacs, formed from fusion of the vaginal labia and filled with nonfunctional fibrous tissue, that looks very much like the scrotum and is located in the same place. The female has no external vagina and must urinate, mate, and deliver young through the urogenital canal traversing the clitoris (Glickman et al. 1992). Erection of the penis or "pseudopenis" apparently is a sign of submission displayed during greeting rituals between

Spotted hyena *(Crocuta crocuta),* photo by E. L. Button.

Spotted hyena *(Crocuta crocuta)*, photo by E. L. Button.

individuals (East, Hofer, and Wickler 1993). Both sexes have two anal scent glands that empty into the rectum. Females usually have a single pair of mammae. The sexual organs of female *Parahyaena* and *Hyaena,* unlike those of female *Crocuta,* do not closely resemble those of males.

The spotted hyena originally occupied nearly all the more open habitats of Africa south of the Sahara (Kingdon 1977). It is especially common in dry acacia bush, open plains, and rocky country. It is found at elevations of up to 4,000 meters. *Crocuta* is probably the most numerous of the large African predators because of its ability quickly to eat and digest entire carcasses, including skin and bones, and because the plasticity of its behavior allows it to function effectively either as a solitary scavenger and predator of small animals or as a group-living hunter of ungulates. Dens vary greatly in size, sometimes accommodating entire communities of hyenas. They are usually in abandoned aardvark holes but may also be in natural caves. Most activity is nocturnal or crepuscular. Up to 80 km may be covered in a night's foraging, though Mills (1990) found the average to be 27 km in the Kalahari. *Crocuta* has keen senses of sight, hearing, and smell.

Except as noted, the information for the remainder of this account was taken from Kruuk's (1972) report of his studies in the Serengeti National Park and the Ngorongoro Crater of northern Tanzania. In these areas the spotted hyena shelters in holes on the plains or in shady places with bushy vegetation on hillsides. Dens may have 12 or more entrances. Activity occurs mainly in the first half of the night, then declines, and then increases again toward dawn. The Ngorongoro population remains in one general area throughout the year, but the Serengeti hyenas may move in response to the seasonal migrations of their prey. During about half of the year Serengeti animals make "commuting trips" from their regular territories, lasting 3–10 days each and averaging about 40 km, to feed on the nearest migratory herds (Hofer and East 1993a, 1993b, 1993c).

The diet consists mainly of medium-sized ungulates, especially wildebeest *(Connochaetes),* and mostly very young, very old, or otherwise inferior animals. One hyena commonly forces a herd of wildebeest to run, watches for a weak individual, and then begins a chase, which is soon joined by other hyenas. Zebras are hunted in a more organized manner by packs of 10–25 hyenas. Chases usually go for less than 2 km, with *Crocuta* averaging 40–50 km/hr. Maximum speed is about 60 km/hr. Roughly one-third of the hunts are successful. Far more animals are killed than are consumed as carrion. Although the spotted hyena is sometimes said to be a scavenger of the lion, most dead prey on which both hyenas and lions were seen feeding had been killed by the hyenas. In the Kalahari, Mills (1990) determined that at least 70 percent of the diet of *Crocuta* results from kills. The specialized teeth and digestive system of *Crocuta* allow it to crush and utilize much more bone than do other predators. *Crocuta* can consume 14.5 kg of food in one meal.

Population densities were estimated at 0.12/sq km in the Serengeti and 0.16–0.24/sq km in Ngorongoro. In the latter area *Crocuta* is clearly organized into large communities, or "clans," each with up to 80 individuals and each occupying a territory of about 30 sq km. Clans are usually divided into smaller hunting packs and individuals, but all members evidently are recognizable to one another. The members mark the borders of their territory with secretions from their anal scent glands and defend the area from other clans. Groups that move about together commonly contain 2–10 individuals, but larger groups form to hunt zebra. In the Serengeti the clan system is less well developed and average group size is smaller. In the Kalahari the average foraging group size is only 3 hyenas and average clan size is 8, but the average territory covers 1,095 sq km (Mills 1990). Within any group of *Crocuta*, females are dominant over males and adults are dominant over young. The animals eat together and may compete by shoving one another, but there generally is no fighting. The spotted hyena is extremely vocal. The well-known laughing sound is emitted by an animal that is being attacked or chased. A whoop or howl is usually given spontaneously by a lone individual with its head held close to the ground; it begins low and deep and increases in volume as it runs to higher pitches. Whoops are given by all members of a group, cubs mainly to request support, adult males generally for sexual advertisement, and adult females in agonistic interaction with other females or outside individuals (East and Hofer 1991*a*, 1991*b*).

Studies in Kenya (Frank 1986*a*, 1986*b*; Holekamp and Smale 1993; Smale, Frank, and Holekamp 1993) have provided further understanding of the internal structure of spotted hyena clans. Each of these groups is a stable community of related females, among which unrelated males reside for varying periods. Whereas females remain in their natal clan for life, males disperse around the time of puberty and temporarily join together in a nomadic group. They subsequently settle within a new clan, sometimes remaining for several years, but then may depart again. Within a clan there is a separate dominance hierarchy for each sex. The highest ranking female and her descendants are dominant over all other animals. By the time they are 6–8 months old the juveniles of a group establish a hierarchy among themselves, and by 12–18 months they generally are able to dominate fully adult animals from matrilines of lower rank than their own and also males that have immigrated into the group. Although all resident males were observed to court females, only the highest-ranking male actually was seen to mate.

There apparently is no permanent pair bonding. Breeding may occur at any time of the year, perhaps with a peak in the wet season. Females are polyestrous and have an estrous cycle of 14 days. Gestation lasts about 110 days. There are usually two young per birth, occasionally one or three. Observations in captivity show that siblings begin to fight violently upon birth and that if both are of the same sex, one will usually be killed (Frank, Glickman, and Light 1991). At birth they weigh about 1.5 kg and their eyes are open. Each clan has a single central denning site where all females bear their cubs, but each mother suckles its own young. There is no cooperative rearing of young as sometimes found in *Parahyaena brunnea* (Mills 1985). Food is not carried to the den or regurgitated for the young. Weaning comes after 12–16 months, when the young are nearly full grown. Sexual maturity is attained at about 2 years by

males and 3 years by females. A captive *Crocuta* lived 41 years and 1 month (Jones 1982).

Human relationships with the spotted hyena have varied (Kingdon 1977; Kruuk 1972). Some native peoples protected it as a valuable scavenger, but others regarded it with superstitious dread. Certain tribes put their dead out for hyenas to consume. Fatal attacks on living humans have occurred. In the twentieth century *Crocuta* has been considered to be a predator of domestic livestock and game and has been widely hunted, trapped, and poisoned. It has declined in numbers over much of its range and has been eliminated in parts of East and South Africa. It now is classified as conservation dependent by the IUCN. The well-studied and nominally protected population in Serengeti National Park may be jeopardized by human hunters who kill both the hyena and its prey when they move beyond the Park boundary (Hofer, East, and Campbell 1993). Kurten (1968) suggested that the disappearance of *Crocuta* in Eurasia in the early postglacial period was associated with the development of agriculture.

CARNIVORA; Family FELIDAE

Cats

This family of 4 Recent genera and 38 species has a natural distribution that includes all land areas of the world except the West Indies, Madagascar, Japan, most of the Philippines, Sulawesi and islands to the east, New Guinea, Australia, New Zealand, Antarctica, and most arctic and oceanic islands. Recognition here of 4 genera—*Felis, Neofelis, Panthera,* and *Acinonyx*—is based on Cabrera (1957), Chasen (1940), Corbet (1978), Corbet and Hill (1991), Ellerman and Morrison-Scott (1966), E. R. Hall (1981), Heptner and Sludskii (1992), Lekagul and McNeely (1977), Medway (1977, 1978), Meester et al. (1986), Skinner and Smithers (1990), Smithers (*in* Meester and Setzer 1977),

Leopard cat *(Felis bengalensis)* and young, photo from U.S. National Zoological Park.

Fishing cat *(Felis viverrina)*, photo from U.S. National Zoological Park.

Stains (1984), and Taylor (1934). There is, however, a great diversity of opinion on how the cats should be classified. Some authorities, such as Romer (1968), divide the family into only 2 genera, *Felis* and *Acinonyx*. Others, including some who recently have done extensive research on the subject, recognize many more genera. The main controversy involves the division of what is here considered to be the genus *Felis* (see account thereof) into genera and subgenera. All of the authorities cited here accept *Acinonyx* as a distinct genus, nearly all accept *Neofelis* (though with some disagreement about its content), and most accept *Panthera*.

The living Felidae are sometimes divided into subfamilies. Stains (1984) recognized the Acinonychinae, for *Acinonyx* (cheetah); the Felinae, for *Felis;* and the Pantherinae, for *Neofelis* and *Panthera*. Wozencraft (*in* Wilson and Reeder 1993) recognized the same three but included *Felis (Pardofelis) marmorata* in the Pantherinae. Leyhausen (1979) accepted the Acinonychinae, for *Acinonyx*, but put all other cats in the Felinae. The cheetah is usually placed last on lists of cats to express its aberrant characters. Actually, however, it seems in some respects to be little changed from the primitive stock that gave rise to all other cats. Locating it at the beginning of a systematic account, as was done by Leyhausen (1979) and Hemmer (1978), is thus fully appropriate. Moreover, placement of *Felis (Puma) concolor* next to *Acinonyx* by Hemmer (1978) and Salles (1992) may be fitting, as Adams (1979) has presented evidence that these two cats evolved from a common ancestor in North America. Analyses by Van Valkenburgh, Grady, and Kurten (1990) also suggest an evolutionary association of the two.

In the family Felidae head and body length is 337–2,800 mm, tail length is 51–1,100 mm, and weight is 1.5–306.0 kg. The color varies from gray to reddish and yellowish brown, and there are often stripes, spots, or rosettes. The pelage is soft and woolly. Its beautiful, glossy appearance is maintained by frequent cleaning with the tongue and paws. The tail is well haired but not bushy, and the whiskers are well developed.

Cats have a lithe, muscular, compact, and deep-chested body. The limbs range from short to long and sinewy. The forefoot has five digits and the hind foot has four. In most species the claws are retractile (to prevent them from becoming blunted), large, compressed, sharp, and strongly curved (to aid in holding living prey). In the cheetah *(Acinonyx)*, however, they are only semiretractile and relatively poorly developed. Except for the naked pads, the feet are well haired to assist in the silent stalking of prey. The baculum is vestigial or absent. Females have two to four pairs of mammae.

The head is rounded and shortened, the ears range from rounded to pointed, and the eyes have pupils that contract vertically. The tongue is suited for laceration and retaining food within the mouth, its surface being covered with sharp-pointed, recurved, horny papillae. The dental formula is: (i 3/3, c 1/1, pm 2–3/2, m 1/1) × 2 = 28 or 30. The incisors are small, unspecialized, and placed in a horizontal line. The canines are elongate, sharp, and slightly recurved. The carnassials, which cut the food, are large and well developed. The upper molar is small.

The dentition of the Felidae, with emphasis on the teeth used for seizing and cutting rather than on those used for grinding, reflects the highly predatory lifestyle of the family. Cats prey on almost any mammal or bird they can overpower and occasionally on fish and reptiles. They stalk their prey, or lie in wait, and then seize the quarry with a short rush. The cheetah *(Acinonyx)*, however, is adapted for a more lengthy pursuit at high speed. Cats walk or trot on their digits, often placing the hind foot in the track of the

A. Lion *(Panthera leo)*, photo by Leonard Lee Rue III. B. Clouded leopard *(Neofelis nebulosa)*, photo by Bernhard Grzimek.
C. Leopard *(Panthera pardus)*, photo by Leonard Lee Rue III.

forefoot. They are agile climbers and good swimmers and have acute hearing and sight. Many are nocturnal, but some are active mainly during daylight. They shelter in trees, hollow logs, caves, crevices, abandoned burrows of other animals, or dense vegetation. They defend themselves with fang and claw or flee, sometimes seeking refuge in trees.

Cats usually are solitary but sometimes are found in pairs or larger groups. The females of most species are polyestrous and give birth once a year, some can have two litters annually, and those of the larger species sometimes breed only every 2–3 years. Gestation periods of 55–119 days have been reported, and litter size is usually 1–6 young. At birth the kittens are usually blind and helpless but are haired and often spotted. They remain with their mother until they can hunt for themselves. Potential longevity is probably at least 15 years for most species, and some individuals have lived more than 30 years.

Some of the larger species of cats have occasionally become a serious menace to human life in localized areas.

Various felids are considered by some people to be a threat to domestic animals. Certain species, especially those with spotted or striped skins, are of considerable value in the fur trade. The big cats, as well as some of the smaller ones, are sought as trophies by hunters. For all of these reasons the Felidae have been extensively hunted and killed by people. As a result, many species and subspecies have become rare or endangered in at least parts of their ranges. The entire family has been placed on appendix 2 of the CITES, except for those species on appendix 1 and the domestic cat (*Felis catus*).

The geological range of the Felidae is late Eocene to Recent in North America and Eurasia, early Eocene to Recent in Africa, and late Pliocene to Recent in South America (Stains 1984). There are several extinct groups, including the saber-toothed cats (subfamily Machairodontinae), which persisted through the end of the Pleistocene. *Smilodon fatalis*, which lived in North America until about 10,000 years ago, was a saber-toothed cat about the size of modern *Panthera leo*. It differed in its shorter and more powerful limbs and shorter tail. Rather than running down antelopes, it apparently preyed on young mammoths and other large, thick-skinned animals, leaping upon them from ambush, dragging them down, and sinking its huge fangs into their underparts (Heald and Shaw 1991).

CARNIVORA; FELIDAE; Genus FELIS
Linnaeus, 1758

Small Cats, Lynxes, and Cougar

There are 16 subgenera and 31 species (Cabrera 1957; Chasen 1940; Corbet 1978; Corbet and Hill 1992; Ellerman and Morrison-Scott 1966; Ewer 1973; Guggisberg 1975; E. R. Hall 1981; Hemmer 1978; Heptner and Sludskii 1992; Lekagul and McNeely 1977; Medway 1977, 1978; Roberts 1977; Smithers *in* Meester and Setzer 1977; Taylor 1934; Werdelin 1981):

subgenus *Felis* Linnaeus, 1758

F. silvestris (wild cat), France and Spain through Kazakhstan and the Arabian Peninsula to north-central China and central India, Great Britain, Balearic Islands, Sardinia, Corsica, Crete, woodland and savannah zones throughout Africa;
F. catus (domestic cat), worldwide in association with people;
F. bieti (Chinese desert cat), southern Mongolia, central China;
F. chaus (jungle cat), Volga River Delta and Egypt to Sinkiang and Indochina, Sri Lanka;
F. margarita (sand cat), desert zone from Morocco and northern Niger to southern Kazakhstan and Pakistan;
F. nigripes (black-footed cat), Namibia, Botswana, South Africa;

subgenus *Otocolobus* Brandt, 1841

F. manul (Pallas's cat), Caspian Sea and Iran to southeastern Siberia and Tibet;

subgenus *Lynx* Kerr, 1792

F. pardina (Spanish lynx), Spain, Portugal;
F. lynx (lynx), western mainland Europe to eastern Siberia and Tibet, possibly Sardinia, Sakhalin;

F. canadensis, Alaska, Canada, northern conterminous United States;
F. rufus (bobcat), southern Canada to Baja California and central Mexico;

subgenus *Caracal* Gray, 1843

F. caracal (caracal), Arabian Peninsula to Aral Sea and northwestern India, most of Africa;

subgenus *Leptailurus* Severtzov, 1858

F. serval (serval), Morocco, Algeria, most of Africa south of the Sahara;

subgenus *Pardofelis* Severtzov, 1858

F. marmorata (marbled cat), Nepal to Indochina and Malay Peninsula, Sumatra, Borneo;

subgenus *Catopuma* Severtzov, 1858

F. temmincki (Asian golden cat), Tibet and Nepal to southeastern China and Malay Peninsula, Sumatra;
F. badia (bay cat), Borneo;

subgenus *Profelis* Severtzov, 1858

F. aurata (African golden cat), Senegal to Kenya and northern Angola;

subgenus *Prionailurus* Severtzov, 1858

F. bengalensis (leopard cat), Ussuri region of southeastern Siberia, Manchuria, Korea, Quelpart and Tsushima islands (between Korea and Japan), most of China east of Tibet, Pakistan to Indochina and Malay Peninsula, Taiwan, Hainan, Sumatra, Java, Bali, Borneo, several islands in the western and central Philippines, Lombok;
F. rubiginosa (rusty-spotted cat), southern India, Sri Lanka;
F. viverrina (fishing cat), Pakistan to Indochina, Sri Lanka, Sumatra, Java;
F. planiceps (flat-headed cat), Malay Peninsula, Sumatra, Borneo;

subgenus *Mayailurus* Imaizumi, 1967

F. iriomotensis (Iriomote cat), Iriomote Island (southern Ryukyu Islands);

subgenus *Oreailurus* Cabrera, 1940

F. jacobita (mountain cat), Andes of southern Peru, southwestern Bolivia, northeastern Chile, and northwestern Argentina;

subgenus *Lynchailurus* Severtzov, 1858

F. colocolo (pampas cat), Ecuador and Mato Grosso region of Brazil to central Chile and Patagonia;

subgenus *Oncifelis* Severtzov, 1858

F. geoffroyi (Geoffroy's cat), Bolivia and extreme southern Brazil to Patagonia;
F. guigna (kodkod), central and southern Chile, southwestern Argentina;

A. Marbled cat *(Felis marmorata)*, photo from *Jour. Bombay Nat. Hist. Soc.* B. Serval *(F. serval)*, photo by Bernhard Grzimek. C. Golden cat *(F. temmincki)*, photo by Ernest P. Walker. D. Pallas's cat *(F. manul)*, photo by Howard E. Uible. E. Pumas *(F. concolor)*, photo by Ernest P. Walker.

Top, black-footed cat *(Felis nigripes)*, photo by John Visser. B. Little spotted cat *(F. tigrina)*. C. Jungle cat *(F. chaus)*, photo by Ernest P. Walker. D. European wild cat *(F. silvestris)*, photo by Bernhard Grzimek. E. Jaguarundi *(F. yagouaroundi)*, photo by Ernest P. Walker.

A. Leopard cat *(Felis bengalensis)*, photo by Lim Boo Liat. B. Margay *(F. wiedii)*, photo by Ernest P. Walker. C. Pampas cat *(F. colocolo)*, photo from San Diego Zoological Garden. D. Fishing cat *(F. viverrina)*, photo from New York Zoological Society. E. Geoffroy's cat *(F. geoffroyi)*, photo by Ernest P. Walker. F. African wild cat *(F. silvestris libyca)*, photo by Bernhard Grzimek.

subgenus *Leopardus* Gray, 1842

F. pardalis (ocelot), Arizona and Texas to northern
 Argentina;
F. wiedii (margay), northern Mexico and possibly southern
 Texas to northern Argentina and Uruguay;
F. tigrina (little spotted cat), Costa Rica to northern
 Argentina;

subgenus *Herpailurus* Severtzov, 1858

F. yagouaroundi (jaguarundi), southern Arizona and
 southern Texas to northern Argentina;

subgenus *Puma* Jardine, 1834

F. concolor (cougar, puma, panther, or mountain lion),
 southern Yukon and Nova Scotia to southern Chile and
 Patagonia.

The sequence, systematic grouping, and individual acceptance of the species in this list is based on the authorities
cited above and also gives some consideration to the detailed morphological analysis of Salles (1992). Most of the
above authorities, as well as Corbet and Hill (1991), Jones
et al. (1992), and Wozencraft (1989a), did not recognize the
various subgenera in the list as being separate genera, except frequently *Lynx*. However, all of the above subgenera
have been treated in the past as full genera, and there has
been an increasing trend toward again accepting them as
such, notably by authorities who have done primary research on the felids (Ewer 1973; Hemmer 1978; Leyhausen
1979; Salles 1992; Wayne et al. 1989; Wozencraft *in* Wilson
and Reeder 1993). There seems to be little disagreement at
the specific level, the main controversy involving the division of what is here considered to be the genus *Felis* into
genera and subgenera. The production in captivity of hybrids between many of the species, including recent cross-
ings of *F. pardalis* and *F. concolor* (Dubost and Royère
1993), has been one argument for not splitting *Felis*.
Wozencraft (*in* Wilson and Reeder 1993) accepted all of the
above subgenera as full genera with content as shown and
also accepted all of the above species as valid, with the following exceptions: *F. catus* was included in *F. silvestris*,
Mayailurus was not recognized and *F. iriomotensis* was included in *F. bengalensis*, and *Lynchailurus* was not recognized and *F. colocolo* was considered part of *Oncifelis*. In a
recent response to Wozencraft, Leyhausen and Pfleiderer
(1994) argued that *F. iriomotensis* is a highly distinctive
and primitive species and that *Mayailurus* may warrant
full generic status.

Corbet and Hill (1991), E. R. Hall (1981), and Jones et al.
(1992) treated *Lynx* as a distinct genus. The preponderance
of available information, however, suggests that there is no
more (or less) justification for recognizing *Lynx* as a full
genus than there is for elevating any other of the subgenera of *Felis* to generic level. The North American *F.
canadensis* and the Iberian *F. pardina* often have been regarded as no more than subspecifically distinct from the
Eurasian *F. lynx*, but studies of the evolution and taxonomy of the group by Werdelin (1981) and García-Perea
(1992) show that all three, as well as *F. rufus*, are distinct
species.

The African and most Asian populations of *F. silvestris*
have been assigned to a separate species, *F. libyca*, by numerous authorities, including Smithers (*in* Meester and
Setzer 1977). There is substantial evidence, however, that *F.
silvestris* and *F. libyca* intergrade in the Middle East. A
multistatistical analysis by Ragni and Randi (1986) showed
broad morphological overlap of *silvestris, libyca,* and *catus,*
indicating that all three should be regarded as conspecific.
There also is much doubt that *F. bieti* of East Asia is distinct
from *F. silvestris* (Corbet 1978). Leyhausen (1979) listed *F.
silvestris* and *F. libyca* as separate species and considered
the populations found from Iran to India to represent still
another species, *F. ornata*. He united all three, however, in
a single superspecies. Leyhausen (1979) also recognized *F.
thinobia,* of Central Asia, and *F. tristis,* of Tibet, as species
distinct from, respectively, *F. margarita* and *F. temmincki.*
Heptner and Sludski (1992) considered *F. euptilura,* found
from southeastern Siberia to central China, a species distinct from *F. bengalensis*. García-Perea (1994) suggested
that *F. (Lynchailurus) pajeros,* of the Andes and Patagonia,
and *F. (L.) braccatus,* of central Brazil, Paraguay, and
Uruguay, are species distinct from *F. (L.) colocolo,* which
would then be restricted to northern and central Chile.
Corbet and Hill (1992) treated *Neofelis* (see account thereof) as a synonym of *Pardofelis*.

For more than a century there have been unconfirmed
reports of a large cat known as the "onza" from the highlands of Mexico. It has been variously associated with *Felis
concolor, Panthera onca,* or an extinct relative of *Acinonyx
jubatus*. In 1986 a specimen of an alleged onza was collected in the Sierra Madre. Published photographs suggested a
slender, long-legged *F. concolor,* and an analysis of mitochondrial DNA (Dratch, Martenson, and O'Brien 1991)
strongly supported referral to that species.

Except for *F. concolor* and some individuals of *F. lynx,* the
members of the genus *Felis* are smaller than those of the
three other genera of cats. Otherwise the characters of *Felis* are the same as those set forth for the family Felidae.
From *Neofelis, Felis* is distinguished by its relatively shorter canine teeth and a smaller gap between the canines and
the cheek teeth. From *Panthera* it is distinguished by a
completely ossified hyoid apparatus without an elastic ligament. From *Acinonyx* it is distinguished by a greater gap
between the canines and the cheek teeth and usually by
shorter limbs and fully retractile claws. Additional information is provided separately for each species. Except as
noted, the information for the following accounts was tak-
en from Guggisberg (1975).

Felis silvestris (wild cat)

Head and body length is usually 500–750 mm, tail length
is 210–350 mm, and weight is usually 3–8 kg. Males are
generally larger than females. The fur is long and dense. In
European populations the ground color is yellowish gray,
the underparts are paler, and the throat is white. Four or
five longitudinal stripes from the forehead to the nape
merge into a dorsal line that ends near the base of the tail.
The tail has several dark encircling marks and a blackish tip.
The legs are transversely striped. In African and Asian
populations the general coloration varies from pale sandy
to gray brown and dark gray; there may be a pattern of distinct spots or stripes. European animals are generally about
one-third larger than domestic cats and have longer legs, a
broader head, and a relatively shorter, more bluntly ending
tail. Females have eight mammae.

The wild cat occupies a variety of forested, open, and
rocky country. It is mainly nocturnal and crepuscular,
spending the day in a hollow tree, thicket, or rock crevice.
It climbs with great agility and seems to enjoy sunning itself on a branch. It normally stays in one area, within which
it has several dens and a system of hunting paths. It may
hunt over a distance of 3–10 km each night (Heptner and
Sludskii 1992; Nowell and Jackson 1996). It usually stalks

its prey, attempting to approach to within a few bounds. The diet consists mainly of rodents and other small mammals and also includes birds, reptiles, and insects.

Average population density under optimal conditions is around 3–5/1,000 ha. (Stahl and Artois 1994). In contrast to *F. catus*, the wild cat is usually solitary, each individual having a well-defined home range. Males defend these areas but may wander outside of them during times of food shortage or to locate estrous females. Some variation in spacing has been reported (Nowell and Jackson 1996). In France the seasonal ranges of males averaged 5.7 sq km and overlapped the ranges of 3–5 females, which averaged 1.8 sq km. In Scotland, however, males and females had equivalent monthly home ranges with an average size of 1.75 sq km and little overlap.

Mating occurs from about January to March in Europe and Central Asia. Females are polyestrous, with heat lasting 2–8 days. Several males collect around a female in heat; there is considerable screeching and other vocalization and sometimes violent fighting. There is usually only a single litter per year, though occasionally a second is produced in the summer. Births in East Africa may occur at any time of year but seem to peak there and in southern Africa during the wet season (Kingdon 1977; Smithers 1971). The gestation period averages 66 days in Europe and about 1 week less in Africa. Litters usually contain two or three young in the wild. At the Berne Zoo, Meyer-Holzapfel (1968) observed that births occurred from March to August and that litter size averaged four and ranged from one to eight. The young weigh about 40 grams at birth, open their eyes after about 10 days, nurse for about 30 days, emerge from the den at 4–5 weeks, begin to hunt with the mother at 12 weeks, probably separate from her at 5 months, and attain sexual maturity at around 1 year. According to Kingdon (1977), captives have lived up to 15 years.

There is some uncertainty about the original range in historical times, partly because *F. silvestris* may be hard to distinguish from, and may even be conspecific with, *F. catus*. Some authorities consider the reported populations on Crete, Corsica, Sardinia, and the Balearic Islands to be distinctive, and now highly endangered, subspecies of *F. silvestris*, but others consider them to be feral domestic cats introduced by people centuries ago (Nowell and Jackson 1996). An interesting compromise position, suggested by Randi and Ragni (1991), is that these island populations do result from introduction but that the introduction took place in prehistoric times and involved the wild African subspecies *F. silvestris libyca*, which was the progenitor of *F. catus*.

The wild cat once occupied most of Europe but had withdrawn from Scandinavia and most of Russia by the Middle Ages because of climatic deterioration. In modern times, especially during the nineteenth century, the species was intensively hunted by persons who considered it to be a threat to game and domestic animals and so was eliminated from much of western and central Europe. Diversion of human activity during World Wars I and II apparently stimulated recovery in such places as Scotland and Germany (Smit and Van Wijngaarden 1981). *F. silvestris* has been utilized in the fur trade and is on appendix 2 of the CITES. Many thousands were taken during some years in the past, but at present there is thought to be little commerce in this species (Nowell and Jackson 1996).

By the 1980s the wild cat had been placed under complete legal protection in most countries of Europe and around the Mediterranean, populations generally seemed to be expanding, and reintroductions had been attempted in several places. Unfortunately, there now has been a re-

versal of this trend, with numbers and distribution again declining in many areas. Populations are mostly small, fragmented, and subject to illegal hunting and deteriorating habitat. There still are substantial numbers in eastern France and the Balkans but fewer than 2,000 in Germany and almost none in Poland. To the east, the wild cat survives only in the Caucasus and small parts of Ukraine and Moldova (Stahl and Artois 1994).

A major threat reported through most of the remaining range is hybridization with domestic *F. catus*. This problem is intensifying in the eastern Mediterranean region and even in sub-Saharan Africa but is of particular concern in Europe, where some authorities fear that very few pure *F. silvestris* remain (Nowell and Jackson 1996; Stahl and Artois 1994). The population in Scotland, the last in the British Isles and numbering a few thousand cats, long has been affected by interbreeding but apparently still includes many genetically distinct animals (Hubbard et al. 1992; Kitchener 1992). The subspecies in Scotland, *F. s. grampia*, is classified as vulnerable by the IUCN. A mysterious black, or "Kellas," cat, recently reported in Scotland, evidently has resulted from hybridization of the wild and domestic species (Kitchener and Easterbee 1992).

Felis catus (domestic cat)

According to the National Geographic Society (1981), there are more than 30 different breeds of domestic cat, and the average measurements of several popular breeds are: head and body length, 460 mm, and tail length, 300 mm. E. Jones (1977) found that feral males on Macquarie Island, south of Australia, averaged 522 mm in head and body length, 269 mm in tail length, and 4.5 kg in weight, while females there averaged 478 mm in head and body length, 252 mm in tail length, and 3.3 kg in weight. Ninety percent of the cats on Macquarie were orange or tabby, and the remainder were black or tortoiseshell. Female *F. catus* have four pairs of mammae.

The domestic cat evidently is descended primarily from the wild cat of Africa and extreme southwestern Asia, *F. silvestris libyca*. The latter may have been present in towns in Palestine as long ago as 7,000 years, and actual domestication occurred in Egypt about 4,000 years ago. Introduction to Europe began around 2,000 years ago, and some interbreeding occurred there with the wild subspecies *F. silvestris silvestris*. Domestication may originally have been associated with the cat's proclivity to prey on the rodents that threatened the stored grain upon which ancient civilizations depended but also seems to have a religious basis (Grzimek 1975; Kingdon 1977; Yurco 1990). The cat was the object of a passionate cult in ancient Egypt, where a city, Bubastis, was dedicated to its worship. The followers of Bastet, the goddess of pleasure, put bronze statues of cats in sanctuaries and carefully mummified the bodies of hundreds of thousands of the animals. The veneration of cats in Egypt intensified about 3,000 years ago and persisted at least into Roman times.

There have been relatively few detailed field studies of *F. catus*, but there is no reason to think that its behavior and ecology under noncaptive conditions differ greatly from what has been found for *F. silvestris*. On Macquarie Island, where the cat population has been feral since 1820, E. Jones (1977) obtained specimens in a variety of habitats by both day and night. The cats sheltered in rabbit burrows, thick vegetation, or piles of rocks. The diet consisted largely of rabbits (also introduced on the island) and also included rats, mice, birds, and carrion. Population density was estimated at 2–7/sq km.

In a rural area of southern Sweden, Liberg (1980) found

Domestic cat *(Felis catus)*, photo by William J. Allen.

a population density of 2.5–3.3/sq km. About 10 percent of the cats were feral, and the rest, including all of the females, were associated with human households. Adult females lived alone or in groups of up to 8 usually closely related individuals. Each member of a group had a home range of 30–40 ha. that overlapped extensively with the ranges of other members of the same group but not with the ranges of the cats in other groups. Most females spent their life in the area in which they were born, seldom wandering more than 600 meters away. Nonferal males remained in their area of birth, along with females, until they were 1.5–3.5 years old but then left and tried to settle somewhere else. Males living in the same group had separate home ranges. There were 6–8 feral males in the study area; their home ranges were 2–4 km across, partly overlapped one another, and sometimes included the areas used by several groups of females. According to Haspel and Calhoon (1989), home ranges of unrestrained urban cats are much smaller than those in rural areas. In Brooklyn, New York, range averaged 2.6 ha. for males and 1.7 ha. for females, the difference evidently being a function of body size. Population density in that area was up to about 5/ha. in habitat characterized by many abandoned buildings and voluminous, poorly contained refuse (Calhoon and Haspel 1989).

F. catus communicates through a variety of vocalizations. Purring differs from the other sounds in that it continues during inspiration as well as expiration and may occur simultaneously with voice production proper. The sound and vibration of purring evidently results primarily from laryngeal modulation of respiratory flow (Sissom, Rice, and Peters 1991).

According to Ewer (1973), the house cat is basically solitary, but individuals in a given area seem to have a social organization and hierarchy. A male newly introduced to an area normally must undergo a series of fights before its position is stabilized in relation to other males. Both males and females sometimes gather within a few meters of one another without evident hostility. A male and female may form a bond that extends beyond the mating process. Females are polyestrous and normally produce two litters annually. They may mate with more than one male per season, and if a litter is lost, they soon enter estrus again. The

gestation period averages 65 days. The number of young per litter averages four and ranges from one to eight. Kittens weigh 85–110 grams at birth, open their eyes after 9–20 days, are weaned at 8 weeks, and attain independence at about 6 months. Hemmer (1976) listed age of sexual maturity in females as 7–12 months.

Although the cat sometimes has been venerated, it also has been associated with evil. Certain superstitions concerning it have persisted to modern times (Grzimek 1975). The species is now generally looked upon with favor by most cultures, but free-ranging cats often are considered to be among the greatest decimators of native wildlife, especially songbirds (Lowery 1974). On Kerguelen Island, a French possession of about 6,200 sq km in the southern Indian Ocean, a pair of cats introduced in 1956 has grown to a population of 10,000 that consumes an estimated 3 million birds annually (Chapuis, Boussès, and Barnaud 1994).

Among domestic animals *F. catus* is the species most commonly reported to be rabid. Although rabies in humans is now exceedingly rare in the United States, with an average of less than 1 indigenously acquired case annually for the past 30 years, efforts to control the disease cost more than $300 million each year. Moreover, 30,000–40,000 people thought to have been exposed to the virus receive the series of anti-rabies vaccinations each year in the United States. In some other countries, such as India, with an estimated 25,000 human deaths annually, rabies remains a serious health problem (Krebs, Wilson, and Child 1995; Squires 1995). In the last several years there has been an increase in human deaths from rabies in the United States, but mainly involving transmission from bats, not cats (see accounts of *Eptesicus* and *Lasionycteris*).

Felis bieti (Chinese desert cat)

Head and body length is 685–840 mm and tail length is 290–350 mm. The build is stocky, the tail relatively short; one male weighed 9.0 kg, a female 6.5 kg (Nowell and Jackson 1996). The general coloration is yellowish gray, the back is somewhat darker, and there are practically no markings on the flanks. The tail is tipped with black and has three or four subterminal blackish rings.

According to Nowell and Jackson (1996), despite its com-

Jungle cat *(Felis chaus)*, photo from San Diego Zoological Society.

mon name, this cat is found mainly in alpine meadows and scrub, and sometimes in steppe and forest edge, at elevations of 2,800–4,100 meters. It is active mostly at night and in the early morning and rests and rears its young in burrows. The diet consists largely of rodents, some of which are dug out of subterranean tunnels. The sexes usually live separately, the mating season is January–March, young often are born in May, litter size is two to four, and age of independence is 7–8 months. *F. bieti* is poorly known but may not be rare, as its pelt is commonly found in local markets. It is on appendix 2 of the CITES.

Felis chaus (jungle cat)

Head and body length is 500–940 mm, tail length is 230–310 mm, and weight is 4–16 kg (Heptner and Sludskii 1992; Lekagul and McNeely 1977; Novikov 1962). The general coloration varies from sandy or yellowish gray to grayish brown and tawny red, usually with no distinct markings on the body. The tail has several dark rings and a black tip. Lekagul and McNeely (1977) noted that the legs are proportionately the longest of any felid's in Thailand and are thus helpful in running down prey.

The jungle cat is found in a variety of open and wooded habitats from sea level to elevations of 2,400 meters but generally is associated with dense vegetative cover and water. It swims well and may dive to catch fish (Nowell and Jackson 1996). It dens in thick vegetation or in the abandoned burrow of a badger, fox, or porcupine. It is active either by day or by night. The diet consists mainly of hares and other small mammals and also includes birds, frogs, and snakes. Mating occurs in Central Asia in February and March, but 3-week-old kittens have been found in Assam in January and February. The gestation period is 66 days, and the usual litter size is three to five young. The newborn weigh 83–161 grams and are weaned after about 3 months (Hayssen, Van Tienhoven, and Van Tienhoven 1993). Sexual maturity comes at 18 months. One specimen lived at the Copenhagen Zoo for 20 years (Marvin L. Jones, Zoological Society of San Diego, pers. comm., 1995).

The jungle cat adapts well to irrigated agriculture and often is found in the vicinity of human settlements (Nowell and Jackson 1996). It thus may be less threatened than most cats. In Central Asia and the Caucasus it is actively hunted because of alleged predation on game birds and poultry and also for its pelt (Heptner and Sludskii 1992). It is on appendix 2 of the CITES.

Felis margarita (sand cat)

Head and body length is 450–572 mm and tail length is 280–348 mm. Weight is 1.3–3.4 kg (Heptner and Sludskii 1992). The general coloration is pale sandy to gray straw ochre, the back is slightly darker, and the belly is white. A fulvous reddish streak runs across each cheek from the corner of the eye, and the tail has two or three subterminal rings and a black tip. The pelage is soft and dense. The soles of the feet are covered with dense hair. The limbs are short, and the ears are set low on the head.

The sand cat is adapted to extremely arid terrain, such as shifting dunes of sand. The padding on its soles facilitates progression over loose, sandy soil. Activity is mainly nocturnal and crepuscular, the day being spent in a shallow burrow. Prey consists of jerboas and other rodents and occasionally hares, birds, and reptiles. The sand cat apparently is able to subsist without drinking free water. A radio-tracking study in Israel determined the presence of at least 22 individuals in an area of 100 sq km; the home range of one adult male was estimated at 16 sq km and overlapped those of neighboring males (Nowell and Jackson 1996).

Births in Central Asia occur in April (Heptner and Sludskii 1992). In Pakistan there may be two litters annually, as kittens have been found in both March–April and October (Roberts 1977). The estrous cycle is 46 days, estrus lasts 5 days, the gestation period is 59–67 days, litters usually contain two to four young, average weight at birth is 39 grams,

Sand cat *(Felis margarita)*, photo by Lothar Schlawe.

and the young open their eyes after 12–16 days (Hemmer 1976; Nowell and Jackson 1996). The young are thought to become independent when still quite small, at perhaps 6–8 months, and they may reach sexual maturity at 9 months; longevity is up to 13 years (Nowell and Jackson 1996).

The sand cat seems to be generally rare, but such a view may reflect a lack of knowledge. Thousands of skins have been taken for the fur trade during some years in Central Asia (Heptner and Sludskii 1992). The species is on appendix 2 of the CITES. The subspecies in Pakistan, *F. m. scheffeli*, was not discovered until 1966 and is now classified as endangered by the USDI and as near threatened by the IUCN. It reportedly declined drastically through uncontrolled exploitation by commercial animal dealers from 1967 to 1972, but recent observations suggest that populations have recovered (Nowell and Jackson 1996).

Felis nigripes (black-footed cat)

This is the smallest cat. Head and body length is 337–500 mm, tail length is 150–200 mm, and weight is 1.5–2.75 kg. The general coloration is dark ochre to pale ochre or sandy, being somewhat darker on the back and paler on the belly. A bold pattern of dark brown to black spots is arranged in rows on the flanks, throat, chest, and belly. There are two streaks across each cheek, two transverse bars on the forelegs, and as many as five transverse bars on the haunches. The tail has a black tip and two or three subterminal bands. The bottoms of the feet are black (since the animal walks on its toes, much of the black is usually visible).

The black-footed cat inhabits dry, open country. It shelters in old termite mounds and the abandoned burrows of other mammals. It is mainly nocturnal, but in captivity it has been found to be more active by day than most other small cats. The diet probably includes rodents, birds, and reptiles. Individuals have been observed to catch birds and hares and to cache them for later use (Nowell and Jackson 1996).

This felid seems to be highly unsocial. Even opposite sexes evidently come together only for 5–10 hours. However, a male's home range of 13 sq km reportedly overlapped a female's range of 12 sq km by about 50 percent (Nowell and Jackson 1996). Gestation lasts 59–68 days, and litters contain one to three young. Newborn weight is about 60–88 grams (Hayssen, Van Tienhoven, and Van Tienhoven 1993). The kittens leave the nest after 28–29 days and take their first solid food a few days later. Captive females have not initially entered heat until 15–21 months (Grzimek 1975).

F. nigripes is on appendix 1 of the CITES and is classified as rare in South Africa (Smithers 1986). It is not thought to be threatened with extinction but may be locally jeopardized by habitat degradation and by traps and poison put out to kill other predators (Nowell and Jackson 1996).

Felis manul (Pallas's cat)

Head and body length is 500–650 mm and tail length is 210–310 mm. Heptner and Sludskii (1992) stated that weight is 2.5–1.5 kg but that some animals may be heavier. The general coloration varies from light gray to yellowish buff and russet; the white tips of the hairs produce a frosted silvery appearance. There are two dark streaks across each side of the head and four rings on the dark-tipped tail. The coat is relatively longer and more dense than that of any other wild species of *Felis*. The fur is especially long near the end of the tail, and on the underparts of the body it is almost twice as long as on the back and sides. Such an

Black-footed cat *(Felis nigripes)*, photo by John Visser.

arrangement provides good insulation for an animal that spends much time lying on frozen ground and snow. The body is massive, the legs are short and stout, the head is short and broad, and the very short, bluntly rounded ears are set low and wide apart.

Pallas's cat inhabits steppes, deserts, and rocky country to elevations of more than 4,000 meters. It dens in a cave, crevice, or burrow dug by another animal. It usually is reported to be nocturnal or crepuscular but may be active by day since its main prey, pikas, are diurnal (Heptner and Sludskii 1992; Nowell and Jackson 1996). The diet also includes small rodents and ground-dwelling birds. According to Stroganov (1969), the young are born in Siberia in late April and May. Newborn weight has been reported as 89 grams (Hayssen, Van Tienhoven, and Van Tienhoven 1993). The estrous cycle lasts 46 days, the gestation period has been variously reported at 60–75 days, litter size is 1–6 young (usually 3 or 4), and females attain sexual maturity at 1 year (Heptner and Sludskii 1992; Nowell and Jackson 1996).

As many as 50,000 pelts of *F. manul* were being taken annually in Mongolia alone during the early twentieth century, but from the 1920s to the 1980s harvests in that country usually were less than 10,000 per year (Heptner and Sludskii 1992; Nowell and Jackson 1996). Broad, Luxmoore, and Jenkins (1988) reported that at least 20,000 skins of *F. manul* still were entering international trade each year and that such activity might be threatening the species. Concern about evident declining populations led to legal protection in Mongolia and China, and commerce in skins has largely ceased. The subspecies *F. m. ferrugineous,* found to the west, from Armenia and Iran to Uzbekistan, is designated as near threatened by the IUCN. The entire species also is on appendix 2 of the CITES.

Felis pardina (Spanish lynx)

Head and body length is 850–1,100 mm, tail length is 125–30 mm, and shoulder height is 600–700 mm. Average weight is 12.9 kg for males and 9.4 kg for females. The upper parts are yellowish red and the underparts are white. There are round black spots on the body, tail, and limbs. The ears have tufts, and the face has a prominent fringe of whiskers (Nowell and Jackson 1996; Smit and Van Wijngaarden 1981; Van Den Brink 1968). There originally seem to have been three different pelage patterns, one characterized by relatively large spots arranged roughly in lines, one with more randomly distributed large spots, and one with very small spots. The latter two patterns evidently have disappeared with the recent decline in the species (Beltrán and Delibes 1993).

According to Nowell and Jackson (1996), the Spanish lynx inhabits open woodland, thickets, and dense scrub. It is primarily nocturnal and travels about 7 km per night. It feeds primarily on rabbits and also takes small ungulates

Pallas's cat *(Felis manul)*, photo by Bernhard Grzimek.

Spanish lynx *(Felis pardina)*, photo by Lothar Schlawe.

and ducks. Reported population densities are about 4.5–16.0/100 sq km. Annual home range averages 18 sq km for males and 10 sq km for females. There is no overlap in the ranges of animals of the same sex but complete overlap between the ranges of opposite sexes. Mating occurs from January to July, peaking in January–February, and parturition peaks in March–April. Gestation lasts about 2 months, litter size is 2–3 young, and independence is attained at 7–10 months, but young remain in their natal territory until they are about 20 months old. Females are capable of breeding in their first winter but may have to wait until they can acquire a territory. Reproductive activity continues until 10 years of age and longevity is up to 13 years.

The Spanish lynx is classified as endangered by the IUCN and the USDI and is on appendix 1 of the CITES. It formerly occurred throughout the Iberian Peninsula, but the range has been reduced to less than 15,000 sq km in scattered mountainous areas and the Guadalquivir Delta. The total number of animals probably does not exceed 1,200, including about 350 breeding females (Nowell and Jackson 1996; Rodriguez and Delibes 1992). There was a major decline during the 1950s and 1960s, when the disease myxomatosis hit the rabbit populations. The decline is continuing as suitable rabbit and lynx habitat is replaced by cereal cultivation and forest plantations.

Felis lynx (Eurasian lynx)

Head and body length is 800–1,300 mm, tail length is 110–245 mm, shoulder height is 600–750 mm, and weight is 8–38 kg (Novikov 1962; Van Den Brink 1968). The average weight is 21.6 kg for males and 18.1 kg for females, about twice that of *F. canadensis* (Nowell and Jackson 1996). The fur is exceedingly dense, and coloration is more variable than in any other cat species (Heptner and Sludskii 1992). The upper parts may be reddish, brown, yellowish, gray, ashy blue, or almost white. Spots, variously darker than the rest of the coat, are almost always present. The underparts are usually white and the terminal part of the tail is black. There is a prominent black tuft on the ear. The summer coat is shorter than the winter coat, and the spots become more prominent. The legs are relatively long and the large feet are thickly haired in winter, producing a "snowshoe effect" for efficient travel through deep snow (Nowell and Jackson 1996).

According to Heptner and Sludskii (1992), the Eurasian lynx is typically associated with forests; it is most common where there is a mixture of spruce and deciduous trees but may sometimes penetrate the forest steppe and steppe zones. In the winter it follows its prey to lower elevations, and if game is scarce it may migrate up to 100 km. It is generally active at night and in the early morning, during which time it hunts over an average distance of about 10 km. Prey is followed for up to several days, then approached by stealth, and at last seized after a final rush. There is some controversy regarding diet. Heptner and Sludskii (1992) indicated that hares are by far the most important food for *F. lynx*, as they are for *F. canadensis*. Data compiled by Nowell and Jackson (1996), however, suggest that *F. lynx* preys primarily on small ungulates, particularly roe deer, chamois, and musk deer, and that smaller animals are eaten only when ungulates are not available. *F. lynx* is capable of killing animals three to four times its own size and in some areas preys mainly on large ungulates (mostly females and young), including red deer, reindeer, and argali *(Ovis ammon)*. The diet also includes rodents, pikas, and birds. Prey is usually dragged several hundred meters before being consumed, and a portion may be cached for later use.

Population densities of around 1–10/100 sq km have been reported in Russia, where numbers are said to fluctuate to some extent in association with the abundance of hares, as in North America. Home range size in Russia is 20–100 sq km, and such an area is traversed every 15–30 days (Heptner and Sludskii 1992). Density in Switzerland, where the species is scarce, is around 1/100 sq km and home ranges are very large. Average size is 264 sq km for males and 168 sq km for females. Female ranges have central core areas, averaging 72 sq km, which do not overlap one another and in which the residents spend most of their time. Male ranges overlap one another to some extent and also encompass most of a female's range, but a male tends to avoid the female's core area (Nowell and Jackson 1996). Heptner and Sludskii (1992) noted that sometimes a mated pair or a mother and young may hunt together, one animal chasing prey in the direction of the other. Mating occurs in Russia from January to March, with births from April to June. The gestation period is 67–74 days. There are 1–4 young, usually 2–3. They weigh about 250 grams at birth, open their eyes after 12 days, take some solid food at 50 days, and are weaned and begin to accompany their mother at about 3 months. They leave her just before the next mating season and attain sexual maturity in their second year. Studies by Kvam (1991) indicate that males in Norway normally reach sexual maturity by 31 months but that some do so at 21 months, and that about 50 percent of females are mature at 9 months, the rest by 21 months. Nowell and Jackson (1996) reported that females are reproductively active until 14 years, males until 16–17 years.

The Eurasian lynx has been intensively hunted and trapped for its valuable fur and because it is considered a threat to game and livestock. It is on appendix 2 of the CITES. The population in Russia has been estimated to number 40,000, and about 5,000–7,000 pelts are taken there annually (Heptner and Sludskii 1992; Nowell and Jackson 1996). The species also still is found in much of China and Central Asia, but it disappeared from most of Europe during the nineteenth century. To the west of Russia, major populations now occur only in Scandinavia and the Carpathian region. Small, isolated groups are present in northeastern Poland, the southern Balkans, and the French Pyrenees. Reintroductions have been carried out recently in parts of Germany, Austria, Switzerland, and Slovenia (Breitenmoser and Breitenmoser-Würsten 1994; Nowell and Jackson 1996; Smit and Van Wijngaarden 1981). The population reestablished in Switzerland now numbers more than 100 animals and is thought to be expanding into northwestern Italy (Guidali, Mingozzi, and Tosi 1990). However, Ragni, Possenti, and Mayr (1993) suggested that the current population in the Alps is different from the original and now extinct subspecies, which they designated *F. lynx alpina*. They also noted that the lynx evidently had disappeared from the Italian peninsula, south of the Alps, by about the beginning of the Bronze Age. The lynx may have occurred in Palestine before that area was largely deforested; in the Arabian region the lynx is now found only in northern Iraq, where it is rare (Harrison 1968).

Felis canadensis (Canada lynx)

Head and body length is about 800–1,000 mm, tail length is 51–138 mm, and weight is 5.1–17.2 kg (Banfield 1974; Burt and Grossenheider 1976). Average weight is 10.7 kg for males and 8.9 kg for females, about half that of the Eurasian *F. lynx* (Nowell and Jackson 1996). The coloration varies but is commonly yellowish brown; the upper parts may have a gray frosted appearance, the underparts are more buffy, and there is often a pattern of dark spots. The markedly short tail may have several dark rings and is tipped with black. The fur is long, lax, and thick. It is especially long on the lower cheeks in winter and gives the impression of a ruff around the neck. The triangular ears are tipped by tufts of black hairs about 40 mm long. The legs are relatively long. The paws are large and densely furred, an adaptation for moving over winter snow. Females have four mammae (Banfield 1974).

The lynx is generally found in tall forests with dense undergrowth but may also enter open forest, rocky areas, or tundra. For shelter it constructs a rough bed under a rock ledge, fallen tree, or shrub (Banfield 1974). It is mainly nocturnal; reports of average nightly movement range from 5 to 19 km (Ewer 1973). The lynx usually keeps to one area but may migrate under adverse conditions. The maximum known movement was by a female that was marked on 5 November 1974 in northern Minnesota and trapped on 20 January 1977 in Ontario, 483 km from the point of release (Mech 1977c). The lynx climbs well and is a good swimmer, sometimes crossing wide rivers. It hunts mainly by eye and also has well-developed hearing. It usually stalks its prey to within a few bounds, or it may wait in ambush for hours (Tumlison 1987). Reports from much of the range of the species indicate that leporids form a major part of the diet. The snowshoe rabbit *(Lepus americanus)* is of particular importance in North America. Deer and other ungulates are utilized heavily in certain areas, especially during the winter. Other foods include rodents, birds, and fish.

In Canada the numbers of lynxes seem to fluctuate, together with those of the snowshoe rabbit, in a regular cycle. Numerical peaks occur at an average interval of 9.6 years, though not at the same time all over Canada. Several authorities have suggested that this phenomenon is actually an artifact, perhaps associated with the intensity of human trapping, but Finerty (1979) upheld the traditional view. Studies in central Alberta (Brand and Keith 1979; Brand, Keith, and Fischer 1976; Nellis, Wetmore, and Keith 1972) indicate that the main direct cause of the cyclical drop in lynx numbers is postpartum mortality of kittens resulting from lack of food. There is a reduced rate of pregnancy in females. Population density is this area varied from a low of 2.3/100 sq km in the winter of 1966–67 to a peak of 10/100 sq km in the winter of 1971–72.

The reported individual home range size is 4–70 sq km for males and 4–25 sq km for females (Nowell and Jackson 1996). The lynx is probably territorial, but the ranges of females may overlap one another to some extent, and the range of a male may include that of a female and her young. Adults tend to avoid one another except during the breeding season. At that time there is an increase in scent marking with urine and also in vocalizations—purring and meowing (Tumlison 1987). Females are monestrous and bear a single litter per year (Banfield 1974; Ewer 1973). Mating occurs mainly in February–March, parturition is usually in May–June, gestation lasts 9–10 weeks, and litters contain 1–8 young, averaging about 4–5 when prey is abundant and 2–3 when prey is scarce (Nowell and Jackson 1996). Banfield (1974) gave birth weight as 197–211 grams. Lactation lasts for 5 months, but some meat is eaten by

Canada lynx *(Felis canadensis)*, photo from San Diego Zoological Garden.

1-month-old kittens. The young usually remain with their mother until the winter mating season, and siblings may stay together for a while afterward. The age of sexual maturity is usually 22–23 months (Nowell and Jackson 1996). One wild female is known to have lived 14 years and 7 months (Chubbs and Phillips 1993), and a captive individual lived 26 years and 9 months (Rich 1983).

The range of *F. canadensis* once extended at least as far south as Oregon, southern Colorado, Nebraska, southern Indiana, and Pennsylvania (E. R. Hall 1981). The only substantial population now remaining in the conterminous United States is that of western Montana and nearby parts of northern Idaho and northeastern Washington. On occasion, especially during periods of cyclically high populations in Canada, the species is found in the states of the northern plains and upper Great Lakes regions. There are small numbers in northern New England and Utah and possibly in Oregon, Wyoming, and Colorado (Armstrong 1972; Cowan 1971; Deems and Pursley 1978; Gunderson 1978; Hunt 1974; Olterman and Verts 1972). From 1988 to 1990, 83 lynxes captured in the Yukon were released in the Adirondacks of New York in an attempt to restore the species (Brocke and Gustafson 1992). In 1994 the U.S. Fish and Wildlife Service rejected a petition to add the lynx in the conterminous United States to the List of Endangered and Threatened Wildlife. This action was taken at higher levels within the agency despite support for listing by the scientific community and by the agency's own field and regional offices with jurisdiction over Montana and other involved states. It may have been a response to political pressure from interests favoring logging, which would disturb lynx habitat and would have become subject to control on federal lands if listing had proceeded (Bourne 1995*b*). After losing a lawsuit on the issue, the Fish and Wildlife Service finally acknowledged that listing was warranted but claimed it could not proceed with such a measure because of higher-priority responsibilities (Knibb 1997).

The lynx has been exterminated in parts of southeastern Canada but still occurs regularly over most of that country. Its numbers there seem to be affected more by the cyclical availability of food than by pressure from human hunting. Populations evidently were relatively high until the early twentieth century, declined until mid-century, and then again increased (Cowan 1971). Subsequent highs and lows in the reported number of pelts taken in Canada have been as follows: 1949–50 trapping season, 3,734; 1954–55, 14,427; 1956–57, 8,748; 1962–63, 51,376; 1966–67, 13,038; 1971–72, 53,589; 1975–76, 13,162; and 1979–80, 34,366. The average price per pelt rose dramatically from $3.62 in 1953 54 to $30.52 in 1968–69 and to $336.36 in 1978–79 (Statistics Canada 1981). At autumn 1984 sales in Ontario the average value of lynx pelts surged to nearly $600.00 (Quinn and Parker 1987). Prices subsequently fell back to around $100 per skin in the early 1990s (data from U.S. Fish and Wildlife Service, 1994). *F. canadensis* is on appendix 2 of the CITES.

Felis rufus (bobcat)

Head and body length is 650–1,050 mm, tail length is 110–90 mm, and shoulder height is 450–580 mm. Banfield

Bobcat *(Felis rufus)*, photo by Lothar Schlawe.

(1974) gave the weight as 4.1–15.3 kg. There are various shades of buff and brown, spotted and lined with dark brown and black. The crown is streaked with black and the backs of the ears are heavily marked with black. The short tail has a black tip, but only on the upper side. *F. rufus* resembles *F. lynx* but is usually smaller and has slenderer legs, smaller feet, shorter fur, and ears that are tufted less conspicuously or not at all. As in *F. lynx*, a ruff of fur extends from the ears to the jowls, giving the impression of sideburns. Fully black individuals occasionally are found in Florida (Regan and Maehr 1990). Females have four mammae (Lowery 1974).

The bobcat is more ubiquitous than the lynx, occurring in forests, mountainous areas, semideserts, and brushland. Its den is usually concealed in a thicket, hollow tree, or rocky crevice. It is mainly nocturnal and terrestrial but climbs with ease. Nightly movements of about 3–11 km have been reported. Prey is usually stalked with great stealth and patience, then seized after a swift leap. The diet consists mainly of small mammals, especially rabbits, and birds. Larger prey, such as deer, is sometimes taken, especially in the winter. A study in Massachusetts found deer to be the most common winter food (McCord 1974).

Reported maximum population densities are 1/18.4 sq km in the western United States and 1/2.6 sq km in the Southeast (Jachowski 1981). Average minimum home range in Louisiana was found to be about 5 sq km for males and 1 sq km for females (Hall and Newsom 1978). In a study in southeastern Idaho, Bailey (1974) found average (and extreme) home range size to be 42.1 (6.5–107.9) sq km for males and 19.3 (9.1–45.3) sq km for females. Female ranges were almost exclusive of one another, but the ranges of males overlapped one another, as well as those of fe-

males. There was a land tenure system, seemingly based on prior right: no neighboring resident or transient permanently settled in an area already occupied by a resident. All observed changes in resident home ranges were attributed to the death of a resident. Territoriality was pronounced, especially in females, and scent marking was accomplished by use of feces, urine, scrapes, and anal gland secretions. Individuals were solitary, avoiding each other even in areas of range overlap, except during the mating season. The bobcat is usually silent but may emit loud screams, hisses, and other sounds during courtship.

Females apparently are seasonally polyestrous; they usually produce a single annual litter during the spring, but there is evidence of a second birth peak in late summer and early autumn, perhaps involving younger females or ones that lost their first litters (Banfield 1974). According to Fritts and Sealander (1978), mating may occur as early as November or as late as August. The estrous cycle is 44 days and estrus lasts 5–10 days (Nowell and Jackson 1996). The gestation period is 60–70 days (Hemmer 1976). The number of young per litter is one to six, commonly three. Weight at birth is 283–368 grams (Banfield 1974). The young open their eyes after 9–10 days, nurse for about 2 months, and begin to travel with the mother at 3–5 months. They separate from the mother in the winter (Jackson 1961), probably in association with the mating season. Females may reach sexual maturity by 1 year, but males do not mate until their second year of life. The oldest individuals captured in a Wyoming study were 12 years of age and still sexually active (Crowe 1975). A captive bobcat lived 32 years and 4 months (Jones 1982).

The bobcat occasionally preys on small domestic mammals and poultry and thus has been hunted and trapped by people. It has been exterminated in much of the Ohio Valley, the upper Mississippi Valley, and the southern Great Lakes region (Deems and Pursley 1978). The bobcat is uncommon in central Mexico (Leopold 1959); the subspecies there, *F. rufus escuinapae,* is listed as endangered by the USDI. It is intensively persecuted by sheepherders, and its habitat is being degraded (Nowell and Jackson 1996). The entire species is on appendix 2 of the CITES.

The value of bobcat fur has varied widely depending on fashion and economic conditions. The average price per pelt rose from about $10 in the 1970–71 season (Deems and Pursley 1978) to $145 in 1978–79 (Jachowski 1981). There was a corresponding increase in the number of bobcats being harvested. The total known annual kill in the United States was approximately 92,000 in the late 1970s. This rate continued or declined slightly in the 1980s. About two-thirds of these animals were being taken primarily for their fur, and most of the pelts were being exported. There was concern that bobcat populations were being seriously reduced in some areas and that little was being done to manage the resource. Jachowski (1981), however, reported substantial improvement in management since *F. rufus* was placed on appendix 2 of the CITES in 1977. Prior to that time few states had a closed season, and many had bounties. Presently there are no bounties, 10 states provide complete protection, and the rest allow a regulated harvest during a limited season. Jachowski estimated that there are between 725,000 and 1,020,000 bobcats in the United States and suggested that current known harvest levels are not jeopardizing overall populations in the country. Export of skins was halted briefly by legal action in 1981 but was subsequently restored. In the 1982–83 season about 75,000 skins were taken and sold for an average price of $142 (Rolley 1987). In 1991–92, 22,077 skins were taken and sold for an average of $63 (Linscombe 1994). Although they did not directly question the above population data, Nowell and Jackson (1996) indicated that there is still concern about

whether commercial trapping is sustainable. They noted also that trade in bobcat fur is declining and that the European Community—the main market for the pelts—announced that after 1995 it would ban all importation of the skins of animals taken in leghold traps.

Felis caracal (caracal)

Head and body length is 600–915 mm, tail length is 230–310 mm, shoulder height is 380–500 mm, and weight is 6–19 kg (Kingdon 1977; Skinner and Smithers 1990). The pelage is dense but relatively short, and there are no side whiskers as in *F. lynx.* The general coloration is reddish brown. There is white on the chin, throat, and belly and a narrow black line from the eye to the nose. The ears are narrow, pointed, black on the outside, and adorned with black tufts up to 45 mm long. Smaller than *F. lynx, F. caracal* has a long and slender body and a tapering tail that is approximately one-third the length of the head and body.

The caracal is found mainly in dry country—woodland, savannah, and scrub—but avoids sandy deserts. Maternal dens are located in porcupine burrows, rocky crevices, or dense vegetation. This cat is largely nocturnal but sometimes is seen by day. It climbs and jumps well. It is mainly terrestrial and is apparently the fastest feline of its size. A radio-tracking study indicated that daily movement averaged 10.4 km for males and 6.6 km for females (Nowell and Jackson 1996). Prey is stalked and then captured after a quick dash or leap. The diet includes birds, rodents, and small antelopes.

In South Africa the home ranges of males were found to measure 5.1–48.0 sq km and to widely overlap one another and also the ranges of females, which measured 3.9–26.7 sq km and overlapped only slightly (Skinner and Smithers 1990). Much larger ranges have been reported in the Negev of Israel, the averages being 221 sq km for males and 57 sq km for females; in the latter area male ranges typically encompassed several female ranges (Nowell and Jackson

Caracals *(Felis caracal)*, photo by Bernhard Grzimek.

1996). The caracal has been reported to be territorial and to mark with urine. Vocalizations include miaows, growls, hisses, and coughing calls (Kingdon 1977). The species is usually seen alone, but Rowe-Rowe (1978b) reported a group of two adults and five young. Various observations suggest that the young may be born at any time of year, though there is a birth peak from October to February in South Africa. The estrous cycle is 14 days, estrus lasts 1–6 days, the gestation period has been reported at 69–81 days, and the number of young per litter is one to six, usually three, young. The kittens open their eyes after 10 days, are weaned at 10–25 weeks, and attain sexual maturity at 12–16 months. One female gave birth at 18 years (Kingdon 1977; Nowell and Jackson 1996; Skinner and Smithers 1990).

The caracal is easily tamed and has been used to assist human hunters in Iran and India. It sometimes raids poultry, however, and thus has been killed by people. It apparently has become scarce in North Africa, South Africa, and central and southwestern Asia. All Asian populations are on appendix 1 of the CITES, and African populations are on appendix 2.

Felis serval (serval)

Head and body length is 670–1,000 mm, tail length is 240–450 mm, shoulder height is 540–620 mm, and weight is 8.7–18.0 kg. Males are generally larger than females (Kingdon 1977). The general coloration of the upper parts ranges from off-white to dark gold, and the underparts are paler, often white (Smithers 1978). The entire pelage is marked either with small, dark spots or with large spots that tend to merge into longitudinal stripes on the head and back. The tail has several rings and a black tip. Melanistic individuals have been widely reported (Nowell and Jackson 1996). The build is light, the legs and neck are long, and the ears are large and rounded.

According to Smithers (1978), the serval is generally a species of the savanna zone and is found in the vicinity of streams with densely vegetated banks. It is primarily nocturnal and may move 3–4 km per night. It is mainly terrestrial and can run or bound swiftly for short distances. Prey is apparently located both by sight and by hearing. Birds up to 3 meters above the ground may be captured by remarkable leaps, but the diet seems to consist mostly of murid rodents. Van Aarde and Skinner (1986a) reported home ranges of 2.1–2.7 sq km in South Africa. In Tanzania, however, a male was found to have a range of at least 11.6 sq km that overlapped the ranges of at least two females (Nowell and Jackson 1996).

The serval is basically solitary. It has a shrill cry and also growls and purrs. There is no definite mating season, but Smithers (1978) reported that births in Zimbabwe occurred mainly in the warm months from September to April. Kingdon (1977) suggested that there are two birth peaks in East Africa, in March–April and September–November. Observations at the Basel Zoo (Wackernagel 1968) show that females can give birth twice a year, with a minimum normal interval of 184 days. Estrus usually lasted only one day, and the average gestation period was determined to be 74 days. The number of young in 20 litters averaged 2.35 and ranged from 1 to 4. Five newborn weighed 230–60 grams each, and one opened its eyes at 9 days. One female at the Basel Zoo gave birth to her last litter at 14 years and died at about 19 years and 9 months.

The serval is on appendix 2 of the CITES. It is hunted for its skin in East Africa and now no longer occurs in areas heavily populated by people (Kingdon 1977). The species has been mercilessly hunted in farming areas of South Africa and is now considered to be rare in that country (Skinner, Fairall, and Bothma 1977). Because of its concentration in riparian vegetation, the serval is particularly vulnerable to both direct hunting and habitat disruption. Such problems may be especially severe in North Africa, where the species seems to survive only in the Atlas Mountains of Morocco (Nowell and Jackson 1996). The IUCN now classifies the subspecies there, F. s. constantinus, as endangered (using the generic name Leptailurus), noting that fewer than 250 mature individuals survive.

Felis marmorata (marbled cat)

Head and body length is 450–530 mm, tail length is 475–550 mm, and weight is 2–5 kg (Lekagul and McNeely 1977). The ground color is brownish gray to bright yellow or rufous brown. The sides of the body are marked with large, irregular, dark blotches, each margined with black. There are solid black dots on the limbs and underparts. The tail is spotted, tipped with black, long, and bushy. The pelage is thick and soft, and the ears are short and rounded. F. marmorata resembles Neofelis nebulosa in the appearance of its coat and in having relatively large canine teeth, and some authorities consider the two species to be closely related. Based on cranial characters and karyology, however, others suggest affinity with Lynx or Panthera (Nowell and Jackson 1996).

The marbled cat is a forest dweller, apparently nocturnal and partly arboreal in habit. It is thought to prey mostly on birds, to some extent on squirrels and rats, and possibly on lizards and frogs. According to Hayssen, Van Tienhoven, and Van Tienhoven (1993), reproduction is not seasonal, litters contain two young, and birth weight is 70–100 grams. Because of human disturbance and habitat destruction, F. marmorata evidently has declined and become very rare in much of its range (IUCN 1978). It is classified as endangered by the USDI and is on appendix 1 of the CITES.

Felis temmincki (Asian golden cat)

Head and body length is 730–1,050 mm and tail length is 430–560 mm. Weight is 8–15 kg (Lekagul and McNeely 1977; Nowell and Jackson 1996). The pelage is of moderate length, dense, and rather harsh. The general coloration varies from golden red to dark brown and gray. Some specimens, especially from the northern parts of the range, have a pattern of spots on the body. The face is marked with white and black streaks, and the underside of the terminal third of the tail is white. The ears are short and rounded.

According to Lekagul and McNeely (1977), the Asian golden cat occurs in deciduous forests, tropical rainforests, and occasionally more open habitats. It is usually terrestrial but is capable of climbing. The diet includes hares, small deer, birds, lizards, and domestic livestock. This cat often hunts in pairs, and the male is said to play an active role in rearing the young. There is no confirmed breeding season. Young weigh about 250 grams at birth. If a litter is lost, the female may produce another within 4 months. Nowell and Jackson (1996) recorded the following: estrous cycle, 39 days; estrus, 6 days; average gestation period, 80 days; litter size, range 1–3, mean, 1.11; age at sexual maturity, 18–24 months; longevity, up to 20 years.

Because of habitat destruction and inability to adjust to the presence of human activity, F. temmincki has declined in much of its range. It is classified as near threatened by the IUCN (under the genus Catopuma) and as endangered by the USDI and is on appendix 1 of the CITES.

Felis badia (bay cat)

Information provided by Nowell and Jackson (1996) and Sunquist et al. (1994) indicates that this species is known only by seven specimens, six of which were collected from 1855 to 1928. Head and body length is 533–670 mm in

Asian golden cat *(Felis temmincki)*, photo from San Diego Zoological Society.

three adults, tail length is 320–91 mm in four adults, and normal weight was estimated at 3–4 kg in a single adult. In most specimens the pelt is bright chestnut above and paler on the belly, with some obscure spots on the underparts and limbs. One specimen is grayish in color. The long and tapering tail has a whitish median streak down the middle of its lower surface and becomes pure white at the tip. *F. badia* closely resembles *F. temmincki* but is much smaller. All collections and precise sightings have been in highland areas, and at least three specimens were collected along rivers. The bay cat has been reported to inhabit dense primary forests, but there have been recent sightings at night in logged dipterocarp forest. Surveys in the 1980s failed to locate the species, which evidently is extremely rare, but an adult female was collected in November 1992. *F. badia* (using the generic name *Catopuma*) is classified as vulnerable by the IUCN, which estimates that fewer than 10,000 mature individuals survive, and is on appendix 2 of the CITES.

Felis aurata (African golden cat)

Head and body length is 616–1,016 mm, tail length is 160–460 mm, shoulder height is 380–510 mm, and weight is 5.3–16.0 kg. Males are generally larger than females

(Kingdon 1977). The overall coloration varies from chestnut through fox red, fawn, gray brown, silver gray, and blue gray to dark slaty. The cheeks, chin, and underparts are white. Some specimens are marked all over with dark brown or dark gray dots; in others the spots are restricted to the belly and insides of the limbs. Black specimens have been recorded. The legs are long, the head is small, and the paws are large.

According to Kingdon (1977), the African golden cat is found mainly in forests, sometimes in mountainous areas. It is said to be active both by day and by night. It is mainly terrestrial but climbs well. Prey is taken by stalking and rushing. The diet includes birds, small ungulates, and domestic animals. *F. aurata* is normally solitary. A pregnant female was taken in Uganda in September. Two litters contained two young each, and two young weighed 195 and 235 grams (Hayssen, Van Tienhoven, and Van Tienhoven 1993). The species is on appendix 2 of the CITES.

Felis bengalensis (leopard cat)

Head and body length is 445–1,070 mm, tail length is 230–440 mm, and weight is 3–7 kg (Lekagul and McNeely 1977; Stroganov 1969). There is much variation in color,

Leopard cat *(Felis bengalensis)*, photo from New York Zoological Society.

but the upper parts are usually pale tawny and the under-parts are white. The body and tail are covered with dark spots. There are usually four longitudinal black bands running from the forehead to behind the neck and breaking into short bands and rows of elongate spots on the shoulders. The tail is indistinctly ringed toward the tip. The head is small, the muzzle is short, and the ears are moderately long and rounded.

The leopard cat is found in many kinds of forested habitat at both high and low elevations. It dens in hollow trees or small caves or under overhangs or large roots. It is mainly nocturnal but often is seen by day. It is an excellent swimmer and has populated many offshore islands (Lekagul and McNeely 1977). It apparently hunts on the ground as well as in trees and feeds on hares, rodents, young deer, birds, reptiles, and fish. Four individuals radio-tracked by Rabinowitz (1990) in Thailand moved about 500–1,000 meters per day. One female used a total area of 6.6 sq km over a 13-month period but had an average monthly home range of 1.8 sq km.

Births have been reported in May in both Siberia and India. According to Lekagul and McNeely (1977), however, breeding continues throughout the year in Southeast Asia. If one litter is lost, the female may mate and produce another within 4–5 months. The gestation period is 65–72 days. The number of young per litter is one to four, usually two or three. The father may participate in rearing the young. The latter open their eyes at about 10 days. Weight at birth is 75–120 grams (Hayssen, Van Tienhoven, and Van Tienhoven 1993). Sexual maturity may be attained at as early as 8 months and longevity is up to 15 years (Nowell and Jackson 1996).

The leopard cat is more adaptable to deforestation and other habitat alteration than most other Asian felids, and it often is found near villages. However, many island populations, notably that inhabiting Panay, Negros, and Cebu in the Philippines, which may be a distinctive subspecies, have declined drastically (Nowell and Jackson 1996). The species is common in most mainland regions and has been intensively exploited for the international fur trade. The annual take in China alone during 1985–88 is believed to have been around 400,000. The European Community, which had received most pelts from China, banned importation in 1988. Japan has since been the main consumer, importing 50,000 skins in 1989. About 1,000–2,000 pelts of the eastern Siberian form, *F. b. euptilura*, were being processed in the 1930s. In recent years the harvest of that subspecies has fallen to 100–300, apparently in association with a serious decline in the population (Heptner and Sludskii 1992). The subspecies *F. b. bengalensis*, of peninsular India and Southeast Asia, is listed as endangered by the USDI. The populations of that subspecies in India, Bangladesh, and Thailand are on appendix 1 of the CITES; all other populations of *F. bengalensis* are on appendix 2.

Felis rubiginosa (rusty-spotted cat)

This is one of the smallest cats. Head and body length is 350–480 mm and tail length is 150–250 mm (Grzimek 1975). Weight is 1.1–1.6 kg (Nowell and Jackson 1996). The upper parts are grizzled gray, with a rufous tinge of varying intensity, and marked with lines of brown, elongate blotches. The belly and insides of the limbs are white with large, dark spots. There are two dark streaks on the face, and four dark streaks run from the top of the head to the nape.

On the mainland of India the rusty-spotted cat seems to be found mostly in scrub, dry grassland, and open country. In Sri Lanka, however, it occurs in humid mountain forests. Nowell and Jackson (1996) suggested that this situation might be explained by the presence on the mainland of the closely related *F. bengalensis*, which is mainly a forest animal, and the presence in Sri Lanka of *F. chaus*, which prefers more open country. The smaller *F. rubiginosa* thus may be forced into whichever habitat is not dominated by the other species. It is nocturnal, frequently enters trees, and preys on birds and small mammals. Births occur in the spring in India (Grzimek 1975). Estrus lasts 5 days, the gestation period is 67 days, and average litter size is 1.5 young (Nowell and Jackson 1996). One specimen was still living after 16 years in captivity (Marvin L. Jones, Zoological Society of San Diego, pers. comm., 1995). The Indian population is on appendix 1 of the CITES and the Sri Lankan population is on appendix 2.

Felis viverrina (fishing cat)

Head and body length is 750–860 mm, tail length is 255–330 mm, shoulder height is 380–406 mm, and weight is 7.7–14.0 kg. The ground color is grizzled gray, sometimes tinged with brown, and there are elongate dark brown spots arranged in longitudinal rows. Six to eight dark lines run from the forehead, over the crown, and along the neck. The fur is short and rather coarse, the head is big and broad, and the tail is short and thick. The forefeet have moderately developed webbing between the digits. The claw sheaths are too small to allow the claws to retract completely.

The fishing cat is found in marshy thickets, mangrove swamps, and densely vegetated areas along creeks. It often wades in shallows and does not hesitate to swim in deep water. It catches prey fish by crouching on a rock or sandbank and using its paw as a scoop. The diet also includes crustaceans, mollusks, frogs, snakes, birds, and small mammals.

In northeastern India mating activity peaks in January–February, births in March–May (Nowell and Jackson 1996). At the Philadelphia Zoo births occurred in March and August (Ulmer 1968). Observations there indicate that the gestation period is 63 days and birth weight is about 170 grams. The young open their eyes by 16 days of age, eat their first meat at 53 days, and attain adult size at 264 days. Litter size averages 2.61 and ranges from 1 to 4, age at independence is 10 months, and average longevity is 12 years (Nowell and Jackson 1996).

There is some question about the original distribution of this species. Van Bree and Khan (1992) reported the first specimen known from mainland Malaysia but suggested that it may have been an escaped captive; they also indicated that presence of the species on Sumatra had not been confirmed and that the known population on Java could have resulted from ancient human introduction. Nowell and Jackson (1996) suggested that the discontinuous distribution may result from the strong association of the fishing cat with major river and coastal floodplains. These wetland systems are being rapidly disrupted by drainage for agriculture and settlement, pollution, and other human activity. *F. viverrina*, unlike *F. chaus*, does not adapt well to cultivated habitats and now is on the verge of extirpation in such areas as the Indus Basin and the southwestern coast of India. It is on appendix 2 of the CITES and is designated as near threatened by the IUCN (under the name *Prionailurus viverrinus*).

Felis planiceps (flat-headed cat)

Head and body length is 410–500 mm and tail length is 130–50 mm (Grzimek 1975). Two specimens from Malaysia weighed 1.6 and 2.1 kg (Muul and Lim 1970). The body is dark brown with a silvery tinge. The underparts are white, generally spotted and splashed with brown. The top of the head is reddish brown. The face is light reddish below the eyes, there are two narrow dark lines running across each cheek, and a yellow line runs from each eye to near the ear. The coat is thick, long, and soft. The pads of the feet are long and narrow, and the claws cannot be completely retracted. The skull is long, narrow, and flat; the nasals are short and narrow, and the orbits are placed well forward and close together (Lekagul and McNeely 1977).

The flat-headed cat is thought to be nocturnal and to hunt for frogs and fish along riverbanks. Observations of a captive kitten (Muul and Lim 1970) suggest that *F. planiceps* is a fishing cat. The kitten seemed to enjoy playing in water, took pieces of fish from the water, and captured live frogs but ignored live birds. Moreover, the long, narrow rostrum and the well-developed first upper premolar of the species would seem to be efficient for seizing slippery prey. Very little additional information is available, but Nowell and Jackson (1996) noted that the species also hunts rodents, that most records are from swamps and riverine forests, and that the gestation period is approximately 56 days. *F. planiceps* appears to be very rare and may be threatened by pollution and clearing of its wetland habitat. It is classified as vulnerable by the IUCN (using the generic name *Prionailurus*) and as endangered by the USDI and is on appendix 1 of the CITES.

Felis iriomotensis (Iriomote cat)

Head and body length is 600 mm and tail length is 200 mm. Average weight is 4.2 kg for males and 3.2 kg for females (Nowell and Jackson 1996). The ground color is dark dusky brown, and spots arranged in longitudinal rows tend to merge into bands. Five to seven lines run from the back of the neck to the shoulders. The body is relatively elongate, the legs and tail are short, and the ears are rounded. Various cranial and karyological characters have been interpreted by different authorities to mean that *F. iriomotensis* represents a monotypic genus or subgenus, a full species, or only a subspecies of *F. bengalensis*.

This cat dwells only on Iriomote, a Japanese-owned island of 292 sq km to the east of Taiwan. Its presence was not known to science until 1965. If it is indeed a full species, it has the smallest population and the most limited range of any species of cat in the world (Yasuma 1988). *F. iriomotensis* typically inhabits lowland, subtropical rainforest along the coast of Iriomote, rather than the interior mountains (Nowell and Jackson 1996). It is partially arboreal, swims well, and shelters in tree holes, on branches, or in rock crevices. Although primarily nocturnal, it sometimes hunts by day for the large skink, *Eumeces kishinouyei*. It also preys on small rodents, bats *(Pteropus)*, birds, amphibians, and crabs. The male's home range averages about 3 sq km and overlaps those of other males and females, and the animal tends to change its range after several months. Female ranges average about 1.75 sq km, seldom overlap one another, and are more stable. Mating apparently occurs in February–March and September–October, births have been observed only in late April and May, gestation lasts approximately 60–70 days, litter size is 1–4 young, and maximum known longevity is more than 10 years.

Although totally protected by Japanese law, the Iriomote cat is losing habitat to deforestation for agriculture and economic development of the island. Its numbers are estimated to be fewer than 100 but are thought to have remained stable since monitoring began in 1982 (Nowell and Jackson 1996). The species is classified as endangered by the IUCN (under the name *Prionailurus bengalensis iriomotensis*) and the USDI and is on appendix 2 of the CITES.

Felis jacobita (mountain cat)

Redford and Eisenberg (1992) reported that head and body length is 577–640 mm, tail length is 413–80 mm, and one

Flat-headed kitten *(Felis planiceps)*, photo from San Diego Zoological Society.

specimen weighed 4 kg. The coat is long, soft, and fine. The upper parts are silvery gray and marked by irregular brown or orange yellow spots and transverse stripes. The underparts are whitish and have blackish spots. The bushy tail is ringed with black or brown and has a light tip.

According to Nowell and Jackson (1996), this cat has been found only in the rocky arid and semiarid zones of the Andes above the timberline, generally at elevations exceeding 3,000–4,000 meters. Its range appears to coincide with the original distribution of viscachas and chinchillas, which probably are its main prey. The recent decline of these large rodents, especially the near extirpation of chinchillas by fur hunters, may have contributed to an evident rarity of *F. jacobita*. The species is classified as vulnerable by the IUCN (using the generic name *Oreailurus*), which estimates that fewer than 10,000 mature individuals survive, and as endangered by the USDI and is on appendix 1 of the CITES.

Felis colocolo (pampas cat)

Head and body length is 435–700 mm, tail length is 220–322 mm, and shoulder height is 300–350 mm. One individual weighed 3 kg (Redford and Eisenberg 1992). The coloration ranges from yellowish white and grayish yellow to brown, gray brown, silvery gray, and light gray. Transverse bands of yellow or brown run obliquely from the back to the flanks. Two bars run from the eyes across the cheeks and meet beneath the throat. The coat is long, the tail is bushy, the face is broad, and the ears are pointed.

The pampas cat inhabits open grassland in some areas but also enters humid forests and mountainous regions. According to Redford and Eisenberg (1992), it has a greater habitat range than any other South American cat, being found from low swamps and marshes to elevations above 5,000 meters. Grimwood (1969) reported it to occur throughout the Andes of Peru. It is mainly terrestrial but may climb a tree if pursued. It hunts at night, killing small mammals, especially guinea pigs *(Cavia)* and ground birds. Litters are said to contain one to three young. *F. colocolo* has been intensively hunted for its fur. It now is on appendix 2 of the CITES.

Felis geoffroyi (Geoffroy's cat)

Head and body length is 422–665 mm, tail length is 240–365 mm, and weight is 2–6 kg (Redford and Eisenberg 1992). The ground color varies widely, from brilliant ochre in the northern parts of the range to silvery gray in the south. The body and limbs are covered with small black

Geoffroy's cat *(Felis geoffroyi)*, photo by K. Rudloff through East Berlin Zoo.

spots. There may be several black streaks on the crown, two on each cheek, and one between the shoulders. The tail is ringed. Melanistic individuals are fairly common (Nowell and Jackson 1996).

Geoffroy's cat inhabits scrubby woodland, open bush country, and pampas grasslands. Ximenez (1975) reported the elevational range to be sea level to 3,300 meters. Data compiled by Nowell and Jackson (1996) indicate that the species is primarily nocturnal, readily enters water, and swims and climbs well. It seems usually to rest by day in the crook of a tree. The diet consists mainly of small mammals but also includes birds, reptiles, amphibians, and fish. Population density in southern Chile is approximately 1/10 sq km. A radio-tracking study in that region determined an average annual home range size of 9.2 sq km for males and 3.7 sq km for females. The ranges of females overlapped, but those of males did not (Johnson and Franklin 1991). The estrous cycle is 20 days and estrus averages 2.5 days (Nowell and Jackson 1996). Hemmer (1976) listed a gestation period of 74–76 days. The single annual litter contains two or three young and in Uruguay is produced from December to May (Ximenez 1975). The young reportedly weigh 65–123 grams at birth and are weaned 8–10 weeks later (Hayssen, Van Tienhoven, and Van Tienhoven 1993). A captive specimen lived to be nearly 21 years old (Marvin L. Jones, Zoological Society of San Diego, pers. comm., 1995).

F. geoffroyi has been heavily exploited for the international fur trade; more than 78,000 skins, mainly from Paraguay, were reported in commerce in 1983 (Broad 1987). The species subsequently was legally protected in most of the countries where it occurs and also placed on ap-pendix 1 of the CITES. Although some exploitation continues, large-scale commercial traffic seems to have stopped (Nowell and Jackson 1996).

Felis guigna (kodkod)

Head and body length is 424–510 mm, tail length is 195–250 mm, and weight is 2.1–2.5 kg (Redford and Eisenberg 1992). The coat is buff or gray brown and is heavily marked with rounded, blackish spots on both the upper and lower parts. The tail has blackish rings. Melanistic individuals are not uncommon.

Nowell and Jackson (1996) indicated that the kodkod is strongly associated with moist forests of the southern Andes but also inhabits secondary forest and shrub country. Although it is usually considered nocturnal, there is evidence that it is also active by day. It is highly arboreal and shelters in trees when inactive. It apparently preys mainly on small rodents and birds and has been reported to raid henhouses. The gestation period is 72–78 days, litter size is one to four young, and longevity is up to 11 years. It may be threatened by deforestation, is classified as vulnerable by the IUCN (using the generic name *Oncifelis*), which estimates that fewer than 10,000 mature individuals survive, and is on appendix 2 of the CITES.

Felis pardalis (ocelot)

Head and body length is 550–1,000 mm and tail length is 300–450 mm (Grzimek 1975; Leopold 1959). Weight is 11.3–15.8 kg. The ground color ranges from whitish or tawny yellow to reddish gray and gray. Dark streaks and spots are arranged in small groups around areas that are darker than the ground color. There are two black stripes on each cheek and one or two transverse bars on the insides of the legs. The tail is either ringed or marked with dark bars on the upper surface.

The ocelot occurs in a great variety of habitats, from humid tropical forests to fairly dry scrub country. A consistent requirement of the species, however, is dense vegetative cover (Nowell and Jackson 1996). It is generally nocturnal, sleeping by day in a hollow tree, in thick vegetation, or on a branch. It is mainly terrestrial but climbs, jumps, and swims well. Mean daily travel distance is 1.8–7.6 km, with males moving up to twice as far as females (Nowell and Jackson 1996). The diet includes rodents, rabbits, young deer and peccaries, birds, snakes, and fish.

Reported population densities are 4 residents/5 sq km in lowland rainforest and 2/5 sq km in savannah, and reported home range size of adults in various areas has been about 1 to 14 sq km (Nowell and Jackson 1996). In a study in Peru, Emmons (1988a) found adult females to occupy exclusive home ranges of about 2 sq km. Male ranges were several times greater and also were exclusive of one another but did overlap a number of female ranges. Individuals generally moved about alone but appeared to make contact frequently and probably maintained a network of social ties. The ocelot communicates by mewing and, during courtship, by yowls not unlike those of *F. catus*.

There is probably no seasonal breeding in the tropics (Grzimek 1975). In Mexico and Texas births are reported to occur in autumn and winter (Leopold 1959). According to Nowell and Jackson (1996), autumn breeding peaks have been noted in Paraguay and Argentina, the estrous cycle averages 25 days, estrus averages 4.6 days, gestation lasts 79–85 days, litter size averages 1.64 and ranges from 1 to 3 young, and age at sexual maturity is 18–22 months in females and 30 months in males. The young weigh 200–340 grams at birth and are weaned 6 weeks later (Hayssen, Van Tienhoven, and Van Tienhoven 1993). One individual lived at the Phoenix Zoo for 21 years and 5 months (Marvin L. Jones, Zoological Society of San Diego, pers. comm., 1995).

The ocelot is classified as endangered by the USDI and is on appendix 1 of the CITES. It was the spotted cat most heavily exploited by the international fur trade from the early 1960s to the mid-1970s, when as many as 200,000 were taken annually (Nowell and Jackson 1996). Prior to 1972, when importation into the United States was prohibited because of the animal's being listed as endangered, enormous numbers of skins were brought into the country—133,069 in 1969 alone. As late as 1975 Great Britain imported 76,838 skins. In the early 1980s ocelot fur coats sold for as high as U.S. $40,000 in West Germany. The ocelot also is reported to be in much demand for use as a pet, a live animal selling for $800 (Thornback and Jenkins 1982). Recent legal protection in most countries where the ocelot occurs and regulation of the international trade in its fur through CITES seem to have contributed to a substantial decline in market hunting and a partial recovery of some of the depleted populations. The species is adaptable and still occurs over most of its original range, though there is concern that its relatively low reproductive rate and its need for dense cover and abundant small prey could make it especially vulnerable to environmental disturbance. Based on a minimum estimated density of 1/5 sq km, there would be at least 800,000 ocelots in forested South America, with total numbers probably 1.5–3.0 million (Nowell and Jackson 1996).

The subspecies *F. p. albescens* is classified as endangered by the IUCN (using the generic name *Leopardus*). It probably once ranged over most of Texas and at least as far east as Arkansas and Louisiana (Lowery 1974). It declined rapidly during the nineteenth century because of hunting and loss of habitat to settlement and is currently restricted to the border region of extreme southern Texas and northeastern Mexico. Of the fewer than 250 mature individuals thought to survive, fewer than 100 are in Texas. The clearing of brush country for agricultural purposes is the main problem in this area.

Felis wiedii (margay)

Head and body length is 463–790 mm and tail length is 331–510 mm. Weight is 2.6–3.9 kg (Redford and Eisenberg 1992). The ground color is yellowish brown above and white below. There are longitudinal rows of dark brown spots, the centers of which are paler than the borders. The margay closely resembles the ocelot but is smaller, has a slenderer build, and has a relatively longer tail.

The margay is mainly, if not exclusively, a forest dweller. It is much more arboreal than the ocelot and is thought to forage in trees. The only specimen known from Texas was taken prior to 1852 and is thought to represent an individual that strayed far from the normal habitat (Leopold 1959). The arboreal acrobatics and effortless climbing of the margay are partly the result of limb structure (Grzimek 1975). The feet are broad and soft and have mobile metatarsals. The hind foot is much more flexible than that of other felids, being able to rotate 180°. The cat thus can hang vertically during descent like a squirrel. It preys on squirrels and other arboreal mammals and also eats birds, some terrestrial rodents, and fruit (Nowell and Jackson 1996; Redford and Eisenberg 1992).

According to Nowell and Jackson (1996), a radio-tracked adult male in Brazil maintained a home range of 16 sq km during a period of 18 months. The estrous cycle is 32–36 days, estrus lasts 4–10 days, the gestation period is 76–84 days, litter size is usually one and sometimes two young, and females first enter estrus at 6–10 months. The young weigh about 170 grams at birth and take solid food 52–57 days later (Hayssen, Van Tienhoven, and Van Tienhoven 1993). A captive margay lived to more than 21 years of age (Marvin L. Jones, Zoological Society of San Diego, pers. comm., 1995).

The margay is classified as endangered by the USDI and is on appendix 1 of the CITES. Leopold (1959) referred to the margay as "an exceedingly rare animal." Grimwood (1969) stated that exports of pelts from Peru were increasing. Paradiso (1972) indicated that at least 6,701 margays, including both live animals and skins, were imported into the United States in 1970. According to Broad, Luxmoore, and Jenkins (1988), the total number of skins known to be in international trade declined from at least 30,000 in 1977 to about 20,000 in 1980 and only 138 in 1985. Although commerce in the fur of the margay has been drastically reduced by legal protection over most of its range and CITES regulation, the species remains threatened by extensive illegal hunting and habitat destruction. Because of its arboreal nature, it probably suffers more than the ocelot from deforestation (Nowell and Jackson 1996; Thornback and Jenkins 1982).

Felis tigrina (little spotted cat)

Head and body length is 400–550 mm, tail length is 250–400 mm, and weight is 1.5–3.0 kg (Leyhausen *in* Grzimek 1990). The upper parts vary in color from light to rich ochre and have rows of large, dark spots. The underparts are paler and less spotted. The tail has 10 or 11 rings and a black tip. One-fifth of all specimens are melanistic.

The little spotted cat lives in forests and seems to prefer montane cloud forest. It generally occurs at higher elevations than the ocelot and the margay, having been found at elevations as high as 4,500 meters (Nowell and Jackson

1996). Its habits in the wild are not known. Captive females have an estrus of several days, a gestation period of 74–76 days, and litters of one or two young. The kittens develop slowly, opening their eyes at 17 days and starting to take solid food at 55 days. Longevity may exceed 17 years (Nowell and Jackson 1996).

Like the ocelot and the margay, the little spotted cat was subject to uncontrolled commerce until recently. At least 3,170 individuals, including both live animals and skins, were imported into the United States in 1970 (Paradiso 1972). The species subsequently was listed as endangered by the USDI, and importation was halted. However, in the early 1980s *F. tigrina* became the leading spotted cat in the international fur trade, with the number of skins in commerce peaking at nearly 84,500 in 1983 (Broad 1987). Eventual legal protection in most countries where *F. tigrina* occurs and addition of the entire species to appendix 1 of the CITES seem to have greatly reduced this market. The species may still be threatened by illegal hunting and loss of restricted forest habitat to agriculture and logging (Nowell and Jackson 1996; Thornback and Jenkins 1982). It is designated as near threatened by the IUCN (under the name *Leopardus tigrinus*).

Felis yagouaroundi (jaguarundi)

Head and body length is 550–770 mm, tail length is 330–600 mm, and weight is 4.5–9.0 kg. There are two color phases: blackish to brownish gray, and fox red to chestnut. The body is slender and elongate, the head is small and flattened, the ears are short and rounded, the legs are short, and the tail is very long. This cat is sometimes said to resemble a weasel or otter in external appearance.

The jaguarundi inhabits lowland forests and thickets. It hunts in the morning and evening and is much less nocturnal than most cats. Recent observations in Central America indicate that it is largely diurnal (McCarthy 1992). It forages mainly on the ground but is an agile climber. The diet includes birds, small mammals, and reptiles. Home range is surprisingly large. The ranges of two males in Belize were 88 and 100 sq km, with overlap less than 25 percent. A female there used a range of 13–20 sq km (Nowell and Jackson 1996). *F. yagouaroundi* has been reported to live in pairs in Paraguay but to be solitary in Mexico. According to Grzimek (1975), there is no definite reproductive season in the tropics, but in Mexico young are produced around March and August. It is not known whether one female gives birth in both seasons. The estrous cycle averages 54 days, estrus averages 3.2 days, gestation lasts 70–75 days, the average litter size is 1.8 and the range is 1–4 young, age at sexual maturity is 2–3 years, and longevity is up to 15 years (Nowell and Jackson 1996).

The pelt is of poor quality and little value (Leopold 1959). The jaguarundi is widespread and not subject to commercial exploitation (Paradiso 1972). Nonetheless, four subspecies—cacomitli, tolteca, fossata, and panamensis—which range from southern Texas and Arizona to Panama, are listed as endangered by the USDI. North American populations, which essentially comprise those four subspecies, are on appendix 1 of the CITES; South American populations are on appendix 2. The IUCN classifies *cacomitli*, of southern Texas and eastern Mexico, as endangered (using the generic name *Herpailurus*), noting that fewer than 250 mature individuals survive. The jaguarundi's status is not well understood, but the animal's numbers have declined in at least the northern portions of its range because of human persecution and habitat destruction (Thornback and Jenkins 1982). The species is not now known to be resident in Arizona, though one was sighted there in 1975 (Jay M. Sheppard, U.S. Fish and Wildlife Ser-

vice, pers. comm., 1985), and only a few individuals survive in Texas. During the late Pleistocene the species occurred as far as Florida, and a small population may have become established there recently through introduction by human agency. There have been sporadic reports of a cat resembling the jaguarundi from other southeastern states.

Felis concolor (cougar, puma, panther, or mountain lion)

The names cougar, puma, panther, and mountain lion are used interchangeably for this species, and various other vernacular terms are applied in certain areas. This is by far the largest species in the genus *Felis*, averaging about the same size as the leopard *(Panthera pardus)*. In males head and body length is 1,050–1,959 mm, tail length is 660–784 mm, and weight is 67–103 kg. In females head and body length is 966–1,517 mm, tail length is 534–815 mm, and weight is usually 36–60 kg. Shoulder height is 600–700 mm. Generally, the smallest animals are in the tropics and the largest are in the far northern and southern parts of the range. There are two variable color phases: one ranges from buff, cinnamon, and tawny to cinnamon rufous and ferruginous; the other ranges from silvery gray to bluish and slaty gray. The body is elongate, the head is small, the face is short, and the neck and tail are long. The limbs are powerfully built; the hind legs are larger than the forelegs. The ears are small, short, and rounded. Females have three pairs of mammae (Banfield 1974).

The cougar has the greatest natural distribution of any mammal in the Western Hemisphere except *Homo sapiens*. It can thrive in montane coniferous forests, lowland tropical forests, swamps, grassland, dry brush country, or any other area with adequate cover and prey. The elevational range extends from sea level to at least 3,350 meters in California and 4,500 meters in Ecuador. There usually is no fixed den except when females are rearing young. Temporary shelter is taken in such places as dense vegetation, rocky crevices, and caves. The cougar is agile and has great jumping power: it may leap from the ground to a height of up to 5.5 meters in a tree. It swims well but commonly prefers not to enter water. Sight is the most acute sense and hearing is also good, but smell is thought to be poorly developed. Activity may be either nocturnal or diurnal. The cougar hunts over a large area, sometimes taking a week to complete a circuit of its home range (Leopold 1959). In Idaho, Seidensticker et al. (1973) found residents occupying fairly distinct but usually contiguous winter–spring and summer–autumn home areas. The latter area was generally larger and at a higher elevation, reflecting the summer movements of ungulate herds.

The cougar carefully stalks its prey and may leap upon the victim's back or seize it after a swift dash. Throughout the range of the cougar the most consistently important food is deer—*Odocoileus* in North America and *Blastoceros*, *Hippocamelus*, and *Mazama* in South America. Estimates of kill frequency vary from about one deer every 3 days for a female with large cubs to one every 16 days for a lone adult (Lindzey 1987). The diet also includes other ungulates, beavers, porcupines, and hares. The most common prey of the current population in Florida is wild hog (Maehr et al. 1990). The kill is usually dragged to a sheltered spot and then partly consumed. The remains are covered with leaves and debris, then visited for additional meals over the next several days.

Detailed studies in central Idaho (Hornocker 1969, 1970; Seidensticker et al. 1973) showed that the cougar population depends almost equally on mule deer *(Odocoileus hemionus)* and elk *(Cervus canadensis)*. The study area contained 1 cougar for every 114 deer and 87 elk. About half of the animals killed were in poor condition. Deer and

Cougar *(Felis concolor)*, photo by Dom Deminick, Colorado Division of Wildlife.

elk populations increased during a four-year study period, evidently being affected more by food availability than by cougar predation; nonetheless, predation was thought to moderate prey oscillations and to remove less fit individuals. During the same period the cougar population remained stable, its numbers being regulated mainly by social factors rather than by food supply. Density was about 1 adult cougar per 35 sq km. The total area used by individuals varied from 31 to 243 sq km in winter–spring and from 106 to 293 sq km in summer–autumn. There was little overlap in the areas occupied by resident adult males, but the areas of resident females often overlapped one another completely and were overlapped by resident male areas. Young, transient individuals of both sexes moved through the areas used by residents. There was a land tenure system based on prior right; transients could not permanently settle in an occupied area unless the resident died. Dispersal and mortality of young individuals unable to establish an area for themselves seemed to limit the size of the cougar population.

According to Lindzey (1987), long-term population densities in other areas have been found to vary from about 1/21 sq km to 1/200 sq km, and reported annual home range has been as great as 1,826 sq km for one animal in Texas. Nowell and Jackson (1996) stated that resident male ranges typically cover several hundred sq km, usually do not overlap one another, and do overlap several resident female ranges, each of which is usually less than 100 sq km. Nowell and Jackson also cited a population density of 7/100 sq km in Patagonia, one of the highest ever reported. The cougar is essentially solitary, with individuals deliberately avoiding one another except during the brief period of courtship. Communication is mostly by visual and olfactory signals, and males regularly make scrapes in the soil or snow, sometimes depositing urine or feces therein. Vocal-

izations include growls, hisses, and birdlike whistles. A very loud scream has been reported on rare occasion, but its function is not known.

There is no specific breeding season, but most births in North America occur in late winter and early spring. In southern Chile all known births have occurred from February to June (Nowell and Jackson 1996). Females are seasonally polyestrous and usually give birth every other year (Banfield 1974). The estrous cycle averages about 23 days (Nowell and Jackson 1996). Estrus lasts about 9 days and the gestation period is 90–96 days. The number of young per litter is one to six, commonly three or four. The kittens weigh 226–453 grams at birth and are spotted until they are about 6 months old. They nurse for 3 months or more but begin to take some meat at 6 weeks. If born in the spring, they are able to accompany the mother by autumn and to make their own kills by the end of winter; nonetheless, they usually remain with their mother for several more months or even another year. Litter mates stay together for 2–3 months after leaving the mother. Females attain sexual maturity at about 2.5 years, but males may not mate until at least 3 years (Banfield 1974). Regular reproductive activity does not begin until a young animal establishes itself in a permanent home area (Seidensticker et al. 1973). Females may remain reproductively active until at least 12 years and males to at least 20 years; captives have lived more than 20 years (Currier 1983). One female killed in the wild was at least 18 years old (Nowell and Jackson 1996).

The cougar is generally considered harmless to people, but attacks have occurred and there is concern about increasing encounters between the two species (Braun 1991). Beier (1991) documented the deaths of nine persons in the United States and Canada from 1890 up to 1990. Two more people were killed in 1991 and 1992, and there have been numerous recent injuries and narrow escapes (Aune 1991; Olson 1991; Rollins and Spencer 1995). Two women were killed in separate attacks in California during 1994 (*New Orleans Times-Picayune*, 14 December 1994, A-10), and a 10-year-old boy died in Colorado in July 1997 (*Washington Post*, 19 July 1997).

The remainder of the account of this species is based largely on Nowak's (1976) report. Information compiled by Nowell and Jackson (1996) indicates that there has been little change in population and legal status in the 20 years since that report. The cougar has long been viewed as a threat to domestic animals, such as horses and sheep, and is also sometimes thought to reduce populations of game. The species has thus been intensively hunted since the arrival of European colonists in the Western Hemisphere. Most successful hunting is done by using dogs to pursue the cat until it seeks refuge in a tree, where it can easily be shot. By the early twentieth century the cougar apparently had been eliminated everywhere to the north of Mexico except in the mountainous parts of the West, in southern Texas, and in Florida. The species still occupies the same regions, and about 16,000 individuals may be present. California provides almost complete protection to the cougar, Texas still allows it to be killed at any time, and all other western states and provinces permit regulated hunting. About 2,100 cougars are killed annually for sport or predator control in the United States and Canada (Green 1991; Tully 1991). Public antipathy seems to have moderated in the last few decades, and the general pattern of decline may have been halted, but loss of habitat and conflict with agricultural interests are still problems.

The subspecies *F. c. coryi*, which formerly occurred from eastern Texas to Florida, and the subspecies *F. c. couguar*, of the northeastern United States and southeastern Canada, are classified as critically endangered by the IUCN (using the generic name *Puma*). These two subspecies, along with *F. c. costaricensis*, of Central America, are also listed as endangered by the USDI and are on appendix 1 of the CITES. All other subspecies are on appendix 2. Since the late 1940s, evidence has accumulated indicating the presence of small surviving cougar populations in south-central and southeastern Canada, the southern Appalachians, and the Ozark region and adjoining forests of Arkansas, southern Missouri, eastern Oklahoma, and northern Louisiana. The most consistent and best-documented reports have come from New Brunswick and Nova Scotia (Cumberland and Dempsey 1994; Stocek 1995).

In southern Florida numerous "panthers" have been killed or live-captured in the last two decades, and a federal-state program of study and conservation is under way (Belden 1986; Belden and Forrester 1980; Shapiro 1981). This program has involved the expenditure of many millions of U.S. dollars to purchase refuge lands, carry out research in the field and captivity, provide vaccinations and medical treatments to the animals, and construct tunnels to allow them to pass safely under a major highway through their range. Nonetheless, the isolated Florida population comprises only 30–50 individuals and is jeopardized by severe inbreeding depression, a reduced reproductive rate, disease and parasites, habitat disruption and fragmentation, and accidental killing on roads (Barone et al. 1994; Nowell and Jackson 1996; Roelke, Martenson, and O'Brien 1993). A controversial project to introduce western cougars to Florida was initiated in 1993; the project is intended to improve genetic viability but could modify the unique characters of the original population. There is some evidence that the population already has experienced genetic introgression as a result of the release of a few individuals

of South American origin into the Everglades several decades ago (O'Brien et al. 1990).

CARNIVORA; FELIDAE; Genus **NEOFELIS**
Gray, 1867

Clouded Leopard

The range of the single species, *N. nebulosa*, extends from central Nepal northeastward to Shaanxi Province in central China and southeastward to Viet Nam and the Malay Peninsula and also includes Taiwan, Hainan, Sumatra, and Borneo. Most authorities, including Ellerman and Morrison-Scott (1966), Ewer (1973), Guggisberg (1975), Hemmer (1978), and Wozencraft (*in* Wilson and Reeder 1993) treat *Neofelis* as a distinct genus with one species. Simpson (1945) considered *Neofelis* to be a subgenus of *Panthera*. Leyhausen (1979) listed *Neofelis* as a full genus but included within it not only the clouded leopard but also the tiger (here referred to as *Panthera tigris*). Based on their shared unique and complex pelage, Corbet and Hill (1992) placed *nebulosa* together with *marmorata* (here referred to as *Felis marmorata*) in the genus *Pardofelis*.

Head and body length is about 616–1,066 mm, tail length is 550–912 mm, and weight is usually 16–23 kg. The coat is grayish or yellowish, with dark markings ("clouds") in such forms as circles, ovals, and rosettes. The markings on the shoulders and back are darker on their posterior margins than on their front margins, suggesting that stripes can be evolved from blotches or spots. The forehead, legs, and base of the tail are spotted, and the remainder of the tail is banded (Medway 1977). Melanistic specimens have been reported (Medway 1977).

The tail is long, the legs are stout, the paws are broad, and the pads are hard. The skull is long, low, and narrow (Guggisberg 1975). The upper canine teeth are relatively longer than those of any other living cat, having a length about three times as great as the basal width at the socket. The first upper premolar is greatly reduced or absent, leaving a wide gap between the canine and the cheek teeth. Unlike that of *Panthera*, the hyoid of *Neofelis* is ossified (Guggisberg 1975).

The clouded leopard inhabits various kinds of forest, perhaps to elevations of up to 2,500 meters. It is usually said to be highly arboreal, to hunt in trees, and to spring on ground prey from overhanging branches. Nowell and Jackson (1996) noted that its arboreal talents rival those of *Felis wiedii* and that it even can run down tree trunks headfirst or move about on horizontal branches with its back to the ground. Information compiled by Guggisberg (1975), however, suggests that *Neofelis* is more terrestrial and diurnal than is generally assumed. Rabinowitz, Andau, and Chai (1987) indicated that it travels mainly on the ground and uses trees primarily as resting sites. A subadult male radio-tracked for eight days was fully terrestrial and even rested in dense patches of grass (Dinerstein and Mehta 1989). *Neofelis* has been reported to feed on birds, monkeys, pigs, cattle, young buffalo, goats, deer, and even porcupines.

Reproduction is known only through observations in captivity (Fellner 1965; Fontaine 1965; Guggisberg 1975; Murphy 1976; Yamada and Durrant 1989). Mating has occurred in all months except June and October, with a peak in December. Cubs have been produced in all months except December, with a peak in March. The estrous cycle averages 30 days, estrus 6 days, and the gestation period 93

Clouded leopard *(Neofelis nebulosa)*, photo from San Diego Zoological Garden.

days. The number of young per litter ranges from one to five but is commonly two. The young weigh about 140–70 grams at birth, open their eyes after 12 days, take some solid food at 10.5 weeks, and nurse for 5 months. The coat is initially black in the patterned areas, and full adult coloration is attained at about 6 months. Both males and females attain sexual maturity at about 26 months. A captive was still living at 19 years and 6 months (Irven 1993). Many captives are gentle and playful and like to be petted by their custodians. Three hundred individuals are known to be in captivity (Millard and Fletchall *in* Bowdoin et al. 1994).

The clouded leopard is classified as vulnerable by the IUCN (1978) and as endangered by the USDI and is on appendix 1 of the CITES. The main problem is loss of forest habitat to agriculture. The genus also has been excessively hunted in some areas for its beautiful pelt, which sells for up to U.S. $2,000 on the black market (Santiapillai 1989). It may already have disappeared from Hainan, and on Taiwan it is restricted to the wildest and most inaccessible parts of the central mountain range. After a survey on Taiwan, Rabinowitz (1988) reported that the most recent sighting was in 1983 but expressed hope that the genus does survive. *Neofelis* still is widespread in mainland China, but suitable habitat is fragmented into small patches and the remaining animals are illegally hunted for their pelts and for other parts that are used in traditional medicines and foods (Nowell and Jackson 1996). *Neofelis* had not been definitely recorded in Nepal since 1863, but a number of individuals were located in 1987–88 and observations suggested that the genus may be more common there than previously thought (Dinerstein and Mehta 1989).

CARNIVORA; FELIDAE; **Genus PANTHERA**
Oken, 1816

Big Cats

There are five subgenera and five species (Cabrera 1957; Corbet 1978; Ellerman and Morrison-Scott 1966; Guggisberg 1975; E. R. Hall 1981; Smithers *in* Meester and Setzer 1977):

subgenus *Uncia* Gray, 1854

P. uncia (snow leopard), mountainous areas from Afghanistan to Lake Baikal and eastern Tibet;

subgenus *Tigris* Frisch, 1775

P. tigris (tiger), eastern Turkey to southeastern Siberia and Malay Peninsula, Sumatra, Java, Bali;

subgenus *Panthera* Oken, 1816

P. pardus (leopard), western Turkey and Arabian Peninsula to southeastern Siberia and Malay Peninsula, Sri Lanka, Java, Kangean Island, most of Africa;

subgenus *Jaguarius* Severtzov, 1858

P. onca (jaguar), southern United States to Argentina;

subgenus *Leo* Brehm, 1829

P. leo (lion), found in historical time from the Balkan and Arabian peninsulas to central India and in almost all of Africa.

The use of the name *Panthera* for this genus is in keeping with Corbet (1978), Ellerman and Morrison-Scott (1966), Hemmer (1978), Leyhausen (1979), Mazak (1981), and most of the other authorities cited herein. A few authorities, such as Cabrera (1957) and Stains (1984), consider *Panthera* to be invalid for technical reasons of nomencla-

curred as far east as 87° E in Bengal, and that the presence of *P. pardus* on Kangean Island may have resulted from introduction.

The members of the genus *Panthera* have an incompletely ossified hyoid apparatus, with an elastic cartilaginous band replacing the bony structure found in other cats (Grzimek 1975). This elastic ligament usually has been considered to be the anatomical feature that allows roaring but limits purring to times of exhaling. At least some other cats can purr both when exhaling and when inhaling. The snow leopard *(P. uncia)* possesses an elastic ligament but does not roar. Detailed morphological studies (Hast 1989; Peters and Hast 1994) now demonstrate that the ability to roar depends on a complex of characters, especially a specialized larynx containing very long vocal folds with a thick pad of fibro-elastic tissue. Such characters are found in the leopard, the jaguar, the lion, and the tiger but not in the snow leopard or any other cat. According to Paul Leyhausen (Max-Planck Institut für Verhaltensphysiologie, pers. comm., c. 1980), certain small cats, such as *Felis nigripes*, produce a full roar but do not sound as formidable as the lion or tiger simply because of the size difference. *Panthera* is also distinguished from *Felis* by having hair that extends to the front edge of the nose (Grzimek 1975). Additional information is provided separately for each species.

Panthera uncia (snow leopard)

Except as noted, the information for the account of this species was taken from Guggisberg (1975), Hemmer (1972), and Schaller (1977). Head and body length is 1,000–1,300 mm, tail length is 800–1,000 mm, shoulder height is about 600 mm, and weight is 25–75 kg. Males usually weigh 45–55 kg, females 35–40 kg (Nowell and Jackson 1996). The ground color varies from pale gray to creamy smoke gray, and the underparts are whitish. On the head, neck, and lower limbs there are solid spots, and on the back, sides, and tail are large rings or rosettes, many enclosing some small spots. The coat is long and thick and the head is relatively small.

The snow leopard is found in the high mountains of Central Asia. In summer it occurs commonly in alpine meadows and rocky areas at elevations of 2,700–6,000 meters. In the winter it may follow its prey down into the forests below 1,800 meters. It sometimes dens in a rocky cavern or crevice. It is often active by day, especially in the early morning and late afternoon. It is graceful and agile and has been reported to leap as far as 15 meters. It tends to remain in a relatively small area for 7–10 days and then shift activity to another part of its home range (Nowell and Jackson 1996). Prey is either stalked or ambushed. The diet includes mountain goats and sheep, deer, boars, marmots, pikas, and domestic livestock.

Population density estimates across the range of *P. uncia* vary from about 0.5 to 10.0 individuals per 100 sq km (Nowell and Jackson 1996). In a study in Nepal, Jackson and Ahlborn (1988) found that an area of 100 sq km supported 5–10 snow leopards. The home ranges of five individuals in this region measured about 12–39 sq km; these ranges overlapped almost entirely both between and within sexes, but the animals kept well apart. Socially the snow leopard is thought to be like the tiger, essentially solitary but not unsociable. The snow leopard does not roar but has several vocalizations, including a loud moaning associated with attraction of a mate.

Births usually occur from April to June both in the wild and in captivity after a gestation period of 90–103 days. The estrous cycle is 15–39 days, estrus 2–12 days (Nowell and Jackson 1996). The young are born in a rocky shelter lined with the mother's fur. The number of young per litter is one to five, usually two or three. The cubs weigh about 450 grams at birth, open their eyes after 7 days, eat their first solid food at 2 months, and follow their mother at 3 months. They hunt with the mother at least through their first winter and attain sexual maturity at about 2 years. Reproductive activity continues until 15 years, and longevity is up to 21 years (Nowell and Jackson 1996).

The snow leopard is classified as endangered by the IUCN, the USDI, and Russia and is on appendix 1 of the CITES. It has declined in numbers through hunting by people because it is considered to be a predator of domestic stock, it is valued as a trophy, and its fur is in demand by commerce. Like those of *P. tigris* (see account thereof), the bones of *P. uncia* are used in certain traditional Oriental medicines (Nowell and Jackson 1996). Although it is protected in China, it continues to be hunted there, and its skin is sold on the open market (Schaller et al. 1988). Contributing to the problem in northwestern India and Nepal, each with perhaps 400 individuals, are a reduction in natural prey and increased use of alpine pastures for livestock,

Snow leopard *(Panthera uncia)*, photo by Ernest P. Walker.

Tiger *(Panthera tigris)*, photo from East Berlin Zoo.

leading to predation on the latter and retaliatory measures by people (Fox et al. 1991; Jackson 1979). It recently has been estimated that a total of about 4,500–7,300 snow leopards remain in the wild and that they occupy 1.9 million sq km of habitat (Nowell and Jackson 1996). There are another 580 in captivity (Wharton and Blomqvist *in* Bowdoin et al. 1994).

Panthera tigris (tiger)

Head and body length is 1,400–2,800 mm, tail length is 600–1,100 mm, and shoulder height is 800–1,100 mm (Leyhausen *in* Grzimek 1990). The subspecies found in southeastern Siberia and Manchuria, *P. t. altaica*, is the largest living cat. The other mainland subspecies are also large, but those of the East Indies are much smaller. In *P. t. altaica* males weigh 180–306 kg, females 100–167 kg; in *P. t. tigris* of India and adjoining countries males weigh 180–258 kg, females 100–160 kg; in *P. t. sumatrae* of Sumatra males weigh 100–140 kg, females 75–110 kg; and in *P. t. balica* of Bali males weigh 90–100 kg, females 65–80 kg (Mazak 1981). The ground color of the upper parts ranges from reddish orange to reddish ochre, and the underparts

are creamy or white. The head, body, tail, and limbs have a series of narrow black, gray, or brown stripes. On the flanks the stripes generally run in a vertical direction; and in some specimens the stripes are much reduced on the shoulders, forelegs, and anterior flanks (Guggisberg 1975). There also is a rare but much publicized variant with a chalky white coat, dark stripes, and icy blue eyes; 103 are now in captivity (Roychoudhury 1987). There have been no records of white tigers in the wild since 1958 (Patnaik and Acharjyo 1990).

Except as noted, the information for the remainder of this account was taken from Mazak (1981). The tiger is tolerant of a wide range of environmental conditions, its only requirements being adequate cover, water, and prey. It is found in such habitats as tropical rainforests, evergreen forests, mangrove swamps, grasslands, savannahs, and rocky country. An individual may have one or more favored dens within its territory in such places as caves, hollow trees, and dense vegetation. The tiger usually does not climb trees but is capable of doing so. It has been reported to cover up to 10 meters in a horizontal leap. It seems to like water and can swim well, easily crossing rivers 6–8 km

Top, Siberian tiger *(Panthera tigris altaica)*. Bottom, Sumatran tiger *(P. t. sumatrae)*. Photos from East Berlin Zoo.

wide and sometimes swimming up to 29 km. It is mainly nocturnal but may be active in daylight, especially in winter in the northern part of its range. Siberian animals have moved up to 60 km per day. In Nepal, Sunquist (1981) determined the usual daily movement to be 10–20 km.

To hunt, the tiger depends more on sight and hearing than on smell. It usually carefully stalks its prey, approaching from the side or rear and attempting to get as close as possible. It then leaps upon the quarry and tries simultaneously to throw it down and grab its throat. Killing is by strangulation or a bite to the back of the neck. The carcass is often dragged to an area within cover or near water. One individual dragged an adult gaur *(Bos gaurus)* 12 meters, and later 13 men tried to pull the carcass but could not move it (Grzimek 1975). After eating its fill, the tiger may cover the remains with grass or debris and then return for additional meals over the next several days. The diet consists mainly of large mammals, such as pigs, deer, antelopes, buffalo, and gaur. Smaller mammals and birds occasionally are taken. A tiger can consume up to 40 kg of meat at one

time, but individuals in zoos are given 5–6 kg per day. Although the tiger is an excellent hunter, it fails in at least 90 percent of its attempts to capture animals. The tiger thus cannot eliminate entire prey populations. Indeed, Sunquist (1981) found that prey numbers, with the exception of one species, were not even being limited by tiger predation.

In Kanha National Park, in central India, Schaller (1967) determined that 10–15 adult tigers were regularly resident in an area of about 320 sq km. In Royal Chitawan National Park, in Nepal, Sunquist (1981) found an overall population density of 1 adult per 36 sq km. Observations in these and other areas indicate much variation in home range size and social behavior, evidently depending on habitat conditions and prey availability. In India individual home range seems usually to be 50–1,000 sq km. In Manchuria and southeastern Siberia the usual size is 500–4,000 sq km and the maximum reported is 10,500 sq km. In Nepal, Smith, McDougal, and Sunquist (1987) found home range size to be 19–151 sq km for males and 10–51 sq km for females. These ranges essentially corresponded to defended territories in that there was no overlap between those of adults of the same sex. A male range, however, overlapped the ranges of several females. Mothers evidently allowed their daughters to establish adjacent territories, but eventually the two generations may become agonistic. Schaller's studies indicate that the same kind of situation may exist in central India but also that females are not territorial there, adults of the same sex sometimes share a home range, and there are transient animals that lack an established range. All of these authorities suggested the presence of a land tenure system based on prior right, by which a resident animal is never replaced on its range until its death. Territorial boundaries are not patrolled, but individuals do visit all parts of their ranges over a period of days or weeks. These areas are marked with urine and feces. Avoidance, rather than fighting, seems to be the rule for tigers; nonetheless, one individual transplanted from its normal range to a different area evidently was killed soon thereafter by another tiger (Seidensticker et al. 1976).

Except for courting pairs and females with young, the tiger is essentially solitary. Even individuals that share a range usually keep 2–5 km apart (Sunquist 1981). The tiger is not unsociable, however, and the animals in a given area (probably close relatives) may know one another and have a generally amicable relationship (Schaller 1967). Several adults may come together briefly, especially to share a kill. Limited evidence suggests that a tiger roars to announce to its associates that it has made a kill. An additional function of roaring seems to be the attraction of the opposite sex. There are a number of other vocalizations, such as purrs and grunts, and the tiger also communicates by marking with urine, feces, and scratches.

Mating may occur at any time but is most frequent from November to April. Females usually give birth every 2–2.5 years and occasionally wait 3–4 years; if all the newborn are lost, however, another litter can be produced within 5 months (Schaller 1967). Females enter estrus at intervals of 3–9 weeks, and receptivity lasts 3–6 days. The gestation period is usually 104–6 days but ranges from 93 to 111 days. Births occur in caves, rocky crevices, or dense vegetation. The number of young per litter is usually two or three and ranges from one to six. The cubs weigh 780–1,600 grams at birth, open their eyes after 6–14 days, nurse for 3–6 months, and begin to travel with the mother at 5–6 months. They are taught how to hunt prey and apparently are capable killers at 11 months (Schaller 1967). They usually separate from the mother at 2 years but may wait another year. Sexual maturity is attained at 3–4 years by females and at 4–5 years by males. About half of all cubs do not survive more than 2 years, but maximum known longevity is about 26 years both in the wild and in captivity.

The tiger probably has been responsible for more human deaths through direct attack than any other wild mammal. In this regard perhaps the most dangerous place in modern history was Singapore and nearby islands, where following extensive settlement during the 1840s more than 1,000 persons were being killed annually (McDougal 1987). About 1,000 more people were reportedly killed each year in India during the early twentieth century. Guggisberg (1975) questioned the accuracy of these statistics, but there seems to be little doubt that some tigers have preyed extensively, or almost exclusively, on people. One individual is said to have killed 430 persons in India. Although such man-eaters have declined with the general reduction in tiger numbers in the twentieth century, the problem persists. In 1972, for example, India's production of honey and beeswax dropped by 50 percent when at least 29 persons who gathered these materials were devoured (Mainstone 1974).

Tigers currently seem to be especially dangerous in the Sundarbans mangrove forest, at the mouth of the Ganges River. Hendrichs (1975c) reported that 129 persons were killed in this area from 1969 to 1971 but noted that only 1 percent of the tigers there actually seem to seek out human prey. P. Jackson (1985) wrote that 429 persons had been killed in the same area during the previous 10 years, but Chakrabarti (1984) noted that unofficial estimates put the average annual toll at 100. A table compiled by Khan (1987) indicates a unique mammalian situation: in some recent years the number of people killed by tigers in the Sundarbans considerably exceeded the number of tigers killed by people.

Because it is considered to be a threat to human life and domestic livestock and also is valued as a big-game trophy, the tiger has been relentlessly hunted, trapped, and poisoned. Some European hunters and Indian maharajahs killed hundreds of tigers each. After World War II, hunting became even more widespread than previously (Guggisberg 1975). The commercial trade in tiger skins intensified in the 1960s, and by 1977 a pelt brought as much as U.S. $4,250 in Great Britain (IUCN 1978). Listing of the tiger as an endangered species by the United States in 1972, comparable protective measures by other countries, and international regulation through CITES contributed to a major decline in the fur market.

However, it subsequently became apparent that an equally devastating problem is posed by the utilization of the tiger's body parts for medicinal preparations and consequent widespread poaching and illegal commerce. Of particular importance are the bones, which are reputed to give strength, relieve pain, and cure numerous diseases, such as rheumatism. They are crushed and processed into various pills, tablets, liquid medicines, and wines. Uses exist for many other parts as well, including even the penis, which is made into a soup and sold for up to U.S. $300 per serving as a supposed aphrodisiac (Norchi and Bolze 1995). Research by Mulliken and Haywood (1994) suggests that there is a demand for such products in Oriental communities throughout the world and that CITES has not been effective in controlling the associated trade. From 1990 to 1992 a minimum of 27 million items containing tiger derivatives were recorded in trade, most of them exported from China and imported by Japan. South Korea and Taiwan also are heavily involved in the trade.

Although tiger products have been in use for centuries, the market long seems to have been supplied mainly by P. tigris amoyensis, the native subspecies of China (see below). The large-scale elimination of that subspecies between the 1950s and 1970s forced the market to go farther afield and also to attract much more attention. Tigers now are being killed illegally throughout the range of the species, and their parts smuggled into China and other places for conversion into consumer items. The consequent alarming decline in the species led to new international efforts at conservation, including U.S. trade sanctions against Taiwan (Bender 1994). That country, South Korea, and China all now legally prohibit commerce in tiger products. However, Peter Jackson (1995), chairman of the IUCN Species Survival Commission Cat Specialist Group, warned that illegal activity is continuing in response to a massive demand from persons who believe in the effectiveness of tiger-based medicines and that responsible countermeasures have been lacking. He suggested that if present trends continue, viable tiger populations could be eliminated by the end of the century.

Even if the tiger had no commercial value, its survival would be threatened by the destruction of its habitat. With the expansion of human populations, the logging of forests, the elimination of natural prey, and the spread of agriculture, there is continuous conflict between people and the tiger, and the latter species is almost always the loser. It is estimated that in 1920 there were about 100,000 tigers in the world. Estimates for wild populations ranged as low as 4,000 in the 1970s (Fisher 1978; Jackson 1978) but subsequently increased to about 6,000–8,000 (Foose 1987b; P. Jackson 1985; Luoma 1987). The apparent increase was thought to have resulted largely from an intensive effort to protect the species and establish reserves in India (Karanth 1987; Panwar 1987), though there were doubts that viable populations could be maintained there for long in the face of growing human encroachment (Ward 1987). The reserves in India are becoming increasingly isolated from one another, and the tiger populations therein are subject to poaching and loss of reproductive viability through inbreeding depression (Norchi and Bolze 1995; Smith and McDougal 1991). More recent analysis has indicated that the higher estimates of the 1980s were overly optimistic and that protective mechanisms have not been as effective as believed (Jackson 1993). This factor, together with the spread of the market for tiger parts, has led to a new estimate of 5,000–7,400 (P. Jackson 1994). The tiger is classified as endangered by the IUCN (except for the three critically endangered subspecies indicated below) and the USDI and is on appendix 1 of the CITES.

Except as noted, the information for the following summaries of the status of the eight subspecies of *P. tigris* was taken from the IUCN (1978), P. Jackson (1993, 1994), Luoma (1987), and Mazak (1981):

P. t. virgata (Caspian tiger) occurred in modern times from eastern Turkey and the Caucasus to the mountains of Kazakhstan and Sinkiang, in the Middle Ages may have reached Ukraine (Heptner and Sludskii 1992), also one report from northern Iraq in 1892 (Kock 1990), a few individuals still present in Turkey in the 1970s, now probably extinct;

P. t. tigris (Bengal tiger), originally found from Pakistan to western Burma, exterminated in Pakistan by 1906 (Roberts 1977), 2,750–3,750 individuals now estimated to survive in India, 300–460 in Bangladesh, 300 in Burma (Seal, Jackson, and Tilson 1987), 150–250 in Nepal, and 50–240 in Bhutan;

P. t. corbetti (Indochinese tiger), still found from eastern Burma to Viet Nam and the Malay Peninsula, 1,050–1,750 individuals estimated to remain but recent studies in Thailand suggest that the lower figure is the more reliable (Rabinowitz 1993);

P. t. amoyensis (Chinese tiger), classified as critically endangered by the IUCN, formerly occurred throughout eastern China, an estimated 4,000 individuals still survived in 1949 (Lu 1987), now confined largely to the Yangtze (Changjiang) Valley and apparently near extinction, with only 30–80 animals left in the wild, another 40 in captivity in China (Tan 1987; Xiang, Tan, and Jia 1987);

P. t. altaica (Siberian tiger), classified as critically endangered by the IUCN, formerly found from Lake Baikal to the Pacific coast and Korea, also sporadically on Sakhalin (Heptner and Sludskii 1992), apparently now very rare or absent in Manchuria and Korea, protection in the Ussuri region of Siberia seemed to result in an increase in numbers and distribution after the 1940s (Heptner and Sludskii 1992; Prynn 1980), only 20–30 survived in 1947 (Prynn 1993) but approximately 500 individuals present in the wild by 1990 (Quigley and Hornocker 1994), subsequent drastic decline as regulatory controls faltered and poaching for Chinese market intensified, now 150–200 in wild, also 632 in captivity (Olney, Ellis, and Fisken 1994);

P. t. sumatrae (Sumatran tiger), classified as critically endangered by the IUCN, population declined rapidly from an estimated 1,000 in the 1970s to about 400 now in the wild, another 194 in captivity (Olney, Ellis, and Fisken 1994);

P. t. sondaica (Javan tiger), almost all suitable habitat destroyed, only 4 or 5 individuals survived in the 1970s, now probably extinct;

P. t. balica (Bali tiger), probably extinct, last known specimen taken in 1937, though Van Den Brink (1980) held out some hope that it survives.

Panthera pardus (leopard)

Head and body length is 910–1,910 mm, tail length is 580–1,100 mm, and shoulder height is 450–780 mm. Males weigh 37–90 kg, and females, 28–60 kg. There is much variation in color and pattern. The ground color ranges from pale straw and gray buff to bright fulvous, deep ochre, and chestnut. The underparts are white. The shoulders, upper arms, back, flanks, and haunches have dark spots arranged in rosettes, which usually enclose an area darker than the ground color. The head, throat, and chest are marked with small black spots, and the belly has large black blotches. Melanistic leopards (black panthers) are common, especially in moist, dense forests (Guggisberg 1975; Kingdon 1977).

The leopard can adapt to almost any habitat that provides it with sufficient food and cover. It occupies lowland forests, mountains, grasslands, brush country, and deserts. A specimen was found at an elevation of 5,638 meters on Kilimanjaro. *P. pardus* is partly sympatric with *P. uncia*, the snow leopard, on the slopes of the Himalayas. The leopard is usually nocturnal, resting by day on the branch of a tree, in dense vegetation, or among rocks. It may move 25 km in a night, or up to 75 km if disturbed. However, typical daily movement for radio-tracked individuals in both South Africa and Thailand was only about 1 or 2 km (Bailey 1993; Rabinowitz 1989). The leopard generally progresses by a slow, silent walk but can briefly run at speeds of more than 60 km/hr. It has been reported to leap more than 6 meters horizontally and more than 3 meters vertically. It climbs with great agility and can descend headfirst. It is a strong swimmer but is not as fond of water as the tiger. Vision and hearing are acute, and the sense of smell seems to be better developed than in the tiger (Guggisberg 1975).

Hunting is accomplished mainly by stalking and stealthily approaching as close as possible to the quarry. Larger animals are seized by the throat and killed by strangulation. Smaller prey may be dispatched by a bite to the back of the neck. The diet is varied but seems to consist mainly of whatever small or medium-sized ungulates are available, such as gazelles, impalas, wildebeests, deer, wild goats and pigs, and domestic livestock. Monkeys and baboons also are commonly taken. If necessary, the leopard can switch to such prey as rodents, rabbits, birds, and even arthropods. Food is frequently stored in trees for later use. Such is the strength of the leopard that it can ascend a tree carrying a carcass larger than itself (Guggisberg 1975; Kingdon 1977). In a study in South Africa, Bailey (1993) found each adult leopard to kill one large prey animal, usually an impala, about every seven days. Leopards themselves are frequently pursued by lions and sometimes by hyenas and wild dogs (*Lycaon*). Large trees or other refuge sites are thus important, especially to a female raising cubs, both for immediate escape and as a place to keep food from the reach of rival species.

Population density is usually about 1/20–30 sq km, but under exceptionally favorable conditions it reportedly has been as high as 1/sq km. Most documented home range sizes are 8–63 sq km. Individuals apparently keep to a re-

Leopard *(Panthera pardus)*, photo from Zoological Society of London.

Jaguar *(Panthera onca)*, photo from Zoological Society of London.

stricted area, which they usually defend against others of the same sex. Territories are marked with urine, and severe intraspecific fighting has been observed. The range of a male may include the range of one or more females. Several males sometimes follow and fight over a female. Apparently, the leopard is normally a solitary species, but there have been reports of males remaining with females after mating and even helping to rear the young. There are a variety of vocalizations, the most common being a coughing grunt and a rasping sound, which seem to function in communication (Grzimek 1975; Guggisberg 1975; Kingdon 1977; Muckenhirn and Eisenberg 1973; Myers 1976; Schaller 1972; R. M. Smith 1977).

In an extended radio-tracking study in Kruger National Park, South Africa, Bailey (1993) determined population density to average 1/28.7 sq km for the whole park and to reach 1/3.3 sq km in an area of especially abundant prey. Social structure was found to be based on a three-tiered land tenure system. First, adult females maintained permanent home ranges of 5.6–29.9 sq km, sometimes exclusive but averaging about 18 percent overlap with one another. These areas were essentially limited territories that assured each resident access to adequate prey, cover, and den sites for raising young. Second, adult males occupied ranges of 16.5–96.1 sq km that overlapped little with ranges of other adult males and seemed to serve mainly to assure access to females. Each adult male range overlapped part or all of up to six female ranges. Third, superimposed on the mosaic of adult ranges were the more fluctuating ranges of young independent animals of both sexes, which essentially were waiting for vacancies in the permanent mosaic. There was extensive overlap of adult and subadult ranges but no contact between the occupants. Indeed, other than mothers with cubs or courting couples (which stayed together for up to five days), leopards were almost always found at least 1 km apart. Spacing was facilitated by vocalizations and by scent marking with ground scrapes and urine. Fighting was almost never recorded.

Breeding occurs throughout the year in most of Africa and India, though there may be peaks in some areas. In the northern portion of the Eurasian range mating apparently takes place mostly from December to February, parturition from March to May (Heptner and Sludskii 1992). In South Africa, Bailey (1993) observed most male-female association in the late dry season, from July to October. A female may give birth every 1–2 years. The estrous cycle averages about 46 days, and heat lasts 6–7 days; the gestation period is 90–105 days. Births occur in caves, crevices, hollow trees, or thickets. The number of young per litter is one to six, usually two or three. The cubs weigh 500–600 grams each at birth, open their eyes after 10 days, are weaned at 3 months, and usually separate from the mother at 18–24 months. Full size and sexual maturity are attained at around 3 years. Maximum longevity in captivity is more than 23 years (Grzimek 1975; Guggisberg 1975; Kingdon 1977).

The leopard seems to be more adaptable to the presence and activities of people than is the tiger and still occurs over a greater portion of its original range. Nonetheless, it is confronted by the same problems—persecution as a predator, value as a trophy, commercial demand for its beautiful fur, and loss of habitat and prey base. Man-eating leopards, though representing only a tiny percentage of the species, have undeniably been a menace in some areas. One in India, for example, is said to have killed more than 200 people during the 1850s (Guggisberg 1975). And from 1982 to 1989, reportedly 170 people were killed by leopards in India (Nowell and Jackson 1996). Intensification of agriculture, along with elimination of natural prey, sets up conflicts between herders and the leopard, usually to the detriment of the latter. The pelt of the leopard has been sought since ancient times, but in the 1960s there was a substantial increase in the worldwide market for the furs of spotted cats. Illegal hunting became rampant, and some leopard populations were decimated. As many as 50,000 leopard skins were marketed annually, of which nearly 10,000 were imported into the United States in some years (Myers 1976; Paradiso 1972). National laws and international agreements subsequently seem to have reduced this traffic. More recently, as in the case of *Panthera tigris* (see account thereof), Asian leopard populations have been subject to intensified poaching to obtain their body parts for use in traditional medicinal preparations (Nowell and Jackson 1996).

The leopard is said to be still relatively common in parts of East and Central Africa. Martin and de Meulenaer (1988) estimated its total numbers in sub-Saharan Africa at about 700,000, including 226,000 in Zaire alone, but these and other such high figures have been questioned (see, e.g., Bailey 1993; Meadows 1991; and Norton 1990). It has been pointed out that Martin and de Meulenaer improperly assumed that all suitable habitat was still occupied, that the leopard always occurred at the maximum density the habitat could support, that relaxation of human killing would result in rapid recovery, and that leopard density was higher in areas with higher rainfall. Stuart and Wilson (1988) estimated numbers at fewer than 15,000 for that part of Africa south of the Zambezi River, a region for which Martin and de Meulenaer had listed figures totaling about 75,000. Norton (1986) reported that the species has been eliminated from most of South Africa. Bailey (1993) determined that there were only about 700 leopards in Kruger National Park, the largest area of high-quality habitat in South Africa, and doubted that there could be even as many as 3,000 leopards in the whole country, whereas Martin and de Meulenaer had set the figure at over 24,000. Bailey anticipated further declines throughout Africa in response to human and livestock encroachment on habitat, with populations in the eastern and southern parts of the continent becoming restricted to isolated protected areas. Myers (1987a) indicated that the leopard populations of Ethiopia, Kenya, Namibia, and Zimbabwe were reduced by 90 percent during the 1970s largely because of poisoning by livestock interests.

Data compiled by Nowell and Jackson (1996) indicate that the leopard also has been greatly reduced in West Africa and completely eliminated in the Sahel zone south of the Sahara. In the entire vast region from North Africa to Central and southwest Asia the species is now rare or absent. Although small, fragmented populations still occur over most of the region, the only country known to have a substantial population is Turkmenistan, with about 150 animals. Numbers also have been greatly reduced in Pakistan and Bangladesh, but there are about 14,000 in India. The leopard still is found in scattered parts of China and has even been taken recently within 50 km of Beijing. Farther to the northeast the species has all but vanished from Manchuria and Korea, and only 25–30 individuals survive in southeastern Siberia. The leopard's status is not well understood in Southeast Asia, but the population on Java is estimated to number 350–700. Santiapillai, Chambers, and Ishwaran (1982) noted that only 400–600 individuals survived in Sri Lanka and that the future of the leopard there looked bleak.

The IUCN no longer has a general designation for *P. pardus* but does classify the following subspecies as indicated: *P. p. orientalis* (southeastern Siberia, Manchuria, Korea), critically endangered; *P. p. japonensis* (northern China), endangered; *P. p. saxicolor* (Turkmenistan, Afghanistan, Iran,

Iraq), endangered; *P. p. melas* (Java), endangered; *P. p. kotiya* (Sri Lanka), endangered; *P. p. tulliana* (Turkey, Caucasus, Syria, Jordan), critically endangered; *P. p. nimr* (Arabian Peninsula), critically endangered; and *P. p. panthera* (northwestern Africa), critically endangered. All of these subspecies have declined drastically, with fewer than 50 *orientalis* and fewer than 250 *tulliana*, *nimr*, and *panthera* thought to survive. Another subspecies sometimes recognized, *P. p. jarvisi*, of the Sinai Peninsula and Negev, apparently now is represented by only about 17 individuals in southern Israel (Nowell and Jackson 1996). In captivity there are 40 *japonensis*, 19 *kotiya*, 120 *orientalis*, and 126 *saxicolor* (Olney, Ellis, and Fisken 1994). The USDI lists the entire species *P. pardus* as endangered except in that part of Africa south of a line corresponding with the southern borders of Equatorial Guinea, Cameroon, Central African Republic, Sudan, Ethiopia, and Somalia. In that region the leopard is listed as threatened, and regulations allow the importation into the United States, under the provisions of the CITES, of trophy leopards taken in this region for purposes of sport hunting; importation for commercial purposes is prohibited. *P. pardus* is on appendix 1 of the CITES.

Panthera onca (jaguar)

Head and body length is 1,120–1,850 mm and tail length is 450–750 mm. Reported weights range from 36 to 158 kg. In Venezuela males usually weigh 90–120 kg, and females, 60–90 kg. The ground color varies from pale yellow through reddish yellow to reddish brown and pales to white or light buff on the underparts. There are black spots on the head, neck, and limbs and large black blotches on the underparts. The shoulders, back, and flanks have spots forming large rosettes that enclose one or more dots in a field darker than the ground color. Along the midline of the back is a row of elongate black spots that may merge into a solid line. Melanistic individuals are common, but the spots on such animals can still be seen in oblique light. *P. onca* averages larger than *P. pardus* and has a relatively shorter tail, a more compact and more powerfully built body, and a larger and broader head (Grzimek 1975; Guggisberg 1975; Mondolfi and Hoogesteijn 1986).

The jaguar is commonly found in forests and savannahs but at the northern extremity of its range may enter scrub country and even deserts. It seems usually to require the presence of much fresh water and is an excellent swimmer. It may den in a cave, canyon, or ruin of a human building. Although it sometimes is said to be nocturnal, radiotelemetry has shown that it is often active in the daytime, with activity peaks around dawn and dusk (Nowell and Jackson 1996). It is not known to migrate, but individuals sometimes shift their range because of seasonal habitat changes, and males have been known to wander for hundreds of kilometers. Daily movement in Belize was found generally to be 2–5 km (Seymour 1989). The jaguar climbs well and is almost as arboreal as the leopard, but most hunting is done on the ground. Prey is stalked or ambushed, and carcasses may be dragged some distance to a sheltered spot. The most important foods are peccaries and capybaras. Tapirs, crocodilians, and fish are also taken (Grzimek 1975; Guggisberg 1975).

In a radio-tracking study in southwestern Brazil, Schaller (1980a, 1980b) found a population density of about 1/25 sq km. There was a land tenure system much like that of the cougar and tiger. Females had home ranges of 25–38 sq km, which overlapped one another. Resident males used areas twice as large, which overlapped the ranges of several females. Studies in Belize indicate that 25–30 jaguars were present in about 250 sq km; male home ranges there averaged 33.4 sq km and overlapped exten-

sively, while two females had nonoverlapping ranges of 10–11 sq km. Estimates of home range in other areas have been as high as 390 sq km (Seymour 1989). The jaguar seems to be basically solitary and territorial. It marks its territory with urine and tree scrapes and has a variety of vocalizations, including roars, grunts, and mews (Guggisberg 1975). A female in estrus may wander far from her regular home range and may sometimes be accompanied briefly by several males. Estrus lasts about 6–17 days (Mondolfi and Hoogesteijn 1986).

Births occur throughout the year in captivity and perhaps also in the wild, especially in tropical areas, but there is evidence for a breeding season in the more northerly and southerly parts of the range. Most births in Paraguay take place in November–December; in Brazil, in December–May; in Argentina, in March–July; in Mexico, in July–September; and in Belize, in June–August, the rainy season (Seymour 1989). The estrous cycle averages 37 days and estrus lasts 6–17 days (Nowell and Jackson 1996). The gestation period commonly is 93–105 days, and litters contain one to four, often two, young. The offspring weigh 700–900 grams each at birth, open their eyes after 3–13 days, suckle for 5–6 months, stay with the mother about 2 years, and attain full size and sexual maturity at 2–4 years (Grzimek 1975; Guggisberg 1975; Mondolfi and Hoogesteijn 1986). A captive specimen at the Wuppertal Zoo in Germany reportedly lived to an age of 24 years (Marvin L. Jones, Zoological Society of San Diego, pers. comm., 1995).

Until the end of the Pleistocene *P. onca* occurred throughout the southern United States, and it seems to have been especially common in Florida. There is some evidence that the species still inhabited the southeastern United States in historical time (Daggett and Henning 1974; Nowak 1975b). Breeding may have continued in Arizona until about 1950 (Brown 1983). Otherwise resident populations had disappeared from the United States by the early twentieth century, though for many years wanderers continued to enter the country from Mexico, whereupon they usually were quickly shot. The most recent confirmed records are from southeastern Arizona and involve an animal killed in December 1986 (Nowak 1994) and jaguars photographed in March and September 1996 (Rabinowitz 1997). The jaguar now has been exterminated in most of Mexico, much of Central America, and, at the other end of its range, in most of eastern Brazil, Uruguay, and all but the northernmost parts of Argentina (Roig 1991; Swank and Teer 1989). Its numbers also have been greatly reduced in the vast Patanal wetland of Brazil, Bolivia, and Paraguay but reportedly are increasing farther south, in the Paraguayan Gran Chaco. Surprisingly, one of the most viable remaining populations, some 600–1,000 cats, is in Belize, the second smallest mainland country in the New World (Nowell and Jackson 1996). There may be another 500 in Guatemala but no more than 500 in all of Mexico (Emmons 1991b).

The jaguar declined because of the same factors that affected other large cats—persecution as a predator, habitat loss, and commercial fur hunting. The jaguar is thought to have killed people on rare occasion, but no individuals are known to have become systematic man-eaters. The species does have a reputation as a serious menace to domestic cattle; indeed, it may actually have increased in colonial times, when livestock was introduced to the savannahs of South America, a continent lacking vast natural herds of ungulates (Guggisberg 1975). As land was cleared and opened to ranching and other human activity, inevitable conflicts arose between the jaguar and people. As in the case of the leopard and other spotted cats, there was a great increase in the commercial demand for jaguar skins in the 1960s. An estimated 15,000 jaguars were then being killed annually in the Amazonian region of Brazil alone. The recorded number of pelts entering the United States reached 13,516 in 1968. Subsequent national and international conservation measures seem to have reduced the kill, but the problem continues (Thornback and Jenkins 1982). *P. onca* is classified as endangered by the USDI and as near threatened by the IUCN and is on appendix 1 of the CITES.

Panthera leo (lion)

Except as noted, the information for the account of this species was taken from Grzimek (1975), Guggisberg (1975), Kingdon (1977), and Schaller (1972). In males head and body length is 1,700–2,500 mm, tail length is 900–1,050 mm, shoulder height is about 1,230 mm, and weight is 150–250 kg. In females head and body length is 1,400–1,750 mm, tail length is 700–1,000 mm, shoulder height is about 1,070 mm, and weight is 120–82 kg. The coloration varies widely, from light buff and silvery gray to yellowish red and dark ochraceous brown. The underparts and insides of the limbs are paler, and the tuft at the end of the tail is black. The male's mane, which apparently serves to protect the neck in intraspecific fighting, is usually yellow, brown, or reddish brown in younger animals but tends to darken with age and may be entirely black.

The preferred habitats of the lion are grassy plains, savannahs, open woodlands, and scrub country. It sometimes enters semideserts and forests and has been recorded in mountains at elevations of up to 5,000 meters. It normally walks at about 4 km/hr and can run for a short distance at 50–60 km/hr. Leaps of up to 12 meters have been reported. The lion readily enters trees by jumping but is not an adept climber. Its senses of sight, hearing, and smell are all thought to be excellent. Activity may occur at any hour but is mainly nocturnal and crepuscular; in places where the lion is protected from human harassment it is commonly seen by day. The average period of inactivity is about 20–21 hours per day. Nightly movements in Nairobi National Park were found to cover 0.5–11.2 km (Rudnai 1973). In the Serengeti most lions remain in a single area throughout the year, but about one-fifth of the animals are nomadic, following the migrations of ungulate herds.

The lion usually hunts by a slow stalk, alternately creeping and freezing, utilizing every available bit of cover; it then makes a final rush and leaps upon the objective. If the intended victim cannot be caught in a chase of 50–100 me-

ters, the lion usually tires and gives up, but pursuits of up to 500 meters have been observed. Small prey may be dispatched by a swipe of the paw; large animals it seizes by the throat and strangles, or it suffocates them by clamping its jaws over the mouth and nostrils. Two lions sometimes approach prey from opposite directions; if one misses, the other tries to capture the victim as it flees by. An entire pride may fan out and then close in on a quarry from all sides. Groups are about twice as likely as lone individuals to capture prey. Most hunts fail: of 61 stalks observed by Rudnai (1973) only 10 were successful. The lion eats anything it can catch and kill, but it depends mostly on animals weighing 50–300 kg. Important prey species are wildebeests, impalas, other antelopes, giraffes, buffalo, wild hogs, and zebras. Carrion is readily taken. An adult male can consume as much as 40 kg of meat at one meal. After making a kill, a lion may rest in the vicinity of the carcass for several days. About 10–20 large animals per lion are killed each year.

Population density in the whole Serengeti ecosystem of East Africa was determined to be 1/10.0–12.7 sq km. Reported densities in most other areas vary from 1/2.6 sq km to 1/50.0 sq km. Nonmigratory lions in the Serengeti live in prides, each with a home range of 20–400 sq km. All or part of the range is a territory, which is vigorously defended against other lions. Nomadic lions have ranges of up to 4,000 sq km; there is much overlap in these areas, and individuals behave amicably. Nomads are commonly found in groups of two to four animals, and membership changes freely.

The basis of a resident pride is a group of related females and their young. These associations may persist for many years, being generally closed to strange females. Most daughters of group members are recruited into the pride, but young males depart as they approach maturity. Several related males frequently leave at about the same time and maintain a nomadic "coalition" until adulthood; sometimes unrelated males join to form a larger and potentially more powerful unit (Packer et al. 1991). Such a group, or occasionally a single male, eventually is able to gain residence with a pride of females and young for an indefinite period. The males cooperatively defend the pride against the approach of outside males. Some males associate with and defend several prides. Males of a coalition, especially if related, show remarkable tolerance in allowing one another to mate (Bertram 1991). Eventually, usually within three years, the pride males are driven off by another group of

African lions *(Panthera leo)*, photo by Bernhard Grzimek.

African lions *(Panthera leo)*, photo by Bernhard Grzimek.

males (Bertram 1975). Studies by Hanby and Bygott (1987) indicate that the factor stimulating departure of subadult males from a pride is the arrival of a new group of adult males and that some subadult females also may depart at such time if they are not yet ready to mate. The newly arrived males generally kill those young that are unable to escape. As a result, females of a pride tend to mate at the same time, to give birth synchronously, and to rear their young communally (Packer, Scheel, and Pusey 1990).

In the Serengeti the average number of lions in a pride was found to be 15, the range was 4–37, and the number of adult males present was 2–4. Prides often were divided into widely scattered smaller groups of about 5 individuals each. Studies by Packer, Scheel, and Pusey (1990) in the Serengeti suggest that such units frequently comprise females that are staying together to forage in order to defend their young against attack by outside males; cooperation to improve hunting success thus would not be the primary function of sociality in lions. In Nairobi National Park, Rudnai (1973) observed a single adult male to be associated with four prides of females. There was a rank order among the females, and a female led each group, even when the male was present, but the male was dominant with respect to access to food. Males living within a pride allow the

females to do almost all of the hunting, but they arrive subsequent to a kill and sometimes drive the others away. Lions appear to behave asocially at a kill, there being much quarreling and snapping and little tolerance shown to subordinates and cubs.

The lion has at least nine distinct vocalizations, including a series of grunts that apparently serve to maintain contact as a pride moves about. The roar, which can be heard by people up to 9 km away, is usually given shortly after sundown for about an hour and then again following a kill and after eating. It apparently has a territorial function. The lion also proclaims its territory by scent marking through urination, defecation, and rubbing its head against a bush.

Breeding occurs throughout the year in India and in Africa south of the Sahara. In any one pride, however, females tend to give birth at about the same time (Bertram 1975). Females are polyestrous, and heat lasts about 4 days. A female normally gives birth every 18–26 months, but if an entire litter is lost, she may mate again within a few days. The gestation period is 100–119 days, and litters contain one to six young, usually three or four. The newborn weigh about 1,300 grams; their eyes may be open at birth, or they may take up to 2 weeks to open. Cubs follow their mother after 3 months, suckle from any lactating female in the pride, and usually are weaned by 6–7 months. They begin to participate in kills at about 11 months, are fully dependent on the adults for food until 16 months, and probably are not capable of surviving on their own until at least 30 months. Sexual maturity is attained at around 3–4 years, but growth continues until about 6 years. The average longevity in zoos is about 13 years, but some captives have lived nearly 30 years.

With the exception of people, their domestic animals, and their commensals, the lion attained the greatest geographical distribution of any terrestrial mammal. Various populations, known from fossils by such names as *Panthera atrox* and *P. spelaea*, are now regarded as conspecific with *P. leo* (Hemmer 1974; Kurten 1985). About 10,000 years ago the lion apparently occurred in most of Africa, in all of Eurasia except probably the southeastern forests, throughout North America, and at least in northern South America. The lion is thought to have disappeared from most of Europe because of the development there of dense forests (Guggisberg 1975). It probably vanished from the Western Hemisphere when many of the large mammals on which it preyed were exterminated through the spread of advanced human hunters at the close of the Pleistocene. It was eliminated on the Balkan Peninsula, its last major stronghold in Europe, about 2,000 years ago and in Palestine at the time of the Crusades.

The continued decline of *P. leo* in modern times has resulted primarily from the expansion of human activity and domestic livestock and the consequent persecution of the lion as a predator. A few lions became regular man-eaters—for example, a pair killed 124 people in Uganda in 1925—and thus gave a sinister reputation to the entire species. Hunting for sport was also a major factor in some areas; for example, one person killed more than 300 lions in India in the mid–nineteenth century. At that time *P. leo* was still common from Asia Minor to central India and in North Africa. The last wild individual of the North African subspecies, *P. leo leo*, was shot in Morocco in 1920, though subsequently animals of the same type were found to exist in captivity (Leyhausen *in* Grzimek 1990). By about 1940 the lion also had been eliminated in Asia except in the Gir Forest, Gujarat State, western India. Vigorous conservation efforts resulted in the number of lions there stabilizing at around 200 for some years and recently increasing to more than 250. Unfortunately, the Gir lions have been under continuous pressure from livestock interests, and conflicts recently have intensified. There have been increasing attacks on people, 20 being killed from 1988 to 1991 (Saberwal et al. 1994). To help relieve these problems, as well as to limit the loss of genetic viability, reintroduction has been proposed at several sites within the former range of the lion in India (Chellam and Johnsingh 1993). There are 119 lions descended from the Gir population in captivity (Olney, Ellis, and Fisken 1994). The subspecies involved, *P. leo persica*, is classified as endangered by the IUCN and the USDI and is on appendix 1 of the CITES. All other lion populations are classified as vulnerable by the IUCN and are on appendix 2.

To the south of the Sahara the lion has become rare in West Africa (Rosevear 1974) and has been exterminated in most of South Africa and much of East Africa. It is still present over a large region but is widely hunted and poisoned by persons owning livestock. Myers (1975b) wrote that since 1950 the number of lions in Africa may have been reduced by half, to 200,000 or fewer; later Myers (1984) set the estimate at only 50,000. He cautioned that the species was rapidly losing ground to agriculture and that by the end of the century it might number only a few thousand individuals and survive only in major parks and

reserves. Nowell and Jackson (1996) agreed that numbers are almost certainly below 100,000, perhaps as low as 50,000, and that the species is becoming increasingly rare outside of protected areas. Stuart and Wilson (1988) suggested that in southern Africa there was little hope for the lion outside of these conservation areas; they estimated numbers south of the Zambezi River to be 6,100–9,100, of which about 4,200 already were in the protected reserves. Myers (1987a) warned, however, that such restriction and fragmentation of the lion population might result in inbreeding and loss of genetic viability. A severe new problem in East Africa is the loss of many lions to an epidemic of canine distemper, evidently spread by *Canis familiaris* (Woodford 1994).

CARNIVORA; FELIDAE; Genus ACINONYX
Brookes, 1828

Cheetah

The single species, *A. jubatus*, originally occurred from Palestine and the Arabian Peninsula to Tajikistan and central India, as well as throughout Africa except in the tropical forest zone and the central Sahara (Ellerman and Morrison-Scott 1966; Guggisberg 1975; Kingdon 1977; Smithers *in* Meester and Setzer 1977).

Except as noted, the information for the remainder of this account was taken from Caro (1994), Eaton (1974), Grzimek (1975), Guggisberg (1975), and Kingdon (1977). Head and body length is 1,120–1,500 mm, tail length is 600–800 mm, shoulder height is 670–940 mm, and weight is 21–72 kg; on the average, males are larger than females. The ground color of the upper parts is tawny or pale buff or grayish white, and the underparts are paler, often white. The pelage is generally marked by round, black spots set closely together and not arranged in rosettes. A black stripe extends from the anterior corner of the eye to the mouth. The last third of the tail has a series of black rings. The coat is coarse. The hair is somewhat longer on the nape than elsewhere, forming a short mane; in young cubs the mane is much more pronounced and extends over the head, neck, and back. *Acinonyx* has a slim body, very long legs, a rounded head, and short ears. The pupil of the eye is round. The paws are very narrow compared with those of other cats and look something like those of dogs. The claws are blunt, only slightly curved, and only partly retractile.

An additional species, *A. rex* (king cheetah), was described in 1927. It was based on specimens that differed from other cheetahs in having longer and softer hair and partial replacement of the normal spots by dark bars. Only 14 skins have been recorded from the wild, most from Zimbabwe and adjacent areas but 1 from Burkina Faso. It is now generally accepted that these specimens represent merely a variety of *A. jubatus* (Frame 1992; Hills and Smithers 1980). Indeed, individuals with the king cheetah markings now have been bred in captivity and recorded within otherwise normal litters (Brand 1983; Skinner and Smithers 1990; Van Aarde and Van Dyk 1986).

The habitat of the cheetah varies widely, from semidesert through open grassland to thick bush. Activity is mostly diurnal, and shelter is sought in dense vegetation. Recorded daily movements are about 3.7 km for a female with cubs and 7.1 km for adult males. The cheetah is capable of climbing and often plays about in trees. It is the fastest terrestrial mammal. Reported estimates of maximum speed range from 80 to 112 km/hr; such velocities, however, cannot be maintained for more than a few hun-

Cheetah *(Acinonyx jubatus)*, photo by Bernhard Grzimek.

dred meters. Unlike most cats, the cheetah does not usually ambush its prey or approach to within springing distance. It stalks an animal and then charges from about 70–100 meters away. It is seldom successful if it attacks from a point more than 200 meters distant, and it can only continue a chase for about 500 meters. Most hunts fail. If an animal is overtaken, it is usually knocked down by the force of the cheetah's charge and then seized by the throat and strangled. The diet consists mainly of gazelles, especially *Gazella thomsoni*, and also includes impalas, other small and medium-sized ungulates, and the calves of large ungulates. A female with cubs may kill such an animal every day, whereas lone adults hunt every two to five days. Hares, other small mammals, and birds are sometimes taken. The cheetah seems to work harder for its living than do the other big cats of Africa and thus may be more vulnerable to environmental changes brought about by human disturbance.

Population density in good habitat varies from about 1/20 sq km to 1/100 sq km. Seasonal concentrations may reach 1/2.5 sq km in some areas. In marginal habitat, however, density may be only 1/250 sq km or less (Myers 1975a). Some home ranges have been reported to measure about 50–130 sq km, but there appears to be considerable spacing variation. The cheetah occurs alone or in small groups. Solitary males and females are usually seminomadic and may occupy large home ranges of 700–1,500 sq km that overlap with one another and with the territories of resident animals. The groups seem usually to be either a female with cubs or two to four related adult males (litter mates that have remained together). The male "coalitions" commonly defend a territory against other males, perhaps thus facilitating access to prey and mates (Caro and Collins 1987). An unrelated male sometimes is accepted into the group. Coalitions persist through the lifetime of the members and seem to function primarily to secure mates. A single male may sometimes acquire a territory if there are no coalitions in the vicinity (Caro 1991). Such a territory is located in an area of high prey density, generally where females concentrate to hunt, and usually measures only about 40 sq km. In contrast to those of most cats, male territories of cheetahs, whether of an individual or a coalition, are usually much smaller than female ranges. Young females normally establish a range overlapping that of their mother, but young males leave their natal range.

Groups avoid one another and mark the area they are using at a given time. Marking is accomplished by regular urination on prominent objects. Such activity also serves to communicate sexual information. The cheetah is normally amicable toward others of its kind, but several males sometimes gather near an estrous female and fight over her. Members of male coalitions, however, usually show little overt competition in such situations. They join forces to control the movements of females, fight together against other males, rest in close proximity, groom one another, and search for their companions if they become separated. There does not seem to be a strong dominance hierarchy. There are a number of antagonistic vocalizations, purrs of contentment, a chirping sound made by a female to her cubs, and an explosive yelp that can be heard by people 2 km away.

Breeding occurs throughout the year, though a birth peak has been reported during the rainy season in East Africa (Nowell and Jackson 1996; Skinner and Smithers 1990). Females are polyestrous, with an average estrous cycle of approximately 12 days and an estrus of 1–3 days. Wild females normally give birth at intervals of 17–20 months; however, if all young are lost, the mother may soon mate and bear another litter. The gestation period is 90–95 days. The number of young per litter is one to eight, usually three to five. The cubs are born in thick vegetation or some other shelter, weigh 150–400 grams at birth, open their eyes after 4–11 days, and are weaned at 3–6 months. Cheetah cubs have an unusually high mortality rate for a large carnivore, especially because of lion predation: only 5 percent may reach independence. They begin to follow their mother after about 6 weeks, and the family may then shift its place of shelter on an almost daily basis. The mother teaches her cubs to hunt. They separate from her at 13–20 months and usually remain together for several months (as noted above, the males sometimes continue to associate for life). Both sexes may be physiologically capable of reproduction at about 18 months, but they seem not to attain full sexual maturity until they are 2–3 years old. A 15-year-old female was still reproductively active. Captives have lived up to 19 years.

People have tamed the cheetah and used it to run down game for at least 4,300 years. It was employed in ancient Egypt, Sumeria, and Assyria and has been used more recently by the royalty of Europe and India. It is usually hooded, like a falcon, when taken out for the chase, then freed when the game is in sight. If the hunt is successful, the cheetah is rewarded with a portion of the kill. If the cheetah should attempt to escape, it soon tires and can be easily caught by persons on horseback. Tame individuals are usually playful and affectionate.

The removal of live cheetahs from the wild has contributed to a decline in the species. Other factors are excessive hunting of both the cheetah and its prey, the spread of people and their livestock, and the fur market. In Namibia and other parts of southern Africa the cheetah is considered a serious predator of domestic sheep, goats, and young cattle, and it is persecuted accordingly (Nowell and Jackson 1996). Some studies also have indicated that *Acinonyx* has an unusually low degree of genetic variation, perhaps reflecting a severe population contraction during evolutionary history, leaving the species especially vulnerable to environmental disruption (Cohn 1986; S. J. O'Brien 1983, 1994; O'Brien et al. 1985; Wayne, Modi, and O'Brien 1986). Analysis of mitochondrial DNA and associated back calculation (Menotti-Raymond and O'Brien 1993) suggest that this hypothetical genetic bottleneck occurred near the end of the Pleistocene, about 10,000 years ago, a period characterized by widespread extinction of large mammals. Nonetheless, Merola (1994) has presented a reasonable argument that genetic constitution in the cheetah is not significantly less variable than in some other carnivores and that the real threat to the species is human habitat disruption. In addition, Caro and Laurenson (1994) reviewed re-

cent studies of the cheetah and found no substantial evidence that genetic problems are contributing to poor reproduction, susceptibility to disease, or mortality of young.

The cheetah seems to be much less adaptable than the leopard to the presence of people. It evidently has disappeared in Asia except in Iran and possibly adjacent parts of Pakistan, Afghanistan, and Turkmenistan. In the mid-1970s the population in Iran was estimated to include more than 250 individuals and was considered to be well protected (IUCN 1976); this population still exists but has declined because of both direct hunting and destruction of its prey (Karami 1992; Nowell and Jackson 1996). In Africa the species remains widely distributed but has become very rare in the northern and western parts of the continent. There still is a population of perhaps 500 animals in the mountains of Mali, Niger, Chad, and southern Algeria (Nowell and Jackson 1996). The species has been extirpated in most of South Africa and much of East Africa, though Hamilton (1986) thought it was holding its own in Kenya and had proved to be more resilient than anticipated. Myers (1975a) suggested that a century earlier the sub-Saharan population comprised about 100,000 individuals. Myers (1987a) estimated the total number remaining at just 10,000–15,000, and Stuart and Wilson (1988) estimated 6,200–8,500 for that part of the continent south of the Zambezi River. According to Grisham and Marker-Kraus (*in* Bowdoin et al. 1994), there are 9,000–12,000 in the wild, including 2,500 in Namibia, which has the largest single national population. Olney, Ellis, and Fisken (1994) reported that there also are 699 in captivity, 555 of which were born in that state. The IUCN classifies the cheetah generally as vulnerable, the northwest African subspecies (*A. j. hecki*) as endangered, and the Asiatic subspecies (*A. j. venaticus*) as critically endangered. The entire species is listed as endangered by the USDI and is on appendix 1 of the CITES.

Index

The scientific names of orders, familes, and genera, which have titled accounts in the text, are in boldface type. The page numbers on which such accounts begin are also in boldface type. Other scientific names, and vernacular names, appear in ordinary type. If the scientific and common names of a mammal are identical, they usually are indexed separately (for example, **Addax** and Addax). If, however, the scientific name does not qualify for boldface type, and the common name is in the singular, the two names are indexed as one (for example, Anoa). Information on an indexed subject sometimes is found in more than one account on the same page. Names in legends of illustrations usually are not indexed independently unless an illustration appears separately from the main text on the genus shown; usually in such a case only the scientific name is indexed. If no illustration of a genus appears on the page indexed for that genus, the reader should look one or two pages either before or after the account. However, not every account has an accompanying illustration (see preface).

GEOLOGICAL TIME

millions of years ago	epoch	period
280		**PERMIAN**
225		**TRIASSIC**
190		**JURASSIC**
135		**CRETACEOUS**
100		
65	Paleocene	**TERTIARY**
55	Eocene	
50		
38	Oligocene	
26	Miocene	
10		
7	Pliocene	
3	Pleistocene	**QUATERNARY**
1		
0.01	Holocene	
0.005	Historical Time	

WEIGHT
scales for comparison of metric and U.S. units of measurement

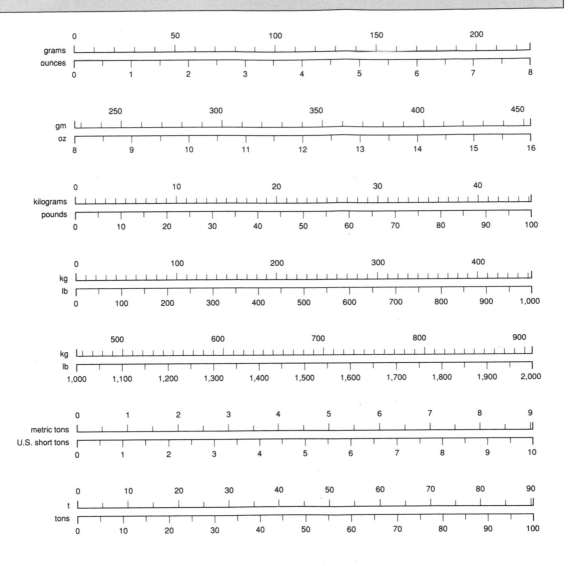

TEMPERATURE
scales for comparison of metric and U.S. units of measurement

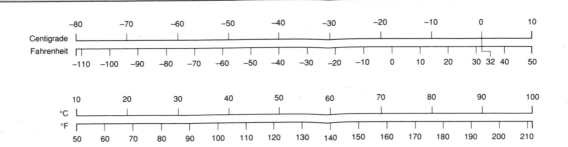